Compendium of Human Anatomic Variation

Compendium of Human Anatomic Variation

Text, Atlas, and World Literature

RONALD A. BERGMAN, PH.D.
Professor of Anatomy

SUE ANN THOMPSON, PH.D.
Assistant Professor of Anatomy

ADEL K. AFIFI, M.D.
Professor of Pediatrics, Neurology, and Anatomy

in the

College of Medicine
The University of Iowa, Iowa City, Iowa

and

FAYSAL A. SAADEH, PH.D.
Associate Professor of Human Morphology

in the

Faculty of Medicine
American University of Beirut, Lebanon

Urban & Schwarzenberg · Baltimore – Munich

Urban & Schwarzenberg, Inc.
7 E. Redwood Street
Baltimore, Maryland 21202
USA

Urban & Schwarzenberg
Pettenkoferstraße 18
D-8000 München 2
West Germany

© Urban & Schwarzenberg, 1988

All rights, including translation, are reserved. No part of this publication may be reproduced, stored in a retrieval system, or transmitted in any other form or by any means, electronic, mechanical, recording, or otherwise without the prior written permission of the publisher.

Printed in Germany

The Publishers have made an extensive effort to trace original copyright holders for permission to use borrowed material. If any has been overlooked, it will be corrected at the first reprint.

Library of Congress Cataloging in Publication Data

Bergman, Ronald A., 1927
  Compendium of human anatomic variation.
  Includes bibliographies and index.
  1. Anatomy, Human–Variation. I. Title.
  QM24.B47 1988   611   87–25511
  ISBN 0-8067-2502-8

ISBN 0-8067-2502-8 Baltimore
ISBN 3-541-72502-8 Munich

Preface

Throughout our careers in teaching anatomy to medical and dental students we have observed many of the variations reported in this catalog. We have also observed that students are often frustrated because the bodies they are dissecting do not conform to atlas or textbook descriptions and, as a result, they may either ignore the variations or inadvertently destroy them in an effort to find conformity.

We find human variations to be a source of great interest because they provide insight into individual development. The same interest must be kindled in students of medicine and the health sciences because individual variation must be considered in diagnosis and treatment. To quote from the preface of the 12th edition of *Morris' Human Anatomy*, edited by Barry Anson and published by the Blakiston Division of McGraw-Hill: "Were the student to remain unaware of the body's prenatal progress toward maturity, he would be baffled by every example of departure from the archetypal pattern or supposedly constant anatomic form. Lacking a genesis, nature's Bible would be an enigmatic book." The loss of *Morris' Human Anatomy* from continued publication is tragic; we are impoverished by its departure.

Most modern textbooks of anatomy are more or less devoid of information on variations that so commonly appear in the dissecting room and, more importantly, in practice. Variations must be considered to be normal and thus must be anticipated and understood. It has been repeatedly stated in the literature that textbook descriptions are accurate or hold in only about 50–70% of individuals. From the standpoint of utilization of anatomic information in a clinical setting, textbooks are not only inadequate but may be dangerously misleading as well. We feel, therefore, that this collection of reported human variations may alleviate, to some extent, profound deficiencies now found in many modern anatomy textbooks. We also see this catalog as a companion to dissecting guides and atlases for use in the dissecting room. It is a reference.

Most of the variations reported here have their investigative origin between 75 and 150 years ago. Their relevance to an understanding of human anatomy and to medicine, surgery, radiology, and other health sciences remains fresh and vital today. References to these will be found at the end of each section of this catalog. We have cited important secondary sources and many of the original references.

Although the value of anatomic information to medicine cannot be questioned, the information is contained in moribund books and journals. Another purpose of this book therefore is to "triage" anatomic facts from as many early sources as possible before they self-destruct from the acid contained in the paper. The problem is very serious.

We do not consider this catalog an exhaustive listing, but rather a beginning, and we hope that interested students of human variation will share their experience with us so that this offering will become increasingly useful. We will also continue our literature searches and report them in subsequent editions.

We are indebted to Tena Perry, Becky Hurt, and Vicki Fagen for their secretarial assistance, Phyllis Bergman for her assistance in bringing this manuscript to a readable and acceptable form, Paul Reimann for his photographic art, David Bergman who assisted in library searches, John Whelan and especially Brenda Robinson for their artistic contributions, Dr. Terence H. Williams for his continuing interest and enthusiastic support, and Braxton Mitchell and Michael Urban of Urban & Schwarzenberg for their encouragement. Linda Powell, our House Editor at Urban & Schwarzenberg, has our great respect and gratitude.

Ronald A. Bergman
Sue Ann Thompson
Adel K. Afifi   Iowa City and Beirut
Faysal A. Saadeh   1988

Dedication

This book is dedicated to our students past and present and to the memory of Buntaro Adachi, Barry Anson, Luigi Calori, Wenzel Gruber, Alexander Macalister, and Leo Testut. This book is also dedicated to our anatomist colleagues worldwide who have served the medical and health science professions and mankind by uncovering the fascinating, clinically invaluable, and vast legacy of variations in the human form.

Table of Contents

Preface	V
Dedication	VI
Acknowledgments	IX
Reference Sources	XI
Anatomists	XIV

Muscles

Alphabetical Listing	1
Muscles with Associated Named and Unnamed Variations	3
Muscles of Upper Extremity	7
Thorax, Shoulder, and Arm	7
Arm and Forearm	10
Forearm and Hand	12
Hand	17
Muscles of Trunk	18
Back, Head, and Neck	18
Thorax	19
Abdominal Wall	20
Lumbar Region	21
Pelvic Diaphragm and Anus	21
Urogenital Diaphragm	21
External Genitalia	22
Muscles of Lower Extremity	22
Hip	22
Thigh	23
Leg and Foot	25
Foot	28
Muscles of Head and Neck (Branchiomeric Muscles)	29
Trigeminal or Craniomandibular	29
Facial	30
Vago-Accessory	31
Orbital	33
Hypoglossal	34
Infrahyoid	34
Scalenes	35
Prevertebral	35
References	36
Glossary of Terms	48
Terminology: Synonyms	51

Cardiovascular System

Alphabetical Listing	57
Heart	60
Arteries	60
Head, Neck, and Thorax	60
Upper Extremity	72
Thorax	75
Abdomen	76
Pelvis	84
Lower Extremity	86
Veins	88
Head, Neck, and Thorax	88
Upper Extremity	90
Abdomen	91
Pelvis	92
Lower Extremity	93
Dural Sinuses	93
Lymphatics	93
References	94
Glossary of Terms	117
Terminology: Synonyms	119

Nervous System

Alphabetical Listing	127
Brain, Brain Stem, and Spinal Cord	129
Cranial Nerves and Ganglia	129
Spinal Nerves and Ganglia	136
Cervical	137
Thoracic	137
Lumbar and Sacral	138
Plexuses	138
Cervical	138
Brachial	139
Lumbosacral	143
Coccygeal	148
Sympathetic Division	148
References	149
Glossary of Terms	159
Terminology: Synonyms	160

Organs

Alphabetical Listing	165
Digestive System and Spleen	166
Respiratory System	167
Urinary System	171

Contents

Male Genital System	172	Sternum	205
Female Genital System	173	Upper Extremity	206
Endocrine System	175	Lower Extremity	207
Miscellaneous	176	References	209
References	177	Glossary of Terms	228
Glossary of Terms	191	Terminology: Synonyms	229
Terminology: Synonyms	192		

Skeleton

Atlas

		Table of Contents	233
Alphabetical Listing	195	Muscles	247
Vertebral Column	197	Cardiovascular System	319
Cranium	199	Nervous System	447
Larynx	204	Organs	509
Ribs	204	Skeleton	539

Acknowledgments

The illustrations by Leonardo da Vinci are reproduced with the gracious permission of **Her Majesty Queen Elizabeth II.**

The assistance of the curatorial staff at the Royal Library, Windsor Castle, is greatly appreciated.

The authors are grateful to:

Dr. R.A. Stockwell, Editor, *Journal of Anatomy*, and Cambridge University Press for permission to reproduce illustrations from the *Journal of Anatomy and Physiology* and the *Journal of Anatomy*.

Longmans Publishing Company for permission to reproduce illustrations from *Quain's Anatomy*, 11th Edition, edited by E.A. Schaefer, J. Symington, and T.H. Bryce.

The Johns Hopkins University Press for permission to quote Dr. Harvey Cushing and to reproduce illustrations from the *Bulletin of the Johns Hopkins Hospital*.

McGraw-Hill Book Company for permission to quote Dr. Barry Anson in the Preface.

VEB Gustav Fischer Verlag, Jena, for permission to reproduce illustrations from *Anatomischer Anzeiger*.

W.B. Saunders Company for permission to reproduce illustrations from *Surgical Anatomy*, 6th Edition, by B.J. Anson and C.B. McVay.

The staff of the Health Sciences Library of The University of Iowa.

Reference Sources

The sources of most of the information to follow have been obtained from original articles from the following journals. Additional information has been obtained from secondary sources (books) listed with the references at the end of each section.

Acta Anatomica
Acta Chirurgica Scandinavica
Acta Dermato-Venereologica
Acta Morphologica Neerlando-Scandinavica
Acta Orthopaedica Scandinavica
Acta Oto-laryngologica
American Family Physician
American Heart Journal
American Journal of Anatomy
American Journal of Cardiology
American Journal of Clinical Pathology
American Journal of Dermatology and Genito-urinary Diseases
American Journal of Diseases of Children
American Journal of Obstetrics and Gynecology
American Journal of Physical Anthropology
Anais Paulistas de Medicina e Cirurgia
Anatomische Hefte
Anatomischer Anzeiger (Centralblatt für die gesamte wissenschaftliche Anatomie)
Angiology
Annales d'Anatomie Pathologique et d'Anatomie Normale Médico-Chirurgicale
Annals of Neurology
Annals of Otology, Rhinology and Laryngology
Annals of Surgery
Annals of the Royal College of Surgeons of England
Archiv für Anatomie (Zeitschrift für Anatomie und Entwicklungsgeschichte)
Archiv für Anatomie, Physiologie und Wissenschaftliche Medizin (Contination of Reil's, Reil's and Autenrieth's, J.F. Meckel's, Joh. Muller's, and Reichert's and Du Bois-Reymond's Archiv)
Archiv für die gesamte Physiologie des Menschen und der Tiere
Archiv für die Physiologie (Reil)
Archiv für Kinderheilkunde
Archiv für Klinische Chirurgie (von Langenbeck's)
Archiv für Ohren-, Nasen- und Kehlkopfheilkunde
Archiv für Orthopädische und Unfall-Chirurgie
Archiv für Pathologische Anatomie und Physiologie und für Klinische Medizin (Virchow's)
Archiv für Psychiatric und Nervenkrankheiten (Berlin)
Archives d'Ophthalmologie
Archives de Physiologie Normale et Pathologique
Archives Générales de Médecine
Archives Italiennes de Biologie
Archives of Dermatology and Syphilology
Archives of Internal Medicine
Archives of Neurology and Psychiatry
Archives of Ophthalmology
Archives of Otolaryngology
Archives of Pathology
Archives of Pediatrics
Archives of Surgery
Archives of the Diseases of Childhood
Archivio di Ortopedia (Milan)
Archivio Italiano di Anatomia e di Embriologia
Australian and New Zealand Journal of Surgery
Beiträge zur Klinischen Chirurgie (Bruns')
Beiträge zur Pathologischen Anatomie und zur Allgemeinen Pathologie (Ziegler)
Berliner Klinische Wochenschrift
Bible
Brain
British Dental Journal
British Heart Journal
British Journal of Plastic Surgery
British Medical Journal
Bulletin de L'Académe Impériale des Sciences de St. Petersbourg
Canadian Journal of Surgery
Canadian Medical Association Journal
Cancer
Centralblatt für Allgemeine Pathologie und Pathologische Anatomie (Ziegler)
Centralblatt für die medicinischen Wissenschaften
Circulation
Clinical Genetics
Clinical Neurosurgery
Clinical Orthopaedics
Clinical Radiology
Correspondenz-Blatt für Schweizer Ärzte
Cutis
Der Nervenarzt
Deutsche Medizinische Wochenschrift
Deutsche Zeitschrift für Chirurgie
Diseases of the Chest
Edinburgh Medical Journal
Excerpta Medica
Folia Morphologica (Warsaw)
Formosan Medical Association Journal
Fortschritte auf dem Gebiete der Röntgenstrahlen und der Nuklearmedizin
Frankfurter Zeitschrift für Pathologie
Gazette des Hôpitaux Civils et Militaires, Lancet Française
Glasgow Medical Journal
Guy's Hospital Reports
Hand
Heart

Reference Sources

Helvetica Paediatrica Acta
Human Biology
Illinois Medical Journal
Industrial Medicine
International Clinics
International College of Surgeons
International Surgery
Internationale Monatsschrift für Anatomie und Physiologie
Jahresberichte über die Fortschritte der Anatomie und Physiologie
Japanese Journal of Medical Sciences
Johns Hopkins Hospital Bulletin
Jordan Medical Journal
Journal de Chirurgie
Journal de l'Anatomie et de la Physiologie Normale et Pathologie de L'Homme et des Animaux
Journal de Médecine, Chirurgie, Pharmacie, etc
Journal de Radiologie d'Electrologie et de Médecine Nucléaire
Journal d'Urologie et de Néphrologie
Journal d'Urologie, Médicale et Chirurgicale
Journal of Anatomy and Physiology, Journal of Anatomy
Journal of Dental Research
Journal of Hand Surgery
Journal of Nervous and Mental Diseases
Journal of Neurosurgery
Journal of Obstetrics and Gynecology
Journal of Pediatrics
Journal of the American Dental Association
Journal of the American Medical Association
Journal of the Anatomical Society of India
Journal of the International College of Surgeons
La Chirurgia Degli Organi di Movimento
Lancet
La Pediatria
La Presse Médicale
La Radiologia Medica
La Sémaine Medicale
La Société de Biologie, Comptes Rendus Hebdomadaires des Séances et Mémoires
L'Association des Anatomistes, Comptes Rendus
Lyon Chirurgical
Lyon Médical
Mayo Clinic, Proceedings of Staff Meetings
Medical Clinics of North America
Medical Times
Medizinische Jahrbücher (Wien)
Memorie della R. Accademia della Scienze dell'Istituto di Bologna
Minerva Pediatrica
Minnesota Medicine
Monatsschrift für Kinderheilkunde
Monatsschrift für Unfallheilkunde und Versicherungsmedizin
Münchener Medizinische Wochenschrift
Muscle and Nerve
Neurologisches Zentralblatt
Neurology (Minneapolis)
Neurosciences Research Program Bulletin
New York State Journal of Medicine
Okajimas' Folia Anatomica Japonica
Oral Surgery, Oral Medicine and Oral Pathology with Quarterly Review of the Literature
Pediatrics
Philadelphia Medical Journal
Philosophical Transactions, Royal Society of London
Physiological Reviews
Plastic and Reconstructive Surgery
Postgraduate Medicine
Prensa Médica Argentina
Proceedings of the Royal Society of Medicine
Quarterly Bulletin of Northwestern University Medical School
Radiologia Clinica
Radiology
Revista da Associacao Paulista de Medicina
Revue Canadienne de Biologie
Revue Neurologique
Revue d'Orthopédie et de Chirurgie de l'Appareil Moteur (Paris)
Rivista di Patologie Nervosa e Mentale
Röntgenpraxis
Scandinavian Journal of Dental Research
Scandinavian Journal of Plastic and Reconstructive Surgery
Schmidt's Jahrbücher der In- und Ausländischen Gesamten Medizin
Schweizerische Medizinische Wochenschrift
Science
Seminars in Roentgenology
Société Médicale des Hôpitaux de Paris, Bulletins et Memoires
Southern Medical Journal
Surgery
Surgery, Gynecology and Obstetrics
Texas State Journal of Medicine
The American Journal of Digestive Diseases
The American Journal of Ophthalmology
The American Journal of Pathology
The American Journal of Roentgenology and Radium Therapy
The American Journal of Surgery
The American Journal of the Medical Sciences
The American Journal of Urology
The American Surgeon, Southern Surgeon
The Anatomical Record
The Boston Medical and Surgical Journal
The British and Foreign Medico-Chirurgical Review *or* Quarterly Journal of Practical Medicine and Surgery
The British Journal of Radiology
The British Journal of Surgery

The British Journal of Tuberculosis and Diseases of the Chest
The British Journal of Urology
The Chinese Medical Journal
The Journal of Bone and Joint Surgery
The Journal of Clinical Endocrinology
The Journal of Comparative Neurology
The Journal of Laryngology, Rhinology and Otology
The Journal of Pediatric Surgery
The Journal of the Kansas Medical Society
The Journal of the Missouri State Medical Association
The Journal of Thoracic and Cardiovascular Surgery
The Journal of Urology
The Medical Journal of Australia
The New England Journal of Medicine
The Ohio State Medical Journal
The Quarterly Journal of Medicine
The Quarterly Review of Biology
The South African Journal of Medical Sciences
The Surgical Clinics of North America
The Transactions of the Royal Irish Academy (Science)
The University Studies of the University of Nebraska
The Western Journal of Surgery, Obstetrics and Gynecology
The Yale Journal of Biology and Medicine
Transactions of the Pathological Society of London
Transactions of the Royal Society of Tropical Medicine and Hygiene
University of Pennsylvania Medical Bulletin
Urological and Cutaneous Review
Urology
Wiener Archiv für Innere Medizin
Wiener Klinische Wochenschrift
Zeitschrift für die Gesamte Anatomie: First part: Zeitschrift für Anatomie und Entwicklungsgeschichte. Second part: Zeitschrift für Konstitutionslehre
Zeitschrift für Geburtshilfe und Gynäkologie
Zeitschrift für Klinische Medizin
Zeitschrift für Menschliche Vererbungs- und Konstitutionslehre
Zeitschrift für Orthopädie und ihre Grenzgebiete
Zeitschrift für Orthopädische Chirurgie
Zeitschrift für Urologische Chirurgie
Zentralblatt für Chirurgie
Zentralblatt für die Gesamte Neurologie und Psychiatrie.

In addition to the primary and secondary references, numerous anatomic variations are also reported that were found or confirmed in the dissecting rooms of The Johns Hopkins Medical School, American University of Beirut Faculty of Medicine, and The University of Iowa College of Medicine by indefatigable first-year medical students, over a period of many years.

Anatomists

The following anatomists have made extensive or significant contributions to our understanding of human variation.

Adachi, Buntaro. Japanese anatomist, 1865–1945.
Andernach, Johann W. (Jean Guinter of Andernach). German physician, 1505–1574.
Andersch, Carolus S. German anatomist, 1732–1777.
Anson, Barry J. American anatomist, 1894–1974.
Arnold, Friedrich. German anatomist, 1803–1890.
Bell, Sir Charles. Scottish anatomist, 1774–1842.
Bertin, Exupère J. French anatomist, 1712–1781.
Boyden, Edward A. American anatomist, 1886–1976.
Breschet, Gilbert. French anatomist, 1784–1845.
Calori, Luigi. Italian anatomist, 1807–1896.
Casserio, Guilio. Italian anatomist, 1556–1616.
Colombo, Matteo Realdo. Italian anatomist, 1516–1559?
Cowper, William. London anatomist, 1666–1709.
Cunningham, Douglas J. Irish anatomist in Scotland, 1850–1909.
Le Double, Anatole Félix. French anatomist, 1848–1913.
Eustachio, Bartolommeo E. Italian anatomist, 1520–1574.
Fallopius, Gabriele. Italian anatomist, 1523–1562.
Glaser, Johann H. Swiss anatomist, 1629–1675.
Gruber, Wenaslaus (Wenzel) L. Russian anatomist, 1814–1890.
Haller, Albrecht von. Swiss physiologist, 1708–1777.
Harvey, William. English physician, 1758–1657.
Henle, Friedrich G.J. German anatomist, 1809–1885.
Herophilus. Greek anatomist, circa 300 B.C.
Heschl, Richard L. Austrian pathologist, 1824–1881.
His, Wilhelm (Sr.). Swiss anatomist in Germany, 1831–1904.
Hyrtl, Joseph. Vienna anatomist, 1810–1894.
Jacobson, Ludwig L. Danish anatomist, 1783–1843.
Leonardo da Vinci. Italian anatomist and artist, 1452–1519.
Luschka, Hubert. German anatomist, 1820–1875.
Macalister, Alexander. Irish anatomist in Cambridge, 1844–1919.
Meckel, Johann F. (Sr.). German anatomist, 1714–1774.
Monro, Alexander (Jr.). Scottish anatomist, 1733–1817.
Morgagni, Giovanni B. Italian anatomist, 1682–1771.
Müller, Johannes P. German anatomist, 1801–1858.
Pecquet, Jean. French anatomist, 1622–1674.
Quain, Jones. English anatomist, 1796–1865.
Quain, Sir Richard. English physician, 1816–1898.
Reichert, Karl B. Berlin anatomist, 1811–1884.
Reil, Johann C. German neurologist, 1759–1813.
Riolan, J. French anatomist, 1577–1657.
Rivinus, August Q. (Bachmann A.Q.). German anatomist, 1652–1723.
Rolando, Luigi. Italian anatomist, 1773–1831.
Santorini, Giovanni D. Italian anatomist, 1681–1737.
Scarpa, Antonio. Venice anatomist, 1747–1832.
Schwalbe, Gustav A. German anatomist, 1844–1916.
Soemmering, Samuel T. von. German anatomist, 1755–1830.
Stensen, Nicholaus (Niels). Danish anatomist, 1638–1686.
Sylvius, Jacobus (Jacques Dubois). French anatomist, 1478–1555.
Testut, Leo. French anatomist, 1849–1925.
Tulpius, Nicholas. Dutch physician, 1593–1674.
Turner, Sir William. English anatomist, 1832–1916.
Verga, Andrea. Italian neurologist, 1811–1895.
Vesalius, Andreas (Andre Wesal). Flemish anatomist, 1514–1564.
Walther, August F. German anatomist, 1688–1746.
Wharton, Thomas. English anatomist, 1614–1673.
Whitnall, Samuel E. English anatomist, 1876–1952.
Willis, Thomas. English physician, 1621–1675.
Winslow Jacob B. Danish anatomist in Paris, 1669–1760.
Wirsung, Johann G. German anatomist in Padua, 1600–1643.
Wood, John. English anatomist, 1875–1891.
Worm, Ole. Danish anatomist, 1588–1654.
Wrisberg, Heinrich A. German anatomist, 1739–1808.

MUSCLES

Alphabetical Listing
(Primary references, in multiple page listings, are in *boldface* type.)

Abducor Accessorius Digiti Minimi (Pedis)   28
Abductor Digiti Minimi (Manus)   13, **18**
Abductor Digiti Minimi (Pedis)   **28**
Abductor Digiti Quinti   14
Abductor Hallucis   **28**
Abductor Manus   15
Abductor Ossis Metatarsi Digiti Quinti   28
Abductor Pollicis Brevis   16, **17**
Abductor Pollicis Longus   14, 15, **16**
Abductor Pollicis Tertius   16
Accessorium Tricipitis   11
Accessorius Ad Abductorem Digiti Minimi (Manus)   18
Accessorius Ad Flexorem Digiti Minimi   13
Accessorius Ad Flexorum Digitorum Profundus   14
Accessorius Ad Quadratum Plantae   27
Accessorius Digitorum Profundus   13
Accessory Abductor Digiti Minimi   18
Accessory Anterior Scalene   35
Accessory Extensor Carpi Radialis   17
Accessory Internal Oblique   20
Acromioclavicularis   33
Adductor Brevis   **24**, 28
Adductor Hallucis Caput Obliquum   **29**
Adductor Hallucis Caput Transversum (Transversus Pedis)   **29**
Adductor Longus   **24**
Adductor Magnus   23, **24**
Adductor Pollicis Brevis   **17**
Anconeus   11, **12**
Anconeus Sextus   11, 13
Anterior Scalene   35
Articularis Genus   **24**
Aryepiglotticus   32
Arymembranaceus   32
Aryvocalis   32
Atlanticomastoideus   35
Atlantobasilaris Internus   35
Atlantomastoid   19
Auricularis Anterior   **31**
Auricularis Inferior   **31**
Auricularis Posterior   **31**
Auricularis Superior   **31**
Axillary Arch   7, 9, 10
Axiobasilaris   35
Azygopharyngeus   32
Basioglossus   34
Biceps Brachii   10, **11**, 12, 15
Biceps Femoris   **23**, 26
Biceps Gracilis   24

Biventer Cervicis   19
Brachialis   9, 10, **11**, 12, 14
Brachiofascialis   11
Brachioradialis   9, 11, 13, **14**, 15
Brachioradialis Brevis   15
Buccinator   **31**
Buccopharyngeus   31
Bulbocavernosus   22
Bulbospongiosus   **22**
Capsularis Subbrachialis   11
Ceratocricoideus   32
Ceratoglossus   34
Ceratopharyngeus   31
Cervicocostohumeralis   34
Chondrocoracoideus   8
Chondroepitrochlearis (Chondrohumeralis)   7
Chondrofascialis   7
Chondroglossus   34
Chondrohumeralis   7
Chondropharyngeus   31
Chondroscapularis   8
Cleidoatlanticus   32
Cleidocervicalis   32
Cleidoepistrophicus   32
Cleidofascialis   34
Cleidohyoideus   34
Cleidomastoideus   32
Cleido-occipital   32
Cleidothyroideae   34
Coccygeofemoral   22
Coccygeus   21
Commes Obliqui Superioris   33
Compressor Narium   30
Compressor Venae Dorsalis   22
Constrictor Pharyngis Inferior   **31**
Constrictor Pharyngis Medius   **31**
Constrictor Pharyngis Superior   **31**
Contrahentes Digitorum   17
Coracobrachialis   7, 9, **10**, 11, 12
Coracobrachialis Inferior (Longus)   11
Coracobrachialis Minor   11
Coracobrachialis Superior (Brevis)   11
Coracoclavicularis Singularis   33
Corrugator Supercilii   **31**
Court Coracobrachialis   11
Court Radial Anterieur   14
Costocoracoideus   8
Costodeltoideus   9
Costoepitrochlearis   7
Cremaster   **20**
Cricoarytenoideus Lateralis   **32**
Cricoarytenoideus Posterior   **32**
Cricohyoideus   32, 34
Cricopharyngeus   31

Cricothyroideus   31, **32**
Cubitocarpeus   14
Curvator Coccygeus Accessorius   21
Deltoideus   7, **9**, 10, 14, 33
Depressor Alae Nasi   30
Depressor Anguli Oris   **30**
Diaphragma   **19**
Digastricus (Anterior Belly)   **29**
Digastricus (Posterior Belly)   **30**, 34
Dilatores Naris Anterior and Posterior   30
Dorsoepitrochlearis   10,11
Dorsofascialis   7, 33
Epistropheobasilaris   35
Epitrochleoanconeus   11, 13
Epitrochleocubital   11
Epitrochleo-olecranonis   11, 13
Extensor Atque Abductor Pollicis Accessorius   16
Extensor Brevis Manus   16
Extensor Carpi Radialis   11
Extensor Carpi Radialis Accessorius   15
Extensor Carpi Radialis Brevis   **15**
Extensor Carpi Radialis Intermedius   15
Extensor Carpi Radialis Longus   14, **15**
Extensor Carpi Ulnaris   12, **15**
Extensor Digiti III   16
Extensor Digiti IV   16
Extensor Digiti Medii   16
Extensor Digiti Minimi (Proprius) (Pedis)   25
Extensor Digiti Minimi (Proprius) (Manus)   15
Extensor Digiti Quinti   15
Extensor Digiti Quinti Brevis (Pedis)   26
Extensor Digitorum   15
Extensor Digitorum Brevis (Manus)   17
Extensor Digitorum Brevis (Pedis)   25
Extensor Digitorum Communis   15
Extensor Digitorum Longus (Pedis)   25
Extensor Hallucis Brevis   25
Extensor Hallucis Longus   25
Extensor Hallucis Longus Minor   25
Extensor Indicis   15, 16
Extensor Indicis Proprius   **16**
Extensor Pollicis Brevis   **16**
Extensor Pollicis Longus   15, **16**
External Oblique   7
Fibulotarsalis   27
Fibulocalcaneus medialis   27
Flexor Accessorius Longus   27
Flexor Carpi Radialis   11, **12**, 18
Flexor Carpi Radialis Brevis   12, 14
Flexor Carpi Radialis Brevis Vel Profundus   14
Flexor Carpi Ulnaris   11, 12, **13**, 18

1

Human Anatomic Variation

Flexor Carpi Ulnaris Brevis   14
Flexor Digiti II Brevis (Pedis)   29
Flexor Digiti Minimi Brevis (Manus)   13, 17, **18**
Flexor Digiti Minimi Brevis (Pedis)   **29**
Flexor Digiti Quinti Brevis   26
Flexor Digitorum Brevis (Pedis)   **28**
Flexor Digitorum Longus (Pedis)   **27**
Flexor Digitorum Profundus   11, **14**
Flexor Digitorum Superficialis (Sublimis)   12, **13**, 14, 17
Flexor Hallucis Brevis   **28**, 29
Flexor Hallucis Longus   26, **27**, 28
Flexor Indicis Profundus   14
Flexor Pollicis Brevis   **17**
Flexor Pollicis Longus   13, **14**, 17
Gastrocnemius   23, **26**, 27
Gastrocnemius Tertius   26
Gemelli   **23**
Genioglossus   **34**
Geniohyoid   **34**
Geniopharyngeus   34
Glenobrachialis   10
Glossopharyngeus   31
Gluteoperinealis   22, 23
Gluteus Maximus   **22**, 23
Gluteus Medius   **23**
Gluteus Medius Accessorius   23
Gluteus Minimus   **23**
Gluteus Quartus   23
Gracilis   **24**
Gracillimus Orbitis   33
Hepatodiaphragmaticus   20
Hyofascialis   34
Hyoglossus   **34**
Iliacus   **22**, 24
Iliacus Minor   22
Iliocapsularis   22
Iliococcygeus   21
Iliocostalis   18
Iliopsoas (Iliacus and Psoas Major)   **22**
Incisurae Mediae Obliquus   32
Incisurae Mediae Obliquus Bicaudatus   32
Inferior Constrictor of Pharynx   35
Inferior Gemellus   23
Inferior Oblique   **33**
Inferior Rectus   33
Infraspinatohumeralis   9
Infraspinatus   **9**
Infraspinatus Minor   9
Infraspinatus Superficialis   9
Interclavicularis Anticus Digastricus   8
Internal Intercostals   20
Internal Oblique   20, 21
Interossei (Manus)   15, **18**
Interossei (Pedis)   25, 26
Ischioaponeuroticus   24
Ischiocavernosus   **22**
Ischiofemoral   22
Lateral Cricoarytenoid   **32**
Lateral Rectus   **33**
Latissimocondyloideus   10, 11
Latissimus Dorsi   7, 9, **10**, 11, 20
Levator Anguli Scapulae   10
Levator Ani   **21**
Levator Claviculae   10, 33
Levator Epiglottidis   34
Levator Glandulae Thyroideae   35
Levator Palpebrae Superioris   **33**
Levator Penis   22

Levator Scapulae   **9**, 18
Longissimus   **19**
Longissimus Capitis   **35**
Longitudinalis Linguae Inferioris Medius   34
Longus Capitis   18, 35
Longus Colli   **35**
Lumbricales (Manus)   13, 14, **17**
Lumbricales (Pedis)   **28**
Mandibuloglossus   34
Mandibulomarginalis   30
Masseter   **29**
Mastoidopharyngeus   32
Medial Rectus   **33**
Mentalis   30
Mentohyoid   29
Myloglossus   34
Mylohyoideus   **29**, 30
Mylopharyngeus   31
Mytiformis   30
Nasalis   **30**
Oblique Arytenoid   32
Obliquus Accessorius Inferior   33
Obliquus Externus   **20**
Obliquus Inferior   **33**
Obliquus Internus   **20**
Obliquus Superior   **33**
Obturator Externus   **24**
Obturator Internus   **23**
Occipital Transverse   31
Occipitalis Minor   **31**
Occipitofrontalis   **31**
Occipitopharyngeus   32
Occipitoscapularis   9
Omocervicalis   33
Omohyoideus   **34**
Omotrachelien   33
Opponens Digiti Minimi   16, 18
Opponens Hallucis   28
Opponens Pollicis   16, **17**
Orbicularis Oculi   **31**
Palatopharyngeus   31
Palmaris Brevis   **17**
Palmaris Longus   **12**, 13, 17, 18, 27
Palmaris Profundus   12
Panniculus Carnosus   **7**, 33
Pectineus   **24**
Pectoralis Intermedius   8
Pectoralis Major   **7**, 9, 10, 11, 20
Pectoralis Minimus   8
Pectoralis Minor   **7**, 9, 10
Pectoralis Quartus (Costocoracoideus)   7
Pectorodorsalis (Axillary Arch)   10
Peroneocalcaneus   26
Peroneocalcaneus Internus   27
Peroneocuboideus   26
Peroneotibialis   26, 27
Peroneus Accessorius   26
Peroneus Brevis   **26**
Peroneus Digiti IV   26
Peroneus Digiti V   26
Peroneus Digiti V Superior or Longus   26
Peroneus Longus   **26**, 29
Peroneus Quartus   **26**
Peroneus Tertius   15, **25**, 26
Petropharyngeus   32
Piriformis   **23**
Pisiannularis   18
Pisimetacarpeus   18
Pisiuncinatus   18

Plantar Interosseous   26
Plantaris   **26**, 27
Platysma   **30**, 31
Popliteus   **26**, 27
Posterior Cricoarytenoid   **32**
Praeclavicularis   8
Praesternoclavicularis Mediale   33
Preclavicularis Lateralis   33
Procerus   30, **31**
Pronator Quadratus   **14**
Pronator Teres   11, **12**, 14, 15, 27
Psoas Major   20, **22**
Psoas Minor   **22**
Pterygoideus Lateralis   **29**
Pterygoideus Medialis   **29**
Pterygoideus Proprius   29
Pterygopharyngeus   31
Pterygospinosus   29
Pubocavernosus   22
Pubococcygeus   21
Puborectalis   21
Pyramidalis   **21**
Quadratus Femoris   **23**, 24
Quadratus Lumborum   20, **21**
Quadratus Plantae   **26**, 28
Quadriceps   **24**
Radialis Internus Brevis Biceps   14
Radiocarpeus   14
Radiocarpien   14
Radiocubitocarpien   14
Radiopalmar   14
Radiopalmaris   13
Rectus Abdominis   7, **21**
Rectus Accessorius   24
Rectus Capitis Anterior   **35**
Rectus Capitis Minimus   35
Rectus Capitis Lateralis   **35**
Rectus Capitis Lateralis Longus   35
Rectus Femoris   23
Rectus Inferior   **33**
Rectus Lateralis   **33**
Rectus Lateralis Accessorius   35
Rectus Medialis   **33**
Rectus Superior   **33**
Retractor Bulbi   33
Retractor Glandulae   33
Retroclavicularis Proprius   33
Rhomboatloideus   18
Rhomboaxoideus   18
Rhomboideus Major   **9**
Rhomboideus Minimus   9
Rhomboideus Minor   **9**, 10, 18
Risorius   **30**
Salpingopharyngeus   31
Saphenous   20
Sartorius   **24**
Scalenus Anterior   **35**
Scalenus Intermedius   35
Scalenus Medius   **35**
Scalenus Minimus   35
Scalenus Pleuralis   35
Scalenus Posterior   **35**
Scapuloclavicularis   8
Scapulocostalis Minor   8
Scapulocostoclavicularis   8
Semimembranosus   23, **24**
Semispinalis Capitis   **19**
Semitendinosus   **23**, 24
Serratus Anterior   7, 8, **9**, 20
Serratus Anterior Posterior   8, 9

Muscles

Serratus Posterior Inferior   18, **19**, 20
Serratus Posterior Superior   18, **19**
Soleus   **26**
Soleus accessorius   26
Sphincter Ani Externus   **21**
Sphincter Colli Profundus   30
Sphincter Urethrae   **22**
Spinalis Cervicis   19
Spinalis Dorsi   19
Spinalis Thoracis   19
Splenius   18
Splenius Capitis Accessorius   10, 18
Splenius Colli Accessorius   18
Stapedius   **31**
Sternalis   7
Sternochondrocoracoideus   8
Sternoclavicularis Anterior (Anticus)   8, 33
Sternocleidomastoideus   **32**, 33, 34
Sternocostalis   **19**
Sternohumeralis   8
Sternohyoideus   21, **34**
Sternohyoideus Azygos   34
Sternomastoideus   7, 32
Sterno-occipital   32
Sternoscapularis or Chondroscapularis   8
Sternothyroideus   21, 31, 32, **34**
Styloglossus   29, **34**
Stylohyoideus   **30**
Stylohyoideus Profundus   30
Stylomandibularis   30
Stylopharyngeus   **31**
Subclavius   **8**
Subclavius Posticus   8
Subcostalis   **19**
Suboccipitales   **19**
Subscapularis   10
Subscapularis Minor or Secundus   10
Superior Constrictor of Pharynx   31

Superior Gemellus   23
Superior Oblique (Orbital)   **33**
Superior Oblique (Suboccipitalis)   19
Superior Rectus   **33**
Superior Thyroarytenoideus   32
Supinator   **15**
Supinator Longus Accessorius   15
Supraclavicularis   8, 30
Supraclavicularis Proprius   9, 33
Supracostalis   7, **8**
Supracostalis Posterior   8
Supraspinatus   **9**
Syndesmopharyngeus   31
Temporalis   29
Temporalis Minor   29
Tensor Bursae Mucosae Tendinum   14
Tensor Capsulae   10
Tensor Capsulae Radiocubitalis (Inferioris)   14
Tensor Capsularis Articulationis Metacarpophalangei Digiti Minimi   18
Tensor Capsuli Tibiotarsalis Anterior   28
Tensor Fasciae Latae   22, **23**
Tensor Fasciae Plantaris   26, 27
Tensor Fasciae Suralis   23, 24
Tensor Laminae Posterioris Vaginae Musculi Recti   21
Tensor Laminae Profundus Fasciae Colli   33
Tensor Laminae Superficialis Fasciae Colli   33
Tensor Semivaginae Articulationis Humeroscapularis   8
Tensor Trochleae   33
Teres Major   9, **10**
Teres Minimus (Scapulae)   9, 10
Teres Minor   9
Thyroarytenoideus   32

Thyrohyoideus   31, 32, **34**
Thyropharyngeus   31
Thyrotrachealis Biceps   32
Tibialis Anterior   **25**
Tibialis Posterior   27, **28**
Tibialis Secundus   28
Tibioastragalus   25
Tibiotarsalis   28
Tibiofascialis Anticus   25
Trachleoclavicularis Imus   33
Transpleuralis   20
Transversalis Cervicis Medius   35
Transversalis Cervicis Posterior Minor   19
Transversalis Dorsi (Thoracis)   19
Transversospinalis   19
Transversus Abdominis   **20**
Transversus Colli   21
Transversus Glabellae   31
Transversus Manus   17
Transversus Menti   30
Transversus Nuchae   **31**
Transversus Pedis   **29**
Transversus Perinei Profundus   **21**
Transversus Perinei Superficialis   **22**
Transversus Thoracis   19
Trapezius   7, 9, 10, 11, 31, **33**
Triceps Brachii   9, **11**, 12
Triticeoglossus   34
Ulnaris Digiti Minimi (Quinti)   15
Ulnaris Externus Brevis   18
Ulnocarpeus   14
Ulnocarpeus Brevis   14
Vastus Intermedius   24
Vastus Lateralis   23
Vocalis   **32**
Zygomaticus Major   **30**
Zygomaticus Minor   **30**

Muscles with Associated Named and Unnamed Variations

(Primary references, in multiple page listings, are in *boldface* type.)

Abductor Digiti Minimi (Manus)   13, **18**
  Accessorius Ad Abductorem Digiti Minimi Manus   18
  Accessory Abductor Digiti Minimi   18
  Pisiannularis   18
  Pisimetacarpeus   18
  Pisiuncinatus   18
Abductor Digiti Minimi (Pedis)   28
  Abductor Ossis Metatarsi Digiti Quinti   28
  Abductor Hallucis   28
  Abductor Manus   15
Abductor Pollicis Longus   14, 15, **16**, 17
  Abductor Pollicis Tertius (Extensor Atque Abductor Pollicis Accessorius)   16

Adductor Brevis   **24**, 28
Adductor Hallucis Caput Obliquum   **29**
  Flexor Digiti II Brevis   29
Adductor Hallucis Caput Transversum (Transversus Pedis)   **29**
Adductor Longus   **24**
Adductor Magnus   23, **24**
Adductor Pollicis Brevis   **17**
  Transversus Manus   17
  Contrahentes Digitorum   17
Anconeus   11, **12**
Auricularis Anterior   **31**
  Auricularis Inferior   31
Auricularis Posterior   **31**
Auricularis Superior   **31**

Biceps Brachii   10, **11**, 12, 15
Biceps Femoris   **23**, 26
  Tensor Fasciae Suralis (Ischio-Aponeuroticus)   23, 24
Brachialis   9, 10, **11**, 12, 14
  Brachiofascialis   11
  Capsularis Subbrachialis   11
Brachioradialis   9, 11, 13, **14**, 15
  Brachioradialis Brevis   15
  Supinator Longus Accessorius   15
Buccinator   **31**
Bulbocavernosus   22
  Compressor Venae Dorsalis   22
Coccygeus   **21**
  Curvator Coccygeus Accessorius   21
Constrictor Pharyngis Inferior   **31**

Human Anatomic Variation

Cricopharyngeus 31
Syndesmopharyngeus 31
Thyropharyngeus 31
Constrictor Pharyngis Medius **31**
   Ceratopharyngeus 31
   Chondropharyngeus 31
Constrictor Pharyngis Superior **31**
   Buccopharyngeus 31
   Glossopharyngeus 31
   Mylopharyngeus 31
   Pterygopharyngeus 31
   Salpingopharyngeus 31
Coracobrachialis 7, 9, **10**, 11, 12
   Coracobrachialis Inferior (Longus) 11
   Coracobrachialis Minor (Court Coracobrachialis) 11
   Coracobrachialis Superior (Brevis) 11
Corrugator Supercilii 31
Cremaster 29
Cricoarytenoideus Lateralis 32
Cricoarytenoideus Posterior 32
   Ceratocricoideus 32
   Cricohyoideus 32, 34
Cricothyroideus 31, **32**
   Incisurae Mediae Obliquus 32
   Incisurae Mediae Obliquus Bicaudatus 32
   Thyrotrachealis Biceps 32
Deltoideus 7, **9**, 10, 14, 33
   Costodeltoideus 9
   Infraspinatohumeralis 9
Depressor Anguli Oris 30
   Transversus Menti 30
Diaphragma **19**
   Hepatodiaphragmaticus 20
Digastricus (Anterior Belly) **29**
   Mentohyoid 29
Digastricus (Posterior Belly) **30**, 34
   Stylomandibularis 30
Epitrochleo-olecranonis 11, 19
Extensor Carpi Radialis Brevis **15**
   Extensor Carpi Radialis Accessorius 15
   Extensor Carpi Radialis Intermedius 15
Extensor Carpi Radialis Longus 14, **15**
   Extensor Carpi Radialis Accessorius 15
   Extensor Carpi Radialis Intermedius 15
Extensor Carpi Ulnaris 12, **15**
   Ulnaris Digiti Minimi (Ulnaris Digiti Quinti) 15
Extensor Digiti Minimi (Proprius) Manus 15
Extensor Digitorum Brevis (Manus) 17
Extensor Digitorum Brevis (Pedis) **25**
Extensor Digitorum Communis **15**
Extensor Digitorum Longus (Pedis) **25**
Extensor Hallucis Longus **25**
   Extensor Hallucis Longus Minor 25
Extensor Indicis Proprius **16**
   Extensor Brevis Manus 16
   Extensor Digiti III 16
   Extensor Digiti IV 16
   Extensor Digiti medii 16
   Extensor Digitorum Brevis (Manus) 17
Extensor Pollicis Brevis **16**
Extensor Pollicis Longus 15, **16**
External Oblique 7

Saphenous 20
Flexor Carpi Radialis 11, **12**, 18
   Flexor Carpi Radialis Brevis (Radiocarpeus) 12, 14
Flexor Carpi Ulnaris 11, 12, **13**, 18
   Epitrochleoanconeus (Epitrochleoolecranonis or Anconeus sextus) 11, 13
Flexor Carpi Ulnaris Brevis (Ulnocarpeus Brevis) 14
Flexor Digiti Minimi Brevis (Manus) 13, 17, **18**
   Tensor Capsularis Articulationis Metacarpophalangei Digiti Minimi 18
   Ulnaris Externus Brevis 18
Flexor Digitorum Brevis (Pedis) **28**
Flexor Digitorum Longus (Pedis) **27**
Flexor Digitorum Profundus 11, **14**
   Accessorius Ad Flexorem Digitorum Profundus 14
   Flexor Iindicis Profundus 14
Flexor Digitorum Superficialis (Sublimis) 12, **13**, 14, 17
   Accessorius Profundus Digitorum 13
   Radiopalmaris 13
Flexor Hallucis Brevis **28**, 29
   Opponens Hallucis 28
Flexor Hallucis Longus 26, **27**, 28
   Fibulotarsalis 27
   Flexor Accessorius Longus 27
   Peroneocalcaneus Internus 27
Flexor Pollicis Brevis **17**
Flexor Pollicis Longus 13, **14**, 17
   Tensor Bursae Mucosae Tendinum 14
Gastrocnemius 23, **26**, 27
   Gastrocnemius Tertius 26
Gemelli **23**
Genioglossus **34**
   Geniopharyngeus 34
   Levator Epiglottidis 34
   Longitudinalis Linguae Inferioris Medius 34
Gluteus Maximus **22**, 23
   Coccygeofemoral 22
   Ischiofemoral 22
Gluteus Medius **23**
   Gluteoperinealis 22, 23
   Gluteus Medius Accessorius 23
   Gluteus Quartus 23
Gluteus Minimus **23**
   Gluteus Quartus 23
Gracilis **24**
   Biceps Gracilis 24
Hyoglossus **34**
   Basioglossus 34
   Ceratoglossus 34
   Chondroglossus 34
   Geniohyoid 34
   Triticeoglossus 34
Iliopsoas (Iliacus and Psoas Major) **22**
   Iliacus Minor (Iliocapsularis) 22
Inferior Oblique **33**
   Obliquus Accessorius Inferior 33
Infraspinatus **9**
   Infraspinatus Minor 9
   Infraspinatus Superficialis 9
Internal Oblique 20, 21
   Accessory Internal Oblique 20
Interossei (Manus) 15, **18**
Ischiocavernosus **22**

Compressor Venae Dorsalis 22
Pubocavernosus (Levator Penis) 22
Lateral Cricoarytenoid **32**
Lateral Rectus **33**
   Retractor Bulbi 33
Latissimus Dorsi 7, 9, **10**, 11, 20
   Axillary Arch (Pectorodorsalis)
   Latissimocondyloideus (Dorsoepitrochlearis) 7, 9, 10
Levator Ani **21**
Levator Palpebrae Superioris **33**
   Retractor Glandulae Lacrimalis 33
   Tensor Trochleae (Gracillimus orbitis) 33
Levator Scapulae **9**, 18
   Levator Claviculae (Omocervicalis or Omotrachelien) 10, 33
Longissimus **19**
   Transversalis Cervicis Posterior Minor 19
   Transversalis Dorsi (Thoracis) 19
Longus Capitis **35**
   Atlantobasilaris Internus (Axiobasilaris) 35
   Epistropheobasilaris 35
Longus Colli **35**
   Atlanticomastoideus (Rectus Lateralis Accessorius) 35
Lumbricales (Manus) 13, 14, **17**
Lumbricales (Pedis) **28**
Masseter **29**
Medial Rectus **33**
   Retractor Bulbi 33
Mylohyoideus **29**, 30
Nasalis **30**
   Compressor Narium 30
   Depressor Alae Nasi (Mytiformis) 30
   Dilatores Naris Anterior and Posterior 30
Obliquus Externus 20
   Saphenous 20
Obliquus Inferior **33**
   Obliquus Accessorius Inferior 33
Obliquus Internus **20**
   Accessory Internal Oblique 20
Obliquus Superior **33**
   Comes Obliqui Superioris 33
Obturator Externus **24**
Obturator Internus **23**
Occipitalis Minor (Transversus Nuchae or Occipital Transverse) 31
Occipitofrontalis **31**
Omohyoideus **34**
   Cervicocostohumeralis 34
   Cleidofascialis 7
   Cleidohyoideus 34
   Cleidothyroideae 34
   Hyofascialis 34
Opponens Digiti Minimi 16, 18
   Ulnaris Externus Brevis 18
Opponens Hallucis 28
Opponens Pollicis 16, **17**
Orbicularis Oculi **31**
   Transversus Glabellae 31
Palmaris Brevis **17**
Palmaris Longus **12**, 13, 17, 18, 27
   Accessorius Ad Flexorem Digiti Minimi (Manus) 13
   Palmaris Profundus 12
Panniculus Carnosus **7**, 33

Muscles

Axillary Arch 7, 9, 10
Dorsofascialis 7, 33
Pectineus 24
Pectoralis Major 7, 9, 10, 11, 20
   Axillary Arch 7, 9, 10
   Chondrofascialis 7
   Costoepitrochlearis 7
   Pectoralis Quartus 7
   Sternalis 7
Pectoralis Minor 7, 9, 10
   Chondrocoracoideus 8
   Pectoralis Minimus 8
   Sternochondrocoracoideus 8
   Sternoclavicularis Anterior 8, 33
   Sternohumeralis 8
   Tensor Semivaginae Articulationis Humeroscapularis 8
Peroneus Brevis 26
   Extensor Digiti Quinti Brevis 26
   Peroneus Accessorius 26
   Peroneus Digiti IV 26
   Peroneus Digiti V 26
Peroneus Longus 26
   Flexor Digiti Quinti Brevis 26
   Peroneus Digiti V Superior or Longus 26
   Plantar Interosseous 26
Peroneus Tertius 15, **25**, 26
Piriformis 23
Plantaris **26**, 27
   Tensor Fasciae Plantaris 26, 27
Platysma **30**, 31
   Mandibulomarginalis 30
   Sphincter Colli Profundus 30
   Supraclavicularis 8, 30
Popliteus 26, **27**
   Peroneotibialis 26, 27
Posterior Cricoarytenoid **32**
   Ceratocricoideus 32
   Cricohyoideus 32, 34
Procerus 30, **31**
Pronator Quadratus **14**
   Court Radial Anterieur 14
   Flexor Carpi Radialis Brevis 12, 14
   Flexor Carpi Radialis Vel Profundus 14
   Radialis Internus Brevis Biceps 14
   Radiocarpeus 14
   Radiocubitocarpien 14
   Radiopalmar 14
   Tensor Capsulae Radiocubitalis (Inferioris) 14
   Ulnocarpeus (Cubitocarpeus) 14
Pronator Teres 11, **12**, 14, 15, 27
Psoas Major 20, **22**
Psoas Minor **22**
Pterygoideus Lateralis **29**
   Pterygoideus Proprius 29
   Pterygospinosus 29
Pterygoideus Medialis **29**
   Styloglossus 29, **34**
Pyramidalis **21**
Quadratus Femoris **23**, 24
Quadratus Lumborum 20, **21**
Quadratus Plantae 26, **28**
   Flexor Accessorius Longus 27
Quadriceps **24**
   Rectus Accessorius 24
Radiocarpeus 14
Rectus Abdominis 7, **21**

Rectus Capitis Anterior **35**
   Rectus Capitis Minimus 35
Rectus Capitis Lateralis **35**
   Rectus Capitis Lateralis Longus 35
Rectus Inferior **33**
Rectus Lateralis **33**
   Retractor Bulbi 33
Rectus Medialis **33**
Rectus Superior **33**
Rhomboideus Major **9**
Rhomboideus Minor **9**, 10, 18
   Occipitoscapularis 9
   Rhomboideus Minimus 9
Risorius **30**
Sartorius **24**
Scalenus Anterior **35**
   Accessory Anterior Scalene 35
   Scalenus Minimus 35
   Transversalis Cervicis Medius 35
Scalenus Medius **35**
Scalenus Posterior **35**
Semimembranosus 23, **24**
Semispinalis Capitis **19**
   Biventer Cervicis 19
Semitendinosus **23**, 24
   Tensor Fasciae Suralis (Ischioaponeuroticus) 23, 24
Serratus Anterior 7, 8, **9**, 20
Serratus Posterior Inferior 18, **19**, 20
Serratus Posterior Superior 18, **19**
Soleus **26**
   Soleus Accessorius 26
   Peroneotibialis 26, 27
   Tensor Fasciae Plantaris 26, 27
Sphincter Ani Externus **21**
Sphincter Urethrae **22**
Splenius **18**
   Rhomboatloideus (Splenius Colli Accessorius) 18
   Rhomboaxoideus 18
   Splenius Capitis Accessorius 10, 18
Sternocleidomastoideus **32**, 33, 34
   Cleidoatlanticus 32
   Cleidocervicalis 32
   Cleidoepistrophicus 32
   Cleidomastoideus 32
   Cleido-occipital 32
   Sternomastoideus 7, 32
   Sterno-occipital 32
Sternohyoideus 21, **34**
   Sternohyoideus Azygos 34
Sternothyroideus 21, 31, 32, **34**
   Levator Glandulae Thyroideae 35
Stylohyoideus **30**
   Stylohyoideus Profundus 30
   Stylomandibularis 30
Stylopharyngeus **31**
   Azygospharyngeus 32
   Mastoidopharyngeus 32
   Occipitopharyngeus 32
   Petropharyngeus 32
Subclavius **8**
   Chondroepitrochlearis (Chondrohumeralis) 7
   Interclavicularis Anticus Digastricus 8
   Pectoralis Intermedius 8
   Pectoralis Minimus 8
   Praeclavicularis 8
   Scapuloclavicularis 8
   Scapulocostalis Minor 8

Scapulocostoclavicularis 8
Sternoscapularis (Chondroscapularis) 8
Subclavius Posticus 8
Supraclavicularis 8, 30
Subcostalis **19**
Suboccipitales **19**
   Atlantomastoid 35
Subscapularis **10**
   Glenobrachialis 34
   Subscapularis Minor (Secundus) 10
   Tensor Capsulae 10
Superior Oblique **33**
   Comes Obliqui Superioris 33
Superior Rectus **33**
   Retractor Bulbi 33
Supinator **15**
Supracostalis 7, **8**
   Supracostalis Posterior 8
Supraspinatus **9**
Temporalis **29**
   Temporalis Minor 29
Tensor Fasciae Latae **22**, 23
Tensor Laminae Posterioris Vaginae Musculi Recti 21
Teres Major 9, **10**
Teres Minor 9
   Teres Minimus 9, 10
Thyroarytenoideus **32**
   Arymembranaceus 32
   Aryvocalis 32
   Superior Thyroarytenoideus 32
Thyrohyoideus **32**
   Cricohyoideus 32, 34
Tibialis Anterior **25**
   Tibioastragalus 25
   Tibiofascialis Anticus 25
Tibialis Posterior 27, **28**
   Tensor Capsuli Tibiotarsalis Anterior 28
   Tibialis Secundus 28
   Tibiotarsalis 28
Transpleuralis **20**
Transversus Abdominis **20**
   Tensor Laminae Posterioris Vaginae Musculi Recti 21
   Transversus Colli 21
Transversus Nuchae **31**
Transversus Perinei Profundus **21**
Transversus Perinei Superficialis **22**
   Gluteoperinealis 22, 23
Transversus Thoracis (Sternocostalis) **19**
Trapezius 7, 9, 10, 11, 31, **33**
   Levator Claviculae (Omocervicalis or Omotrachelien) 10, 33
   Dorsofascialis 7, 33
   Retroclavicularis Proprius 33
   Sternoclavicularis Anticus 8, 33
   Supraclavicularis Proprius 9, 33
   Trachleoclavicularis Imus 33
Triceps Brachii 9, **11**, 12
   Epitrochleo-olecranonis (Epitrochleoanconeus or Epitrochleocubital or Anconeus Sextus 11, 13
   Latissimocondyloideus (Accessorium Tricipitis or Dorsoepitrochlearis) 10, 11
Ulnocarpeus Brevis 14
Zygomaticus Minor **30**

Muscles of Upper Extremity

Thorax, Shoulder, and Arm

M. Panniculus Carnosus

The embryologic origin of these muscles probably accounts for the special degree of variability not only in extent and kinds of attachments but in the variety of separate aberrant slips that are occasionally present. In lower mammals, a great sheet of muscle arises from the early pectoral group and spreads beyond this group. This sheet of muscle, the panniculus carnosus, regresses in higher primates and humans, and its functional significance is diminished since the arm is so free that almost any portion of the body surface may be reached by the hand. The panniculus is represented only by vestigial remnants. Remnants may be found where the pars abdominalis of the *pectoralis major* and *pectoralis minor* come into close relationship. Hence, when found, remnants of the panniculus are seen as extra muscular slips from the abdominal aponeurosis which spreads forward on the rectus sheath, or are attached to the fascia over the *serratus anterior*. Other examples include separate bundles, all of which arise below the *pectoralis major* to be inserted at various points, e.g., the axillary fascia or short of it, the pectoral ridge of the humerus behind the upper part of the pectoral tendon, the fascia between the *coracobrachialis* and *pectoralis minor,* or the coracoid process.

Other slips may be attached to the fascia of the *serratus anterior*. Still others, arising from the border of the *latissimus dorsi*, extend to the fascia of the arm or pectoral insertion. In many cases marginal slips of the *latissimus* are bound up with tendinous remnants of the panniculus and so find secondary attachments to fascia, the humerus, or the coracoid process. The common variety of *axillary arch* is a fleshy slip of varying dimensions, often divided by a tendinous intersection, which extends from *latissimus dorsi* in the axillary fascia to *pectoralis major,* to the short head of biceps brachii, or to the coracoid process. Another sheet of the panniculus, termed *dorsofascialis*, has been reported in a position superficial to the *trapezius* muscle.

References 135, 178, 390, 436, 444, 453, 475, 511, 513, 520, 536, 638, 665, 687, 688, 690–698.

M. Pectoralis Major
M. Pectoralis Minor

From the surface of pectoralis major, some superficial fibers arise from the fascia of the clavicular head and are inserted into the fascia of the deltoid muscle. *M. sternalis* (in 3–5% of individuals) is a fleshy band of longitudinal fibers, often bilateral, of varying length and width, located close to and generally parallel with the sternum. When typical, it arises from the sheath of the *rectus abdominis*, fascia of the *external oblique,* pectoralis major, or costal cartilages (3–7) and ends above in fascia, upper costal cartilages, or manubrium, or may be joined to the sternal or medial head of the *sternomastoideus*. The muscle may exhibit tendinous inscriptions. It is considered by some authors to be a vestige of the *panniculus carnosus*. In a study of 1000 cadavers (438 females and 562 males), M. sternalis was found more frequently in females (8.7%) than males (6.4%). Its reported incidence in the white population is about 4–7%, the black population about 8.4%, and the Oriental population about 11.5%. (See also *M. Supracostalis*.)

The pectoral sheet is divided into a superficial and a deep layer. The outer sheet (pectoralis major) is divided into clavicular, manubrial, sternal, and abdominal portions, all of which may be more or less separable. The clavicular head may extend laterally on the clavicle as far as the *deltoideus* and be fused with it, or its origin may be confined to its sternal end. It is occasionally divided into two distinct parts; it may also decussate across the midline. Its sternal and costal heads may be absent and its complete absence unilaterally and bilaterally has been recorded. *Pectoralis quartus* usually arises near the costochondral junction of the fifth and sixth ribs. It extends laterally along the border of pectoralis major, but is entirely separate from it. It crosses the axilla to be inserted on or near the deep surface of the tendon of pectoralis major. The quartus forms a long flat band with an average width of 1 cm. The origin and insertion of the quartus are frequently joined by a common band of connective tissue to an *axillary arch* and/or to a *sternalis* muscle when these muscles are present. *M. chondrofascialis* (Macalister) extends from the insertion of pectoralis major to the medial intermuscular septum of the arm. Sometimes the level of insertion of pectoralis major on the fascia of the arm is lower than usual. A muscular slip, *costoepitrochlearis, chondroepitrochlearis,* or *chondrohumeralis,* has been described arising from one or more ribs, crossing the axilla, and inserting into the median intermuscular septum or medial epicondyle of the humerus. It may occur in 12–20% of bodies (Le Double).

The deep layer of the pectoral sheet consists of the pectoralis minor and certain muscles that appear only occasionally (see M. Subclavius). Pectoralis minor reportedly originates from ribs 2, 3, and 4 or 2, 3, 4, and 5 in 67% of subjects examined; in 24%, it arises from ribs 3, 4, and 5. It rarely arises from the first or sixth rib. The pectoralis minor sometimes has its insertion extended beyond the coracoid process. The tendon, in such cases, passes over the process, from which it is separated by a bursa, through the coracoacromial ligament to the capsule of the shoulder joint and/or greater tubercle of the humerus. Pectoralis minor may send slips to the *subclavius* and pectoralis major. The unilateral and bilateral absence of pectoralis minor has been reported. *Chondrocoracoideus* (Wood) or *costocoracoideus* arises by one or more slips from ribs 6 to 8 and the rectus sheath to insert with the short head of the biceps.

The manubrial part of the deep layer is normally absent, but may be represented in rare cases by anomalous slips. The *sternoclavicularis anterior* muscles (which Gruber reported in about 3% of bodies) arise from the border of the manubrium and are inserted into the anteroinferior border of the clavicle as far as, or even beyond, the middle of the bone. *Pectoralis minimus* (Gruber) arises from the first rib cartilage and is inserted into the coracoid process. The *tensor semivaginae articulationis humeroscapularis* arises from the first, first and second, second and third, or third and fourth rib cartilages and from the border of the sternum, to be inserted by an aponeurosis into the capsule of the shoulder joint or into the wall of the subacromial bursa. Another slip, *sternohumeralis*, has been observed passing from the first rib cartilage and border of the sternum to the fascia of the arm. Finally, the *sternochondrocoracoideus* arises from the third rib cartilage and/or border of the sternum and is inserted into the coracoid process.

References 9, 13, 16, 18, 36–38, 50, 73, 79, 83, 92, 112, 115, 124, 125, 146, 155, 157, 164, 174, 175, 180, 213, 235, 251, 267, 298, 326, 352, 358, 366, 376–378, 385, 389, 399, 413, 415, 418, 424, 426, 453, 457, 462, 471, 473, 475, 479, 481, 499, 505–507, 510, 519, 526, 556, 560, 568, 569, 581, 582, 595, 620, 626, 638, 640, 646, 649, 658, 672, 674, 682, 685–688, 690–699.

M. Supracostalis

Slips are occasionally found that extend from the ventral end of the first rib to the fourth. Other slips may terminate on the second or third rib. M. supracostalis has also been found attached to the deep fascia of the neck or to the scalene muscles. Still other slips, e.g., *supracostalis posterior*, may be placed dorsal to the origin of *serratus anterior*. They may occur singly or in multiples, bilaterally or unilaterally. They are classed along with the *serratus posterior* muscles as remnants of the superfical lamellae of the primitive external oblique muscle.

References 64, 395, 432, 453, 521, 687, 692, 693, 695, 697, 698.

M. Subclavius

The subclavius may be regarded as a derivative of the deep lamella of the pectoral sheet. Its insertion may spread from the clavicle to the coracoclavicular ligament, coracoid process, transverse ligament of the scapula, or superior border of the scapula and the humerus. It has been described as being replaced by a ligament and by *pectoralis minimus*. It may also be doubled. The subclavius may arise from the second rib and insert on the coracoclavicular ligament.

Another muscle, the *sternoscapularis*, *chondroscapularis*, or *subclavius posticus*, separate from the subclavius and placed behind it, may be regarded as the deepest derivative of the pectoral sheet. It arises from the first rib cartilage, separate from and at a right-angle to the subclavius, or from the edge of the presternum. It passes behind and beneath the clavicle and normal subclavius (above or beneath the subclavian vessels and brachial plexus) to attach to the upper or cranial border of the scapula or to the fascia over the supraspinatus.

Other variations include the *scapuloclavicularis*, a small muscle passing from the root of the coracoid and transverse scapular ligament to the back of the clavicle, and *pectoralis intermedius*, a fleshy slip that arises from the third and fourth ribs between pectoralis major and minor and is inserted into the coracoid process. All these muscles belong to the fifth cervical to first thoracic myotomes. *M. scapulocostalis minor* arises from the scapular notch and inserts on the first rib, and *scapulocostoclavicularis*, a variant of *scapuloclavicularis* cited above, arises from the upper border of the scapula and inserts partly on the clavicle and partly on the first costal cartilage.

A subclavius muscle, originating normally from the junction of the first rib and its costal cartilage, has been found divided into two slips. The upper slip inserted normally onto the middle third of the clavicle. The lower slip, after crossing the cords of the brachial plexus and axillary artery, passed deep to coracobrachialis and terminated by fusing with the tendons of latissimus dorsi and teres major in the bicipital groove.

Several other muscles associated with the clavicle have been described. *M. praeclavicularis* originates from the sternoclavicular joint, courses anterior to the clavicle and across the acromion to insert into the skin over the deltoid, or into the anterior part of the middle of the clavicle. *M. interclavicularis anticus digastricus* has two fleshy bellies joining the two clavicles bilaterally across the manubrium, on which it is inserted by a tendinous intersection covered only by fascia and skin. *M. supraclavicularis* runs over the clavicle, arising from the lateral one-third of the clavicle and/or coracoclavicular ligament to insert on the medial end of the clavicle and/or the interclavicular ligament. (See also the notation of *Supraclavicularis Proprius* listed under M. Trapezius.)

References   12, 34, 35, 99, 123, 131, 162, 166, 167, 209, 210, 213, 215, 238, 250, 265, 271, 272, 281, 395, 416, 444, 445, 453, 476, 513, 530, 541, 552, 591, 630, 687, 690–698.

M. Serratus Anterior

This muscle may receive a slip from the tenth rib. On the other hand, the highest digitation often has no attachment to the first rib, or the lower slips may be absent and the muscle may not descend below the seventh rib. In one study it was found that the number of digitations varied between seven and 12. The last digitation stopped at the seventh rib in 1%, eighth in 40%, ninth in 38%, tenth in 10%, and eleventh rib in 0.5% of cases. The first digitation attached to the first rib in 75%, or to fascia of the first rib in 21% of cases. In 3%, the attachment of the first digitation was the second rib. The entire muscle may be absent. The muscle sheet is composed of three parts, the middle of which may be defective, and the gap between upper and lower parts may be filled with fibrous connective tissue. Rarely, the muscle may be continuous with *levator scapulae*. Some slips passing from the first and second ribs to the vertebral border of the scapula and supplied by the dorsal scapular nerve may be looked upon as belonging to the rhomboid sheet and therefore as variations of those muscles. Other slips may extend to adjacent muscles. Slips arising from digitations on ribs 6 and 7 and joining *pectoralis minor* and *coracobrachialis* have been described as *axillary arch* muscles (see listing for Axillary Arch).

References   430, 433, 453, 514, 687, 690–698.

M. Deltoideus

The clavicular and even the acromial part of this muscle may be missing. Rarely, the muscular sheet may be split into three main parts that are more or less separated from each other; the independence of the acromial part was reported by Macalister. One or more of the parts may be deficient or absent. The blending of *trapezius* with deltoideus may occur. Accessory slips may join the deltoid with *infraspinatus*, *latissimus dorsi*, or *teres major*. Also, additional slips are seen arising from the vertebral border of the scapula or infraspinatus fascia or from the axillary border of the scapula and joining the dorsal part of the muscle or expanding into the fascia of the arm, e.g., *infraspinatohumeralis*. Another head, *costodeltoideus* (Calori), may arise from the lateral edge of the scapula between *teres minor* and *infraspinatus*. Well-developed accessory slips have been observed arising from the spine of the scapula or the infraspinatus fascia and inserting into the humerus between the deltoid and *triceps* or into the fascia of the arm. The deltoid may be joined to *pectoralis major*. It may be connected distally with *brachialis* and *brachioradialis*. The connection with the latter may be partly fleshy and partly tendinous. The acromial portion has been found fused with *brachioradialis*, so that the muscle extended onto the radial border of the forearm. Occasionally some fibers arising from the scapular spine extend superficially over the surface of the muscle and end in the skin of the arm.

References   208, 227, 453, 568, 686, 687, 690–698.

M. Supraspinatus
M. Infraspinatus

The supraspinatus is very constant in form and attachments. Macalister called the muscle singularly invariable. Rare variations, however, include (a) a division of its belly into two parts, (b) an additional slip that arises from the suprascapular ligament, (c) a connection between its tendon and that of *pectoralis minor*, (d) a slip passing from its tendon to that of *pectoralis major* or the pectoral ridge of the humerus, and (e) attachment to a tendinous lamella below pectoralis major.

The infraspinatus is sometimes inseparably united with *teres minor*. *Infraspinatus minor* is a superior fascicle that is distinctly separated from the main muscle mass. *Infraspinatus superficialis* is a fascicle extending from the border of the spine of the scapula to the greater tuberosity of the humerus. A slip connecting this muscle with the posterior border of the *deltoid* (see *M. Deltoideus*) has also been described.

References   453, 687, 690–698.

M. Rhomboideus Minor
M. Rhomboideus Major

The rhomboideus minor has been reported as absent more frequently than rhomboideus major. These two muscles may be inseparable and appear as a single muscle or they may be divided in several fascicles making their identification as discrete entities difficult.

Several slips of muscle have been identified as associated with these muscles: (a) an obliquely placed bundle deep to rhomboideus major, (b) slips to *latissimus dorsi* and *teres major*, (c) a slip, *occipitoscapularis*, passing between the *trapezius* and *splenius* and extending to the occipital bone of the skull, and (d) a slip at the level of rhomboideus minor, termed *rhomboideus minimus*, extending from the scapula to thoracic or lower cervical vertebral spines. (See listing for M. Serratus Anterior.)

References   453, 572, 687, 690–698.

M. Levator Scapulae

The following variations have been described: (a) a variable origin from the first to seventh cervical vertebrae, most commonly, however, from the first and second; (b) slips extending to the temporal or occipital bones; (c) slips to the *trapezius*, *serratus anterior*, and *serratus anterior posterior*; (d) slips to the clavicle, first and second ribs, and spinous process of the first thoracic vertebra; (e) parts of the muscle separated from each other, arising from a vertebral origin to insert on the angle of the scapula (*levator anguli scapulae*); (f) a detached slip from various cervical vertebrae to the lateral end of the clavicle and acromion, corresponding to the *levator claviculae* found normally in other vertebrates; and (g) insertion with *rhomboideus minor*, or bundles to *splenius capitis* or the *scalenes*.

References   453, 687, 690–698.

M. Teres Minor

Absence of this muscle has been reported. The partial separation into two slips has also been recorded: one inserted into the greater tubercle and the other into the neck of the humerus (*teres minimus* of Gruber). The origin of the muscle may be extended so that it entirely covers the infraspinatus and replaces the fascial sheet covering that muscle. Detached fleshy slips sometimes seen extending from the vertebral border of the scapula to the greater tubercle of the humerus are in all probability parts of this sheet, and the fascial sheet may represent it in tendinous form. Various slips described with *deltoideus* appear to be related, suggesting an associated developmental origin of these two muscles.

References   246, 453, 661, 687, 690–698.

M. Teres Major

This muscle is closely related in development to the *latissimus dorsi*. It is part of the same muscle sheet and is often fused with a fasciculus of *latissimus dorsi* at the inferior angle of the scapula. A muscular slip has been observed joining the scapular head of the *triceps*, and also one descending into the fascia of the upper arm on its lateral aspect. An accessory slip originating from the vertebral border of the scapula just below the spine has also been described. In rare cases, teres major may be absent.

References   186, 453, 687, 690–698.

M. Subscapularis

The chief variation is an additional muscle, *subscapularis minor* or *secundus*, which arises from the upper part of the axillary border of the scapula and is inserted into the capsule of the joint, onto the crest of the lesser tubercle of the humerus, or somewhat below the lesser tubercle. It was found by Gruber in 10 of 200 limbs. A rarer variation is a slip arising from the subscapular tendon and passing to the axillary fascia and skin at the base of the axilla. A slip connecting the *subscapularis* to the *pectoralis major* has been observed; sometimes the tendon is connected to the short head of the *biceps* muscle. Other variants include *tensor capsulae*, which arises from an accessory process below the lesser tubercle, crosses the tendon of subscapularis, and inserts into the joint capsule; and *glenobrachialis* (Gruber), which arises with the long head of the biceps from the supraglenoid tubercle of the scapula and inserts onto the surgical neck of the humerus.

References   453, 504, 687, 690–698.

M. Latissimus Dorsi

Variations in the extent of this muscle and its attachments are considerable. It may arise only from the ribs rather than from the vertebral column, ribs, scapula, and crest of the ilium as is most common. The costal origin may be reduced to rib 12. The muscle may be divided into separate fascicles, frequently with one arising from the scapula. The muscle may be continuous with *teres major*. A slip extending from the *latissimus dorsi* may cross the axillary space to the tendon of *pectoralis major* and has been termed the axillary arch, *achselbogen*, or *pectorodorsalis* muscle. This muscle may also be fused with *teres major*. It may extend to the brachial fascia of *coracobrachialis*, to the tendon of the long head of the *biceps*, to the axillary fascia, to *pectoralis minor*, or to the coracoid process. *Axillary arch* muscles were discovered by Ramsay in 1795[1] and have a frequency of about 7% in dissecting room specimens. *Latissimocondyloideus* or *dorsoepitrochlearis*, a muscle found in about 5% of bodies, may pass from the tendon of latissimus dorsi to the brachial and forearm fascia, to the humerus, to the lateral epicondyle and olecranon, or to the long head of the *triceps*. Latissimus dorsi may be absent bilaterally.

References   155, 186, 453, 525, 687, 688, 690–698.

[1] Original reference: *Edinburgh Med. Surg. J.* 8:281, 1812.

Arm and Forearm

M. Coracobrachialis

This muscle is chiefly a flexor muscle, although in the arm it corresponds to the adductor mass seen in the thigh. It develops from a tissue mass forming the *biceps* and *brachialis* and is supplied by the musculocutaneous nerve.

The muscle actually consists of three parts: (a) proximal part arising from the coracoid process and inserted into the humerus close to the lesser tubercle; (b) a middle part of intermediate size; and (c) a distal part, which is the largest and most superficially placed. The muscle may extend to the medial epicondyle or near it, or it may be inserted into a supracondylar process. In humans, the middle part is the most constant, but it is generally accompanied by a part of the distal portion, with the musculocutaneous nerve passing between them. In 3.5–6.5% of arms the muscle is not traversed by the musculocutaneous nerve. The distal part is occasionally found quite separate form the other parts. The proximal part is represented occasionally by a proximal extension of insertion onto the surgical neck of the humerus or capsule of the shoulder joint, or by an accessory head *(coracobrachialis superior* or *brevis)*. The distal portion is sometimes present in the form of the *coracobrachialis inferior* or *longus* (Wood), and may be attached to the humerus, to a fibrous band of the medial intermuscular septum (ligament of Struthers), or even to the medial epicondyle. The muscle is sometimes connected to the *brachialis* muscle, and a fasciculus has been observed joining the medial head of the *triceps*. *Coracobrachialis minor* (*le court coracobrachialis* of Cruveilhier) is an accessory muscle originating from the coracoid process; it crosses the radial nerve in the axilla and inserts into the tendinous part of *latissimus dorsi*. Coracobrachialis may be absent.

References 48, 453, 686, 687, 690–698.

M. Biceps Brachii

This muscle is very variable. In about 12% of arms, a humeral head is found in addition to those normally arising from the coracoid process (medial or short head) and glenoid lip of the scapula (lateral or long head). A biceps with more than two heads is found in about 8% of Chinese, 10% of white Europeans, 12% of black Africans, and 18% of Japanese. Three additional accessory slips may occasionally be found, and they may appear singly or in combination so that the muscle may appear to have three, four, or five heads. The most common slip is the one arising from the humerus at the insertion of the *coracobrachialis,* extending between it and the *brachialis* muscle. It joins the short head, but most of its fibers pass into the semilunar fascia. It may be isolated and end entirely in the fascia. The two other accessory heads are rare and take several different forms. When fully developed, they arise close together from the neck of the humerus, below the lesser tubercle and behind the pectoral tendon to which they may be more or less united. The more lateral of the two slips joins the long head of the muscle, whereas the medial joins the short head.

In other instances, the two heads of the biceps muscle may be totally separate or fused and either head may be absent. In the absence of the long head, the tendon may be found arising from the bicipital groove, one of the tubercles, the capsule of the joint, or the tendon of *pectoralis major*. The tendon may be doubled or it may be represented only by the lacertus. An accessory slip may arise from the deltoid muscle. At the distal end of the muscle, various muscular or tendinous slips have been described connecting it with the lower end of the humerus, ulna, radius, antebrachial fascia, or neighboring muscles. Hence supernumerary heads may also arise from *brachialis, brachioradialis,* or *pronator teres* muscles. Thus a fleshy slip occasionally arises from the medial border of the biceps muscle and passes to the medial intermuscular septum or medial epicondyle over the brachial artery. The brachial artery may pass through this muscle slip or its tendon. Fasciculi have also been seen passing to *brachialis, pronator teres, flexor carpi radialis, flexor digitorum profundus,* or the lateral side of *brachioradialis*. The absence of biceps brachii has been reported.

References 138, 207, 211, 232, 410, 422, 453, 574, 604, 614, 626, 685–687, 690–698.

M. Brachialis

The most frequent varieties of the muscle consist of its subdivision into two or more parts. When the muscle is divided, the distal insertions are irregular and variable. The parts may be attached to the (a) coronoid process of the ulna, (b) radius on or below the tuberosity, (c) radius and ulna, (d) radius and a tendinous band between it and the coronoid process, (e) fascia of the forearm (the part is then named *brachiofascialis* of Wood), or (f) muscles of the forearm arising from the medial epicondyle. A slip from m. brachialis may pass to the semilunar fascia, and a deep detached bundle has been observed attached to the capsule of the elbow joint (*capsularis subbrachialis*). The brachialis is sometimes closely connected or even fused with *brachioradialis* and *extensor carpi radialis*; it may also be absent.

References 199, 453, 686, 687, 690–698.

M. Triceps Brachii

Variations in the triceps are neither numerous nor common. The fibers of the long head may be found extending to the capsule of the joint above or distally for a greater extent than usual on the axillary border of the scapula. A slip (present in 5% of bodies) may join *latissimus dorsi* and corresponds to *latissimocondyloideus* (*dorsoepitrochlearis* or *accessorium tricipitis*), which is common among anthropoids. An accessory or fourth head has been reported arising from the medial aspect of the humerus above or near the deep head, or from the axillary border of the scapula, from the coracoid process, or even from the capsule of the shoulder joint. Occasionally the medial head of the triceps extends to form an arch across the ulnar groove. The tendon of insertion may contain a sesamoid bone, the patella cubiti (sesamum cubiti or "elbow disk").

Epitrochleo-olecranonis (*epitrochleoanconeus, epitrochleocubital,* or *anconeus sextus* [Gruber]), a muscle distinct from the triceps, extends from the medial epicondyle of the humerus and arches across the groove for the ulnar nerve to insert on the olecranon process. It occurs in 25% of bodies, and takes the place of the fibrous arch normally passing between the epicondylar and ulnar heads of *flexor carpi ulnaris*. Triceps brachii and epitrochleo-olecranonis may be fused with *anconeus*.

References 453, 686, 687, 690–698.

M. Anconeus

The anconeus may be united, in varying degrees, with the *triceps, extensor carpi ulnaris,* or epitrochleoanconeus (if present). It may represent a portion of the triceps that has been segmented off in this position. It may also be missing.

References 453, 686, 687, 690–698.

Forearm and Hand

M. Pronator Teres

The most interesting variation of this muscle is the extension of its origin, in a proximal direction, to a supracondylar process or to a ligament that connects it to the medial epicondyle. In these cases the course of the brachial artery is usually changed: accompanied by the median nerve, it passes behind the process, deep to the accessory portion of the muscle, and reaches the antecubital space. In cases of high bifurcation of the brachial artery it is usually the ulnar artery that passes deep to the process or ligament (of Struthers). Even in the absence of a supracondylar process, an accessory bundle or head may arise from the median intermuscular septum, humerus, fascia of the arm, or adjacent muscle. Additional heads may arise from the *biceps, brachialis,* or humerus near the insertion of *coracobrachialis.* The coronoid head is often absent or rudimentary. Pronator teres has also been seen reinforced by fibers from the front of the ulna. Accessory fasciculi may also connect it with *flexor carpi radialis, flexor digitorum superficialis,* or *brachialis.* The two heads of the muscle may be completely separated, and the coronoid and the humeral heads may also be divided (doubled). The radial insertion may be more extensive than usual. Fasciculi may extend to the long flexor of the thumb.

A sesamoid bone may be found in the thick humeral tendon of pronator teres.

References 31, 40, 187, 450, 453, 686, 687, 690–698.

M. Flexor Carpi Radialis

At its origin, the flexor carpi radialis has been observed to receive an additional slip from the tendon of insertion of the *biceps* or its tendinous expansion, *brachialis,* coronoid process of the ulna, or from the anterior oblique line on the radius. Its insertion is subject to frequent variations, the most common being partial or total attachment to the trapezium. It may also be partly inserted into the flexor retinaculum or navicular bone, and a supplementary slip to the third metacarpal may be extended to the fourth. *Flexor carpi radialis brevis,* a small muscle arising from the radius, usually inserts on the fibrous sheath of the tendon of flexor carpi radialis. It was reported present in one of 400 limbs (Gruber) and in six of 70 limbs (Wood). (See also M. Pronator Quadratus.)

Flexor carpi radialis may be absent.

References 233, 453, 687, 689, 690–698.

M. Palmaris Longus

This is probably the most variable muscle in the body. It is absent in 11.2% of bodies (all sources) but has been reported absent in only 3.4% of Japanese and 2.2% of Chinese. Its absence was, reportedly, recorded as early as 1559, by Colombos (*De Re Anatomica Libri*). In a study of 800 living subjects it was absent on both sides in 7.7% of cases, absent on the right in 4.5%, and absent on the left in 5.2%. M. palmaris longus is absent more often in the female and on the left side in both sexes. It may be digastric or fleshy throughout its entire length. It may have a proximal tendon as well as a distal one; it may be fleshy distally and tendinous proximally (palmaris longus inversus); or it may be reduced to a mere tendinous band. Occasionally there are two palmaris longus muscles, one having the normal form and the other having one of the forms referred to previously, or one of the following additional modes of origin. The muscle may arise from the medial intermuscular septum, the *biceps* or *brachialis,* the fascia of the forearm proximally or one of the neighboring muscles, the coronoid process, or the radius. The muscle may be doubled at its proximal end, the additional slip arising from one of the sites just indicated. The insertion is equally variable. It may be attached to the fascia of the forearm, tendon of the *flexor carpi ulnaris,* flexor retinaculum, pisiform bone, scaphoid bone, short abductor of the thumb, fascia and muscles of the hypothenar eminence, one of the flexor tendons, or near the metacarpophalangeal joint. It may enter Guyon's canal and compress the ulnar nerve and artery. Its tendon of insertion may be split into two or more bundles, in which case the accessory tendon or tendons may be inserted into one of the structures named above. The tendon of palmaris longus may enter and insert within the carpal tunnel. It may also substitute for the ring finger slip of flexor digitorum superficialis.

Palmaris longus variations have been classified as (a) complete agenesis, (b) variation in location and form of its fleshy part, (c) aberrancy of attachment at its origin or insertion, (d) duplication and triplication, (e) accessory slips, and (f) replacing elements of similar form or position.

Palmaris profundus is a muscle arising from the lateral edge of the radius, in its middle third, external to *flexor digitorum superficialis* and deep to *pronator teres.* Its tendon passes beneath the flexor retinaculum (to the radial side of the median nerve) and broadens in the palm to insert into the deep side of the palmar aponeurosis. The tendon of palmaris longus may give rise to an additional muscle, *accessorius ad flexorem digiti minimi*, which usually inserts on the body and head of the fifth metacarpal between *abductor digiti minimi* and *flexor digiti minimi brevis*. It has been reported that an accessorius may have two bellies, one arising from the tendon of *flexor carpi ulnaris* and the other arising from the tendon of palmaris longus. A forearm with three palmaris longus muscles has been recorded.

References   2, 29, 75, 96, 163, 182, 185, 223, 231, 233, 384, 397, 407, 453, 480, 491, 531, 540, 561, 592, 599, 601, 626, 631, 685–698.
Additional reference: Madsen, E. Musculus palmaris longus inversus. *Nordisk Medicin* 37:855–856, 1947.

M. Flexor Carpi Ulnaris

In one case, the anterior fibers of this muscle formed an accessory muscle that joined the main belly below the middle of the forearm; it was supplied not by the ulnar but by the median nerve. Because its two heads are completely independent and usually supplied by the ulnar nerve, it has been suggested that the muscle is formed of two fused elements. An additional slip of origin from the medial side of the coronoid process is often present. Partial insertion into the flexor retinaculum has been recorded and its attachment to the fifth metacarpal may extend to the fourth or even to the third metacarpal bone by the radial expansion of a pisimetacarpal band of fibers. Extension of the fibers to the capsule of the fifth metacarpophalangeal joint has been observed. The occurrence of a slip connecting the tendon with the abductor of the little finger has also been reported.

The *epitrochleoanconeus* (*epitrochleo-olecranonis* or *anconeus sextus* [Gruber]) is a small anomalous muscle closely associated with flexor carpi ulnaris. It arises from the posterior surface of the medial epicondyle of the humerus and is inserted into the olecranon process. It is superficial to the ulnar nerve from which it receives its innervation, and takes the place of the fibrous arch of deep fascia usually found in the same location. The muscle has been reported present, however, in about 25% of bodies.

References   183, 293, 365, 453, 687, 689–698.

M. Flexor Digitorum Superficialis (Sublimis)

Some variations in this muscle have been termed retrogressive in that they represent remnants of the connections between two sheets of muscle. Thus, a slip very frequently connects the superficialis with the long flexor of the thumb (*flexor pollicis longus*); muscular slips occasionally pass between the superficial and deep flexors of the fingers. Sometimes a muscular slip derived but separated from superficialis, or one arising directly from the coronoid process, ends on a tendon that joins one of the tendons of the deep flexor (*accessorius profundus digitorum* of Gantzer). Wood reported the muscle present in five of 36 subjects. Tendinous slips frequently connect the tendons of the two muscles. On the other hand, the occasional separation (up to their origins) of individual muscle bellies has been considered a progressive variation. The extensive origin of the muscle from the radius is a human feature. The radial head shows much variability; complete absence has been recorded. One or more of the muscle's distal divisions, e.g., the radial head of the index finger, may be deficient or absent. The belly to the middle finger may be absent or reinforced by a slip from the radial tuberosity. The most common example of this is the absence of the belly for the little finger. Normally, this is the smallest division of the muscle, and it arises from the side of the intermediate tendon. In the absence of the fourth belly of the muscle, its place may be taken by a muscle arising from the medial collateral ligament of the ulna, the *deep flexor*, the transverse carpal ligament and palmar aponeurosis or the ligament alone, or the *fourth lumbrical*. A slip has been seen passing from the superficial aspect of the muscle to the *palmaris longus*, or replacing it if absent, and ending in the fascia of the forearm or palmar aponeurosis. A slip (*radiopalmaris*) may arise directly from the radius beneath the sublimis and attach to the palmar aponeurosis or the common sheath of the flexor tendons.

Some variation occurs in the origin of the tendons from the flexor digitorum superficialis: a slip from the deep part of the muscle sometimes forms the chief part of the ring finger division or there may be a slip from the deep part to the middle finger tendon. The radial head has been seen supplying the tendon to the fourth digit. The tendon for the third digit arises from the extreme radial side of the radial head only and is reinforced by a tendon arising from the epicondyle. An accessory slip is sometimes present passing from the tuberosity of the ulna to the index and middle finger portions. Another slip may arise from the muscle and attach to the tendon of *brachioradialis*. Very rarely, the whole muscle is digastric. Cases in which the medial and deep tendons are replaced by a sheet of muscle (representing a second superficial flexor) and in which the tendon has failed to divide within the flexor sheath have been reported. An anomalous muscle (*palmar flexor digitorum superficialis accessorius*) may arise from the palmar fascia and distal border of the transverse carpal ligament and end in a tendon that joins the flexor tendon of the index finger at the level of the metacarpophalangeal joint. The muscle is innervated by the median nerve.

References   111, 204, 340, 394, 453, 456, 584, 601, 650, 667, 686–688, 690–698.

M. Flexor Digitorum Profundus

The various ways in which the superficial and deep flexors may be united have already been noted under *m. flexor digitorum superficialis*. In addition, a connection with the *flexor pollicis longus* is frequent, generally in the form of a slip passing to the index finger tendon of the deep flexor. The number of bellies of the muscle may be increased. The origin of the belly to the index finger may not be limited to the radius but may reach the interosseus membrane. The separation of the index finger tendon from the others is a distinctly human characteristic, correlated with the specialization of the index finger. The degree of the muscle's freedom varies. It may be quite independent, forming a *flexor indicis profundus*. The radial origin may extend a considerable distance on the bone alongside the long flexor of the thumb. A tendon to the middle finger may be absent. A muscular slip (*accessorius ad flexorem digitorum profundus*) arising from the coronoid process and ending on one or another (usually the index or middle) of the tendons of the muscle has been reported in about 20% of bodies.

References   244, 319, 391, 428, 453, 673, 687, 689–698.

M. Flexor Pollicis Longus

Flexor pollicis longus is uniquely human. In other primates there is only one common deep flexor that provides a tendon to the thumb. In some cases, this condition is exactly reproduced in humans; in other cases, the muscle has been found fused with the index tendon of the *flexor digitorum profundus* (as in the gorilla and chimpanzee). The index tendon may split and two tendons may insert on the index finger; it also may give a tendinous slip to the *first lumbrical*. Another slip may attach to the synovial bursa forming a *tensor bursae mucosae tendinum*. Fleshy slips are sometimes present uniting the belly of *flexor pollicis longus* with *flexor digitorum superficialis* or *profundus*. These, as well as humeral and coronoid heads, are indications of the original connection between the divisions of the deep flexor sheet of muscles. The accessory coronoid head is subject to some variation, occasionally being divided distally into two or three slips passing to *flexor pollicis longus, flexor digitorum profundus, flexor digitorum superficialis*, or *pronator teres*. Pronator teres may give some fasciculi to the flexor of the thumb. Flexor pollicis longus may arise from *brachialis*.

References   152, 353, 428, 453, 458, 656, 673, 686, 687, 689–698.

M. Pronator Quadratus

In very rare instances, even when the radius is normal, the absence of a pronator quadratus has been noted. It is occasionally laminated into two or more layers consisting of fasciculi running in different directions. It may be continued down on the carpus or metacarpus, in some cases as a *radiocarpal* fasciculus, in other cases as an *ulnocarpal* fasciculus. Aberrant fibers arising from the distal edge of the muscle may insert into the thenar muscle mass and act as an accessory adductor of the thumb. Pronator quadratus may also extend in a proximal direction on the radius and may be joined to *pronator teres* or *flexor carpi radialis brevis*.

The *radiocarpeus* (*radiocarpien* [Fano], *court radial anterieur* [Le Double], *flexor carpi radialis brevis vel profundus* [Wood], *flexor carpi radialis brevis*) is a small penniform muscle, present in about 16% of individuals (all sources), which arises from the radius (usually from its volar border and surface) proximal to pronator quadratus. It may extend proximally to the ulna or medial epicondyle and have additional attachment sites. It is inserted by a tendon on one or more carpal bones or on one or more metacarpals, or on both. The *ulnocarpeus brevis*, (*flexor carpi ulnaris brevis* or *cubitocarpeus*) is a small muscle, of rare occurrence, corresponding to and resembling the previously described radial muscle. It arises from the ulna, on its medial or volar surface, and is inserted into the pisiform bone, the hamulus, fifth metacarpal, the capsule of the carpal articulation, or *abductor digiti quinti*. *Radiocubitocarpien* (Calori) (*radialis internus brevis biceps*) arises from two heads, from the radius and ulna, and terminates distally on the carpus. *Radiopalmar* arises from the anterior face of the radius and superior border of the flexor retinaculum and terminates on the deep surface of the palmar aponeurosis.

M. tensor capsulae radiocubitalis (*inferioris*) is a small muscular slip that arises from the volar aspect of the distal radius, crosses superficial to pronator quadratus in a distal direction, and inserts into the radioulnar capsule.

References   95, 228, 234, 293, 307, 314, 320, 384, 452, 453, 687, 689–698.

M. Brachioradialis

The brachioradialis is often united with the *brachialis* at its origin, and its origin may extend proximally as far as the insertion of the *deltoideus*, with which it may be connected. It may also be joined to *extensor carpi radialis longus* and *abductor pollicis longus*. The muscle may be divided, with the second belly attached distally to the radius (in the neighborhood of the tuberosity) or even to the ulna. In about 7% of cases, the tendon is split into two or three slips, occasionally allowing the radial nerve to pass through, and a slip may also extend to the fascia of the forearm. The insertion may ascend on the lateral face of the radius or may be extended distally to the navicular, trapezium, or base of the third metacarpal bone. Brachioradialis may be joined to the *supinator*. An accessory brachioradialis termed *supinator longus accessorius* or *brachioradialis brevis* (Gruber) arises adjacent to brachioradialis, inserts on the radial tuberosity, and acts as a true supinator. It may also insert into the *supinator*, the tendon of *pronator teres*, or onto the ulna. Brachioradialis may be doubled or absent.

References 327, 392, 453, 687, 690–698.

M. Abductor Manus

This rare muscle arises from the common extensor origin, and courses between *extensor carpi radialis longus* and *brachioradialis* to insert on the trapezium.

References 453, 687, 692–694, 696–698.

M. Extensor Carpi Radialis Longus
M. Extensor Carpi Radialis Brevis

In lower mammals the two extensores carpi radiales are represented by one muscle; human cases have been reported in which the muscles were fused, the common belly having two or three tendons inserted into the metacarpal bones. A trigastric radialis longus has been reported inserting partially on the base of the first metacarpal and partially on the base of the second metacarpal. Incomplete fusion frequently occurs, effected by cross slips from one muscle to the other. The longus and brevis are noted for exchange of fleshy and tendinous slips. These connecting fasciculi may be replaced by distinct muscles connected with one or both of the radial extensors. *Extensor carpi radialis intermedius* is a rare muscle that arises independently from the epicondyle of the humerus between the two usual muscles and is inserted into the second and third metacarpals. More frequently, the accessory muscle is represented by a slip from one or both of the radial extensors and is inserted, in the same way, onto the second or third metacarpal or both of these bones. The *extensor carpi radialis accessorius* is an additional muscle sometimes seen arising from the humerus, with or below the long radial extensor, and inserting most frequently onto the metacarpal bone of the thumb. It is represented at times by a slip from the tendon of the extensor longus.

The muscle of the radial extensors may be divided, but more commonly it is one of the tendons that is split into two or sometimes three tendons prior to insertion. The tendons may be attached to the second and third metacarpal bones. Occasionally, a slip also passes to the fourth metacarpal or to the trapezium. Extensor carpi radialis longus may be joined to *abductor pollicis longus* or to some of the *interosseous* muscles.

References 43, 263, 311, 392, 453, 687, 690–698.

M. Supinator

The division of this muscle into two lamellae may be more pronounced than is normally the case. An accessory fasciculus from the lateral epicondyle has been observed, as well as fibers inserted into the tendon of insertion of the *biceps,* the bursa under the tendon, and the tuberosity of the radius.

References 453, 687, 692, 693, 695–698.

M. Extensor Digitorum Communis

Variations of extensor digitorum communis include the occasional deficiency of one or more of the tendons of insertion or an increase in their number to five, supplying the five digits. If one of the tendons is absent it is usually that to the fifth digit. More frequently, however, the tendons are limited to the index or middle finger alone, although an additional slip to the thumb is occasionally seen. Any digit may receive two tendons. The fleshy part of the muscle may consist of two to four bellies, one for each tendon of insertion. Tendinous slips may leave the muscle and join those of *extensor indicis* or *extensor pollicis longus*.

References 43, 243, 273, 282, 299, 300, 345, 453, 686, 687, 689–698.

M. Extensor Digiti Minimi (Proprius) (Manus)

A thin fibrous slip from the lateral epicondyle of the humerus and continuous with the common tendon is occasionally present. More rarely, there is a supplementary origin from the dorsal surface of the ulna. The belly of the muscle may be doubled. The tendon of insertion is sometimes divided and gives a slip to the fourth digit. An ulnar slip has been observed to end on the base of the fifth metacarpal, thus exactly reproducing the arrangement of the serially homologous slip known as *peroneus tertius* in the leg. The muscle may be absent in rare cases, but fusion with *extensor digitorum* is frequent, in which case the tendon comes from the common extensor.

References 43, 309, 310, 315, 328, 342, 442, 453, 686, 687, 689–698.

M. Extensor Carpi Ulnaris

The extensor carpi ulnaris, being the ulnar marginal element of the primitive extensor sheet, corresponds to the two peroneal muscles in the leg. Several of the variations observed are of interest in this connection: (a) the muscle is occasionally doubled throughout, (b) there may be a partial attachment of its tendon to the base of the third or fourth metacarpal bone, (c) an additional tendon may replace an absent *extensor digiti quinti*, and (d) an independent muscle *ulnaris digiti minimi* or *ulnaris digiti quinti* has been observed arising from the lower part of the dorsal surface of the ulna and inserting into the base of the first phalanx of the little finger. This muscle may be represented by a fasciculus from the belly of the extensor carpi ulnaris, or more frequently (44%) by a dorsal slip from its tendon, which may insert into the metacarpal bone, first phalanx, or extensor tendon of the little finger. In many cases (52%), a slip extends from the tendon anteriorly over the *opponens digiti minimi* and inserts into fascia covering the metacarpal bone, the capsule of the metacarpophalangeal joint, or the first phalanx of the little finger. The slip is sometimes joined or replaced by a muscular fasciculus arising from, or in the neighborhood of, the pisiform bone. Agenesis of extensor carpi ulnaris is rare (0.55%).

References   43, 197, 308, 453, 687, 690–698.

M. Abductor Pollicis Longus
M. Extensor Pollicis Brevis
M. Extensor Pollicis Longus

Variations in the muscles of the thumb have been reported to occur in one out of every six subjects dissected. The most common variations occur in the long abductor and consist of an extensive cleavage of its tendon, or even of the whole muscle, into separate parts. On the other hand, it may be reduced to a small slip arising only from the radius. The slips into which its tendon is divided may find attachment to the flexor retinaculum, scaphoid, trapezium, *abductor pollicis brevis*, *opponens pollicis*, or first phalanx, or the tendon may be inserted onto the base of the first metacarpal bone. An additional tendon may give origin to *opponens pollicis*. In one study of 38 forearms, only seven abductor pollicis longus muscles had a single ("normal") tendon, 19 had one aberrant tendon, 20 had two, and six had three anomalous tendons. In the 19 cases of single aberrant tendon, the tendon inserted into the tendon of *abductor pollicis brevis* in 12 cases, inserted onto the trapezium in six, and joined the volar carpal ligament in one. In the 20 cases of double aberrant tendon, 10 inserted into the *abductor pollicis brevis*, eight inserted onto the trapezium, and two inserted into the *opponens pollicis*. In six cases of triple aberrant tendon, two tendons inserted into the *abductor pollicis brevis*, two inserted onto the trapezium, and two inserted into the tendon of *opponens pollicis*.

In another study of 127 dissected forearms, 10.2% of the abductor pollicis longus muscles had a single tendon that inserted upon the radial side of the first metacarpal only. In the remaining 89.8%, the abductor pollicis longus had a dual insertion: into the abductor pollicis brevis, 59%; onto the trapezium, 58%; into the *opponens pollicis*, 15.7%; onto the first phalanx of the thumb, 7.1%; and into the carpal-metacarpal joint, 2.3%.

The extensor pollicis brevis, which is a muscle peculiar to (or individualized to a greater degree in) humans, is sometimes absent, being fused with the abductor longus as in the Simiidae. On the other hand, its tendon is often united with that of the long extensor and inserted with it, or it may be continued as an independent slip to the base of the distal phalanx. In a study of 118 forearms the extensor pollicis brevis inserted onto the first phalanx in 72% of cases, both phalanges in 21.2%, and the terminal phalanx in only 6.8%. The extensor pollicis has been reported absent in 6.3% of cases. In another study, the muscle was absent in five of 120 limbs. The muscle may terminate in two tendons, one of which attaches to the base of the first metacarpal.

Doubling of the extensor longus is not infrequent, and the ulnar portion of the muscle may pass beneath the dorsal annular ligament with the common extensor. It has been reported that extensor pollicis longus is absent in about 1.5% of individuals. A slip from the tendon of the long extensor to the *extensor indicis* is occasionally seen. A rarer variation is an additional extensor between *extensor indicis* and *extensor pollicis longus*, with a double tendon and insertion into both digits. This additional extensor may replace *extensor pollicis longus* or *extensor indicis*.

Abductor pollicis tertius (*extensor atque abductor pollicis accessorius*), a rare muscle, arises from the dorsal aspect of the radius with abductor pollicis longus and inserts, after fusion with abductor pollicis brevis, into the first metacarpal.

References   28, 43, 107, 117, 139, 296, 301, 302, 306, 311, 406, 411, 453, 508, 554, 593, 652, 686, 687, 690–698.

M. Extensor Indicis Proprius

The muscle of the index finger is rarely absent. The origin of the muscle may shift from the ulna to the radius, carpus, or interosseous membrane. An anomalous extensor indicis muscle has been found arising from the lunate, navicular, and capitate bones and terminating on the head of the proximal phalanx of the index finger. It may have two heads or the muscle may be doubled completely. Its tendon is occasionally double and one of the slips may pass, although rarely, to the thumb or ring finger, or more commonly to the middle finger. This last slip, forming an *extensor digiti III*, may occur as a separate muscle (2–5%) arising from the ulna, or from the posterior ligament of the wrist joint below the indicator. An *extensor digiti IV* is a rarer variation. These slips of the deep or short extensor appear to be reversions to a primitive arrangement in which the muscle provides tendons to the whole series of digits. The tendons may also be poorly developed. An *extensor brevis manus* is also found in rare cases. It may appear in two forms: slips arising from the back of the wrist, and slips arising from the carpus or metacarpus and connected to the tendons of interosseous muscles, as well as the tendons of the extensor.

Extensor digiti medii (present in 10% of bodies) arises from the ulna beneath the extensor indicis, with which it may be fused. It sends a tendon to the extensor aponeurosis of the middle finger or sends slips to the middle finger and index finger.

Extensor digitorum brevis or *extensor brevis digitorum manus* resembles the muscle of the same name on the dorsum of the foot. It may have from one to four fasciculi. The most common fasciculus is one that provides a tendon to the extensor of the index finger. This arrangement for the middle finger is nearly as frequent. The fasciculi usually arise from the bones of the ulnar half of the carpus (lunate, triquetrum, hamatum, and capitatum) and from the dorsal ligaments joining the bones. The tendons are inserted either into the corresponding extensor tendons or onto the metacarpals. The muscle is found in about 9% of bodies.

References   43, 53, 59, 60, 97, 149, 196, 296, 301, 302, 357, 386, 442, 453, 460, 508, 545, 583, 589, 601, 622, 629, 652, 684, 686, 687, 689–698.

Hand

Mm. Lumbricales

The lumbricals show frequent variations (20% of individuals). They may be reduced to three or even two, and in rare instances all four are absent. Rarely, the number may be increased by the development of accessory slips. In the case of the first lumbrical, an accessory slip may arise from the tendon of *flexor pollicis longus*, the tendon of *flexor digitorum superficialis*, the first metacarpal, *opponens pollicis*, or the palmar carpal ligament. Cases have also been observed in which a fasciculus arose from the muscular belly of the superficial or deep flexor and joined the first lumbrical. The second lumbrical may arise from the two tendons between which it lies. On the other hand, the third and fourth may arise from one tendon only. Variations of insertion frequently occur with the third and fourth lumbricals. The origin of one or another of the lumbricals may be displaced proximally, arising from the flexor retinaculum, from a special tendon from the deep or superficial flexor, or, in the case of the first lumbrical, from the tendon of the *flexor pollicis longus*. They may be attached in the palm to the superficial as well as to the deep flexor tendons. Frequently, the lumbricals may be bifid distally and insert into two fingers. This occurs in the case of the third lumbrical in about 40% of individuals.

The destination of one or more of the lumbricals may be changed, and one finger sometimes has two lumbricals inserted into it. They may be inserted into the first phalanx instead of or in addition to the normal attachment to the extensor tendon. The fourth lumbrical has been observed to replace the fourth tendon of *flexor superficialis*.

References   42, 43, 86, 150, 200, 339, 453, 464, 488, 535, 550, 563, 601, 687, 689–698.

M. Palmaris Brevis

This muscle is rarely absent (about 2% of individuals), but varies greatly in its size and strength and in its length and direction. It is sometimes found joining *flexor digiti minimi brevis*. It may insert onto the pisiform bone. The palmaris brevis may be doubled and may simulate a soft-tissue tumor of the hand.

References   429, 453, 592, 629, 686, 687, 689–698.

M. Abductor Pollicis Brevis

The abductor pollicis is often divided into an outer and an inner part, or it may receive slips from the navicular bone or styloid process of the radius. A third head arising from *opponens pollicis* has been reported. An accessory slip is frequently found joining the muscle with *abductor pollicis longus*. More rarely, accessory slips connect abductor pollicis with the *palmaris longus*, *extensor carpi radialis longus* (*accessory extensor carpi radialis*), *opponens*, or *flexor pollicis brevis*. A cutaneous slip is frequently present attached to the skin over the proximal part of the thenar eminence, and inserted below on the radial side of the first phalanx. The absence of abductor pollicis has been reported.

References   453, 567, 686, 687, 689–698.

M. Opponens Pollicis

The opponens pollicis is sometimes divided into two parts; its absence has been reported.

References   453, 686, 689, 690, 692, 693, 696, 698.

M. Flexor Pollicis Brevis

The flexor pollicis brevis has been reported absent or incorporated with the *opponens*. The belly of the muscle may be divided.

An interesting pattern of innervation has been reported for this muscle. The superficial head may be innervated by the median or ulnar nerve or both with a frequency ratio of 13:10:7, while the deep head is innervated primarily by the ulnar (16 of 24 cases), by the median (three of 24 cases), or by both nerves (five of 24 cases).

References   134, 353, 453, 458, 554, 687, 689, 690, 692–698.

M. Adductor Pollicis Brevis

The adductor pollicis is a triangular muscle that shows a tendency to split into radial parts. The relative extent of the two parts and the closeness of their connection are subject to variation. Occasionally, fasciculi are seen arising from the palmar metacarpophalangeal ligaments, forming a *transversus manus*. A *contrahentes digitorum* has been reported, which is normally present in amphibia and most mammals but absent from the gorilla, orangutan, and humans. When present in humans, it arises from the carpals or the base of the metacarpals and inserts on the head of the fourth and fifth metacarpals and phalanges. It adducts the fourth and fifth digits toward the midaxial line (third or middle digit). Its embryologic development has been described. Although the muscle as such normally disappears, a part usually persists and is incorpoated into adductor pollicis.

References   43, 360, 381, 453, 592, 687, 689, 690, 692–698.

M. Abductor Digiti Minimi

The abductor digiti minimi is occasionally divided into two or even three slips; in other cases it is united with *flexor digiti minimi brevis*. Its absence has been reported. An accessory head, *accessorius ad abductorem digiti minimi manus*, is frequently present. It may arise from the tendon of the *flexor carpi ulnaris*, from the flexor retinaculum, the tendon of *palmaris longus* (if present), fascia of the forearm, the ulna, or even the radius. In some cases the additional head arises a considerable distance above the wrist from the intermuscular fascia, beneath either *flexor carpi ulnaris* or *flexor carpi radialis*. Passing distally, it may cover the ulnar artery and end in the *abductor* or *flexor digiti minimi brevis*. A portion of the muscle is occasionally inserted into the metacarpal bone or into the palmar ligament of the metacarpophalangeal joint. An *accessory abductor digiti minimi* was found which compressed the deep palmar branch of the ulnar nerve. A small separate slip may extend from the pisiform bone to a metacarpal bone, constituting a *pisimetacarpeus*. The *pisiuncinatus* is a small muscle occurring in 2–5% of hands, extending between the pisiform bone and the hook of the hamate bone. A similar slip, *pisiannularis*, may extend from the pisiform to the flexor retinaculum.

References   43, 81, 242, 294, 343, 453, 592, 687, 689, 690–698.

M. Flexor Digiti Minimi Brevis

This muscle is extremely variable and frequently united with, or replaced by, the *abductor* and sometimes the *opponens digiti minimi*. It may also be replaced by a tendinous band that arises from *flexor carpi ulnaris* and inserts onto the base of the fifth proximal phalanx and hook of the hamate. The absence of *flexor digiti minimi brevis* has been reported. *Tensor capsularis articulationis metacarpophalangei digiti minimi* is a small muscle that arises from the ligaments joining the pisiform and hamate bones. It inserts into the palmar surface of the metacarpophalangeal joint of the fifth digit.

Although not an intrinsic hand muscle, the rare *ulnaris externus brevis* arises 6.5 cm proximal to the distal end of the ulna. It inserts on the fourth and fifth metacarpals not unlike *opponens digiti minimi*, which inserts on the fifth metacarpal, to which it may be functionally related.

References   96, 198, 453, 687, 689–698.

Mm. Interossei

The interossei are occasionally doubled in one or more spaces or they may be absent in one or more spaces.

The palmar interossei insert into the extensor expansion. The first dorsal interosseus usually inserts into the proximal phalanx. The three other dorsal interossei are variable. Most frequently they insert into the extensor expansion and proximal phalanx. The second dorsal interosseus inserts into the proximal phalanx more frequently than into the fourth, and into the fourth more frequently than into the third. In about 3% of cases the first dorsal interosseus is innervated by the median nerve. The second dorsal interosseus may have three heads.

References   43, 78, 381, 409, 414, 453, 555, 687, 689–698.

Muscles of Trunk

Back, Head, and Neck

M. Splenius

The absence of this muscle has been recorded in at least two cases. Its origins may be displaced proximally by one or two vertebrae, and frequently the origins do not correspond on both sides. The spinal origin may extend to the cranial end of ligamentum nuchae. The muscle is sometimes separated into two parts, the superior part overlapping the inferior at its origin. Slips joining the *longissimus capitis*, *iliocostalis*, or *levator scapulae* are sometimes found. The sheet may be broken up into separate parts forming a mastoid and an occipital division. Accessory slips, in rare cases, pass superficial to *serratus posterior superior* and may be more or less connected with splenius. One such slip, *rhomboatloideus* or *splenius colli accessorius*, arises from the lower cervical and upper thoracic spines with *serratus posterior superior*, passes superficial to that muscle along the border of the splenius, and inserts on the transverse process of the atlas. Its reported frequency is about 8% (Le Double). When this slip inserts onto the occipital bone or mastoid process it is named *splenius capitis accessorius*. In another report, *splenius colli accessorius* or *rhomboaxoideus* was found arising from the fascia of *rhomboideus minor* and *serratus posterior inferior*.

The splenius represents a superficial prolongation of the lateral column of muscles to the head. Its attachment to the spinous processes of the vertebrae is probably due to a dorsal extension of the muscle, associated with a change in the direction of its fibers. This is supported by the observation that the nerve supply of splenius is derived from lateral branches of the posterior division of spinal nerves, which are limited to cervical muscles.

References   226, 401, 453, 687, 690–698.

M. Longissimus

When the tendinous slips attached to the lumbar spines are cut, the deep medial surface of the thoracolumbar part of longissimus is exposed, revealing a variable number of slender accessory slips of origin. These come from the transverse processes of the lower thoracic vertebrae, mamillary processes of the lumbar vertebrae, or aponeurosis of the underlying *transversospinalis*. These may unite to form an independent muscle, *transversalis dorsi* or *thoracis*, with a separate insertion onto the upper two or three thoracic vertebrae. Similarly, accessory slips may be seen in the cervical section of the column arising from the roots of the transverse processes of the cervical vertebrae and forming, when more or less independent, the *transversalis cervicis posterior minor*.

The attachments of the components of the longissimus are subject to considerable variation. The costal attachments may be reduced either at the cranial or caudal end of the series; the longissimus cervicis may extend to the tenth or eleventh transverse process, or its slips may be reduced in number. The cranial section may be absent; it has been found divided into two parts. Its slips may be limited to the thorax or to the cervical vertebrae. Slips detached from longissimus capitis have been observed originating from the transverse process of the atlas, or from the tendon of the cervical part of *longissimus* going to the atlas, and inserting onto the mastoid process.

References   453, 687, 690–698.

M. Semispinalis Capitis

The slips of attachment vary considerably in number. The muscle may receive accessory slips from *longissimus* or *spinalis (dorsi) thoracis*. A rare variation is the presence of a second lamella under the medial portion of the muscle, extending from one or more of the upper thoracic spines to the occipital bone. Another slip of muscle, *biventer cervicis*, is superficial and located medially, may appear to have a tendinous intersection, but remains fleshy; it may have two distinct bellies, or two distinct bellies joined only by a strong tendon (several centimeters in length). It arises from lower cervical and upper thoracic vertebrae and spines of transverse processes, and inserts onto the occipital bone and inion.

References   453, 687, 690–698.

Mm. Suboccipitales

Division or doubling of the rectus muscles has been reported as frequent, although doubling of the oblique muscles is rare. Accessory slips may be found as a continuation of *spinalis cervicis*. They extend from the spines of one or two cervical vertebrae and ligamentum nuchae to the occipital bone. A small muscle has been found passing from the spine of the axis to the mastoid process along the lateral border of the inferior oblique. The *atlantomastoid* is a small muscle arising from the transverse process of the atlas, passing between the superior oblique and rectus lateralis, and inserting onto the mastoid process.

References   253, 261, 453, 686, 687, 690–698.

Thorax

M. Serratus Posterior Inferior
M. Serratus Posterior Superior

There is considerable variation in the number of digitations of these muscles. The number of slips of the serratus superior may be reduced to three; less frequently the number may be increased to five or six slips. The serratus inferior may be reduced to two digitations, the first and the last being absent. When increased to six, they arise from ribs 1–6. Absence of the superior and the inferior muscle has been recorded but this is extremely rare; they are replaced by fibrous tissue. Cases have been described in which the serratus inferior showed irregular arrangements of its digitations as a result of associated accessory muscle tissue.

References   514, 651, 687, 690–698.

M. Subcostalis
M. Transversus Thoracis or Sternocostalis

The subcostals are usually better developed on the inner surface of the thorax and are variable in number.

Transversus thoracis arises from the posterior surface of the sternum and xiphoid (to the level of the third costal cartilage) and extends onto the internal surface of the second (third) to sixth costal cartilages. Transversus thoracis may extend to the first rib. Absence on one or both sides has been recorded. It is sometimes divided into separate fasciculi which may number from two to six, and its division into two lamellae has also been observed. This muscle was Hyrtl's candidate for the most variable in the body.

References   453, 514, 687, 690–698.

M. Diaphragma

The sternal portion of the diaphragm may be absent,

and the costal origins extend to the sixth or even to the fifth rib. The slip from the twelfth rib may be absent; if this is associated with an absence of the fibers from the lateral arched band, the trigonum lumbocostale becomes an extensive gap. This condition in exaggerated form occurs in congenital diaphragmatic hernia. In such cases, it may be noted that the lumbar portion of the diaphragm extends anteroposteriorly on the medial side of the aperture. The extension of muscle fibers onto the fascia of the *psoas* or *quadratus lumborum* occasionally occurs. Connections with the *transversus abdominis* have been reported. Various detached or aberrant bundles of muscle fibers have also been described. They may compress the renal artery. From the right crus, a slip may extend in the suspensory ligament of Treitz to the back of the duodenum near the duodenojejunal junction, with an extension to the mesentery along blood vessels. From the medial borders of the crura, a band of connective tissue containing muscle fibers may be joined to the surface of the esophagus. Frequently, detached bundles of muscle fibers are found in the tendinous center, located on the surface or between the layers of the aponeurosis. Partial duplication of the diaphragm has been reported. Transverse muscle bundles have been found running anteroposteriorly to the aorta or central tendon. A muscle, *hepatodiaphragmaticus*, has been reported arising from the lateral aspect of the central tendon, anterior to the esophagus and to the right of and beneath the liver, to form a tendon that blended with the obliterated ductus venosus and umbilical vein.

References 348, 349, 441, 453, 558, 585, 627, 687, 688, 690–698.

M. Transpleuralis

Muscular bands have been reported in the oblique fissure of the right lung, which originate from the sixth rib and extend to the central part of the diaphragm just behind the caval opening.

References 8, 201, 453.

Abdominal Wall

M. Obliquus Externus (External Oblique)

The external oblique varies in the number of its attachments to the ribs. Absence of the highest and lowest digitation is common. On the other hand, one or more slips may be doubled, most frequently those from the eighth or ninth ribs, or an additional slip may arise from the lumbar fascia below the last rib. In addition to the connection with the *pectoralis major*, there may be union, frequently by tendinous intersection or more rarely by direct continuity, between the slips of origin of the external oblique and *serratus anterior* muscles. A similar connection may exist between external oblique and *latissimus dorsi* or *serratus posterior inferior*. In some cases, deep accessory bundles arise from the ribs and are separated by connective tissue from the main sheet of muscle. These slips arise from one or more of the lower ribs and are attached independently at their distal end to the iliac crest, anterior superior iliac spine, inguinal ligament, or sheath of the rectus. Detached slips may arise beneath the digitations from the fourth to the seventh ribs and from the fascia over the fifth/sixth intercostal space, and end in the upper part of the rectus sheath. A detached slip, the *saphenous* muscle, has been described, attached at both ends to the inguinal ligament and looping around the saphenous vein. The presence of a tendinous intersection in the external oblique muscle is interesting in view of its segmental character, as evidenced by the regularly arranged tendinous intersections in the muscle found in many mammals.

References 249, 341, 453, 641, 687, 690–698.

M. Obliquus Internus (Internal Oblique)

A fibrous band or inscription frequently interrupts the internal oblique muscle opposite the tenth or eleventh rib. In rare instances a similar inscription has been seen opposite the twelfth or even the eighth rib. They apparently represent tendinous intersections between the internal oblique and the *internal intercostal* muscles. An additional slip occasionally arises from the eighth costal cartilage. It has been reported that the internal oblique inserts onto the eleventh rib in 1% of cases, the tenth in 66.5%, the ninth in 31%, and the eighth rib in 1.5% of cases. Absence of the inguinal portion or of the anterior superior part of the muscle has been recorded. Fusion of the lower part of the muscle with the *transversus abdominis* may occur. A deep slip was found inserted on the transverse process of the second lumbar vertebra in one case. The part of the internal oblique muscle that arises from the iliac crest is often thick and may be divided into an anterior and a posterior part (*accessory internal oblique*). Macalister reported a highly developed internal oblique that was traversed by the spermatic cord.

References 239, 453, 687, 690–698.

M. Cremaster

In the female this muscle is represented by a few fasciculi on the round ligament. It may arise from the transversalis fascia or may be fused with the *transversus abdominis*, which may then take part in the formation of the muscle. The muscle may arise from two heads.

References 453, 687, 690–698.

M. Transversus Abdominis

The number of costal slips varies: it may be reduced to five or increased to seven. The muscle has been found fused with the *internal oblique*; it has also been reported absent. Absence of the inguinal bundles has been recorded; on the other hand, extension of the transversus to the pubic crest has been noted, and the spermatic cord has been seen piercing its lower border. The insertion onto the seventh rib may be absent, but in some cases there may be a supernumerary digitation for the sixth rib. The muscle bundles may extend as far inferiorly as the spermatic cord, they may remain superior to the anterior superior spine, or may have an intermediate position. Beneath the transversus, supernumerary muscle slips, e.g., *tensor laminae posterioris vaginae musculi recti*, have been reported.

Based on a study of 200 body-halves, it has been reported that, while the *internal oblique* arises from almost the entire inguinal ligament, the transversus abdominis seldom (3% of cases) arises from the entire inguinal ligament. The inferior edge of the transversus seldom forms an arch but it is not the roof of the inguinal canal. The *internal oblique* and transversus unite to form a conjoined aponeurosis rather than a conjoined tendon. The conjoined tendon has been defined (Cunningham) as the union of the aponeuroses of internal oblique and transversus muscles, which arise from the inguinal ligament and insert medially into the linea alba and inferiorly onto the pubic crest and pectineal line. The term conjoined tendon is a misnomer in most cases (about 95%). When the fused aponeuroses of the two muscles are unusually thickened (about 5% of cases), the fibers curve downward in front of the spermatic cord and insert onto the iliopectineal line.

A curious muscle, *transversus colli*, described by Luschka as a cranial extension of transversus abdominis (and thoracis), arises from the upper edge of the first costal cartilage, courses between *sternohyoideus* and *sternothyroideus*, and inserts into the interclavicular ligament or capsule of the sternoclavicular joint and surrounding fascia.

References   103, 104, 237, 287, 453, 683, 687, 690–698.

M. Rectus Abdominis

The rectus muscle of lower mammals extends further over the thorax than that of the human subject. In many of the lower primates, it reaches the first rib and the clavicle. It may rise to the fourth or even the third rib in humans, and in certain very rare cases even higher. On the other hand, the slip attached to the fifth rib may be absent. The muscle usually inserts onto two or three ribs. In one study, six types of insertion were found. In 67%, the insertions were ribs 4 and 5; 4, 5, and 6; 5 and 6; or 5, 6, and 7. In 20%, the insertions were ribs 6 and 7; and in 10%, ribs 6, 7, and 8. Slender muscular slips are sometimes found between the rectus and the deep inguinal ring. Absence of a part or of even the whole muscle has been reported. It may also be doubled.

References   110, 146, 268, 283, 286, 432, 453, 469, 588, 687, 692, 693, 695, 697, 698.

M. Pyramidalis

This muscle varies greatly in length (20–138 mm) and size. It is often absent on one or both sides, in which case the size of the lower part of the *rectus* is often increased. It is rarely symmetric bilaterally. It is absent in about 8% of all subjects studied, more frequently in females and more often on the right side. Reported frequencies are 82% in white populations, 90% in blacks, 95% in Japanese, and "always" present in Chinese. In some instances, the muscle has been found doubled and may extend upward as far as the umbilicus. It is usually supplied by the subcostal nerve and occasionally by the iliohypogastric.

References   15, 22, 44, 453, 687, 690, 692, 693, 695–698.

Lumbar Region

M. Quadratus Lumborum

The muscle varies according to the degree of development of its vertebral slips. The insertion of the muscle has been found to extend to the eleventh rib.

References   453, 687, 690, 692, 693, 695–698.

Pelvic Diaphragm and Anus

M. Coccygeus

The coccygeal muscle varies greatly in the extent of its fleshy development. It may be doubled or may be partially fused with *levator ani*. Occasionally it is absent. *M. curvator coccygeus accessorius* arises from the lower anterior part of the transverse process of the fourth sacral vertebra and inserts onto the edge of the third and fourth coccygeal vertebrae.

References   333, 453, 662, 687, 690–698.

M. Levator Ani

The muscle is divisible into three parts: iliococcygeus, pubococcygeus, and puborectalis. Because it shows great individual variation in structure, this muscle is described differently by different authors and these descriptions may differ substantially from a dissected specimen.

References   27, 453, 687, 688, 690–698.

M. Sphincter Ani Externus

This muscle shows considerable individual variations in structure.

References   453, 687, 690–698.

Urogenital Diaphragm

M. Transversus Perinei Profundus

This muscle is variable in structure; it may also be absent, more frequently in females.

References   453, 687, 690–695, 697, 698.

M. Sphincter Urethrae

There is considerable variation in the muscles composing the urethral sphincter.

References 453, 687, 690–698.

M. Transversus Perinei Superficialis

This muscle is frequently absent or poorly developed. Sometimes it consists of several separate fasciculi, and it may be inserted wholly or in large part into the *bulbocavernosus* or *external sphincter*. A muscular slip is occasionally found arising from the fascia at the lower border of the *gluteus maximus* and running into the transversus perinei.

A muscular slip is described (*gluteoperinealis*) that arises from the aponeurosis of *gluteus maximus* or upper fascia lata and inserts under the origin of *bulbocavernosus* in the albuginea of the bulbous urethra.

References 260, 262, 453, 687, 690–698.

External Genitalia

M. Bulbospongiosus

This muscle is variable in structure judging from the descriptions given by various authors.

References 453, 687, 690–698.

M. Ischiocavernosus

This muscle is obviously larger in the male. Some of the more anterior fiber bundles may extend to the dorsal surface of the penis and form a *pubocavernosus* or *levator penis* muscle. A *compressor venae dorsalis* has been described, composed of a few fasciculi that arise from the sheath of the corpus cavernosum urethrae and from the median raphe, and unite with those of the opposite side by a tendon that passes over the dorsal vein.

References 368, 453, 687, 690, 696, 698.

Muscles of Lower Extremity

Hip

M. Iliopsoas (Iliacus and Psoas Major)

The psoas and iliacus are occasionally completely independent muscles. The psoas may be divided longitudinally into fascicles. An accessory slip has been described lateral to the muscle and separated from it by the femoral nerve. Iliacus may be pierced by the femoral nerve. The *iliacus minor* or *iliocapsularis* (third head of iliacus), a small detached portion of the iliacus, is frequently present. It arises from the anterior inferior spine of the iliac bone and inserts into the lower part of the intertrochanteric line of the femur, or into the iliofemoral ligament. A slip from iliacus may run medial to the psoas muscle.

References 69, 116, 453, 687, 690–698.

M. Psoas Minor

This muscle is not constant in humans. In a report of 182 subjects it was present on both sides in 70 subjects, on the right side in 12, on the left side in eight, and absent on both sides in 92. Various sources report that it is present in about 56% of bodies. In one study of 2627 cadavers, bilateral agenesis occurred in 54.5%. The muscle was absent in 49.9% of 722 body-halves in Orientals, 57% of 4507 body-halves in whites, and 66.6% of 674 body-halves in blacks. When present, it varies considerably in its site of origin. It may be connected only with the first lumbar vertebra, or with the second lumbar and the intervertebral disc above it; it has been observed to arise from two heads, separated from each other, or to be divided throughout. At its insertion it may end on the fascia iliaca, inguinal ligament, neck of the femur, or lesser trochanter with the *psoas major*. The tendon of insertion may bifurcate, with one terminating as usual and the second attaching to the synchondrosis between the fifth lumbar vertebra and the sacrum.

References 453, 571, 687, 690–698.

M. Tensor Fasciae Latae

This muscle is occasionally found divided longitudinally, partially or entirely. Accessory slips have been described proceeding from the iliac crest, inguinal ligament, or aponeurosis of the abdominal wall. Union with *gluteus maximus* has been observed. This muscle is considered by some authors to be an atrophied relic of a "deltoid" of the hip. It may be absent.

References 295, 453, 687, 690–698.

M. Gluteus Maximus

Variations are rare. The most medial fibers may separate to insert on the lateral lip of the linea aspera. The muscle may receive an additional fasciculus of origin from the lumbar aponeurosis or ischial tuberosity. This fasciculus may be independent (*ischiofemoral* muscle). A distinct slip (*coccygeofemoral*) at the lower border, arising from the coccyx and attached to the femur, may also be found and represents a caudal head. The fibers arising from the sacrotuberous ligament and the margin of the sacrum are normally separated from the superficial part by a layer of areolar tissue. A very rare variation is the fusion of gluteus maximus and *tensor fasciae latae*.

References 333, 425, 453, 687, 690–698.

M. Gluteus Medius

This muscle may be divided into two lamellae, or the deeper fibers may end on a separate tendon that inserts into the upper border of the greater trochanter. The muscle may supply slips that join the tendon of *piriformis*. A *gluteus medius accessorius* has been described as arising from the iliac crest under *tensor fasciae latae* and *gluteus medius* and inserting on the greater trochanter partly with medius and partly independently.

Gluteus quartus is a rare muscle arising between gluteus medius and minimus and inserting on the tip of the greater trochanter. A *gluteoperinealis* muscle has also been reported.

References 279, 383, 453, 643, 687, 690–698.

M. Gluteus Minimus

Detached slips may connect the muscle with the *tensor fasciae latae, superior gemellus,* upper part of the origin of the *vastus lateralis,* or the capsule of the hip joint. This last form may appear as an accessory muscle under cover of the main mass. The anterior and posterior parts may be separate, or the anterior fibers may detach from the rest and form an independent muscle, the *gluteus quartus*. Gluteus quartus may be homologous to invertor femoris of apes.

References 329, 453, 687, 690–698.

M. Piriformis

The most frequent variation is the division of the muscle into two parts. This is usually associated with a high division of the sciatic nerve, with the peroneal portion of the sciatic emerging between the two parts. The muscle is rarely divided by the tibial or by the entire sciatic nerve. The cleavage generally affects only the muscle belly, but may involve the entire muscle. Subdivision into several fascicles has also been observed. The muscle has been reported absent. Fusion frequently takes place between *piriformis* and *gluteus medius, gluteus minimus, superior gemellus,* or in rare cases even the *internal obturator* muscle. Its digitations may be reduced to two or increased to four. It may insert onto the capsule of the hip joint.

References 102, 177, 453, 477, 687, 690–698.

M. Obturator Internus
Mm. Gemelli

The obturator internus muscle is not subject to much variation. Complete division into two parts, as well as accessory slips, have been noted. The gemelli vary in size. The superior is small or absent, or may be doubled and may insert onto the capsule of the hip joint. The inferior gemellus may also be doubled; its absence has also been reported, but more rarely than that of the superior muscle. In very rare cases both gemelli may be absent. The superior gemellus may be fused with *piriformis* or *gluteus minimus*, and the inferior gemellus with the *quadratus femoris*. Both are frequently more or less continuous with the obturator internus.

References 277, 304, 453, 624, 687, 690–698.

M. Quadratus Femoris

The quadratus may be fused with the *inferior gemellus* or *adductor magnus,* replaced by a large inferior gemellus or *obturator internus,* or absent altogether (1–2%). It may be divided at its femoral insertion into two parts: one posterior, with the normal attachment, and the other anterior, with insertion into the intertrochanteric crest.

References 49, 284, 453, 600, 687, 690–698.

Thigh

M. Biceps Femoris

There is support for the suggestion that the two parts of the biceps represent two separate muscles that have been united in humans. In rare cases the parts remain separate either partially or entirely, or the short head may be absent. The normal attachment of the muscle to the crural fascia represents an extension to the leg that is found in some mammals. The crural extension may be seen as muscular slips (*tensor fasciae suralis*) attached to fascia, *gastrocnemius*, or tendo achillis. Accessory slips have been noted, which arise from various sources including the sacrum, coccyx, sacrotuberous ligament, ischial tuberosity, *gluteus maximus*, gluteal ridge of the femur or linea aspera at its upper extremity, fascia lata, *gastrocnemius* and lateral condyle of the femur, *adductor magnus,* or *vastus lateralis*. In 41% of 62 limbs, the long head took origin from the sacrotuberous ligament.

References 30, 41, 195, 230, 240, 321, 361, 364, 367, 400, 453, 477, 490, 687, 690–698.

M. Semitendinosus

Complete separation of the semitendinosus from the *biceps* has been reported. The muscle is sometimes partly fused with *semimembranosus* and cases have been reported in which the *biceps* and semitendinosus were joined by a muscular slip. Accessory slips have been observed arising from the coccyx, sacrotuberous ligament, or ischial tuberosity and joining the belly of the muscle. In 39% of 62 limbs, the semitendinosus took origin from the sacrotuberous ligament. A rare variation is one in which a slip arises from the linea aspera below the semitendinosus and ends on the medial condyle and back of the knee joint. *Tensor fasciae suralis* or *ischioaponeuroticus* is a slip that leaves the belly of the semitendinosus (it may also be one of the other hamstrings) and ends in a tendon that joins the fascia of the leg.

References   205, 230, 240, 322, 453, 687, 690–698.

M. Semimembranosus

Reduction and complete absence of the semimembranosus have been recorded. It may be doubled or split along its length. The second belly may have the form of a small fleshy slip with proximal and distal tendons, or it may arise from the sacrotuberous ligament. The distal tendon may be inserted in various ways. In one case, it was attached by one slip to the main tendon, with a larger slip to the oblique popliteal ligament and a delicate reflected slip inserted onto the medial supracondylar ridge. Sometimes the tendon joins that of the *adductor magnus*. The muscle may consist of up to four bellies with the extra bellies inserting into the *adductor magnus*, planum popliteum, and tibia. Partial fusion with the *semitendinosus* was mentioned under *M. semitendinosus*. As in the case of the other hamstrings, extension of the proximal attachments to the coccyx has been reported.

References   100, 194, 205, 323, 324, 453, 687, 690–698.

M. Sartorius

In rare cases, this muscle may be completely divided longitudinally into two parts, the accessory portion being inserted into the fascia lata, the femur above the medial condyle, the ligament of the patella, the tendon of *semitendinosus*, or the tendon of its twin. Separate or accessory heads of origin occasionally arise from the notch below the anterior superior iliac spine, the iliopectineal line, the pubic bone close to the symphysis, or the inguinal ligament. The tendon of insertion has been found to end in the fascia lata, capsule of the knee joint, and fascia of the leg. The absence of the muscle has been reported.

References   57, 74, 453, 687, 690–698.

Mm. Quadriceps

The deep parts of the quadriceps are not subject to many variations. The vasti are sometimes bilaminar. This, in the case of the vastus lateralis, is an accentuation of a normal feature. The vasti may be continuous with the rectus femoris, and all three parts may be united into a common fleshy mass. The absence of rectus femoris, as well as the entire muscle mass, has been reported. Rarely, a slip (*rectus accessorius*) may arise from a tendon at the edge of the acetabulum and insert into the ventral edge of vastus lateralis. The quadriceps of the thigh corresponds in a general way to the triceps group (including the anconeus) in the arm.

References   147, 276, 292, 344, 453, 675, 687, 690–698.

M. Articularius Genus

This muscle arises deep to *vastus intermedius*, from the distal third of the femur, and inserts into the fibrous portion of the articular capsule of the knee joint at the level of the suprapatellar bursa.

References   145, 453, 687, 696, 698.

M. Pectineus

The pectineus may be divided into two parts, each of which receives its nerve supply from different sources: the medial from the obturator and the lateral from the femoral. This division occurs normally in many mammals including some primates. Pectineus may be supplied entirely by an accessory obturator nerve. A common variation is union with *adductor longus*. Accessory fasciculi may unite the pectineus with the *obturator externus*, the *iliacus*, the capsule of the hip, or the lesser trochanter.

References   453, 687, 690–698.

M. Gracilis

Occasionally, the gracilis supplies fibers to the fascia lata and to the fascia of the leg. A *biceps gracilis* has been reported with a normal upper head and a lower head arising from the ramus of the ischium.

References   334, 453, 633, 687, 690–698.

M. Adductor Longus
M. Adductor Brevis
M. Adductor Magnus

In humans, these three muscles may be subdivided into a greater number of parts than is usual. Thus, the adductor longus may be doubled, the adductor brevis is frequently found divided into two or three parts, and the adductor magnus is a complex mass. The adductor magnus may be fused with the *quadratus femoris* or with the long or the short adductor. Irregular slips from both adductor magnus and biceps femoris insert into the tendon of adductor magnus and onto the adductor tubercle. Adductor longus may extend as far as the knee, with its tendon of insertion inseparable from that of adductor magnus. Adductor brevis may provide a slip to *obturator externus*.

References   364, 453, 687, 690–698.

M. Obturator Externus

The external obturator muscle shows little variation. Sometimes a part of the muscle is isolated above the point of exit of the obturator nerve. The muscle may be divided into three parts; this subdivision has been described by some authors as the normal arrangement. It may be connected with *adductor brevis*.

References   453, 654, 687, 690–698.

Leg and Foot

M. Tibialis Anterior

The tendon is frequently found divided into a cuneal and metatarsal slip; in very rare instances, this division may involve the whole muscle. In a few cases, a deep portion has been found inserted into the talus (m. *tibioastragalus* of Gruber), and the tendon also supplies a slip to the head of the first metatarsal or base of the first phalanx. A separate muscular slip, the *tibiofascialis anticus,* sometimes occurs. It arises from the lower end of the tibia and ends on the annular ligament or deep fascia; it may be represented only by a tendinous slip from the tibialis anterior itself.

References 143, 330, 453, 482, 553, 598, 687, 690–698.
Additional references: *Bahnsen, G. Zeitschrift für rationelle Medicin, dritte Reihe, Bd. 3, S. 56, 1868. Leipzig und Heidelberg.
Seelaus, H. K. On certain muscle anomalies of the lower extremity. *Anat. Rec.* 35:185–191, 1927.

M. Extensor Hallucis Longus

This muscle is occasionally united at its origin with *extensor digitorum longus*. The tendon may be divided into three slips confined to the great toe; a slip to the second toe has also been observed. Occasionally this muscle, which is usually unipennate, has three bellies and three tendons. When one of these three bellies represents a separate muscle, it is named *extensor hallucis longus minor*. Doubling of the muscle may occur, with the extra muscle running lateral to the main muscle. It may join *extensor hallucis brevis* distally or insert independently onto the first metatarsal, the proximal part of the first phalanx, or both phalanges of the great toe.

References 43, 245, 256, 453, 687, 690–698.

M. Extensor Digitorum Longus (Pedis)

Variations in this muscle relate mainly to the origin and arrangement of its various tendons. They may be doubled, or extra slips may arise from one or more tendons to their corresponding metatarsal bones, to the *short extensor,* or to one of the *interosseous* muscles. A slip to the great toe from the innermost tendon has also been seen. In contrast to the corresponding tendons in the hand, all are independent of their neighbors; there are no connecting bands between them. The muscle belly supplying the fifth toe may be reinforced by *peroneus tertius* and may also contribute to *extensor hallucis longus* or *extensor hallucis brevis*.

References 43, 54, 243, 252, 346, 453, 687, 690–698.

M. Peroneus Tertius

The peroneus tertius varies in size: it may be as large as the *extensor digitorum longus* or it may be absent. This muscle is peculiar to humans, and represents the *extensor digiti minimi (proprius) (pedis)* with its insertion displaced to the base of the fifth metatarsal. It often inserts on the base of the fourth as well. It is not so completely separated from the common extensor at its origin as is the corresponding muscle of the forearm. A slip occasionally joins the extensor tendon of the fourth and little toe or fourth interosseus, and represents the radial tendon of the *extensor digiti minimi (proprius)*; its main tendon corresponds to the ulnar slip in the hand.

References 54, 453, 586, 596, 687, 690–698.

M. Extensor Digitorum Brevis (Pedis)

The short extensor of the toes is both ontogenetically and phylogenetically a fibular muscle. Its position on the dorsum of the foot is secondary. The muscle may have three distinct heads. An accessory fasciculus originating from the talus and navicular bone or from the third cuneiform and third metatarsal bones has been found inserting into the second slip of the muscle, as well as a fasciculus from the cuboid to the third slip. The tendons vary in number. They may be reduced to two or even one, and cases have been

Accessory Insertions of the Tibialis Anterior Tendon

Investigator	Year	No. of specimens studied	Base of proximal phalanx of big toe or metatar-sophal. articul.	Distal part of metatarsal I	Fascia dorsalis pedis	Plantar aponeurosis	Ankle joint	Total
Wood	1867	?	50%	–	–	–	–	>50%
Knott	1877	30	1 (3%)	2 (6.6%)	2 (6.6%)	–	–	5 (16.6%)
Le Double	1897	42	–	–	1 (2.4%)	–	–	1 (2.4%)
Hallisy	1930	290	9 (3.1%)	3 (1%)	–	–	–	13 (4.5%)
Musiet	1965	122	4 (3.3%)	–	2 (1.7%)	–	–	6 (5.2%)
Stepien	1967	116	4 (3.4%)	–	2 (1.7%)	–	–	6 (5.2%)
Luchansky, Paz	1984	124	10 (8%)	8 (6.5%)	5 (4%)	2 (1.6%)	2 (1.6%)	27 (21.7%)

Source: Luchansky, E. and Z. Paz. Variations in the insertion of tibialis anterior muscle. *Anat. Anz. 162:*129–136, 1986.
Tibialis anterior usually inserts onto the medial surface of the first cuneiform and the base of the first metatarsal.

described in which the muscle was absent. In very rare instances, a slip may join the long extensor tendon of the little toe, increasing the slips to five. When five tendons are present, it is usually due to the presence of supernumerary fascicles arising between the normal slips and ending in various ways. Such slips are not infrequent, especially in the interval between the great and second toes. These slips may join a dorsal *interosseous* muscle.

References   453, 687, 690–698.

M. Peroneus Longus
M. Peroneus Brevis

In rare cases, fasciculi have been seen arising from the lateral femoral condyle, and it has been suggested that the fibular collateral ligament represents a primitive femoral attachment of the muscle. Peroneus longus in the plantar foot may serve as the origin of *flexor digiti quinti brevis* or *the plantar interosseous* muscles. The tendon of *peroneus brevis* may deviate to insert on the dorsum of the fifth metatarsal or into the fourth dorsal interosseous muscle. It may also attach to flexor digiti quinti. Its tendon may bifurcate, with one part joining *peroneus tertius*. The *peroneus accessorius* (peroneus quartus) is a slip arising high up on the fibula between the long and short peroneal muscles and joining the tendon of the longus in the sole of the foot. Another detached slip, which arises from the proximal part of the fibula and attaches to the fibular malleolus, the peroneal process of the calcaneus, or the lateral ligament of the ankle, is probably a separated portion of peroneus longus. However, it may represent a *peroneus digiti V superior* or *longus*, such as is seen in catarrhine monkeys. There are other slips that arise from the lower part of the fibula behind peroneus brevis and are attached to the lateral aspect of the calcaneus, to the cuboid, or into the dorsal digital expansion of the fifth toe. These are vestiges of a muscle, the *extensor digiti quinti brevis*, which appears in complete form in very rare cases. A case has been reported of concomitant peroneus brevis, *peroneus digiti IV* and *peroneus digiti V*.

Peroneus longus and brevis are probably represented in the forearm by the extensor carpi ulnaris.

References   7, 126, 191, 303, 309, 331, 332, 335, 336, 356, 359, 382, 414, 435, 453, 483, 533, 625, 681, 687, 690–698.

M. Peroneus Quartus

Gruber found this muscle to be present in about 13% of his specimens. The muscle arises from the posterior intermuscular septum and inserts on the fibular aspect of the calcaneus. Hence it has also been termed *peroneocalcaneus*. (See also the discussion of M. Flexor Hallucis Longus.) This muscle may also exist as a *peroneocuboideus* when it inserts on the cuboid. Peroneus quartus muscles are located in the pre-Achilles fat pad of the lower calf and may appear radiologically as a soft-tissue mass.

References   453, 509, 533, 587, 615, 687, 690, 692, 696, 698.

M. Gastrocnemius
M. Soleus

In view of its phylogenetic history and its development in humans, the gastrocnemius has been considered a muscle of the fibular side of the leg. This may also be true of the lateral head of the soleus. The soleus appears first in lower mammals and may be a derivative of the lateral head of the gastrocnemius. The medial head is probably a medial extension from the fibular head. The gastrocnemius is frequently joined by a separate fasciculus or head (*gastrocnemius tertius*) arising from some part of the popliteal surface of the femur. Caput tertium may also arise from the long head of the *biceps femoris*, lateral epicondyle, knee joint capsule, midfibula, and crural fascia. The *plantaris* muscle itself can be the third head. It may merge with the two heads at their junction, or separately with one or the other head, or with the fascia under the muscle; it has also been seen passing between the popliteal artery and nerve. Absence of the lateral head and its reduction to a fibrous cord have been observed. In some cases, the two heads are separated close to their insertions on the calcaneus, with the muscle remaining separate from the soleus. Another variation is the absence of the tibial head of the soleus. An accessory muscular slip (*soleus accessorius*) may arise from the soleal line with the soleus or be detached from the muscle distally. It lies between the soleus tendo achillis (externally) and *flexor hallucis longus* (internally). It ends on the tendo achillis, or it may have a separate attachment on the medial surface of the calcaneus or tibial collateral ligament of the ankle. Additional slips are (a) *tensor fasciae plantaris,* which inserts into the plantar fascia or *quadratus plantae,* and (b) *peroneotibialis,* which is a tensor of the tendinous arch of the soleus originating from the posterior aspect of the medial surface of the upper tibia and inserting into the arcus tendineus of soleus. Peroneotibialis was found in 128 of 860 limbs; it usually occurs bilaterally, but, when unilateral, is more frequently found in the right limb.

Either or both heads of the gastrocnemius may be doubled; the soleus may be absent or doubled.

References   140, 159, 170, 264, 269, 437, 453, 467, 623, 687, 690–698.

M. Plantaris

The plantaris varies in its mode of origin and more frequently in its insertion. An accessory head may arise along with the usual one from its normal origin, making the muscle bicipital. The muscle may arise from the fascia of the *popliteus*, the origin of the lateral head of the *gastrocnemius*, the fibula between peroneus longus and flexor hallucis longus, or fascia of the leg. Distally, it frequently joins the tendo achillis or ends in the fascia of the leg or tibial collateral ligament of the ankle. Its tendon is sometimes enclosed in the lower part of the tendo achillis; it may send a slip to the plantar fascia. A variant, *tensor fasciae plantaris* (Wood), arises from the oblique line of the tibia and inserts on the fascia overlying the calcaneus.

Like *palmaris longus,* plantaris, which is poorly developed in humans, is a remnant of the superficial layer of the original common flexor of the digits. It is separated, by the calcaneous into proximal and distal parts. It is frequently absent or represented as a fibrous band. In a series of 392 limbs, the plantaris was absent in about 8%, more commonly from the left limb. In another series of 552 limbs, it was absent in 5.4%, also predominantly from the left limb. In yet another study of 750 limbs, plantaris was recorded absent in 6.6% of limbs. Its tendon is commonly used (as a spare part) for repairs when a tendon from *palmaris longus* is inadequate or absent.

References   132, 185, 241, 453, 687, 690–698.

M. Popliteus

The popliteus represents the pronator element of the flexor-pronator mass and corresponds to the upper head of the *pronator teres* of the arm. A rare variation is the presence of a second head, or accessory slip, which may arise from the upper and posterior part of the lateral condyle of the femur on the medial side of *plantaris*, from the lateral head of the *gastrocnemius*, or in some cases from the posterior ligament of the knee joint. It is inserted into the medial side of the oblique line and the medial border of the tibia along with the medial fibers of popliteus. Crossing the popliteal space, this accessory slip may lie in front of or behind the blood vessels. The *peroneotibialis* is a small muscle that arises from the medial side of the head of the fibula, crosses the interosseous space deep to popliteus, and inserts into the upper end of the soleal line or into the tendinous arch of the soleus.

Accessory slips of popliteus may radiate both proximally and distally: to the posterior wall of the knee joint capsule, to the head of the fibula, and to the tibia at the level of insertion of the posterior cruciate ligament.

References   247, 270, 350, 363, 419, 440, 453, 502, 657, 687, 690–698.

M. Flexor Digitorum Longus

The deep portion of this muscle may provide slips to a variable number of digital tendons of the long flexor, most frequently to those of the second, third, and fourth toes. The tendon to the second toe may be absent and replaced by a tendon from *flexor hallucis longus*. A slip is sometimes provided to the fifth toe, replacing the tendon from the short flexor. Another slip passes to the tendinous connection between the tendon of *flexor hallucis longus* and that of *flexor digitorum longus*. Rarely, a slip joins the tendon of the great toe itself, and the muscle has been seen to end entirely on this tendon. The lateral head is often absent and sometimes the whole muscle is missing. The muscle may have an accessory head arising from the tibia, fibula, calcaneus, *tibialis posterior*, or *flexor hallucis longus*. An accessory head arising from the crural fascia may split to join both flexor digitorum longus and *flexor hallucis longus*.

References   39, 172, 192, 337, 338, 453, 670, 687, 690–698.

M. Flexor Hallucis Longus

The connection between the tendon of the long flexor of the great toe and that of *flexor digitorum longus* is constant, but the mode of union is variable. A slip is commonly sent from the flexor hallucis to the *flexor digitorum*, but frequently an additional slip passes, in reverse, from flexor digitorum to flexor hallucis. Complete division of the tendons is very rare. A connecting slip from flexor hallucis is distributed, in more than 50% of cases, to the tendons of the second and third toes; less frequently, it passes to the second toe alone, or to the second, third, and fourth, or, rarely, to all five toes. A doubled *flexor hallucis longus* has been reported, with both tendons inserting on the terminal phalanx of the great toe.

A muscle closely related to *flexor hallucis longus* is *m. fibulotarsalis,* which arises from the fibula and inserts, via the sustentaculum, to the underside of calcaneus and navicular bones.

In the normal arrangement of the long flexor muscles, several features indicate that their degree of differentiation is not the same as that between the corresponding muscles in the arm. In the leg, both flexors are closely connected; in the foot, the flexors of the great toe do not act on the great toe alone, but also send slips to other toes. In a few instances, however, progressive variations have been observed in the form of special muscular slips for the second and fifth toes. Supplementary fasciculi of varying size have been described arising from the tibia, *tibialis posterior*, fibula, or fascia of the leg and joining the tendon of the *flexor digitorum longus* at the ankle. A slip has also been observed arising from the fascia over *flexor digitorum* and joining the tendon of the *flexor hallucis longus* in the sole of the foot.

A more frequent variation (1–5% of bodies) is *flexor accessorius longus,* also known as *fibulocalcaneus medialis, peroneocalcaneus internus* (Macalister), *flexor accessorius digiti longus, accessorius ad quadratum plantae, accessorius ad flexor accessorium, accessorius longus ad flexor digitorum longum, accessorius flexori hallucis longi superior* (Rein-

hart), and *accessorius ad accessorium* (*Turner*). Its origin in the leg can vary widely (from the tibia or fibula or from both bones, the deep fascia, or one of the muscles); it ends in a tendon which, after passing beneath the flexor retinaculum of the foot, joins the *accessorius* (*quadratus plantae*) or *flexor digitorum longus* in the sole.

This muscle, when present, is located in the pre-Achilles fat pad of the lower calf and may appear radiographically as a soft-tissue mass.

References   172, 225, 403, 453, 534, 687, 690–698.

M. Tibialis Posterior

Variations of this muscle are few and rare. At least two cases have been reported in which the muscle consisted of two parts; the accessory part was inserted separately in the sole of the foot. The insertion onto the first cuneiform may be absent. Another muscle, *tibialis secundus,* arises from the posterior aspect of the tibia below *tibialis posterior* and inserts into the capsule of the ankle joint. A similar slip (*tensor capsuli tibiotarsalis anterior*), ending on the anterior wall of the capsule of the ankle joint, is considered to be a variant of the long accessory muscle just described.

Tibiotarsalis arises from the popliteal line and back of the tibia and inserts in the plantar fascia. This muscle is considered to be a tensor fasciae plantaris.

References   11, 453, 482, 687, 690–698.

Foot

M. Quadratus Plantae (Flexor Accessorius)

This muscle is commonly reduced in size: the lateral head may be missing, as may, very rarely, the medial head or the whole muscle. Heads may be duplicated or divided into superficial or deep parts. The manner of attachment to the tendon of flexor digitorum longus varies. The muscle may be inserted in part or wholly into the long flexor of the great toe.

M. quadratus plantae may receive (in about 5% of cases) an accessory slip from the fibula or tibia, one of the muscles of the leg, the fascia of the leg or foot, or the medial surface of the calcaneus or related areas. (See also M. Flexor Hallucis Longus.)

References   264, 448, 509, 676, 686, 693, 697.

Additional references: Winckler, G. La morphogénèse de la chair carée de Sylvius. *Arch. Anat. Hist. Emb.* 9:311–341, 1928–29.

Winckler, G. and G. Gianoli. La véritable terminasion de la chair carrée de Sylvius (musc. quadratus plantae). *Arch. Anat. Hist. Emb.* 38:46–66, 1955.

M. Lumbricales

One or more of the lumbrical muscles has been reported absent; absence of the first lumbrical is extremely rare. Doubling of the second, third, and fourth has also been reported. Various attachments have been noted: the second, third, or fourth lumbrical may arise from the short instead of the long flexor tendon, and the first may arise from the tendon of the *tibialis posterior*. A frequent observation is the contribution of fleshy or tendinous fibers to the lumbricals from *flexor digitorum longus* and *flexor hallucis longus*.

References   448, 453, 564, 687, 690–698.

M. Flexor Digitorum Brevis

The slip from flexor brevis to the little toe may be defective or absent. It may be present but fail to reach the toe, ending in fascia. It may arise separately from the fibular band of the plantar aponeurosis. It is absent in about 21% of cases (Le Double), but may be replaced by a small fusiform muscle arising from the long flexor tendon, with occasional additional attachments to the medial tubercle of the calcaneus, lateral intermuscular septum, or *accessorius* (*quadratus plantae*). The muscle or its tendons may be more or less joined to the tendons of *flexor digitorum longus*. Variation in this muscle occurs in 63% of all limbs.

References   453, 486, 487, 687, 690–698.

M. Abductor Hallucis

The abductor hallucis occasionally sends a slip to the base of the first phalanx of the second and third toe. Its tendon of insertion is sometimes joined by a muscular slip arising from the skin on the medial border of the foot, somewhat in front of the ankle.

References   43, 453, 687, 690–698.

M. Abductor Digiti Minimi (Pedis)

An additional slip originating from the base of the fifth metatarsal bone may be present, part of which may be inserted into the base of the metatarsal bone. These fibers may form a small discrete muscle, *abductor ossis metatarsi digiti quinti*, which arises from the lateral tubercle of the calcaneus and is inserted into the tuberosity of the fifth metatarsal bone. It may be adherent to the abductor of the little toe and has been seen to extend to the middle or even the anterior part of the metatarsal bone. This occurs in more than 40% of cases. *Abductor accessorius digiti minimi* is a rare muscle arising from the lateral process of the tuber calcanei and inserting into the lateral surface of the base of the proximal phalanx of the little toe.

References   43, 325, 453, 687, 690–698.

M. Flexor Hallucis Brevis

The flexor brevis is often fused with the *adductor*. It may be covered by the *abductor hallucis* and *short flexor* of the toes. A slip (*opponens hallucis*) which is attached to the first metatarsal bone has been described. Sometimes this slip is separated from the rest of the muscle and is inserted on the length of the metatarsal shaft. Occasionally, another slip of muscle is observed arising from the cuneiform bone and joining the medial head of the muscle at its insertion. *Flexor hallucis brevis* occasionally sends a slip to the proximal phalanx of the second toe.

References   453, 687, 690–698.

M. Adductor Hallucis Caput Obliquum

A slip has been observed passing from this muscle to the base of the first phalanx of the second or even the third toe. This is of theoretical interest when considered as a separate muscular fasciculus because it may represent a vestige of a plantar adductor of the second toe. A slip from the adductor is occasionally found ending on the metatarsal bone of the great toe.

M. flexor brevis digiti II pedis arises from the plantar aspect of the base of the fourth metatarsal adjacent to adductor hallucis and inserts into the capsule of the second metatarsophalangeal joint.

References   43, 360, 453, 687, 690–698.

M. Adductor Hallucis Caput Transversum (Transversus Pedis)

The slip from the little toe is frequently absent; more rarely, other slips or the entire muscle may be missing. In a study of 91 bodies it was absent in 67% of limbs. Muscle arising from the base of the second metatarsal may receive substantial contributions from *peroneus longus* prior to the latter's usual insertion. This head may appear as a separate fascicle.

References   43, 122, 360, 453, 687, 690–698.

M. Flexor Digiti Minimi Brevis (Pedis)

In the great majority of cases, some of the deeper fibers end on the lateral border of the shaft of the fifth metatarsal bone. This part of the muscle is occasionally found as a separate and independent slip.

References   453, 687, 695–698.

Muscles of Head and Neck (Branchiomeric Muscles)

The branchiomeric muscles are those skeletal muscles derived from the mesoderm associated with the branchial arches. They are supplied by the following cranial nerves: trigeminus, facialis, and glossopharyngeo-vago-accessorius groups.

Trigeminal or Craniomandibular

M. Masseter
M. Temporalis
M. Pterygoideus Lateralis
M. Pterygoideus Medialis

The muscles of mastication are derivatives of a single muscular mass; indications of their common origin can be seen in partial unions of the various muscles. Thus, fibers from the posterior portion of the deeper head of the masseter may join the temporal, fibers from both the temporal and masseter sometimes pass to the anterior border of the fibrocartilage of the mandibular articulation, and the temporal and lateral pterygoid have been observed in some cases to be connected. Fibers may arise from the posterior edge of temporalis to form a distinct muscle termed *temporalis minor* (Henle). It inserts onto the mandibular notch.

Additional muscles apparently belonging to this group have been noted in some cases; *pterygoideus proprius* (Henle), which extends from the infratemporal crest of the sphenoid to the posterior edge of the lateral pterygoid plate, and *pterygospinosus*, which has attachments to the spine of the sphenoid and the posterior border of the lateral pterygoid plate. The significance of these muscles passing between immovable parts is obscure. The masseter may be divided into two layers. The lateral pterygoid may exhibit two distinct fasciculi; it may join the *temporalis* and *digastricus*. The medial pterygoid may send a fascicle to the *masseter* and give rise to the *styloglossus*.

References   51, 168, 181, 203, 289, 451, 453, 454, 655.

M. Mylohyoideus

The mylohyoid may not extend to the hyoid bone; instead it may fuse with neighboring muscles. It is rarely absent. The mylohyoid may be broken up into distinct bundles by lobes of the submandibular gland.

References   297, 453, 459, 687, 690–698.

M. Digastricus (Anterior Belly)

A close relationship exists between the *mylohyoid* and the anterior belly of the digastric. There is usually some exchange of fibers, sometimes amounting to a complete fusion, between the two muscles. Doubling of the anterior belly is a rather common variation, and the anterior bellies of opposite sides may be united by conversion of the fascia that typically passes between them into muscular tissue. An independent muscle, the *mentohyoid* (*Macalister*), extending between the body of the hyoid and the symphysis of the mandible, may be found along the medial border of the anterior belly, and may be considered as a separate part of the digastric. The anterior belly may be absent. Tri- and quadrigastric muscles have been reported.

References   70, 148, 171, 289, 312, 453, 512, 529, 547, 594, 607, 628, 668, 687, 690–698.

Facial

M. Digastricus (Posterior Belly)
M. Stylohyoideus

A close relationship exists between the stylohyoid and the posterior belly of the digastric; one or the other occasionally fails to separate from the common mass from which they are derived. A bundle of muscle fibers sometimes passes from the tip of the styloid process to the angle of the mandible, forming what may be termed *stylomandibularis*.

The stylohyoid may be absent uni- or bilaterally; duplication of the stylohyoid also has been observed. The second slip, *stylohyoideus profundus*, varies in its insertion, sometimes accompanying the stylohyoid proper, sometimes inserting into the lesser cornu of the hyoid, and in some cases replacing the stylohyoid ligament.

If the stylohyoid fails to divide near its insertion (which typically permits passage of the intermediate tendon of the digastric), the insertion may be a single head passing in front of or behind the digastric tendon.

The posterior belly of the digastric may originate from any part of the mastoid groove or even from the outer portion of the superior nuchal line. Occasionally, it fuses completely with the stylohyoid muscle. In certain cases in which the anterior belly fails to differentiate from the *mylohyoid*, the posterior belly inserts into the angle of the mandible instead of into the hyoid bone. Fasciculi may also pass from either belly to neighboring structures. *Digastricus* is frequently divided by a tendinous inscription.

A case of a "digastric" muscle with no connection between the anterior and posterior bellies has been reported. The posterior belly arose from the incisura mastoidea and inserted on the styloid process, while the anterior belly arose from the digastric fossa and inserted on the side of the hyoid body.

A trigastric muscle with the third head arising from the inferomedial aspect of the mandible and inserting on the intertendon has been reported. Another form of trigastric involved the splitting of stylohyoideus, permitting passage of the external carotid with its facial and occipital branches. The more superficial portion of the muscle accompanied the posterior belly of the digastric and inserted into the intertendon of the digastric muscle.

References   70, 148, 171, 288, 387, 451, 453, 512, 628, 663, 668, 687, 690–698.

M. Platysma

There is usually decussation of the two contralateral muscles, across the median line – especially in their upper parts, where a certain amount of decussation may be considered normal. Muscle development is subject to considerable variation, sometimes forming a very thin, pale layer largely interspersed among connective tissue, and at other times forming strong, deeply colored bundles with much less intermixture with connective tissue. Its extension upon the face may also vary considerably, sometimes being traceable as high as the zygoma and, dorsally, to behind the ear. On the other hand, the muscle may be considerably reduced in size, especially distally; complete absence of the lower half has been observed. Rarely, a deep transverse layer is found. *M. platysma* is intimately but variably associated with *mentalis*.

A rare case of a muscle described as a *sphincter colli profundus* has been reported. This muscle sheet arose deep to platysma from the medial part of the clavicle and inserted into the fascia of the pinna. In this particular case the platysma was reduced in size and *supraclavicularis* and *mandibulomarginalis* muscles were present.

References   153, 371, 453, 557, 687, 690–698.

M. Nasalis

Fibers from the nasalis may pass upward upon the nasal bones and enter into the formation of *procerus*. Frequently, the pars alaris and pars transversa are recognized as distinct muscles, *depressor alae nasi* (*mytiformis*) and *compressor narium*, respectively. Thin muscular slips on the outer margin of the nostrils are sometimes described as distinct muscles, *dilatores naris anterior* and *posterior*.

References   370, 453, 687, 690–698.

M. Risorius

The risorius (Santorini) is frequently absent or may be represented by some scattered muscular bands. Its intimate association with the *depressor anguli oris* suggests derivation from that muscle.

References   66, 427, 453, 687, 690–698.

M. Depressor Anguli Oris

A bundle of fibers may arise from the inferior border of the depressor and pass obliquely downward and inward toward the median line beneath the chin, either terminating in superficial fascia of that region or uniting with a counterpart from the opposite side. This slip has been regarded as a distinct muscle, the *transversus menti*. The depressor may be divided into three distinct fascicles.

References   93, 289, 453, 687, 690–698.

M. Zygomaticus Major
M. Zygomaticus Minor

Zygomaticus major may be absent or doubled. Zygomaticus minor may be absent or fused with zygomaticus major. The muscle may be doubled throughout or only at one end.

References   453, 687, 690–698.

M. Buccinator

This muscle may be composed of two laminae and may be continuous with the *superior constrictor* of the pharynx.

References   453, 687, 690–698.

M. Orbicularis Oculi
M. Corrugator Supercilii
M. Procerus

These muscles vary in their differentiation and extent and may be fused with other muscles in the region. The two orbicularis oculi muscles may be united by a *transversus glabellae* muscle. An upward extension of *platysma* has been seen to reach the lower eyelid. The palpebral and orbital portions of these muscles may be completely separate. The orbital portion of orbicularis oculi may be absent, and a rudimentary palpebral portion of the same muscle has been reported. The absence of the orbital portion of corrugator and procerus has also been reported.

References   686, 692, 693, 695, 696, 698, 701, 702.

M. Occipitofrontalis

The occipital belly of this muscle may be absent, and the muscles of the two sides may be fused medially. The frontal belly is rarely absent and may supply slips to the arch of the frontal bone, maxilla, or nasal bones. Fibers of the frontal part may interdigitate across the midline.

References   453, 687, 690–698, 700.

M. Occipitalis Minor (Transversus Nuchae)

This small muscle is frequently absent. It is located between the occipital protuberances and the posterior auricular region, and is frequently fused with the *auricularis posterior*. It may be found either deep or superficial to *trapezius*. It is also known as the *occipital transverse* muscle.

Reported frequencies are remarkable. *Occipitalis minor* is "always" present in Malays, is found in 56% of blacks, 50% of Japanese, and 36% of Europeans, and is "nonexistent" in Hottentots and Melanesians.

Variations include origination from the external occipital protuberance and insertion into the parotid fascia and the angle of the mouth interdigitating with platysma. Insertion into the angle of the mouth allows it to act as a "risorius" muscle.

References   285, 453, 687, 690–698.

M. Auricularis Anterior
M. Auricularis Superior
M. Auricularis Posterior

The most constant of these three ear muscles is the superior. The posterior may be joined with *occipitofrontalis*, *transversus nuchae*, or *platysma*. It is uncommon for any of these three ear muscles to be absent; the most frequently absent is the anterior. *Auricularis inferior* is rarely found in humans. Connections between the superior muscle and *transversus nuchae* have been reported.

References   26, 52, 453, 605, 687, 690–698.

M. Stapedius

This muscle may be doubled or absent.

References   453, 687, 690, 695–698.

Vago-Accessory

M. Constrictor Pharyngis Superior

Because a considerable degree of independence may exist between the bundles of fibers coming from different areas of the line of origin, this muscle has been described as consisting of various parts designated *pterygopharyngeus*, *buccopharyngeus*, *mylopharyngeus*, and *glossopharyngeus*.

Frequently, a bundle of fibers arises from the basilar portion of the occipital bone, or even from the inferior surface of the petrous portion of the temporal or the spine of the sphenoid, and passes downward to be inserted along with the *palatopharyngeus*. The *salpingopharyngeus* is a bundle that arises from the cartilaginous portion of the eustachian tube and inserts with the *palatopharyngeus*.

References   10, 142, 453, 687, 690–698.

M. Constrictor Pharyngis Medius

As in the case of the superior constrictor, the fibers from different points of origin may have considerable independence. Thus, the fibers from the greater cornu of the hyoid are referred to as *ceratopharyngeus* and the remaining fibers as *chondropharyngeus*.

References   453, 687, 690–698.

M. Constrictor Pharyngeus Inferior

The portions of this muscle arising from each of the two laryngeal cartilages may be more or less distinct and have been named *thyropharyngeus* and *cricopharyngeus*. The inferior constrictor may be more or less united with fibers from *cricothyroid*, *sternothyroid*, and *thyrohyoid*. A slip (*syndesmopharyngeus*) from the lateral thyrohyoid ligament occurs frequently.

References   453, 687, 690–698.

M. Stylopharyngeus

This muscle is frequently divided into two or three slips. Supernumerary muscles of the pharynx are common. They frequently originate from the petrous part of the temporal bone in front of the carotid canal, the vaginal process of the temporal spine (*petropharyngeus*), the basilar process (*occipitopharyngeus*), and, very rarely, the mastoid process (*mastoidopharyngeus*). Occasionally a small median slip arises from the pharyngeal tubercle of the occipital bone and inserts on the raphe or posterior wall of the pharynx (*azygopharyngis*).

The stylopharyngeus is innervated by the glossopharyngeal nerve, while the constrictors above are innervated by the pharyngeal plexus from the pharyngeal branch of the vagus (thought to consist primarily of nerve fibers from the bulbar portion of the accessory nerve).

References 7, 316, 347, 453, 687, 690–698.

M. Cricoarytenoideus Lateralis (Lateral Cricoarytenoid)
M. Cricoarytenoideus Posterior (Posterior Cricoarytenoid)

In 50% of cases the lateral cricoarytenoid is inseparable from *thyroarytenoideus*.

An occasional small slip may be found in contact with the lower border of the posterior cricoarytenoid. The *ceratocricoideus* (of Merkel) is a short, slender bundle arising from the cricoid cartilage near its lower border, a little behind the inferior cornu of the thyroid cartilage, and passing obliquely outward and upward to be inserted into that process. It usually exists on one side only and was found in seven of 32 bodies.

A muscle, termed *cricohyoideus*, may be found arising from the anterior aspect of the cricoid arch and inserting on the posterior aspect of the hyoid body.

References 313, 453, 687, 690–695, 697, 698.

M. Cricothyroideus

The right and left cricothyroid may cross the midline and interdigitate before inserting onto the cricoid cartilage. A very small band of muscle may be found on the inferior lateral aspect of the thyroid cartilage. It does not have any association with the cricoid, hence its function in this location is difficult to understand. It has, however, the elegant name of *incisurae mediae obliquus* (Gruber), and may be found unilaterally or bilaterally. The muscle may also present with two heads, one located on the thyroid cartilage and the second arising from the *sternothyroid*. It is then named *incisurae mediae obliquus bicaudatus* (Gruber).

Another muscle having two heads arises from the thyroid cartilage immediately below *thyrohyoideus*. Its heads join and the muscle crosses the thyroid gland (coursing from superior lateral to the inferior medial position) to insert by a thin band of connective tissue on the midline of the trachea. This rare muscle is named *thyreotrachealis biceps* (Gruber).

References 217–222.

M. Thyroarytenoideus
M. Vocalis

A *superior thyroarytenoideus* muscle, in the form of a thin band, arises from the inner, anterior superior surface of the thyroid cartilage and crosses the cartilage and *thyroarytenoideus*, inserting on the inferior lateral aspect of the arytenoid cartilage of the larynx.

The portion of vocalis that inserts onto the vocal ligament has been designated as the *aryvocalis* muscle (of Ludwig).

In the same region, a small, somewhat square-shaped muscle extends superiorly from the *aryepiglottic* muscle (which arises from the *oblique arytenoid*) to insert into the connective tissue of the aryepiglottic fold. It is named *arymembranaceus*.

Reference 700.

M. Sternocleidomastoideus

The amount of fusion of the two heads of this muscle varies considerably. They are frequently separated into cleidomastoid and sternomastoid parts; this has been regarded as normal by some authors. In addition, the muscle frequently separates into other parts, which are arranged in two layers: a superficial layer consisting of a superficial sternomastoid, sterno-occipital, and cleido-occipital part, and a deep layer consisting of a deep sternomastoid and cleidomastoid part. The names indicate the attachments of the various parts. A supernumerary cleido-occipital (Wood), more or less separate from the sternocleidomastoid, has a reported frequency of 33%.

Occasionally, the lower portion of the muscle is traversed by a tendinous intersection, a peculiarity of interest in connection with the formation of the muscle by the fusion of parts derived from different myotomes.

In many animals, the cleidomastoid portion of the muscle is quite distinct from the sternomastoid portion; this condition is frequently found in humans. Other variants of cleidomastoideus include *cleidoepistrophicus*, which originates from the second cervical vertebra and inserts on the clavicle adjacent to the sternocleidomastoid; *cleidocervicalis*, which originates from the carotid tubercle of the fourth cervical vertebra and inserts on the clavicle adjacent to the sternocleidomastoid; and *cleidoatlanticus*, which originates from the transverse process of the atlas and inserts on the clavicle about 2 cm away from cleidomastoideus. The width of the clavicular portion of the muscle varies greatly. The origin of sternocleidomastoideus may spread from the acromial to the sternal end of the clavicle. The sternal head has been seen to extend as far distally as the juncton of the fifth rib. Slips from the muscle may pass to various neighboring structures, e.g., the parotid fascia, thyroid cartilage, angle of the mandible, and contralateral clavicle, and slips have been found arising from the anterior tubercles of the transverse process of the upper three cervical vertebrae and from the hyoid bone. The entire muscle may be doubled.

References   14, 254, 257–259, 317, 404, 453, 687, 690–698.

M. Trapezius

Like the sternomastoid, the trapezius is a compound muscle consisting of three distinct parts, as indicated by a more or less distinct separation of the clavicular portion of the muscle from the rest and, less frequently, by a separation of the lower from the middle portion. Occasionally, bundles pass from the anterior border of the clavicular portion to join the cleido-occipitalis part of *sternocleidomastoideus*, indicating perhaps the common origin of the two muscles. Variations likewise occur in the extent of the spinal attachment of the trapezius, owing to the reduction of one or another of its parts; this attachment usually extends lower in the muscle on the right side than in that on the left. The trapezius may send a fascicle to the sternum and to the *deltoid*. A rare muscle, *supraclavicularis proprius*, may be found adjacent and superficial to the trapezius. *Supraclavicularis proprius* consists of an anterior sheet inserting on the clavicle with the *trapezius* and a posterior sheet inserting into fascia colli superficialis. This muscle arises from the sternum and sternal end of the clavicle, and may insert into the fascia of the dorsal scapular region as well as into those areas cited above.

Supernumerary clavicular muscles are numerous but uncommon. They include *sternoclavicularis anticus* (also referred to as *praesternoclavicularis mediale, coracoclavicularis singularis, acromioclavicularis* or *preclavicularis lateralis*), *supraclavicularis proprius* (*tensor laminae superficialis fasciale colli*), and *retroclavicularis proprius* (*tensor laminae profundus fasciale colli*). *Dorsofascialis,* another sheet of the *panniculus carnosus*, has been reported superficial to *trapezius*.

A bundle of fibers (*omocervicalis, omotrachelien,* or *levator claviculae*) may lie beneath the cervical portion of the trapezius. It attaches distally to the lateral end of the clavicle or to the acromion, and originates above from transverse processes of some of the upper cervical vertebrae, usually the atlas and axis. It occurs in 2% of cases.

Trachleoclavicularis imus (Gruber) is a small supernumerary muscle extending between the clavicle and the transverse process of the sixth cervical vertebra.

The bilateral and unilateral absence of trapezius has been noted. The muscle's origin may not ascend above the axis or descend below the third thoracic vertebra.

References   45, 120, 238, 305, 351, 412, 453, 511, 552, 572, 577, 638, 648, 687, 690–698.

Orbital

M. Levator Palpebrae Superioris

Absence of the levator palpebrae superioris has been noted. When the muscle is present, a slip (*tensor trochleae* [Budge] or *m. gracillimus orbitis* [Bochdalek]) from its belly or proximal part may pass to the sclera, trochlea of the *superior oblique*, and neighboring tissues. The reported frequency of 15 of 20 cases (75%) is probably much too high. The rarity of this variation has been recorded. Fleshy bundles may pass from the front and upper part of the planum of the ethmoid bone across the orbit to the levator. A fasiculus that sometimes passes from the lateral border of the muscle to the lacrimal gland has an effective action as a *retractor glandulae lacrimalis*. Its presence replaces the normal fascial connection between the levator and lacrimal gland.

References   453, 687, 690, 692, 695–698.

M. Rectus Inferior (Inferior Rectus)
M. Rectus Lateralis (Lateral Rectus)
M. Rectus Medialis (Medial Rectus)
M. Rectus Superior (Superior Rectus)
M. Obliquus Inferior (Inferior Oblique)
M. Obliquus Superior (Superior Oblique)

The four recti may be poorly developed and may be connected to each other by muscular slips. The inferior rectus may provide a large muscular bundle passing lateral to the optic nerve to join the superior rectus. A fasciculus of the lateral rectus may pass to the inferior rectus or to the lateral wall of the orbit. Two additional fasciculi sometimes occur: one passing forward to end on the inferior tarsal plate and another inserting on the lateral wall of the orbit. The lateral rectus may also provide a fascicle that passes across the posterior third of the orbit, beneath the optic nerve, to fuse with the belly of the medial rectus. A bifid insertion of the medical rectus has been reported. The superior rectus may provide a muscular slip which may also arise from the annulus of Zinn; it passes downward and forward across the lateral aspect of the optic nerve, to join the inferior rectus. An anomalous muscle bundle from the inferior oblique (*m. obliquus accessorius inferior*) has been found passing from the apex of the orbit to join the inferior rectus. The superior oblique may be closely accompanied by an offshoot from the levator palpebrae superioris, the *comes obliqui superioris*. The lateral and medial recti may be absent.

Occasionally, slips corresponding to a *retractor bulbi* muscle are found; these may be supplied by the abducens and oculomotor nerves. In its simplest form the retractor may arise from the lateral rectus. In some reported instances it forms a muscular funnel around the optic nerve and is inserted onto the back of the eyeball. It may be found divided into four slips that attach to the posterior hemisphere of the globe behind and between the four recti. It may also be represented by a single muscle bundle between the lateral and superior recti. A fibrous ring with four slips joined to the deep surface of the four recti has been reported. The inferior oblique and the lateral rectus may be doubled.

References 179, 379, 453, 455, 538, 608, 671, 687, 690–698, 701, 702.

Hypoglossal

M. Genioglossus

The genioglossus may send a slip to the epiglottis *levator epiglottidis*. It may send some fibers to the superior constrictor of the pharynx (*geniopharyngeus*) or to the stylohyoid ligament. Parts of the muscle may become isolated; for instance, a fasciculus might extend from the mental spine to the tip of the tongue (*longitudinalis linguae inferior medius*).

References 446, 453, 687, 690–698.

M. Hyoglossus

Fibers arising from the lesser cornu of the hyoid bone and inserting on the dorsum of the tongue are frequently separate from the hyoglossus and have been named *chondroglossus*. The fibers arising from the body of the hyoid are frequently separated by a distinct interval from those arising from the greater cornu, the former constituting the *basioglossus* and the latter the *ceratoglossus*. A bundle of fibers, forming the *triticeoglossus*, may arise from the cartilago triticea, passing proximally and forward to insert along with the posterior fibers of the hyoglossus. *Geniohyoid*, which may be doubled bilaterally, may originate from hyoglossus fibers.

References 446, 453, 687, 690–698.

M. Styloglossus

The styloglossus may be doubled or it may be absent on one or both sides. In the latter two cases, it may be replaced by *myloglossus* (Wood) (or *mandibuloglossus*), which arises from the inner surface of the angle of the mandible or from the stylomandibular ligament and is inserted into the sides of the tongue. This muscle is usually in the form of small bundles of fibers. The stylohyoid ligament may pass in front of or behind the *digastric* muscle.

References 290, 453, 645, 687, 690–698.

Infrahyoid

M. Omohyoideus
M. Sternohyoideus
M. Thyrohyoideus
M. Sternothyroideus

These muscles vary considerably in the extent of their development. They may be more or less continuous; the sternal attachment of the sternohyoid is absent more frequently than the clavicular attachment. A straplike muscle, *sternohyoideus azygos*, may be found in the midline of the neck from the posterior surface of the manubrium to the hyoid bone. The region between the omohyoid and sternohyoid may be filled with muscle instead of fascia. Each infrahyoid muscle may be divided longitudinally into two distinct fasciculi, or may send fasciculi to each other or to the pretracheal layer of cervical fascia. The infrahyoid muscles may vary in their origin or insertion.

The omohyoid is the most frequently absent of this group. One belly is absent more frequently than both bellies; the inferior belly may be doubled, with the second belly possibly arising from the coracoid process. The inferior belly may attach along the entire length of the clavicle. When the inferior belly is absent, the superior belly arises from the clavicle and the resulting muscle is termed *cleidohyoideus*. The intermediate tendon of the omohyoid may be reduced to a tendinous inscription; in about 10% of cases it is absent. The inferior attachment may take place on the scapular spine, acromion, coracoid process, or even the first rib or clavicle. An extra fasciculus from the clavicle is found in 3% of cases. Occasionally, a muscle is found extending from the sternum to the clavicle, behind the origin of *sternocleidomastoideus*. Another fascicle, usually arising behind the clavicular head of the sternocleidomastoid, may extend in various directions upward toward the head. It frequently inserts onto the thyroid cartilage (*cleidothyroideae*) or onto the hyoid bone (*cleidohyoideus*). The insertion of the superior belly may overshoot the hyoid bone and attach alongside the digastric. A case has been reported of the fascial sling being muscular, resulting in a T-shaped, three-bellied omohyoid.

Unusual omohyoid forms include: *cleidofascialis*, which originates from the medial third of the clavicle and inserts into fascia colli; *m. cleidohyoideus*, which originates behind the origin of the cleidomastoid part of *sternocleidomastoideus* and inserts onto the body of the hyoid cartilage; and *m. hyofascialis*, which originates from the hyoid and inserts into the omosternoclavicular fascia. *M. cervicocostohumeralis* is considered as a case of extreme displacement of the omohyoid. Its origin is the lesser tubercle of the humerus and its insertions are the transverse process of the sixth cervical vertebra and cartilage of the first rib.

Thyrohyoideus may insert onto the cricoid cartilage. A *cricohyoid* muscle is described as a variant of thyrohyoid. The thyrohyoid is often continuous with the sternothyroid.

The medial fibers on both sides of sternothyroideus may form a cruciate pattern. The muscle may exist in two strata, or it may be divided longitudinally into bundles; the lateral bundle may terminate in cervical fascia.

A bundle of fibers, *levator glandulae thyroideae*, is sometimes found passing either from the lower border of the hyoid or from the thyroid cartilage to the lobe, isthmus, or pyramid of the thyroid gland. Included are those fibers joining the *inferior constrictor* of the pharynx to the thyroid gland.

References   101, 114, 154, 217–222, 257, 266, 274, 275, 278, 280, 388, 423, 453, 551, 576, 595, 617, 621, 637, 660, 687, 690–698.

Scalenes

M. Scalenus Anterior
M. Scalenus Medius
M. Scalenus Posterior

The scalenes vary in the degree of their costal and vertebral attachments, as well as in the interrelationship caused by the fusion of numerous fasciculi. The problem of definition has been compounded by authors who ascribe various muscles to different parts of the scalenus mass.

Frequently (about 30% of cases) a poorly defined muscle mass is present. A *scalenus minimus* (s.* *intermedius* [Testut], s. *pleuralis* [Le Double]) may originate from the anterior tubercle of the sixth and seventh cervical vertebrae and insert on the first rib (behind the subclavian artery groove). It joins the pleural cupula through a suprapleural membrane. When the muscle is not present, a connective tissue band is found. Another variation, *m. transversalis cervicis medius*, arises from the transverse processes of the second and fourth cervical vertebrae and courses between the middle and posterior scalenes to insert onto the seventh and eighth cervical vertebrae.

Typically, the anterior scalene arises from the fourth, fifth, and sixth cervical vertebrae (anterior tubercles), occasionally from the third, and rarely from the seventh. When the anterior scalene is doubled, one of the two muscles is termed *accessory anterior scalene*.

The middle scalene arises from the costotransverse lamellae of the lower five cervical vertebrae. It arises, in about 80% of cases, from the axis, in about 60% from the atlas, and in about 10% from the seventh cervical vertebra.

The posterior scalene may arise from the third, fourth, and the seventh vertebrae, as well as from the usual fifth and sixth; it inserts on the second and occasionally the third rib. The posterior scalene is frequently fused with the first external intercostal muscle.

References   65, 71, 98, 160, 193, 396, 421, 434, 453, 485, 570, 579, 606, 687, 690–698.

Prevertebral

M. Longus Colli
M. Longus Capitis

The most significant and common variations in prevertebral muscles concern the number of vertebrae to which the tendons of origin and insertion are attached. In addition there may be fusion with the *anterior scalene*.

In about 4% of cases a muscle, *atlantobasilaris internus* (Gruber) or *axiobasilaris*, extends along the medial border of longus capitis from the anterior tubercle of the atlas to the base of the skull. Another muscle, *epistropheobasilaris*, is similar but arises from the second cervical vertebra. *Atlanticomastoideus* (Gruber) or *rectus lateralis accessorius* (Winslow) is a cylindrical, ribbon-like, or fusiform muscle extending from the transverse process of the atlas to the mastoid process.

The longus capitis is occasionally inseparable from longus colli.

References   291, 453, 687, 690–698.

M. Rectus Capitis Anterior

This small flat muscle, partly covered by *longus capitis*, arises from the transverse process of the atlas and is inserted into the basilar process of the occipital bone. It may be absent, or may fail to reach the occipital bone, inserting into the occipitoatlantal ligament. A medial part may separate from the primary muscle mass and is termed *rectus capitis minimus*.

References   255, 291, 453, 687, 690–698.

M. Rectus Capitis Lateralis

An accessory bundle, *rectus capitis lateralis longus*, arising from the transverse process of the axis and inserting on the occipital bone has been reported.

References   453, 687, 692, 695, 697, 698.

---

* s. = Synonymus

References

+ The authors have not had access to these references. They are included because of their historic and/or scientific importance.
* These journal articles are notable for their content, literature review, and/or illustrations.

Journals

1* ADACHI, B. Anatomische Untersuchungen an Japanern. *Z. Morphol. Anthropol.* 2:200–222, 1900.

2* ADACHI, B. Beiträge zur Anatomie der Japaner. XII. Die Statistik der Muskelvarietäten. *Z. Morphol. Anthropol.* 12:261–312, 1909.

3+ ADACHI, B. Muscles of the face – Microscopic examination of their origin and insertion in the skin. (in Japanese.) *Kyoto J. Med. Sci.* 15:123–133, 1918.

4 ADRIAO, M. Note sur quelques variations musculaires de l'épaule. *L'Assoc. Anatomistes, Comptes Rendus.* 28:445–449, 1933.

5 ADRIAO, M. Note sur un muscle ptérygo-épineau. *L'Assoc. Anatomistes, Comptes Rendus* 28:450–453, 1933.

6 AGARWAL, I.P. A supernumerary stylopharyngeus muscle. *J. Anat. Soc. India* 4:45, 1955.

7 ALKEN, R. and G. ARNOLD. Ein doppelseitiger, akzessorischer Muskel der Peroneusgruppe. *Z. Orthop.* 108:38–42, 1970.

8 ALLEN, L. Transpleural muscles. *J. Thoracic Surg.* 19:290–291, 1950.

9* ALMSTEAD, S. Congenital defects of the pectoral muscles. *Lancet* 1:1179, 1933.

10 ANDERSON, R.J. A variety of mylopharyngeus and other unusual muscular abnormalities. *J. Anat. Physiol.* 14:357–359, 1897.

11 ANDERSON, R.J. Vorkommen eines Musculus tibio-tarsalis sive tensor fasciae plantaris. *Arch. Pathol. Anat. Physiol. Klin. Med.* 81:574–575, 1880.

12 ANDERSON, R.J. Über einige Varietäten des M. subclavius. *Arch. Pathol. Anat. Physiol. Klin. Med.* 81:575, 1880.

13* ANONYMOUS. Anatomie du vivant. Deformation du thorax par anomalie musculaire. *La Presse Médicale No. 14* 30:278, 1922.

14* ANONYMOUS. Anatomie du vivant. Attitudes vicieuses du cou d'origine musculare. *La Presse Médicale No. 76* 30:1594, 1922.

15* ANSON, B.J., BEATON, L.E. and C.B. MCVAY. The pyramidalis muscle. *Anat. Rec.* 72:405–411, 1938.

16* ANSON, B.J., BEATON, L.E. and J.J. MCDONALD. The origin of the m. pectoralis minor. *Am. J. Anat.* 72:629–630, 1937–38.

17* ANSON, B.J. and C.B. MCVAY. Inguinal hernia. The anatomy of the region. *Surg. Gynecol. Obstet.* 66:186–191, 1938.

18* ANSON B.J., JAMIESON, R.W., O'CONOR, V.J. JR. and L.E. BEATON. The pectoral muscles. An anatomical study of 400 body-halves. *Q. Bull. Northwestern Univ. Med. School* 27:211–218, 1953.

19 APPLETON, A.B. On the morphology of the cervico-costo-humeralis muscle of Gruber. *J. Anat.* 63:437–446, 1929.

20 APPLETON, A.B. and P.G.R. GHEY. An example of the cervico-costo-humeral muscle of Gruber. *J. Anat.* 63:434–436, 1929.

21 AREY, L.B. and M.J. TREMAINE. The muscle content of the lower esophagus of man. *Anat. Rec.* 56:315–320, 1933.

22 ASHLEY-MONTAGU, M.F. Anthropological significance of the musculus pyramidalis and its variability in man. *Am. J. Phys. Anthropol.* 25:435–490, 1939.

23 AUGUSTYNIAK, E., KOCON, T. and S. ZAJAC. Rare cases of symmetric muscular variations of the upper extremities. *Folia Morphol.* 26:39–42, 1967.

24* AUGUSTYNIAK, E. and S. ZAJAC. An accessory muscle of the shoulder girdle. *Folia Morphol.* 27:471–473, 1968.

25 AUMÜLLER, G. Über Bau und Funktion des Musculus adductor minimus. *Anat. Anz.* 126:337–342, 1970.

26* AUSTONI, A. Muscoli auricolari estrinseci dell'uomo. *Archiv. Ital. Anat. Embriol.* 7:193–243, 1908.

27 AYOUB, S.F. The anterior fibres of the levator ani muscle in man. *J. Anat.* 128:571–580, 1979.

28 BABA, M.A. The accessory tendon of the abductor pollicis longus muscle. *Anat. Rec.* 119:541–547, 1954.

29 BACKHOUSE, K.M. and D. CHURCHILL-DAVIDSON. Anomalous palmaris longus muscle producing carpal tunnel-like compression. *Hand* 7:22–24, 1975.

30 BAJAJ, I.D. and S. SOOD. Anomalous insertion of the biceps femoris muscle. *J. Anat. Soc. India* 11:92, 1962.

31 BANCHI, A. Faisceau accessoire du m. pronator teres du nerf médian et de l'artère humérale à la région du coude. *Arch. Ital. Biol.* 45:283–284, 1906.

32 BANCHI, A. Un muscle manieux à deux faisceaux. *Arch. Ital. Biol.* 45:284, 1906.

33 BANKART, J., PYE-SMITH, P.H. and J.J. PHILLIPS. Note of abnormalities observed in the dissecting room during sessons of 1866–7, 1867–8. *Guy's Hosp. Rep.* 14:436–455, 1869.

34 BANT, A. A propos de la question morphologique des muscles périclaviculares surnuméraires chez l'homme et les Mammifères. *Soc. Biol. Comptes Rendus Hebdomadaires des Séances et Mémoires* 90:900–902, 1368, 1924.

35 BANT, A. Deux cas de muscle acromio-claviculaire. *Soc. Biol. Comptes Rendus Hebdomadaires des Séances et Mémoires* 90:902–903, 1924.

36 BARDELEBEN, K. VON. Der Musculus "sternalis". *Centralblatt Med. Wissen.* 27:433–435, 1875.

37* BARGE, J.A.J. Ein Fall von vollständigem Defekt der beiden Brustmuskeln. *Anat. Anz.* 64:102–119, 1927–28.

38 BARLOW, R.N. The sternalis muscle in American whites and negroes. *Anat. Rec.* 61:413–426, 1935.

39 BARLOW, T.E. An unusual anomaly of m. flexor digitorum longus. *J. Anat.* 83:224–226, 1949.

40 BARRETT, J.H. An additional (third and separate) head of the pronator teres muscle. *J. Anat.* 70:577–578, 1935–36.

41 BARRY, D. and J.S. BOTHROYD. Tensor fasciae suralis. *J. Anat.* 58:382–383, 1924.

42* BASU, S.S. and S. HAZARY. Variations of the lumbrical muscles of the hand. *Anat. Rec.* 136:501–503, 1960.

43 BAUMANN, J.A. Valeur, variation, et équivalences des muscles extenseurs, interosseux, adducteurs et abducterurs de la main et du pied chez l'homme. *Acta Anat.* 4:10–16, 1947.

44* BEATON, L.E. and B.J. ANSON. The pyramidalis muscle: Its occurrence and size in American Whites and Negroes. *Am. J. Phys. Anthrop.* 25:261–269, 1939.

45* BEATON, L.E. and B.J. ANSON. Variations in the origin of the m. trapezius. *Anat. Rec.* 83:41–46, 1942.

46 BEATTIE, J. and F.L. HORSFALL. An anomalous facial muscle. *J. Anat.* 65:145–148, 1931.

47 BECCARI, N. Le muscle transverse du thorax et les insertions sternocostales. *Arch. Ital. Biol.* 62:283–285, 1914.

48 BECKER, A.E. Variation of the M. coracobrachialis. *Acta Morphol. Neerl. Scand.* 5:217–220, 1963.

49 BELLAMY, E. Note on the absence of the quadratus femoris muscle and on a spine possessing a sixth lumbar vertebra, the first rib being rudimentary. *J. Anat. Physiol.* 9:185–186, 1875.

50 BERGER, O. Angeborener Defect der Brustmuskeln. *Arch. Pathol. Anat. Physiol. Klin. Med.* 72:438–442, 1878.

51 BERTELLI, D. Le muscle temporal superficiel. *Arch. Ital. Biol.* 11:420, 1889.

52 BERTELLI, D. Le muscle auriculaire anterior. *Arch. Ital. Biol.* 13:354, 1890.

53 BHADKAMKAR, A.R. and V.R. MYSOREKAR. Bilateral extensor digitorum brevis muscle in the hand. *J. Anat. Soc. India* 9:104–105, 1960.

54* BHADKAMKAR, A.R. and V.R. MYSOREKAR. Anomalous peroneus tertius and extensor digitorum longus muscles. *J. Anat. Soc. India* 10:37–38, 1961.

55 BHARGAVA, I. An extra muscle in forearm. *J. Anat. Soc. India* 5:81–82, 1956.

56 BHARGAVA, K.N., SANYAL, K.P. and S.N. BHARGAVA. Lateral musculature of the leg as seen in hundred Indian cadavers. *Indian J. Med. Sci.* 15:181–185, 1961.

57 BHATNAGAR, B.N.S. and D. NARAYAN. Bicipital sartorius. *J. Anat. Soc. India* 8:32–33, 1959.

58* BING, R. Über angeborene Muskeldefecte. *Arch. Pathol. Anat. Physiol. Klin. Med.* 170:175–228, 1902.

59 BINGOLD, A.C. An extensor indicis brevis. *Br. J. Surg.* 51:236–237, 1964.

60 BINNS, J.H. Two cases of extensor digitorum brevis manus. *Hand* 4:263–264, 1972.

61 BLEICHER, M. and F. MATHIEU. Sur une formation musculaire

reliant la pyramide de lalouette a l'os hyoide: le muscle hyopyramidal. *L'Assoc. Anatomistes, Comptes Rendus.* 27:52–59, 1932.
62 BLEICHER, M. and E. LEGAIT. L'espace sous-quadricipital et ses éléments anatomiques. *L'Assoc. Anatomistes, Comptes Rendus.* 28:72–84, 1933.
63 BLUNT, M.J. The posterior wall of the inguinal canal. *Br. J. Surg.* 39:230–232, 1951.
64 BOCHDALEK, JR. Ein anomaler Musculus supracostalis anterior. *Arch. Pathol. Anat. Physiol. Klin. Med.* 41:257–258, 1867.
65 BOCHDALEK, JR. Vor dem Musculus scalenus anticus ihren Verlauf nehmende Art. subclavia dextra. *Arch. Pathol. Anat. Physiol. Klin. Med.* 41:261–263, 1867.
66 BOHDANOWICZ, F. Statistique de l'absence du muscle risorius chez les Polonais. *L'Assoc. Anatomistes, Comptes Rendus.* 26:73–76, 1931.
67 BOIS-REYMOND, R. Beschreibung einer Anzahl Muskelvarietäten an einem Individuum. *Anat. Anz.* 9:451–455, 1894.
68* BÖKEN, H. Über einige Varietäten mit Defektbildung der platten Rückenmuskulatur. *Anat. Anz.* 46:515–522, 1914.
69 BONIN, G. VON. Bau und Variationen des Iliopsoas. *Anat. Anz.* 69:476–485, 1930.
70 BOVERO, A. Sur les muscles digastriques de l'os hyoïde. *Arch. Ital. Biol.* 26:158, 1896.
71 BOYD, G.I. Abnormality of subclavian artery associated with presence of the scalenus minimus. *J. Anat.* 68:280–281, 1933–34.
72 BRADLEY, S.M. Notes on myological peculiarities. *J. Anat. Physiol.* 6:420–421, 1872.
73 BRETTSCHNEIDER, H. Über selten vorkommende tiefe Brustmuskeln. *Anat. Anz.* 99:67–71, 1952–53.
74 BROCK, G.S. A two-headed sartorius. *J. Anat. Physiol.* 13:578, 1879.
75 BRONES, M.F. Anatomical variations of the palmaris longus, causing carpal tunnel syndrome. Case reports. *Plast. Reconstr. Surg.* 62:798–800, 1978.
76 BROWN, J.M. Variations in myology. *J. Anat. Physiol.* 14:512–513, 1897.
77 BRUNI, A.C. Recherches sur les muscles surnuméraires du dos de la main. *Arch. Ital. Biol.* 42:483–484, 1907.
78 BRUNN, A. VON. Varietät des Musc. interosseus dorsal. manus II. *Arch. Anat. Physiol. Wissen. Med.* 1873: p. 126.
79 BRYCE, T.H. Note on the varieties of the pectoral sheet of muscle. *J. Anat. Physiol.* 34:75–78, 1899.
80 BUCHER, O. Eine seltene Varietät des Musculus biventer mandibulae. *Anat. Anz.* 91:81–87, 1941.
81 BUCHER, O. Über die Varietäten des M. abductor digiti quinti. *Anat. Anz.* 94:317–324, 1943.
82 BURKARD, O. Zwei seltene Anomalien an der Musculatur der tiefen Lage der Hinterseite des Unterschenkels. *Arch. Anat. Entwickelungsgesch.* 1902, 344–346.
83 BURKE, C.V. Congenital absence of pectoral muscles. *Med. Rec.* 62:976, 1902.
84 BURKITT A.N. and G.H.S. LIGHTOLLER. The facial musculature of the Australian Aboriginal. Part I. *J. Anat.* 61:14–39, 1926.
85 BURKITT, A.N. and G.H.S. LIGHTOLLER. The facial musculature of the Australian Aboriginal. Part II. *J. Anat.* 62:33–57, 1927.
86 BUTLER, B. JR. and E.C. BIGLEY, JR. Aberrant index (first) lumbrical tendinous origin associated with carpal-tunnel syndrome: A case report. *J. Bone Joint Surg.* 53A:160–162, 1971.
87* CALORI, L. Di alcuni nuovi muscoli soprannumerarii degli arti. *Mem. R. Accad. Sci. Istituto di Bologna S.2*, 6:137–147, 1866.
88* CALORI, L. Intorno ad alcune varieta incontrate nella muscolatura degli arti superiori annotazioni anatomische. *Mem. R. Accad. Sci. Istituto di Bologna S.2*, 6:157–174, 1866.
89* CALORI, L. Di alcune varieta muscolari dell' avambraccio e dell' eminenza ipothenar. *Mem. Accad. R. Sci. Istituto di Bologna S.2*, 7:359–381, 1867.
90* CALORI, L. Varieta dei muscoli del trunco e descrizione di una pettorina di fanciullo singolare per varie anom. *Mem. R. Accad. Sci. Istituto di Bologna S.2*, 7:383–393, 1867.
91* CALORI, L. Delle anomalie più importanti di ossa, vasi, nervi, e muscoli occorse nell'ultimo biennio facendo anatomia del corpo umano. *Mem. R. Accad. Sci. Istituto di Bologna S.2*, 8:417–482, 1868.
92* CALORI, L. Sopra il muscolo episternale e le sue anatomische interpretazioni. *Mem. R. Accad. Sci. Istituto di Bologna S.4*, 9:131–140, 1888.
93 CAMPIONI, B. The transversus menti. Its innervation and variations. *L'Assoc. Anatomistes, Comptes Rendus.* 26:47–53, 1931.
94 CAREY, J.M. and W.H. HOLLINSHEAD. An anatomic study of the esophageal hiatus. *Surg. Gynecol. Obstet.* 100:196–200, 1955.
95 CARLETON, A. Flexor carpi radialis brevis vel profundus. *J. Anat.* 69:292–293, 1934–35.
96 CARR, N.D., O'CALLAGHAN, J.D. and R. VAUGHN. An unusual flexor of the fifth finger. *Acta Anat.* 98:376–379, 1977.
97 CAULDWELL, E.W., ANSON, B.J. and R.R. WRIGHT. The extensor indicis proprius muscle. A study of 263 consecutive specimens. *Q. Bull. Northwestern Univ. Med. School.* 17:267–279, 1943.
98 Cave, A.J.E. A note on the origin of the m. scalenus medius. *J. Anat.* 67: 480–483, 1933.
99 CAVE, A.J.E. and R.W. BROWN. On the tendon of the subclavius muscle. *J. Bone Joint Surg.* 34B: 466–469, 1952.
100 CAVE, A.J.E. and C.J. PORTEOUS. A note on the semimembranosus muscle. *Ann. R. Coll. Surg. Engl.* 24:251–256, 1959.
101 CECCHERELLI, G. Sur quelques anomalies des muscles peaussiers de la face et de muscle omo-hyoïdien. *Arch. Ital. Biol.* 43:317–318, 1905.
102 CELLI, E. Sur la morphologie du M. piriformis. *Arch. Ital. Biol.* 41:132–133, 1914.
103 CHANDLER, S.B. and M. SCHADEWALD. The conjoined aponeurosis versus the conjoined tendon. *Anat. Rec.* 82:404, 1942.
104 CHANDLER, S.B. and M. SCHADEWALD. Studies on the inguinal region. I. The conjoined aponeurosis versus the conjoined tendon. *Anat. Rec.* 89:339–343, 1944.
105 CHELMER-FAÏNSILBER, R.B. Über die Sehnen der langen Beuger der Zehen. *Anat. Anz.* 71:185–194, 1930–31.
106 CHI, T.K. M. tensor ligamenti cruciati cruris (?). *J. Anat.* 72:132–133, 1937–38.
107 CHOPRA, R.P., SINGH, S. and S.S. MAKHNI. The insertion of the abductor pollicis longus and extensor pollicis brevis. *J. Anat. Soc. India* 7:105–109, 1954–56.
108* CHOUCHKOV, H. and D. DSCHEROV. Über eine seltene symmetrische Muskelvarietät der unteren Extremitäten. *Anat. Anz.* 120:91–96, 1967.
109 CHOUKE, K.S. Some interesting anomalies of the human body. *Anat. Rec.* 36:389–394, 1927.
110 CHOUKE, K.S. The constitution of the sheath of the rectus abdominis muscle. *Anat. Rec.* 61:341–349, 1935.
111* CHOWDHARY, D.S. A rare anomaly of m. flexor digitorum sublimis. *J. Anat.* 85:100–101, 1951.
112* CHRISTOPHER, F. Congenital absence of the pectoral muscles. *J. Bone Joint Surg.* 10:350–351, 1928.
113 CIHÁK, R. The occurrence of Mm. contrahentes and their changes during the development of the human hand. *Folia Morphol.* 15:197–205, 1967.
114 CIVALLERI, A. Contribution à l'etude des Musculi levatores glandulae thyreoidae et à l'innervation des Musculi sternothyreoidus et thyreohyoideus. *Arch. Ital. Biol.* 45:281–282, 1906.
115 CLARK, E. Congenital variation of the pectoral muscles, with a report of a case. *J. Anat. Physiol.* 49:155–164, 1915.
116 CLARKSON, R.D. and H. RAINY. Unusual arrangement of the psoas muscle. *J. Anat. Physiol.* 23:504–506, 1889.
117* COLEMAN, S.S., MACAFEE, D.K. and B.J. ANSON. The insertion of the abductor pollicis longus muscle. An anatomical study of 175 specimens. *Q. Bull. Northwest. Univ. Med. Sch.* 27:117–122, 1953.
118 COOPER, G.W. Fascial variants of the trigonum lumbale (Petiti). *Anat. Rec.* 114:1–7, 1952.
119 Cornejo, R.H. Sur la signification et fonction du ligament acromiocoracoidien. *Ann. Anat. Pathol.* 13:383–393, 1936.
120 CORNER, E.M. Proceedings of the Anatomical Society of Great Britain and Ireland, June 1900. A communication of a case of omo-trachelian muscle in a living subject. *J. Anat. Physiol.* 35:iii, 1901
121 CORSY, F. Un nouveau cas d'appareil hyoïdien. Anomalies multiples des muscles du deuxième arc. *Soc. Biol. Comptes Rendus Hebdomadaires des Séances et Mémoires.* 83:366–368, 1920.
122 CRALLEY, J.C. and J.M. SCHUBERTH. The transverse head of adductor hallucis. *Anat. Anz.* 146:400–409, 1979.
123 CRERAR, J.W. Note on the absence of the subclavius muscle. *J. Anat. Physiol.* 26:554, 1892.
124 CUNNINGHAM, D.J. The musculus sternalis. *J. Anat. Physiol.* 18:208–210, 1884.
125 CUNNINGHAM, D.J. The musculus sternalis. *J. Anat. Physiol.* 22:391–407, 1888.
126 CUNNINGHAM, D.J. and J. BROOKS. The peroneus quinti digiti. *Proc. R. Irish Acad, Ser. 3.* 1:78–81, 1889.

127 CURNOW, J. Notes on some irregularities in muscles and nerves. *J. Anat Physiol.* 7:304–309, 1873.
128 CURNOW, J. Notes on some muscular irregularities. *J. Anat. Physiol.* 8:377–379, 1874.
129 CURNOW, J. Variations in the arrangement of the extensor muscles of the forearm. *J. Anat. Physiol.* 10:595–601, 1876.
130 DAHLGARD, D.L. and G.E. KAUTH. An anomalous arrangement of the flexor musculature of the forearm and hand. *Anat. Rec.* 152:251–255, 1965.
131 DATTA, K.N. and S.K. BASU. An abnormal slip of left subclavius muscle having an additional nerve supply. *J. Anat. Soc. India.* 3:42, 1954.
132* DASELER, E.H. and B.J. ANSON. The plantaris muscle. An anatomical study of 750 specimens. *J. Bone Joint Surg.* 25:822–827, 1943.
133 DAVIES-COLLEY, J.N.C., TAYLOR, F. and B.N. DALTON. Notes of abnormalities observed in the dissecting room from October, 1870 to June, 1872. *Guy's Hosp. Rep.* 28:389–400, 1873.
134* DAY, M.H. and J.R. NAPIER. The two heads of flexor pollicis brevis. *J. Anat.* 95:123–130, 1961.
135* DBALY, J. Ein Fall von beiderseitigem Vorkommen des muskulären Achselbogens. *Anat. Anz.* 137:75–78, 1975.
136 DEBIERRE, C. Anomalies des muscles et des nerfs. *Soc. Biol. Comptes Rendus Hebdomadaires des Séances et Mémoires.* 38:176–177, 1886.
137 DEBIERRE. Des anomalies des muscles coexistantes et corrélatives des anomalies des nerfs. *Soc. Bio. Comptes Rendus Hebdomadaires des Séances et Mémoires.* 40:86–90, 1888.
138* DEBIERRE, C. Sur le biceps brachial a trois chefts. *Soc. Biol. Comptes Rendus Hebdomadaires des Séances et Mémoires.* 40:486–489, 1888.
139* DEBIERRE, C. Anomalie des muscles radiaux externes et du long abducteur du pouce de l'homme. *Soc. Biol. Comptes Rendus Hebdomadaires des Séances et Mémoires.* 40:547–550, 1888.
140* DELITZIN, S.N. Über einen supernumerären Muskel des Unterschenkels (Musculus soleus accessorius?), welcher den Nervus tibialis durchbohrt. *Int. Monat. Anat. Physiol.* 19:349–354, 1901.
141 DENTICI, S. Sur un faisceau accessoire, non encore décrit, du muscle petit oblique de l'abdomen. *Arch. Ital. Biol.* 44:383, 1915.
142 DE PAULA ASSIS, J.E. Observacoes anâtomicas sôbre o musculus salpingopharyngeus. *Rev. Bras. Oto-Rino-Laringo.* 15:169–181, 1947. Cited in Excerpta Medica, Sec. 1, Vol. 3, abstract 363, 1949.
143 D'EVANT, T. Sur la signification morphologique du m. tibialis anticus. *Arch. Ital. Biol.* 32:457, 1899.
144 D'EVANT, T. Les muscles tenseurs de la synoviale radio-bicipitale. *Arch. Ital. Biol.* 37:479, 1902.
145 DIDIO, L.J.A., ZAPPALÁ, A., CARDOSO, A.D. and R.A. DIAZ. Musculus articularis genu in human fetuses, newborn and young individuals. *Anat. Anz.* 124:121–132, 1969.
146 DOBSON, G.E. Note on the rectus abdominis et sternalis muscle. *J. Anat. Physiol.* 17:84–85, 1883.
147* DRACHMANN, A.G. Trans. by J.W. MOORE. Case of congenital absence of the quadriceps extensor cruris muscle. *J. Anat. Physiol.* 7:310–311, 1873.
148* DRATSCH, S. Ein Fall von Variationen des M. digastricus mandibulae. *Anat. Anz.* 69:81–82, 1930.
149 DUNN, A.W. and C.M. EVARTS. The extensor digitorum brevis manus muscle. *Clin. Ortho.* 28:210–212, 1963.
150* DURKSEN, F. Anomalous lumbrical muscles in the hand: A case report. *J. Hand Surg.* 3:550–551, 1978.
151 DWIGHT, J. Muscular abnormalities. *J. Anat. Physiol.* 22:96–102, 1888.
152* DYKES, J. and B.J. ANSON. The accessory tendon of the flexor pollicis longus muscle. *Anat. Rec.* 90:83–87, 1944.
153 EHRENBERG, G. Eine seltene Abnormität des Platysma. *Anat. Anz.* 26:343–347, 1905.
154 EISLER, P. Der M. levator glandulae thyreoidea und verwandte praelaryngeale Muskelbildungen. *Anat. Anz.* 17:183–196, 1900.
155 ENDRES, H. Über ein Zwischenmuskelbündel im Gebiete des M. pectoralis major und latissimus dorsi. *Anat. Anz.* 8:387–397, 1893.
156 ENTIN, M., BARSKY, A. and A. SWANSON. Classification of the congenital malformations of the hand and upper extremity. *Hand* 4:215–219, 1972.
157 EULENBURG, A. Ein Fall von angeborener Anomalie der Brustmuskeln. *Deutsche Med. Wochenschr.* 3(35):413–415, 1877.
158 EYLER, D.L. and J.E. MARKEE. The anatomy and function of the intrinsic musculature of the fingers. *J. Bone Joint Surg.* 36A:1–9, 1954.

159 FALLER, A. Zur Deutung der akzessorischen Köpfe des Schollenmuskels. *Anat. Anz.* 93:161–179, 1942.
160 FAWCETT, E. What is Sibson's muscle (scalenus pleuralis)? *J. Anat. Physiol.* 30:433–436, 1896.
161 FEDOROFF, D.N. Zur Morphologie der Muskelvarianten des Lateralgebietes des Halses. *Anatomischer Anzeiger* 62:338–346, 1926–27.
162 FERET, J. A rare case of "costoepitrochlear muscle." *Folia Morphol.* 28:204–208, 1969.
163* FERET, J., JAGIELLO, W. and A. WIŚNIEWSKI. Morphologic variation of the long palmar muscle (M. palmaris longus) in man. *Folia Morphol.* 24:64–69, 1965.
164 FICK, R. Drei Fälle von Musculus sternalis. *Anat. Anz.* 6:601–606, 1891.
165 FISCHER, H. Note a propus d'un muscle surnuméraire de la tête. *L'Assoc. Anatomistes, Comptes Rendus.* 21:206–209, 1926.
166 FISCHER, H. A propos d'une curieuse disposition musculaire. Existence d'un muscle petit – pectoro – epitrochléen. *L'Assoc. Anatomistes, Comptes Rendus.* 22:99–101, 1927.
167* FITZGERALD, R.R. A case showing the chondro-epitrochlearis muscle. *J. Anat.* 70:273–274, 1935–36.
168 FLEISSIG, J. Eine Varietät des Musculus masseter und der Mandibula. *Anat. Anz.* 36:505–510, 1910.
169 FRORIEP, A. Über den Hautmuskel des Halses und seine Beziehung zu den unteren Gesichtsmuskeln. *Arch. Anat. Entwicklungsgesch.* 1877:46–62.
170 FLOWER, J.M. Note on an accessory soleus muscle. *J. Anat.* 65:548–549, 1931.
171 FOCACCI, M. Contribution à l'étude du Muscle interdigastrique de Bianchi. *Arch. Ital. Biol.* 36:476, 1901.
172 FOCACCI, M. Etude morphologique sur les fléchisseurs longs du pied. *Arch. Ital. Biol.* 39:486–487, 1903.
173* FONTES, V. Note sur le muscle manieux. *L'Assoc. Anatomistes, Comptes Rendus.* 28:289–294, 1933.
174 FORSYTH, A. Absence of the pectorales. *Lancet* 1:618, 1873.
175 FRANK, J. Über einen im Leben beobachteten M. sternalis. *Anat. Anz.* 46:648–652, 1914.
176 FRAZER, J.E. The derivation of the human hypothenar muscles. *J. Anat. Physiol.* 42:326–344. 1908.
177 FREDERICQ, H. Un cas de division complète du nerf grand sciatique accompagnée d'une duplicité complete du muscle pyramidal. *L'Assoc. Anatomistes, Comptes Rendus.* 10:168–170, 1908.
178 FRITSCH, G. Abnorme Muskelbündel der Achselhöhle. *Arch. Anat. Physiol. Wissen. Med.* 1869:367–371.
179 FUJITA, T. Eine Muskelanomalie in der Orbita. *Folia Anat. Jpn.* 16:301–302, 1938.
180 FUKUYAMA, U. der Musculus sternalis bei den Nordchinesen. *Folia Anat. Jpn.* 19:69–92, 1940.
181 FUSARI, R. Contribution à la connaissance morphologique du muscle temporal. *Arch. Ital. Biol.* 28:471–472, 1897.
182 GAILLARD, J. Fréquence du muscle petit palmaire (M. Palmaris longus) chez les Mélano-Indiens. *L'Assoc. Anatomistes, Comptes Rendus.* 45:371–377, 1958.
183 GALTON, J.C. On the epitrochleo-anconeus or anconeus sextus (Gruber). *J. Anat. Physiol.* 9:169–175, 1875.
184 GEGENBAUR, C. Ein Fall von mehrfachen Muskelanomalien an der oberen Extremität. *Arch. Path. Anat. Physiol. Klin. Med.* 21:376–385, 1861.
185 GEORGE, R. Co-incidence of palmaris longus and plantaris muscles. *Anat. Rec.* 116:521–523, 1953.
186 GÉRARD, G. and P. CORDIER. Au sujet d'un cas de fusion des tendons des muscles grand dorsal et grand rond. *Soc. Biol., Comptes Rendus Hebdomadaires des Séances et Mémoires.* 83:927, 1920.
187 GIANNELLI, L. Sur une anomalie peu commune du muscle rond pronateur. *Arch. Ital. Biol.* 31:186, 1899.
188 GIANNELLI, L. Sur quelques anomalies musculaires. *Arch. Ital. Biol.* 31:186, 1899.
189 GIANNELLI, L., LAGHI, A. and V. QUERCIOLI. Sur les variétés anatomiques les plus importantes observées durant l'année scolastique. *Arch. Ital Biol.* 32:474, 1899.
190 GIANNELLI, L. Multiples anomalies musculaires chez un même individu. *Arch. Ital. Biol.* 47:481, 1907.
191 GIANNELLI, L. Vestiges de M. Peronaeus digiti V (superior et posterior) et de M. extensor proprius digit V pedis chez l'homme. *Arch. Ital. Biol.* 47:474–475, 1913.
192 GIES, T. Der Flexor digitorum pedis communis longus und seine Varietäten. *Arch. Anat. Physiol. Wissen. Med.* 1868: pp. 231–239.
193* GILIS, P. Note sur l'anatomie des muscles scalènes. *Soc. Biol.,*

194. GILIS, P. Anomalíe et absence re'elle du muscle demi-membraneaux sur le même subjet. *Soc. Biol., Comptes Rendus Hebdomadaires des Sèances et Mémoires.* 47:642–644, 1895.
195* GLADYKOWSKA-RZECZYCZKA, J. Accessory bundle of the short head of the biceps femoris muscle. *Folia Morphol.* 23:75–78, 1964.
196. GLASGOW, E.F. Bilateral extensor digitorum brevis manus. *Med. J. Aust.* 54:24–25, 1967.
197. GLOOBE, H. and S. LIBERTY. Bilateral agenesis of extensor carpi ulnaris. *Hand* 5:175–176, 1973.
198. GLOOBE, H. and P. PECKET. An anomalous muscle in the canal of Guyon (A possible ulnar nerve compression). *Anat. Anz.* 133:477–479, 1973.
199. GODLEWSKI, E. Note sur la constitution des insertions inférieures du muscle brachial antérieur. *L'Assoc. Anatomistes, Comptes Rendus.* 6:146–148, 1904.
200. GOLDBERG, S. The origin of the lumbrical muscles in the hand of the South African native. *Hand* 2:168–171, 1970.
201. GOSZCZYŃSKA, I. Normal position and mobility of the diaphragm in women. *Folia Morphol.* 31:289–300, 1972.
202. GRÄFENBERG, E. Die Entwicklung der Knochen, Muskeln und Nerven der Hand und die Frage von den Bewegungen der Hand bestimmten Muskeln des Unterarms. *Anat. Hefte* 30:1–154, 1906.
203. GRANT, P.G. Lateral pterygoid: Two muscles? *Am. J. Anat.* 138:1–9, 1973.
204. GRAPER, L. Eine sehr seltene Varietät des M. flexor digitorum sublimis. *Anat. Anz.* 50:80–84, 1917–18.
205. GRAY, D.J. Some anomalous hamstring muscles. *Anat. Rec.* 91:33–38, 1945.
206. GREEN, H.L.H.H. The occurrence of a tenuissimus muscle in a human adult. *J. Anat.* 65:266–271, 1931.
207* GREIG, H.W., ANSON, B.J. and J.M. BUDINGER. Variations in the form and attachments of the biceps brachii muscle. *Q. Bull. Northwestern Med. School* 26:241–244, 1952.
208. GREINERT, E. Muskelvarietät: Hautmuskel über dem M. deltoideus. *Anat. Anz.* 36:643–645, 1910.
209* GRIFFITH, T.W. Anatomical notes and queries. Bilateral chondro-epitrochlearis in the living subject. *J. Anat. Physiol.* 33:501–502, 1899.
210. GRIFFITH, T.W. Note on a case of muscular abnormality observed during life. *J. Anat. Physiol.* 36:387–388, 1902.
211* GROENEVELD, G. Über einen M. biceps brachii mit fehlender Sehne. *Anat. Anz.* 63:358–361, 1927.
212* GRUBER, W. Die eigenen Spanner des Ringbandes des Radius – Musculi tensores proprii ligamenti annularis radii – bei dem Menschen. *Arch. Anat. Physiol. Wissen. Med.* 1865:377–391.
213* GRUBER, W. Neue supernumeräre Schlüsselbeinmuskeln. *Arch. Anat. Physiol. Wissen. Med.* 1865: pp. 703–718.
214* GRUBER, W. Die eigenen Spanner des Ringbandes des Radius – Musculi tensores proprii ligamenti annularis radii – bei dem Menschen. *Arch. Anat. Physiol. Wissen. Med.* 1865: pp. 377–391.
215* GRUBER, W. Neue supernumeräre Schlüsselbeinmuskeln. *Arch. Anat. Physiol. Wissen. Med.* 1865: pp. 703–718.
216. GRUBER, W. Über das Zungenbein-Schildknorpel-Hilfsband (Ligamentum hyothyreoideum accessorium). *Arch. Anat. Physiol. Wissen. Med.* 1868: pp. 633–634.
217. GRUBER, W. Über die Muskeln des unteren Schildknorpelrandes (Musculi thyroeoidei marginales inferiores). *Arch. Anat. Physiol. Wissen. Med.* 1868: pp. 635–639.
218. GRUBER, W. Über den seltenen Schildknorpelhorn-Gießbeckenknorpelmuskel (Musculus keratoarytaenoideus). *Arch. Anat. Physiol. Wissen. Med.* 1868: pp. 640–641.
219. GRUBER, W. Über eine neue Variante des Musculus thyreotrachealis und über den Musculus hyo-trachealis. *Arch. Anat. Physiol. Wissen. Med.* 1868: pp. 642–645.
220. GRUBER, W. Über die Muskeln des unteren Schildknorpelrandes – Musculi thyreoidei marginalis inferiores. *Arch. Anat. Physiol. Wissen. Med.* 1868: pp. 635–639.
221* GRUBER, W. Über den seltenen Schildknorpelhorn-Gießbeckenknorpelmuskel. *Arch. Anat. Physiol. Wissen. Med.* 1868: pp. 640–641.
222. GRUBER, W. Über eine neue Variante des Musculus thyreotrachealis und über den Musculus hyotrachealis. *Arch. Anat. Physiol. Wissen. Med.* 1868: pp. 642–645.
223* GRUBER, W. Über die Varietäten des Musculus palmaris longus. *Mém. l'Acad. Imp. Sci. St. Petersbourg S.* 7 11:1–26, 1868.
224. GRUBER, W. Über das neue Anheftungsbündel des Oesophagus an den Bronchus dexter-Musculus bronchooesophageus dexter. *Arch. Anat. Physiol. Wissen. Med.* 1869: pp. 519–520.
225. GRUBER, W. Über einen Musculus tibio-astragaleus anticus des Menschen. *Arch. Anat. Physiol. Wissen. Medizin* 1871: pp. 663–668.
226. GRUBER, W. Über den Musculus und über die neue Bursa mucosa ilio-costocervicalis am Tuberculum der ersten Rippe, und über accidentelle Bursae mucosae am Rücken. *Arch. Anat. Physiol. Wissen. Med.* 1871: pp. 669–693.
227* GRUBER, W. Mangel der mittleren Portion des Musculus deltoideus. *Arch. Path. Anat. Physiol. Klin. Med.* 54:184–185, 1872.
228* GRUBER, W. Ein Spanner der unteren Radio-Cubitalkapsel: Tensor capsulae radiocubitalis inferioris. *Arch. Path. Anat. Physiol. Klin. Med.* 54:186–187, 1872.
229* GRUBER, W. Dissection d'un bras droit ayant le pouce double. *Bull. l'Acad. Imp. Sci. de St. Petersbourg* 17:24–31, 1872.
230* GRUBER, W. Sur un muscle tenseur de l'aponeurose surale partant du demitendineux. *Bull. l'Acad. Imp. Sci. St. Petersbourg* 17:289–291, 1872.
231* GRUBER, W. Sur quelques varietes du muscle palmaire grele. *Bull. l'Acad. Imp. Sci. St. Petersbourg* 17:291–294, 1872.
232. GRUBER, W. Sur un muscle biceps humeral ayant une portion coracoidienne et une portion humeral anomale à la place de la portion glenoidienne. *Bull. l'Acad. Imp. Sci. St. Petersbourg* 17:314–318, 1872.
233. GRUBER, W. Un muscle radial interne long bicaude chez l'homme, remplacant par une portion surnumeraire le palmaire grele, qui manque. *Bull. l'Acad. Imp. Sci. St. Petersbourg* 17:318–319, 1872.
234. GRUBER, W. Sur les varietes du muscle radial interne court. *Bull. l'Acad. Imp. Sci. St. Petersbourg* 17:379–38, 1872.
235. GRUBER, W. Sur un muscle costo-coracoidien surnumeraire chez l'homme. *Bull. l'Acad. Imp. Sci. St. Petersbourg* 17:408–413, 1872.
236. GRUBER, W. Sur le muscle sterno-fascial chez l'homme. *Bull. l'Acad. Imp. Sci. St. Petersbourg* 17:497–498, 1872.
237. GRUBER, W. Sur quelques muscles surnumeraires de l'abdomen, chez l'homme. *Bull. l'Acad. Imp. Sci. St. Petersbourg* 18:142–147, 1873.
238. GRUBER W., Sur un muscle cleido-hyoidien et sur un muscle supraclaviculaire singulier chez l'homme. *Bull. d. l'Acad. Imp. des Sci. de St. Petersbourg.* 18:154–157, 1873.
239. GRUBER, W. Un cas de muscle oblique interne de l'abdomen, prive completement de sa portion inguinale. *Bull. l'Acad. Imp. Sci. St. Petersbourg* 18:157–158, 1873.
240. GRUBER, W. Sur une variante du muscle tenseur de l'aponevrose surale, partant du muscle demi-tendineux. *Bull. l'Acad. Imp. Sci. St. Petersbourg* 18:184–186, 1873.
241* GRUBER, W. Über den Musculus plantaris bicaudatus mit Endigung seines supernumerären Schwanzes im Ligamentum popliteum. *Arch. Anat. Physiol. Wissen. Med.* 1874: pp. 467–473.
242* GRUBER, W. Ein Musculus piso-hamatus beim Menschen. *Arch. Anat. Physiol. Wissen. Med.* 1875: pp. 202–203.
243* GRUBER, W. Über den Musculus extensor digitorum communis manus anomalus mit 5 Sehnen zu allen Fingern, und über den Musculus extensor digitorum longus pedis anomalus mit 5 Sehnen zu allen Zehen. *Arch. Anat. Physiol. Wissen. Med.* 1875: pp. 204–210.
244. GRUBER, W. Ein Fall des Vorkommens des Musculus flexor pollicis longus beim Menschen: als Tensor bursae mucosae tendinum m. flexorum, oder als Kopf des M. flexor digitorum profundus manus. *Arch. Anat. Physiol. Wissen. Med.* 1875: pp. 211–214.
245* GRUBER, W. Über die Varietäten des Musculus extensor hallucis longus. *Arch. Anat. Physiol. Wissen. Med.* 1875: pp. 565–589.
246. GRUBER, W. Ein Musculus teres minimus scapulae. *Arch. Anat. Physiol. Wissen. Med.* 1875: pp. 593–598.
247* GRUBER, W. Über den Musculus popliteus biceps. *Arch. Anat. Physiol. Wissen. Med.* 1875: pp. 599–605.
248. GRUBER, W. Über einige seltene Zungenbein- und Kehlkopfmuskeln. *Arch. Anat. Physiol. Wissen. Med.* 1875: pp. 606–609.
249. GRUBER, W. Zwei neue Fälle eines rudimentären Musculus obliquus externus abdominis II. *Arch. path. Anat. Physiol. Klin. Med.* 65:16–17, 1875.
250. GRUBER, W. Ein Musculus scapulo-clavicularis. *Arch. Path. Anat. Physiol. Klin. Med.* 65:18–21, 1875.
251. GRUBER, Ein neuer Fall des Musculus tensor semivaginae articulationis humeroscapularis. *Arch. Path. Anat. Physiol. Klin. Med.* 65:21–23, 1875.

252 GRUBER, W. Ein Musculus extensor digiti II pedis longus. *Arch. Path. Anat. Physiol. Klin. Med.* 65:23–25, 1875.

253* GRUBER, W. Über den Musculus atlantico-mastoideus. *Arch. Anat. Physiol. Wissen. Med.* 1876: pp. 733–738.

254* GRUBER, W. Ein Musculus cleido-epistrophicus bei Existenz des Musculus cleidomastoideus der Norm. *Arch. Anat. Physiol. Wissen. Med.* 1876: pp. 739–745.

255* GRUBER, W. Über den Musculus rectus capitis anticus medius s. minimus. *Arch. Anat. Physiol. Wissen. Med.* 1876: pp. 746–749.

256* GRUBER, W. Ein neuer Fall von Musculus extensor hallucis longus tricaudatus. *Arch. Anat. Physiol. Wissen. Med.* 1876: pp. 750–752.

257 GRUBER, W. Ein Musculus cleido-cervicalis s. tracheloclavicularis imus. *Arch. Anat. Physiol. Wissen. Med.* 1876: pp. 757–758.

258 GRUBER, W. Vorkommen des Musculus cleido-mastoideus als Musculus cleido-epistrophicus. *Arch. Anat. Physiol. Wissen. Med.* 1876: pp. 759–760.

259 GRUBER, W. Ein Musculus cleido-atlanticus. *Arch. Anat. Physiol. Wissen. Med.* 1876: p. 761.

260 GRUBER, W. Über einen anomalen Musculus transversus perinei superficialis. *Arch. Path. Anat. Phys. Klin. Med.* 67:353–358, 1876.

261 GRUBER, W. Verlängerung des Musculus spinalis und semispinalis cervicis durch ein gemeinschaftliches Fleischbündel auf das Hinterhaupt. *Arch. Path. Anat. Physiol. Klin. Med.* 67:358–359, 1876.

262 GRUBER, W. Neuer Fall eines außerhalb des Regio perinealis entsprungenen Musculus transversus perinei superficialis. *Arch. Path. Anat. Physiol. Klin. Med.* 68:287–290, 1876.

263* GRUBER, W. Über den Musculus radialis externus accessorius. *Arch. Anat. Physiol. Wissen. Med.* 1877: pp. 388–398.

264* GRUBER, W. Über den Musculus peroneo-tibialis beim Menschen. *Arch. Anat. Physiol. Wissen. Med.* 1877: pp. 401–410.

265* GRUBER, W. Über einen besonderen Canal für den Nervus medianus in Sulcus bicipitalis internus bei Vorkommen eines Musculus costoepitrochlearis. *Arch. Anat. Physiol. Wissen. Med.* 1877: pp. 411–423.

266* GRUBER, W. Über das neue Anheftungsbündel des Oesophagus an die Glandula thyroidea-Musculus thyreo-oesophageus. *Arch. Path. Anat. Physiol. Klin. Med.* 69:396–398, 1877.

267 GRUBER, W. Ein von der Sternocostalportion des Musculus pectoralis major abgegebener besonderer Tensor fasciae brachialis. *Arch. Path. Anat. Physiol. Klin. Med.* 69:398–400, 1877.

268 GRUBER, W. Tensor laminae posterioris vaginae musculi recti abdominis. *Arch. Path. Anat. Physiol. Klin. Med.* 69:400–403, 1877.

269 GRUBER, W. Über den Sehnenbogen des Musculus soleus und seine ungewöhnlichen Spannmuskeln. *Arch. Anat. Physiol. Wissen. Med.* 1878: pp. 474–480.

270 GRUBER, W. Nachträge über den Musculus peroneotibialis. *Arch. Anat. Physiol. Wissen. Med.* 1878: pp. 481–485.

271 GRUBER, W. Ein Musculus praeclavicularis subcutaneous. *Arch. Path. Anat. Physiol. Klin. Med.* 72:496–497, 1878.

272 GRUBER, W. Ein Nachtrag zum Vorkommen des Musculus interclavicularis anticus digastricus. *Arch. Path. Anat. Physiol. Klin. Med.* 72:497–500, 1878.

273 GRUBER, W. Musculus extensor digitorum communis manus anomalus mit 5 Sehnen zu allen Fingern. *Arch. Path. Anat. Physiol. Klin. Med.* 72:500–501, 1878.

274 GRUBER, W. Über einen Fall einseitigen Vorkommens zweier den Musculus omohyoideus substituirender Musculi cleidohyoidei. *Arch. Path. Anat. Physiol. Klin. Med.* 73:342–345, 1878.

275 GRUBER, W. Beobachtungen über den Mangel des Musculus omohyoideus. *Arch. Path. Anat. Physiol. Klin. Med.* 73:345–346, 1878.

276 GRUBER, W. Beobachtungen über den Mangel des Musculus quadratus femoris. *Arch. Path. Anat. Physiol. Klin. Med.* 73:346–348, 1878.

277 GRUBER, W. Ein Musculus obturator internus biceps. *Arch. Path. Anat. Physiol. Klin. Med.* 73:348–350, 1878.

278 GRUBER, W. Ein den oberen Bauch des Omohyoideus (bei Mangel des unteren Bauches des letzteren) repräsentirender Musculus hyofacialis. *Arch. Path. Anat. Physiol. Klin. Med.* 74:454–456, 1878.

279 GRUBER, W. Über den Gesäß-Mittelfleischmuskel (Musculus gluteo-perinealis). *Arch. Path. Anat. Physiol. Klin. Med.* 74:456–460, 1878.

280 GRUBER, W. Musculus hyo-facialis. *Arch. Path. Anat. Physiol. Klin. Med.* 77:123, 1879.

281 GRUBER, W. Nachträge zum Vorkommen des Musculus scapulo-costalis minor und neuer Musculus scapulo-costoclavicularis. *Arch. Path. Anat. Physiol. Klin. Med.* 77:123–128, 1879.

282 GRUBER, W. Nachtrag zum Vorkommen des Musculus extensor digitorum communis manus mit 5 Sehnen zu allen Fingern. *Arch Path. Anat. Physiol. Klin. Med.* 77:129–130, 1879.

283 GRUBER, W. Nachtrag zum Vorkommen des Musculus tensor laminae posterioris vaginae musculi recti abdominis. *Arch. Path. Anat. Physiol. Klin. Med.* 77:130–131, 1879.

284 GRUBER, W. Nachtrag zu den Beobachtungen über den Mangel des Musculus quadratus femoris. *Arch. Path. Anat. Physiol. Klin. Med.* 77:131, 1879.

285 GRUBER, W. Ein in der Haut oder Fasciae des Gesichtes und mit seiner Endzacke am Mundwinkel als Musculus risorius endender Musculus occipitalis minor-Santorini. *Arch. Path. Anat. Physiol. Klin. Med.* 80:83–86, 1880.

286 GRUBER, W. Tensor laminae posterioris vaginae musculi recti abdominis. *Arch. Path. Anat. Physiol. Klin. Med.* 80:87–88, 1880.

287 GRUBER, W. Ein mit seiner inguinalen Portion durch die ganze Regio inguinalis sich herab erstreckender Musculus transversus abdominis. *Arch. Path. Anat. Physiol. Klin. Med.* 80:88–91, 1880.

288 GRUBER, W. Über den Musculus trigastricus maxillae inferioris. *Arch. Path. Anat. Physiol. Klin. Med.* 81:445–449, 1880.

289 GRUBER, W. Musculus digastricus maxillae inferioris mit Ursprung seines vorderen Bauches an und hinter der Mitte des Seitentheiles der Maxilla im Bereiche der Strecke zwischen dem Ansatze des M. masseter und dem Ursprunge des M. depressor anguli oris. *Arch. Path. Anat. Physiol. Klin. Med.* 81:449–453, 1880.

290* GRUBER, W. Über den Musculus myloglossus bei Mangel und Vorkommen des M. styloglossus. *Arch. Path. Anat. Physiol. Klin. Med.* 81:453–457, 1880.

291* GRUBER, W. Über den Musculus atlantico-basilaris, M. epistropheobasilaris und andere von den obersten Halswirbeln zwischen den Mm. recti capitis antici majores vorkommende Muskelvarietäten. *Arch. Path. Anat. Physiol. Klin. Med.* 82:465–473, 1880.

292 GRUBER, W. Ein bilaminärer Musculus vastus externus biceps. *Arch. Path. Anat. Physiol. Klin. Med.* 82:473–474, 1880.

293 GRUBER, W. Ein Musculus ulnaris externus brevis beim Menschen. *Arch. Path. Anat. Physiol. Klin. Med.* 86:15–19, 1881.

294* GRUBER, W. Über den Arcus tendineus piso-hamatus, und dessen Substituten, den Musculus pisohamatus. *Arch. Path. Anat. Physiol. Klin. Med.* 86:19–25, 1881.

295 GRUBER, W. Vollständiger Mangel des Musculus tensor fasciae latae beim Menschen. *Arch. Path. Anat. Physiol. Klin. Med.* 86:25–27, 1881.

296 GRUBER, W. Über den dem constanten Musculus extensor pollicis et indicis gewisser Säugetiere homologen supernumerären Muskel beim Menschen. *Arch. Path. Anat. Physiol. Klin. Med.* 86:471–491, 1881.

297 GRUBER, W. Supernumeräre, der Glandula submaxilaris zur Stütze dienende Schicht des Musculus mylohyoideus. *Arch. Path. Anat. Physiol. Klin. Med.* 86:491–492, 1881.

298 GRUBER, W. Musculus sterno-fascialis. *Arch. Path. Anat. Physiol. Klin. Med.* 86:492–493, 1881.

299 GRUBER, W. Über einen anomalen, den constanten Musculus extensor digitorum communis manus zu allen 5 Fingern bei den Säugetieren Genera Myogale und Fiber homologen Muskel beim Menschen. *Arch. Path. Anat. Physiol. Klin. Med.* 90:88–97, 1882.

300 GRUBER, W. Über anomale, dem constanten Musculus extensor digitorum communis manus zum 2.–4. Finger bei Aspalax und Dasypus, und dem constanten Musculus extensor digitorum communis manus zum 1.–4. Finger bei Echidna homologe Muskeln beim Menschen. *Arch. Path. Anat. Physiol. Klin. Med.* 90:97–103, 1882.

301 GRUBER, W. Über den zum Extensor pollicis et indicis singularis gewordenen Extensor pollicis longus beim Menschen. *Arch. Path. Anat. Physiol. Klin. Med.* 90:103–108, 1882.

302 GRUBER, W. Über einen zum Extensor indicis et pollicis singularis gewordenen Extensor indicis proprius beim Menschen. – Dasypeus Bildung. *Arch. Path. Anat. Physiol. Klin. Med.* 90:108–110, 1882.

303 GRUBER, W. Ein neuer Musculus peroneo-calcaneus externus anterior. *Arch. Path. Anat. Physiol. Klin. Med.* 95:177–179, 1884.

304 GRUBER, W. Mangel beider Musculi gemelli bei Anwesenheit des Obturator internus. *Arch. Path. Anat. Physiol. Klin. Med.* 95:180–183, 1884.

305 GRUBER, W. Neuer Musculus retro-clavicularis proprius-Tensor laminae profundae fascia colli. *Arch. Path. Anat. Physiol. Klin. Med.* 98:416–422, 1884.

306* GRUBER, W. Neuer Musculus radialis digiti I. s. pollicis. *Arch. Path. Anat. Physiol. Klin. Med.* 98:422–424, 1884.

307 GRUBER, W. Der musculus ulno-carpeus externus (eine neue Variante des Musculus ulnaris externus brevis) und sein Homologen, der Musculus peroneo-cuboideus. *Arch. Path. Anat. Physiol. Klin. Med.* 99:475–478, 1885.

308 GRUBER, W. Mangel des Musculus ulnaris externus bis auf einen Sehnenstreifen. *Arch. Path. Anat. Physiol. Klin. Med.* 99:478–480, 1885.

309 GRUBER, W. Zweibäuchiger Extensor digiti V proprius manus mit Insertion eines seiner Bäuche an die Basis des Metacarpale V, und die ihm homologe Variante des Peroneus III. *Arch. Path. Anat. Physiol. Klin. Med.* 99:480–483, 1885

310 GRUBER, W. Absoluter Mangel des Extensor digiti quinti proprius manus bei Mangel seiner Vagina im Ligamentum carpi dorsale: ganz ausnahmsweise und nur beim Menschen (2. Fall) bei Vorkommen des Muskels dennoch Mangel der Vagina constant bei den Prosimiae. *Arch. Path. Anat. Physiol. Klin. Med.* 99:484–488, 1885.

311 GRUBER, W. Über die im Bereiche des Lig. carpi dorsale schon congenital und in der Norm vorkommende Communication der Sehnenscheide für den Extensor pollicis longus mit der für die Radialis externi (nach 500 Untersuchungen), und über einen Fall von Hygroma proliferum dieser mit einander communicirender Sehnenscheiden. *Arch. Path. Anat. Physiol. Klin. Med.* 101:245–253, 1885.

312 GRUBER, W. Auftreten der Bäuche des Musculus digastricus maxillae inferioris als selbständige Muskeln Musculus mentohyoideus et M. mastoideo-maxillaris. *Arch. Path. Anat. Physiol. Klin. Med.* 101:253–255, 1885.

313 GRUBER, W. Über den seltenen Musculus crico-hyoideus medianus oder den wahren supernumerären Musculus crico-hyoideus. *Arch. Path. Anat. Physiol. Klin. Med.* 101:256–259, 1885.

314 GRUBER, W. Ein Musculus tensor ligamenti carpi dorsalis. *Arch. Path. Anat. Physiol. Klin. Med.* 101:259–260, 1885.

315* GRUBER, W. Duplicität des Musculus extensor digiti quinti et quarti proprius manus. *Arch. Path. Anat. Physiol. Klin. Med.* 101:260–262, 1885.

316 GRUBER, W. Ein Musculus stylo-pharyngeus biceps durch Auftreten eines vom Processus mastoides entsprungenen Caput accessorium. *Arch. Path. Anat. Physiol. Klin. Med.* 102:536–538, 1885.

317 GRUBER, W. Supernumerärer Bauch des Musculus sternocleidomastoideus in der Richtung des hinteren Bauches des M. digastricus maxillae inferioris und abwärts von diesem zum Os hyoides. *Arch. Path. Anat. Physiol. Klin. Med.* 102:538–539, 1885.

318 GRUBER, W. Ein Musculus cleido-occipitalis mit enorm breiter Endaponeurose. *Arch. Path. Anat. Physiol. Klin. Med.* 102:539–540, 1885.

319 GRUBER, W. Musculus flexor digitorum profundus mit Mangel einer Sehne zum Mittelfinger. *Arch. Path. Anat. Physiol. Klin. Med.* 103:472–474, 1886.

320 GRUBER, W. Ein Tensor capsulae radio-carpalis dorsalis. *Arch. Path. Anat. Physiol. Klin. Med.* 103:474–475, 1886.

321 GRUBER, W. Ein Tensor fasciae femoris posterior digastricus. *Arch. Path. Anat. Physiol. Klin. Med.* 103:475–477, 1886.

322 GRUBER, W. Wahrer Musculus semitendinosus biceps. *Arch. Path. Anat. Physiol. Klin. Med.* 103:477–479, 1886.

323 GRUBER, W. Musculus semimembranosus mit zwei bis vier Bäuchen. *Arch. Path. Anat. Physiol. Klin. Med.* 103:480–483, 1886.

324 GRUBER, W. Vollständiger Mangel des Musculus semimembranosus an der einen Seite und rudimentäres Vorkommen an der anderen Seite. *Arch. Path. Anat. Physiol. Klin. Med.* 103:483–484, 1886.

325 GRUBER, W. Über den anomalen Musculus abductor metatarsi quinti, seine Substitution durch einen Sehnenstrang (neu) und sein Auftreten als M. abductor metatarsi quinti circumflexus (neu) beim Menschen, sowie über analoge Homologien dafür bei Säugethieren. *Arch. Path. Anat. Physiol. Klin. Med.* 106:489–501, 1886.

326 GRUBER, W. Mangel der Portio sterno-costalis des Musculus pectoralis major und gleichzeitiger Mangel der Mamma an der entsprechenden Seite bei einer Jung-frau. *Arch. Path. Anat. Physiol. Klin. Med.* 106:501–502, 1886.

327 GRUBER, W. Auftreten des Musculus radialis externus accessorius als Musculus supinator longus II. s. accessorius. *Arch. Path. Anat. Physiol. Klin. Med.* 107:476–478, 1887.

328 GRUBER, W. Seltener Extensor proprius digiti quinti manus bei dem Menschen und bei Säugethieren. *Arch. Path. Anat. Physiol. Klin. Med.* 107:478–480, 1887.

329 GRUBER, W. Über einen Musculus glutaeus quartus bei dem Menschen und einen homologen Muskel bei Säugethieren. *Arch. Path. Anat. Physiol. Klin. Med.* 107:480–484, 1887.

330 GRUBER, W. Ein dreibäuchiger Musculus tibialis anticus, zugleich ein Tensor ligamenti cruciati tarsi. *Arch. Path. Anat. Physiol. Klin. Med.* 107:485, 1887.

331 GRUBER, W. Musculus peroneus longus und brevis zu einem Muskel verschmolzen. *Arch. Path. Anat. Physiol. Klin. Med.* 107:485–487, 1887.

332 GRUBER, W. Ein Musculus peroneus brevis mit Insertion an den Calcaneus, bei Abgabe einer mit einem Fleischbauche versehenen Fussrückensehne zur 5. Zehe (eines auf den Fussrücken verkürzten Musculus peroneus digiti quinti). *Arch. Path. Anat. Physiol. Klin. Med.* 107:487–489, 1887.

333 GRUBER, W. Ein seltener Curvator coccygis accessorius beim Menschen, homolog dem constanten Depressor caudae longus bei gewissen Säugethieren. *Arch. Path. Anat. Physiol. Klin. Med.* 109:1–4, 1887.

334 GRUBER, W. Ein Musculus gracilis biceps. *Arch. Path. Anat. Physiol. Klin. Med.* 109:4, 1887.

335 GRUBER, W. Dreibäuchiger Musculus peroneus longus. *Arch. Path. Anat. Physiol. Klin. Med.* 109:5, 1887.

336 GRUBER, W. Ein Musculus peroneo-malleolaris als Tensor des Ligamentum intermusculare externum posterius fasciae cruris. *Arch. Path. Anat. Physiol. Klin. Med.* 109:5–6, 1887.

337 GRUBER, W. Ein Musculus flexor brevis digiti II pedis. *Arch. Path. Anat. Physiol. Klin. Med.* 109:6–8, 1887.

338 GRUBER, W. Ein Musculus flexor brevis digiti IV pedis. *Arch. Path. Anat. Physiol. Klin. Med.* 109:8, 1887.

339 GRUBER, W. Eine Reihe neuer Varietäten des Musculus lumbricalis I manus. *Arch. Path. Anat. Physiol. Klin. Med.* 110:555–559, 1887.

340 GRUBER, W. Auftreten des Zeigefingerbauches des Flexor digitorum sublimis als Venter bifissus digastricus biceps. *Arch. Path. Anat. Physiol. Klin. Med.* 110:559–561, 1887.

341 GRUBER, W. Ein rudimentärer Musculus obliquus abdominis externus accessorius. *Arch. Path. Anat. Physiol. Klin. Med.* 110:561, 1887.

342 GRUBER, W. Reduction des Ulnaris internus digiti V auf einen Muskel in der Vola manus. *Arch. Path. Anat. Physiol. Klin. Med.* 336–365, 1888.

343 GRUBER, W. Ein Musculus piso-metacarpeus. *Arch. Path. Anat. Physiol. Klin. Med.* 114:365–367, 1888.

344 GRUBER, W. Ein Musculus rectus femoris accessorius. *Arch. Path. Anat. Physiol. Klin. Med.* 114:367–368, 1888.

345 GRUBER, W. Vagina propria im Ligamentum carpi dorsale für den Bauch des Extensor digitorum longus manus zum Zeigefinger beim Menschen – und dieselbe analoge Vagina für den Bauch desselben Muskels für die 2. Zehe des Vorderfusses bei Phascolarctos cinereus. *Arch. Path. Anat. Physiol. Klin. Med.* 114:369–371, 1888.

346 GRUBER, W. Theilung der Vagina am Ligamentum der Fascia cruralis in der Fussbeuge in zwei secundäre Vaginae für den Musculus extensor digitorium longus pedis bei dem Menschen und bei Säugethieren. *Arch. Path. Anat. Physiol. Klin. Med.* 114:371–374, 1888.

347 GUERRIER, Y. and NGO-VAN-HEIN. L'appareil musculaire du segment pharyngo-oesophagien ètude statistique. *L'Assoc. Anatomistes, Comptes Rendus.* 39:279–288, 1952.

348 GUERRIER, Y. and A. THÉVENET. La région diaphragmatique antérieure. *L'Assoc. Anatomistes, Comptes Rendus.* 42:620–633, 1955.

349 GUINANE, F.R. Two anomalies in the construction of the diaphragm. *J. Anat.* 59:83–86, 1924.

350 HADZISELIMOVIC, V. and N. GLUHBEGOVIC. Vergleichende anatomische Untersuchungen des Musculus popliteus. *Anat. Anz.* 125:388–399, 1969.

351* HAFFNER, HERBERT VON. Eine seltene doppelseitige Anomalie des Trapezius. *Int. Monatsschr. Anat. Physiol.* 20:313–318, 1903.

352* HALL, G.A.M. Congenital absence of pectoralis major muscle. *Chin. Med. J.* 39:224–225, 1925.

353 HALL, H.S. Complete absence of the superficial flexors of the thumb and concurrent muscular anomalies. *J. Anat. Physiol.* 38:287–289, 1903.

354* HALLISY, J.E. The muscular variations in the human foot. A quantitative study. *Am. J. Anat.* 45:411–442, 1930.

355 HARBESON, A.E. The origin of the first dorsal interosseus muscle of the foot. *J. Anat.* 68:116–118, 1933–34.

356 HARBESON, A.E. Further studies on the origin of the first dorsal

interosseus muscle of the foot from the tendon of the peroneus longus. *J. Anat.* 72:463–464, 1937–38.
357 HART, J.A.L. Extensor digitorum brevis manus. *Hand* 4:265–267, 1972.
358 HASAN, M. Pectoralis minimus muscle. *J. Anat. Soc. India* 11:89, 1962.
359 HECKLER, P. Study of the peroneus of the tarsus. *Anat. Rec.* 26:79–82, 1923.
360 HEPBURN, D. The adductor muscles of the thumb and great toe. *J. Anat. Physiol.* 27:408–410, 1893.
361 HEPBURN, D. Complete separation of two heads of biceps flexor cruris muscle. *J. Anat. Physiol.* 27:282–284, 1893.
362 HEPBURN, D. Abnormalities of muscles, nerves, heart, vessels and ligaments. *J. Anat. Physiol.* 30:570–583, 1896.
363 HIGGINS, H. The popliteus muscle. *J. Anat. Physiol.* 29:569–573, 1895.
364 HILDEBRAND, R. Discovery of a variant in the region of the adductor magnus and the short head of the biceps femoris *Anat. Anz.* 144:48–50, 1978.
365 HIRASAWA, Y., SAWAMURA, H. and K. SAKAKIDA. Entrapment neuropathy due to bilateral epitrochleoanconeus muscles. A case report. *J. Hand Surg.* 4:181–184, 1979.
366 HOFFMAN. Ein Fall von angeborenem Brustmuskeldefect mit Atrophie des Armes und Schwimmhautbildung. *Arch. Path. Anat. Physiol. Klin. Med.* 146:163–172, 1869.
367 HOLMGREN, H. Ein Fall von bilateraler Doppelbildung des Caput longum Musculi bicipitis femoris. *Anatomischer Anzeiger* 86:269–271, 1938.
368 HOUSTON. Two newly-discovered muscles for compressing the dorsal vein of the penis in man and other animals. *Am. J. Med. Sci.* 8:477–478, 1831.
369 HOWELL, A.B. Anatomy of the inguinal region. *Surgery* 6:653–662, 1939.
370* HUBER, E. Über die Morphologie des M. procerus nasi des Menschen. *Anat. Anz.* 51:302–308, 1918–19.
371 HUBER, E. Überreste des Sphincter colli profundus beim Menschen. *Anat. Anz.* 51:480–492, 1918–19.
372 HUMPHRY, G.M. Lectures on the varieties in the muscles of man. *Br. Med. J.* 2:33–37, 51–54, 78–80, 108–111, 1873.
373 HUMPHRY, G.M. Abstract of lectures on varieties in the muscles of man. *Lancet* 1:834–835, 1873.
374 HUMPHRY, G.M. Abstract of lectures on varieties in the muscles of man. *Lancet* 1:871 872, 1873.
375 HUNTINGTON, G.S. Present problems of myological research and the significance and classification of muscular variations. *Am. J. Anat.* 2:157–175, 1902–03.
376* HUNTINGTON, G.S. The derivation and significance of certain supernumerary muscles of the pectoral region. *J. Anat. Physiol.* 39:1–54, 1905.
377* INGALLS, N.W. Musculi sternales and infraclavicularis. *Anat. Rec.* 7:203–206, 1913.
378 IRVINE, E.D. and J.B. TILLEY. Congenital absence of the pectoral muscles. *Arch. Dis. Child.* 12:123–126, 1937.
379* ISOMURA, G. Nerve supply for anomalous ocular muscle in man. *Anat. Anz.* 142:255–265, 1977.
380 JACOBI, F.H. Beitrag zur Anatomie der Steissbeinmusculatur des Menschen. *Arch. Anat. Physiol. Wissen. Med.* 1888: pp. 353–364.
381 JACQUART, H. Exemple d'insertion anormale du muscle adducteur du pouce sur une main d'homme, qui prouve que ce muscle n'est en réalité que le premier interosseux palmaire. *Soc. Biol. Comptes Rendus des Séances et Mémoires* 11:252–253, 1859.
382 JAVADEKAR, B.S. Peroneus digiti minimi. *J. Anat. Soc. India* 10:44, 1961.
383* JAZUTA, K.M. Glutaeus medius accessorius. *Anat. Anz.* 72:10–11, 1931.
384 JEFFERY, A.K. Compression of the deep palmar branch of the ulnar nerve by an anomalous muscle: Case report and review. *J. Bone Joint Surg.* 53B:718–723, 1971.
385 JOESSEL. Ein besonderer Fall von Musculus sternalis. *Arch. Anat. Physiol. Wissen. Med.* 1878: pp. 429–432.
386 JONES, B.V. An anomalous extensor indicis muscle. *J. Bone Joint Surg.* 41B:763–765, 1959.
387* JORDON, J. On the deep stylohyoid muscle in man. *Folia Morphol.* 22:7–10, 1963.
388 JUNG, A. Eine noch nicht beschriebene Anomalie des Musculus omo-hyoideus. *Anat. Anz.* 7:582–584, 1892.
389* KACKER, G.N. Sternalis muscle in U.P. Indian subjects. *J. Anat. Soc. India* 9:101–103, 1960.
390* KAMEDA, Y. An anomalous muscle (accessory subscapularis-teres-latissimus muscle) in the axilla penetrating the brachial plexus in man. *Acta Anat.* 96:513–533, 1976.
391 KANEFF, A. Über eine sehr seltene Kombination von Varietäten des tiefen Fingerstreckers, M. extensor digitorum profundus, beim Menschen. *Anat. Anz.* 107:424–428, 1959.
392 KANEFF, A. Sehr seltene Kombination von Muskelvarietäten der radialen und dorsalen Muskelgruppen eines menschlichen Vorderarmes. *Anat. Anz.* 125:535–538, 1969.
393 KANEFF, A. and K. KOITSCHEFF. Eine seltene Kombination von Muskelvarietäten am Rücken des menschlichen Fußes. *Anat. Anz.* 114:425–429, 1964.
394 KATER, N.W. Two cases of supernumerary radio-palmar muscle – muscle surnumeraire radio-palmaire of Testut. *J. Anat. Physiol.* 36:76, 1902.
395 KEITH, A.M. Note on supracostalis anterior. *J. Anat. Physiol.* 27:333–334, 1894.
396 KELLER, A. Ein Musculus fibulotarsalis beim Menschen. *Anat. Anz.* 56:481–488, 1923.
397* KING, T.S. and R. O'RAHILLY. M. palmaris accessorius and duplication of m. palmaris longus. *Acta Anat.* 10:327–331, 1950.
398* KIRGIS, H.D. and A.F. REED. Significant anatomic relations in the syndrome of the scalene muscles. *Ann. Surg.* 127:1182–1201, 1948.
399* KIRK, T.S. Sternalis muscle (in the living). *J. Anat.* 59:192, 1924–25.
400 KLAATSCH, H. Über eine dem Tenuissimus ähnliche Variation am Biceps femoris des Menschen. *Anat. Anz.* 38:305–310, 1911.
401 KNAPP, E. Über eine seltene Varietät der Nackenmuskulatur (M. splenius accessorius). *Anat. Anz.* 118:178–180, 1966.
402 KNOTT, J.F. Muscular anomalies. *J. Anat. Physiol.* 15:139–140, 1881.
403 KÖRNER, F. Eine Verdoppelung des Musculus flexor hallucis longus. *Anat. Anz.* 97:16–22, 1949–50.
404 KRAUSE, W. Der M. sternocleidomastoideus. *Centralblat. Med. Wissen.* 1876, pp. 433–435.
405 KRAUSE, W. Myologische Bemerkungen. *Arch. Anat. Physiol. Wissen. Med.* 1881: pp. 419–422.
406 KRISHNAMURTI, A. Anomalous insertion of extensor pollicis longus. *J. Anat. Soc. India* 11:90–91, 1962.
407 KRUEG, G. Ein accessorischer palmaris longus mit doppelter Endsehne. *Wiener med. Wochenschr.* 49:1229, 1872.
408* KUCIŃSKI, P., OKRASZEWSKA, E. and W. PISZCZ. Variability of the course of the cricothyroid muscle in humans. *Folia Morphol.* 38:391–396, 1979.
409 KUCZYNSKI, K. The variations in the insertion of the first dorsal interosseous muscle and their significance in the rheumatoid arthritis. *Hand* 4:37–39, 1972.
410 KUDO, K. and K. BANDO. Varietäten-Statistik des M. biceps brachii beim nordchinesischen Weibe. *Folia Anat. Jpn.* 28:299–301, 1956.
411 LACY, T. III, GOLDSTEIN, L.A. and C.E. TOBIN. Anatomical and clinical study of the variations in the insertions of the abductor pollicis longus tendon, associated with stenosing tendovaginitis. *J. Bone Joint Surg.* 33A:347–350, 1951.
412 LAIDLAW, P.P. A supraclavicularis proprius (Gruber). *J. Anat. Physiol.* 36:417–418, 1902.
413 LAKE, G.B. Congenital absence of the right pectoralis major muscle. *Med. J. Rec.* 119:552, 1924.
414 LAMONT, J.C. Note on a tendon found in association with the insertion of the peroneus longus, and the origin of the first dorsal interosseus muscles. *J. Anat. Physiol.* 42:236, 1908.
415 LANDER, K.F. The pectoralis minor: A morphological study. *J. Anat.* 52:292–318, 1918.
416 LANDRY, S.O. The phylogenetic significance of the chondro-epitrochlearis muscle and its accompanying pectoral abnormalities. *J. Anat.* 92:57–61, 1958.
417 LANE, W.A. Abnormal muscles of the hand., *J. Anat. Physiol.* 21:673, 1887.
418 LANE, W.A. A coraco-clavicular sternal muscle. *J. Anat. Physiol.* 21:673–674, 1887.
419 LAST, R.J. The popliteus muscle and the lateral meniscus. *J. Bone Joint Surg.* 32B:93–99, 1950.
420 LAURENT, H. Über einige Muskelvarietäten. *Anat. Anz.* 12:168–172, 1896.
421 LAZORTHES, G. and A. HAUMONT. La boutonnière scalénique (scalène antérieur, petit scalène). *L'Assoc. Anatomistes, Comptes Rendus* 39:312–315, 1952.
422* LECCO, T.M. Ein Fall von vollständigem Fehlen des langen Kopfes des M. biceps brachii und die damit in Zusammenhang stehen-

423 LEHR, R.P. Musculus levator glandulae thyroidea: An observation. *Anat. Anz.* 146:494–496, 1979.

424* LENGSFELDER, M. Über einen angeborenen Defekt des Musculus pectoralis. *Wiener Klin. Wochenschr.* 15:1306–1309, 1902.

425 LEVADOUX, M. and LEVÊQUE. Note sur les insertions inférieures du muscle grand fessier. *L'Assoc. Anatomistes, Comptes Rendus* 9:128–129, 1907.

426 LEWIS, W.H. Observations on the pectoralis major muscle in man. *John Hopkins Hosp. Bull.* 12:172–177, 1901.

427 LIGHTOLLER, G.H.S. Facial muscles. The modiolus and muscles surrounding the rima oris with some remarks about the panniculus adiposus. *J. Anat.* 60:1–85, 1925.

428* LINBURG, R.M. and B.E. COMSTOCK. Anomalous tendon slips from the flexor pollicis longus to the flexor digitorum profundus. *J. Hand. Surg.* 4:79–83, 1979.

429 LIPSCOMB, P.R. Duplication of hypothenar muscles simulating soft-tissue tumor of the hand: Report of a case. *J. Bone Joint Surg.* 42A:1058–1061, 1960.

430 LIVINI, F. Contribution à la morphologie du M. serratus anterior chez l'homme. *Arch. Ital. Biol.* 44:115–116, 1905.

431 LIVINI, F. Contribution à la morphologie du m. rectus abdominis et du m. supracostalus chez l'homme. *Arch. Ital. Biol.* 44:116–117, 1905.

432 LIVINI, F. Contribuzione alla morfologia des m. rectus abdominis e del m. supra-costalis nell'uomo. *Arch. Ital. Anat. Embriol.* 4:81–115, 1905.

433 LIVINI, F. Morfologia del M. Serratus anterior nell'uomo. *Archiv. Ital. Anat. Embriol.* 6:429–441, 1907.

434 LIVINI, F. Osservazioni anatomische e considerazioni critiche intorno al M. Scaleno nell'uomo. *Archiv. Ital. Anat. Embriol.* 7:1–17, 1908.

435 LO-CASCIO, G. Sur un cas, non encore décrit, relatif au mode anormale de se comporter des muscles péronés latéraux de l'homme. *Arch. Ital. Biol.* 59:474, 1913.

436 LO-CASCIO, G. Contribution à la morphologie de l'arc axillaire de Langer. *Arch. Ital. Biol.* 62:285, 1915.

437 LOETZKE, H.H. and K. TRZENSCHIK. Beitrag zur Frage der Varianten des M. soleus beim Menschen. *Anat. Anz.* 124:28–36, 1969.

438* LOOMIS, L.K. Variations of stenosing tenosynovitis at the radial styloid process. *J. Bone Joint Surg.* 33A:340–346, 1951.

439 LOTOSCHNIKOW, A.K. Die Besonderheiten der Plastik und Topographie des Halses bei einigen Muskelanomalien. *Anat. Anz.* 68:151–156, 1929–30.

440 LOVEJOY, F.J. JR. and T.P. HARDEN. Popliteus muscle in man. *Anat. Rec.* 169:727–730, 1971.

441 LOW, A. A note on the crura of the diaphragm and the muscle of Treitz. *J. Anat. Physiol.* 42:93–96, 1908.

442 LUCIEN, M. Le muscle court extenséur du cinquieme orteil chez l'homme. *Soc. Biol. Comptes Rendus Hebdomadaires des Séances et Mémoires* 67:67–68, 1909.

443 LUCIEN, M. L'indépendance des faisceaux constitutifs du muscle pedieux. *Soc. de Biol. Comptes Rendus Hebdomadaires des Séances et Mémoires* 67:376–377, 1909.

444 LUNGHETTI, C. Sur un muscle surnuméraire axillo-épitrochléaire et sue d'autres anomalies musculaires. *Arch. Ital. Biol.* 47:481–482, 1907.

445* LUSCHKA, H. VON. Ein Musculus supraclavicularis beim Menschen. *Arch. Anat. Physiol. Wissen. Med.* 1856: pp. 282–285.

446 LUSCHKA, H. VON. Der Musc. hyo- und genio-epiglotticus. *Arch. Anat. Physiol. Wissen. Med.* 1868: pp. 224–230.

447 LUSCHKA, H. VON. Der Musc. pubo-transversalis des Menschen. *Arch. Anat. Physiol. Wissen. Med.* 1870: pp. 227–231.

448 MACALISTER, A. On muscular anomalies in human anatomy, and their bearing upon homotypical myology. *Proc. R. Irish Acad.* 10:126–164, 1866–69.

449 MACALISTER, A. Notes on an instance of irregularity in the muscles around the shoulder joint. *J. Anatomy Physiol.* 1:316–319, 1868.

450 MACALISTER, A. On the nature of the coronoid portion of pronator radii teres. *J. Anat. Physiol.* 2:8–12, 1868.

451 MACALISTER, A. The varieties of the styloid muscles. *J. Anat. Physiol.* 5:28–31, 1871.

452 MACALISTER, A. The varieties of the pronator quadratus. *J. Anat. Physiol.* 5:32–34, 1871.

453 MACALISTER, A. Observations on muscular anomalies in the human anatomy. (Third series with a catalogue of the principal muscular variations hitherto published.) *Trans. R. Irish Acad. Sci.* 25:1–130, 1875.

454 MACDOUGALL, J.D.B. The attachments of the masseter muscle. *Brit. Dent. J.* 98:193–199, 1955.

455* MACEWEN, E.M. Orbital variations. *Anat. Rec.* 46:223–232, 1930.

456 MAINLAND, D. An uncommon abnormality of the flexor digitorum sublimis muscle. *J. Anat.* 62:86–89, 1927.

457 MAKHNI, S.S. A short note on the origin of pectoralis minor muscle. *J. Anat. Soc. India* 4:100, 1955.

458 MALHOTRA, V.K., SING, N.P. and S.P. TEWARI. The accessory head of the flexor pollicis muscle and its nerve supply. *Anat. Anz.* 151:503–505, 1982.

459 MALPAS, P. Anomalies of the mylohyoid muscle. *J. Anat.* 61:64–67, 1927.

460 MANNO, A. Uncas de M. extensor digitorum brevis de la main. *Arch. Ital. Biol.* 47:482, 1907.

461* MARTIN, B.F. The oblique cord of the forearm. *J. Anat.* 92:609–615, 1958.

462* MARTIRENÉ. Absence congénitate des muscles pectoraux. *Rev. d'Orthopéd.* 14:209–217, 1903.

463 MEEK, M.O. Some rare muscular anomalies. *J. Anat Physiol.* 69:376–377, 1915.

464* MEHTA, H.J. and W.U. GARDNER. A study of lumbrical muscles in the human hand. *Am. J.Anat.* 109:227–238, 1961.

465 MEYER, A.W. Spolia anatomica addenda I. *Anat. Rec.* 9:483–527, 1914.

466 MEYER, A.W. Spolia anatomica. *J. Anat. Physiol.* 48:107–173, 1914.

467 MIAŚKIEWICZ, C. Studies on the morphology of the gastrocnemius muscle in the black race. *Folia Morphol.* 29:266–272, 1970.

468 MILIANITCH, N. and R. SPIRIDONOVITCH. Variations de la morphologie et de l'innervation des muscles de l'epaule et du bassin chez les Serbes (Etude basée sur 200 cas). *L'Assoc. Anatomistes, Comptes Rendus* 21:385–395, 1926.

469* MILLOY, F.J., ANSON, B.J. and D.K. MCAFFEE. The rectus abdominis muscle and the epigastric arteries. *Surg. Gynecol. Obstet.* 110:293–302, 1960.

470 MINNE, DEPREUX, and FONTAINE. Structure du tendon d'achille importance réciproque de ses constituants. *L'Assoc. Anatomistes, Comptes Rendus* 39:663–666, 1952.

471 MISRA, B.D. The sternalis muscle. *J. Anat. Soc. India.* 3:47–48, 1954.

472 MISRA, B.D. Rotator humeri. *J. Anat. Soc. India* 6:115, 1957.

473 MIURA, T. Findings of hand anomaly associated with pectoral muscle anomaly. *Hand* 10:205–212, 1978.

474* MOGI, E. Muskelvarietäten der unteren Extremitäten bei den japanischen Zwillingsfeten. *Folia Anat. Jpn.* 19:93–95, 1940.

475* MONTEIRO, H. L'arc axillaire musculaire et ses relations avec les faisceaus pectorauc aberrants. *L'Assoc. Antomistes, Comptes Rendus* 21:262–269, 1926.

476 MOODY, R.O. A note on the occurrence of the scapulo-clavicular muscle. *Trans. First Pan-Am. Med. Congress* 1893, pp. 1165–1166.

477 MOORE, A.T. An anomalous connection of the piriformis and biceps femoris muscles. *Anat. Rec.* 23:307–309, 1922.

478 MOORE, W.D. Notices of recent Dutch and Scandinavian contributions to anatomical and physiological science. *J. Anat. Physiol.* 5:227–232, 1871.

479 MORLEY, E.B. Congenital defect of the pectoralis muscles. *Lancet* 1:1101–1102, 1923.

480 MORRISON, J.T. A palmaris longus muscle with a reversed belly, forming an accessory flexor muscle of the little finger. *J. Anat. Physiol.* 50:324–326, 1916.

481 MÜLLER, G. Beschreibung eines M. sternalis, der in Afghanistan beobachtet wurde. *Anat. Anz.* 108:96–98, 1960.

482* MUSIAL, W.W. Variations of the terminal insertions of the anterior and posterior tibial muscles in man. *Folia Morphol.* 22:237–247, 1963.

483* MUSIAL, W.W. Variations of the terminal insertions of the peroneus longus and peroneus brevis muscles in man. *Folia Morphol.* 22:294–302, 1963.

484 NAKANO, T. Beiträge zur Anatomie der Chinesen. Die Statistik der Muskelvarietäten. *Folia Anat. Jpn.* 1:273–282, 1923.

485 NAT, B.S. The scalenus muscle. *J. Anat.* 58:268–270, 1924.

486* NATHAN, H. and H. GLOOBE. Flexor digitorum brevis – anatomical variations. *Anat. Anz.* 135:295–301, 1974.

487 NATHANIEL, D. A note on the variation of the flexor digitorum brevis. *J. Anat. Soc. India* 3:103–105, 1954.

488 NICOLA, B. Sur les insertions distales des musculi lumbricales dans la main de l'homme. *Arch. Ital. Biol.* 45:282–283, 1906.
489 NICOLAS, A. Transmission héréditaire d'une anomalie musculaire. *Soc. Biol. Comptes Rendus Hebdomadaires des Séances et Mémoires.* 42:656–657, 1890.
490 NIIZIMA, M. Ein Fall von überzähligem Kopf des M. biceps femoris beim Menschen. *Folia Anat. Jpn.* 16:187–192, 1938.
491 NIKOLAJEW, P.W. Zur Frage über die funktionale Bedeutsamkeit der Mm. palmaris longi beim Menschen. *Anat. Anz.* 75:145–160, 1932-33.
492 NISHI, S. Über einige bisher unbekannte oder wenig berücksichtigte, abnorme Muskelbündel im Gebiete der Brustwirbelsäule. – Fasciculi costoarcuales, interarcualis thoracici, transversoarticulares, articulo-costales und transversocostales beim Menschen. *Anat. Anz.* 85:110–115, 1937-38.
493* OCCHIPINTI, G. Studi di morfologia musculare comparata. Nota I. – I muscoli della spalla in una centuria di Siciliani. *Archiv. Ital. Anat. Embriol.* 31:345–369, 1933.
494* OCCHIPINTI, G. Studi di morfologia musculare comparata. Nota II. – I muscoli dell Braccio in una centuria di Siciliani. *Archiv. Ital. Anat. Embriol.* 31:370–395, 1933.
495 OCHILTREE, A.B. Some muscular anomalies of the lower limb. *J. Anat. Physiol.* 47:31–34, 1913.
496 OISHI, K. Observations on the laryngeal muscles of Madam Tamaki Miura, primadonna of the "Madam Butterfly." *Folia Anat. Jpn.* 28:581–594, 1956.
497 PAN, N. Some variations in the musculature of the superior extremity. *J. Anat.* 69:530–533, 1934–35.
498 PARDI, I. Variétés anatomiques. *Arch. Ital. Biol.* 45:284, 1906.
499 PARSONS, F.G. On the morphology of the musculus sternalis. *J. Anat. Physiol.* 27:505–507, 1893.
500 PARSONS, F.G. The muscles of mammals, with special relation to human myology. *J. Anat. Physiol.* 32:428–450, 1898.
501 PARSONS, F.G. The muscles of mammals, with special relation to human myology. *J. Anat. Physiol.* 32:721–752, 1898.
502 PARSONS, F.G. Note on abnormal muscle in popliteal space. *J. Anat.* 54:170, 1919.
503 PARSONS, F.G. and A. ROBINSON. Eighth report of the committee of collective investigation of the Anatomical Society of Great Britain and Ireland for the year 1897–1898. *J. Anat. Physiol.* 33:189–203, 1899.
504 PATERSON, A.M. The subscapularis muscle. *J. Anat. Physiol.* 46:11, 1916.
505* PATTEN, C.J. Right sternalis muscle with expanded fenestrated tendon. *J. Anat.* 68:424–425, 1933-34.
506 PATTEN, C.J. Right sternalis muscle, narrow and spindle-shaped. *J. Anat.* 68:426, 1933-34.
507 PATTEN, C.J. Bilateral capsular insertion of part of the tendon of the pectoralis minor. *J. Anat.* 68:427, 1933-34.
508* PEELING, W.B. Short extensor muscles of the hand. *Br. J. Surg.* 53:359–360, 1966.
509 PERKINS, J.D. An anomalous muscle of the leg: peronaeo-calcaneus internus. *Anat. Rec.* 8:21–25, 1914.
510 PERRIN, J.B. Notes on some variations of the pectoralis major, with its associate muscles seen during sessions 1868–69, 69–70, at King's College, London. *J. Anat. Physiol.* 5:233–240, 1871.
511 PERRIN, J.B. A rudiment of the dorsal portion of the panniculus carnosus, superficial to the trapezius. *J. Anat. Physiol.* 5:241, 1871.
512 PERRIN, J.B. A peculiar additional digastric muscle. *J. Anat. Physiol.* 5:251–256, 1871.
513* PICHLER, K. Achselbogen und M. chondro-epitrochlearis bei demselben Träger. *Anat. Anz.* 49:383–384, 1916.
514 DE PINA, L. Le muscle petit dentelé postérieur et supérieur chez l'homme et les primates. *L'Assoc. Anatomistes, Comptes Rendus* 28:523–531, 1933.
515 PIRES DE LIMA, J.A. Sur la fréquence de quelques anomalies musculaires chez les Portugais. *Soc. Biol. Comptes Rendus Hebdomadaires des Séances et Mémoires* 88:1329–1330, 1923.
516 PITZORNO, M. Musculus interflexorius. *Arch. Ital. Biol.* 46:291–292, 1906.
517 PITZORNO, M. Contributo alla morfologia dell'arco ascellare muscolare di Länger. *Archiv. Ital. Anat. Embriologia* 10:129–144, 1911.
518 POHLMAN, A.G. Multiple anomalies in the upper extremity of one cadaver. *J. Anat. Physiol.* 42:433–437, 1908.
519* POLAND, A. Deficiency of the pectoral muscles. *Guy's Hosp. Rep.* 6:191–193, 1841.
520* PRINCETEAU. Note pour servir a l'histoire des anomalies musculaires du creux de l'aisselle. *Soc. de Biol. Comptes Rendus Hebdomadaires des Séances et Mémoires* 44:202–206, 1892.
521 PYE-SMITH. Ein zweiter Fall von Musculus supracostalis anterior anomalus. *Arch. Path. Anat. Physiol. Klin. Med.* 43:142, 1868.
522 PYE-SMITH, P.H., HOWSE, H.G. and J.N.C. DAVIES-COLLEY. Notes of abnormalities observed in the dissecting room during the winter sessions of 1868–9 and 1869–70. *Guy's Hosp. Rep.* 16:147–164, 1871.
523 DE QUERVAIN, F. Über eine Form von chronischer Tendovaginitis. *Correspondenz-Blatt Schweizer Aerzte.* 25:389–394, 1895.
524 RANSON, W.B. Notes on some variations of the shoulder-muscles. *J. Anat. Physiol.* 19:508–509, 1885.
525 RAO, G.R.K.H. and V.R. RAO. Dorso-epitrochlearis brachii. *J. Anat. Soc. India.* 5:33–34, 1956.
526 RAO, V.S. and G.R.K.H. RAO. The sternalis muscle. *J. Anat. Soc. India.* 3:49–51, 1954.
527 RECTOR, J.M. Congenital muscular defects with special reference to deficiencies of the pectoral muscles. *J. Pediatr.* 7:625–630, 1935.
528 REGNAULT, F. Les causes des anomalies musculaires. *L'Assoc. Anatomistes, Comptes Rendus* 4:19–20, 1902.
529 REID, R.W. Anterior belly of the digastric muscle on both sides, arising from the lower border of the horizontal ramus of the inferior maxillary bone. *J. Anat. Physiol.* 21:78, 1887.
530 REID, R.W. Supra-clavicularis muscle. *J. Anat. Physiol.* 23:353, 1889.
531* REIMANN, A.F., DASELER, E.H., ANSON, B.J. and L.E. BEATON. The palmaris longus muscle and tendon. A study of 1600 extremities. *Anat. Rec.* 89:495–505, 1944.
532* REIMANN, R. Überzählige Musculi peronaei beim Menschen. *Z. Morphol. Anthropol.* 69:16–31, 1978.
533* REIMANN, R. Vier Musculi peronei in einem menschlichen Unterschenkel. *Anat. Anz.* 145:205–207, 1979.
534 REINHARDT. Beobachtung eines Musculus accessorius flexoris hallucis longi superior. *Arch. Anat. Physiol. Wissen. Med.* 1846: pp. 298–299.
535 REINHARDT, E. Über den Ansatz der Musculi lumbaricales an der Hand des Menschen. *Anat. Anz.* 20:129–134, 1902.
536 RENVALL, G. Ein Fall von doppelseitigen Turner-Perrinschem Musculus dorsofascialis beim Menschen. *Anat. Anz.* 31:545–554, 1907.
537 RENVALL, G. Eine ungewöhnliche Varietät des M. pectoralis major, zugleich ein Beitrag zur Sternalisfrage. *Anat. Anz.* 35:401–407, 1910.
538 REX, H. Über einen abnormen Augenmuskel (Musc. obliquus accessorius inferior). *Anat. Anz.* 2:625–630, 1887.
539* ROBACKI, R. and T. OANCEA. Am Lebenden aufgefundene Muskelvarietäten. *Anat. Anz.* 117:124–128, 1965.
540 ROBERTS, P.H. An anomalous accessory palmaris longus muscle. *Hand* 4:40–41, 1972.
541 ROEGHOLT, M.N. Musculus supraclavicularis proprius. *Anat. Anz.* 45:474–477, 1913.
542 ROESCH, W. Ein Gefäßscheidenmuskel am Halse. *Anat. Anz.* 46:366–368, 1914.
543 ROHAN, R.F. and L. TURNER. The levator palati muscle. *J. Anat.* 90:153–154, 1956.
544 ROLLESTON, H.D. Some abnormalities of the muscles of the upper limb. *J. Anat. Physiol.* 21:328–330, 1887.
545 ROSS, A.J. and C.A. TROY. The clinical significance of the extensor digitorum brevis manus. *J. Bone Joint Surg.* 51B:473–478, 1969.
546 ROTHER, P., LUSCHNITZ, E., BEAU, S. and P. LOHMANN. Der Ursprung der ischiokruralen Muskelgruppe des Menschen. *Anat. Anz.* 135:64–71, 1974.
547 RUFFINI, A. Sul muscolo interdigastrico di Bianchi e sull'aponeurosi soprajoidea od intermediojoidea. Rarissimo caso di mancanza bilaterale del ventre anteriore del m. digastrico della mandibola. *Archiv. Ital. Anat. Embriol.* 7:578–601, 1908.
548 RUGE, G. Der Hautrumpfmuskel des Menschen. *Gegenbaur's Morphol. Jahrb.* 47:677–682, 1914.
549* RUGH, J.T. JR. An anomalous muscle of the forearm inserting into the sheath of the median nerve. *Anat. Rec.* 41:299–300, 1929.
550* RUSSELL, K.F. and S. SUNDERLAND. Abnormalities of the lumbrical muscles of the hand. *J. Anat.* 72:306–307, 1937-38.
551 RUSSUE, I.G. Überzähliges Bündel des Musculus omohyoideus. *Anat. Anz.* 88:420–423, 1939.
552 RUTHERFORD, N. A curious arrangement of the retro-clavicular musculature. *Anat. Anz.* 37:148–150, 1910.

553 Sala, L. Sur un muscle tibio-péronéo-astragalien. *Arch. Ital. Biol.* 26:489, 1896.
554 Salama, R. and S.L. Weissman. Congenital bilateral anaomalous band between flexor and extensor pollicis longus tendons. *Hand* 7:25–26, 1975.
555 Salsbury, C.R. The interosseus muscles of the hand. *J. Anat.* 71:395–403, 1936–37.
556 Samuel, E. Congenital absence of the pectoralis major. *Br. J. Radiol.* 18:20–21, 1945.
557 Sanchez & Sanchez, M. Quelques variations morphologiques des muscles peauciers de la tête. *L'Assoc. Anatomistes, Comptes Rendus* 28:586–592, 1933.
558 Sappington, T.B. and R.A. Daniel. Accessory diaphragm. *J. Thoracic Surg.* 21:212–216, 1951.
559* Sauer, M.E. The cricoesophageal tendon. A recommendation for its inclusion in official anatomical nomenclature. *Anat. Rec.* 109:691–697, 1951.
560 Sauser, G. Beobachtung zweier Muskelvarietäten an Lebenden. *Wien. Klin. Wochenschr.* 48:430–432, 1935.
561 Schaeffer, J.P. On the variations of the palmaris longus muscle. *Anat. Rec.* 3:275–278, 1909.
562 Schmidt, H.M. Transversus nuchae muscle: Two observations of hitherto unknown deviations of its nerve supply. *Anat. Anz.* 151:144–150, 1982.
563* Schmidt, R., Heinrichs, H.-J. and D. Reissig. Die Mm. lumbricales an der Hand des Menschen, ihre Variationen in Ursprung und Ansatz. *Anat. Anz.* 113:414–449, 1963.
564* Schmidt, R., Reissig, D. and H.-J. Heinrichs. Die Mm. lumbricales am Fuß des Menschen. *Anat. Anzeiger* 113:450–453, 1963.
565* Schmidlmüller. Beschreibung eines seltenen Halsmuskels. *Arch. Physiol.* 8:269–270, 1807–08.
566 Schramm, U. and D.G. von Keyserlingk. Studien über Latissimusbögen des Oberarmes. *Anat. Anz.* 156:75–78, 1984.
567 Schuldt, H.H. Ein zweibäuchiger Musculus abductor tertius des rechten Daumens. *Anat. Anz.* 96:418, 1947–48.
568 Schultz, A.H. The position of the insertion of the pectoralis major and deltoid muscles on the humerus of man. *Am J. Anat.* 23:155–174, 1918.
569 Schultz, J. Zwei Musculi sternalis. *Anat. Anz.* 3:228–234, 1888.
570* Sebileau, P. Le muscle scalène. *Soc. Biol. Comptes Rendus Hebdomadaires des Séances et Mémoires* 43:201–221, 1891.
571 Seib, G.A. Incidence of the m. psoas minor in man. *Am. J. Phys. Anthropol.* 19:229–246, 1934.
572 Selden, B.R. Congenital absence of trapezius and rhomboideus major muscles. *J. Bone Joint Surg.* 17:1058–1059, 1935.
573 Sewell, R.B.S. The small or superficial thyro-arytenoideus muscle. *J. Anat. Physiol.* 39:301–307, 1905.
574 Shah, M.A. and M. Shah. Quadricipital M. biceps brachii. *J. Anat.* 80:54, 1946.
575 Shattock, G.S. Note on the anatomy of the thyro-arytenoid muscle in the human larynx. *J. Anat. Physiol.* 16:485, 1882.
576 Shattock, G.S. A "kerato-thyro-hyoid" muscle as a variation in human anatomy. *J. Anat. Physiol.* 17:124–125, 1883.
577 Sheehan, D. Bilateral absence of trapezius. *J. Anat.* 67:180–181, 1932.
578 Shepard, F.J. Anomalous muscle of the thorax connected with the diaphragm. *J. Anat. Physiol.* 30:168, 1896.
579 Shore, L.R. An example of the muscle scalenus minimus. *J. Anat.* 60:418–419, 1926.
580 Siddiqi, M. A.H. and A.N. Mullick. On the anatomy of intercostal spaces in man and certain other mammals. *J. Anat.* 69:350–355, 1935.
581* Sklodowski, J. Über einen Fall von angeborenem rechtseitigem Mangel der Musculi pectoralis major et minor mit gleichzeitigen Missbildungen der rechten Hand. *Arch. Path. Anat. Physiol. Klin. Med.* 121:600–604, 1890.
582 Slobodin, S.G. Zur Frage über den Musculus sternalis. *Anat. Anz.* 79:207–211, 1934–35.
583 Smith, E.B. Some points in the anatomy of the dorsum of the hand, with special reference to the morphology of the extensor brevis digitorium manus. *J. Anat. Physiol.* 31:45–58, 1897.
584 Smith, R.J. Anomalous muscle belly of the flexor digitorum superficialis causing carpal-tunnel syndrome. Report of a case. *J. Bone Joint Surg.* 53A:1215–1216, 1971.
585 Sokolowska-Pituchowa, J., Goszczyński, M. and I. Goszczyńska. Position of the diaphragm in pregnant women. *Folia Morphol.* 31:448–450, 1972.
586* Sokolowska-Pituchowa, J., Miaśkiewicz, C., Skawina, A. and K. Makos. Morphology and some measurements of the peroneus tertius muscle in man. *Folia Morphologica* 33:91–103, 1974.
587 Sokolowska-Pituchowa, J., Miaśkiewicz, C., Skawina, A. and K. Makos. Morphologic types of the third peroneal muscle and some indices of the foot in man. *Folia Morphol.* 34:361–366, 1975.
588 Sosa, G.C., Liotta, D. and S. Liotta. La gaine des muscles grands droits de l'abdomen considérations anatomiques dans les 3/4 supérieurs. *L'Assoc. Anatomistes, Comptes Rendus* 39:329–338, 1952.
589 Souter, W.A. The extensor digitorum brevis manus. *Br. J. Surg.* 53:821–823, 1966.
590 Sperino, G. and R. Balli. Sur de nombreuses variétés musculaires recontrées dans les membres supérieurs d'un adulte. *Arch. Ital. Biol.* 44:117, 1905.
591 Stadler, H. Über einen Musculus supraclavicularis proprius. *Anat. Anz.* 69:284–286, 1930.
592* Stark, H.H., Otter, T.A., Boyes, J.H. and T.A. Richard. "Atavistic contrahentes digitorum" and associated muscle abnormalities of the hand: A cause of symptoms. *J. Bone Joint Surg.* 61A:286–289, 1979.
593 Stein, A.H. Variations of the tendons of insertion of the abductor pollicis longus and the extensor pollicis brevis. *Anat. Rec.* 110:49–55, 1951.
594 Stein, M. Über einen Fall von vollkommenem Mangel des vorderen Digastricusbauches. *Anat. Anz.* 47:345–352, 1914–15.
595 Steinbach, K. Über Varietäten der Unterzungenbein- und Brustmuskulatur. *Anat. Anz.* 56:488–506, 1923.
596 Stephenson, P.H. On an unusual anomaly of the peroneus tertius in a Chinese. *Anat. Rec.* 22:81–83, 1921.
597 Stephenson, P.H. On an anomalous digastric muscle in the thigh of a Chinese. *Anat. Rec.* 23:281–290, 1922.
598 Stepień, E. Accessory insertions of the tendon of the anterior tibialis muscle in man. *Folia Morphol.* 26:216–219, 1967.
599 Stewart, S.F. Reverse palmaris longus. *J. Bone Joint Surg.* 36B:689–690, 1954.
600 Stibbe, E.P. Anatomical notes: Complete absence of the quadratus femoris. *J. Anat.* 64:97, 1930.
601* Still, J.M. Jr. and H.E. Kleinert. Anomalous muscles of nerve entrapment in the wrist and hand. *J. Plastic Reconstr. Surg.* 52:394–400, 1973.
602 Stintzing, R. Der angeborene und erworbene Defect der Brustmuskeln. Zugleich ein klinischer Beitrag zur progressiven Muskelatrophie. *Deutsches Arch. Klin. Med.* 45:205–232, 1889.
603 Stirling, W. The trachealis muscle of man and animals. *J. Anat. Physiol.* 17:204–206, 1883.
604 Stolowsky, A. Drei seltene Anomalien des M. biceps brachii. *Anat. Hefte.* 12:299–336, 1899.
605 Stopczyk, J. La morphologie du muscle auriculaire antérieur. *L'Assoc. Anatomistes, Comptes Rendus* 26:496–498, 1931.
606 Stott, C.F. A note on the scalenus minimus muscle. *J. Anat.* 62:359–361, 1928.
607 Stracker, O. Die Häufigkeit interponierter Muskelkörper zwischen den vorderen Bäuchen des M. digastricus. *Anat. Anz.* 33:227–236, 1908.
608 Strangeways, T. On a supernumerary oblique muscle of the eyeball. *J. Anat. Physiol.* 2:245–246, 1868.
609 Straus, W.L. Jr. The phylogeny of the human forearm extensors. *Hum. Biol.* 13:23–50, 203–238, 1941.
610 von Streckeisen, A. Beiträge zur Morphologie der Schilddrüse. *Arch. Path. Anat. Physiol. Klin. Med.* 103:131–186, 1885.
611 Sunderland, S. The actions of the extensor digitorum communis, interosseus and lumbrical muscles. *Am. J. Anat.* 77:189–217, 1945.
612 Sunderland, S. The innervation of the flexor digitorum profundus and lumbrical muscles. *Anat. Rec.* 93:317–321, 1945.
613* Sutton, J.B. On the nature of ligaments. Part V. *J. Anat. Physiol.* 22:542–553, 1888.
614* Swieter, M.G. and S.W. Carmichael. Bilateral three-headed biceps brachii muscles. *Anat. Anz.* 148:346–349, 1980.
615* Szabados, A. Über einen Fall von M. peroneaus quartus. *Anat. Anz.* 118:162–163, 1966.
616 Tait, L. Notes on unusual accessory muscles. *J. Anat. Physiol.* 4:236–238, 1870.
617 Tamega, O.J., Garcia, P.J., Soaves, J.C. and N.L. Zorzetto. About a case of absence of the superior belly of the omohyoid muscle. *Anat. Anz.* 154:39–42, 1983.
618 Tataroff, D. Über die Muskel der Ohrmuschel und einige Be-

sonderheiten des Ohrknorpels. *Arch. Anat. Physiol. Wissen. Med.* 1887: pp. 35–58.

619* TAVARES, A.S. L'innervation des muscles pectoraux. *Acta Anat.* 21:132–141, 1954.

620 TAYLOR, A.E. Case of clavicular insertion of the pectoralis minor. *J. Anat. Physiol.* 32:218, 1898.

621 TAYLOR, J. An unusual variation of the omo-hyoid muscle. *J. Anat.* 59:331–332, 1925.

622 TENCHINI, L. Sur un nouveau muscle surnuméraire de la région postérieure de l'avant-bras humain (M. extensor digiti indicis et medii) associé à un faisceau manieux. *Arch. Ital. Biol.* 39:481, 1903.

623 TERNOWSKY, W.N. and M. SADIKOWA. Ein akzessorischer M. soleus. *Anat. Anz.* 61:280–284, 1926.

624 TERRY, R.J. Absence of superior gemellus muscle in American Whites and Negroes. *Am. J. Phys. Anthropol.* 29:47–56, 1942.

625 TESTUT, L. Le muscle péronier du cinquième orteil chez l'homme. *Bull. mém. Soc. Anat. Paris.* 1884, pp. 352–363.

626 TESTUT, L. Les anomalies muscularies chez les Nègres et chez les Blancs. *Int. Monatschr. Anat. Physiol.* 1:285– 291, 1884.

627 THÉVENET, A. and J.B. PRIOTON. Les insertions postérieures du diaphragme. *L'Assoc. Anatomistes, Comptes Rendus* 44:722–737, 1957.

628 THIEL, W. Eine seltene Varietät des Venter mastoideus m. biventeris und ihre phylogenetische Deutung. *Anat. Anz.* 101:16–23, 1954–55.

629 THOMAS, C.G. JR. Clinical manifestations of an accessory palmaris muscle. *J. Bone Joint Surg.* 40A:929–930, 1958.

630* THOMPSON, I.M. M. chondro-epitrochlearis bilateralis. *J. Anat.* 71:130, 1936–37.

631 THOMPSON, J.W., MCBATTS, J. and C.H. DANFORTH. Hereditary and racial variations in the musculus palmaris longus. *Am. J. Phys. Anthropol.* 4:205–218, 1921.

632 THOMSON, A. Notes on some unusual variations in human anatomy. *J. Anat. Physiol.* 19:328–332, 1885.

633 TRICOMI-ALLEGRA, G. Musculus gracilis biceps, faisceau surnuméraire naissant de la rotule. *Arch. Ital. Biol.* 53:153, 1910.

634 TROJANOWSKI, A. Sur le muscles mandibulo-marginal. *L'Assoc. Anatomistes, Comptes Rendus* 26:511–513, 1931.

635 TSUGE, K. Congenital aplasia or hypoplasia of the finger extensors. *Hand* 7:15–21, 1975.

636 TURKEWITSCH, N. Zur Frage über Anomalien der Nackenmuskeln. *Anat. Anz.* 74:123–126, 1932.

637 TURKEWITSCH, N. Musculus thyreotrachialis. *Folia Anat. Jpn.* 12:155–158, 1934.

638 TURNER, W. On the musculus sternalis. *J. Anat. Physiol.* 1:246–253, 1867.

639 TURNER, W. A rudiment of the panniculus carnosus superficial to the trapezius. *J. Anat. Physiol.* 5:116–117, 1871.

640 TWEEDY, J. On a case of absence of the thoracic portion of the pectoralis major and the whole of the pectoralis minor muscle. *Lancet* 1:443, 1873.

641 TYRIE, C.C.B. Musculus saphenous. *J. Anat. Physiol.* 28:288–290, 1894.

642 URBANOWICZ, Z. A combination of congenital anomalies of the upper extremity. *Folia Morphol.* 27:162–171, 1968.

643 VALENTI, G. Sur un muscle fessier surnuméraire. *Arch. Ital. Biol.* 53:153, 1910.

644 VALENTI, G. Sur un muscle radial accessoire. *Arch. Ital. Biol.* 57:302–303, 1912.

645 VALENTI, G. Sur un muscle mandibulo-glosse (M. Mylo-Glossus Wood). *Arch. Ital. Biol.* 75:77, 1925.

646 VARAGLIA, S. Sur la signification d'un prolongement fibreux qui va du tendon du m. pectoralis major à la capsule d l'articulatio humeri. *Arch. Ital. Biol.* 36:340–341, 1901.

647 VARAGLIA, S. Sur quelques dispositions myologiques peu connues de la région poplitée chez l'homme. *Arch. Ital. Biol.* 39:486, 1903.

648 VASTARINI-CRESI, G. Nouvelle variété de musculus supraclavicularis chez l'homme et sa signification probable. *Arch. Ital. Biol.* 46:290–291, 1906.

649* VELLUDA, C.C. and V. DAGHIE. Über einen costo-coracoiden oder WOOD'schen Muskel. *Anat. Anz.* 65:139–142, 1928.

650 VICHARE, N.A. Anomalous muscle belly of the flexor digitorum superficialis. Report of a case. *J. Bone Joint Surg.* 52B:757–759, 1970.

651 VILDE, J. Über eine bisher unbekannte Ursprungsmöglichkeit des Musculus serratus posterior inferior. *Anat. Anz.* 59:464–467, 1924–25.

652 DE VILHENA, H. Le muscle surnuméraire extenseur commun du pouce et de l'index. *L'Assoc. Anatomistes, Comptes Rendus* 28:685–702, 1933.

653* VIRCHOW, H. Die Anomalie des Mittelgesichts. *Z. Anat. Entwicklungsgesch.* 84:555–596, 1927.

654* VRIJMAN, L.H. Eine seltene Abart des M. obturator externus. *Anat. Anz.* 59:246–250, 1924–25.

655 WAGSTAFFE, W.W. Two cases showing a peculiar arrangement in the fibres of the external pterygoid muscle in man. *J. Anat. Physiol.* 5:281–284, 1871.

656 WAGSTAFFE, W.W. Partial deficiency of the tendon of the long flexor of the thumb. *J. Anat. Physiol.* 6:212–214, 1872.

657 WAGSTAFFE, W.W. Description of an accessory muscle in connection with the popliteus. *J. Anat. Physiol.* 6:214–215, 1872.

658 WALLACE-JONES, H. Congenital absence of the pectoral muscle. *Br. Med. J.* 2:59–60, 1926.

659 WALMSLEY, T. The costal musculature. *J. Anat. Physiol.* 50:165–171, 1916.

660 WALMSLEY, T. Observations on the omohyoid muscle. *J. Anat.* 52:319–325, 1918.

661 WATERSTON, D. Variations in the teres minor muscle. *Anat. Anz.* 32:331–333, 1908.

662 WATSON, M. The curvatores coccygis muscles of man. *J. Anat. Physiol.* 14:407–412, 1879.

663 WEBER, A. Signification du muscle stylo-hyoïdien profound. *Soc. Biol. Comptes Rendus Hebdomadaires des Séances et Mémoires* 81:816–817, 1918.

664 WEBER, A. and R. COLLIN. Variations des insertions musculaires sur la tubérosité ischiatique chez l'homme. *L'Assoc. Anatomistes, Comptes Rendus* 6:42–46, 1904.

665 WEISSBERG, H. Über einen Fall von muskulösem Achselbogen. *Anat. Anz.* 74:105–117, 1932.

666 WELLS, L.H. and E.A. THOMAS. A note on two abnormal laryngeal muscles in a Zulu. *J. Anat.* 61:340–343, 1927.

667* WESSER, D.R., CALOSTYPIS, F. and S. HOFFMAN. The evolutionary significance of an aberrant flexor superficialis muscle in the human palm. *J. Bone Joint Surg.* 51A:396–398, 1969.

668 WEST, S.H. A peculiar digastric muscle – a variety of occipito-hyoid. *J. Anat. Physiol.* 8:150–151, 1874.

669 WHILLIS, J. A note on the muscles of the palate and the superior constrictor. *J. Anat.* 65:92–95, 1930.

670 WHITE, S. Note respecting the course of the flexor longus digitorum pedis. *J. Anat. Physiol.* 18:118–119, 1884.

671 WHITENALL, S.E. An instance of the retractor bulbi muscle in man. *J. Anat. Physiol.* 46:36–40, 1912.

672* WIDAL and LEMIERRE. Absence congénitale du muscle petit pectoral et du chef sternocostal du grand pectoral. *Bull. Mém. Soc. Méd. Hôpitaux Paris.* 19(3): 880–884, 1902.

673 WILKINSON, J.L. The insertions of the flexores pollicis longus et digitorum profundus. *J. Anat.* 87:75–88, 1953.

674* WILLIAMS, G.A. Pectoral muscle defects. Cases illustrating three varieties. *J. Bone Joint Surg.* 12:417–418, 1930.

675 WILLIAMS, W.R. The anatomy of the quadriceps extensor cruris. *J. Anat. Physiol.* 13:204–218, 1879.

676 WOOD, J. On some varieties in human myology. *Proc. R. Soc. Lond.* 13:299–303, 1864.

677 WOOD, J. Additional varieties in human myology. *Proc. R. Soc. Lond.* 14:378–392, 1865.

678 WOOD, J. On human muscular variations in their relation to comparative anatomy. *J. Anat. Physiol.* 1:44–59, 1867.

679 WOOD, J. Variations in human myology observed during the winter session of 1867-68 at King's College, London. *Proc. R. Soc. Lond.* 17:483–525, 1868.

680* WOOD, J. On a group of varieties of the muscles of the human neck, shoulder, and chest, and their transitional forms and homologies in the mammalia. *Phil. Trans. R. Soc. (Lond.)* 160:83–116, 1870.

681* WRIGHT, R.R. GREIG, W. and B.J. ANSON. Accessory tendinous (peroneal) origin of the first dorsal interosseous muscle. A study of 125 specimens of lower extremity. *Q. Bull. Northwestern Univ. Med. School* 20:339–341, 1946.

682 YAP, S.E. Musculus sternalis in Filipinos. *Anat. Rec.* 21:353–366, 1921.

683 ZIEMAN, S.A. The fallacy of the conjoined tendon. *Am. J. Surg.* 50:17–21, 1940.

684* ZIENTARSKI, B. A rare variation of the extensor indicis muscle. *Folia Morphol.* 35:401–493, 1976.

Books

685 AASAR, Y.H. *Anatomical Anomalies*, Fouad I University Press, Cairo, 1947.
686 ANSON, B.J. *Atlas of Human Anatomy*. W.B. Saunders Co., Philadelphia, 1950.
687 ANSON, B.J. Ed. *Morris' Human Anatomy*, 12th Ed. The Blakiston Division, McGraw-Hill Book Company, New York, 1966.
688 BENNINGHOFF, A. *Lehrbuch der Anatomie des Menschen*. Urban & Schwarzenberg, Munich, 1952.
689 BUNNELL, S. *Surgery of the Hand*, 2nd Ed. J.B. LIPPINCOTT Co., Philadelphia, 1948.
690 GARDNER, E., GRAY, D.J. and R. O'RAHILLY, *Anatomy*, 3rd Ed. W.B. Saunders Company, Philadelphia, 1969.
691 GEGENBAUR, C. *Lehrbuch der Anatomie des Menschen*, 3rd Ed. V. Wilhelm Engelmann, Leipzig, 1888.
692 HENLE, J. *Handbuch der Systematischen Anatomie des Menschen*, 3 volumes. von Friedrich Vieweg und Sohn, Braunschweig, 1868.
693 HUBER, C.G., Ed. *Piersol's Human Anatomy*, 9th Ed. L.B. Lippincott Company, Philadelphia, 1930.
694 HYRTL, J. *Lehrbuch der Anatomie des Menschen*, 19th Ed. Wilhelm Braumüller, Wien, 1887.
695 JACKSON, C.M., Ed. *Morris' Human Anatomy*, 9th Ed. P. Blakiston's Son & Co., Inc., Philadelphia, 1933.
696 KOPSCH, F. *Rauber's Lehrbuch der Anatomie des Menschen*. Georg Thieme, Leipzig, 1908.
697 LATARJET, A. *Testut's Traite D'Anatomie Humaine*, 9th Ed. G. Doin & Cie., Paris, 1948.
698 SCHAEFER, E.A., SYMINGTON, J. and T.H. Bryce, Eds. *Quain's Anatomy*, 11th Ed. Longmans, Green, and Co., London, 1923.
699+ TESTUT, L. *Les Anomalies Muscularies Chez l'Homme Expliques par l'Anatomie Comparée*. Masson, Paris, 1884.
700 TOLDT, C. *An Atlas of Human Anatomy for Students and Physicians*. The Macmillian Company, New York, 1928.
701 WHITNALL, S.E. *The Anatomy of the Human Orbit*, 2nd Ed. Oxford University Press, London, 1932.
702 WOLFF, E. *The Anatomy of the Eye and Orbit*, 4th Ed. Lewis, London, 1954.

Glossary of Terms

Their origin and definition related to Muscles.

The following abbreviations are used: F., French; G., Greek; L., Latin; and ME., Middle English.

Abductor (L. *abducere*, to move away). A muscle that draws a structure away from the axis of the body or one of its parts.

Accessorius (L. *accessorius*, to move toward). Accessory or supernumerary. Also denoting specific muscles.

Accessory (L. *accessorius*, to move toward). Supernumerary, adjuvant.

Adductor (L. *adducere*, to bring forward). A muscle that draws a structure toward the axis of the body or one of its parts.

Alae (L. *ala*, wing). Relating to the ala of the nose, etc.

Anconeus (G. *ankon*, elbow). Musculus anconeus.

Ani (L. *anus*, anal orifice). Pertaining to a muscle that is attached to the anus.

Anticus (L. *anticus*, anterior). Designating an anterior muscle.

Arch (L. *arcus*, a bow). Any structure resembling a bent bow or an arch.

Articulationis (L. *articulationes*, the forming of new joints of a vine). Pertaining to two muscles of the hand and shoulder inserted into a joint capsule.

Arytenoid (G. *arytenoideus*, ladle-shaped). Pertaining to a muscle attached to this laryngeal cartilage.

Atlanto- (G. *Atlas*, in Greek mythology a Titan who supported the world on his shoulders). Relating to the atlas (first cervical vertebra).

Atloideus. See Atlanto-.

Auricularis (L. *auricularis*, the external ear). Pertaining to a muscle of the external ear. Also referring to the fifth digit of the hand because of its use in cleaning the external auditory meatus.

Axillary (L. *axilla*, armpit). Pertaining to a muscle in the region of the axilla or armpit, e. g., axillary arch muscle.

Azygos (G. *a*, without + *zygon*, yoke). Unpaired structure.

Basilaris (L., G., *basis*, base). Pertaining to the base, body, or lower part of a structure, e. g., base of the skull.

Biceps (L. *bi*, two + *caput*, head). Two heads, pertaining to certain muscles, e. g., biceps brachii.

Biventer (L. *bi*, two + *venter*, belly). Muscle having two bellies.

Brachialis (G. *brachion*, arm; L. *brachium*, arm). Relating to the arm.

Brachii (G. *brachion*, arm). Of the arm.

Brachio- (G. *brachion*, arm). Relating to the arm.

Brevis (L. *brevis*, short). A short muscle or short head, e. g., short head of biceps brachii muscle.

Buccinator (L. *buccinator*, trumpeter). A muscle of the cheek.

Bucco- (L. *bucca*, cheek). Pertaining to the cheek.

Bulbo- (L. *bulbus*, a bulbus root). Any globular or fusiform structure. A muscle covering a bulbar structure.

Capitis (L. *caput*, head). Pertaining to the head.

Capsularis (L. *capsa*, a chest or box). A muscle joined to a capsule as, for example, of a joint. Any structure so designated as a capsule.

Carnosus (L. *carnis*, flesh or muscle). Pertaining to muscular tissue.

Carpi (G. *karpos*, wrist). Pertaining to the eight bones of the wrist.

Cavernosus (L. *caverna*, a grotto or hollow). Pertaining to the cavernous tissue of the reproductive organs.

Cerato- (G. *keras*, horn). Relating to a muscle that arises from the greater horn of the hyoid bone.

Chondro- (G. *chondros*, cartilage). Pertaining to muscles that arise from costal cartilage.

Cilii (L. *cilium*, eyelid). Pertaining to eyebrow, e. g., m. corrugator supercilii.

Clavicularis (L. *clavicula*, small key). Pertaining to the clavicle.

Cleido- (G. *kleis*, clavicle). Relating to the clavicle as the origin of muscle.

Coccygeus (G. *kokkyx*, a cuckoo). Musculus coccygeus.

Colli (L. *collum*, neck). Pertaining to the neck or to the neck of a structure.

Communis (L. *communis*, in common). Relating to more than one structure working as a single unit.

Compressor (L. *compressus*, to press together). A muscle that, when contracted, causes pressure on a structure.

Condyloideus (G. *kondylos*, knuckle). Pertaining to muscles attached to the outer edge of a joint.

Constrictor (L. *constringere*, to draw together). A muscle that, upon contraction, reduces the size of a canal. A sphincter.

Coraco- (G. *korakodes*, a crow's beak). Denoting a muscle that arises from the coracoid process of the scapula.

Cornu (L. *cornu*, horn). Any structure resembling a horn in shape.

Corrugator (L. *con*, together + *ruga*, wrinkle). A muscle that wrinkles the skin.

Costalis (L. *costa*, rib). Pertaining to muscles attached to the ribs.

Cremaster (G. *kremaster*, a suspender). Musculus cremaster, the muscle by which the testicles are suspended.

Crico- (G. *krikos*, a ring). Denoting a muscle that arises from the cricoid cartilage.

Crural (L. *crus*, leg). Pertaining to the leg (from knee to ankle) or to any other structure designated as a crus.

Deltoideus (G. *deltoeides*, shaped like the letter delta). Musculus deltoideus.

Dentate (L. *dentatus*, toothed). Notched muscles, e. g., the serrati.

Diaphragm (G. *diaphragma*, a partition). Muscles diaphragma.

Digastricus (G. *di*, two + *gaster*, belly). Denoting a muscle with two fleshy parts separated by a tendinous intersection. Musculus digastricus.

Dilatores (ME. *dilaten*, to dilate). Denoting a muscle that opens an orifice.

Dorso- (L. *dorsum*, back). Related to the dorsal surface.

Epi- (G. *epi*, upon). Denoting a muscle attached to a particular structure.

Epistropheus (G. *epistropheus*, the pivot). Second cervical vertebra.

Epitrochlearis (L. *trochlea*, pulley or block). Pertaining to a muscle inserted on the humeral epicondyle.

Extensor (L. *ex-tendere*, to stretch out). A muscle that, upon contraction, tends to straighten a limb; the antagonist of a flexor.

Femoris (L. *femur*, thigh). Pertaining to the femur or thigh.

Flexor (L. *flectere*, to bend). A muscle that, upon contraction, tends to bend a joint; the antagonist of an extensor.

Gastrocnemius (G. *gaster*, belly + *kneme*, leg). The belly of the leg., i.e., musculus gastrocnemius.

Gemelli (L. *geminus*, twin). Musculi gemelli.

Genio- (G. *geneion*, chin). Pertaining to a muscle arising from the chin (mandible).

Glosso- (G. *glossa*, tongue). Pertaining to a muscle that arises from or inserts on the tongue.

Gluteus (G. *gloutos*, buttock). Pertaining to the muscles of the buttocks.

Muscles

Gracilis (L. *gracilis*, slender or delicate). Musculus gracilis.
Hallucis (L. *hallex*, great toe). First digit of the foot.
Humero- (G. *homos*, shoulder). Pertaining to the bone of the arm.
Hyo- (G. *hyoeides*, hyoid). Relating to the U-shaped hyoid bone.
Hyoideus. See Hyo-.
Iliacus (L. *ilium*, groin). Musculus iliacus.
Ilio- (L. *ilium*, groin). Pertaining to the iliac region.
Indicis (L. *index*, one that points). The forefinger.
Inferior (L. *inferior*, lower). Lower, caudal.
Infra- (L. *infra*, below). Pertaining to a position below a named structure, e. g., infraspinatus.
Internal (L. *internus*, interior). Deep or away from the surface.
Inter- (L. *inter*, between). Between or among.
Ischio- (G. *ischion*, hip). Pertaining to the ischium.
Lateral (L. *lateralis*, lateral). On the outer side away from the midline.
Latissimo- (L. *latus*, broad). A term applied to some broad flat muscles. Musculus latissimus dorsi.
Levator (L. *levare*, to lift). One of several muscles whose function is to lift the structure into which it is inserted.
Linguae (L. *lingua*, tongue). Pertaining to, or toward, the tongue.
Longissimus (L. *longus*, long). A name given to certain long muscles.
Lumborum (L. *lumbus*, a loin). Pertaining to the back and sides between the pelvis and ribs.
Lumbricales (L. *lumbricus*, an earthworm). Muscles resembling an earthworm.
Mandibulo (L. *mandere*, to chew). Pertaining to a muscle arising from the mandible.
Manus (L. *manus*, hand). Pertaining to the hand.
Masseter (G. *maseter*, masticator). Musculus masseter.
Mastoideus (G. *mastos*, breast + *eidos*, resemblance). Resembling a mamma or a breast-shaped structure.
Medial (L. *medialis*, middle). Relating to a muscle nearer to the median or midsagittal plane.
Mentalis (L. *mentum*, chin). Relating to the chin. Musculus mentalis.
Mento- (L. *mentum*, chin). See Mentalis.
Metacarpo- (G. *meta*, after + *carpus*, wrist). Pertaining to the bone adjacent to the wrist.
Musculus (L. *mus*, a mouse). A muscle.
Myo- (G. *mys*, a muscle). Relating to muscle.
Mytiformis (G. *mytilos*, mussel + *forma*, shape). Musculus mytiformis.
Naris (L. *naris*, nostril). Pertaining to the nostril.
Nasalis (L. *nasus*, nose). Pertaining to the nose.
Nuchae (F. *nuque*, back of the neck). Pertaining to back of the neck.
Obturator (L. *obturare*, to occlude). Pertaining to a muscle associated with the obturator membrane and obturator foramen.
Occipitalis (L. *ob*, before or against + *caput*, head). Pertaining to the occipital bone.
Omo- (L. *omo*, shoulder). Pertaining to a muscle attached to the scapula.
Opponens (L. *opponere*, to place against). A name given to several adductor muscles of the fingers and toes.
Oris (L. *os*, mouth). Relating to the entrance to the digestive system, the mouth.
Os (L. *os*, bone). A bone.
Palato- (L. *palatum*, palate). Relating to the palate.
Palmaris (L. *palma*, palm of the hand). Pertaining to a muscle of the forearm which may insert into the palmar aponeurosis.
Palpebrae (L. *palpebra*, eyelid). Relating to the eyelid.

Panniculus (L. *pannus*, cloth). Pertaining to a thin sheet of muscle.
Pectineus (L. *pecten*, a comb). Pertaining to the os pubis or any ridged structure. Musculus pectineus.
Pectoro- (L. *pecuts, pector-*, chest). Pertaining to the muscles of the chest wall.
Pedis (L. *pes*, foot). Relating to the foot.
Penis (L. *penis*, tail). Relating to the penis.
Peroneus (G. *perone*, brooch, fibula). Pertaining to one of several muscles on the lateral (or fibular) side of the leg.
Phalangei (L., F., G. *phalanx*, line of soldiers). Pertaining to the bones of the fingers.
Pharyngeus (G. *pharynx*, throat). Pertaining to the pharynx.
Piriformis (L. *pirum*, pear + *forma*, form). Pear-shaped.
Pisiform (L. *pisum*, pea + *forma*, form). Pea-shaped or pea-sized.
Plantaris (L. *plantaris*, sole of foot). Pertaining to a muscle of the foot. Musculus plantaris.
Platysma (G. *platys*, flat or broad). A flat, broad dermal muscle of the thorax and neck.
Pollicis (L. *pollex*, thumb). Relating to the thumb.
Popliteus (L. *poples*, the ham of the knee). Pertaining to the popliteal space. Musculus popliteus.
Procerus (L. *procerus*, long or stretched-out). Musculus procerus.
Pronator (L. *pronare*, to bend forward). Muscle that, on contraction, rotates the hand so that the palm of the hand faces backward when the arm is in the anatomical position.
Psoas (G. *psoa*, muscles of the loins). Pertaining to two muscles of the lumbosacral regions. The "tenderloin."
Pterygoideus (G. *pteryx*, or *pteryg-*, wing + *eidos*, resemblance). Wing-shaped. Applied to muscles associated with the pterygoid process of the sphenoid bone.
Pubo- (L. *pubes*, genitalis). Pertaining to the attachment of muscles to the os pubis.
Pyramidalis (G. *pyramis*, pyramid). Pertaining to muscles having, more or less, a pyramidal shape.
Quadratus (L. *quadratus*, square). A name given to muscles having, more or less, a square shape.
Quadriceps (L. *quadri-*, four + *caput*, head). A name given to a muscle having four heads, e. g., quadriceps femoris of the thigh.
Quinti (L. *quintus*, fifth). Fifth.
Radio- (L. *radius*, ray). Pertaining to muscles associated with the os radius of the forearm.
Rectalis (L. *rectus*, straight). Pertaining to muscles associated with the distal segment of the large intestine.
Rhombo- (G. *rhombos*, a rhomb). Resembling a rhomb, an oblique parallelogram of unequal sides. Relating to two superficial muscles of the back.
Risorius (L. *risor*, laughter). Pertaining to a facial muscle, musculus risorius.
Salpingo- (G. *salpinx*, trumpet). Pertaining to a muscle attached to the eustachian tube and pharynx.
Saphenous (G. *saphenes*, visible). Relating to the saphenous vein.
Sartorius (L. *sartor*, a tailor). Musculus sartorius.
Scalenus (G. *skalenos*, uneven). Pertaining to muscles having uneven sides or length.
Scapulo- (L. *scapulae*, shoulder blades). Pertaining to a muscle associated with the scapula.
Semi- (L. *semis*, half). Prefix denoting half or partly.
Serratus (L. *serra*, saw). Pertaining to muscles that are serrated, notched, or dentate.
Soleus (L. *solea*, a sandal). Musculus soleus.
Spinosus (L. *spina*, thorn). Related to the spinous processes of the vertebral column.
Splenius (G. *splenion*, a bandage). Musculus splenius and others.

Human Anatomic Variation

Stapedius (L. *stapes*, stirrup). A muscle inserted into the stapes. Musculus stapedius.

Sterno- (G. *sternon*, the chest). Pertaining to muscles attached to the sternum.

Stylo- (G. *stylos*, pillar or post). Pertaining to muscles attached to the styloid process of the temporal bone.

Sub- (L. *sub*, under). Denoting muscles that are beneath or inferior to a named structure, e. g., subclavius.

Superior (L. *superus*, above). Denoting a muscle located above a muscle in an inferior position or structure to which it is attached.

Supinator (L. *supinare*, to place on back). Denoting a muscle that, upon contraction, rotates the forearm and hand with the palm facing anteriorly when the hand and forearm are in the anatomical position.

Supra- (L. *supra*, above). Prefix to note the position of a muscle above a named structure, e. g., supracostalis.

Suralis (L. *sura*, calf of the leg). Relating to the calf.

Temporalis (L. *tempus*, time, temple). Relating to the temple, musculus temporalis.

Tensor (L. *tendere*, to stretch). Pertaining to a muscle whose function is to make a structure, to which it is attached, firm and tense.

Teres (L. *tero*, round or smooth). Denoting certain muscles that are round and long.

Thyro- (G. *thyreos*, an oblong shield). Denoting certain muscles attached to the thyroid cartilage.

Tibialis (L. *tibia*, a pipe or flute). Pertaining to muscles attached to the tibia.

Trachelian (G. *trachelos*, neck). Cervical.

Transversus (L. *trans*, across + *vertare*, to turn). Denoting muscles that lie across the long axis of an organ or part.

Trapezius (G. *trapezoin*, a table). A four-sided muscle having no two sides parallel. A superficial muscle of the back. Musculus trapezius.

Triangularis (L. *tri-*, three + *angulus*, angle). A muscle that has, more or less, a triangular shape. Musculus triangularis.

Triceps (L. *tri*, three + *caput*, head). Denoting a muscle with three heads. Musculus triceps.

Triticeo- (L. *triticum*, a grain of wheat). Pertaining to a muscle attached in part to the cartilago triticea.

Ulnaris (L. *ulna*, elbow, arm). Pertaining to the larger and more medial bone of the forearm.

Uncinatus (L. *uncus*, hook). Os hamatum or unciform bone.

Urethrae (G. *ourethra*, urethra). Relating to the urethra.

Vaginae (L. *vagina*, sheath). Pertaining to muscles attached to a joint capsule.

Vastus (L. *vastus*, huge). A large muscle of the thigh. Musculus vastus lateralis.

Zygomaticus (G. *zygoma*, a bar or bolt). Pertaining to the zygomatic bone. Musculus zygomaticus.

Terminology: Synonyms

Terminology for muscles from the older English, French, and German literature related to more modern usage.

Old Terminology	Modern Usage
M. abductor externus.	M. abductor pollicis brevis.
M. abductor indicis.	M. extensor indicis proprius.
M. abductor ossis metatarsi quinti.	M. abductor digiti quinti (pedis).
M. abductor pollicis bicornis.	M. abductor pollicis longus.
M. accelator urinae.	M. bulbocavernosus.
M. accessoire du long fléchisseur.	M. quadratus plantae.
M. accessor buccinatoris.	M. incisivus labii inferioris.
Mm. accessores orbiculares.	Mm. incisivi.
M. accessorium tricipitis.	M. latissimocondyloideus.
M. accessorius ad flexorem digiti minimi.	M. flexor digiti quinti accessorius.
M. accessorius perforantis.	M. quadratus plantae.
M. Achselbogen.	M. axillary arch.
M. acromioclavicularis.	M. praeclavicularis lateralis.
M. adducteur hallucis.	M. adductor hallucis.
M. adductor digiti quinti.	M. opponens digiti quinti.
M. adductor hallucis, caput transversum s.* breve s. parvum.	M. adductor hallucis, transverse head.
M. adductor ossis metacarpi digiti quinti.	M. opponens digiti quinti.
M. adductor transversus.	M. adductor hallucis, transverse head.
Mm. adductores anguli oris.	M. incisivi.
Aeusserer Ellenmuskel.	M. extensor carpi ulnaris.
Aeusserer oder kleiner Flügelmuskel.	M. pterygoideus lateralis.
Aeusserer schiefer Bauchmuskel.	M. obliquus externus (external oblique).
Afterschliesser.	M. sphincter ani externus.
M. anconeus brevis.	M. triceps brachii, lateral head.
M. anconeus brevis, caput externum s. magnum s. secundum.	M. triceps brachii, lateral head.
M. anconeus internus.	M. triceps brachii, medial (deep) head.
M. anconeus internus, caput internum s. tertium s. parvum.	M. triceps brachii, medial (deep) head.
M. anconeus longus.	M. triceps brachii, long head (caput longum s. primum).
M. anconeus parvus.	M. anconeus.
M. anconeus quartus.	M. anconeus.
M. anconeus sextus.	M. epitrochleoanconeus.
M. angulaire.	M. levator scapulae.
M. anitersor s. aniscalptor.	M. latissimus dorsi.
M. ano-cavernosus.	M. bulbocavernosus.
Mm. arcuum transversales.	Mm. rotatores brevis.
Armheber.	M. deltoideus.
M. Arschkratzer.	M. latissimus dorsi.
M. atlantobasilaris internus.	M. axiobasilaris.
M. atollens humerum.	M. deltoideus.
Aufheber des Augenlids.	M. levator palpebrae superioris.
Mm. auriculares posterior s. retrahentes auriculae.	M. auricularis posterior.
M. auricularis anterier s. attrahens auriculae.	M. temporalis.

Old Terminology	Modern Usage
M. auricularis anterior s. attrahens auriculae.	M. auricularis superior.
M. auricularis superior s. attollens auriculae.	M. auricularis superior.
M. auriculo-temporalis.	M. auricularis superior.
M. azygos.	M. levator glandulae thyroideae.
M. azygos uvulae.	M. uvulae.
Backenmuskel.	M. buccinator.
M. basioglossus.	M. hyoglossus.
Bauschmuskeln.	Mm. splenius.
Beimuskel.	M. quadratus plantae.
M. biceps cruris.	M. biceps femoris.
M. biceps flexor cubiti.	M. biceps brachii.
M. biceps huméral.	M. biceps brachii.
Birnmuskel.	M. piriformis.
M. biventer.	M. digastricus.
M. biventer cervicis.	M. semispinalis capitis, medial head.
M. biventer mandibulare.	M. digastricus.
M. brachiaeus internus.	M. brachialis.
M. brachialis s. brachieus posterieur.	M. triceps brachii.
M. brachialis anterieur.	M. brachialis.
M. brachialis anticus.	M. brachialis.
M. brachialis exterieur.	M. triceps brachii, medial (deep) head.
Breiter Halsmuskel.	M. platysma.
Breiter Rückenmuskel.	M. latissimus dorsi.
Brustbeinschildmuskel.	M. sternothyroideus.
Brustbeinschildknorpelmuskel.	M. sternothyroideus.
Brustbeinzungenbeinmuskel.	M. sternohyoideus.
Brustschildmuskel.	M. sternothyroideus.
M. buccinato-labial, portion buccale.	M. buccinator.
M. buccinato-labial, portion labial.	M. orbicularis oris.
M. bulbo-urethralis.	M. bulbocavernosus.
M. caninus.	M. levator anguli oris.
M. caro quadrata Sylvii.	M. quadratus plantae.
M. carré pronateur.	M. pronator quadratus.
M. catenae.	M. tibialis anterior.
M. cavo quadrata manus.	M. palmaris brevis.
M. cephalo-humeralis.	M. cleido-occipitalis (Wood).
M. cephalopharyngeus.	M. constrictor pharyngis-superior (superior constructor).
M. ceratoglossus.	M. hyoglossus.
M. cervicalis descendens s. adcendens.	M. iliocostalis cervicis.
M. chair carrée de Sylvius.	M. quadratus plantae.
M. chondroepitrochlearis.	M. chondrohumeralis.
M. chondroglossus.	M. hyoglossus.
M. circumflexus palati.	M. tensor veli palatini.
M. cleido-hyoidien.	M. sternohyoideus.
M. cleidomastoideus, caput claviculare.	M. cleidomastoideus, clavicular head.
M. cleidooccipitalis (Wood).	M. cephalohumeralis.
M. complexus.	M. semispinalis capitis.
M. complexus major.	M. semispinalis capitis.
M. complexus minor s. parvus.	M. longissimus capitis.

* s. = Synonymous

Human Anatomic Variation

M. compressor.	M. bulbocavernosus.	M. epicranius temporalis.	M. temporalis.
M. compressor bulbi.	M. bulbocavernosus.	M. episternalis.	M. sternalis.
Conjoined tendon.	Conjoined aponeurosis.	M. epitrochleo-olecranonis.	M. epitrochleoanconeus.
M. constrictor s. orbicularis ani.	M. sphincter ani externus.	M. erector clitoridis.	M. ischiocavernosus.
M. constrictor cunni.	M. bulbocavernosus (in the female).	M. erector penis.	M. ischiocavernosus.
	M. bulbocavernosus.	M. erythroides.	M. cremaster.
M. constrictor cunni superficialis.		M. extensor cruris.	M. rectus femoris.
M. constrictor isthmi faucium.	M. palatoglossus.	M. extensor cruris medialis superficialis.	M. rectus femoris.
M. constrictor labiorum.	M. orbicularis oris.	M. extensor cubiti.	M. triceps brachii.
M. constrictor pharyngeus medius.	M. constrictor pharyngis medius (middle constrictor).	M. extensor digitorum pedis longus.	M. extensor digitorum communis longus.
M. constrictor prolabii superieur et inferieur.	M. orbicularis oris.	M. extensor dorsi communis.	M. sacrospinalis.
		M. extensor ossis metacarpi pollicis.	M. abductor pollicis longus.
M. coracohyoideus.	M. omohyoideus.	M. extensor pedis.	Mm. gastrocnemius and soleus.
M. coracopectoralis.	M. pectoralis minor.		
M. coracoradialis.	M. biceps brachii, short head (caput breve).	M. extensor peroneus of fifth toe.	M. peroneus accessorius inferior.
M. costalis dorsi.	M. iliocostalis thoracis.	Extensor peroneus of the fifth toe.	M. peroneus accessorius medius.
Mm. costarum depressores proprii Cowperi.	Mm. subcostales.	M. extensor pollicis major.	M. extensor pollicis longus.
		M. extensor pollicis minor.	M. extensor pollicis brevis.
M. costocoracoideus.	M. pectoralis quartus.	M. extensor primi internodii pollicis.	M. extensor pollicis brevis.
M. costohyoideus.	M. omohyoideus.		
M. court radial anterieur.	M. flexor carpi radialis brevis.	M. extensor quadriceps.	M. quadriceps.
M. cranii cutaneus.	M. epicranius or occipitofrontalis.	M. extensor secundi internodii pollicis.	M. extensor pollicis longus.
M. crotaphites.	M. temporalis.		
M. cubital antérieur.	M. flexor carpi ulnaris.	M. extensor triceps.	M. quadriceps (less the rectus femoris).
M. cubital postérieur.	M. extensor carpi ulnaris.		
M. cubitocarpus.	M. ulnocarpeus.	M. extensor triceps.	M. triceps brachii.
M. cucullaris.	M. trapezius.	M. faisceau péronéo-calcanéen externe.	M. peroneus accessorius inferior.
Deltamuskel.	M. deltoideus.		
M. demi-membraneux.	M. semimembranosus.	M. faisceau péronéo-cuboiden.	M. peroneus accessorius inferior.
M. demi-tendineux.	M. semitendinosus.		
M. depressor anguli oris s. depressor labiorum communis.	M. triangularis (menti).	M. faisceau péronéo-malléolaire.	M. peroneus accessorius inferior.
M. depressor clitoridis.	M. ischiocavernosus.	M. faisceau péronéo-metatarsien.	M. peroneus accessorius inferior.
M. depressor septi mobilis narium.	M. nasalis labii superioris.	M. fibularis accessorius.	M. peroneus accessorius medius.
M. diaphragma pelvis.	M. levator ani.	M. fibularis quartus.	M. peroneus accessorius inferior.
M. digastrique.	M. digastricus.		
M. digastricus maxillae interioris.	M. digastricus.	M. fléchisseur superficiel ou sublime.	M. flexor digitorum superficialis or sublimis.
M. digastricus ossis hyoidei.	M. digastricus.	Fléchisseurs tibiaux s. Mm. demi-tendineaux et demi-membraneaux.	Mm. semitendinosus and semimembranosus.
M. director s. erector penis.	M. ischiocavernosus.		
Dornmuskeln.	Mm. spinales.	M. flexor accessorius.	M. quadratus plantae.
M. dorsoepitrochlearis.	M. latissimocondyloideus.	M. flexor antibrachii radialis.	M. biceps brachii.
Dreieckiger Armmuskel.	M. deltoideus.		
Dreieckiger Brustmuskel.	M. transversus thoracis (anterior).	M. flexor antibrachii ulnaris.	M. brachialis.
Dreiköpfiger Armmuskel.	M. triceps brachii.	M. flexor carpi radialis brevis s. profundus (Wood).	M. radialis internus brevis.
M. droit antérieur.	M. rectus femoris.	M. flexor carpi radialis brevis vel profundus.	M. flexor (carpi) radialis (internus) brevis.
Droit interne.	M. gracilis.		
Dünner Sohlenmuskel.	M. plantaris.	M. flexor cruris fibularis s. externus.	M. biceps femoris.
M. ejaculator seminis.	M. bulbocavernosus (in the male).	M. flexor digitorum communis brevis s. sublimis s. perforatus.	M. flexor digitorum brevis (plantae).
Ellenbogenbeuger.	M. brachialis.		
M. Ellenbogenstrecker der Hand.	M. extensor carpi ulnaris.	M. flexor digitorum communis longus s. perforans s. profundus.	M. flexor digitorum longus (pedis).
M. epicranius auricularis posterior.	M. auricularis posterior.	M. flexor digitorum perforans.	M. flexor digitorum profundus.
M. epicranius auricularis superioris.	M. auricularis superior.	M. flexor digitorum perforatus.	M. flexor digitorum superficialis.
M. epicranius frontalis.	M. frontalis.		
M. epicranius occipitalis.	M. occipitalis.	M. flexor femoris.	M. iliopsoas.

Muscles

M. flexor perforatus.

M. flexor pollicis proprius longus.
M. flexor radii.
M. fourth peroneus.

M. gastrocnemius internus.
Gerader Bauchmuskel.
Gerader Schenkelmuskel.
M. glenoradialis.

M. glossopalatinus.
M. glossostaphylinus.
M. gluteus magnus.
M. gluteus major.
M. gluteus minor s. tertius.
M. gluteus secundus s. iliacus externus.
M. gluteus tertius.
M. gnathopharyngeus.

M. gracillimus orbitis.
M. grand adducteur profundus.
M. grand complexus.
M. le grand dentelé.

M. grand dorsal.
M. grand droit.
M. grand fessier.
M. grand oblique.

M. grand palmaire.
M. grand rond.
Griffelzungenbeinmuskel.
Großer Brustmuskel.
Großer Flügelmuskel.
Großer Gesäßmuskel.
Großer Rautenmuskel.
Großer runder Armmuskel.
Großer Sägemuskel.

Haken-Armmuskel.
Hakenmuskel.
Halbdornmuskeln.
Halbhäutiger Muskel.
Halbsehniger oder halbflechsiger Muskel.
Handsehnenspanner.
Hauthalsmuskel.
Hebemuskel.
Hintere Sägemuskeln.
M. hippicus.
Hodenmuskel.
Hohlhandmuskel.
M. hyopharyngeus.

M. hyothyreoideus.
M. hyothyreoidien de Duverney.

M. iliacus externus.
M. iliocapsularis.
M. iliocostalis dorsi.
M. immersus.
M. indicator s. indicatoris.
Mm. infracostales.
M. infrascapularis.

M. flexor digitorum superficialis.
M. flexor pollicis longus.

M. biceps brachii.
M. peroneus accessorius inferior.
M. soleus.
M. rectus abdominis.
M. rectus femoris.
M. biceps brachii, long head (caput longum).
M. palatoglossus.
M. palatoglossus.
M. gluteus maximus.
M. gluteus maximus.
M. gluteus minimus.
M. gluteus medius.

M. gluteus minimus.
M. constrictor pharyngis superior (superior constrictor).
M. tensor trochleae.
M. adductor magnus (femoris).
M. semispinalis capitis.
M. serratus anterior or anticus.
M. latissimus dorsi.
M. rectus abdominis.
M. gluteus maximus.
M. external obliquus externus (external oblique).
M. flexor carpi radialis.
M. teres major.
M. stylohyoideus.
M. pectoralis major.
M. pterygoideus medialis.
M. gluteus maximus.
M. rhomboideus major.
M. teres major.
M. serratus anterior or anticus.

M. coracobrachialis.
M. coracobrachialis.
Mm. semispinales.
M. semimembranosus.
M. semitendinosus.

M. palmaris longus.
M. platysma.
M. cremaster.
Mm. serratus posterior.
M. tibialis anterior.
M. cremaster.
M. palmaris longus.
M. constrictor pharyngis medius (middle constrictor).
M. thyrohyoideus.
M. levator glandulae thyroideae.

M. piriformis.
M. iliacus minor.
M. iliocostalis thoracis.
M. subscapularis.
M. extensor indicis proprius.
Mm. subcostales.
M. subscapularis.

Innerer Armmuskel.
Innerer Brustmuskel.

Innerer Ellenbogenmuskel.
Innerer schiefer Bauchmuskel.
Innerer oder großer Flügelmuskel.
M. internus.
M. interspinales supernumerarii.
M. intertransversaire antérieur.
Mm. intertransversales.
Mm. intracostales.
M. ischioaponeuroticus.

Kappenmuskel.
Kaumuskel.
Kiefermuskel.
Kieferzungenbeinmuskel.
Kinnzungenbeinmuskel.
Kleiner Gesäßmuskel.
Kleiner Rautenmuskel.
Kleiner runder Armmuskel.
Kleinerer Peroneus.

Kniekehlenmuskel.
Knorrenmuskel.
Kuhmuskel.
Kurzer äußerer Speichenmuskel.
Kurzer Rückwärtswender.
Kurzer Speichenstrecker.

Langer äußerer Speichenmuskel.
Langer Rückgratsstrecker.
Langer Rückwärtswender.
Langer Sohlenmuskel.
Langer Speichenstrecker.

Langer's arch (muscle).
M. latissimus colli.
M. levator s. dilator pharyngeus.
M. levator anguli scapulae.
M. levator claviculae.
M. levator humeri internus.
M. levator labii interior s. incisivus inferior.
M. levator labiorum communis.
M. levator menti.
M. levator pharyngeus externus.
M. levator pharyngis internus.
M. levator uvulae.
M. lividus.
M. long ou rond pronateur.
M. longissimus dorsi.
M. longus atlantis.

M. mallei internus.
M. mandibularis externus.
M. manducatorius.
M. mastoideus colli.
M. mesothenar.

M. brachialis.
M. transversus thoracis (anterior).
M. flexor carpi ulnaris.
M. obliquus internus (internal oblique).
M. pterygoideus medialis.

M. flexor carpi ulnaris.
M. spinalis cervicis.

M. rectus capitis anterior.

Mm. intertransversarii.
Mm. subcostales.
M. tensor fasciae suralis.

M. trapezius.
M. masseter.
M. masseter.
M. mylohyoideus.
M. geniohyoideus.
M. gluteus minimus.
M. rhomboideus minor.
M. teres minor.
M. peroneus accessorius inferior.

M. popliteus.
M. anconeus.
M. masseter.
M. extensor carpi radialis brevis s. secundus.
M. supinator.
M. extensor carpi radialis brevis s. secundus.

M. extensor carpi radialis longus.
M. sacrospinalis.
M. brachioradialis.
M. plantaris.
M. extensor carpi radialis longus.

Axillary arch (muscle).
M. platysma.
M. stylopharyngeus.

M. levator scapulae.
M. cleido-occipitalis (Wood).
M. coracobrachialis.
M. mentalis.

M. levator anguli oris.

M. mentalis.
M. stylopharyngeus.

M. palatopharyngeus.

M. uvulae.
M. pectineus.
M. pronator teres.
M. longissimus thoracis.
M. longus colli.

M. tensor tympani.
M. masseter.
M. masseter.
M. sternocleidomastoideus.
M. adductor pollicis brevis.

53

Human Anatomic Variation

Mittlerer Gesäßmuskel.	M. gluteus medius.	M. peronaeo-metatarsalis.	M. peroneus accessorius inferior.
Mönchskappenmuskel.	M. trapezius.		
M. motor uvulae.	M. uvulae.	M. peronaeus accessorius.	M. peroneus accessorius medius.
Moyen fessier.	M. gluteus medius.		
Moyen péronier.	M. peroneus brevis.	M. peronaeus brevis secundus.	M. peroneus accessorius medius.
M. multifidus spinae.	M. multifidus.		
M. musculorum lumbricalium s. lumbricales pedis propriae s. vaginulae tendinum muskel lumbricalium pedis synoviales.	M. lumbricales (pedis).	M. peronaeus brevis II.	M. peroneus accessorius inferior.
		M. peronaeus digiti V posterior.	M. peroneus accessorius inferior.
		M. peronaeus digiti V superior.	M. peroneus accessorius medius.
Muscle de la houppe du menton.	M. mentalis.	M. peronaeus intermedius.	M. peroneus accessorius inferior.
M. mytiformis.	M. depressor alae nasi.	M. peronaeus digiti minimi pedis.	M. peroneus accessorius medius.
Niederzieher des Penis.	M. ischiocavernosus.	M. peronaeus parvus.	M. peroneus accessorius medius.
Novus musculi conchae proprius.	M. temporalis.		
M. nutator capitis.	M. sternocleidomastoideus.	M. peronaeus quartus.	M. peroneus accessorius inferior.
M. nutator capitis anterior.	M. sternomastoideus, sternal head.	M. peronaeus quartus.	M. peroneus accessorius superior.
M. nutator capitis exterior.	M. cleidomastoideus, clavicular head.	M. peronaeus quinti digiti.	M. peroneus accessorius medius.
M. nutator capitis interior.	M. sternomastoideus, sternal head.	M. peronaeus quintus.	M. peroneus accessorius imus.
N. nutator posterior.	M. cleidomastoideus, clavicular head.	Peronaeus sextus.	M. peroneus accessorius inferior.
Obergrätenmuskel.	M. supraspinatus.	M. péronéo-calcanéen externe.	M. peroneus accessorius inferior.
Oberschädelmuskel.	M. epicranius or occipitofrontalis.	M. péronéo-calcaneus externus.	M. peroneus accessorius inferior.
M. obliquus abdominis externus s. descendens s. oblique descendens.	M. obliquus externus (external oblique).	M. péronéo-cuboidien.	M. peroneus accessorius inferior.
M. obliquus abdominal internus s. ascendens s. oblique ascendens.	M. obliquus internus (internal oblique).	M. peroneus accessorius.	M. peroneus accessorius medius.
		M. peroneus anticus s. secundus s. medius.	M. peroneus brevis.
M. obliquus capitis major.	M. obliquus capitis inferior (inferior capitis oblique).	M. peroneus digiti minimi.	M. peroneus accessorius inferior.
M. obliquus capitis minimus.	M. obliquus capitis superior (superior capitis oblique).	M. peroneus digiti quinti.	M. peroneus accessorius inferior.
M. obliquus capitis superioris.	M. obliquus capitis superior (superior capitis oblique).	M. peroneus digiti quinti.	M. peroneus medius.
M. obliquus longus.	M. obliquus superior (superior oblique).	Peroneus muscle of tarsus.	M. peroneus accessorius inferior.
M. omocervicalis s. omotrachelien.	M. levator claviculae.	M. peroneus primus s. posticus.	M. peroneus longus.
M. omoplathyoiden.	M. omohyoideus.	M. peroneus quinti.	M. peroneus accessorius inferior.
M. opistothenar.	M. sacrospinalis.	M. péronier antérieur.	M. peroneus tertius.
M. orbicularis ani.	M. sphincter ani externus.	M. péronier accessorie.	M. peroneus accessorius inferior.
M. orbicularis palpebrae.	M. orbicularis oculi.	M. péronier accessoire.	M. peroneus accessorius medius.
M. orbito-palpebralis.	M. levator palpebrae superioris.	M. péronier du cinquième orteil.	M. peroneus accessorius inferior.
M. palatostaphylinus.	M. uvulae.	M. péronier latéraux.	Mm. peroneus longus and brevis.
M. palmaire grêle.	M. palmaris longus.		
M. palmaris cutaneus.	M. palmaris brevis.	Pes anserinus.	Aponeurotic expansion of tendon of insertion of sartorius muscle.
M. patientiae.	M. levator scapulae.		
M. le peaucier.	M. platysma.		
Peaucier de la main.	M. palmaris brevis.	M. petit adducteur profundus.	M. adductor brevis (femoris).
M. peaucier du crâne.	M. epicranius or occipitofrontalis.		
M. pectinalis.	M. pectineus.	M. petit dentalé anterior.	M. transversus thoracis (anterior).
M. pectoralis.	M. pectoralis major.	M. petit fessier.	M. gluteus minimus.
M. pectorodorsalis.	M. axillary arch.	M. petit oblique.	M. obliquus internus (internal oblique).
M. pediaeus internus.	M. flexor digitorum brevis (plantae).		
M. perforatus Casserii s. coracoideus.	M. coracobrachialis.	M. petit péronier.	M. peroneus tertius.
M. peristaphylinus internus.	M. levator veli palatini.	M. petit rond.	M. teres minor.

Muscles

Petits dentelés postérieurs.
M. petro-malléal.
M. petrosalpingostaphylinus.
M. petrostaphylinus.
M. pharyngostaphylinus.
M. phrenicus.
Caput plantare flexoris digitorum plantae longi.
M. platysma myoides.
Pozzi's muscle.

M. premier adducteur superficial.
Premier radial externe.

M. presternalis.
M. productor labii inferiores.
M. pronator inferior.
M. pronator rotundus.
Mm. protractores anguli oris.
M. psoas-iliaque.
M. psoas magnus.
M. psoas parvus.
M. pterygoideus externus.
M. pterygoideus internus.
M. pterygoideus major.
M. pterygoideus minimus.
M. pterygostaphylinus internus.
M. pterystaphylinus internus.
M. pyramidalis.
M. pyramidalis abdominis.
M. pyramidalis menti.
M. pyramido-stapéal.
M. pyriformis.
M. quadratus genae.
M. quadratus menti.
M. quadratus plantae pedis.
M. quatriéme péronier.

M. quatrième péronier d'Otto.

M. radialis anticus.
M. radialis extensor brevis s. secundus.
M. radialis extensor longus.
M. radialis internus.
M. radialis internus minor.
M. radio-carpeus s. radiocarpo-metacarpeus.
M. radiocarpien.
Raben-Armmuskel.
M. rectus abdominis posticus.
M. rectus anterior.
M. rectus capitis anterior.
M. rectus capitis anticus major.
M. rectus capitis anticus minor.
M. rectus capitis internus major.
M. rectus capitis internus minor.
M. rectus capitis posterior major.
M. rectus capitis posterior minor.

Mm. serratus posterior.
M. tensor tympani.
M. levator veli palatini.
M. levator veli palatini.
M. palatopharyngeus.
M. diaphragma (diaphragm).
M. quadratus plantae.

M. platysma.
M. extensor digitorum brevis manus.
M. pectineus.

M. extensor carpi radialis longus.
M. sternalis.
M. incisivus labii inferioris.
M. pronator quadratus.
M. pronator teres.
Mm. incisivi.

M. iliopsoas.
M. psoas major.
M. psoas minor.
M. pterygoideus lateralis.
M. pterygoideus medialis.
M. pterygoideus medialis.
M. pterygoideus lateralis.
M. levator veli palatini.

M. levator veli palatini.
M. piriformis.
M. pyramidalis.
M. triangularis menti.
M. stapedius.
M. piriformis.
M. platysma.
M. depressor labii inferioris.
M. quadratus plantae.
M. peroneus accessorius inferior.
M. peroneus accessorius inferior.

M. flexor carpi radialis.
M. extensor carpi radialis brevis.
M. extensor carpi radialis longus.
M. flexor carpi radialis.
M. radialis internus brevis.
M. flexor (carpi) radialis (internus) brevis.
M. flexor carpi radialis brevis.
M. coracobrachialis.
M. quadratus lumborum.
M. rectus femoris.
M. longus capitis.
M. longus capitis.

M. rectus capitis anterior.

M. longus capitis.

M. rectus capitis anterior.

M. rectus capitis major.

M. rectus capitis minor.

M. rectus capitis posterior profundus.
M. rectus capitis posterior superficialis s. inferior.
M. rectus internus.
M. rectus sternalis.
M. rectus thoracicus.
M. rectus thoracicus s. sternocostalis.
M. rhomboatloideus.
M. rhomboideus inferior.
M. rhomboideus superior.
Riemenmuskeln.
Rippenhalter.
Mm. rotatores dorsi.
Ruthensteifer.

M. sacrolumbaris s. lumbocostalis.
Mm. scalène postérieur.

M. scalenus intermedius s. pleuralis.
M. scalenus lumborum and ileolumbalis.
M. scalenus primus.

M. scalenus prior.

M. scalenus secundus.
M. scalenus tertius.

Schädelmuskel.

Schambeinmuskel.
Schildzungenbeinmuskel.
Schlafmuskel.
Schläfenmuskel.
Schlanker Schenkelmuskel.
Schlüsselbeinmuskel.
Schollenmuskel.
Schulterblattzungenbeinmuskel.
Schulterheber.
Schulterzungenbeinmuskel.
Schwimmuskel.

M. second adducteur superficiel.
M. second peroneus.

M. second radial externe.

M. semifibularis.
M. seminervosus.
M. semispinalis dorsi und colli.
M. septum transversum.
M. serratus anterior.
M. serratus anterior major.

M. serratus anterior minor.
Mm. serratus interior.
M. serratus magnus.

Sixth peroneal muscle.

M. soléaire.
M. sousclavier.
Mm. sous-costaux.

M. rectus capitis minor.

M. rectus capitis major.

M. gracilis.
M. sternalis.
M. sternocostalis.
M. supracostalis.

M. splenius colli accessorius.
M. rhomboideus major.
M. rhomboideus minor.
Mm. splenius.
Mm. scaleni.
Mm. rotatores brevis.
M. ischiocavernosus.

M. iliocostalis.

Mm. scalenus medius and posticus.
M. scalenus minimus.

M. quadratus lumborum.

M. scalenus anterior or anticus.
M. scalenus anterior or anticus.
M. scalenus medius.
M. scalenus posterior or posticus.
M. epicranius or occipitofrontalis.
M. pectineus.
M. thyrohyoideus.
M. temporalis.
M. temporalis.
M. gracilis.
M. subclavius.
M. soleus.
M. omohyoideus.

M. levator scapulae.
M. omohyoideus.
M. tibialis posterior or posticus.
M. adductor longus (femoris).

M. peroneus accessorius medius.
M. extensor carpi radialis brevis s. secundus.
M. peroneus brevis.
M. semitendinosus.
M. semispinalis thoracis and cervicis.
M. diaphragma (diaphragm).
M. pectoralis minor.
M. serratus anterior or anticus.
M. pectoralis minor.
Mm. subcostales.
M. serratus anterior or anticus.
M. peroneus accessorius inferior.
M. soleus.
M. subclavius.
Mm. subcostales.

55

Human Anatomic Variation

M. sous-épineux.
M. sous-scapulaire.
Faisceau sous-symphysien.
M. sphincter s. constrictor labiorum.
M. sphincter ani cutaneous.
M. sphincter oris.
M. sphincter palpebrae.
M. spinalis cervicis.
Mm. spinotransversales brevissimi.
M. splenius colli.
Steigbügelmuskel.
M. sternalis brutorum.
M. sternoabdominalis.
M. sternoclavicularis.
M. sternoclavicularis posticus.
M. sternoclavicularis superior.
M. sternoclavicularis superioris.
M. sternocleidohyoideus.
M. sternocostalis.

M. sternocostalis.

M. sternocostalis.
M. sternomastoideus, caput sternale.
Stirnmuskel.
M. stylochondrohyoideus.
M. stylohyoideus novus s. alter s. parvus.
M. subscapularis secundus.
M. subcutaneus colli.
M. subpopliteus.
Mm. superspinales.
M. superspinalis colli.
M. supinator brevis.
M. supinator longus.
M. supinator radii brevis.
M. supinator radii longus.
M. supraclavicularis medialis.

Mm. supracostales.
M. supracostalis anterior.
Mm. surales.

M. surcostaux.
Sus-épineux.
M. sustentator penis.
M. tensor fasciae.
M. tensor membranae synovialis tarsi.
M. tensor vaginae femoris.
M. thoracicus.
M. thyreopharyngopalatinus.
M. thyreopharyngostaphylinus.
M. tibiaeus posterior s. nauticus.
M. tibialis anticus.
M. trachelomastoideus.
M. trachelomastoideus minor s. accessorius.
M. transversaire épineux.
M. transversaire épineux antérieur.
M. transversalis capitis.

M. infraspinatus.
M. subscapularis.
M. transversus menti.
M. orbicularis oris.

M. sphincter ani externus.
M. orbicularis oris.
M. orbicularis oculi.
M. semispinalis cervicis.
Mm. rotatores brevis.

M. splenius cervicis.
M. stapedius.
M. sternalis.
M. transversus abdominis.
M. praeclavicularis medialis.
M. retroclavicularis medialis.

M. supraclavicularis medialis.

M. supraclavicularis medialis.

M. sternohyoideus.
M. transversus thoracis (anterior).
M. rectus thoracis s. M. sternalis.
M. supracostalis.
M. sternomastoideus, sternal head.
M. frontalis.
M. stylohyoideus profundus.
M. stylohyoideus profundus.

M. subscapularis minor.
M. platysma.
M. popliteus.
M. spinalis cervicis.
M. spinalis cervicis.
M. supinator.
M. brachioradialis.
M. supinator.
M. brachioradialis.
M. sternoclavicularis superior.
Mm. levatores costarum.
M. supracostalis (Wood).
Mm. gastrocnemius and soleus.
Mm. levatores costarum.
M. supraspinatus.
M. ischiocavernosus.
M. tensor fasciae latae.
M. peroneus accessorius inferior.
M. tensor fasciae latae.
M. sternalis.
M. palatopharyngeus.
M. palatopharyngeus.

M. tibialis posterior or posticus.
M. tibialis anterior.
M. longissimus capitis.
M. longissimus cervicis.

M. transversospinalis.
M. longus capitis.

M. longissimus capitis.

M. transversalis cervicis.
M. transversalis cervicis posterior major.
M. transversalis cervicis posterior minor.
M. transversalis dorsi.
M. transverse vaginae.

M. transverso-analis.

M. transversus.
M. transversus pectoris.
M. transversus perinei medius.

M. transversus perinei posterior inferior.

M. transversus perinei superficialis.

M. transversus thoracis posterior medialis.
M. transversus s. transversalis pedis s. plantae.
M. transversus s. triangularis.

Mm. triangulares.
M. triangularis coccygis.
M. triangularis sterni.

M. triceps extensor cubiti.
M. triceps surae.
M. trochlearis.

Trompetermuskel.
M. tunica carnae.
M. ulnaris externus.
M. ulnaris internus.
M. ulnocarpeus brevis.
Untergrätenmuskel.
Unterschlüsselbeinmuskel.
Unterschulterblattmuskel.
M. uvula palatina.
M. vastus internus.

Verdoppelung des M. peronaeus brevis.
Vieltheiliger Rückgratsmuskel.
Viereckiger Vorwärtswender.
Vorderarmstrecker.
Vorderer Schienenmuskel.
Vorderschulterblattmuskel.
Wadenbeinbeuger.
Zungenbeinschildknorpelmuskel.
Zweibäuchiger Unterkiefermuskel.
Zweiköpfiger Armmuskel.
Zweiköpfiger Beuger.
Zwerchmuskel.
Zwischendornmuskeln.
Zwischenquerfortsatzmuskeln.
M. zygomaticus.

M. longissimus cervicis.
M. longissimus cervicis.

M. longissimus cervicis.

M. longissimus thoracis.
M. transversus perinei profundus (deep transverse perinei).
M. transversus perinei superficialis (superficial transverse perinei).
M. mylohyoideus.
M. transversus thoracis (anterior).
M. transversus perinei superficialis (superficial transverse perinei).
M. transversus perinei superficialis (superficial transverse perinei).
M. transversus perinei superficialis (superficial transverse perinei).
M. subcostalis.

M. adductor hallucis, transverse head.
M. transversus perinei superficialis (superficial transverse perinei).
Mm. scaleni.
M. coccygeus.
M. transversus thoracis (anterior).
M. triceps brachii.
Mm. gastrocnemius and soleus.
M. obliquus superior (superior oblique).
M. buccinator.
M. cremaster.
M. extensor carpi ulnaris.
M. flexor carpi ulnaris.
M. flexor carpi ulnaris brevis.
M. infraspinatus.
M. subclavius.
M. subscapularis.
M. uvula.
M. triceps brachii, medial (deep) head.
M. peroneus accessorius inferior.
M. multifidus.

M. pronator quadratus.
M. triceps brachii.
M. tibialis anterior.
M. subscapularis.
M. biceps femoris.
M. thyrohyoideus.

M. digastricus.

M. biceps brachii.
M. biceps femoris.
M. diaphragma (diaphragm).
Mm. interspinales.
Mm. intertransversarii.

M. zygomaticus major.

56

CARDIOVASCULAR SYSTEM

Alphabetical Listing

(Primary references, in multiple page listings, are in *boldface* type.)

Arteries

Accessory
  Ascending Pharyngeal 64
  Femoral Circumflex 86
  Gastroduodenal 77
  Gonadal 76
  Hepatic 76, 77, 78, 80, 83
  Inferior Epigastric 85
  Inferior Phrenic 76
  Inferior Thyroid 70
  Internal Thoracic 76
  Interosseous 74
  Lateral Thoracic 72
  Lumbar 76
  Median Sacral 76
  Meningeal (Pterygomeningeus) 66
  Middle Colic 77, 80
  Middle Rectal 83
  Obturator 84
  Pectoral 73
  Perforating 86
  Peroneal 88
  Posterior Communicating 68
  Posterior Interosseous 75
  Posterior Tibial 87
  Radial Recurrent 74
  Renal 76, 81, 82, 83, 84
  Splenic 77
  Superior Mesenteric 76
  Superior Thyroid 64, 65
  Thoracoacromial 72
  Transverse Facial 64
  Ulnar Recurrent 74
  Umbilical 76
  Vertebral 71
Alar Thoracic 72
Angular 64, 65
Antebrachial 73
Aorta **60**, 72, 76, 77, 80, 81, 82, 83
  Abdominal 71, **76**
  Ascending **60**, 71
  Descending **60**, 62
  Thoracic 62, 69, 71, 76
Aortic Arch **61**, 63, 64, 68, 70, 71, 94
  Branches of
    Doubled 61
  Arch 61
    Deep Palmar **75**
    Gonadal, of Luschka 83
    Plantar **88**

Superficial Palmar **74**, 75
Superficial Plantar 88
Arteria Aberrans 69, 72, 73, 74, 76, 78, 79, 86, 87
  Subclavia Dextra Lusoria 61
Arteria Femoris Magna 86
Arteria Lusoria 61, 69
Arteria Penis 84
Arteria Saphena Parva 81
Articular, Medial Superior 86
Auricular, Posterior 65
Axillary 69, 71, **72**, 75
Axis (Embryonic) 84
Basilar 66, **72**
Brachial 72, **73**, 74
  Superficial 73
Brachiocephalic 60, **63**, 69, 70
Bronchial 63, 70, **71**, **75**, 76
  Inferior 76
Buccal 66
Bulbus Arteriosus 61
Carotid 62
  Common 61, 63, **64**, 72
  External **64**, 65
  Internal **66**, 67, 68
Celiac 80
Celiac Trunk 76, **77**, 79
Celiacocolic Trunk 79
Celiacomesenteric Trunk 79
Central, of Retina 66
Cerebellar
  Anterior Inferior **67**, 68
  Middle 72
  Posterior Inferior **67**, 72
Cerebral
  Anterior **67**, 68
  Azygos Anterior 67
  Common Anterior 67
  Deep 67, 68
  Median Anterior 67
  Middle 66, 67, **68**
  Posterior **68**, 72
  Recurrent of Heubner **68**
Cervical
  Ascending 65, 70, **71**, 72
  Deep **71**, 72
  Superficial **71**, 72
  Transverse 69, **70**, 72
Cervicoscapular Trunk 70
Choroidal, Anterior 66
Circle of Willis 64, **66**
Circumflex Humeral
  Anterior 72, **73**, 74

Posterior 72, **73**, 74
Circumflex Iliac 84, 85, **86**
Circumflex, Lateral **86**
Circumflex Scapular, Posterior 72
Colic
  Left 79, 80, 83
  Middle 77, 80, 83, 84
  Right 80
Communicating
  Anterior **67**, 68
  Posterior 67, **68**
Conus Arteriosus 61, 63
Coronary 62, **63**
Costocervical Trunk 64, 69, **71**, 72
Cricothyroid 65
Cystic 78, 79, 80
Deep Cervical (Cervicalis Profundus) 70, **71**, 72
Deep Temporal 66
  Anterior 66
  Middle 66
Digital 75, 87
Dorsalis Pedis 81
Dorsalis Penis/clitoris 84, 85, 86
Ductus Arteriosus 60, 61, 69
Epigastric 80, 86
  Inferior 84, **85**, 86
  Superficial 85, 86
Esophageal, Superior 70
Facial 64, **65**
  Transverse 65, 66
Femoral 84, **86**
Femoral Circumflex
  Lateral 81, **86**
  Medial 85, **86**, 87
Gastric **77**
  Left 76, **77**, 78
  Right 77
Gastroduodenal 77, 78, 80
Gastroepiploic 77, 79, 80
Gluteal
  Inferior **84**, 93
  Superior 84
Gonadal 77, 80, 81, 82, **83**, 84
Great Saphenous 86, 87
Hemorrhoidal, Superior 79
Hepatic 76, **77**, 78, 79, 80
  Common 77, 80
  Left 77, 78
  Proper 78
  Right 80
Hepatogastric Trunk 79
Hepatosplenic Trunk 79

Hepatosplenogastric Trunk 79
Hepatosplenomesenteric Trunk 79
Hepatosuperiormesenteric Trunk 77
Hyoid 65
Ileocolic 80
Iliac
   Circumflex **86**
   Common 81, 82, 83, **84**, 85
   External 82, **85**
   Internal 76, 81, 82, 83, **84**, 85
Iliolumbar **84**, 86
Inguinal 85
Intercostal **76**
   Superior (Supreme) 63, 69, 70, **71**, 72, 76
Interosseous 72
   Anterior 75
   Common 73, **75**
   Palmar 75
   Posterior 75
Ischiadic (Sciatic) 84, 87, 93
Labial 65
   Inferior 65
   Superior 65
Lacrimal 66, 67
Laryngeal, Superior 65
Ligamentum Teres Femoris 86, 87
Lingual **64**, 65, 66
   Dorsal 64
Linguofacial Trunk 64
Lumbar **77**, 83
Masseteric **65**
Maxillary 64, **65**
Median (Comes Nervi Mediani) **75**
Median Basilar 72
Median Cubital 74
Meningeal
   Accessory **66**
   Middle **66**, 67
Mesenteric
   Inferior 79, 80, 81, 82, **83**
   Superior 76, 77, 78, **79**, 80, 82
Metacarpal 75
Nasal, Lateral (Alar Nasi) 65
Obturator **84**, 85, 86, 87
Occipital 64, **65**, 66, 72
Olfactory 67
Omphalomesenteric 80
Ophthalmic 64, 65, 66, **67**
Ovarian 81, 82, **83**
Palatine, Ascending 65
Palmar Interosseus 75
Palmar, Superficial **74**, 75
Pancreatic 81
   Dorsal 77, 79, 80
   Inferior 79, 80
   Supreme 76, 77
Pancreatoduodenal, Superior 78
Parotid 65, 70
Penis, Dorsal 84
Perforating 86
Pericardiacophrenic 63, 70
Pericardial 63, 70
Perineal 84
Peroneal 87, **88**
Pharyngeal, Ascending 64, **65**, 66
Phrenic
   Inferior **76**, 77, 80, 81
Plantar, Deep 87
   Lateral 87

Pleural Cupula 70
Popliteal 84, **87**
Princeps Pollicis 75
Profunda
   Brachii 72, 73, **74**
   Femoris 84, 85, **86**, 87
   Penis 84
Prostatic 83, 84, 85
Pudendal
   External 84, 85, 86
   Internal 84, **85**
Pulmonary **60**, 63, 71
Radial 69, 70, 72, 73, 74, **75**
Radialis Indicis 75
Rectal, Middle 84, **85**
Renal 63, 76, 77, 80, **81**, 83
   Anterior Superior Capsular Adipose 81, 82
   Hilar 81, 82
   Polar 80, 81, **82**
Retinacular
   Inferior 87
   Superior 87
Sacral
   Middle 77, 81, 82, **83**, 84
   Lateral **84**, 85
Saphena Parva 87
Saphenous
   Great 87
   Small 87
Scapular
   Dorsal 69, **70**, 73
   Transverse 63, 64
Sciatic 84, 87, 93
Seminal Vesical 85
Spermatic 81, 82, **83**
Spinal 72
   Anterior 72
   Posterior 61, 72
Splenic 76, 77, **79**, 80
Splenogastric Trunk 77, 79
Sternocleidomastoid 64, **65**
Stylomastoid 65
Subclavian 60, 62, 64, **68**, 69, 70, 71, 72
   Branches of **69**
Sublingual 65
Submental 65
Subscapular 65, 70, 71, 72, **73**
Supraorbital 67
Suprarenal 76, **80**
   Inferior 76, 80, 81, 82, 83
   Middle 77, 80, 81, 83
   Superior 80
Suprascapular 69, **70**, 72
Temporal 64
   Superficial **65**
   Temporobuccal Trunk 65
   Temporomasseteric Trunk 66
Testicular 81, 82, **83**
Thoracic
   Internal 62, 63, 64, 69, 70, **71**, 76
   Lateral 70, 71, 72, **73**
   Supreme **79**
Thoracoacromial 70, **73**
Thoracodorsal 72, **73**
Thymic 62, 63, 70, **71**
Thyrocervical Trunk 63, 69, **70**, 71, 72
Thyroid
   Inferior 62, 63, 64, 65, 69, **70**, 71, 72

   Superior 64, **65**
Thyroidea Ima 62, **63**, 64, 70
Thyrolinguofacial Trunk 64
Thyroscapular Trunk 70
Thyrovertebral Trunk 70
Tibial
   Anterior 86, **87**, 88
   Posterior 86, **87**, 88
Tonsillar 65
Tracheoesophageal 70
Truncus Arteriosus 61
Tympanic 65
Ulnar 69, 70, 72, 73, **74**
   Collateral 72
   Inferior **74**
   Superior **74**
Umbilical 76, 83, 84, **85**
Ureteric 81, 84
Uterine 83, **84**, 85
Vaginal 83, **84**, 85
Vertebral 62, 64, 65, 66, 69, 70, **71**, 72
Vesical 83
   Inferior 84, **85**
   Middle **85**
   Superior 84, **85**
Vesicodeferential **85**
Vesicoprostatic 84

Heart

Heart **60**
   Apex **60**
   Atrioventricular Node **60**, 63
   Sinoatrial Node **60**, 63
   Valves **60**

Veins

Accessory
   Cephalic 90
   External Jugular 89
   External Thoracic 89
   Hemiazygos 88
   Renal 92
   Splenic 92
   Subclavian 89
   Vertebral 89
   Angular 89
Anterior Jugular 89
Arcus Venosus Dorsalis 90
Ascending 91
Auricular, Posterior 89
Axillary **90**
Azygos 88, **89**
Basilic 90
Brachial Artery Venae Comites 90
Brachiocephalic 88, **89**
Brachiocephalic Trunk **88**
Bronchial 89
Cardiac
   Great 88
   Left **88**
   Middle 88
   Right **88**
   Small Cardinal 88
Cardinal (Embryonic) 88

Cephalic   89, **90**
Cerebral   **89**
Cervical
   Superficial   **90**
   Inferior   **90**
   Transverse   89
Coronary Sinus   **88**
Duodenal   92
Dural Sinuses   **93**
   Cavernous   93
   Petrosal   93
   Superior Sagittal   93
   Tentorial   93
   Transverse   93
Epigastric, Inferior   **92**
Facial   **89**
Femoral   86, **93**
Gastric, Left   **92**
Gastroepiploic   92
Gluteal   92
Gonadal   89, 92
Hemiazygos   **89**
Hepatic   **91**
Iliac
   Common   **92**
   External   **92**
   Internal   **92**
Innominate   88
Intercostal
   Superior (Supreme)   88, 89, **90**
Jugular

Anterior   **89**
External   **89**
Internal   88, **90**
Laryngeal, Superior   90
Lingual   89, **90**
Lumbar   89, 92
Median
   Antebrachial   90
   Basilic   90
   Cephalic   90
   Cubital   90
Mesenteric
   Inferior   **92**
   Superior   **92**
Oblique   90
Obturator   92
Ovarian   89, 92
Palatine   90
Pancreaticoduodenal   92
Petrosal, Inferior   90
Pharyngeal   89
Phrenic
   Inferior   92
   Superior   89
Popliteal   **93**
Portal   77, **92**
Pudendal, Internal   92
Pulmonary   **89**
Rectal, Superior   **92**
Renal   83, 89, **91**, 92
Sacral Trunk   92

Cardiovascular System

Scapular, Transverse   90
Sciatic   92
Spermatic   89, 92
Splenic   92
Subclavian   88, 89, **90**
Submental   89
Suprarenal   89, **92**
Suprascapular   90
Thoracic
   Internal   88, 89, **90**
   Lateral   90
Thymic   90
Thyroid
   Inferior   **90**
   Superior   **90**
Thyroidea Ima   89
Uterovaginal   92
Vena Cava   76
   Inferior   76, 82, 89, **91**, 92
   Superior   60, **88**, 89
   Left (Persistent)   88
Venae Comites   90
Vertebral   **90**

Lymphatics

Thoracic duct   69, **93**

Human Anatomic Variation

Among the most important considerations in the study of the vascular system are its significant variations. Although many of them cause no disturbance in the functions of the body, they may be of great importance to the surgeon. One group of variations represents persistent fetal forms of circulation. Another group represents individual variations, certain of which may be explained by a study of usual anastomoses. Variations may be rare, or so common as to make it difficult to determine the normal.

Only the major vessels and some of their primary branches are considered in this review.

Heart

Common cardiac anomalies are (a) atresia or stenosis of the pulmonary trunk or aorta, (b) defective closure of septal openings, (c) defective closure of the ductus arteriosus, (d) reverse rotation of the heart, and (e) mirror imaging. Situs inversus viscerum and dextrocardia are examples of (d) and (e). Incomplete closure of the foramen ovale, "probe patency," is common (25% of adult hearts) and may be regarded as a normal variant rather than an abnormality. The foramen is closed physiologically but patent anatomically. In one study of 2648 subjects, probe patency was found in 23.1%. There is virtually no sex difference (25.5% occurrence in females and 21.4% in males). A heart with a more or less bifid apex and three-chambered hearts in adult subjects have been reported. Cor triatriatum, a heart with a doubled left atrium, has also been reported.

The heart valves may be composed of accessory cusps which vary in size. The posterior (inferior) cusp may be divided in the right atrioventricular valve, and the bicuspid valve of the left ventricle may have an anterior or posterior cusp or both. The aortic and pulmonary valves may present with two cusps in front and one behind. The aortic valve may be composed of two or four cusps rather than the usual three. The mitral valve may be doubled.

The (right) superior vena cava may be joined to the left atrium. Other anomalous vascular connections of the heart will be found in subsequent sections.

The heart normally weighs from 280 to 389 grams in males and 250 to 384 grams in females.

The pericardium may be partially or completely absent.

References 6, 52, 74, 95, 114, 118, 135, 141, 150, 166, 170, 177, 190, 195, 196, 216, 232, 233, 260, 297, 301, 324, 346, 353, 424, 446, 449, 475, 477, 480, 498, 534, 545, 583, 605, 614, 628, 644, 675–677, 693, 715, 729, 735, 745, 749, 754, 757, 773, 793, 813, 826, 828, 853, 858, 870, 887, 891, 892, 900, 925, 927, 930, 950, 957, 958, 962, 966, 970, 976, 977, 1017, 1027, 1054, 1073, 1081, 1090, 1095, 1105, 1154, 1156, 1170, 1181, 1204, 1217, 1224, 1235, 1243, 1258, 1286, 1294, 1313, 1315, 1316, 1320, 1327, 1348–1350, 1354–1359, 1361, 1362, 1365, 1366.

Arteries

Head, Neck, and Thorax

Pulmonary

The pulmonary artery (carrying deoxygenated blood) may appear as a branch at the origin of the arch of the ascending aorta. Conversely, the aorta may be a branch of the pulmonary trunk (subject surviving 20 years) or even a branch of the right ventricle (subject surviving 13 years). The right branch of the pulmonary artery is occasionally doubled, with the second branch arising from the right ventricle. The right and left branches may be doubled. The left branch may provide a branch to the left subclavian, and the right a branch to the brachiocephalic trunk. In some cases, the ductus arteriosus arises from the right ventricle.

References 19, 23, 37, 47, 48, 89, 114, 123, 144–147, 166, 169, 190, 198, 232, 295, 321, 324, 361, 369, 399, 402, 423, 465, 551, 554, 575, 583, 613, 614, 622, 628, 653, 655, 664, 754, 757, 793, 798, 807, 808, 826, 838, 840, 843, 853, 976, 1011, 1073, 1084, 1087, 1088, 1151, 1162, 1170, 1217, 1218, 1228, 1252, 1258, 1319, 1349, 1350, 1354–1359, 1361, 1362, 1364–1367.

Aorta

The ascending aorta is occasionally doubled. In one case the vessel bifurcated 8 mm after its origin into a right and left trunk, each providing the vessels of that side before reuniting to form the descending aorta.

The variations or absence of the arch of the aorta are usually based on persistent fetal conditions and may be associated with abnormalities of the heart. Many of the variations are due to different modes of transformation of the primary vessels of the brachial arches, especially the fourth. Since the aorta and pulmonary artery develop from a common conus arteriosus, irregular and imperfect development of the septum between them may also produce variations.

At one stage of development, two aortic arches, a right and a left, are present, and such a condition occasionally persists in the adult. In such cases, since the portion of the aorta derived from the bulbus arteriosus is directed upward and to the right and the descending aorta is lying to the left side of the vertebral column, a right arch passes from right to left behind the esophagus. A left arch may curve over the right bronchus. Rarely, the aortic arch is doubled, forming a ring around the trachea. The ring closes as a single descending aorta.

Another variation is a single aortic arch curving to the right instead of to the left. This may be due to the persistence of the lower portion of the right dorsal aorta and the disappearance of the left, and may be associated with a complete inversion of all the viscera, situs inversus.

If the inferior portion of the right dorsal aorta involutes, and the part that normally forms the proximal portion of the right subclavian persists, the right subclavian arises from the descending portion of the aortic arch. In such cases the subclavian passes either behind the esophagus or between the trachea and esophagus (arteria subclavia dextra lusoria) to reach the right side of the body. It was found in 22 of 2169 bodies (all sources). An aortic arch (right) having the left subclavian as the first branch has been reported.

Another group of variations is based on the persistence of the ductus arteriosus, which is derived from the sixth branchial arch vessel. In some rare cases that are accompanied by a partial or complete closure of the aorta (coarctation), life may be maintained after birth if a collateral circulation for the aorta is established. Included in this group of variations are cases in which the pulmonary artery arises from the aorta, and the blood of the pulmonary arteries passes from the aorta through the ductus arteriosus. In truncus arteriosus, a single vessel leaves the heart and supplies systemic, pulmonary, and coronary arteries. The incidence of common aorticopulmonary trunk is about 0.7% in cases of congenital heart disease and it occurs in about 0.0086% of live births. In most cases the truncus arises over both ventricles, which have large ventricular septal defects. There are numerous variations, including the following: a single pulmonary trunk and ascending aorta arising from the truncus; left and right pulmonary arteries arising independently from the truncus; associated aortic arch atresia or hypoplasia; unilateral absence of one pulmonary artery (usually on the side of the arch), with the second pulmonary artery arising from the truncus; and so on.

A ductus arteriosus was found draining into the left subclavian artery in a 25-year-old man.

Many variations occur in the number and position of the vessels arising from the aortic arch. There may be from one to six branches. The formation of a single branch involves the fusion of the two aortic stems and a shortening of the arch (no true arch) so that the ascending aorta gives rise to the left subclavian, left common carotid, and right brachiocephalic. A vertically descending branch of the aorta enters the abdominal cavity. A more common form is that found in most apes, in which the brachiocepahlic and left common carotid unite to form one branch.

The avian form, with two arterial branches, is extremely rare. The following combinations have been reported: (a) the right common carotid and right subclavian arising from the arch, and the left common carotid and left subclavian with other branches arising from the descending aorta; (b) two brachiocehpalics (bi-innominate), with side branches arising symmetrically; (c) two brachiocephalics, with the right giving both common carotids and the left both subclavians; (d) one case of right common carotid and left brachiocephalic with the right subclavian arising from the descending aorta; (f) one case of both carotids arising from a common stem and a left subclavian, and a right subclavian arising from the descending aorta; and (g) a right arch replacing the left, one trunk arising from the arch proper which divides into the left common carotid, right common carotid, and right subclavian, and the left subclavian arising from low on the arch, i.e., the descending part. In rare instances, variations of the usual three branches and their arrangement are found. With the aorta arching (normally) to the left, the order of branching may be as follows: (a) right subclavian, right carotid, left brachiocephalic; (b) right brachiocephalic, left subclavian, left carotid; (c) right subclavian, a bicarotid trunk, left subclavian; (d) a bicarotid trunk, left subclavian, right subclavian; (e) a bicarotid trunk, right subclavian, left subclavian; (f) the brachiocepahlic providing the left carotid, the left vertebral arising from aortic arch; and (g) the brachiocephalic providing the left carotid, the left subclavian providing the left vertebral.

With the aorta arching to the right (no concomitant situs inversus), the order of branching may be as follows: (a) left brachiocephalic, right common carotid, right subclavian; (b) left carotid, right carotid, right subclavian, the left subclavian arising form a patent ductus arteriosus; and (c) right subclavian, right carotid, left carotid, the left subclavian arising from a patent ductus arteriosus.

When there are more than three branches, the vertebral arteries are usually added or the extra branch may be the thyroidea ima, which ascends in front of the trachea to the thyroid gland. A common form, with four vessels, is one in which the left vertebral arises between the left carotid and left subclavian. The following orders of branching have also been reported:

(a) Right carotid, right subclavian, left carotid, left subclavian.
(b) Right carotid, left carotid, right subclavian, left subclavian.
(c) Right carotid, left carotid, left subclavian, right subclavian (from descending aorta). This variation has an incidence of about 0.4%.
(d) Left carotid, right carotid, right subclavian, left subclavian.
(e) Left carotid, right carotid, left subclavian, right subclavian.
(f) Right subclavian, right carotid, left carotid, left subclavian (absence of innominate).
(g) Right brachiocephalic, left carotid, left vertebral, left subclavian.
(h) The usual branches (80% of cases) – right brachiocephalic, left common carotid, left subclavian – plus a vertebral (the left vertebral more often than the right vertebral). This variation has an incidence of about 5%.
(i) Right brachiocephalic, left carotid, left subclavian, left vertebral.
(j) Right brachiocephalic, left carotid, left subclavian, right subclavian.
(k) Right brachiocephalic, left vertebral, left subclavian, right subclavian.
(l) Right subclavian, bicarotid stem, left vertebral, left subclavian.
(m) Right innominate (with right carotid and right internal thoracic branches), left carotid, left subclavian, right subclavian.
(n) Right brachiocephalic, left carotid, left subclavian, right vertebral.
(o) Right brachiocephalic, left internal carotid, left external carotid, left subclavian.
(p) In cases of right aortic arch: left carotid, right carotid, right subclavian, left subclavian.
(q) Right brachiocephalic, right internal thoracic, left carotid, left subclavian.
(r) Right brachiocephalic, right inferior thyroid, left carotid, left subclavian.
(s) Right brachiocephalic, left carotid, left inferior thyroid, left subclavian.
(t) Right brachiocephalic, left carotid, left superior intercostal, left subclavian.
(u) Right brachiocephalic, left thymic, left carotid, left subclavian.

When there are five arteries, the extra branches are usually the right subclavian and left vertebral. However, additions to the usual complement of three may include an internal and an external carotid in place of a common carotid, or two vertebrals. In the absence of the brachicephalic, the subclavian, common carotid, and one vertebral are branches of the arch. The reported order of branching follows:

(a) Right brachiocephalic, right vertebral, left carotid, left vertebral, left subclavian.
(b) Right brachiocephalic, left carotid, left vertebral, left subclavian, right subclavian (from descending aorta).
(c) Right subclavian, right carotid, left carotid, left vertebral, left subclavian.
(d) Right carotid, left carotid, left vertebral, left subclavian, right subclavian (from descending aorta).
(e) Right carotid, left carotid, left subclavian, left vertebral, right subclavian (from descending aorta).
(f) Right brachiocephalic, right internal thoracic, left carotid, left vertebral, left subclavian.
(g) In cases of right arch: left carotid, right carotid, right vertebral, right subclavian, left subclavian.
(h) Also in cases of right arch: innominate with left carotid and left vertebral branches, right carotid, right vertebral, right subclavian, left subclavian.

Cases of six branches result from the separate origin for both vertebrals and both subclavians. Additional examples of branching order include the following: (a) with a double aortic arch, the subclavian, external carotid, and internal carotid arise successively on the right and left side; (b) right subclavian, right vertebral, right carotid, left carotid, left vertebral, left subclavian; and (c) innominate with a thyroidea ima branch, two left vertebrals, left inferior thyroid, left subclavian.

In about 80% of persons, the branching order is right brachiocephalic, left common carotid, and left subclavian; in 11%, a common stem exists for the brachiocephalic and left common carotid, with the left subclavian arising independently from the arch.

Uncommonly, the three branches are accompanied by less significant branches: (a) a left thymic, (b) a right internal thoracic, (c) a right inferior thyroid, (d) a thyroidea ima, and (e) a left ventricular coronary.

The following arteries have been found arising from the aortic arch: external carotid, internal carotid, right vertebral, inferior thyroid, thymic, thyroidea ima, and left coronary. In rare instances, the lower parts of the lungs may receive anomalous arteries from the thoracic aorta (or the upper part of the abdominal aorta). The thoracic aorta may give rise to a right subclavian (level: T_2, T_3, or T_4), a supreme intercostal, and a renal artery which descends through the aortic hiatus.

References 5, 49, 50, 54, 69, 72, 80, 97, 110, 114, 116, 117, 122, 127, 140, 173, 190, 201, 206, 214, 263, 267, 270, 294, 324, 354–356, 360, 367, 370, 374, 375, 402, 405, 418, 423, 447, 457, 460, 461, 484, 485, 488–493, 496, 498, 541, 548–552, 568, 570, 577, 578, 580, 581, 583, 613–615, 624, 628, 637, 644, 651, 661, 664, 683, 686, 689, 700, 708, 718, 755, 757, 766, 767, 769, 771, 782, 784–786, 825, 826, 830, 838, 842, 845, 848, 853, 887, 919–921, 943, 960, 964, 975, 976, 980, 995, 1001, 1007, 1012, 1026, 1028, 1030, 1047, 1056, 1062, 1073, 1081, 1093, 1098, 1109, 1136, 1147, 1155, 1157, 1160, 1163, 1168, 1170, 1190, 1196, 1199, 1217, 1228, 1246, 1258, 1266, 1276, 1282, 1285, 1291, 1299, 1301, 1307, 1308, 1322, 1347–1350, 1354–1359, 1361, 1362, 1364–1368.

Additional reference: Webb, W. R., Gamsu, G., Speckman, J. M., Kaiser, J. A., Federle, M. P. and M. J. Lipton. CT demonstration of mediastinal aortic arch anomalies. *J. Comput. Assist. Tomogr.* 6:445–451, 1982.

Coronary

Branches of the coronary arteries may vary in distribution, number, and size. The left coronary artery is more variable than the right. The branches may arise as a common trunk, or both arise from the same aortic sinus. The anterior interventricular and terminal branches of the left coronary sometimes arise separately from the aortic sinus. In some cases, a single coronary artery serves the entire heart; either the right or the left coronary is absent. One coronary artery may be larger than usual and the other correspondingly smaller. Very rarely, an extra coronary artery arises from the pulmonary artery. Occasionally, there are two interventricular branches, or two or more posterior interventricular branches. There may be three or even four independent coronary arteries which are generally quite small. A conus artery arising from the aorta was found in 50% of 651 subjects. Coronary artery preponderance occurs in about 30% of cases; left coronary is preponderant in 12% of cases and right coronary in about 18%. Left coronary artery dominance is eight times more frequent in males (18.2%) than in females (2.6%), while right coronary artery preponderance is almost twice as common in females (23.1%) as in males (14.6%).

The right coronary artery occasionally arises from the pulmonary trunk, usually without adverse consequences. Cases have also been reported of the left coronary arising from the pulmonary trunk, but this is usually associated with myocardial ischemia, with patients dying at 13 months or younger. In one case, however, a female patient survived 60 years; in another case, a male survived 34 years (cause of death unrelated to heart disease). In 14,600 consecutive autopsies, this condition was found three times. When both coronary arteries arise from the pulmonary trunk, death occurs shortly after birth.

It has been reported that in 50% of hearts, the sinoatrial and atrioventricular nodes were supplied by the right coronary, and in 7% of hearts, these nodes were supplied by the left coronary. In the remainder (43%), one artery supplied one node and the other artery supplied the other, in either combination. Spalteholz, however, reported the sinoatrial node to be supplied in 68% of cases by the right coronary, in about 32% by the left coronary, and very rarely by an extracardiac artery.

An infrequent branch of the right coronary is a septal branch that supplies the middle part of the septum and both limbs of the conduction system. Occasionally, a coronary artery arises from the common carotid, later giving rise to right and left branches. "Large and direct" anastomoses between the right and left coronary arteries have been reported in 9% of individuals. Smaller anastomoses between the two vessels occur in most hearts. The absence of right/left coronary anastomoses was said to occur in 3% of hearts.

References 12, 17, 46, 66, 71, 124, 131, 160, 194, 220, 238, 260, 314, 315, 339, 357, 361, 371, 417, 419, 465, 487, 543, 576, 617, 618, 631, 634, 635, 653, 664, 671, 721, 741, 798, 804, 888, 900, 901, 985, 997, 1034, 1040, 1082–1084, 1122, 1143, 1148, 1151, 1165, 1189, 1227, 1233, 1283, 1303, 1349, 1354, 1356–1359, 1362, 1364–1368.

Brachiocephalic

The variations in the brachiocephalic artery are of surgical interest. If it divides lower than normal, this increases the length of the first portion of the right subclavian artery. If it divides higher than usual, it may incline to the left rising in front of the trachea above the sternum. Under these circumstances it poses a danger during tracheotomy. When abnormally long and inclining to the left, the brachiocephalic may pass behind the trachea or the esophagus to reach the right side. It may give rise to the thyroidea ima artery, and more rarely the vertebral, inferior thyroid, internal thoracic (or a smaller branch such as a bronchial, thymic, pericardiac, tracheal, mediastinal, and diaphragmatic), and muscular branches to the sternohyoid and sternothyroid. The brachiocephalic trunk is sometimes absent (0.44% of cases).

References 14, 326, 627, 1347, 1349, 1354–1359, 1361, 1362, 1364–1368.

Thyroidea Ima (of Neubauer)

This artery may occur in 4–10% of individuals. Adachi, however, reported only a single case in 271 (0.4%) subjects studied. Thyroidea ima arises from any one of the following sources: aortic arch, brachiocephalic trunk, common carotid, internal thoracic, pericardiacophrenic, subclavian, thyrocervical trunk, inferior thyroid, or transverse scapular artery. It usually but not always arises from the right side.

Gruber analyzed the origin of this artery in 90 reported cases: in 12 cases it arose from the aortic arch, in nine from between the brachiocephalic and left carotid, in two from between the right subclavian and right carotid, in one from between the left carotid and subclavian, in 39 from the right brachiocephalic, in 16 from the carotid, in six from the internal thoracic, in three from the right subclavian, in one from the right inferior thyroid, and in one from the transverse scapular artery.

References 331, 508, 518, 1008, 1010, 1041, 1049, 1167, 1241, 1347, 1349, 1354–1359, 1361, 1362, 1364–1367.

Common Carotids

A number of variations are of interest. The artery may obliquely cross the lower part of the trachea above the level of the sternum. This occurs on the right side when the brachiocephalic is situated to the left of the midline, when the right common carotid arises as the second branch of the aortic arch, or when the right and left common carotids arise as a common stem from the aorta. It occurs on the left side when the left common carotid arises from the brachiocephalic. The left common carotid varies more than the right, arising either from the brachiocephalic or from a common stem with the right common carotid.

The right common carotid, when arising from the aorta, on occasion runs behind the trachea and esophagus to the right side of the neck. The origin of the right common carotid may be above or below the usual point, depending on whether the brachiocephalic bifurcates higher or lower than usual. A low bifurcation of the brachiocephalic is somewhat more common. The common carotid sometimes follows a very tortuous course, forming one or more distinct loops in the neck.

The artery may bifurcate higher or lower than normal; a high bifurcation is more common. The bifurcation can occur as high as the hyoid bone (Quain, 5/295; Poynter, 3/400) or even the styloid process, or as low as the cricoid cartilage (Quain, 10/295; Poynter, 18/400) or within 3.7 cm of its origin. The artery may not bifurcate but provide branches usually derived from the external carotid as it ascends in the neck. A common carotid may be absent, the external and internal carotids arising directly from the aorta.

The common carotid may provide one or more of the branches usually derived from the external carotid (e.g., superior thyroid). It may give rise to a thyroidea ima. Cases have also been reported in which a vertebral, inferior thyroid, or laryngeal artery is a branch of the common carotid. The vagus nerve may run in front of the common carotid artery instead of behind it.

Collateral circulation after ligature of the common carotid is maintained by the following anastomoses: the two internal carotids through the circle of Willis; the two vertebrals; the inferior thyroid with the superior thyroid; the deep cervical branch of the costocervical trunk (superior intercostal) with the descending branch of the occipital; the superior thyroid, lingual, facial, occipital, and temporal with the corresponding arteries of the opposite side; and the ophthalmic with the angular. The anastomosis between the deep cervical branch of the costocervical trunk and the descending branch of the occipital is important; it is located deep at the back of the neck between the semispinalis capitis and cervicis muscles.

References 142, 662, 725, 834, 1089, 1316, 1347–1349, 1354–1359, 1361, 1362, 1364–1367.

External Carotid

Variations in the origin of this artery are discussed under Common Carotids. It may be absent unilaterally or bilaterally. When unilaterally absent, the branches usually derived from it arise form the upward continuation of the common trunk or from the contralateral vessel. The artery is sometimes located superficially, and runs lateral to the stylohyoid muscle or between the posterior belly of the digastric muscle and the stylohyoid.

The branches of the external carotid may arise irregularly or be diminished or increased in number. When increased in number (by two or more), they arise as a common stem or by the addition of branches not usually derived from this artery, such as the sternomastoid branch of either the superior thyroid or occipital artery. On occasion, all of the branches arise close together from a common point just above the origin of the artery from the common carotid. The external carotid may provide an accessory superior thyroid, a dorsal lingual, an accessory transverse facial, and an accessory ascending pharyngeal.

A curious variation is an external carotid composed of two separate trunks that unite behind the condylar process of the mandible forming an annulus from which the various branches arise.

Variations in some of the branches of the external carotid are as follows. The ascending pharyngeal may arise from the external carotid or the bifurcation of the common carotid (65–80%), the occipital (14–20%), or the common carotid (7–9%). The lingual arises from a common trunk with the facial (linguofacial trunk) in 10–20% of cases; a rare combination branch of the external carotid is a thyrolinguofacial trunk. The sternocleidomastoid, usually a branch of the superior thyroid, may arise as an independent branch of the external carotid.

References 273, 514, 650, 780, 923, 949, 998, 1230, 1316, 1347–1349, 1355–1359, 1361, 1362, 1365–1367.

Lingual. This artery may arise from the common carotid. It may also arise in common with the facial (Quain's series, 20% of cases) or superior thyroid or both, with the maxillary (Haller, 14%; Quain, 20%; Livini, 25%; Dall'Acqua, 20%; Poynter, 15%) or with the ascending pharyngeal (Livini, 0.5%). The lingual sometimes branches from the facial. It may also give rise to both the facial and the maxillary, and to the submental, ascending palatine, superior laryngeal, or accessory superior thyroid. Its hyoid branch may be absent. The lingual is occasionally replaced, totally or partially, by a branch of the maxillary.

References   712, 923, 1349, 1355–1359, 1361, 1362, 1365–1367.

Facial.   The facial artery is frequently rudimentary. It may terminate as a submental artery, i.e., not reaching the face, or as a labial or alar nasi (lateral nasal) artery and not as the angular (43% of cases). In its absence it may be replaced by the transverse facial. The facial can replace the frontal branch of the ophthalmic. Unusual branches include an ascending pharyngeal, superior laryngeal, tonsillar, sternocleidomastoid, maxillary, or sublingual.

Its inferior and superior labial branches are sometimes poorly developed or absent, in which case they are replaced by the contralateral vessel.

References   276–278, 923, 1349, 1355, 1357–1359, 1361, 1362, 1365–1367.

Sternocleidomastoid.   When absent this artery may be replaced by branches from the occipital, parotid, posterior auricular, lingual, or suprascapular.

References   1349, 1355–1359, 1361, 1362, 1365–1367.

Superior Thyroid.   This artery is generally considered to be present in 100% of cases, and its absence has only been reported once. The superior thyroid has been reported to give rise to a sizable branch that descends anterior to the clavicle and supplies pectoralis major. An unusually large superior thyroid may replace the contralateral vessel or the inferior thyroid. It can also be so small that it is represented only by muscular branches and the superior laryngeal. It may arise from the subclavian or the common carotid (Quain, 41/292; Livini, 9%; Poynter, 7% of cases), or share a common origin with the facial and with the lingual and facial. It is occasionally doubled (Quain, 3/292). The superior laryngeal branch sometimes arises from a common trunk with the lingual (Livini, 1.5%), facial, or external or even common carotid. The superior laryngeal may traverse the thyroid cartilage through an anomalous foramen. The cricothyroid branch may arise from the inferior thyroid or the thyroid branch of the superior thyroid. In some cases, enlarged right and left cricothyroid arteries anastomose in the midline and give rise to a median descending vessel that supplies the middle lobe of the thyroid gland, if present.

References   22, 29, 296, 331, 379, 549, 725, 923, 1008, 1041, 1049, 1145, 1167, 1347–1349, 1354–1362, 1365–1367.

Occipital.   The occipital may be a branch of the internal carotid, thyrocervical trunk, or ascending cervical. The posterior auricular (Livini, 12%), ascending pharyngeal, or stylomastoid may be branches of the occipital. It sometimes arises in common with the ascending pharyngeal (Quain, 13%; Livini, 16%), or is continuous with a vertebral artery through a large anastomosis.

References   523, 1349, 1354–1362, 1365–1367.

Posterior Auricular.   This artery is sometimes poorly developed or even terminates as the stylomastoid artery. It may arise in common with the ascending pharyngeal (Livini, 1%) or with the occipital; the occipital may replace an absent posterior auricular. An infrequent branch is the transverse facial.

References   1349, 1354–1362, 1365–1367.

Ascending Pharyngeal.   When absent (infrequent), this artery is replaced by a branch from the facial. It may be a branch of the occipital (Livini, 13%; Quain, 14%; Poynter, 15%), internal carotid (6%), or even the common carotid. It is occasionally doubled, and may give rise to the ascending palatine or superior laryngeal.

References   1349, 1354–1362, 1365–1367.

Sternocleidomastoid.   This artery may arise from the external carotid (Livini, 42%; Poynter, 37%).

Reference   1365.

Superficial Temporal.   The superficial temporal occasionally gives rise to the maxillary above the zygomatic arch, or to a tympanic artery. It may be very small, in which case it is replaced by the posterior auricular artery. When enlarged, it may give rise to branches normally provided by the facial: the lateral nasal and superior and inferior labial. Its transverse facial branch is sometimes doubled.

References   275, 1349, 1354–1362, 1365–1367.

Masseteric.   The masseteric artery is frequently absent (Poynter, 25%; Livini, 60%; Dall'Acqua, 25%).

Reference   1365.

Maxillary.   The origin of this artery is usually constant, but very rarely a branch of the facial artery. Quain reported a case in which the maxillary provided two vessels that entered the cranial cavity (one branch through the foramen rotundum, the other through the foramen ovale) to compensate for the absence of the internal carotid. The maxillary may be doubled.

Of surgical importance is the relationship of the maxillary artery to the lateral pterygoid muscle. The maxillary artery was found medial to the lateral pterygoid muscle in 30.5% of 180 dissections and lateral to this muscle in 69.5%.

The anterior deep temporal branch of the maxillary sometimes exists as a temporobuccal trunk, which divides into the anterior deep temporal (behind the nerve of the same name) and buccal arteries. The anterior deep temporal may substitute for the lacrimal artery.

The middle deep temporal branch of the maxillary may exist (16% of cases) as a temporomasseteric trunk, which gives rise to the middle deep temporal and a branch to the masseter muscle.

In a study involving 447 specimens, the maxillary artery was found passing superficial to the lateral pterygoid in 54% and deep to the lateral pterygoid in 46%.

References   111, 486, 650, 744, 801, 915, 1006, 1135, 1216, 1347–1349, 1354–1359, 1361, 1362, 1366, 1367.

Middle Meningeal. The middle meningeal artery may be absent. In 99% of cases it enters the foramen spinosum, but when this foramen is absent, the artery enters the foramen ovale along with the mandibular division of the trigeminal nerve. The middle meningeal gives rise in about 50% of cases to an accessory meningeal artery, which enters the skull via the foramen ovale or the emissary sphenoidal foramen (20% of cases).

The artery may or may not supply the trigeminal ganglion, which usually gets its blood supply from the internal carotid or, less often, the accessory meningeal artery. The anterior division of the middle meningeal artery occasionally gives rise to a medial branch (0.8% of cases). This enters the skull through the superior orbital fissure or a small foramen in the greater wing of the sphenoid to anastomose with the ophthalmic artery. The entire ophthalmic system of arteries occasionally arises from the middle meningeal artery. On the other hand, the middle meningeal, if absent, may be replaced by branches from the lacrimal or ophthalmic artery.

References   27, 30, 222, 268, 440, 510, 559, 589, 612, 792, 907, 1347–1349, 1354–1359, 1361, 1362, 1366, 1367, 1371.

Accessory Meningeal. An accessory meningeal artery is present in about 96% of individuals and originates from either the middle meningeal or maxillary artery with almost equal frequency. When the maxillary passes behind the lateral pterygoid muscle, the accessory meningeal arises from the maxillary artery; when the maxillary passes in front of the lateral pterygoid muscle, the accessory meningeal arises from the middle meningeal. The accessory meningeal may be doubled.

The artery distributes branches to the medial pterygoid muscle, superior head of the lateral pterygoid, tensor veli palatini, parts of the sphenoid bone, middle cranial fossa, root of the mandibular branch of the trigeminal nerve, and the otic ganglion. It has also been reported to supply the trigeminal ganglion in about 50% of cases. Some have suggested that this artery would be better named pterygomeningeus, for historic and distribution reasons.

References   1, 91, 1347, 1349, 1354–1359, 1361, 1362, 1366, 1367.

Internal Carotid

The internal carotid rarely provides branches in the neck. However, the ascending pharyngeal, occipital, transverse facial, and lingual arteries, as well as a laryngeal and a meningeal artery, have all been reported to arise from the internal carotid prior to its entry into the carotid canal. The internal carotid may take a sinuous course with a pronounced curve or a horseshoe shape, or may even form a complete loop in the upper cervical part of the vessel. This artery is occasionally absent unilaterally or bilaterally. In a review of carotid angiograms from 1407 patients, one case of bilateral agenesis of the internal carotid arteries was found (0.07%). When absent, the internal carotid may be replaced by branches of the maxillary artery. The carotid canal(s) may also be absent.

In one case, the internal carotid gave rise to a "vertebral" artery that entered the cranial cavity through the hypoglossal canal. The internal carotid occasionally provides an aberrant vessel in the cavernous sinus, which passes posteriorly to anastomose with the basilar artery. There have also been reports of the internal carotid giving rise to the basilar artery.

Unusual branches of the internal carotid include the ascending pharyngeal, occipital, transverse facial, accessory meningeal, and central artery of the retina.

In a dissection study of 200 half-necks (or sides), the internal carotid ran dorsolateral to the external carotid in 38%, dorsal to the external carotid in 38.5%, and dorsomedial to the external carotid in 23.5% of cases.

References   183, 373, 396, 397, 593, 604, 794, 852, 949, 956, 963, 1013, 1141, 1354–1362, 1366, 1367.

Circle of Willis. This circle of arterial vessels is extremely variable and illustrated in the atlas section.

Stehbens regards a normal circle of Willis to be one in which there is a complete anastomotic polygon, which may show considerable variability not only of its components but also its branches. The so-called normal "textbook" description holds in about 34.5% of cases based on a composite review of 1413 brains. Not even Willis's original illustration conforms to the usual textbook description.

References   75, 98, 115, 241, 337, 382, 687, 790, 794, 841, 1037, 1137, 1304, 1354–1362, 1366, 1367.

Anterior Choroidal. This artery arises from the internal carotid in 85% of cases, middle cerebral in 8%, and the junction of these two vessels in 7%.

It may be diminished or absent, or replaced by a branch of the deep cerebral or posterior communicating artery.

References 199, 579, 902, 1129, 1349, 1354–1362, 1366, 1367, 1369.

Ophthalmic. In a series of 170 specimens, this artery arose from the internal carotid in 164 cases. In four cases it arose from the internal carotid and middle meningeal artery through an enlargment of a normal anastomosis between the orbital branch of the middle meningeal and recurrent meningeal branch of the lacrimal artery. This artery occasionally arises from the posterior communicating artery (circle of Willis), or from the middle cerebral when the internal carotid is absent. The middle meningeal artery may arise from the ophthalmic instead of the maxillary artery.

The ophthalmic artery may enter the orbit via a special bony canal separate from the optic canal (noted in eight of 170 specimens). In these rare cases the term "double optic foramen" has been applied. In some cases the artery enters the orbit through the most medial part of the superior orbital fissure. In a study of 56 specimens, the artery's origin was found to be inferior and medial to the optic nerve in 23 cases, under the center of the nerve in 21 cases, inferior and lateral to the nerve in nine cases, lateral to the nerve in one case, under the medial border of the nerve in one case, and medial to the nerve in one case.

The ophthalmic artery may be associated with the development of a branch of the lacrimal artery that passes through the sphenoidal fissure to anastomose with the middle meningeal artery. This branch may be large, forming the main stem of the lacrimal artery. The connection with the ophthalmic sometimes disappears entirely. The entire ophthalmic system of arteries occasionally arises from the frontal branch of the middle meningeal artery.

The supraorbital branch of the ophthalmic may arise from the lacrimal branch of the ophthalmic.

References 4, 73, 224, 268, 380, 407, 408, 440, 536, 559, 564, 565, 589, 612, 658, 665, 907, 1078, 1110, 1125, 1126, 1318, 1340, 1349, 1354–1362, 1366, 1367, 1369–1371.

Anterior Inferior Cerebellar and Posterior Inferior Cerebellar. One of the anterior or posterior inferior cerebellar arteries may be missing. Occasionally, one of the anterior inferior cerebellar arteries arises from the posterior cerebral artery. The posterior inferior cerebellar may provide the posterior spinal artery.

References 55, 98, 535, 537, 681, 682, 751, 752, 1306, 1349, 1354–1362, 1366, 1367, 1369.

Anterior Cerebral. Both the right and left anterior cerebral arteries may run as one vessel to divide distally, or one may be a branch of the contralateral artery. The arteries are sometimes doubled. A persistent olfactory artery may be a branch of the anterior cerebral; it courses through the cribriform plate to reach the nasal cavity.

In a study of 50 brains, 10% had an A-1 segment of the anterior cerebral artery (i.e., the initial segment between its point of origin from the internal carotid and the point of origin of the anterior communicating artery from the anterior cerebral) with a diameter of 1.5 mm or less. In only one of 50 brains was the diameter less than 1.0 mm. In a study of 1647 circles of Willis, the A-1 segment was hypoplastic in 7%.

Hypoplasia of the A-1 segment may be associated with aneurysms of the anterior communicating artery. Infrequent variations are duplication of the A-1 segment, and a third or median anterior cerebral artery arising from the anterior communicating artery.

The anterior cerebrals may join to form a median common anterior cerebral.

References 24, 74, 98, 410, 688, 841, 898, 1046, 1127, 1137, 1304, 1349, 1354–1362, 1366, 1367, 1369.

Anterior Communicating. This artery is occasionally doubled, and may exist in one of several patterns. In 50 brains, the anterior communicating artery had a diameter of 1.5 mm or less in 44%, and 1.0 mm or less in 16%. In one study of 1647 brains, the anterior communicating artery was so small in 6% of cases that it restricted circulation (from one side to the other) between the two anterior cerebrals.

The anterior communicating artery, contrary to many textbook descriptions, may not be oriented in a strictly transverse plane. It can be oriented in an oblique or anterior-posterior plane if one anterior cerebral artery passes between the hemispheres behind the other, as occurs in 80% of brains.

The anterior communicating artery is usually between 2 and 3 mm in length, but may vary from 0.3 to 7.0 mm. It may be curved, kinked, or tortuous rather than straight.

One anterior communicating artery is found in 60% of brains; the other 40% of brains may have two or three arteries. The artery is rarely absent. When doubled or tripled, the different arteries may be similar in diameter or may vary markedly. Commonly one is large and the other(s) small.

In one review (seven investigators) of variations of 1813 anterior communicating arteries, it was absent in four cases (0.22%); there were small areas of fusion in 17 cases (0.94%), an azygos anterior cerebral in 21 (1.16%), a double V or Y form in 225 (12.4%), and a treble or plexiform vessel in 56 (3.09%). The artery was single in 1473 cases (about 81%).

References 25, 174, 483, 1349, 1354–1362, 1366, 1367, 1369.

Recurrent Artery of Heubner. Descriptions of the origin of the recurrent artery are highly inconsistent. Most arise from the A-2 segment of the anterior cerebral artery (i.e., the segment in the interhemispheric fissure, distal to the anterior communicating artery) within 2 or 3 mm of the anterior communicating artery. However, it may arise as far as 1.0 cm proximal to the anterior communicating artery. The recurrent artery is rarely absent and is rarely doubled.

References   1349, 1360, 1361, 1367, 1369.

Posterior Communicating. This artery may be absent or replaced by a branch from the middle cerebral. In one review of nine studies of 2362 brains, the posterior communicating artery was absent unilaterally in 3.09% and bilaterally in 0.3%.

References   1349, 1360, 1361, 1367, 1369.

Middle Cerebral. In a study of 25 brains (50 middle cerebral arteries), 10 forms of division were found. The principal forms of variation included (a) a single vessel arising proximal to the sylvian fissure (34%), (b) more than one major vessel arising proximal to the sylvian fissure (26%), (c) a trifurcation at the mouth of the sylvian fissure (34%), and (d) a bifurcation at the mouth of the sylvian fissure (16%). The middle cerebral may be absent or replaced by a branch from the deep cerebral. On occasion it gives rise to an accessory posterior communicating artery.

References   98, 205, 335, 411, 412, 414, 630, 1038, 1349, 1360, 1361, 1367, 1369.

Posterior Cerebral. Occasionally, the proximal part of one of the posterior cerebral arteries is greatly diminished in size. The blood reaches the terminal region of the vessel from the internal carotid via the posterior communicating artery. The posterior cerebral may course below, rather than over, the oculomotor nerve (cranial nerve III); it may be absent and replaced by an accessory contralateral vessel. The posterior cerebral occasionally provides the anterior inferior cerebellar artery.

References   82, 98, 413, 458, 595, 1297, 1349, 1360, 1361, 1367, 1369.

Subclavian Artery

Variations of origin of the subclavian arteries are also considered along with variations of the aortic arch (Aorta).

The right subclavian artery may arise directly from the arch of the aorta, or as the first, second, third, or fourth branch of that vessel. When it arises as the first branch, it takes the place usually occupied by the brachiocephalic. When it arises as the distal or last branch, it may course behind both the trachea and esophagus (retroesophageal) to reach the groove on the first rib. In four studies on a total of 1783 bodies, the frequency of this variation was about 0.6%. In other cases in which the right subclavian is the last branch of the aortic arch it may pass between the trachea and esophagus or it may follow a pretracheal course to reach the right side. Sometimes when the subclavian artery passes between the trachea and the esophagus it compresses both the artery and the esophagus, giving rise to a "disease" named "dysphagia lusoria" by Bayford, from the term "lusus naturae," meaning "a sport of nature" (Bayford, D. An account of a singular case of obstructed deglutition. *Mem. Med. Soc. Lond.* 2:271–282, 1794). An excerpt from Bayford's report follows: "Jane Fordham, was born in Bassenbourn, near Royston, in Hertfordshire, in the last year of the last century. From her infancy she was observed to have some difficulty in swallowing, but it was not much attended to, till she entered into her thirteenth year, when she first experienced those symptoms which commonly precede the erruption of the menses. At this time it became so considerable as induced her to have recourse to medicine for relief; but she received no sensible benefit from anything, except repeated bleeding... She went on in this way, with the disease gradually increasing for many years, during which time she constantly observed, that her difficulty of swallowing was increased by violent exercise, and as she expressed herself, by everything that heated her blood; so that she was frequently obliged to lose blood once or twice in the intervals of the month... For the last twenty years of her life this poor creature could scarcely, from day to day, muster up resolution to force food down to prevent her starving, so much was the difficulty now increased. This difficulty she described as arising from an obstruction opposite the first bone of the sternum. The food did not return when it came to that place; but seemed to make a momentary stop: and in the instant she felt an inexpressible something approaching to strangulation or suffocation, which she could only compare to what she conceived of the agonies of death... Different kinds of foods made no sensible difference in the effects, except that solids gave her less uneasiness than fluids: for which reason she took very sparingly of the latter. As her complaints were continually aggravated, rather than diminished by time, she became at length unable to struggle; and being worn out with fatigue and famine (for it is asserted that she scarcely swallowed a single morsel for the last three weeks of her life), she sunk into her grave in the beginning of February, 1761. I felt myself interested in making an inquiry into the cause of the obstruction by an examination of the dead body. It was not thought necessay to open the head, as there was no reason to expect any mischief in that part. The cavities of the thorax and abdomen were fully examined. ... At length by mere accident I discovered an extraordinary lusus naturae in the disposition of the right subclavian artery. The right subclavian arising (arose) from the posterior part of the aorta, behind on the left hand of the left sublcavian artery. In crossing from the left to right it ... insinuated itself between the trachea and esophagus. It should seem that this peculiar origin and course of the right subclavian artery has hitherto escaped the observation of anatomists, otherwise in all probability it would have been recorded... It is impossible to suppose the act of deglutition performed in this case, without a certain degree of pressure upon the artery, where it passes between the trachea and oesophagus. Hence must arise an interruption of the circulation, which in so large an artery, and so near the heart, could scarcely fail of exciting very disagreeable sensation... The act of deglutition will not be accompanied with pain or soreness, but extreme anxiety and violent palpitations of the heart."

The subclavian is rarely the second or third branch of the aortic arch, but when it is, it runs behind the right common carotid.

The right subclavian arises as the distal or last branch of the aortic arch when the right aortic arch remains intact and the normal root of the subclavian artery has become obliterated. An arteria aberrans, arising from the right subclavian (or superior intercostal), can generally be traced to the third thoracic vertebra behind the esophagus, and in a number of such cases can be followed across the vertebral column where it anastomoses with a branch of the thoracic aorta arising below the ductus arteriosus. It is the enlargement of this anastomosis (which is the remnant of the primitive right dorsal aorta in the embryo) that gives rise to the variation. The recurrent laryngeal nerve in such cases follows a direct course to the larynx instead of winding recurrently around the subclavian artery, because the right fourth arch that forms the first part of the subclavian is obliterated.

The right subclavian may arise higher or lower in the neck than usual, depending on whether the brachiocephalic divides above or below its normal position. On occasion the subclavian arises from the thoracic aorta.

The subclavian may perforate or pass in front of the scalenus anterior; when it perforates scalenus anterior, arterial rings may be formed around muscle slips. The artery can ascend above or remain below the level of the clavicle. The third part of the artery may be covered by the trapezius or sternomastoid, or by a clavicular origin of the omohyoid. The subclavian vein occasionally accompanies the artery behind the scalenus anterior; the artery may also pass between scalenus medius and posterior. The right subclavian artery may receive the thoracic duct; the left subclavian may receive a patent ductus arteriosus.

The subclavian artery divides in some cases at the medial border of scalenus anterior, with the two branches continuing through the axilla and down the arm to become the radial and ulnar arteries. Anomalous right subclavian arteries occur in 0.8–1% of individuals (all sources). If a cervical rib is present, the artery may run above the anomalous rib.

References 33, 63, 64, 92, 93, 101, 107, 113, 121, 132, 139, 149, 185, 189, 204, 247, 254, 262, 269, 274, 282, 291, 292, 317, 420, 421, 447, 452, 455, 464, 489, 490, 492, 549, 561, 562, 581, 599, 600, 700, 719, 789, 848, 914, 926, 948, 960, 974, 979, 1004, 1019, 1030, 1061, 1093, 1123, 1133–1136, 1159, 1160, 1182, 1215, 1305, 1316, 1321, 1322, 1347–1349, 1354–1359, 1362, 1364–1367.

Additional references: Mears, J.E. Description of an anomalous origin of the right subclavian artery, associated with anomalies of origin of the branches of both subclavian arteries; with remarks. *Am. J. Med. Sci.* 62:401–409, 1871.
Proto, A.V., Cuthbert, N.W. and L. Raider. Aberrant right subclavian artery: Further observations. *Am. J. Roentgenol.* 148:253–257, 1987.
Wood, J. Two specimens of abnormal origin of the right subclavian artery. *Trans. Path. Soc. Lond.* 10:119–128, 1859.
Raphael, R.L., Schnabel, T.G. Jr. and S.S. Leopold. A new method for demonstrating aberrant right subclavian artery. *Radiology* 58:89–93, 1952.

Variations in Branches of Subclavian Artery.

There is considerable variation in the branching of the subclavian artery, and the branches may be arranged differently on the two sides of the body. The usual branching order on the right side is for the vertebral, thyrocervical trunk (common trunk of inferior thyroid, transverse cervical, and suprascapular arteries in 50% of cases), and internal thoracic to arise from the first part of the right subclavian. When the ascending cervical is a branch of the inferior thyroid, the transverse cervical artery and costocervical trunk arise from the second part of the subclavian. The costocervical trunk may arise from the third part, on the right side in 75% of cases and on the left in 30%. There may be one or two branches from the third part (costocervical and dorsal scapular artery in about 35% of cases). When the dorsal scapular artery is absent the costocervical trunk arises from the transverse cervical at the angle of the scapula. The transverse cervical may arise from the dorsal scapular when the dorsal scapular arises from the third part of the subclavian.

There are three additional types of variations. (a) The vertebral, internal thoracic, costocervical, and inferior thyroid arise from the first part of the subclavian, while the transverse cervical arises from the second part and the suprascapular from either the third part or from the axillary artery. (b) The inferior thyroid, suprascapular, and transverse cervical arise from a common stem from the first part. (c) In the rarest form, the inferior thyroid and superficial cervical arteries arise from a common trunk from the first part of the subclavian, while the suprascapular artery arises from the internal thoracic.

The subclavian may give rise to the inferior thyroid (27 of 273 cases), ascending cervical, suprascapular, supreme intercostal, deep cervical, accessory root to the vertebral, thyroidea ima, accessory inferior thyroid, and the radial and ulnar arteries (without the axillary intervening). It sometimes provides a stem that gives rise to the internal thoracic, inferior thyroid, suprascapular, superior esophageal, an accessory or lateral internal thoracic, a vessel to the brachial plexus, a pericardial branch, a bronchial, or a pleural cupula branch.

References 101, 137, 204, 282, 291, 518, 606, 608, 926, 948, 960, 1050, 1347–1349, 1354–1359, 1361, 1362, 1366, 1367.

Thyrocervical Trunk. In 83% of cases, there is an inconstant pattern of branching in this trunk. The most frequent variations include the following. (a) The inferior thyroid has a common stem (of varying length) for the suprascapular and transverse cervical (30%), or both of the latter arise independently from the thyrocervical trunk (28%). (b) The trunk may bifurcate into an inferior thyroid with a suprascapular branch (15%) or with a transverse cervical branch (8%); the alternate branch has a different source. (c) A simple trifurcation of the thyrocervical trunk occurred in 46.75% of 770 half-bodies or sides. (d) The trunk occasionally provides a tracheoesophageal branch that is either very small or up to about 2 mm in diameter.

When the thyrocervical trunk is absent, the inferior thyroid is absent or arises from another source, usually as a branch from the first part of the subclavian. The other branches form a combined "cervicoscapular trunk" (10%), giving rise to transverse cervical and suprascapular arteries. A "thyrovertebral trunk" has also been reported. The internal thoracic may arise from the thyrocervical trunk (9%).

References 137, 282, 291, 299, 509, 518, 606, 660, 800, 926, 1008, 1050, 1051, 1347–1349, 1354–1359, 1361, 1362, 1366, 1367.

Inferior Thyroid. In a study of 200 cadavers, the inferior thyroid arose from the thyrocervical trunk in 90.5%, subclavian in 7.5%, and very rarely from the common carotid, aortic arch, brachiocephalic, internal thoracic, pericardiacophrenic or vertebral, or arose as a common stem with the contralateral inferior thyroid. The inferior thyroid may also arise as a common trunk with the dorsal scapular and suprascapular arteries (thyroscapular trunk). It may course behind the common carotid. This artery has been reported absent and doubled.

The inferior thyroid and its branches may lie anterior or posterior to the recurrent laryngeal nerve. It occasionally gives rise to branches to the anterior mediastinum, thymus, and bronchi, and may anastomose with the vertebral, superior intercostal, and bronchial arteries.

The inferior thyroid, when usually small or absent, may be replaced or supplemented by thyroidea ima.

When the vertebral artery ascends to a level above the sixth cervical vertebra before entering the foramen in the transverse process, the inferior thyroid will pass beneath (or dorsal to) the anomalous vertebral artery in about 50% of cases.

References 21, 29, 68, 282, 291, 331, 379, 517, 549, 642, 737, 756, 799, 926, 1008, 1041, 1049, 1167, 1241, 1347–1349, 1354–1359, 1361, 1362, 1365–1367.

Transverse Cervical. The transverse cervical artery arises from the subclavian in about 61% of cases and from the thyrocervical trunk in about 38%. When it arises from the subclavian, it is usually behind or lateral to scalenus anterior. When absent, it is compensated for by branches from the dorsal scapular, superficial cervical, costocervical trunk, or the distal or third part of the subclavian or even the axillary artery. If the artery arises medial to scalenus anterior it is, without exception, a branch of the thyrocervical trunk. It may arise from the dorsal scapular (possibly as high as 20%). In addition, it sometimes arises from the inferior thyroid, internal thoracic, costocervical trunk, or first part of the subclavian.

References 282, 291, 606, 608, 660, 926, 1022, 1050, 1347, 1349, 1354–1359, 1361, 1362, 1365–1367.

Suprascapular. In a study of 400 cadavers, the suprascapular artery arose from the thyrocervical trunk in about 92%, internal thoracic in 4.1%, and transverse cervical in 2.5%, independently from the subclavian in about 2%, and rarely from the axillary, costocervical trunk, or dorsal scapular artery. It may arise from the subclavian or axillary artery (reported as high as 10%), or from an accessory inferior thyroid, the subscapular, or the thoracoacromial artery. When it arises from the second or third part of the subclavian, it usually passes through, rather than over, the brachial plexus. A thyroidea ima may be a branch of the suprascapular.

References 282, 291, 606, 608, 926, 1022, 1065, 1347, 1349, 1354–1359, 1361, 1362, 1365–1367.

Dorsal Scapular. In 70% of individuals, this artery arises from the second or third part of the subclavian as an independent branch. In 30%, it arises from the transverse cervical artery. If the artery arises from the subclavian, it usually passes dorsally between the trunks of the brachial plexus (middle and superior in 46%, middle and inferior in 44%). It may perforate the middle and posterior scalene muscles. If the dorsal scapular artery is a branch of the transverse cervical, it arises at the superior angle of the scapula and descends vertically along the medial border of the scapula.

References   1347, 1365.

Costocervical Trunk.   This trunk is absent in 8–10% of cases, and thus its main branches may arise independently, frequently from the same (second) part of the subclavian.

References   282, 291, 926, 1347, 1349, 1354–1359, 1361, 1362, 1365–1367.

Internal Thoracic.   The origins of this artery may vary as follows: (a) arising from the subclavian, behind or in front of scalenus anterior; (b) the right artery arising from the ascending aorta: or (c) arising in common with the inferior thyroid. The internal thoracic may give rise to a lateral internal thoracic artery as it enters the thorax. The lateral internal thoracic runs about midway between the spine and sternum or somewhat forward, spans up to six intercostal spaces, and anastomoses with intercostal arteries of those spaces. A lateral internal thoracic artery may arise from the costocervical trunk, thyrocervical trunk, or ascending cervical. The existence of this vessel must be remembered when paracentesis is performed.

Accessory internal thoracic vessels have been reported, which may give rise to superficial or deep cervical branches.

References   18, 137, 282, 291, 509, 515, 711, 811, 926, 1347, 1349, 1354–1359, 1362, 1364, 1366, 1367.

Thymic.   The thymic usually arises from the anterior mediastinal branch of the internal thoracic, but it may also arise as a sizable branch of the ascending aorta or aortic arch, or from the brachiocephalic, common carotid, inferior thyroid, pericardiacophrenic, or internal thoracic directly.

In one study, the blood supply to the thymus was reported as follows. Of 77 subjects studied, the thymus was supplied by the internal thoracic in 59 (the right internal thoracic in 28 and the left in 29), the pericardiacophrenic in 13 (the right side in eight and the left in five), the inferior thyroid in six (the right side in two and the left in four), and the brachiocephalic in two.

References   740, 1347, 1349, 1354–1359, 1362, 1364, 1366, 1367. (See also Organs reference 67.)

Supreme (Superior) Intercostal.   This artery may be a branch of the vertebral, in which case it passes through the transverse foramen of the seventh cervical vertebra and between the ribs and transverse processes of the upper three intercostal spaces. The supreme intercostal may arise from the thyrocervical trunk, deep cervical (cervicalis profundus), inferior thyroid, second intercostal, or axillary. The left artery has been reported as a branch of the aortic arch in cases having four arch branches. In one case it arose from a lateral internal thoracic artery. It may be absent, multiple, or diminished in size; in the latter case, it supplies only the first intercostal space.

References   533, 926, 1259, 1349, 1354–1359, 1362, 1365–1367.

Ascending Cervical.   When this artery is absent it is replaced by branches from the superficial cervical, suprascapular, or vertebral artery.

References   1347, 1354–1359, 1362, 1366, 1367.

Deep Cervical.   There are several variations in this artery. It may be a branch of the supreme intercostal or internal thoracic, or it may join the vertebral above the atlas. When smaller than usual, its area of supply may be supplemented by branches of other regional vessels. The artery occasionally provides an accessory vertebral. It may be a branch of the vertebral, thyrocervical trunk, inferior thyroid, suprascapular, transverse cervical, or superior (supreme) intercostal.

References   1349, 1354–1359, 1362, 1366, 1367.

Superficial Cervical.   This artery may arise from a common stem with the suprascapular or from the transverse cervical. In some cases it provides an accessory vertebral artery.

References   1349, 1354–1359, 1362, 1366, 1367.

Bronchial.   The bronchial may arise from the internal thoracic, subclavian, inferior thyroid, supreme thoracic, thoracic aorta, abdominal aorta, costocervical trunk, or intercostal. It occasionally replaces a pulmonary artery.

References   209, 232, 582, 668, 770, 788, 848, 860, 948, 1214, 1349, 1354–1359, 1362, 1366, 1367.

Vertebral Artery.   The right vertebral artery may arise (a) from the first part of the subclavian, nearer than normal to the brachiocephalic (1% of cases) or to the anterior scalene muscle; (b) directly from the arch of the aorta (3% of cases); (c) from the right common carotid, when the right subclavian is given off from the aorta beyond the left subclavian; or (d) from the brachiocephalic trunk. The right artery may pass behind the esophagus.

The left vertebral artery may arise directly from the left common carotid, or the root of the subclavian, close to the aortic arch. It may arise from the arch of the aorta. There are occasionally two left vertebral arteries. In one such case, one arose from the arch of the aorta, the other from the left subclavian artery in the normal position; the latter was smaller than normal. The left and right vertebrals may be doubled, with each vessel entering a different transverse foramen. In some cases, a left vertebral arises from the arch of the aorta between the origins of the left common carotid and left subclavian; in one subject, the vertebral entered the foramen of the transverse process of the fifth cervical vertebra. The left vertebral in one case originated from the aorta at the upper angle of the junction of the left subclavian with the arch of the aorta; the vessel entered the foramen of the transverse process of the fifth cervical vertebra. Either vertebral artery may enter the foramen in the second through seventh cervical vertebrae. When entering one of the higher vertebral foramina, the artery may lie behind the common carotid. The vertebrals enter the sixth cervical vertebral foramen in 88% of cases and the seventh cervical foramen in 5%; they ascend to enter the fifth cervical in 7% of cases. Either vertebral artery may give rise to, or arise from, the inferior thyroid, and may give rise to the superior intercostal, deep cervical, or occipital artery. Either artery may be much increased or diminished in size. In 3% of cases, either artery may arise from the thyrocervical trunk or from the costocervical trunk.

The two vertebral arteries are usually unequal in size, the right being smaller than the left. The rule is not absolute, since Cruveilhier has reported seeing the left vertebral reduced to a small twig with the right being of normal caliber. The inferior thyroid may lie behind (dorsal to) the vertebral artery. Very rarely, the two vertebrals fail to unite to form the median basilar artery, hence the basilar may appear to be doubled. The two longitudinal trunks may be united by anastomoses between them. The vertebral artery may exist as several vessels that eventually unite, or it may form an arterial ring traversed by the hypoglossal nerve. One of the anterior spinal branches (usually the left) is occasionally absent, in which case it is replaced by branches from the contralateral vertebral. Occasionally, a spinal branch arises to supply the cervical enlargement of the spinal cord.

The following branches of the vertebral have been reported missing: the inferior cerebellar, middle cerebellar, and left anterior spinal, one of the posterior spinal arteries, and even the posterior cerebral artery.

References 28, 99, 103, 219, 439, 517, 558, 626, 679, 760, 821, 914, 926, 967, 1255, 1256, 1305, 1338, 1349, 1354–1359, 1362, 1364–1367.

Basilar. The basilar may exist as two longitudinal trunks that may be united by anastomoses. Occasionally, the basilar splits into two vessels that reunite. When the two vessels "fuse," the resulting single vessel may have a median sagittal partition. In some subjects, the basilar passes through a foramen in the dorsum sellae. An aberrant branch may arise that pierces the dorsum sellae and joins the internal carotid.

In one case of a rudimentary vertebral artery that stopped short of the skull, the basilar arose as a branch of the internal carotid in the neck and entered the cranial cavity through the hypoglossal canal. The vertebral artery on the contralateral side ended as a posterior inferior cerebellar artery.

References 90, 103, 672, 1072, 1141, 1349, 1354–1359, 1362, 1366, 1367, 1369.

Upper Extremity

Axillary

Discussed here are the major variations in the axillary artery. Occasionally it gives rise to the radial artery or, more rarely, the ulnar artery. Still more rarely, it gives rise to the interosseous artery or a vas aberrans. It may give rise to a common trunk, from which may arise the subscapular, anterior and posterior circumflex humeral, profunda brachii, and ulnar collateral arteries. The branches of the brachial plexus may surround this common trunk, but not the main brachial artery. In some cases the brachial artery appears to be an enlarged "vas aberrans," and the common trunk to be the main brachial artery, the lower portion of which has been obliterated. A vas aberrans, according to Quain, is a long, usually slender vessel that originates either from the axillary or the brachial and unites with forearm arteries or their branches. The vas aberrans may extend into the hand to join the superficial palmar arch. In eight of nine cases, the aberrant vessel joined the radial artery; hence it rarely joined with the ulnar. The axillary artery may be represented by two parallel vessels that arise from the first portion of the subclavian and continue distally as the ulnar and radial arteries. The first part of the axillary may also provide an accessory thoracoacromial artery. The third part of the axillary artery is occasionally covered by a muscular slip (the axillary arch) derived from the upper part of the tendon of latissimus dorsi; this slip is always present in early fetal life, but usually atrophies later.

Any of the following branches that normally arise from the subclavian artery may also arise instead from the axillary, either from a common trunk or in groups: the inferior thyroid, ascending cervical, superficial cervical, deep cervical, internal thoracic, transverse cervical, or suprascapular.

Unusual branches of the axillary include a glandular artery to lymph nodes and skin of the axilla (so-called alar thoracic artery), an accessory lateral thoracic, anterior and posterior circumflex humeral, circumflex scapular, thoracodorsal, and a branch to the mammary gland.

References 38, 108, 137, 211, 293, 588, 607, 633, 872, 876, 979, 1132, 1223, 1349, 1354–1359, 1362, 1366, 1367.

Variations in Branches of Axillary Artery.

The following variations have been recorded in the

branches of the axillary artery. The first part of the axillary may give rise to the dorsal scapular artery (Poynter, 0.5%).

Supreme Thoracic. The supreme throacic is occasionally missing (Poynter, 10%). It may replace the pectoral branch of the thoracoacromial and may itself arise as a branch of the thoracoacromial artery (Poynter, 2.5%).

References 137, 293, 588, 607, 876, 1223, 1349, 1354–1359, 1362, 1365–1367.

Thoracoacromial. The thoracoacromial may be missing. In some cases it has four branches: the supreme thoracic, thoracic branch of the thoracoacromial, accessory pectoral branches, and lateral thoracic.

The first part of the axillary may, in rare cases, give rise to the subscapular artery or supply a branch to the subscapular muscle.

References 137, 293, 588, 607, 876, 1223, 1349, 1354–1359, 1362, 1366, 1367.

Lateral Thoracic. This artery may arise from the thoracoacromial (Poynter, 12%). The lateral thoracic may be replaced by or it may replace the thoracodorsal. On occasion it gives rise to the ulnar artery, and may itself be a branch of the subscapular (Poynter, 8%). In about 24% of cases it is doubled (Poynter).

References 293, 588, 607, 720, 876, 1223, 1349, 1354–1359, 1362, 1365–1367.

Anterior and Posterior Circumflex Humeral. The branches of the third part of the axillary artery are subject to great variation. The two circumflex arteries may arise from a common trunk (Poynter, 20%; Quain, 6%; Pellegrini, 22%; Hitzrot, 16%), usually alone or rarely together with the profunda brachii and muscular branches. In a study of 610 limbs, the anterior artery had a common origin with a neighboring vessel in 20.2% of cases, most frequently with the posterior artery (13.4%). It is usually small, is doubled in 8% of cases, and is the last branch of the axillary in 35% (Poynter). The posterior artery has an origin in common with other vessels in 47% of cases, most frequently with the subscapular (Poynter, 10%). The two circumflex arteries together or the anterior circumflex alone may be the only branch of the third part of the axillary. In some cases the posterior circumflex humeral arises from the profunda brachii or from the subscapular arteries. The anterior and posterior circumflex humerals may be doubled.

References 293, 588, 607, 876, 1223, 1349, 1354–1359, 1362, 1365–1367.

Thoracodorsal. This artery may be supplemented by an accessory vessel.

References 1066, 1223, 1349, 1354–1359, 1362, 1366, 1367.

Subscapular. Occasionally the subscapular arises from the second part of the axillary, and it may also arise from the brachial. Quite often it arises from a common trunk with one or both circumflex arteries. It may arise as a branch from the lateral thoracic, posterior circumflex humeral, or profunda brachii. The posterior circumflex humeral artery may arise from the subscapular artery. In one documented case, the subscapular arose as a branch of the inferior thyroid.

References 293, 588, 607, 876, 1223, 1349, 1354–1359, 1362, 1365–1367.

Brachial

Major variations are present in about 25% of subjects studied. In one study of 610 limbs, the usual textbook description of the axillary continuing as the brachial was observed in only 80% of specimens. A superficial brachial, a vessel that has a course anterior to that of the median nerve, was found in about 13%. The superficial branch continued as the radial twice as frequently as it continued as the ulnar artery. Less frequently it continued as both arteries. A communication at the elbow between the superficial brachial and the brachial was occasionally observed. One of the major variations is a high proximal division into terminal branches (radial (15%), ulnar (2%), common interosseous, vas aberrans, and superficial median antebrachial artery). This may occur at any point in the normal course of the vessel, but is most common in the upper third of the arm and least common in the middle third. The two vessels into which the brachial divides run parallel to each other to the bend of the elbow, in the usual position of the brachial. From this point one branch follows the normal course of the radial artery through the forearm and the other takes the normal course of the ulnar artery, supplying as usual the common interosseous artery. This arrangement is considered a simple high division of the brachial. In the following cases, the disposition of the two vessels is different. (a) The two arteries communicate at the elbow by a cross branch, or reunite and then divide again in the usual manner. (b) One vessel follows the course of the ulnar artery in the forearm, and the other divides into the radial and common interosseous. This condition is considered a high origin of the ulnar. (c) One artery divides into the radial and ulnar as usual, and the other takes the course of the common interosseous and divides into the anterior and posterior interosseous arteries or, much more rarely, takes the course of the posterior interosseous artery. The anterior interosseous comes from the ulnar. (d) The vessels follow a course in the upper arm different from that of the normal brachial.

The branch representing the radial may (a) cross over or under the other branch, (b) perforate the deep fascia above the elbow and run beneath the skin to its place in the forearm, or (c) pass behind the tendon of the biceps.

The branch representing the ulnar may (a) run to the front of the medial epicondyle with the median nerve, reaching its usual position by descending beneath the fascia and pronator teres or, more rarely, beneath some of the flexor muscles or merely beneath the skin, or (b) pass with the ulnar nerve behind the medial epicondyle and beneath the muscles to its usual place in the forearm.

An enlarged vas aberrans may be present. This long vessel arises from the brachial, usually near the origin of the profunda, and joins, most commonly, the radial artery or, more rarely, one of its branches or the ulnar. It is normally quite small, and descends over the median nerve to the biceps muscle. At times this vessel takes the place of the brachial; the median nerve will then be found behind the artery.

The brachial may run with the median nerve toward the medial epicondyle, where it may turn around a supracondyloid process if present (Gruber, 2.7%) and then descend to its normal position beneath the pronator teres.

Latarjet suggests five possibilities for branching from the premature division of the brachial artery:
(1) Radial and an ulnar-common interosseous trunk.
(2) Ulnar and a radial-common interosseous trunk.
(3) Common interosseous or presistent median artery and a radioulnar trunk.
(4) Radial, ulnar, and a common interosseous artery.
(5) A vas aberrans and normal brachial artery.

The variations are usually unilateral. Quain reported 61 cases, with 43 involving one side only and 18 involving both sides. In five of the 18 cases the same variation was bilateral, and in 13 cases the variation differed on the two sides.

The brachial may give rise to a median cubital, a median articular cubital artery, or a branch to the mammary gland. It may be covered by various muscular slips derived from the adjacent muscles.

The frequency of brachial variations based on studies by various authors (3337 arms) is about 20%.

References   105, 109, 227, 228, 400, 597, 633, 787, 829, 954, 1005, 1131, 1132, 1271, 1349, 1354–1359, 1362, 1365–1367.

Variations in Branches of Brachial Artery.

The following variations have been recorded in the branches of the brachial artery.

Profunda Brachii.   The profunda may arise from the third part of the axillary artery in common with one or more branches of that vessel, e.g., subscapular (Poynter, 19%), or arise as a common trunk with the superior ulnar collateral or anterior and/or posterior circumflex humeral. It may give rise to the posterior circumflex, which then runs upward behind the teres major to reach the back of the shoulder. In some cases, the deltoid branch of the profunda brachii arises from the brachial or from the superior ulnar collateral. The profunda may be reduced in size, ending in muscle without giving rise to the radial and medial collaterals.

References   227, 536, 1271, 1349, 1354–1359, 1362, 1365–1367.

Superior Ulnar Collateral.   The superior ulnar collateral may arise with the profunda or from a trunk common to several other branches of the axillary and brachial arteries. When absent, the inferior ulnar collateral takes its place.

References   1349, 1354–1359, 1362, 1366, 1367.

Inferior Ulnar Collateral.   The inferior ulnar collateral may take the place of the superior ulnar collateral. It may be very small, with the superior taking its place.

References   1349, 1354–1359, 1362, 1366, 1367.

Ulnar

The ulnar artery may arise from the brachial above the usual point of division or from the axillary, in which case it usually passes over the flexor muscles, but beneath the fascia, to reach its usual position in the forearm. The recurrent arteries and the common interosseous are usually derived from the trunk vessel from which the ulnar arises. It may run beneath the muscles or merely beneath the skin, where it may be vulnerable to iatrogenic problems. In cases of high division of the brachial, the ulnar may run in a subcutaneous position or beneath the fascia throughout its extent in the forearm. In some cases of high origin and superficial course, it gives rise to the subscapular and profunda brachii but not the common interosseous. It may also provide an accessory ulnar recurrent, radial recurrent, or accessory interosseous branch, which may enter the palm and provide any of the branches of the superficial palmar arch. In some cases in which origin from the brachial is normal, the ulnar takes a superficial course, covered by fascia in the forearm, and the recurrent branches and the common interosseous arise from the radial. The ulnar may be a branch of the profunda brachii. The ulnar is sometimes doubled; in some of these cases, one of the two arteries may take a superficial course in the forearm.

References   109, 500, 504, 536, 829, 954, 1016, 1268, 1342, 1349, 1354–1359, 1362, 1365–1367.

Superficial Palmar Arch.   The superficial palmar arch is subject to significant variations. The completion of the arch on the radial side is extremely variable. In a study of 200 subjects, a superficial arch was lacking in 32% (Jaschtschinski). It may be formed by the superficial branch of the ulnar anastomosing with the superficial palmar of the radial. The superficial palmar branch may be larger than usual, and have a greater role than the ulnar in the formation of the arch. In some cases, the arch is reinforced by a large median artery, arising frequently from the common interosseous, or by an enlarged metacarpal artery. The arch may be doubled, with both the superficial branch of the ulnar and the superficial palmar dividing into two branches that anastomose across the palm. The arteries of the thumb and radial side of the index finger occasionally arise from the arch. The arch may be incomplete, the medial digital branches arising from the ulnar and the lateral branches arising from the superficial palmar, the radial in the palm, or an enlarged median artery. When the arch is absent, the digital arteries arise from enlarged metacarpal arteries from the deep palmar arch or from enlarged dorsal metacarpal arteries.

References   249, 638, 829, 1005, 1194, 1268, 1349, 1355–1359, 1362, 1365–1367.

Median

A significant median artery (comes nervi mediani) may persist in the adult. The main supply of the hand in the embryo is the median artery. This is followed, in development, by the anterior interosseous which is finally replaced by the ulnar and radial arteries as the principal supply of the hand. Either the anterior interosseous or the median may enter into the formation of the superficial palmar arch if the ulnar is deficient or absent. In one study, a median artery entered the hand through the carpal tunnel: bilaterally in four and unilaterally in three cases (66 bodies, 132 hand dissections). In five cases, median arteries participated in the superficial palmar arch, either by joining the radial (three cases) or the ulnar (two cases). In six cases, there was no palmar arch and the median artery supplied the fingers directly. The median frequently arises from the common interosseous or anterior interosseous. In some cases the vessel runs a superficial course through the forearm. The median may end in the radial artery.

References   340, 435, 466, 500, 829, 881, 954, 1121, 1194, 1349, 1354–1359, 1362, 1365–1367.

Common Interosseous

The palmar and dorsal (posterior) vessels may arise separately from the ulnar. A high origin of the common interosseous has been described as arising from the axillary. Palmar interosseous arteries may be so well developed that they substitute for branches of the ulnar and/or radial arteries. The common interosseous may be a branch of a high radial artery and it may provide the radial deep to pronator teres. A normal anterior interosseous may be absent and replaced by branches of the radial. At times it joins the radial artery at the wrist, or if the radial stops short of the carpus the interosseous may replace it. The recurrent interosseous branch of the posterior interosseous is frequently missing. The posterior interosseus may give rise to an accessory posterior interosseous; it may also give rise to a "superficial external interosseous artery" that courses superficial to flexor carpi ulnaris to the wrist.

References   1194, 1349, 1354–1359, 1362, 1365–1367.

Radial

The radial artery may arise from the brachial more proximally than usual, from the axillary artery, or from the brachial lower than the bend of the elbow, but this low division of the brachial is rare. It may run superficial to the fascia of the forearm. In some cases it crosses over, instead of under, the extensors of the thumb. The radial may terminate in the forearm or be absent, with its place in the forearm and hand being supplied by the ulnar, palmar interosseous, or enlarged median artery. It may be joined by a vas aberrans from the brachial or axillary artery. The radial artery may be doubled or tripled, with the three divisions arising from the axillary, median, and anterior interosseous.

References   81, 231, 479, 494, 501–503, 536, 638, 903, 1005, 1016, 1076, 1077, 1208, 1349, 1354–1359, 1362, 1365–1367.

Deep Palmar Arch.   The deep palmar arch may be larger than usual, and its metacarpal branches may replace one or more of the common digital arteries by dividing at the cleft of the fingers into digital branches. The arch may be reinforced by enlarged posterior perforating branches from the radial and its branches on the back of the hand or by a large palmar interosseous. In some cases the radial joins the deep arch by passing through the second, instead of the first interosseous space. The princeps pollicis and radialis indicis may arise from the superficial palmar, or from a separate branch of the radial that passes through the first interosseous space. When the radialis indicis originates from the superficial arch it is distributed to the radial side of the index finger. The radialis indicis originates from the deep arch in 45% of cases, and the superficial arch in 13%; both arches contribute in 42%. The deep palmar arch is statistically less variable than the superficial arch. When absent (0.25%, Jaschtschinski), it is replaced by branches of the superficial palmar arch or by the dorsal system of arteries.

References   249, 638, 851, 903, 1194, 1349, 1354–1359, 1362, 1365–1367.

Thorax

Bronchial

The bronchial arteries are the nutrient arteries of the substance of the lung; they also supply bronchial glands, and in part the pericardium and esophagus. These vessels vary frequently in number and in their mode of origin. On the right side there is usually one bronchial artery, which arises from the first aortic intercostal artery or by a common trunk with the upper left bronchial artery from the descending thoracic aorta. On the left side there are generally two arteries, both of which arise from the descending thoracic aorta, one near the origin of that trunk, and the other, the inferior bronchial, lower down. Each artery is directed to the back part of the corresponding bronchus, along which it runs, dividing and subdividing with the successive bronchial ramifications in the substance of the lung.

The place of origin of the bronchial arteries is subject to considerable variation. The artery of the right sides has been found to arise singly from the aorta, from the internal thoracic, or from the inferior thyroid. The bronchial arteries of the two sides have been seen to arise by a common trunk from the subclavian (Haller). In one case, two common trunks, each furnishing a branch to the right and left lungs, descended into the thorax; one trunk arose from the internal thoracic and the other from the superior intercostal artery (R. Quain). In some cases the bronchials arise from the inferior surface of the arch instead of from the descending thoracic aorta. Instances also occur of two distinct bronchial arteries for each lung.

References 209, 232, 582, 668, 770, 778, 848, 860, 948, 1214, 1349, 1354–1359, 1362 1364, 1365, 1367.

Intercostal

The arteries of the first and second intercostal spaces usually arise from the superior intercostal branch of the subclavian, but occasionally the artery of the second space, and more rarely that of the first, may arise from the thoracic aorta. The arteries of the third and fourth intercostal spaces, as well as those of the first and second, may arise from the superior intercostal, the aortic intercostals being correspondingly reduced in number.

Occasionally the second intercostal is formed by a branch that arises from the first aortic intercostal and runs upward to the second space over the neck of the third rib; a similar condition may be found in the lower arteries, two or more intercostal spaces being supplied from a common stem. Finally, the right and left arteries of one or all of the intercostal pairs may arise from a common stem, which arises from the posterior median line of the aorta.

References 1357–1359, 1367.

Abdomen

Abdominal Aorta

Variations in the abdominal aorta are not common. According to Quain, in 10 out of every 13 subjects examined, the bifurcation of the aorta occurred within 1.2 cm above or below the level of the highest part of the iliac crest. The most common point for its bifurcation is opposite the lower border of the body of the fourth lumbar, but it may divide opposite the disc between the L4 and L5 or rarely opposite the fifth lumbar vertebra. A higher division is less common. The artery has been found dividing as high as the origin of the renal arteries, or even as high as the second lumbar vertebra.

The following rare variations have been described. The aorta may pass through the esophageal opening in the diaphragm. It may lie on the right side of the vena cava, the vein passing over the upper part of the aorta to gain the caval opening. The aorta may have a vena cava on each side, the left vein passing across the upper part of the artery to open into the right vein just below the caval opening. In some cases the aorta gives rise to a pulmonary branch close to the origin of the celiac artery; the abnormal vessel then passes through the esophageal opening and supplies a branch to the lower lobe of each lung. The aorta may run to the right of the inferior vena cava. Direct branches of the abdominal aorta sometimes include an accessory inferior phrenic, splenic, hepatic, accessory hepatic, supreme pancreatic, accessory superior mesenteric, inferior suprarenals, accessory gonadal, accessory and fifth lumbar, internal iliac, umbilical, and/or accessory median sacral. The aorta may have a marked sinuous course throughout the abdomen.

References 15, 16, 32, 159, 208, 248, 707, 1229, 1231, 1232, 1349, 1354–1359, 1362, 1366, 1367.

Variations in Branches of Abdominal Aorta.

The following variations have been recorded in the branches of the abdominal aorta.

Inferior Phrenic. Variations in origin of the phrenic arteries are numerous and supplementary phrenic vessels are common. The inferior phrenic arteries arise from a common trunk (55%, Poynter) from the aorta or celiac trunk in 18–30% of cases, or as independent branches from these same sources in 62%. Other sources may be hepatic, left gastric, renal, suprarenal, or superior mesenteric arteries in about 8% of cases. When independent, the right and left arteries usually arise from separate sources, in which case they may arise asymmetrically. Occasionally, the inferior phrenic may supply an accessory renal artery and suprarenal branches.

In one study (with added data from five other sources) of 848 bodies, the origin of the inferior phrenic was as follows: right and left separately from celiac trunk, 20.3% of cases; as a common trunk from the aorta, 19.7%; right from the aorta, left from the celiac, 14.2%; common trunk from the celiac, 13.6%; separately from the aorta, 13.2%; right from celiac, left from aorta, 6.8%; right from renal, left from aorta, 3.7%; right from renal, left from celiac, 3.5%; right and left from left gastric, 0.7%; right from left gastric, left from aorta, 0.5%; right from aorta, left from renal, 0.5%; right from celiac, left from left gastric, 0.5%; right from aorta, left from left gastric, 0.4%; right and left from renal, 0.4%; and all other sources and combinations, 1.9%. In some cases in which the right inferior gastric arose from a renal artery, it was from a superior aortic renal artery (four cases). When both the right and left inferior phrenics arose from the left gastric, it was from a common trunk in three cases. When the arteries arose from the aorta as a common trunk, it was usually from the left side of the aorta.

References   35, 478, 707, 850, 987, 1349, 1354–1359, 1362, 1365–1367.

Lumbar.   One or more pairs of lumbar arteries may arise as a common stem from the back of the aorta. The first lumbar may be joined at its origin with the lowest intercostal artery; the third and fourth lumbars, or less often the second and third lumbars, may arise from the aorta as a common stem. The fifth pair is sometimes absent. The first lumbar may provide the inferior phrenic or middle suprarenal. On occasion, one of the lumbar arteries provides the gonadal. The fourth lumbar on either side may provide the middle sacral, or both arteries arise as a common stem with the middle sacral.

References   35, 707, 1326, 1349, 1354–1359, 1362, 1365–1367.

Celiac Trunk

A typical three-branched celiac trunk occurred in 86% of 756 cases studied.

The celiac trunk may be absent (0.4% Poynter); the branches usually arising from it then arise separately from the aorta. In one case, the splenic gave rise to the left gastric, and the common hepatic originated from the superior mesenteric. The trunk may be shorter or longer than usual. In the latter case, the branches commonly arise separately from the trunk instead of radiating from one point. The trunk may provide only two branches, usually the splenic and hepatic, more rarely the gastric and splenic, or it may have more than three branches. The additional branch may be one of the inferior phrenics, a trunk common for the two inferior phrenics or for the inferior phrenic and left gastric, a gastroduodenal (or an accessory gastroduodenal), a second gastric or accessory splenic artery, a superior mesenteric (1.4%, Poynter), a middle or accessory middle colic, a supreme pancreatic, or a dorsal pancreatic. One of the normal branches of the celiac artery may be absent, and may be replaced by a stem common to the inferior phrenics, by the right middle suprarenal and the right gastroepiploic, or more rarely by some other branch. The branches of the celiac trunk sometimes arise from the superior mesenteric artery. The celiac, superior mesenteric, and inferior mesenteric may be joined together by one continuous longitudinal anastomosis. Cases of "celiac trunk" in the form of the splenogastric trunk or a hepatosuperior mesenteric trunk have been reported. A branch of the celiac trunk may pass through the esophageal hiatus of the diaphragm to supply the lower medial quarter of the right lung.

References   61, 102, 186, 215, 237, 319, 341, 345, 358, 376, 426, 442, 478, 567, 704, 707, 775, 850, 879, 934, 946, 984, 993, 1000, 1043, 1059, 1060, 1091, 1191, 1193, 1195, 1203, 1293, 1302, 1309, 1332, 1337, 1344, 1349, 1354–1359, 1362, 1365–1367.

Variations in Branches of Celiac Trunk.

The following variations have been recorded in the branches of the celiac trunk.

Left Gastric.   The left gastric may arise directly from the abdominal aorta (5%, Rossi; 6.7%, Eaton), and may provide one or both of the inferior phrenics or a trunk common for the two. There may be two left gastric arteries instead of one. The left gastric sometimes provides an enlarged branch to the hepatic artery (18–25% of individuals) or an accessory hepatic, and sometimes entirely replaces the left hepatic artery (11–12% of individuals).

References   79, 215, 707, 759, 775, 1349, 1354–1359, 1362, 1365–1367.

Hepatic.   The hepatic blood supply is said to fit the standard textbook description in slightly more than one-half of subjects. This may be a generous estimate. The common hepatic artery usually arises from the celiac trunk (85% of cases), but it may also arise directly from the aorta or from the left gastric, superior mesenteric (3% of cases), gastroduodenal, or right renal artery. Together with a normal artery, an accessory hepatic may arise from one of the above-named or neighboring branches. The common hepatic artery may be doubled, tripled, or missing (12% of cases), in which case one or more accessory arteries derived from one of the above sources may take its place. In one case of three hepatic arteries, a right and a left hepatic arose independently from the celiac trunk, and one branch arose from the superior mesenteric which supplied the gallbladder (Poynter). The common hepatic is almost always located between the bile duct and portal vein. The course of the right hepatic artery observed in 165 specimens was as follows: anterior to the bile duct, 24% of cases; posterior to the bile duct, 64%; anterior to the portal vein, 0.1%; and posterior to the portal vein, 9%. It arose from the superior mesenteric in 12%. The cystic artery arises from the right hepatic in about 89%

of cases. Although the cystic normally arises from the hepatic arterial system, it may also arise from the gastroduodenal artery. The left hepatic artery arises from the common hepatic in 89% of cases; in 11% of cases the vessel is aberrant and may arise from the left gastric. Supernumerary or "accessory" vessels to the left lobe of the liver occur in 23% of cases. In one study of 257 cadavers, 21% had one accessory hepatic artery, 9.3% had two, 7% had three, and 3.1% had four. "Accessory hepatic arteries and ducts" must not be considered simply as accessory or extra, but always as an essential, tissue-sustaining blood supply and as mandatory biliary drainage ducts (Michels). The hepatic artery may supply branches to the diaphragm. The gallbladder may be supplied by one (79% of cases), two (19.9%), or three cystic arteries (1.8%).

The hepatic pedicle is of surgical importance, hence the following description of the variations of the right hepatic, cystic, and gastroduodenal arteries in relation to the biliary system. The relations of the right hepatic artery to the hepatic duct have already been described. In 70% of individuals, the right hepatic artery is behind the hepatic bile duct, and the cystic artery arises from the right hepatic artery at a point between this artery and the cystic duct. In 12%, the hepatic artery passes across the front of the hepatic duct before entering the right lobe of the liver. The cystic artery arises as in the 70% of cases described above. In 10%, the right hepatic artery runs parallel, and in close proximity, to the cystic duct and neck of the gallbladder throughout its course. At the level of the neck of the gallbladder, it enters the right lobe of the liver. In 8% of individuals, the right hepatic artery passes across the right edge of the hepatic duct and either enters the liver or loops around the hepatic duct to pass behind it. Before entering the liver, the hepatic artery gives rise to the cystic artery.

The variations of the gastroduodenal artery in relation to the common bile duct may be described as follows. In 42% of individuals, a branch of the gastroduodenal artery (superior pancreatoduodenal artery) crosses the part of the common bile duct that lies behind the duodenum. In 38%, the gastroduodenal artery projects over the left edge of the common bile duct in its retroduodenal segment. In 20%, the gastroduodenal artery crosses the anterior aspect of the retroduodenal part of the common bile duct. The right hepatic artery may pass in front or behind the common bile duct. The gastroduodenal artery may give rise to the cystic artery.

Variations in origin of a single cystic artery have been reported as follows. There is only one cystic artery in 88% of individuals. Of these individuals, the cystic artery arises from the right hepatic artery in 82%; from the proper hepatic artery in 3%; from the left hepatic artery, distal to the origin of the right hepatic artery, in 2%; and from the gastroduodenal in 1%, crossing the common bile duct to reach the neck of the gallbladder. The relations of the single cystic artery to the hepatic bile duct have been described as follows. The cystic artery arises close to the right edge of the hepatic duct in 71% of individuals; arises on the left side of the hepatic duct or common bile duct in 27%, crossing one of these structures to reach the neck of the gallbladder; and arises behind the hepatic bile duct in 2%.

In cases of two cystic arteries, their origins have been reported as follows. Two cystic arteries occur in 12% of individuals according to one author and in 18% according to another; we assume here a frequency close to 15%. Of these individuals, both cystic arteries arise from the right hepatic artery in 9%; one cystic artery arises from the right hepatic and the second from the gastroduodenal artery in about 3%; one artery arises from the right hepatic and the second from the proper hepatic artery in 1.5%; and both cystic arteries arise from the left hepatic artery in 1.5%.

Thompson[1] provided the following listing of probable or estimated frequencies of what he described as the most important anatomic arrangements in the hepatic pedicle.

5–9%	Hepatic artery absent.
	Left hepatic from celiac.
	Right hepatic artery from superior mesenteric.
	Right hepatic artery dorsal to portal vein.
	Two cystic arteries from right hepatic.
	Common hepatic duct formed within liver.
10–19%	Left hepatic artery from left gastric.
	Accessory right hepatic arteries.
	Right hepatic artery ventral to common hepatic duct.
	Right hepatic artery dorsal to cystic duct.
	Two cystic arteries.
	Cystic artery ventral to common hepatic duct.
	Common hepatic duct entirely to right of portal vein.
	Cystic duct presenting dorsal arterial relations.
20–29%	Right hepatic artery presenting peculiar relationships to biliary ducts.
	Cystic duct presenting ventral arterial relations.

---

[1] Thompson, I. M. *On the Arteries and Ducts in the Hepatic Pedicle.* University of California Publications in Anatomy, Vol.1, No. 4, pp. 55–160. Berkeley, 1933.

30–39%	Aberrant arteries to the liver.
	Right hepatic artery close to left or upper aspect of cystic duct.
	"High opening" of cystic and common hepatic ducts.
	Common hepatic duct presenting ventral arterial relations.
40–49%	Cystic artery closely related to cystic duct.
	Either common hepatic or common bile duct (supraduodenal portion) presenting reasonably large ventral arterial relations.
50–59%	Right hepatic artery close to cystic duct (frequency not satisfactorily established).
60–69%	"Low opening" of cystic and common hepatic ducts.
70–79%	Right hepatic artery dorsal to common hepatic duct.
	Cystic artery crossing neither common bile nor common hepatic duct.
	Angular junction of cystic and common hepatic ducts.
	Common hepatic duct ventral to portal vein.
80–89%	Right hepatic artery from hepatic.
	Right hepatic artery dorsal to common or common bile duct.
	Single cystic artery.
	Supraduodenal portion of common bile duct entirely to right of portal vein.
90–100%	Hepatic artery from celiac.
	Hepatic artery entirely to left of common bile and common hepatic ducts.
	Hepatic artery ventral to portal vein.
	Gastroduodenal artery from hepatic.
	Gastroduodenal artery entirely to left of supraduodenal portion of common bile duct.
	Left hepatic artery from hepatic.
	Left hepatic artery entirely to left to common bile and common hepatic ducts.
	Right hepatic artery ventral to portal vein.
	Cystic artery from right hepatic.
	Definite right and left hepatic ducts outside liver.
	Cystic duct ventral to portal vein.

References 13, 34, 161, 162, 164, 167, 171, 180, 181, 215, 243, 264, 330, 341, 363, 381, 383, 384, 398, 425, 434, 566, 646, 666, 673, 685, 707, 733, 775, 805, 863–865, 875, 880, 883, 893, 908, 928, 933, 935, 941, 951, 1036, 1052, 1094, 1192, 1193, 1237, 1239, 1333, 1334, 1349, 1354–1359, 1362, 1365–1367.

Splenic. The variations of the splenic artery are numerous. It may divide into two branches that reunite, the splenic vein passing through the loop thus formed. It may give rise to branches normally derived from other vessels, such as the left gastric, middle colic, and left hepatic.

The dorsal pancreatic artery arises from the splenic in 37% of cases, the celiac in 33%, the superior mesenteric in 21%, and the common hepatic artery in 8%, typically close to the aortic source.

As indicated in a previous section (Celiac Trunk), the celiac trunk may lack one or more of its main branches. In such cases, the branch in question may arise from the aorta or superior mesenteric, either independently or in conjunction with another branch. Variations have been reported to occur with the following frequencies (approximate).

(1) Hepatosplenogastric trunk; normal complement 90%.
(2) Hepatosplenic trunk; hepatic and splenic, 3.5%.
(3) Hepatosplenomesenteric trunk; hepatic, splenic, and superior mesenteric, 0.5%.
(4) Hepatogastric trunk; hepatic and left gastric, 1.5%.
(5) Splenogastric trunk; splenic and left gastric, 5.5%.
(6) Celiacomesenteric trunk; superior mesenteric in conjunction with hepatosplenogastric trunk, 1–2.5%.
(7) Celiacocolic trunk; middle or accessory middle colic arising from the celiac trunk is extremely rare.

The splenic and hepatic arteries have been described arising from a common trunk from the superior mesenteric artery. The inferior pancreatic artery may also supply the spleen.

The tortuous course of the splenic artery is considered so variable that no two arteries are alike. Splenic arteries are end arteries in the strictest sense, and interference with the blood supply to the spleen will result in necrotic areas in the organ.

In congenital absence of the spleen, the splenic artery terminates in the pancreas.

References 3, 197, 215, 243, 707, 775, 862, 866, 1158, 1213, 1349, 1354–1359, 1362, 1365–1367.

Superior Mesenteric

The variations in this artery are numerous. It may be doubled, arising from the aorta as two trunks. On occasion it provides accessory branches to the liver, stomach, pancreas, spleen, and gallbladder. The artery may give rise to branches normally derived from other sources: the hepatic or its right or left branch, the cystic, the gastroduodenal or its right gastroepiploic branch, the left gastric, or a larger pancreatic branch. In some cases it gives rise to the left colic and superior hemorrhoidal, thus wholly or partly replacing the inferior mesenteric. The colic and intestinal branches of this artery may vary considerably in their origin and course, and in the number of primary and secondary loops that they form. A rare variation is the persistence of an omphalomesenteric artery running near the umbilicus and providing a branch to the urachus, to the liver through the falciform ligament, or to the rectus abdominis muscle anastomosing with the epigastric artery. The superior mesenteric also sometimes arises in common with the celiac. As many as nine to 18 of its branches supply the jejunum and ileum, although an 11 to 14 branch supply is most common. The minimum/maximum range for the jejunum and ileum is six to 21 branches.

References   167, 179, 237, 426, 433, 629, 645, 667, 690, 691, 704, 707, 805, 854, 855, 859, 861, 866, 867, 929, 1015, 1058, 1106, 1116, 1118, 1119, 1152, 1153, 1166, 1238, 1248, 1302, 1314, 1344, 1349, 1354–1359, 1362, 1366, 1367.

Additional reference: Kitamura, S., Nishiguchi, T., Sakai, A. and K. Kumamoto. Rare case of the inferior mesenteric artery arising from the superior mesenteric artery. *Anat. Rec.* 217:99–102, 1987.

Variations in Branches of Superior Mesenteric Artery.

Serveral variations in the branches of the superior mesenteric artery have been recorded.

Inconstant branches have been noted. These are the dorsal pancreatic (21% of cases), inferior pancreatic, right hepatic (accessory or replacing, 14%), common hepatic, and an accessory middle colic. In some cases, the inferior mesenteric, splenic, gastroduodenal, right gastroepiploic, or even the cystic artery arise from the superior mesenteric.

Of the more constant branches, a single middle colic is most common (88% of cases); it may arise independently from the superior mesenteric (60%) or from a common stem with the right colic (25%). Accessory middle colics arise from the superior mesenteric in about 9% of individuals. In 27%, an additional large left branch courses to the left colic flexure, ramifying to supply that area. The dorsal pancreatic and inferior pancreatic sometimes arise from the middle colic. The middle colic is absent in 3% of cases.

The right colic is a single vessel in 78% of individuals, and arises idependently from the superior mesenteric in only 28% of the population. Most frequently, it arises with, or as a branch of, the middle colic, ileocolic, or left colic arteries. It is absent in 13% of individuals.

The ileocolic arises independently from the superior mesenteric in 63% of individuals; in the remainder, it arises as a common trunk, usually with the right colic to the ascending colon.

References   136, 167, 629, 667, 690, 691, 707, 805, 866, 867, 1045, 1058, 1106, 1118, 1152, 1153, 1166, 1174, 1238, 1249, 1314, 1349, 1354–1359, 1362, 1366, 1367.

Suprarenal

In a classic study of the arteries of the suprarenal gland (111 cadaver dissections), it was concluded that the patterns (or course) of that structure's blood supply vary to such an extent that no two are ever alike.

In another study (50 dissections), the suprarenal arteries were classified according to the region of the gland they supply. The superior suprarenal arteries supply the superomedial part of the gland, the middle suprarenal supplies the anteromedial part of the gland, and the inferior suprarenal supplies the posterior and the inferolateral aspect of the organ. In approximately 96% of the 50 cases studied, superior suprarenal arteries arose primarily from the inferior phrenic. In only one case did the supply come from the aorta; in four cases, the superior area of the left suprarenal gland was supplied by both the aorta and the inferior phrenic arteries.

The middle suprarenal artery is usually single but can originate in any of a number of sites. The right and left middle suprarenal arose from the inferior phrenic in 24 cases, from the renal artery in four cases, and from the celiac trunk in three cases. The middle suprarenal was replaced (on the right or left side) by the superior suprarenal artery in nine cases, and by the inferior suprarenal artery in 41 cases. In one case, both the inferior suprarenal artery and middle suprarenal artery supplied the anteromedial part of the organ.

The inferior suprarenal arteries are considered to be very important because they supply all or most of the gland. They may be grouped, according to their point of origin from the renal artery, as proximal, middle, or distal. They are often multiple; they are larger and always branch before entering the organ, unlike the superior suprarenal artery. The inferior suprarenal artery arose from the renal artery in 46% of the 50 cases. In 30% of cases, the artery arose from the aorta; in 17 cases, both the aorta and the renal artery supplied the posterior and inferolateral parts of the suprarenal gland. The gonadal artery also sometimes gives rise to the inferior suprarenal artery.

The superior suprarenal arteries can consist of anywhere from three to 30 small branches from the inferior phrenic artery. They can also originate as direct branches from the aorta, celiac, or superior renal polar, or from a second inferior phrenic.

The middle suprarenal is usually a single vessel (85% of cases in one study). It often arises from the aorta, but it may also originate from the inferior phrenic, renal, aortic superior renal polar, or celiac trunk. The artery may be doubled, in which case the accessory artery may arise from one of the sources just mentioned. If absent, it is replaced by a superior or inferior suprarenal artery.

In 66% of cases, the inferior suprarenal is a single vessel arising from the aorta (above or below the origin of the renal); renal, in cases of doubled renal from the lower vessel; superior renal polar; branch of the renal, e.g., gonadal, ureteral, or inferior phrenic artery; or a common trunk with the middle suprarenal arising from the aorta, renal, or inferior phrenic. In 23% of cases, the inferior suprarenal is doubled, with one branch arising from the aorta and the second from the renal near the hilus of the kidney or from an aortic superior renal polar or one of its branches. In about 11% of cases, the inferior suprarenal is multiple, its branches arising from sources listed for the single vessel variety. In a study of 200 glands, 12% had no inferior suprarenal, 75% had one to seven, and 13% had eight to 12 arteries. The mean was 2.75 arteries. The inferior suprarenal contributes less than the superior suprarenals, but more than the middle suprarenal arteries, to the total vascularization of the gland.

The anterior superior capsular adipose branch of the renal provides three to 10 small branches to the inferior and lateral areas of the suprarenal gland. These branches may arise separately from the aorta, middle suprarenal, superior or uppermost branch of doubled renals, or inferior phrenic, or from a common trunk from the gonadal artery. A rare pattern of blood supply is one in which an inferior phrenic (arising from the aorta) supplies almost all the superior, middle, and inferior portions of the suprarenals as well as an aortic renal polar, kidney capsular branch, and gonadal artery.

The middle suprarenal may be absent, multiple, or may not supply the suprarenal gland at all.

References 288, 316, 415, 416,, 429, 431, 610, 707, 764, 797, 827, 849, 850, 988, 1349, 1354–1359, 1362, 1365–1367.

Renal

Variations in the number, source, and course of the renal arteries are common. The right and left renal arteries may arise from the aorta by a common stem. They may arise from the aorta at a lower point than usual, in which case the kidneys lie below their usual position. There may be several renal arteries on each side, or the renal artery may divide, close to its origin, into several branches. The renal artery on one or both sides may arise from the bifurcation of the aorta or from the common iliac, internal iliac, inferior mesenteric, or middle sacral artery. The right artery may cross in front of, instead of behind, the vena cava. Branches of the renal artery may "perforate" the substance of the kidney instead of entering at the hilus. The renal artery may give rise to branches normally derived from other vessels, such as the inferior phrenic, hepatic, right renal, middle and inferior suprarenals, gonadal, pancreatic, some of the colic arteries, and one or more of the lumbar arteries. "Accessory renal arteries," varying in size and generally derived from the aorta, are common (25% of all reported cases), and may enter the kidney at almost any point. It is a misnomer to call such structures "accessory renal arteries"; they are essential, tissue-sustaining vessels. Frequently, the renal arteries provide two to five slender branches to the suprarenal gland, whereas the inferior suprarenal arteries coming from the aorta or the renal arteries often supply one to four branches (capsular branches) to the superior pole of the kidney and its fat body.

The inferior phrenic artery sometimes also supplies an "accessory renal artery." A study of the arterial supply of the kidney (composite of 45 authors and 10,967 kidneys) demonstrated the following: single artery to each kidney (72% (SD 2.51 ±) of cases); one artery with an upper polar branch (412 of 3269 cases or 12.6%); two arteries (530 of 4901 cases or 11%); one hilar and one upper aortic polar branch (211 of 3384 cases or 6.0%); one hilar and one lower aortic polar branch (115 of 3674 cases or 3%); three arteries (70 of 4019 cases or 1.7%); and two arteries, one with an upper polar branch (37 of 1370 cases or 2.7%).

The inferior suprarenal arteries (one to three), which occur bilaterally, are constant branches of the single renal artery. An inconstant, but frequent (from 16 to 22% of cases depending on author), branch is the gonadal (testicular or ovarian) artery, a component of the renal pedicle. The inferior phrenic is another inconstant branch sometimes arising from the right renal artery or from an aortic superior renal polar artery. Because the inferior phrenic often supplies most or all of the blood to the suprarenal gland it must be identified and saved when surgery is performed on the renal pedicle.

In one study, the renals were located between the lower third of the first lumbar level and the cranial third of the second lumbar vertebra. The right and left renals were reported to be at the same level in about 30% of cases, while the right was higher in 47% and the left in 23% of cases. A single renal on one side and multiple (two, three, or four) renals on the other is not unusual. There are no reported sex- or race-related differences.

The right renal artery is longer than the left renal and normally courses behind the inferior vena cava. When multiple renal arteries are present, the more caudal vessels often take a precaval course. The right renal artery, from aortic origin to its division point, can measure 0.5–8 cm long; the length of the left can vary from 0.5 to 6 cm. In most cases, a single renal artery divides into anterior and posterior trunks; that division might occur anywhere along the artery's course to the kidney hilum. Precocious (near the origin) division may be interpreted erroneously as dual (or multiple) arteries. Most commonly, there are four branches, but there can be two to five entering the parenchyma of the organ. Based upon their relationship to the renal pelvis (that is, on whether they represent anterior or posterior division of the artery), these branches have been termed the anterior superior, anterior medial, anterior inferior, and a fourth branch that usually runs posterior to the renal pelvis, the retropyelic rami. The kidney has been shown to be composed of five segments, i.e., apical, upper, middle, lower, and posterior. The three terminal branches of the anterior division are generally referred to as the upper, middle, and lower segmental (terminal) branches; the fourth (posterior) branch supplies the posterior segment. The apical segment of the organ is usually supplied by the anterior division, but in 10% of cases the posterior branch may also supply an apical artery.

So-called "accessory" or "supernumerary" renal arteries are present in 30% of cases (based on a composite of 45 studies involving 10,967 kidneys). These vary in number from two to four, although there may be, rarely, five or six, arranged either unilaterally or bilaterally. In 16% of subjects, there was variation in the number of renal arteries between kidneys in the same subject. Accessory renal arteries can arise from the aorta as high (superiorly) as the diaphragm or as low (inferiorly) as the internal iliac artery.

Doubled renal arteries occur in 10% of cases (based on 10,967 kidneys studied). They may be side by side, one in front of the other, or spaced so that they enter the kidney at opposite ends of the hilum. In cases of doubled renal arteries, there may be a primary aortic hilar renal and an anterior or posterior division (vessel) or a renal polar artery.

Very rarely, the lower renal arises near the aortic bifurcation or from the common iliac. Sometimes one of the arteries passes in front of or behind the renal pelvis; the upper may cross the lower renal and the ureteropelvic junction to enter the inferior pole of the kidney.

In 1–2% of cases, three hilar renal arteries are derived from the aorta. The typical pattern for triple renal arteries is that two are hilar (of aortic origin) and the third is a superior or lower renal polar artery. The superior or the intermediate of the three renals may supply a superior renal polar or a suprarenal branch. The inferior vessel may provide a gonadal (testicular or ovarian) or inferior polar.

The presence of four renal hilar arteries derived from the aorta is rare. This pattern arises from the aorta between the superior and inferior mesenteric arteries. The upper two vessels may have a postcaval course and the two lower may pass in front of the inferior vena cava. The upper two may pass in front of the renal pelvis and the lower two behind it. Quadruple renals may exist as two hilar and two polar, three hilar and one polar, or one hilar and three polar renal arteries. Usually only one of these is large and the others are smaller and distributed to the superior or inferior extremities of the kidney.

Superior renal polar arteries are usually single. They arise as separate branches from the aorta in 7% of cases, or as branches of the renal artery in 12% of cases. There may be two or three unilaterally or bilaterally with different aortic or renal sources. They often arise from an inferior suprarenal or from a branch of an inferior suprarenal, i.e., an adipose renal capsular branch to the kidney. Superior renal polar arteries can also arise from the inferior phrenic or from the superior mesenteric artery. Some superior renal polar arteries provide serveral branches to the suprarenal gland, dividing before entering the superior pole of the kidney. Because they are often concealed or take a high course to the kidney, superior polar arteries can cause surgical problems. When such vessels are torn, the consequences may be a massive fatal hemorrhage; when they are ligated, devascularized and necrotic areas will develop in the kidney. All renal arteries, including accessory or supernumerary renals, are end arteries in the strictest sense; there is no collateral arterial supply (Michels).

References 20, 36, 40–45, 59, 60, 157, 172, 202, 212, 217, 265, 288, 319, 322, 347, 362, 364–366, 428, 431, 438, 456, 470–472, 526, 540, 560, 573, 601, 610, 669, 707, 765, 778, 806, 870, 885, 894–897, 911, 937, 953, 983, 989, 999, 1003, 1029, 1055, 1057, 1063, 1064, 1097, 1142, 1149, 1150, 1187, 1274, 1280, 1287, 1289, 1312, 1328, 1349, 1354–1359, 1362, 1365–1367.

Inferior Renal Polar Arteries. Inferior renal polar arteries are usually single and arise from the aorta (5.5% of cases) or the renal artery (1.4%). They have also occasionally been reported arising from a suprarenal, common iliac, or superior mesenteric artery. They may have a high aortic origin, and hence may cross the lower renal arteries to supply the inferior pole of the kidney. The inferior polar arteries are sometimes doubled, with one arising from the aorta and the other from the renal, or the pair from either source. They may or may not be accompanied by a vein.

Inferior polar renal arteries have been implicated as an etiologic factor in a form of hydronephrosis correctable by surgery. Aberrant renal arteries are common in fused kidneys (horseshoe- and dumbbell- or disc-shaped): two to six such arteries are characteristically longer or shorter than usual. The anomalous vessels may originate from the aorta, gonadal, common iliac, middle sacral, external or internal iliac, or superior or inferior mesenteric arteries. There are reports in the literature of seven, eight, and 10 arteries supplying horseshoe-shaped kidneys, but such findings, are very rare. In one study, the vascularization pattern of a dumbbell-shaped kidney had seven arteries: four from the aorta, one from each common iliac artery, and one from a left testicular artery. Studies indicate that 84% of cases of fused kidneys occur in males.

The blood supply of the renal pelvis is usually from the renal arteries which form serveral anastomotic branches. The blood supply of the ureter may arise at its proximal end from the aorta or the renal, common iliac, or internal iliac arteries and reaches the ureter on its medial side, or it may arise from its distal end from a testicular or ovarian, vesical, uterine, or vaginal artery on its lateral aspect. In another study, it was noted that the blood supply to the right and left ureters was usually asymmetric and there was marked variation in individual cases as to the source of the blood supply.

References 20, 36, 40–45, 59, 60, 279, 281, 553, 573, 643, 707, 738, 836, 850, 870, 911, 953, 983, 989, 1063, 1064, 1097, 1128, 1187, 1349, 1354–1359, 1362, 1366, 1367.

Gonadal (Testicular and Ovarian)

If one or both testicular arteries are missing, the testes are supplied by branches from the vesical or prostatic arteries passing under the arch of the pubis. One or both gonadal arteries may arise from the renal artery; more rarely, they arise from the middle suprarenal or lumbar arteries. One may branch off higher than the other. They may arise from a common stem, and one or both may be doubled, tripled, or quadrupled throughout or in a particular part of their course. The right spermatic may run behind, instead of in front of, the inferior vena cava.

The gonadal arteries have been found arching over the left renal vein, and have been identified as the origin of inferior suprarenal branches. One or both arteries might pass through a cleft formed by partially doubled gonadal veins. No sex-related differences in origin of these arteries have been reported.

The gonadal arteries usually arise from the anterior surface of the aorta below the renal vein, although cases have been reported of origins behind or above the renal vein. In 15% of cases, the gonadal arteries arise from the renal artery, from one of the branches of the renal artery, or from a supernumerary renal. The right gonadal may arise from a renal artery, the left from the aorta or vice versa. More rarely, a gonadal artery can originate from a suprarenal, phrenic, superior mesenteric, lumbar, common iliac, or internal iliac artery.

In some cases (17%), the gonadal arteries are doubled on one side; less commonly, they are doubled on both sides. The two gonadals may have an aortic and a renal origin or both may have the same origin. The inferior of doubled gonadal arteries is usually of aortic origin, whereas the one arising superiorly can be from either the renal artery or the aorta. Occasionally gonadal arteries arising from the aorta have two or three roots which subsequently merge into one. Their origins are generally between the first and third lumbar vertebrae.

Gonadal arteries have been classified into three types according to their relationship to the renal vein. One clinically and surgically important fact is that gonadal arteries, by crossing the renal vein, are frequently components of the renal pedicle (18% of all cases, on the left side about 34% and on the right about 16%). Testicular or ovarian arteries may arise from the aorta or renal artery and take the following course: arising behind, below, or above the renal artery and descending to the pelvis; arising behind or below the renal vein with the left artery ascending from behind the renal vein to curve over its upper border (lateral to the suprarenal vein) and then descending (anterior to the renal vein) to the pelvis; and as an arched artery passing in front of the renal vein when that vein passes behind the aorta. When gonadal arteries arise behind or below the renal vein, as just described, the right artery usually descends to the pelvis behind (posterior to) the renal vein. The arched testicular or ovarian artery (of Luschka) occurs in 22% of these cases.

References 288, 343, 344, 430, 521, 555, 556, 707, 726, 791, 824, 850, 913, 931, 953, 959, 1057, 1349, 1354–1359, 1362, 1366, 1367.

Inferior Mesenteric

Duplication of this artery has been reported. The artery may arise from the left common iliac artery. Its left colic branch may be absent. The inferior mesenteric occasionally encroaches on the territory of the superior mesenteric artery, or is replaced by the superior mesenteric. The inferior mesenteric artery may be absent entirely, or just absent from its usual location. This artery sometimes arises from the superior mesenteric by a common trunk with the middle colic. On the other hand, the inferior mesenteric may give rise to the middle colic or to an accessory right hepatic, accessory renal artery, or a common artery for both umbilical arteries when it arises from its usual position from the aorta.

References 8, 84, 85, 100, 138, 178, 237, 323, 426, 453, 474, 480, 704, 707, 833, 867, 868, 1035, 1058, 1059, 1117–1120, 1177, 1290, 1314, 1331, 1344, 1349, 1354–1359, 1362, 1366, 1367.

Middle Sacral

The origin of the middle sacral artery is variable. It may arise from one of the common iliacs or from the fourth lumbar. It may give rise to an accessory renal artery or to an accessory middle rectal artery or a renal artery in cases of pelvic kidney.

References 1325, 1349, 1354–1359, 1362, 1366, 1367.

Pelvis

Common Iliac

The common iliac arteries are as short as 1.2 cm or as long as 11 cm; the usual length is 3.7–7.5 cm. These variations in length may depend on whether the aorta bifurcates above or below the usual point, or on whether the common iliac arteries divide higher or lower than usual. A low aortic bifurcation is more common than a high bifurcation, as is the case with the common iliacs. The common iliacs may be absent; the external and internal iliacs then arise together from the end of the aorta. Either artery may provide a large branch, such as the iliolumbar or lateral or middle sacral, sometimes a lumbar, or occasionally an accessory renal or gonadal artery. In some cases right and left common iliacs are joined by a transverse anastomotic vessel. The embryonic axis (ischiadic) artery may persist and be of significant size and importance to give rise to the popliteal artery. Occasionally the common iliac fails to bifurcate, sinking into the pelvis to provide the usual pelvic branches and then emerging to continue as the external iliac. Unusual branches of the common iliac include a middle colic, umbilical, obturator, and circumflex iliac.

References 120, 521, 762, 763, 828, 968, 1254, 1349, 1354–1359, 1362, 1366, 1367.

Internal Iliac

The internal iliac may be longer or shorter than usual. It is seldom less than 2.5 cm in length, but may be as short as 1.2 cm or as long as 7.5 cm. Length variation generally depends upon the length of the common iliacs; when they bifurcate higher than usual, the internal iliac is longer and may lie initially above the brim of the true pelvis. The length may also depend upon the internal iliac itself dividing higher or lower than usual into its branches. This division may occur anywhere between the brim of the pelvis and the upper border of the sacrosciatic foramen. In some cases the branches arise without the artery dividing into an anterior and a posterior division, or one or more branches arise above the division. The internal iliac may give rise to the following: superior mesenteric, vesicoprostatic (which supplies the testes), prostatic (which also supplies the rectum), a common trunk for a superior vesical and a profunda penis, an independent arteria penis supplying the profunda penis arteries when the dorsal penis is a branch of the internal pudendal, and/or an inferior epigastric artery. Branches of the anterior and posterior divisions or the internal iliac may exchange origins.

References 53, 153, 217, 271, 328, 427, 462, 776, 922, 955, 968, 1042, 1101, 1349, 1354–1359, 1362, 1366, 1367.

Iliolumbar. This artery may be absent, reduced, or partially replaced by one of the lumbars. It occasionally gives rise to a lateral superior sacral artery.

References 217, 271, 776, 955, 1349, 1354–1359, 1362, 1366, 1367.

Lateral Sacral. The vessels from both sides may arise in common. Superior branches are sometimes absent and may be replaced by branches from the common iliac, internal iliac, iliolumbar, or sciatic arteries. When inferior branches are absent, they are sometimes replaced by branches from the middle rectal, gluteal, and sciatic arteries. The lateral sacrals may provide the inferior vesical and middle rectal arteries.

References 217, 271, 776, 955, 1349, 1354–1359, 1362, 1366, 1367.

Inferior Gluteal. This artery may form a common trunk with the superior gluteal (Jaschtschinski, 25%; Lipschutz, 43%). It has been reported doubled and it may be joined to the obturator.

The ischiadic (sciatic) artery, the embryonic axis artery, may persist as the primary artery of the thigh and continue as the popliteal.

References 120, 271, 722, 776, 818, 955, 982, 1349, 1354–1359, 1362, 1365–1367.

Obturator. This artery is variable in origin. It may arise from the common iliac, anterior division of the internal iliac (41.4%), inferior epigastric (25% based on observations in 4044 bodies), superior gluteal (10%), inferior gluteal-internal pudendal trunk (10%), inferior gluteal (4.7%), internal pudendal (3.8%), or external iliac (1.1%). The obturator has also been found arising from the femoral artery adjacent to its profunda branch. In only 23% of cases is a similar origin found on both sides of the body. The obturator artery gives rise to the ligamentum teres femoris artery in about 54% of subjects. The obturator may have two roots, one arising normally and the other from the external iliac. Accessory obturators may be found. Unusual branches of the obturator include an iliolumbar, inferior vesical, uterine, vaginal, dorsal penis, ureteric, perineal, external pudendal, and inferior epigastric. Its acetabular branch may be absent. The artery may be absent and its distribution replaced by branches from profunda femoris.

References 88, 221, 256, 266, 271, 298, 325, 603, 776, 955, 987, 990, 1349, 1354–1359, 1362, 1365–1367.

Uterine. The uterine artery usually arises from the internal iliac but may arise in common with the vaginal artery or middle rectal. The artery may be doubled or composed of multiple parts. It occasionally supplies branches that may be designated as superior vesical, inferior vesical, and ureteral arteries.

References 134, 258, 271, 448, 776, 955, 1349, 1354–1359, 1362, 1366, 1367.

Vaginal. This artery usually arises from the uterine artery, sometimes as several branches, and sometimes from the internal iliac in common with the uterine artery. The vaginal may also arise from the middle rectal or superior vesical artery.

References 258, 271, 448, 776, 955, 1349, 1354, 1356–1359, 1362, 1366, 1367.

Middle Rectal. This artery is occasionally absent. It usually arises from the internal iliac; however, it has been reported as arising from the inferior vesical or internal pudendal. Unusual branches include vaginal, seminal vesical, prostatic, and inferior lateral sacral arteries.

References 244, 278, 371, 776, 955, 1349, 1354–1359, 1362, 1366, 1367.

Inferior Vesical. Anastomoses between the inferior vesical and uterine arteries were found in 60 of 70 specimens.

References 151, 152, 244, 271, 448, 463, 776, 955, 1112, 1349, 1354–1357, 1359, 1362, 1366, 1367.

Middle Vesical. This artery arose from the umbilical in 99 of 132 cases. It was bilateral in 33 of 66 and unilateral in 33 of 66 bodies. When unilateral, it occurs with almost equal frequency from the right (18 of 33 cases) and left (15 of 33) sides.

References 151, 152, 244, 271, 448, 463, 776, 1112, 1349, 1354–1357, 1359, 1362, 1366, 1367.

Superior Vesical. The number of superior vesicals varies from one to four. They commonly arise from the umbilical artery, but were also found arising from the uterine (six of 66 sides), the vesicodeferential (six of 66 sides), and the obturator (six of 132 sides). A large anastomosis between the superior and inferior vesical arteries was found in 60 of 70 specimens studied.

References 151, 152, 244, 271, 448, 463, 776, 955, 1112, 1349, 1354, 1355, 1357, 1359, 1362, 1366, 1367.

Vesicodeferential. The vesicodeferential is usually a branch of the umbilical and may also arise from the prostatic and the pudendal. This artery (which supplies the trigone of the bladder) occurred in 48 of 66 bodies. It may give rise to a superior vesical artery.

References 244, 271, 448, 463, 776, 1112, 1349, 1354, 1355, 1357, 1359, 1362, 1365–1367.

Umbilical. This artery may give rise to the vesicodeferential, middle rectal, vaginal, accessory inferior epigastric, and an inguinal branch.

References 271, 378, 776, 955, 1349, 1354–1359, 1362, 1365–1367.

Internal Pudendal. The internal pudendal may arise in common with the obturator or the umbilical. It may end as a perineal artery, with the artery to the penis arising from the internal iliac. In some cases it splits into two vessels, one of which remains in the pelvis to supply pelvic viscera. The dorsalis penis artery may arise from profunda femoris.

References 271, 325, 776, 955, 1349, 1354–1359, 1362, 1366, 1367.

External Iliac

The external iliac may be longer or shorter than usual, resulting from the bifurcation of the common iliac above or below its usual point. When longer, it often takes a very tortuous course, making a partial loop or bend that may dip below the brim of the pelvis. The artery may be much smaller in diameter than usual; this may be the case when the femoral or main vessel of the lower limb arises from the inferior gluteal or other branch of the internal iliac. It then often ends in profunda femoris. The external iliac may give rise to a large branch, such as the circumflex iliac or inferior epigastric, at a point higher than usual. On occasion it gives rise to the obturator artery (1.1% of cases). The artery may terminate as the deep femoral in cases in which the anastomoses between the inferior gluteal and popliteal are extensive. In some cases it gives rise to branches usually arising from the femoral, i.e., superficial external pudendal or profunda femoris.

References 290, 342, 499, 592, 762, 763, 968, 1349, 1354, 1355, 1357–1359, 1362, 1366, 1367.

Inferior Epigastric. The epigastric may arise from the external iliac at a higher point than usual; it has been seen arising as much as 6.2 cm above the inguinal ligament. It may arise from the femoral below the inguinal ligament or even from the profunda femoris artery, or as a common trunk with the circumflex iliac. It is sometimes doubled. The epigastric may arise from, or conversely give rise to, the obturator. This variation is due to the enlargement of the normal anastomosis of the epigastric and obturator through their pubic branches. It is of considerable importance to the surgeon, since the obturator artery, when derived from the inferior epigastric, may run either lateral or medial to the femoral ring to reach the obturator foramen. This origin of the obturator occurs in about 20% of individuals. The artery courses around the inner side of the femoral ring where it is vulnerable to injury during surgical correction of femoral hernia. The inferior epigastric occasionally provides the dorsal artery of the penis/clitoris, circumflex iliac, superficial epigastric, and medial femoral circumflex.

References 298, 591, 873, 1207, 1349, 1354–1359, 1362, 1365–1367.

Lower Extremity

Deep Circumflex Iliac

The circumflex iliac may arise (a) like the epigastric, from the external iliac higher than normal, although seldom if ever as high as the epigastric; (b) more rarely, from the femoral below the inguinal ligament, or (c) as a common trunk with the epigastric. It may be absent, doubled, or tripled. The deep circumflex iliac occasionally arises in common with the inferior epigastric or from the femoral artery. The artery may give rise to the external pudendal and medial femoral circumflex, and may provide muscular branches in place of the lateral femoral circumflex.

References   223, 499, 1169, 1298, 1349, 1354–1359, 1362, 1366, 1367.

Femoral Artery

The following are important variations in the femoral artery. It may arise from the inferior gluteal or internal iliac, leave the pelvis, and traverse the back of the thigh with the sciatic nerve to the popliteal space. The external iliac under these circumstances ends in the deep (profunda) femoral or lateral circumflex, or some other branch of the femoral. A doubling of the femoral artery may occur below the origin of the deep femoral; the vessel may reunite lower down the thigh. On occasion, a vas aberrans arises from the medial side of the femoral or external iliac and joins the femoral lower down. The femoral vein may remain on the medial side of the artery throughout its course in the thigh, or it may be doubled, especially in the adductor canal. There is often a plexiform arrangement of the vein around the artery in this situation.

The femoral may give rise to the inferior epigastric, deep circumflex iliac, lateral femoral circumflex, medial femoral circumflex, iliolumbar, accessory femoral circumflex (medial and lateral), descending branch of the lateral femoral circumflex, and fourth or lowest perforating artery (great femoral nutrient or arteria femoris magna). The fourth perforating artery sometimes splits distally into an anterior and posterior tibial; it may also provide the medial superior articular artery to the knee joint, the dorsal artery of the penis, the obturator, and/or a great saphenous artery. The great saphenous courses between adductor magnus and vastus medialis, pierces the crural fascia below the knee, and runs with the saphenous vein to the medial malleolus or ends at the knee.

References   2, 65, 211, 678, 777, 822, 857, 905, 906, 938, 1075, 1099, 1100, 1224, 1298, 1330, 1349, 1354–1359, 1362, 1366, 1367.

Profunda Femoris. This artery courses lateral, medial, or anterior to the femoral. If it arises from the medial aspect of the femoral, the femoral may appear to split into three vessels of almost equal caliber: femoral, profunda, and lateral circumflex. In 200 limbs, the usual origin of the medial and lateral circumflex was found in 123 specimens, the medial arose from the femoral in 41, and the lateral arose from the femoral in 26. The profunda may be underdeveloped and terminate as the second or third perforating. If overdeveloped, it courses medial to biceps femoris in the popliteal fossa to supply the popliteal region.

The variations of the profunda and its branches are numerous, and, to a considerable extent, largely associated with one another. The various vessels of the profunda complex may more or less dissociate, one or another of them having an independent origin from the femoral; this may occur to such an extent that a profunda femoris as a definite vessel may not exist.

The point of origin of the profunda from the femoral is usually stated to be about 4 cm distant from the inguinal ligament, but the figure must be taken as an average from which there may be wide departures. Thus, in 430 thighs, Quain found that the distance from the inguinal ligament of the origin of the profunda was between 2.5 and 5.1 cm in 68%; of these it was between 2.5 and 3.8 cm in 42.6%. This distance was less than 2.5 cm in 24.6% of the thighs and more than 5.1 cm in only 7.4%. Quain's data are as follows: origin at the inguinal ligament, seven cases; 0–1.3 cm below the inguinal ligament, 13; 1.3–2.5 cm below, 86; 2.5–3.8 cm below, 183; 3.8–5.1 cm below, 109; 5.1–6.3 cm below, 19; 6.3–7.6 cm below, 12; and 11.6 cm below, one case.

The median number of perforating artery branches has been reported to be three, although two to six may be found (not including the termination of profunda femoris). Rare variations of profunda femoris include absence of the artery, perforating arteries arising from the femoral and from the medial circumflex femoral, and the profunda passing in front of the femoral artery and vein. Unusual branches include the inferior epigastric, obturator, dorsal artery of the penis, circumflex iliac, external pudendal, and accessory perforating arteries. No side or sex differences have been reported.

References   128, 648, 1099, 1100, 1324, 1329, 1349, 1354–1359, 1362, 1366, 1367.

Medial and Lateral Femoral Circumflex. Variations of the medial femoral circumflex include the following. It may arise (a) from the profunda femoris before the lateral circumflex; (b) from the femoral artery, external iliac or one of its branches, or superficial epigastric; or (c) from the inferior epigastric artery. The artery of ligamentum teres femoris arises from the medial femoral circumflex in about 15% of cases, from both the obturator and medial circumflex via an anastomotic connection in about 24%, and from both arteries independently in about 7%. The artery of ligamentum teres may supply the head of the femur; however, in most subjects it is absent or less important than the superior retinacular arteries, which are the most important blood supply to the head of the femur. The inferior retinaculars are of subsidiary importance. The medial femoral circumflex may give rise to a perforating artery. Variations of the lateral femoral circumflex include the following. It may arise from the femoral proximal to the origin of profunda femoris, or in common with the obturator. It may be doubled: one branch from the femoral and one from the profunda, both from the profunda, or both from the femoral proximal to profunda.

As indicated above, one of the circumflex arteries (more frequently the medial femoral circumflex) may arise independently from the femoral, and the point of origin of the independent vessel may be either above or below that of the profunda. When the medial femoral circumflex is the independent vessel, its origin is most frequently above that of the profunda (with an independent medial femoral circumflex, the origin of the profunda is apt to be somewhat below the typical point). With a high origin of the profunda, the lateral femoral circumflex may be represented by two vessels, one arising from the profunda, the other (accessory) arising from the femoral distally. Occasionally, both circumflexes arise independently from the femoral, the profunda in such cases usually having a low origin, and one of the perforating arteries may arise from the circumflexes. An extreme case of this, representing an almost complete dissociation of the profunda, has been described by Ruge. In this case, the superior perforating arises from the medial femoral circumflex and the middle perforating from the lateral femoral circumflex; what may be termed the profunda arises 9.7 cm below the inguinal ligament and provides only the inferior perforating artery.

The medial femoral circumflex may be very much reduced in size or even absent, its territory being supplied by branches from the obturator artery. Occasionally, although rarely, one of the perforating branches arises directly from the femoral, and a similar origin has also been observed for the descending branch of the lateral femoral circumflex.

No side or sex differences have been reported for either vessel.

References 57, 221, 223, 226, 325, 499, 1099, 1100, 1207, 1298, 1300, 1349, 1354–1359, 1362, 1366, 1367.

Popliteal. The principal variations are as follows. The popliteal may divide higher or, more rarely, lower than usual. It may divide into the anterior tibial, posterior tibial, and peroneal, or into a posterior tibial and a common trunk for the anterior tibial and peroneal. The popliteal vein may be deeper than the artery or separated from it by a slip of the gastrocnemius muscle. The popliteal is sometimes a continuation not of the femoral but of a sizable ischiadic (sciatic) artery. Unusual branches of the popliteal include an ascending branch to hamstring muscles, anastomosis with one of the perforating femorals, an aberrant artery that descends to the popliteal ligament and then reunites with the popliteal artery, an accessory posterior tibial, and a branch to the gastrocnemius (where it may form an arterial ring around the tibial nerve). The articular branches to the knee are variable. The popliteal occasionally provides a small saphenous artery (arteria saphena parva) that courses with the vein of the same name and the sural nerve to the tarsus. The popliteal artery usually passes over the popliteal muscle and between the lateral and medial heads of gastrocnemius. Entrapment and compression of the popliteal artery may occur whenever its course is altered, for example, when it passes beneath the popliteus muscle or some part of the gastrocnemius, usually the medial head, which may be anomalous.

References 120, 511, 519, 722, 818, 982, 1102, 1173, 1202, 1349, 1354–1359, 1362, 1366, 1367.

Anterior Tibial. The anterior tibial may be absent, rudimentary, or highly developed. It may terminate at the lower limit of the leg and fail to join the dorsalis pedis artery, in which case the dorsalis pedis arises from perforating branches of the peroneal. If the anterior tibial is absent, it is replaced by perforating branches from the posterior tibial in the leg, and by perforating branches from the peroneal in the foot. In cases of a high origin, the anterior tibial may lie deep to the popliteus muscle (hence compressed against the tibia) or between this muscle and the oblique popliteal ligament. With high origin it gives rise to branches normally originating from the popliteal and the peroneal artery. An enlarged anterior tibial may supply both the dorsum of the foot and the plantar arch. The anterior tibial may become plexiform on the dorsum of the foot, giving rise to the dorsal intermetatarsals.

References 130, 336, 519, 1349, 1354–1359, 1362, 1365–1367.

Posterior Tibial. The posterior tibial may be absent, rudimentary, or replaced by the peroneal artery. It may supply all the common digital arteries. On occasion it pierces the interosseous membrane to join the anterior tibial. It may turn anteriorly (in the distal quarter of the leg) to replace the anterior tibial. The peroneal in such cases provides the medial and lateral plantars. The peroneal is sometimes absent or enhanced by a branch from the posterior tibial. The great saphenous artery may arise from the posterior tibial. The posterior tibial may, in conjunction with the lateral plantar and deep plantar, form a superficial plantar arch, analogous to the superficial palmar arch of the hand.

References 519, 1349, 1354–1359, 1362, 1366, 1367.

Peroneal. The peroneal may arise (unusually) in the distal leg. It may be absent or replaced by the posterior tibial. On occasion it gives rise to the anterior tibial, or rejoins the posterior tibial. An accessory peroneal has been noted. Quain reported this artery unusually well-developed in 15% of 209 cadavers.

References 519, 817, 819, 1349, 1354–1359, 1362, 1366, 1367.

Plantar Arch. The arch may be doubled or absent, in which case the plantar foot is supplied by a single vessel, the posterior tibial.

References 519, 1349, 1354–1359, 1362, 1366, 1367.

Veins

Head, Neck, and Thorax

Coronary Sinus

The coronary sinus may be obliterated. The great cardiac vein then drains into the superior vena cava or left brachiocephalic vein. Several veins, including the middle cardiac, may converge to empty into the right atrium.

In the absence of an ostium for the coronary sinus, the blood carried in the cardiac veins may reach the right atrium of the heart by passing successively through a left (persistent) superior vena cava, left innominate vein, and right superior vena cava.

The coronary sinus may open into the left atrium. In one study, two of 150 (1.33%) hearts lacked a valve. The coronary sinus has been reported to be as long as 5.4 cm but is usually about 2.5 cm in length.

References 572, 695, 810, 978, 1014, 1044, 1349, 1354–1359, 1362, 1365–1367.

Superior Vena Cava and Brachiocephalic Trunk

The variations in the vena cava and brachiocephalic veins arise from the embryonic development of the great veins. They may be classified in the following way.

Variations Due to Persistence of Left Common Cardinal Vein.

The left subclavian may join the left internal jugular vein to form a trunk that continues almost vertically downward over the arch of the aorta in front of the root of the left lung and opens into the coronary sinus of the heart. This variation is known as the persistent left superior vena cava and results from retention of the embryonic arrangement. It occurs (all sources) in about 0.4% of cases. A persistent left superior vena cava may be associated with the presence of two azygos veins. When a left superior vena cava is present, the accessory hemiazygos vein (the remnant of the anterior part of the left cardinal vein) may open into it by arching over the root of the left lung, in a way similar to that in which the azygos (the right cardinal vein) opens into the superior vena cava over the root of the right lung. The normal left brachiocephalic vein may be absent, or quite small or rudimentary. The left superior vena cava may open into the left auricle (atrium) of the heart, coronary sinus, innominate vein, or right superior vena cava. On occasion, a left superior vena cava communicates with a normal inferior vena cava by means of a channel that is a remnant of a persistent sinus venosus. In another variation, a vein runs from the left brachiocephalic or from the left superior intercostal vein through the vestigial fold of the pericardium to the coronary sinus.

Variations Due to Persistence of Left and Suppression of Right Common Cardinal Veins.

The right brachiocephalic vein may cross the arch of the aorta to join a vertical left brachiocephalic vein, thus forming a left superior vena cava. The right superior vena cava is absent. The arrangement of the azygos veins under these conditions may be reversed: a left azygos opens (over the root of the left lung) into the left superior vena cava, and the right hemiazygos and accessory veins are organized in a manner similar to the normal arrangement of left hemiazygos veins. This arrangement of veins may occur independently of any general transposition of the viscera. There are other variations in the development of the great veins from the common cardinals, primitive jugular, and anterior and posterior cardinal veins.

The right superior vena cava is usually connected to the left atrium, but may be joined to the coronary sinus by a vein of substantial length and size. Doubled superior vena cava has an incidence of 0.16% (3000 reported cases). Unusual tributaries of the superior vena cava include the internal thoracic on the right side, the right supreme intercostal, and the thyroidea ima vein.

References 31, 56, 87, 94, 191, 192, 230, 240, 253, 259, 318, 403, 404, 447, 467, 476, 497, 516, 522, 523, 527, 531, 532, 546, 547, 575, 602, 609, 615, 619, 649, 680, 696, 717, 724, 796, 839, 858, 889, 910, 942, 961, 962, 978, 1014, 1024, 1061, 1074, 1092, 1108, 1124, 1144, 1146, 1161, 1206, 1224, 1234–1236, 1247, 1251, 1257, 1273, 1277, 1339, 1349, 1353–1359, 1362, 1365–1367.

Brachiocephalic

The right and left vessels occasionally open into the right atrium separately. The left vein may pierce the thymus (2% of individuals, Gruber). Unusual tributaries include the thyroidea ima, internal thoracic, accessory vertebral, accessory external jugular, accessory subclavian, and supreme intercostal. The brachiocephalic trunk may cross over the thymus gland.

References 236, 283, 406, 513, 525, 557, 713, 714, 1224, 1349, 1353–1359, 1362, 1365–1367.

Azygos

In one clinically significant variation, the azygos curves over and grooves the right lung before entering the superior vena cava just before it enters the heart. In one study of 200 bodies, the azygos showed variation in 26%. The cross connection of the hemiazygos to the azygos occurred with almost equal frequency at the eighth or the ninth vertebra. At least two cross connections were present in 100%; three cross connections occurred in 4% and four or more in 6%.

Rarely, the azygos vein passes through the aortic hiatus of the diaphragm. It may open directly into the right atrium (ascribed to the embryonic disappearance of the right common cardinal vein). The azygos occasionally opens into the right brachiocephalic or even the right subclavian (believed to be caused by an elongated embryonic right common cardinal vein). If the hemiazygos is underdeveloped, its branches open directly into the azygos. The azygos may receive either or both supreme intercostals and any combination of intercostals and lumbars. The azygos may drain into the hemiazygos, which in turn drains into the right atrium (Gruber). Unusual tributaries include the right internal thoracic, superior phrenic, bronchial, and gonadal veins.

References 31, 39, 86, 176, 200, 253, 257, 312, 395, 468, 523, 527, 569, 598, 652, 896, 910, 912, 1023, 1096, 1188, 1197, 1234, 1317, 1349, 1353–1359, 1362, 1365–1367.

Hemiazygos

The hemiazygos veins have been found passing to the right side of the body ventral to (over) the aorta. The vein may open into the right atrium, brachiocephalic, internal thoracic, left subclavian (ascribed to an elongated embryonic left common cardinal vein), or left brachiocephalic (after receiving abnormally the azygos vein). Unusual tributaries include the superior phrenic, gonadal, renal, and suprarenal veins.

References 31, 86, 285, 496, 498, 544, 774, 899, 1197, 1349, 1353–1359, 1362, 1366, 1367.

Pulmonary

Pulmonary veins vary in number and size. There are reports of one or more pulmonary veins entering either the right atrium or the inferior or superior vena cava. In one case, a left pulmonary vein opened into the left brachiocephalic vein.

References 47, 58, 144–147, 154, 158, 175, 250–252, 255, 259, 308, 321, 436, 482, 516, 532, 554, 583, 611, 622, 649, 698, 909, 916, 947, 1090, 1162, 1224, 1236, 1250, 1349, 1353, 1355–1357, 1359, 1362, 1366, 1367.

Common Facial

The common facial vein may (a) cross over the sternomastoid and open into the external jugular vein; (b) open into the anterior jugular vein; (c) run beneath the posterior belly of the digastricus and stylohyoid muscles; and (d) receive the lingual vein, pharyngeal vein, or both, and the submental. (e) The anterior facial may receive the angular, posterior facial, and lingual.

References 168, 1347, 1354–1359, 1362, 1366, 1367.

Anterior Jugular

The anterior jugular may exist as two parallel vessels, one opening into the transverse cervical and one into the external jugular. Two anterior jugulars are also sometimes present, with one median vessel opening into the external jugular and the other into the subclavian vein.

References 168, 505, 529, 1347, 1354–1359, 1362, 1366, 1367.

External Jugular

The external jugular veins are so variable that it is not possible to determine a "normal" pattern (Poynter). Both veins may be very small, or one may be much larger than the other. The external jugular may be missing on one or both sides; the veins that normally form it then open into the internal jugular. It may be formed merely by the posterior auricular vein. On occasion it receives the facial, lingual, and cephalic veins. It may pass over the clavicle and open into the cephalic, subclavian, or internal jugular. The vein is sometimes doubled. The external jugular may descend posterior to the inferior belly of the omohyoid. In some cases it forms an annulus around the clavicle. The vein occasionally receives a lingual vein or an accessory external thoracic that ascends subcutaneously from the nipple.

References 163, 168, 1349, 1355–1359, 1362, 1365–1367.

Cerebral

Variations in cerebral veins have been reported, none of which are considered especially noteworthy.

References 302–306, 736, 1349, 1355–1359, 1362, 1365–1367.

Internal Jugular

This vein may course in front of, or rarely, behind the common carotid. It may be rudimentary, its blood return taken over by the external jugular. Unusual tributaries include the supreme thoracic, bronchial, superior laryngeal, palatine, superficial cervical, vertebral, suprascapular, and transverse cervical veins.

References   1270, 1349, 1355–1359, 1362, 1365–1367.

Inferior Thyroid

The two veins may form a transverse venous arch across the trachea. The inferior thyroid may receive bronchial or thymic veins.

References   1349, 1355, 1356, 1358, 1359, 1362, 1366, 1367.

Superior Thyroid

This vein may receive the inferior petrosal vein.

References   1349, 1355–1359, 1362, 1366, 1367.

Vertebral

The vertebral may have two roots that encircle the corresponding subclavian. It may receive the upper intercostal veins as tributaries. It has been found passing in front of the clavicle.

References   1349, 1353,–1359, 1362, 1365–1367.

Lingual

This vein may receive the submental vein.

References   1349, 1353–1359, 1362, 1366, 1367.

Inferior Superficial Cervical

This vein may be small or absent. It may receive the inferior thyroid or thyroidea ima vein.

References   168, 1349, 1354–1359, 1362, 1366, 1367.

Internal Thoracic

On occasion this vein receives the lateral thoracic veins. It may drain into the vena cava or the azygos.

References   533, 1349, 1354–1359, 1362, 1365–1367.

Supreme Intercostal

The supreme intercostal may drain into the azygos on the right side and the hemiazygos on the left.

References   533, 896, 1349, 1354–1359, 1362, 1366, 1367.

Cephalic

This vein is occasionally accompanied by an accessory cephalic vein. The cephalic vein may be small or absent.

References   1349, 1354–1359, 1362, 1366, 1367.

Axillary

Variations in the axillary vein have been reported. It may be doubled, usually forming two trunks with frequent interconnections

References   311, 359, 530, 1053, 1270, 1349, 1354–1359, 1362, 1365–1367.

Subclavian

The subclavian vein may run on a higher plane than usual, even lying above the subclavian artery. It may pass with the artery behind the scalenus anterior, or run behind the scalenus anterior with the artery lying in front of that muscle. The vein may split forming a venous circle and enclosing the scalenus anterior. It occasionally passes between the clavicle and the subclavius. The subclavian may directly receive an inferior superficial cervical, transverse cervical, transverse scapular, suprascapular, anterior jugular, bronchial, cephalic, or venae comites of the brachial artery.

References   447, 512, 1270, 1349, 1354–1359, 1362, 1365–1367.

Upper Extremity

In a study of 300 arms, a dorsal venous arch (arcus venosus dorsalis (manus)) was present in 100% of cases. It was found at the level of the middle of the metacarpals and was convex distally. In 82% of cases, a large ascending vein passed from the center of the concavity to terminate in 65% of cases in the cephalic and in 17% in the basilic vein. Judging from this study the origin of the basilic vein is the ulnar side of the dorsal venous arch, and the basilic is the main venous outlet in 53% of cases. In 47% of cases, a large vein, the oblique, leaves the origin of the basilic to join the cephalic. An accessory cephalic arises proximally from the basilic (parallel with the oblique vein) to join the cephalic in 28% of cases.

On the volar surface, a venous rete is best marked on the thenar, hypothenar, and volar spects of the wrist; its outlet in 99% of cases is the median antebrachial vein.

The cephalic vein, defined as the vein originating in the radial end of the dorsal venous arch, winds around the radial border of the forearm and passes proximally along the arm to the shoulder region.

In one study, the cephalic was found to be of two types. In most cases, a level just distal to the bend of the elbow it gives rise to the median cubital, which continues in the bicipital sulcus and deltopectoral triangle to its termination. In 16% of cases, it receives the median cubital from the median antebrachial vein.

The termination of the cephalic is almost constant, but two variations were reported. In one case, the cephalic crossed the distal third of the arm to join the basilic. In the other case, the cephalic became continuous with the median cubital and thus terminated in the basilic vein.

The median antebrachial may terminate in the basilic (43% of cases). A case of bilateral absence of the cephalic vein has been reported.

In 35 of 44 cases studied, the basilic was found to terminate in the axillary vein without any specific line of demarcation.

The median antebrachial is the main outlet of the volar venous rete. In the arms studied (300), it terminated in the vicinity of the elbow joint as follows: in 43% of cases it terminated in the basilic and in 43% in the median cubital; in 13% it divided to terminate in both the basilic (vena mediana basilica) and cephalic (vena mediana cephalica). In four cases, the median antebrachial vein was absent.

The accessory cephalic (82% of 300 cases) is a vessel running parallel to the median antebrachial. In 39% of cases, this vessel originated from the distal part of the dorsal forearm, in 28% from the ulnar end of the dorsal venous arch, and in 16% from the cephalic at a point where the cephalic turns around the radial border of the forearm. In 17% of cases, it was either not present or extremely small.

The median cubital is remarkably constant (present in 84% of cases). In 43% of cases studied, it received the median antebrachial as a tributary; in 4% it arose from the cephalic. Its function is to transfer blood from the radial to the ulnar side of the forearm. The median cubital may be doubled (4% of cases), with the extra vessel lying proximal to the main trunk. In four of 300 cases, the cephalic did not give rise to the median cubital vein. In four of 300 cases, only the cephalic and basilic were present, with the median cubital, median antebrachial, and accessory cephalic being absent. In six of 300 cases, the anastomosis between the cephalic and basilic by way of a median basilic and median cephalic was doubled.

The main venous return in the upper limb is by the basilic vein, with all others converging upon it. In 83% of cases, the general arrangement is two longitudinally directed veins, the cephalic and basilic, connected by the median cubital. In 16% of cases, the cephalic and basilic are connected by the "M-shaped" median antebrachial, median basilic, and median cephalic. In 1% of cases, there are no connections between the cephalic and basilic in the vicinity of the elbow.

References   255, 856, 1349, 1354–1359, 1362, 1366, 1367.

Abdomen

Inferior Vena Cava

The inferior vena cava, in cases of transposition of the viscera, may lie on the left side of the aorta. It may also lie to the left of the aorta, without visceral transposition, crossing to the right to gain the caval opening immediately below the diaphragm. It may also cross over after receiving the left renal vein. When the inferior vena cava is doubled, the left cava usually passes across the aorta into the right after receiving the left renal vein. In this case, there may be a communication between the right and left veins in the position of the normal left common iliac vein. The inferior vena cava may join with the azygos vein to enter the superior vena cava. In this case, the hepatic veins open independently into the right atrium. If the inferior vena cava is absent, the blood from the lower extremities may pass through the diaphragm into the superior vena cava by way of a large vein in the location of the ascending lumbar and azygos veins. As a result, the hepatic veins drain directly into the right atrium through the normal caval opening in the diaphragm.

The inferior vena cava may receive the left spermatic vein. On occasion it receives a left accessory renal vein that passes behind the aorta. It is into this vessel that the usual tributaries of the left renal vein may open. The right ureter has been found passing behind the vena cava; it may receive several accessory renal veins, and as many as seven on each side have been seen. The lumbar veins may enter the inferior vena cava on one or both sides as a common trunk.

It should be noted that any venous channel returning blood via the aortic hiatus of the diaphragm is called a "persistent posterior cardinal vein."

The inferior vena cava may be unusually short (about 6 cm) and may be formed from three roots.

References   9, 26, 31, 96, 125, 143, 148, 182, 200, 242, 261, 348, 401, 406, 423, 443–445, 507, 571, 616, 647, 659, 699, 706, 717, 742, 743, 749, 757, 761, 781, 795, 809, 832, 835, 874, 918, 932, 953, 971, 1009, 1018, 1023, 1029, 1039, 1079, 1086, 1096, 1130, 1262, 1264, 1284, 1288, 1335, 1349, 1353, 1355–1359, 1362, 1365–1367.

Hepatic

The usual number of hepatic veins is three. The right hepatic vein may pass through the caval diaphragmatic opening before joining the vena cava. One of the hepatic veins has been demonstrated to drain directly into the right ventricle. A case has been described in which an accessory vein arising from three roots from the right lobe of the liver coursed adjacent to the inferior vena cava to enter the right ventricle. The vessel contained valves. A common hepatic may be formed that empties into the right atrium.

References   709, 733, 747, 1094, 1139, 1336, 1349, 1354–1359, 1362, 1366, 1367.

Renal

The renal veins show less variation than do the renal arteries. One study found multiple renal veins to be rare on the left side (1%) and common on the right (28%). In another study, the right renal vein was found to have one to three accessories (18% of 203 cases) and the left renal vein to have one or two accessories (9% of 203 cases).

In about 2% of cases, the left renal vein may be retroaortic or circumaortic.

The right renal vein may be doubled, even though the left renal vein is usually single. The left renal may anastomose with the splenic vein and receive lumbar veins.

References 210, 287, 322, 438, 705, 724, 874, 897, 1029, 1142, 1274, 1275, 1280, 1311, 1323, 1349, 1354–1359, 1362, 1365–1367.

Left Gastric

In a study of 220 cadavers, the left gastric was an affluent of the portal vein in about 67% of cases and of the splenic (lienal) vein in about 33%. A literature review of the studies of 15 authors showed that of 1383 patients, the left gastric vein joined the portal in 54%, the splenoportal junction in 16%, and the splenic vein in 29.0%. In the remainder the site of junction was not reported. There were no ethnic or sex differences noted.

The left gastric vein may terminate intrahepatically.

References 79, 313, 1349, 1354–1359, 1362, 1366, 1367.

Suprarenal

This vein may drain into the inferior phrenic and it may receive a gonadal.

References 287, 316, 952, 1349, 1354–1359, 1362, 1366, 1367.

Superior Mesenteric

This vein is extremely variable in origin, although its area of drainage and course are reasonably constant. It may have between 10 and 15 tributaries. The superior mesenteric may receive an accessory splenic and sometimes drains into a renal vein.

References 287, 441, 1349, 1354–1359, 1362, 1365–1367.

Inferior Mesenteric

This vein usually, but not always, joins the superior mesenteric and splenic to form the portal vein. In many cases, the inferior mesenteric vein joins the splenic or superior mesenteric in a highly variable manner.

References 287, 1349, 1354–1359, 1362, 1366, 1367.

Portal

The portal vein is rarely variable. It often receives the left gastric vein, and may also receive an accessory splenic, pancreaticoduodenal, or right gastroepiploic. The portal vein may lie in front of instead of behind the duodenum, common bile duct, and hepatic artery.

References 187, 188, 218, 287, 320, 329, 403, 437, 539, 594, 694, 723, 939, 940, 1279, 1349, 1354–1359, 1362, 1366, 1367.

Gonadal

Both spermatic veins were found draining into a left renal. In a case of doubled inferior vena cava, a left ovarian was found draining into the left inferior vena cava. The gonadal vein may be represented by several vessels or may form a plexus. The veins may be partially divided into two parts permitting passage of the gonadal arteries through the cleft vessels. The gonadal may receive a duodenal or a suprarenal vein.

References 287, 716, 1349, 1354–1359, 1362, 1366, 1367.

Pelvis

Common Iliac

Either common iliac vein may be doubled in part or throughout its length. One of the veins may be absent, in which case the external and internal iliac veins join the opposite common iliac to form the vena cava. The right and left internal iliac veins occasionally unite and open as a common trunk at the confluence of the right and left external iliac veins to form the vena cava. The middle sacral trunk vein may divide, and one branch may open into the right common iliac vein and the other into the left. The common iliacs occasionally receive an accessory renal vein or even the entire renal drainage. A connection between the portal and right common iliac has been reported.

References 83, 528, 724, 1349, 1354–1359, 1362, 1365–1367.

Internal Iliac

A venous annulus may form around the internal iliac artery. It may receive an aberrant vessel representing a confluence of the internal pudendal, obturator, gluteal, and sciatic.

References 507, 1349, 1354–1359, 1362, 1366, 1367.

External Iliac

The vein may be doubled or lie lateral to the external iliac artery. It may receive a gonadal or obturator vein.

References 83, 218, 300, 1349, 1354–1359, 1362, 1366, 1367.

Inferior Epigastric

The inferior epigastric vein may be doubled, run through the femoral ring, or connect with the obturator or pudendal.

Reference 1365.

Superior Rectal

This vein may receive a uterovaginal vein.

References 1349, 1354–1359, 1362, 1366, 1367.

Lower Extremity

Femoral

The femoral vein may be doubled in part of or throughout its length. It may split into two and encircle the femoral artery. It occasionally passes through the adductor magnus above the femoral artery and runs independently from the artery until it joins the profunda vein to form the common femoral vein. The femoral may run with the sciatic nerve and pierce the adductor magnus at the level of the lesser trochanter. When this occurs, the femoral artery is usually small, and the inferior gluteal (sciatic) artery is the primary vessel of the thigh. A plexus-forming femoral vein has been reported. The obturator may communicate with the femoral vein

References   83, 120, 226, 280, 352, 506, 710, 938, 1349, 1354–1359, 1362, 1366, 1367.

Popliteal. The popliteal may lie between the artery and the bone. It is sometimes doubled in part of or throughout the popliteal space. The doubled veins may unite in front and behind the artery to form a plexus around the vessel.

References   284, 1349, 1354–1359, 1362, 1366, 1367.

Dural Sinuses

The confluences of the venous sinuses of the dura mater are variable. The superior sagittal and straight sinuses may drain into a common opening, or they may bifurcate before joining the transverse sinuses. The straight sinus occasionally bifurcates and joins one of the transverse sinuses. The superior sagittal sinus may enter one of the transverse sinuses.

A "typical" confluence of the superior sagittal, straight, and two transverse sinuses was found in four of 50 cases in one study and in two of 42 cases in another study. In 11 of 42 cases, the superior sagittal sinus joined a transverse sinus. In nine cases, it joined the right transverse sinus; in two others, it joined the left transverse sinus. In these 11 cases, the straight sinus bifurcated into right and left limbs, each of which opened into the corresponding transverse sinus.

In the remaining 29 cases, the superior sagittal also bifurcated so that it too, as well as the straight sinus, opened into the transverse sinus of the corresponding side.

References   70, 165, 239, 728, 820, 1347, 1354–1359, 1362, 1366, 1367.

Transverse Sinus

The confluence or horizontal segment of the transverse is sometimes absent, as is, on occasion, the left transverse sinus. The transverse sometimes communicates via an aberrant vein with the cavernous sinus or ophthalmic vein.

References   70, 165, 563, 1347, 1354, 1356–1359, 1362, 1365.

Superior Sagittal

The superior sagittal sometimes exists as two separate parallel channels because of a median dural partition. Its is also sometimes absent.

References   70, 165, 1347, 1354–1359, 1362, 1366, 1367.

Petrosal

A middle petrosal sinus might be present. The superior petrosal sinus occasionally communicates via an aberrant vein with the superior ophthalmic vein.

References   70, 165, 245, 621, 1347, 1354–1359, 1362, 1366, 1367.

Tentorial

The tentorial might be absent, and sometimes has as unusual tributaries the occipital sinus and an accessory common internal cerebral vein.

References   70, 165, 307, 1347, 1354–1359, 1362, 1366, 1367.

Cavernous

This sinus might be absent. It sometimes receives a vein via the foramen rotundum.

References   70, 165, 193, 1347, 1354–1359, 1362, 1366, 1367.

Lymphatics

Thoracic Duct

The right lymphatic duct forms various combinations with the right subclavian vein and jugular trunks. It is rare to find a simple right lymphatic duct that enters directly into the junction of the internal jugular and subclavian veins.

The thoracic duct, which drains most of the body, may end by joining the left internal jugular, the junction of the left internal jugular and left subclavian veins, or the brachiocephalic vein. It is common for the duct to have several terminal divisions; it might be doubled or even tripled over a variable distance. When doubled it can open on both the right and left sides, at the venous angle. Persistence of paired ducts has been reported in cases of doubled aortic arch and in low origin (from the descending aorta) of the right subclavian. Persistence of the (embryonic) right duct, with obliteration of usual left duct, has been noted in a case of right aortic arch of the type presenting a left brachiocephalic trunk. The size, shape, and location of the cysterna chyli is extremely variable; the structure is absent in about 50% of cases examined.

The thoracic duct has been found in a normal subject, without other vascular anomalies, to terminate on the right side at the junction of the right suclavian and internal jugular veins. The thoracic duct has also been found draining into the azygos vein and inferior vena cava. In one study, the thoracic duct opened into the following vessels (in order of decreasing frequency): (a) left venous angle, (b) internal jugular, (c) left subclavian, (d) vertebral, and (e) others, including the azygos and brachiocephalic. The thoracic duct has also been found draining into the subclavian artery.

The length of the thoracic duct varies from 38 to 45 cm in the adult. It usually begins at the vertebral level of L_2 but can begin at L_3, L_1, or even T_{12}.

References 189, 286, 368, 455, 459, 544, 632, 639, 654, 784, 877, 914, 1048, 1183, 1190, 1209, 1240, 1265, 1317, 1347, 1354–1359, 1362, 1364, 1366, 1367.

References

* These journal articles are notable for their content, literature review, and/or illustrations.
+ The authors have not had access to these references. They are included because of their historic and/or scientific importance.

Journals

1* AARON, C. and A.R. CHAWAF. Variations de la carotide externe et de ses branches. *L'Assoc. Anatomistes, Comptes Rendus.* 52:125–133, 1967.
2 AASAR, Y.H. The saphenous artery. *J. Anat.* 73:194, 1938–1939.
3 ABADIA-FENOLL, F. Über eine äußerst stark geschlängelte und verlängerte Arteria lienalis. *Anat. Anz.* 115:339–344, 1964.
4 ABBIE, A.A. The blood supply of the visual pathways. *Med. J. Aust.* 2:199–202, 1938.
5 ABBOTT, F.C. Specimen of right aortic arch. Proceedings of the Anatomical Society of Great Britain and Ireland. *J. Anat. Physiol.* 26:581–582, 1892.
6 ABBOTT, H.K. Case of abnormal arrangement of aortic valves. *J. Anat. Physiol.* 28:103, 1904.
7* ADACHI, B. *Das Arteriensystem der Japaner.* Kenkyusha Press, Tokyo, 1928.
8 ADACHI, B. Das Fehlen der A. mesenterica inferior bei einem Japaner. *Anat. Anz.* 69:431–433, 1930.
9 ADACHI, B. Statistik der Varietät der V. cava caudalis bei den Japanern. *Anat. Anz.* 85:215–223, 1937–1938.
10* ADACHI, B. *Venensystem der Japaner.* Kenkyusha Press, Tokyo, 1940.
11 AEBY, C. Ein seltener Fall von Gefässanomalie. *Correspondenz-Blatt Schweizer Aerzte.* 2:120–123, 1872.
12* AHMED, S.H., EL-RAKHAWY, M.T., ABDALLA, A. and R.G. HARRISON. A new conception of coronary artery preponderance. *Acta Anat.* 83:87–94, 1973.
13 ALEKSANDROWICZ, R. On the vascular segments of the liver. *Folia Morphol.* 23:26–40, 1964.
14* ALEKSANDROWICZ, R. Two cases of absence of the brachiocephalic trunk. *Folia Morphol.* 26:220–222, 1967.
15 ALEKSANDROWICZ, R., GOSEK, M. and M. PROROK. Normal and pathologic dimensions of the abdominal aorta. *Folia Morphol.* 33:309–315, 1974.
16 ALEKSANDROWICZ, R., GOSEK, M. and M. PROROK. Dimensions of the lumen of the visceral branches of the abdominal aorta under normal and pathologic conditions. *Folia Morphol.* 34:77–84, 1975.
17* ALEXANDER, R.W. and G.C. GRIFFITH. Anomalies of the coronary arteries and their clinical significance. *Circulation* 14:800–805, 1956.
18 ALEXANDER, W.F. The course and incidence of the lateral costal branch of the internal mammary artery. *Anat. Rec.* 94:446, 1946.
19* ALEXANDER, S.C., FIEGEL, S.J. and R.N. CLASS. Congenital absence of the left pulmonary artery. *Am. Heart J.* 50:465–470, 1955.
20 ALEZAIS. Le rein en fér a cheval et les anomalies des artères rénales. *Soc. de Biol., Comptes Rendus Hebdomadaires des Séances et Mémoires.* 60:889–891, 1906.
21 ALLAN, F.D. An accessory or superficial inferior thyroid artery in a full term infant. *Anat. Rec.* 112:539–542, 1952.
22 ALLMER, K. Ein Fall einer abnormen oberen Schilddrüsenarterie. *Anat. Anz.* 94:310–313, 1943.
23 ALM, L. Aorta und Arteria pulmonalis aus der rechten Kammer bei einem dreizehnjährigen Mädchen. *Arch. Pathol. Anat. Physiol. Klin. Med.* 298:229–238, 1937.
24 ALMEIDA, F. DE. Notes sur l'artère cérébrale anterieure. *L'Assoc. Anatomistes, Comptes Rendus.* 28:8–10, 1933.
25 ALMEIDA, F. DE. Notes sur l'artère communicante antérieure. *L'Assoc. Anatomistes Comptes Rendus.* 28:14–15, 1933.
26 ALTEN, H. VON. Über linksseitige Lage der Vena cava inferior. *Anat. Anz.* 43:337–348, 1913.
27 ALTMANN, F. Über zwei seltene Anomalien der Arteria meningea media. *Z. Gesamte Anat.* 97:509–514, 1932.
28 ANDERSON, R.J. A new abnormality in connection with the vertebral artery. *J. Anat. Physiol.* 14:248–249, 1879.
29 ANDERSON, R.J. Abnormal arrangement of the thyroid arteries. *J. Anat. Physiol.* 14:353–356, 1879.
30 ANDERSON, R.J. A palatine branch from the middle meningeal artery. *J. Anat. Physiol.* 15:136–138, 1881.
31 ANDREASSI, G. Über die Verbindung zwischen dem oberen und unteren Hohlvenensystem in Hinblick auf die laterale und mediale Wurzel der Vena azygos und der Vena hemiazygos. *Anat. Anz.* 92:240–245, 1941–42.
32 ANNAN, J.L. Case of an abnormal sinuous aorta. *J. Anat. Physiol.* 44:241–243, 1910.
33* ANSON, B.J. The anomalous right subclavian artery. *Surg. Gynecol. Obstet.* 62:708–711, 1936.
34* ANSON, B.J. Anatomical considerations in surgery of the gallbladder. *Q. Bull. Northwestern University Medical School* 30:250–259, 1956.
35 ANSON, B.J. and C.B. MCVAY. The topographical positions and the mutual relations of the visceral branches of the abdominal aorta. A study of 100 consecutive cadavers. *Anat. Rec.* 67:7–15, 1936.
36* ANSON, B.J., RICHARDSON, G.A. and W.L. MINEAR. Variations in the number and arrangement of the renal vessels. A study of the blood supply of four hundred kidneys. *J. Urol.* 36:211–219, 1936.
37* ANSON, B.J. and C.B. MCVAY. The fossa ovalis, and related blood vessels. *Anat. Rec.* 72:399–404, 1938.

38* ANSON, B.J., WRIGHT, R.R. and J.A. WOLFER. Blood supply of the mammary gland. *Surg. Gynecol. Obstet.* 69:468–473, 1939.
39* ANSON, B.J., SIEKERT, R.G., RICHMOND, T.E. and W.E. BISHOP. The accessory pulmonary lobe of the azygos vein. *Q. Bull. Northwestern University Medical School* 24:285–290, 1940.
40* ANSON, B.J., PICK, J.W. and E.W. CAULDWELL. The anatomy of the commoner renal anomalies. Ectopic and horseshoe kidneys. *J. Urol.* 47:112–132, 1942.
41* ANSON, B.J. and E.W. CAULDWELL. The pararenal vascular system. A study of 425 anatomical specimens. *Q. Bull. Northwestern University Medical School* 21:320–328, 1947.
42* ANSON, B.J., CAULDWELL, E.W., PICK, J.R. and L.E. BEATON. The blood supply of the kidney, suprarenal gland and associated structures. *Surg. Gynecol. Obstet.* 84:313–320, 1947.
43* ANSON, B.J., CAULDWELL, E.W., PICK, J.W. and L.E. BEATON. The anatomy of the pararenal system of veins with comments on the renal arteries. *J. Urol.* 60:714–737, 1948.
44* ANSON, B.J. and L. KURTH. Common variations in the renal blood supply. *Surg. Gynecol. Obstet.* 100:156–162, 1955.
45* ANSON, B.J. and E.H. DASELER. Common variations in renal anatomy, affecting blood supply, form, and topography. *Surg. Gynecol. Obstet.* 112:439–449, 1961.
46* ANTOPOL, W. and M.A. KUGEL. Anomalous origin of the left circumflex coronary artery. *Am. Heart J.* 8:802–806, 1933.
47 APPLETON, A.B. The arteries and veins of the lungs. I. Right upper lobe. *J. Anat.* 79:97–120, 1945.
48 ARCE, J. Accessory pulmonary artery probably arising from abdominal aorta. *J. Thoracic Surg.* 12:548–552, 1943.
49 ARKIN, A. Totale Persistenz des rechten Aortenbogens im Röntgenbild. *Wien. Arch. Inn. Med.* 12:385–416, 1926.
50 ARKIN, A. Double aortic arch with total persistence of the right and isthmus stenosis of the left arch: A new clinical and x-ray picture. *Am. Heart J.* 11:444–474, 1936.
51* ARMSTRONG, W.G. and J.W. HINTON. Multiple divisions of the recurrent laryngeal nerve. An anatomic study. *Arch. Surg.* 62:532–539, 1951.
52 ARNAUD, H. Présentation d'un cas d'anomalie de position du coeur. *Soc. Biol. Comptes Rendus Hebdomadaires des Séances et Memoirès* 43:4–8, 1891.
53 ASHLEY, F.L. and B.J. ANSON. The hypogastric artery in American Whites and Negroes. *Am. J. Phys. Anthropol.* 28:381–395, 1941.
54 ATANASIU, I., OANCEA, T. and V. PANAITESCU. Arteria subclavia dextra als letzter Ast des Aortenbogens. *Anat. Anz.* 118:114–116, 1966.
55* ATKINSON, W.J. The anterior inferior cerebellar artery. Its variations, pontine distribution, and significance in the surgery of cerebello-pontine angle tumours. *J. Neurol. Neurosurg. Psychiatry* 12:137–151, 1949.
56* ATWELL, W.J. and P. ZOLTOWSKI. A case of a left superior vena cava without a corresponding vessel on the right side. *Anat. Rec.* 70:525–532, 1938.
57* AUBURTIN, G. Die beiden Arteriae circumflexae femoris des Menschen. *Anat. Anz.* 27:247–269, 1905.
58 AÜER, J. The development of the human pulmonary vein and its major variations. *Anat. Rec.* 101:581–594, 1948.
59 AUGUSTYN, M. Arterial vascularization of the double kidney. *Folia Morphol.* 40:159–166, 1981.
60* AUGUSTYN, M. Supernumerary segmental branches of the renal artery. *Folia Morphol.* 41:441–447, 1982.
61* AUGUSTYNIAK, E. A rare case of absence of the celiac artery. *Folia Morphol.* 24:411–412, 1965.
62 AUGUSTYNIAK, E. and S. ZAJAC. A rare case of multiple vascular variation. *Folia Morphol.* 27:427–431, 1968.
63 AUTENRIETH. Zusatz zu der Abhandlung: De dysphagia lusoria. *Arch. Physiol.* 8:264–268, 1807–08.
64 AUTENRIETH and PFLEIDERER. Disseratatio inauguralis. De Dysphagia lusoria. *Arch. Physiol.* 7:145–188, 1807.
65 AUZIAS-TURENNE. Anomalie de l'artère fémorale. *Soc. Biol. Comptes Rendus des Séances et Memoires* 6:71–72, 1854.
66 AYER, A.A. and Y.G. RAO. A radiographic investigation of the coronary arterial pattern in human hearts. *J. Anat. Soc. India* 6:63–66, 1957.
67 BACHHAMMER, J. Über einige Varietäten des menschlichen Körpers. *Arch. Anat. Physiol.* 1879, pp. 139–166.
68 BACHHUBER, C.A. Complications of thyroid surgery: Anatomy of the recurrent laryngeal nerve, middle thyroid vein and inferior thyroid artery. *Am. J. Surg.* 60:96–99, 1943.
69 BAHNSON, H.T. and A. BLALOCK. Aortic vascular rings encountered in the surgical treatment of congenital pulmonic stenosis. *Ann. Surg.* 131:356–362, 1950.
70 BALÓ, J. The dural venous sinuses. *Anat. Rec.* 106:319–325, 1950.
71 BANCHI, A. Morfologia delle arteriae coronariae cordis. *Arch. Ital. Anat. Embriol.* 3:87–164, 1904.
72 BANCHI, A. Il V arco aortico branchiale nella interpretazione di alcune varietà dell'arco aorta e dei suoi rami, con osservazioni originali. *Arch. Ital. Anat. Embriol.* 6:389–427, 1907.
73* BANSAL, P.C. A case of abnormal ophthalmic arteries. *J. Anat. Soc. India* 10:41–43, 1961.
74* BAPTISTA, A.G. Studies on the arteries of the brain. II. The anterior cerebral artery: Some anatomic features and their clinical significance. *Neurology (Minn.)* 13:825–835, 1963.
75* BAPTISTA, A.G. Studies on the arteries of the brain. III. Circle of Willis: Morphologic features. *Acta Neurol. Scand.* 40:398–414, 1964.
76 BAPTISTA, B.V. Des valvules sigmoides aortiques. *Assoc. Anatomistes Comptes Rendus* 26:42–46, 1931.
77 BARBOSA SUEIRO, M.B. Observation de quelques artères avec un trajet superficiel anormal chez quelques membres d'une famille. *Assoc. Anatomistes Comptes Rendus* 28:25–27, 1933.
78 BARDELEBEN, H. VON. Abdominalanastomose der Nabelarterien. *Anat. Anz.* 10:725–728, 1895.
79 BARLOW, T.E., BENTLEY, F.H. and D.N. WALDER. Arteries, veins and arteriovenous anastomoses in the human stomach. *Surg. Gynecol. Obstet.* 93:657–671, 1951.
80 BARRY, A. The aortic arch derivations in the human adult. *Anat. Rec.* 111:221–238, 1951.
81 BARTELS, P. Über eine Ösenbildung der Arteria recurrens radialis für den Nervus radialis profundus, kombiniert mit anderen Abnormitäten. *Anat. Hefte* 15:203–212, 1900.
82 BARTOSIAK, P., BOROWSKI, S. and B. GOLAB. Variations of the origin, the course and branching of the posterior cerebral arteries in man. *Folia Morphol.* 42:165–173, 1983.
83 BASMAJIAN, J.V. The distribution of valves in the femoral, external iliac and common iliac veins and their relationship to varicose veins. *Surg. Gynecol. Obstet.* 95:537–542, 1952.
84* BASMAJIAN, J.V. The marginal anastomoses of the arteries to the large intestine. *Surg. Gynecol. Obstet.* 99:614–616, 1954.
85 BASMAJIAN, J.V. The main arteries of the large intestine. *Surg. Gynecol. Obstet.* 101:585–591, 1955.
86 BASTIEN. Anomalie de la veine azygos droite. *Soc. Biol. Comptes Rendus des Séances et Memoires* 11:204, 1859.
87 BASU, B.N. Anatomical note. Persistent "left superior vena cava," "left duct of Cuvier" and left horn of the sinus venosus. *J. Anat.* 66:268–270, 1932.
88* BATTEUR, M. Un cas exceptionnel de variation de l'artère obturatrice. *Assoc. Anatomistes Comptes Rendus* 38:116, 1951.
89* BATTS, M. JR. The pulmonary artery arising from the aorta. *J. Thoracic Surg.* 8:565–569, 1939.
90 BATUJEFF, N. Eine seltene Arterienanomalie (Ursprung der A. basilaris aus der A. carotis interna). *Anat. Anz.* 4:282–285, 1889.
91* BAUMEL, J.J. and D.Y. BEARD. The accessory meningeal artery of man. *J. Anat.* 95:386–402, 1961.
92 BEAN, R.B. Observations on a study of the subclavian artery in man. *Johns Hopkins Hosp. Bull.* 15:203–205, 1904.
93* BEAN, R.B. A composite study of the subclavian artery in man. *Am. J. Anat.* 4:303–328, 1905.
94* BEATTIE, J. The importance of anomalies of the superior vena cava in man. *Can. Med. Assoc. J.* 25:281–284, 1931.
95 BECK, C.S. Congenital deficiency of the pericardium. The function of the pericardium. *Arch. Surg.* 22:282–288, 1931.
96* BECKER, F.F. A singular left sided inferior vena cava. *Anat. Rec.* 143:117–120, 1962.
97 BEDFORD, D.E. and J. PARKINSON. Right-sided aortic arch (situs inversus arcus aortae). *Br. J. Radiol.* 9:776–798, 1936.
98* BEEVOR, C.E. Cerebral blood supply. *Brain* 30:403–425, 1908.
99* BELL, R.H., SWIGART, L.L. and B.J. ANSON. The relation of the vertebral artery to the cervical vertebrae. Based on a study of 200 specimens. *Q. Bull. Northwestern University Medical School* 24:184–185, 1950.
100* BENTON, R.S. and W.B. COTTER. A hitherto undocumented variation of the inferior mesenteric artery in man. *Anat. Rec.* 145:171–173, 1963.
101* BEREZA, S. A case of aberrant subclavian artery and other neurovascular anomalies in the neck. *Folia Morphol.* 25:589–599, 1966.
102* BERGMANN, L. Über einige seltene Anomalien der Arteria coeliaca. *Z. Anat. Entwicklunggesch.* 101:525–533, 1933.
103 BERRY, R. and J. ANDERSON. A case of nonunion of the verte-

brales with consequent abnormal origin of the basilaris. *Anat. Anz.* 35:54–65, 1910.

104 BERRY, R. and H. NEWTON. A study of the superficial veins of the superior extremity in 300 living subjects. *Anat. Anz.* 33:591–602, 1908.

105 BERTACCHINI, P. Un cas d'anomalie artérielle du bras et d'apophyse susépitrochléenne de l'humérus associés. *Arch. Ital. Biol.* 26:159, 1896.

106 BERTELLI, R. Il sistema arterioso degli arti studiato con metodo radiografico in rapporto alla cinetica. *Arch. Ital. Anat. Embriol.* 36:215–287, 289–343, 1936.

107 BEVIER, G. An anomalous origin of the subclavian artery. *Anat. Rec.* 9:785–789, 1915.

108 BHARGAVA, I. Anomalous branching of axillary artery. *J. Anat. Soc. India* 5:78–80, 1956.

109 BIANCHI, L. Considerazioni sopra un caso di arteria brachio-ulnare superficiale. *Anat. Anz.* 94:73–88, 1943.

110 BIEDERMANN, F. Der rechtsseitige Aortenbogen im Röntgenbild. *Fortschr. Röntgenstr.* 43:168–187, 1931.

111 BIEN, G. Eine seltene Varietät der Arteria maxillaris interna. *Anat. Anz.* 30:421–426, 1907.

112 BIERMANN, H. Die chirurgische Bedeutung der Lagevariationen der Arteria maxillaris. *Anat. Anz.* 94:289–309, 1943.

113 BILLET, H. Suite de l'étude des rapports des branches collatérales de la sousclavière avec les troncs primaires du plexus brachial. *Assoc. Anatomistes Comptes Rendus* 32:49–52, 1937.

114 BIRMINGHAM, A. Extreme anomaly of the heart and great vessels. *J. Anat. Physiol.* 27:139–150, 1893.

115* BLACKBURN, J.W. Anomalies of the encephalic arteries among the insane. A study of the arteries at the base of the encephalon in two hundred and twenty consecutive cases of mental disease, with special reference to anomalies of the circle of Willis. *J. Comp. Neurol.* 17:493–509, 1907.

116 BLACKFORD, L.M. Coarctation of the aorta. *Arch. Int. Med.* 41:702–735, 1928.

117 BLACKFORD, L.M., DAVENPORT, T.F. and R.H. BAYLEY. Right aortic arch. Clinical report of a case with associated anomalies. *Am. J. Dis. Child.* 44:823–844, 1932.

118 BLACKWELL-MORISON, A. Cardiac malformation with vascular anomalies. *J. Anat.* 62:227–231, 1928.

119 BLAIR, C.B. JR., NANDY, K. and G.H. BOURNE. Vascular anomalies of the face and neck. *Anat. Rec.* 144:251–257, 1962.

120* BLAIR, C.B. JR. and K. NANDY. Persistence of the axis artery of the lower limb. *Anat. Rec.* 152:161–172, 1965.

121 BLAIR, D.M., DAVIES, F. and W. MCKISSOCK. The etiology of the vascular symptoms of cervical rib. *Br. J. Surg.* 22:406–414, 1934–35.

122* BLAKE, H.A., MANION, W.C. and F.C. SPENCER. Atresia or absence of the aortic isthmus. *J. Thorac. Cardiovasc. Surg.* 43:607–614, 1962.

123 BLALOCK, A. Surgical procedures employed and anatomical variations encountered in the treatment of congenital pulmonic stenosis. *Surg. Gynecol. Obstet.* 87:385–409, 1948.

124 BLAND, E.F., WHITE, P.D. and J. GARLAND. Congenital anomalies of the coronary arteries. Report of an unusual case associated with cardiac hypertrophy. *Am. Heart J.* 8:787–801, 1933.

125* BLASINGAME, F.J.L. and C.H. BURGE. A case of left postrenal inferior vena cava without transposition of the viscera. *Anat. Rec.* 69:465–470, 1937.

126 BLEICHER, M. and P. WEBER. De la répartition des valvules veineuses dans la saphène interne. *Assoc. Anatomistes Comptes Rendus* 27:60–65, 1932.

127 BLINCOE, H., LOWANCE, M.I. and J. VENABLE. A double aortic arch in man. *Anat. Rec.* 66:505–517, 1936.

128 BLODA, E., SIEROCIŃSKI, W. and A. KLING. Variation of the arteria profunda femoris in man. *Folia Morphol.* 41:123–131, 1982.

129* BLUM, L., MEDL, T. and E.B.C. KEEFER. Aortic homograft substitution for the postrenal inferior vena cava. *Arch. Surg.* 72:567–570, 1956.

130 BLUME, W. Eine seltene Varietät im Bereich der Arteria tibialis anterior. *Anat. Anz.* 66:140–141, 1928–29.

131 BOCHDALEK, JR. Anomaler Verlauf der Kranzarterien des Herzens. *Arch. Pathol. Anat. Physiol. Klin. Med.* 41:260, 1867.

132 BOCHDALEK, JR. Einsenkung des Lig. arteriosum (Ductus arteriosus Botalli) in die linke Arteria subclavia. *Arch. Pathol. Anat. Physiol. Klin. Med.* 41:259–260, 1867.

133* BONNEY, C.W. Studies in the anatomy of the kidney. *Am. J. Urol.* 8:607–615, 1912.

134* BORELL, U. and I. FERNSTRÖM. Uterine artery adnexal branches. Arteriography study in human subjects. *Acta Radiol.* 40:561–582, 1953.

135 BORST. Cor triatratum. *Zentralbl. Allg. Pathol. Pathol. Anat.* 16:812, 1905.

136* BOURRET, P. Les artères de l'appendice. *Assoc. Anatomistes Comptes Rendus* 38:221–232, 1951.

137 BOVERO, A. Morphologie des artères de la glande mammaire. *Arch. Ital. Biol.* 48:486–487, 1907.

138 BOXALL, T.A., SMART, P.J.G. and J.D. GRIFFITHS. The blood supply of the distal segment of the rectum in anterior resection. *Br. J. Surg.* 50:399–404, 1963.

139 BOYD, G.I. Abnormality of subclavian artery associated with presence of the scalenus minimus. *J. Anat.* 68:280–281, 1934.

140* BOYD, G.I. Abnormal bifurcation of aorta. *J. Anat.* 71:417–418, 1936–37.

141 BOYD, G.I. Anomalous heart. *J. Anat.* 72:299–300, 1937–38.

142 BOYD, J.D. Absence of the right common carotid artery. *J. Anat.* 68:551–557, 1933–34.

143* BOYDEN, E.A. Description of a horseshoe kidney associated with left inferior vena cava and disc-shaped suprarenal glands, together with a note on the occurrence of horseshoe kidneys in human embryo. *Anat. Rec.* 51:187–211, 1932.

144* BOYDEN, E.A. The anatomical hazards of lingulectomy. *Surgery* 20:828–829, 1946.

145* BOYDEN, E.A. The nomenclature of the bronchopulmonary segments and their blood supply. *Dis. Chest* 39:1–6, 1961.

146* BOYDEN, E.A. and J.G. SCANNELL. An analysis of variations in the bronchovascular pattern of the right upper lobe of fifty lungs. *Am. J. Anat.* 82:27–73, 1948.

147* BOYDEN, E.A. and C.J. HAMRE. An analysis of variations in the broncho-vascular patterns of the middle lobe in fifty dissected and twenty injected lungs. *J. Thorac. Surg.* 21:172–188, 1951.

148* BOYER, C.C. Anomalous inferior vena cava formed from supracardinal complexes. *Anat. Rec.* 117:829–839, 1953.

149 BRADLEY, S.M. Note on a peculiar origin of the right subclavian artery. *J. Anat. Physiol.* 5:341–342, 1871.

150 BRADLEY, S.M. Tricoelian human heart. *Br. Med. J.* 1:33–34, 1873.

151 BRAITHWAITE, J.L. The arterial supply of the urinary bladder. *J. Anat.* 85:413P, 1951.

152* BRAITHWAITE, J.L. The arterial supply of the male urinary bladder. *Br. J. Urol.* 24:64–71, 1952.

153* BRAITHWAITE, J.L. Variations in origin of the parietal branches of the internal iliac artery. *J. Anat.* 86:423–430, 1952.

154 BRANTIGAN, O.C. Anomalies of the pulmonary veins. Their surgical significance. *Surg. Gynecol. Obstet.* 84:653–658, 1947.

155 BREMER, J.L. The origin of the renal artery in mammals and its anomalies. *Am. J. Anat.* 18:179–200, 1915.

156 BROCKIS, J.G. and D.B. MOFFATT. The intrinsic blood vessels of the pelvic colon. *J. Anat.* 92:52–56, 1958.

157* BRÖDEL, M. The intrinsic blood vessels of the kidney and their significance in nephrotomy. *Johns Hopkins Hosp. Bull.* 12:10–13, 1901.

158* BRODY, H. Drainage of the pulmonary veins into the right side of the heart. *Arch. Pathol.* 33:221–240, 1942.

159 BROMAN, I. Über die Entwicklung und „Wanderung" der Zweige der Aorta abdominalis beim Menschen. *Anat. Hefte.* 36:407–549, 1908.

160 BROOKS, H.ST.J. Two cases of an abnormal coronary artery of the heart arising from the pulmonary artery: With some remarks upon the effect of this anomaly in producing cirsoid dilatation of the vessels. *J. Anat. Physiol.* 20:26–29, 1886.

161* BROWN, E.Z. Variations in origin and course in hepatic artery. *South. Med. J.* 28:599–603, 1935.

162* BROWN, J.M. Abnormal cystic artery. *J. Anat. Physiol.* 14:373, 1879.

163 BROWN, S. The external jugular vein in American Whites and Negroes. *Am. J. Phys. Anthropol.* 28:213–226, 1941.

164* BROWNE, E.Z. Variations in origin and course of the hepatic artery and its branches. *Surgery* 8:424–445, 1940.

165* BROWNING, H. The confluence of dural venous sinuses. *Am. J. Anat.* 92:307–329, 1953.

166 BRÜNINGHAUSEN, A. Über zwei Fälle von Anomalie der Pulmonalarterienklappen. *Arch. Pathol. Anat. Physiol. Klin. Med.* 82:200–202, 1880.

167 BRUNIN, A. Sur un cas d'amastomose retro-pancreatique entre l'artère hepatique et l'artère mésentérique supérieure. *Anat. Anz.* 27:90–96, 1905.

168. BRUNO, G. Ricerche col metodo statistico sella morfologia delle vene della faccia e del collo nell'Uomo. *Arch. Ital. Anat. Embriol.* 21:273–293, 1924.
169. BRUWER, A., CLAGETT, O.T. and J.R. MCDONALD. Anomalous arteries to the lung associated with congenital pulmonary abnormality. *J. Thoracic Surg.* 19:957–972, 1950.
170. BUCHANAN, G. Malformation of the heart. Cyanosis. *Trans. Pathol. Soc. Lond.* 8:149–150, 1856.
171*. BUDDE. Beiträge zur Kenntnis der Topographie der normalen A. hepatica und ihrer Varietäten sowie der Blutversorgung der Leber. *Dtsch. Z. Chir.* 86:18–40, 1906.
172. BUSCH, W. Die arterielle Gefäßversorgung der Nebennieren. (Zugleich ein Beitrag zur Anatomie der Nierenarterien). *Arch. Pathol. Anat. Physiol. Klin. Med.* 324:688–699, 1954.
173. BUSCHENDORFF, C. Beitrag zur Kenntnis der Persistenz eines doppelten Aortenbogens und seine klinische Bedeutung. *Zentralbl. Herz Gefäßkr.* 9:165–171, 1917.
174*. BUSSE, O. Aneurysmen und Bildungsfehler der Arteria communicans anterior. *Arch. Pathol. Anat. Physiol. Klin. Med.* 229:184–206, 1921.
175. BUTLER, H. Abnormal disposition of the pulmonary veins. *J. Anat.* 86:485–486, 1952.
176*. BUTLER, H. and K. BALANKURA. Preaortic thoracic duct and azygos veins. *Anat. Rec.* 113:409–419, 1952.
177. BYRON, F. Ectopia cordis. Report of a case with attempted operative correction. *J. Thoracic Surg.* 17:717–722, 1948.
178. CABANIÉ and SOUTOUL. Essai de systématisation du mode de ramescence de l'artère mésentérique inférieure. *Assoc. Anatomistes Comptes Rendus* 40:930–936, 1953.
179. CABANIÉ and SOUTOUL. L'arcade artérielle intermésentérique (à propos de 8 cas). *Assoc. Anatomistes Comptes Rendus* 40:937–950, 1953.
180*. CABANIÉ, H., SENTENAC and J. DEJUSSIEU. A propos de 9 cas de triplicité de l'artère hépatique. *Assoc. Anatomistes Comptes Rendus* 45:212–215, 1958.
181. CABANIÉ, H., SENTENAC and J. DEJUSSIEU. Note sur le mode d'origine de l'artère cystique. *Assoc. Anatomistes Comptes Rendus* 45:216–218, 1958.
182*. CABANIÉ, H., PERRUCHIO, P., SOUTOUL, J. and J. DEJUSSIEU. Sur un cas d'absence du segment sous-rénal de la veine cave inférieure. *Assoc. Anatomistes Comptes Rendus* 45:219–225, 1958.
183. CADARSO, J.J. and J.J. GOYANES. A rare case of tortuosity of the internal carotid artery. *J. Anat.* 60:119–120, 1926.
184*. CAIRNEY, J. Tortuosity of the cervical segment of the internal carotid artery. *J. Anat.* 59:87–96, 1925.
185*. CAIRNEY, J. The anomalous right subclavian artery considered in the light of recent findings in arterial development; with a note on two cases of an unusual relation of the innominate artery to the trachea. *J. Anat.* 59:265–296, 1925.
186. CALAS, F., MARTIN, R., BOUCHET, Y. and D. POLLIAK. Les artères de la tête du pancréas. *Assoc. Anatomistes Comptes Rendus* 42:362–367, 1955.
187. CALAS, F., COUPPIE, G., MARTIN, R. and Y. BOUCHET. Etude des affluents et de la formation de la veine porte. *Assoc. Anatomistes Comptes Rendus* 45:254–271, 1958.
188*. CALORI, L. Sulle comunicazioni della vena porta con le vene generali del corpo. *Mem. R. Accad. Sci. Istituto di Bologna S. 4* 9:461–469, 1888.
189*. CALORI, L. Sopra un caso d'inversione dei condotti toracici accompagnato da inversa origine dell'arteria succlavia destra e sulla genesi delle due anomalie. *Mem. R. Accad. Sci. Istituto di Bologna S. 5* 1:189–196, 1890.
190. CAMERON, A.H.F. Note of a case of peculiar malformation of the heart and great vessels. *J. Anat. Physiol.* 5:339–341, 1871.
191. CAMERON, J. Persistence of the left posterior cardinal vein. *J. Anat. Physiol.* 45:416–419, 1911.
192. CAMERON, J. A specimen showing complete remains of the left superior vena cava; with a description of a venule, which represents the "lost" extrapericardiac portion of the left duct of Cuvier. *J. Anat. Physiol.* 49:140–147, 1915.
193. CAMPBELL, E.H. The cavernous sinus. Anatomical and clinical considerations. *Annales of Otol. Rhinol. Laryngol.* 42:51–63, 1933.
194+. CAMPBELL, J. The coronary arteries of the heart. Thesis. Queen's University, Belfast, 1928.
195. CAMPBELL, K. Accessory segment in the pulmonary valve. *J. Anat. Physiol.* 30:347–348, 1896.
196*. CAPITAN, L. Un cas d'inversion du coeur exclusivement. *Soc. Biol. Comptes Rendus Hebdomadaires des Séances et Mémoires* 50:1104–1105, 1898.
197. CARMEL, A.G. The tortuous splenic artery. *Anat. Rec.* 29:352, 1925.
198. CARO, C., LERMANDA, V.C. and H.A. LYONS. Aortic origin of the right pulmonary artery. *Br. Heart J.* 19:345–352, 1957.
199. CARPENTER, M.B., NOBACK, C.R. and M.L. MOSS. The anterior choroidal artery. Its origin, course, distribution and variations. *Arch. Neurol. Psychiatry* 71:714–722, 1954.
200*. CARPENTIER and BERTAUX. Anomalie des systèms de la veine cave inférieure et des veines azygos, observée chez un homme adulte. *Soc. Biol. Comptes Rendus Hebdomadaires de Séances et Memoires* 40:339–340, 1888.
201. CARSON, M.J. and J. GOODFRIEND. Constricting vascular rings. *J. Pediatr.* 34:155–165, 1949.
202. CARSON, W.J. Anomalous renal vessels and their surgical significance. *West. J. Surg. Obstet. Gynecol.* 40:597–601, 1932.
203. CARVER, E. Irregularities in the arteries and muscles of an idiot. *J. Anat. Physiol.* 3:257–261, 1869.
204. CASALI, R. Di tre rami sopranumerari non ancora descritti dell' A. succlavia. Arteria del muscolo succlavio, Arteria della cupola pleurica, Arteria esofogea superiore. *Arch. Ital. Anat. Embriol.* 10:122–128, 1911.
205. CASTRO-RODRIGUES, M. Contribution a l'etude de l'artère sylvienne. *Assoc. Anatomistes Comptes Rendus* 42:375–381, 1955.
206. CATRITSIS, E. and M. JUSTER. Sur la position et l'orientation de la crosse aortique. *Assoc. Anatomistes Comptes Rendus* 37:44–50, 1950.
207*. CAULDWELL, W.E. and H.C. MOLOY. Anatomical variations in the female pelvis and their effect on labor with a suggested clarification. *Am. J. Obstet. Gynecol.* 26:479–504, 1933.
208. CAULDWELL, E.W. and B.J. ANSON. The visceral branches of the abdominal aorta: Topographical relationships. *Am. J. Anat.* 73:27–57, 1943.
209*. CAULDWELL, E.W., SIEKERT, R.G., LININGER, R.E. and B.J. ANSON. The bronchial arteries. An anatomic study of 150 human cadavers. *Surg. Gynecol. Obstet.* 86:395–412, 1948.
210. CAYOTTE, J. and P. HAHN. A propos de trois observations de veines rénales gauches rétro-aortiques. *Assoc. Anatomistes Comptes Rendus* 39:55–58, 1952.
211. CAYOTTE, J. and P. QUEREUX. Ramescence anormale des artères axillaire et fémorale. *Assoc. Anatomistes Comptes Rendus* 40:968–972, 1953.
212. CAYOTTE, J. and A. BRULÉ. La vascularisation du pole supérieur de rein. *Assoc. Anatomistes Comptes Rendus* 41:1019–1027, 1954.
213. CELLI, E. Les trous anormaux vasculaires ef nerveux alvéolo-patatins. *Arch. Ital. Biol.* 54:282, 1915.
214*. CELORIA, G.C. and R.B. PATTON. Congenital absence of the aortic arch. *Am. Heart J.* 58:407–413, 1959.
215*. CHADZYPANAGIOTIS, D. and L. AMERSKI. A rare case of the celiac trunk anomaly. *Folia Morphol.* 37:401–405, 1978.
216. CHAFFEY, W.C. Congenital malformation of heart, with persistent left duct of Cuvier (card specimen). *Trans. Pathol. Soc. Lond.* 36:175–176, 1885.
217. CHALIER, A. Anomalies des artèrer rénales et du bassinet. *J. Urol. (Paris)* 1:103–104, 1912.
218. CHAMPNEYS, F. Communication between the external iliac and portal veins. *J. Anat. Physiol.* 6:417–419, 1872.
219. CHANAMIRJAN, A. Versuch eines systematischen Studiums der Variationen der Wirbelarterie. *Anat. Anz.* 68:163–178, 1929–30.
220. CHANDER, S. and I. JIT. Single coronary artery. *J. Anat. Soc. India.* 6:116–118, 1957.
221. CHANDLER, S.B. and P.H. KREUSCHER. A study of the blood supply of the ligamentum teres and its relation to the circulation of the head of the femur. *J. Bone Surg.* 14:834–846, 1932.
222. CHANDLER, S.B. and C.E. DEREZINSKI. The variations of the middle meningeal artery within the middle cranial fossa. *Anat. Rec.* 62:309–319, 1935.
223. CHANG, S.C., CHEN, E.Y. and W.L. SUNG. Observations on the deep and circumflex femoral arteries in Chinese. *Acta Anat. Sinica* 8:282–292, 1963.
224*. CHANMUGAM, P.K. Note on an unusual ophthalmic artery associated with other abnormalities. *J. Anat.* 70:580–582, 1935–36.
225*. CHARLES, C.M. On the arrangement of the superficial veins of the cubital fossa in American white and American negro males. *Anat. Rec.* 54:9–14, 1932.
226*. CHARLES, C.M., FINLEY, T.L., BAIRD, R.D. and J.S. COPE. On the termination of the circumflex veins of the thigh. *Anat. Rec.* 46:125–132, 1930.

227* CHARLES, C.M., PENN, L., HOLDEN, H.F., MILLER, R.A. and E.B. ALVIS. The origin of the deep brachial artery in American white and in American negro males. *Anat. Rec.* 50:299–302, 1931.

228 CHARLES, J.J. Notes on some cases of abnormal arrangement of arteries of the upper extremity. *J. Anat. Physiol.* 7:300–301, 1873.

229 CHARLES, J.J. Abnormalities in the arteries of the upper extremity. *J. Anat. Physiol.* 9:180–181, 1875.

230 CHARLES, J.J. Notes of a case of persistent left superior vena cava, the right superior vena cava being in great part a fibrous cord. *J. Anat. Physiol.* 23:649–650, 1889.

231 CHARLES, J.J. A case of absence of the radial artery. *J. Anat. Physiol.* 28:449–450, 1894.

232 CHARRIN and LENOIR. Maladie bleue; communication interventriculaire; absence d'artère pulmonaire; quatre artères bronchiques. *Soc. Biol. Comptes Rendus Hebdomadaires des Séances et Memoires* 42:598–599, 1890.

233 CHASE, M.R. Congenital deficiency of the pericardium. *J. Anat. Physiol.* 50:299–302, 1916.

234 CHAVATZAS, D. Revision of the incidence of congenital absence of dorsalis pedis artery by an ultrasonic technique. *Anat. Rec.* 178:289–290, 1974.

235* CHÉRIÈ-LIGNIÈRE, M. Le vene del collo nell'uomo. *Int. Monatsschr. Anat. Physiol.* 31:63–115, 1914–15.

236 CHIENE, J. On a case in which the innominate veins opened separately into the right auricle, and in which the intestines were misplaced; with remarks on the development of the parts. *J. Anat. Physiol.* 2:13–18, 1868.

237 CHIENE, J. Complete obliteration of celiac and mesenteric arteries; the viscera receiving their blood supply through extra-peritoneal system of vessels. *J. Anat. Physiol.* 3:65–72, 1869.

238 CHINN, J. and M.A. CHINN. Report of an accessory coronary artery arising from the pulmonary artery. *Anat. Rec.* 139:23–28, 1961.

239 CHIRIAC, V., FRASIN, G., CHIRIAC, R. and W. ZILLMANN. Varietät des Sinus durae matris der oberen aboralen Gruppe. *Anat. Anz.* 130:526–527, 1972.

240* CHOUKÉ, K.S. A case of bilateral superior vena cava in an adult. *Anat. Rec.* 74:151–157, 1939.

241* CHRZANOWSKA, G. and J. MIERŻWA. Variation of the cerebral arterial circle (of Willis) in man. *Folia Morphol.* 38:365–373, 1979.

242 CIVALLERI, A. Note sur un cas d'oblitération de la veine cave inférieure avec circulation collaterale donnée par les azygos. *Arch. Ital. Biol.* 32:459, 1899.

243* CLAUSEN, H.J. An unusual variation in origin of hepatic and splenic arteries. *Anat. Rec.* 123:335–340, 1955.

244 CLEGG, E.J. The arterial supply of the human prostate and seminal vesicles. *J. Anat.* 89:209–216, 1955.

245 COATES, A.E. A note on the superior petrosal sinus and its relation to the sensory root of the trigeminal nerve. *J. Anat.* 68:428, 1933–34.

246 COBB, S.T. The cerebral circulation; the question of "end-arteries" of brain and mechanism of infarction. *Arch. Neurol. Psychiatry* 25:273–280, 1931.

247 COBEY, J.F. An anomalous right subclavian artery. *Anat. Rec.* 8:15–19, 1914.

248 COLE, F.H., ALLEY, F.H. and R.S. JONES. Aberrant systemic arteries to the lower lung. *Surg. Gynecol. Obstet.* 93:589–596, 1951.

249* COLEMAN, S.S. and B.J. ANSON. Arterial patterns in the hand based upon a study of 650 specimens. *Surg. Gynecol. Obstet.* 113:409–424, 1961.

250 COMPERE, D.E. and H.F. FORSYTH. Anomalous pulmonary veins. *J. Thoracic Surg.* 13:63–66, 1944.

251 CONN, L.C., CALDER, J., MACGREGOR, J.W. and R.F. SHANER. Report of a case in which all the pulmonary veins from both lungs drain into the superior vena cava. *Anat. Rec.* 83:335–340, 1942.

252 COOKE, F.N., EVANS, J.M., KISTIN, A.D. and B. BLADES. An anomaly of the pulmonary veins. *J. Thoracic Surg.* 21:452–459, 1951.

253 COOPER, C.M. A pericardial sac in which the large azygos vein pierced the sac before opening into the superior vena cava. *J. Anat. Physiol.* 34:426, 1900.

254* COPLEMANN, B. Anomalous right subclavian artery. *Am. J. Roentgenol.* 54:270–275, 1945.

255 CORDIER, G. and C. CABROL. Les veines du poumon gauche. Caractères généraux et systématisation. *Assoc. Anatomistes Comptes Rendus* 40:627–638, 1953.

256 CORDIER, P. and L. PARDOEN. Deux variétés d'origine de l'artère obturatrice. *Soc. Biol. Comptes Rendus Hebdomadaires des Séances et Mémoires* 84:896–897, 1921.

257* CORDIER, P., DEVOS, L. and A. DELCROIX. Essai de classification des variations du système azygos intrathoracique (interprétation embryologique). *Assoc. Anatomistes Comptes Rendus* 33:100–118, 1938.

258* CORDIER, P., DEVOS, L. and A. DELCROIX. L'anastomose artérielle uteroovarienne. *Assoc. Anatomistes Comptes Rendus* 1939, pp. 48–79.

259 CORDS, E. Ein Fall von teilweiser Erhaltung der V. cava sup. sin. zusammen mit einer Anomalie der Vv. pulmonales. *Anat. Anz.* 54:491–495, 1921.

260 CORREIA, M. Contribution a l'etude de la circulation artérielle du coeur. *Assoc. Anatomistes Comptes Rendus* 28:200–204, 1933.

261 CORSY, F. Anomalie dans de système veine cave inférieure. Faut-il maintenir le schéma classique du développement de ce système? *Soc. Biol. Comptes Rendus Hebdomadaires des Séances et Mémoires* 83:365–366, 1920.

262 COSTA, A.J. A proposito de una anomalia arterial. La arteria subclavia derecha saliendo del cayado aortico en su porción final. *Rev. Assoc. Méd. Argentina* 60:740–743, 1946. Cited in *Excerpta Medica*, Sec. 1, Vol. 2, abstract 525, p. 250, 1948.

263 COTRIM, E. Os diâmetros aórticos. *Ann. Paulistas Med. Cir.* 37:809–839, 1939.

264 COUINAUD, C. Distribution de l'artère hépatique dans le foie. *Acta Anat.* 22:49–81, 1954.

265 CRELIN, E.S. JR. An unusual anomalous blood vessel connecting the renal and internal spermatic arteries. *Anat. Rec.* 102:205–211, 1948.

266 CROCK, H.V. A revision of the anatomy of the arteries supplying the upper end of the human femur. *J. Anat.* 99:77–88, 1965.

267* CRYSTAL, D.K., EDMONS, H.W. and P.F. BETZOLD. Symmetrical double aortic arch. *West. J. Surg.* 55:389–392, 1947.

268* CURNOW, J. Two instances of irregular ophthalmic and middle meningeal arteries. *J. Anat. Physiol.* 8:155–156, 1874.

269* CURNOW. Right subclavian artery passing in front of the scalenus anticus, Proc. Anat. Soc. of Great Britain and Ireland. *J. Anat. Physiol.* 22:v–vi, 1888.

270* CURNOW, J. Double aortic arch enclosing trachea and oesophagus. *Trans. Pathol. Soc. Lond.* 26:33–38, 1875.

271* CURTIS, A.H., ANSON, B.J., ASHLEY, F.L. and T. JONES. The blood vessels of the female pelvis in relation to gynecological surgery. *Surg. Gynecol. Obstet.* 75:421–423, 1942.

272* CURTIS, J.B. Cerebral angiography. *Br. J. Surg.* 38:295–331, 1951.

273* CZERWIŃSKI, F. Variability of the course of external carotid artery and its rami in man in the light of anatomical and radiological studies. *Folia Morphol.* 40:447–453, 1981.

274 DAHM, M. Zur Eindellung der Speiseröhre bei links entspringender Arteria subclavia dextra. *Fortschr. Röntgenstr.* 62:108–114, 1940.

275 DALL'ACQUA, U. L'artère temporale superficielle de l'homme. *Arch. Ital. Biol.* 36:347–348, 1901.

276 DALL'ACQUA, U. and A. MENEGHETTI. Ricerche di anatomia comparata sulle arteria della faccia. *Arch. Ital. Anat. Embriol.* 4:161–182, 306–365, 1905.

277 DALL'ACQUA, U. and A. MENEGHETTI. Recherches d'anatomie comparée sur les artères de la face. *Arch. Ital. Biol.* 46:302–304, 1906.

278 DALL'ACQUA, U. and A. MENEGHETTI. Recherches d'anatomie comparée sur les artères de la face. *Arch. Ital. Biol.* 47:484–486, 1907.

279 DANIEL, O. and R. SHACKMAN. The blood supply of the human ureter in relation to ureterocolic anastomosis. *Br. J. Urol.* 24:334–343, 1952.

280* DASELER, E.H., ANSON, B.J., REIMANN, A.F. and L.E. BEATON. The saphenous venous tributaries and related structures in relation to the technique of high ligation. *Surg. Gynecol. Obstet.* 82:53–63, 1946.

281* DASELER, E.H., ANSON, B.J., HAMBLEY, W.C. and A.F. REIMANN. The cystic artery and constitutents of the hepatic pedicle. A study of 500 specimens. *Surg. Gynecol. Obstet.* 85:47–63, 1947.

282* DASELER, E.H. and B.J. ANSON. Surgical anatomy of the subclavian artery and its branches. *Surg. Gynecol. Obstet.* 108:149–174, 1959.

283 DASER, P. Über eine seltene Lage-Anomalie der Vena anonyma sinistra. *Anat. Anz.* 20:553–555, 1902.

284 DAVIDSON, D.G. Abnormal arrangement of the veins about the popliteal space. *J. Anat. Physiol.* 25:154, 1891.

285 DAVIS, F.A.G. Unusual course of vena hemiazygos. *J. Anat.* 70:321, 1935–36.
286* DAVIS, H.K. A statistical study of the thoracic duct in man. *Am. J. Anat.* 17:211–244, 1915.
287* DAVIS, R.A., MILLOY, F.J. and B.J. ANSON. Lumbar, renal, and associated parietal and visceral veins based upon a study of 100 specimens. *Surg. Gynecol. Obstet.* 107:1–22, 1958.
288 DAWSON, A.B. and J.H. RIES. An anomalous arterial supply to suprarenal, kidney and ovary. *Anat. Rec.* 23:161–168, 1922.
289 DECKER, K. and E. HIPP. Der basale Gefäßkranz. Morphologie und Angiographie. *Anat. Anz.* 105:100–116, 1958.
290 DEES, J.E. Anomalous relationship between ureter and external iliac artery. *J. Urol.* 44:207–215, 1940.
291* DE GARIS, C.F. Patterns of branching of the subclavian artery in White and Negro stocks. *Am. J. Phys. Anthropol.* 7:95–107, 1924.
292* DE GARIS, C.F. Aortic axillary collaterals and the pattern of arm arteries in anomalous right subclavian artery. *Am. J. Anat.* 51:189–213, 1932.
293 DE GARIS, C.F. and W.B. SWARTLEY. The axillary artery in white and negro stocks. *Am. J. Anat.* 41:353–397, 1928.
294 DE GARIS, C.F., BLACK, I.H. and E.A. RIEMENSCHNEIDER. Patterns of the aortic arch in American White and Negro stocks, with comparative notes on certain other mammals. *J. Anat.* 67:599–619, 1933.
295 DELBUONO, M.S. and A. MELIK. Ein Fall von Hypoplasie der linken A. pulmonalis mit Schrumpfung der linken Lunge. *Radiol. Clin.* 28:150–155, 1959.
296* DELITZIN, S. Über eine Varietät des Verlaufes der Arteria thyreoidea superior. *Arch. Anat. Physiol. Wissen. Med.* 1892, pp. 105–106.
297 DELITZIN, S. Beobachtungen über die vierte Halbmondklappe in der Arteria pulmonalis. *Arch. Anat. Physiol. Wissen. Med.* 1892, pp. 107–112.
298 DELITZIN, S. Über eine complicierte Anomalie im Gebiete der Arteria obturatoria und epigastrica inferior. *Arch. Anat. Physiol. Wissen. Med.* 1896, pp. 413–422.
299 DELITZIN, S. Ein Fall von Durchbohrung des M. scalenus anterior durch den Truncus thyreo-cervicalis. *Arch. Anat. Physiol. Wissen. Med.* 1899, pp. 124–129.
300 DELITZIN, S.B. Ein Fall von Inselbildung an der Vena iliaca externa dextra. *Int. Monatsschr. Anat. Physiol.* 19:355–358, 1901.
301 DELMAS, A. and E. CATRITSIS. Le poids du coeur. *Assoc. Anatomistes Comptes Rendus* 36:176–181, 1949.
302* DELMAS, A. and B. PERTUISET. Les veines du cortex cérébral. Distribution générale, variations types veineux de distribution. *Assoc. Anatomistes Comptes Rendus* 36:185–193, 1949.
303* DELMAS, A. and B. PERTUISET. Les veines du cortex cérébral. 2) Topographic régionale. *Assoc. Anatomistes Comptes Rendus* 36:194–203, 1949.
304* DELMAS, A. and G. BERTRAND. La veine retro-Rolandique. *Assoc. Anatomistes Comptes Rendus* 37:116–122, 1950.
305* DELMAS, A. and J. CHIFFLET. Le pressoir d'Herophile. *Assoc. Anatomistes Comptes Rendus* 37:123–131, 1950.
306 DELMAS, A. and G. BERTRAND. Les veines frontales externes ou préfrontales. *Assoc. Anatomistes Comptes Rendus* 38:378–381, 1951.
307 DELMAS, A. and H. KOWSARIAN. Lacunes veineuses de la tente du cervelet. *Assoc. Anatomistes Comptes Rendus* 38:382–387, 1951.
308 DELMAS, A. and A. HEFFEZ. Sur l'abouchement des veines pulmonaires dans l'oreillette gauche (nombre et calibre). *Assoc. Anatomistes Comptes Rendus* 39:777–783, 1952.
309 DEPREUX and FONTAINE. Contribution a l'étude de la vascularisation artérielle du sinus maxillaire. *Assoc. Anatomistes Comptes Rendus* 37:132–141, 1950.
310 DEPREUX, R. and P. HERLEMONT. Les anastomoses entre les systèmes saphène interne et externe. *Assoc. Anatomistes Comptes Rendus* 45:349–354, 1958.
311 D'EVANT, T. Rare anomalie de la veine axillaire. *Arch. Ital. Biol.* 22:155–156, 1895.
312 DÉVÉ, M.F. Le lobule de la veine azygos ou "lobule de Wrisberg." *Bull. Mém. Soc. Anat.* 74:489–514, 1899.
313* DIDIO, L.J.A. The termination of the vena gastrica sinistra in 220 cadavers. *Anat. Rec.* 141:141–144, 1961.
314* DIDIO, L.J.A. and T.W. WAKEFIELD. Coronary arterial predominance or balance on the surface of the human cardiac ventricles. *Anat. Anz.* 137:147–158, 1975.
315* DIETRICH, W. Ursprung der vorderen Kranzarterie aus der Lungenschlagader mit ungewöhnlichen Veräderungen des Herzmuskels und der Gefäßwände. *Arch. Pathol. Anat. Physiol. Klin. Med.* 303:436–456, 1939.
316 DOBBIE, J.W. and T. SYMINGTON. The human adrenal gland with special reference to the vasculature. *J. Endocrinol.* 34:479–489, 1966.
317 DOLGO-SABUROFF, B. Zur Frage über die Lagebeziehungen zwischen der Arteria subclavia und der Scalenusmuskulatur beim Menschen. *Anat. Anz.* 76:97–113, 1933.
318 DONADIO, N. Ein Fall von Verdoppelung der Vena cava superior. *Anat. Anz.* 59:321–327, 1924–25.
319 DORSHE, H.H. VAN and A. DORN. Varietäten im Bereiche der A. coeliaca und an den Nierengefäßen. *Anat. Anz.* 118:297–304, 1966.
320* DOUGLASS, B.E., BAGGENSTROSS, A.H. and W.H. HOLLINSHEAD. The anatomy of the portal vein and its tributaries. *Surg. Gynecol. Obstet.* 91:562–575, 1950.
321 DOUGLASS, R. Anomalous pulmonary vessels. *J. Thoracic Surg.* 17:712–716, 1948.
322 DOUVILLE, E. and W.H. HOLLINSHEAD. The blood supply of the normal renal pelvis. *J. Urol.* 73:906–912, 1955.
323 DRUMMOND, H. The arterial supply of the rectum and pelvic colon. *Br. J. Surg.* 1:677–685, 1913–14.
324 DRY, T.J., EDWARDS, J.E., PARKER, R.L., BURCHELL, H.B., ROGERS, H.M. and A.H. BULBULIAN. Congenital anomalies of the heart and great vessels. Clinico-pathologic study of 132 cases. *Postgrad. Med.* 4:231–263, 327–359, 1948.
325* DSCHAU, F. Eine bisher unbekannte Varietät der Arteria obturatoria und pudenda accessoria mit einer Varietät der A. circumflexa femoris medialis. *Anat. Anz.* 83:25–29, 1936–37.
326 DUBRUEIL, M.A. Anomalie du tronc artériel brachio-céphalique. *Société Biol. Comptes Rendus des Séances et Memoires* 14:189, 1862.
327* DUDZIAK, M., SZOSTAKIEWICZ-SAWICKA, H. and M. URBANOWICZ. Rare variant of venous stems of the retroperitoneal space in man. *Folia Morphol.* 39:171–180, 1980.
328 DUFOUR, RIGAUD and GOUAZÉ. Variations du mode de division du tronc de l'artère hypogastrique en fonction due sexe. *Assoc. Anatomistes Comptes Rendus* 38:449–456, 1951.
329 DUFOUR, M., RIGAUD, CABANIE and GOUAZÉ. Note sur le confluent portal et le mode l'abouchement des veines coronaire, stomachique et gastro-colique. *Assoc. Anatomistes Comptes Rendus* 39:195–200, 1952.
330* DUFOUR, M., RIGAUD, CABANIE and GOUAZÉ. Sur deux dispositions exceptionnelles des artères hépatiques accessoires. *Assoc. Anatomistes Comptes Rendus* 39:201–207, 1952.
331 DUFOUR, RIGAUD, CABANIÉ and LAFFONT. Contribution a l'étude des artères thyroidiennes. *Assoc. Anatomistes Comptes Rendus* 41:713–720, 1954.
332 DUFOUR, M., RIGAUD, A., GOUAZÉ, A. and J. PELLEGRINO. Relations entre le système artériel antérieur et la morphologie des sillon de la face interne du cerveau. *Assoc. Anatomistes Comptes Rendus* 44:258–264, 1957.
333 DURET, H. Revue critique de quelques recherches récentes sur la circulation cérébrale. *L'encephale* 1:7–27, 1910.
334 DURET, M. Recherches anatomiques sur la circulation de l'encephale. *Arch. Physiol. Norm. Pathol.* 1:316–353, 1874.
335 DUROUX, P.E., DUJOL, H. and J. BOSSY. Contribution on l'etude de l'artère sylvienne et de ses territoires vasculaires. *Assoc. Anatomistes Comptes Rendus* 39:135–145, 1952.
336 DUROUX, DUJOL and C. GABRIELLE. A propos des branches de l'artère tibiale antérieure sitot après son passage au-dessus de la membrane interosseuse. *Assoc. Anatomistes Comptes Rendus* 39:151–152, 1952.
337 DUROUX, P.E., DUJOL, H., AVET, J. and C. GABRIELLE. A propos du polygone de Willis. *Assoc. Anatomistes Comptes Rendus* 40:328–338, 1953.
338 DUROZIEZ, P. Mémoire sur la persistance du canal artériel sans autra communication anormale. *Soc. Biol. Comptes Rendus des Séances et Memoires* 14:279–292, 1862.
339 DUTRA, F.R. Anomalies of coronary arteries; report of two cases, with comment on the dynamics of development of the coronary circulation. *Arch. Intern. Med.* 85:955–965, 1950.
340 DUVAL, J.M. and J.P. LOESCHER. L'artére du nerf médian et ses variations. *Assoc. Anatomistes Comptes Rendus* 46:231–236, 1959.
341 DUVAL, J.M., PREVOT, J. and J.P. LOESCHER. Un cas de tronc coeliaco-mésentérique donnant deux artères hépatiques. *Assoc. Anatomistes Comptes Rendus* 46:237–242, 1959.
342 DVORAK, H.J. Anomaly of external iliac artery found during a

herniotomy. *Rozhledy Chirurgii* 26:235–236, 1947. Cited in *Excerpta Medica*, Sec. 1, Vol. 3, abstract 522, 1949.
343 DZIALLAS, P. Über eine bisher unbekannte Variation der Vena spermatica dextra. *Anat. Anz.* 96:348–352, 1947–48.
344 DZIALLAS, P. Über die Klappenverhältnisse der Venae spermaticae des Menschen. *Anat. Anz.* 97:57–63, 1949–50.
345* EATON, P.B. The coeliac axis. *Anat. Rec.* 13:369–374, 1917.
346 EBSTEIN, W. Über einen sehr seltenen Fall von Insufficienz der Valvula tricuspidalis bedingt durch eine angeborene hochgradige Missbildung derselben. *Arch. Anat Physiol.* 1866, pp. 238–258.
347 EDSMAN, G. Accessory vessels of the kidney and their diagnosis in hydronephrosis. *Acta. Radiol.* 42:26–32, 1954.
348* EDWARDS, E.A. Clinical anatomy of lesser variations of the inferior vena cava; and a proposal for classifying the anomalies of this vessel. *Angiology* 2:85–99, 1951.
349 EDWARDS, E.A. The anatomic basis for ischemia localized in certain muscles of the lower limb. *Surg. Gynecol. Obstet.* 97:87–94, 1953.
350* EDWARDS, E.A. The anatomy of collateral circulation. *Surg. Gynecol. Obstet.* 108:183–194, 1958.
351 EDWARDS, E.A. Organization of the small arteries of the hand and digits. *Am. J. Surg.* 99:837–846, 1960.
352* EDWARDS, E.A. and J.D. ROEBUCK. Applied anatomy of the femoral vein and its tributaries. *Surg. Obstet. Gynecol.* 85:547–557, 1947.
353 EDWARDS, F.R., FARQUHAR, H.G., HAY, J.D. and G.J. REES. Surgical treatment of atrial septal defects. *Br. Med. J.* 2:1463–1469, 1955.
354* EDWARDS, J.E. "Vascular rings" related to anomalies of the aortic arches. *Mod. Concepts Cardiovasc. Dis.* 17:19–20, 1948.
355* EDWARDS, J.E. Retro-esophageal segment of the left aortic arch, right ligamentum arteriosum and right descending aorta causing a congenital vascular ring about the trachea and esophagus. *Proc. Staff Meeting Mayo Clin.* 23:108–116, 1948.
356* EDWARDS, J.E. Anomalies of the derivatives of aortic arch system. *Med. Clin. North Am.* 32:925–949, 1948.
357 EDWARDS, J.E. Editorial. Anomalous coronary arteries with special reference to arteriovenous-like communications. *Circulation* 17:1001–1006, 1958.
358* EDWARDS, L.F. The retroduodenal artery. *Anat. Rec.* 81:351–355, 1941.
359 E'EVANT, T. Rare anomalie de la veine axillaire. *Arch. Ital. Biol.* 22:155–156, 1895.
360 EHLER, E. and T. HACKENBERG. Gefäßabnormitäten im Retroperitonealsitus eines 68jährigen Mannes. *Anat. Anz.* 107:375–383, 1959.
361 EIDLOW, S. and E.R. MACKENZIE. Anomalous origin of the left coronary artery from the pulmonary artery. Report of a case diagnosed clinically and confirmed by necropsy. *Am. Heart J.* 32:243–249, 1946.
362 EISENDRATH, D.N. Surgical importance of supernumerary arteries to the kidney. *Ann. Surg.* 68:53–54, 1918.
363* EISENDRATH, D.N. Anomalies of the bile ducts and blood vessels as the cause of accidents in biliary surgery. *JAMA* 71:864–867, 1918.
364* EISENDRATH, D.N. The relation of variations in the renal vessels to pyelotomy and nephrectomy. *Ann. Surg.* 71:726–743, 1920.
365* EISENDRATH, D.N. and D.S. STRAUSS. The surgical importance of accessory renal arteries. *JAMA* 55:1375–1377, 1910.
366 EISLER. Anomalie der Art. renalis bei Verlagerung der Niere. *Anat. Anz.* 4:465–467, 1889.
367* EKSTRÖM, G. and P. SANDBLOM. Double aortic arch. *Acta Chir. Scand.* 102:183–202, 1951.
368* ELIZAROWSKI, S.I. Die chirurgische Anatomie des Halsabschnittes des Ductus thoracicus. *Anat. Anz.* 91:248–258, 1941.
369 ELLIOTT, F.M. and L. REID. Some new facts about the pulmonary artery and its branching pattern. *Clin. Radiol.* 16:193–198, 1965.
370 ENDO, M. Topographische Untersuchung über die Abgangsstelle der großen Bauchaortenäste bei den japanischen Kindern. *Folia Anat. Jpn.* 16:395–408, 1938.
371 ENGELMAN, G. Ein Fall von Mangel einer Coronararterie. *Anat. Anz.* 14:348–350, 1898.
372* ESPERANCA PINA, J.A. Morphological study of the human anterior cardiac veins, venae cardis anteriores. *Acta Anat.* 92:145–159, 1975.
373 EVANS, T.H. Carotid canal anomaly. Other instances of absent internal carotid artery. *Med. Times* 84:1069–1072, 1956.
374* EVANS, W. Congenital stenosis (coarctation), atresia and interruption of aortic arch (a study of 28 cases). *Q. J. Med.* 2:1–31, 1933.
375 EWALD, W. Einige Fälle von Arcus aortae dexter. *Frankfurter Zeitschr. Pathol.* 34:87–97, 1926.
376* FALCONER, C.W.A. and E. GRIFFITHS. The anatomy of the blood-vessels in the region of the pancreas. *Br. J. Surg.* 37:334–344, 1949–50.
377* FALLER, A. Zur Kenntnis der Gefässverhältnisse der Carotisteilungsstelle. *Schweiz. Med. Wochenschr.* 76:1156–1158, 1946.
378 FALLER, A. Das Verhalten der arteriellen Gefässe in den verschiedenen Wandschichten des Rectum beim Menschen. *Acta Anat.* 30:275–286, 1957.
379 FALLER, A. and O. SCHÄRER. Über die Variabilität der Arteriae thyreoideae. *Acta Anat.* 4:119–122, 1947–48.
380 FAWCETT, E. The origin and intra-cranial course of the ophthalmic artery, and the relationship they bear to the optic nerve. *J. Anat. Physiol.* 30:49–52, 1895.
381 FAWCETT, E. Interesting abnormality of the hepatic artery, with an explanation of the condition. *J. Anat. Physiol.* 30:206–208, 1896.
382 FAWCETT, E. and J.V. BLACHFORD. The circle of Willis: An examination of 700 specimens. *Journal Anat. Physiol.* 40:63–70, 1906.
383 FEIGL, W., W. FIRBAS and H. SINZINGER. Über das Vorkommen akzessorischer Leberarterien und ihre Genese sowie ein Diskussion ihrer Nomenklatur. *Anat. Anz.* 134:139–171, 1973.
384 FERREIRA, A.D.S., MENDES, J.C. and J.T. DA COSTA. Sur l'anatomie chirurgicale de l'artère cystique. *Assoc. Anatomistes Comptes Rendus* 52:521–525, 1967.
385* FERREIRA, A.D.S., PEREIRA, J.P. and M. ANDREA. Segmentation artérielle du rein. *Assoc. Anatomistes Comptes Rendus* 52:526–533, 1967.
386* FIANDRA, O., BARCIA, A., CORTES, R. and J. STANHAM. Partial anomalous pulmonary venous drainage into the inferior vena cava. *Acta Radiol.* 57:301–310, 1962.
387* FIELDSTEIN, L.E. and J. PICK. Drainage of the coronary sinus into the left auricle. Report of a rare congenital cardiac anomaly. *Am. J. Clin. Pathol.* 12:66–69, 1942.
388* FINDLAY, C.W. and H.C. MAIER. Anomalies of the pulmonary vessels and their surgical significance. *Surgery* 29:604–641, 1951.
389 FINE, H. and E.N. KEEN. The arteries of the human kidney. *J. Anat.* 100:881–894, 1966.
390* FINERTY, J.C. Persistent ischiadic artery. *Anat. Rec.* 98:587–595, 1947.
391 FINERTY, J.C. and M. TROTTER. Two unusual anomalies in the same cadaver (Negro male, age 85 years). *Anat. Rec.* 97:415, 1947.
392 FINGERLAND, A. Trois cas de trajet anormal de l'artère sousclavière droite. *Assoc. Anatomistes Comptes Rendus* 23:133–139, 1928.
393* FISCHER, W. Anatomisch-röntgenologische Untersuchungen über die Verteilung und Anastomosenbildung der viszeralen und parietalen weiblichen Beckenarterien. *Z. Geburtsh. Gynäk.* 154:321–340, 1959.
394* FISHER, A.G.T. A case of complete absence of both internal carotid arteries, with a preliminary note on the developmental history of the stapedial artery. *J. Anat. Physiol.* 48:37–46, 1913.
395 FISHER, E. Seltener Verlauf der Vena azygos (Abspaltung eines Lungenlappens). *Anat. Anz.* 15:476–481, 1899.
396 FLEMMING, E.E. Absence of left internal carotid. *J. Anat. Physiol.* 29:xxiii–xxiv, 1895.
397 FLESCH, M. Ein weiterer Fall von Theilung der Arteria carotis interna in der Schädelhöhle. *Arch. Anat. Physiol.* 1886, pp. 151–155.
398* FLINT, E.R. Abnormalities of the right hepatic, cystic, and gastroduodenal arteries, and of the bile ducts. *Br. J. Surg.* 10:509–519, 1923.
399 FLYNN, J.E., SIEBENS, A.A. and S.F. WILLIAM. Congenital absence of a main branch of pulmonary artery. *Am. J. Med. Sci.* 228:673–679, 1954.
400* FORSTER, A. Über zwei Fälle einer seltenen Arterienvarietät an der Ansatzsehne des Biceps brachii. *Anat. Anz.* 50:193–194, 1917–18.
401 FRAENKEL, W. Linksseitige Vena cava inferior. *Anat. Anz.* 37:240–241, 1910.
402 FRAENTZEL, O. Ein Fall von abnormer Communication der Aorta mit der Arteria pulmonalis. *Arch. Path. Anat. Physiol. Klin. Med.* 43:420–426, 1868.
403 FRASER, J. and A.K. BROWN. A clinical syndrome associated

with a rare anomaly of the vena portal system. *Surg. Gynecol. Obstet.* 78:520–524, 1944.
404* FRASER, R.S., DVORKIN, J., ROSSALL, R.E. and R. EIDEM. Left superior vena cava. A review of associated congenital heart lesions, catheterization data and roentgenologic findings. *Am. J. Med.* 31:711–716, 1961.
405 FRAY, W.W. Right aortic arch. *Radiology* 26:27–36, 1936.
406* FRIEDMAN, S.M. Report of two unusual venous anomalies. (Left postrenal inferior vena cava; post-aortic left innominate vein). *Anat. Rec.* 92:71–76, 1945.
407 FRYCZKOWSKI, A. Collateral branch of the intracranial segment of the ophthalmic artery in man. *Folia Morphol.* 30:518–523, 1971.
408 FRYCZKOWSKI, A. Collateral branches of the ophthalmic artery running to the intracanalicular segment of the optic nerve in man. *Folia Morphol.* 31:285–288, 1972.
409* GABRIELLE, H., LATARJET, M., LECUIRE, J. and M. MEUNIER. Anatomie macroscopique des artères de lécorce cérébrale humaine. *Assoc. Anatomistes Comptes Rendus* 36:230–276, 1949.
410* GABRIELLE, H., LATARJET, M., LECUIRE, J., SAUTOT, J. and M. CHARPIN. Contribution a l'etude macroscopique de l'artère cerebrale antérieure et de la vascularisation arterielle du lobe frontal chez l'homme. *Assoc. Anatomistes Comptes Rendus* 36:277–290, 1949.
411* GABRIELLE, H., LATARJET, M., LECUIRE, J. and P.E. DUROUX. Contribution a l'étude de tronc de l'artère sylvienne et de la vascularisation artérielle du lobe de l'insula chez l'homme. *Assoc. Anatomistes Comptes Rendus* 36:291–311, 1949.
412* GABRIELLE, H., LATARJET, M., LECUIRE, J., REPLUMAZ, P.R. and G. MARET. Contribution a l'étude macroscopique des artères de la région Rolandique chez l'homme. *Assoc. Anatomistes Comptes Rendus* 36:313–325, 1949.
413* GABRIELLE, H., LATARJET, M., LECUIRE, J. and L. EICHOLZ. Contribution a l'étude de l'anatomie macroscopique de l'artère cérébrale postérieure et de la vascularisation artérielle du lobe occipital chez l'homme. *Assoc. Anatomistes Comptes Rendus* 36:326–338, 1949.
414* GABRIELLE, H., LATARJET, M., LECUIRE, J. and P. MILLERET. Contribution a l'étude macroscopique de la vascularisation du lobe temporal chez l'homme. *Assoc. Anatomistes Comptes Rendus* 36:339–352, 1949.
415* GAGNON, R. The arterial supply of the human adrenal gland. *Rev. Can. Biol.* 16:421–433, 1957.
416* GAGNON, R. Les artères surrénales inférieures chez l'homme. *Rev. Can. Biol.* 25:135–144, 1966.
417* GALLAVARDIN, L. and P. RAVAULT. Anomalie d'origine de la coronaire antérieure. *Lyon Méd.* 135:270–272, 1925.
418 GARLAND, L.H. Persistent right-sided aortic arch. *Am. J. Roentgenol.* 39:713–719, 1938.
419 GASUL, B.M. and E. LOEFFLER. Anomalous origin of the left coronary artery from the pulmonary artery (Bland-White-Garland Syndrome). Report of four cases. *Pediatrics* 4:498–507, 1949.
420 GEDDES, A.C. A note on abnormal right subclavian artery. *J. Anat. Physiol.* 45:197–198, 1911.
421 GEDDES, A.C. Note on the reputed passage of abnormal right subclavian artery (A) in front of the trachea, (B) between the trachea and oesophagus. *J. Anat. Physiol.* 45:199–201, 1911.
422 GEDDES, A.C. Abnormal superior vena cava. *Anat. Anz.* 41:449–453, 1912.
423* GEIPEL, P. Weitere Beiträge zum Situs transversus und zur Lehre von den Transpositionen der großen Gefäße des Herzens. *Arch. Kinderheilk.* 35:112–145, 222–259, 1903.
424 GEIPEL, P. Mißbildungen der Tricuspidalis. *Arch. Pathol. Anat. Physiol. Klin. Med.* 171:298–334, 1903.
425 GENTES and PHILIP. L'artère hepatique gauche. Sa signification. Ses rapports avec l'indépendance des lobes du foie. *Soc. Biol. Comptes Rendus des Sëances et Memoires* 61:640–642, 1906.
426* GEORGE, R. Topography of the unpaired visceral branches of the abdominal aorta. *J. Anat.* 69:196–205, 1935.
427 GÉRARD, G. Variabilité des rapports vasculaires du bassinet. *Assoc. Anatomistes Comptes Rendus* 3:147–154, 1901.
428 GÉRARD, G. Les artères rénales. (Note statistique d'après l'étude de 150 paires de reins). *J. Anat. Physiol. Norm. Pathol. Homme Animaux* 47:531–534, 1911.
429 GÉRARD, G. Contribution à l'étude morphologique des artères des capsules surrénales de l'homme. *J. Anat. Physiol. Norm. Pathol. Homme Animaux* 49:269–303, 1913.
430* GÉRARD, G. Sur les variations d'origine et de nombre des artères génitales, spermatiques ou ovariennes, de l'homme. *Soc. Biol. Comptes Rendus Hebdomadaires des Séances et Mémoires* 74:778, 1913.
431 GÉRARD, G. Sur un cas de solidarité artérielle entre le rein et la surrénale gauche chez l'homme. *Soc. Biol. Comptes Rendus Hebdomadaires des Séances et Mémoires* 74:857–858, 1913.
432 GÉRARD, G. Mention d'une anastomose veineuse rénocave rétroaortique obliquement desendante. *Soc. Biol. Comptes Rendus Hebdomadaires des Séances et Mémoires* 83:185–186, 1920.
433 GÉRARD, G. Théorie des artères segmentaires abdominales marginales. *Soc. Biol. Comptes Rendus Hebdomadaires des Séances et Mémoires* 83:822–824, 1920.
434* GÉRARD, G. and D. CORDONNIER. Un cas-type de triplicité de l'artère hépatique. *Soc. Biol. Comptes Rendus Hebdomadaires des Séances et Mémoires* 76:619–621, 1914.
435 GÉRARD, G. and P. CORDIER. Deux nouveaux cas d'anomalies de l'artère du nerf médian. *Soc. Biol. Comptes Rendus Hebdomadaires des Séances et Mémoires* 83:1108–1110, 1920.
436 GHON, A. Ein Beitrag zu den Anomalien der Pulmonalvenen. *Beiträge Pathol. Anat. Allg. Pathol.* 62:175–193, 1916.
437* GILFILLAN, R.S. Anatomic study of the portal vein and its main branches. *Arch. Surg.* 61:449–461, 1950.
438 GILLASPIE, C., MILLER, L. and M. BASKIN. Anomalous renal vessels and their surgical significance. *Anat. Rec.* 11:77–86, 1916.
439* GILLILAN, L.A. The arterial blood supply of the human spinal cord. *J. Comp. Neurol.* 110:75–103, 1958.
440* GILLILAN, L.A. The collateral circulation of the human orbit. *Arch. Ophthalmol.* 65:684–694, 1961.
441* GILLOT, C., HUREAU, J., AARON, C., MARTINI, R. and G. THALER. The superior mesenteric vein. An anatomic and surgical study of eighty-one subjects. *J. Int. Coll. Surg.* 41:339–369, 1964.
442* GISEL, A. Eine seltene Variation im Verästelungstypus der Arteria coeliaca. Ein Beitrag zur Frage der „ventralen Längsanastomose". *Anat. Anz.* 94:208–220, 1943.
443* GLADSTONE, R.J. A case in which the right ureter passed behind the inferior vena cava: With a short note upon a case in which the left renal vein passed behind the abdominal aorta, and the bearing of these abnormalities on the development of abdominal veins. *J. Anat. Physiol.* 45:225–231, 1910.
444 GLADSTONE, R.J. A case of left inferior vena cava occurring in a female subject in whom the left superior intercostal vein joined the vena azygos major, and the twelfth ribs were absent. *J. Anat. Physiol.* 46:220–227, 1912.
445* GLADSTONE, R.J. Development of the inferior vena cava in the light of recent research, with especial reference to certain abnormalities, and current descriptions of the ascending lumbar and azygos veins. *J. Anat.* 64:70–93, 1929.
446 GLADSTONE, R.J. and C.H. REISMANN. Two examples of cardiac malformation. *J. Anat. Physiol.* 50:228–238, 1916.
447 GLADSTONE, R.J. and C.P.G. WAKELEY. Two cases, considered from the developmental standpoint, in which the right subclavian artery arose from the arch of the aorta beyond the origin of the left subclavian artery; with a note on the relation of the subclavian veins to the cardinal system. *J. Anat. Physiol.* 49:362–374, 1915.
448* GLADYKOWKSA-RZECZYCKA, J. and M. MIESZKOWSKA. Vascular variation in the genito-urinary system. *Folia Morphol.* 36:113–116, 1977.
449* GLASSER, S.T. An anatomic study of venous variations at the fossa ovalis. The significance of recurrences following ligation. *Arch. Surg.* 46:289–295, 1943.
450 GODINOV, V.M. The arterial system of the brain. *Am. J. Phys. Anthropol.* 13:359–388, 1929.
451* GOETZEN, B. The internal cerebral veins in man. Part I. *Folia Morphol.* 23:44–69, 1964.
452* GOLDBLOOM, A.A. The anomalous right subclavian artery and its possible clinical significance. *Surg. Gynecol. Obstet.* 34:378–384, 1922.
453* GOLIGHER, J.C. The blood supply to the sigmoid colon and rectum with reference to the technique of rectal resection with restoration of continuity. *Br. J. Surg.* 37:157–162, 1949.
454* GOLIGHER, J.C., LEACOCK, A.G. and J.J. BROSSY. The surgical anatomy of the anal canal. *Br. J. Surg.* 43:51–61, 1955.
455 GOLUB, D.M. Ein Fall eines anomalen Ursprungs der A. subcl. dextra unterhalb der A. subcl. sin. kombiniert mit dem Tr. bicaroticus und einem rechtsseitigen Münden des Ductus thoracicus. *Anat. Anz.* 67:387–392, 1929.
456 GOLUBEV, A. Zur chirurgischen Anatomie der Nierenarterien. *Z. Urol. Chir.* 21:307, 1927.
457 GORDON, S.J. Double aortic arch. *Journal Pediatr.* 30:428–437, 1947.

458 GORDON-SHAW, C. Two cases of reduplication of the arteria cerebri posterior. *J. Anat. Physiol.* 44:244–248, 1910.
459 GORTAT-ZALEWSKA, W. Variation of origin and course of the abdominal portion of the thoracic duct in man. *Folia Morphol.* 34:53–58, 1975.
460 GOTTSCHAU, M. Zwei seltene Varietäten der Stämme des Aortenbogens. *Arch. Anat. Physiol.* 1885, pp. 245–252.
461 GOTTSCHAU, M. Eine seltene Aorten-Anomalie. *Anat. Anz.* 2:37–39, 1887.
462 GOUAZÉ, A., SENTENAE, M. and R. ODANO. Contribution a l'étude de l'irrigation artérielle de la prostate. *Assoc. Anatomistes Comptes Rendus* 43:374–381, 1956.
463* GOUAZÉ, A., ODANO, R. and G. CHANTEPIE. Contribution a l'étude de la vascularisation artérielle de la vessie. *Assoc. Anatomistes Comptes Rendus* 44:331–340, 1957.
464 GOULD, E.L.P. A case of abnormal right subclavian artery. *J. Anat. Physiol.* 43:324–328, 1909.
465 GOULEY, B.A. Anomalous left coronary artery arising from the pulmonary artery (adult type). *Am. Heart J.* 40:630–637, 1950.
466* GOŹDZIEWSKI, S. Variation of the axial arteries of the upper extremity in man. *Folia Morphol.* 37:91–97, 1978.
467+ GOZULOFF, G.I. Left superior vena cava, the right being absent (In Russian). *Universitatskiya Izviestiya, Kiyev* 50:1–29, 1910.
468 GRAF, G. Infarto emorragico di un lobo azigos polmonare in un neonato di tre settimane, portatore di una complessa cardiopatia congenita (transposizione corretta die grossi; vasi, pervietà die setti interventricolare ed interatriale). *Pediatria* 46:740–751, 1938.
469 GRANT, R.T. and L.E. VIKO. Observations on the anatomy of the Thebesian vessels of the heart. *Heart* 15:103–123, 1929.
470* GRAVES, F.T. The anatomy of the intrarenal arteries and its application to segmental resection of the kidney. *Br. J. Surg.* 42:132–139, 1954–55.
471* GRAVES, F.T. The aberrant renal artery. *J. Anat.* 90:553–558, 1956.
472 GRAY, G.M. Multiple renal arteries. *Anat. Anz.* 29:266–270, 1906.
473 GREEN, H.T. The venous drainage of the human hypophysis cerebri. *Am. J. Anat.* 100:435–462, 1957.
474 GREENBERG, M.W. Blood supply of the rectosigmoid and rectum. *Ann. Surg.* 131:100–108, 1950.
475 GREENFIELD, W.S. Congenital malformation of the aortic valves. *Trans. Pathol. Soc. Lond.* 27:110–113, 1876.
476 GREENFIELD, W.S. Persistence of left vena cava superior, with absence of right. *Trans. Pathol. Soc. Lond.* 27:120–124, 1876.
477* GREENFIELD, W.S. Double mitral valve. *Trans. Pathol. Soc. Lond.* 27:128–129, 1876.
478* GRIEG, H.W., ANSON, B.J. and S.S. COLEMAN. The inferior phrenic artery. Types of origin in 850 body-halves and diaphragmatic relationship. *Q. Bull. Northwestern University Medical School* 25:345–350, 1951.
479 GRIFFITH, T.W. An abnormal muscle of the hand, with remarks on the course of the radial artery. *J. Anat. Physiol.* 31:283–287, 1897.
480 GRIFFITH, T.W. An example of a peculiar malformation of the tricuspid valve. *J. Anat. Physiol.* 37:251–254, 1903.
481 GRIFFITHS, J.D. Extramural and intramural blood-supply of colon. *Br. Med. J.* 1:323–326, 1961.
482 GRISHMAN, A., POPPEL, M.H., SIMPSON, R.S. and M.L. SUSSMAN. The roentgenographic and angiocardiographic aspects of (1) aberrant insertion of pulmonary veins associated with interatrial defect and (2) congenital arteriovenous aneurysm of the lung. *Am. J. Roentgenol.* 62:500–508, 1949.
483 GRISOLI, J., PIGANIOL, G. and R. SEDAN. L'artère cérébrale communicante antérieure. *Assoc. Anatomistes Comptes Rendus* 44:349–355, 1957.
484* GRISWOLD, H.E. and M.D. YOUNG. Double aortic arch. Report of two cases and review of the literature. *Pediatrics* 4:751–768, 1949.
485* GROB, M. Über Anomalien des Aortenbogens und ihre entwicklungsgeschichtliche Genese. *Helv. Pediatr. Acta* 4:274–293, 1949.
486 GRÖNROOS, H. Eine seltene Anordnung der Arteria maxillaris externa bei einem Erwachsenen. *Anat. Anz.* 20:9–16, 1902.
487 GROSS, L. and M.A. KUGEL. The arterial blood vascular distribution to the left and right ventricles of the human heart. *Am. Heart J.* 9:165–177, 1933.
488* GROSS, R.E. Surgical relief for tracheal obstruction from a vascular ring. *N. Engl. J. Med.* 233:586–590, 1945.
489* GROSS, R.E. Surgical treatment for dysphagia lusoria. *Ann. Surg.* 124:532–534, 1946.
490 GROSS, R.E. Arterial malformations which cause compression of the trachea or esophagus. *Circulation* 11:124–134, 1955.
491* GROSS, R.E. and P.F. WARE. The surgical significance of aortic arch anomalies. *Surg. Gynecol. Obstet.* 83:435–448, 1946.
492* GROSS, R.E. and E.B.D. NEUHAUSER. Compression of the trachea or esophagus by vascular anomalies. Surgical therapy in 40 cases. *Pediatrics* 7:69–88, 1951.
493* GRUBER, G.B. Zwei Fälle von Dextropositio des Aortenbogens. *Frankfurter Z. Pathol.* 10:375–382, 1912.
494 GRUBER, W. Zur Anatomie der Arteria radialis. *Arch. Anat. Physiol. Wissen. Med.* 1864, pp. 434–455.
495 GRUBER, W. Zu den Anomalien der Arteria pediaea. *Arch. Anat. Physiol. Wissen. Med.* 1864, pp. 512–515.
496 GRUBER, W. Über einen Fall von Einmündung der Vena hemiazyga in das Atrium dextrum cordis beim Menschen. *Arch. Anat. Physiol. Wissen. Med.* 1864, pp. 729–740.
497 GRUBER, W. Rudimentäre Vena cava superior sinistra bei einem Erwachsenen. *Arch. Pathol. Anat. Physiol. Klin. Med.* 32:114–117, 1865.
498 GRUBER, W. Weitere Fälle von Einmündung der Vena hemiazyga in das Atrium dextrum cordis beim Menschen. *Arch. Anat. Physiol. Wissen. Med.* 1866, pp. 224–230.
499 GRUBER, W. Enorm hoher Ursprung einer supernumerären Arteria circumflexa ilei interna von der Arteria iliaca externa. *Arch. Anat. Physiol. Wissen. Med.* 1867, pp. 547–551.
500 GRUBER, W. Über die Arteria mediana antibrachii superficialis, Arteria ulnaris antibrachii superficialis und Duplicität der Arteria ulnaris. *Arch. Anat. Physiol. Wissen. Med.* 1867, pp. 668–687.
501* GRUBER, W. Dreiwurzelige Arteria radialis. *Arch. Anat. Physiol. Wissen. Med.* 1870, pp. 180–188.
502 GRUBER, W. Rudimentäre Arteria radialis. *Arch. Anat. Physiol. Wissen. Med.* 1870, pp. 189–196.
503* GRUBER, W. Subcutaner Verlauf des Ramus dorsalis der Arteria radialis am Unterarm- und Handwurzelrücken. *Arch. Anat. Physiol. Wissen. Med.* 1870, pp. 484–489.
504* GRUBER, W. Duplicität der Arteria ulnaris – neuer Fall. *Arch. Anat. Physiol. Wissen. Med.* 1871, pp. 286–296.
505 GRUBER, W. Anomalie der Vena jugularis anterior. *Arch. Pathol. Anat. Physiol. Klin. Med.* 54:188, 1872.
506 GRUBER, W. Anomale Mündung der Vena saphena parva in die Vena cruralis. *Arch. Pathol. Anat. Physiol. Klin. Med.* 54:189, 1872.
507 GRUBER, W. Bildung der Vena cava inferior durch Zusammenfluß von 3 Stämmen, und Bildung von Inseln zwischen der Vena iliaca externa und interna der linken Seite. *Arch. Pathol. Anat. Physiol. Klin. Med.* 54:190, 1872.
508 GRUBER, W. Über die Arteria thyreoidea ima. *Arch. Pathol. Anat. Physiol. Klin. Med.* 54:445–484, 1872.
509 GRUBER, W. Über die Varianten des ungewöhnlichen Ursprunges der Arteria mammaria interna und des Truncus thyreo-cervicalis. *Arch. Pathol. Anat. Physiol. Klin. Med.* 54:485–491, 1872.
510 GRUBER, W. Über einen anomalen Kanal für eine aus der Arteria meningea media innerhalb der Schädelhöhle entspringende Arteria temporalis profunda. *Arch. Pathol. Anat. Physiol. Klin. Med.* 63:100–103, 1875.
511 GRUBER, W. Anomaler Verlauf der Arteria poplitea durch den Sulcus popliteus internus und Obliteration derselben auf diesem Umwege. *Arch. Pathol. Anat. Physiol. Klin. Med.* 65:262–270, 1875.
512 GRUBER, W. Über den anomalen Intraclavicularkanal zum Durchgange für Venen. *Arch. Pathol. Anat. Klin. Med.* 66:457–461, 1876.
513 GRUBER, W. Verlauf der Vena anonyma sinistra vor der Thymus. *Arch. Pathol. Anat. Physiol. Klin. Med.* 66:462–463, 1876.
514 GRUBER, W. Ein Fall des Verlaufes der Art. carotis externa zwischen dem Musc. digastricus und Musc. stylohyoideus. *Arch. Pathol. Anat. Physiol. Klin. Med.* 66:464, 1876.
515 GRUBER, W. Ein Nachtrag zu den Anomalien des Ursprunges der Arteria mammaria interna. *Arch. Pathol. Anat. Physiol. Klin. Med.* 66:464–465, 1876.
516 GRUBER, W. Ein Fall von Einmündung der Vena pulmonalis dextra superior in die Vena cava superior. *Arch. Pathol. Anat. Physiol. Klin. Med.* 68:284, 1876.
517 GRUBER, W. Ursprung der Arteria vertebralis dextra von der Subclavia knapp neben der Carotis, mit Kreuzung der Thyreoidea inferior von vorn während ihres Verlaufes. *Arch. Pathol. Anat. Physiol. Klin. Med.* 74:433–434, 1878.
518* GRUBER, W. Zweiwurzlige Arteria vertebralis dextra bei Ursprung der accessorischen Wurzel von einem vom Anfange der

Subclavia entstandenen Truncus thyreovertebralis (und mit Vorkommen einer Arteria thyreoidea ima). *Arch. Pathol. Anat. Physiol. Klin. Med.* 74:435–438, 1878.
519 GRUBER, W. Hohe Theilung der Arteria poplitea in die A. tibialis postica und in den Truncus communis für die A. peronea und die A. tibialis antica, mit Endigung der A. tibialis postica als A. plantaris interna und der A. peronea als A. plantaris externa. *Arch. Pathol. Anat. Physiol. Klin. Med.* 74:438–443, 1878.
520 GRUBER, W. Beide Venae faciales anteriores als Aeste einer abnorm starken Vena superficialis colli anterior dextra. *Arch. Pathol. Anat. Physiol. Klin. Med.* 74:444–447, 1878.
521 GRUBER, W. Duplicität der Arteria spermatica interna bei Ursprung der supernumerären Arterie aus der Arteria iliaca communis. *Arch. Pathol. Anat. Physiol. Klin. Med.* 81:457–459, 1880.
522* GRUBER, W. Vorkommen einer Vena cava superior sinistra (bei Abwesenheit der V. cava superior der Norm). *Arch. Pathol. Anat. Physiol. Klin. Med.* 81:458–462, 1880.
523 GRUBER, W. Duplicität der Vena cava superior, mit Vorkommen zweier Venae azygae und einer sufficienten Valvula an der Mündung der Vena azygos sinistra. *Arch. Pathol. Anat. Physiol. Klin. Med.* 81:462–465, 1880.
524 GRUBER, W. Duplicität der Arteria occipitalis. *Arch. Pathol. Anat. Physiol. Klin. Med.* 82:474–475, 1880.
525 GRUBER, W. Verlauf der Vena anonyma sinistra vor der Thymusdrüse. *Arch. Pathol. Anat. Physiol. Klin. Med.* 82:475, 1880.
526 GRUBER, W. Vorkommen einer Zwischennierenarterie (Arteria inter-renalis beim Menschen). *Arch. Pathol. Anat. Physiol. Klin. Med.* 86:35–38, 1881.
527 GRUBER, W. Duplicität der Vena cava superior mit Vorkommen zweier transversaler Communicationsäste und zweier Venae azygae. *Arch. Pathol. Anat. Physiol. Klin. Med.* 86:38–41, 1881.
528 GRUBER, W. In Bildungshemmung begründetes, anscheinend bis über den 1. Lendenwirbel verlängertes und mit einem Ramus communicans vor dem 5. Lendenwirbel versehenes Auftreten der Venae ilacae communis. *Arch. Pathol. Anat. Physiol. Klin. Med.* 86:493–495, 1881.
529 GRUBER, W. Vena mediana colli vera unica von enormer Dicke. *Arch. Pathol. Anat. Physiol. Klin. Med.* 98:431–435, 1884.
530 GRUBER, W. Weitere Nachträge in Bezug auf den anomalen Infraclavicularkanal – 11.–14. Fall (Verlauf der Vena axillaris durch denselben – vorher nicht gesehen). *Arch. Pathol. Anat. Physiol. Klin. Med.* 98:435–440, 1884.
531 GRUBER, W. Duplicität der Vena cava superior (11. Fall eigener Beobachtung) bei Verschluß der Mündung der Vena cava superior sinistra in das Atrium dextrum und deren Auftreten als Abführungskanal der Herzvenen in die Vena anonyma sinistra (sicherer 1. Fall). *Arch. Pathol. Anat. Physiol. Klin. Med.* 99:492–497, 1885.
532 GRUBER, W. Einmündung der Vena pulmonalis dextra superior in die Vena cava superior. *Arch. Pathol. Anat. Physiol. Klin. Med.* 102:3–5, 1885.
533 GRZYBIAK, M., SZOSTAKIEWICZ-SAWICKA, H. and A. TREDER. Remarks on pathways of drainage from the left upper intercostal spaces in man. *Folia Morphol.* 34:301–313, 1975.
534 GUBLER, M. Coeur d'adulte avec persistance du trou de Botal et communication des deux ventricules a travers la cloison interventriculaire. *Soc. Biol. Comptes Rendus des Séances et Memoires* 13:279–281, 1861.
535* GUERRIER, Y. and G. VILLACEQUE. Origine et comportement des artères cérébelleuse moyenne et auditive interne. *Assoc. Anatomistes Comptes Rendus* 36:377–382, 1949.
536 GUERRIER, Y. and R. PALEIRAC. Le tronc artériel antibrachial. *Assoc. Anatomistes Comptes Rendus* 38:566–575, 1951.
537 GUERRIER, Y. and J. OLIVIER. A propos de l'origine des artères cérébelleuses. *Assoc. Anatomistes Comptes Rendus* 38:576–579, 1951.
538 GUERRIER, Y. and C. BOUDET. Variation de calibre de l'artère centrale de la rétine. *Assoc. Anatomistes Comptes Rendus* 39:889, 1952.
539 GUERRIER, Y. and P.E. RAPP. La veine porte extra et intra-hépatique. I. Le mode de division extra-hépatique. *Assoc. Anatomistes Comptes Rendus* 40:272–274, 1953.
540 GUGGEMOS, E., NYSTROM, J., PEPPY, S.J., SINATRA, C. and H. BRODY. A rare case of an arterial connection between the left and right kidneys. *Ann. Surg.* 156:940–943, 1962.
541* GUGLIELMO, L.D. and M. GUTTADAURO. Anatomic variations in the coronary arteries. An arteriographic study in living subjects. *Acta Radiol.* 41:303–416, 1954.
542 GULISANO, M., ZECCHI, S., PACINI, P. and G.E. ORLANDINI. The behavior of some human arteries as regards the correct circumference: A statistical research. *Anat. Anz.* 152:341–357, 1982.
543 HACKENSELLNER, H.A. Koronaranomalien unter 1000 auslesefrei untersuchten Herzen. *Anat. Anz.* 101:123–130, 1954–55.
544 HAFFERL, A. Ein Fall von abnormem Verlauf des Ductus thoracicus und der V. hemiazygos. *Anat. Anz.* 57:373–378, 1923–24.
545 HALE. Case of a heart with a single ventricle, with observations on the production of cyanosis. *Trans. Pathol. Soc. Lond.* 4:87–96, 1853.
546 HALPERT, B. Complete situs inversus of the vena cava superior. *Anat. Rec.* 35:38, 1927.
547 HALPERT, B. and F. COMAN. Complete situs inversus of the vena cava superior. *Am. J. Pathol.* 6:191–197, 1930.
548 HALPERT, B., SNODDY, W.T., BOHAN, K.E. and C.L. FREEDE. Right aortic arch with a vascular ring constricting esophagus and trachea. Report of two cases. *Arch. Pathol.* 47:429–434, 1949.
549 HAMMER, D.L. and A.M. MEIS. Thyroid arteries and anomalous subclavian in the White and the Negro. *Am. J. Phys. Anthropol.* 28:227–237, 1941.
550 HAMMER, G. Situs inversus arcus aortae (Hohe Rechtslage der Aorta). *Fortschr. Gebiete Röntgenstrahlen.* 34:517–523, 1926.
551* HANLON, C.R. and A. BLALOCK. Complete transposition of the aorta and the pulmonary artery. *Ann. Surg.* 127:385–397, 1948.
552 HARLEY, H.R.S. The development and anomalies of the aortic arch and its branches. *Br. J. Surg.* 46:561–573, 1958–59.
553 HARPER, W.F. Observations on the blood supply of the human ureter. *Br. J. Urol.* 14:63–72, 1942.
554 HARRIS, H.A. and I. LEWIS. Anomalies of the lungs, with special reference to the dangers of abnormal vessels in lobectomy. *J. Thoracic Surg.* 9:667–671, 1940.
555 HARRISON, R.G. The distribution of the vasal and cremasteric arteries to the testis and their functional importance. *J. Anat.* 83:267–281, 1949.
556 HARRISON, R.G. and A.E. BARCLAY. Distribution of testicular artery (internal spermatic artery) to human testis. *Br. J. Urol.* 20:57–66, 1948.
557* HART-SMITH. Abnormal left innominate vein. *J. Anat. Physiol.* 29:xxiv–xxvi, 1895.
558 HASSLER, O. Blood supply to human spinal cord. A microangiographic study. *Neurol.* 15:302–307, 1966.
559 HARVEY, J.C. and L.M. HOWARD. A rare type of anomalous ophthalmic artery in a negro. *Anat. Rec.* 92:87–90, 1945.
560 HARVEY, R.A. A case of multiple renal arteries. *Anat. Rec.* 8:333–339, 1914.
561 HARVEY, R.W. Notes on two cases of anomalous right subclavian artery. *Anat. Rec.* 12:329–330, 1917.
562* HAUG, H. and U. MERKEL. Abnormer Ursprung und Verlauf der A. subclavia dextra. *Anat. Anz.* 117:319–321, 1965.
563 HAYNER, J.C. Variations of the torcular Herophili and transverse sinuses. *Anat. Rec.* 103:542, 1949.
564* HAYREH, S.S. Arteries of the orbit in the human being. *Br. J. Surg.* 50:938–953, 1963.
565 HAYREH, S.S. and R. DASS. The ophthalmic artery. I. Origin and intra-cranial and intra-canalicular course. *Br. J. Ophthalmol.* 46:65–98, 1962.
566* HEALEY, J.E. JR. and P.C. SCHROY. Anatomy of the biliary ducts within the human liver: Analysis of the prevailing pattern of branchings and the major variations of the biliary ducts. *Arch. Surg.* 66:599–616, 1953.
567* HECHT, V. Über einen Fall von Kollateralkreislauf im Gebiet der Arteria coeliaca. *Anat. Anz.* 26:570–576, 1905.
568 HEIDSIECK, E. Zur Skeletotopie der großen Äste der Bauchaorta. *Anat. Anz.* 66:6–24, 1928.
569 HEITZMAN, E.R., SCRIVANI, J.V., MARTINO, J. and J. MORO. The azygos vein and its pleural reflections. I. Normal roentgen anatomy. *Radiology* 101:249–266, 1971.
570 HELLER, A. Doppelter Aortenbogen. *Münchener Med. Wochenschr.* 51:1666, 1904.
571 HELLER, A. Mangel der Vena cava inferior. *Münchener Med. Wochenschr.* 51:1666, 1904.
572 HELLERSTEIN, H.K. and J.L. ORBISON. Anatomic variations of the orifice of the human coronary sinus. *Circulation* 3:514–523, 1951.
573* HELLSTRÖM, J. A contribution to the knowledge of the relation of abnormally running renal vessels to hydronephrosis and an investigation of the arterial conditions in 50 kidneys. *Acta Chir. Scand.* 61:289–330, 1927.
574 HENSMAN, A. On the relations of the dorsal artery of the foot to the cuneiform bones. *J. Anat. Physiol.* 18:60–61, 1884.

575 HEPBURN, D. Double superior vena cava, right pulmonary veins opening into the right auricle, and a special inter-auricular foramen. *J. Anat. Physiol.* 21:438–443, 1887.

576 HEPBURN, D. Rare abnormal arrangement of the coronary arteries. *J. Anat. Physiol.* 29:459–462, 1895.

577* HERBUT, P.A. Anomalies of the aortic arch. *Arch. Pathol.* 35:717–729, 1943.

578 HERBUT, P.A. and T.T. SMITH. Constricting double aortic arch. *Arch. Otolaryngol.* 37:558–562, 1943.

579* HERMAN, L.H., FERNANDO, O.U. and E.S. GURDJIAN. The anterior choroidal artery: An anatomical study of its area of distribution. *Anat. Rec.* 154:95–101, 1966.

580 HERMANN, W.W. Double aortic arch. *Arch. Pathol.* 6:418–425, 1928.

581 HERRINGHAM, W.P. Right aorta with persistent left aortic root giving origin to the left subclavian, Proc. Anat. Soc. *J. Anat. Physiol.* 25:vi–vii, 1891.

582 HEWITT, R.W. Anatomical notes. An abnormal left bronchial artery. *J. Anat.* 64:363, 1930.

583 HICKMAN. Malformation of Heart; 1) Transposition of auricles and of aorta; absence of pulmonary artery; patent foramen ovale; communicating ventricles; with lateral transposition of principal viscera; cyanosis. 2) Transposition of viscera. Malformation of Heart; pulmonary veins from right lung entering left auricle and from left lung entering left auricle and from left lung entering right auricle. *Transactions Pathol. Soc. Lond.* 20:88–92, 93–98, 1869.

584 HILL, R.M. Vascular anomalies of the upper limbs associated with cervical ribs. *Br. J. Surg.* 27:100–110, 1939–40.

585 HILL, W.C.O. Atrial arteries in a human heart. *J. Anat.* 79:41–43, 1945.

586 HINDZE, B. Die Hirnarterien einiger hervorragender Persönlichkeiten. *Anat. Anz.* 62:1–24, 1926–27.

587 HIRAKÓ, G. Über das Verhältnis der zuführenden zu den abführenden Gefäßen des Organs (artero-venöser Index). *Folia Anat. Jpn.* 28:229–236, 1956.

588* HITZROT, J.M. A composite study of the axillary artery in man. *Johns Hopkins Hosp. Bull.* 12:136–145, 1901.

589 HIURA, A. An anomalous ophthalmic artery arising from the middle meningeal artery. *Anat. Anz.* 147:473–476, 1980.

590* HOCHSTETTER, A. VON. Eine einzigartige abnorme Anordnung der Arterien des Gesichtes. *Anat. Anz.* 113:221–223, 1963.

591 HOCHSTETTER, A. VON. Eine Schleife der Arteria epigastrica inferior im Canalis inguinalis. *Anat. Anz.* 109:221–224, 1961.

592 HOCHSTETTER, A. VON. Ein Venenring um die Arteria iliaca externa dextra. *Anat. Anz.* 99:351–353, 1952–53.

593* HOCHSTETTER, F. Über zwei Fälle einer seltenen Varietät der A. carotis interna. *Arch. Anat. Physiol. Wissen. Med.* 1886, pp. 396–400.

594 HOCHSTETTER, F. Über das normale Vorkommen von Klappen in den Magenverzweigungen der Pfortader beim Menschen und einigen Säugethieren. *Arch. Anat. Physiol. Wissen. Med.* 1887, pp. 137–142.

595* HOCHSTETTER, F. Über einige Fälle einer bisher anscheinend noch nicht beobachteten Varietät der A. cerebralis posterior des Menschen. *Z. Anat. Entwicklungsgesch.* 107:633–646, 1937.

596* HODES, P.J., CAMPOY, F., RIGGS, H.E. and P. BLY. Cerebral angiography. Fundamentals in anatomy and physiology. *Am. J. Roentgenol.* 70:61–82, 1953.

597 HOFER, K. and G. HOFER. Über den Verlauf der Arteria brachialis mit dem Nervus medianus zwischen den beiden Köpfen des Musculus pronator teres. *Anat. Anz.* 36:510–514, 1910.

598 HOLTBY, J.R.D. A pleural fold containing the vena azygos. *J. Anat. Physiol.* 69:236–239, 1915.

599 HOLTZAPFEL, G. Ungewöhnlicher Ursprung und Verlauf der Arteria subclavia dextra. *Anat. Hefte* 12:369–523, 1899.

600 HOPKINSON, J.P. Aneurism of the aorta, and unusual origin of right subclavian. *Am. J. Med. Sci.* 7:556–557, 1831.

601 HOU-JENSEN, H.M. Die Verästelung der Arteria renalis in der Niere des Menschen. *Z. Anat. Entwicklungsgesch.* 91:1–129, 1930.

602 HOWDEN, R. Case of double superior vena cava with left-sided arrangement of the azygos veins. *J. Anat. Physiol.* 21:72–75, 1887.

603 HOWE, W.W. JR., LACEY, T. and R.P. SCHWARTZ. A study of the gross anatomy of the arteries supplying the proximal portion of the femur and the acetabulum. *J. Bone Joint Surg.* 32A:856–866, 1950.

604 HÜBNER, H.J. Zum Verlauf der Arteria carotis interna im Bereich des Halses. *Anat. Anz.* 121:489–496, 1967.

605 HUECK, O. Eine seltene Tricuspidalismißbildung. *Arch. Pathol. Anat. Physiol. Klin. Med.* 319:247–264, 1950.

606* HUELKE, D.F. A study of the transverse cervical and dorsal scapular arteries. *Anat. Rec.* 132:233–245, 1958.

607* HUELKE, D.F. Variation in the origins of the branches of the axillary artery. *Anat. Rec.* 135:33–41, 1959.

608* HUELKE, D.F. The dorsal scapular artery. A proposed term for the artery to the rhomboids. *Anat. Rec.* 142:57–61, 1962.

609 HUFFMIRE, A.P. and G.C. BOWER. A case of persistence of the left superior vena cava in an aged adult. *Anat. Rec.* 17:127–129, 1919–20.

610 HUGHES, A.W. Abnormal arrangement of arteries in the region of the kidney and supra-renal bodies. *J. Anat. Physiol.* 26:305–307, 1892.

611 HUGHES, C.W. and P.C. RUMORE. Anomalous pulmonary veins. *Arch. Pathol.* 37:364–366, 1944.

612 HUIRA, A. An anomalous ophthalmic artery arising from the middle meningeal artery. *Anat. Anz.* 147:473–476, 1980.

613 HÜLSE, W. Beitrag zur Kenntnis der totalen Persistenz des Truncus arteriosus communis. *Arch. Pathol. Anat. Physiol. Klin. Med.* 225:16–23, 1918.

614* HUMPHREYS, G.H. II. The surgery of congenital heart disease. *Surg. Clin. North Am.* 1948, pp. 353–365.

615 HURLEY, L.E. and A.E. COATES. A case of right-sided aortic arch and persistent left superior vena cava. *J. Anat.* 61:333–339, 1927.

616* HUSEBY, R.A. and E.A. BOYDEN. Absence of the hepatic portion of the inferior vena cava with bilateral retention of the supracardinal system. *Anat. Rec.* 81:537–544, 1941.

617* HUTCHINSON, M.C.E. A study of the atrial arteries in man. *J. Anat.* 125:39–54, 1978.

618 HUTTON, W.K. Some anomalous coronary sinus. *J. Anat. Physiol.* 69:407–413, 1915.

619* HUU, N. Veine cave supérieure gauche et pathologie cardio-vasculaire moderne. *Assoc. Anatomistes Comptes Rendus* 43:619–635, 1956.

620 HUZLY, A. Zur Anatomie (Bildung und Rückbildung) des Ductus venosus Arantii (Beschreibung einer ungewöhnlichen Variation). *Anat. Anz.* 93:1–15, 1942.

621 HYRTL. Der Sinus ophthalmo-petrosus. *Wien. Med. Wochenschr.* N. 19 12:290–291, 1862.

622* INGALLS, N.W. Anomalous pulmonary vessels. *Anat. Rec.* 53:269–281, 1932.

623 IONESCU, M., MIHAIL, N. and C. IONESCU. Blutversorgung der normalen Niere des Menschen durch mehrfache Nierenschlagadern. *Anat. Anz.* 111:398–406, 1962.

624 ISSAJEW, P.O. Der doppelte Aortenbogen. *Anat. Anz.* 73:153–158, 1931–32.

625 IWANOW, G. Über einige Varietäten der Oberschenkelarterien beim Menschen. *Anat. Anz.* 64:269–275, 1927–28.

626 IYER, A.A. Some anomalies of origin of the vertebral artery. *J. Anat.* 62:121–122, 1927.

627 JAENSCH, P.A. Eine seltene Abweichung im Verlauf der Arteria anonyma. *Anat. Anz.* 55:138–142, 1922.

628 JAFFE, K. Fall von Mißbildung des Herzens und der Gefäße. *Z. Anat. Entwicklungsgesch.* 60:411–437, 1921.

629 JAGODZIŃSKA, K., KOHMANN, A. and S. KOHMANN. Variations of the arterial system in the region of the cecum and vermiform appendix. *Folia Morphol.* 30:198–201, 1971.

630* JAIN, K.K. Some observations on the anatomy of the middle cerebral artery. *Can. J. Surg.* 7:134–139, 1964.

631 JAIN, S.P. and S. HAZARY. Coronary arterial pattern in man and some other animals. *J. Anat. Soc. India.* 7:1–4, 1958.

632 JAKUBCZIK, I. and H. ZIEGLER. Seltene Anomalien der großen intrathoracalen Blut- und Lymphgefäßstämme. *Anat. Anz.* 112:257–268, 1963.

633* JAKUBOWICZ, M. and M. BRUSKA. A case of variations in arterial vascularization of the upper limb. *Folia Morphologica* 39:55–59, 1980.

634* JAMES, T.N. The arteries of the free ventricular walls in man. *Anat. Rec.* 136:371–384, 1960.

635 JAMES, T.N. and G.E. BURCH. The atrial coronary arteries in man. *Circulation* 17:90–98, 1958.

636 JANSKÝ, M., PLUCNAR, B. and Z. SVOBODA. Beitrag zum Studium von Varietäten der subkutanen Halsvenen des Menschen. *Acta Anat.* 37:298–310, 1959.

637 JARVIS, J.F. Asymptomatic right aortic arch in the adult: A case report and discussion. *Anat. Anz.* 119:305–311, 1966.

638 JASCHTSCHINSKI, S.N. Morphologie und Topographie des Arcus

volaris sublimis und profundus des Menschen. *Anat. Hefte* 7:161–188, 1897.
639* JDANOV, D.A. Anatomie du thoracique et des principaux collecteurs lymphatiques du tronc chez l'homme. *Acta Anat.* 37:20–47, 1959.
640* JEFFERSON, G. and D. STEWART. On the veins of the diploë. *Br. J. Surg.* 16:70–88, 1928–29.
641 JEIDELL, H. A note on the source and character of the early blood vessels of the kidney. *Anat. Rec.* 5:47–54, 1911.
642 JENNY, H. Abnorme einseitige Verdoppelung der Arteria thyreoidea inferior. *Anat. Anz.* 40:623–640, 1912.
643 JEWETT, H.J. Accessory renal vessels: Their influence in certain cases of hydronephrosis. *Surg. Gynecol. Obstet.* 68:666–676, 1939.
644 JEX-BLAKE, A.J. Obstruction of the oesophagus caused by a persistent ductus arteriosus. *Lancet* 2:542–544, 1926.
645 JOHANOWICZ-WIŚNIEWSKA, H. and E. RATAJCZYK-PAKALSKA. Arteries of the jejunum and ileum. *Folia Morphol.* 35:159–168, 1976.
646* JOHNSTON, E.V. and B.J. ANSON. Variations in the formation and vascular relationships of the bile ducts. *Surg. Gynecol. Obstet.* 94:669–686, 1952.
647 JOHNSTON, T.B. Anomaly of the vena cava inferior; with a note on the relationships which help to determine the nature of anomalies of some of the abdominal veins. *J. Anat. Physiol.* 47:235–245, 1913.
648 JOHNSTON, T.B. A rare anomaly of the arteria profunda fermoris. *Anat. Anz.* 42:269–272, 1912.
649 JOHNSTON, T.B. A rare vascular anomaly – opening of the upper left pulmonary vein into a persistent left superior vena cava. *J. Anat. Physiol.* 69:182–186, 1915.
650* JOESSEL. Neue Anomalien der Carotis externa und der Maxillaris interna. *Arch. Anat. Physiol.* 1878, pp. 433–437.
651* JORDAN, F.L.J. and T.TH. TER HAAR. Ein Fall vollkommenen Verschlusses der Aorta thoracica. *Anat. Anz.* 66:24–30, 1928–29.
652* JORDAN, J. Contribution to the variations of the azygos vein system in man. *Folia Morphol.* 23:41–43, 1964.
653* JORDAN, R.A., DRY, T.J. and J.E. EDWARDS. Anomalous origin of the right coronary artery from the pulmonary trunk. *Proc. Staff Meeting Mayo Clin.* 25:673–678, 1950.
654* JOSSIFOW, G.M. Der Anfang des Ductus thoracicus und dessen Erweiterung. *Arch. Anat. Physiol. Wissen. Med.* 1905, pp. 68–76.
655* JUE, K.L., RAGHIB, G., AMPLATZ, K., ADAMS, P. JR., and J.E. EDWARDS. Anomalous origin of the left pulmonary artery from the right pulmonary artery. *Am. J. Roentgenol.* 95:598–610, 1965.
656 JUSTER, M. and E. CATRITSIS. Valeur fonctionnelle des gros troncs artériels de l'avantbras. *Assoc. Anatomistes Comptes Rendus* 37:236–243, 1950.
657* KADANOFF, D. and G. BALKANSKY. Zwei Fälle mit seltenen Varietäten der Arterien der oberen Extremität. *Anat. Anz.* 118:289–296, 1966.
658 KADASNE, D.K. A case of abnormal ophthalmic artery. *J. Anat. Soc. India* 13:108–109, 1964.
659* KAESTNER, S. Eintreten der hinteren Cardinalvenen für die fehlende Vena cava inferior beim erwachsenen Menschen. *Arch. Anat. Physiol. Wissen. Med.* 1900, pp. 271–280.
660* KÄGI, J. Beitrag zur Topographie der Arteria transversa colli. Eine Untersuchung an 134 Halshälften. *Anat. Anz.* 107:168–186, 1959.
661 KAMATH, C.R. and E. THOMAS. Report of a case of solitary aortic trunk with a suggested classification of the anomaly on an embryological basis. *J. Anat. Soc. India* 9:24–32, 1960.
662 KANTOR, H. Tiefe Teilung der Arteria carotis communis. *Anat. Anz.* 26:492–496, 1905.
663 KAPLAN, H.A. Arteries of the brain: An anatomic study. *Acta Radiol.* 46:364–370, 1956.
664 KARCZ, R., KOHMANN, A. and S. KOHMANN. Varieties of the pulmonary trunk, aorta and coronary arteries of the heart. *Folia Morphol.* 30:341–345, 1971.
665 KARPOWICZ, S. Anomalie rare de l'art. ophthalamique chez l'homme. *Assoc. Anatomistes Comptes Rendus* 1939, pp. 167–168.
666 KARSTEN TALLMADGE, G. Un cas d'anomalie rare de l'artere hépatique. *Anat. Anz.* 85:433–435, 1937–38.
667 KARSTEN TALLMADGE, G. On the fixity of the gastroduodenal artery. *Anat. Anz.* 88:161–167, 1939.
668 KASAI, T. and S. CHIBA. Microscopic anatomy of the bronchial arteries. *Anat. Anz.* 145:166–181, 1979.
669 KATER, M.W. Case of multiple renal arteries. *J. Anat. Physiol.* 36:77, 1902.

670 KATÔ, H. Eine seltene Variation der A. transversa scapulae. *Folia Anat. Jpn.* 18:223–228, 1939.
671* KAUNITZ, P.L. Origin of left coronary artery from pulmonary artery. Review of the literature and report of two cases. *Am. Heart J.* 33:182–206, 1947.
672* KEDZIA, A. Morphologic variation of area drained by the basilar vein. *Folia Morphol.* 36:203–210, 1977.
673 KEDZIOR, E. and J. KUŚ. The blood vessels of the gallbladder. *Folia Morphol.* 24:357–365, 1965.
674* KEEN, J.A. A study of the arterial variations in the limbs, with special reference to symmetry of vascular patterns. *Am. J. Anat.* 108:245–261, 1961.
675 KEITH, A. Partial deficiency in the pericardium. *J. Anat. Physiol.* 41:6–7, 1907.
676* KEITH, A. Malformations of the heart. *Lancet* 2:359–363, 433–435, 519–523, 1909.
677 KEITH, A. Six specimens of abnormal heart. *J. Anat. Physiol.* 46:211–214, 1912.
678 KELLY, H.H. Some rare and new anomalies in man: with three cases of double femoral artery. *Am. J. Med. Sci.* 83:138–142, 1882.
679 KEMMETMÜLLER, H. Über eine seltene Varietät der Art. vertebralis. *Anat. Hefte* 44:305–362, 1911.
680 KEYES, D.C. and H.C. KEYES. A case of persistent left superior vena cava with reversed azygos system. *Anat. Rec.* 31:23–26, 1925.
681* KIELBASIŃSKI, G. Arteries of the inferior part of the vermis cerebelli in man. *Folia Morphol.* 35:149–157, 1976.
682 KIELBASIŃSKI, G. Course and area supplied by the anterior inferior cerebellar arteries in man. *Folia Morphol.* 36:181–190, 1977.
683* KIRKLIN, J.W. and O.T. CLAGETT. Vascular "rings" producing respiratory obstruction in infants. *Proc. Staff Meeting Mayo Clin.* 25:360–367, 1950.
684 KISS, A., PARTILLA, H. and E. PERNKOPF. Die Transposition der Venenmündungen. *Arch. Pathol. Anat. Physiol. Klin. Med.* 324:707–748, 1954.
685* KISS, F. Über einige Varietäten der Arteria hepatica und Arteria cystica. *Z. Anat. Entwicklungsgesch.* 81:601–617, 1926.
686 KLEINERMAN, J., YANG, W., HACKEL, D.B. and N. KAUFMAN. Absence of the transverse aortic arch. *Arch. Pathol.* 65:490–498, 1958.
687 KLEISS, E. Die verschiedenen Formen des Circulus arteriosus Willisi. *Anat. Anz.* 92:216–230, 1941–42.
688 KLEISS, E. Die Arteria cerebralis anterior. *Anat. Anz.* 95:353–372, 1944–45.
689 KLIMA, M. and J. RICHTER. Gefäßvarianten im Retroperitonealsitus. *Anat. Anz.* 138:56–61, 1975.
690* KLING, A. Variations of the arteries supplying the terminal portion of ileum and initial portion of ascending colon. *Folia Morphol.* 37:389–400, 1978.
691 KLING, A. and M. MICHNIEWICZ-NOWAK. Anastomoses between the colic branch of the ileocolic artery and the right colic artery. *Folia Morphol.* 35:307–312, 1976.
692 KLOPPER, Z.D. The normal kidney and its anomalies. *Am. J. Derm. Genit. Urin. Dis.* 16:61–63, 1912.
693 KNAPE. Eine seltene Herzmißbildung bei Situs inversus abdominis. *Arch. Pathol. Anat. Physiol. Klin. Med.* 209:473–476, 1912.
694 KNIGHT, H.O. An anomalous portal vein and its surgical dangers. *Ann. Surg.* 74:697–699, 1921.
695* KOCON, T. A rare variation of the cardiac coronary sinus. *Folia Morphol.* 22:11–14, 1963.
696* KODAMA, K. Fortbestand der V. supracardinalis dextra und sinistra bei einem Japaner. *Folia Anat. Jpn.* 9:72–76, 1930.
697 KODAMA, K. Persistenz des Ductus Cuvieri sinister bei einem Japaner. *Folia Anat. Jpn.* 9:365–369, 1931.
698 KOLESNIKOW, N. Ein seltener Fall der Mündung der Vena pulmonalis in die Vena cava superior. *Anat. Anz.* 74:233–237, 1932.
699 KOLLMANN, J. Abnormitäten im Bereich der Vena cava inferior. *Anat. Anz.* 8:75–80, 97–116, 1893.
700* KOMMERELL, B. Verlagerung des Ösophagus durch eine abnorm verlaufende Arteria subclavia dextra (Arteria lusoria). *Fortschr. Röntgenstr.* 54:590–595, 1936.
701* KORFEL, Z. and M. PANEK. A rare variation of the segmental arteries of the right lung in man observed *in vivo*. *Folia Morphol.* 26:223–225, 1967.
702* KORMANO, M. and H. SUORANTA. An angiographic study of the arterial pattern of the human testis. *Anat. Anz.* 128:69–76, 1970.
703 KOSIŃSKI, C. Observations on the superficial venous system of the lower extremity. *J. Anat.* 60:131–141, 1925.

704 KOSIŃSKI, C. Quelques observations sur les rameaux du tronc coeliaque et des artères mésentériques chez l'homme basée sur l'examen de 55 pièces anatomiques à Wilno. *Assoc. Anatomistes Comptes Rendus* 23:241–260, 1928.

705 KOSIŃSKI, H. The veins of the kidneys. *Folia Morphol.* 35:277–285, 1976.

706 KOSIR, V. Persistierende Kardinalvenen und fehlende V. cava inferior. *Anat. Anz.* 55:365–368, 1922.

707* KOSTINOVITCH, L.I. A case of simultaneous occurrence of a number of variations of the visceral branches of the abdominal aorta. *Anat. Rec.* 67:399–403, 1937.

708 KOTLJARTSCHUK, P.S. Ein Fall von doppeltem Aortenbogen. *Zentralbl. Allg. Pathol. Pathol. Anat.* 60:116–118, 1934.

709 KOWALSKI, E. and J. PAWLACZYK. Venous blood vessels of rare shape in the liver. *Folia Morphol.* 28:497–501, 1969.

710 KRAHN, V. and H. LUZIUS. Besondere Lagebeziehungen kleinerer Arterien zur Vena saphena magna im Bereich ihrer Einmündung in die Vena femoralis. *Anat. Anz.* 141:285–291, 1977.

711* KRECHOWIECKI, A., DANIEL, B. and S. WIECHOWSKI. Variation of the internal thoracic artery. *Folia Morphologica* 32:173–184, 1973.

712 KREN, O. Über eine seltene Anomalie der Arteria lingualis nebst einer kurzen Zusammenfassung ihrer häufigsten topographischen Varietäten. *Anat. Anz.* 73:107–110, 1931-32.

713 KRETSCHNY, F. Verschliessung der Vena anonyma dextra durch eine Struma substernalis. *Wien. Med. Wochenschr.* 27:3–6, 1877.

714 KRISHNAMURTI, A. Double left innominate vein in a Chinese cadaver. *J. Anat. Soc. India.* 13:100–102, 1964.

715* KRISTIN, A.D., EVANS, J.M. and A.E. BRIGULIO. Ebstein's anomaly of the tricuspid valve: Angiocardiographic diagnosis. *Am. Heart J.* 50:634–640, 1955.

716* KRIZAN, Z. Über eine ungewöhnliche Mündung der Vena ovarica dextra beim Menschen. *Anat. Anz.* 105:328–331, 1958.

717 KRIZAN, Z., HERMAN, O. and V. DZIDROV. Teilweiser Fortbestand des Supracardinalsystems neben der normalen Vena cava inferior beim Menschen. *Acta Anat.* 34:312–325, 1958.

718 KROHN, I. Über eine Anomalie der Aorta thoracica. *Anat. Anz.* 126:211–215, 1970.

719* KROKOWSKI, E. Arteria lusoria. *Fortschr. Röntgenstr.* 93:376–377, 1960.

720* KROPP, B.N. The lateral costal branch of the internal mammary artery. *J. Thoracic Surg.* 21:421–425, 1951.

721* KRUMBHAAR, E.B. and W.E. EHRICH. Varieties of single coronary artery in man, occurring as isolated cardiac anomalies. *Am. J. Med. Sci.* 196:407–413, 1938.

722 KUBOTA, K., HAIBARA, T. and M. NOGUCHI. A case of bilateral existence of ischiadic artery in man. *Folia Anat. Jpn.* 30:339–340, 1958.

723* KUBOTA, K., NOGUCHI, I., NARITA, M., NAKANO, T. and R. NAKAGAWA. A case of collateral circulation between the portal and systemic veins through the left para-umbilical and left inferior epigastric veins in man. *Folia Anat. Jpn.* 30:341–342, 1958.

724 KUHLENBECK, H. Teilweise erhaltene linke Vena supracardinalis, Fensterbildung im Bereiche der rechten Vena iliaca communis und akzessorische Nierengefäße bei einem erwachsenen Menschen. *Anat. Anz.* 70:117–122, 1930.

725* KUKWA, A. and A. ZBRODOWSKI. A rare case of origin of the superior thyroid artery from the left common carotid artery. *Folia Morphol.* 25:600–602, 1966.

726* KURRAT, H.J. and M. HESSE. Doppelbildung der Arteria testicularis sinistra und Verlaufsvarianten – Eine entwicklungsgeschichtliche Betrachtung. *Anat. Anz.* 145:303–307, 1979.

727 KURYLCIO, L., METERA, A., MODRZEWSKI, Z. and A. WILGOS-ZYŃSKI. Arteries supplying the gastric fundus in man. *Folia Morphol.* 29:154–159, 1970.

728* LABBÉ, C. Anomalies des sinus de la dure-mère. Dévelopment de ces sinus. Considérations sur la suppléance reciproque de ces canaux veineux dans les cas d'absence de l'un d'eux description de quelques sinus peu connus. *Arch. Physiol. S. 3* 1:1–27, 1883.

729* LABUNOWA, T. Cor biloculare in man. *Folia Morphol.* 26:154–160, 1967.

730 LAING, P.G. The blood supply of the femoral shaft. *J. Bone Joint Surg.* 35B:462–466, 1953.

731 LAING, P.G. The arterial supply of the adult humerus. *J. Bone Joint Surg.* 38A:1105–1116, 1956.

732 LAING, P.G. The blood supply of the bones of the forearm and hand. *Surg. Clin. North Am.* 40:311–320, 1960.

733 LANDER, H.H., LYMAN, R.Y. and B.J. ANSON. An anatomical consideration of the structure in the hepatic pedicle. A study of 100 consecutive cadavers. *Q. Bull. Northwestern University Medical School* 15:103–109, 1941.

734 LANE, W.A. Rare abnormality of the large arteries of the heart. *J. Anat. Physiol.* 21:97–100, 1887.

735 LANG, F.J. Zur Kenntnis der angeborenen Herzbeuteldefekte. *Arch. Pathol. Anat. Physiol. Klin. Med.* 230:608–621, 1921.

736 LANG, J., KÖTH, R. and G. REISS. Über die Bildung, die Zuflüsse und den Verlauf der V. basalis und der V. cerebri interna. *Anat. Anz.* 150:385–423, 1981.

737 LA ROCCA, C. Rameau présternal, non encore décrit, de l'artère thyréoïdienne inférieure droite. *Arch. Ital. Biol.* 47:486–487, 1907.

738 LATARJET and LAROYENNE. Les artères de l'uretée. *Assoc. Anatomistes Comptes Rendus* 10:109–116, 1908.

739 LATARJET, A. Anatomie de la veine infra-pylorique (Pyloric vein). *Lyon Chir.* 6:377–388, 1911.

740 LATARJET, A. and J. MURARD. La vascularisation artérielle du thymus. *Lyon Chir.* 6:493–501, 1911.

741* LATARJET, M. and P.E. DUROUX. Les branches auriculaires des artères coronaires du coeur. *Assoc. Anatomistes Comptes Rendus* 43:450–465, 1956.

742* LATIMER, H.B. and H.N. VIRDEN. A case of complete absence of the inferior vena cava. *J. Kansas Med. Soc.* 45:346–347, 353, 1944.

743 LAUBER, H. Ein Fall von teilweiser Persistenz der hinteren Cardinalvenen beim Menschen. *Anat. Anz.* 19:590–594, 1901.

744 LAUBER, H. Über einige Varietäten im Verlaufe der Arteria maxillaris interna. *Anat. Anz.* 19:444–448, 1901.

745 LAUENSTEIN, C. Varietät der Klappen des rechten Atrium. *Arch. Pathol. Anat. Klin. Med.* 68:632–633, 1876.

746 LAUX, G., GUERRIER, Y. and G. MARCHAL. La bourse retrocardiaque. *Assoc. Anatomistes Comptes Rendus* 37:271–278, 1950.

747 LAUX, G. and P.E. RAPP. Le dispositif veineux du lobe de Spigel. *Assoc. Anatomistes Comptes Rendus* 40:264–271, 1953.

748 LAUX, G. and J.B. PRIOTON. Gaine des muscles spinaux son système artériel intra-aponeurotique. *Assoc. Anatomistes Comptes Rendus* 41:817–830, 1954.

749 LAWRENCE, T.W.P. and D. NABARRO. A case of congenital malformation of the heart, with abnormalities of the abdominal viscera: Absence of spleen, absence of hepatic section of inferior vena cava. *J. Anat. Physiol.* 36:63–75, 1902.

750* LAZORTHES, G. and J. POULHÈS. Les amarres veineuses du cerveau et du cervelet. *Assoc. Anatomistes Comptes Rendus* 35:203–209, 1948.

751 LAZORTHES, G., POULHÈS J. and J. ESPAGNO. Les artères du cervelet. *Association Anatomistes Comptes Rendus* 37:279–288, 1950.

752 LAZORTHES, G., POULHÈS J. and J. ESPAGNO. Les territoires vasculaires du cortex cérébelleux. *Assoc. Anatomistes Comptes Rendus* 37:289–297, 1950.

753 LAZORTHES, G., POULHÈS J. and J. ESPAGNO. La vascularisation artérielle des noyaux de cervelet. *Assoc. Anatomistes Comptes Rendus* 38:649–653, 1951.

754 LEBERT. Über einen Fall von Ursprung der Aorta aus dem rechten Ventrikel neben der A. pulmonalis mit Offenbleiben des Eirundenloches, sowie der Kammerscheidewand, ohne Cyanose, und Erhaltung des Lebens bis zum zwanzigsten Jahre. *Arch. Pathol. Anat. Physiol. Klin. Med.* 28:405–414, 1863.

755 LE DOUBLE. Des variations des troncs de la convexité de la crosse de l'aorte de l'homme et principalement de la reproduction chez lui de la formule aortique de l'orang, du gibbon, des singes quadropèdes et des carnassiers. *Assoc. Anatomistes Comptes Rendus* 3:242–246, 1901.

756* LEE, I.N. Anomalous relationship of the inferior thyroid artery. *Anat. Rec.* 122:499–506, 1955.

757 LEES, D.B. Case of malformation of the heart, with transposition of the aorta and pulmonary artery. *Trans. Pathol. Soc. Lond.* 31:58–62, 1880.

758 LE GENDRE. Anomalie de la veine cave inférieure. *Soc. Biol. Comptes Rendus des Séances et Mémoires* 11:205, 1859.

759 LEIRCHE, R. and F. VILLEMIN. Le rameau hepatique de l'artère coronaire stomachique. *Soc. Biol. Comptes Rendus Hebdomadaires des Séances et Mémoires* 61:721–722, 1906.

760 LEROUX, M.A. Anastomose de l'artère vertébrale avec la cervicale profunde. *Soc. Biol. Comptes Rendus des Séances et Mémoires* 2:196–198, 1850.

761 LEUDET. Anomalie de la veine cave. *Soc. Biol. Comptes Rendus Hebdomadaires des Séances et Mémoires* 4:180, 1852.

762 LEVI, G. Ovservations sur les variations des artères iliaques. *Arch. Ital. Biol.* 37:489, 1902.
763 LEVI, G. Morfologia delle arterie iliache. *Arch. Ital. Anat. Embriol.* 1:120–172, 295–346, 523–605, 1902.
764 LEVI, G. Les variations des artères surrénales et des artères rénales étudiées avec la méthode statistique sériale. *Arch. Ital. Biol.* 52:148, 1909.
765 LEVI, G. Le variazione delle arterie surrenali e renali studiate col metodo statistico seriale. *Archiv. Ital. Anat. Embriol.* 8:35–71, 1909.
766* LEWIS, C.W.D and J.N.M. PARRY. Double aortic arch. *Anat. Rec.* 101:613–615, 1948.
767 LEYDEN, E. Ein Bemerkenswerther Fall von Stenose des Ostium aorticum. *Arch. Pathol. Anat. Physiol. Klin. Med.* 29:197–198, 1864.
768* LIBERSA and VOISIN. Vascularisation artérielle du ganglion de Gasser. *Assoc. Anatomistes Comptes Rendus* 38:663–670, 1951.
769* LIE, J.T. The malformation complex of the absence of the arch of the aorta – Steidele's complex. *Am. Heart J.* 73:615–625, 1967.
770* LIEBOW, A.A. Patterns of origin and distribution of the major bronchial arteries in man. *Am. J. Anat.* 117:19–32, 1965.
771* LIECHTY, J.D., SHIELDS, T.W. and B.J. ANSON. Variations pertaining to the aortic arches and their branches. *Q. Bull. Northwestern University Medical School* 31:136–143, 1957.
772 LIÉGEOIS. Anomalies des reins. *Soc. Biol. Comptes Rendus des Séances et Mémoires* 11:255–256, 1859.
773 LIND, J. Heart volume in normal infants. *Acta Radiol. [Suppl.]* 82:1–127, 1950.
774 LINDSAY, H.C. An abnormal vena hemiazygos. *J. Anat.* 59:438, 1925.
775* LIPSHUTZ, B. A composite study of the coeliac axis artery. *Ann. Surg.* 65:159–169, 1917.
776* LIPSHUTZ, B. A composite study of the hypogastric artery and its branches. *Ann. Surg.* 67:584–608, 1918.
777* LIPSHUTZ, B.B. Studies on the blood vascular tree. I. A composite study of the femoral artery. *Anat. Rec.* 10:361–370, 1916.
778 LIPSHUTZ, B. and C. HOFFMAN. Renal arterial variations and extra-peritoneal abdominal nephrectomy. *Ann. Surg.* 84:525–532, 1926.
779* LITTLE, J.M., ZYLSTRA, P.L., WEST, J. and J. MAY. Circulatory patterns in the normal hand. *Br. J. Surg.* 60:652–655, 1973.
780 LIVINI, F. Le type normal et les variations de l'A. carotis externa. *Arch. Ital. Biol.* 39:487, 1903.
781 LIZE, I. Double inferior vena cava. *Folia Morphol.* 28:241–243, 1969.
782 LIZE, I. Abnormal origin of the great blood vessels from the aortic arch. *Folia Morphol.* 29:355–357, 1970.
783* LJUBOMUDROFF, A.P. Zur Morphologie der Arterienanastomosen in der Fossa cubiti. *Z. Anat. Entwicklungsgesch.* 84:795–813, 1928.
784 LLORCA, O. Über den rechtsseitigen Verlauf des Ductus thoracicus bei erhaltener rechter Aortenwurzel. *Anat. Anz.* 81:283–290, 1935–36.
785* LOCKHART, R.D. Complete double aortic arch. *J. Anat.* 64:189–193, 1930.
786 LOCKWOOD, C.B. Right aortic arch. *Trans. Pathol. Soc. Lond.* 35:132, 1884.
787* LOETZKE, H.H. and W. KLEINAU. Gleichzeitiges Vorkommen der Aa. brachialis superficialis, radialis und antebrachialis dors. superfic. sowie deren Aufzweigungen. *Anat. Anz.* 122:137–141, 1968.
788 LOHMAN, A.H.M. and E.J. DEKORT. Eine Abnormität des Ursprunges und Verlaufes der Bronchialarerien der linken Lunge. *Anat. Anz.* 112:12–18, 1963.
789* LOMBARDI, G. Il quadro radiologico della arteria lusoria (arteria succlavia destra anomala). *Radiol. Med.* 37:809–816, 1936.
790* LONGO, L. Le anomalie del poligono di Willis nell'uomo studiate comparativamente in alcuni mammiferi ed uccelli. *Anat. Anz.* 27:170–176, 1905.
791 LOONEY, W.W. An unusual aberrant right internal spermatic vein. *Anat. Rec.* 23:333–334, 1922.
792 LOW, F.N. An anomalous middle meningeal artery. *Anat. Rec.* 95:347–351, 1946.
793 LÖWENTHAL, H. Über angeborene Dextrocardie ohne Situs viscerum inversus; Fehlen der Arteria pulmonalis, mit Ductus Botalli als arterielles Gefäss. Lücke im Septum ventriculorum. Kleiner linker Ventrikel. *Z. Klin. Med.* 41:130–136, 1900.
794* LOWREY, L.G. Anomaly in the circle of Willis, due to absence of the right internal carotid artery. *Anat. Rec.* 10:221–222, 1916.

795 LUCAS, M.F. A case of double inferior vena cava. *J. Anat. Physiol.* 51:69–70, 1917.
796 LUCIEN, M. Considérations sur le développement de la veine cave supérieure chez l'homme. *Assoc. Anatomistes Comptes Rendus* 41:977–981, 1954.
797 LUNA, E. Sur l'irrigation artérielle des glandes surrénales chez l'homme. Sur la classification des artères surrénales. Le développement des artères surrénales chez l'homme. *Arch. Ital. Biol.* 53:159–160, 1910.
798 LUNDQUIST, C. and K. AMPLATZ. Anomalous origin of the left coronary artery from the pulmonary artery. *Am. J. Roentgenol.* 95:611–620, 1965.
799 LURJE, A.S. Über einige Eigentümlichkeiten der Topographie der A. thyreoid. infer. im Zusammenhang mit den sie umgebenden Bildungen. *Anat. Anz.* 79:113–128, 1934–35.
800 LURJE, A.S. Einige topographisch-morphologische Angaben über ein aus dem Truncus thyreocervicalis in medialer Richtung abgehendes Arterienzweiglein (Ramus tracheooesophageus). *Anat. Anz.* 79:200–202, 1934–35.
801* LURJE, A.S. On the topographical anatomy of the internal maxillary artery. *Acta Anat.* 2:219–231, 1947.
802 LÜTKEN, P. Investigation into the position of the nutrient foramina and the direction of the vessel canals in the shafts of the humerus and femur in man. *Acta Anat.* 9:57–68, 1950.
803* LYLE, D.J. Annual Reviews: Neuro-ophthalmology. *AMA Arch. Ophthal* 56:768–689, 1956.
804 LYON, R.A., JOHANSMANN, R.J. and K. DODD. Anomalous origin of the left coronary artery. *Am. J. Dis. Child.* 72:675–690, 1946.
805 LYSSENKOW, N.K. Fall einer seltenen Variation des Ursprunges der A. hepatica (Truncus hepato-mesentericus Adachi). *Anat. Anz.* 75:184–187, 1932–33.
806 MACALISTER, A. Multiple renal arteries. *J. Anat. Physiol.* 17:250–252, 1883.
807 MADOFF, I.M., GAENSLER, E.A. and J. STRIEDER. Congenital absence of the right pulmonary artery; diagnosis by angiocardiography with cardiorespiratory studies. *N. Engl. J. Med.* 247:149–157, 1952.
808* MAIER, H.C. Absence or hypoplasia of pulmonary artery with anomalous systemic arteries to the lung. *J. Thoracic Surg.* 28:145–162, 1954.
809 MAKOWSKI, A. and W. SIEROCIŃSKI. A variation of the inferior vena cava in man. *Folia Morphol.* 34:103–108, 1975.
810 MALHOTRA, V.K., TEWARI, S.P., TEWARI, P.S. and S.K. AGARWAL. Coronary sinus and its tributaries. *Anat. Anz.* 148:331–332, 1980.
811 MALINIAC, J.W. Arterial blood supply of the breast. Revised anatomic data relating to reconstructive surgery. *Arch. Surg.* 47:329–343, 1943.
812 MALL, F.P. On the development of the blood vessel of the brain in the human embryo. *Am. J. Anat.* 4:1–18, 1905.
813 MANN, M. Cor triloculare biatriatum. Eine entwicklungsgeschichtliche Studie. *Beitr. Pathol. Anat. Allg. Pathol.* 6:485–510, 1889.
814 MANNERS-SMITH, T. The limb arteries of primates. *J. Anat. Physiol.* 46:95–172, 1912.
815 MANNO, A. Arteriac plantares pedis mammalium. *Int. Monatsschr. Anat. Physiol.* 22:293–359, 1905.
816 MANNO, A. Arteriae plantares pedis mammalium. *Arch. Ital. Biol.* 46:305–307, 1906.
817 MANNO, A. Arteria peronea communis, arteria peronea profunda, arteria peronea superficialis. Contributo alla morfologia della circolazione arteriosa nell'arto abdominale. *Int. Monatsschr. Anat. Physiol.* 23:272–334, 1906.
818 MANNO, A. Sur une variété d'arteria ischiatica chez l'homme. *Arch. Ital. Biol.* 47:487, 1907.
819 MANNO, A. Arteria peronea communis, arteria peronea profunda, arteria peronea superficialis. *Arch. Ital. Biol.* 47:487–489, 1907.
820* MANNU, A. Il confluente dei seni della dura madre, la variazioni e il suo significato. *Int. Monatsschr. Anat. Physiol.* 24:304–397, 1908.
821 MANNU, A. Variazioni dell'arteria vertebralis nell'uomo e nei mammiferi. *Arch. Italiano Anatomia Embriologia* 13:79–113, 1914–15.
822* MANSBERGER, A.R. Saphenofemoral junction anomalies. *Surg. Obstet. Gynecol.* 91:533–536, 1950.
823* MANSFIELD, A.O. and J.M. HOWARD. Absence of both common iliac arteries. A case report. *Anat. Rec.* 150:363–381, 1964.

824 MARSH, T. Abnormal arrangement of spermatic arteries and of the right cord of the testis. *J. Anat. Physiol.* 32:216–217, 1898.

825 MARTENS, G. Zwei Fälle von Aortenatresie. *Arch. Pathol. Anat. Physiol. Klin. Med.* 121:322–335, 1890.

826 MARTIN, C.P. Anatomical notes. A case of congenital abnormality of the heart with apparently an unusual abnormality of the great vessels. *J. Anat.* 65:395–398, 1931.

827 MARTIN, K. Ungewöhnlicher Verlauf einer Arteria suprarenalis. *Acta Anat.* 24:48–50, 1955.

828 MARTINOTTI, G. and G. SPERINO. Des anomalies numériques des valvules sémi-lunaires, aortiques et pulmonaires. *Arch. Ital. Biol.* 6:181–193, 1884.

829 MARUYAMA, K. Seltene Varietät der Arterien der oberen Extremität bei einem Japaner (A. brachialis superficialis lateralis inferior, A. antebrachialis superficialis mediano-ulnaris, Arcus volaris superficialis vom Typus mediano-ulnaris). *Folia Anat. Jpn.* 22:551–567, 1944.

830 MAUTNER, H. and M. LÖWY. Transposition der Aorta oder Persistenz einer rechtskammerigen Aorta. *Arch. Pathol. Anat. Physiol. Klin. Med.* 229:337–344, 1921.

831 MAXIMEMKOV, A.N. Structural and functional peculiarities in some parts of the venous system. *Anat. Rec.* 136:239, 1960.

832* MAXWELL, E.V. and G.S. ERWIN. Four cases of anomalous inferior vena cava with an explanation of their developmental origin. *J. Anat.* 62:184–197, 1928.

833 MAYO, C.W. Blood supply of the colon: Surgical consideration. *Surg. Clin. N. Am.* 35:1117–1122, 1955.

834* McAFEE, D.K., ANSON, B.J. and J.J. McDONALD. Variation in the point of bifurcation of the common carotid artery. *Q. Bull. Northwestern University Medical School* 27:226–229, 1953.

835* McCLURE, C.F.W. and E.G. BUTLER. The development of the vena cava in man. *Am. J. Anat.* 35:331–383, 1925.

836* McCORMACK, L.J. and B.J. ANSON. The arterial supply of the ureter. *Q. Bull. Northwestern University Medical School* 24:291–294, 1950.

837* McCORMACK, L.J., CAULDWELL, E.W. and B.J. ANSON. Brachial and antebrachial arterial patterns; a study of 750 extremities. *Surg. Gynecol. Obstet.* 96:43–54, 1953.

838 McCOTTER, R.E. On the occurrence of pulmonary arteries arising from the thoracic aorta. *Anat. Rec.* 4:291–298, 1910.

839* McCOTTER, R.E. Three cases of the persistence of the left superior vena cava. *Anat. Rec.* 10:371–383, 1915–16.

840 McCRAE, J. A case of congenital atresia of pulmonary artery, with transposition of viscera; a second case of transposition. *J. Anat.* 40:28–33, 1905.

841* McCULLOUGH, A.W. Some anomalies of the cerebral arterial circle (of Willis) and related vessels. *Anat. Rec.* 142:537–543, 1962.

842* McDONALD, J.J and B.J. ANSON. Variations in the origin of arteries derived from the aortic arch, in American Whites and Negroes. *Am. J. Phys. Anthropol.* 27:91–107, 1940.

843* McKIM, J.S. and F.W. WIGLESWORTH. Absence of left pulmonary artery. A report of six cases with autopsy findings in three. *Am. Heart J.* 47:845–859, 1954.

844 MEARS, W.P. Abnormal distribution of arteries and veins in the neck. *J. Anat. Physiol.* 22:304–306, 1888.

845* MEHNERT, E. Über die klinische Bedeutung der Oesophagus- und Aortenvariationen. *Arch. Klin. Chir.* 58:183–245, 1899.

846 MEILLÈRE, J. Etude de la vascularization des tuniques du segment gauche de colon ses applications chirurgicales. *Ann. Anat. Pathol. (Paris)* 4:867–888, 1927.

847* MEILLIÈRE, D. Variations in the abdominal arteries applied to radical surgery and transplantation procedure. *Rev. Surg.* 25:305–313, 1968.

848* MENKE, J.F. An anomalous a. bronchialis dextra from the a. subclavia dextra, secondarily connected to the aorta thoracalis. *Anat. Rec.* 65:55–58, 1936.

849 MERKLIN, R.J. Arterial supply of the suprarenal gland. *Anat. Rec.* 144:359–371, 1962.

850* MERKLIN, R.J. and N.A. MICHELS. The variant renal and suprarenal blood supply with data on the inferior phrenic, ureteral and gonadal arteries. A statistical analysis based on 185 dissections and review of the literature. *J. Int. Coll. Surg.* 29:41–76, 1958.

851 MESTDAGH, H. Anatomie des Arcus volaris profundus beim Menschen. *Anat. Anz.* 147:180–187, 1980.

852* METZ, H., MURRAY-LESLIE, R.M., BANNISTER, R.G., BULL, J.W.D. and J. MARSHALL. Kinking of the internal carotid artery in relation to cerebrovascular disease. *Lancet* 1:424–426, 1961.

853 MEYER, H. Über die Transposition der aus dem Herzen hervortretenden grossen Arterienstämme. *Arch. Pathol. Anat. Physiol. Klin. Med.* 12:364–385, 1857.

854* MEYER, P. Contribution à l'étude de la première artère jéjunale. *Assoc. Anatomistes Comptes Rendus* 39:250–257, 1952.

855 MEYER, P. Etude de tronc artériel pancréatico-duodéno-jéjunal. *Assoc. Anatomistes Comptes Rendus* 40:99–119, 1953.

856 MIAŚKIEWICZ, C. and M. AUGUSTYN. The system of superficial veins of the anterior ulnar region in negroes. *Folia Morphol.* 29:15–19, 1970.

857* MIAŚKIEWICZ, C., PATYK, K. and A. WIŚNIEWSKI. Variation of the course of the great saphenous vein on the leg. *Folia Morphol.* 29:374–380, 1970.

858* MICHAELSOHN, A. Einmündung aller Lungenvenen in die persistierende Vena cava superior sinistra und Cor biloculare bei einem 21jährigen Manne. *Frankfurter Z. Pathol.* 23:222–246, 1920.

859* MICHALEWSKI, K. A case of origin of the cystic artery from the superior mesenteric artery. *Folia Morphol.* 24:366–368, 1965.

860* MICHALEWSKI, K. Topography of the bronchial branches of the aorta. *Folia Morphol.* 28:417–441, 1969.

861 MICHNIEWICZ-NOWAK, M. Arteries supplying the ascending colon in man. *Folia Morphol.* 34:293–300, 1975.

862* MICHELS, N.A. The variational anatomy of the spleen and splenic artery. *Am. J. Anat.* 70:21–72, 1942.

863* MICHELS, N.A. The hepatic, cystic and retroduodenal arteries and their relations to the biliary ducts. *Ann. Surg.* 133:503–524, 1951.

864* MICHELS, N.A. Variational anatomy of the hepatic, cystic and retroduodenal arteries. *Arch. Surg.* 66:20–34, 1953.

865 MICHELS, N.A. Newer anatomy of the liver – Variant blood supply and collateral circulation. *JAMA* 172:125–132, 1960.

866* MICHELS, N.A. The anatomic variations of the arterial pancreaticoduodenal arcades. Their import in regional resection involving the gall-bladder, bile ducts, liver, pancreas and part of the small and large intestine. *J. Int. Coll. Surg.* 37:13–40, 1962.

867* MICHELS, N.A., SIDDHARTH, P., KORNBLITH, P.L. and W.W. PARKE. The variant blood supply to the small and large intestines. Its import in regional resections. A new anatomic study based on four hundred dissections with a complete review of the literature. *J. Int. Coll. Surg.* 39:127–170, 1963.

868* MICHELS, N.A., SIDDHARTH, P., KORNBLITH, P.L. and W.W. PARKE. The variant blood supply to the descending colon, rectosigmoid and rectum based on 400 dissections. *Dis. Colon Rectum* 8:251–278, 1965.

869 MICHELSSON, G. Über eine obere Extremität mit mehrfachen Arterienvarietäten. *Anat. Anz.* 53:241–258, 1920–21.

870* MIERZWA, J. and T. KOZIELEC. Variation of the anterior cardiac veins and their orifices in the right atrium in man. *Folia Morphol.* 34:125–133, 1975.

871* MIERZWA, J. and G. CHRZANOWSKA. Arteries of horseshoe kidney with fused lower poles. *Folia Morphol.* 37:381–388, 1978.

872 MILLER, R.A. Observations upon arrangement of the axillary artery and brachial plexus. *Am. J. Anat.* 64:143–163, 1939.

873* MILLOY, F.J., ANSON, B.J. and D.K. McAFFEE. The rectus abdominis muscle and the epigastric arterias. *Surg. Gynecol. Obstet.* 110:293–302, 1960.

874* MILLOY, F.J., ANSON, B.J. and E.W. CAULDWELL. Variations in the inferior caval veins and in their renal and lumbar communications. *Surg. Gynecol. Obstet.* 115:131–142, 1962.

875 MILLS, C.P. An unusual combination of aberrant hepatic arteries. *Br. J. Surg.* 41:181–182, 1953–54.

876 MINGTZU, P'AN. The origin of branches of the axillary artery in the Chinese. *Am. J. Phys. Anthropol.* 27:269–279, 1940.

877 MINKIN, A. Zur Frage des rechtsseitigen Verlaufes des Ductus thoracicus. *Anat. Anz.* 60:314–318, 1925–26.

878 MINNE, J. and C. LIBERSA. Les artères du bulbe olfactif. *Assoc. Anatomistes Comptes Rendus* 38:723–730, 1951.

879* MINNE, J., VOISID, C. and C. LIBERSA. A propos de l'irrigation artérielle de l'estomac et du bulbe duodenal. *Assoc. Anatomistes Comptes Rendus* 40:396–408, 1953.

880 MIRRIZZI, P.L. Beitrag zur Abnormität des rechten Astes der Leberarterie. *Zentralbl. Chir.* 60:1171–1174, 1933.

881 MISRA, B.D. The arteria mediana. *J. Anat. Soc. India* 4:48, 1955.

882 MITCHELL, S.W. Aneurism of an anomalous artery causing antero-posterior division of the chiasm of the optic nerves and producing bitemporal hemianopsia. *J. Nerv. Ment. Dis.* 16:44–62, 1889.

883 MITCHELL, G. A rare abnormality of the hepatic artery. *Br. J. Surg.* 22:178–180, 1934–35.

884* MITTERWALLNER, F. VON. Variationsstatistische Untersuchungen an den basalen Hirngefäßen. *Acta Anat.* 24:51–88, 1955.

885* MLYNARCZYK, L., WOŹNIAK, W. and A. KIERSZ. Varianten in der Anzahl und im Verlauf der Nierenarterien. *Anat. Anz.* 118:67–81, 1966.

886 MOFFAT, D.B. A case of persistence of the primitive olfactory artery. *Anat. Anz.* 121:477–479, 1967.

887* MOLLER, J.H. and J.E. EDWARDS. Interruption of aortic arch. Anatomic patterns and associated cardiac malformations. *Am. J. Roentgenol.* 95:557–572, 1965.

888 MÖNCKEBERG, J.G. Über eine seltene Anomalie des Koronararterienabgangs. *Zentralbl. Herzen Gefäßkr.* 6:441–445, 1914.

889 MÖNCKEBERG, J.G. Das Verhalten des Sinusknotens bei Fehlen der Vena cava superior dextra und Persistenz der Vena cava superior sinistra. *Beitr. Pathol. Anat. Allg. Pathol.* 69:537–548, 1921.

890 MONIZ, E. L'angiographie cérébrale chez le vivant son importance anatomique. *Assoc. Anatomistes Comptes Rendus* 28:454–464, 1933.

891 MONTEIRO, H.B. Multiple anomalies in a human heart. *J. Anat.* 52:335–342, 1918.

892 MOORE, R.L. Congenital deficiency of the pericardium. *Arch. Surg.* 11:765–777, 1925.

893 MOOSMAN, D.A. Where and how to find the cystic artery during cholecystectomy. *Surg. Gynecol. Obstet.* 141:769–772, 1975.

894 MORE, R.H. and G.L. DUFF. The renal arterial vasculature in man. *Am. J. Pathol.* 27:95–117, 1951.

895 MÖRIKE, K.D. Der Verlauf der Nierenarterien und ihr möglicher Einfluß auf die Lage der Nieren. *Anat. Anz.* 116:485–502, 1965.

896 MORISON, B.G. The arrangement of the azygos and superior intercostal veins in the thorax. *J. Anat. Physiol.* 13:346–349, 1879.

897 MORISON, D.M. A study of the renal circulation with special reference to its finer distribution. *Am. J. Anat.* 37:53–93, 1926.

898* MORRIS, A.A. and C.M. PECK. Roentgenographic study of the variations in the normal anterior cerebral artery. One hundred cases studied in the lateral plane. *Am. J. Roentgenol.* 74:818–826, 1955.

899* MORTON, W.R.M. Pre-aortic drainage of the hemiazygos veins. Report of two cases. *Anat. Rec.* 101:187–191, 1948.

900 MOUCHET, A. and A. NOUREDDINE. L'artère du pilier antérieur du ventricule droit ou artère de la branche droite du faisceau de His. *Assoc. Anatomistes Comptes Rendus* 21:415–421, 1926.

901 MOUCHET, A. and A. NOUREDDINE. Note sur les artères auriculaires du coeur chez l'homme. *Assoc. Anatomistes Comptes Rendus* 24:367–376, 1929.

902* MOUNIEUR-KUHN, A. and A. BOUCHEF. L'artère choroïdienne antèrieure. *Assoc. Anatomistes Comptes Rendus* 43:598–618, 1956.

903* MOZERSKY, D.J., BUCKLEY, C.J., HAGOOD, C.O., JR., CAPPS, W.F. JR., and F.J. DANNEMILLER, JR. Ultrasonic evaluation of the palmar circulation. A useful adjunct to radial artery cannulation. *Am. J. Surg.* 126:810–812, 1973.

904* MÜLLER, E. Beiträge zur Morphologie des Gefäßsystems. Die Armarterien des Menschen. *Anat. Hefte* 22:376–574, 1903.

905 MUSGROVE, J. Bifurcation of the femoral artery with subsequent reunion. *J. Anat. Physiol.* 26:239–244, 1892.

906 MUSGROVE, J. Additional note on bifurcation of the femoral artery. *J. Anat. Physiol.* 26:555–556, 1892.

907 MUSGROVE, J. Origin of ophthalmic artery from the middle meningeal. *J. Anat. Physiol.* 27:279–281, 1893.

908 MUTEL. Triplicité de l'artère hépatique. *Soc. Biol. Comptes Rendus Hebdomadaires des Séances et Mémoires* 83:81–83, 1920.

909 NABARRO, D. Two hearts showing peculiarities of the great veins. *J. Anat. Physiol.* 37:382–391, 1903.

910 NANDY, K. and C.B. BLAIR, JR. Double superior venae cavae with completely paired azygos veins. *Anat. Rec.* 15:1–9, 1965.

911* NATHAN, H. Observations on aberrant renal arteries curving around and compressing the renal vein. *Circulation* 18:1131–1134, 1958.

912 NATHAN, H. Anatomical observations on the course of the azygos vein. *Thorax* 15:229–232, 1960.

913 NATHAN, H., TOBIAS, P.V. and M.D. WELLSTED. An unusual case of right and left testicular arteries arching over the left renal vein. *Br. J. Urol.* 48:135–138, 1976.

914* NATHAN, H. and M.R. SEIDEL. The association of a retroesophageal right subclavian artery, a right-sided terminating thoracic duct, and a left vertebral artery of aortic origin: Anatomical and clinical considerations. *Acta Anat.* 117:362–373, 1983.

915 NAVARRO, J.A.C., FILHO, J.L.T. and N.L. ZORZETTO. Anatomy of the maxillary artery in (and) to the pterygomaxillopalatine fossa. *Anat. Anz.* 152:413–433, 1982.

916 NEIL, C.A. Development of the pulmonary veins with reference to the embryology of anomalies of pulmonary venous return. *Pediatrics* 18:880–887, 1956.

917 NEUBERGER, H. Über einige Arterienvarietäten am Hals. *Anat. Anz.* 41:618–625, 1912.

918 NEUBERGER, H. Ein Fall von vollkommener Persistenz der linken Vena cardinalis posterior bei fehlender Vena cava inferior. *Anat. Anz.* 43:65–80, 1913.

919 NEUHAUSER, E.B.D. Roentgen diagnosis of double aortic arch and other anomalies of the great vessels. *Am. J. Roentgenol.* 56:1–12, 1946.

920 NEUHAUSER, E.B.D. Tracheo-esophageal constriction produced by right aortic arch and left ligamentum arteriosum. *Am. J. Roentgenol.* 62:493–499, 1949.

921* NIKOLAJEW, P.W. Einige Fälle seltener Variationen in den Abzweigungen der Äste von der Aorta. *Anat. Anzeiger* 76:53–61, 1933.

922 NITSCHKE, T. and F. PREUSS. Die Hauptäste der A. iliaca int. bei Mensch und Haussäugetieren in vergleichend-anatomisch häufigster Reihenfolge. *Anat. Anz.* 128:439–453, 1971.

923* NIZANKOWSKI, C. Seltener Fall eines gemeinsamen Stammes der Schilddrüsenzungengesichtsschlagader beim Menschen. (Truncus thyrolinguofacialis). *Anat. Anz.* 132:530–534, 1972.

924* NIZANKOWSKI, C., RAJCHEL, Z. and M. ZIÓŁKOWSKI. Abnormal origin of arteries from the aorta arch in man. *Folia Morphol.* 34:109–116, 1975.

925* NIZANKOWSKI, C., RAJCHEL, Z. and M. ZIÓŁKOWSKI. Cor biloculare in man. *Folia Morphol.* 35:55–65, 1976.

926* NIZANKOWSKI, C., NOCZYŃSKI, L. and E. SUDER. Variability of the origin of ramifications of the subclavian artery in humans (studies on the Polish population). *Folia Morphol.* 41:281–294, 1982.

927 NIZANKOWSKI, C., ZIÓŁKOWSKI, M., MAREK, J. and J. RABCZYŃSKI. Fossa ovalis of the interatrial septum in humans. *Folia Morphol.* 41:413–418, 1982.

928* NOCZYŃSKI, L. Vascular anomalies of the hepatic arteries and their practical importance. *Folia Morphol.* 35:85–93, 1976.

929 NOER, R.J. The blood vessels of the jejunum and ileum: A comparative study of man and certain laboratory animals. *Am. J. Anat.* 73:293–334, 1943.

930 NOTES, M. Adult heart without ventricular septum. *Br. Med. J.* 2:190, 1870.

931* NOTKOVICH, H. Variations of the testicular and ovarian arteries in relation to the renal pedicle. *Surg. Gynecol. Obstet.* 103:487–495, 1956.

932* NOWAK, M. A rare variation of the tributaries of the inferior vena cava in man. *Folia Morphol.* 22:144–152, 1963.

933* NOWAK, M. Variation of the arteries supplying the liver in man. *Folia Morphol.* 25:263–270, 1966.

934* NOWAK, M. Variation of the celiac trunk in man. *Folia Morphol.* 27:51–69, 1968.

935* NOWAK, M. Variation of the cystic artery in man. *Folia Morphol.* 36:89–98, 1977.

936 NUTZEL, H. Beitrag zur Kenntnis der Mißbildungen im Bereiche der oberen Hohlvene. *Frankfurter Z. Pathol.* 15:1–19, 1914.

937 NUZUM, F. Retro-aortic left renal veins. *JAMA* 62:1238–1241, 1914.

938 NUZZI, O. Studio anatomica su particolari rapporti fra vene e arteria femorale nel settore medio della coscia. *Archiv. Ital. Anat. Embriol.* 48:1:21, 1943.

939 OBEL, W. Morphology of the trunk of the portal vein in man in various periods of life. *Folia Morphol.* 34:85–101, 1975.

940* OBEL, W. Initial branches of the portal vein in the developmental aspect in man. *Folia Morphol.* 39:327–340, 1980.

941 OBREBOWSKI, A. A rare case of blood supply of the liver through a set of accessory arteries. *Folia Morphol.* 26:43–46, 1967.

942 ODGERS, P.N.B. A case of bilateral superior vena cava in the adult. *J. Anat.* 62:221–223, 1928.

943 ODGERS, P.N.B. Circum-aortic venous rings. *J. Anat.* 66:98–99, 1931.

944* OEFFINGER, H. Eine bemerkenswerthe Varietät der Vorderarmarterien. *Arch. Pathol. Anat. Physiol. Klin. Med.* 39:424–426, 1867.

945 OLIVIERI, L. Le arterie del mesencefalo. *Arch. Ital. Anat. Embriol.* 51:243–283, 1946.

946* OLSEN, L.L. and R.T. WOODBURNE. The vascular relations of the pancreas. *Surg. Gynecol. Obstet.* 99:713–719, 1954.

947 O'MALLEY, T.S. An anomalous vena pulmonalis within the lung. *Anat. Rec.* 12:173–175, 1917.

948 O'RAHILLY, R., DEBSON, H. and T.S. KING. Subclavian origin of bronchial arteries. *Anat. Rec.* 108:227–238, 1950.

949 ORR, A.E. A rare anomaly of the carotid arteries (internal and external). *J. Anat. Physiol.* 41:51, 1907.

950* OSEMLAK, J. and M. STELMASIAK. Variation of shape and size of the human heart in various periods of life. *Folia Morphol.* 34:383–394, 1975.

951 OSLER, G. and R. DOW. Variations and anomalies of the biliary duct system and its associated blood supply. *West. J. Surg.* 53:316–321, 1945.

952 OTTAVIANI, G. Sulle vascolarizzazione venosa delle ghiandole surrenali dell'uomo. *Arch. Italiano Anat. Embriol.* 36:173–214, 1936.

953 OTULAKOWSKI, B. and W. WOŹNIAK. A case of origin of the testicular artery from the renal artery, and course of the testicular artery behind the inferior vena cava. *Folia Morphol.* 34:349–351, 1975.

954 PABST, R. and H. LIPPERT. Beiderseitiges Vorkommen von A. brachialis superficialis, A. ulnaris superficialis und A. mediana. *Anat. Anz.* 123:223–226, 1968.

955 PÁC, L., HAMPLOVÁ, M. and O. PELCOVÁ. An atypical case of arising of some parietal branches of the arteria iliaca interna in man. *Anat. Anz.* 141:450–454, 1977.

956 PAKULA, H. and J. ZAJGNER. The small branches of the cavernous portion of the internal carotid artery. Anatomic and angiographic studies. *Folia Morphol.* 29:9–14, 1970.

957* PALTAUF, R. Dextrocardie und Dextroversio cordis. *Wien. Klin. Wochenschr.* 14:1032–1036, 1901.

958 PAN, N. Remnants of the venous valves in the adult heart. *J. Anat.* 70:435–438, 1935–36.

959 PANEK-MIKULA, J. and K.A. GÓRSKI. Developmental and clinical aspects of altered course of the left testicular artery in man. *Folia Morphol.* 36:323–328, 1977.

960 PAPADATOS, D. Trois observations anatomiques sur l'origine de l'Arteria subclavia dextra comme dernière branche de l'arcus aortae (Arteria lusoria). *Anat. Anz.* 140:100–117, 1976.

961* PAPEZ, J.W. Two cases of persistent left superior vena cava in man. *Anat. Rec.* 70:191–198, 1937–38.

962 PAPILIAN, V. and R. RUBIN. Der Zusammenhang des Sinus venosus mit der Vena cava sinistra. *Anat. Anz.* 57:276–277, 1923–24.

963 PARKINSON, D. Collateral circulation of cavernous carotid artery: Anatomy. *Can. J. Surg.* 7:251–268, 1964.

964 PÄRNÄNEN, P.O. and W. SIPILÄ. The role of compensation in the formation of anomalies of the vascular system illustrated by two rare anomalies of the arch of the aorta. *Ann. Med. Exper. Biol. Fenniae.* 26:25–33, 1948. Cited in Excerpta Medica, Sec. 1, Vol. 3, abstract 663, 1949.

965 PARSONS, F.G. and A. ROBINSON. Eighth report of the committee of collective investigation of the Anatomical Society of Great Britain and Ireland for the year 1897–98. *J. Anat. Physiol.* 33:189–203, 1899.

966 PASCHELES. Ein Fall von Dextrocardie. *Wien. Klin. Wochenschr.* 10:618–619, 1897.

967 PATERSON, A.M. Notes on abnormalities, with special reference to the vertebral arteries. *J. Anat. Physiol.* 18:295–303, 1884.

968 PATERSON, A.M. Obliteration of the left common, external, and internal iliac arteries. *J. Anat. Physiol.* 44:56, 1910.

969* PATEY, D.H. and I. RANGER. Some points in the surgical anatomy of the parotid gland. *Br. J. Surg.* 45:250–258, 1957.

970* PATTEN, B. The closure of the foramen ovale. *Am. J. Anat.* 48:19–44, 1931.

971 PATTEN, C. Persistence of the embryonic arrangement of the post-renal part of the cardinal veins. *Anat. Anz.* 34:189–191, 1909.

972 PATTERSON, E.L. Sources of arterial blood supply to the superior and middle cervical sympathetic ganglia and the ganglion intermédiaire. *J. Anat.* 84:329–341, 1950.

973 PATTERSON, E.L. The arterial supply to the stellate ganglion. *J. Anat.* 87:219–227, 1953.

974 PATTISON, J.N. Anomalous right subclavian artery. *Br. Heart J.* 15:150–158, 1953.

975 PAUL, R.N. A new anomaly of the aorta. Left aortic arch with right descending aorta. *J. Pediatr.* 32:19–29, 1948.

976 PEACOCK. Malformation of the heart; absence of ductus arteriosus; small size of the pulmonary artery; aorta arising from both ventricles; irregular course of the aorta, etc. *Trans. Pathol. Soc. Lond.* 11:40–43, 1860.

977 PEACOCK. Two specimens of malformation of the heart. *Br. Med. J.* 2:473, 1870.

978* PEELE, T.L. A case of closed coronary sinus and left superior vena cava. *Anat. Rec.* 54:83–86, 1932.

979 PELLEGRINI, A. Le arteriae subclavia e axillaris nell'uomo studiate col metodo statistico. *Arch. Ital. Anat. Embriol.* 5:205–255, 466–505, 1906.

980* PENSE, G. Ein Fall von rechtsseitigen Aortenbogen und seine entwicklungsgeschichtliche Deutung. *Anat. Anz.* 70:257–274, 1930.

981* PERLIŃSKI, L. Variation of the course and the division of the dorsalis pedis artery in man. *Folia Morphol.* 40:141–148, 1981.

982* PERNKOPF, E. Über einen Fall von beiderseitiger Persistenz der Arteria ischiadica. *Anat. Anz.* 55:536–542, 1922.

983 PETRÉN, G. Beiträge zur Frage von der klinischen Bedeutung der akzessorischen Nierengefäße. *Beitr. Klin. Chir.* 125:493–518, 1921.

984 PETRÉN, T. Die Arterien und Venen des Duodenums und des Pancreaskopfes beim Menschen. *Z. Anat. Entwicklungsgesch.* 90:234–277, 1929.

985 PETRÉN, T. Ein Fall von Mangel der A. coronaria dextra. *Arch. Pathol. Anat. Physiol. Klin. Med.* 278:158–164, 1930.

986 PEYGOT, M. Anomaly of the venous system. *Am. J. Med. Sci.* 14:218, 1834.

987* PFITZNER, W. Über die Ursprungsverhältnisse der Arteria obturatoria. *Anat. Anz.* 4:504–514, 1889.

988* PICK, J.W. and B.J. ANSON. Inferior phrenic artery: Origin and suprarenal branches. *Anat. Rec.* 78:413–427, 1940.

989* PICK, J.W. and B.J. ANSON. The renal vascular pedicle. An anatomical study of 430 body-halves. *J. Urol.* 44:411–434, 1940.

990* PICK, J.W., B.J. ANSON and F.L. ASHLEY. The origin of the obturator artery. A study of 640 body-halves. *Am. J. Anat.* 70:317–344, 1942.

991 PICK, J.W., ANSON, B.J. and H.W. BURNETT, JR. Communication between lymphatic and venous system at renal level in man. *Q. Bull. Northwestern University Medical School* 18:307–316, 1944.

992 PIERSON, H.H. Seven arterial anomalies of the human leg and foot. *Anat. Rec.* 30:139–145, 1925.

993* PIERSON, J.M. The arterial blood supply of the pancreas. *Surg. Gynecol. Obstet.* 77:426–432, 1943.

994* PIKKIEFF, E. On subcutaneous veins of the neck. *J. Anat.* 72:119–127, 1937–38.

995* PILLSBURY, R.C., LOWER, R.R. and N.E. SHUMWAY. Atresia of the aortic arch. *Circulation* 30:749–754, 1964.

996* PIRES DE LIMA, J.A. Anatomie du vivant. Anomalies artérielles. *La Presse Médicales* (No. 99) 30:2078, 1922.

997 PIRINGER-KUCHINKA, A. Typische Varietät der Koronararterien. *Anat. Anz.* 98:97–104, 1951–52.

998 PISK, W. Über eine seltene Varietät im Verlaufe der Arteria carotis externa beim Menschen und beim Hunde. *Anat. Anz.* 45:373–378, 1914.

999 PITZORNO. Rare anomalie des artères et des veines émulgentes. *Arch. Ital. Biol.* 23:98, 1895.

1000 PITZORNO, M. Morfologia delle arterie del Pancreas. *Arch. Ital. Anat. Embriol.* 18:1–48, 1920–21.

1001 PIZZORNO, A. Sur une intéressante variété des rameaux de l'arc de l'aorte. *Arch. Ital. Biol.* 50:154, 1908.

1002 PLATZER, W. Zwei Fälle von Transpositionen mit funktioneller Korrektur. *Arch. Pathol. Anat. Physiol. Klin. Med.* 327:400–418, 1955.

1003* POISEL, S. and H.P. SPÄNGLER. Über aberrante und akzessorische Nierenarterien bei Nieren in typischer Lage. *Anat. Anz.* 124:244–259, 1969.

1004* POKER, N., FINBY, N. and I. STEINBERG. The subclavian arteries: Roentgen study in health and disease. *Am. J. Roentgenol.* 80:193–216, 1958.

1005 POLONSKAJA, R. Zur Frage der Arterienanastomosen im Gebiete der Ellenbogenbeuge des Menschen. *Anat. Anz.* 74:303–317, 1932.

1006 POLONSKAJA, R. Zur Frage der Morphologie der Art. maxillaris externa bei Vertretern der weißen und der gelben Rasse. *Anat. Anz.* 80:331–339, 1935.

1007* POTTS, W.J., GIBSON, S. and R. ROTHWELL. Double aortic arch. *Arch. Surg.* 57:227–233, 1948.

1008* POULHES, J., HEMOUS, G. and P. METREAU. La distribution des artères thyroidennes. *Assoc. Anatomistes Comptes Rendus* 41:478–488, 1954.

1009 POWELL, E.D.U. and J.M. MULLANEY. The Chiari network and the value of the inferior vena cava. *British Heart J.* 22:579–584, 1960.

1010 PRATT, G.W. The thyroidea ima artery. *J. Anat. Physiol.* 50:239–240, 1916.
1011 PREISZ, H. Beiträge zur Lehre von den angeborenen Herzanomalien. *Beitr. Pathol. Anat. Allg. Pathol.* 7:245–298, 1890.
1012* PRIMAN, J. Notes on the anomalies of the aortic arch and of its large branches. *Anat. Rec.* 42:335–347, 1929.
1013* PRIMAN, J. and D.H. CHRISTIE. A case of abnormal internal carotid artery and associated vascular anomalies. *Anat. Rec.* 134:87–95, 1959.
1014* PROWS, M.S. Two cases of bilateral superior venae cavae, one draining a closed coronary sinus. *Anat. Rec.* 87:99–106, 1942.
1015 QUÉNU, L. Les artères du jéjunum – Types et variétés. *Assoc. Anatomistes Comptes Rendus* 38:833–841, 1951.
1016 RADZIKOWSKI, A. and P. SZULCZYK. The radial recurrent and ulnar recurrent arteries. *Folia Morphol.* 31:109–116, 1972.
1017 RANIER, F.J. Ein Fall von Mißbildung der Aortenklappen. *Int. Monatsschr. Anat. Physiol.* 24:246, 1908.
1018 RAO, G.R.K.H. Double inferior vena cava. *J. Anat. Soc. India* 8:30–31, 1959.
1019 RAO, V.R. and G.R.K.H. RAO. Anomalous right subclavian artery. *J. Anat. Soc. India* 5:74–77, 1956.
1020* RATAJCZYK-PAKALSKA, E. Variation of the smallest cardiac veins. *Folia Morphol.* 37:415–416, 1978.
1021 RATAJCZYK-PAKALSKA, E. and W.J. KOLFF. Types of the smallest cardiac veins. *Folia Morphol.* 40:345–355, 1981.
1022 READ, W.T. and M. TROTTER. The origins of transverse cervical and of transverse scapular arteries in American Whites and Negroes. *Am. J. Phys. Anthropol.* 28:239–247, 1941.
1023 REAGEN, F.P. The supposed homology of vena azygos and vena cava inferior considered in the light of new facts concerning their development. *Anat. Rec.* 35:129–148, 1927.
1024* REED, A.F. A left superior vena cava draining the blood from a closed coronary sinus. *J. Anat.* 73:195–197, 1938–39.
1025* REICHARDT, W. and R. CAMERON. Anatomy of the pancreatic veins. A postmortem and clinical phlebographic investigation. *Acta Radiol. Diagn.* 21:33–41, 1980.
1026 REICHER, M. Contribution à l'étude anatomo-topographique du système de l'aorte chez l'homme adulte et le nouveau-né. *Assoc. Anatomistes Comptes Rendus* 26:445–461, 1931.
1027 REID, D.G. A case of multiple heart anomalies. *J. Anat. Physiol.* 46:86–88, 1912.
1028 REID, D.G. Three examples of a right aortic arch. *J. Anat. Physiol.* 47:174–188, 1914.
1029* REIS, R.H. and G. ESENTHER. Variations in the pattern of renal vessels and their relations to the type of posterior vena cava in man. *Am. J. Anat.* 104:295–318, 1959.
1030 RENANDER, A. Roentgen-diagnosed anomaly of oesophagus and arcus aortae. Dysphagia Lusoria. *Acta Radiol.* 7:298–308, 1926.
1031 RENDALL, S.M. Unusual abnormality of the arteries at the base of the brain. *J. Anat. Physiol.* 13:397, 1879.
1032 RHOTON, A.L. and D. PERLMUTTER. Microsurgical anatomy of anterior communicating artery aneurysms. *Neurol. Res.* 2:217–251, 1980.
1033 RICHMOND, R.S. New abnormalities of the arteries of the upper extremity. *J. Anat. Physiol.* 14:351–352, 1879.
1034 RICHTER, O. Über das Fehlen einer Kranzarterie. *Arch. Pathol. Anat. Physiol. Klin. Med.* 299:637–642, 1937.
1035 RIGAUD and CABANIÉ. A propos d'un cas d'artère mésentérique moyenne considérations embryologiques. *Assoc. Anatomistes Comptes Rendus* 39:208–213, 1952.
1036* RIGAUD, A., CABANIE, H., GOVAZC, A. and J. DEJUSSIEU. Considérations sur les artères hépatiques accessoires (a propos de 120 dissections). *Assoc. Anatomistes Comptes Rendus* 42:1197–1205, 1955.
1037 RIGGS, H. and C. RUPP. Variations in form of the circle of Willis: The relation of the variations to collateral circulation, anatomic analysis. *Arch. Neurol.* 8:24–30, 1963.
1038* RING, B.A. Middle cerebral artery: Anatomical and radiographic study. *Acta Radiol.* 57:289–300, 1962.
1039 RISCHBIETH, H. Anomaly of the inferior vena cava: Duplication of the post-renal segment. *J. Anat. Physiol.* 48:287–292, 1914.
1040* ROBERTS, J.T. and S.D. LOUBE. Congenital single coronary artery in man: Report of 9 new cases, one having thrombosis with right ventricular and atrial (auricular) infarction. *Am. Heart J.* 34:188–208, 1947.
1041 ROBERTS, L. The thyroid arteries considered in relation to their surgical importance. *J. Anat.* 64:50–61, 1930.
1042* ROBERTS, W.H. and G.L. KRISHINGNER. Comparative study of human internal iliac artery based on Adachi classification. *Anat. Rec.* 158:191–196, 1967.
1043 ROBERTS, W.H. and P. ENGEN. Anomalous blood supply to the upper abdominal organs secondary to atresia of the celiac trunk. *Anat. Anz.* 143:396–398, 1978.
1044* ROBERTS, W.H., ENGEN, P. and E. FUJIMOTO. Anomalous opening of coronary sinus into the left atrium. *Anat. Anz.* 147:471–472, 1980.
1045* ROBILLARD, G.L. and A.L. SHAPIRO. Variational anatomy of middle colic artery; its significance in gastric and colonic surgery. *J. Int. Coll. Surg.* 10:157–169, 1947.
1046 ROBINSON, L.R. An unusual human anterior cerebral artery. *J. Anat.* 93:131–133, 1959.
1047 ROCHE, U.J., STEINBERG, I. and G.P. ROBB. Right-sided aorta with descending aorta simulating aneurysm. *Arch. Intern. Med.* 67:995–1007, 1941.
1048 RODRIGUEZ, A.L. and J.M. GENIS-GALVEZ. Varieté du ductus thoracicus non décrite chez l'homme: canal thoracique traversant l'anse de Vieussens. *Acta Anat.* 31:61–65, 1957.
1049 ROGERS, L. The thyroid arteries considered in relation to their surgical importance. *J. Anat.* 64:50–61, 1929.
1050 RÖHLICH, K. Über die Arteria transversa colli des Menschen. *Anat. Anz.* 79:37–52, 1934–35.
1051* RÖHLICH, K. Über den Truncus thyreocervicalis des Menschen. *Anat. Anz.* 90:129–148, 1940–41.
1052 ROLLESTON, H.D. Abnormal vascular supply to the liver. *J. Anat. Physiol.* 24:132–133, 1890.
1053 ROMINGER, C.J. The normal axillary venogram. *Am. J. Roentgenol.* 80:217–224, 1958.
1054 RONKA, E.K.F. and C.F. TESSMER. Congenital absence of the pericardium. *Am. J. Pathol.* 20:137–140, 1944.
1055 RONSTROM, G.N. Incidence of single and multiple renal arteries in Negroes. *Am. J. Phys. Anthropol.* 5:485–490, 1947.
1056 ROSCHDESTWENSKIJ, K.G. Der doppelte Aortenbogen. *Anat. Anz.* 68:145–151, 1929–30.
1057 ROSENBAUER, K.A. Beitrag zur Variation der Vasa renalia. Abnormer Verlauf einer rechten A. renalis unter der Mündung der V. spermatica und Einmündung einer rechten Nierenvene in die V. spermatica. *Anat. Anz.* 107:209–215, 1959.
1058 ROSS, J.A. Vascular patterns of small and large intestine compared. *Br. J. Surg.* 39:330–333, 1950.
1059* ROSS, J.A. The surgical significance of the intestinal vascular patterns. *Edinburgh Med. J.* 57:572–579, 1950.
1060* ROSSI, G. and E. COVA. Studio morfologico delle Arterie dello stomaco. *Arch. Italiano Anat. Embriol.* 3:485–526, 566–657, 1904.
1061 ROSZEL, A. Über die Anomalie der Arteria subclavia dextra und die Persistenz der Vena cava superior sinistra. *Anat. Anz.* 118:348–367, 1966.
1062* RUIZ VILLALOBOS, M.C., BALDERRAMA, D.P., LOPEZ, J.L. and M. COSTELLANOS. Complete interruption of the aorta. *Am. J. Cardiol.* 8:664–669, 1961.
1063* RUPERT, R.R. Irregular kidney vessels found in fifty cadavers. *Surg. Gynecol. Obstet.* 17:580–585, 1913.
1064* RUPERT, R.R. Further study of irregular kidney vessels as found in one hundred eighteen cadavers. *Surg. Gynecol. Obstet.* 21:471–480, 1915.
1065* SAADEH, F.A. The suprascapular artery: Case report of an unusual origin. *Anat. Anz.* 145:83–86, 1979.
1066* SAADEH, F.A. Accessory thoracodorsal artery. *Anat. Anz.* 157:319–321, 1984.
1067 SABIN, F.R. The lymphatic system in human embryos, with a consideration of the morphology of the system as a whole. *Am. J. Anat.* 9:43–91, 1909.
1068* SALASSA, J.R., PEARSON, B.W. and W.S. PAYNE. Gross and microscopical blood supply of the trachea. *Ann. Thorac. Surg.* 24:100–107, 1977.
1069 SALVI, G. Artère dorsale du pied. *Arch. Ital. Biol.* 31:188–193, 1899.
1070 SALVI, G. Arteriae superficiales et arteriae comitantes de l'extrémité. *Arch. Ital. Biol.* 32:457–459, 1899.
1071 SALVI, G. Arteriae dorsales carpi contribution à la morphologie de la circulation dans le membre thoracique. *Arch. Ital. Biol.* 33:473–474, 1900.
1072 SANDER, W. Auffälliges Verhalten der Basilararterie. *Arch. Pathol. Anat. Physiol. Klin. Med.* 72:284–285, 1878.
1073 SANDERS, G. Case of congenital malformation of the heart; with transposition of the aorta and pulmonary artery. *J. Anat. Physiol.* 27:464–473, 1893.

1074* SANDERS, J.M. Bilateral superior vena cavae. *Anat. Rec.* 94:657–662, 1946.

1075 SANDERS, R.J. Relationships of the common femoral artery. *Anat. Rec.* 145:169–170, 1963.

1076 SANKOTT, A.M. Über eine neue Varietät der Art. radialis. *Anat. Anz.* 52:502–511, 1919–20.

1077 SANKOTT, A.M. Zweiter Beitrag zur Kasuistik der Varietäten der Art. radialis. *Anat. Anz.* 53:397–398, 1920–21.

1078 SANTIS, M. DE, ANDERSON, K.J., KING, D.W. and J. NIELSEN. Variability in relationships of arteries and nerves in the human orbit. *Anat. Anz.* 157:227–231, 1984.

1079 SAPPEY, C. and DUMONTPALLIER. Note sur un cas d'oblitération de la veine cave inférieure avec circulation collatérale, suivie de faits analogues démontrant qu'il existe trois principales variétes d'oblitération de cette veine. *Soc. Biol. Comptes Rendus des Séances et Mémoires* 13:135–155, 1861.

1080 SCHALL, K. Über einige Anomalien des Gefäßsystems. *Anat. Anz.* 62:94–95, 1926–27.

1081 SCHEIBER, S.H. Über einen Fall von Obliteration der Aorta, entsprechend dem Ductus arteriosus Botalli. *Arch. Pathol. Anat. Physiol. Klin. Med.* 26:218–220, 1863.

1082 SCHLESINGER, M.J. Relation of anatomic pattern to pathologic conditions of the coronary arteries. *Arch. Pathol.* 30:403–415, 1940.

1083 SCHLESINGER, M.J., ZOLL, P.M. and S. WESSLER. The conus artery; a third coronary artery. *Am. Heart J.* 38:823–836, 1949.

1084 SCHLEY, J. Abnormer Ursprung der rechten Kranzarterie aus der Pulmonalis bei einem 61jährigen Mann. *Frankfurter Z. Pathol.* 32:1–7, 1925.

1085 SCHILLING, W. Doppelter Aortenbogen in Verbindung mit Verkürzung des Ösophagus und teilweiser Verlagerung des Magens in die Brusthöhle. *Anat. Anz.* 74:344–357, 1932.

1086 SCHMIDT, G.P. Über eine besondere Form der doppelseitigen V. cava inferior. *Anat. Anz.* 137:200–206, 1975.

1087* SCHMITZ, H. and P. THURN. Zur Asymmetrie der Lungenarterien. *Fortschr. Röntgenstr.* 88:133–145, 1958.

1088 SCHNEIDERMAN, L.J. Isolated congenital absence of the right pulmonary artery: A caution as to its diagnosis and proposal for its embryogenesis. Report of a case with review. *Am. Heart J.* 55:772–780, 1958.

1089 SCHREIBER, H. Seltene Varietät der Art. carotis. *Anat. Anz.* 91:269–271, 1941.

1090 SCHRÖDER, R. Über Anomalien der Pulmonalvenen, zugleich ein Beitrag zum Cor biloculare. *Arch. Pathol. Anat. Physiol.* 205:122–138, 1911.

1091 SCHUBEL, A.L. Abnormer Verlauf der 10. Interkostalarterie in Verbindung mit seltenen Gefäßvarietäten im Bereich der A. coeliaca. *Anat. Anz.* 102:279–286, 1955–56.

1092 SCHÜTZ, H. Einige Fälle von Entwicklungsanomalie der Vena cava superior (Persistenz des linken Ductus Cuvieri.) *Arch. Pathol. Anat. Physiol. Klin. Med.* 216:35–45, 1914.

1093* SEALY, W.C. A report of two cases of the anomalous origin of the right subclavian artery from the descending aorta. *J. Thoracic Surg.* 21:319–324, 1951.

1094 SEGALL, H. The blood and bile channels of the liver. An experimental anatomical investigation of the blood and bile channels of the liver with special reference to the compensatory arterial circulation of the liver in its relation to surgical ligation of the hepatic artery. Report of a case of arteriosclerotic aneurism of the gastroduodenal artery. *Surg. Gynecol. Obstet.* 37:152–178, 1923.

1095* SEIB, G.A. Incidences of the patent foramen ovale cordis in adult American whites and American negroes. *Am. J. Anat.* 55:511–525, 1934.

1096 SEIB, G.A. The azygos system of veins in American whites and American negroes, including observations on the inferior caval venous system. *Am. J. Phys. Anthropol.* 19:39–163, 1934.

1097* SELDOWITSCH, J.B. Über die Multiplicität der Nierenarterie und deren chirurgische Bedeutung. *Arch. Klin. Chir.* 89:1071–1112, 1909.

1098 SEN, P.K. and K.K. DATEY. Incomplete aortic vascular rings. *Br. J. Surg.* 44:175–178, 1956–57.

1099 SENIOR, H.D. The development of the arteries of the lower extremity. *Am. J. Anat.* 25:55–95, 1919.

1100 SENIOR, H.D. The description of the larger direct or indirect muscular branches of the human femoral artery: A morphogenetic study. *Am. J. Anat.* 33:243–266, 1924.

1101* SENIOR, H.D. An interpretation of the recorded arterial anomalies of the human pelvis and thigh. *Am. J. Anat.* 36:1–46, 1925.

1102* SENIOR, H.D. Abnormal branching of the human popliteal artery. *Am. J. Anat.* 44:111–119, 1929.

1103* SENNEVILLE, LIBERSA, VOISIN, and GUIOT. Étude anatomo-radiologique des anomalies artérielles du membre supérieur chez l'adulte et chez le foetus. *Assoc. Anatomistes Comptes Rendus* 37:454–461, 1950.

1104 SEVITT, S. and R.G. THOMPSON. The distribution and anastomoses of arteries supplying the head and neck of the femur. *J. Bone Joint Surg.* 47B:560–573, 1965.

1105 SHAFIROFF, R.G.P. Congenital absence of the pericardium. *J. Thoracic Surg.* 21:30–33, 1951.

1106 SHAH, M.A. and M. SHAH. The arterial supply of the vermiform appendix. *Anat. Rec.* 95:457–460, 1946.

1107* SHAPIRO, A.L. and G.L. ROBILLARD. The esophageal arteries. *Ann. Surg.* 131:171–185, 1950.

1108 SHATTOCK, S.G. Specimen of double vena cava (persistent left duct of Cuvier) the vessel of the left side being the larger. *Trans. Pathol. Soc. Lond.* 35:130–131, 1885.

1109 SHAW, D.L. An aorta with a double arch. *JAMA* 28:538–540, 1897.

1110 SHEA, M. Carotid-ophthalmic anastomoses. Frequency of external carotid and ophthalmic artery anastomoses. *Br. J. Ophthalmol.* 40:497–501, 1956.

1111* SHEEHAN, D. Some observations on the course and distribution of the circumflex arteries in man. *Anat. Anz.* 75:129–145, 1932–33.

1112* SHEHATA, R. The arterial supply of the urinary bladder. *Acta Anat.* 96:128–134, 1976.

1113* SHELLSHEAR, J.L. A contribution to our knowledge of the arterial supply of the cerebral cortex in man. *Brain* 50:236–253, 1927.

1114* SHENKIN, H.A., SPITZ, E.B., GRANT, F.C. and S.S. KETY. Physiologic studies of arterio-venous anomalies of the brain. *J. Neurosurg.* 5:165–172, 1948.

1115 SHEPARD, F.J. Some vascular anomalies observed during the session 1888–89. *J. Anat. Physiol.* 24:69–71, 1890.

1116* SIEROCIŃSKI, W. Arteries supplying the left colic flexure in man. *Folia Morphol.* 34:117–124, 1975.

1117 SIEROCIŃSKI, W. Studies on the arteries supplying the descending and sigmoid colon in man. *Folia Morphol.* 35:287–306, 1976.

1118 SIEROCIŃSKI, W. Variations of colic veins in man. *Folia Morphol.* 40:149–157, 1981.

1119* SIEROCIŃSKI, W. and M. MICHNIEWICZ-NOWAK. Variation of the arising and the course of arteries vascularizing the transverse colon in man. *Folia Morphol.* 39:271–281, 1980.

1120 SIEROCIŃSKI, W., KLING, A. and M. MICHNIEWICZ-NOWAK. The occurrence and the course of "the marginal artery" as the anastomosis of arteries vascularizing the cecum and the colon in man. *Folia Morphol.* 40:167–179, 1981.

1121* SINGER, E. Embryological pattern persisting in the arteries of the arm. *Anat. Rec.* 55:403–409, 1933.

1122 SINGER, R. The coronary arteries of the Bantu heart. *S. Afr. Med. J.* 33:310–315, 1959.

1123 SINGH, B., MUKERJEC, R.N. and S.S. MAKHNI. Anomalous right subclavian artery. *J. Anat. Soc. India* 8:34–35, 1959.

1124* SINGH, B. and V.K. AHUJA. Double superior vena cava with double vena azygos. *J. Anat. Soc. India* 12:33–35, 1963.

1125* SINGH, S. and R. DASS. The central artery of the retina. I. Origin and course. *Br. J. Ophthalmol.* 44:193–211, 1960.

1126* SINGH, S. and R. DASS. The central artery of the retina. II. A study of its distribution and anastomoses. *Br. J. Ophthalmol.* 44:280–299, 1960.

1127 SINZINGER, H. et al. Über eine zusätzliche linke Arteria cerebri anterior beim Menschen. *Anat. Anz.* 133:377–381, 1973.

1128 SIRCA, A. The arterial supply of the pelvic ureter in the female. *Acta Anat.* 33:252, 1958.

1129 SJÖGREN, S.E. The anterior choroidal artery. *Acta Radiol.* 46:143–157, 1956.

1130 SKAMNAKIS, S.N. Eine Anomalie der Vena cava inf. und abnormer Verlauf des rechten Ureters. *Anat. Anz.* 73:50–56, 1931–32.

1131* SKOPAKOFF, C. Über die Variabilität der Ab- und Verzweigung der A. brachialis superficialis. *Anat. Anz.* 106:356–368, 1959.

1132* SKOPAKOFF, C. Über die Variabilität der Abzweigung der Hauptäste der Aa. axillaris et brachialis. *Anat. Anz.* 107:294–304, 1959.

1133* SKOPAKOFF, C. Über die Variabilität der Abzweigung der A. subclavia und ihrer Hauptäste. *Anat. Anz.* 115:393–402, 1964.

1134* SKOPAKOFF, C. Über die Variabilität der Abzweigung der A.

subclavia und ihrer Hauptäste. II. Abzweigung der Hauptäste der A. subclavia. *Anat. Anz.* 116:73–91, 1965.
1135* SKOPAKOFF, C. Über die Variabilität im Verlauf der A. maxillaris. *Anat. Anz.* 123:534–546, 1968.
1136* SLAGER, R.F. and K.P. KLASSEN. Anomalous right subclavian artery arising distal to a coarctation of the aorta. *Ann. Surg.* 147:93–97, 1958.
1137* SLANY, A. Anomalien des Circulus arteriosus Willisi in ihrer Beziehung zur Aneurysmenbildung an der Hirnbasis. *Arch. Pathol. Anat. Physiol. Klin. Med.* 301:62–71, 1938.
1138 SLEDZIEWSKI, H.G. Trajet des vasseaux efférents des ganglions lymphatiques diàphragmatiques dans les médiastins. (Note résumée). *Assoc. Anatomistes Comptes Rendus* 26:467–477, 1931.
1139 SLEDZIŃSKI, Z. and T. TYSZKIEWICZ. Hepatic veins of the right part of the liver in man. *Folia Morphol.* 34:315–322, 1975.
1140 SMITH, G.E. A peculiar enlargement of the supra-scapular vein. *J. Anat. Physiol.* 29:99, 1895.
1141 SMITH, G.E. Note on an anomalous anastomosis between the internal carotid and the basilar artery. *J. Anat. Physiol.* 43:310–311, 1909.
1142* SMITH, G.T. The renal vascular patterns in man. *J. Urol.* 89:275–288, 1963.
1143 SMITH, J.C. Review of single coronary artery with report of 2 cases. *Circulation* 1:1168–1175, 1950.
1144 SMITH, J.H. Description of a case of persistent left duct of Cuvier in man. *Anat. Rec.* 17:131–133, 1919–20.
1145* SMITH, S.D. and R.S. BENTON. A rare origin of the superior thyroid artery. *Acta Anat.* 101:91–93, 1978.
1146* SMITH, W.C. A case of left superior vena cava without a corresponding vessel on the right side. *Anat. Rec.* 11:191–198, 1916.
1147 SNELLING, C.E. and I.H. ERB. Double aortic arch. *Arch. Dis. Child.* 8:401–408, 1933.
1148 SNOW, P.J.D. A case of single coronary artery with stereographic demonstration of the arterial distribution. *Br. Heart J.* 15:261–263, 1953.
1149 SOHIER, H.M.L., GOUAZÉ, A. and M. TORLOIS. La ramescence de l'artère rénale en fonction de la destinée lobaire de ses branches (d'aprés 200 reins). *Assoc. Anatomistes Comptes Rendus* 43:783–787, 1956.
1150 SOHIER, H.M.L., GOUAZÉ, A., SENTENAC, M. and M. TORLOIS. Recherches sur le systéme pyramidal du rein humain et sur les relations de ce système avec la morphologie du hile et avec la disposition des branches de l'artère rénale. *Assoc. Anatomistes Comptes Rendus* 43:788–795, 1956.
1151 SOLOFF, L.A. Anomalous coronary arteries arising from the pulmonary artery. Report of a case in which the left artery arose from the pulmonary artery. *Am. Heart J.* 24:118–127, 1942.
1152* SONNELAND, J., ANSON, B.J. and L.E. BEATON. Surgical anatomy of the arterial supply of the colon from the superior mesenteric artery based upon a study of 600 specimens. *Surg. Gynecol. Obstet.* 106:385–398, 1958.
1153 SOULIGOUX, C. and L. LAGANE. Note sur le mode de terminaison fonctionnellement anastomotique des branches de l'artère mésentérique supérieure. *Soc. Biol. Comptes Rendus des Séances et Mémoires* 69:612–613, 1920.
1154* SOUTHWORTH, H. and C.S. STEPHENSON. Congenital defects of the pericardium. *Arch. Intern. Med.* 61:223–240, 1938.
1155 SPENCER, J. and R. DRESSER. Right-sided aorta. *Am. J. Roentgenol.* 36:183–187, 1936.
1156* SPITZER, A. Über den Bauplan des normalen und mißbildeten Herzens. Versuch einer phylogenetischen Theorie. *Arch. Pathol. Anat. Physiol. Klin. Med.* 243:81–272, 1923.
1157 SPRONG, D.H. JR. and N.L. CUTLER. A case of human right aorta. *Anat. Rec.* 45:365–375, 1930.
1158* SSOSON-JAROSCHEWITSCH, J. Zur chirurgischen Anatomie des Milzhilus. *Z. Anat. Entwicklungsges.* 84:218–237, 1927.
1159* STAHEL, H. Zur Anatomie und Chirurgie der Art. subclavia. *Arch. Anat. Physiol. Wissen. Med.* 1886, pp. 211–235.
1160 STAUFFER, H.M. and H.H. POTE. Anomalous right subclavian artery originating on the left as the last branch of the aortic arch. *Am. J. Roentgenol.* 56:13–17, 1946.
1161* STEINBERG, I., DUBILIER, W. JR. and D.S. LUKAS. Persistence of left superior vena cava. *Dis. Chest* 24:479–488, 1953.
1162* STEINBERG, I. and A. FINBY. Clinical and angiocardiographic features of congenital anomalies of the pulmonary circulation: A classification and review. *Angiology* 7:378–395, 1956.
1163 STEINER, P. and M.J. FINEGOLD. Truncus arteriosus with atresia of the aortic arch. *Arch. Pathol.* 79:518–523, 1965.
1164 STELMASIAK, M. and S. BRYC. The anterior artery of the falx cerebri and its roentgendiagnostic significance. *Folia Morphol.* 29:369–373, 1970.
1165* STELMASIAK, M. and J. OSEMLAK. Relation of the coronary arteries to the areas supplied by them in man. *Folia Morphol.* 31:457–467, 1972.
1166 STEWARD, J.A. and F.W. RANKIN. Blood supply of the large intestine. Its surgical considerations. *Arch. Surg.* 26:843–891, 1933.
1167 STEWART, J.D. Circulation of the human thyroid. *Arch. Surg.* 25:1157–1165, 1932.
1168 STIBBE, E.P. True congenital diverticulum of the trachea in a subject showing also right aortic arch. *Journal Anat.* 64:62–66, 1929.
1169 STIEDA, H. Über die Arteria circumflexa ilium. *Anat. Anz.* 7:232–246, 1892.
1170 STOKES, A. Abnormal position of the heart and great vessels associated with the transposition of the viscera. *J. Anat. Physiol.* 43:301–307, 1909.
1171 STOPFORD, J.S.B. The arteries of the pons and medulla oblongata. *J. Anat. Physiol.* 50:133–164, 255–280, 1916.
1172* STRICKLAND, B. and W. URQUHART. Digital arteriography. *Br. J. Radiol.* 36:465–476, 1963.
1173 STUART, T.P.A. Note on a variation in the course of the popliteal artery. *J. Anat. Physiol.* 13:162, 1879.
1174* SUAREZ, A.R. Une variation anatomique dans la distribution de l'artère appendiculaire d'intérét chirurgical. *Assoc. Anatomistes Comptes Rendus* 36:659–661, 1949.
1175* SUDAKEVITCH, T. The variations in the system of the trunks of the posterior ciliary arteries. *Br. J. Ophthalmol.* 31:738–760, 1947.
1176 SUGAR, O., HOLDEN, L.B. and C.B. POWELL. Vertebral angiography. *Am. J. Roentgenol.* 61:166–182, 1949.
1177* SUNDERLAND, S. Blood supply of the distal colon. *Aust. N.Z. J. Surg.* 11:253–263, 1942.
1178 SUNDERLAND, S. Blood supply of the nerves of the upper limb in man. *Arch. Neurol. Psychiatry* 53:91–115, 1945.
1179 SUNDERLAND, S. Blood supply of the sciatic nerve and its popliteal divisions in man. *Arch. Neurol. Psychiatry* 54:283–289, 1945.
1180 SUNDERLAND, S. Neurovascular relations and anomalies at the base of the brain. *J. Neurol. Neurosurg. Psychiatry* 11:243–257, 1948.
1181 SUNDERLAND, S. and R.J. WRIGHT-SMITH. Congenital pericardial defects. *Br. Heart J.* 6:167–175, 1944.
1182 SUNDERLAND, S. and G.M. BEDBROOK. Narrowing of the second part of the subclavian artery. *Anat. Rec.* 104:299–307, 1949.
1183 SVITZER. Beobachtung einer Theilung des Ductus thoracicus. *Arch. Anat. Physiol. Wissen. Med.* 1845, p. 21.
1184 SWATKO, A. and M. MATUSEWIEZ. Variation of the place of origin and course of posterior and anterior spinal arteries in man. *Folia Morphol.* 38:77–84, 1979.
1185 SWEET, R.H., FINDLAY, C.W. JR., and G.C. REYERSBACH. The diagnosis and treatment of tracheal and esophageal obstruction due to congenital vascular ring. *J. Pediatr.* 30:1–17, 1947.
1186* SWIGART, LaV.L., SIEKERT, R.G., HAMBLEY, W.C. and B.J. ANSON. The esophageal arteries. An anatomic study of 150 specimens. *Surg. Gynecol. Obstet.* 90:234–243, 1950.
1187 SYKES, D. The arterial supply of the human kidney with special reference to accessory arteries. *Br. J. Surg.* 50:368–374, 1963.
1188 SYLWANOWICZ, W. Sur les valvules du système des azygos. *Assoc. Anatomistes Comptes Rendus* 26:503–506, 1931.
1189 SYMMERS, W.St.C. Note on accessory coronary arteries. *J. Anat. Physiol.* 41:141–142, 1907.
1190 SZAWLOWSKI, J. Über das Verhalten des Ductus thoracicus bei Persistenz der rechten absteigenden Aortenwurzel. *Anat. Anz.* 3:839–849, 1888.
1191* SZKUDLAREK, R. Pancreatoduodenal veins in man. *Folia Morphol.* 39:15–28, 1980.
1192* TALLMADGE, G.K. Un cas d'anomalie rare de l'artère hépatique. *Anat. Anz.* 85:433–435, 1938.
1193 TALLMADGE, G.K. On the fixity of the gastroduodenal artery. *Anat. Anz.* 88:161–167, 1939.
1194 TANDLER, J. Zur Anatomie der Arterien der Hand. *Anat. Hefte* 7:263–282, 1897.
1195* TANDLER, J. Über die Varietäten der Arteria coeliaca und deren Entwicklung. *Anat. Hefte* 25:473–499, 1904.
1196 TANIGUCHI, T. Beitrag zur Topographie der großen Äste der Bauchaorta. *Folia Anat. Jpn.* 9:201–214, 1931.

1197* TARUFFI, C. Sulle anomalie delle vene azigos ed emiazigos. *Mem. Accad. Sci. Istituto di Bologna S. 4* 2:675–681, 1880.

1198 TAVARES, A. Trou sous-jugulaire de Serrano. *Assoc. Anatomistes Comptes Rendus* 22:212–215, 1927.

1199 TAYLOR, R.S. and J.M.P. GRELL. Rare anomaly of the aortic arch. *J. Anat. Physiol.* 36:288–289, 1902.

1200 TELFORD, E.D. and J.S.B. STOPFORD. The vascular complications of cervical rib. *Br. J. Surg.* 18:557–564, 1930–31.

1201 TENCHINI, L. La bulbe jugulaire inferieure de l'homme. *Arch. Ital. Biol.* 34:474, 1900.

1202 TERRIER, M. Disposition anormale des jumeaux de al jambe et de l'artère poplitée. *Gazette des Hôpitaux* 46:317, 1873.

1203 THANE. Obliteration of Coeliac axis, Proc. Anat. Soc. Great Britain and Ireland. *J. Anat. Physiol.* 22:xvii–xviii, 1888.

1204 THÉVENET, A. and J.B. PRIOTON. Le cloisonnement du coelome péricardique ses anomalies. *Assoc. Anatomistes Comptes Rendus* 43:796–806, 1956.

1205 THOMAS, L. and L. CHRISTIAENS. Recherches sur les rapports vasculaires de l'amygdale. *Assoc. Anatomistes Comptes Rendus* 27:481–499, 1932.

1206 THOMPSON, I.M. Anatomical note. Venae cavae superiores dextra et sinistra of equal size in an adult. *J. Anat.* 63:496–497, 1929.

1207 THOMSON, A. Origin of the internal circumflex from the deep epigastric artery. *J. Anat. Physiol.* 18:379–383, 1883.

1208 THOMSON, A. Note on two instances of abnormality in the course and distribution of the radial artery. *J. Anat. Physiol.* 18:265–269, 1884.

1209 THOMSON, A. Variations of the thoracic duct associated with abnormal arterial distribution. *J. Anat. Physiol.* 18:416–425, 1884.

1210* THOMSON, A. Report of the committee of collective investigation of the Anatomical Society of Great Britain and Ireland for the year 1889–90. *J. Anat. Physiol.* 25:89–101, 1891.

1211* THOMSON, A. Second annual report of the committee of collective investigation of the Anatomical Society of Great Britain and Ireland for the year 1890–91. *J. Anat. Physiol.* 26:74–93, 1892.

1212 THOMSON, A. Third annual report of the committee of collective investigation of the Anatomical Society of Great Britain and Ireland for the year 1891–92. *J. Anat. Physiol.* 27:189–194, 1893.

1213 TISCHENDORF, F. Zum Problem der Milzarterie, II. Eine abnorm stark und unregelmäßig geschlängelte A. lienalis mit zweiteiligem Truncus coeliacus. *Anat. Anz.* 134:108–119, 1973.

1214 TOBIN, C.E. The bronchial arteries and their connections with other vessels in the human lung. *Surg. Gynecol. Obstet.* 95:741–750, 1952.

1215 TODD, T.W. The arterial lesion in cases of cervical rib. *J. Anat. Physiol.* 47:250–253, 1913.

1216 TOKARSKI, S. Les variations de l'artère maxillaire interne chez l'homme expliquées par les variations chez les primates. *Assoc. Anatomistes Comptes Rendus* 26:507–510, 1931.

1217 TOOTH, H.H. Case of origin of the aorta from the infundibulum, and of the pulmonary artery from the sinus of the right ventricle. *Trans. Pathol. Soc. Lond.* 35:127–129, 1884.

1218* TORI, G. and G. GARUSI. Congenital absence of right pulmonary artery with patent ductus arteriosus and auricular septal defects. *Radiol. Clin.* 28:228–250, 1959.

1219 TREDER, A. Muscular trabeculae at the base of the right ventricle in the human heart. *Folia Morphol.* 31:468–477, 1972.

1220 TRICOME-ALLEGRA, G. Arcade plantaire superficielle. *Arch. Ital. Biol.* 47:489, 1907.

1221 TRIMINGHAM, H.L. and J.A. MCDONALD. Congenital anomalies in the region of the umbilicus. *Surg. Gynecol. Obstet.* 80:152–163, 1945.

1222 TROTTER, M. The level of termination of the popliteal artery in the White and the Negro. *Am. J. Phys. Anthropol.* 27:109–118, 1940.

1223 TROTTER, M., HENDERSON, J.L., GASS, H., BRUA, R.S., WEISMAN, S., AGRESS, H., CURTIS, G.H. and E.R. WESTBROOK. The origins of branches of the axillary artery in Whites and in American Negroes. *Anat. Rec.* 46:133–137, 1930.

1224 TROYER, J.R. A multiple anomaly of the human heart and great veins. *Anat. Rec.* 139:509–513, 1961.

1225* TSCHERNYSCHEFF, V. and I. GRIOGOROWSKY. Über die arterielle Versorgung des Kleinhirns. *Arch. Psychiatry (Berlin)* 89:482–569, 1929–30.

1226* TURNBULL, I.M., BRIEG, A. and O. HASSLER. Blood supply of cervical spinal cord in man. A microangiographic cadaver study. *J. Neurosurg.* 24:951–965, 1966.

1227* TURNER, F.C. Case of abnormal development of the coronary arteries of the heart. *J. Anat. Physiol.* 19:119, 1885.

1228 TURNER, W. On irregularities of the pulmonary artery, arch of the aorta, with an attempt to illustrate their mode of origin by a reference to development. *Br. Foreign Med. Rev.* 30:173–189, 461–482, 1862.

1229 TURNER, W. On the existence of a system of anastomosing arteries between and connecting the visceral and parietal branches of the abdominal aorta. *Br. Foreign Med. Rev.* 32:222–227, 1863.

1230 TURNER, W. A loop-like bifurcation of the external carotid artery. *J. Anat. Physiol.* 13:399, 1879.

1231 TYRIE, C.C.B. Axial rotation of abdominal aorta, with associated abnormalities of the branches. *J. Anat. Physiol.* 28:281–287, 1894.

1232 UNGER, K. Beitrag zur Kenntnis der Venae cordis minimae (Thebesii) des menschlichen Herzens. *Z. Anat. Entwicklungsgesch.* 108:356–375, 1938.

1233 URWITZ, S. Arteria septi cordis dextra. *Anat. Anz.* 85:97–110, 1937–38.

1234* VAKAET, L., POPPELIER, G. and P. VERMEIRE. Sur un case d'anomalie combinée de la veine cave supérieure et du système azygos. *Acta Anat.* 32:235–239, 1958.

1235 VALLEIX. Transposition irrégulière des organes, de droite à gauche; absence de la cloison inter-auriculaire du coeur; ventricle pulmonaire rudimentaire, et ne communiquant pas avec les oreillettes; cloison inter-ventriculaires incomplète deux veines caves supérieures... *Arch. Gén. Méd.* 8:78–91, 1835.

1236 VAN CLEAVE, C.D. A multiple anomaly of the great veins and interatrial septum in a human heart. *Anat. Rec.* 50:45–51, 1931.

1237* VANDAMME, J.P.J., BONTE, J. and G. VAN DER SCHUEREN. A reevaluation of hepatic and cystic arteries. The importance of the aberrant hepatic branches. *Acta Anat.* 73:192–209, 1969.

1238* VANDAMME, J.P.J. and G. VAN DER SCHUEREN. Re-evaluation of the colic irrigation from the superior mesenteric artery. *Acta Anat.* 95:578–588, 1976.

1239 VAN GENT, I. and M.A. LUYENDIJK-ELSHAUT. Anomalies du patron vasculaire hépatique dan des cas de défauts cardiaques congénitaux. *Assoc. Anatomistes Comptes Rendus* 44:913–922, 1957.

1240 VAN PERNIS, P.A. Variations of the thoracic duct. *Surgery* 26:806–809, 1949.

1241 VARAGLIA, S. Sur une artère thyréoïdienne inférieur accessoire. *Arch. Ital. Biol.* 36:348, 1901.

1242* VÉBER, K. Beitrag zur Frage der Gangrän des oralen Darmstumpfes nach Mastdarmresection. *Arch. Klin. Chir.* 98:579–607, 1912.

1243 VEHSEMEYER. Ein Fall von congenitaler Dexiokardie zugleich ein Beitrag zur Verwerthung der Röntgenstrahlen im Gebiete der inneren Medicin. *Deutsche Med. Wochenschr.* 23:180–181, 1897.

1244 VENKAMMA, D. and Y. APPAJAE. A case of unusual blood supply to the mammary gland. *Indian J. Surg.* 9:210–212, 1947. Cited in *Excerpta Medica*, Sec. 1, Vol. 3, abstract 664, 1949.

1245 VERNEUIL, M. Anomalies artérielles. *Soc. Biol. Comptes Rendus des Séances et Mémoires* 1:197–198, 1849.

1246 VERSARI, R. Rare anomalie de l'aorte et considérations sur le trajet typique de nerf laryngien droit dans les cas d'arc aortique à droite. *Arch. Ital. Biol.* 28:169–170, 1897.

1247 VERSARI, R. Rare anomalie de la valvule d'Eustache chez un homme adulte avec double veine cave supérieure. *Arch. Ital. Biol.* 46:298, 1906.

1248 VILLEMIN, DUFOUR and RIGAUD. Les branches intestinales de la mésentérique supérieure. (Etude faite en vue des possibilités d'ascension chirurgicale des premières anses jejunales.) *Assoc. Anatomistes Comptes Rendus* 37:506–514, 1950.

1249 VILLEMIN, DUFOUR, RIGAUD and CABANIE. Considérations anatomochirurgicales sur la vascularisation due colon droit et da la dernière anse iléale ches l'adulte. (Leur ultilisation dan l'élévation de ce segment intestinal dans l'oesophagoplastie prethoracique.) *Assoc. Anatomistes Comptes Rendus* 38:984–994, 1951.

1250 VILLEMIN, G., DUFOUR, R., RIGAUD, A. and A. GOUAZE. Le système veineux pulmonaire selon les types morphologiques. *Assoc. Anatomistes Comptes Rendus* 39:174–177, 1952.

1251 VILLEMIN, G., DUFOUR, R., RIGAUD, A. and A. GOUAZE. A propos d'une disposition peu fréquente de persistance de la veine cave supérieure gauche chez l'homme. *Assoc. Anatomistes Comptes Rendus* 39:178–182, 1952.

1252* VILLEMIN, G., DUFOUR, R., RIGAUD, A. and A. GOUAZE. Influence des types morphologiques sur le systeme artériel pulmonaire. *Assoc. Anatomistes Comptes Rendus* 39:183–194, 1952.

1253* VOISIN and DEVAMBEZ. Contribution a l'etude de la vascularisation artérielle du bulbe doudénal. *Assoc. Anatomistes Comptes Rendus* 36:691–696, 1949.

1254 VONWILLER, P. Eine seltene Varietät der Arteria iliaca communis sinistra. *Anat. Anz.* 50:155–158, 1917–18.

1255 WALDEYER, N.W. Sur la situation de l'artère vertébrale. *Assoc. Anatomistes Comptes Rendus* 8:83–84, 1906.

1256 WALDEYER, N.W. Weitere Bemerkungen über die Lage der Arteria vertebralis. *Assoc. Anatomistes Comptes Rendus* 11:181, 1909.

1257 WALLRAFF, J. Ein Fall von linker Vena cava superior ohne entsprechende Vene auf der rechten Seite. *Anat. Anz.* 87:305–311, 1938–39.

1258 WALMSLEY, T. Transposition of the ventricles and the arterial stems. *J. Anat.* 65:528–540, 1931.

1259 WALSHAM, W.J. Abnormal origin and distribution of the upper seven right intercostal arteries, with remarks. *J. Anat. Physiol.* 16:444–445, 1882.

1260* WALTNER, J.G. Anatomic variations of the lateral and sigmoid sinuses. *Arch. Otolaryngol.* 39:307–312, 1944.

1261 WANKOFF, W. Über einige Gesetzmäßigkeiten bei der Variabilität der Arterien der oberen Extremität. *Anat. Anz.* 111:216–240, 1962.

1262 WARING, H.F. Left vena cava inferior. *J. Anat. Physiol.* 28:46–50, 1894.

1263 WASIŃSKA-SIEROCIŃSKA, K. Variation of the origin of the lienal vein in man. *Folia Morphol.* 37:251–260, 1978.

1264 WATERSTON, D. Duplication of the post-renal segment of the vena cava inferior. *J. Anat. Physiol.* 47:433–435, 1913.

1265 WATSON, M. Note on the termination of the thoracic duct on the right side. *J. Anat. Physiol.* 6:427–428, 1872.

1266 WATSON, M. Notes on a case of double aortic arch. *J. Anat. Physiol.* 11:229–234, 1877.

1267* WATT, J.C. and A.N. MCKELLOP. Relation of arteries to roots of nerves in posterior cranial fossa in man. *Arch. Surg.* 30:336–345, 1935.

1268* WEATHERSBY, H.T. The artery of the index finger. *Anat. Rec.* 122:57–64, 1955.

1269* WEATHERSBY, H.T. Unusual variation of the ulnar artery. *Anat. Rec.* 124:245–248, 1956.

1270 WEATHERSBY, H.T. The valves of the axillary, subclavian and internal jugular veins. *Anat. Rec.* 124:379–380, 1956.

1271* WEATHERSBY, H.T. Anomalies of brachial and antebrachial arteries of surgical significance. *South Med. Journal* 49:46–49, 1956.

1272* WEATHERSBY, H.T. The origin of the artery of the ligamentum teres femoris. *J. Bone Joint Surg.* 41A:261–263, 1959.

1273 WEIGERT, C. Über einen Fall von links verlaufender Vena cava superior, mutmaßlich bedingt durch frühzeitige Synostose der Sutura mastoidea dextra. *Arch. Pathol. Anat. Physiol. Klin. Med.* 84:184–188, 1881.

1274 WEINSTEIN, B.B., COUNTISS, E.H. and V.J. DERBES. The renal vessels in 203 cadavers. *Urol. Cutan. Rev.* 44:137–139, 1940.

1275 WEINSTEIN, B.B., COUNTISS, E.H. and V.J. DERBES. Retroaortic renal vein. *Urol. Cutan. Rev.* 46:17–19, 1942.

1276+ WELCH, W.H. Duplicature of arch of aorta with aneurysm. *Johns Hopkins Hosp. Bull.* 2:142, 1891.

1277 WENNERSTRAND, J.R. A case with a vena cava superior on both sides. *Anat. Anz.* 112:338–343, 1963.

1278 WERLIN and DOLGOPOL. Case No. 1489, Clinical Conference, Willard Parker Hospital, New York. *Arch. Pediatr.* 56:539–541, 1939.

1279 WERMUTH, E.G. Anastomoses between the rectal and uterine veins forming a connexion between the somatic and portal venous system in the recto-uterine pouch. *J. Anat.* 74:116–126, 1939.

1280* WESTERBORN, A. Über Nierentorsion und abnorm verlaufende Nierengefäße. *Arch. Chir. Scand.* 68:25–36, 1931.

1281 WHARTON, G.K. The blood supply of the pancreas, with special reference to that of the islands of Langerhans. *Anat. Rec.* 53:55–81, 1932.

1282 WHEELER, C. and M.E. ABBOTT. Double aortic arch and pulmonary atresia, with pulmonic circulation maintained through a persistent left aortic root, in a man aged twenty-nine. *Can. Med. Assoc. J.* 19:297–303, 1928.

1283* WHITE, N.K. and J.E. EDWARDS. Anomalies of the coronary arteries: Report of four cases. *Archives Pathol.* 45:766–771, 1948.

1284 WHITE, W.H. Left inferior vena cava. *Trans. Pathol. Soc. Lond.* 35:131, 1884.

1285 WHITE, W.H. A case of co-arctation of the aorta. *Trans. Pathol. Soc. Lond.* 36:178–182, 1885.

1286 WHITE, W.H. Patent ductus arteriosus. *Trans. Pathol. Soc. Lond.* 36:182–183, 1885.

1287 WHITTAKER, C.R. A curious abnormality of the renal arteries. *J. Anat. Physiol.* 29:493, 1905.

1288 WICKE, A. Über einen Fall von Inselbildung im Bereiche der hinteren Hohlvene (Vena cava posterior) und Durchtrift des rechten Harnleiters beim Erwachsenen. *Z. Anat. Entwicklungsges.* 84:524–533, 1927.

1289 WICKE, L., SPÄNGLER, H.P., DIMOPOULOS, J., FIRBAS, W. and F. OLBERT. Zur Dignität der Variationen der Nierenarterienabgänge in Angiogramm. *Anat. Anz.* 135:140–150, 1974.

1290 WIDMER, O. Die Rectalarterien des Menschen. *Z. Anat. Entwicklungsgesch.* 118:398–416, 1955.

1291* WIEDEMANN, G. Ein neuer Fall von rechtsseitigem Aortenbogen (unter besonderer Berücksichtigung der Röntgenliteratur). *Anat. Anz.* 91:195–206, 1941.

1292* WILHELM, E. Über einen Fall von Collateralkreislauf nach Verlegung der unteren Hohlvene. Ein Beitrag zur Lehre von der funktionellen Anpassung des Venensystems. *Z. Anat. Entwicklungsgesch.* 92:93–132, 1930.

1293 WILKIE, D.P.D. The blood supply of the duodenum, with special reference to the supraduodenal artery. *Surg. Gynecol. Obstet.* 13:399–405, 1911.

1294 WILLIAM, N. and A. ABRIKOSSOFF. Ein Herz mit linkem Doppelvorhof. *Arch. Pathol. Anat. Physiol. Klin. Med.* 203:404–420, 1911.

1295* WILLIAMS, A.F. The formation of the popliteal vein. *Surg. Gynecol. Obstet.* 97:769–772, 1953.

1296 WILLIAMS, D.B. and W.S. PAYNE. Observations on esophageal blood supply. *Mayo Clin. Proc.* 57:448–453, 1982.

1297 WILLIAMS, D.J. The origin of the posterior cerebral artery. *Brain* 59:175–180, 1936.

1298* WILLIAMS, G.D., SINDELAR, R.J., PEART, J.C., MARTIN, N.A., MCINTIRE, L.R. and C.H. MARTIN. Origin of the deep and circumflex femoral group of arteries. *Anat. Rec.* 46:273–279, 1930.

1299 WILLIAMS, G.D., AFF, H.M., SCHMECKEBIER, M., EDMONDS, H.W. and E.G. GRAUL. Variations in the arrangement of the branches arising from the aortic arch in American Whites and Negroes. *Anat. Rec.* 54:247–251, 1932.

1300* WILLIAMS, G.D., MARTIN, C.H. and L.R. MCINTIRE. Origin of the deep and circumflex femoral group of arteries. *Anat. Rec.* 60:189–196, 1934.

1301* WILLIAMS, G.D. and H.W. EDMONDS. Variations in the arrangement of the branches arising from the aortic arch in American whites and negroes (a second study). *Anat. Rec.* 62:139–146, 1935.

1302* WILMER, H.A. The blood supply of the first part of the duodenum. With description of the gastroduodenal plexus. *Surgery* 9:679–687, 1941.

1303 WILSON, P.M. An unusual variation of the coronary arteries. *Anat. Anz.* 116:299–302, 1965.

1304 WINDLE, B.C.A. The arteries forming the circle of Willis. *J. Anat. Physiol.* 22:289–293, 1888.

1305 WINDLE, W.F., ZEISS, F.R. and M.S. ADAMSKI. Note on a case of anomalous right vertebral and subclavian arteries. *J. Anat.* 62:512–514, 1928.

1306 WOLF, B.S., NEWMAN, C.M. and M.T. KHILNANI. The posterior inferior cerebellar artery on vertebral angiography. *Am. J. Roentgenol.* 87:322–337, 1962.

1307 WOLMAN, I.J. Syndrome of constricting double aortic arch in infancy. Report of a case. *J. Pediatr.* 14:527–533, 1939.

1308* WOODBURNE, R.T. A case of right aortic arch and associated venous anomalies. *Anat. Rec.* 111:617–627, 1951.

1309* WOODBURNE, R.T. and L.L. OLSEN. The arteries of the pancreas. *Anat. Rec.* 111:255–270, 1951.

1310* WOODHALL, B. Anatomy of the cranial blood sinuses with particular reference to the lateral. *Laryngoscope* 49:966–1010, 1939.

1311 WOŹNIAK, W. A case of termination of the ileocolic vein in the right renal vein associated with accessory renal blood vessels. *Folia Morphol.* 24:195–200, 1965.

1312 WOŹNIAK, W., KIERZ, A. and S. WAWRZYNIAK. The question of the renal arterial segments. *Anatomischer Anz.* 132:332–340, 1972.

1313 WRBA, H. Ein primitives Herz bei einem erwachsenen Menschen (Persistenz frühembryonaler Herzformen). *Arch. Pathol. Anat. Physiol. Klin. Med.* 324:662–670, 1954.

1314* WRIGHT, C.W. Aberrant blood vessels and nerves in a case of left-sided vermiform appendix, cecum, and anomalous colon. *Anat. Rec.* 133:187–202, 1959.

1315* WRIGHT, R.R., ANSON, B.J. and H.C. CLEVELAND. The vestigial valves and the interatrial foramen of the adult human heart. *Anat. Rec.* 100:331–355, 1948.

1316 WURTZ, K.G. and N.B. POWELL. Two unusual vascular and cardiac anomalies. I. Vascular ring of the esophagus with patent ductus arteriosus. Origin of the left subclavian and carotid arteries. II. Persistent atrioventricular communis and aortic dextroposition with mongolism. *J. Pediatr.* 33:722–733, 1948.

1317 WUTZER, C.W. Einmündung des Ductus thoracicus in die Vena azygos. *Arch. Anat. Physiol.* 311–318, 1834.

1318* WYBAR, K.C. Anastomoses between the retinal and ciliary arterial circulations. *Br. J. Ophthalmol.* 40:65–81, 1956.

1319 WYMAN, S.M. Congenital absence of a pulmonary artery: Its demonstration by roentgenography. *Radiology* 62:321–328, 1954.

1320 YATER, W.M. Variations and anomalies of the venous valves of the right atrium of the human heart. *Arch. Pathol.* 7:418–441, 1929.

1321* YAZUTA, K. Zur Varietätenstatistik des Ursprunges und der Lage einiger Äste der A. subclavia. *Anat. Anz.* 63:139–143, 1927.

1322* YEAGER, V.L. and N. LERTPRAPAI. Heart and aortic arch anomalies: Anomalous right subclavian artery, complete interruption of the aortic arch, and associated anomalies of the heart. *J. Med. Assoc. Thailand* 53:279–288, 1970.

1323* YELIN, G. Retroaortic renal vein. *J. Urol.* 44:406–410, 1940.

1324 YOUNG, A.H. Abnormal arrangement of the branches of the femoral artery. Note on the absence of profunda femoris. *J. Anat. Physiol.* 13:154–156, 1879.

1325 YOUNG, A.H. Abnormalities of the middle sacral artery and their morphological significance. *J. Anat. Physiol.* 31:169–175, 1897.

1326 YOUNG, A.H. Observations on the lumbar arteries. *J. Anat. Physiol.* 39:295–300, 1905.

1327 YOUNG, A.H. Rare anomaly of the heart – a three chambered heart in an adult aged thirty-five years. *J. Anat. Physiol.* 41:190–197, 1907.

1328* YOUNG, A.H. and P. THOMPSON. Abnormalities of the renal arteries, with remarks on their development and morphology. *J. Anat. Physiol.* 38:1–14, 1903.

1329 ZAAIJER, T. Seltene Abweichung (Schlingenbildung um die Vena cruralis) der Arteria profunda femoris. *Anat. Anz.* 9:502–508, 1894.

1330* ZALUSKA, S. and Z. UROANOWICZ. Variation of origins of the branches of the femoral artery in man and macaca. *Folia Morphol.* 29:273–283, 1970.

1331* ZEBROWSKI, W., AUGUSTYNIAK, E. and S. ZAJAC. Variation of origin and branches of the inferior mesenteric artery and its anastomoses. *Folia Morphol.* 30:510–517, 1971.

1332 ZIEGLER, H.R. Excision of the head of the pancreas for carcinoma with studies of its blood supply. *Surg. Gynecol. Obstet.* 74:137–145, 1942.

1333* ZIENTARSKI, B. Blood vessels of the caudate lobe of the human liver. *Folia Morphol.* 35:95–103, 1976.

1334 ZIENTARSKI, B. Accessory hepatic veins in man. *Folia Morphol.* 38:85–92, 1979.

1335 ZIENTARSKI, B. Dimensions of the inferior vena cava in humans. *Folia Morphol.* 41:307–314, 1982.

1336 ZIENTARSKI, B. Topography of hepatic veins in man. *Folia Morphol.* 41:419–428, 1982.

1337* ZIENTARSKI, B. and A. KASZUBA. A rare variation of the celiac trunk in the form of a gastrolienal and hepatomesenterial trunk. *Folia Morphol.* 34:403–406, 1975.

1338 ZOLNAI, B. Die zwischen der Arteria vertebralis und den vertebralen und zerebralen Venen bestehende Verbindung am atlantookzipitalen Abschnitt beim Menschen. *Anat. Anz.* 114:400–407, 1964.

1339 ZOLTOWSKY, P. and W. ATWELL. A case of a left superior vena cava without a corresponding vessel on the right side. *Anat. Rec.* 70:525–532, 1938.

1340 ZUCKERKANDL, E. Zur Anatomie der Orbitalarterien. *Med. Jahrb. (Wien)* 348–350, 1876.

1341 ZUCKERKANDL, E. Über ein abnormes Verhalten der Zungenschlagadern. *Wien. Med. Wochenschr.* Nr. 29:834–835, 1881.

1342 ZUCKERKANDL, E. Über die tiefen Hohlhandäste der Arteria ulnaris. *Anat. Hefte* 6:533–560, 1896.

1343 ZUMSTEIN, J. Zur Anatomie und Entwickelung des Venensystems des Menschen. *Anat. Hefte* 6:571–608, 1896.

1344 ZWERINA, H. and S. POISEL. Über eine Anastomose zwischen dem Truncus coeliacus, der Arteria mesenterica superior und der Arteria mesenterica inferior mit anderen Varietäten der unpaarigen und paarigen Äste der Bauchaorta bei einem Individuum. *Anat. Anz.* 119:427–435, 1966.

Books

1345* ADACHI, B. *Das Arteriensystem der Japaner*. Kenkyusha Press, Kyoto, 1928.

1346* ADACHI, B. *Das Venensystem der Japaner*. Kenkyusha Press, Tokyo, 1933, 1940.

1347* ANSON, B.J. *Atlas of Human Anatomy*. W.B. Saunders Co., Philadelphia, 1950.

1348 ANSON, B.J. and W.G. MADDOCK. *Callander's Surgical Anatomy*. W.B. Saunders Co., Philadelphia and London, 1952.

1349* ANSON, B.J. Ed. *Morris' Human Anatomy*, 12th Ed. The Blakiston Division, McGraw-Hill Book Company, New York, 1966.

1350 AREY, L.B. *Developmental Anatomy*, 7th Ed, Revised. W.B. Saunders Co., Philadelphia, 1974.

1351+ DUBREVIL-CHAMBARDEL, L. *Variations des Artéres du Membre Supérieur*. Masson, Paris, 1926.

1352* DUBRUEIL, J.M. *Des Anomalies Arterielles*. Bailliere, Paris, 1847.

1353 FRANKLIN, K.J. *A Monograph on Veins*. Charles C Thomas, Springfield, Ill., 1937.

1354 GARDNER, E., GRAY, D.J. and R. O'RAHILLY. *Anatomy*, 3rd Ed. W.B. Saunders Co., Philadelphia, 1969.

1355 GEGENBAUR, C. *Lehrbuch der Anatomie des Menschen*, 3rd. Ed. V. Wilhelm Engelmann, Leipzig, 1888.

1356 HENLE, J. *Handbuch der Systematischen Anatomie des Menschen*, 3 volumes. Von Friedrich Vieweg und Sohn, Braunschweig, 1868.

1357 HUBER, G.C., Ed. *Piersol's Human Anatomy*, 9th Ed. L.B. Lippincott Co., Philadelphia, 1930.

1358 HYRTL, J. *Lehrbuch der Anatomie des Menschen*, 19th Ed. Wilhelm Braumüller, Wien, 1887.

1359 JACKSON, C.M., Ed. *Morris' Human Anatomy*, 9th Ed. P. Blakiston's Son & Co., Inc., Philadelphia, 1933.

1360 KAPLAN, H.A. and D.H. FORD. *The Brain Vascular System*. Elsevier, Amsterdam, 1966.

1361 KOPSCH, F. *Rauber's Lehrbuch und Atlas der Anatomie des Menschen*. Georg Thieme, Leipzig, 1908.

1362 LATARJET, A. *Testut's Traite D'Anatomie Humaine*, 9th Ed. G. Doin & Cie., Paris, 1948.

1363 MICHELS, N.A. *Blood Supply and Anatomy of the Upper Abdominal Organs, with a Descriptive Atlas*. J.B. Lippincott Co., Philadelphia, 1955.

1364* POYNTER, C.W.M. *Arterial Anomalies Pertaining to the Aortic Arches and the Branches Arising from Them*. The University Studies of the University of Nebraska. 16:229–345, Lincoln, 1916.

1365* POYNTER, C.W.M. *Congenital Anomalies of the Arteries and Veins of the Human Body with Bibliography*. The University Studies of the University of Nebraska. 22:1–106, Lincoln, 1922.

1366+ QUAIN, R. *Anatomy of the Arteries of the Human Body*. Taylor and Walton, London, 1844.

1367 SCHAEFER, E.A., SYMINGTON, J. and T.H. BRYCE, Eds. *Quain's Anatomy*, 11th Ed. Longmans, Green, and Co., London, 1915.

1368 TAUSSIG, H.B. *Congenital Malformations of the Heart*, 2nd Ed. Harvard University Press, Cambridge, 1960.

1369 WADDINGTON, M.M. *Atlas of Cerebral Angiography with Anatomic Correlation*. Little, Brown and Co., Boston, 1974.

1370 WHITNALL, S.E. *The Anatomy of the Human Orbit*, 2nd Ed. Oxford University Press, London, 1932.

1371 WOLFF, E. *The Anatomy of the Eye and Orbit*, 4th Ed. Lewis, London, 1954.

Glossary of Terms

Their origin and definition related to the Cardiovascular System.

The following abreviations are used: Ar., Arabic; AS., Anglo-Saxon; G., Greek; and L., Latin.

Abdomen (L. *abdomen*).
Abdominal (L. *abdominalis*, pertaining to the abdomen).
Aberrans (L. *ab*, from + *errare*, to wander). Wandering or deviating from the usual or normal course.
Accessory (L. *accedere*, to approach). Supplementary or affording aid to another similar and generally more important vessel. Important exceptions are accessory hepatic, renal, and splenic arteries, which are essential, tissue-sustaining, and arteries without anastomotic connections.
Acromial (G. *akron*, extremity + *omos*, shoulder). Pertaining to the acromion, the point of the shoulder, e. g., acromial artery.
Aorta (L. from G. *aorte*, from *aeiro*, to lift up).
Apex (L. *apex*, summit or tip). Referred originally to the small swollen tuft on the top of the cap of the chief priest, thus by extension, the top or summit of any part. Apex of the heart.
Artery (G. *arteria*, from *aer*, air + *terein*, to keep). The arteries were believed to contain air.
Auricular (L. *auricula*, little ear). Pertaining to the auricles of the heart or to the little finger.
Azygos (G. *a* + *zygon*, yoke or pair). Unpaired, e. g., vena azygos.
Brachial (L. *brachium*, G. *brachion*, arm). Pertaining to the arm. Brachial blood vessels.
Brachium. See Brachial.
Bulbus (L. *bulbus*, G. *bolbos*, a rounded mass or enlargement), e.g., bulbus arteriosus.
Cardinal (L. *cardo*, hinge). Of primary or preeminent importance. Cardinal vein.
Carotid (G. *karos*, deep sleep, from the word *kar* or head, from where sleep is induced). The ancients called the arteries of the neck carotid because they believed that when they were pressed hard, the animal became sleepy.
Celiac (G. *koilia*, belly). Pertaining to the celiac trunk.
Cephalic (G. *kephale*, head). Pertaining to the head.
Cervical (L. *cervix*, neck), e.g., cervical artery.
Circumflex (L. *circumflexus*, bent around). Curved like a bow.
Cleido- (G. *kleis*, bolt, hook). Pertaining to the clavicle.
Collateral (L. *com*, together + *latus*, side). Small side branch.
Comites (L. *comes*, companion). Companion veins.
Conus (G. *konos*, a cone).
Coronary (L. *corona*, G. *korone*, encircling in the manner of a crown). Arteries of the heart; also formerly left and right gastric arteries and veins.
Cystic (G. *kystis*, bladder). Pertaining to the gallbladder.
Digital (L. *digitus*, a finger or toe). Pertaining to a finger or toe.
Ductus (L. *ducere*, to lead or conduct), e.g., ductus arteriosus.
Duodenal (L. *duodeni*, twelve). Pertaining to the duodenum. The duodenum was so named because it is "twelve fingerbreadths" in length.
Dural (L. *dura*, hard). Pertaining to the dura mater.
Epigastric (G. *epi*, on + *gaster*, belly). Pertaining to the epigastrium.
Epiploic (G. *epiploon*, omentum). Pertaining to the omentum.
Facial (L. *facies*, face), e.g., facial vein.
Femoral (L. *femur*, thigh). Pertaining to the femur or thigh.
Gastric (G. *gaster*, belly or stomach), e.g., gastric artery.
Gluteal (G. *gloutos*, buttock). Pertaining to the buttocks.

Gonadal (L. *gonas*, G. *gone*, seed). Pertaining to the gonad.
Heart (AS. *heorte*, originally from Sanskrit *hrid*, source of heat).
Hemorrhoidal (G. *haima*, blood + *rhoia*, flow). Vessels likely to bleed. Applied to arteries and veins in the region of the rectum and anus.
Hepatic (G. *hepatos*, liver). Pertaining to the liver.
Humeral (L. *humerus*, shoulder). Pertaining to the humerus.
Hypogastric (G. *hypo*, under + *gaster*, stomach). Internal iliac artery.
Iliac (L. *ilium*, flank, soft part), e.g., common iliac artery.
Ima (L. *lowest*), e.g., thyroidea ima.
Indicis (L. *indicare*, to point out). Pertaining to the indicator or the first finger of the hand.
Intercostal (L. *inter*, between + *costa*, rib). Situated between the ribs.
Interosseus (L. *inter*, between + *os*, bone). Situated between bone.
Jugular (L. *jugulum*, throat, from *jugum*, a yoke, because the yoke is carried on the neck). Pertaining to the throat or neck.
Lien- (L. *lien*, G. *splen*, spleen). Pertaining to the spleen.
Lingual (L. *lingua*, tongue). Pertaining to the tongue.
Lumbar (L. *lumbus*, loin), e.g., lumbar artery.
Median (L. *medius*, middle), e.g., median basilar artery.
Mesenteric (G. *meso*, middle + *enteron*, intestine). Pertaining to the mesentery.
Metacarpal (G. *meta*, after, beyond + *karpos*, wrist). Pertaining to the part of the hand between the wrist and fingers.
Obturator (L. *obturare*, to occlude), e.g., obturator vein.
Occipital (L. *ob*, before or against + *caput*, head). Pertaining to the occiput.
Omphalo- (G. *omphalos*, the navel). Pertaining to the umbilicus.
Ovarian (L. *ovarium*, egg receptacle). Pertaining to the ovary.
Palmar (L. *palma*, palm of the hand), e.g., superficial palmar artery.
Pancreatic (G. *pan*, all + *kreas*, flesh). Pertaining to the pancreas.
Peroneal (G. *perone*, brooch, anything pointed for piercing or pinning, fibula). Blood vessels of the leg.
Pollicis (L. *pollex*, thumb, from *polleo*, strong). The thumb is so named because it is stronger than the other fingers.
Popliteal (L. *poples*, ham, back of the knee). Pertaining to the posterior surface of the knee. It is possible that this area had a special name because the Roman soldier wore a short skirt or kilt with the knees exposed, and a favorite stroke in fighting was a cut behind the knee to "hamstring" the opponent.
Portal (L. *porta*, gate). An entrance or gateway. The portal vein carries nutrients to the liver.
Princeps (L. *princeps*, principal chief).
Profunda (L. *profundus*, deep).
Radial (L. *radius*, stroke of a wheel). The radial artery is on the lateral side of the forearm.
Radialis. See Radial.
Rectal (L. *rectum*, straight). Pertaining to the rectum.
Renal (L. *ren*, kidney). Pertaining to the kidney.
Sacral (L. *sacrum*, sacred). Pertaining to the sacrum.
Saphenous (Ar. *safin*, standing). Saphenous vein.
Scapular (L. *scapula*, shoulder), e.g., transverse scapular vein.

117

Septal (L. *septum*, dividing wall). Pertaining to a septum in the heart.
Spermatic (G. *sperma*, seed), e.g., spermatic vein.
Splenic (L. *splen*, spleen), e.g., splenic artery.
Sternum (G. *sternon*, breast or chest).
Subclavian (L. *sub*, under + *clavis*, key). Blood vessels under the clavicle.
Testicular (L. *testiculus*, from *testis*, witness), e.g., testicular artery.
Thoracic (G. *thorax*, chest). Pertaining to the chest.
Thyroid (G. *thyreoeides*, from *thyreos*, oblong shield + *eidos*, form). Resembling a shield.

Tibial (L. *tibia*, pipe or flute). Pertaining to the tibia.
Ulnar (L. *ulna*, elbow, arm). Pertaining to the ulna.
Uterine (L. *uterus*, womb). Pertaining to the uterus.
Vaginal (L. *vagina*, sheath). Pertaining to the vagina.
Valves (L. *valvae*, folding doors). Heart valves.
Vein (L. *vena*, vein, from *venire*, to come to). So named because blood comes to the heart through these vessels.
Vertebral (L. *vertebra*, vertebra), e.g., vertebral vein.
Vesical (L. *vesica*, bladder). Pertaining to the urinary bladder.

Terminology: Synonyms

Terminology for the cardiovascular system from the older English, French, and German literature related to more modern usage.

Arteries

Old Terminology	Modern Usage	Old Terminology	Modern Usage
A. abdominalis.	A. circumflex iliac.	A. brachialis profunda.	A. profunda brachii.
A. abdominalis externus.	A. inferior epigastric.	Brustschulterpulsader.	A. thoracoacromial.
A. abdominalis subcutanea.	A. inferior epigastric.	A. buccinatoria.	A. buccalis (buccal).
A. accessory superior colic artery.	A. colica media (middle colic).	A. bulbi penis.	A. bulbi urethrae.
Achselpulsader.	A. axillary.	A. bulbi urethrae.	A. bulbi penis.
A. acromiothoracic.	A. thoracoacromial.	A. bulbina.	A. bulbocavernosa.
Aeussere Hüftpulsader.	A. external iliac.	A. bulbosa.	A. bulbocavernosa.
Aeussere Kieferpulsader.	A. external maxillary (facial artery).	Canal artériel.	Ductus arteriosus.
A. anastomotica.	A. inferior ulnar collateral.	A. capsularis.	A. suprarenal.
A. anonyma iliaca.	A. common iliac.	Aa. cardiacae.	Aa. coronary.
Aa. anterior, media, posterior cerebri.	Aa. cerebri (cerebral).	Rr. cardiaci.	A. right gastric branch.
A. anterior dental.	A. alveolares superiores anteriores.	A. carotid facialis.	A. external carotid.
		A. carotis cerebralis.	A. internal carotid.
A. anterior iliac.	A. iliaca externa (external iliac).	A. carotis communis.	A. common carotid.
		A. carotis externa.	A. external carotid.
Antlitzpulsader.	A. external maxillary.	A. carotis primitiva.	A. common carotid.
A. aorta sacrococcygea.	A. sacralis mediana (middle sacral).	A. carpea dorsalis.	A. dorsal metacarpal.
		A. carpea dorsalis (rad.).	A. dorsal radial, carpal branch.
A. aortic suprarenal.	A. suprarenalis media (middle suprarenal).	A. carpea volaris (rad.).	A. palmar radial, carpal branch.
A. aortic uterine.	A. ovarica (ovarian).	A. caudal or coccygeal or sacrococcygeal.	A. sacralis mediana (middle sacral).
A. appendicalis.	A. appendicular.		
A. appendicularis.	A. appendicular.	A. cavi tympani.	A. tympanica (tympanic).
A. arciformes renis.	A. arcuatae renis.	Cavum pericardii.	Pericardial cavity.
A. arcus dorsalis pedis.	A. arcuata pedis.	A. cephalic.	A. carotis communis (common carotid).
Armpulsader.	A. brachial.		
Armspindelpulsader.	A. radial.	A. cerebellar inferior magna.	A. posterior inferior cerebellar.
Arteriae palatinae minores.	A. accessory palatine.		
V. arteriosa.	A. pulmonary.	A. cerebelli inferior.	A. anterior inferior cerebellar.
A. articularis genu azygos.	A. superior lateral genicular.		
A. articularis genu superior interna.	A. superior medial genicular.	A. cerebelli inferior anterior.	A. anterior inferior cerebellar.
A. articularis genu superioris interna secunda.	A. superior medial genicular.	A. cerebelli inferior posterior.	A. posterior inferior cerebellar.
A. articularis genu superioris medialis.	A. superior medial genicular.	A. cerebelli superior.	A. superior cerebellar.
		A. cerebelli superior anterior and posterior.	A. superior cerebellar.
A. articularis genu superficialis.	A. descending genicular.	A. cerebri media.	A. middle cerebral.
		A. cerebri posterior.	A. posterior cerebral.
A. articularis genu superioris profunda.	A. superior medial genicular.	A. cerebri transversa.	A. middle cerebral.
		A. cervicalis.	A. deep cervical.
A. articularis genu suprema.	A. descending genicular.	A. cervicalis adscendens.	A. ascending cervical.
A. atrabiliaria.	A. suprarenal.	A. cervicalis posticus.	A. deep cervical.
A. auditiva externus.	A. tympanic.	A. cervicalis superficialis.	A. superficial cervical.
A. auditiva interna.	A. labyrinthi.	Chorda ductus arteriosi.	Ligamentum arteriosum.
Auricle.	Atrium and the auricular appendix.	A. choroidea.	A. choroidea anterior (anterior choroidal).
Auricle.	Auricular appendix of atrium.	A. choroidea anterior.	A. choroidea (choroidal).
A. azygos, of vagina.	Vertical branch (posterior) of superior and inferior vaginal arteries.	A. circumflexa humeri anterior.	A. anterior circumflex humeral.
		A. circumflexa humeri major.	A. posterior circumflex humeral.
Balkenpulsader.	A. anterior cerebral.	A. circumflexa humeri minor.	A. anterior circumflex humeral.
Bauchspeicheldrüsenzwölffingerdarmpulsader.	A. superior pancreaticoduodenal.		
		A. circumflexa humeri posterior.	A. posterior circumflex humeral.
Bauchstück der A. cruralis.	A. external iliac.		
Beckenpulsader.	A. internal iliac.	A. circumflexa ilium.	A. circumflex iliac.

119

Human Anatomic Variation

A. circumflexa musculi sternocleidomastoideus inferior.
Aa. colica dextra.
Aa. colica dextra inferioris
Aa. colica dextra media.
Aa. colica media.
Aa. colicae dextra.
Aa. collateral digital arteries.

A. collateralis externa s.* magna.
A. collateralis internus.
A. collateralis radialis.
A. collateralis radialis superior.
A. collateralis ulnaris inferior.
A. collateralis ulnaris prima.
A. collateralis ulnaris secunda.
A. collateralis ulnaris superioris.
Comes nervi mediani.
Comes nervi phrenici.
Comes nervi ischiadici.
A. common carotid.

A. common femoral.

Aa. common volar digital.

Aa. common digital arteries of foot.
A. communicans Willisii.
A. coronaria dextra.

A. coronaria sinistra.

A. coronaria ventriculi dextra.
A. coronaria ventriculi inferioris dextra.
A. coronaria ventriculi inferioris sinistra.
A. coronaria ventriculi sinistra.
A. coronaria ventriculi superior sinistra.
Aa. coronariae cordis.
A. coronary anterior s. interna.
A. coronary posterior s. externa.
A. corporis callosi.
A. corporis cavernosa urethrae.
R. costalis lateralis.

A. cricothyreoidea.
A. cruralis.
A. cruralis iliaca.
A. cubitalis.

A. deep brachial.
A. deep facial.
A. deferentialis.
Aa. dentales superioris.
R. descendens, A. subscapularis.

* s. = synonymous

A. sternocleidmastoid.

Aa. right colic.
Aa. right colic.
Aa. right middle colic.
Aa. right middle colic.
Aa. right colic.
Aa. digitales palmares propriae.
A. profunda brachii.

A. superior ulnar collateral.
A. profunda brachii.
A. profunda brachii, deltoid branch.

A. inferior ulnar collateral.
A. superior ulnar collateral.
A. inferior ulnar collateral.

A. superior ulnar collateral.

A. median.
A. pericardiacphrenic.
A. ischiadic or sciatic.
A. carotis communis (common carotid).
A. femoral (above origin of profunda (deep) femoral).
Aa. arteriae digitales palmares communes.
Aa. metatarseae plantares.

A. posterior communicating.
A. coronaria (cordis) dextra (right coronary).
A. coronaria (cordis) sinistra (left coronary).
A. right gastric.

A. right gastroepiploic.

A. left gastroepiploic.

A. left gastric.

A. left gastric.

Aa. coronary.
A. right coronary.

A. left coronary.

A. anterior cerebral.
A. bulbocavernosa.

Lateral thoracic branch of internal thoracic.
A. inferior thyroid.
A. femoral.
A. external iliac.
A. ulnar.

A. profunda brachii.
A. maxillaris (maxillary).
A. ductus deferentis.
Aa. superior alveolar.
A. thoracodorsal.

Aa. diaphragmatic.

Aa. diaphragmaticae.
A. diaphragmatica superior.
Aa. digitales palmares communes.
Aa. digitales volares communis.
Aa. digitales volares propriae.
A. digitalis communis volaris prima.
A. dorsal (or posterior) interosseous (of forearm).
A. dorsalis scapulae.
A. dorsalis suprema.
Ductus arteriosus Botalli.
Aa. duodenal.

A. duodenalis inferioris.

Eingeweidepulsader
Ellenbogenpulsader
A. emulgent
A. emulgens
A. epigastrica inferioris externis
A. epigastrica interna
Aa. ethmoideae
Aa. etmoidales
A. external carotid

A. external epigastric

A. external mammary

A. external maxillary
A. external plantar

A. external spermatic artery

A. facial

A. facialis anterior
A. fallopian
A. femoralis superficialis
A. fibular
A. fibularis
A. fifth lumbar
A. Flügelgaumenpulsader
A. fossa Sylvii
A. frontal
A. frontalis
A. funiculi spermatici
A. funicular
Fussohlenpulsader

Gallenblasenarterie.
Aa. gastrica breves.
A. gastrica inferioris.
A. gastrica inferioris.
A. gastrica superioris dextra.
A. gastrica superioris sinistra.
Aa. gastricae breves.
A. gastro-epiploica sinistra.
A. genu descendens.
A. genu inferior lateralis.
A. genu inferior medialis.
A. genu superior lateralis.
A. genu superior medialis.

Aa. phrenica inferiores (inferior phrenic).
Aa. inferior phrenic.
A. pericardiacophrenic.
Aa. digitales volares communes.
Aa. digitalis palmares communes.
Aa. digitales palmares propriae.
A. princeps pollicis and indicis.
A. interossea posterior (posterior interosseous).
A. transverse cervical.
A. ascending cervical.
Ductus arteriosus.
Aa. inferior pancreaticoduodenal.
A. posterior inferior pancreaticoduodenal.
A. celiac.
A. ulnar.
A. renal.
A. renal.
A. circumflex iliac.

A. inferior epigastric.
Aa. ethmoidal.
Aa. ethmoidal.
A. carotis externa (external carotid).
A. circumflexa ilium profunda (deep circumflex iliac).
A. thoracica lateralis (lateral thoracic).
A. facialis (facial).
A. plantaris lateralis (lateral plantar).
A. cremasterica.

A. carotis externa (external carotid).
A. external maxillary
A. uterina (uterine).
A. femoral.
A. peronea.
A. peroneal.
A. lumbalis ima.
A. pterygopalatine.
A. middle cerebral.
A. supratrochlearis.
A. supratrochlearis.
A. cremasteric.
A. testicularis (testicular).
A. medial plantar.

A. cystic.
Aa. short gastric.
A. left gastroepiploic.
A. right gastroepiploic.
A. right gastric.
A. left gastric.
Aa. short gastric.
A. left gastroepiploic.
A. genus suprema.
A. genus inferior lateralis.
A. genus inferior medialis.
A. genus superior lateralis.
A. genus superior medialis.

Cardiovascular System

A. glutea.
A. gluteae inferior.
A. gluteae superior.

Aa. great phrenic.
Große Kranzpulsader des Magens.
Grundpulsader.

A. haemorrhoidalis interna.
A. haemorrhoidalis media.

Heiligbeinpulsader.
A. hepatic funiculus of Rauber.
A. hepatica communis.
A. hepatica propria.
Herzkegel.
Herzkrone s. corona cordis.
Herzohr.
A. highest thoracic.

Hüftbeinlochpulsader.
Hüftpulsader.
A. humeraria.
A. hypogastric.

A. hypogastrica.

Aa. ileae.
A. iliaca anterior.
A. iliaca parva.
A. iliaca posterior.
A. iliaca primitiva.
Aa. iliocolica.
A. iliolumbalis.
A. inferior capsular.

A. inferior dental.
A. inferior right colic.
A. infrascapularis.
Innere Brustpulsader.
Innere Hüftpulsader.
A. intercostalis s. costalis prima.
A. internal auditory.
A. internal mammary.
A. internal maxillary.
A. interossea dorsalis.

A. interossea internus.
A. interossea palmaris.
A. interossea perforans superioris.
A. interossea superficialis.
A. interosseus externa.
A. interosseus volaris.
A. ischiadica.

Kniekehlenpulsader.
Kopfpulsader.
Kranzarterien.
Kreuzpulsader.
Kurze Bauchpulsader.

Rr. labiales anteriores aa. femoralis.

A. superior gluteal.
A. glutea inferior (inferior gluteal).
A. glutea superior (superior gluteal).
Aa. phrenicae inferiores (inferior phrenic).
A. left gastric.

A. basilaris (basilar).

A. superior hemorrhoidal.
A. rectalis media (middle rectal).
A. lateral sacral.
A. hepatica propria (proper hepatic).
A. common hepatic.
A. common hepatic.
Ventricular cone of the heart.
Atrioventricular groove.
Auricle (atrium) of heart.
A. thoracica suprema (supreme thoracic).
A. obturator.
A. common iliac.
A. brachial.
A. iliaca interna (internal iliac).
A. internal iliac.

A. ilei.
A. external iliac.
A. iliolumbar.
A. superior gluteal.
A. common iliac.
Aa. right colic.
A. iliolumbar.
A. suprarenalis inferior (inferior suprarenal).
A. alveolaris inferior.
A. ileocolica (iliocolic).
A. subscapular.
A. internal thoracic.
A. internal iliac.
A. supreme intercostal.

A. labyrinthi.
A. thoracica interna.
A. maxillaris (maxillary).
A. interossea posterior (posterior interosseous).
A. anterior interosseous.
A. anterior interosseous.
A. posterior interosseous.

A. median.
A. posterior interosseous.
A. anterior interosseous.
A. inferior gluteal.

A. popliteal.
A. carotis (carotid).
Aa. coronary.
A. lateral sacral.
A. celiac.

Aa. labiales anteriores vulvae.

Rr. labiales posteriores aa. pudendae internae.
R. lateral infracostal.

A. left auricular.

A. left coronary (gastric).

A. left inferior gastric.

A. lienalis.
Linke Kranzpulsader des Magens.
A. long thoracic.
Aa. lumbales.
Aa. lumbares.

Magennetzpulsader.
Magenzwölffingerdarmpulsader.
A. mammaria externa.
A. mammaria internus.
A. mandibularis.
Mastdarmpulsader.
A. maxillaris s. dentalis inferior.
A. maxillaris externa.
A. maxillaris interna.
A. maxima Galeni.
A. meningea magna.
A. meningea medial.
Aa. metacarpeae dorsales (rad.).
Aa. metacarpeae volares.
A. metacarpea volaris sublimis (rad.).
A. metacarpea volaris sublimis.
A. middle capsular.

A. muscularis profunda.
Aa. musculares inferiores.
Aa. of Müller.

Aa. nasalis media.
A. nasopalatine.
A. nervi mediani.
A. Neubauer's.

Oberarmpulsader.
Obere Bauchpulsader.
Obere Schilddrüsenpulsader.
Oberflächliche Bauchdeckenpulsader.
Ostium venosum dextrum.

Ostium venosum sinistrum.

Rr. ovarici.
A. ovarii.

A. palatina descendens s. superior.
A. pancreatico-duodenalis superioris.
A. pediaea.
A. perforans prima femoris.
A. perforans secunda femoris.
A. perforans tertia femoris.

A. labiales posteriores vulvae.
Lateral thoracic branch of internal thoracic.
A. coronaria sinistra (left coronary).
A. gastrica sinistra (left gastric).
A. gastroepiploica sinistra (left gastroepiploic).
A. splenic.
A. left gastric.

A. lateral thoracic.
Aa. lumbar.
Aa. lumbar.

A. right gastroepiploic.
A. gastroduodenal.

A. lateral internal thoracic.
A. internal thoracic.
A. inferior alveolar.
A. superior hemorrhoidal.
A. inferior alveolar.

A. facialis (facial).
A. maxillaris (maxillary).
Aorta.
A. middle meningeal.
A. middle meningeal.
Aa. dorsal metacarpal.

Aa. metacarpeae palmares.
A. superficial palmar metacarpal branch of radial.
A. superficial palmar metacarpal.
A. suprarenalis media (middle suprarenal).
A. profunda femoris.
Aa. sural, medial, and lateral.
Aa. helicinae penis.

A. posterior ethmoidal.
A. sphenopalatina.
A. median.
A. thyroidea ima.

A. brachial.
A. inferior epigastric.
A. superior thyroid.
A. inferior epigastric.

Right auriculoventricular aperture.
Left auriculoventricular aperture.
A. ovarian.
A. ovarian.

A. pterygopalatine.

A. superior pancreaticoduodenal.
A. dorsalis pedis.
Aa. perforantae primae.
Aa. perforantae secundae.

Aa. perforantae tertiae.

Human Anatomic Variation

A. pelvica.
Aa. pericardiace posticus.
A. perinei.
A. peronea.
A. pharyngea ascendens.
A. pharyngea inferioris s. pharyngobasilaris.
A. pharyngomeningea.
Aa. phrenicae inferioris.
Aa. phrenicae magna.
A. phrenicocostalis.
A. posterior meningeal.

A. posterior pelvic.

A. profunda cerebri.
A. profunda humeri.
A. profunda inferior.
A. profunda superioris.
R. profundus arteriae transversae colli.
Aa. proper volar digital.

A. pterygoidea.
A. pudenda.
A. pudenda communis.
Aa. pudendae externae.
A. pulmonalis communis.
A. pylorica.

A. radial volaris sublimis.

A. radial volaris superficialis.

A. radialis.
A. radio-palmaris.

A. ranina.
A. ranine.
A. rectalis inferior.
A. rectalis superior.
Aa. recurrens anterior and posterior (ulnar).
A. recurrens radialis.
A. recurrens ulnaris.
A. renocapsularis.
A. right coronary (gastric).

A. right inferior gastric.

Rippenpulsadern.
Rückenschlagader des Fußes.
Ruthenpulsader.

A. sacra lateralis.
A. sacra media.
A. sacralis media.

A. scapularis inferior.
A. scapularis posticus.
A. scapularis propria.
A. scapularis superior.
Schampulsader.
Schlüsselbeinpulsader.
Rr. scrotales anteriores aa. femoralis.
Rr. scrotales posteriores aa. pudendae internae.

A. internal iliac.
A. bronchial.
A. perinealis (perineal).
A. fibularis (fibular).
A. ascending pharyngeal.
A. ascending pharyngeal.

A. ascending pharyngeal.
Aa. inferior phrenic.
A. inferior phrenic.
A. musculophrenic.
A. meningeal branch of vertebral. Also meningeal branch of ascending pharyngeal.
A. iliaca interna (internal iliac).
A. posterior cerebral.
A. profunda brachii.
A. superior ulnar collateral.
A. profunda brachii.
A. scapularis descendens (dorsalis) (dorsal scapular).
Aa. digitales palmares propriae.
A. pterygopalatine.
A. internal pudendal.
A. internal pudendal.
Aa. external pudendal.
A. pulmonary.
A. gastrica dextra (right gastric).

A. superficial palmar branch of radial.
A. superficial palmar branch of radial.
A. radial.
A. superficial palmar branch of radial.
A. lingualis (lingual).
A. profunda linguae.
A. hemorrhoidal.
A. hemorrhoidal.
Aa. anterior and posterior ulnar recurrent.
A. radial recurrent.
A. ulnar recurrent.
A. suprarenal.
A. gastrica dextra (right gastric).
A. gastroepiploica dextra (right gastroepiploic).
Aa. intercostal.
A. dorsalis pedis.
A. bulbi penis.

A. lateral sacral.
A. medial sacral.
A. sacralis mediana (middle sacral).
A. subscapular.
A. transverse cervical.
A. circumflex scapular.
A. suprascapular.
A. internal pudendal.
A. subclavia (subclavian).
Aa. scrotales posteriores.

Aa. scrotales posteriores.

A. short thoracic.
Sitzbeinpulsader.
A. small iliac.
Speichenpulsader.
A. spermatica deferentialis.
A. spermatica externa.
A. spermatica interna.
A. spinose.
Rr. sternocleidomastoidei aa. occipitalis.
A. superficial femoral.

A. superficialis arteriae transversae colli.
A. superficialis perinei.
A. superficialis volae (rad.).

A. superior dental.

Aa. superior diaphragmatic.
A. superior phrenic.
A. superior thoracic.
A. supramaxillaris.
A. suprarenalis aortica.
A. suprarenalis media.
Aa. surales medialis and lateralis.
A. sylvian.

A. tarsea internus.
A. temporalis.
A. thoracalis lateralis.

A. thoracalis suprema.

A. thoracic axis.

A. thoracica.
A. thoracia acromialis.
A. thoracica humeraria.
A. thoracica inferior.
A. thoracica longa.
A. thoracica major.
A. thoracica posterior.
A. thoracica prima s. minor.
A. thoracica secunda.
A. thoracica s. infrascapularis.
A. thoracica tertia.
Aa. thoracicae.
Rr. thymici aa. thoracicae internea.
A. thyreoidea superior.
A. tibialis anticus.

A. transversa carpi anterior.

A. transversa carpi dorsalis.

A. transversa carpi posterior.

A. transversa carpi volaris.

A. transversa colli.
A. transversa perinei.
A. transversa scapulae.

A. transversalis cervicis.

A. supreme thoracic.
A. inferior gluteal.
A. iliolumbal.
A. radial.
A. deferential.
A. cremasteric.
A. testicular.
A. middle meningeal.
A. sternocleidomastoid.

A. femoral (below origin of profunda (deep) femoral).
A. cervicalis superficialis (superficial cervical).
A. perineal.
A. superficialis palmar branch of radial.
A. alveolaris superior posterior.
Aa. phrenicae superiores.
A. pericardiacophrenic.
A. supreme thoracic.
A. superior alveolar.
A. suprarenal.
A. suprarenal.
Aa. sural, medial, and lateral.

A. cerebri media (middle cerebral).

A. dorsalis pedis.
A. superficial temporal.
A. thoracica lateralis (lateral thoracic).
A. thoracica suprema (supreme thoracic).
A. thoracoacromialis (thoracoacromial).
A. thoracodorsal.
A. thoracoacromial.
A. thoracoacromial.
A. long thoracic.
A. lateral thoracic.
A. long thoracic.
A. thoracodorsal.
A. supreme thoracic.

A. thoracoacromial.
A. thoracodorsal.

A. long thoracic.
A. thoracoacromial.
Aa. thymicae (thymic).

A. superior thyroid.
A. tibialis anterior (anterior tibial).
A. palmar radial, carpal branch.
A. dorsal carpal branch of radial.
A. dorsal carpal branch of radial.
A. palmar radial, carpal branch.
A. transverse cervical.
A. perineal.
A. suprascapularis (suprascapular).
A. superficial cervical.

Cardiovascular System

Old Terminology	Modern Usage
A. transverse scapular.	A. suprascapularis (suprascapular).
A. tripus.	A. celiac trunk.
Truncus coeliacus.	A. celiac trunk.
A. tubo-ovarian.	A. ovarica (ovarian).
Aa. ulnar metacarpal.	Aa. digitales palmares communes.
Untere Bauchdeckenpulsader.	A. inferor epigastric.
Unterschlüsselbeinpulsader.	A. subclavian.
A. utero-ovarica.	A. ovarian.
Vagina vasorum.	Carotid sheath.
Aa. venoeae.	Vv. pulmonary.
A. vermiform.	A. appendicularis.
A. vidian.	A. canalis pterygoidei.
A. volar interosseous.	A. interossea anterior (anterior interosseous).
Aa. volar metacarpal.	Aa. metacarpeae palmares (palmar metacarpal).
A. volar radial of index finger.	A. radialis indicis.
A. volaris indicis radialis.	A. radialis indicis.
R. volaris superficialis A. ulnaris.	A. superficial palmar, metacarpal branch.
Vordere Schienbeinpulsader.	A. tibialis anterior (anterior tibial).
Vordere untere Rumpfarterie.	A. inferior epigastric.
Wadenbeinpulsader.	A. peroneal.
Wirbelpulsader.	A. vertebralis (vertebral).
Wurmfortsatzpulsader.	A. appendicular.
Zapfenpulsader.	A. basilar.
A. Zinn's.	A. centralis retinae.

Veins

Old Terminology	Modern Usage
Angulus venosus s. venous angle (right and left).	Junction of internal jugular and subclavian veins.
V. anonyma iliaca.	V. common iliac.
Vv. anonymae dextra et sinistra.	Vv. brachiocephalicae (dextra et sinistra) (right and left brachiocephalic).
Venosus artery.	Pulmonary trunk.
Vv. articulares mandibulae.	Vv. articulares temporomandibulares.
V. ascendens.	V. anterior cerebral.
Vv. auditiva internae.	Vv. labyrinthi.
V. azygos minor.	V. hemiazygos.
V. azygos sinistra.	V. hemiazygos.
V. azyga magna.	V. azygos.
V. basalis Rosenthali.	V. basalis.
V. basilaris.	V. anterior cerebral.
V. canalis pterygoidei Vidii.	V. canalis pterygoidei.
Vv. cardiac veins.	Vv. cordis.
V. carotis externa.	V. external facial.
V. cava ascendens.	V. inferior vena cava.
V. cava descendens.	V. superior vena cava.
V. cephalica internus.	V. internal jugular.
V. cephalica posterior.	V. internal jugular.
V. cerebralis magna.	V. great cerebral.
V. cerebralis magna Galeni.	V. great cerebral.
V. cerebri anterior.	V. anterior cerebral.
Vv. cerebri internus communis.	V. great cerebral.
V. cerebri magna Galeni.	V. cerebri magna (great cerebral).
V. cerebri media.	V. deep middle cerebral.
V. cerebri media.	V. cerebri media superficialis (superficial middle cerebral).
V. cerebri mediana inferior.	V. deep middle cerebral.
Vv. cerebri posterior inferior.	V. deep middle cerebral.
V. cervicalis profunda.	V. deep cervical.
V. choroidea.	V. choroid.
V. choroidea lateralis.	V. choroid.
V. choroidea lateralis externa.	V. choroid.
Vv. choroideae oculi.	Vv. vorticosae (vorticose).
Vv. ciliares posterior.	Vv. vorticose (from posterior ciliary).
V. common facial.	V. facialis (facial).
V. coronaria dextra.	V. small cardiac.
V. coronaria magna.	V. great coronary.
V. coronaria parva.	V. small cardiac.
V. coronaria sinistra.	V. great cardiac.
V. coronaria ventriculi.	V. left gastric.
V. coronaria ventriculi.	V. gastrica sinistra (left gastric) or dextra (right or dextra gastric).
V. coronaria ventriculi sinistra.	V. gastrica sinistra (left gastric).
V. corporis callosi.	V. anterior cerebral branch to corpus callosum.
V. corporis striati.	V. thalamostriate.
V. cutanea colli majoris.	V. external jugular.
V. cutanea colli posterioris.	V. external jugular.
V. cutanea radialis.	V. cephalic.
V. cutanea ulnaris.	V. basilic.
Vv. digitales palmares.	Vv. digitales volares communes.
Vv. digitales volares propriae.	Vv. digitales palmares.
Vv. dorsales penis subcutaneae.	Vv. dorsales penis superficiales (superficial dorsal veins of penis).
V. dorsalis penis.	V. dorsal of penis.
V. dorsalis penis mediana.	V. dorsal of penis.
V. dorsi nasi superioris.	V. nasalis (external nasal).
Drosselblutader.	V. jugular.
V. emissarium.	V. emissaria.
V. emissarium condyloideum.	V. emissaria condylaris.
V. facialis anterior.	V. facial.
V. facialis communis.	V. facial.
V. facialis cutanea communis.	V. facial.
V. facialis externis.	V. external facial.
V. facialis interna.	V. facial.
V. facialis posterior.	V. retromandibularis (retromandibular).
Vv. fibulares.	Vv. peroneae (peroneal).
Vv. fibulares or fibular.	Vv. peroneae (peroneal).
V. fossa Sylvii.	V. deep middle cerebral.
V. Galeni.	V. great cerebral.
V. gastrica dextra.	V. coronaria ventriculi.
V. grande trachélienne externa.	V. deep cervical.
Vv. haemorrhoidales inferiores.	Vv. rectales inferiores (inferior rectal).
V. hypogastrica.	V. iliaca interna (internal iliac).

123

Human Anatomic Variation

V. hypogastrica.	V. internal iliac.	Sinus basilaris posterior.	Occipital sinus.
		Sinus caroticus.	Cavernous sinus.
V. iliaca.	V. common iliac.	Sinus communis venarum cardiacarum.	Coronary sinus.
V. iliaca communis.	V. common iliac.		
V. iliaca primitiva.	V. common iliac.	Sinus falciformis superioris.	Superior sagittal sinus.
		Sinus lateralis.	Transverse sinus.
V. jugularis cephalica.	V. internal jugular.	Sinus longitudinalis.	Superior sagittal sinus.
V. jugularis communis.	V. brachiocephalic.	Sinus obliquus.	Straight sinus.
V. jugularis communis.	V. internal jugular.	Sinus perpendicularis.	Straight sinus.
V. jugularis externa posterioris.	V. external jugular.	Sinus petrobasilaris.	Superior petrosal sinus.
		Sinus petrosal profundus.	Inferior petrosal sinus.
		Sinus petrosal superficialis.	Superior petrosal sinus.
V. large azygos.	V. azygos.		
V. left marginal.	V. posterior cardiac.	Sinus quartus.	Straight sinus.
V. lienalis.	V. gastrolienal or splenic.	Sinus rectus.	Straight sinus.
V. longitudinalis inferioris.	Inferior sagittal sinus.	Sinus sagittalis minor.	Inferior sagittal sinus.
Vv. lumbales.	Vv. lumbar.	Sinus tentorii.	Straight sinus.
Vv. lumbales transversae.	Vv. lumbar.	Sinus tentorii lateralis.	Superior petrosal sinus.
		Sinus tentorii medius.	Straight sinus.
V. magna cordis.	V. great cardiac.	Sinus tentorii posterior and sinus sigmoideus.	Transverse sinus.
Vv. majores cerebrales periphericae superioris.	Vv. superior cerebral.		
		Sinus triangularis.	Superior sagittal sinus.
V. mammaria interna.	V. thoracica interna (internal thoracic).	Sinus venae coronarie.	Coronary sinus.
		Sinus venae coronarie magnae proprius.	V. great coronary.
V. maxillaris externa.	V. facial.		
V. mediani.	V. median cubital.	Sinus venarum (cavarum).	Portion of right atrium lying to left of sulcus terminalis.
V. mediana antibrachii.	V. median cubital.		
V. mesenteric magna.	V. superior mesenteric.		
V. mesenterica.	V. superior mesenteric.	Sinus venarum s. saccus reuniens.	Right atrium lying to left of sulcus terminalis.
V. mesenterica superioris.	V. superior mesenteric.		
Vv. metacarpeae volares.	Vv. metacarpeae palmares (palmar metacarpal).	Sinus venosus.	Right atrium.
		V. subrenalis.	V. renal capsular.
		Vv. sus-hepatique.	Vv. hepatic.
V. obliqua atrii sinistri Marshalli.	V. obliqua atrii sinistri.		
		V. temporalis communis.	V. external facial.
V. ophthalmic cerebralis.	V. superior ophthalmic.	V. temporalis profundus.	V. middle temporal.
V. ophthalmica externa.	V. inferior ophthalmic.	V. temporalis superficialis anterior.	V. superficial temporal.
V. ophthalmica facialis.	V. inferior ophthalmic.		
V. ophthalmica inferioris.	V. inferior ophthalmic.	V. temporo-maxillaris.	V. external facial.
V. ophthalmica internus.	V. superior ophthalmic.	V. terminalis.	V. thalamostriate.
V. ophthalmica superioris.	V. superior ophthalmic.	Torcular Herophili.	Confluens of dural sinuses.
V. palatina.	V. palatina externa (external palatine).	V. trachélienne interieur.	V. vertebral.
		V. transversa scapulae.	V. suprascapularis (suprascapular).
Vv. parotideae posteriores.	Vv. parotideae.		
Vv. parumbilicales Sappeyi.	Vv. paraumbilicales.	V. transverse colic.	V. middle colic.
Vv. perforantes femoris.	Vv. perforantes.	Truncus anonymus.	V. brachiocephalic.
V. Pfortader.	V. portal.		
Vv. phrenicae.	Vv. inferior phrenic.	Vagina Vasorum.	Carotid sheath.
Vv. phrenicae superioris.	Vv. pericardiacophrenic.	V. velata.	V. thalamostriate.
Plexus pharyngeus.	Pharyngeal plexus.	Vena anonyma brachiocephalica.	V. brachiocephalic.
Plexus pharyngeus superficialis.	Pharyngeal plexus.	R. venae cephalicae anterioris anticus.	V. facial.
Plexus spermaticus.	Plexus pampiniformis.		
Plica venae cavae sinistrae.	Vestigial fold (Marshall).	V. vertebralis externa.	V. deep cervical.
V. portae.	V. portal.	V. vertebralis interna.	V. vertebral.
V. portarum.	V. portal.	V. vertebralis lateralis.	V. vertebral.
V. posterior interventricular.	V. middle cardiac.	V. vertebralis posterior.	V. deep cervical.
V. prepylorica.	V. coronaria ventriculi.	V. vertebralis profunda.	V. vertebral.
Vv. pulmonales sinistrae.	Vv. pulmonalis inferior sinistra and pulmonalis superior sinistra.	V. vertebralis superficialis.	V. deep cervical.
		Vv. vertebro-costales.	Vv. intercostal.
		Vv. vertebro-lumbares.	Vv. lumbar.
		Vestigial fold (Marshall).	Pericardial fold between left pulmonary artery and subjacent pulmonary vein. Contains fibrous vestige of left superior vena cava (duct of Cuvier).
Receptaculum.	Cavernous sinus.		
V. reflexa.	V. thalamostriate.		
V. right azygos.	V. azygos.		
V. satellite du nerf lingual.	V. dorsal lingual.		
V. septi lucidi.	V. septi pellucidi.		
Sinus alae parvae.	Sphenoparietal sinus.	Zeltblutleiter.	Straight sinus.

Lymphatics

Old Terminology	Modern Usage		
Bruströhre.	Thoracic duct.	Milchbrustgang.	Thoracic duct.
		Milchsaftgang.	Thoracic duct.
L. cysterna s. ampulla s. receptaculum chyli.	Cisterna chyli.		
		Receptaculum pecqueti.	Cisterna chyli.
L. ductus chyliferus s. lumbothoracicus s. pecquetianus.	Thoracic duct.	L. saccus acteus.	Cisterna chyli.
		Speisesaftröhre.	Thoracic duct.

NERVOUS SYSTEM

Alphabetical Listing

(Primary references, in multiple page listings, are in *boldface* type.)

Abducens (CN VI) **134**
Ansa Cervicalis 137
Brachial Plexus **139**
   Axillary N. 140
   Dorsal Scapular N. 139
   Lateral Brachial Cutaneous N. 140
   Long Thoracic N. 140
   Medial Antebrachial Cutaneous N. 140
   Medial Brachial Cutaneous N. 140
   Musculocutaneous N. 141
   Median N. 141
   N. to M. Subclavius 140
   Pectoral Nn., Lateral and Medial 140
   Posterior Interosseus N. 142
   Radial N. 142
   Subscapular Nn. 140
   Supraclavicular Nn. 139
   Suprascapular N. 139
   Thoracodorsal N. 140
   Ulnar N. 142
Brain **129**
   Anterior Commissure 129
   Brain Stem 131
      Pyramids 131
   Cavum
      Septi Pellucidi 129
      Veli Interpositi 129
      Vergae 129
   Cerebral Hemispheres 131
   Corpus Callosum 129
   Fissures and Sulci 129
      Angular 130
      Calcarine 131
      Central 130
      Cingulate 131
      Collateral 131
      Frontal
         Inferior 131
         Middle 131
         Superior 131
      Intermedius
         Posterior 130
         Primus (Anterior) 130
      Intraparietal 130
      Lateral Frontomarginal 131
      Olfactory 130
      Orbital 130
         Anterior 130
         Posterior 130
      Parieto-occipital 131
      Precentral 131
         Inferior 131
         Medius 131
         Superior 131

Postcentral 130
   Inferior 130
   Superior 130
Subcentral
   Anterior 130
   Posterior 130
Sylvian 129
Temporal
   Middle 130
   Superior 130
Transverse Retrocentral (Postcentral) 130
Fornix 129
Gyrus
   Angular 130
   Frontal 130
   Rolandic 131
   Supramarginal 130
Insula 129
Nucleus, Midline 129
Septum Pellucidum 129
Weight 129
Cardiac Plexus 148
Cervical Plexus **138**
   Communicating Cervical Branches 138
   Great Auricular N. 138
   Occipital N.
      Greater 138
      Lesser 138
   Phrenic N. 138
   Superficial Cervical N. 138
   Supraclavicular N. 139
Cervical Spinal Nerves **137**
   Eighth Cervical 137
   Fifth and Sixth Cervical 137
      Upper Subscapular N. 137
   First Cervical 137
      Suboccipital N. 137
   Fourth Cervical 137
   Second Cervical 137
   Seventh Cervical 139
   Sixth Cervical 137
   Third Cervical 135
Coccygeal Plexus 148
Descendens
   Cervicalis 138
Descendens
   Hypoglossi 137
Facial (CN VII) **134**
   Chorda Tympani 135
Furcal Nerve 143
Glossopharyngeal (CN IX) **135**
Hypoglossal (CN XII) **136**
Lumbar Spinal Nerves 138

Fifth Lumbar 143
First Lumbar **143**
Fourth Lumbar 143
Second Lumbar 143
Third Lumbar 143
Lumbar Plexus **143**
   Accessory Obturator N. 144
   Femoral N. 145
   First Lumbar N. 143
   Genitofemoral N. 144
   Iliohypogastric N. 143
   Ilioinguinal N. 143
   Lateral Femoral Cutaneous N. 144
   Obturator N. 144
      Accessory Obturator N. 144
   Saphenous N. 145
Lumbosacral Plexus 143
Lumbosacral Trunk 143
Oculomotor (CN III) **132**
   Ciliary Ganglion 132
Olfactory (CN I) **132**
Optic (CN II) **132**
Plexuses
   Brachial 134
   Cardiac 148
   Cervical 138
   Coccygeal 148
   Lumbar 143
   Lumbosacral 143
   Pudendal 141
   Sacral 145
Pudendal Plexus **147**
   N. to External Sphincter Ani M. 147
   Perforating Cutaneous N. 147
   Posterior Femoral Cutaneous N. 147
   Pudendal N. 147
      Inferior Rectal N. 148
Sacral Plexus **145**
   Gluteal N.
      Inferior 145
      Superior 145
   N. to M. Quadratus Femoris 145
   Peroneal N.
      Common 146
      Deep 146
      Superficial 146
   Plantar Nn., Lateral and Medial 147
   Posterior Femoral Cutaneous N. 146
   Sciatic N. 145
   Sural N. 146
   Tibial N. 146
Sacral Spinal Nerves 138
   Fifth Sacral 147
   First Sacral 147

Human Anatomic Variation

- Fourth Sacral **147**
- Second Sacral 147
- Third Sacral **147**
- Spinal Accessory (CN XI) **136**
- Spinal Cord **131**
- Spinal Nerves and Ganglia **136**
  - Ansa Cervicalis **137**
  - Ganglia Aberrantia 136
- Sympathetic Division **148**
  - Cardiac Plexus **148**
  - Cervical Ganglia
    - Inferior **148**
    - Middle **148**
    - Superior **148**
  - Cervical Rami Communicans **148**
  - Coccygeal Ganglion (Ganglion Impar) 149
  - First Thoracic Ganglion **148**
  - Lumbar Ganglia **149**
  - Sacral Ganglia **149**
  - Splanchnic N.
    - Greater **149**
    - Least **149**
    - Lesser **149**
    - Supreme **149**
  - Splanchnicus Imus 149
  - Stellate Ganglion (Cervicothoracic) 148
  - Superior Cardiac N. **148**
  - Temporal Ganglion 148
- Tibial Nerve **146**
  - Lateral and Medial Plantar Nn. **147**
- Thoracic Spinal Nerves **137**
  - Eighth 137
  - Eleventh 143
  - First 137
  - Intercostals 137
  - Second 137
    - Intercostobrachial N. 137
  - Twelfth **143**
- Trigeminal (CN V) **132**
  - Alveolar N.
    - Inferior **134**
    - Posterior Superior **133**
  - Auriculotemporal N. **134**
  - Buccal N. **134**
  - Ethmoidal N. **133**
  - Frontal N. **133**
  - Infraorbital N. **133**
  - Infratrochlear N. **133**
  - Lacrimal N. **132**
  - Lingual N. **134**
  - Mandibular N. **134**
  - Maxillary N. **133**
  - Nasal (Nasociliary) N. **133**
  - Ophthalamic **132**
  - Pterygopalatine (Sphenopalatine) Ganglion **133**
  - Sublingual Ganglion 134
  - Supraorbital N. **133**
  - Zygomatic N. **133**
- Trochlear (CN IV) **132**
- Vagus (CN X) **135**
  - Auricular N. **135**
  - Laryngeal N.
    - External 135
    - Internal **135**
    - Superior **135**
  - Pharyngeal Plexus **135**
  - Recurrent Laryngeal N. **135**
    - Inferior Laryngeal N. 135

Brain, Brain Stem, and Spinal Cord

Brain

Brain Weight

The brain usually weighs between 1200 and 1400 grams. It has been reported that the brain of Anatole France (French writer and Nobel laureate) weighed 1040 gm. The reported brain weights (in grams) of some other notable individuals include: Ivan Turgenieff, 2012; Cuvier, 1830; John Abercrombie, 1786; Thackeray, 1658; Kant, 1600; Spurzheim, 1559; Dirichlet, 1520; Daniel Webster, 1518; Fuchs, 1500; Louis Agassiz, 1495; Gauss, 1491; Helmholz, 1440; Dupuytren, 1437; Dante, 1420; Goltz, 1395; Liebig, 1352; Walt Whitman, 1282; Tiedemann, 1253; Hausmann, 1225; and Gall, 1198.

Sir William Osler is reported to have said that his brain would not appear to be different from the average. He was subsequently proved to be correct in this assumption.

References 215, 262, 518, 520.

Cavum Septi Pellucidi

The septum pellucidum normally has a cavity of 1 to 2 mm in width. When this cavity is larger, it is called cavum septi pellucidi or fifth ventricle. The term fifth ventricle is a misnomer because unlike ventricular cavities it does not contain cerebrospinal fluid nor is it lined by ependyma. The incidence of cavum septi pellucidi was 20.3% in 1032 brains studied.

References 86, 385, 432, 522.

Additional reference: Thompson, I.M. On certain abnormal conditions of the septum pellucidum. *Univ. Calif. Pubs. Anat.* 1 (3):21–54, 1932.

Cavum Vergae

The cavum vergae is the posterior extension of the cavum septi pellucidi. It is also called the sixth ventricle, which is a misnomer because the cavum does not contain cerebrospinal fluid nor is it lined by ependyma. The cavity was first described by the Italian anatomist, Andrea Verga, in 1851. It may exist as a separate cavity rather than communicating with the cavum septi pellucidi. The incidence of the cavum vergae was 2.3% in 1032 brains.

References 86, 385, 499, 522.

Additional reference: *Backman, E. Septum pellucidum and Verga's ventrikel. Upsala Läkareforen. Forhandl. N.F., 29:215, 1924.

Cavum Veli Interpositi

The cavum veli interpositi is located in front of the quadrigeminal (superior) cistern above the roof of the third ventricle. It develops as a result of abnormal separation of the limbs of the fornix. In one study, the cavum veli interpositi was found in 13 of 1200 (1.1%) pneumoencephalograms, and in another study, in 53 of 158 (30%) patients less than two years old.

References 86, 217, 334, 499, 522.

Corpus Callosum

The corpus callosum may be absent.

References 15, 50, 62, 116, 190, 244, 341, 456, 522.

Septum Pellucidum

The septum pellucidum may be absent. Agenesis of the septum pellucidum is often associated with agenesis of the corpus callosum, but either condition may occur without the other. When agenesis of the septum pellucidum is associated with hypoplasia of the optic disk, the condition is referred to as septo-optic dysplasia. The septum may also be perforated.

References 23, 50, 148, 279, 299, 498, 522.

Insula

In a study of 200 brains, including a few mentally retarded individuals, paralytics, and a series of young fetuses, it was found that the left insula was more deeply marked by sulci than the right, and its length, on average, was greater by 11 mm. The central sulcus of the insula is usually the first to appear developmentally, and it is more pronounced in males.

References 359, 389.

Anterior Commissure

The anterior commissure may be absent.

Reference 50.

Fornix

The fornix may be absent.

Reference 50.

Midline Nucleus

In 70% of brains, the massa intermedia (interthalamic adhesion) forms a bridge of gray matter, which crosses the third ventricle and constitutes part of a bilateral nuclear complex.

Reference 50.

Fissures and Sulci

Considerable variations have been reported in the course and development of the fissures and convolutions, not only in individuals but also on opposite sides of the same brain. These variations have not shown any constant relationship to age, sex, occupation, race, or intelligence.

Sylvian Fissure. The caudal extremity of the sylvian (lateral) fissure may be forked (41%) forming ascending and descending branches. The ascending branch is the most constant and is regarded as the true continuation of the fissure. The ascending branch forms the axis of the supramarginal gyrus. Triradiate termination has been reported.

According to Cunningham, the anterior rami of the sylvian fissure may assume different configurations: they may be Y-shaped where the two rami, enclosing the triangular frontal gyrus, join the sylvian fissure by a common stem (32% of hemispheres); the two rami may join the sylvian fissure separately, and they may be V- or U-shaped (37% of hemispheres); or a single ramus may join the fissure (30% of hemispheres). Less frequently, there are more than two rami, or the two rami are reduced to one (taking the position of either one of them) or are intermediate in position. In one study, two rami were observed in 85% of brains, one ramus in 12%, and three rami in 3%. The length of the anterior ascending ramus varies from 10 to 30 mm, and of the anterior horizontal, from 10 to 25 mm.

The sylvian fissure is frequently longer on the left hemisphere than on the right.

Postcentral Sulcus. This sulcus may be formed by up to four different segments.

Intraparietal Sulcus. The intraparietal sulcus may be formed by various combinations of its three phylogenetic components: a rostral vertical component, the postcentral inferior and superior sulci; a middle horizontal component, the intraparietal proper; and a caudal component, the paroccipital.

The inferior and superior postcentral components may separate from the intraparietal sulcus proper and join the postcentral sulcus. The inferior postcentral sulcus may terminate in a forked fashion with the anterior branch directed toward the lower concavity of the central sulcus. Occasionally the inferior postcentral sulcus gives rise below the bifurcation to the sulcus retrocentralis (postcentralis) transversus. The development of the transverse retrocentral (postcentral) sulcus bears an inverse relationship to the development of the subcentral posterior sulcus situated close to the sylvian fissure. The transverse postcentral and the subcentral sulci may join the sylvian fissure.

The relationship of the horizontal component (intraparietal sulcus proper) to the superior and inferior postcentral sulci has been described under five categories: in category I, all three sulci are separate; in category II, the horizontal component is continuous with the inferior postcentral component; in category III, the superior and inferior postcentral sulci are confluent and are separate from the intraparietal sulcus proper; in category IV, all three components are continuous with each other; and in category V, the horizontal component is continuous with the superior postcentral component.

The intraparietal sulcus proper may have one or two dorsal branches extending into the superior parietal lobule and continuous with one or more of its sulci. One or more short ventral branches may also extend into the inferior parietal lobule. One of these branches, the intermedius primus or anterior, marks the boundary between the supramarginal and angular gyri. The intermedius primus or anterior may be continuous with the superior temporal sulcus. Another branch, the sulcus angularis, may be axial to the angular gyrus, and it may be continuous with the superior temporal sulcus. The sulcus intermedius posterior is very variable and represents another inferior branch of the intraparietal sulcus proper.

Superior Temporal Sulcus. The superior temporal sulcus is usually a continuous sulcus but is occasionally separated into two or three sections, especially in the left hemisphere. The sulcus may encroach on the territory of the middle temporal sulcus. The terminal portion of the sulcus may be forked.

Middle Temporal Sulcus. This sulcus may separate into two to six parts. It may interdigitate with the superior temporal sulcus.

Orbital Sulcus. In its simplest form, this sulcus is triradiate with an anterior branch (orbitalis anterior) and two posterior branches (orbitalis posterior and lateralis). The two posterior branches form a U-shaped furrow with a caudally directed concavity. A second anterior branch (orbitalis anterior lateralis) may be present lateral to the orbitalis anterior branch. The orbitalis anterior lateralis may join with the U-shaped furrow and with the orbitalis anterior forming an H-shaped configuration. The two longitudinal arms of the H-shaped pattern may separate to form two distinct sagittal furrows.

Olfactory Sulcus. This sulcus usually bends medially at its rostral extremity, but it occasionally bends laterally. Its caudal extremity usually forks: the medial branch is short, while the lateral is usually long and bends anteriorly to hook around or anastomose with the posterior orbital sulcus.

Central Sulcus (Fissure). Usually, the central sulcus presents two anterior convexities, but occasionally, four or five may occur. The upper half of the sulcus may turn sharply backward. The upper end of the sulcus usually cuts the dorsal border of the brain and turns caudally; rarely, it ends as much as 1 cm from the dorsal border. The upper third of the sulcus is frequently separated from the rest of the sulcus. The lower part of the sulcus may be directed forward and may unite with the subcentral anterior sulcus. In some cases, the central sulcus may be interrupted, doubled, or absent; when doubled, the sulci are separated by the rolandic gyrus. The name rolandic was first applied by Leuret in 1839 (*Anatomie Comparée du Système Nerveux*) because the Italian anatomist Rolando directed attention to the sulcus. However, this sulcus was recorded by Vicq d'Azyr (*Traité d'Anatomie et de Physiologie*) in 1796.

Precentral Sulcus. The precentral sulcus may be composed of two or three parts: precentral inferior, precentral medius, and precentral superior. In about 17% of brains, the precentral sulcus does not show discontinuity into separate parts.

The upper extremity of the precentral inferior sulcus may be located anterior to the lower end of the precentral superior sulcus or may unite with the latter forming a continuous sulcus. Occasionally, a horizontal ramus extends from the precentral inferior sulcus above the inferior frontal sulcus; this horizontal ramus may be confluent with the middle or superior frontal sulci.

A separate precentral medius sulcus may be present. Most frequently, it is derived from the upper part of the precentral inferior sulcus, or the posterior part of the horizontal ramus. The precentral superior sulcus may give rise to the superior frontal sulcus. A posterior ramus of the precentral superior sulcus is sometimes present and may appear as an extension of the superior frontal sulcus. A superior precentral sulcus may cut (12%) or reach (14%) the superomedial border of the frontal lobe, but in most cases (74%) it fails to reach that border.

Inferior Frontal Sulcus. The inferior frontal sulcus may be attached to the precentral inferior sulcus, or it may be interrupted. It sometimes terminates in a bifurcation. The anterior part of the sulcus may be separated from the rest of the sulcus; the separated part is referred to as the lateral frontomarginal sulcus.

Superior Frontal Sulcus. This may form a continuous furrow or be broken up into several parts. It is more constant than the inferior frontal sulcus. The posterior end of the sulcus may (64%) be connected with the superior precentral sulcus.

Middle Frontal Sulcus. This sulcus is rarely uninterrupted. Its posterior end may be free or may join the middle or lower segment of the precentral sulcus.

Cingulate Sulcus. The cingulate sulcus may show partial or complete interruption in 28–50% of brains. The subcallosal portion of the sulcus may be doubled or tripled. In some cases the marginal ramus passes upward over the superomedial border to the dorsolateral surface.

Parieto-Occipital Fissure. This fissure may take a straight course or may assume a simple or an S-shaped curve.

Calcarine Fissure. This fissure may be hook-like. Its anterior part may show one or several side branches pointing backward and downward, and its posterior part may be interrupted or may extend to the dorsolateral surface of the hemisphere.

Collateral Fissure. The anterior end of the collateral fissure may be continuous with the rhinal fissure or with the anterior segment of the inferior temporal sulcus. The posterior end of the collateral fissure occasionally bifurcates into Y- or T-shaped branches.

References (Fissures and Sulci) 67, 70, 71, 81, 94, 144, 264, 268, 388, 407, 434, 439, 468.

Cerebral Hemispheres

The cerebral hemispheres may be asymmetric. The brain may be incompletely or imperfectly divided into two hemispheres.

Reference 50.

Brain Stem

Heterotopias of white matter within the medulla oblongata have been described.

The left pyramid decussates first in about 73% of individuals. Interesting variations in the pyramidal tracts have been reported. In about 71% of 144 specimens, both uncrossed and crossed pyramidal tracts were present bilaterally. In 13%, only crossed tracts were present; in 12%, both crossed tracts and one uncrossed tract were present; in 2.8%, only uncrossed tracts were found; and in 0.7%, one pyramid was totally crossed, and the other pyramid was totally uncrossed.

References 95, 179, 426, 457.

Spinal Cord

Diastematomyelia is a condition in which a portion of or the entire spinal cord is divided into two lateral halves. Each half of the cord is enveloped in its own membranes and gives rise to its own spinal roots. Hydromyelia or hydrorrhachis interna is a defective closure or arrangement of the divisions of the primary fetal central canal. Its presence may be indicated by subsequent development of syringomyelia or by its association with spina bifida. Hydrorrhachis externa is the abnormal congenital accumulation of fluid between the meninges of the cord causing a diminution of cord volume. Heterotopias (misplacement of parts) of the spinal cord include gray matter islands located in white matter; these may be isolated or partially joined to the gray horns by slender strands of gray matter. Malposition of bundles of white matter or passage of bundles of white matter (nerve fibers) through unusual sites to reach their destination has been described. Additional variations include (a) doubling of the spinal cord in some of its parts; (b) the spinal cord (in an adult) extending to the level of the third sacral vertebra; and (c) doubling of the central canal. The denticulate ligaments vary between 18 and 22 in normal individuals.

In one study of 129 adults, the point of termination of the spinal cord varied from the lower third of the twelfth thoracic vertebra to the middle third of the third lumbar. In 95% of these 129 cases, the cord terminated at the level of the first and second lumbar vertebrae.

References   180, 203, 208, 232, 258, 360, 382, 438, 457, 463, 464, 471.

Cranial Nerves and Ganglia

The present system for enumeration of cranial nerves (CN) was proposed by Soemmering in 1778.

Olfactory (CN I) Nerve

The following structures have been reported absent: olfactory nerve, bulb, tract, and lobes.

References   182, 193, 289, 298, 453, 512.

Optic (CN II) Nerve

In one study of 50 brains, the intracranial length of the nerve varied between 4.8 and 15.1 mm; the majority were 8–10 mm. The optic chiasm may be absent. The optic nerves and tracts may be unequal on either side, or the nerve may be absent on either side. The optic nerves are occasionally doubled and "accessory" optic nerves have been reported. There are reports of the optic tract being directly joined to the pedunculus cerebri and to the brain stem. The entire visual system has been reported absent in an adult.

References   280, 408, 473, 527.

Oculomotor (CN III) Nerve

Reported variations in the oculomotor nerve involve the occurrence of unusual branches or deviations in the course of its branches. The oculomotor may give rise to a branch that supplies the lateral rectus muscle, either in addition to or replacing its usual supply from the abducens nerve, which may be absent. It may provide a branch to the superior oblique muscle. The oculomotor has been reported to communicate with the sixth nerve (abducens) as they pass through the cavernous sinus. The upper division of the oculomotor may communicate with the nasociliary nerve. In a minor variation in the orbital course of the oculomotor nerve to the inferior oblique muscle, some of its branches pierce the inferior rectus muscle or the ciliary ganglion. The nerve has been found perforated, near its origin, by the posterior cerebral artery.

References   54, 72, 192, 512, 516, 519, 521, 525.

Ciliary Ganglion. The following variations in the size and roots of this ganglion have been reported. It may be absent or very small (grossly unidentifiable), in which case its constituent neurons are scattered among the nerves connected with the ganglion (oculomotor and ophthalmic division of the trigeminal). The ganglion may be doubled or even tripled. The motor root (branch of the oculomotor nerve to the inferior oblique muscle) sometimes bifurcates before reaching the ganglion. The sympathetic root may enter the ganglion separately or as part of the sensory root, or it may accompany the oculomotor nerve. Additional roots entering the ganglion have been described: these arise from the superior division of the oculomotor, trochlear, lacrimal, and abducens nerves, and from the pterygopalatine ganglion. The sensory root (branch of the ophthalmic nerve) may (a) be absent or multiple, in which case the sensory fibers from the nasociliary nerve are probably carried by the long ciliary nerves to the eyeball; or (b) arise directly from the semilunar (trigeminal) ganglion, the trunk of the ophthalmic nerve, or the supraorbital or lacrimal nerves. There may be one or more accessory ganglia; an accessory ganglion has been reported to lie on the medial side of the optic nerve. The ganglion has been found perforated by a ciliary artery.

References   502, 507, 510, 512, 516, 518, 519, 525, 527.

Trochlear (CN IV) Nerve

Variations in the branches and orbital course of the trochlear nerve have been described. The nerve has been reported to send a branch to M. orbicularis palpebrarum (orbicularis oculi) and to join sensory nerves (frontal, supratrochlear, infratrochlear, and nasal). These branches probably represent an aberrant course of sensory fibers from the trigeminal nerve. One variation in the orbital course consists of the trochlear nerve piercing (instead of coursing over) M. levator palpebrae superioris on its way to the superior oblique muscle.

References   72, 507, 510, 512, 516, 518, 519, 525, 527.

Trigeminal (CN V) Nerve
Ophthalmic Division (V$_1$)

Lacrimal Nerve. This nerve may appear to be derived from the trochlear nerve; however, the probable source in such cases is the ophthalmic nerve through its communicating branch to the trochlear in the cavernous sinus. The lacrimal nerve may be small at its origin, increasing in size later in its course by the addition of fibers derived from the temporal branch of the maxillary division of the trigeminal nerve. The lacrimal may be absent and replaced by the temporal branch of the maxillary division of the trigeminal nerve, or it may replace this nerve. It occasionally gives rise to a ciliary nerve, or it receives a branch from the long ciliary nerve of the ciliary ganglion or from the ganglion directly. It may receive accessory roots from the supraorbital or nasociliary nerves. The bifurcation of the lacrimal into its terminal branches may occur on the posterior wall of the orbital cavity. A branch of the lacrimal has been noted piercing the sclera. The lacrimal may exchange fibers with the ciliary ganglion.

References   83, 516, 525.

Frontal Nerve.   A variation has been reported in which the frontal nerve divides at a variable point before leaving the orbit to form supratrochlear and supraorbital branches. In such cases, the supraorbital branch passes through the supraorbital foramen through which the undivided nerve ordinarily passes. When the foramen is absent, it may have a special groove, the frontal notch (Henle).

References   83, 516, 525.

Infratrochlear Nerve.   This nerve may anastomose with a branch from the upper division of the oculomotor. In one case, a branch from a communication between the supra- and infratrochlear nerves pierced the orbital plate, ran beneath the dura, pierced the frontal bone, and supplied the overlying skin.

Reference   83.

Ethmoidal Nerve.   This nerve may be limited to the nasal cavity. It may also traverse the posterior ethmoidal foramen to gain entrance into the cranial cavity.

References   83, 516, 525.

Nasal (Nasociliary) Nerve.   Several variations in the branches of this nerve have been reported. It may send branches to the superior rectus, medial rectus, and superior levator palpebrae muscles. Branches emanating from a small ganglion connected to the nasal have been followed to the oculomotor and abducens nerves. The infratrochlear branch of the nasal nerve may be missing, in which case the areas normally supplied by this branch (skin of upper eyelid, root of nose, conjunctiva, and lacrimal caruncle and sac) receive their supply from the supratrochlear branch of the frontal nerve. Branches of the nasal nerve have been described passing to the frontal, ethmoidal, and sphenoidal sinuses. The branches to the frontal and anterior ethmoidal sinuses arise in the anterior ethmoidal foramen; branches to the sphenoidal and posterior ethmoidal sinuses arise in the posterior ethmoidal foramen. The branches to the sphenoidal sinuses are known as sphenoidal branches, whereas the branches to the posterior ethmoidal sinuses are known as sphenoethmoidal or posterior ethmoidal branches. An anastomosis between the nasal and lacrimal nerves has been reported.

References   83, 516, 525.

Trigeminal (CN V) Nerve
Maxillary Division (V$_2$)

The maxillary nerve may split into two trunks, each entering the skull through a separate foramen.

References   83, 118, 486, 525.

Infraorbital Nerve.   In about 54% of cases, a middle alveolar nerve arises from the infraorbital, 28% show significant variation, and in 18% the middle alveolar nerve is absent. The branch supplying the upper lip may run independently in the floor of the orbital cavity; other branches of the nerve arise earlier and enter another canal medial to the main channel. The nerve may be doubled.

References   83, 209, 516, 525.

Posterior Superior Alveolar Nerve.   In the absence of the buccal nerve, the posterior superior alveolar nerve distributes branches to the areas normally supplied by this nerve (mucous membrane and skin of the cheek).

References   83, 127, 516, 525.

Zygomatic Nerve.   The following variations have been reported in this nerve or its two branches (the temporal and facial or malar). The nerve may pass through the zygomatic bone before it divides into two branches, or the two branches may pass separately through foramina in the zygomatic bone instead of passing through a common foramen (sphenozygomatic foramen). The temporal branch in some cases passes through the sphenomaxillary fissure into the temporal fossa. Either branch of the zygomatic may be absent or smaller than normal, in which case the other branch compensates by carrying the additional nerve fibers. The area usually supplied by the zygomatic branch (skin of the zygomatic region) may be supplied instead by the infraorbital nerve. The area usually supplied by the temporal branch (skin of the anterior temporal region) may be supplied solely or additionally by the lacrimal nerve.

References   83, 516, 525.

Pterygopalatine (Sphenopalatine) Ganglion.   Several variations in the branches of this ganglion have been reported. The lesser palatine nerve, which is ordinarily one of the descending (palatine) branches of the ganglion and supplies the mucous membrane of the soft palate and tonsils, may be absent. There may be a reversal in the course of the left and right nasopalatine nerves through the foramina of Scarpa (greater and lesser palatine): the left nerve passes through the posterior (instead of the anterior) foramen and the right nerve passes through the anterior (instead of the posterior) foramen. Both nerves belong to the internal branches of the sphenopalatine ganglion. In addition to the usual ascending, descending, and internal branches of the ganglion, branches have been described passing to the abducens nerve, ciliary ganglion, or optic nerve, or through the sheath of the optic nerve.

References   83, 516, 525.

Trigeminal (CN V) Nerve
Mandibular Division (V$_3$)

Mandibular Nerve.   Some of the branches of the smaller, superior division of this nerve may arise directly from the main trunk before it divides in two. In some cases, the two divisions of the nerve are separated by a small fibrous band (pterygosphenoid ligament), which extends from the root of the external pterygoid plate to a point on the greater wing of the sphenoid in front of the foramen spinosum. This ligament is sometimes replaced by bone, and the smaller division of the nerve passes through a special foramen beneath the greater wing of the sphenoid, just lateral to the foramen ovale.

References   83, 516, 525.

Lingual Nerve.   A minute sublingual ganglion has been described arising from the lingual nerve or submandibular ganglion and supplying the sublingual gland. This nerve may pierce the lateral pterygoid rather than passing between the two pterygoid muscles. It occasionally provides motor branches to the medial and lateral pterygoids and to the palatoglossus muscles.

References   128, 134, 220, 345, 525.

Buccal Nerve (N. Buccinator).   The following variations in the course and/or branches of this nerve have been reported. Instead of its usual course on the medial side of the temporal muscle, the nerve may pierce the muscle. When absent, the nerve is replaced by the posterior superior alveolar nerves. The buccal nerve may also arise from the inferior alveolar nerve, in which case it emerges from the inferior dental canal through a small foramen in the alveolar border of the mandible, just anterior to the ramus. The nerve may arise directly from the gasserian (trigeminal, semilunar) ganglion, in which case it exits from the cranial cavity through a special foramen between the foramina ovale and rotundum.

References   53, 83, 516, 525.

Auriculotemporal Nerve.   This nerve usually arises by two roots from the posterior division of the mandibular nerve. The two roots normally surround the middle meningeal artery before joining to form a single trunk. A variation in this relationship has been described in which the middle meningeal artery pierces the anterior root instead of passing between the two roots.

References   83, 516, 525.

Inferior Alveolar Nerve.   The inferior alveolar nerve may form a single trunk with the lingual nerve, extending as far as the mandibular (dental) foramen. It may be separated from the lingual nerve by an accessory ligament extending from the lateral pterygoid plate and the spine of the sphenoid to the lateral side of the pterygospinous ligament. The inferior alveolar nerve is sometimes perforated by the internal maxillary artery. It may have accessory roots from other divisions of the mandibular nerve. In some cases, the mylohyoid branch of the inferior alveolar nerve gives rise to a branch that pierces the mylohyoid muscle and joins the lingual nerve. Branches have been described arising from the mylohyoid branch and supplying the mm. depressor anguli oris and platysma, the skin below the chin, and the submandibular (submaxillary) gland. The inferior alveolar may form connections with the auriculotemporal nerve. In one case, the roots of the third lower molar tooth were found surrounding the inferior alveolar nerve.

References   83, 220, 227, 258, 346, 412.
References (Trigeminal (CN V) Nerve)   53, 83, 118, 127, 128, 134, 209, 220, 227, 345, 346, 412, 486, 502–505, 507, 509–512, 516–519, 521, 524, 525.

Abducens (CN VI) Nerve

Several variations have been reported in the course and/or distribution of this nerve. It may be absent on one side, in which case the extraocular muscle normally innervated by the abducens (lateral rectus) is supplied by the oculomotor nerve. The abducens may leave the brain stem not as a single bundle but as several widely separated fascicles emerging from the lower border of the pons or between fibers of the pyramid. The abducens occasionally furnishes a branch to the superior rectus, or to the nasociliary nerve or ciliary ganglion.

References   72, 336, 504, 507, 509–512, 516–519, 521, 524, 525.

Facial (CN VII) Nerve

A pattern of variation has been established for the facial nerve (see Atlas). However, the clear separation of the nerve into five main terminal branches (temporal, zygomatic, buccal, mandibular, and cervical) may be difficult.

The posterior auricular and occipital branches may arise before their exit from the stylomastoid foramen. Branches from the facial nerve have been observed to replace those from the glossopharyngeal (CN IX) in supplying the root of the tongue and palate. The nerve to the digastric muscle may also supply the styloglossus and, in part, the sternocleidomastoid. The facial nerve may furnish branches to CN X (vagus) and CN XI (spinal accessory).

In a study of 78 preparations, the chorda tympani originated from the facial nerve within the facial canal in 74, and extracranially in four. The chorda tympani may run independently of the lingual nerve after its exit from the iter chordae anterius and the petrotympanic fissure. It has been observed attached to the inferior alveolar nerve until the latter entered the alveolar canal; the chorda tympani then supplied the salivary glands and joined the lingual to supply the tongue.

References   2, 13, 88, 92, 156, 224, 236, 276, 277, 314, 326, 502–505, 507, 509, 511, 512, 516–519, 521, 524, 525.

Glossopharyngeal (CN IX) Nerve

This nerve may supply the mylohyoid and anterior belly of the digastric muscles. A communicating branch sometimes arises between the glossopharyngeal and facial nerves. The tympanic branch of the glossopharyngeal may arise from two roots, one of which may come from the vagus. The glossopharyngeal may receive fibers from the hypoglossal just distal to the hypoglossal foramen.

References   137, 323, 356, 357, 371, 504, 507, 512, 516–519, 521, 525.

Vagus (CN X) Nerve

The right and left vagus may differ significantly in size. The nerves may lie in front of both vessels in the carotid sheath. The left vagus has been found passing anterior to the common carotid artery and thyroid gland. Descendens hypoglossi (C1 and C2) may be carried partly or completely by the vagus instead of by the hypoglossal nerve. The distribution of the right and left vagus nerves around the esophageal hiatus is variable. The vagus may contain accessory parathyroid tissue.

References   65, 103, 105, 145, 189, 205, 238, 239, 240, 241, 269, 290, 371, 427, 477, 500, 504, 507, 509, 511, 512, 516–519, 525.

Auricular Nerve. This nerve may be absent, or may join with the main trunk of the facial nerve reaching its destination through the posterior auricular nerve. Communicating branches between the auricular and facial nerves or the auricular nerve that traverses the temporal bone may be absent.

References   512, 517, 519, 525.

Pharyngeal Plexus. The pharyngeal plexus, formed by the pharyngeal branches of the vagus and glossopharyngeal nerves and branches from the superior cervical ganglion, has been reported to give rise to a slender branch, the middle laryngeal nerve. It supplies the cricothyroid muscle, then pierces the cricothyroid membrane to supply the mucous membrane of the lower part of the larynx.

References   461, 475.

Superior Laryngeal Nerve. Instead of its usual course on the medial side of the internal carotid artery, this nerve may pass lateral to the artery. The external laryngeal branch often arises independently from the main trunk. In addition to its usual territory of supply (cricothyroid muscle, inferior laryngeal constrictor, and cardiac sympathetics), this nerve, through its external laryngeal branch, has been reported to send fibers to the following structures: thyroid gland; pharyngeal plexus; sternohyoid, sternothyroid, thyrohyoid, and lateral cricoarytenoid muscles; mucous membrane of the true vocal cords; and the lower portion of the larynx. The superior laryngeal nerve may arise by two roots; one of these may come from CN IX (glossopharyngeal).

References   32, 101, 265, 274, 461, 475.

Internal Laryngeal Branch of Superior Laryngeal Nerve. Instead of piercing the thyrohyoid membrane to reach the larynx, this nerve may pass through a small foramen in the thyroid cartilage. The arytenoid muscle receives its primary innervation from the internal laryngeal nerve and occasionally receives an additional supply from the recurrent laryngeal nerve.

References   32, 98, 101, 265, 347, 461, 475.

Recurrent Laryngeal Nerve. There are several reported variations in the course and branches of this nerve. The right recurrent laryngeal nerve normally passes beneath and behind the subclavian artery and then ascends to enter the larynx. In cases in which the subclavian artery arises dorsally either between the esophagus and trachea (arteria lusoria) or retroesophageal, the right "recurrent" laryngeal nerve passes directly downward and medially from the vagus nerve to the larynx. In these cases there is no recurrent laryngeal nerve. The recurrent (inferior) laryngeal nerve, more commonly the right, tends to lie anterior to the inferior thyroid artery (50% of cases). The left recurrent laryngeal, three times more frequently than the right, lies behind the inferior thyroid artery. The nerve may be posterior to the lower branch of the inferior thyroid and anterior to the upper branch, or vice versa. In addition to supplying the inferior constrictor muscle of the pharynx, the nerve may supply the cricothyroid muscle. The recurrent laryngeal nerve may exist as two bundles on one side.

> References 3, 6, 32, 45, 47, 101, 225, 241, 265, 274, 306, 318, 328, 354, 428, 458, 461, 475.
>
> Additional references: Stedman, G. W. Singular distribution of some of the nerves and arteries in the neck and the top of the thorax. *Edin. Med. Surg. J.* 19:564, 1823.
>
> * Hart, J. A case of irregular origin and course of the right subclavian artery and right inferior laryngeal nerve; with remarks. *Edin. Med. Surg. J.* 25:286, 1826.
>
> Hooper, F. H. The anatomy and physiology of the recurrent laryngeal nerves. *N. Y. Med. J.* 46:29–34, 63–66, 99–103, 150–152, 188–190, 1887.
>
> Work, W. P. Unusual position of the right recurrent laryngeal nerve. *Ann. Otol. Rhinol. Laryngol.* 50:769–775, 1941.

Spinal Accessory (CN XI) Nerve

The following variations in the origin, course, and branches of this nerve have been reported. The lower limit of origin from the anterior horn of the spinal cord may be as high as the third cervical segment (instead of the fifth or sixth segment) or as low as the first thoracic segment. The nerve may exit from the subdural space below the first cervical nerve and reenter the space at a higher level. It may pass beneath (but never pierce) the sternocleidomastoid muscle. The nerve sometimes supplies only the sternocleidomastoid muscle, in which case the trapezius is supplied by the third and fourth cervical nerves. The nerve may provide a branch to the descendens cervicalis. In about 70% of cases, the accessory nerve crosses anterior to the internal jugular vein, and in about 30% of cases, posterior to the vein. It may pass through an annulus of the internal jugular vein.

> References 14, 202, 219, 252, 290, 327, 337, 417, 472, 504, 510, 512, 516, 517, 519, 525.

Hypoglossal (CN XII) Nerve

A number of variations in the course and branches of this nerve have been reported. The nerve may have a posterior (dorsal) root bearing a ganglion reminiscent of its embryologic origin (like a spinal nerve). The hypoglossal nerve has been reported to exit from the posterior surface of the medulla oblongata instead of its usual exit anteriorly. In some cases, instead of coursing dorsal to the vertebral artery, the rootlets of the nerve, upon leaving the medulla oblongata, lie on either side of the artery (artery passing between the rootlets), lie anterior to the artery, or pass through an annulus of the vertebral artery. The right and left nerves may be connected by crossing fibers situated between the genioglossus and geniohyoid muscles or in the substance of the geniohyoid (ansa suprahyoidea hypoglossi of Hyrtl). The reported frequency is 8–10% of individuals. The hypoglossal may send branches to one or more of the following muscles: mylohyoid, digastric, and stylohyoid.

In some cases, the descendens hypoglossi (C1 and C2) is carried partly or totally by the vagus nerve. The origin of fibers, in either case, remains the first and second cervical nerve. The descendens hypoglossi may send branches to the thorax, where it joins either the vagus nerve or the sympathetics. The origin of these branches is probably either vagal or sympathetic fibers that have joined the descendens hypoglossi. The descendens hypoglossi occasionally sends a branch to the sternal head of the sternocleidomastoid muscle.

Branches from the hypoglossal have been traced to cardiac nerves, but these have been interpreted as consisting of aberrant vagal nerve fibers.

> References 28, 35, 68, 128, 137, 196, 235, 269, 339, 367, 444, 477, 482–484, 504, 510, 512, 516, 517, 519, 525.

Spinal Nerves and Ganglia

There are a number of variations in the posterior (dorsal or sensory) and anterior (ventral or motor) roots and spinal ganglia. The first cervical nerve may either have no dorsal root or share a dorsal root with, or derive a dorsal root from, the spinal accessory (CN XI) nerve. The dorsal root and ganglion of the first cervical nerve may be rudimentary or entirely absent. With this singular exception, the sensory or dorsal root of spinal nerves is always larger than the motor or ventral root. The anterior or posterior root (or both) are sometimes absent in one or more thoracic spinal nerves. The dorsal root ganglia associated with any one of the lumbar or upper sacral spinal nerves may be doubled. Ganglia aberrantia (small detached portions of spinal ganglia) may be found along the posterior roots of the upper cervical, lumbar, and sacral spinal nerves.

In the coccygeal region, anterior roots may be completely absent or increased in number (one, two, or more). This is usually associated with an anomalous number or arrangement of vertebrae. The spinal ganglion and sensory root of the coccygeal nerves may be absent; the same may occur to a lesser degree in other spinal nerves. Occasionally, rudimentary anterior roots are found in the filum terminale. One or two additional coccygeal nerves may be present bilaterally or only on one side.

Communicating branches between roots (dorsal or ventral) of adjacent spinal nerves may occur.

The majority of spinal ganglia are located in the intervertebral foramina, closely ensheathed but entirely outside the cavity or sac of the dura mater. The ganglia of the fifth lumbar and first four sacral nerves lie within the vertebral canal, but these are also extradural. The fifth sacral and the coccygeal nerves (when present) are located within the dural cavity. This may also be true of the first cervical ganglion when it is present.

References   504, 512, 516–519, 525.

Cervical Spinal Nerves

Ansa Cervicalis

The ansa cervicalis may arise from the first, second, and third or only from the second and third cervical nerves. The ramus descendens hypoglossi may be replaced by the vagus (CN X). When no ansa is present, the innervation of the infrahyoid muscles is by branches arising directly from C2 and C3. The nerve to the thyrohyoid may arise as a branch of the ramus descendens hypoglossi. The phrenic nerve may also receive a contribution from the descendens hypoglossi. Variable connections with the cervical sympathetic trunk have been reported.

References   245, 269, 367, 444, 483, 484, 504, 507, 509, 510–512, 516–519, 525.

First Cervical Nerve

The dorsal root of the first cervical nerve is sometimes absent (8%). The root may arise partly or wholly from, or in common with, the accessory nerve. The dorsal branch of the first cervical nerve (the suboccipital nerve) occasionally supplies a cutaneous branch to the back of the head. It accompanies the occipital artery and is joined beneath the skin with the greater and lesser occipital nerves. (See also Vagus (CN X) and Hypoglossal (CN XII) Nerves.)

References   245, 307, 504, 507, 509–512, 516–518, 525.

Second Cervical Nerve

The greater occipital branch of the second cervical nerve has been reported to send a branch to the auricle. The greater and lesser occipital nerves complement each other, with each one compensating for any deficiency in nerve supply of the other. The external (lateral) branch of the second cervical nerve may have a cutaneous branch in addition to its usual muscular branches; it may also supply the suboccipital superior oblique muscle in addition to its usual supply to the semispinalis capitis, inferior oblique, semispinalis cervicis, and multifidus spinae muscles. (See also Vagus (CN X) and Hypoglossal (CN XII) Nerves.)

References   245, 307, 504, 507, 510–512, 516–519, 525.

Third Cervical Nerve

When the ansa cervicalis is absent, the third cervical participates in the innervation of the infrahyoid muscles.

References   245, 504, 507, 510–512, 516, 517, 519, 525.

Fifth and Sixth Cervical Nerves

The cutaneous branches of the fifth and sixth cervical nerves may be very small or completely absent. The upper subscapular nerve arises from the fifth and sixth, but in 50% of cases it also receives a contribution from the fourth.

References   504, 507, 510–512, 516, 517, 519, 525.

Eighth Cervical Nerve

The eighth cervical nerve may give rise to a cutaneous branch of considerable size.

References   504, 512, 516, 518, 526.

Thoracic Spinal Nerves

Defects in roots of thoracic nerves have been reported. In one study of 16 spinal cords, only three had the full complement of thoracic roots. In three cases, one thoracic spinal nerve (dorsal and ventral roots) was absent; in three, a dorsal root was absent; and in seven, a ventral root was absent.

The first thoracic may fail to contribute to the formation of the posterior cord of the brachial plexus or it may be the sole source of the medial cord of the plexus.

The first and second thoracic nerves are occasionally joined by a connecting branch.

The lateral branch of the second thoracic, the intercostobrachial (intercostohumeral) nerve, may divide into an anterior and posterior division. The anterior division may, however, be absent. In either case, the lateral branch anastomoses with the medial brachial cutaneous, usually with the lateral branch of the third intercostal, and rarely with the lateral of the first intercostal if it is present.

The size of the intercostobrachial cutaneous nerve and the extent of its distrubution are reportedly inversely proportional to the size of the other cutaneous nerves of the upper arm, especially the medial brachial cutaneous nerve. When the medial brachial cutaneous nerve is absent, it is replaced by the intercostobrachial.

The first thoracic nerve may have no cutaneous branch. The sixth, seventh, and eighth nerves may provide cutaneous branches from both their lateral and medial branches. Thoracic spinal nerves supply the sternalis muscle, when present (3–5% of individuals), and the axillary arch muscle, when present.

References   40, 64, 66, 79, 90, 243, 287, 478, 481, 504, 512, 516, 519, 525.

Lumbar and Sacral Spinal Nerves

The ganglia of lumbar and upper sacral nerves may appear to be doubled, with two distinct swellings on the dorsal root.

References   89, 469, 504, 512, 516, 519, 525.

Plexuses

Cervical Plexus

The cervical plexus consists of three loops. A large part of the anterior primary division of the first cervical nerve is given to the hypoglossal or cervical loop. The remainder contributes to the cervical plexus by joining the second cervical nerve, thus forming the first loop. From the first loop of the plexus, two branches of the first cervical nerve usually enter the sheath of the hypoglossal nerve (or may instead enter the sheath of the vagus nerve) and descend with it to contribute to the hypoglossal or cervical loop. The first cervical nerve leaves the hypoglossal (or vagus) as the descendens cervicalis or hypoglossi at the level of the occipital artery and later joins the communicans cervicalis, which arises from the second and third cervical nerves, completing the hypoglossal or cervical loop. The location of this loop is variable. It is usually found between the sheaths of the sternocleidomastoid and carotid artery superficial to the internal jugular vein, but may instead be located between the carotid artery and the internal jugular or, rarely, dorsal to both artery and vein. It may be long, terminating below the level of the thyroid cartilage, or short, terminating at the level of the hyoid bone. The descendens hypoglossi (or cervicalis) runs distally on the sheaths of the great vessels or may be located within one of these sheaths.

The second cervical nerve contributes to the first and second loop of the cervical plexus. The third cervical unites with the second and fourth cervical to complete the lower loops of the plexus.

Lesser Occipital Nerve

The lesser occipital nerve may be very small and distributed only to the skin of the neck. In such cases, the greater occipital nerve supplies the areas usually supplied by the lesser occipital. In some cases instead of its usual dorsal and upward course beneath the deep fascia along the posterior border of the sternocleidomastoid muscle, the lesser occipital nerve passes directly backward, piercing the trapezius muscle near its upper border before reaching the scalp. The auricular branch of the lesser occipital nerve may be derived from the greater occipital. It usually arises from the second and third cervical nerves or from the loop between them.

References   74, 504, 512, 516, 519, 525.

Great Auricular Nerve

The mastoid branch of the great auricular nerve may arise independently from the cervical plexus, in which case it passes upward between the lesser occipital and great auricular nerves to reach its destination. The great auricular nerve may arise solely from the third or from the third and fourth cervical nerves.

References   74, 504, 512, 516, 519, 525.

Superficial Cervical Nerve

Instead of a single nerve, the superficial cervical may have two or more branches arising from the cervical plexus. The nerve may arise solely from the third or from the third and fourth cervical nerves. In its course in the neck it sometimes passes through an annulus in the external jugular.

References   25, 74, 504, 512, 519, 525.

Communicating Cervical Branches

An additional communicating branch from the fourth cervical nerve may join the descending cervical nerve in the ansa cervicalis.

References   74, 504, 512, 516, 519, 525.

Phrenic Nerve

The following variations in origin, course, and distribution of the phrenic have been reported. The phrenic nerve may receive additional roots from one or more of the following nerves: nerve to subclavius; nerve to sternohyoid; second or rarely sixth cervical nerve; descendens cervicalis; ansa cervicalis; and brachial plexus. It may receive a branch from CN XII (hypoglossal) and may communicate with CN XI (spinal accessory). The phrenic may arise exclusively from the nerve to the subclavius. On occasion it supplies a branch to the subclavius muscle. The phrenic may pass along the lateral border of or pierce the anterior scalene muscle. The consolidation of the phrenic into a single trunk may not occur until it enters the thorax. The size of the nerve may vary bilaterally.

Instead of descending behind the subclavian vein,

the phrenic may pass anterior to it or even through a ring or annulus formed by the vein. An accessory phrenic nerve may also arise from the fifth or fifth and sixth cervical nerves and pass in front or behind the subclavian vein to join the main trunk of the phrenic nerve at the root of the neck or within the thorax. The incidence of an accessory phrenic nerve was reported to be 75.7% in 309 cases studied. The phrenic nerve may supply a branch to scalenus anterior. The phrenic may descend in contact with the posterior surface of the sternocleidomastoid muscle, cross in front of the subclavian and right brachiocephalic veins, and lie about 1.5 cm in front of the scalenus anterior muscle and subclavian artery. In one case, the right phrenic was found passing through an annulus of the subclavian vein.

References   34, 120, 121, 130, 133, 151, 202, 216, 223, 247, 286, 313, 316, 343, 374, 380, 444, 455, 492, 504, 510–512, 516–519, 521, 524, 525.

Brachial Plexus

The brachial plexus is formed by the anterior primary divisions of the lower four cervical nerves and the greater part of the anterior primary division of the first thoracic nerve. It may be joined by branches from the fourth cervical and the second thoracic nerves.

In one study of 175 plexuses, more than 60% received a communication from the fourth cervical nerve. Of those that did not receive a contribution from the fourth, almost 30% of the 175 plexuses studied received the entire anterior primary division of the fifth cervical. In the remaining 10%, a portion of the fifth cervical contributed to the cervical plexus.

Brachial plexuses receiving a branch from the fourth cervical nerve are more cephalic with reference to the vertebral column and have been designated as high or prefixed. Those receiving the largest part of the fifth cervical nerve are more caudal with reference to the vertebral column and have been designated as low or postfixed brachial plexuses.

The following variations in the organization or course of components of this plexus have been reported. The fifth cervical nerve may pass in front of or through the anterior scalene muscle. The sixth cervical nerve may cross the anterior scalene. The seventh cervical nerve (as the middle trunk) may break up into three branches (instead of the usual two), one branch going to each of the three cords. Occasionally the fibers of the posterior cord may arise from only the seventh and eighth cervical nerves. The first thoracic nerve may fail to contribute to the posterior cord or may be the sole source of the medial cord. Instead of the usual three cords, only two cords, a smaller and larger, may be present. In such cases, the larger cord will replace either the medial and lateral or the medial and posterior cords. Cases of a single cord have been reported.

A prefixed (high) type of brachial plexus exists in which the fourth cervical nerve contributes a fairly large branch, the dorsal branch of the first thoracic nerve is small, and there is no ascending branch from the second thoracic nerve. A postfixed (low) type of plexus is occasionally encountered in which the branch from the fourth cervical nerve is absent, the dorsal branch of the first thoracic nerve is large, and the second thoracic nerve supplies a branch to the plexus. In humans the brachial plexus is prefixed when compared to that in all primates below the anthropoids. The variations in this plexus may be appreciated by examining the illustrations in the Atlas.

References   36, 37, 49, 66, 104, 122, 172, 173, 218, 267, 284, 386, 396, 462, 467, 504, 509–513, 516–519, 521, 524, 525.

Supraclavicular Nerves

The following variations have been reported in these nerves. A branch to the phrenic nerve may pass down into the thorax over the subclavian artery or vein before joining the phrenic nerve. An additional bundle from the fifth cervical nerve joining this branch has been reported. A branch to the rhomboid muscles may pass through the trapezius muscle on its way to the skin close to the midline, at the level of the spinous processes of the fifth and sixth thoracic vertebrae.

In 10 of 254 cases, the clavicle was perforated by the middle branch of these nerves. They traversed the bone in intraclavicular canals and were visible on x-ray films.

References   313, 319, 504, 512, 518, 519, 525, 528.

Suprascapular Nerve

Several variations in the formation, course, and distribution of this nerve have been reported. It may receive fibers from the fourth cervical nerve (50% of cases), in addition to those usually supplied by the fifth and sixth cervical nerves. The suprascapular may arise entirely from the fifth cervical nerve. In some cases, it divides into two parts, an upper and a lower; the upper passes through or above the suprascapular notch (transverse ligament) and the lower passes through a foramen below the notch. The entire nerve may also pass over the transverse scapular ligament. The suprascapular may give branches to the following structures: teres minor and the upper part of the subscapular muscle, the scapula and its periosteum, and the acromioclavicular articulation.

References   504, 510–512, 516, 519, 525.

Dorsal (Posterior) Scapular Nerve

Instead of its usual location on the deep surface of

the levator scapulae muscle, the nerve may pierce the muscle.

References   102, 504, 510–512, 516, 519, 521, 525.

Long (Posterior) Thoracic Nerve (External Respiratory Nerve of Bell)

The following variations occur in the origin of this nerve. A contribution has been reported from the fourth cervical and from the fifth cervical nerve, which may not join the main trunk of the nerve but pass directly to the upper digitations of m. serratus anterior. The root from the seventh cervical nerve may be missing. An additional root from the eighth cervical nerve has been reported. Infrequently, the long thoracic nerve does not pass through scalenus medius and the nerve trunk is formed over the first rib. The nerve is usually formed of the fifth, sixth, and seventh cervical nerves; the first two pierce m. scalenus medius while the seventh passes in front of that muscle.

References   174, 504, 510–512, 516, 519, 521, 525.

Nerve to M. Subclavius

The nerve to the subclavius may supply a branch to the phrenic nerve (accessory phrenic nerve). It joins the phrenic nerve either at the root of the neck or less frequently in the thorax. Rarely, the entire phrenic nerve arises from this branch, in which case it crosses the third part of the subclavian artery. A branch may communicate with the lateral and medial pectoral nerves, and another may supply the clavicular head of the sternocleidomastoid muscle.

References   313, 504, 510–512, 516, 519, 521, 525.

Lateral and Medial Pectoral Nerves

The medial pectoral (from the medial cord of the brachial plexus) may supply twigs to the clavicular head of the deltoid muscle and the acromioclavicular articulation. The lateral pectoral (from the lateral cord of the brachial plexus) may not receive fibers from the first thoracic nerve. Cutaneous branches from the lateral pectoral nerve have been described supplying the skin of the mammary region and axilla. Fibers from the lateral pectoral that supply m. pectoralis major may wind around the lower border of pectoralis minor instead of piercing it.

The lateral and medial pectoral nerves together may form an ansa. Either nerve may form a loop around a neighboring vessel.

References   39, 504, 510–512, 516, 519, 521, 525.

Subscapular Nerves

The upper trunk of these nerves (upper subscapular) may arise from either the fifth or sixth cervical nerve instead of from both. The fourth cervical may also contribute. The middle (thoracodorsal) may arise solely from the fifth or seventh, or from the seventh and eighth, cervical nerves. The lower subscapular nerve may arise from the fifth or sixth, or from the sixth and seventh cervical nerves.

The nerve supply to the lower part of mm. subscapularis and teres major (ordinarily from the lower subscapular) sometimes proceeds separately from the brachial plexus. The branch to m. teres major may also arise from the axillary (circumflex) nerve.

The lower subscapular and axillary nerves have been found to traverse m. subscapularis beneath an accessory subscapular muscle (subscapularis minor). This arrangement is considered to be a possible nerve entrapment site.

In 50% of cases studied, the upper and lower subscapular nerves receive contributions from the fourth cervical nerve.

References   504, 510–512, 516, 519, 521, 525.

Thoracodorsal Nerve

This nerve may arise solely from the seventh cervical nerve, or from the seventh and eighth. It may also receive an additional branch from the cervical nerve. This nerve is also known as the middle subscapular nerve.

References   333, 504, 510–512, 516, 519, 521, 525.

Axillary Nerve

The axillary nerve may be derived solely from the fifth cervical nerve; the contribution from the sixth cervical may be insignificant or absent altogether. It may pierce and supply a branch to m. subscapularis. Branches to the long head of the triceps and infraspinatus muscles have been described. The nerve to teres major may arise from the axillary nerve instead of the lower subscapular nerve.

The axillary and lower subscapular nerves have been found to traversing subscapularis beneath an accessory subscapular muscle. Breisch considered the arrangement a possible nerve entrapment site.

References   143, 410, 504, 511, 512, 516, 519, 521, 525.

Additional reference: Breisch E. A. *Anat. Rec.* 216: 440–442, 1986.

Lateral Brachial Cutaneous Nerve

This nerve usually arises from the radial, but it may also arise from the posterior brachial cutaneous nerve.

References   504, 510–512, 516, 519, 521, 525.

Medial Antebrachial Cutaneous Nerve

The posterior division of this nerve may (a) arise separately from the posterior branch of the lower trunk of the brachial plexus; (b) be joined with or replaced by a branch of the ulnar nerve; or (c) extend to the hand, replacing the dorsal branch of the ulnar nerve.

References   451, 504, 510–512, 516, 519, 521, 525.

Medial Brachial Cutaneous Nerve

This nerve may receive fibers from the eighth cervical to the second thoracic nerves. The connection with the intercostobrachial nerve may be multiple (two or more communicating branches), forming a plexus at the posterior boundary of the axillary space. The nerve sometimes communicates with the lateral cutaneous branch of the third intercostal nerve. When absent, the medial brachial cutaneous is replaced (rarely) by a lateral cutaneous branch from the first thoracic nerve, which traverses the first intercostal space, or more commonly by the intercostobrachial nerve. The medial cutaneous nerve usually leaves the brachial plexus as a single branch, or it may be combined with the medial antebrachial cutaneous nerve.

References   497, 504, 510–512, 516, 519, 521, 525.

Musculocutaneous Nerve

This nerve arises from the lateral cord (90.5% of 75 arms), from the lateral and posterior cord (4%), from the median nerve (2%), as two separate bundles from the medial and lateral cords (1.4%), or from the posterior cord (1.4%).

A number of variations in the course and distribution of the musculocutaneous nerve have been reported. Instead of piercing the coracobrachialis muscle, the nerve may adhere to the median nerve for some distance down the arm and then, either as a single trunk or as several branches, pass between the biceps and brachialis muscles to supply all three muscles (about 22% of arms). Sometimes only a part of the nerve follows this course; this part then rejoins the main trunk after it transits through and supplies coracobrachialis. In some cases, instead of the whole trunk of the nerve piercing coracobrachialis, only its muscular branch or only its cutaneous branch pierces the muscle. The musculocutaneous nerve may be accompanied by fibers from the median nerve as it transits coracobrachialis; a communicating branch passes from the musculocutaneous to the median nerve. Instead of penetrating coracobrachialis, the nerve may pass behind it or between it and the short head of the biceps muscle. Occasionally, the nerve perforates, not only coracobrachialis, but also the brachialis or the short head of the biceps. Very rarely the lateral cord may pierce coracobrachialis and then divide into the musculocutaneous and lateral head of the median nerve.

Branches to the following structures have been reported: brachioradialis, pronator teres, the skin of the dorsum of the hand over and adjacent to the first metacarpal bone, the dorsum of the thumb (the radial nerve being absent), both sides of the ring finger, and the radial side of the little finger. The musculocutaneous may receive a root from the seventh cervical nerve. Anastomoses with the median nerve usually occur in the lower third of the arm, and may form an ansa; the frequency of anastomoses in this region is 8% of cases. The musculocutaneous nerve may be doubled, unusually short, or absent.

References   5, 12, 16, 24, 51, 52, 56, 57, 87, 99, 124, 126, 160, 212, 229, 250, 267, 282, 349, 351, 415, 459, 460, 504, 510–513, 519, 525.

Median Nerve

The following variations in the formation, course, and distribution of this nerve have been reported. The contribution from the first thoracic nerve to the formation of the median nerve may be missing. Either the outer or the inner branch (from the lateral and medial cords) of the nerve may consist of two nerve bundles. The site of union between the lateral and medial branches is quite variable and has been found as far down as the elbow. The two branches may enclose the axillary vein as well as the axillary artery. The median nerve or either of its branches may pass behind the axillary artery instead of in front of it; the nerve may also pass behind the brachial artery. The lateral branch of the nerve sometimes arises in the middle of the arm, behind the axillary artery, and joins the medial branch.

The median nerve usually passes between the two heads of pronator teres. It may also pass superficial or deep to the two heads or may pierce the humeral head. The nerve may lie on the superficial surface of flexor digitorum superficialis instead of deep to it. In some cases, the median splits, forming a cleft in the forearm and allowing the ulnar artery or one of its branches, the superficial long head of the flexor pollicis, or an extra palmaris longus muscle to pass through.

Communications have been described between the median and ulnar nerves in the arm, forearm, or hand. Up to 90% of all subjects studied are reported to have median-ulnar nerve anastomoses, 10% of which are bilateral. They are more frequently reported, however, on the right side. These median-ulnar anastomoses may carry nerve fibers to muscles of the hand that are commonly supplied by the ulnar nerve. Opponens pollicis may be supplied by both the ulnar and median nerves (30% of cases). The superficial head of the short flexor of the thumb (flexor pollicis brevis) is very frequently (83% of cases) supplied by both the median and ulnar nerves, and in about 25% of cases by the ulnar alone.

The anterior interosseous branch of the median nerve may receive a bundle of fibers from the radial nerve through the interosseous membrane. Branches from the median nerve to m. abductor indicis and to the first dorsal interosseous muscles (3% of individuals) have been described. The third, fourth, and fifth common digital branches of the median may arise in the proximal forearm and enter the hand independently. The anterior and posterior interosseous nerves may communicate at the lower end of the interosseous space.

The median may communicate with the musculocutaneous nerve. In 50% of individuals, the part of flexor digitorum profundus to the index and middle fingers is supplied by the median nerve while the part to the ring and little finger is supplied by the ulnar. The median nerve may encroach on the part of the muscle supplied by the ulnar nerve. Only rarely does the ulnar nerve encroach to supply the middle finger part of flexor digitorum profundus.

The median nerve may innervate biceps brachii. Occasionally, the median nerve (and brachial artery) passes through a channel formed by a supracondylar process and ligament, where it (they) may become compressed, producing clinical symptoms.

A median nerve was found entering the belly of flexor digitorum superficialis, on its lateral side, and emerging from its tendon (to the index finger) about 25 mm above the wrist joint.

The median nerve may pass into the hand within the flexor retinaculum, where it may also give rise to the motor recurrent branch. The recurrent branch may also arise above or pass through the retinaculum to reach the thenar muscles.

The median nerve may be perforated by the median artery, when present.

References 12, 27, 41, 42, 48, 51, 52, 57–59, 78, 87, 100, 106, 107, 113, 123, 126, 131, 153, 160, 161, 170, 177, 195, 200, 206, 246, 267, 271, 282, 301, 303, 311, 321, 332, 349, 365, 373, 376, 383, 403, 409, 414, 415, 419–421, 425, 429–431, 433, 437, 466, 480, 487, 504, 510–513, 516, 519, 525.

Ulnar Nerve

There are a number of variations in the formation, course, and distribution of this nerve. It may have a root from the seventh cervical nerve by way of the lateral cord. It may be derived from the eighth cervical nerve only or from both the seventh and eighth cervical nerves. The ulnar may pass in front of the medial epicondyle of the humerus or lie behind the condyle.

Connecting branches with the median nerve have been reported in the arm and forearm. Other connecting branches with the medial brachial cutaneous and radial nerve have also been described. Muscular branches to the medial head of the triceps, anconeus, flexor digitorum superficialis, first and second lumbricals, and superficial head of the flexor pollicis brevis have been reported. When an epitrochleoanconeus muscle is present, it receives a branch from the ulnar nerve.

The branch to the dorsum of the hand may be deficient; this is compensated for by the radial, the dorsal lateral antebrachial cutaneous branch of the radial, or the medial brachial cutaneous nerves. The ulnar may supply all four fingers when the radial nerve is deficient. The dorsal terminal branches may encroach on the radial side of the hand, often reaching the radial side of the second metacarpal bone and even the dorsum of the first phalanx of the thumb. Opponens pollicis is supplied by the median and ulnar nerves in about 30% of cases. The superficial head of the short flexor of the thumb is supplied by the median and ulnar in about 83% of cases and by the ulnar alone in 25%. Much anomalous innervation is due to the interchange of axons between the ulnar and median nerves, either in the forearm (most commonly) or in the hand.

References 12, 48, 87, 126, 131, 155, 159, 162, 163, 166, 171, 195, 229, 233, 267, 271, 275, 282, 301, 303, 338, 348, 349, 373, 376, 401, 405, 414, 415, 420, 433, 480, 501, 504, 510–513, 516, 519, 525.

Radial Nerve

The radial nerve may accompany the axillary (circumflex) nerve through the quadrilateral space (bounded by mm. subscapularis and teres minor above, teres major below, the humoral head of the triceps medially, and the humerus laterally). The radial may communicate with the ulnar nerve in the arm.

The radial may (a) supply a small branch to the lateral portion of the brachialis; (b) be the motor supply to extensor carpi radialis brevis (56% of cases); (c) supply dorsal digital nerves to the little finger and ulnar side of the third finger (instead of the ulnar nerve); and (d) supply the middle and index fingers through its posterior interosseous branch. The dorsal lateral antebrachial cutaneous branch of the nerve may extend to the first phalanx of the third finger and the second phalanx of the little finger; it may send a branch to the palm. The radial nerve may supply the entire dorsum of the hand including the dorsal aspect of all the fingers.

Cases have been reported in which the radial nerve is absent, its area of supply receiving branches from the musculocutaneous nerve (lateral antebrachial cutaneous for the thumb) and the ulnar nerve (remainder of the digits).

The radial branch to anconeus may run in the sheath of the ulnar; leaving the sheath it may reach the muscle in company with the superior ulnar collateral artery. The superficial branch of the radial may be doubled. When a dorsoepitrochlearis muscle is present (18–20% of cases) it is innervated by the radial nerve.

References 12, 16, 17, 43, 87, 93, 126, 159, 162, 178, 188, 191, 195, 229, 233, 260, 267, 282, 303, 338, 349, 369, 373, 378, 401, 415, 420, 422, 459, 504, 510–513, 516, 519, 525.

Posterior Interosseous Nerve

The posterior interosseous nerve may pass over instead of perforating the supinator muscle. It may supply the anconeus muscle, extensor carpi radialis brevis (36% of cases), and the opposed surfaces of the middle and index fingers. The posterior and anterior interosseous nerves sometimes communicate.

References 87, 229, 267, 282, 369, 378, 415, 420, 504, 510–512, 519, 525.

Lumbosacral Plexus

The lumbosacral plexus is formed by the union of the anterior primary divisions of the lumbar, sacral, and coccygeal nerves. In about 50% of cases it receives a branch from the twelfth thoracic nerve. Its components are distributed to the lower extremity in a manner homologous and similar to the distribution of the parts of the brachial plexus to the upper extremity; the lumbar nerves are distributed similarly to the nerves from the anterior (medial and lateral) cords of the brachial plexus, and the sacral nerves are distributed similarly to the nerves from the posterior cord of the brachial plexus.

Partly for convenience of description and partly because of the differences in position and course of some of the nerves arising from it, the lumbosacral plexus is subdivided into four parts: the lumbar, sacral, pudendal, and coccygeal plexuses. It must be remembered that these plexuses overlap and that there is no definite line of demarcation between their origins and distributions.

References 19, 114, 184, 197, 273, 392, 448, 504, 510–512, 516–519, 524, 525.

Lumbar Plexus

In one study of 64 lumbar plexuses, 47 were formed by four spinal nerves, 14 by five, and three plexuses by six spinal nerves.

In its usual form, the first three lumbar nerves are entirely distributed to the lumbar plexus. The fourth lumbar is divided between the lumbar and sacral plexuses, hence it is called the furcal (meaning forked) nerve. The name furcal is also applied to any nerve or nerves that enter into the formation of both the lumbar and sacral plexuses. The proportion of the fourth lumbar contributing to the sacral plexus, however, is variable. In its usual form, the lateral femoral cutaneous nerve arises from L_2 and L_3, but it may arise in "a high form" (prefixed) from L_1 and L_2 or in a "low form" (postfixed) from L_3 and L_4. The obturator usually arises from L_2, L_3, and L_4, but may arise in a "high form" from L_1, L_2, L_3, and L_4, (very rarely from L_1, L_2, and L_3); or in a "low form" from L_2, L_3, L_4, and L_5. The femoral usually arises from L_1, L_2, L_3, and L_4; or in a "high form" from T_{12}, L_1, L_2, L_3, and L_4 or in a "low form" from L_1, L_2, L_3, L_4, and L_5. Furcal nerves usually arise from L_4, or in a "high form" from L_3 and L_4 or in a "low form" from L_4 and L_5 or L_5 alone.

The lumbosacral plexus is frequently (usually) asymmetric bilaterally.

References 21, 114, 197, 363, 448, 449, 470, 504, 510–512, 516–519, 524, 525.

Range of Variation of Lumbar Plexus (from Bardeen)

Nerve	High form of plexus	Usual form	Low form of plexus
Lateral femoral cutaneous	L_1–L_2	L_2–L_3	L_3–L_4
Femoral	T_{12}–L_4	L_1–L_4	L_1–L_5
Obturator	L_1–L_4	L_2–L_3	L_2–L_5
Furcal	L_3–L_4	L_4	L_4–L_5

Twelfth Thoracic Nerve. In 50% of cases studied, the twelfth thoracic participates in the lumbosacral plexus.

References 19, 20, 152, 504, 510–512, 516–519, 524, 525.

First Lumbar Nerve. This nerve usually has variable connections with both the subcostal (T_{12}) nerve and the second lumbar nerve. This contributes to the right/left asymmetry of the lumbosacral plexuses.

References 19, 20, 152, 504, 510–512, 516–519, 524, 525.

Lumbosacral Trunk. The fourth lumbar (or furcal) nerve usually contributes to both the lumbar plexus and (through the lumbosacral trunk) the sacral plexus. Contribution to the lumbosacral trunk may come from the third or the fifth lumbar nerve. Sometimes the branch from the fourth lumbar nerve to the lumbosacral plexus is absent. In such cases, the fifth lumbar nerve supplies branches to both the lumbar and sacral plexuses. Thus, contributions to both plexuses may come from the fourth lumbar nerve, the third and fourth, the fourth and fifth, or the fifth lumbar nerve alone.

References 21, 197, 448, 510–512, 516–519, 524, 525.

Iliohypogastric Nerve. The iliac branch of the iliohypogastric nerve may be absent, replaced by the lateral cutaneous branch of the twelfth thoracic nerve. The hypogastric branch may supply the pyramidalis muscle and may be joined with the twelfth thoracic nerve. The iliohypogastric nerve is sometimes derived from the twelfth thoracic nerve and may also receive a root from the eleventh thoracic nerve.

References 19, 20, 198, 320, 504, 510–512, 516–519, 524, 525.

Ilioinguinal Nerve. This nerve may be small and terminate near the iliac crest by joining the iliohypogastric nerve, which sends branches to replace the absent terminal part of the ilioinguinal nerve. When sent, the nerve is replaced by either the genital (more commonly) or the femoral branches of the genitofemoral nerve. The ilioinguinal may provide a lateral cutaneous or iliac branch to supply the skin in the region of the anterior superior iliac spine. It may partially or completely replace the genital branch of the genitofemoral nerve or the lateral femoral cutaneous nerve. The ilioinguinal and iliohypogastric nerves sometimes arise as a common trunk, usually separating between the transversus and internal oblique muscles. The ilioinguinal nerve may be derived from the last thoracic nerve (T_{12}), from a loop between the first and second lumbar nerves, or even from the second and third lumbar nerves. The rectus abdominis muscle may receive a branch from the ilioinguinal.

In one study of 200 bodies, the ilioinguinal nerve arose from the lumbar plexus in 72.5% and by a common trunk with the iliohypogastric nerve in 25%; it was absent in 2.5%. The ilioinguinal nerve was formed from one root in 92.5% and from two roots in about 5% of cases. In 86%, the ilioinguinal carried fibers from one spinal nerve (primarily from L_1), and in 11%, from two spinal nerves (T_{12}, L_1; L_1, L_2; or L_2, L_3).

Within the inguinal canal, the nerve usually lies ventral to the spermatic cord (60% of cases) but it may lie beneath (dorsal) the cord and/or within it. This nerve may also leave the superficial inguinal ring at its lateral aspect instead of its usual medial position. In some cases, the nerve runs outside the inguinal ring.

References 108, 198, 309, 320, 495, 504, 510–512, 516–519, 524, 525.

Genitofemoral Nerve. This is considered by some authors to be the most variable nerve of the lumbar plexus. The genital and femoral (crural) branches of the nerve may arise as separate offshoots of the lumbar plexus. Either of them may be derived entirely from the first or second lumbar nerves and occasionally from the third. The genital branch sometimes receives fibers from the twelfth thoracic nerve.

The genitofemoral or either of its branches (genital or femoral) may be absent. In such cases, the ilioinguinal nerve replaces the genital branch, while the lateral cutaneous or the anterior femoral nerve replaces the femoral branch. The branches of the genitofemoral may replace or join the ilioinguinal nerve.

The genital branch may bypass the deep inguinal ring running superficial to it in the aponeurosis of the external abdominal oblique muscle. The femoral branch may replace or join the lateral or middle cutaneous nerve. On occasion, the femoral branch has an extensive distribution to the skin of the thigh (upper two-thirds). The genital branch may supply the lower fibers of the internal oblique and transversalis muscles. Occasionally, the nerve divides within the substance of the psoas muscle and the two terminal branches emerge separately from the anterior surface of the muscle.

In a study of 200 bodies, the genitofemoral nerve was a single trunk in 80%, and two separate branches, genital and femoral, in 20%. The single trunk may arise from L_1, L_2 or L_2, L_3 and the two trunks from L_1, L_2 or $L_1, L_2,$ and L_3. The third lumbar was represented in 0.75% of cases, the second lumbar in every case. The level of division into terminal branches was highly variable.

References 228, 447, 504, 510–512, 516–519, 525.

Lateral Femoral Cutaneous Nerve. This nerve may arise from the femoral nerve or as an independent branch of the lumbar plexus (L_2, L_3). Several variations in the formation, course, and branches of this nerve have been reported. The nerve may be absent on one side and may be replaced by a branch of the anterior femoral cutaneous nerve or by the ilioinguinal nerve. The lateral cutaneous is sometimes associated with the anterior femoral cutaneous nerve until it passes distal to the inguinal ligament. The posterior branch of the lateral cutaneous nerve may be replaced by a branch of the genitofemoral nerve.

The nerve may pass beneath the inguinal ligament at a point midway between the anterior superior iliac spine and the femoral artery. In some cases, the posterior branch emerges from beneath the inguinal ligament about 5 cm medial to the anterior superior iliac spine. The nerve may course anterior or posterior to, or pass through, the sartorius muscle. Its subsequent division into anterior and posterior branches provides the sensory innervation of the anterior and lateral sides of the thigh.

References 228, 450, 504, 510–512, 516–519, 525.

Obturator Nerve. There are reports of several variations in the formation, course, and distribution of this nerve. The contribution to the nerve from the second lumbar nerve may be absent. Branches from the obturator to the following structures have been reported: mm. obturator internus and pectineus, the obturator artery, and the periosteum of the pelvic surface of os pubis. Both the anterior and posterior branches of the nerve may pass posterior to adductor brevis. The branch from the main trunk of the nerve to obturator externus may pass to the lateral (instead of the medial) side of the obturator nerve. The obturator nerve may have additonal roots from the first or fifth lumbar nerves. The cutaneous branch is frequently absent.

An accessory obturator nerve has a reported incidence of 29%, 13%, and 8% according to different authors. It arises from the third or more commonly the third and fourth lumbar nerves between the roots of the femoral and obturator nerves. The accessory nerve may also arise from $L_2, L_3,$ and L_4; L_2, L_3; or

L$_3$; or it may arise from the obturator nerve. Although it may be closely related to the femoral nerve, it usually courses with the obturator to the level of the brim of the pelvis, but instead of passing through the obturator foramen, it descends along the medial border of the psoas muscle, crosses the anterior brim of the pelvis, passes beneath m. pectineus, and terminates in three branches, which are also variable. These branches supply the pectineus, the hip joint, and, by rejoining the obturator, the adductor muscles. The terminal branches, individually, usually replace the femoral branch to the pectineus and usually supply the hip joint. They may, however, only supply m. pectineus or they may make a significant contribution to the innervation of the adductor muscles.

References 210, 213, 294, 325, 340, 418, 489, 494, 504, 510–512, 516–519, 524, 525.

Femoral Nerve. In its course to the thigh, the femoral may pierce m. iliacus. The nerve has been reported to enter the thigh between the femoral artery and vein.

The anterior cutaneous branch of the femoral nerve may arise from the beginning of the femoral nerve or directly from the lumbar plexus. It may partly or completely replace the femoral branch of the genitofemoral nerve. The posterior branch of the medial cutaneous nerve may be very small or absent, in which case its usual area of supply is provided by the obturator or saphenous nerve.

The saphenous nerve may end at the knee and may be replaced in the leg by a branch of the tibial nerve. The patellar branch of the saphenous nerve may arise from the nerve to the vastus medialis. The saphenous sometimes provides the medial dorsal digital nerve to the great toe.

Branches from the femoral nerve to mm. tensor fasciae latae and adductor longus have been reported. A branch was also found passing behind the femoral artery and vein, joining an accessory obturator nerve, and supplying a part of the obturator muscle. The portion of the nerve arising from L$_4$ may run a separate course. Leaving the pelvis with the superior gluteal nerve, it may pass under the fascia lata to supply rectus femoris and vastus lateralis. The femoral nerve may run buried in a cleft in the iliacus under the psoas.

References 111, 125, 146, 181, 263, 310, 317, 325, 340, 344, 397, 436, 504, 510–512, 516–519, 524, 525.

Sacral Plexus

The origins of nerves of the sacral plexus are variable. The range of these variations, primarily from the work of Eisler, is tabulated below.

References 114, 324, 448, 504, 510–512, 516–519, 524, 525.

Range of Variation of Sacral Plexus (from Bardeen)

Nerve	High form of plexus	Usual form	Low form of plexus
Composition			
Lumbosacral trunk	L$_4$	L$_4$	L$_4$
N. to quadratus femoris	L$_4$, L$_5$	L$_4$, L$_5$, L$_1$	L$_5$, L$_1$
N. to obturator internus	L$_4$, L$_5$, S$_1$, S$_2$	L$_5$, S$_1$, S$_2$	S$_1$, S$_2$, S$_3$
Tibial	L$_3$, L$_4$, L$_5$, S$_1$, S$_2$	L$_4$, L$_5$, S$_1$, S$_2$, S$_3$	L$_5$, S$_1$, S$_2$, S$_3$, S$_4$
Superior gluteal	L$_4$, L$_5$, S$_1$	L$_4$, L$_5$, S$_1$, S$_2$	L$_5$, S$_1$, S$_2$
Inferior gluteal	L$_4$, L$_5$, S$_1$	L$_5$, L$_1$, S$_2$	L$_5$, S$_1$, S$_2$
N. to piriformis	L$_5$, S$_1$, S$_2$	S$_1$, S$_2$	S$_1$, S$_2$, S$_3$
Common peroneal	L$_3$, L$_4$, L$_5$, S$_1$	L$_4$, L$_5$, S$_1$, S$_2$	L$_5$, S$_1$, S$_2$, S$_3$
Posterior femoral cutaneous	L$_5$, S$_1$, S$_2$	S$_1$, S$_2$, S$_3$	S$_2$, S$_3$, S$_4$
Pudendal	L$_5$, S$_1$, S$_2$, S$_3$	S$_1$, S$_2$, S$_3$, S$_4$	S$_1$, S$_2$, S$_3$, S$_4$
Extreme Range of Variation			
Lumbosacral trunk	L$_3$ or L$_3$, L$_4$	L$_4$	L$_4$, L$_5$ or L$_5$
Common peroneal	L$_3$, L$_4$, L$_5$, S$_1$, S$_2$	L$_4$, L$_5$, S$_1$, S$_2$	L$_4$, L$_5$, S$_1$, S$_2$, S$_3$, S$_4$
Tibial	L$_3$, L$_4$, L$_5$, S$_1$, S$_2$	L$_4$, L$_5$, S$_1$, S$_2$, S$_3$,	L$_4$, L$_5$, S$_1$, S$_2$, S$_3$, S$_4$
Posterior femoral cutaneous	L$_5$, S$_1$, S$_2$, S$_3$	L$_5$, S$_1$, S$_2$, S$_3$, S$_4$	L$_5$, S$_1$, S$_2$, S$_3$, S$_4$

Nerve to M. Quadratus Femoris. The nerve to the quadratus femoris has been reported to supply (in addition to the quadratus femoris, the inferior gemellus, and the hip joint) branches to the upper part of the adductor magnus and superior gemellus. The branch to the superior gemellus may be this muscle's only nerve supply.

References 479, 504, 512, 519, 525.

Superior Gluteal Nerve. This nerve may receive fibers from the second sacral. A branch to m. piriformis may arise from the lowest root of this nerve.

References 504, 512, 518, 519, 525.

Inferior Gluteal Nerve. This nerve is frequently joined in the early part of its course with the posterior femoral cutaneous nerve. It may also be joined with the nerve to the short head of the biceps. The inferior gluteal frequently provides a communicating branch that joins the posterior femoral cutaneous nerve.

References 435, 504, 512, 516, 518, 519.

Sciatic Nerve. A number of variations in the course and distribution of the nerve have been reported. Bifurcation into its two major divisions (common peroneal and tibial) may occur anywhere between the sacral plexus and the lower part of the thigh. The two terminal branches of the sciatic may arise directly from the sacral plexus. The division of the nerve into its terminal branches has been reported to occur below the popliteal space. The nerve to the short head of the biceps femoris sometimes arises directly from the sacral plexus. In one study of 420 limbs, the sciatic nerve passed beneath the piriformis in 87.5%, through the piriformis in 12% (peroneal division), and above the piriformis in 0.5%. In another study of 138 subjects, the sciatic passed beneath the piriformis in 118 cases (85.5%), the peroneal division passed through the piriformis in 17 (12.3%), and the entire sciatic passed through the piriformis in three (2.2%).

References   26, 60, 84, 135, 181, 228, 285, 291, 305, 322, 325, 424, 440, 454, 502, 504, 507, 510, 512, 516, 518, 519, 524, 525.

Posterior Femoral Cutaneous Nerve. In cases of separate origin of the peroneal and tibial nerves, the posterior femoral cutaneous arises from the sacral plexus in two parts. The ventral part descends with the tibial below the piriformis and gives rise to the inferior pudendal and medial femoral branches. The dorsal part passes through the piriformis with the peroneal nerve, and supplies the gluteal and lateral femoral branches. The posterior femoral cutaneous nerve may be joined by a branch from the sciatic on the back of the thigh.

References   228, 302, 504, 512, 518, 519, 525.

Common Peroneal. The sciatic usually divides into the common peroneal and tibial nerves at the level of the lower thigh. Very often these two nerves arise separately from the sacral plexus. They may be separated in the greater sciatic foramen by the piriformis and pass into the thigh as contiguous but separate structures.

Huelke (1958) reported that the peroneal communicating nerve arose directly from the common peroneal nerve in 54.7%, usually as a branch separate from the lateral sural cutaneous nerve (41.5%). The peroneal communicating nerve gave rise to the lateral sural cutaneous branches in 13.2% of sides studied (198 adult lower extremities). The peroneal communicating nerve was a terminal branch of the lateral sural cutaneous nerve in one-third of the sides, and arose from a trunk common to it and to the lateral sural cutaneous nerve in 12%. The peroneal communicating nerve was absent in 19.7% of the 198 sides, and, due to its absence, no sural nerve was formed in these cases. When this occurs, it is usually the medial sural cutaneous nerve that passes onto the dorsum of the foot as the lateral dorsal cutaneous nerve. Only 58.6% of the cadavers had the same type of origin of the peroneal communicating nerve in both legs. No significant differences were found between the right and left sides, between sexes, or the place where the peroneal communicating or sural nerves arose. The union between the peroneal communicating and medial sural cutaneous nerves was seen on 159 sides (80.3%). This union took place more often in the lower half of the leg (75%).

References   26, 60, 84, 135, 285, 291, 305, 322, 440, 454, 502, 504, 507, 510, 512, 516, 518, 519, 521, 524, 525.

Deep Peroneal (Anterior Tibial) Nerve. The following variations in the digital distribution of the nerve have been reported. The nerve may have no digital branches whatsoever. It may supply the medial side of the great toe, adjacent sides of the second and third toes, or the lateral three-and-one-half toes. Absence of the cutaneous part of the superficial peroneal and the deep peroneal nerve and its branch to extensor digitorum brevis has been reported. In this case the nerves were replaced by the saphenous and sural nerves.

References   4, 80, 158, 228, 242, 283, 413, 476, 504, 512, 518, 519, 525.

Superficial Peroneal Nerve. The medial branch may be absent and may be replaced by the deep peroneal (anterior tibial) nerve. Deficiencies in the lateral branch are satisfied by the sural nerve. The nerve has been described as arising from the nerve to peroneus brevis, and extending downward to supply a rare peroneus quartus (extensor digiti quinti brevis), after which it continued on the dorsum of the foot to furnish a nerve to extensor digitorum brevis (Bryce).

References   4, 63, 158, 283, 413, 476, 504, 512, 518, 519, 525.

Tibial and Sural Nerves. Several variations have been reported in the formation and distribution of branches of these nerves. The point of union of the two branches of the sural nerve (tibial), which connects the medial sural cutaneous with the peroneal communicating branch (arising from the common peroneal nerve), is subject to wide variation. It may be high in the popliteal space or there may be no union at all. The union may occur 3 cm below the origin of the peroneal communicating nerve. The two branches of the sural nerve may arise 3 cm apart about 10 cm above the knee and pierce the medial head of the gastrocnemius muscle before joining the peroneal communicating nerve. The two branches may not rejoin. The sural nerve sometimes supplies the dorsal cutaneous area of the lateral two-and-one-half toes. The sural may terminate at the lateral border of the foot without providing any digital branches.

When the medial sural cutaneous nerve is joined by the peroneal communicating nerve the combined nerve is termed the sural nerve. Huelke (1958) used his studies and others to construct the following table.

Occurrence of Sural Nerve

Author	Present	Absent	
		Medial sural cutaneous n. becomes the lateral dorsal cutaneous n.	Lateral sural cutaneous n. becomes the lateral dorsal cutaneous n.
	%	%	%
Huelke ('58) 198 limbs	80.3	19.2	0.5
Huelke ('57) 352 limbs	80.7	19.0	0.3
Williams ('54) 257 limbs	83.7	16.0	0.4
P'an ('39) 286 limbs	81.5	13.3	5.2
Mogi ('38) 180 limbs	83.3	16.7	–
Ssokolow ('33) 500 limbs	52.6	43.8	3.6
Andreassi ('31) 144 limbs	63.9	34.7	1.4
Kosinski ('26) 118 limbs	40.2	53.8	6.0
Catania ('24) 100 limbs	65.0	29.0	6.0
Bardeen ('06) 76 limbs	59.2	39.5	1.3

References   1, 18, 186, 187, 228, 283, 410, 496, 504, 512, 518, 519, 525.

Lateral and Medial Plantar Nerves. The first and second lumbricals may receive branches from both the lateral and medial plantars. The branch of the lateral nerve to the second lumbrical courses forward, beneath the transversus (adductor hallucis), then turns backward over the transversus to reach the lumbrical muscle. Cruveilhier described a branch of the lateral plantar that pierced the transversus to reach the third lumbrical. The lateral nerve rarely provides a branch to the lateral head of flexor hallucis brevis.

References   18, 413, 504, 512, 518, 519, 525.

Third and Fourth Sacral Nerves. The levator ani may be supplied by either the third or the fourth sacral nerves. In one study of 30 subjects, the levator muscle was supplied by the third sacral in 17 and by the fourth sacral in 13 subjects.

References   504, 512, 516, 518, 519, 525.

Pudendal Plexus

The pudendal plexus, like the parts of the lumbosacral plexus described previously, varies in its formation. The following table shows the extreme range of variation and the usual form of the largest nerve (pudendal or pudic) of this plexus.

Range of Variation of Pudendal Plexus (from Bardeen)

Nerve	High form of plexus	Usual form	Low form of plexus
	Composition		
Pudendal	S_2, S_3	S_2, S_3, S_4	S_3, S_4
	Extreme Range of Variation		
Pudendal	$L_5, S_1, S_2, S_3, S_4, S_5$	S_1, S_2, S_3, S_4	S_2, S_3, S_4, S_5

Nerve to External M. Sphincter Ani. The nerve to the external sphincter ani (muscular branch of the pudendal plexus) may pass between mm. coccygeus and levator ani instead of piercing the coccygeus.

References   504, 512, 516, 518, 519, 525.

Perforating Cutaneous Nerve. This nerve, when present (64% of cases), arises from the second and third sacral nerves. Instead of piercing the sacrosciatic ligament, the nerve may accompany the pudendal nerve or pass between the ligament and gluteus maximus. It may be replaced by a branch of the posterior femoral cutaneous nerve or by the perforans coccygeus major, which arises from the third and fourth or the fourth and fifth sacral nerves. It is sometimes joined at its origin with the pudendal nerve.

References   504, 512, 516, 518, 519, 525.

Posterior Femoral Cutaneous Nerve. This nerve may originate from two trunks: from the common peroneal and from the tibial nerves after they separate from their sciatic origin. In such cases, the ventral trunk accompanies the tibial nerve below the piriformis and provides the perineal and medial femoral branches of the posterior femoral cutaneous nerve; the dorsal trunk accompanies the common peroneal nerve beneath the piriformis and provides the gluteal and femoral branches of the posterior femoral cutaneous. The posterior femoral cutaneous nerve may be joined in the back of the thigh by a branch from the sciatic. A perineal branch may pierce the sacroiliac ligament. In some cases, the posterior femoral cutaneous ends behind the knee and may be replaced in the leg by a branch from the sural nerve.

References   504, 512, 516, 518, 519, 525.

Pudendal Nerve. Several variations have been reported in the formation of the nerve and course of its branches. The nerve may have roots arising from both the fifth lumbar and fifth sacral nerves. The inferior rectal branch may pierce the sacrospinous ligament on its course to the perineum. The lateral superficial perineal branch of the pudendal sometimes pierces the sacrotuberous ligament.

References   315, 504, 512, 516, 518, 519, 525.

Inferior Rectal (Hemorrhoidal) Nerve.
This nerve arises from S_3, S_4 and occasionally S_2. It may arise from the sacral plexus directly rather than being one of the three terminal branches (inferior rectal, perineal, and dorsal nerve of the clitoris or penis) of the pudendal.

References   366, 504, 512, 516, 518, 519, 525.

Coccygeal Plexus

Although usually regarded as a subdivision of the pudendal plexus, this plexus has been described with the coccygeal nerves. It is usually formed by the anterior division of the fifth sacral and coccygeal nerves; it usually receives contributions from the fourth sacral as well.

References   504, 512, 516, 518, 519, 524, 525.

Sympathetic Division

Cervical Rami Communicantes

Communicating branches between the cervical sympathetic trunk and descendens cervicalis are variable. In one study, the following patterns were noted: communicating branches were found (a) between the superior cervical ganglion and descendens cervicalis; (b) joining C_2 and C_3; and (c) variably joining the connecting loop between C_2 and C_3.

Reference   367.

Superior Cervical Ganglion

The superior cervical ganglion may be doubled, with one part superficial to and the other posterior to the carotid sheath. A variable number of fibers from the superior cervical ganglion join the inferior ganglion of the vagus, hypoglossal, and descendens cervicalis nerves. Ganglion cells, intermingled with nerve fibers from the sympathetic trunk, have been reported within the cranium.

References   199, 296, 367, 504, 512, 516, 518, 519, 524, 525.

Superior Cardiac Nerve

The superior cardiac nerve (a branch from the superior cervical ganglion) may be absent, particularly on the right side. In such cases, it is replaced by a branch from the vagus or from the external laryngeal nerve. The nerve may not have a separate course, but may be part of another cardiac nerve from one of the other cervical ganglia. Variations in the course of the other cardiac nerves are common. A pretracheal branch arising from the communication between the superior cardiac nerve and the recurrent laryngeal nerve has been reported. It descends on the trachea and terminates in the pericardium and anterior pulmonary plexus.

A ganglion (the so-called temporal ganglion) is sometimes found as a part of the external carotid (artery) plexus near the point of origin of the posterior auricular artery. It may be about 2 mm in length. It has been reported to receive nerve fibers from the stylohyoid branch of the facial nerve.

References   115, 199, 296, 504, 512, 516, 518, 519, 524, 525.

Middle Cervical Ganglion

The sympathetic plexus in the region of the middle cervical ganglion may be posterior to the inferior thyroid artery or may be bifurcated, with the artery lying between the two trunks. The middle cervical ganglion may be absent, in which case the middle cardiac nerves are carried in the interganglionic chain. When there are two ganglia, the upper (usually smaller) part is called the middle cervical ganglion of Arnold (or Luschka) or the thyroid ganglion of Krause, and the larger lower part is called the middle cervical ganglion of Swan (or Krause) or the inferior cervical ganglion of Arnold (or Luschka or Rüdinger).

The middle cervical ganglion may communicate with the phrenic nerve. According to Allan (1958), in cases of retroesophageal right subclavian artery the sympathetic (middle and inferior) cardiac nerves follow the vessel in its anomalous course to reach the cardiac plexus.

References   3, 115, 199, 296, 504, 512, 516, 518, 519, 525.

Inferior Cervical Ganglion

The inferior cervical ganglion is irregular in shape and usually larger than the middle ganglion. It is frequently combined with the first thoracic and is then named the stellate (cervicothoracic) ganglion. A direct communication to the phrenic nerve is often derived from the inferior ganglion, less frequently from the middle cervical ganglion.

The sympathetic chain on the left side may split to encircle the subclavian artery.

References   3, 199, 254, 296, 330, 504, 512, 516, 518, 519, 525.

First Thoracic Ganglion

The first thoracic is usually larger than the others and is variable in shape (ovoid or semilunar). It may be fused with the inferior cervical (stellate) or second thoracic ganglion.

References   112, 199, 204, 504, 512, 516, 518, 519, 525.

Cardiac Plexus

The nerves that supply the heart usually converge

anteriorly toward the lower part of the trachea but may be followed separately to the heart. Cardiac ganglia, varying in size and number, may occur along the cervicothoracic nerves that supply the aorta, pulmonary trunk, and arteries, as well as the heart.

References 112, 115, 293, 402, 504, 512, 516, 518, 519, 525.

Supreme Splanchnic Nerve

Wrisberg observed a supreme splanchnic nerve eight times in a "large number" of bodies. It was described as being formed by branches from cardiac nerves, and from lower cervical and upper thoracic ganglia.

References 119, 234, 270, 355.

Greater Splanchnic Nerve

The greater splanchnic is usually formed from the seventh, eighth, and ninth thoracic ganglia, but the upper and lower limits are variable. The nerve arises symmetrically in only 12% of bodies (100 studied). So-called normal nerves (T_7–T_9) occurred in 23%; only 9% of these were symmetric bilaterally.

References 110, 234, 270, 329, 355, 504, 512, 516, 518, 519, 525.

Lesser Splanchnic Nerve

This nerve may be absent (7% on the right side, 2% on the left, and 2% bilaterally). It is usually formed from the tenth and eleventh thoracic ganglia. The nerve arises symmetrically in 35% of bodies. "Normal" nerves (T_{10}–T_{11}) occurred in 14% of subjects; only 1% of these were symmetric bilaterally. The greater and lesser nerves may be carried in the same trunk.

References 110, 234, 270, 329, 355, 504, 512, 516, 518, 519, 525.

Least (Lowest) Splanchnic Nerve

This nerve may be absent (11% on the right side and 4% on the left). The splanchnicus imus usually arises from the twelfth thoracic ganglion. It may be absent bilaterally in 2% of bodies. When absent, the lesser splanchnic supplies the renal plexus. The nerve arises symmetrically in 79% of bodies. "Normal" nerves were found in 92% of bodies and were symmetric in 77%.

References 110, 234, 329, 355, 504, 512, 516, 518, 519, 525.

Lumbar Ganglia

Although there are usually four lumbar ganglia, the number may vary from three to eight. When there are fewer than four ganglia they tend to be larger than usual.

References 77, 109, 330, 504, 512, 516, 518, 519, 525.

Sacral Ganglia

These vary greatly in size and number, but there are usually four. A single median terminal ganglion (ganglion impar or coccygeal ganglion) is often found at the distal end of the sympathetic chain of ganglia on the proximal part of the coccyx. Variations in size and number of ganglia are greater in the sacral region than in the thoracic and lumbar regions.

The sympathetic chain of ganglia has been found terminating at the level of the second sacral nerve.

References 82, 119, 504, 512, 516, 518, 519, 525.
References (Sympathetic Division) 3, 7–9, 29, 30, 44, 46, 76, 77, 82, 109, 110, 112, 115, 119, 169, 194, 204, 222, 226, 237, 296, 342, 350, 352, 353, 368, 390, 398, 399, 491, 493, 504, 511, 512, 516–519, 521, 524, 525.

References

+ The authors have not had access to these references. They are included because of their historic and/or scientific importance.
* These journal articles are notable for their content, literature review, and/or illustrations.

Journals

1 AASAR, Y.H. A peculiar relation of the femoral head of the biceps femoris to the common peroneal nerve. *J. Anat.* 72:305, 1937–38.
2 ADAMS, S.B., WHEELER, J.F.G. and F.H. EDGEWORTH. On the innervation of the platysma and mandibulo-auricularis. *J. Anat.* 63:242–252, 1929.
3* ALLAN, F.D. An analysis of the cervicothoracic visceral branches of the vagus and the sympathetic trunk in the presence of an anomalous right subclavian artery. *Anat. Rec.* 132:71–80, 1958.
4 ANDREASSI, G. Observations sur l'origine, le comportement et la distribution du nerf saphène externe, de l'accessoire du saphène externe et du nerf cutané péronier chez l'homme. *Assoc. Anatomistes Comptes Rendus* 40:558–564, 1953.
5 APPLETON, A.B. A case of abnormal distribution of the n. musculocutaneus, with complete absence of the ramus cutaneus n. radialis. *J. Anat. Physiol.* 46:89–94, 1912.
6* ARMSTRONG, W.G. and J.W. HINTON. Multiple divisions of the recurrent laryngeal nerve: An anatomic study. *Arch. Surg.* 62:532–539, 1951.
7* ASHLEY, F.L. and B.J. ANSON. The pelvic autonomic nerves in the male. *Surg. Gynecol. Obstet.* 82:598–608, 1946.
8 ATLAS, L.N. A modified form of lumbar sympathectomy for denervating the blood vessels of the leg and foot; anatomic considerations. *Ann. Surg.* 111:117–125, 1940.
9 ATLAS, L.N. Sympathetic denervation limited to the blood vessels of the leg and foot, II. *Ann. Surg.* 116:476–479, 1942.
10 ATUMI, H. Über den Verlauf, insbesondere die Spaltung und die sogenannte Unterbrechung des Brustgrenzstranges bei Japanern. *Folia Anat. Jpn.* 22:511–522, 1944.
11 ATUMI, H. Über die Brustganglien bei Japanern. *Folia Anat. Jpn.* 22:523–567, 1944.
12 BACKHOUSE, K.M. Innervation of the hand. *Hand* 7:107–114, 1975.
13 BAILEY, H. The treatment of tumors of the parotid gland with special reference to total parotidectomy. *Br. J. Surg.* 28:337–346, 1941.
14 BALAGURA, S. and R.G. KATZ. Undecussated innervation to the sternocleidomastoid muscle. A reinstatement. *Ann. Neurol.* 7:84–85, 1980.

15. BANCHI, A. Étude anatomique d'un cerveau sans corps calleux. *Arch. Ital. Biol.* 44:126–127, 1905.
16. BANT, A.J. Sur un cas très rare de l'innervation anormal du membre supérieur. Absence de la branche antérieure du nerf radial suppléé par le nerf musculocutané. *Soc. Biol. Comptes Rendus des Séances et Mémoires* 92:377–378, 1925.
17. BANT, A.J. Quelques remarques sur la question de l'innervation de la face dorsale de la main et un cas très rare d'innervation anormale chez l'homme. *Assoc. Anatomistes Comptes Rendus* 27:29–36, 1932.
18*. BARBOUR, J.R. Puzzling foot symptoms from abnormal nerve distribution. *Med. J. Aust.* 1:275–276, 1947.
19*. BARDEEN, C.R. A statistical study of the abdominal and border nerves in man. *Am. J. Anat.* 1:203–228, 1901–02.
20*. BARDEEN, C.R. Development and variation of the nerves and the musculature of the inferior extremity and of the neighboring regions of the trunk in man. *Am. J. Anat.* 6:259–390, 1906–07.
21. BARDEEN, C.R. and A.W. ELTING. A statistical study of the variations in the formation and position of the lumbo-sacral plexus in man. *Anat. Anz.* 19:124–128, 209–232, 1901.
22*. BARNES, S. Degenerations in hemiplegia: With special reference to a ventro-lateral pyramidal tract, the accessory fillet and Pick's bundle. *Brain* 24:463–501, 1901.
23. BASU, B.N. Complete absence of septum pellucidum. *J. Anat.* 69:394–395, 1934–35.
24. BAUR, R. Innervation des M. brachioradialis aus dem N. musculocutaneous als seltene Varietät. *Anat. Anz.* 128:108–110, 1971.
25. BAUTZMANN, H. Verlauf des Nervus cutaneus colli durch eine Insel der Vena jugularis externa. *Anat. Anz.* 70:516–519, 1930.
26*. BEATON, L.E. and B.J. ANSON. The relation of the sciatic nerve and of its subdivisions to the piriformis muscle. *Anat. Rec.* 70:1–5, 1938.
27*. BEATON, L.E. and B.J. ANSON. The relation of the median nerve to the pronator teres muscle. *Anat. Rec.* 75:23–26, 1939.
28. BEAU, A. and R. GRANDPIERRE. L'innervation du muscle géniohyoidien. *Assoc. Anatomistes Comptes Rendus* 33:32–38, 1938.
29. BECK, T.S. On the structure and functions of the sympathetic nervous system. *Lancet* 1:615–617, 1847.
30*. BECKER, R.F. and J.A. GRUNT. The cervical sympathetic ganglia. *Anat. Rec.* 127:1–14, 1957.
31*. BEREZA, S. A case of aberrant subclavian artery and other neurovascular anomalies in the neck. *Folia Morphol.* 25:589–599, 1966.
32*. BERLIN, D.D. and F.H. LAHEY. Dissections of the recurrent and superior laryngeal nerves. *Surg. Gynecol. Obstet.* 49:102–104, 1929.
33. BERTACCHINI, P. Sur une remarqable anomalie du cerveau humain. *Arch. Ital. Biol.* 26:160, 1896.
34. BERTELLI, D. Distribuzione dei nervi frenici nel diaframma dei Mammiferi. *Arch. Ital. Anat. Embriol.* 32:110–148, 1933–34.
35. BETTI, U.A. Des connexions du nerf hypoglosse avec les nerfs cervicaux. *Arch. Ital. Biol.* 28:156–159, 1897.
36. BILLET, H. Les troncs primaires du plexus brachial. *Assoc. Anatomistes Comptes Rendus* 28:63–71, 1933.
37. BILLET, H. Suite de l'étude des rapports des branches collaterales de la souclaviere avec les troncs primaires de plexus brachial. *Assoc. Anatomistes Comptes Rendus* 32:49–52, 1937.
38. BINSWANGER, O. Über eine Mißbildung des Gehirns. *Arch. Pathol. Anat. Physiol. Klin. Med.* 87:427–476, 1882.
39. BIRMINGHAM, A. Homology and innervation of the achselbogen and pectoralis quartus, and the nature of the lateral cutaneous nerve of the thorax. *J. Anat. Physiol.* 23:206–223, 1889.
40. BISHOP, W.E., CARR, B.W., ANSON, B.J. and F.L. ASHLEY. The parietal intermuscular plexus of the thoracic nerves. *Q. Bull. Northwestern University Medical School* 17:209–216, 1943.
41. BLAIR, W.F. and K. JOOS. The innervation of flexor carpi radialis. An interfasicular dissection. *Arch. Neurol.* 39:647–649, 1982.
42. BORCHARDT, M. and WJASMENSKI. Der Nervus medianus. *Beitr. Klin. Chir.* 107:553–582, 1917.
43. BORCHARDT, M. and WJASMENSKI. Der Nervus radialis. *Beitr. Klin. Chir.* 117:475–513, 1919.
44*. BOTAR, J. Nouvelle orientation dans la systématisation du sympathique. *Assoc. Anatomistes Comptes Rendus* 28:85–101, 1933.
45*. BOWDEN, R.E.M. The surgical anatomy of the recurrent laryngeal nerve. *Br. J. Surg.* 43:153–163, 1955–56.
46. BOYD, J.D. and P.A.G. MONRO. Partial retention of autonomic function after paravertebral sympathectomy. *Lancet* 2:892–895, 1949.
47. BRENNER, A. Über das Verhältnis des N. laryngeus inferior vagi zu einigen Aortenvarietäten und zu dem Aortensystem der durch Lungen atmenden Wirbeltiere überhaupt. *Arch. Anat. Entwickgsch.*, 1883, pp. 373–396.
48*. BROOKS, H.ST.J. Variations in the nerve supply of the lumbrical muscles in the hand and foot, with some observations on the innervation of the perforating flexors. *J. Anat. Physiol.* 21:575–585, 1887.
49. BROOKS, W.T. The brachial plexus of the macaque monkey and its analogy with that of man. *J. Anat. Physiol.* 17:329–332, 1883.
50*. BRUCE, A. On the absence of the corpus callosum in the human brain with a description of a new case. *Brain* 12:171–190, 1890.
51*. BUCH, C. Zur Variation der Innervationsweise des M. biceps brachii unter Beachtung der Astabgabe vom N. musculocutaneous und vom N. medianus. *Anat. Anz.* 114:131–140, 1964.
52*. BUCH-HANSEN, K. Über Varietäten des Nervus medianus und des Nervus musculo cutaneus und deren Beziehungen. *Anat. Anz.* 102:187–203, 1955–56.
53. BURR, H.S. and G.B. ROBINSON. An anatomical study of the gasserian ganglion, with particular reference to the nature and extent of Meckel's cave. *Anat. Rec.* 29:269–282, 1925.
54. BUSACCA, A. Les anastomoses des nerfs moteurs de l'oeil dan la région du sinus caverneux. *Arch. Ital. Biol.* 75:145–146, 1925.
55*. CALIOT, P., CABANIÉ, P., BOUSQUET, V. and D. MIDY. A contribution to the study of the innervation of the sternocleidomastoid muscle. *Anat. Clin.* 6:21–28, 1984.
56. CALORI, L. Delle corrispondenze del nervo muscolo-cutaneo con il capo soprannumerario del bicipite bracchiale e col bracchiale interno. *Mem. R. Accad. Sci. Istituto di Bologna S.2* 6:149–155, 1866.
57*. CALORI, L. Delle anomalie più importanti di ossa, vasi, nervi, e muscoli occorse nell'ultimo biennio facendo anatomia del corpo umano. *Mem. R. Accad. Sci. Istituto di Bologna S.2* 8:417–482, 1868.
58. CALORI, L. Sul variato rapporto vascolare delle due radici del nervo mediano e sulle condizioni atte a produrlo. *Mem. R. Accad. Sci. Istituto di Bologna S.3* 8:443–456, 1877.
59*. CALORI, L. Intorno al canale sopracondiloideo dell'omero nell'uomo. *Mem. R. Accad. Sci. Istituto di Bologna S.4* 2:37–46, 1880.
60*. CALORI, L. Sull'alta divisione del nervo grande ischiatico considerata come differenza nazionale e sulle varietà del muscolo piriforme. *Mem. R. Accad. Sci. Istituto di Bologna S.4* 2:623–633, 1880.
61*. CALORI, L. Sui nervi di un rene a ferro di cavallo con ectopia del rene sinistro. *Mem. R. Accad. Sci. Istituto di Bologna S.4* 9:451–460, 1888.
62. CASS, A.B. and D.L. REEVES. Partial agenesis of the corpus callosum. Diagnosis by ventriculographic examination. *Arch. Surg.* 39:667–681, 1939.
63*. CATANIA, V. Il comportamento dei nervi cutanei dorsali piede. *Arch. Ital. Anat. Embriol.* 21:295–331, 1924.
64. CAVE, A.J.E. The distribution of the first intercostal nerve and its relations to the first rib. *J. Anat.* 63:367–379, 1929.
65*. CHAMBERLIN, J.A. and T. WINSHIP. Anatomic variations of the vagus nerves – their significance in vagus neurectomy. *Surgery* 22:1–19, 1947.
66. CHETRICK, A. and L.R.M. DEL GUERCIO. Unusual brachial plexus formation. *Yale J. Biol. Med.* 23:395–398, 1951.
67. CHI, T.K. and C. CHANG. The sulcal pattern of the Chinese brain. *Am. J. Phys. Anthropol.* 28:167–207, 1941.
68. CHIARUGI, G. Sur l'existence d'une racine dorsale rudimentaire avec ganglion pour le nerf hypoglosse chez l'homme. *Arch. Ital. Biol.* 11:421–422, 1889.
69. CITELLI, S. Sur la fréquence et sur la signification d'un sillon glottidien chez l'homme. Sur la valeur de l'angle vocal. *Arch. Ital. Biol.* 47:499–500, 1907.
70. CONNOLLY, C.J. The fissural pattern in the brain of Negroes and Whites. The frontal lobe. *Am. J. Phys. Anthropol.* 28:133–165, 1941.
71. CONNOLLY, C.J. The fissural pattern in the brain of Negroes and Whites. The parietal and temporal lobes. *Am. J. Phys. Anthropol.* 29:225–265, 1942.
72. CORDIER and COULOUMA. Recherches sur les nerfs du sinus carotidien et sur leurs variations. *Assoc. Anatomistes Comptes Rendus* 27:161–178, 1932.
73*. CORDIER and COULOUMA. Les nerfs érecteurs. *Assoc. Anatomistes Comptes Rendus* 28:143–199, 1933.
74*. CORDIER, COULOUMA, DEVOS and DELCROIX. Contribution a

l'étude de la constitution du plexus cervical chez l'homme et quelques primates. *Assoc. Anatomistes Comptes Rendus* 31:114–123, 1936.
75. CORDIER, COULOUMA and VAN VARSEVELD. Contribution a l'étude de la constitution du plexus sacré chez l'homme et quelques primates. *Assoc. Anatomistes Comptes Rendus* 31:124–132, 1936.
76. COTTE, G. and R. NOËL. Sur l'existence de ganglion sympathiques dans le nerf présacré. *Assoc. Anatomistes Comptes Rendus* 23:99, 1928.
77* COWLEY, R.A. and G.H. YEAGER. Observations on the lumbar sympathetic nervous system. *Surgery* 25:880–890, 1949.
78. CRANDALL, R.C. and A.L. HAMEL. Bipartite median nerve at the wrist. *J. Bone Joint Surg. (Am.)* 61:311, 1979.
79. CUNNINGHAM, D.J. Note on a connecting twig between the anterior divisions of the first and second dorsal nerves. *J. Anat. Physiol.* 11:539–540, 1877.
80. CUNNINGHAM, D.J. Note on the distribution of the anterior tibial nerve on the dorsum of the foot. *J. Anat. Physiol.* 13:389–399, 1879.
81. CUNNINGHAM, D.J. The fissure of Rolando. *J. Anat. Physiol.* 25:1–23, 1890.
82* CURTIS, A.H., ANSON, B.J., ASHLEY, F.L. and T. JONES. The anatomy of the pelvic autonomic nerves in relation to gynecology. *Surg. Gynecol. Obstet.* 75:743–750, 1942.
83* CUSHING, H. The sensory distribution of the fifth cranial nerve. *Johns Hopkins Hosp. Bull.* 15:213–232, 1904.
84. CUTORE, G. La division du grand nerf sciatique chez l'homme. *Arch. Ital. Biol.* 37:491, 1902.
85. CUTORE, G. and G. FICHERA. Variétés anatomiques. *Arch. Ital. Biol.* 36:329, 1901.
86* DANDY, W.E. Congenital cerebral cysts of the cavum septi pellucidi (fifth ventricle) and cavum vergae (sixth ventricle). *Arch. Neurol. Psychiatry* 25:44–66, 1931.
87* DANKMEIJER, J. and J.M. WALTMAN. Sur l'innervation de la face dorsale des doigts humains. *Acta Anat.* 10:377–384, 1950.
88. DARGENT, M. and P.E. DUROUX. Données anatomiques concernant la morphologie et certains rapports du facial intra-parotidien. *La Presse Médicale* No. 37, pp. 523–524, 1946.
89. DAVIES, F. A note on the first lumbar nerve (anterior ramus). *J. Anat.* 70:177–178, 1935.
90* DAVIES, F., GLADSTONE, R.J. and E.P. STIBBE. The anatomy of the intercostal nerves. *J. Anat.* 66:323–333, 1932.
91. DAVIES, F., FRANCIS, E.T.B. and T.S. KING. Neurological studies of the cardiac ventricles of mammals. *J. Anat.* 86:130–142, 1952.
92* DAVIS, R.A., ANSON, B.J., BUDINGER, J.M. and LeR.E. KURTH. Surgical anatomy of the facial nerve and parotid gland based upon a study of 350 cervicofacial halves. *Surg. Gynecol. Obstet.* 102:385–412, 1956.
93. DEBIERRE, C. De quelques anomalies nervuses et en particulier d'une anomalie dans la distribution du nerf radial. *Soc. Biol. Comptes Rendus Hebdomadaires des Séances et Mémoires* 37:706–707, 1885.
94. DEBIERRE, C. Sur les anomalies des circonvolutions du cervéau de l'homme. Duplicilé de la scissure de Rolando, absence de la scissure de Rolando. *Soc. Biol. Comptes Rendus Hebdomadaires des Séances et Mémoires* 43:369–372, 1891.
95* DEJERINE and DEJERINE. Le faisceau pyramidal direct. *Rev. Neurol.* 12:253–274, 1904.
96. DELMAS, A. and S.H. KOWSARIAN. Portion transverse du sinus lateral. *Association Anatomistes Comptes Rendus* 39:788–791, 1952.
97. DELMAS, A. and B. PERTUISET. La topomètrie cranio-encéphalique variations internes du complexe striato-thalamique. *Assoc. Anatomistes Comptes Rendus* 42:428–430, 1955.
98. DEMMEL, U. Über einen Nerven im Foramen thyroideum. *Anat. Anz.* 143:456–465, 1978.
99. D'EVANT, T. Observations sur une anomalie de nerf perforant de Casserio. *Arch. Ital. Biol.* 22:162–163, 1895.
100. DIDIO, L.J.A. and J.G. D'ANGELO. Nervus medianus piercing the caput humerale of the m. pronator teres. *Anat. Anz.* 112:385–388, 1963.
101. DILWORTH, T.F.M. The nerves of the human larynx. *J. Anat.* 56:48–52, 1921.
102. DIXON, A.F. Abnormal distribution of the nervus dorsalis scapulae and of certain intercostal nerves. *J. Anat. Physiol.* 30:209–210, 1896.
103* DOUBILET, H., SHAFIROFF, B.G.P. and J.H. MULHOLLAND. The anatomy of the periesophageal vagi. *Ann. Surg.* 127:128–135, 1948.
104. DOW, D.R. The anatomy of rudimentary first thoracic ribs, with special reference to the arrangement of the brachial plexus. *J. Anat.* 59:166–179, 1925.
105* DRAGSTEDT, L.R., FOURNIER, H.J., WOODWARD, E.R., TOVEE, E.B. and P.V. HARPER, JR. Transabdominal gastric vagotomy. A study of the anatomy and surgery of the vagus nerves at the lower portion of the esophagus. *Surg. Gynecol. Obstet.* 85:461–466, 1947.
106. DUROUX, P.E., BARRY, P. and P. LACOSTE. Le rameau thenarien du médian. *Assoc. Anatomistes Comptes Rendus* 39:146–150, 1952.
107. DUROUX, DUJOL, GABRIELLE, C. and GARDE. Le anomalies du nerf médian. *Assoc. Anatomistes Comptes Rendus* 39:153–154, 1952.
108. ECKMANN, I. Ein Beitrag zur Kenntnis des terminalen Verlaufs des N. ilioinguinalis. *Anat. Anz.* 140:15–30, 1976.
109* EDWARDS, E.A. Operative anatomy of the lumbar sympathetic chain. *Angiology* 2:184–198, 1951.
110* EDWARDS, L.F. and R.C. BAKER. Variations in the formation of the splanchnic nerves in man. *Anatomical Record* 77:335–342, 1940.
111. EHLER, E. et al. Zur Frage der Verteilung der Femoralisäste im M. vastus medialis mit besonderer Berücksichtigung der Kniegelenkäste. *Anat. Anz.* 107:414–423, 1959.
112* EHRLICH, E. JR. and W.F. ALEXANDER. Surgical implications of upper thoracic independent sympathetic pathways. *Arch. Surg.* 62:609–614, 1951.
113. EIKEN, O., CARSTAM, N. and A. EDDELAND. Anomalous distal branching of the median nerve. Case reports. *Scand. J. Plast. Reconstr. Surg.* 5:149–152, 1971.
114* EISLER, P. *Der Plexus lumbosacralis des Menschen.* Halle, 1892.
115* ELLISON, J.P. and T.H. WILLIAMS. Sympathetic nerve pathways to the human heart and their variations. *Am. J. Anat.* 124:149–162, 1969.
116* VAN EPPS, E.F. Agenesis of the corpus callosum with concomitant malformations, including atresia of the foramens of Luschka and Magendie. *Am. J. Roentgenol.* 70:47–60, 1953.
117* ERALP, I. A propos d'un cas de trajet anormal unilatéral de la branche externe du nerf spinal. *Assoc. Anatomistes Comptes Rendus* 45:362–365, 1958.
118. EVANS, T.H. Unlisted nerve branches of the maxillary division of the trigeminal nerve (fifth cranial) which are related to nerve association of the lacrimal and salivary systems. *Am. J. Ophthalmol.* 47:225–226, 1959.
119. FAWCETT, E. An unusual mode and site of termination of the right sacral sympathetic cord. *J. Anat. Physiol.* 29:329, 1895.
120* FELIX, W. Anatomische, experimentelle und klinische Untersuchungen über den N. phrenicus und über die Zwerchfellinnervation. *Deutsche Z. Chir.* 171:283–397, 1922.
121. FELIX, W. Über den Nervus phrenicus und die Zwerchfellinnervation. *Zentralbl. Chir.* 49:1832, 1922.
122. FENART, R. La morphogenese du plexus brachial, ses rapports avec la formation du cou et due membre superieur. *Acta Anat.* 32:322–360, 1958.
123. FERNER, H. Ein abnormer Verlauf des Nervus medianus vor dem M. pronator teres. *Anat. Anz.* 84:151–156, 1937.
124. FERNER, H. Der Nervus musculocutaneus, seine Verlaufsvariäten am Oberarm und deren Beziehungen zur Entwicklung eines Caput tertium bicipitis. *Z. Anat. Entwicklungsgesch.* 108:567–586, 1938.
125. FERNER, H. Ein abnormer Verlauf des Nervus femoralis. *Anat. Anz.* 90:124–127, 1940–41.
126. FETROW, K.O. Practical and important variations in sensory nerve supply to the hand. *Hand* 2:178–184, 1970.
127* FITZGERALD, M.J.T. The occurrence of a middle superior alveolar nerve in man. *J. Anat.* 90:520–522, 1956.
128. FITZGERALD, M.J.T. and M.E. LAW. The peripheral connexions between the lingual and hypoglossal nerves. *J. Anat.* 92:178–188, 1958.
129* FOERSTER, O. The dermatomes in man. *Brain* 56:1–39, 1933.
130. FONTES, V. Les origines du nerf phrénique. *Assoc. Anatomistes Comptes Rendus* 42:518–526, 1955.
131. FORREST, W.J. Motor innervation of human thenar and hypothenar muscles in 25 hands: A study combining electromyography and percutaneous nerve stimulation. *Can. J. Surg.* 10:196–199, 1967.
132. FÖRSTER. Ein Fall von Defect beider Augenäpfel bei einem Neu-

geborenen. *Arch. Pathol. Anat. Physiol. Klin. Med.* 13:53–58, 1858.
133. FRANK, E. Fünf, teils seltene Sonderfälle des Nervus phrenicus. *Anat. Anz.* 103:177–186, 1956.
134. FRANK, E. Zwei Sonderfälle des Nervus lingualis. *Anat. Anz.* 103:187–191, 1956.
135. FREDERICQ, H. Un cas de division complète du nerf grand sciatique accompagnée d'une duplicité complète du muscle pyramidal. *Assoc. Anatomistes Comptes Rendus* 10:168–170, 1908.
136* FREEMAN, W. The radix spinalis trigemini and the principle of usurpation. *Arch. Neurol. Psychiatry* 15:607–612, 1926.
137. FRÓES DA FONSECA, A. Beobachtung einer Anastomose zwischen Glossopharyngeus und Hypoglossus. *Anat. Anz.* 55:551–553, 1922.
138* GALIBERT, P., DELANDTSHEER, J.M. and P. HERLEMONT. Le variations topographiques du trou de Monro. *Assoc. Anatomistes Comptes Rendus* 46:279–287, 1959.
139* GARDNER, E. The innervation of the knee joint. *Anat. Rec.* 101:109–130, 1948.
140. GARDNER, E. The innervation of the hip joint. *Anat. Rec.* 101:353–371, 1948.
141. GARLING-PALMER, R. Une anomalie de situation du nerf médian du bras. *Ann. Anat. Pathol.* 5:332–333, 1928.
142* GEHWOLF, S. Ein Fall außergewöhnlicher Nervenverzweigung in der Hohlhand. *Anat. Anz.* 54:1–8, 1921.
143. GÉRARD, G. and P. CORDIER. Anomalies dans la distribution du nerf circonflexe. *Soc. Biol. Comptes Rendus Hebdomadaires des Séances et Mémoires* 83:927–929, 1920.
144* GERLACH, E.J. and H. WEBER. Über ein menschliches Gehirn mit beiderseitiger Verdoppelung der Zentralfurche. *Anat. Anz.* 67:440–452, 1929.
145. GIBSON, A. Bilateral abnormal relationship of the vagus nerve in its cervical portion. *J. Anat. Physiol.* 69:389–392, 1915.
146. GISEL, A. Ein Nervus femoralis accessorius. *Anat. Anz.* 96:371–376, 1947–48.
147* GRÄFENBERG, E. Die Entwicklung der Knochen, Muskeln und Nerven der Hand und für die Bewegungen der Hand bestimmten Muskeln des Unterarms. *Anat. Hefte* 30:1–154, 1906.
148. GRANT, J.C.B. Perforation of the septum pellucidum of the brain. *Anat. Rec.* 30:153–154, 1925.
149. GRAY, D.J. The intrinsic nerves of the testis. *Anat. Rec.* 98:325–331, 1947.
150* GRAY, D.J. and E. GARDNER. The innervation of the joints of the wrist and hand. *Anat. Rec.* 151:261–266, 1965.
151. GREENFIELD, J. and G.M. CURTIS. The "sniff test" in thoracic surgery. With a review of 119 phrenic nerve interruptions. *J. Thorac. Surg.* 12:79–85, 1942.
152* GRIFFIN, M. Some varieties of the last dorsal and first lumbar nerves. *J. Anat. Physiol.* 26:48–55, 1891.
153. GRUBER, W. Anomaler Verlauf des Nervus medianus vor dem Musculus pronator teres, bei Durchbohrung des letzteren durch die hoch am Oberarm von der Arteria brachialis entsprungene Arteria interossea. *Arch. Anat. Physiol. Wissen. Med.* 1867, pp. 552–559.
154. GRUBER, W. Über den anomalen Verlauf des Nervus ulnaris vor dem Epitrochleus. *Arch. Anat. Physiol. Wissen. Med.* 1867, pp. 560–564.
155* GRUBER, W. Über die Verbindung des Nervus medianus mit dem Nervus ulnaris am Unterarme des Menschen und der Säugetiere. *Arch. Anat. Physiol. Wissen. Med.* 1870, pp. 501–522.
156. GRUBER, W. Ungewöhnliche Lage des Nervus facialis in der Parotis zu den Gefäßen. *Arch. Pathol. Anat. Physiol. Klin. Med.* 54:190, 1872.
157* GRUBER, W. Verlauf des Nervus phrenicus durch eine sehr enge Insel der Vena subclavia. *Arch. Pathol. Anat. Physiol. Klin. Med.* 56:436, 1872.
158. GRUBER, W. Zu den Varietäten des Nervus peroneus profundus et superficialis. *Arch. Pathol. Anat. Physiol. Klin. Med.* 81:471–475, 1880.
159. GRUBER, W. Der Nervus radio-cutaneous externus als Substitut des nervus ulnaris am Rücken der Hand und der Finger. *Arch. Pathol. Anat. Physiol. Klin. Med.* 81:27–29, 1881.
160. GRUBER, W. Über eine durch den Theilungswinkel der Arteria brachialis tretende Ansa recurrens zwischen dem Nervus musculocutaneous brachii und dem Nervus medianus beim Menschen. *Arch. Pathol. Anat. Physiol. Klin. Med.* 86:29–33, 1881.
161. GRUBER, W. Abgang des Ramus volaris digitorum communis III des Nervus medianus in verschiedener Höhe am Unterarme. *Arch. Pathol. Anat. Physiol. Klin. Med.* 86:33–35, 1881.

162. GRUBER, W. Verästelung des Ramus superficialis des Nervus radialis in der Haut des Rückens der ganzen Hand und aller Finger bei nur verkümmert vorhandenem Ramus dorsalis des Nervus ulnaris. *Arch. Pathol. Anat. Physiol. Klin. Med.* 102:5–7, 1885.
163. GRUBER, W. Abgang eines supernumerären Nervus cutaneous ulnaris antibrachii vom Nervus ulnaris am Oberarm über dem Epitrochleus humeri, der einen Kanal an der Spitze des Letzteren passiert. *Arch. Pathologische Anat. Physiol. Klin. Med.* 102:7–9, 1885.
164* GRZYBOWSKI, J. Innervation de la dure-mère cranienne chez l'homme. *Assoc. Anatomistes Comptes Rendus* 26:233–238, 1931.
165. GUERRIER, Y. and L. GESP. L'innervation du sinus maxillaire. *Assoc. Anatomistes Comptes Rendus* 38:580–588, 1951.
166. HABERER, H. VON. Eine sehr seltene Varietät des Nervus ulnaris. *Anat. Anz.* 47:596–602, 1914–15.
167. HAJÓS, E. Über ein scheinbar abnormes Bündel der menschlichen Oblongata. Eine direkte cerebro-bulbo-cerebellare Pyramidenbahn. *Z. Ges. Neurol. Psychiatry* 21:182–194, 1914.
168. HANOT, V. Quatre observations de dédoublement de la deuxiéme circonvolution frontale chez des malfaiteurs. *Soc. Biol. Comptes Rendus des Séances et Mémoires* 31:365, 1879.
169* HARMAN, N.B. The pelvic splanchnic nerves: An examination into their range and character. *J. Anat. Physiol.* 33:386–399, 1899.
170. HARNESS, D. and E. SEKELES. The double anastomotic innervation of thenar muscles. *J. Anat.* 109:461–466, 1971.
171* HARRIS, C.M. Peripheral entrapment neuropathies of upper extremity. *N.Y. State J. Med.* 72:717–724, 1972.
172. HARRIS, W. The true form of the brachial plexus, and its motor distribution. *J. Anat. Physiol.* 38:399–422, 1904.
173* HASAN, M. and D. NARAYAN. A single cord human brachial plexus. *J. Anat. Soc. India* 13:103–104, 1964.
174* HAYEK, H. VON. Variationsstatistische Untersuchungen über die erste Verlaufsstrecke des Nervus thoracicus longus. *Anat. Anz.* 98:105–115, 1951–52.
175. HEPBURN, D. Some variations in the arrangement of the nerves of the human body. *J. Anat. Physiol.* 21:511–513, 1887.
176. HEPBURN, D. Abnormalities of muscles, nerves, heart, vessels and ligaments. *J. Anat. Physiol.* 30:570–583, 1896.
177. HIGHET, W.B. Innervation and function of the thenar muscles. *Lancet* 1:227–230, 1943.
178. HIRASAWA, K. and Y. UKITA. Über eine seltene Varietät des Ramus superficialis n. radialis und des N. cutaneus antebrachii radialis, nebst einer allgemeinen, phylogenetischen Betrachtung der Handrückeninnervation. *Folia Anat. Jpn.* 21:519–536, 1941–42.
179* HOCHE, A. Über Variationen im Verlauf der Pyramidenbahn. *Neurol. Centralbl.* 16:993–997, 1897.
180* HOLLIDAY, M. and R.P. CORMACK. Notes on a case of partial duplication of the spinal cord in the lumbar region. *Trans. R. Soc. Trop. Med. Hygiene* 21:215–218, 1927.
181. HOLLINSHEAD, W.H. and J.E. MARKEE. The multiple innervation of limb muscles in man. *J. Bone Joint Surg.* 28:721–731, 1946.
182. HOROWITZ, E. Über einen Fall von einseitiger Aplasie des Nervus olfactorius. *Anat. Anz.* 81:353–359, 1935–36.
183. HORWITZ, M.T. Normal anatomy and variations of the peripheral nerves of the leg and foot. Application in operations for vascular diseases: Study of one hundred specimens. *Arch. Surg.* 36:626–636, 1938.
184* HORWITZ, M.T. The anatomy of (A) the lumbosacral nerve plexus – its relation to variations of vertebral segmentation, and (B) the posterior sacral nerve plexus. *Anat. Rec.* 74:91–107, 1939.
185* HOWDEN, R. Variations in the hippocampus major and eminentia collateralis in the human brain. *J. Anat. Physiol.* 22:283–288, 1888.
186* HUELKE, D.F. A study of the formation of the sural nerve in adult man. *Am. J. Phys. Anthropol.* 15:137–145, 1957.
187* HUELKE, D.F. The origin of the peroneal communicating nerve in adult man. *Anat. Rec.* 132:81–92, 1958.
188. HUTTON, W.K. Remarks on the innervation of the dorsum manus, with special reference to certain rare abnormalities. *J. Anat. Physiol.* 40:326–331, 1906.
189. HUU, N. Les nerfs de Telford, Stopford et Mitchell et la problème de la terminaison du pneumogastrique. Première note: Les arguments anatomiques. *Assoc. Anatomistes Comptes Rendus* 41:821–830, 1954.
190. HYNDMAN, O.R. and W. PENFIELD. Agenesis of the corpus callosum. Its recognition by ventriculography. *Arch. Neurol. Psychiatry* 37:1251–1270, 1937.

191 IP, M.C. and K.S.F. CHANG. A study on the radial supply of the human brachialis muscle. *Anat. Rec.* 162:363–371, 1968.

192 ISOMURA, G. Nerve supply for anomalous ocular muscle in man. *Anat. Anz.* 142:255–265, 1977.

193* ISSAJEW, P.O. Ein Fall der Abwesenheit des N. olfactorius. *Anat. Anz.* 74:398–400, 1932.

194 IWANOW, G. Zur Frage der Topographie der Paraganglien beim Menschen. *Z. Anat. Entwicklungsgesch.* 84:544–547, 1927.

195 JABALEY, M.E., WALLACE, W.H. and F.R. HECKLER. Internal topography of major nerves of the forearm and hand: A current view. *J. Hand Surg.* 5:1–18, 1980.

196 JAMESON, R.M. and W.O. REINHARDT. The supra-hyoid hypoglossal ansa (of Hyrtl): A nerve crossing the midline of the body. *Anat. Rec.* 97:392–393, 1947.

197* JAMIESON, E.B. Some anomalies in nerves arising from the lumbar plexus, and a bilaminar musculus pectineus in a foetus; and on variations in the nerve supply in man and in some other mammals. *J. Anat. Physiol.* 37:266–286, 1903.

198* JAMIESON, R.W., SWIGART, LaV.L. and B.J. ANSON. Points of parietal perforation of the ilioinguinal and iliohypogastric nerves in relation to optimal sites for local anesthesia. *Q. Bull. Northwestern University Medical School* 26:22–26, 1952.

199* JAMIESON, R.W., SMITH, D.B. and B.J. ANSON. The cervical sympathetic ganglia. An anatomical study of 100 cervicothoracic dissections. *Q. Bull. Northwestern University Medical School* 26:219–227, 1952.

200* JAMIESON, R.W. and B.J. ANSON. The relation of the median nerve to the heads of origin of the pronator teres muscle. A study of 300 specimens. *Q. Bull. Northwestern University Medical School* 26:34–35, 1952.

201* JAMIESON, R.W., SMITH, D.B. and B.J. ANSON. The cervical sympathetic ganglia. An anatomical study of 100 cervicothoracic dissections. *Q. Bull. Northwestern University School* 26:219–227, 1952.

202 JAYA, Y. Liver as a content of the right sided diaphragmatic hernia – A case report. *J. Anat. Soc. India* 9:37–38, 1960.

203 JIT, I. and V.M. CHARNALIA. The vertebral level of the termination of the spinal cord. *J. Anat. Soc. India* 8:93–101, 1959.

204 JIT, I. and R.N. MUKERJEE. Observations on the anatomy of the human thoracic sympathetic chain and its branches; with an anatomical assessment of operations of hypertension. *J. Anat. Soc. India* 9:55–82, 1960.

205 JO, A. Eine seltene Anastomose des Nervus vagus. *Anat. Anz.* 129:180–182, 1971.

206* JOHNSON, R.K. and M.M. SHREWSBURY. Anatomical course of the thenar branch of the median nerve. *J. Bone Joint Surg. (Am.)* 52:269–273, 1970.

207 JOHNSTON, H.M. The cutaneous branches of the posterior primary division of the spinal nerves and their distribution in the skin. *J. Anat. Physiol.* 42:80–92, 1909.

208 JOLYET. Sur un cas d'anomalie du canal central de la moelle épinière, observée sur la moelle de Mme V., agée de 80 ans, morte à la Salpêtrière. *Soc. Biol. Comptes Rendus des Séances et Mémoires* 19:7–8, 1867.

209* JORDAN, J. Double infraorbital nerve in man. *Folia Morphol.* 26:405–407, 1967.

210* KAISER, R.A. Obturator neurectomy for coxalgia. An anatomical study of the obturator and the accessory obturator nerve. *J. Bone Joint Surg. (Am.)* 31:815–819, 1949.

211* KARMAŃSKA, W., KARMAŃSKA, J. and E. STOKLOSA. Segmental innervation of the skin by the dorsal branches of spinal nerves in newborns. *Folia Morphol.* 32:1–9, 1973.

212* KASAI, T. About the N. cutaneus brachii lateralis inferior. *Am. J. Anat.* 112:305–309, 1963.

213 KATRITSIS, E., ANAGNOSTOPOULOU, S. and N. PAPADOPOULOS. Anatomical observations on the accessory obturator nerve (Based on 1000 specimens). *Anat. Anz.* 148:440–445, 1980.

214* KEEGAN, J.J. and F.D. GARRETT. The segmental distribution of the cutaneous nerves in the limbs of man. *Anat. Rec.* 102:409–437, 1948.

215 KEITH, A. The brain of Anatole France. *Br. Med. J.* 2:1048–1049, 1927.

216 KELLEY, W.O. Phrenic nerve paralysis. Special considerations of the accessory phrenic nerve. *Journal Thoracic Surg.* 19:923–928, 1950.

217 KEMPE, L.G. and E. BUSCH. Clinical significance of cisterna veli interpositi. *Arch. Neurochir.* 16:241–248, 1967.

218* KERR, A.T. The brachial plexus of nerves in man, the variations in its formation and branches. *Am. J. Anat.* 23:285–395, 1918.

219 KESSEL, F. Verlauf des Nervus accessorius durch eine Insel der Vena jugularis interna. *Anat. Anz.* 65:162–164, 1928.

220 KIDO, S. and K. KAMEDA. Über die Ursprungsweise des N. mylophyoideus und die Anastomosenbildung zwischen dem N. alveolaris mandibularis und dem N. lingualis. *Folia Anat. Jpn.* 19:597–605, 1940.

221 KING, T.S. and J.B. COAKLEY. The intrinsic nerve cells of the cardiac atria of mammals and man. *J. Anat.* 92:353–376, 1958.

222 KIRGIS, H.D. and A. KUNTZ. Inconstant sympathetic neural pathways. Their relations to sympathetic denervation of upper extremity. *Arch. Surg.* 44:95–102, 1942.

223 KISS, F. and H.C. BALLON. Contribution to the nerve supply of the diaphragm. *Anat. Rec.* 41:285–298, 1925.

224 KLEIN, M. Surgical anatomy of the facial nerve. *Br. J. Ophthalmology* 30:668–675, 1946.

225 KLUYSKENS, P. Sur la localisation de la division du nerf laryngé inférieur. *Assoc. Anatomistes Comptes Rendus* 39:36–41, 1952.

226 KOCHANOWSKI, J. The vertebral loop in man. *Folia Morphol.* 37:353–355, 1978.

227 KOCH-LANGENTREU, J. Abnormer Verlauf des Nervus alveolaris inferior. *Z. Stomatologie* 44:317–318, 1947. Cited in *Excerpta Medica*, Sec. 1, Vol. 2, abstract 1305, 1948.

228 KOSINSKI, C. La valeur anthropologique des variations dans l'innervation cutanée des membres inférieurs. *Assoc. Anatomistes Comptes Rendus* 21:291–295, 1926.

229 KOSINSKI, C. L'innervation cutanée de la face drosale de la main, basée sur l'examen de 300 pièces anatomiques, avec quelques notions d'anatomie comparée. *Assoc. Anatomistes Comptes Rendus* 22:121–133, 1927.

230* KOSINSKI, C. The course, mutual relations and distribution of the cutaneous nerves of the metazonal region of leg and foot. *J. Anat.* 60:274–297, 1926.

231 KOWSARIAN, S.H. Le sinus pétreux supérieur. *Assoc. Anatomistes Comptes Rendus* 39:793–796, 1952.

232 KRAUS, W. and A. WEIL. Duplicité de la Moelle Epinière. *Rev. Neurol.* 33:317–323, 1926.

233 KRAUSE, W. Über den Ramus collateralis ulnaris Nervi radialis. *Arch. Anat. Physiol. Wissen. Med.* 1868, pp. 134–136.

234 KRÓLIKOWSKA, M. and R. ZAWODNIAK. Variation of origin of major, minor and the lowest splanchnic nerves in man. *Folia Morphol.* 38:341–346, 1979.

235 KUBIK, W. Ansa suprahyoidea hypoglossi. *Folia Morphol.* 22:253–256, 1963.

236 KULCZYŃSKI, B. and W. WOŹNIAK. Variation of the origin and course of the chorda tympani. *Folia Morphol.* 37:237–241, 1978.

237 KUNTZ, A. and W.F. ALEXANDER. Surgical implications of lower thoracic and lumbar independent sympathetic pathways. *Arch. Surg.* 61:1007–1018, 1950.

238* KURYLCIO, L. Forms of the vagus trunk and their distribution in the esophageal hiatus in man. *Folia Morphol.* 30:73–85, 1971.

239* KURYLCIO, L. Anatomic varieties of the vagus trunks from the point of view of selective vagotomy. *Folia Morphol.* 31:85–108, 1972.

240* KURYLCIO, L. The terminal branches of the vagus trunks in man. *Folia Morphol.* 31:231–259, 1972.

241* LAHEY, F.H. Exposure of the recurrent laryngeal nerves in thyroid operations. *Surg. Gynecol. Obstet.* 78:239–244, 1944.

242 LAMBERT, E.H. The accessory deep peroneal nerve: A common variation in innervation of extensor digitorum brevis. *Neurology* 19:1169–1176, 1969.

243 LAMONT, J.C. Notes on the nervous supply of the musculus sternalis. *J. Anat. Physiol.* 21:514–515, 1887.

244* DE LANGE, C. On brains with total and partial lack of the corpus callosum and on the nature of the longitudinal callosal bundle. *J. Nerv. Mental Dis.* 62:449–476, 1925.

245 LANGSAM, C.L.M. Omohyoideus in American Whites and Negroes. *Am. J. Phys. Anthropol.* 28:249–259, 1941.

246* LANZ, U. Anatomical variations of the median nerve in the carpal tunnel. *J. Hand Surg.* 2:44–53, 1977.

247 LARKIN, F.C. Accessory phrenic nerve. *J. Anat. Physiol.* 23:340, 1889.

248* LAST, R.J. Innervation of the limbs. *J. Bone Joint Surg. (Br.)* 31:452–464, 1949.

249 LATARJET, A. and DEGROAT. Note preliminaire a l'étude des nerfs du gros intestin. *Assoc. Anatomistes Comptes Rendus* 21:334–337, 1926.

250 LATARJET, M., NEIDHART, J.H., MORRIN, A. and J.M. AUTISSIER. L'entrée du nerf musculo-cutané dans le muscle coracobrachial. *Assoc. Anatomistes Comptes Rendus* 138:755–765, 1967.

251 LATTES, L. Asymétries cérébrales. *Arch. Ital. Biol.* 48:475–476, 1907.
252 LAUX, G., GUERRIER, Y. and J. OLIVIER. Les rapports du spinal médullaire avec les deux premiers nerfs cervicaux. *Assoc. Anatomistes Comptes Rendus* 37:262–270, 1950.
253 LAUX, G., GUERRIER, Y. and G. MARCHAL. Ansa supra-hyoidienne de Hyrtl. *Assoc. Anatomistes Comptes Rendus* 38:648, 1951.
254 LAZORTHES, G. and L. CASSAN. Essai de schématisation des ganglions étiolé et intermédiaire (ganglion cervico-thoracique) d'après 80 dissections chez l'adulte et chez le foetus. *Assoc. Anatomistes Comptes Rendus* 1939, pp. 193–210.
255 LAZORTHES, G. and J. POULHÈS. Les variations de la scissure calcarine et de la corne occipitale des ventricules lateraux. *Assoc. Anatomistes Comptes Rendus* 35:210–218, 1948.
256 LAZORTHES and J. POULHÈS. Les ventricules latéraux de l'homme. *Assoc. Anatomistes Comptes Rendus* 36:432–436, 1949.
257 LAZORTHES, G., DE LA SOUJEOLE, A. and J. ESPAGNO. Note sur les vaisseaux de l'angle pontocérébelleux. Variations et rapports avec la racine du trijumeau. *Assoc. Anatomistes Comptes Rendus* 36:437–438, 1949.
258 LAZORTHES, G., POULHÈS, J. and S. GAUBERT. La signification et la hauteur des arcades du ligament dentelé. *Assoc. Anatomistes Comptes Rendus* 39:303–311, 1952.
259 LAZORTHES, G., BIMES, C., PLANEL, H. and J. GAUBERT. Les variations régionales de l'épaisseur de la dure-mère. *Assoc. Anatomistes Comptes Rendus* 40:159–160, 1953.
260 LEARMONTH, J.R. A variation in the distribution of the radial branch of the musculospiral nerve. *J. Anat.* 53:371–372, 1919.
261 LEBOUCQ, G. La mensuration de la surface des hémisphères cérébraux. *Assoc. Anatomistes Comptes Rendus* 21:338–343, 1926.
262 LEBOUCQ, G. Rapport entre le poids du cerveau humain et son éncore. *Assoc. Anatomistes Comptes Rendus* 22:139–144, 1927.
263 LEGGETT, T.H. JR. and J. LINITZ. Eine Varietät eines Teiles des N. femoralis. *Anat. Anz.* 43:232–233, 1913.
264 LEGGIARDÍ-LAURA, C. Sur la signification de la "duplicité de la scissure de Rolando" et sur un rapport constant de la scissure post-rolandique. *Arch. Ital. Biol.* 36:355, 1901.
265 LEMERE, F. Innervation of the larynx: I. Innervation of laryngeal muscles. *Am. J. Anat.* 51:417–432, 1932.
266 LEONTOWITSCH, A. Die Innervation der menschlichen Haut. *Int. Monatsschr. Anat. Physiol.* 18:142–310, 1901.
267 LINELL, E.A. The distribution of nerves in the upper limb, with reference to variabilities and their clinical significance. *J. Anat.* 55:79–112, 1921.
268 LINELL, E.A. and B.M. KOSTER. A developmental anomaly of the cerebral cortex. *Arch. Neurol. Psychiatry* 12:167–172, 1924.
269 LIPPMANN, R. VON Abnormer Ursprung des Ramus descendens n. hypoglossi aus dem N. vagus. *Anat. Anz.* 37:1–4, 1910.
270 LOEWENECK, H., STORK, H.T. and P. LOEWENECK. Variationsstatistische Untersuchung zum Zwerchfelldurchtritt der Nervi splanchnici. *Anat. Anz.* 126:531–547, 1970.
271 LUNA, E. Fréquente anastomose entre le nerf médian et le rameau profond du nerf cubital. *Arch. Ital. Biol.* 55:141, 1911.
272 LUYS. Anomalies des fibres arciformes du bulbe. *Soc. Biol. Comptes Rendus des Séances et Mémoires* 32:229–231, 1880.
273* MATWEJEW, W.P. Zur Typenanatomie des Plexus lumbosacralis. *Anat. Anz.* 83:422–436, 1937.
274 MAYET, A. Zur Innervation des M. cricothyreoideus. *Anat. Anz.* 103:340–343, 1956.
275 MCCARTHY, R.E. and E.A. NALEBUFF. Anomalous volar branch of the dorsal cutaneous ulnar nerve: A case report. *J. Hand Surg.* 5:19–20, 1980.
276* MCCORMACK, L.J., CAULDWELL, E.W. and B.J. ANSON. The surgical anatomy of facial nerve with special reference to the parotid gland. *Surg. Gynecol. Obstet.* 80:620–630, 1945.
277* MCKENZIE, J. The parotid gland in relation to the facial nerve. *J. Anat.* 82:183–186, 1948.
278 MEURMANN, O.H. Theories of vocal cord paralysis. *Acta Otolaryngol.* 38:460–472, 1950.
279 MEYER, E. Die Erweiterung des Ventriculus septi pellucidi. *Arch. Psychiatry* 92:9–36, 1930.
280* MEYER, H. Eine historische Notiz über eine Varietät des N. opticus. *Arch. Anat. Physiol. Wissen. Med.* 1870, pp. 523–524.
281 MILIANITCH, N. and B. CHLYIVITCH. Variations de la morphologie et de l'innervation des muscles de lépaule et du bassin chez les Serbes (Etude basée sur 200 cas). *Assoc. Anatomistes Comptes Rendus* 21:396–402, 1926.

282 MILLER, M.R., RALSTON, H.J. III and M. KASAHARA. The pattern of cutaneous innervation of the human hand. *Am. J. Anat.* 102:183–217, 1958.
283 MILLER, M.R. and M. KASAHARA. The pattern of cutaneous innervation of the human foot. *Am. J. Anat.* 105:233–255, 1959.
284* MILLER, R.A. Observations upon the arrangement of the axillary artery and the brachial plexus. *American J. Anat.* 64:143–163, 1939.
285 MING-TZU, P'AN. The relation of the sciatic nerve to the piriformis muscle in the chinese. *Am. J. Phys. Anthropol.* 28:375–380, 1941.
286 MINNE, J., SENNEVILLE, A. and Y. GUYOT. Remarques sur la division et les branches terminales du nerf phrenique droit. *Assoc. Anatomistes Comptes Rendus* 36:484–491, 1949.
287* MINNE, SENNEVILLE and LIBERSA. Sur quelques particularités concernant la trajet et la distribution des nerfs intercostaux. *Assoc. Anatomistes Comptes Rendus* 36:492–496, 1949.
288* MINNE, DEPREUS and FONTAINE. Innervation de la région talonnière parties molles et plans osseux. *Assoc. Anatomistes Comptes Rendus* 38:704–709, 1951.
289* MIRSALIS, T. Ein neuer Fall von Arhinencephalie. *Anat. Anz.* 67:353–360, 1929.
290 MIRTO, D. and E. PUSATERI. Sur les rapports anastomotiques entre le nerf accessoire et le vague. *Arch. Ital. Biol.* 1896, pp. 25–348.
291 MISRA, B.D. The relations of the sciatic nerve to the piriformis in Indian cadavers. *J. Anat. Soc. India* 3:44, 1954.
292 MITCHELL, G.A.G. The innervation of the kidney, ureter, testicle and epididymis. *J. Anat.* 70:10–32, 1935.
293* MIZERES, N.J. The cardiac plexus in man. *Am. J. Anat.* 112:141–151, 1963.
294* MOGHADDAM, T.C. Variationen des N. obturatorius und N. obturatorius accessorius. *Anat. Anz.* 113:1–18, 1963.
295 MONGKOLLUGSANA, D. and L.F. EDWARDS. The extra-osseous innervation of the gingivae. *J. Dent. Res.* 36:516–524, 1957.
296 MONTEIRO, H. and A. RODRIGUES. Sur les variations du nerf vertebral. *Assoc. Anatomistes Comptes Rendus* 26:406–419, 1931.
297 MORRIS, H.H. and B.H. PETERS. Pronator syndrome: Clinical and electrophysiological features in seven cases. *J. Neurol. Neurosurg. Psychiatry* 39:461–464, 1976.
298 MORTON, W.R.M. Arhinencephaly and multiple anomalies occurring in a human full term foetus. *Anat. Rec.* 98:45–58, 1947.
299 DE MOSSIER, G. Agenesie du septum pellucidum avec malformation du tractus optique: la dysplasie septo-optique. *Schweiz. Arch. Neurol. Psychiatry* 77:267–292, 1956.
300 MURAKAMI, T. On the distribution of the cutaneous nerves of the upper extremities in Macacus cyclopsis. *Folia Anat. Jpn.* 30:343–367, 1958.
301 MURPHEY, F., KIRKLIN, J.W. and A.I. FINLAYSON. Anomalous innervation of the intrinsic muscles of the hand. *Surg. Gynecol. Obstet.* 83:15–23, 1946.
302 NAKANISHI, T., KANNO, Y. and T. KANESHIGE. Comparative morphological remarks on the origin of the posterior femoral cutaneous nerve. *Anat. Anz.* 139:8–23, 1976.
303* NAKANO, K.K. The entrapment neuropathies. *Muscle Nerve* 1:264–279, 1978.
304 NEGUS, V.E. Certain anatomical and physiological considerations in paralysis of the larynx. *Proc. R. Soc. Med.* 40:849–853, 1947.
305* NIZANKOWSKI, C., SLOCIAK, J. and J. SZYBEJKO. Variation of the course of the sciatic nerve in man. *Folia Morphol.* 31:451–456, 1972.
306* NORDLAND, M. The larynx as related to surgery of the thyroid based on an anatomical study. *Surg. Gynecol. Obstet.* 51:449–459, 1930.
307* OBERLÄNDER, W. Beobachtung eines sensiblen Astes aus dem Ramus dorsalis des I. Cervicalsegmentes. *Anat. Anz.* 138:26–28, 1975.
308 OCCHIPINTI, G. and S. ANZALONE. Confronto encefalometrico fra Siciliani e Sardi. *Arch. Ital. Anat. Embriol.* 42:78–127, 1939.
309 OELRICH, T.M. and D.A. MOOSMAN. The aberrant course of the cutaneous component of the ilioinguinal nerve. *Anat. Rec.* 189:233–236, 1977.
310 OERTEL, O. Abnormer Verlauf des Nervus femoralis. *Anat. Anz.* 56:550–553, 1923.
311* OGDEN, J.A. An unusual branch of the median nerve. *J. Bone Joint Surg. (Am.)* 54:1779–1781, 1972.
312 OGLE, J.W. A nerve piercing the walls of an artery. *J. Anat. Physiol.* 29:20, 1895.
313 OLIVIER, E., MINNE and ROCQUET. La traversée diaphragmati-

que du nerf phrénique droit et ses variations. *Assoc. Anatomistes Comptes Rendus* 21:441–444, 1926.
314* OLIVIER, G., ABOULKER, P. and C. EYRIES. Les différents types de nerf facial intra-parotidien. *Assoc. Anatomistes Comptes Rendus* 40:859–869, 1953.
315 OLEZEWSKI, J. Variations of the pudendal nerve in man. *Folia Morphol.* 41:245–252, 1982.
316 ONO, N. Untersuchungen und Studien über die Ursprungszellen des N. phrenicus. *Jpn. J. Med. Sci.* 5:1–34, 1934–35.
317 OUTI, H. Pri la interrilato de la femura nervo kaj la psoasmuskolo, precipe pri la t.n. akcesora femura nerve kaj la akcesora fasko de la psoas-muskolo. *Folia Anat. Jpn.* 28:384–412, 1956.
318 PAPADATOS, D. Deux observations anatomiques d'un nerf récurrent non-récurrent. *Anat. Anz.* 144:97–106, 1978.
319 PAPADATOS, D. Supraclavicular nerves piercing the clavicle: A study of 10 cases. *Anat. Anz.* 147:371–381, 1980.
320 PAPADOPOULOS, N.J. and E.D. KATRITSIS. Some observations on the course and relations of the iliohypogastric and ilioinguinal nerves (based on 348 specimens). *Anat. Anzeiger* 149:357–364, 1981.
321 PAPATHANASSIOU, B.T. A variant of the motor branch of the median nerve in the hand. *J. Bone Joint Surg. (Br.)* 50:156–157, 1968.
322* PARSONS, F.G. and A. KEITH. Sixth annual report of the Committee of Collective Investigation of the Anatomical Society of Great Britain and Ireland. *J. Anat. Physiol.* 31:31–44, 1896–97.
323 PARTRIDGE, E.J. The relations of the glossopharyngeal nerve at its exit from the cranial cavity. *J. Anat.* 52:332–334, 1918.
324 PATERSON, A.M. Morphology of the sacral plexus in man. *J. Anat. Physiol.* 21:407–412, 1887.
325 PATERSON, A.M. The origin and distribution of the nerves to the lower limb. *J. Anat. Physiol.* 28:84–95, 169–193, 1894.
326* PATEY, D.H. and I. RANGER. Some points in the surgical anatomy of the parotid gland. *Br. J. Surg.* 45:250–258, 1957.
327* PEARSON, A.A., SAUTER, R.W. and G.R. HERRIN. The accessory nerve and its relation to the upper spinal nerves. *Am. J. Anat.* 114:371–391, 1964.
328* PEMBERTON, J. DE J. and J.M. MILLER. Anomaly of the right inferior laryngeal nerve. *Arch. Surg.* 42:712–718, 1941.
329 PEREIRA, S. Sur les rapports des nerfs splanchniques avec les piliers du diaphragme chez l'homme. *Assoc. Anatomistes Comptes Rendus* 25:252–266, 1930.
330* PERLOW, S. and K.L. VEHE. Gross anatomy of the stellate and the lumbar sympathetic ganglia. *Am. J. Surg.* 30:454–458, 1935.
331 PFISTER, H. Zur Anthropologie des Rückenmarks. *Neurol. Centralbl.* 22:757–762, 819–824, 1903.
332 PHALEN, G.S. The carpal tunnel syndrome. Seventeen year's experience in diagnosis and treatment of six hundred fifty-four hands. *J. Bone Joint Surg. (Am.)* 48:211–228, 1966.
333 PHILLIPS, W.F.R. Innervation of an axillary arch muscle. *Anat. Rec.* 7:131–132, 1913.
334 PICARD, L., LEYMARIE, F., ROLAND, J., SIGIEL, M., MASSON, J.P., ANDRÉ, J.M. and M. RENARD. Cavum veli interpositi. Roentgen anatomy – pathology and physiology. *Neuroradiology* 10:215–220, 1976.
335 PICK, A. Über ein abnormes Faserbündel in der menschlichen Medulla oblongata. *Arch. Psychiatry* 21:636–640, 1890.
336 PIFFER, C.R. and N.L. ZORZETTO, Course and relations of the abducens nerve. *Anat. Anz.* 147:42–46, 1980.
337 PIFFER, C.R., GARCIA, P.J. and J.C. SOARES. Course of the external branch of the accessory nerve and its relations with the first portion of the internal jugular vein. *Anat. Anz.* 148:252–257, 1980.
338 PILATI, L. Variazioni della distribuzione dei rami del radiale e dell'ulnare nell'innervazione cutanea de dorso della dita dello mano. *Boll. Soc. Ital. Biol. Sperimentale* 23:446–448, 1974. Cited in Excerpta Medica, Sec. 1, Vol. 3, abstract 45, 1949.
339* PLATZER, W. Eine transmediane Anastomose der beiden Nn. hypoglossi. *Arch. Psychiatr. Nervenkr.* 199:372–376, 1959.
340 POLACEK, P. Ein Beitrag zur Frage des N. femoralis accessorius und N. obturatorius accessorius. *Anat. Anz.* 105:141–148, 1958.
341 POTERIN-DUMOTEL, Absence congénitale du corps calleux sans troubles fonctionnels durant la vie. *Soc. Biol. Comptes Rendus des Séances et Mémoires* 14:94–99, 1862.
342 POTTS, T.K. The main peripheral connections of the human sympathetic nervous system. *J. Anat.* 59:129–135, 1925.
343 PRIOTON, J.B. and A. THEVENET. La distribution intra-diaphragmatique des nerfs phréniques. *Assoc. Anatomistes Comptes Rendus* 44:635:645, 1957.

344 PÜRNER, J. Über den peripheren Verlauf des N. saphenous. *Anat. Anz.* 129:114–132, 1971.
345 RÁCZ, L. and T. MAROS, The anatomic variants of the lingual nerve in humans. *Anat. Anz.* 149:64–71, 1981.
346 RÁCZ, L., MAROS, T. and L. SEVES-STRUM. Anatomical variations of the nervus alveolaris inferior and their importance for the practice. *Anat. Anz.* 149:329–332, 1981.
347 RAMASWAMY, S. A ganglion on the internal laryngeal nerve. A preliminary report. *J. Anat. Soc. India* 12:1–7, 1963.
348 RAMSTROM, M. Untersuchungen über die Innervation des Caput mediale tricipitis brachii. *Anat. Anz.* 51:420–431, 1918–19.
349 RANSCHBURG, P. Über die Anastomosen der Nerven der oberen Extremität des Menschen mit Rücksicht auf ihren neurologische und nervenchirurgische Bedeutung. *Neurol. Centralbl.* 36:521–534, 1917.
350 RANSON, S. Anatomy of the sympathetic nervous system with reference to sympathectomy and ramisection. *JAMA* 86:1886–1890, 1926.
351 RAO, G.R.K.H. and V.R. RAO. Musculocutaneous nerve of the arm. *J. Anat. Soc. India.* 4:48, 1955.
352 RAY, B.S. and A.D. CONSOLE. Residual sympathetic pathways after paravertebral sympathectomy. *Journal Neurosurg.* 5:23–50, 1948.
353 REAL, K. Eine Abnormität im Verhalten des Grenzstranges des Nervus sympathicus. *Arch. Anat. Physiol. Wissen. Med.* 1871, pp. 180–183.
354* REED, A.F. The relations of the inferior laryngeal nerve to the inferior thyroid artery. *Anat. Rec.* 85:17–23, 1943.
355 REED, A.F. The origins of the splanchnic nerves. *Anat. Rec.* 109:341, 1951.
356 REICHERT, F.L. Neuralgias of the glossopharyngeal nerve. With particular reference to the sensory, gustatory and secretory functions of the nerve. *Arch. Neurol. Psychiatry.* 32:1030–1037, 1934.
357 REICHERT, F.L. and E.J. POTH. Recent knowledge regarding the physiology of the glossopharyngeal nerve in man with an analysis of its sensory, motor, gustatory and secretory functions. *Johns Hopkins Hosp. Bull.* 53:131–139, 1933.
358* REID, R.W. The relations between the superficial origins of the spinal nerves from the spinal cord and the spinous processes of vertebrae. *J. Anat. Physiol.* 23:341–353, 1889.
359 REIL, Mangel des mittleren und freyen theils des Balkens im Menschengehirn. *Arch. Physiol.* 11:341–344, 1812.
360* REIMANN, A.F. and B.J. ANSON. Vertebral level of termination of the spinal cord with report of a case of sacral cord. *Anat. Rec.* 88:127–138, 1944.
361 REMAK, E. Zur vicariierenden Funktion peripherer Nerven des Menschen. *Berl. Klin. Wochenschr.* 48:601–604, 1874.
362 RICH, A.R. The innervation of the tensor veli palatini and levator veli palatini muscles. *Johns Hopkins Hosp. Bull.* 31:305–310, 1920.
363 RIGAUD, A., CABANIE, H., DEJUSSIEU, J. and J. PLESSIS. Note sur la fréquente asymétrie des plexus lombaires. *Assoc. Anatomistes Comptes Rendus* 42:1206–1214, 1955.
364 RIGAUD, A., CABANIÉ, H., DEJUSSIEU, J. and J. PELLEGRINO. Note sur l'origine et la signification des nerfs abdomino-génitaux. *Assoc. Anatomistes Comptes Rendus* 43:736–741, 1956.
365* ROBBINS, H. Anatomical study of the median nerve in the carpal tunnel and etiologies of the carpal-tunnel syndrome. *J. Bone Joint Surg. (Am.)* 45:953–966, 1963.
366 ROBERTS, W.H.B. and W.H. TAYLOR. Inferior rectal nerve variations as it relates to pudendal block. *Anat. Rec.* 177:461–463, 1973.
367 RODRIGUES, A. Le descendens cervicalis chez l'homme et chez les mammifères (quelques notes sur son évolution phylogénique. *Assoc. Anatomistes Comptes Rendus* 25:267–282, 1930.
368 RODRIGUES, A. Communicating branches between the cervical sympathetic and the descendens cervicalis. *J. Anat.* 64:308–318, 1930.
369* ROLES, N.C. and R.H. MAUDSLEY. Radial tunnel syndrome. *J. Bone Joint Surg. (Br.)* 54:499–508, 1972.
370 ROMANES, G.J. The spinal cord in a case of congenital absence of the right limb below the knee. *J. Anat.* 77:1–5, 1942–43.
371 ROMANES, G.J. The distribution of taste buds on the human epiglottis. *J. Anat.* 87:464, 1953.
372* ROVERE, D.D. and B. DE VECCHI. Anomalie del cervelletto, prima osservazione di scissione in due lobi distinti del verme. *Riv. Patol. Nerv. Mentale* 7:241–254, 1902.
373 ROWNTREE, T. Anomalous innervation of the hand muscles. *J. Bone Joint Surg. (Br.)* 31:505–510, 1949.

374. RUHEMANN, E. Die Topographie des Nervus phrenicus unter abnormen Verhältnissen. *Arch. Klin. Chirurgie* 139:557–562, 1926.
375.* RUSSELL, J.S.R. Contributions to the study of the afferent and efferent tracts in the spinal cord. *Brain* 21:145–179, 1898.
376. SADR, A.R. Cross nerve anastomosis in man. *Ann. Surg.* 124:599–603, 1946.
377. SAKA, H. Über die Innervation von dem Periost und der Gelenkkapsel der oberen Extremität des Menschen. *Folia Anat. Jpn.* 30:291–321, 1958.
378. SALSBURY, C.R. The nerve to the extensor carpi radialis brevis. *Br. J. Surg.* 26:95–97, 1938–39.
379.* SALTER, H. Congenital absence of the middle portions of left hemisphere of the cerebellum, in a case of chronic hydrocephalus. *Trans. Pathol. Soc. Lond.* 4:31–34, 1853.
380. SAUERBRUCH, F. Die Beeinflussung von Lungenerkrankungen durch künstliche Lähmung des Zwerchfells (Phrenikotomie). *Münch. Med. Wochenschr.* 60:625–626, 1913.
381. SAUERBRUCH, F. Kritische Bemerkungen zur Behandlung von Lungenerkrankungen durch künstliche Lähmung des Zwerchfells. *Münch. Med. Wochenschr.* 70:693–695, 1923.
382. SCHNEIDERLING, W. Unvollkommene dorso-ventrale Verdoppelung des Rückenmarkes. *Arch. Pathol. Anat Physiol. Klin. Med.* 301:479–489, 1938.
383.* SCHULTZ, R.J., ENDLER, P.M. and H.D. HUDDLESTON. Anomalous median nerve and an anomalous muscle belly of the first lumbrical associated with carpal-tunnel syndrome. Case report. *J. Bone Joint Surg. (Am.)* 55:1744–1746, 1973.
384.* SCHWARTZ, H.G. and G. WEDDELL. Observations on the pathways transmitting the sensation of taste. *Brain* 61:99–115, 1938.
385. SCHWIDDE, J.T. Incidence of cavum septi pellucidi and cavum vergae in 1032 human brains. *Arch. Neurol. Psychiatry* 67:625–632, 1952.
386. SCOTT, S. A record of the decussations of the brachial plexus in man. *J. Anat. Physiol.* 40:412–415, 1906.
387.* SEELAUS, H.K. Some notes on a muscle which takes origin from the sheath of a nerve. *Anat. Rec.* 26:307–308, 1923.
388. SERGI, S. Variations de developpement du lobe frontal chez l'homme. *Arch. Ital. Biol.* 55:138–139, 1911.
389. SERGI, S. Variations des sillons de l'insula dans le cerveau humain. *Arch. Ital. Biol.* 55:139–140, 1911.
390. SHEEHAN, D. Spinal autonomic outflows in man and monkey. *J. Comp. Neurol.* 75:341–370, 1941.
391. SHEPARD, F.J. Musculus sternalis and its nerve supply. *J. Anat. Physiol.* 23:303–307, 1889.
392. SHERRINGTON, C.S. Notes on the arrangement of some motor nerve fibers in the lumbosacral plexus. *J. Physiol.* 13:621–772, 1892.
393. SHIMADA, K. Beiträge zur Anatomie des Zentralnervensystems der Japaner. VI. Ventriculus lateralis. *Folia Anat. Jpn.* 9:429–486, 1931.
394. SHIMADA, K. Contributions to anatomy of the central nervous system of the Japanese. XI. Upon the vermal arbor vitae. *Folia Anat. Jpn.* 28:207–227, 1956.
395. SHIMADA, K. and K. KODAMA. Beiträge zur Anatomie des Zentralnervensystems der Japaner. VIII. Insula Reili. *Folia Anat. Jpn.* 12:423–444, 1934.
396. SINGER, E. Human brachial plexus united into a single cord. Description and interpretation. *Anat. Rec.* 55:411–419, 1933.
397. SIRANG, H. Ursprung, Verlauf und Äste des N. saphenous. *Anat. Anz.* 130:158–169, 1972.
398. SIWE, S. A. The cervical part of the gangliated cord, with special reference to its connections with the spinal nerves and certain cerebral nerves. *Am. J. Anat.* 48:479–497, 1931.
399. SKOOG, T. Ganglia in the communicating rami of the cervical sympathetic trunk. *Lancet* 2:457–460, 1947.
400. SMITH, G.E. Some rare nerve and muscle anomalies, with remarks on their significance. *J. Anat. Physiol.* 29:84–98, 1895.
401. SMITH, G.E. Note on the communication between musculospiral and ulnar nerves. *J. Anat. Physiol.* 38:162–163, 1904.
402.* SMITH, R.B. The occurrence and location of intrinsic cardiac ganglia and nerve plexuses in the human neonate. *Anat. Rec.* 169:33–40, 1971.
403. SOKOLOW, P.A. Abweichungen in dem Verlaufe der Gefäße und Nerven im Zusammenhang mit dem Vorhandensein des Proc. supracondyloideus am Oberarmbein. *Anat. Anz.* 68:156–163, 1929–30.
404. SOUTHAM, J.A. The inferior mesenteric ganglion. *J. Anat.* 93:304–308, 1959.
405. SPANNER, R. Eine seltene Varietät des Ramus palmaris n. ulnaris. *Anat. Anz.* 56:222–225, 1923.
406. SPERINO, G. L'encéphale de l'anatomiste Carlo Giacomini. *Arch. Ital. Biol.* 36:354–355, 1901.
407.* SPERINO, G. Descrizione morfologica dell'encefalo del Prof. Carlo Giacomini. *Int. Monatsschr. Anatomie Physiol.* 18:313–386, 1901.
408. SPILLER, W.G. A case of complete absence of the visual system in an adult. *University Penn. Med. Bull.* 14:448–453, 1902.
409. SPINNER, M. The anterior interosseous nerve syndrome with special attention to its variations. *J. Bone Joint Surg. (Am.)* 52:84–94, 1970.
410. SSOKOLOW, P. Zur Anatomie des N. suralis beim Menschen und Affen. *Z. Ges. Anat.* 100:194–217, 1933.
411. STÁNCZYK, J.L. Variation of the gyrus and sulcus cinguli in phylo- and onto-genesis. *Folia Morphol.* 42:243–269, 1983.
412. STARKIE, C. and D. STEWART. The intra-mandibular course of the inferior dental nerve. *J. Anat.* 65:319–323, 1931.
413. STILWELL, Jr., D.L. The innervation of deep structures of the human foot. *Am. J. Anat.* 101:59–73, 1957.
414. STILWELL, D.L. JR. The innervation of deep structures of the hand. *Am. J. Anat.* 101:75–99, 1957.
415.* STOPFORD, J.S.B. The variations in distribution of the cutaneous nerves of the hand and digits. *J. Anat.* 53:14–25, 1919.
416.* STOPFORD, J.B.S. The nerve supply of the inter-phalangeal and metacarpophalangeal joints. *J. Anat.* 56:1–11, 1921.
417. STRAUS, W.L. JR. and A.B. HOWELL. The spinal accessory nerve and its musculature. *Q. Rev. Biol.* 11:387–402, 1936.
418. SUGIHARA, T. M. pectineus kaj gia nervo. *Folia Anat. Jpn.* 28:377–388, 1956.
419. SUNDERLAND, S. The innervation of the flexor digitorum profundus and lumbrical muscles. *Anat. Rec.* 93:317–321, 1945.
420. SUNDERLAND, S. The intraneural topography of the radial, median, and ulnar nerves. *Brain* 68:243–298, 1945.
421. SUNDERLAND, S. The innervation of the first dorsal interosseous muscle of the hand. *Anat. Rec.* 95:7–10, 1946.
422. SUNDERLAND, S. Metrical and non-metrical features of the muscular branches of the radial nerve. *J. Comp. Neurol.* 85:93–112, 1946.
423. SUNDERLAND, S. and E.S.R. HUGHES. Metrical and non-metrical features of the muscular branches of the ulnar nerve. *J. Comp. Neurol.* 85:113–126, 1946.
424. SUNDERLAND, S. and E.S.R. HUGHES. Metrical and non-metrical features of muscular branches of the sciatic nerve and its medial and lateral popliteal divisions. *J. Comp. Neurol.* 85:205–222, 1946.
425. SUNDERLAND, S. and L.J. RAY. Metrical and non-metrical features of the muscular branches of the median nerve. *J. Comp. Neurol.* 85:191–203, 1946.
426.* SWEET, W.H. Recent observations pertinent to improving anterolateral cordotomy. *Clin. Neurosurg.* 23:80–95, 1976.
427.* TAGUCHI, K. Über eine seltene Anomalie des Verlaufes des Vagusstammes und eines seiner Äste. *Arch. Anat. Physiol. Wissen. Med.* 1888, pp. 365–368.
428. TAGUCHI, K. Die Lage des Nervus recurrens nervi vagi zur Arteria thyreoidea inferior. *Arch. Anat. Physiol. Wissen. Med.* 1889, pp. 309–325.
429. TALEISNIK, J. The palmar cutaneous branch of the median nerve and the approach to the carpal tunnel. An anatomical study. *J. Bone Joint Surg. (Am.)* 55:1212–1217, 1973.
430. TANZER, R.C. The carpal-tunnel syndrome. A clinical and anatomical study. *J. Bone Joint Surg. (Am.)* 41:626–634, 1959.
431. THOMPSON, I.M. Anomaly of the median nerve and flexor digitorum sublimis muscle. *Anat. Rec.* 23:375–379, 1922.
432. THOMPSON, I.M. On the cavum septi pellucidi. *J. Anat.* 67:59–77, 1932.
433. THOMPSON, I.M. The diagnostic application of our knowledge of the normal variability of cutaneous nerve areas, exemplified by the median and ulnar nerves. *J. Anat.* 69:159–164, 1934–35.
434. THOMSON, R.B. Complete unilateral interruption of the fissure of Rolando. *J. Anat. Physiol.* 45:433–437, 1911–12.
435.* TILLMANN, B. Verlaufsvarianten des N. gluteus inferior. *Anat. Anz.* 145:293–302, 1979.
436. TINDARO RENDA, D. Variazione insolita nel decorso del nervo femorale, nell' uomo. *Anat. Anz.* 121:214–219, 1967.
437. TOUNTAS, C.P., MACDONALD, C.J., MEYERHOFF, J.D. and D.M. BIHRLE. Carpal tunnel syndrome. A review of 507 patients. *Minn. Med.* 66:479–482, 1983.
438. TRÉTIAKOFF, C. and V. RAMOS. Etude anatomique d'un cas de

moelle bifide (double moelle lombaire). *Rev. Neurol.* 28:768–772, 1921.
439 TRICOMI, G. Deux cas de duplicité du sulcus Rolandi. *Arch. Ital. Biol.* 37:490, 1902.
440 TROTTER, M. The relation of the sciatic nerve to the piriformis muscle. *Anat. Rec.* 52:321–323, 1932.
441 TRUMBLE, H.C. The parasympathetic nerve supply to the distal colon. *Med. J. Aust.* 2:149–151, 1934.
442 TURNER, W. Variations in the distribution of the nerves of the human body. *J. Anat. Physiol.* 6:101–106, 1872.
443 TURNER, W. Further examples of variations in the arrangement of the nerves of the human body. *J. Anat. Physiol.* 8:297–299, 1874.
444 TURNER, W. A phrenic nerve receiving a root of origin from the descendens hypoglossi. *J. Anat. Physiol.* 27:427, 1893.
445 UGOLOTTI, F. Nuove ricerche sulle vie piramidali nell'uomo. (A proposito di une recente publicazione di P. MARIE e G. GUILLIAN). *Riv. Patol. Nerv. Mentale* 8:145–154, 1903.
446 UKITA, Y. Über das Innervationsbild des Fußrückens bei Japanern. *Folia Anat. Jpn.* 733–775, 1944.
447* URBANOWICZ, Z. External structure of the genitofemoral nerve in postfetal life in man. *Folia Morphol.* 34:425–434, 1975.
448* URBANOWICZ, Z. Connections between the lumbar and the sacral plexus in man. *Folia Morphol.* 40:271–279, 1981.
449 URBANOWICZ, Z. and S. ZALUSKA. Formation of the lumbar plexus in man and macaca. *Folia Morphol.* 28:256–271, 1969.
450* URBANOWICZ, Z. and S. ZALUSKA. External structure of the cutaneous femoral lateral nerve in postfetal life in man. *Folia Morphol.* 36:265–271, 1977.
451* URBANOWICZ, Z. and S. ZALUSKA. External structure of the medial cutaneous nerve of the forearm in the extrauterine life of man. *Folia Morphol.* 39:61–67, 1980.
452 UTUNO, K. Über die Innervation des Zehenrückens bei Japanern. *Folia Anat. Jpn.* 22:215–271, 1943.
453* VALENTI, G. Un cas d'absence unilatéral de l'appareil olfactif. *Arch. Ital. Biol.* 54:457, 1910.
454 VALLOIS, H.V. Le niveau de bifurcation du nerf grand sciatique chez l'homme. *Assoc. Anatomistes Comptes Rendus* 24:519–529, 1929.
455 VAN DER STRICHT, J. A propos d'une anomalie du nerf phrenique. *Assoc. Anatomistes Comptes Rendus* 37:478–484, 1950.
456* VAN EPPS, E.F. Agenesis of the corpus callosum with concomitant malformations, including atresia of the foramens of Luschka and Magendie. *Am. J. Roentgenol.* 70:47–60, 1953.
457* VAN GIESON, I. A study of the artifacts of the nervous system. *N.Y. Med. J.* 56:337–346, 437–447, 1892.
458 VELEANU, C. Vagus nerve passing in front of the left lobe of the thyroid gland. *Anat. Anz.* 141:84–85, 1977.
459 VERSARI, R. Un cas d'absence du rameau superficiel ou antérieur du nerf radial remplacé par le nerf musculo-cutané. *Arch. Ital. Biol.* 30:324, 1898.
460 VITALI, G. Varietà bilaterale del nerf musculo-cutané et des artères du membre supérieur; biceps à trois chefs et arc axillaire chez le même individu. *Arch. Ital Biol.* 57:313–314, 1912.
461 VOGEL, P.H. The innervation of the larynx of man and the dog. *Am. J. Anat.* 90:427–440, 1952.
462 WACHTLER, F. Eine seltene Varietät im Gefäß-Nervenbündel der Axilla. *Anat. Anz.* 143:450–455, 1978.
463* WAGNER, J. Notiz über einen teilweise doppelten Centralkanal im Rückenmark des Menschen. *Arch. Anat. Physiol. Wissen. Med.* 1861, pp. 735–747.
464 WALKER, A.E. Dilatation of the vertebral canal associated with congenital anomalies of the spinal cord. *Am. J. Roentgenol.* 52:571–582, 1944.
465 WALLACE D. Nerve supply of musculus sternalis. *J. Anat. Physiol.* 21:153–154, 1887.
466* WALLACE, W.A. and P.A.M. WESTON. The arrangement of the digital nerves within the human thumb and index fingers. *J. Anat.* 118:381–382, 1974.
467 WALSH, J.F. The anatomy of the brachial plexus. *Am. J. Med. Sci.* 74:388–399, 1877.
468 WATERSTON, D. Complete bilateral interruption of the fissure of Rolando. *J. Anat. Physiol.* 41:143–146, 1907.
469 WEBBER, R.H. The lumbar nerves in a body with six lumbar vertebrae. *Anat. Rec.* 126:123–126, 1956.
470* WEBBER, R.H. Some variations in the lumbar plexus of nerves in man. *Acta Anat.* 44:336–345, 1961.
471 WEBER, F.P. Duplication of the spinal cord. *Br. Med. J.* 1:1106, 1928.

472* WEIGNER, K. Bezeichnungen des Nervus accessorius zu den proximalen Spinalnerven. *Anat. Hefte* 17:549–587, 1894.
473* WEINBERG, R. Ungekreuzte Sehnervenfasern. *Z. Anat. Entwicklungsgesch.* 79:433–446, 1926.
474 WERTHEIMER, L.G. The sensory nerves of the hip joint. *J. Bone Joint Surg. (Am.)* 34:477–487, 1952.
475 WILLIAMS, A.F. The nerve supply of the laryngeal muscles. *J. Laryngol.* 65:343–348, 1951.
476* WILLIAMS, D.D. A study of the human fibular communicating nerve. *Anat. Rec.* 120:533–543, 1954.
477 WILLM, A. Absence totale de l'anse de l'hypoglosse, les nerfs des muscles soushyoidiens naissant en apparence de pneumogastrique. *Ann. Anat. Pathol.* 5:580–582, 1928.
478 WILSON, J.T. The innervation of axillary muscular arches in man, with remarks on its significance. *J. Anat. Physiol.* 22:294–299, 1888.
479 WILSON, J.T. Abnormal distribution of the nerve to quadratus femoris in man, with remarks on its significance. *J. Anat. Physiol.* 23:354–357, 1889.
480 WILSON, J.T. Two cases of variation in the nerve supply of the first lumbrical muscle in the hand. *J. Anat. Physiol.* 24:22–26, 1890.
481 WILSON, J.T. The innervation of the achselbogen muscle. *J. Anat. Physiol.* 47:8–17, 1913.
482* WIMMER, A. Einige, darunter seltene Nervenvariationen des Halsgebietes A. Vier Variationen der Ansa nervi hypoglossi B. Verlauf des Nervus accessorius durch ein "Knopfloch" der Vena jugularis interna sinistra. *Anat. Anz.* 105:391–399, 1958.
483 WINCKLER, G. A propos des relations que existent entre le plexus cervical et le nerf grand hypoglosse. *Assoc. Anatomistes Comptes Rendus* 42:1415–1419, 1955.
484* WISCHNEWSKY, A.S. Die Aufbautypen des Ramus descendens nervi hypoglossi. *Z. Anat. Entwicklungsgeschichte* 92:551–564, 1930.
485 WOLLARD, H.H. and E.A. CARMICHAEL. The testis and referred pain. *Brain* 56:293–303, 1933.
486* WOLLMANN, F. Zum Verlauf des Nervus maxillaris auf Grund einer Varietätbeobachtung. *Anat. Anz.* 93:129–132, 1942.
487* WOOD, V.E. and G.K. FRYKMAN. Unusual branching of the median nerve at the wrist: A case report. *J. Bone Joint Surg. (Am.)* 60:267–268, 1978.
488 WOODBURNE, R.T. The sacral parasympathetic innervation of the colon. *Anat. Rec.* 124:67–76, 1956.
489* WOODBURNE, R.T. The accessory obturator nerve and the innervation of the pectineus muscle. *Anat. Rec.* 136:367–369, 1960.
490 WOOD-JONES, F.W. On the relation of the limb plexuses to the ribs and vertebral column. *J. Anat. Physiol.* 44:377–393, 1910.
491* WRETE, M. The anatomy of the sympathetic trunks in man. *J. Anat.* 93:448–459, 1959.
492 YANO, K. Zur Anatomie und Histologie des Nervus phrenicus und sogenannten Nebenphrenicus, nebst Bemerkungen über ihre Verbindung mit Sympathicus. *Folia Anat. Jpn.* 6:247–290, 1928.
493* YEAGER, G.H. and R.A. COWLEY. Anatomical observations on the lumbar sympathetics with evaluation of sympathectomies in organic peripheral vascular disease. *Ann. Surg.* 127:953–967, 1948.
494* ZALUSKA, S. The obturator nerve in man and macaca. *Folia Morphol.* 30:89–96, 1971.
495* ZALUSKA, S. External structure of the ilioinguinal nerve in postfetal life in man. *Folia Morphol.* 34:419–424, 1975.
496* ZALUSKA, S. and Z. URBANOWICZ. The tibial and common peroneal nerves in man and macaca. *Folia Morphologica* 30:167–175, 1971.
497* ZALUSKA, S., URBANOWICZ, Z., SZCZEPIŃSKA, J. and H. STEFANIAK-WOJTASIK. External structure of the medial cutaneous nerve of the arm in postfetal life in man. *Folia Morphol.* 37:359–365, 1978.
498 ZELLWEGER, H. and G. VON MURALT. Zur Pathologie des Septum pellucidum im Pneumoencephalogramm. *Helv. Paediatr. Acta* 7:229–252, 1952.
499 ZELLWEGER, H. and E.F. VAN EPPS. The cavum veli interpositi and its differentiation from cavum vergae. *Am. J. Roentgenol.* 87:793–805, 1959.
500* ZHU, T., TAN, J., ZHANG, Z., ZHANG, B. and Y. MA. Vagus nerve anatomy at the lower esophagus and stomach. A study of 100 cadavers. *Chin. Med. J.* 93:629–636, 1980.
501 ZUCKERKANDL, E. Über das Gleiten des Ulnarnerven auf die volare Seite des Epicondylus internus während der Flexion im Ellbogengelenke. *Med. Jahrb. (Wien)* 1880, pp. 135–140.

Human Anatomic Variation

Books

502 ANSON, B.J. *Atlas of Human Anatomy*, W.B. Saunders Co., Philadelphia, 1950.

503 ANSON, B.J. and W.G. MADDOCK, *Callander's Surgical Anatomy*, W.B. Saunders Co., Philadelphia, 1952.

504 ANSON, B.J., Ed. *Morris' Human Anatomy*, 12th Ed. The Blakiston Division, McGraw-Hill Book Company, New York, 1966.

505 ANSON, B.J. and C.B. McVAY. *Surgical Anatomy*, 5th Ed. W.B. Saunders Co., Philadelphia, 1971. (See also 6th Ed., 1984.)

506 BARR, M.L. and J.A. KIERNAN, *The Human Nervous System*, 4th Ed. Harper & Row, Philadelphia, 1983.

507 BENNINGHOFF, A. *Lehrbuch der Anatomie des Menschen*. Urban & Schwarzenberg, Munich, 1952.

508 BUNNELL, S. *Surgery of the Hand*, 2nd Ed. J.B. Lippincott Co., Philadelphia, 1948.

509 CRUVEILHIER, J. *Traité d'Anatomie Descriptive*, 3rd Ed. G. Doin & Cie., Paris, 1851.

510 GARDNER, E., GRAY, D.J. and R. O'RAHILLY. *Anatomy*, 3rd Ed. W.B. Saunders Co., Philadelphia, 1969.

511 GEGENBAUR, C. *Lehrbuch der Anatomie des Menschen*, 3rd Ed. V. Wilhelm Engelmann, Leipzig, 1888.

512 HENLE, J. *Handbuch der Systematischen Anatomie des Menschen*, 3 volumes. von Friedrich Vieweg und Sohn, Braunschweig, 1868.

513 HIRASAWA, K. *Arbeiten aus der dritten Abteilung des Anatomischen Institutes der Kaiserlichen Universität Kyoto*. Series A: Untersuchungen über das periphere Nervensystem. Book 2: Plexus brachialis und die Nerven der oberen Extremität. Kyoto, 1931.

514+ HIRSCHFELD, L. *Traité et Iconographic du Système Nerveux et des Organes des Sens de l'Homme, Atlas*, 2nd Ed. Victor Masson et Fils, Paris, 1866.

515+ HOVELACQUE, A. *Anatomie des nerfs craniens et rachidiens et du systeme grand sympathique chez l'homme*. G. Doin & Cie., Paris, 1927.

516 HUBER, G.C., Ed. *Piersol's Human Anatomy*, 9th Ed. L.B. Lippincott Co., Philadelphia, 1930.

517 HYRTL, J. *Lehrbuch der Anatomie des Menschen*, 19th Ed. Wilhelm Braumüller, Wien, 1887.

518 JACKSON, C.M., Ed. *Morris' Human Anatomy*, 9th Ed. P. Blakiston's Son & Co., Inc., Philadelphia, 1933.

519 KOPSCH, F. *Rauber's Lehrbuch der Anatomie des Menschen*. Georg Thieme, Leipzig, 1908.

520 LAGOS, J.C. *Differential Diagnosis in Pediatric Neurology*. Little, Brown and Co., Boston, 1971.

521 LATARJET, A. *Testut's Traité d'Anatomie Humaine*, 9th Ed. G. Doin & Cie., Paris, 1948.

522 MORI, K. *Anomalies of the Central Nervous System. Neuroradiology and Neurosurgery*. Thieme-Stratton, Inc., New York, 1985.

523+ PITRES, A. and L. TESTUT. *Le Nerfs en Schémas*, G. Doin & Cie., Paris, 1925.

524 ROMANES, G.J. *Cunningham's Textbook of Anatomy*, 11th Ed. Oxford University Press, London, 1972.

525 SCHAEFER, E.A., SYMINGTON, J. and T.H. BRYCE, Eds. *Quain's Anatomy*, 11th Ed. Longmans, Green, and Co., London, 1915.

526+ SWAN, J. *A Demonstration of the Nerves of the Human Body*. Longman, Rees, Orme, Brown, Green, and Longman, London, 1834.

527 WHITNALL, S.E. *The Anatomy of the Human Orbit*, 2nd Ed. Oxford University Press, London, 1932.

528 ZIMMER, E.A. *A. Köhler/E.A. Zimmer Borderlands of the Normal and Early Pathologic in Skeletal Roentgenology*, 11th Ed., translated and edited by S. P. Wilk, Grune & Stratton, New York, 1968.

Glossary of Terms

Their origin and definition related to the Nervous System.

The following abbreviations are used: G., Greek; H., Hebrew; and L., Latin.

Abducens (L. *abducere*, to move away). The abducens nerve supplies the lateral rectus muscle which draws the pupil of the eye away from the midline.

Alveolar (L. *alveolaris*, hollow), e.g., anterior superior alveolar nerve.

Ani (L. *anus*, anal orifice, from Sanskrit, to sit). e.g., nerve to external sphincter ani.

Ansa (L. *ansa*, handle) Shaped like a loop or arc, e.g., ansa cervicalis.

Auricular (L. *auricula*, little ear). Pertaining to the external ear.

Auriculotemporal (L. *auricula*, little ear + *tempora*, temples). Pertaining to the nerve supply to the external ear and temporal region.

Axillary (L. *axis alae*, axle of the wing). Axillary nerve (named by Winslow).

Brachial (L. *brachium*, G. *brachion*, arm). Pertaining to the nerves of the arm.

Buccal (L. *bucca*, cheek; H. *bukkah*, empty or hollow). Pertaining to nerves of the cheek.

Cervical (L. *cervix*, neck). The nerves of the neck.

Ciliary (L. *cilium*, eyelash). The ciliary ganglion and ciliary nerves.

Femoral (L. *femor*, thigh). Also suggested is the origin from *feo*, to be fruitful, due to some relationship between the thighs and sex or the bearing of children. Femoral nerve.

Furcal (L. *furca*, fork). Fourth lumbar nerve, so named because of its division, with one part contributing to the lumbar plexus and the other to the lumbosacral plexus. By definition, the third and fifth lumbar nerves may also be furcal nerves.

Lacrimal (L. *lacrimal*, tear). A nerve supplying the lacrimal gland.

Laryngeal (G. *larynx*, upper part of the windpipe). Pertaining to the nerves of the larynx.

Lumbar (L. *lumbus*, loin). Lumbar plexus and lumbar spinal nerves.

Lumbosacral (L. *lumbus*, loin + *sacrum*, sacred). The lumbosacral plexus of nerves.

Mandibular (L. *mandibula*, lower jaw, from *mandere*, to chew). Pertaining to the lower jaw.

Median (L. *medianus*, middle). Used in anatomy for structures in the middle. The nerve of the ventral forearm and ventrolateral aspect of the hand.

Musculocutaneous (L. *musculus*, muscle + *cutis*, skin). A nerve supplying muscle and skin on the ventral aspect of the arm and forearm.

Nasal (L. *nasalis*, from Sanskrit *nasa*, nose). Pertaining to the nose.

Nerve (L. *nervus*, from Sanskrit *nauree*, string).

Obturator (L. *obturare*, to occlude). Obturator nerve.

Occipital (L. *ob*, before or against + *caput*, head). Occipital nerve.

Oculomotor (L. *oculus*, eye + *motor*, mover). Oculomotor nerve.

Orbital (L. *orbis*, circle). Pertaining to the orbit. The use of the word for the socket of the eye first appeared in a translation of the *Canon of Avicenna* by Gerard of Cremona.

Palatine (L. *palatum*, palate). The palatine nerves.

Pectoral (L. *pectus*, breast or chest). An old term for an ornamental plate, cloth, or other decoration worn on the breast. It was also used for medicines that were food for affections of the chest. In this case, referring to the nerves supplying the pectoral region.

Peroneal (G. *perone*, brooch, anything pointed for piercing or pinning, fibula). Pertaining to nerves of the leg.

Phrenic (L. *phrenicus*, diaphragm; G. *phren*, mind). Pertaining to the mind; pertaining to the diaphragm and to the nerves supplying the diaphragm.

Plexus (L. *plexus*, braid). Fallopius described the brachial plexus as a "tangle of nerves" in the arm.

Pudendal (L. *pudere*, to be ashamed). Pudendal nerve is the nerve of the "shameful parts," which modesty demands should be covered, i.e., the genitals.

Radial (L. *radius*, spoke of a wheel). The radial nerve is a motor and sensory nerve that supplies the dorsal aspect of the arm, forearm, and hand.

Scapular (L. *scapula*, shoulder blade). Originally derived from Greek, to dig, because it somewhat resembles a digging instrument such as a spade.

Sciatic (L. *sciaticus*, subject to pain in the loins; from G. *ischiadikos*, hip joint, hip, or loins). Sciatic nerve. The largest peripheral nerve of the body.

Sphenoid (G. *sphen*, wedge + G. *eidos* form). Wedge-shaped.

Sphincter (G. *sphinkter*, that which binds tight), e.g., sphincter ani.

Splanchnic (G. *splanchnon*, viscus). The splanchnic nerves or nerves to viscera.

Stellate (L. *stella*, a star). Stellate ganglion.

Subclavian (L. *sub*, under + *clavis*, key). Pertaining to any structure beneath the clavicle.

Sura (L. *sura*, calf of leg). Sural nerve.

Thoracic (G. *thorax*, chest). Thoracic spinal nerves.

Tibial (L. *tibia*, pipe or flute). Tibial nerve.

Trigeminal (L. *tri*, three + *geminus*, twin). The name was coined by Winslow because the nerve has three divisions.

Trochlear (G. *trochilia*, pulley or block). Resembling a pulley. The superior oblique muscle supplied by the trochlear nerve was originally called the trochlear muscle. This muscle has a pulley through which it operates.

Ulnar (L. *ulna*, elbow). The ulnar nerve is a motor and sensory nerve that supplies the medial side of the forearm and hand.

Vagus (L. *vagare*, to wander). The vagus nerve is a "wanderer" because this parasympathetic nerve supplies structures in the head, neck, thorax, and abdomen. Pertaining to its wide distribution to many organs.

Zygomatic (G. *zygoma*, bolt or bar). Pertaining to the nerves in the region of the zygoma. The zygomatic bone was named by Galen.

Terminology: Synonyms

Terminology for the nervous system from the older English, French, and German literature related to more modern usage.

Old Terminology	Modern Usage
Old Terminology	Modern Usage
N. accessorius cutanei internus.	N. medial brachial cutaneous.
N. accessorius willisii.	N. accessory (CN XI).
N. acusticus.	N. vestibulocochlear.
Nn. alveolo-dentales.	Nn. superior alveolar.
N. alveolar superior anterior major.	N. anterior superior alveolar.
N. alveolaris mandibularis.	N. inferior alveolar.
N. alveolaris maxillae inferioris.	N. inferior alveolar.
N. alveolaris superiores anterior minor.	N. middle superior alveolar.
N. alveolaris superior medialis.	N. middle superior alveolar.
Ammon's horn.	Hippocampus.
N. anastomoticus.	N. deep petrosal or nerve of the pterygoid canal.
N. branche de l'angulaire et du rhomboide.	N. dorsal scapular.
Ansa hypoglossi.	Ansa cervicalis.
Ansa infrahyoidea.	Ansa cervicalis.
Ansa Vieusenii.	Ansa subclavia (fibers connecting the middle cervical sympathetic ganglia forming a loop around the subclavian artery).
N. anterior crural.	N. femoral.
N. anterior tibial.	N. deep peroneal.
Aquaeductus Sylvii.	Aquaeductus cerebri.
Armspindelnerv.	N. radial.
N. articularis.	N. axillary.
Ascending frontal convolution.	Precentral gyrus.
N. auditorius.	N. vestibulocochlear.
Augenhöhlenast.	N. ophthalmic division (V_1), trigeminal nerve.
Augennerv.	N. ophthalmic division (V_1), trigeminal nerve.
N. auricularis s.* auricularis anterior.	N. auriculotemporal.
R. auricularis Arnoldi.	R. auricular, vagus nerve.
N. auricularis cervicalis s. posterior.	N. great auricular.
N. auricularis magnus.	N. great auricular.
N. axillaris.	N. axillary.
N. bigeminus.	N. third sacral.
Rr. bucco-labiales superiores.	Rr. buccal, facial nerve.
Bulbus cinerus.	Bulbus olfactorius, olfactory bulb.
Canal aqueduc.	Cavum Vergae.
N. cardiacus s. cordis supremus.	N. superior cardiac.
N. cardiacus externus and internus.	N. middle cardiac.
N. cardiacus magnus s. profundus.	N. middle cardiac.
N. cardiacus superficialis.	N. superior cardiac.
N. cardiacus minor.	N. inferior cardiac.
N. cardiacus quartus s. imus.	N. inferior cardiac.
N. cardiacus tertius s. parvus.	N. inferior cardiac.
Central lobe of brain.	Insula.
Centralfurche.	Fissure of Rolando.
N. cervicalis superficialis.	N. transverse cervical.
Tela choroidea ventriculi quarti.	Chorioid tela of fourth ventricle.
Nn. cilliares longi interni.	Nn. long ciliary.
Circulus arteriosus cerebri.	Circle of Willis.
N. circumflexus.	N. axillary.
N. circumflexus humeri.	N. axillary.
Cisterna ambiens.	Cisterna venae magnae cerebri.
Cisterna cerebellomedullaris.	Cisterna magna.
Cisterna venae magnae cerebri.	Cisterna superior.
Nn. costales s. intercostales.	Nn. thoracic (spinal).
N. of Cotunnius.	N. nasopalatine.
R. cricothyroideus.	R. external laryngeal, vagus nerve.
N. crotaphitico-buccinatoris.	N. mandibular division (V_3), trigeminal nerve.
N. cruralis.	N. femoral.
N. cruralis anterior.	N. femoral.
N. cruralis internus.	N. obturator.
N. cruralis posterior.	N. obturator.
N. cubitalis.	N. ulnar.
N. cutaneous colli.	N. superficial cervical.
N. cutaneous palmaris antibrachii.	N. palmar cutaneous.
N. cutaneous palmaris longus.	N. palmar cutaneous.
N. cutaneus anterior external.	N. lateral femoral cutaneous.
N. cutaneus brachii externus s. lateralis.	N. musculocutaneus.
N. cutaneus externus.	N. lateral femoral cutaneous.
N. cutaneus internus.	N. medial brachial cutaneous.
N. cutaneus internus femoris major.	N. saphenous.
N. cutaneus internus major.	N. medial antebrachial cutaneous.
N. cutaneus internus minor.	N. medial brachial cutaneous.
N. cutaneus medialis.	N. medial brachial cutaneous.
N. cutaneus medius.	N. medial antebrachial cutaneous.
N. deep radial.	N. posterior interosseous.
Nn. dentales.	Nn. superior alveolar.
N. descendens cervicis.	N. descendens hypoglossi.
N. descendens colli internus.	N. descendens cervicalis.
N. descendens hypoglossi.	N. descendens cervicalis.
N. descendens noni.	N. descending cervical branch of ansa hypoglossi.
N. descendens vagi.	N. descendens cervicalis.
Rr. descendentes.	Nn. superior labial.
N. diaphragmaticus s. respiratorius internus.	N. phrenic.
Nn. dorsales.	Nn. thoracic (spinal).
N. dorsalis scapulae.	N. to rhomboid nn.
Ellenbogennerv.	N. ulnar.
N. external respiratory, of Bell.	N. long thoracic.
N. external saphenous.	N. sural.
R. externus s. femoralis cutaneus N. inguinalis interni.	N. genitofemoral.

* s. = Synonymous

Nervous System

R. externus, N. laryngeus superioris.	R. external laryngeal, vagus nerve.	Ganglion sublinguale s. submaxillare.	Submandibular ganglion.
N. facial, sensory root.	N. intermedius.	Ganglion superius N. vagi.	Superior ganglion, vagus nerve.
Rr. faciales.	Rr. buccal, facial nerve.	Ganglion thoracicum primum.	Inferior cervical ganglion.
Rr. faciales medii.	Rr. buccal, facial nerve.		
Rr. faciales temporales.	Rr. temporal, facial nerve.	Ganglion thyreoideum.	Middle cervical ganglion.
Fasciola cinerea.	Fasciolar gyrus.	Ganglion vertebrale.	Inferior cervical ganglion.
R. femoralis nervi genito-cruralis.	N. genitofemoral.	Ganglion of Wrisberg.	Cardiac ganglion.
N. femorocutaneus.	N. lateral femoral cutaneous.	N. gangliosus.	N. sympathetic.
N. fibularis.	N. peroneal.	Gänsefussgeflecht.	Parotid plexus, facial nerve.
Fifth ventricle.	Cavum septi pellucidi.	Gaumenkeilbeinknoten.	Pterygopalatine ganglion.
N. first cervical, dorsal primary ramus.	N. suboccipital.	Gaumennerven.	Nn. palatine.
		R. genitalis n. genitocruralis.	N. external spermatic, genital branch of genitofemoral.
Fissure callosomarginal.	Cingulate sulcus.		
Fissure chorioid.	Choroidal sulcus.	Geschmacksnerv.	N. lingual.
Fissure of Rolando.	Central sulcus.	N. grand ischiatico.	N. sciatic.
Fissure of Sylvius.	Lateral cerebral fissure.	N. occipitalis major.	N. greater occipital.
Foramen of Monro.	Interventricular foramen.	N. splanchnicus major.	N. greater splanchnic.
Fossa Sylvian.	Fossa cerebri lateralis.	Grosshirnschenkel s. brain stem.	Truncus encephali (NA).
N. fossae jugularis.	R. auricular, vagus nerve.		
Ventriculus quartus.	Fourth ventricle.	N. gustatorius.	N. lingual.
N. frontalis.	N. supraorbital.	Gyrus anterior central.	Precentral gyrus.
N. frontonasalis.	N. supratrochlear.	Gyrus fornicatus.	Limbic lobe.
N. furcal.	Fourth lumbar nerve (usually). The third and fifth lumbar nerves may also be called furcal nerves if they contribute by branching to both the lumbar and sacral plexuses.	Gyrus, Heschl's.	Superior transverse temporal gyrus.
		Hippocampus major.	Hippocampus.
		Hörnerv.	N. vestibulocochlear.
		Hüft-Beckennerv.	N. iliohypogastric.
		Hüftleistennerv.	N. ilioinguinal.
		Hüftlochnerv.	N. obturator.
		Hüftnerv.	N. sciatic.
Ganglion Anderschii s. inferior.	Inferior (petrosal) ganglion, glossopharyngeal nerve.	Hyrtl's loop.	Anastomosis between right and left hypoglossal nerves lying between geniohyoid and genioglossus muscles or within the substance of geniohyoid.
Ganglion of Arnold.	Otic ganglion.		
Ganglion auriculare.	Otic ganglion.		
Ganglion cervicale magnum.	Superior cervical ganglion.		
Ganglion cervicale medium.	Middle cervical ganglion.		
Ganglion ciliary.	Ciliary ganglion.		
Ganglion coccygeum.	Coccygeal ganglion.	N. impar.	Filum terminale.
Ganglion Ehrenritteri.	Superior (petrosal) ganglion, glossopharyngeal nerve.	Indusium griseum.	Supracallosal gyrus.
		N. branche inférieure du sous-scapulaire.	N. lower subscapular.
Ganglion fusiforme s. olivare.	Superior cervical ganglion.		
Ganglion geniculi.	Geniculate ganglion.	N. inferior clunial.	N. inferior gluteal.
Ganglion Geniculum.	Geniculate ganglion.	N. inferior dental.	N. inferior alveolar.
Ganglion impar s. Walteri.	Coccygeal ganglion.	N. inferior maxillary.	N. mandibular.
Ganglion jugulare.	Superior ganglion, vagus nerve.	N. inframaxillaris.	N. mandibular division (V_3), trigeminal nerve.
Ganglion jugulare s. jugulare superior.	Superior (petrosal) ganglion, glossopharyngeal nerve.	Rr. infraorbitales.	Rr. buccal, facial nerve.
		N. inguinalis.	N. genitofemoral.
Ganglion lenticulare.	Ciliary ganglion.	N. inguinalis externus.	N. lateral femoral cutaneous.
Ganglion linguale.	Submandibular ganglion.	N. intercostalis Willisii s. intercostalis magnos.	N. sympathetic.
Ganglion maxillare.	Submandibular ganglion.		
Ganglion Meckellii minoris.	Submandibular ganglion.	N. intercostohumeral.	N. intercostobrachial.
Ganglion Meckel's.	Pterygopalatine ganglion.	N. internal cutaneous.	N. medial antebrachial cutaneous.
Ganglion Mülleri.	Superior (petrosal) ganglion, glossopharyngeal nerve.	R. internus s. scrotalis n. inguinalis interni.	N. external spermatic, genital branch of genitofemoral.
Ganglion nasale.	Pterygopalatine ganglion.	R. internus, N. laryngeus superioris.	R. internal laryngeal (superior laryngeal), vagus nerve.
Ganglion nodose.	Inferior ganglion, vagus nerve.		
Ganglion ophthalmicum.	Ciliary ganglion.	R. interosseus s. volaris.	N. anterior interosseous.
Ganglion oticum.	Otic ganglion.	Intumescentia ganglioformis.	Geniculate ganglion.
Ganglion rhinicum.	Pterygopalatine ganglion.	N. ischiadicus.	N. sciatic.
Ganglion of Scarpa.	Vestibular ganglion.	N. ischiadicus magnus.	N. sciatic.
Ganglion semilunare s. abdominalie s. solare.	Celiac ganglion.	N. ischiadic minor.	N. inferior gluteal.
		Island of Reil.	Insula.
		Isthmus of gyrus fornicatus.	Isthmus of hippocampal fissure.
Ganglion sphenopalatinum s. Meckelii.	Pterygopalatine ganglion.		
Ganglion sphlanchnicum.	Celiac ganglion.		

Human Anatomic Variation

N. of Jacobson.	N. tympanic.	N. naso-ocularis s. oculo-nasalis s. nasalis.	N. nasociliary.
Knieknoten.	Geniculate ganglion.	N. nasopalatinus Scarpae.	N. nasopalatine.
		Nerf dorsal de l'omoplate.	N. dorsal scapular.
Nn. labiales superiores.	Nn. superior labial.	Nerf du grand dentelè.	N. long thoracic.
N. lacrymalis.	N. lacrimal.	Nerf du grand dorsale.	N. thoracodorsal.
N. lacrymopalpebralis.	N. lacrimal.	Nerf du grand pectoral.	N. lateral pectoral.
N. laryngeal inferioris.	N. inferior laryngeal (continuation of the recurrent laryngeal), vagus nerve.	Nerf grand thoracique.	N. lateral pectoral.
		Nerf marginal de l'omoplate.	N. thoracodorsal.
		Nerf medio-digital.	N. median.
R. laryngeus superior internus.	R. internal laryngeal, vagus nerve.	Nerf perforans du coracobrachial.	N. musculocutaneous.
N. lateral anterior thoracic.	N. lateral pectoral.	Nerf du petit pectoral.	N. medial pectoral.
N. lateral cutaneous nerve of forearm.	N. lateral antebrachial cutaneous.	Nerf radio-cutane.	N. musculocutaneous.
		Nerf radio-digital.	N. radial.
N. lateraler Hautnerv des Armes.	N. musculocutaneous.	Nerf sous-scapulaire inférieur.	N. lower subscapular.
Nn. lateralis narium.	Nn. external nasal.	Nerf du (muscle) sous-clavier.	N. to subclavius.
N. latissimus dorsi.	N. thoracodorsal.	Nerf subscapulaire.	N. suprascapular.
Lenden-Leistennerv.	N. genitofemoral.	Nerf thoracique antérieur.	N. lateral pectoral.
N. lesser internal cutaneous.	N. medial brachial cutaneous.	Nerf thoracique postérieur ou inférieur.	N. medial pectoral.
N. occipitalis minor.	N. lesser occipital.		
N. splanchnicus minor.	N. lesser splanchnic.	Nerf thoracique supérieur.	N. to subclavius.
Linsen- oder Blendungsnervenknoten.	Ciliary ganglion.	N. noni (Willis).	N. hypoglossal (CN XII, of Soemmerring). Nervus nonus means ninth nerve (in the enumeration of Willis).
N. long subscapular.	N. thoracodorsal.		
N. lowest splanchnic.	N. least splanchnic.		
N. lumbo-inguinalis.	N. genitofemoral.		
		Oberflächlicher Schläfennerv.	N. auriculotemporal.
R. magnus n. mediani.	N. musculocutaneous.	Oberkiefernerv.	N. maxillary division (V$_2$), trigeminal nerve.
N. major splanchnic.	N. greater splanchnic.		
Rr. malares s. faciales superiores.	Rr. zygomatic, facial nerve.	Oberrollnerv.	N. supratrochlear.
		N. occipitalis internus.	N. greater occipital.
N. mandibularis.	N. inferior alveolar.	N. occipitalis magnus s. maximus.	N. greater occipital.
N. marginalis scapulae.	N. thoracodorsal.		
N. maxillaris inferior.	N. inferior alveolar.	N. occipitalis major.	N. greater occipital.
N. maxillaris inferioris.	N. mandibular division (V$_3$), trigeminal nerve.	N. occipitalis minor.	N. lesser occipital.
		N. occipitalis parvus s. externus s. anterior.	N. lesser occipital.
N. maxillaris superioris.	N. maxillary division (V$_2$), trigeminal nerve.		
		N. octavus.	N. vestibulocochlear.
N. medial anterior thoracic.	N. medial pectoral.	Rr. orbitales.	Rr. zygomatic, facial nerve.
N. medial cutaneous nerve of the forearm.	N. medial antebrachial cutaneous.	N. orbitalis.	N. ophthalmic division (V$_1$), trigeminal nerve.
N. meningeus posterior.	N. meningeal, vagus nerve.		
N. middle subscapular.	N. thoracodorsal.	Granulationes arachnoideales.	Pacchionian bodies.
N. minor splanchnic.	N. lesser splanchnic.		
Mittelarmnerv.	N. median.	Nn. palatini.	Nn. palatine.
R. muscularis regionis infrahyoideae.	R. descendens cervicalis.	Nn. palatini descendentes.	Nn. palatine.
		N. palatinus lateralis.	N. medial or external palatine.
N. musculi mallei interne.	N. tensor tympani.		
N. musculo-cutaneous.	N. superficial peroneal.	N. palatinus medius.	N. posterior or small palatine.
N. musculo-cutaneus peroneus externus s. ischiadicus popliteus externus.	N. lateral superficial peroneal.	N. palatinus minimus s. externus.	N. medial or external palatine.
		N. palatinus minor s. internal.	N. posterior or small palatine.
N. musculospiral.	N. radial.	N. palatinus posteriores.	N. posterior or small palatine.
Muskelhautnerv.	N. musculocutaneous.	N. palatinus posterior minor.	N. posterior or small palatine.
		Rr. palpebrales.	Rr. zygomatic, facial nerve.
N. nasal s. oculonasal.	N. nasociliary.	Nn. palpebrales inferiores.	Nn. inferior palpebral.
Rr. nasales s. infraorbitales.	Rr. buccal, facial nerve.	N. pars intermedia, of facial.	N. intermedius, facial nerve.
Nn. nasales laterales s. superficiales nasi.	Nn. external nasal.	N. patheticus.	N. trochlear.
		N. pectoralis posterior.	N. long thoracic.
Nn. nasales superiores.	Nn. pterygopalatine.	N. perforans Gasseri s. coracobrachialis.	N. musculocutaneous.
Nn. nasales subcutanei.	Nn. external nasal.		
N. nasalis.	N. ethmoidal.	N. peroneus communis.	N. fibularis communis (common fibular).
N. nasalis anterior.	N. ethmoidal.		
N. nasalis exterior.	N. infratrochlear.	N. peroneus profundus.	N. fibularis profundus (deep fibular).
N. nasalis inferior.	N. ethmoidal.		
Nasenaugennerv.	N. nasociliary.	N. peroneus superficialis.	N. fibularis superficialis (superficial fibular).
Nasenknoten.	Pterygopalatine ganglion.		
N. nasodentalis.	N. anterior superior alveolar.	Petit nerf thoracique.	N. medial pectoral.

162

Nervous System

Plexus axillaris.	Brachial plexus.	Stria Lancisii.	Medial longitudinal stria.
Plexus cruralis.	Lumbar plexus.	N. subclavian.	N. to subclavius.
Plexus epigastric.	Celiac plexus.	N. subcutaneus colli.	N. transverse colli.
Plexus parotideus.	Parotid plexus, facial nerve.	N. subcutaneus colli inferior.	N. transverse cervical.
Plexus pes anserinus.	Parotid plexus, facial nerve.	N. subcutaneus colli superior.	N. transverse colli.
Plexus pes anserinus major.	Parotid plexus, facial nerve.	N. subcutaneus malae.	N. zygomatic.
Plexus pudendus nervosus.	N. pudendal.	N. subscapularis.	N. upper subscapular.
Plexus sacralis posterior.	Sacral plexus.	N. subscapularis inferior.	N. lower subscapular.
Plexus solar.	Celiac plexus.	N. subscapularis inferior.	N. thoracodorsal.
N. pneumogastric.	N. vagus.	N. subscapularis longus.	N. thoracodorsal.
Pons of Varolius.	Pons.	N. subscapularis medius.	N. lower subscapular.
N. popliteus s. popliteus internus.	N. tibial.	Substantia perforata intercruralis.	Posterior perforated substance.
N. popliteus externus.	N. peroneal.	Sulcus, limiting of Reil.	Circular sulcus.
N. portio minor, trigeminal.	N. motor root, trigeminal.	Sulcus of Monro.	Hypothalamic sulcus.
R. profundus n. mediani.	N. anterior interosseous.	N. superficial radial.	N. radial.
R. profundus nervi pudendi.	N. dorsal of penis (clitoris).	N. superficialis colli s. profundus subcutaneus colli medius.	N. transverse cervical.
N. pronatoris quadrati.	N. anterior interosseous.	N. superior clunial.	N. superior gluteal.
N. pterygoideus.	N. deep petrosal or nerve of the pterygoid canal.	N. superior maxillary.	N. maxillary.
Nn. pterygopalatini.	Nn. palatine.	N. supramaxillaris.	N. maxillary division (V_2), trigeminal nerve.
N. pterygopalatinus s. nasopalatinus.	N. pterygopalatine.	N. sympathicus magnus.	Nn. sympathetic.
N. pudendus externus.	N. external spermatic, genital branch of genital femoral.	N. temporalis cutaneus.	N. auriculotemporal.
N. pudendus superior s. externus.	N. dorsal of penis (clitoris).	N. temporalis superficialis.	N. auriculotemporal.
N. pudic.	N. pudendal.	N. temporomalar.	N. zygomatic.
		Rr. temporo-frontales.	Rr. temporal, facial nerve.
Quadrigeminal plate.	Tectum.	N. third occipital.	N. occipitalis tertius.
Quadrale lobule.	Precuneus.	Nn. thoracales anterior.	Nn. lateral and medial pectoral.
N. quinti recurrens.	N. deep petrosal or nerve of the pterygoid canal.	N. thoracalis anterior externus.	N. lateral pectoral.
N. radial s. musculospiral trunk (divides into two branches): (1) r. superficialis nervi radialis s. radial nerve; (2) r. profundus nervi radialis s. posterior interosseous nerve.		N. thoracalis anterior externus (primus).	N. lateral pectoral.
		N. thoracalis anterior internus s. secundus.	N. medial pectoral.
		Nn. thoracici anteriores, inferiores s. ventrales.	Nn. lateral and medial pectoral.
N. recurrens.	N. meningeal, vagus nerve.	N. thoracicodorsalis.	N. dorsal scapular.
N. recurrens s. adscendens.	N. recurrent laryngeal, vagus nerve.	N. thoracicus s. pectoralis posterior.	N. dorsal scapular.
N. recurrens internus.	N. deep petrosal or nerve of the pterygoid canal.	N. thoracicus lateralis s. medius s. longus.	N. long thoracic.
Riechbeinnerv.	N. ethmoidal.	N. thoracicus longus.	N. long thoracic.
N. renalis posterior (Walter).	N. least splanchnic.	N. thoracicus longus.	N. thoracodorsal.
N. respiratorius externus (of Bell).	N. long thoracic.	N. thoracicus posterior.	N. long thoracic.
Rosennerv.	N. saphenous.	N. branche thoracique postérieure.	N. medial pectoral.
		Thränendrüsenast.	N. lacrimal.
N. saphenus internus s. major.	N. saphenous.	N. tibialis posterior.	N. tibial.
N. scapularis.	N. suprascapular.	N. trifacialis.	N. trigeminal.
N. scapularis medius.	N. suprascapular.	N. trisplanchnicus.	N. sympathetic (splanchnic nerves).
N. scapularis medius Lussana.	N. suprascapular.	N. tympanico-lingualis.	N. chorda tympani.
N. scapularis superior.	N. suprascapular.		
N. of Scarpa.	N. nasopalatine.	Unterkiefernerv.	N. inferior alveolar.
Septum posticum of Schwalbe.	Subarachnoid septum.	Unterkieferknoten.	Submandibular ganglion.
N. short saphenous.		Unterrollnerv.	N. infratrochlear.
Sixth ventricle.	N. sural.		
N. small sciatic.	Cavum Vergae.	Vallecula Sylvii.	Vallecula cerebri lateralis.
	N. posterior femoral cutaneous, of the thigh.	Velum interpositum.	Choroid tela of third ventricle.
Speichennerv.	N. radial.	Ventricle of Strambio.	Cavum Vergae.
N. sphenopalatinus.	N. pterygopalatine.	Ventriculus fornicis.	Cavum Vergae.
N. sphenopalatinus internus.	N. nasopalatine.	Ventriculus triangularis.	Cavum Vergae.
N. spiral.	N. radial.	Verga's ventricle.	Cavum Vergae.
N. splanchnicus imus.	N. least splanchnic.	Verstopfungsnerv.	N. obturator.
Stirnnerv.	N. supraorbital.	N. vidian.	N. of pterygoid canal.

N. vidianus.

N. of Wrisberg.
N. Wrisberg'scher.

Zahnhöhlennerven.

N. deep petrosal or nerve of
 the pterygoid canal.

N. intermedius, facial nerve.
N. medial brachial cutaneous.

Nn. superior alveolar.

Zahnhöhlennerv.
Zahnnerven.
Zungenknoten.
Zungennerv.
Zurücklaufender Flügelnerv.

N. inferior alveolar.
Nn. superior alveolar.
Submandibular ganglion.
N. lingual.
N. deep petrosal or nerve of
 the pterygoid canal.

ORGANS

Alphabetical Listing

(Primary references, in multiple page listings, are in *boldface* type.)

Bulbourethral Glands **173**
Digestive System and Spleen **166**
Ductus Deferens 122
Duodenum **167**
Endocrine System **175**
Esophagus **166**
Eyelids **176**
Female Genital System **173**
Gallbladder **168**
Hermaphroditism **175**
Ileum **167**
    Meckel's Diverticulum 167
Iris **176**
Jejunum **167**
Kidneys **171**
    Ectopic 171
    Horseshoe 172
    Pelvic 172
Lacrimal Gland **176**
Large Intestine **167**
    Cecum 167
    Colon 167
    Imperforate Anus 167
Larynx 169
Lip **166**
Liver **167**
    Accessory 168
    Appendix Fibrosa Hepatis 168

Hepar Succenturiatum 168
Lobus Posterior 167
Lobus Venae Cavae 168
Pons Hepatis 168
Lungs **169**
    Blood Supply **169**
    Lobes and Fissures **169**
Male Genital System **172**
Mammary Gland **174**
    Polymastia 174
    Polythelia 174
Mesenteries and Omenta **167**
Nasolacrimal Duct **176**
Nose 169
Ovaries **173**
Pancreas **168**
    Accessory 168
    Annular 168
Parathyroid Glands **175**
Parotid Gland **166**
Penis **172**
Pharynx **166**
Pineal Gland **176**
Prostate **173**
Respiratory System **169**
Seminal Vesicles **173**
Situs Inversus 166
Skin **176**

Small Intestine **166**
Spleen **168**
    Accessory 169
Stomach **166**
Sublingual Gland **166**
Submandibular Gland **166**
Suprarenal (Adrenal) Gland **175**
    Accessory 176
Testes **172**
Thymus **176**
Thyroid Gland **175**
    Accessory 175
    Lobus Pyramidalis 175
Tonsils (Palatine) **166**
Trachea **169**
Ureters **171**
Urethra **171**
Urinary Bladder **171**
Urinary System **171**
Uterine (Fallopian) Tubes **173**
Uterus **173**
    Uterus Bicornis 174
    Uterus Eptus 174
Vagina **174**
Vas Deferens **172**
Vermiform Process (Appendix) **167**

Digestive System and Spleen

Lip

A case of double lip has been reported.

Reference 527.

Parotid Gland

A case has been reported of displacement of the parotid gland on one side, with the entire gland being located on the masseter muscle. The gland may extend to the mandibular ramus to cover the buccinator muscle. Its absence has also been recorded. The parotid duct may be doubled.

References 89, 124, 276, 286, 318, 535, 574, 577, 601, 856–859, 861–864, 866, 868–870, 873–875.

Submandibular and Sublingual Glands

The submandibular gland has been found intimately associated or fused with the sublingual gland; its bilateral absence has also been reported. The sublingual gland has a variable relation to the mandible.

References 336, 577, 599, 630, 732, 784, 866, 868, 869, 872–875.

Palatine Tonsils

Palatine tonsils exist as two types: pedunculated or buried. The pedunculated type is nearly spherical and attached by a small base to a shallow tonsillar sinus. The buried type is attached by a broad base to a deep sinus. If the buried type is abnormally enlarged the condition may be obscured by tonsillar folds.

References 856, 858, 866, 872, 874, 875.

Pharynx

The principal variations of the pharynx are due to the persistence of the visceral clefts or to irregularities in their position. Various cases of this type have been described as pharyngeal diverticula or cervical fistulae. A pituitary nodule has been found in the pharyngeal wall.

References 162, 426, 441, 442, 541, 858, 860, 862–864, 866, 868–870, 872–875.

Esophagus

The upper part of the esophagus may communicate with the trachea by a congenital fistula, the result of arrested development in which this part of the foregut becomes divided into an alimentary and a respiratory tube. This condition may be associated with an atresia of the upper end of the esophagus. Diverticula occasionally occur, especially on the posterior wall, and the esophagus may become considerably dilated just above the point where it pierces the diaphragm. Patches of the mucous membrane of the upper part of the esophagus are frequently found whose structure resembles the cardiac part of the stomach. The right subclavian artery (*arteria lusoria*) may pass from left to right behind or in front of the esophagus. Rarely, skeletal muscle may be found in the distal segment of the esophagus. An anomalous lobe of the lung occasionally arises from the esophagus.

References 311, 426, 427, 483, 505, 514, 828, 858, 859, 860, 862, 866, 868–870, 872, 874, 875.

Stomach

In complete transposition of the viscera, the stomach will lie on the right side. Congenital hourglass constriction of the stomach is rare (46 of 50 cases reported, however, were in women). Congenital stenosis of the pylorus occasionally occurs. There may also be a doubled or tripled pyloric antrum, or the antrum may be replaced by a tube with thickened walls. The fundus of the stomach sometimes has a diverticulum.

The stomach may be partially retroperitoneal (along with the spleen and left lobe of the liver). This is usually associated with defective mesenteries and omenta. Heterotopic islands of pancreatic tissue have been found in the stomach wall.

Situs inversus has an incidence of 1:4100 in Japan, 1:5000 in Israel, and 1:10,000–15,000 in the United States and Northern Europe. Situs inversus was recognized by Fabricus in 1600 when he described a case of reversed liver and spleen. Petrius Servius in 1615 described total transposition of viscera. Another similar observation was made by Riolan in 1668 in a case of an executed criminal. Kuchenmeister in 1824 recognized situs inversus in a living person for the first time.

In a study of 18,480 operated patients, 25 children and four adults had variations of position of abdominal viscera. In 15 cases there was one complete absence of intestinal rotation, three cases of incomplete rotation, nine cases of abnormal rotation, and two cases of situs inversus. Situs inversus has been associated developmentally with nonmotile cilia (Afzelius) and is transmitted as an autosomal recessive trait (Kartagener's syndrome).

References 1, 4, 6–8, 29, 35, 37, 43, 57, 66, 108, 123, 138, 147, 151, 160, 161, 165, 168, 173, 210, 212, 227, 229, 243, 248, 250, 256, 270, 273, 278, 283, 301, 311, 316, 347, 354, 363, 425, 437, 440, 446, 473, 484, 488, 493, 530, 543, 554, 571, 591, 597, 600, 606, 607, 621, 636, 638, 641, 643, 647, 652, 656, 659, 662, 663, 665, 668, 676, 692, 697, 703, 708, 714, 727, 765, 774, 777, 789, 793, 840, 854, 858, 860, 864, 866, 868–870, 872, 874, 875.

Small Intestine

The small intestine, which measures 7 m immediately post mortem, may measure 2–4 m (or less) in the living. Postmortem lengths vary between 5 and 8 m, and length can be correlated with stature.

References 29, 141, 168, 311, 317, 363, 434, 440, 565, 700, 786, 858.

Duodenum. The position of the duodenum does not vary greatly. The third part of the duodenum generally crosses in front of the third lumbar vertebra, and not infrequently it is as low as the fourth; it rarely comes into relation with the fifth lumbar (three of 337 subjects). In four cases the duodenum ended on the right side of the vertebral column.

Duodenal diverticula are not uncommon. They are usually single, consist of all the coats of the intestine, and project from the concave side of the duodenum and, consequently, toward the pancreas.

The duodenum is one part of the alimentary canal that, during embryonic life, is liable to physiologic occlusion by proliferation of its epithelial lining. This may persist as an atresia or become the focus of a stenosis. Megaloduodenum has been reported. Anomalous pancreatic nodules have been found in the wall of the duodenum.

References 23, 48, 87, 114, 152, 168, 221, 252, 267, 309, 347, 357, 391, 405, 428, 576, 582, 583, 675, 684, 773, 813, 858–860, 863, 864, 866, 868–870, 872, 874, 875.

Jejunum. Heteropic islands of pancreatic tissue have been found in the wall of the jejunum.

References 53, 168, 221, 309, 357.

Ileum. In about one in 100 subjects (367 found in 40,439 patients), a tubular or conical diverticulum (Meckel's diverticulum) arises from the ileum. Its average position is about 52 cm (varying from 15 to 100 cm) proximal to the ileocolic junction, but it has been observed anywhere from 2.7 to 300 cm proximal to the junction. The diverticulum usually arises from the ileum on the side opposite the attachment of the mesentery. As a rule, it varies between 1 and 15 cm (average, 4.1 cm) in length, and is about the same caliber as the bowel, from which it arises. This diverticulum results from the persistence of a part of the vitelline duct of early fetal life. It is very rarely attached to the umbilicus. The diverticulum may possess aberrant islands of gastric mucosa, pancreatic, and duodenal glands. Aberrant tissue has been found in 29.5% of cases, with gastric mucosa accounting for 24.6%. The diverticulum was without aberrant tissue in 47.5% of cases.

The ileocecal valve is variable.

References 53, 119, 126, 127, 168, 184, 221, 238, 309, 395, 429, 555, 568, 769, 847, 852, 858, 860, 866, 868, 870, 872, 874, 875.

Large Intestine

The principal variations are due to irregular or defective development. In cases of complete transposition of the viscera, the cecum and appendix lie in the left iliac fossa, and the sigmoid colon in the right. The ascending colon may not be completely retroperitoneal, resulting in a condition in which the transverse colon falls obliquely across the abdomen, covering the cecum and appendix, and extends from the right pelvic inlet to the left splenic flexure. The proximal portion of the large intestine, as far as the left part of the transverse colon, may remain attached, along with the jejunoileum, to the posterior abdominal wall by a common mesentery. The rotation of the intestinal loop, associated with the descent of the cecum and appendix and the formation of a descending colon, may be permanently arrested in any part of its course. In some cases, the descending colon remains near the median plane and is connected to the front of the vertebral column by a mesentery. The descending colon may cross the posterior abdominal wall, with the sigmoid colon descending on the right side. The sigmoid may lie in the abdomen just inferior to the splenic flexure of the transverse colon, or in the right iliac fossa. Occasionally, the rectum opens into the bladder or urethra, or it may terminate blindly, giving rise to the condition known as imperforate anus. It may also be doubled.

The large intestine is about 2 m in length. Partial duplication of the large intestine has been reported.

References 22, 29, 76, 167, 168, 208, 221, 223, 225, 275, 317, 350, 434, 440, 459, 498, 506, 510, 532, 558, 569, 575, 581, 616, 637, 638, 648, 670, 700, 713, 723, 748, 778, 786, 843, 848, 854, 858, 860, 862–864, 866, 867–870, 872, 875.

Vermiform Process (Appendix). The appendix may be found almost anywhere in the abdominal cavity depending upon its development and rotation of the gut. In *situs inversus* it will be found abnormally in the left lower quadrant of the abdomen. Depending on its size, it may lie in the true pelvis. It varies in size from less than 5 mm (¼ inch) to 230 mm (9¼ inch) in length. The appendix may be absent or duplicated. Three appendices in one individual have been reported. Appendicular diverticula may occur.

References 18, 29, 74, 142, 148, 163, 168, 217, 221, 251, 285, 290, 293, 367, 381, 409, 509, 516, 608, 671, 682, 696, 719, 730, 798, 807, 811, 817, 843, 858, 866, 872, 875.

Mesenteries and Omenta. A common mesentery for the jejunum, ileum, ascending colon, and part or all of the transverse colon has been reported. The greater omentum may be defective with consequent continuity between the abdominal cavity and omental bursa.

References 29, 36, 315, 316, 333, 389, 457, 775, 802, 818.

Liver

The liver is not subject to great or frequent deviation from its ordinary form and relation. It has been found without any division into lobes. On the other hand, a case has been recorded in which the adult liver was divided into 12 lobes, and similar cases of excess lobes have been observed. A detached portion, forming a sort of accessory liver, is occasionally found appended to the left extremity of the gland by a fold of peritoneum containing blood vessels. Two additional lobes may occur, the *lobus posterior,* projecting through the epiploic foramen to lie behind the stomach, and the *lobus venae cavae*, projecting downward along the course of the inferior vena cava.

At the left extremity of the left lobe of the adult liver a fibrous band (sometimes containing liver substance) of variable length may attach to the diaphragm or posterior abdominal wall. It represents the atrophied remnant of the larger organ in fetal life and is termed *appendix fibrosa hepatis*.

The fossa venae umbilicalis is often bridged by a piece of liver, the *pons hepatis*. It unites the quadrate and left lobes, and may vary considerably in size. The upper surface sometimes shows longitudinal furrows which are occupied, when the organ is in situ, by folds of the diaphragm.

There is a report of an accessory liver adherent to the pancreas. Its bile duct drained into the pancreatic duct system. Other isolated masses of hepatic tissue (hepar succenturiatum) have been found on the wall of the gallbladder, ligamentum teres, spleen, and greater omentum.

References   15, 29, 53, 83, 134, 174, 199, 241, 325, 370, 422, 461, 471, 508, 609, 751, 766–768, 804, 805, 858, 860, 866, 868–870, 872, 875.

Gallbladder. The gallbladder may be absent, in which case the hepatic duct is dilated within the liver or in some part of its course. In three studies, the congenital absence of the gallbladder was reported as follows. In 18,350 autopsies the incidence of absence was 0.065%; in 7,919 autopsies, 0.075%; and in 3,661 autopsies, 0.03%. The gallbaldder and cystic duct may be absent. Sometimes the gallbladder is irregular in form or constricted across its middle; more rarely, it is partially divided in a longitudinal direction. Cases have been recorded in which there were two distinct gallbladders, each having a cystic duct that joined the hepatic duct. The cystic duct may itself be doubled. Doubled gallbladders have been found in 5 of 19,000 patients. The gallbladder has been found on the left side (to the left of ligamentum teres) in subjects in whom there was no general transposition of the thoracic and abdominal viscera.

The right and left divisions of the hepatic duct sometimes continue separately for some distance within the lesser omentum. The common bile duct may open separately from the pancreatic duct into the duodenum. There may be one or two distinct duodenal papillae. Anomalous pancreatic tissue is occasionally found in the wall of the gallblader.

References   21, 27, 44, 60, 80, 81, 91, 92, 94, 175, 194, 219, 239, 240, 244, 257, 266, 268, 282, 298, 299, 303, 310, 324, 362, 364, 368, 382, 386, 390, 411, 430, 439, 444, 461, 465, 494, 503, 550, 566, 593, 626, 642, 660, 679, 687, 699, 718, 720, 735, 736, 739, 747, 803, 840, 856–859, 866, 868–870, 872, 874, 875.

Pancreas

One of the most interesting but rare variations of the pancreas is the condition known as the *annular pancreas*. The duodenum (usually its descending part) is surrounded by a ring of pancreatic tissue continuous with the head of the gland. There may be a constriction of the duodenum at the location of the ring, with dilations above and below. It is generally considered that the ventral rudiment of the pancreas consists of a right and left part. The left part usually atrophies but may persist, or the right half may exhibit excessive growth and surround the pancreas. The duct contained in the ring joins the ductus pancreaticus near the duodenum. The tail of the pancreas is sometimes bifid, and the part of the head lying behind the mesenteric vessels may be separate from the remainder of the gland, forming what is called the lesser (accessory) pancreas.

Various examples of an accessory pancreas have been described. They may occur in the mesentery of the small intestine, the wall of the duodenum, the upper part of the jejunum, or more rarely, in the wall of the stomach, ileum, gallbladder, or spleen. They are generally situated either in the submucous or the subserous coat, but may extend into the muscular layer. An accessory pancreas was found near the origin of the inferior mesenteric artery in an 11-year-old boy. The size of these masses varies in diameter, from a few millimeters to 5 cm. The masses may contain pancreatic islet cells. In a few cases, some pancreatic tissue has been found at the blind end of intestinal (Meckel's) diverticula.

Studies of variations in the pancreatic ducts indicate that in approximately 40% of adults the accessory duct is not functionally joined to the chief or primary pancreatic duct. In 15% of cases there is a dual drainage from both ducts, in 7% there is a dual drainage and a functional connection between the two ducts, and in 2% the accessory duct does not drain into the duodenum but into the primary duct.

References   24, 29, 40, 49, 53, 56, 57, 77, 106, 132, 152, 172, 183, 200, 207, 252, 273, 278, 313, 346, 368, 390, 392, 411, 432, 452, 469, 479, 550, 631, 655, 660, 664, 665, 679, 688, 718, 735, 737, 739, 750, 755, 756, 769, 820, 852, 856–859.

Spleen

Notches on the anterior, posterior, and inferior borders, and fissures on the diaphragmatic surface of the spleen have been reported. The anterior border was free of notches in eight of 113 specimens examined; most commonly there were two notches, but in one case there were seven. Notches on the posterior border occurred in 32% of cases and on the inferior border in 8%. The fissures on the diaphragmatic surface do not appear to represent the persistence of a primate lobulation.

When the spleen is absent, the splenic artery terminates in the pancreas.

Small, detached, roundish nodules are occasionally found in the neighborhood of the spleen and are similar to it in substance. These are commonly called accessory or supplementary spleens (*splenculi, lienculi*). One or two most commonly occur, but up to 23 have been found. The small masses vary from the size of a pea (0.5 cm) to that of a walnut (3.75 cm). They may be found along the borders of the spleen, especially its anterior aspect, and in the hilum. They may also occur in the scrotum, greater omentum, transverse mesocolon, and gastrolineal ligament, behind the left lobe of the liver, near the pancreas, and embedded in the connective tissue surrounding the splenic vessels. They occur most frequently, however, in the tail of the pancreas (about one of every six cases of accessory spleen). The incidence of accessory spleen is 10% or more.

Heterotopic islands of pancreatic tissue have been found in the spleen.

Human adult spleens have been reported to weigh between 80 and 300 grams.

References   38, 53, 70, 85, 182, 189, 345, 347, 348, 349, 388, 448, 531, 572, 621, 668, 705, 707, 726, 740, 856–859, 866, 868–870, 872, 874, 875.

Respiratory System

Nose

The nose may be divided in the midline, duplicated, or reduced to a half nose.

Reference   758.

Larynx

A foramen is commonly found near the upper lateral angle, a little below the superior tubercle of the thyroid cartilage, which transmits the superior laryngeal artery and rarely some fibers of the external branch of the superior laryngeal nerve. The foramen represents a cleft between the fourth and fifth branchial bars. It is common for one of the superior horns to be shorter than the other, and also common for one to be absent, more frequently on the left side. The thyroid cartilage may articulate with the hyoid bone. Accessory thyrohyoid ligaments and additional or unusual cricothyroid articulations have been reported.

Saccular diverticula of the laryngeal ventricle are reported often. Additional bursae may be associated with muscles originating or inserting on the larynx. The epiglottic cartilage may be rudimentary, with duplication of the mucosa substituting for cartilage.

References   75, 236, 320–323, 328, 330, 334, 483, 858, 859, 862–864, 866,–870, 872, 874, 875.

Lungs and Trachea

Variations of the Lobes and Fissures of the Lungs.

An accessory inferior lobe is frequently found on the undersurface, extending up onto the medial surface in front of the pulmonary ligament. This lobe may be indicated by shallow fissures or be clearly defined. It may present a tongue-like projection inward or may comprise the entire medial portion of the base, but usually consists of one-fifth to one-third of the base. The lobe may occur on either side or on both; it is larger and more frequently well-defined on the right, but more often present, or at least indicated, on the left. It was found in 47% of 210 lungs.

Occasional absence of a fissure marking off the middle lobe has been reported. An irregular fissure may subdivide the left lung into three lobes, and both lungs may be further subdivided, especially the right. In one study, a single pair of lungs was divided into 11 clearly defined lobes with fully developed pleural fissures. However, one more frequently finds the commonly described five lobes not separated by fissures.

The most detailed study of the human lung has been performed by Boyden and associates. The brief summary that follows is from some of that work. The lung may show various degrees of developmental arrest: (a) agenesis, absence of one or both lungs; (b) aplasia, formation of rudimentary bronchi; and (c) hypoplasia, incomplete development of the lung. Thomas and Boyden reported three cases of agenesis of the right lung with numerous variations of the left lung. In two cases, the left lung had three lobes and in the third it had no lobulation. In addition, there are cases of variations in bronchopulmonary segments, left eparterial bronchus, and vascular anomalies including coarctation of the aorta, persistent ductus arteriosus, and left posttracheal or bronchial pulmonary artery.

Hypoplasia of only a single lobe or of the entire lung has been described. In one case, the right lung appeared to have two lobes, but within the oblique fissure a small bronchus was found to supply a small piece of lung tissue; the bronchus originated at the expected site of the middle bronchus.

Rudimentary lung tissue may be found arising in ectopic positions such as (most commonly) the trachea. Examples of these so-called supernumerary lungs are described by Boyden; they arise from tracheal diverticula, and are not to be confused with displaced lobar bronchi. In one study, five specimens of this type were found in 6000 autopsies (0.08%). Another type of supernumerary lung is the Rokitansky lobes, isolated structures located in the lower thorax. They occur most frequently on the left side, usually above the diaphragm. In one study, in 27 of 41 cases the lobes were located above the diaphragm and in five below the diaphragm. These isolated masses of lung tissue are usually separated from both the tracheobronchial tree and the pulmonary vascular system. It has been reported that these isolated masses are subject to the same diseases that affect lungs. Anomalous lung tissue has been reported as an outgrowth of the wall of the stomach.

Aeby (1880) first characterized the human lungs as asymmetrical because of the presence of a right eparterial bronchus. Boyden has reported, however, that in very rare instances, a left eparterial bronchus may also occur (five cases, with only one case in 100 consecutive dissections). Although this anomaly has an incidence of "no more" than 1%, it is of surgical importance that it be recognized when present.

Lobar bronchi may be displaced to a point higher or lower than usual on the primary lung sacs. Cases are known in which the whole right upper lobe bronchus originates immediately above the middle lobe bronchus. The right upper and middle lobe bronchi may arise from a single stem located at the usual site of the middle lobe. The middle lobe bronchus may arise from a normally placed right upper lobe bronchus, either from the main stem or from the proximal part of the anterior segmental bronchus. The right upper lobe bronchus sometimes originates from the right lateral wall of the trachea.

Variations in the patterns of the bronchial trees are, for the most part, due to displacement of segmental and subsegmental bronchi. These variations are important considerations in resection of bronchopulmonary segments. One such variant, which supplies all or part of the apical segment, arises on the right side of the trachea or from the lateral wall of the right primary bronchus before the origin of the lobar bronchus. Boyden has termed this the "preeparterial bronchus." In addition, one of the upper lobe bronchi may originate at a lower level. The right anterior bronchus, so-called B^2, may arise from the middle-lobe stem. Boyden has termed this the "postarterial bronchus." Usually the left medial basal (B^7) and the left anterior basal (B^8) bronchi arise from a common stem. In 7% of 100 specimens studied, Boyden found that B^7 arises independently, at a higher level on the tree. In the left lower lobe, the posterior ramus of the anterior segment (B^2a) may be absent (35% of cases studied). When present, it occupies the site at which the absence of the left upper lobe most frequently occurs. Other examples and a detailed analysis of variations in the lung can be found in the numerous detailed works of Boyden and co-workers.

The most common anomaly of the lungs is the presence of supernumerary fissures. These fissures do not always separate segments but may enter subsegmental or interbronchial planes. The most common site for a supernumerary fissure is the plane between the medial basal (B^7) and anterior basal (B^8) segments of the right lower lobe. This partially segregates a cardiac or infracardiac lobe. In one study it was found in 35% of 180 human lungs, and in another study it was found in 38% of 50 injected right lungs. These variations would have been found in surgical exploration and, most of them, in radiologic examinations. A comparable fissure occurs, infrequently, in the left lung. Another common fissure segregates, more or less, the superior segment of the right or left lower lobe (so-called "dorsal lobe of Nelson"). In a study by Dévé it was found bilaterally in 12 cases, in the right lower lobe in 40, and in the left lower lobe in 14. In this study, the separated part was called "the posterior lobe."

Yet another common cleavage is the left horizontal fissure. This subdivides the left upper lobe into two almost equal parts, the lower part being called the "left middle lobe." Four types of left middle lobe have been described by Boyden: (a) a true middle lobe separating normal upper and lower division segments, (b) a compressed lingular, (c) an expanded lingular, and (d) an ectopic pulmonary type, which always separates the sector of an eparterial bronchus from the left upper lobe.

Another common form of lung variation includes the absence of fissures. In a study of 277 lungs, the horizontal fissure was absent in 21% and incomplete in 67%. Incomplete oblique fissures occur in about 30% of both right and left lungs.

Certain segments of the upper and middle lobes have a predilection for disease processes. In the right lung, subsegment B^3a is the site of secondary (Assmann's) focus of pulmonary tuberculosis. Segment B^3 on the right side and subsegment B^2a on both the right and left sides are common sites of lung abscesses. Segments B^4 and B^5 are common sites of bronchiectasis. In the lower lobes, the superior segment (B^6), the subsuperior bronchus (B^*), and accessory subsuperior bronchi from B^9 and B^{10} are common sites of lung abscess.

A small process of the right lung just above the base, behind the termination of the inferior vena cava, rarely becomes isolated as the lobus cava. The vena azygos may be displaced outward, so that, instead of curving over the root of the lung, it makes a deep fissure in the upper part of the right lung, called the azygos lobe. This variation may be recognized radiographically and has the shape of an inverted "teardrop" or a comma. In one study its incidence was about 0.5% (7 of 1400 specimens examined); in another study of 50,000 roentgen examinations its incidence was about 0.26%. All sources combined (323,641 examinations) indicate the incidence to be

0.57%. The lobe of the azygos vein has been known since 1777; it was first described by Wrisberg in the cadaver of a 3-year-old boy. In this first case the lobe appeared bilaterally. At least 11 other cases of left-sided lobes have been reported. The bronchial supply of the azygos lobe has been identified as the apical segment B^1a or B^1b. Larger lobes may include both segments or B^1a and B^3a, the apical rami of the apical and posterior segments.

Rarely a segmental bronchus arises from the trachea. An anomalous lobe of the lung may also arise from the esophagus. The absence of a lung has been reported. The right lung is usually heavier than the left by about 50–100 gm. It is also usually shorter and wider.

The trachea usually divides at the level of the fifth or sixth thoracic vertebra and sometimes at the level of the seventh thoracic. The shape of the trachea is variable.

Variations in Pulmonary Blood Supply.

Although vascular variations are cataloged in another section (Cardiovascular System), the inclusion here of the pulmonary vasculature is appropriate. Accessory bronchial arteries may arise from the left aorta (embryologic), subclavian arteries, highest intercostal artery, and right internal thoracic artery. Accessory pulmonary arteries may arise from the thoracic aorta below the hilum of the lung and atypically from the derivatives of the embryonic aortic arches. In the 25 recorded cases, they arise from the ascending aorta and arch of the aorta in 10, from the innominate artery in four, from the right subclavian in two, from the left subclavian in five, and from the descending aorta below the hilum of each lung in four. These aberrancies are usually associated with malformations of the heart and developmental anomalies of the aortic arch.

The pulmonary veins may also retain connections with the derivatives of the anterior cardinal veins or other embryonic vessels. In one study summarizing 106 cases, the drainage from the lungs passed entirely into the right atrium or its tributaries in 36% and partially in the remaining 64%. Of the 64%, the site of drainage was (in order of frequency) superior vena cava, right atrium, and left innominate vein. Pulmonary venous anomalies occur about twice as frequently on the right as on the left. Other unusual sites of drainage include the coronary sinus, inferior vena cava, azygos vein, left subclavian vein, portal vein, and left persistent superior vena cava.

Variations in the mode of entrance of pulmonary veins into the left atrium are of surgical importance. In one study, in 3% of 148 specimens, the right lung drained into a common right pulmonary vein before emptying into the left atrium. On the left side, the frequency may be as high as 25%. The right pulmonary vein may also be represented by three veins, which arise from the usual three lobes of the right lung and empty into the left atrium. In one reported case, however, the middle of the three veins drained a portion of the right middle lobe and also a part of the lower lobe. One may not assume therefore that each of these three veins is entirely restricted to a particular lobe.

References   3–5, 12, 13, 28, 32, 34, 41, 63–65, 68, 72, 73, 95–98, 100–102, 111, 121, 124, 131, 139, 140, 157, 164, 176, 179, 180, 193, 203, 205, 214, 224, 226, 229, 246, 250, 254, 262, 272, 277, 280, 287, 335, 337, 361, 378, 384, 399, 404, 410, 425, 447, 458, 480, 485, 501, 507, 517, 528, 578, 580, 612, 617, 622, 644, 649, 661, 667, 685, 686, 689, 693, 694, 717, 721, 722, 743, 744, 749, 761, 763, 783, 787, 791, 792, 797, 812, 841, 844, 851, 856, 858, 859, 860, 862, 863, 866, 867, 869, 870, 872, 874, 875.

Urinary System

Kidneys, Ureters, Bladder, and Urethra

The kidneys sometimes vary from their normal form, being either longer and narrower or shorter and more rounded. The characteristic fetal lobulation may persist in the adult (7%). Occasionally, one kidney is very small, while the other is proportionately enlarged. Numerous cases are recorded of absence of one kidney, most frequently the right. Congenital absence of a kidney has a reported frequency of 0.24% (6500 autopsies). Ectopic kidney has a frequency of 1:500 to 1:110; solitary kidney, 1:1000, solitary pelvic kidney, 1:22,000; and one normal and one pelvic kidney, 1:3000. The single kidney is usually enlarged, but not invariably. The left kidney is usually larger than the right. On average, in males the right is 12.7 × 6.3 cm while the left is 13.2 × 6.4 cm. In females, the right is 12.4 × 5.9 cm and the left is 12.8 × 6.1 cm. These figures are based on a radiologic study. The occurrence of an additional kidney is rare. In one case, the right kidney was normal, but on the left side there were two kidneys (upper and lower) separated from one another by a distance of 3 cm. The ureter from the upper kidney was joined by that from the lower, so that there were only two ureteric openings into the bladder.

The two kidneys may be joined by their lower ends across the front of the great blood vessels and vertebral column. The conjoined organ usually has the form of a horseshoe, with the convexity downward due to fusion of the caudal ends of the two kidneys.

Human Anatomic Variation

Horseshoe kidney was found in 0.21% of 6500 autopsies in one study and in 0.12% in 68,000 autopsies (1:850) in another. Cumulative data from 192 authors (202,777 autopsies and x-ray studies) indicated a total of 378 horseshoe kidneys. The incidence therefore is about 1:536 or 0.19%; 84% of cases occur in males and about 17% in females. In one study of 2424 cases, 15 horseshoe kidneys were found in males and three in females.

Sometimes two united kidneys are situated in the lumbar region on one side of the vertebral column, or more rarely in the cavity of the pelvis. *Pelvic kidney* was found in 0.04% of 6500 autopsies. "Dumbbell-shaped" organs or two disk-shaped kidneys joined together at their midpoint have also been described.

The kidney may be movable, owing to the laxity of its areolar and adipose capsule. These conditions occur much more frequently with the right kidney.

The *calices minores* may vary in number from six to 14, the smaller number indicating that some of the calices minores embrace the apices of several pyramids. Variations in the calices majores and pelvis are more marked. The *calices majores* may pass downward for some distance beyond the hilum, and end by joining to form the ureter without undergoing any obvious expansion. In such cases, the pelvis is absent; if the calices dilate, one or two pelves may be present.

Bilateral anephrogenesis has an incidence of 0.0017%.

The principal variation in the ureter is a more or less complete division into two. As a rule, the two ureters unite a little above the bladder, so that there is only one vesical orifice. When the division is complete there are two separate openings into the bladder. In one study of 26,480 autopsies, 136 double or bifid ureters were found (0.5%).

In rare cases, three or four ureters may be found. Several instances are recorded in which a supernumerary ureter, proceeding from the upper part of the kidney, opened directly into the urethra. Ureters have also been reported to open into the vagina, the seminal vesicle, or the prostatic urethra.

The right ureter has been found passing behind the inferior vena cava (postcaval or retrocaval ureter) between that vessel and the aorta.

Urinary bladders of variable shape (e.g., hourglass bladder) have been reported. The bladder may be doubled. Agenesis has a reported incidence of 1:600,000 (0.00017%).

The urethra may be doubled.

References   2, 11, 16, 17, 25, 30, 31, 42, 50, 61, 69, 88, 90, 93, 103, 104, 118, 120, 128, 136, 137, 144, 145, 167, 169, 171, 182, 190, 196–198, 201, 202, 206, 211, 215, 218, 228, 234, 237, 242, 259, 269, 274, 279, 281, 289, 291, 294, 295, 300, 305, 306, 308, 312, 314, 331, 338–342, 344, 352, 353, 358, 359, 365, 371, 372, 373, 376, 380, 383, 385, 387, 393, 394, 397, 398, 400–403, 412, 416, 419, 420, 423, 435, 436, 445, 464, 466, 468, 481, 491, 513–515, 519, 521, 544, 551, 557, 561, 563, 564, 567, 585–589, 594–596, 610, 613, 618, 620, 627–629, 639, 645, 653, 658, 672, 702, 710, 711, 715, 724, 725, 729, 738, 741, 742, 753, 762, 776, 782, 785, 790, 794, 806, 819, 821, 830, 831, 836, 838, 845, 856–859, 860, 862, 864, 866–868, 870, 872, 874, 875.

Male Genital System

Testes

In addition to irregularities in position, the testes may vary in number. A case of three and of four testes has been verified. One or both testicles may be absent or may be very rudimentary. Various cases of supernumerary testicles have been reported during operations or in physical examination (without histologic proof, however). In their descent, the testes may become lodged in the perineum, inguinal canal, or abdomen. Two testes may lie in the same serous sac on one side of the scrotum. The position of the testicles has been found to be a diagnostic sign of situs inversus totalis.

References   84, 160, 177, 178, 296, 297, 332, 433, 470, 472, 477, 495, 496, 522, 529, 573, 734, 770, 829, 835, 858, 866, 868, 871, 872, 874, 875.

Additional references: Al-Habbal, Z. and A. Y. Izzidian. Polyorchidism: Case report and review of the literature. *J. Ped. Surg. 19*:212, 1984.

Baker, L. L., Hajek, P. C., Burkhard, T. K. and R. F. Mattrey. Case report. Polyorchidism: evaluation by MR. *Am. J. Roentgenol. 148*:305–306, 1987.

Rifkind, M. D., Kurtz, A. B., Pasto, M. E. and B. B. Goldberg. Polyorchidism diagnosed preoperatively by ultrasonography. *J. Ultrasound Med. 2*:93–94, 1985.

Snow, B. W., Tarry, W. F. and J. W. Duckett. Polyorchidism: an unusual case. *J. Urol. 133*:483, 1985.

Ductus Deferens (Vas Deferens)

Congenital bilateral absence of the vas deferens without associated kidney and urinary tract anomalies has been reported.

Reference   297.

Penis

Bifid conditions and complete duplication (diphallus) may occur. The penis has been reported absent, in which case the urethra drained into the rectum. Congenital absence has a reported frequency of 1:30,000,000 (documentation unavailable). An os penis has been reported.

References 20, 33, 42, 45, 167, 216, 222, 249, 260, 288, 360, 413, 431, 453, 456,462, 490, 492, 526, 615, 619, 650, 657, 674, 681, 704, 728, 799.
Additional reference: Agarwal, S. and S.K. Sogani. Diphallus. *J. Ped. Surg.* 19:213, 1984.

Seminal Vesicles

Absence of the seminal vesicles is very rare. They are often bilaterally unequal in size, shape, and form.

References 297, 858, 874.

Prostate

Apart from abnormalities in size, the prostate is subject to few variations. Persistence of the original independence of the lateral lobes, absence of the middle lobe, and presence of a fourth lobe have been reported. Variations in the relations and mode of ending of the ejaculatory ducts (fusion into a single canal, or termination in the prostatic utricle or by a special canal below the crest) or in the prostatic utricle (absence, enlarged size, or unusual opening) are related to deviations in the development of the generative tract.

References 297, 307, 504, 858, 866, 868, 869, 872, 874, 875.

Bulbourethral Glands

In addition to abnormalities in size, the two glands may be fused into a single mass, or one or both may be absent. Sometimes their absence is questionable since the organs may be represented by rudimentary glands embedded entirely within the substance of the corpus cavernosum urethrae.

References 858, 866, 868, 869, 872, 874, 875.

Female Genital System

Ovaries

Abnormalities in the sex glands of the female are, for the most part, due to developmental deviations. Incompleteness or modification of its descent from the abdomen affect the position of the ovary; it may retain its original suprapelvic position and lie above or upon the psoas major muscle, or it may follow the round ligament and pass partly or entirely through the inguinal canal into the labium majus. The adult ovary may present marked deviations from its typical form, sometimes being unusually long, spheroidal, flattened, triangular, crescentic, or otherwise irregular.

Supernumerary ovaries, varying in size, are not infrequent, occurring in 2–4% of women. Their usual location is along the line marking the transition of the peritoneum into the germinal epithelium. Isolation of a portion of the ovarian anlage, probably by a peritoneal band, may be responsible for these bodies, which consist of normal follicle-bearing ovarian tissue. The ovaries may themselves be the site of anomalous adrenal tissue.

Absence of one or both ovaries is extremely rare.

An ovary may become involved in inguinal hernia; the displacement of the ovaries into the rectouterine pouch (of Douglas) has also been reported.

References 82, 105, 129, 204, 640, 646, 677, 832, 858, 860, 863, 864, 866, 868, 869, 872, 874, 875.

Uterine (Fallopian) Tubes

Apart from anomalous locations that accompany malposition of the uterus and ovary, variations of the uterine tube usually depend on developmental faults traceable to imperfect or aberrant formation of the müllerian ducts. Retention of the fetal tortuosity, stunted development, or complete absence may involve one or both tubes. Complete doubling of the uterine tubes may occur in association with supernumerary ovaries. Occasionally, partial duplication of the tube is observed, consisting of a short canal ending in a diminutive fimbriated extremity in the vicinity of the infundibulum. Such accessory tubes are probably due to a repetition of the invagination that normally produces the infundibulum. The uterine "tubes" may be uncanalized. Quite frequently, the tube has from one to three fringed accessory openings that may lie close to the fimbriated end or at some distance along the tube. Congenital absence of the uterine tubes is rare. A uterine tube has been found in a femoral hernia. Tubal variations are not necessarily associated with ovarian anomalies.

References 82, 129, 185, 356, 666, 677, 712, 858, 860, 862, 866, 868–872, 874, 875.

Uterus

The chief anomalous conditions of the uterus are due to defective development or imperfect fusion of the müllerian ducts, which unite to form the normal organ. Arrested development of the lower part of these fetal canals accounts for the complete absence of the uterus and vagina. Depending on the extent to which failure of fusion occurs, all degrees of doubling are produced. In the most pronounced cases, in which the müllerian ducts remain separate throughout their entire length, two completely distinct uteri and vaginae may result, each set being capable of performing the functions of the normal organs. On the other hand, a slight indentation of the fundus may be the only evidence of imperfect union. Between these extremes all gradations occur: (a) the body may be completely cleft (*uterus bicornis*), with or without divided cervix; (b) the doubling may be partial and limited to branching of the fundus; or (c) the faulty fusion may be manifested by only a partition, more or less complete, that divides the uterine cavity into two compartments (*uterus eptus*), although the external form of the organ is almost or quite normal. When, in conjunction with any of the foregoing variations, one of the component müllerian ducts fails to keep pace in its growth, all degrees of asymmetrical development may result, from complete suppression of one of the tubes in a bicornuate uterus to unilateral diminution of the fundus. Subsequent arrest of what began as normal development may result in permanent retention of a fetal or infantile type of uterus.

Cases of congenital absence of the os uteri, tripartite uteri, uterine diverticula, and uterus have been reported.

References 47, 61, 82, 133, 156, 209, 230, 232, 356, 377, 408, 450, 451, 463, 482, 512, 546, 584, 590, 592, 677, 816, 839, 858, 860, 863, 864, 866, 867–872, 874, 875.

Additional reference: Mintz, M. C., Thickman, D. I., Gussmann, D. and H. Y. Kressel. MR evaluation of uterine anomalies. *Am J. Roentgenol. 148*:287–290, 1987.

Vagina

The most important variations are related to defective development and imperfect fusion of the component müllerian ducts, which are often associated with anomalies of the uterus. When these ducts fail to reach the urogenital sinus, the vagina (and often the uterus) may be absent. Vaginal agenesis was first described by Realdus Columbus in 1572. Duplication, more or less complete, follows persistence of separate or imperfectly fused müllerian ducts. The doubling may not extend throughout the length of the vagina, but may be represented by an imperfect and partial septum, isolated bands, or a twin hymen. Unequal development of the müllerian ducts accounts for the marked asymmetry occasionally seen, notably in double vaginae, in which one canal may be very rudimentary, end blindly, or be essentially normal but doubled throughout.

A doubled vagina may be associated with a single uterus, although this condition is usually found in cases of doubled or bipartite uteri. In some instances the cervices are also doubled.

The vagina may be markedly foreshortened.

Reports of the vagina opening into the urethra, bladder, or rectum, or onto the abdominal wall in adults are very rare.

Pregnancy in a case of an anovaginal orifice (i.e., lack of a normal external vaginal opening) has been reported.

References 14, 61, 125, 143, 170, 209, 210, 230, 356, 377, 449, 454, 455, 590, 801, 816, 824, 842, 858, 860, 864, 866–869, 871, 872, 874, 875.

Mammary Gland

The mammae frequently develop asymmetrically, the right often being larger and lower than the left. While absence of one or both mammae is very rare, with or without associated absence of the nipple, an increase in their number is a relatively common occurrence. The supernumerary mammae vary greatly in the extent to which they are developed, being represented sometimes by well-formed accessory glands (*polymastia*) that may become functioning organs, but more often, particularly in the male subject, only by rudimentary nipples (*polythelia*) or even pigmented areas suggesting areolae. In women, polythelia may be associated with greater or less development of glandular tissue. Doubled nipples have been reported frequently. The recorded frequency of polythelia in men (14%) is questionable. A better estimate in healthy individuals is about 0.4–5% with no sex differences reported. The occurrence of rudimentary supernumerary nipples in males is undoubtedly more common than usually recognized. Renal anomalies accompanied polythelia in nine of 37 (27%) subjects studied.

The usual position of the accessory mammae is below and somewhat medial to the normal glands and corresponds to the mammary line in other animals. In some cases, they may be found above and laterally (especially in Japanese subjects).

The number of accessory glands varies: three pairs in one case, five milk-secreting organs in another, and eight glands in both sexes (1–2% of females and males) have been recorded. They are often asymmetrically placed and not uniformly developed. Comparative studies of the mammae in lower animals and the disposition of the supernumerary organs in the human subject suggest the probability that remote human ancestors normally possessed more than two; the occasional occurrence of the anomalous mammae perhaps indicates a reversion to the primitive condition. In addition to the supernumerary mammae in positions anticipated by the milk-ridges, rudimentary organs sometimes occupy very unusual locations, including the back, lateral thorax, neck, shoulder, inner aspect of the arm, axilla, buttocks, hip, thigh, and labium majus.

Massive hypertrophy has been reported, with a combined, bilateral weight of 124 lbs, with one gland weighing 63 lbs in a 13-year-old girl whose remaining body weight was less then the combined weight of the two glands. Micromastia has also been reported.

References 10, 52, 55, 58, 107, 115, 122, 154, 159, 166, 213, 220, 247, 261, 351, 355, 419, 424, 460, 474, 500, 533, 536, 537, 598, 602, 604, 605, 611, 635, 709, 752, 794, 795, 822, 833, 834, 846, 858, 874, 875.

Hermaphroditism

Cases of individuals with a functional ovary and uterus (with monthly bleeding) and with a penis, testicle (with spermatogenesis), epididymis, seminal vesicle, and prostate have been reported.

References   86, 109, 233, 245, 255, 263, 379, 415, 467, 487, 489, 520, 545, 547, 549, 559, 570, 579, 683, 690, 695, 698, 716, 745, 746, 759, 760, 771, 778, 827, 850, 871.

Endocrine System

Thyroid Gland

Most of the variations in the thyroid gland are due to a partial persistence of the median or thyroglossal duct. The most obvious example of this persistence is the *lobus pyramidalis*. In one study it was found in 104 of 153 cases (nearly 68%) while in another study it was found in only 24 of 60 cases (40%). It is usually attached to the isthmus on the left side of the median plane, but it may join either of the lateral lobes. It is usually connected above with the hyoid bone. Of the 104 specimens in which the pyramidal process was found to be present, 55 were glandular up to the hyoid bone; in 12 the process was connected to the hyoid by fibrous tissue, and in two by muscle. The muscular fasciculi, which are occasionally found to descend from the hyoid bone to the thyroid gland or its pyramidal process, are known as the *levator glandulae thyroidea*. The fibers are most frequently derived from the thyrohyoid muscle, but occasionally they are independent. In one case there were two pyramidal processes; in another, a single process divided into two parts, one for each lateral lobe. Accessory thyroids may be formed by transverse division of the pyramidal process into several separate masses; more rarely they are associated with the lateral lobes. The small glandular masses resemble the thyroid in structure. They are frequently found in front of or above the hyoid bone, and are derived from the upper part of the thyroglossal duct. Thyroid tissue may also be found in the thorax in association with the aortic arch, and in the tongue around the foramen cecum (three cases in 800,000 patients).

As a rule, the two lateral lobes are almost equally well-developed, but occasionally they are very unequal in size, and in rare cases one lobe may be absent. The isthmus varies greatly in size and is frequently absent.

The thyroid gland may be absent.

References   9, 46, 62, 78, 130, 191, 192, 302, 319, 326, 327, 329, 406, 414, 417, 418, 421, 511, 524, 525, 623, 706, 814, 815, 837, 857, 858, 860, 863, 864, 866–870, 872, 874, 875.

Parathyroid Glands

The usual text description of the parathyroid holds true in only about 25% of cases studied; their number and location is highly variable. They may be located within the thorax embedded in thymic tissue, in the posterior mediastinum, and in the areolar tissue associated with the pericardium.

In one study of 86 glands, 76 were found dorsal to the thyroid gland, one at the ventrocaudal pole and nine caudal to it. Of the latter nine, four were in contact with the thyroid capsule and five were found in the loose connective tissue ventral and lateral to the trachea, or in the tissue extending from the thyroid capsule to the thymus. Of the 76 glands found on the dorsum of the thyroid, six were associated with the cranial third of the thyroid, and the remainder were distributed over the caudal two-thirds with a concentration near the cricoid cartilage and first tracheal ring. Exceptional locations were as follows: between the isthmus of the thyroid and trachea, one case; in the triangular cleft formed by the thyroid, trachea, and esophagus, three cases; and near the midline between esophagus and trachea, one case. In one study of 25 larynges, 86 glands were found (averaging 3.2 per specimen). The number per specimen varied from two to six. The glands were distributed as follows: 7% were opposite the cranial third of the thyroid, 57% opposite the middle third, 25.6% opposite the caudal third, and 10.4% caudal to the thyroid. One gland was found (partly) on the ventral surface of the thyroid. None was found embedded in the thyroid.

The parathyroid glands may be located on the anterior surface of the thyroid gland. A parathyroid gland was found on the anterior surface of the thyroid isthmus. In a study of 42 larynges, 30 glands were identified, of which 35.9% were found on the ventrolateral thyroid capsule and 9.5% on the ventral thyroid capsule. An accessory parathyroid was found in the vagus nerve.

References   71, 369, 552, 553, 623, 825, 826, 853, 858, 866, 872, 874, 875.

Suprarenal (Adrenal) Gland

These glands vary considerably in size and shape: they may be pyramidal, quadrilateral, foliate, or spherical. The right gland is usually quadrilateral, the left foliate (leaf-like). In one study, the gland varied in length from 3.5 to 7.3 cm, in width from 1.5 to 4.0 cm, and in thickness from 0.3 to 1.8 cm. The right and left glands, on average, vary only slightly.

One or both glands may be absent in their normal

position. Accessory suprarenal glands are occasionally found, varying in size from a pinhead (1 mm) to a large pea (5 mm). The smaller glands apparently have no medullary substance, but the larger ones may possess a medulla. The accessory glands may be found near, upon, or under the capsule of the kidney, and adhered to it by connective tissue. Sometimes they are partially embedded in the substance of the kidney or liver, or they may be found near organs that were in the neighborhood of the suprarenal glands during their embryonic origin and have moved away in the process of development. These include the genital glands (ovary and testis), broad ligament of the uterus, and spermatic cord. Accessory cortical suprarenal tissue may be found at any point along the descending path of the gonads, including the pelvis, inguinal canal, and scrotum.

Accessory medullary tissue is common and may be found in a variety of places, including the heart and along the aorta. Histologic confirmation of an intracranial intradural adrenal gland has been reported.

References   59, 116, 181, 204, 271, 497, 614, 673, 754, 823, 858, 866–870, 872, 874, 875.

Pancreas

See the discussion under Digestive System and Spleen.

Pineal Gland

The pineal body is calcified in 40% of subjects over 20 years of age, but rarely in individuals under this age. Because the calcium is visible on x-ray it is a significant and useful aid in diagnosis of space-occupying lesions within the skull. Choroid plexus and habenular calcification have also been useful in diagnosis of lesions of the nervous system. Calcified choroid plexus has a reported incidence of 6.1–28% in all persons over 20 years of age.

References   235, 264, 265, 486, 499, 518, 733, 796.

Miscellaneous

Skin

The thickness of skin (epidermis and dermis) averages between 1 and 2 mm. It is 3 to 6 mm in the interscapular region, back of the neck, and flexor surfaces of hand and foot. It is less than 0.5 mm over the tympanic membrane and eyelid. The integument is thicker on the posterior or extensor surfaces than on corresponding regions of the anterior or flexor surfaces, with the exception of the hand and foot.

The color of skin is dependent upon intrinsic pigmentation and blood flow. Pigment may be absent (albinism) or generally reduced (partial albinism).

There are variations in color, types, and amount (hypertrichosis) and distribution of hair. Cutaneous horns have been reported.

References   26, 153, 155, 231, 603, 633, 634, 678, 731, 757, 764, 780, 781, 861–864, 866, 870, 872, 874, 875.

Eyelids

The eyelids may be absent. The margin of either lid is sometimes notched or fissured. The palpebral fissure may be wider or narrower than usual, or the lids may be united by vascularized tissue. The muscles of the lid occasionally fail to develop. Supernumerary eyelids may occur.

References   632, 876, 877.

Lacrimal Gland and Nasolacrimal Duct

There may be supernumerary (two to four) puncta. Duplication of the canaliculus is rare (1:60,000). The puncta are occasionally imperforate. Diverticula of sinuses in the lacrimal duct may occur, or the nasolacrimal duct may be doubled.

Reference   877.

Iris

The iris may be partially or totally absent (coloboma). In some cases, the irides differ in color (congenital heterochromia). The embryonic pupillary membrane may persist; the pupillary membrane was described by Waschendorff in 1738.

References   343, 375, 874, 876, 877.

Thymus

The size and shape of the thymus is variable. It may be located behind the brachiocephalic vein. It occasionally contains anomalous parathyroid tissue.

References   67, 859, 874.

References

* These journal articles are notable for their content, literature review, and/or illustrations.
+ The authors have not had access to these references. They are included because of their historic and/or scientific importance.

Journals

1. AASAR, Y.H. A case of an unusual arrangement of the gut. *J. Anat.* 72:579–583, 1937–38.
2* ABESHOUSE, B.S. Surgery of the congenital anomalous kidney. *Surg. Gynecol. Obstet.* 78:288–303, 1944.
3. ADACHI, B. Selteneres Vorkommen des sog. "Lobus venae azygos" bei den Japanern. *Anat. Anz.* 89:214–216, 1940.
4. ADAMS, R. and E.D. CHURCHILL. Situs inversus, sinusitis, bronchiectasis. Report of five cases, including frequency statistics. *J. Thoracic Surg.* 7:206–217, 1937.
5. AEBY, D. Der Bronchialbaum des Menschen bei Situs inversus. *Arch. Anat. Physiol. Wissen. Med.* 1882, pp. 31–32.
6* AFZELIUS, B.A. A human syndrome caused by immotile cilia. *Science* 193:317–319, 1976.
7. AFZELIUS, B.A. Genetical and ultrastructural aspects of the immotile-cilia syndrome. *Am. J. Hum. Genet.* 33:852–864, 1981.
8* AFZELIUS, B.A., CARLSTEN, J. and S. KARLSSON. Clinical, pathologic, and ultrastructural features of situs inversus and the immotile-cilia syndrome in a dog. *J. Am. Vet. Med. Assoc.* 184:560–563, 1984.
9* D'AJUTOLO, G. Delle strume tiroidee accessorie ed in particolare di una mediastinica e di due cervicali nello stesso individuo. *Mem. R. Accad. Sci. Istituto di Bologna S. 4* 10:773–782, 1889.
10* ALBERT, H. Diffuse idiopathic hypertrophy of the mammary glands of the female. *JAMA* 55:1339–1343, 1910.
11. ALBU, I., VAIDA, A. and F. GRIGORESCU-SIDO. Ein Fall von mehrfacher beidseitiger Nierenanomalie beim Menschen. *Anat. Anz.* 133:207–213, 1973.
12. ALEZAIS. Anomalie de division du poumon droit. *Soc. Biol. Comptes Rendus Hebdomadaires des Séances et Mémoires* 55:144–145, 1903.
13. ALLEN, W. A variety of pulmonary lobation and its relations to the thoracic parietes, as illustrated by comparative anatomy and abnormalities in the human subject. *J. Anat. Physiol.* 16:605–614, 1882.
14. ALMEIDA, A. DE. La macronymphie chez les femmes indigènes de l'Angola. *Assoc. Anatomistes Comptes Rendus* 43:131–150, 1956.
15. ALMEIDA, F. DE. Une anomalie rare du foie. *Assoc. Anatomistes Comptes Rendus* 28:11–13, 1933.
16. ALMEIDA, F. DE. Sur l'irrigation d'un rein en fer-à-cheval. *Assoc. Anatomistes Comptes Rendus* 28:16–18, 1933.
17. ALMQUIST, P.O. Unilateral double ureter with partly vaginal, partly normal outlet. *Acta Chir. Scand.* 93:495–502, 1946.
18. ALTUCHOFF, N. Ungewöhnlich langer Wurmfortsatz, Positio mesenterica. *Anat. Anz.* 22:206–210, 1903.
19. AMIN, M. Persistent opening of anterior cloacal depression. *J. Anat.* 73:192–193, 1938–39.
20* AMR, S.S., MUGHAIREH, A.M., URI, F.I., BUSTAMI, F.M.F. and A.M. SHAMAYLEH. Agenesis of the penis: Report of two cases and review of the literature. *Jordan Med. J.* 18:221–230, 1984.
21. AMUSSAT, M. Congenital absence of the gall-bladder. *Am. J. Med. Sci.* 8:478, 1831.
22. ANCEL, P. and P. CAVAILLON. Sur les mésocôlons ascendant et descendant et leur mode de formation chez l'homme. *Assoc. Anatomistes Comptes Rendus* 9:1–11, 1907.
23. ANDERSON, J.H. Abnormalities of the duodenum. *Br. J. Surg.* 10:316–321, 1922–23.
24. ANDERSON, J.R. and H. WAPSHAW. Annular pancreas. *Br. J. Surg.* 39:43–49, 1951–52.
25. ANILE, A. A propos d'un rein en fer à Cheval. *Arch. Ital. Biol.* 32:468, 1899.
26* ANONYMOUS. Anatomie du vivant. Corne cutanée du cuir chevelu. *La Presse Médicale* No. 104. 30:2183, 1922.
27* ANSON, B.J. Anatomical considerations in surgery of the gallbladder. *Q. Bull. Northwestern University Medical School* 30:250–259, 1956.
28* ANSON, B.J. and H.V. SMITH. The accessory pulmonary lobe of the azygos vein. *Am. J. Roentgenol.* 35:630–634, 1936.
29* ANSON, B.J., LYMAN, R.Y. and H.H. LANDER. The abdominal viscera *in situ*. A study of 125 consecutive cadavers. *Anat. Rec.* 67:17–21, 1936.
30* ANSON, B.J. and L.W. RIBA. The anatomical and surgical features of ectopic kidney. *Surg. Gynecol. Obstet.* 68:37–44, 1939.
31* ANSON, B.J., PICK, J.W. and E.W. CAULDWELL. The anatomy of commoner renal anomalies. Ectopic and horseshoe kidneys. *J. Urol.* 47:112–132, 1942.
32* ANSON, B.J., SIEKERT, R.G., RICHMOND, T.E. and W.E. BISHOP. The accessory pulmonary lobe of the azygos vein. An anatomical report, with a record of incidence. *Q. Bull. Northwestern University Medical School* 24:285–290, 1950.
33* ANTONY, J. and M.K. CHANDRASEKHARA. Penile agenesis. *Int. Surgery* 58:650–652, 1973.
34. APPLETON, A.B. Segments and blood vessels of the lungs. *Lancet* 2:592–594, 1944.
35. ARGE, E. Transposition of the viscera and sterility in men. *Lancet* 1:412–414, 1960.
36. ARMSTRONG, G.E. Deficiency of the mesentery over the lower ileum. *Br. J. Surg.* 9:287–289, 1921–22.
37. ARNEILLE, J.R. Clinical observations on congenital and acquired transposition of the viscera. *Am. J. Med. Sci.* 124:885 898, 1902.
38* ARNOLD, J. Ein Fall von Cor triloculare biatriatum, Communication der Lungenvenen mit der Pfortader und Mangel der Milz. *Arch. Pathol. Anat. Physiol. Klin. Med.* 42:449–472, 1868.
39. ARVIDSON, K. Location and variation in number of taste buds in human fungiform papilla. *Scand. J. Dent. Res.* 87:435–442, 1979.
40. ASKANAZY, M. Zur Pathogenese der Magenkrebse und über ihren gelegentlichen Ursprung aus angeborenen epithelialen Keimen in der Magenwand. *Dtsch. Med. Wochenschr.* 49:3–6, 49–51, 1923.
41. ATKIN, E.E. A specimen of the accessory lobe of the azygos vein. *Lancet* 2:1221–1222, 1934.
42. ATTIE, J. Congenital absence of the penis: A report of a case with congenital concealed penile agenesis and congenital absence of the left kidney and ureter. *J. Urol.* 86:343–345, 1961.
43. AUGIER, M. and M. BOPPE. Sur quelques formes de l'estomac humain adulte. *Assoc. Anatomistes Comptes Rendus* 15:66–77, 1913.
44* AUGUSTYNIAK, E., KOCON, K. and S. ZAJAC. A rare case of double cystic duct. *Folia Morphol.* 27:70–72, 1968.
45* AZPIROZ, J. Agenesis of the penis. *J. Pediatr. Surg.* 6:74, 1971.
46. BAILEY, H. Thyroglossal cysts and fistulae. *Br. J. Surg.* 12:579–589, 1924–25.
47. BAINBRIDGE, W.S. Duplex uterus with multiple pregnancy: Report of cases. *Am. J. Obstet. Gynecol.* 7:285–288, 1924.
48. BAKER, A.H. and J. KIRK. A case of megalo-duodenum with various vascular abnormalities. *J. Anat.* 68:350–353, 1933–34.
49. BALDWIN, W.M. A specimen of annular pancreas. *Anat. Rec.* 4:299–204, 1910.
50. BALLOWITZ, E. Über angeborenen, einseitigen, vollkommenen Nierenmangel. *Arch. Pathol. Anat. Physiol. Klin. Med.* 141:309–390, 1895.
51. BANH, D.B. and HOANG. Recherches anthropologiques sur les organes génitaux de la Vietnamienne. *Assoc. Anatomistes Comptes Rendus* 38:457–464, 1951.
52. BANNER, E.R. Unilateral micromastia associated with abnormalities of musculature of chest wall on same side. *Br. Med. J.* 2:420, 1946.
53. BARBOSA, J.J. DE C., DOCKERTY, M.B. and J.M. WAUGH. Pancreatic heterotopia: Review of literature and report of 41 authenticated surgical cases, of which 25 were clinically significant. *Surg. Gynecol. Obstet.* 82:527–542, 1946.
54. BARCLAY-SMITH, E. A liver exhibiting multiple anomalies. *J. Anat. Physiol.* 42:346–348, 1909.
55. BARDELEBEN, K. VON. Weitere Untersuchungen über die Hyperthelie bei Männern. *Anat. Anz.* 7:87–92, 1892.
56. BARGE, J.A.J. Eine seltene Variation des menschlichen Pankreas. *Anat. Anz.* 56:417–427, 1923.
57. BARON. Vice de conformation. Transposition complète des viscères abdominaux et thorachiques. *Arch. Gén. Méd.* 10:131, 1826.

58* BARTELS, M. Überzahl der Brustwarzen. *Arch. Anat. Physiol. Wissen. Med.* 1872, pp. 304–306.
59 BARTLETT, C.J. Direct union between adrenals and kidneys (subcapsular location of adrenals). *Anat. Rec.* 10:67–77, 1915.
60 BARTONE, N.F. and R.V. GRIECO. Absent gallbladder and cystic duct. *Am. J. Roentgenol.* 110:252–255, 1970.
61 BASTIEN and LE GENDRE. Absence complète de l'utérus de trompes, du vagin et du rein droit. *Soc. Biol. Comptes Rendus des Séances et Mémoires* 11:96–99, 1859.
62* BEACH, F. Trachea, showing absence of thyroid gland, and fatty tumours, from a case of sporadic cretinism. *Trans. Pathol. Soc. Lond.* 27:316–321, 1876.
63 BEAU, CAYOTTE, LUX and STREIFF. La parabronche externe du lobe superieur du poumon humain. *Assoc. Anatomistes Comptes Rendus* 38:128–133, 1951.
64 BEAU, CAYOTTE, PRÉVOT and DELESTRE. Systématisation du lobe moyen du poumon humain. *Assoc. Anatomistes Comptes Rendus* 38:134–139, 1951.
65 BEAU, CAYOTTE, GILLE and BRULÉ. Systématisation du lobe inférieur du poumon humain. *Assoc. Anatomistes Comptes Rendus* 38:140–146, 1951.
66 BECK, R. Ein Fall von Situs viscerum inversus totalis. *Münchener Med. Wochenschr.* 63:122, 1916.
67* BELL, R.H., KNAPP, B.I., ANSON, B.J. and S.J. LARSON. Form, size, blood supply and relations of the adult thymus. *Q. Bull. Northwestern University Medical School* 28:156–164, 1954.
68 BENEDICK, A.J. and H. WESSLER. The azygos lobe of the lung. *Am. J. Roentgenol.* 20:1–6, 1928.
69 BENJAMIN, J.A. and C.E. TOBIN. Abnormalities of the kidneys, ureters, and perinephric fascia: Anatomic and clinical study. *J. Urol.* 65:715–733, 1951.
70 BENNETT-JONES, M.J. and C.A. ST. HILL. Accessory spleen in the scrotum. *Br. J. Surg.* 40:259–262, 1952.
71 BÉRARD, L. and H. ALAMARTINE. Les parathyroides externes de l'homme. *Soc. Biol. Comptes Rendus Hebdomadaires des Séances et Mémoires* 66:619–621, 1909.
72* BÉRARD, M., SOURNIA, J. and L. GRÉZARD. Difficultés inhérentes à l'existence d'un lobe azygos lors des interventions pulmonaires endothoraciques. *Lyon Chir.* 45:98–102, 1950.
73* BERG, R.M., BOYDEN, E.A. and F.R. SMITH. An analysis of variations of segmental bronchi of the left lower lobe of fifty dissected, and ten injected lungs. *J. Thorac. Surg.* 18:216–236, 1949.
74 BERRY, R.J.A. The anatomy of the vermiform appendix. *Anat. Anz.* 10:761–769, 1895.
75 BETTI, U.A. Sur les rapports du larynx avec la colonne vertébrale chez l'homme. *Arch. Ital. Biol.* 32:467, 1899.
76 BHATTACHAYYA, B. Three cases of right-sided sigmoid colon. *J. Anat.* 60:229–232, 1926.
77 BICKFORD, B.J. and J.C.F.L. WILLIAMSON. Annular pancreas. *Br. J. Surg.* 39:49–52, 1951–52.
78* BILL, A.H. Cysts and sinuses of the neck of thyroglossal and branchial origin. *Surg. Clin. N. Am.* 36:1599–1611, 1956.
79 BIONDI, D. Lippenspalte und deren Complicationen. *Arch. Pathol. Anat. Physiol. Klin. Med.* 111:125–176, 1888.
80 BLANTON, D.E., BREAM, C.A. and S.R. MANDEL. Gallbladder ectopia: A review of anomalies of position. *Am. J. Roentgenol.* 121:396–400, 1974.
81 BLECHSCHMIDT, C.M. Agenesis of the gallbladder – borderline: Case of normality. *Anat. Anz.* 151:281–285, 1982.
82 BLOT, M. Atrophie ou plutôt état rudimentaire d'un des ovaries, avec absence de la trompe du même côte chez une femme d'une quarantaine d'années. Nouvelle description des ligaments ronds. *Soc. Biol. Comptes Rendus des Séances et Mémoires* 8:176–179, 1856.
83 BOETTCHER, A. Seltene angeborene Formanomalie der Leber. *Arch. Pathol. Anat. Physiol. Klin. Med.* 34:103–106, 1865.
84 BOGGON, R.H. Polyorchidism. *Br. J. Surg.* 20:630–639, 1932–33.
85 BOGGS, J.D. and W. REED. Congenital absence of the spleen. *Q. Bull. Northwestern University Medical School* 27:289–293, 1953.
86* BOLK, L. Pseudohermaphroditismus masculinus occultus. *Anat. Anz.* 32:129–137, 1908.
87* BONIN, G. VON. Partial inversion of the duodenum. *Anat. Rec.* 89:71–73, 1944.
88* BOTEZ, G. Considérations sur la pathologie et la chirurgie du rein en fer a cheval. *J. Urol. (Paris)* 1:193–204, 373–392, 502–516, 633–654, 1912.
89 BOULGAKOW, B. A case of maldevelopment of the right parotid gland. *J. Anat.* 60:341, 1926.
90* BOULGAKOW, B. and N. FAHMY. A case of unilateral duplication of the ureter. *J. Anat.* 60:342–344, 1926.
91 BOWER, J.O. Congenital absence of the gall bladder. *Ann. Surg.* 88:80–90, 1928.
92* BOYDEN, E.A. The accessory gallbladder – an embryological and comparative study of aberrant vesicles occurring in man and the domestic mammals. *Am. J. Anat.* 38:177–231, 1927.
93* BOYDEN, E.A. Description of a horseshoe kidney associated with left inferior vena cava and disc-shaped suprarenal glands, together with a note on the occurrence of horseshoe kidneys in human embryos. *Anat. Rec.* 51:187–211, 1932.
94* BOYDEN, E.A. The problem of the double ductus choledochus: An interpretation of an accessory bile duct found attached to the pars superior of the duodenum. *Anatomical Rec.* 55:71–93, 1932.
95* BOYDEN, E.A. A synthesis of the prevailing patterns of the bronchopulmonary segments in the light of their variations. *Dis. Chest* 15:657–668, 1949.
96* BOYDEN, E.A. The distribution of bronchi in gross anomalies of the right upper lobe, particularly lobes subdivided by the azygos vein and those containing pre-eparterial bronchi. *Radiology* 58:797–807, 1952.
97* BOYDEN, E.A. A critique of the international nomenclature on bronchopulmonary segments. *Dis. Chest* 23:266–269, 1953.
98 BOYDEN, E.A. Developmental anomalies of the lungs. *Am. J. Surg.* 89:79–89, 1955.
99 BOYDEN, E.A. The anatomy of the choledochoduodenal junctions in man. *Surg. Gynecol. Obstet.* 104:641–652, 1957.
100* BOYDEN, E.A. The nomenclature of the bronchopulmonary segments and their blood supply. *Dis. Chest* 39:1–6, 1961.
101 BOYDEN, E.A. and J.F. HARTMANN. An analysis of variations in the bronchopulmonary segments of the left upper lobes of fifty lungs. *Am. J. Anat.* 79:321–360, 1946.
102* BOYDEN, E.A. and J.G. SCANNELL. An analysis of variations in the bronchovascular patterns of the right upper lobe of fifty lungs. *Am. J. Anat.* 82:27–74, 1948.
103 BRAASCH, W.F. Anomalous renal rotation and associated anomalies. *J. Urol.* 25:9–21, 1931.
104* BRAASCH, W.F. and A.J. SCHOLL, JR. Pathologic complications with duplication of the renal pelvis and ureter (double kidney). *J. Urol.* 8:507–558, 1922.
105 BRANCA, A. Note sur l'ovaire ectopique. *Assoc. Anatomistes Comptes Rendus* 3:253–254, 1901.
106* BRANDT, G. Die Gefäßarchitektur des humanen kaudalen Pancreas. *Anat. Anz.* 157:73–81, 1984.
107 BRAQUEHAYE, J. and REMLINGER. Mamelle surnuméraire audessous de l'ombilic chez un homme. *Soc. Biol. Comptes Rendus Hebdomadaires des Séances et Mémoires* 51:598–599, 1899.
108* BRASH, J.C. and M.J. STEWART. A case of partial transposition of mesogastric viscera. *J. Anat.* 54:276–286, 1920.
109 BRAUN, H. Ein Fall von Pseudohermaphroditismus masculinus externus. *Z. Geburtshilfe Gynäk.* 28:375–382, 1894.
110 BRAY, E. Grandezze ponderali di alcuni visceri umani in rapporto alle dimensioni esterne dei corrispondenti segmenti corporei. Nota II. *Arch. Ital. Anat. Embriol.* 30:198–214, 1932.
111 BRAY, E. Grandezze ponderali di alcuni visceri umani in rapporto alle dimensioni esterne dei corrispondenti segmenti coporel. Elaborazione statistica del materiale ruccolta a Firenze da Castaldi e Vannucci. Nota III: Cuore, Pulmoni. *Arch. Ital. Anat. Embriol.* 32:257–274, 1933–34.
112 BRAY, E. Grandezze ponderali di alcuni visceri umani in rapporto alle dimensioni esterne dei corrispondenti segmenti corporel. Nota I. *Monit. Zool. Ital.* 42:101–113, 1931.
113 BRAZE, A. Bicornate uterus with pregnancy in each horn. *JAMA* 123:474–476, 1943.
114 BRETON, M. Anomalies duodénales. *La Presse Médicale* 42:627, 1934.
115* BRIGHTMORE, T. Bilateral double nipples. *Br. J. Surg.* 59:55–57, 1972.
116 BRITES, G. Glandes surrénales, accessoires, sous-capsulaires dans le rein de l'homme. *Assoc. Anatomistes Comptes Rendus* 28:102–109, 1933.
117 BROEK, A. Ein Fall vollkommener Agenesie des rechten Urogenitalapparates. *Anat. Anz.* 31:417–423, 1907.
118* BROESIKE, G. Ein Fall von congenitaler S-förmiger Verwachsung beider Nieren. *Arch. Pathol. Anat. Physiol. Klin. Med.* 98:338–341, 1884.
119 BROOKES, V.S. Meckel's diverticulum in children. A report of 43 cases. *Br. J. Surg.* 42:57–66, 1954–55.

120 BROWN, M. Variations in the position and development of the kidneys. *J. Anat. Physiol.* 28:193–194, 1894.

121 BROWN, S. and M. BRAVERMAN. Azygos lobe of the right lung. *Radiology* 17:575–576, 1931.

122 BRUCE, J.M. Supernumerary nipples and mammae. With an account of sixty-five instances observed. *J. Anat. Physiol.* 13:425–448, 1879.

123 BRÜGGEMANN. Fall einer vollständigen Transposition aller Eingeweide. *Schmidt's Jahrb. Med.* 10:219–220, 1836.

124 BRÜNNER, H. Zur makroskopischen Form der Glandula parotis. *Anat. Anz.* 110:327–335, 1962.

125 BRYAN, A.L., NIGRO, J.A. and V.S. COUNSELLER. One hundred cases of congenital absence of the vagina. *Surg. Gynecol. Obstet.* 88:79–86, 1949.

126 BUCHANAN, A.M. Diverticulum (Meckel's) of small intestine. *J. Anat. Physiol.* 27:559–563, 1893.

127* BUIRGE, R.E. Gross variations in the ileocecal valve. A study of the factors underlying incompetency. *Anat. Rec.* 86:373–385, 1943.

128 BUMPUS, JR., H.C. Unusual duplication of the renal pelvis. *J. Urol.* 25:39–41, 1931.

129 BURGE, E.S. Absence of left ovary and a portion of left fallopian tube in a 19-year-old student. *Q. Bull Northwestern University Medical School* 32:4–5, 1958.

130 BUTLIN. Tumeurs glandulaires de la racine de la langue. *La Sémaine Méd.* 10:87, 1890.

131 CAIRNEY, J. The lobe of the azygos vein. Note on two additional cases. *J. Anat.* 58:54–58, 1924.

132* CALAS, F. BOUCHET, Y. and R. MARTIN. Etude morphologique des canaux de la tête du pancreas. *Assoc. Anatomistes Comptes Rendus* 41:866–876, 1954.

133 CALORI, L. Di alcuni particolari intorno le parti genitali muliebri. *Mem. R. Accad. Scienze Istituto di Bologna S. 2* 1:171–201, 1852.

134* CALORI, L. Sulla coesistenza di un'eccessiva divisione del fegato e di qualche dito soprannumerairo nelle mani o nei piedi. *Mem. R. Accad. Sci. Istituto di Bologna S. 4* 2:335–343, 1880.

135* CALORI, L. Di una inversione splancnica generale nell'uomo accompagnata da alcuni notabili del capo con esso lei convenienti e da estranee anomalie. *Mem. R. Accad. Sci. Istituto di Bologna S. 4* 2:597–622, 1880.

136 CALORI, L. Caso di un rene a ferro di cavallo. *Mem. R. Accad. Sci. Istituto di Bologna S. 4* 6:73–83, 1884.

137* CALORI, L. Sui nervi di un rene a ferro di cavallo con ectopia del rene sinistro. *Mem. R. Accad. Sci. Istituto di Bologna S. 4* 9:451–460, 1888.

138 CAMERON, J.L. and A. NICHOLLS. Two rare abnormalities occurring in the same subject. Partial absence of the corpus callosum. Stomach situated entirely within the thorax. *Can. Med. Assoc. J.* 11:448–453, 1921.

139* CAMPBELL, A.H. and A.G. LIDDELOW. Significant variations in the shape of the trachea and large bronchi. *Med. J. Aust.* 1:1017–1020, 1967.

140 CAMPBELL, J.M. Bilateral anomaly of the laryngeal saccule. *J. Anat.* 72:465–466, 1937–38.

141 CAMPBELL, R.T. Intestinal obstruction with congenital absence of the left diaphragm. *Br. J. Surg.* 41:56–60, 1953–54.

142 CAPONE, A.J. and H. MILLER. Left-sided appendicitis in a dextro cardiac patient. *Am. J. Surg.* 71:282–283, 1946.

143* CAPRARO, V.J. and M.B. GALLEGO. Vaginal agenesis. *Am. J. Obstet. Gynecol.* 124:98–107, 1976.

144 CARLETON, A. Crossed ectopia of the kidney and its possible cause. *J. Anat.* 71:292–298, 1936–37.

145 CARSON, W.J. Congenital solitary kidney. *Wis. Med. J.* 29:154–156, 1930.

146 CARTER, J.P., ISA, N.N., HASHEM, N. and F.O. RAASCH, JR. Congenital absence of the penis. A case report. *J. Urol.* 99:766–768, 1968.

147 CATON, R. Case of complete transposition of viscera. *J. Anat. Physiol.* 21:446, 1897.

148* CAVE, A.J.E. Appendix vermiformis duplex. *J. Anat.* 70:283–292, 1935–36.

149 CEVIDALLI, A. Sur les lignes papillaires des doigts de la main. *Arch. Ital. Biol.* 48:482, 1907.

150 CEVIDALLI, A. and G. BENASSI. Recherches sur les plis palmares. *Arch. Ital. Biol.* 48:482–483, 1907.

151 CHAPLIN, T. Report of a case of the transposition of the viscera. *Lancet* 2:478, 1854.

152 CHAPMAN, J.L. and H.W. MOSSMAN. Annular pancreas accompanied by aberrant pancreatic nodule in duodenum. *Am. J. Surg.* 60:286–288, 1943.

153* CHARACHE, H. Cutaneous horn of the scalp. *Am. J. Surg.* 29:297–298, 1935.

154 CHARCOT and LE GENDRE. Deux cas mamelons surnuméraires observés chez la femme. *Soc. Biol. Comptes Rendus des Séances et Mémoires* 11:164–165, 1859.

155 CHASE, H.B. Growth of hair. *Physiological Rev.* 34:113–126, 1954.

156 CHEN, H.Y. and Y.P. CHEN. A case of uterus didelphys with pregnancy in one of the uteri. *Formosan Med. Assoc. J.* 54:26–27, 1955.

157 CHIENE, J. Note on a supernumerary lobe to the right lung. *J. Anat. Physiol.* 4:89–91, 1870.

158 CHILDE, A.E. Calcification of the choroid plexus and its displacement by expanding intracranial lesions. *Am. J. Roentgenol.* 45:523–536, 1941.

159 DE CHOLNOKY, T. Supernumerary breast. *Arch. Surg.* 39:926–941, 1939.

160 CHOLST, M.R. Discrepancies in pain and symptom distribution. Position of the testicles as a diagnostic sign in situs inversus totalis. *Am. J. Surg.* 73:104–107, 1947.

161 CHUDNOFF, J. and H. SHAPIRO. Two cases of complete situs inversus. *Anat. Rec.* 74:189–194, 1939.

162 CIVALLERI, A. L'hypophyse pharyngienne chez l'homme. *Assoc. Anatomistes Comptes Rendus* 10:128–133, 1908.

163* CLADO. Appendice caecal. *Soc. Biol. Comptes Rendus Hebdomadaires des Séances et Mémoires* 44:133–172, 1892.

164 CLELAND, J. Case of the supernumerary lobe of the right lung. *J. Anat. Physiol.* 4:200, 1870.

165* CLEVELAND, M. Situs inversus viscerum. An anatomic study. *Arch. Surg.* 13:343–368, 1926.

166 COE, A.H. Case of supernumery nipple associated with maternal impression. *Med. Rec.* 34:479, 1888.

167 COHEN, S.J. Diphallus with duplication of the colon and bladder. *Proc. R. Soc. Med.* 61:305–306, 1968.

168* COLTON, E.J., ANSON, B.J., GIBBS, E.W., MCCORMACK, L.J., REIMANN, A.F. and E.H. DASELER. Positions of abdominal viscera. *Q. Bull. Northwestern University Medical School* 21:154–155, 1947.

169 COPPRIDGE, W.M. Bilateral pelvic kidneys. *J. Urol.* 32:231–235, 1934.

170 CORDIER, G. and MALINAS. Quelques précisions sur l'absence congénitale du vagin (à propos de dix cas personnels). *Assoc. Anatomistes Comptes Rendus* 38:285–293, 1951.

171* CORDS, E. Über eine Anomalie des Nierenbeckens bei normaler Lage des Organes. *Anat. Anz.* 38:549–554, 1911.

172 CORDS, E. Ein Fall von ringförmigem Pankreas. *Anat. Anz.* 39:33–40, 1911.

173 CORNAZ. Inversion splanchnique complète. *Gaz. Hôp.* 32:315–316, 1859.

174 CORSY, F. Lobe surnuméraire du foie, implanté sur la face inférieure de la vésicule biliaire. *Soc. Biol. Comptes Rendus Hebdomadaires des Séances et Mémoires* 86:695–696, 1922.

175 COUINAUD, C. Mode d'origine de la voie biliaire principàle. *Assoc. Anatomistes Comptes Rendus* 40:603–610, 1953.

176 COULOUMA and DEVOS. Les scissures pulmonaires et leurs variations chez l'homme et les mammifères. *Assoc. Anatomistes Comptes Rendus* 32:112–139, 1937.

177 COUNSELLER, V.S. and M.A. WALKER. Congenital absence of testes (anorchia). *Ann. Surg.* 98:104–109, 1933.

178 COUNSELLER, V.S., NICHOLS, D.R. and H.L. SMITH. Congenital absence of testis. A report of seven cases of monorchidism. *J. Urol.* 44:237–241, 1940.

179 CRAWFORD, J.H. Tomographic appearance of the azygos lobe with a description of two cases and a report of seven cases. *Br. J. Radiol.* 17:319–322, 1944.

180 CRIVELLARI, C.A.L. DE, NEWTON, E.F. and C.A. CRIVELLARI. Lóbulo de la azigos normal y patológico. *Prensa Méd. Argent.* 30:2210–2214, 1943.

181 CULP, O.S. Adrenal heterotopia. A survey of the literature and report of a case. *J. Urol.* 41:303–309, 1939.

182* CUNNINGHAM, D.J. The form of the spleen and the kidneys. *J. Anat. Physiol.* 29:501–517, 1895.

183 CUNNINGHAM, G.J. Annular pancreas. *Br. J. Surg.* 27:678–681, 1939–40.

184 CURD, H.H. A histologic study of Meckel's diverticulum with special reference to heterotopic tissues. *Arch. Surg.* 32:506–523, 1936.

185* CURTIS, A.H. and B.J. ANSON. Bilateral double infundibulum of the uterine tube. *Anat. Rec.* 71:177–179, 1938.
186 CURTIS, A.H., ANSON, B.J. and C.B. McVAY. The anatomy of the pelvic and urogenital diaphragms, in relation to urethrocele and cystocele. *Surg. Gynecol. Obstet.* 68:161–166, 1939.
187* CURTIS, A.H., ANSON, B.J. and L.E. BEATON. The anatomy of the subperitoneal tissues and ligamentous structures in relation to surgery of the female pelvic viscera. *Surg. Gynecol. Obstet.* 70:643–656, 1940.
188* CURTIS, A.H., ANSON, B.J. and F.L. ASHLEY. Further studies in gynecological anatomy and related clinical problems. *Surg. Gynecol. Obstet.* 74:709–727, 1942.
189 CURTIS, G.M. and D. MOVITZ. The surgical significance of the accessory spleen. *Ann. Surg.* 123:276–298, 1946.
190 CUTORE, G. Rein unique ectopique et autres anomalies de développement dans le cadavre d'une petite fille. *Arch. Ital. Biol.* 60:309–310, 1913.
191 DALGAARD, J.B. and P. WETTELAND. Lateral cervical thyroid metastases. A follow-up study of 39 cases. (Aberrant thyroid tissue I). *Acta Chir. Scand.* 111:431–443, 1956.
192 DALGAARD, J.B. and P. WETTELAND. Thyroglossal anomalies. A follow-up of 58 cases. (Aberrant thyroid tissue II). *Acta Chir. Scand.* 111:444–455, 1956.
193 DALLA ROSA, L. Beitrag zur Casuistik und Morphologie der Varietäten des menschlichen Bronchialbaumes. *Wien. Klin. Wochenschr.* 2:437–438, 461–463, 483–487, 1889.
194* DANZIS, M. Congenital absence of the gall bladder. *Am. J. Surg.* 29:202–207, 308, 1935.
195* DARDINSKI, V.J. The anatomy of the major duodenal papilla of man, with special reference to its musculature. *J. Anat.* 69:469–478, 1934–35.
196* DARNER, H.L. Bilateral ectopic kidneys. *J. Urol.* 12:193–214, 1924.
197 DASELER, E.H. and B.J. ANSON. Anatomical relations of ectopic iliolumbar kidneys, bilateral in adult, unilateral in fetus. *J. Urol.* 49:789–802, 1943.
198 DAVIDSON, A. Notice of a case of malposition of the right kidney. *J. Anat. Physiol.* 2:282, 1868.
199 DAVIES, J.N.P. An accessory liver in an African. *Br. Med. J.* 2:736–737, 1946.
200* DAWSON, W. and J. LANGMAN. An anatomical-radiological study on the pancreatic duct pattern in man. *Anat. Rec.* 139:59–68, 1961.
201 DAY, R.V. Some rare anomalies of the kidney and ureter with case reports. *Surg. Gynecol. Obstet.* 38:51–57, 1924.
202 DE, M.N. and S.C. SINHA. A case of displaced and malformed kidney. *J. Anat.* 68:357–360, 1933–34.
203* DEBAKEY, M., AREY, J.B. and R. BRUNAZZI. Successful removal of lower accessory lung. *J. Thorac. Surg.* 19:304–311, 1950.
204 DEBEYRE, A. and O. RICHE. Surrénale accessoire dans l'ovarie. *Soc. Biol. Comptes Rendus Hebdomadaires des Séances et Mémoires* 63:733–734, 1907.
205 DEBEYRE, A., COULOUMA and DEVOS. Les scissures et les cloisons interzonaires du poumon. *Assoc. Anatomistes Comptes Rendus* 33:156–164, 1938.
206 DEES, J.E. Anomalous relationship between ureter and external iliac artery. *J. Urol.* 44:207–215, 1940.
207 DELHOUGNE, F. Über Pankreaskeime im Magen. *Arch. Klin. Chir.* 129:116–123, 1924.
208* DELMAS, J. Sur la forme du caecum. *Assoc. Anatomistes Comptes Rendus* 8:94–96, 1906.
209* DEPAUL. Deux observations tératologiques (avec deux planches), portant sur des examples remarquables de vices de conformation, l'un de utérus et du vagin, l'autre des organes genito-urinaires. *Soc. Biol. Comptes Rendus des Séances et Mémoires* 5:271–284, 1853.
210 DEPAUL, M. Imperforation congénitale du vagin; dilatation considérable de ce conduit avec accumulation de liquide dans son intérieur; estomac situé à droite. *Soc. Biol. Comptes Rendus des Séances et Mémoires* 9:46–48, 1857.
211 DERBES, V.J. and M.J. LANASA. Postcaval ureter and hydronephrosis. Case report. *Urol. Cutan. Rev.* 41:172–175, 1937.
212 DEROW, J.R. and A.L. STEARNS. Report of two cases of complete transposition of the viscera. *Military Surg.* 101:139–142, 1947. Cited in *Excerpta Medica*, Sec. 1, Vol. 3, abstract 897, 1949.
213 DESLONGCHAMPS and RAYER. Mamelles surnuméraires chez la femme. *Soc. Biol. Comptes Rendus des Séances et Mémoires* 4:9–10, 1852.
214 D'HOUR, H. and J. CRINQUETTE. Orifices bronchoscopiques et anatomie bronchique du lobe moyen. *Assoc. Anatomistes Comptes Rendus* 38:425–429, 1951.
215 DIAL, W.A. Retrocaval ureter and right aorta. *Anat. Rec.* 65:239–245, 1936.
216* DI BATTISTA, C., LAUDIZI, L. and G. TAMBORINO. Focomelia ed agenesis pene in neonato. *Minerva Pediatr.* 27:675–679, 1975.
217 DIXON, A.F. A rare condition of the vermiform appendix. *J. Anat. Physiol.* 31:442–445, 1897.
218 DIXON, A.F. Supernumerary kidney: The occurrence of three kidneys in an adult male subject. *J. Anat. Physiol.* 45:117–121, 1911.
219 DIXON, C.F. and A.L. LICHTMAN. Congenital absence of the gall bladder. *Surgery* 17:11–21, 1945.
220 DOMANSKY, K.Z. Hyperthelie-hypermastie. *Rozhledy v Chirurgii* 26:88–91, 1947. Cited in *Excerpta Medica*, Sec. 1, Vol. 3, abstract 384, 1949.
221 DONALD, C. Volvulus of small gut, caecum, and ascending colon, associated with congenital reversed rotation of intestine and with pregnancy. *Br. J. Surg.* 15:269–272, 1927–28.
222 DONALD, C. A case of human diphallus. *J. Anat.* 64:523–526, 1930.
223* DOTT, N.M. Anomalies of intestinal rotation: Their embryology and surgical aspects: with a report of five cases. *Br. J. Surg.* 11:251–286, 1923–24.
224+ DOVAY, E. Lobe pulmonaire accessoire par anomalie de l'azygos (lobule de Wrisberg). Le sillon creusé dan le poumon a-t-il une conséquence pathologique? *Bull. Mem. Soc. Anat.* 89:26–31, 1926.
225 DUBEY, P.N. Anomalous descending colon. *J. Anat. Soc. India* 7:112, 1907.
226 DUBEY, P.N., ROBERT, S.L. and K.P. HARDAS. Costomediastinal border of the left pleura. *J. Anat. Soc. India* 11:87–88, 1962.
227 DUBLED, M. Cas de transposition des viscères; recueilli á l'hôpital Cochin (service de M. Bertin). *Arch. Gén. Méd.* 6:573–577, 1824.
228 DUFF, P.A. Retrocaval ureter: Case report. *J. Urol.* 63:496–499, 1950.
229 DUGUET. Transposition complète des viscères thoraciques et abdominaux. *Soc. Biol. Comptes Rendus des Séances et Mémoires* 33:55–58, 1881.
230 DUNCAN, J.M. On cases of vagina duplex et uterus simplex and of saccated uterus. *J. Anat. Physiol.* 1:269–274, 1868.
231* DUPERTUIS, C.W., ATKINSON, W.B. and H. ELFTMAN. Sex differences in pubic hair distribution. *Hum. Biol.* 17:137–142, 1945.
232 DUPONT, R. and SIMARD. Absence congénitale de l'uterus et cancer de l'ovarie à type Séminome. *Ann. Anat. Pathol.* 2:529–540, 1925.
233* DVORAK, R. Über einen Fall von Pseudohermaphroditismus masculinus internus (tubularis). *Virchows Arch. Pathol. Anat. Physiol. Klin. Med.* 251:616–627, 1924.
234 DWIGHT, T. Case of absence of the right kidney. *J. Anat. Physiol.* 29:18–19, 1895.
235 DYKE, C.G. Indirect signs of brain tumor as noted in routine roentgen examinations. Displacement of the pineal shadow. (A survey of 3,000 consecutive skull examinations). *Am. J. Roentgenol.* 23:598–606, 1930.
236 EBERTH, C.J. Hochgradiger Defect der Epiglottis. *Arch. Pathol. Anat. Physiol. Klin. Med.* 43:135–136, 1868.
237 EDWARDS, H. Aberrant accessory ureter associated with hydronephrosis. *J. Anat.* 64:94–96, 1929.
238 EDWARDS, H. Congenital diverticula of the intestine: With the report of a case exhibiting heterotopia. *Br. J. Surg.* 17:7–21, 1929–30.
239 EISENDRATH, D.N. Anomalies of the bile ducts and blood vessels as the cause of accidents in biliary surgery. *JAMA* 71:864–866, 1918.
240* EISENDRATH, D.N. The clinical importance of anatomical anomalies in biliary surgery. *Boston Med. Surg. J.* 182:573–578, 1920.
241 EISERTH, P. Beiträge zur Kenntnis der Nebenlebern. *Arch. Pathol. Anat. Physiol. Klin. Med.* 307:307–313, 1941.
242 EKEHORN, G. Zur Behandlung durch anormale Nierengefäße verursachte Hydronephrose. *Z. Urol. Chir.* 10:122–124, 1922.
243 ELIASSON, R., MOSSBERG, B., CAMNER, P. and B.A. AFZELIUS. The immotile-cilia syndrome. A congenital ciliary abnormality as an etiologic factor in chronic airway infections and male sterility. *N. Engl. J. Med.* 297:1–6, 1977.
244 ENGEL, E. Nouvelles recherches sur les vaisseaux biliaires aberrants. *Arch. Ital. Biol.* 60:303–304, 1913.
245* ENGLE, E.T., YEAW, R.C. and J.K. LATTIMER. True hermaphroditism. Supplementary report of a case. *J. Urol.* 56:731–736, 1946.

246 ETTER, L.E. Variations in the position of the azygos septum and its incidence in fifty thousand roentgen examinations. *Am. J. Roentgenol.* 58:726–729, 1947.

247* EVANS, W. Polythelia in cardio-arterial disease. *Br. Heart J.* 21:130–136, 1959.

248 EWART, G.A. A case of hour-glass stomach. *Br. J. Surg.* 9:42–45, 1921–22.

249 FARAH, R. and G. RENO. Congenital absence of the penis. *J. Urol.* 107:154–155, 1972.

250 FAVARO, G. Les cavités pleurales rétrocardiaques de l'homme dans la transposition des viscères. *Arch. Ital. Biol.* 60:307, 1913.

251 FAWCETT, E. An unusually large terminal vermiform appendix, with recurved small conical caecum; with some remarks on the peritoneal pouches. *J. Anat. Physiol.* 29:498–500, 1895.

252 FELDMAN, M. and T. WEINBERG. Aberrant pancreas: A cause of duodenal syndrome. *JAMA* 148:893–898, 1952.

253 FERRANDU, S. Le alterazioni congenite del tubo digerente e di altri visceri dell'asimmetria interna del corpo e alle vedute di E. Pernkopf. *Arch. Ital. Anat. Embriol.* 36:1–34, 1936.

254* FERRY, R.M. JR. and E.A. BOYDEN. Variations in the bronchovascular patterns of the right lower lobe of fifty lungs. *J. Thorac. Surg.* 22:188–201, 1951.

255 FIBIGER, J. Beiträge zur Kenntnis des weiblichen Scheinzwittertums. *Virchows Arch. Pathol. Anat. Physiol. Klin. Med.* 181:1–51, 1905.

256 FLEINER, W. Situs viscerum inversus abdominalis mit Eventration des rechtsgelagerten Magens und Stauungsektasie der Speiseröhre. *Münchener Med. Wochenschr.* 63:113–117, 1916.

257* FLINT, E.R. Abnormalities of the right hepatic, cystic, and gastroduodenal arteries, and of the bile ducts. *Br. J. Surg.* 10:509–519, 1922–23.

258 FOCACCI, M. Contribution à l'étude des anomalies de développement. *Arch. Ital. Biol.* 39:167, 1903.

259 FORBES, G. Pelvic ectopic kidney. *Br. J. Surg.* 33:139–142, 1945–46.

260 FORSHALL, I. and P.P. RICKHAM. Transposition of the penis and scrotum. *Br. J. Urol.* 28:250–252, 1956.

261* FORSTER, A. Ein Fall überzähliger rudimentärer Mammabildung an der Innenseite des Oberschenkels eines Mannes. *Anat. Anz.* 49:529–535, 1916–17.

262 FOSTER-CARTER, A.F. Broncho-pulmonary abnormalities. *Br. J. Tuberc.* 40:111–124, 1946.

263 FRAENCKEL, P. Ein Fall von Pseudohermaphroditismus femininus externus. *Arch. Pathol. Anat. Physiol. Klin. Med.* 215:378–403, 1914.

264 FRAY, W.W. Roentgenologic study of orientation of the pineal body. *Arch. Neurol. Psychiatry* 38:1199–1207, 1937.

265 FRAY, W.W. A roentgenological study of pineal orientation. *Am. J. Roentgenol.* 39:899–907, 1938.

266* FREY, C., BIZER, L. and C. ERNST. Agenesis of the gallbladder. *Am. J. Surg.* 114:917–926, 1967.

267 FRIEDMAN, S. The position and mobility of the duodenum in the living subject. *Am. J. Anat.* 79:147–165, 1946.

268* FRIEND, E. Abnormalities of the bile ducts and their blood vessels and their surgical significance. *Illinois Med. J.* 56:169–180, 1929.

269* FURNISS, H.D. Supernumerary ureters with extravesical openings. *J. Urol.* 8:495–505, 1922.

270 GACHET, M. Observation de transposition du coeur et des autres viscéres chez un sujet de vingt et un ans du sexe féminin. *Gaz. Hôp.* 34:406, 1861.

271 GANFINI, C. Quelques particularités morphologiques et topographiques des glandulae suprarenales de l'homme. *Arch. Ital. Biol.* 46:323–324, 1906.

272 GANS, S.L. and W.J. POTTS. Anomalous lobe of lung arising from the esophagus. *J. Thorac. Surg.* 21:313–318, 1951.

273 GARDINER, J.P. A case of hour-glass stomach with accessory pancreas. *JAMA* 49:1598–1600, 1907.

274 GARDINI. Absence unilatérale du rein. *J. Urol. (Paris)* 1:105, 1912.

275* GARDNER, C.E. JR. The surgical significance of anomalies of intestinal rotation. *Ann. Surg.* 131:879–898, 1950.

276 GASSER, R.F. The early development of the parotid gland around the facial nerve and its branches in man. *Anat. Rec.* 167:63–77, 1970.

277 GEDDES, A.C. Apparent triplication of the apex of the right lung. *J. Anat. Physiol.* 45:11–15, 1910.

278 GEGENBAUER, C. Ein Fall von Nebenpankreas in der Magenwand. *Arch. Anat. Physiol. Wissen. Med.* 1863, pp. 163–165.

279 GELDEREN, C. VAN. Drei Fälle doppelseitiger, vollständig doppelter Ureteren (mit 4 Blasenostien). *Arch. Pathol. Anat. Physiol. Klin. Med.* 263:89–98, 1927.

280* GENNADIEW, A.N. Zur Frage der Morphologie des Lungenläppchens der Vena azygos. *Z. Anat. Entwicklungsgesch.* 92:178–212, 1930.

281* GÉRARD, G. Les anomalies congénitales du rein chez l'homme. Essai de classification d'après 527 cas. *J. Anat. Physiol. Norm. Pathol. Homme Animaux* 41:411–439, 1905.

282* GERWIG, W.H., COUNTRYMAN, L.K. and A.C. GOMEZ. Congenital absence of the gallbladder and cystic duct. *Ann. Surg.* 153:113–125, 1961.

283 GÉRY, L.H. De l'inversion complète des viscères; observation suivie de considérations généales sur ce sujet. *Arch. Gén. Méd.* 1:62–75, 1843.

284 GIACICH. Ein Fall geschlechtlicher Anomalie. *Wien. Med. Wochenschr.* 27:679–680, 1877.

285 GIANNELLI, L. Notes anatomiques sur l'appendice caecal. *Arch. Ital. Biol.* 41:474, 1904.

286 GIANNELLI, L., LAGHI A. and V. QUÉRCIOLI. Recherches sur le canal de sténon. *Arch. Ital. Biol. 32:460*, 1899.

287 GILLASPIE, C., MILLER, L.I. and M. BASKIN. Anomalies in lobation of lungs with review of literature. *Anat. Rec.* 11:65–75, 1916.

288* GILLIES, H. and R.J. HARRISON. Congenital absence of the penis. *Br. J. Plast. Surg.* 1:8–28, 1948.

289* GLADSTONE, R.J. A case in which the right ureter passed behind the inferior vena cava: With a short note upon a case in which the left renal vein passed behind the abdominal aorta, and the bearing of these abnormalities on the development of the abdominal veins. *J. Anat. Physiol.* 45:225–231, 1911.

290 GLADSTONE, R.J. Congenital absence of the appendix of the caecum. *J. Anat. Physiol.* 69:414–417, 1915.

291 GLADSTONE, R.J. A case of congenital absence of the left kidney and ureter. *J. Anat. Physiol.* 69:418–428, 1915.

292 GLADSTONE, R.J. Axillary mammae in a man. *J. Anat.* 69:239–246, 1930.

293 GLADSTONE, R.J. and C.P.G. WAKELEY. The relative frequency of the various positions of the vermiform appendix: As ascertained by an analysis of 3,000 cases: with an account of its development. *Br. J. Surg.* 11:503–520, 1923–24.

294 GLENN, J.F. Agenesis of the bladder. *JAMA* 169:2016–2018, 1959.

295 GLENN, J.F. Analysis of 51 patients with horseshoe kidney. *N. Engl. J. Med.* 261:684–687, 1959.

296 GODARD, E. Sur l'absence congénitale du testicule. *Soc. Biol. Comptes Rendus des Séances et Mémoires* 11:311–327, 1859.

297 GODARD, E. Note sur l'absence congénitale du canal excréteur et du réservoir de la semence, le testicule existant. *Soc. Biol. Comptes Rendus des Séances et Mémoires* 11:329–342, 1859.

298* GOODWIN, T.W. Anomalous additional cystic ducts: Report of two cases in which operation was performed. *Proc. Staff Meeting Mayo Clin.* 9:440–442, 1934.

299 GOOR, D.A. and P.A. EBERT. Anomalies of the biliary tree. Report of a repair of an accessory bile duct and review of literature. *Arch. Surg.* 104:302–309, 1972.

300 GORDON-TAYLOR, G. On horseshoes and on horseshoe kidney, concave downwards. *Br. J. Urol.* 8:112–118, 1936.

301* GORHAM, G.W. and J.G. MARSELIS. Kartagener's triad, a family study. *Johns Hopkins Hosp. Bull.* 104:11–16, 1959.

302 GOW, W.J. Total absence of the left lobe of the thyroid body. *J. Anat. Physiol.* 18:118, 1884.

303 GRAY, H.K. and F.B. WHITESELL. Anatomic relationship of the cystic duct to the cystic artery in 100 consecutive cases of cholecystectomy. *Surg. Clin. N. Am.* 30:1001–1004, 1950.

304* GRAY, J.H. The hair tracts of the Australian Aboriginal. *J. Anat.* 69:206–225, 1935.

305 GRAVES, R.C. and L.M. DAVIDOFF. Anomalous relationship of the right ureter to the inferior vena cava. *J. Urol.* 8:75–79, 1922.

306* GREISINGER, J.F. Supernumerary kidney. *J. Urol.* 38:331–356, 1937.

307 GRIFFITHS, J. Observations on the anatomy of the prostate. *J. Anat. Physiol.* 23:374–386, 1889.

308 GRIM, K.B. Bilateral renal and ureteral agenesis. *J. Urol.* 44:397–400, 1940.

309 GRÖNROOS, H. Über einen Fall abnormer Lagerung des Darmkanals beim Erwachsenen. *Anat. Anz.* 9:89–103, 1894.

310* GROSS, R.E. Congenital anomalies of the gallbladder. A review of 148 cases, with report of double gallbladder. *Arch. Surg.* 32:131–162, 1936.

311 GROSS, R.E., NEUHAUSER, E.B.D. and L.A. LONGINO. Thora-

cic diverticula which originate from the intestine. *Ann. Surg.* 131:363–375, 1950.

312* GROSS, R.E. and T.C. MOORE. Duplication of the urethra. Report of two cases and summary of the literature. *Arch. Surg.* 60:749–761, 1950.

313 GRUBER, G.B. Über das Ringpankreas (Pancreas anulare). *Münchener Med. Wochenschr.* 67:676, 1920.

314 GRUBER, G.B. Entwicklungsstörungen der Nieren und Harnleiter. *Münchener Med. Wochenschr.* 67:1001, 1920.

315 GRUBER, W. Weitere Beiträge zu den Bildungshemmungen der Mesenterien. *Arch. Anat. Physiol. Wissen. Med.* 1864, pp. 478–490.

316 GRUBER, W. Über das Vorkommen eines Mesenterium commune für das Jejuno-Ileum und die größere Anfangshälfte des Dickdarmes bei seitlicher Transposition der Viscera aller Rumpfhöhlen. Resultate aus den bis jetzt gemachten Beobachtungen seitlicher Transposition der Viscera aller Rumpfhöhlen zugleich, oder jener der Bauch- und Beckenhöhle allein. *Arch. Anat. Physiol. Wissen. Med.* 1865, pp. 558–569.

317 GRUBER, W. Weitere Beiträge zu den durch Bildungsfehler bedingten Lagerungsanomalien des Darmes. *Arch. Pathol. Anat. Physiol. Klin. Med.* 32:94–104, 1865.

318 GRUBER, W. Congenitaler Mangel der Parotis der rechten Seite in der Excavatio parotidea und Ersatz derselben durch eine in der Regio massetrico-buccinatoria gelagerte, voluminöse Drüse bei einem Manne. *Arch. Pathol. Anat. Physiol. Klin. Med.* 32:228–231, 1865.

319 GRUBER, W. Über das Zungenbein-Schildknorpel-Hilfsband (Ligamentum hyothyreoideum accessorium). *Arch. Anat. Physiol. Wissen. Med.* 1868, pp. 633–634.

320 GRUBER, W. Über Kehlköpfe mit supernumerären Articulationes crico-thyroideae. *Arch. Anat. Physiol. Wissen. Med.* 1874, pp. 454–462.

321 GRUBER, W. Über Kehlköpfe mit einem supernumerären Processus medianus; und über andere mit supernumerärem Tubercula lateralia am mittleren, zwischen den Gelenkwülsten gelagerten Abschnitte des oberen Randes der Lamina der Cartilago cricoidea. *Arch. Anat. Physiol. Wissen. Med.* 1874, pp. 463–466.

322 GRUBER, W. Über einen Kehlkopf des Menschen mit teilweise außerhalb desselben gelagerten seitlichen Ventrikelsäcken – Sacci ventriculares extralaryngei laterales. *Arch. Anat. Physiol. Wissen. Med.* 1874, pp. 606–632.

323 GRUBER, W. Eine accidentelle Bursa mucosa des Musculus laryngo-pharyngeus am Cornu majus der Cartilago thyreoidea. *Arch. Anat. Physiol. Wissen. Med.* 1875, pp. 590–592.

324 GRUBER, W. Ein gabelförmig gespaltener Gallenblasengang – Ductus cysticus bifurcatus. *Arch. Pathol. Anat. Physiol. Klin. Med.* 63:97, 1875.

325 GRUBER, W. Eine hinter einer zungenförmigen Verlängerung des linken Leberlappens versteckt gelagerte Milz mit anomaler Form und Anheftung. *Arch. Pathol. Anat. Physiol. Klin. Med.* 65:397–398, 1875.

326 GRUBER, W. Über die Glandula thyreoidea ohne Isthmus beim Menschen. *Arch. Anat. Physiol. Wissenschaftliche Med.* 1876, pp. 208–220.

327 GRUBER, W. Über die Glandula thyreoidea accessoria. *Arch. Pathol. Anat. Physiol. Klin. Med.* 66:447–454, 1876.

328 GRUBER, W. Über eine congenitale Articulatio hyothyreoidea anomalo. *Arch. Anat. Physiol. Wissen. Med.* 1876, pp. 753–756.

329 GRUBER, W. Eine Glandula thyreoidea bipartita bei Anwesenheit eines rudimentären Isthmus. *Arch. Pathol. Anat. Physiol. Klin. Med.* 66:454–455, 1876.

330 GRUBER, W. Kehlkopf mit theilweise außerhalb desselben gelagertem seitlichem Ventrikelsacke – Saccus ventricularis extralaryngeus lateralis – an der linken Seite. *Arch. Pathol. Anat. Physiol. Klin. Med.* 67:361–364, 1876.

331* GRUBER, W. Tiefe Lage der linken Niere und Lage der congenital enorm vergrößerten rechten Niere in ihrem unteren Viertel in der Fovea fasciae iliacae der Fossa iliaca dextra. *Arch. Anat. Physiol. Klin. Med.* 68:276–284, 1876.

332 GRUBER, W. Ein Fall von rechtsseitiger Cryptorchie mit Lagerung des Testikels und seiner Anhänge zwischen der die vordere Wand des Canalis inguinalis bildenden Muskelschichten. *Arch. Pathol. Anat. Physiol. Klin. Med.* 73:332–336, 1878.

333 GRUBER, W. Mesenterium commune für das Colon ascendens und Jejuno-Ileum. *Arch. Pathol. Anat. Physiol. Klin. Med.* 77:132–133, 1879.

334 GRUBER, W. Kehlkopf mit teilweise außerhalb desselben seitlich gelagertem Ventrikelsacke – Saccus ventricularis extralaryngeus lateralis dexter. *Arch. Pathol. Anat. Physiol. Klin. Med.* 78:106–107, 1879.

335 GRUBER, W. Vierlappige rechte Lunge eines Erwachsenen, in Folge Auftretens eines Spitzenlappens durch einen supernumerären verticalen Einschnitt-Verlauf des Bogens der Vena azygos in diesem Einschnitte. *Arch. Pathol. Anat. Physiol. Klin. Med.* 81:475–477, 1880.

336 GRUBER, W. Congenitaler Mangel beider Glandulae submaxillares bei einem wohlgebildeten, erwachsenen Subjecte. *Arch. Pathol. Anat. Physiol. Klin. Med.* 102:9–11, 1885.

337 GRUBER, W. Vierlappige rechte Lunge eines Erwachsenen in Folge Auftretens eines Spitzenlappens durch einen supernumerären-tiefen Einschnitt-Verlauf des Bogens der Vena azygos am Boden dieses Einschnittes. *Arch. Pathol. Anat. Physiol. Klin. Med.* 103:484–487, 1886.

338 GRUBER, W. Durch rechtwinklige Verschmelzung der Nieren an deren unteren Enden entstandene zweischenklige Niere mit Lagerung in der rechten Hälfte der Bauchhöhle und im hinteren Abschnitte der rechten Fossa iliaca. *Arch. Pathol. Anat. Physiol. Klin. Med.* 107:489–492, 1887.

339 GRUENWALD, P. and S. N. SURKS. Pre-ureteric vena cava and its embryological explanation. *J. Urol.* 49:195–201, 1943.

340 GUIZZETTI, P. and F. PARISET. Beziehungen zwischen Mißbildungen der Nieren und der Geschlechtsorgane. *Virchows Arch. Pathol. Anat. Physiol. Klin. Med.* 204:372–392, 1911.

341 GUTTMAN, P. Über zwei Fälle von congenitalem Mangel der rechten Niere, compliciert mit Anomalien der Genitalorgane. *Arch. Pathol. Anat. Physiol. Klin. Med.* 92:187–191, 1883.

342 GUYOT, J. and G. JEANNENEY. Étude physio-pathologique d'un vein a uretère double. *J. Urol. (Paris)* 15:81–88, 1923.

343 HAEBERLIN, C. A congenital anomaly of the iris. *Arch. Ophthalmol.* 37:717–718, 1908.

344 HALASZ, H. Niere mit doppeltem Ureter. *Anat. Anz.* 9:631–632, 1894.

345* HALBERT, B. and W.L. EATON. Accessory spleens: A pilot study of 600 necropsies. *Anat. Rec.* 109:371, 1951.

346 HALE, K. A study of the accessory pancreas. *Ann. Surg.* 83:774–781, 1926.

347 HALFF, J. Ein Fall von Situs inversus des Magens, des Duodenums und der Milz bei einem 63jährigen, weiblichen Individuum. *Münchener Med. Wochenschr.* 2:2287–2289, 1904.

348 HALPERT, B. and F. GYÖRKEY. Lesions observed in the accessory spleens of 311 patients. *Am. J. Clin. Pathol.* 32:165–168, 1959.

349 HALPERT, B. and F. GYÖRKEY. Accessory spleens: A survey of 3,000 necropsies. *Anat. Rec.* 133:389, 1959.

350 HAMDY, M. and M.F. SOROUR. On a case of displacement of the descending colon. *Pathol. Klin. Med.* 43:242–243, 1909.

351 HANDYSIDE, P.D. Notice of quadruple mammae, – the two lower rudimentary – in two adult brothers. *J. Anat. Physiol.* 7:56–59, 1873.

352 HANLEY, H.G. A horseshoe and a supernumerary kidney. A triple kidney with a horseshoe component. *Br. J. Surg.* 30:165–168, 1942–43.

353 HANLEY, H.G. and W.A. STEEL. The solitary ectopic pelvic kidney with reports of two further cases. *Br. J. Surg.* 34:402–407, 1946–47.

354 HARDY, W. Transposition congénitale des viscères. *Arch. Gén. Méd.* 3:277–278, 1833.

355 HARE, C.J. On a case of secretion of milk from the right axilla. *Lancet* 2:405–407, 1860.

356 HARMON, N.B. Anomalous female urogenital organs. *J. Anat. Physiol.* 32:211–215, 1898.

357 HARMON, N.B. The duodenal-jejunal flexure: Its variations, and their significance. *J. Anat. Physiol.* 32:665–674, 1898.

358* HARPSTER, C.M., BROWN, T.H. and H.A. DELCHER. Abnormalities of the kidney and ureter. A case of double kidney and double ureter with a review of the literature. *J. Urol.* 8:459–490, 1922.

359 HARRILL, H.C. Retrocaval ureter. Report of a case with operative correction of the defect. *J. Urol.* 44:450–457, 1940.

360 HARRIS, R.P. Congenital absence of the penis, the urethra making its exit into or below the rectum, and emptying the bladder by, or exterior to, the anus. *Philadelphia Med. Journal* 1:71–77, 1898.

361 HASSE, C. Ein seltener Fall von Lungenschnürung. *Anat. Anz.* 32:385–388, 1908.

362* HATFIELD, P.M. and R.E. WISE. Anatomic variation in the gallbladder and bile ducts. *Semin. Roentgenol.* 11:157–164, 1976.

363* HAUSCHILD, M.W. Über Situs inversus des Dünndarms infolge

364* HAYES, M.A., GOLDENBERG, I.S. and C.C. BISHOP. The developmental basis for bile duct anomalies. *Surg. Gynecol. Obstet.* 107:447–456, 1958.

abnormer Drehung des Dünndarmgekröses. *Anat. Anz.* 55:110–122, 1922.

365 HEADINGS, D.M. and R.E. PALMER. Supernumerary kidney with ureter opening into vagina. *Am. J. Surg.* 71:370–373, 1946.

366* HEALEY, J.E. JR. and P.C. SCHROY. Anatomy of the biliary ducts within the human liver. Analysis of the prevailing pattern of branchings and the major variations of the biliary ducts. *Arch. Surg.* 66:599–616, 1953.

367 HEDINGER, E. Kongenitale Divertikelbildung im Processus vermiformis. *Arch. Pathol. Anat. Physiol. Klin. Med.* 178:25–43, 1904.

368 HEDRY, N. VON. Nebenpankreas in der Gallenblasenwand. *Beitr. Klin. Chir.* 132:570–572, 1924.

369* HEINBACH, W.F. JR. A study of the number and location of the parathyroid glands in man. *Anat. Rec.* 56:251–261, 1933.

370 HELLER, A. Mangelhafte Entwicklung des rechten Leberlappens. *Arch. Pathol. Anat. Physiol. Klin. Med.* 51:355–357, 1870.

371* HELLSTRÖM, J. Zur Kenntnis der Doppelnieren. *Z. Urol. Chir.* 23:31–56, 1927.

372* HELLSTRÖM, J. Über die Varianten der Nierengefäße. *Z. Urol. Chir.* 24:253–272, 1928.

373 HELM, F. Zur Topographie der menschlichen Niere. *Anat. Anz.* 11:97–104, 1896.

374 HENRICHSEN, K. Pseudohermaphroditismus masculinus externus completus. *Arch. Pathol. Anat. Physiol. Klin. Med.* 94:211–217, 1883.

375 HENTZSCHEL. Congenital absence of the iris. *Am. J. Med. Sci.* 8:215, 1831.

376 HEPBURN, D. Floating kidney. *J. Anat. Physiol.* 19:178–185, 1885.

377 HEPBURN, D. Malformation of the female organs of generation – vagina duplex et uterus duplex – associated with the displacement of the right kidney. *J. Anat. Physiol.* 25:24–30, 1891.

378 HEPBURN, D. Note on a right lung which resembled a left lung in presenting only apical and basal lobes. *Journal Anat.* 59:326–327, 1925.

379 HEPPNER, C.L. Über den wahren Hermaphroditismus beim Menschen. *Arch. Anat. Physiol. Wissen. Med.* 1870, pp. 679–717.

380* HERMAN, J.L. and G. FETTEROLF. "Dumb-bell" kidney. *Ann. Surg.* 57:868–878, 1913.

381* HERRLIGKOFFER, K.M. Abnorme Lage des Processus vermiformis bei Hochstand des Caecum. *Anat. Anz.* 94:313–317, 1943.

382* HICKEN, N.F., CORAY, Q.B. and B. FRANZ. Anatomic variations of the extrahepatic biliary system as seen by cholangiographic studies. *Surg. Gynecol. Obstet.* 88:577–584, 1949.

383 HIGGINS, C.C. Horseshoe kidney with report of eighteen cases. *Int. Clin.* 4:198–206, 1928.

384 HJELM, R. and O. HULTÉN. Röntgenologische Studien über den Lobus der Vena azygos. *Acta Radiol.* 9:126–135, 1928.

385 HJORT, E.F. A case of double bladder. *Acta Chir. Scand.* 94:262–270, 1946.

386 HOCHSTETTER, F. Anomalien der Pfortader und der Nabelvene in Verbindung mit Defect oder Linkslage der Gallenblase. *Arch. Anat. Physiol. Wissen. Med.* 1886, pp. 369–384.

387 HOCHSTETTER, F. Über den Einfluß der Entwickelung der bleibenden Nieren auf die Lage des Urnierenabschnittes der hinteren Cardinalvenen. *Anat. Anz.* 3:938–940, 1888.

388 HODENPYL, E. A case of apparent absence of the spleen, with general compensatory lymphatic hyperplasia. *Med. Rec.* 54:695–698, 1898.

389 HOFFMANN, K. Über eine Spaltbildung im großen Netz, verbunden mit Mesenterium commune und Hernia inguinalis. *Anat. Anz.* 81:51–55, 1935–36.

390 HOLZAPFEL, R. Die Mündung von Gallen- und Pankreasgang beim Menschen. *Anat. Anz.* 69:449–453, 1930.

391 HOLZWEISSIG. Ein Pankreasdivertikel im Dünndarm. *Beitr. Pathol. Anat. Allg. Pathol.* 71:702–704, 1922.

392* HOWARD, J. and R. JONES. The anatomy of the pancreatic ducts. The etiology of acute pancreatitis. *Am. J. Med. Sci.* 214:617–622, 1947.

393 HOWARTH, V.S. Solitary kidney. Report of three cases. *Br. J. Surg.* 37:327–328, 1949–50.

394 HOWDEN, R. Case of misplaced kidney, with undescended testicle and rudimentary vas deferens on the same side. *J. Anat. Physiol.* 21:551–557, 1887.

395 HOWELL, L.M. Meckel's diverticulum. A consideration of the anomaly with a review of sixty-one cases. *Am. J. Dis. Child.* 71:365–371, 1946.

396 HUECK, W. Über das Mesenchym. Die Bedeutung seiner Entwicklung und seines Baues für die Pathologie. *Beitr. Pathol. Anat. Allg. Pathol.* 66:330–376, 1919.

397* HUFFMAN, L.F. Congenital displacements of the kidney. *J. Urol.* 12:363–377, 1924.

398 HUFFMAN, L.F. Unilateral renal aplasia, ureter opening into vas. *J. Urol.* 12:379–382, 1924.

399 HUMPHRY, L. Accessory lobe to the left lung. *J. Anat. Physiol.* 19:345–346, 1885.

400 HUNTINGTON, G.S. The genetic interpretation and surgical significance of some variations of the genitourinary tract. *Harvey Lect.* 2:222–267, 1908.

401 HÜTER, C. Großes angeborenes Divertikel der Urethra mit klappenförmigem Verschluß der Urethra und Incontinentia urinae. *Arch. Pathol. Anat. Physiol. Klin. Med.* 46:32–36, 1869.

402 IRMISCH, G.W. and E.N. COOK. Double and accessory urethra. *Minn. Med.* 29:999–1002, 1946.

403 IWANOW, G. Über die Lagebeziehungen der Nieren und Nebennieren beim Menschen. *Anat. Anz.* 64:163–173, 1927.

404* JACKSON, C.L. and J.F. HUBER. Correlated applied anatomy of the bronchial tree and lungs with a system of nomenclature. *Dis. Chest* 9:319–326, 1943.

405 JACKSON, C.M. An unusual duodenal diverticulum. *J. Anat. Physiol.* 42:219–220, 1908.

406 JACQUES. Deux cas d'ectopie thyroïdienne. *Soc. Biol. Comptes Rendus Hebdomadaires des Séances et Mémoires* 59:714–715, 1905.

407* JAIN, S.R., SEPAHA, G.C. and G.D. KHANDELWAL. A case of unilateral amastia. *J. Anat. Soc. India* 10:45–46, 1961.

408* JARCHO, J. Malformations of the uterus. Review of the subject, including embryology, comparative anatomy, diagnosis and report of cases. *Am. J. Surg.* 71:106–166, 1946.

409 JELESIJEVIC, V. Diverticulum processus vermiformis. *Anat. Anz.* 135:226–234, 1974.

410 JIT, I. Broncho-vascular anatomy of the azygos lobe of the lung. *J. Anat. Soc. India* 7:19–29, 1958.

411 JOB, T.T. Relationship of common bile and pancreatic ducts. *Anat. Rec.* 32:212–213, 1926.

412 JOHNSON, F.P. Urethral anomalies. *J. Urol.* 23:693–699, 1930.

413 JOHNSTON, W.G. JR. YEATMAN, G.W. and J.W. WEIGEL. Congenital absence of the penis. *J. Urol.* 117:508–511, 1977.

414 JONES, T. Lingual thyroid gland in a cretin of 78 years. *Anat. Rec.* 63:315–316, 1935.

415 JORDAN, H.E. The histology of a testis from a case of human hermaphroditism with a consideration of the significance of hermaphroditism in relation to the question of sex differentiation. *Am. J. Anat.* 31:27–53, 1922.

416 JUDD, E.S. Partial resection of the kidney. *Ann. Surg.* 82:458–471, 1925.

417 JURASZ, A. Eine seltene Anomalie des Schildknorpels und dessen Verbindung mit dem Zungenbeine. *Arch. Anat. Physiol. Wissen. Med.* 1877, pp. 399–400.

418 KADYI, H. Über accessorische Schilddrüsenläppchen in der Zungenbeingegend (Glandula praehyoides et suprahyoides). *Arch. Anat. Physiol. Wissen. Med.* 1879, pp. 312–320.

419 KAHN, S.A. and R.F. WAGNER, Jr. Polythelia and unilateral renal agenesis. *Cutis* 30:225–226, 1982.

420 KALBERG, W. Eine einseitige rechte verwachsene Niere. *Anat. Anz.* 74:117–123, 1932.

421 KANTHACK, A.A. The thyreo-glossal duct. *J. Anat. Physiol.* 25:155–165, 1891.

422* KANTOR, H. Zwei Fälle von Lebermißbildung. *Arch. Pathol. Anat. Physiol. Klin. Med.* 174:571–576, 1903.

423* KARLTREU, A.R.B. VON. Ein Fall von linksseitiger Doppelniere. *Arch. Pathol. Anat. Physiol. Klin. Med.* 152:545–550, 1898.

424* KARSNER, H.T. Gynecomastia. *Am. J. Pathol.* 22:235–315, 1946.

425 KARTAGENER, M. and A. HORLACHER. Bronchiektasien bei Situs viscerum inversus. *Schweiz. Med. Wochenschr.* 16:782–784, 1935.

426 KAUFMANN, C. Die Struma retropharyngo-oesophagea. *Dtsch. Z. Chir.* 18:255–292, 1883.

427 KEITH, A. and J.E. SPICER. Three cases of malformation of the tracheo-oesophageal septum. *J. Anat. Physiol.* 41:52–55, 1907.

428 KELLOGG, E.L. Abnormalities in the shape and position of the duodenum. *Am. J. Surg.* 12:462–465, 1931.

429 KELYNACK, T.N. Cases of Meckel's diverticulum. *J. Anat. Physiol.* 26:554–555, 1892.

430 KENNON, R. Double gallbladder opening by two cystic ducts into the common bile duct. *Br. J. Surg.* 20:522, 1933.
431 KESSLER, W. and A.P. MCLAUGHLIN. Agenesis of the penis. Embryology and management. *Urology* 1:226–229, 1973.
432 KEYL, R. Ein Fall von Ringpankreas. *Anat. Anz.* 58:209–218, 1924.
433 KING, E.S.J. An anatomical analysis of an example of polyorchidism. *J. Anat.* 65:427–437, 1931.
434 KOCH, W. Die angeborenen ungewöhnlichen Lagen und Gestaltungen des menschlichen Darmes. *Dtsch. Z. Chir.* 50:1–63, 1899.
435 KOENIG, U.D. Besondere Fehlbildungen bei Nebennieren. *Anat. Anz.* 132:303–309, 1972.
436* KOLISKO, F. Ein Fall abnormen Verlaufes des rechten Ureters. *Anat. Anz.* 34:520–524, 1909.
437* KOLLER, A. Ein Fall von Situs inversus und seine Deutung. *Arch. Pathol. Anat. Physiol. Klin. Med.* 156:115–150, 1899.
438 KORENCHEVSKY, V. The longest span of life based on records of centenarians in England and Wales. *Br. Med. J.* 2:14–16, 1947.
439 KORFEL, Z. Two cases of rare variations of the extrahepatic bile ducts observed *in vivo*. *Folia Morphol.* 25:216–218, 1966.
440* KORFEL, Z., CHOJNACKI, M., CHILIMONIUK, M. and I. SZEWKO-SZWAYKOWSKA. Variations of position of the abdominal viscera in man. *Folia Morphol.* 32:71–78, 1973.
441 KOSLOWSKY, B. Ein Fall von angeborener Hals-Schlundkopffistel (Fistula colli congenita). *Arch. Pathol. Anat. Physiol. Klin. Med.* 117:108–150, 1889.
442 KOSTANECKI, K. VON. Zur Kenntnis der Pharynxdivertikel des Menschen mit besonderer Berücksichtigung der Divertikelbildungen im Nasenrachenraum. *Arch. Pathol. Anat. Physiol. Klin. Med.* 117:108–150, 1889.
443 KOSTANECKI, K. VON and A.V. MIELECKI. Die angeborenen Kiemenfisteln des Menschen. Ihre anatomische Bedeutung und ihr Verhältnis zu verwandten branchiogenen Mißbildungen. *Arch. Pathol. Anat. Physiol. Klin. Med.* 120:385–436; 121:55–247, 1890.
444 KREILKAMP, B.L. and E.A. BOYDEN. Variability in the composition of the sphincter of Oddi. A possible factor in the pathologic physiology of the biliary tract. *Anat. Rec.* 76:485–497, 1940.
445* KREUTZMANN, H.A.R. Unusual ureteral anomaly. With a discussion of the embryology. *J. Urol.* 38:67–73, 1937.
446 KRIEG, E.G. Heterotopic pancreatic tissue producing pyloric obstruction. *Ann. Surg.* 113:364–370, 1941.
447 KRIVINKA, R. Über einen Fall von linksseitigem Vorkommen des Lobus Wrisbergi. *Röntgenpraxis* 11:234–237, 1939.
448 KRUMBHAAR, E.B. and S.W. LIPPINCOTT. Postmortem weight of "normal" human spleen of different ages. *Am. J. Med. Sci.* 197:344–358, 1939.
449* KRYSINSKI, S. Eine seltene Hymenanomalie. *Arch. Pathol. Anat. Physiol. Klin. Med.* 111:386, 1888.
450* KUBASSOW, P. Beitrag zur Lehre von der doppelten Gebärmutter (Uterus didelphys) nebst besonderer Würdigung der Aetiologie dieser Difformität. *Arch. Pathol. Anat. Physiol. Klin. Med.* 92:35–81, 1883.
451* KUBASSOW, P. Noch ein Fall von doppelter Gebärmutter und Scheide (Uterus et Vagina duplices), intra vitam diagnosticirt. *Arch. Pathol. Anat. Physiol. Klin. Med.* 92:81–86, 1883.
452 KUBOTA, K. A case of rare type of an accessory pancreas. *Folia Anat. Jpn.* 27:193–196, 1955.
453 KUKSTINAT, G.J. and R.J. HASTERLIK. Congenital absence of penis. *Arch. Pathol.* 27:984–993, 1939.
454* KULIGOWSKA, S. La longueur du vagin et son rapport a la stature de la femme. *Assoc. Anatomistes Comptes Rendus* 26:330–333, 1931.
455 KULIGOWSKA, S. Dimensions des grandes et des petites lèvres. *Assoc. Anatomistes Comptes Rendus* 26:334–339, 1931.
456 KÜTTNER, H. Über angeborene Verdoppelung des Penis. *Beitr. Klin. Chir.* 15:364–374, 1896.
457 LABBOK, A.J. Partielle retroperitoneale Lage des Magens, der Milz und des linken Leberlappens mit einer Hypoplasie der Mesenterien und mit Fehlen der großen und kleinen Omentum. *Anat. Anz.* 76:409–415, 1933.
458 LACHMAN, E. A comparison of the posterior boundaries of lungs and pleura as demonstrated on the cadaver and on the roentgenogram of the living. *Anat. Rec.* 83:521–542, 1942.
459* LADD, W.E. and R.E. GROSS. Congenital malformations of anus and rectum: Report of 162 cases. *Am. J. Surg.* 23:167–183, 1934.
460 LANDAU, T. Über einige Anomalien der Brustdrüsensecretion. *Dtsch. Med. Wochenschr.* 16:745–747, 1890.
461* LANDER, H.H., LYMAN, R.Y. and B.J. ANSON. An anatomical consideration of the structure in the hepatic pedicle. *Q. Bull. Northwestern University Medical School* 15:103–109, 1941.
462* LANGE, M. Über komplette Verdoppelung des Penis, kombiniert mit rudimentärer Verdoppelung der Harnblase und Atresia ani. *Beitr. Pathol. Anat. Allg. Pathol.* 24:223–230, 1898.
463 LANGER, C. Ein neuer Fall von Uterus masculinus bei Erwachsenen. *Arch. Anat. Physiol. Wissen. Med.* 1881, pp. 392–397.
464 LANGREDER, W. Die weibliche Urethra, funktionelle Anatomie, Pathologie und Therapie des Verschlußmechanismus. *Zentralb. Gynäkol.* 78:561–609, 1956.
465 LATIMER, E.O., MENDEZ, F.L. JR. and W.J. HAGE. Congenital absence of gallbladder. Report of a case. *Ann. Surg.* 126:229–242, 1947.
466 LAWRENCE, T.W.P. Cases of anomalous kidney and ureter. *J. Anat. Physiol.* 31:559–606, 1931.
467 LAYCOCK, H.T. and D.V. DAVIES. A case of true hermaphroditism. *Br. J. Surg.* 41:79–82, 1953–54.
468 LAZARUS, J.A. Horseshoe kidney. A report of five cases. *J. Urol.* 27:471–487, 1932.
469 LECCO, T.M. Pancrease anulare. *Wien. Klin. Wochenschr.* 23:875, 1910.
470 LECÈNE, P. Un cas de Triorchidie vraie chez l'homme. *Ann. Anat. Pathol.* 1:71–75, 1924.
471 LEDENYI, J. Vier Fälle von Leberanomalie. *Anat. Anz.* 72:281–294, 1931.
472 LEGENDRE, E.Q. Observation de monorchide chez l'homme. *Soc. Biol. Comptes Rendus des Séances et Mémoires* 8:216–218, 1856.
473 LEGROUX. Cancer du pylore. – Diagnostic rendu douteux par une transposition générale des vicères. *Gaz. Hôp.* 29:466, 1856.
474* LEICHTENSTERN. Über das Vorkommen und die Bedeutung supernumerärer (accessorischer) Brüste und Brustwarzen. Auf Grund 13 eigener und 19 aus der Literatur gesammelter Beobachtungen. *Arch. Pathol. Anat. Physiol. Klin. Med.* 73:222–256, 1878.
475 LEICHTENSTERN. Über das Vorkommen und die Bedeutung supernumerärer (accessorischer) nebst besonderer Würdigung der Aetiologie dieser Difformität. *Arch. Pathol. Anat. Physiol. Klin. Med.* 92:35–81, 1883.
476 LENHOSSEK, J. VON. Knorpelähnliche und wahre Knochenbildung im männlichen Gliede eines Erwachsenen. *Arch. Pathol. Anat. Physiol. Klin. Med.* 60:1–14, 1874.
477 LENHOSSEK, M. VON. Ectopia testis transversa. *Anat. Anz.* 1:376–381, 1886.
478* LESSER, E. Ein Fall von Hypertrichosis universalis und frühzeitiger Geschlechtsreife. *Z. Klin. Med.* 41:122–129, 1900.
479 LETULLE, M. Pancréas surnuméraires. *Société Biol. Comptes Rendus Hebdomadaires des Séances et Mémoires* 52:233–235, 1900.
480 LEUDET, E. Trois bronches naissant de la trachee. *Soc. Biol. Comptes Rendus des Séances et Mémoires* 8:54, 1856.
481 LEUDET, E. Rein gauche placé au niveau de l'angle sacro-vertébral; anomalie des artères et veines rénales. *Soc. Biol. Comptes Rendus des Séances et Mémoires* 8:55, 1856.
482 LEUDET, E. Remarques sur une variété d'utérus bicorne rudimentaire décrite par quelques auteurs comme des cas d'absence de l'utérus. *Soc. Biol. Comptes Rendus des Séances et Mémoires* 13:121–133, 1861.
483 LEVEN, L. Blinde Endigung des Halstheiles der Speiseröhre und directe Fortsetzung ihrer Pars thoracica in die Luftröhre. *Arch. Pathol. Anat. Physiol. Klin. Med.* 114:553–556, 1888.
484* LEWALD, L.T. Complete transposition of the viscera. A report of twenty-nine cases, with remarks on etiology. *JAMA* 84:261–268, 1925.
485 LIANG, L.S. and C.A.R. SNELL. The bronchial tree in Indonesians of East-Java (the right upper lobe bronchus). *Assoc. Anatomistes Comptes Rendus* 42:924–932, 1955.
486 LILJA, B. Displacement of the calcified pineal body in roentgen pictures as an aid in diagnosing intracranial tumors. *Acta Radiol. [Suppl.]* 37:1–182, 1939.
487* LINDVALL, S. and F. WAHLGREN. Ein Fall von Hermaphroditismus verus, Gynandromorphismus. *Virchows Arch. Pathol. Anat. Physiol. Klin. Med.* 297:1–15, 1936.
488 LINEBACK, P.E. An extraordinary case of situs inversus totalis. *JAMA* 75:1775–1778, 1920.
489 LINSS, W. Über einen Fall von Pseudohermaphroditismus. *Anat. Anz.* 113:131–135, 1963.
490 LIPSHUTZ, A. A note on a case of bifid penis. *J. Anat.* 58:254–255, 1924.

491 LIPSHUTZ, B. and C. HOFFMAN. A contribution to the knowledge of fused kidneys. *Ann. Surg.* 68:39–52, 1918.

492* LISA, L., HANAK, J., CERNY, M., FAFLOVA, H., KAFKA, H. and J. BRAZA. Agenesis of the penis. *J. Pediatr. Surg.* 8:327–328, 1973.

493 LOCHTE. Ein Fall von Situs viscerum irregularis, nebst einem Beitrag zur Lehre von der Transposition der arteriellen großen Gefäßstämme des Herzens. *Beitr. Pathol. Anat. Allg. Pathol.* 24:187–222, 1898.

494 LOCKHART, R.D. Congenital absence of the gall bladder. *J. Anat. Physiol.* 57:108–109, 1928.

495 LOCKWOOD, C.B. Development and transition of the testis, normal and abnormal. *J. Anat. Physiol.* 21:635–664, 1887.

496* LOCKWOOD, C.B. Development and transition of the testis, normal and abnormal. *J. Anat. Physiol.* 22:38–77, 1888.

497 LOCKWOOD, C.B. On the presence of adrenal structures in the inguinal canal. *J. Anat. Physiol.* 34:79–83, 1899.

498 LOHMAN, A.J.M. and ST. J. ZIEKENHUIS. Partieele verdubbeling van den dikken darm (colon duplex) met unipolaire atresie. Nederlandsch. *Tijdschrift voor Geneeskunde* 90:1735–1738, 1946. Cited in *Excerpta Medica*, Sec. 1, Vol. 2: abstract 200, p. 101, 1948.

499 LORENZ, R. Zur Lagebestimmung der verkalkten Glandula pinealis im Röntgenbild. *Fortschritte Röntgenstr.* 61:338–348, 1940.

500 LOURIA, H.W. Case of unilateral amastia. *Am. J. Obstet. Gynecol.* 8:364–365, 1924.

501* LUCIEN, M. and A. BEAU. La systématisation pulmonaire. Ses bases morphologiques et ses modalitiés. *Assoc. Anatomistes, Comptes Rendus* 38:3–92, 1951.

502 LUNN, H.F. The anatomy of inguinal hernia. *Ann. R. Coll. Surg.* 2:285–298, 1948.

503 LURJE, A. Topography of extrahepatic biliary passages with reference to dangers of surgical technique. *Ann. Surg.* 105:161–168, 1937.

504 LUSCHKA, H. VON. Das vordere Mittelstück der Prostata und die Aberration desselben. *Arch. Pathol. Anat. Physiol. Klin. Med.* 34:592–597, 1865.

505 LUSCHKA, H. VON. Blinde Endigung des Halstheiles der Speiseröhre und Communication ihrer Pars thoracica mit der Luftröhre. *Arch. Path. Anat. Physiol. Klin. Med.* 47:378–381, 1869.

506 LYONS, A.S. The sigmoid as a source of right-sided symptoms. *Ann. Surg.* 127:399–409, 1948.

507* MACKENZIE, C.F., MCASLAN, T.C., SHIN, B., SCHELLINGER, D. and M. HELRICH. The shape of the human adult trachea. *Anesthesiology* 49:48–50, 1978.

508 MACMAHON, H.E. Congenital anomalies of the liver. *Am. J. Pathol.* 5:499–507, 1929.

509 MACPHAIL, S.R. Note on length of vermiform process in 220 consecutive post-mortem examinations. *J. Anat.* 51:308, 1917.

510 MADDOX, E.E. Right-sided sigmoid flexure and rectum. *J. Anat. Physiol.* 17:403, 1883.

511* MADELUNG. Anatomisches und Chirurgisches über die Glandula thyreoidea accessoria. *Von Langenbeck's Arch. Klin. Chir.* 24:71–107, 1879.

512 MAGNON, L.A. Anomalies et malformations génitales de la femme: étude clinique et radiologique. *J. Chir.* 34:674–676, 1929.

513 MAGNUS, H. Über eine Anomalie der männlichen Harnröhre. *Arch. Pathol. Anat. Physiol. Klin. Med.* 47:307–308, 1869.

514 MAGOUN, J.A.H. Renal anomalies. *J. Urol.* 27:435–469, 1932.

515 MAHON, R.B. Abnormal arrangement of the kidney and its vessels. *J. Anat. Physiol.* 23:339–340, 1889.

516* MAISEL, H. The position of the human vermiform appendix in fetal and adult age groups. *Anat. Rec.* 136:385–389, 1960.

517 MAKHNI, S.S. Some variations of the human lingular bronchi. *J. Anat. Soc. India* 6:40–43, 1957.

518 MALBIN, M. Mobile calcified choroid plexus. *Radiology* 51:383–386, 1948.

519* MARCHAND. Über accessorische Nebennieren im Ligamentum latum. *Arch. Pathol. Anat. Physiol. Klin. Med.* 92:11–19, 1883.

520 MARCHAND. Ein neuer Fall von Hermaphroditismus. (H. spurius masculinus?) *Arch. Pathol. Anat. Physiol. Klin. Med.* 92:286–295, 1883.

521 MARION, G. Un cas d'urètre double chez l'homme. *J. Urol. (Paris)* 1:235–236, 1912.

522 MARSH, F. Two testicles on one side. *Br. Med. J.* 2:1354, 1911.

523 MARSH, J.P. Congenital absence of the entire oesophagus with report of a case. *Am. J. Med. Sci.* 124:304–309, 1902.

524 MARSHALL, C.F. Thyro-glossal duct or canal of His. *J. Anat. Physiol.* 26:94–99, 1892.

525* MARSHALL, C.F. Variations in the form of the thyroid gland in man. *J. Anat. Physiol.* 29:234–239, 1895.

526 MARTEN, A. Die angeborene Verwachsung des Penis und Scrotum. *Arch. Pathol. Anat. Physiol. Klin. Med.* 28:555–556, 1863.

527* MASON, M.L., ANSON, B.J. and L.E. BEATON. The surgical and anatomical aspects of a case of double lip. *Surg. Gynecol. Obstet.* 70:12–17, 1940.

528 MATHER, J.H. and R. COOPE. Accessory lobe of the azygos vein. *Br. J. Radiol.* 1:481–485, 1928.

529 MATTHEW. On eunuchs. *Bible, New Testament* Matthew 19:12, Circa 80 A.D.

530 MAYO, C.W. and R.G. RICE. Situs inversus totalis. A statistical review of data on seventy-six cases, with special reference to diseases of the biliary tract. *Arch. Surg.* 58:724–730, 1949.

531 MCCORMICK, W.F. and M. KASHGARIAN. The weight of the adult human spleen. *Am. J. Clin. Pathol.* 43:332–333, 1965.

532 MCCOY, W.N. Imperforate anus; the rectum opening into the vulva; successful operation. *American J. Med. Sci.* 74:287–288, 1877.

533* MCGILLICUDDY, T.J. The mammae and their anomalies. *Med. Rec.* 40:446–448, 1891.

534* MCLAUGHLIN, C.R. Reduplication of mouth, tongue and mandible. *Br. J. Plast. Surg.* 1:89–95, 1948.

535 MCWHORTER, G.L. The relations of the superficial and deep lobes of the parotid gland to the ducts and to the facial nerve. *Anat. Rec.* 12:149–154, 1917.

536 MÉHES, K. Association of supernumerary nipples with other anomalies. *J. Pediatr.* 95:274–275, 1979.

537 MÉHES, K. Association of supernumerary nipples with other anomalies. *J. Pediatr.* 102:161, 1983.

538 MEINERTZ, J. Ein ungewöhnlicher Fall von angeborener Mißbildung des Herzens. *Arch. Pathol. Anat. Physiol. Klin. Med.* 166:385–403, 1901.

539* MEITNER, E.R. Über ein monoventrikuläres Herz. *Anat. Anz.* 105:301–304, 1958.

540* MEITNER, E.R. Über ein Cor triatriatum. *Anat. Anz.* 105:305–308, 1958.

541 MELCHIONNA, R.H. and R.A. MOORE. The pharyngeal pituitary gland. *Am. J. Pathol.* 14:763–773, 1938.

542 MELISSINOS, K. Beckenniere mit persistierender Vena cardinalis dextra. *Anat. Anz.* 39:149–155, 1911.

543 MELSOME, W.S. Hour-glass stomach, with peculiar pyloric orifice. *J. Anat. Physiol.* 29:xxvi–xxvii, 1895.

544 MENZIES, W.F. Two cases of single kidney. *J. Anat. Physiol.* 21:510, 1887.

545 MESSNER. Ein neuer Fall von Hermaphroditismus verus (Hermaphroditismus verus unilateralis?). *Arch. Pathol. Anat. Physiol. Klin. Med.* 129:203–213, 1892.

546 MEYER. Different species of double uterus; uterus bicameratus, uterus bilocularis; uterus bicornis and uterus bipartitus. *Am. J. Med. Sci.* 7:222, 1830.

547 MEYER, H. Ein Fall von Hermaphroditismus lateralis. *Arch. Pathol. Anat. Physiol. Klin. Med.* 11:420–427, 1857.

548 MEYER, R. Zum Mangel der Geschlechtsdrüsen mit und ohne zwittrige Erscheinungen. *Arch. Pathol. Anat. Physiol. Klin. Med.* 255:33–46, 1925.

549* MIKULICZ-RADECKI, V. VON. Die Varianten des Hermaphroditismus. *Z. Geburtsh. Gynäkol.* 152:1–23, 1959.

550* MILLBOURN, E. On the excretory ducts of the pancreas in man, with special reference to their relations to each other, to the common bile duct and to the duodenum. *Acta Anat.* 9:1–34, 1950.

551 MILLS, W.M. Two cases of supernumerary kidney. *J. Anat. Physiol.* 46:313–318, 1912.

552 MILLZNER, R.J. The occurrence of parathyroids on the anterior surface of the thyroid gland. *JAMA* 88:1053–1055, 1927.

553* MILLZNER, R.J. The normal variations in the position of the human parathyroid glands. *Anat. Rec.* 48:399–405, 1931.

554 MINNE, J., SENNEVILLE, A., LIBERSA, C. and C. VOISIN. Estomac bilocule congénital. *Assoc. Anatomistes Comptes Rendus* 38:731–735, 1951.

555 MITCHELL, L.J. Notes on a series of thirty-nine cases of Meckel's diverticulum. *J. Anat. Physiol.* 32:675–678, 1898.

556 MIURA. Ein Fall mit angeborenen Herzanomalien. *Arch. Pathol. Anat. Physiol. Klin. Med.* 115:353–355, 1889.

557 MOËLL, H. Size of normal kidneys. *Acta Radiol.* 46:640–645, 1956.

558 MOLE, R.H. Congenital non-rotation of the intestine. *Br. J. Surg.* 17:670–672, 1929–30.

559 MOLLER, P. Ein Fall von komplettem Pseudohermaphroditismus

masculinus. *Arch. Pathol. Anat. Physiol. Klin. Med.* 223:363–379, 1917.
560 MOLLWO, C. Zwei Fälle von angeborenen Herzfehlern. *Arch. Pathol. Anat. Physiol. Klin. Med.* 19:438–444, 1860.
561 MONTEIRO, H., RODRIGUES, A. and O. RIBEIRO. Quelques cas de variations de l'appareil urinaire supérieur. *Ann. Anat. Pathol.* 5:222–229, 1928.
562 MOODY, R.O., CHAMBERLAIN, W.E. and R.G. VAN NUYS. Visceral anatomy of healthy adults. *Am. J. Anat.* 37:273–288, 1926.
563* MOODY, R.O. and R.G. VAN NUYS. The position and mobility of the kidneys in healthy young men and women. *Anat. Rec.* 76:111–133, 1940.
564 MOORE, F.C. Unilateral renal aplasia: "Unsymmetrical kidney." *J. Anat. Physiol.* 33:400–412, 1899.
565 MOORE, T.C. and J.S. BATTERSBY. Congenital duplications of the small intestine. Report of 11 cases. *Surg. Gynecol. Obstet.* 95:557–567, 1952.
566* MOOSMAN, D.A. and F.A. COLLER. Prevention of traumatic injury to the bile ducts. A study of the structures of the cystohepatic angle encountered in cholecystectomy and supraduodenal choledochostomy. *Am. J. Surg.* 82:132–143, 1951.
567 MORGAN, E.K. and C.M. STONE. Bilateral renal ectopia. *J. Urol.* 38:427–429, 1937.
568* MORIAN. Über das offene Meckel'sche Divertikel. *Arch. Klin. Chir.* 58:306–316, 1899.
569 MORITO, Y. Über eine seltene Varietät des Blinddarmes und Colon ascendens bei einem Japaner. *Folia Anat. Jpn.* 21:541–551, 1941–42.
570* MOURA, A.C. DE, and L.P. BASTO. True hermaphroditism. *J. Urol.* 56:725–730, 1946.
571 MÜLLER, H. Über Situs inversus partialis. *Beitr. Pathol. Anat. Allg. Pathol.* 51:632–647, 1911.
572 MURPHY, J.W. and W.A. MITCHELL. Congenital absence of the spleen. *Pediatrics* 20:253–256, 1957.
573 MUSCHAT, M. Perineal testicle. *J. Urol.* 30:353–355, 1933.
574 NAJWER, K. and J. JASZKIN-GRUDZIŃSKA. A case of two-lobar parotid gland. *Folia Morphol.* 41:335–337, 1982.
575 NAKAZIMA, T. Eine Riesenschlingenbildung des Colon transversum bei einem Japaner. *Folia Anat. Jpn.* 18:219–222, 1939.
576 NAT, B.S., MOOKERJI, I.M.S. and P.D. MOOKERJI. Abnormal duodenum. *J. Anat.* 64:250–253, 1930.
577 NAYAK, U.V. and A.A. AYYAR. Anomalies in the development of salivary glands. *J. Anat.* 61:261–262, 1927.
578 NEIL, J.H. Azygos lobes of the lung, and terminology of the bronchopulmonary segments of the bronchial tree. *Ann Otol. Rhinol. Laryngol.* 59:409–413, 1950.
579 NEILL, J. Case of hermaphroditism. *Am. J. Med. Sci.* 22:558, 1851.
580 NELSON, H.P. and G. SIMON. The accessory lobe of the azygos vein. *Br. Med. J.* 1:9–11, 1931.
581 NICOLA, N.E. Abnormal position of the iliac and pelvic colons. *J. Anat.* 56:53, 1921.
582* NIMEH, W. Les anomalies du duodénum. *La Presse Médicale* No. 69. 30:746, 1922.
583 NOER, R.J. and C.G. JOHNSTON. Decompression of the small bowel in intestinal obstruction. *Am. J. Dig. Dis.* 6:46–49, 1939.
584 NOTICES, A. Bilobate uterus. *Am. J. Med. Sci.* 7:222, 1830.
585 OBICI, A. Les anomalies d'embonchure des Uretères. *Arch. Ital. Biol.* 26:486–488, 1896.
586 ODDONO, E. Sur un rein en ectopie pelvienne congénitale et sur la segmentation du rein. *Arch. Ital. Biol.* 32:468, 1899.
587 OGDEN, H.D. and E. MALTRY. Pelvic single kidney. Report of an additional case. *J. Urol.* 44:13–18, 1940.
588* OGNEW, B.W. Die Hufeisenniere des Menschen im Zusammenhang mit Varianten im Bau anderer Organe. *Anat. Anz.* 69:330–341, 1930.
589* O'HEERON, M.K. and F.A. WEBSTER. Report of two cases of double urethra in the male. *J. Urol.* 55:391–396, 1946.
590* OLLIVIER, A. Note sur un cas d'utérus et de vagin doubles. *Soc. Biol. Comptes Rendus des Séances et Memoires* 22:35–38, 1870.
591 ORR, A.E. Hour-glass stomach. *J. Anat. Physiol.* 16:49–50, 1907.
592* OSEMLAK, J. Variation of the shape and size of the human uterus in different periods of life. *Folia Morphol.* 30:442–456, 1971.
593* OSEMLAK, J. A combination of congenital anomalies of the bile ducts not described hitherto in man. *Folia Morphol.* 31:123–126, 1972.
594 OTIS, F.N. Remarks on the urethral calibre and its proportionale relation to the size of the penis. *Arch. Dermatol.* 2:210–212, 1876.
595 PAISECKI, Z., JUGOWSKI, F. and J. PIOTROWSKI. Forms of renal pelvises in man in casts and radiograms. *Folia Morphol.* 28:161–171, 1969.
596 PAN, N. Misplaced kidney. *J. Anat.* 58:271–272, 1924.
597 PAN, N. Transposition of abdominal viscera. *J. Anat.* 60:202–206, 1926.
598 PARAVICINI, G. Cas de polymastie, ou de pléomazie, chez des individus de sexe masculin. *Arch. Ital. Biol.* 40:325, 1903.
599 PARDI, F. Sur une rare variété de la glandula sublingualis dans l'espèce humaine. Le ductus sublingualis major s. Bartholini et la glandula sublingualis monostomatica s. Bartholini de l'homme. *Arch. Ital. Biol.* 47:493–494, 1907.
600 PARISOT. Transposition complète des viscères. *Arch Gén. Méd.* 5:222, 1839.
601* PATEY, D.H. and I. RANGER. Some points in the surgical anatomy of the parotid gland. *Br. J. Surg.* 45:250–258, 1957.
602 PAULL, F. Total absence of the left mammary gland from non-development. *Lancet* 1:648, 1862.
603 PEDERSEN, J. Hypertrichosis in women. *Acta Dermato-Venereol.* 23:1–13, 1942.
604 PELLEGRINI, J.R. and R.F. WAGNER, JR. Polythelia and associated conditions. *Am. Fam. Physician.* 28:129–132, 1983.
605 PERKINS, R.A. Supernumerary breast on buttock. *JAMA* 76:792, 1921.
606* PERNKOPF, E. Der partielle Situs inversus der Eingeweide beim Menschen. Gedanken zum Problem der Asymmetrie und zum Phänomen der Inversion. *Z. Anat. Entwicklungsgesch.* 79:577–752, 1926.
607 PERNKOPF, E. Eine divertikelartige Bildung am Magenfundus beim Menschen. *Anat. Anz.* 66:257–262, 1928–29.
608 PERRUCHIO, P., MOLLARET, L., SOUTOUL, J. and G. BAUD. Note à propos d'un cas de duplicité de l'appendice vermiforme et d'un diverticule coecal pseudoappendiculaire. *Assoc. Anatomistes Comptes Rendus* 45:643–646, 1958.
609 PETREN, T. Abweichende Leberlobierung. *Anat. Anz.* 63:362–364, 1927.
610 PICK, J.W. and B.J. ANSON. Retrocaval ureter: Report of a case, with a discussion of its clinical significance. *J. Urol.* 43:672–685, 1940.
611 PIERACCINI, G. Asimmetria morfologica, differenze anatomica e fisiologica tra le mammelle nella donna. *Arch. Ital. Anat. Embriol.* 33:71–94, 1934.
612 PIERGROSSI, A.R. Sul lobo medio del polmone sinistro. *Radiol. Med.* 32:154–165, 1946. Cited in *Excerpta Medica*, Sec. 1, Vol 2, abstract 529, p. 251, 1948.
613 PIERSON, L.E. Unilateral fused kidney. *J. Urol.* 28:217–231, 1932.
614 PILLIET, A.H. and V. VEAU. Capsule surrénale aberrante de ligament large. *Soc. Biol. Comptes Rendus Hebdomadaires des Séances et Mémoires* 49:64–68, 1897.
615 PIRES DE LIMA, J.A. Note on a case of bifid penis, with penial hypospadia. *J. Anat. Physiol.* 44:85–89, 1915.
616 PIRES DE LIMA, J.A. La duplicité de l'anus et sa signification morphologique. *Assoc. Anatomistes Comptes Rendus* 22:203–208, 1927.
617* PITEL, M. and E.A. BOYDEN. Variations in the bronchovascular patterns of the left lower lobe of fifty lungs. *J. Thorac. Surg.* 26:633–653, 1953.
618 PITZORNO, M. Rare anomalie rénal gauche. *Arch. Ital. Biol.* 26:160, 1896.
619* POHLANDT, F., KÜHN, H., TELLER, W. and H. THOMÄ. Penisagenesie. Weibliche Geschlechtszuweisung unter psychotherapeutischer Betreuung der Eltern. *Dtsch. Med. Wochenschr.* 99:2166–2172, 1974.
620* POHLMAN, A. Abnormalities in the form of the kidney and ureter dependent on the development of the renal bud. *Bull. Johns Hopkins Hosp.* 16:51–60, 1905.
621 POLHEMUS, D.W. and W.B. SCHAFER. Congenital absence of the spleen: Syndrome with atrioventricularis communis and situs inversus. *Pediatrics* 9:696–718, 1952.
622 PONFICK. Ein Fall von angeborener primärer Atrophie der rechten Lunge. *Arch. Pathol. Anat. Physiol. Klin. Med.* 50:633–637, 1870.
623 POOL, E.H. and H.C. FALK. Concerning the surgical anatomy of the thyroid, with special reference to the parathyroid glands. *Ann. Surg.* 63:71–77, 1916.
624 PORTER, E.C. Measurement of the cervical spinal cord in pantopaque myelography. *Am. J. Roentgenol.* 76:270–272, 1956.
625 POTEL, G. and P. CORDIER. Pourquoi le vein mobile est-il plus fréquent a droite qu'a gauche? *J. Urol. (Paris)* 15:161–180, 1923.

626* PRIESEL, A. Verdoppelung der Gallenblase beim Menschen. *Arch. Pathol. Anat. Physiol. Klin. Med.* 265:76–83, 1927.
627 PRIESEL, A. Angeborenes Fehlen beider Samenleiter bei normalen Harnwegen. *Arch. Pathol. Anat. Physiol. Klin. Med.* 286:24–41, 1932.
628 PRIMAN, J. Uretère double et loi de Weigert-Meyer. *Soc. Biol. Comptes Rendus Hebdomadaires des Séances et Mémoires* 90:1125, 1924.
629 PRIMAN, J. A consideration of normal and abnormal positions of the hilum of the kidney. *Anat. Rec.* 42:355–363, 1929.
630 PROSBA-MACKIEWICZ, M. and H. SZOSTAKIEWICA-SAWICKA. Remarks on the position of the sublingual salivary gland in relation to the mandible. *Folia Morphol.* 32:149–164, 1973.
631 PRZYSTASZ, T. Additional pancreas. *Folia Morphol.* 39:341–345, 1980.
632 PTASIŃSKA-URBAŃSKA, M. Variation of the structure of the tarsal discs in man. *Folia Morphol.* 24:273–279, 1965.
633* PTASIŃSKA-URBAŃSKA, M. A rare case of more than fourteen centimeter long cutaneous horn of the hair-covered skin of the head. *Folia Morphol.* 29:478–479, 1970.
634* PTASIŃSKA-URBAŃSKA, M. A rare case of *cutis verticis gyrata*. *Folia Morphol.* 30:127–128, 1971.
635 PURVES, R. and J.A. HADLEY. Accessory breasts in the labia majora. *Br. J. Surg.* 15:279–281, 1927.
636 PYE-SMITH, P.H. Lateral transposition of the viscera. *Trans. Pathol. Soc. Lond.* 19:447, 1868.
637 QUÉNU, L., CHABROL, J. and P. HERLEMONT. Le colon, ses variations, ses artères. *Assoc. Anatomistes Comptes Rendus* 41:760–769, 1954.
638 QUERVAIN, F. DE. Über Rechtslagerung des ganzen Dickdarms und partiellen Situs inversus. *Arch. Klin. Chir.* 65:256–265, 1902.
639 QUINN, W.P. A case of bilateral renal ectopy. *J. Urol.* 44:10–12, 1940.
640 RADWIN, L.S., MICHELSON, J.P. and J. SHORE. Primary ovarian agenesis. *J. Pediatr.* 34:143–154, 1949.
641 RAHMAN, M.A.A. A subject with complete transposition of viscera. *J. Anat.* 51:304–307, 1917.
642 RAINS, A.J.H. Congenital absence of the gall-bladder: Cholangiography: gall stone solvents. *Br. J. Surg.* 39:37–39, 1951–52.
643 RALEIGH. Ein Fall von verkehrter Lage der Baucheingeweide. *Schmidt's Jahrb. Med.* 14:21–22, 1837.
644 RAMAN, T.K. Azygos lobe of lung. *J. Indian Med. Assoc.* 21:98–99, 1944.
645* RANDALL, A. and E.W. CAMPBELL. Anomalous relationship of the right ureter to the vena cava. *J. Urol.* 34:565–583, 1935.
646 RANDERATH, E. Über einen Fall von angeborenem Mangel beider Eierstöcke. *Arch. Pathol. Anat. Physiol. Klin. Med.* 254:798–810, 1925.
647 RANIER, F.J. Vier Fälle von topographischen Anomalien des Darmes. *Int. Monatsschr. Anat. Physiol.* 24:247–252, 1908.
648 RANNSHOFF, J. Considerations of the anatomy, physiology and pathology of the caecum and appendix. *JAMA* 11:40–46, 1888.
649 RAO, D.S. The azygos lobe of the lung. *J. Indian Med. Assoc.* 21:69, 1951.
650* RÄUBER. Angeborener Mangel des männlichen Gliedes. *Arch. Pathol. Anat. Physiol. Klin. Med.* 121:604, 1890.
651* REGMUNT-SOBIESZCZANSKI, L. Nouvelle modification du tubercule de Darwin complétant la classification de G. Schwalbe. *Assoc. Anatomistes Comptes Rendus* 26:443–444, 1931.
652 REID, D.G. Imperfect torsion of the intestinal loop. *J. Anat. Physiol.* 42:320–325, 1908.
653 REID, D.G. Bilateral duplication of the ureter. *J. Anat.* 51:30–45, 1917.
654 REINER, L., MAZZOLENI, A., RODRIGUEZ, F.L. and R.R. FREUDENTHAL. The weight of the human heart. *Arch. Pathol.* 68:58–73, 1959.
655 REITMANN, K. Zwei Fälle von accessorischem Pankreas. *Anat. Anz.* 23:155–157, 1903.
656 RICHARDS, W.F. Situs inversus viscerum, absent frontal sinuses with ethmoid and maxillary infection and bronchiectasis. Kartagener's triad. *Tubercle* 25:27–29, 1944. Cited in *Excerpta Medica*, Sec. 1, Vol. 2, abstract 527, p. 251, 1948.
657* RICHART, R. and K. BENIRSCHKE. Penile agenesis. Report of a case, review of the world literature, and discussion of pertinent embryology. *Arch. Pathol.* 70:252–260, 1960.
658 RICHMOND, W.S. Abnormal ureters. *J. Anat. Physiol.* 19:120, 1885.
659 RIECHELMANN, W. Über Situs viscerum inversus. *Dtsch. Z. Chir.* 74:345–350, 1904.
660* RIENHOFF, W.F. JR. and K.L. PICKRELL. Pancreatitis. An anatomic study of the pancreatic and extrahepatic biliary systems. *Arch. Surg.* 51:205–219, 1945.
661 RIGLER, L.G. and L.G. ERICKSEN. The inferior accessory lobe of the lung. *Am. J. Roentgenol.* 29:384–392, 1933.
662 RISEL. Zwei Fälle von partiellem Situs inversus der Bauchorgane. *Zentralbl. Allg. Pathol. Pathol. Anat.* 20:511, 1909.
663* RISEL, W. Die Literatur der partiellen Situs inversus der Bauchorgane. *Zentralbl. Allg. Pathol. Pathol. Anat.* 20:673–731, 1909.
664* RITTER, L. Zum klinischen Bilde und Sitz versprengter Pankreaskeime. *Beitr. Klin. Chir.* 124:157–172, 1921.
665 ROACH, J.F. and M.H. POPPEL. The roentgen demonstration of an aberrant pancreatic nodule in the stomach. Report of three cases. *Am. J. Roentgenol.* 56:586–589, 1946.
666 ROBACZYŃSKI, J. and M. ZIÓŁKOWSKI. Supernumerary oviducts and abdominal ostia of oviducts in women. *Folia Morphol.* 22:162–167, 1963.
667 ROBBINS, R.H. Abnormal azygos pleural fold. *J. Anat.* 70:579, 1935–36.
668 ROBERT, H.L.F. Hemmungsbildung des Magens, Mangel der Milz und des Netzes. *Arch. Anat. Physiol. Wissen. Med.* 1842, pp. 57–60.
669* ROBERTS, W.H., HABENICHT, J. and G. KRISHINGER. The pelvic and perineal fasciae and their neural and vascular relationships. *Anat. Rec.* 149:707–720, 1964.
670 ROBERTS, W.H. and P. ENGEN. The blood supply of the abdominally placed vertical sigmoid colon. *Anat. Anz.* 142:165–167, 1977.
671 ROBINSON, J.O. Congenital absence of vermiform appendix. *Br. J. Surg.* 39:344–345, 1951–52.
672* ROBSON, M.C. and E.B. RUTH. Bilocular bladder: An anatomical study of a case; with considerations of urinary tract anomalies. *Anat. Rec.* 142:62–68, 1962.
673 ROLLESTON, H.D. Note on the anatomy of the suprarenal bodies. *J. Anat. Physiol.* 26:548–553, 1892.
674 ROMANOVSKY, R. Dystopia testis transversa. *Anat. Anz.* 26:635–639, 1905.
675 ROTH, M. Über Divertikelbildung am Duodenum. *Arch. Pathol. Anat. Physiol. Klin. Med.* 56:197–201, 1872.
676* ROTT, H.D. Kartagener's syndrome and the syndrome of immotile cilia. *Hum. Genet.* 46:249–261, 1979.
677 ROWLEY, W.N. Uterine anomaly: Duplication of uterus, three tubes and three ovaries. *Ann. Surg.* 127:676–680, 1948.
678* ROYS, C.K. Cutaneous horns. A report of three cases. *Ann. Surg.* 46:674–677, 1907.
679 RUGE, E. Beiträge zur chirurgischen Anatomie der großen Gallenwege (Ductus hepaticus, cysticus, choledochus und pancreaticus). *Arch. Klin. Chir.* 87:47–78, 1908.
680* RUGE, H. Über Defekte der Vorhofsscheidewand des Herzens. *Arch. Pathol. Anat. Physiol. Klin. Med.* 126:323–375, 1891.
681* RUKSTINAT, G.T. and R.J. HASTERLIK. Congenital absence of the penis. *Arch. Pathol.* 27:984–993, 1939.
682 SAARE, J.J. Absence of the vermiform appendix (report of a case of necropsy). *Acta Anat.* 23:327–329, 1955.
683 SAND, K. Hermaphrodisme (vrai) glandulaire alternant chez un individu de dix ans. *J. Urol. (Paris)* 15:181–194, 1923.
684 SAUNDERS, J.B. DE C.M. and H.H. LINDNER. Congenital anomalies of the duodenum. *Ann. Surg.* 112:321–338, 1940.
685* SCANNELL, J.G. A study of variations of the bronchopulmonary segments in the left upper lobe. *J. Thorac. Surg.* 16:530–537, 1947.
686* SCANNELL, J.G. and E.A. BOYDEN. A study of the variations of the bronchopulmonary segments of the right upper lobe. *J. Thorac. Surg.* 17:232–237, 1948.
687* SCHACHNER, A. Anomalies of the gall-bladder and bile passages. With the report of a double gall-bladder and a floating gallbladder. *Ann. Surg.* 64:419–433, 1916.
688 SCHAEFER, J.H. The normal weight of the pancreas in the adult human being: A biometric study. *Anat. Rec.* 32:119–132, 1926.
689* SCHAFFNER, G. Über den Lobus inferior accessorius der menschlichen Lunge. *Arch. Pathol. Anat. Physiol. Klin. Med.* 152:1–25, 1898.
690 SCHAPIRO, G. Zur Frage des Hermaphroditismus. *Arch. Pathol. Anat. Physiol. Klin. Med.* 266:392–406, 1927.
691 SCHEIBER, S.H. Einige angeborene Anomalien, beobachtet im Pathologisch-Anatomischen Institute zu Bukarest. *Med. Jahrb.* 1875, pp. 257–264.
692 SCHELENZ, C. Ein neuer Fall von Situs inversus partialis. *Zentralbl. Allg. Pathol. Pathol. Anat.* 21:489–492, 1910.

693. SCHMITT, H. Mittellappen der linken Lunge und seltene Form des Lobus venae azygos. *Röntgenpraxis* 10:568–571, 1938.
694. SCHMITZ-CLIEVER, E. Über das Vorkommen des Lobus venae azygos der linken Lungenseite. *Fortschr. Röntgenstr.* 72:728–731, 1950.
695.* SCHMORL, G. Ein Fall von Hermaphroditismus. *Arch. Pathol. Anat. Physiol. Klin. Med.* 113:229–244, 1888.
696. SCHRIDDE, H. Über den angeborenen Mangel des Processus vermiformis. *Arch. Pathol. Anat. Physiol. Klin. Med.* 177:150–166, 1904.
697. SCHULTZE, B.S. Ein Fall von Heterotaxie der Bauch- und Brusteingeweide. *Arch. Pathol. Anat. Physiol. Klin. Med.* 22:209–229, 1861.
698.* SCHULTZE, B.S. Der Hermaphrodit Katharina Hohmann aus Melrichstadt. *Arch. Pathol. Anat. Physiol. Klin. Med.* 43:329–335, 1868.
699.* SCHUMACHER, O. Über eine Dystopie der Gallenblase bei abnormer Aufteilung der V. portae. *Anat. Anz.* 69:453–459, 1930.
700. SEIFERT, F. Lageanomalien des Darmes bei einem Erwachsenen. *Anat. Anz.* 47:209–217, 1914–15.
701. SENNEVILLE, A. and P. GALIBERT. Etude comparative des variations scissurales chez les foetus et chez l'adulte. *Assoc. Anatomistes Comptes Rendus* 38:894–917, 1951.
702. SERBENY, A. Ein Fall von Hufeisenniere mit drei getrennt in die Blase mündenden Ureteren. *Anat. Anz.* 63:131–138, 1927.
703. SERRES. Transposition des viscères. *Arch. Gén. Méd.* 16:124–125, 1828.
704.* SETH, R.E. and A.H. PEACOCK. Double penis. *Urol. Cutan. Rev.* 36:590–592, 1932.
705. SETTLE, E.B. The surgical importance of accessory spleens with report of two cases. *Am. J. Surg.* 50:22–26, 1940.
706. SHARER, R.F. Substernal thyroid. *Am. J. Surg.* 32:56–62, 99, 107, 1936.
707. SHEPARD, R.K. The form of the human spleen. *J. Anat. Physiol.* 37:50–69, 1903.
708. SHERK, H.H. Total transpositions of viscera. *Surg. Obstet. Gynecol.* 35:53–57, 1922.
709.* SHEWMAKE, S.W. and G.T. IZUNO. Supernumerary areolae. *Arch. Dermatol.* 113:823–825, 1977.
710. SHIH, H.E. Postcaval ureter. *J. Urol.* 38:61–66, 1937.
711.* SHORE, L.R. A report on a specimen of fused pelvic kidney found in a Bantu native of South Africa. *J. Anat.* 64:344–352, 1930.
712. SHUTE, W.B. Congenital absence of the vagina. Report of a case with normal menstruation following operation. *Obstet. Gynecol.* 12:310–316, 1958.
713. SIEGLBAUER, F. Verlagerung des Colon sigmoideum mit Tiefstand der linken Niere. *Anat. Anz.* 35:33–47, 1910.
714. SIEWERT, A.K. Über einen Fall von Bronchiectasie bei einem Patienten mit Situs inversus viscerum. *Berliner Klin. Wochenschr.* 41:139–141, 1904.
715. SIMON, H.E. and N.A. BRANDEBERRY. Anomalies of the urachus: Persistent fetal bladder. *J. Urol.* 55:401–408, 1946.
716. SIMON, W. Hermaphroditismus verus. *Arch. Pathol. Anat. Physiol. Klin. Med.* 172:1–29, 1903.
717. SIMPSON, G.C.E. A case of accessory lobe of the right lung. *J. Anat. Physiol.* 42:221–225, 1908.
718. SINGH, I. Observations on the mode of termination of the bile and pancreatic ducts: Anatomical factors in pancreatitis. *J. Anat. Soc. India* 5:54–60, 1956.
719. SINGLETON, A.O. and W.B. KING. Persistent vitelline duct continuous with the appendix. *Surgery* 29:278–280, 1951.
720. SINNER, B.L. Unusual anomaly of the gallbladder and bile duct. *J. Missouri State Med. Assoc.* 45:265–266, 1948. Cited in *Excerpta Medica*, Sec. 1, Vol. 3, abstract 665, 1949.
721. SMART, J. Complete congenital agenesis of a lung. *Q. J. Med.* 15:125–140, 1946. Cited in *Excerpta Medica*, Sec. 1, Vol. 2, abstract 1182, p. 426, 1948.
722.* SMITH, F.R. and E.A. BOYDEN. An analysis of variations of the segmental bronchi of the right lower lobe of fifty injected lungs. *J. Thorac. Surg.* 18:195–215, 1949.
723. SMITH, G.E. Note on an abnormal colon. *J. Anat. Physiol.* 38:32–33, 1904.
724.* SMITH, I. Triplicate ureter. *Br. J. Surg.* 34:182–185, 1946–47.
725. SMITH, S. and B. BOULGAKOW. Two cases of misplaced fixed kidneys. *J. Anat.* 59:188–191, 1925.
726. SNEATH, W.A. An apparent third testicle consisting of a scrotal spleen. *J. Anat. Physiol.* 47:340–342, 1913.
727. SNOWDEN, H. Vollkommene Transposition der Eingeweide. *Schmidt's Jahrb. Med.* 14:313, 1837.

728. SODERDAHL, D.W., BROSMAN, S.A. and W.E. GOODWIN. Penile agenesis. *J. Urol.* 108:496–499, 1972.
729. SOKOLOW, B.M. Über die hufeisenförmige Niere. *Anat. Anz.* 68:449–495, 1929–30.
730. SOLANKE, T.F. The position, length and content of the vermiform appendix in Nigerians. *Br. J. Surg.* 57:100–102, 1970.
731.* SOUTHWOOD, W.F.W. The thickness of the skin. *Plast. Reconstr. Surg.* 15:423–429, 1955.
732. SPERINO, G. Absence congénitale de la glande submaxillaris dans sa place normale; sa transposition au-dessus du m. mylo-hyoideus; fusion partielle de cette glande avec la glandula sublingualis. *Arch. Ital. Biol.* 40:318, 1903.
733. STAUFFER, H.M., SNOW, L.E. and A.B. ADAMS. Roentgenologic recognition of habenular calcification as distinct from calcification in the pineal body. Its application in cerebral localization. *Am. J. Roentgenol.* 70:83–92, 1953.
734. STEARNS, L.M. Congenital anorchia with report of a case of unilateral anorchia. *Nebr. State Med. J.* 20:227–228, 1935.
735.* STELMASIAK, M. and J. KARSKI. Morphology of the orifices of the common bile duct and pancreatic ducts in man. *Folia Morphol.* 32:287–295, 1973.
736.* STELMASIAK, M., KURYLCIO, L. and Z. WÓJCIK. The anatomic basis of drainage of extrahepatic bile ducts. *Folia Morphol.* 29:1–8, 1970.
737.* STELMASIAK, M. and J. KARSKI. Relation of the accessory pancreatic duct to the duodenal muscularis in man. *Folia Morphol.* 32:407–411, 1973.
738. STEPHENS, F.D. Anatomical vagaries of double ureters. *Aust. N.Z. J. Surg.* 38:27–33, 1958.
739.* STERLING, J.A. The common channel for bile and pancreatic ducts. *Surg. Gynecol. Obstet.* 98:420–424, 1954.
740. STERNBERG, C. Ein Fall von Agenesie der Milz. *Arch. Pathol. Anat. Physiol. Klin. Med.* 173:571–575, 1903.
741. STEVENS, A.R. Pelvic single kidneys. *J. Urol.* 37:610–618, 1937.
742. STEWART, M.J. and S.D. LODGE. On unilateral fused kidney and allied renal malformations. *Br. J. Surg.* 11:27–37, 1923–24.
743. STIBBE, E.P. The accessory pulmonary lobe of the vena azygos. *J. Anat.* 53:305–314, 1919.
744. STIBBE, E.P. True congenital diverticulum of the trachea in a subject showing also right aortic arch. *Journal Anat.* 64:62–66, 1929.
745.* STIRLING, W.C. Report of a case of true hermaphroditism. A discussion of 37 cases previously reported. *J. Urol.* 56:720–724, 1946.
746. STOJALOWSKI, K. and J. DEBSKI. Ein neuer Fall von Hermaphroditismus verus beim Menschen. *Arch. Pathol. Anat. Physiol. Klin. Med.* 290:358–372, 1933.
747. STOLKIND, E. Double gall-bladder. Report of a case and review of 38 cases. *Br. J. Surg.* 27:760–766, 1939–40.
748. STREHL, H. Congenitale Retroposition des Dickdarmes. *Arch. Klin. Chir.* 87:8–19, 1908.
749. SUESS, E. Zur Röntgenologie des Lobus venae azygae. *Med. Klin.* 24:1790–1792, 1928.
750. SUGAWARA, K. and I. SHIBATA. Ein Fall von Pancreas annulare. *Folia Anat. Jpn.* 3:239–242, 1925.
751. SUTTON, J.B. The ligamentum teres. *J. Anat. Physiol.* 17:191–193, 1883.
752.* SUTTON, J.B. Supernumerary mammae and nipples in man, monkeys, cows, etc. *Am. J. Med. Sci.* 97:247–257, 1889.
753. SWICK, M. A quadruple kidney with a horseshoe component, demonstrated with the aid of excretion urography. *J. Mt. Sinai Hosp.* 4:7–11, 1937–38.
754. SWINYARD, C.A. Volume and cortico-medullary ratio of the adult human suprarenal gland. *Anat. Rec.* 76:69–79, 1940.
755. SYBIRSKA-SZCZEPAŃSKA, B. Very rare localization of an accessory pancreas. *Folia Morphol.* 23:432–435, 1964.
756. SYMINGTON, J. A rare abnormality of the pancreas. *J. Anat. Physiol.* 19:292, 1885.
757. SZCZEŚNIAK-ZARZYCKA, N. Studies on the color of eyes and hair in male and female students in the Lublin region. *Folia Morphol.* 26:208–215, 1967.
758.* TARUFFI, C. Nuovo caso di meso-rino-schisi nell'umomo. *Mem. R. Accad. Sci. Istituto di Bologna S. 5* 1:227–233, 1890.
759.* TARUFFI, C. Sur l'hermaphrodisme. *Arch. Ital. Biol.* 32:477–480, 1899.
760.* TARUFFI, C. L'Ermafroditismo. *Med. R. Accad. Sci. Istituto di Bologna S. 5* 7:697–751, 1898.
761.* TEGGIHALLI, M.M. The azygos lobe. *J. Anat. Soc. India* 4:49, 1955.

762 THEILMANN, C.H. Drei Nieren im Leichname eines Menschen. *Arch. Pathol. Anat. Physiol. Klin. Med.* 5:511–515, 1835.
763 THOMAS, L.B. and E.A. BOYDEN. Agenesis of the right lung. *Surgery* 31:429–435, 1952.
764 THOMAS, P.K. and D.G. FERRIMAN. Variation in facial and pubic hair growth in white women. *Am. J. Phys. Anthropol.* 15:171–180, 1957.
765 THOMAS, W.T. Observations on fifty cases of hour-glass stomach subjected to operation. *Br. J. Surg.* 9:37–41, 1921–22.
766 THOMPSON, P. A note on the development of the septum transversum and the liver. *J. Anat. Physiol.* 42:170–175, 1908.
767 THOMPSON, P. The development of the lobus quadratus of the liver, with special reference to an unusual anomaly of this lobe in the adult. *J. Anat. Physiol.* 48:222–237, 1914.
768 THOMSON, A. Some variations in the anatomy of the human liver. *J. Anat. Physiol.* 19:303–306, 1885.
769* THOMSON, A. Diverticula of the alimentary tract, with particular reference to those met within the ileum resulting from an accessory pancreas and from tuberculosis. *Medico-Chir. Trans. Edinb.* 27:118–129, 1908.
770 THOREK, M. and P. THOREK. Anorchidism (absence of the testicle) with case report. *J. Urol.* 30:345–351, 1933.
771* THURSTON, E.W. Über einen eigenartigen Fall von "Intersex" beim Menschen. *Arch. Pathol. Anat. Physiol. Klin. Med.* 292:220–232, 1934.
772 TOBIN, C.E. and J.A. BENJAMIN. Anatomic and clinical re-evaluation of Camper's, Scarpa's, and Colles' fasciae. *Surg. Gynecol. Obstet.* 88:545–559, 1949.
773 TÖNDURY, G. Ueber Situs inversus partialis des Duodenum. *Z. Anat. Entwicklungsgesch.* 106:251–270, 1936–37.
774 TÖNDURY, G. Ueber Situs inversus partialis beim Menschen. *Z. Anat. Entwicklungsgesch.* 106:648–662, 1936–37.
775 TÖNDURY, G. Mesenterium commune. *Schweizerische Med. Wochenschr.* 77:1127–1129, 1947. Cited in *Excerpta Medica*, Sec. 1, Vol. 2, abstract 1409, pp. 492–493, 1948.
776 TONKOFF, W. Beitrag zu den Nierenanomalien. *Int. Monatsschr. Anat. Physiol.* 20:449–461, 1902.
777* TORGERSEN, J. Genic factors in visceral asymmetry and in the development and pathologic changes of lungs, heart and abdominal organs. *Arch. Pathol.* 47:566–593, 1949.
778 TRICOMI-ALLEGRA, G. Cas d'absence complète congénitale du rectum. *Arch. Ital. Biol.* 46:320, 1906.
779 TRICOMI, G. and L. DE GAETANI. Anomalies d'un appareil uro-génital mâle. *Arch. Ital. Biol.* 30:310–311, 1898.
780 TROTTER, M. A review of the classification of hair. *Am. J. Phys. Anthropol.* 24:105–126, 1938.
781 TROTTER, M. Classifications of hair color. *Am. J. Phys. Anthropol.* 25:237–260, 1939.
782 TROTTER, M. and J.C. FINERTY. An anomalous urinary bladder. *Anat. Anz.* 100:259–269, 1948.
783 TURNER, R.S. A note on the geometry of the tracheal bifurcation. *Anat. Rec.* 143:189–194, 1962.
784 TURNER, W. Note on a case of displacement of the submaxillary glands. *J. Anat. Physiol.* 4:147, 1870.
785 TWEEDIE, H.C. Case of a single unilateral kidney. *J. Anat. Physiol.* 28:340–341, 1894.
786* UNDERHILL, B.M.L. Intestinal length in man. *Br. Med. J.* 2:1243–1246, 1955.
787 UNDERWOOD, F.A. and N. TATTERSALL. The accessory lobe of the azygos vein: A record of fourteen cases, with special reference to heredity as an aetiological factor, and to pathological features of the condition. *Tubercle* 15:1–12, 1933.
788 UNSHELM, E. Ein klinisch bemerkenswerter Fall von Pseudohermaphroditismus masculinus completus. *Arch. Pathol. Anat. Physiol. Klin. Med.* 265:318–329, 1927.
789 UPSON, W.S. Transposed viscera. *Am. J. Roentgenol.* 8:385–388, 1930.
790 VALENTI, G. Sur un cas d'ectopie rénale congenitale. *Arch. Ital. Biol.* 28:171–172, 1897.
791 VALLE, A.R. Agenesis of the lung. *American J. Surg.* 89:90–100, 1955.
792 VALLE, A.R. and M.L. WHITE, Jr. Subdiaphragmatic aberrant pulmonary tissue. *Dis. Chest* 13:63–68, 1947.
793* VARANO, N.R. and R.J. MERKLIN. Situs inversus. Review of the literature, report of four cases and an analysis of the clinical implications. *J. Int. Coll. Surg.* 33:131–148, 1960.
794 VARSANO, I.B., JABER, L., GARTY, B., MUKAMEL, M.M. and M. GRÜNEBAUM. Urinary tract abnormalities in children with supernumerary nipples. *Pediatrics* 73:103–105, 1984.
795 VASSAL, P.A. and P. VOIZARD. L'hyperthélie en Afrique du Nord. *Assoc. Anatomistes Comptes Rendus* 42:1371–1380, 1955.
796* VASTINE, J.H. and K.K. KINNEY. The pineal shadow as an aid in localization of brain tumors. *Am. J. Roentgenol.* 17:320–324, 1927.
797 VELDE, G. Ein eigentümlicher Schattenstreifen in der rechten Lungenspitze. *Fortschr. Röntgenstr.* 36:315–318, 1927.
798 VELLUDA, C.C. Absence du caecum et de l'appendice iléo-caecal, de la valvule iléo-caecale et des bandelettes muscularies du côlon ascendant chez l'homme. *Ann. Anat. Pathol.* 13:1030–1033, 1936.
799 VERA-ROMAN, J.M. Robinow dwarfing syndrome accompanied by penile agenesis and hemivertebrae. *Am. J. Dis. Child.* 126:206–208, 1973.
800 VERNEUIL. Recherches anatomiques pour servir a l'histoire des kystes de la partie supérieure et médiane du cou. *Arch. Gén. Méd. Ser. 5* 1:185–191, 450–468, 1853.
801 VERNEUIL. Vice de conformation du vagin; brièveté rétrécissement probablement congénital de ce conduit. *Soc. Biol. Comptes Rendus des Séances et Mémoires* 8:201–202, 1856.
802 VILLAR, F. Disposition anormale du péritoine pariétal. Diaphragme péritonéal divisant en deux loges la grande cavité abdominale. *Assoc. Anatomistes Comptes Rendus* 8:56–58, 1906.
803* VILLAREAL, L. Congenital absence of the gallbladder with case report. *Ann. Surg.* 127:745–748, 1948.
804 VILLEMIN, DUFOUR and RIGAUD. Variations concomitantes de la morphologie du foie et de la région thoraco-abdominale. *Assoc. Anatomistes Comptes Rendus* 36:685–690, 1949.
805 WAKEFIELD, P.A. Congenital malformation of the liver – absence of the lobus Spigelli, rudimentary quadrate and left lobes, with enlargement of the right lobe. *J. Anat. Physiol.* 33:50–53, 1899.
806* WAKELEY, C.P.G. A case of duplication of the ureters. *J. Anat. Physiol.* 49:148–154, 1915.
807* WAKELEY, C.P.G. The position of the vermiform appendix as ascertained by an analysis of 10,000 cases. *J. Anat.* 67:277–283, 1933.
808 WALDEYER, A. Ein menschlicher Kehlkopf mit beiderseitigen seitlichen Ventrikelsäcken. *Anat. Anz.* 66:354–359, 1928–29.
809 WALLRAFF, J. and A. ROSZEL. Angeborene Zwerchfellhernie. *Anat. Anz.* 132:149–153, 1972.
810 WALLRAFF, J. and A. ROSZEL. Linksseitiger Herzbeuteldefekt und Verlauf des linken Nervus phrenicus in einer tiefen senkrechten Furche mitten auf der vorderen Herzfläche. *Anat. Anz.* 131:481–486, 1972.
811 WALMSLEY, T. A diverticulum of the appendix. *J. Anat.* 64:47–49, 1929.
812 WANG, K.P. and H.P. TAI. An analysis of variations of the segmental vessels of the right lower lobe in 50 Chinese lungs. *Acta Anat. Sin.* 8:408–423, 1965.
813 WANKE, R. Duodenalanomalien im Röntgenbild. *Fortschr. Röntgenstr.* 39:249–262, 1929.
814 WAPSHAW, H. Lingual thyroid. *British J. Surg.* 30:160–165, 1942–43.
815* WARD, G.E., CANTRELL, J.R. and W.B. ALLAN. The surgical treatment of lingual thyroid. *Ann. Surg.* 139:536–546, 1954.
816 WARREN, J.M. Complete absence of vagina and uterus. *Boston Med. Surg. J.* 56:297, 1857.
817 WATT, J.K. Appendix duplex. *Br. J. Surg.* 46:472–473, 1958–59.
818 WAUGH, G.E. Congenital malformations of the mesentery: A clinical entity. *Br. J. Surg.* 15:438–449, 1927–28.
819 WEDENSKY, K. A rare anomaly of the right kidney and ureter. *Am. J. Urol.* 8:25–26, 1912.
820 WEIDMAN, F.D. Aberrant pancreas in the splenic capsule. *Anat. Rec.* 7:133–139, 1913.
821 WEIGERT, C. Über einige Bildungsfehler der Ureteren. *Arch. Pathol. Anat. Physiol. Klin. Med.* 70:490–501, 1877.
822 WEINBERG, S.K. and A.G. MOLUSKY. Aberrant axillary breast tissue. A report of a family with six affected women in two generations. *Clin. Genet.* 10:325–328, 1976.
823 WEINLECHNER. Über einen bemerkenswerten Fall von Schilddrüsenknoten. *Centralbl. Chir.* 10:140–141, 1883.
824* WEINTROB, M. Abnormalities of the female genitalia. *J. Int. Coll. Surgeons* 7:381–396, 1944.
825 WELLBROCK, W.C.A. The occurrence of accessory parathyroid glands. *JAMA* 92:1821–1822, 1929.
826 WELSH, D.A. Concerning the parathyroid glands: A critical anatomical, and experimental study. *J. Anat. Physiol.* 32:380–402, 1898.
827* WERMANN. Ein Fall von Pseudohermaphroditismus masculinus

completus. *Arch. Pathol. Anat. Physiol. Klin. Med.* 104:81–91, 1886.
828 WHITE, H. Roentgenology of congenital anomalies of the esophagus. *Q. Bull. Northwestern University Medical School* 31:52–57, 1957.
829 WILDBOLZ, H. Ein Fall von kongenitaler Anorchie. *Correspondenz-Blatt Schweizer Aerzte* 47:1307–1314, 1917.
830* WILHELMJ, C.M. A case of cross displacement of the left kidney with fusion. *Anat. Rec.* 18:167–177, 1920.
831 WILHELMJ, C.M. A case of double ureter in man with failure of development of the kidney about the aberrant ureter. *Anat. Rec.* 18:179–192, 1920.
832* WILKINS, L. and W. FLEISCHMANN. Ovarian agenesis. Pathology, associated clinical symptoms and the bearing on the theories of sex differentiation. *J. Clin. Endocrinol.* 4:351–375, 1944.
833 WILLIAMS, W.R. Polymastism, with special reference to mammae erraticae and development of neoplasms from supernumerary mammary structures. *J. Anat. Physiol.* 25:225–255, 1891.
834 WILLIAMS, W.R. Mammary variations per defectum. *J. Anat. Physiol.* 25:304–315, 1891.
835 WILSON, W.A. and J. LITTLER. Polyorchidism. A report of two cases with torsion. *Br. J. Surg.* 41:302–306, 1953.
836 WIMMER, H. Doppelbildungen an den Nieren und ein Versuch ihrer entwicklungsgeschichtlichen Deutung. *Arch. Pathol. Anat. Physiol. Klin. Med.* 200:487–522, 1910.
837 WOLF, R. Fall von accessorischer Schilddrüse in der Zunge. *Dtsch. Med. Wochenschr.* 15:638, 1889.
838 WÖLFLER, A. Zur chirurgischen Anatomie und Pathologie des Kropfes und Nebenkropfes. *Arch. Klin. Chir.* 40:169–243, 1890.
839 WONG, A.I.H. Pregnancy in a double uterus. *Chin. Med. J.* 47:61–65, 1933.
840 WOOD, G.O. and A. BLALOCK. Situs inversus totalis and disease of the biliary tract. *Arch. Surg.* 40:885–896, 1940.
841 WOODBURNE, R.T. The costomediastinal border of the left pleura in the precordial area. *Anat. Rec.* 97:197–210, 1947.
842 WORD, B. The construction of an artificial vagina, with ten year follow up studies. *South. Med. J.* 44:375–382, 1951.
843 WRIGHT, C.W. A case of left-sided vermiform appendix, cecum and anomalous colon. *Anat. Rec.* 123:291–298, 1955.
844+ WRISBERG, H.A. Observationes anatomicae de vena azyga duplici, aliisque hujus venae varietatibus. *Novis Commentariis Societatis Reg. Sci. Gottingen* 8:14, 1777.
845 WWEDENSKY. Un cas rare d'anomalie du rein droit et des uretères. *J. Urol. (Paris)* 1:106, 1912.
846 WYLIE, W. Case of entire absence of both mammae in a female, age 21 years. *Br. Med. J.* 1885 (2), p. 235.
847 YATES, H.B. A remarkable Meckel's diverticulum. *Br. J. Surg.* 17:456–462, 1929–30.
848 YOUNG, R.B. Abnormal disposition of the colon. *J. Anat. Physiol.* 19:98–108, 1885.
849 ZAHN, W. Ueber congenitale Knorpelreste am Halse. *Arch. Pathol. Anat. Physiol. Klin. Med.* 115:47–52, 1889.
850 ZANDER, R. Ein Fall von echtem Hermaphroditismus beim Menschen. *Anat. Anz.* 23:27–28, 1903.
851 ZAWADOWSKI, W. Le lobule de la veine azygos (lobule de Wrisberg) sa visibilité sur les radiographies pulmonaires. *J. Radiol. Electrol.* 14:273–282, 1930.
852 ZENKER, F.A. Nebenpancreas in der Darmwand. *Arch. Pathol. Anat. Physiol. Klin. Med.* 21:369–376, 1861.
853* ZEREN, Z. Eine Lagevarietät der Glandula parathyreoidea. *Acta Anat.* 2:358–360, 1946.

Books

854 AASAR, Y.H. *Anatomical Anomalies.* Fouad I University Press, Cairo, 1947.
855+ AHLFELD, F. *Die Mißbildungen des Menschen.* Grunow, Leipzig, 1880.
856 ANSON, B.J. *Atlas of Human Anatomy.* W.B. Saunders Co., Philadelphia, 1950.
857 ANSON, B.J. and W.G. MADDOCK. *Callander's Surgical Anatomy.* W.B. Saunders Co., Philadelphia and London, 1952.
858 ANSON, B.J., Ed. *Morris' Human Anatomy*, 12th Ed. The Blakiston Division, McGraw-Hill Book Company, New York, 1966.
859 ANSON, B.J. and C.B. McVAY. *Surgical Anatomy*, 5th Ed. W.B. Saunders Co., Philadelphia, 1971. (Also 6th Ed., 1984.)
860 AREY, L.B. *Developmental Anatomy*, 7th Ed, Revised. W.B. Saunders Co., Philadelphia, 1974.
861 BENNINGHOFF, A. *Lehrbuch der Anatomie des Menschen.* Urban & Schwarzenberg, Munich, 1952.
862 CRUVEILHIER, J. *Traité d'Anatomie Descriptive*, 3rd Ed. Paris, 1851.
863 GARDNER, E., GRAY, D.J. and R. O'RAHILLY. *Anatomy*, 3rd Ed. W.B. Saunders Co., Philadelphia, 1969.
864 GEGENBAUR, C. *Lehrbuch der Anatomie des Menschen*, 3rd Ed. V. Wilhelm Engelmann, Leipzig, 1888.
865 GOULD, G.M. and W.L. PYLE. *Anomalies and Curiosities of Medicine.* W.B. Saunders Co., Philadelphia, 1897.
866 HENLE, J. *Handbuch der Systematischen Anatomie des Menschen*, 3 volumes. Von Friedrich Vieweg und Sohn, Braunschweig, 1868.
867 HOLLINSHEAD, W.H. *Anatomy for Surgeons.* Hoeber Medical Division, Harper and Row, New York, 1969.
868 HUBER, G.C., Ed. *Piersol's Human Anatomy*, 9th Ed. L.B. Lippincott Co., Philadelphia, 1930.
869 HYRTL, J. *Lehrbuch der Anatomie des Menschen*, 19th Ed. Wilhelm Braumüller, Wien, 1887.
870 JACKSON, C.M., Ed. *Morris' Human Anatomy*, 9th Ed. P. Blakiston's Son & Co., Inc., Philadelphia, 1933.
871 JONES, H.W. JR. and W.W. SCOTT. *Hermaphroditism, Genital Anomalies and Related Endocrine Disorders.* Williams & Wilkins, Baltimore 1958.
872 KOPSCH, F. *Rauber's Lehrbuch der Anatomie des Menschen.* Georg Thieme, Leipzig, 1908.
873 KOPSCH, F. *Rauber-Kopsch Lehrbuch und Atlas der Anatomie des Menschen.* Georg Thieme, Stuttgart, 1955.
874 LATARJET, A. *Testut's Traité d'Anatomie Humaine*, 9th Ed. G. Doin & Cie., Paris, 1948.
875 SCHAEFER, E.A., SYMINGTON, J. and T.H. BRYCE, Ed. *Quain's Anatomy*, 11th Ed. Longmans, Green & Co., London, 1915.
876 WHITNALL, S.E. *The Anatomy of the Human Orbit*, 2nd Ed. Oxford University Press, London, 1932.
877 WOLFF, E. *The Anatomy of the Eye and Orbit*, 4th Ed. Lewis, London, 1954.

Glossary of Terms
Their origin and definition related to Organs.

The following abbreviations are used: Ar., Arabic; AS., Anglo-Saxon; F., French; G., Greek; H. Hebrew; L. Latin; and ME., Middle English.

Appendix (L. *appendere*, to hang upon). Something hung from or attached to something else.

Bulbourethral (L. *bulbus*, a bulb; G. *urethra*, to make water).

Duodenum (L. *duodeni*, twelve). So called because it is twelve fingerbreadths in length.

Esophagus (G. *oisophagos*, from *oisein*, to carry, and *phagma*, food). To bear or carry food.

Genital (L. *gentialis*, belonging to birth).

Gall (L. *galla*, bile).

Ileum (G. *twisted*). The twisted small intestine.

Intestine (L. *intestinus*, internal).

Jejunum (L. *jejunus*, empty).

Liver (L. *Jecur*; G. *hepar*).

Lung (L. *pulmo*; G. *pneumon* or *pleumon*). Anglo-Saxon *lunge*, light, thus named for their lightness.

Mammary (L. *mammarius*, pertaining to the mamma, breast).

Ovary (L. *ovarium*, egg receptacle). The female gonad.

Pancreas (G. *pan*, all, and *kreas*, flesh). Described by Herophilus and named from its meaty or fleshy character.

Parotid (G. *para*, beside, and *ous*, ear). Situated near the ear.

Penis (L. *penis*, tail).

Prostate (G. *pro*, before, and *histanai*, to stand). The term means "standing before" and applied to one who was a chief or a leader.

Seminal (L. *seminalis*, pertaining to semen or seed).

Spleen (L. *lien*; G. *splen*).

Stomach (G. *stomachos*; L. *stomachus*, a mouth or opening).

Suprarenal (L. *supra*, above, *ren*, kidney).

Testis (L. *testis*, a testicle). The male gonad. Testis also meant witness. The testes are so named because they are evidence (witness) of virility. Under Roman law, no man was admissible as a witness unless his testicles were present.

Thyroid (G. *thyreoeides* from *thyreos*, shield, and *eidos*, form). Resembling a shield.

Urinary (L. *urina*; G. *ouron*, urine).

Uterine (L. *uterinus*, womb). Pertaining to the uterus.

Uterus (L., G. *hystera*, womb). Derived from uter (utris), a leather wineskin or water bottle made of leather or hide.

Vagina (L., a sheath). Vagina was a sheath for a sword (L. *gladius*) and gladius was a common term for the penis.

Vermiform (L. *vermiformis* from *vermis*, worm, and *forma*, shape). Shaped like a worm.

Vesicle (L. *vesicula*, little bladder).

Terminology: Synonyms

Terminology for organs from the older English, French, and German literature related to more modern usage.

Old Terminology	Modern Usage	Old Terminology	Modern Usage
Adamsapfel.	Laryngeal prominence, thyroid cartilage.	Cornu majus s. longum.	Superior horn, thyroid cartilage.
Amygdala.	Palatine tonsils.	Cornu minus s. breve.	Inferior horn, thyroid cartilage.
Appendix glandulae thyreoidea.	Pyramidal lobe, thyroid gland.	Cornu superius.	Superior horn, thyroid cartilage.
Appendix vermicularis.	Vermiform process, appendix.	Corpora fibrospongiosa.	Corpora cavernosa penis.
Arcus glossopalatinus.	Arcus palatoglossus (palatoglossal arch).	Corpora spongiosa s. nervosa penis.	Corpora cavernosa penis.
Arcus pharyngopalatinus.	Arcus palatopharyngeus (palatopharyngeal arch).	Corpus cavernosum s. spongiosum inferius.	Corpus cavernosum urethrae.
Arteria aspera.	Trachea.	corpus thymicum s. thymianum.	Thymus.
Aspera arteria.	Trachea.	Cutis.	Skin.
Augenzähne.	Teeth, canine.	Cystis s. vesicula bilis.	Gallbladder.
Aureola.	Areola.	Cystis fellea.	Gallbladder.
Aveolus s.* sinus laryngis s. Morgagni.	Laryngeal ventricle.	Dens sapientiae.	Teeth, molar, wisdom.
Balanus.	Glans penis.	Dentes cuspidati s. laniarii s. angulares.	Teeth, canine.
Bauchspeicheldrüse.	Pancreas.	Dentes decidui.	Teeth, deciduous.
Brustdrüse.	Mamma.	Dentes incisive.	Teeth, incisor.
Canal de Sténon.	Stensen's duct or parotid duct.	Dentes incisores s. primores.	Teeth, incisor.
Canal of Itis.	Thyroglossal duct.	Dentes infantiles.	Teeth, deciduous.
Canalis pancreaticus azygos.	Accessory pancreatic duct.	Dentes lactei.	Teeth, deciduous.
Canalus gutturalis.	Auditory tube.	Dentes molares anteriores s. minores s. bicuspidati.	Teeth, premolars.
Canalus palatinus tympani.	Auditory tube.	Dentes molares posteriores s. majores.	Teeth, molars.
Capitula arytenoidea.	Corniculate cartilage.	Dentes multicuspidati.	Teeth, molars.
Capsulae suprarenales.	Suprarenal glands, capsules of.	Dentes praemolares.	Teeth, premolars.
Caput coli.	Cecum.	Ductus Bartholinianus.	Duct of sublingual gland.
Caput pancreatis s. extremitas dextra s. obtusa s. duodenalis.	Head of pancreas.	Ductus choledochus.	Bile duct, common.
		Ductus pancreaticus accessorius.	Accessory pancreatic duct.
Caput penis.	Glans penis.	Ductus pancreaticus directus.	Pancreatic duct.
Cartilago s. corpuscula Wrisbergii.	Cuneiform cartilage.	Ductus pancreaticus recurrens.	Accessory pancreatic duct.
Cartilago annularis.	Cricoid cartilage.	Ductus pancreaticus superiorius.	Accessory pancreatic duct.
Cartilago cuneiformes.	Cuneiform cartilage.	Ductus Santorini.	Accessory pancreatic duct.
Cartilago gutturales.	Arytenoid cartilage.	Ductus Stenonianus.	Parotid duct.
Cartilago pyramidales.	Arytenoid cartilage.	Ductus Stensonianus.	Parotid duct.
Cartilago Santorinianae.	Corniculate cartilage.	Ductus venosus Arantii.	Ductus venosus.
Cartilago scutiformis.	Thyroid cartilage.	Ductus Whartonianus.	Submandibular duct.
Cartilago triquetrae.	Arytenoid cartilage.	Ductus Wirsungianus.	Pancreatic duct.
Cauda pancreatis s. extremitas sinistra s. acuta s. lienalis.	Tail of pancreas.	Eckzähne.	Teeth, canines.
Cavum buccale.	Vestibule of mouth.	Eierstöcke.	Ovaries.
Cavum oris.	Mouth.	Eileiter.	Oviduct.
Cavum oris externus.	Vestibule of mouth.	Epiploon.	Greater omentum.
Cholecystis.	Gallbladder.	Eustachische Röhre.	Auditory tube.
Columna media.	Pyramidal lobe, thyroid gland.	Eustachische Trompete.	Auditory tube.
Conduit accessoire ou supplémentaire.	Accessory pancreatic duct.	Excisura cartilago thyreoideae.	Superior thyroid notch.
Conduit excréteur principal.	Pancreatic duct.	Flexura iliaca s. ramanum.	Sigmoid flexure, of rectum.
Cornea opaca.	Sclera.	Foramen cecum or Morgagni.	Foramen cecum linguae.
Cornea pellucida.	Cornea.	Fourth tonsil.	Lingual tonsil.
Cornicula s. capitula s. corpuscula Santorini.	Corniculate cartilage.	Fruchthalter.	Uterus.
Cornu inferius.	Inferior horn, thyroid cartilage.	Fundus ventriculi.	Fundus of stomach.
		Gargareon.	Uvula.
		Gaster.	Stomach.

* s. = synonymus

Organs

Gastrocolic omentum.	Greater omentum.	Nebennieren.	Suprarenal glands.
Gebärmutter.	Uterus.	Netzhaut.	Retina.
Gießbeckenknorpel.	Arytenoid cartilage.	Nieren.	Kidneys.
Glandula maxillaris s. angularis.	Submandibular gland.	Nierendrüsen.	Suprarenal glands.
Glandula salivalis externa.	Parotid gland.	Nodus gutturis.	Laryngeal prominence.
Glandula salivalis interna.	Submandibular and sublingual glands.	Obernieren.	Suprarenal glands.
Glandula thymus.	Thymus.	Ohrendrüse.	Parotid gland.
Glandulae s. capsulae atrabiliariae.	Suprarenal glands.	Ohrspeicheldrüse.	Parotid gland.
Glandulae genales.	Buccal glands.	Ohrtrompete.	Auditory tube.
Glandulae morales.	Buccal glands.	Omentula.	Appendices epiploicae.
Glottis vera.	True vocal fold.	Omentum majus.	Greater omentum.
Great cul-de-sac.	Fundus of stomach.	Orchides.	Testicles.
Grimmdarmklappe.	Ileocolic valve.		
Grundknorpel.	Cricoid cartilage.	Palatum molle.	Soft palate.
		Pancreas of Winslow.	Uncinate process of pancreas.
Harngang.	Ureters.	Pancreas Winslowi.	Processus uncinatus of pancreas.
Harnleiter.	Ureters.	Papillae arcuatae.	Papillae filiformes.
Harte Augenhaut.	Sclera.	Papillae caliciformes.	Papillae vallate.
Hautorgan.	Skin.	Papillae capitatae.	Papillae vallate.
Hepar.	Liver.	Papillae circumvallate.	Papillae vallate.
Hintere Backenzähne.	Teeth, molar.	Papillae claratae.	Papillae fungiformes.
Hintere Mundhöhle.	Mouth.	Papillae conicae.	Papillae vallate.
Hinterste Backenzähne.	Teeth, molar, wisdom.	Papillae conoideae.	Papillae filiformes.
Hoden.	Testicles.	Papillae corolliformes.	Papillae filiformes.
Hodensack.	Scrotum.	Papillae lenticulares.	Papillae fungiformes.
Hornhaut.	Cornea (or epidermis).	Papillae magnae.	Papillae vallate.
Hundszähne.	Teeth, canine.	Papillae maximae.	Papillae vallate.
		Papillae mediae.	Papillae fungiformes.
Incisura cartilago thyreoideae superioris.	Superior thyroid notch.	Papillae minores.	Papillae filiformes.
Integumentum commune.	Skin.	Papillae obtusae.	Papillae fungiformes.
Intestinum angustum.	Small intestine.	Papillae petiolatae.	Papillae vallate.
Intestinum erassum.	Large intestine.	Papillae truncatae.	Papillae vallate.
Intestinum gracile.	Small intestine.	Papillae villosae.	Papillae filiformes.
Intestinum mesenteriale.	Jejunum and ileum.	Paukenfell.	Tympanic membrane.
Intestinum tenue.	Small intestine.	Plicae thyreo-arytaenoideae inferior.	True vocal fold.
Jecur.	Liver.	Plicae thyreo-arytaenoideae superior.	False vocal fold.
Jejuno-ileum.	Jejunum and ileum.	Portio duodenalis s. verticalis.	Head of pancreas.
Jungfernhäutchen.	Hymen.	Portio lienalis.	Fundus of stomach.
		Porus biliarius.	Bile duct, common.
Kehldeckelknorpel.	Epiglottic cartilage.	Pouch of Douglas.	Rectovaginal pouch.
Kindzähne.	Teeth, deciduous.	Priapus.	Penis.
Kinnbackendrüse.	Submandibular gland.	Processes pyramidalis.	Pyramidal lobes, thyroid gland.
Kropfdrüse.	Thyroid gland.	Prominentia laryngis.	Laryngeal prominence, thyroid cartilage.
Lien.	Spleen.	Pyramis.	Pyramidal lobe, thyroid gland.
Ligament, large.	Broad ligament.		
Lingual cartilage.	Epiglottic cartilage.	Radix linguae.	Root of tongue.
Lobus anonymus.	Lobus quadratus of liver.	Regenbogenhaut.	Iris.
Lobus anterior.	Lobus quadratus of liver.	Reißzähne.	Teeth, canine.
Luette.	Uvula.	Renes.	Kidneys.
Luschka's gland.	Pharyngeal tonsil.	Renes succenturiati.	Suprarenal glands.
Luschka's tonsil.	Pharyngeal tonsil.	Rima vocalis.	True vocal fold.
		Ringknorpel.	Cricoid cartilage.
Magendrüse.	Pancreas.	Ruthe.	Penis.
Mammila.	Nipple.	Ruthenzellkörper.	Corpora cavernosa penis.
Matrice.	Uterus.		
Membrana hyothyreoidea.	Thyrohyoid membrane.	Saccus coecus.	Fundus of stomach.
Membrana obturatoria laryngis.	Thyrohyoid membrane.	Salpinx.	Tube.
Milchfleisch.		Scheide.	Vagina.
Milz.	Thymus.	Scheidenklappe.	Hymen.
Molares.	Spleen.	Schildknorpel.	Thyroid cartilage.
Mutterscheide.	Teeth, molar.	Schnepfknorpel.	Arytenoid cartilage.
Muttertrompete.	Vagina.	Schneppenknorpel.	Arytenoid cartilage.
	Oviduct.	Schwammkörper.	Corpora cavernosa penis.

193

Human Anatomic Variation

Schwellbares.	Cavernous tissue.	Tuba.	Uterine tube (oviduct).
Schwellgewebe.	Cavernous tissue.	Tube uterina s. Fallopiae.	Uterine tube (oviduct).
Schwellkörper.	Corpora cavernosa penis.	Tube.	Auditory tube.
Septum membranaceum auris.	Tympanic membrane.	Tunica amphiblestroides.	Retina.
Small cul-de-sac.	Pyloric antrum of stomach.	Tunica nervea oculi.	Retina.
Spigelian lobe of liver.	Caudate lobe of liver.		
Spitzzähne.	Teeth, canine.	Valvula Bauhini.	Ileocolic valve.
Splen.	Spleen.	Valvula coli.	Ileocolic valve.
Staphyle.	Uvula.	Valvula Fallopiae.	Ileocolic valve.
Stellknorpel.	Arytenoid cartilage.	Valvula Tulpii.	Ileocolic valve.
Stockzähne.	Teeth, molar.	Valvula vaginae.	Hymen.
Stomachus.	Stomach.	Velum palatinum.	Soft palate.
Submaxillary gland.	Submandibular gland.	Velum pendulum palati.	Soft palate.
		Ventriculus.	Stomach.
Tenia libera.	Tenia coli, one of three tenia (topographically unrelated to mesentery or omentum).	Ventriculus laryngis.	Laryngeal ventricle.
		Ventriculus Morgagni.	Laryngeal ventricle.
		Vesicula fellea.	Gallbladder.
Tenia mesocolica.	Tenia coli, one of three tenia (position of attachment of embryonic dorsal mesentery).	Vesica fellea.	Gallbladder.
		Vestibulum oris.	Vestibule of mouth.
		Vordere Backenzähne.	Premolars.
Tenia omentalis.	Tenia coli, one of three tenia (position of attachment of greater omentum to transverse colon).	Vordere Mundhöhle.	Vestibule of mouth.
		Vorderzähne.	Teeth, incisor.
		Wechselzähne.	Teeth, deciduous.
		Weisheitszahn.	Teeth, molar, wisdom.
Teste muliebres.	Ovaries.	Weiße Augenhaut.	Sclera.
Tonsilla pharyngea.	Pharyngeal tonsil.		
Trommelfell.	Tympanic membrane.	Zäpfchen.	Uvula.
Tuba acustica.	Auditory tube.	Zitze.	Nipple.
Tuba Eustachii.	Auditory tube.		

SKELETON

Alphabetical Listing

(Primary references, in multiple page listings, are in *boldface* type.)

Arm **206**
Bifid Vertebrae 197
Bone
  Accessory 207, 208
  Atlas 199
    Proatlas 197
  Axis 197
  Calcaneus 208
  Carpal 206
    Accessory 207
  Clavicle **206**
  Cranium **199**
  Cuboid 208
  Cuneiform 208
  Ethmoid **202**
  Femur 208
  Fibula 208
  Frontal **200**
  Humerus **206**
  Hyoid 201
  Ilium 207
  Inca 200
  Intercuneiform 209
  Interparietal 200
  Ischium 207
  Lacrimal **202**
  Lunate 201
  Mandible **204**
  Maxillary **203**
  Metacarpal 207
  Metatarsal 208
  Nasal **202**
  Navicular 209
  Occipital **199**
  Orbital 203
  Palatine **203**
  Parietal 199, **200**
  Patella 208
  Pelvis **207**
  Phalanges 207
  Preinterparietal 200
  Radius 206
  Scaphoid 207
  Scapula **206**
  Sesamoid 207, 209
  Sphenoid **200**
  Sternum **205**
  Talus 209
  Tarsal 208
    Accessory 208
  Temporal **201**
  Tibia 208
  Trapezoid 207
  Triquetrum 207
    Accessory 207

  Ulna 206
  Vomer **202**
  Wormian (Sutural) 199, 201
  Zygomatic **203**
Canal
  Accessory Condyloid 199
  Carotid 201
  Condyloid 199
  Craniopharyngeus Lateralis 201
  Hypoglossal 199
  Incisive 203
  Infraorbital 203
    Supplementary 203
  Innominatus 201
  Lesser Palatine 203
  Mylohyoid 204
  Optic 201
  Porus crotaphytico-buccinatorius 200
  Posterior Ethmoidal 202
  Retrotransverse 197
  Zygomaticofacial 203
  Zygomaticotemporal 203
Carpus 206
Cartilage
  Medial Meniscus 208
  Sesamoid 204
  Thyroid **204**
Cervical Rib 197, 205
Cleft Palate 203
Cleidocranial Dysostosis 200, 206
Concha Santorini 202
Condyle
  Atlantoaxial 199
  Occipital 199
  Occipitoatlantal 199
  Third (Tertiary) 197, 199
Costoclavicular Ligament 206
Cranium **199**
Ectrocheiry 207
Fissure
  Orbital
    Inferior 203
    Superior 201
Fontanelle
  Sagittal 200
  Sphenoidal 200
Foot **208**
  Accessory Bones 208
Foramen
  Accessory 203
  Caroticoclinoid 201
  Greater Sciatic 207
  Hypoglossal 199
  Infraorbital 203
  Jugular 202

Jugulare Spurium 202
Magnum 199
Mental 204
Obturator 207
Optic 201
Ovale 201
Parietal 200
Pterygoalar 200
Pterygospinous 200
Rotundum 201
Sphenoidal Emissary 201
Sphenopalatine 203
Spinosum 201
Supratrochlear 206
Transversarium 197
Vesalii 200
Vertebrarterial 197
Forearm **206**
Fossa
  Canine 203
  Hypophyseal 201
  Hypotrochanteric 208
  Iliac 207
  Middle Cerebellar of Verga (Vermian) 200
  Nasal 203
  Pterygoid 201
  Subarcuate 201
  Torcular (Herophili) 200
Groove
  Infraorbital 203
  Mylohyoid 204
  Retrotransverse 197
  Sinus of Breschet 200
Half-Vertebrae 198
Hand **206**
  Accessory Bones 207
Hemilumbarization 198
Hemisacralization 198
Hyperostosis Frontalis Interna 200
Knee **208**
  Menisci 208
Lamina
  Cribrosa 202
  Orbitalis 202
  Papyracea 202
Larynx **204**
  Thyroid Cartilage **204**
Leg **208**
Ligament
  Costoclavicular 206
  Cruciate 208
  Meniscofemoral 208
  Pterygopetrosal 201
  of Struthers 206

Human Anatomic Variation

Transverse 208
Lumbarization 198
Lumbar Rib 198
Occipital Condyles 199
Occipital Vertebrae 199
Os
   Acetabuli Persistens 207
   Acromion 206
   Ainoicum 203
   Bregmaticum 200
   Calcaneum Secundarium 209
   Centrale 207
   Coronoides 206
   Coxae 207
   Cuneiforme Mediale Bipartum 208
   Gruberi 207
   Incisivum 203
   Intercuneiforme 209
   Intermetatarsum 208
   Japonicum 203
   Lunatotriquetrum 207
   Odontoideum 197
   Paracuneiforme 208
   Pterion 200
   Pubis 208
   Radiale Externum 207
   Subcapitatum 207
   Styloideum 207
   Sustentaculi Proprium 209
   Tibiale Externum 208
   Trapezium Secundarium 207
   Trapezoideum Secundarium 207
   Triangulare 207
   Trigonium 208
   Ulnostyloideum 206
   Vesalianum 208
   Vesalianum Carpi 207
Ossicle (Ossiculum)
   Ethmolacrimale Inferius 202
   Ethmolacrimale Superius 202
   Perilacrimal 202
   Scaphotrapezial 207
   Suprasternal 205
   Tarsal 208
Pelvis **207**
Platycnemia 208
Process (Processus)
   Alveolar 203
   Clinoid 200
   Condylar 204
   Coracoid 206

Coronoid 204
Deltoid 206
Frontal 203
Intrajugular 199
Jugular 199
Lemurinicus 204
Mamillary 198
Marginalis 203
Mastoid 202
Maxillary 203
Middle Clinoid 201
Orbital 203
Palatine 203
Paramastoid 199
Posterior Clinoid 200
Postglenoid 202
Sphenoidal 201
Styloid 201
Supracondylar 206
Vaginalis 201
Xiphoid 205
Zygomatic 203
Ribs **204**
   Bicipital 205
   Bifid 197
   Bifurcated 205
   Cervical 197, 205
   Coccygeal 205
   Floating 205
   Lumbar 205
   Thoracic 197
Sinus
   of Breschet 200
   Frontal 200
   Inferior Petrosal 202
   Maxillary 203
   Sphenoidal 201
   Superior Petrosal 202
Sternum **205**
   Cleft 205
   Ectopia Cordis 205
Styloid Process 201
Sulcus
   Arteriae Vertebralis 197
   Obturator 201
   Occipital 199
   Sagittal 199
   Transverse 199
Supracondylar Process 206
Suprasternal Ossicles 205
Suture

Anteroposterior 200
Frontal (Metopic) 200
Frontosquamous 201
Frontozygomatic 203
Intermaxillary 202
Internasal 202
Median Palatine 203
Mendosa 199
Palatomaxillary 203
Petrosquamous 201
Subsagittal 200
Subcondyloid 204
Subcoronoid 204
Transverse Palatine 203
Tail 198
Talonavicular Synostosis 209
Teeth 204
   Canine 204
   Deciduous 204
   Dens in Dente 204
   Incisors 204
   Molar 204
   Premolar 204
   Pulp 204
   Supernumerary 204
   Wisdom 204
Thigh **208**
   Third Trochanter 208
Thoracic Ribs 197
Thyroid Cartilage **204**
Torus Occipitalis 199
Transitional Vertebrae 198
Tubercle
   Crista Tentori 202
   Orbital 203
   Precondylar 199
Vertebra
   Bifid 197
   Cervical **197**
   Coccygeal **198**
   Half- 198
   Lumbar **198**
   Occipital 199
   Prominens 197
   Sacral **198**
   Thoracic **197**
   Transitional 198
Vertebral Column **197**
   Numerical Variation **199**
Wrist and Hand **206**
Zygomatic Process 203

Vertebral Column

Cervical Vertebrae

The cervical vertebrae exhibit considerable variation in their spinous processes. As a rule, the second, third, fourth, and fifth vertebrae possess bifid spines. The sixth and seventh demonstrate a tendency to be bifid with two small lateral tubercles. The sixth occasionally has a bifid spine, the seventh rarely. Occasionally all the cervical spines, with the exception of the second, are nonbifid.

The laminae of the inferior cervical vertrebrae frequently exhibit dorsally distinct tubercles from which fasciculi of the multifidus muscle arise. The tubercles may be found on the fourth and fifth vertebrae, but more frequently on the sixth and seventh. Anterior tubercles are bony projections that may be seen, usually, on the sixth cervical in lateral views of cervical spines.

A *foramen transversarium*, typical of the foramina of the cervical vertebral transverse processes, has been found in the first lumbar, fifth lumbar, and first sacral vertebrae.

An accessory transverse foramen, posterior to and smaller than the primary foramen, may be found in the sixth vertebra, less frequently in adjacent vertebrae.

The transverse foramen of the seventh cervical vertebra is normally small and may be absent. It does not normally transmit the vertebral artery but frequently allows passage of a vein.

The atlas may show incomplete ossification of the anterior and posterior arch. The absence of the anterior arch has been reported in an adult. The posterior arch may possess facets, which articulate with facets on the edge of the foramen magnum. The anterior arch may also have facets articulating with projections on the occipital bone. On occasion a groove for the vertebral artery is converted to a foramen. Paired ossicles have been reported in the posterior atlanto-occipital membrane, and have been interpreted as vestiges of a *proatlas*. Fusion, either partial or complete, of the atlas with the occipital bone may occur (incidence about 1%). The sulcus *arteriae vertebralis* of the posterior arch may be bridged over by a bony process from the inferior articular process, producing a bony canal for the vertebral artery (10% of cases). A *retrotransverse groove* (canal) may occasionally be present on the lateral mass of the atlas, posterior to the root of its transverse process. It houses an anastomotic vein. The posterior tubercle of the atlas is rarely bifid but may occasionally present two tubercles.

The atlas may be fused with the occipital bone in varying degrees.

The dens of the axis may (a) develop as a separate bone, the os *odontoideum*; (b) arise from the anterior arch; (c) articulate with a third occipital condyle; or (d) be absent. The axis (epistropheus) may be fused with either the atlas or with the third vertebra.

The vertebrarterial foramen associated with the sixth cervical vertebra occasionally appears as a notch and may also be associated with a secondary foramen for the vertebral vein.

The seventh cervical vertebra (*vertebra prominens*) is responsible for the palpable prominence at the dorsum of the neck in 70% of individuals; the sixth cervical is the next most frequent in forming this prominence (20%), and the spinous process of the first thoracic is the least frequent (10%).

The transverse process of the seventh vertebra may be bifid, and occasionally the costal process is replaced by a *cervical rib*. The body of the seventh sometimes bears bilaterally, near its caudal border, a costal pit for the head of the first rib. When a costal pit is present there is usually a well-developed cervical rib. Cervical ribs associated with the seventh vertebra appear in a variety of forms. They may be fully formed or merely enlargements of costal processes. In very rare cases there is a complete pair of ribs, which articulate with the sternum. They may be rudimentary, appearing as a fibrous band that joins the usual first thoracic rib or first rib cartilage. The cervical rib may be unattached at its ventral (distal) end.

Rudimentary cervical ribs have been observed in one case on the sixth cervical vertebra, in two cases on the fourth, but never on the fifth. Cervical ribs are normally present in reptiles and birds.

References   1, 8, 17, 18, 25, 49, 61, 68, 71, 94, 95, 103, 105, 115, 117, 119, 120, 128, 136, 148, 161, 163, 164, 166, 167, 232, 233, 252, 263, 269, 280, 296–298, 306, 307, 316, 317, 322, 324, 329, 338, 341, 347, 353, 360, 369, 416, 434, 435, 471, 476, 504, 512, 525, 529, 533, 536, 538, 555, 561, 563, 566, 589, 592, 597, 618, 619, 627, 647, 655, 656, 688, 692, 700, 704, 706, 722, 742, 757, 760, 768, 794, 795, 801, 807, 822, 828, 834, 836, 846, 859, 868, 882, 891, 897, 901, 902, 914, 946, 951, 959, 978, 1000, 1012–1014, 1023, 1027, 1035, 1056, 1063, 1080, 1102, 1112, 1117, 1126–1130, 1136, 1137, 1139, 1140, 1141, 1146, 1148.

Thoracic Vertebrae

The body of the first thoracic vertebra may bear, on one or both sides, two costal pits resembling the usual cervical type. The articulation of the head of the first rib is completed by a costal pit on the body of the seventh cervical vertebra. In rare cases the first thoracic rib may be rudimentary and similar in appearance to cervical ribs. The body of the second thoracic vertebra usually exhibits an identifying tubercle marking the attachment of a fasciculus of the medial portion of the longus colli muscle. The twelfth thoracic vertebra, in the absence of the twelfth pair of ribs, commonly conforms to the lumbar type. The transverse process of the tenth may lack a facet for costal articulation. The lumbar form of transverse process may be present in the eleventh thoracic vertebra. In addition, the transition from the thoracic to the lumbar type of articular process may occur in the eleventh thoracic instead of in the twelfth thoracic vertebra. Articular processes may be of the lumbar type on one side and thoracic on the other side, resulting in an asymmetry in articulation. Occasionally, a *half-vertebra* is found in the thoracic and lumbar regions and very rarely in cervical and sacral vertebrae. Such specimens have a wedge-shaped half-body, a half-arch with a transverse process, superior and inferior articular processes, and half a spinous process. The half-vertebra is, as a rule, ankylosed to adjacent vertebrae.

References 25, 61, 69, 117, 119, 165, 191, 203, 211, 233, 238, 241, 248, 249, 270, 277, 338, 417, 418, 447, 761, 817, 861, 900, 991, 1013, 1040, 1069, 1082, 1128, 1129, 1134, 1136–1138, 1140, 1141, 1146, 1148.

Lumbar Vertebrae

The first lumbar, instead of the twelfth thoracic vertebra, may present the transitional form of articular process. A lumbar rib may be present, united with the ventral surface of the transverse process or the tip of a short transverse process. The accessory rib may be elongated but is usually rudimentary. Occasionally, the mamillary and accessory processes are united by a bony bridge forming a foramen behind the transverse process.

The transverse process of the fourth lumbar has been reported arising from the side of the vertebral body and having no connection with either the arch or pedicle.

The fifth lumbar vertebra is subject to numerous variations of several sorts, of which the following two are of special interest. In the first, the pedicles of the arch show a remarkable deviation, a complete dissolution of the continuity of the arch immediately caudal to the superior articular processes, differing from the conditions found in parts of the column. In a large series of skeletons, the incidence was about 4%. In such specimens the anterior part consists of the vertebral body carrying the pedicles and the transverse and superior articular processes, whereas the posterior segment is composed of the laminae and the spinous and inferior articular processes. This condition is occasionally present, either unilaterally or bilaterally, in the other lumbar vertebrae and even in the thoracic vertebrae. In the second type, the fifth lumbar shows a tendency, on one or both sides, to conform to the type of the first sacral vertebra, with which it may become fused. However, there may only be four lumbar vertebrae. Failure of the union of the halves of the vertebral arch and defective laminae, in this and other regions of the column, are found in spina bifida.

Lumbosacral facets are variable. In an x-ray study of 3000 subjects, the facets faced backward in 57%, pointed inward in 12%, and were mixed or asymmetric in 31% of cases.

References 25, 27, 35, 41, 61, 64, 67, 112, 118, 119, 187, 191, 198, 205, 211, 233, 254, 328, 336, 346, 469, 487, 571, 617, 635, 665, 702, 718, 721, 733, 737, 783, 791, 836, 848, 852, 912, 936, 986, 988, 1022, 1029, 1030, 1049, 1058, 1077, 1087, 1110, 1119, 1128–1130, 1134–1141, 1146, 1148.

Sacrum and Coccyx

The number of vertebrae in the sacrum may be increased by fusion of the first coccygeal, by (less often) addition of the last (fifth) lumbar, or by addition of both the last lumbar and first coccygeal vertebrae. The number may be reduced to four by the "lumbarization" of the first sacral vertebra. In one study of 631 bodies, the sacrum was composed of five vertebrae in 77%, six in 21.7%, four in 1%, and seven in 0.2%. The lumbosacral junction is occasionally composed of a vertebra with characteristics of a lumbar vertebra on one side and sacral on the other (so-called *hemilumbarization* or *hemisacralization*).

The articular surface of the sacrum may extend variably, over no more than the first two vertebrae or to the fourth sacral vertebra. Accessory articular facets may occur on the lateral sacral crest at the level of the first or second dorsal sacral foramen; these have been found to be unpaired or bilateral and to articulate with appropriate facets on the ilium.

The sacral canal may be open dorsally because of failure of laminar fusion. Coalescence of the coccyx and sacrum occurs more often and earlier in males.

The subdural and subarachnoid spaces extend into the sacral canal as far as the middle third of the body of the second sacral vertebra; however, in 46% of 56 cadavers, these spaces extended caudal to this level.

The female sacrum is usually broader (in proportion to length), less curved, and directed more obliquely dorsally than the male sacrum. The curvature of the female sacrum occurs in the caudal part of the bone; in the male, the curvature is distributed over the entire length of the bone. Sex characteristics, however, are variable.

Partial or complete absence of the sacrum and/or coccyx has been reported.

The presence of a tail in humans has been reported on several occasions. In one case, the tail was 23 cm in length.

References 25, 26, 35, 41, 61, 64, 92, 101, 112, 114, 118, 119, 187, 198, 205, 207, 224, 225, 233, 234, 242, 266, 275, 276, 303, 328, 331, 345, 346, 364, 477, 483, 487, 488, 508, 532, 537, 560, 633, 635, 636, 673, 702, 718, 721, 729, 783, 788, 791, 819, 837, 848, 849, 852, 871, 912, 919, 936, 988, 1022, 1029–1032, 1058, 1060 1077, 1087, 1110, 1115, 1119, 1128–1130, 1136–1141, 1146, 1148.

Numerical Variation in Vertebral Column

The usual grouping formula of seven cervical, 12 thoracic, five lumbar, five sacral, and four coccygeal vertebrae is found in only about 20% of cases studied. The cervical region is reported to be the most constant, the coccygeal the most variable.

The number of elements of the vertebral column has been reported to vary between 32 and 35. Addition to a group is frequently seen, which occurs through the reduction in the number of vertebrae of an adjacent group, the total number being unchanged. In this variation, the vertebra added is usually intermediate in form between the adjacent groups. The location of such a vertebra is predominantly at the ends of the column and at the levels of transition between its regions. Thus, sacralization of the fifth lumbar, lumbar-like articular processes in the eleventh thoracic, and thoracic costal facets on the seventh cervical are observed.

The levels of transition may be shifted cephalad, resulting in 23 mobile vertebrae, or shifted caudad, resulting in 25 presacral vertebrae. Such variations may occur in 2–11% of the population.

In one study of 748 vertebral columns, 717 had 17 thoracolumbar vertebrae, 26 had 18, and five had 16. Of the 717 columns, four had a lumbar increase and a thoracic decrease while five had a lumbar decrease with a compensating thoracic increase.

References 20, 21, 47, 50, 54, 56, 82, 102, 116, 146, 189, 204, 336, 645, 646, 899, 943, 991, 992, 1128, 1130, 1136, 1137, 1139–1141, 1146, 1148.

Cranium

Occipital Bone

The articular surface of the occipital condyles may possess a transverse ridge of cartilage that corresponds to a groove on the articular surface of the atlas, or notches at the margin may partly divide the condyle in two parts.

A shallow fossa is occasionally found on the ventral surface of the basilar portion anterior to the pharyngeal tubercle, which has been interpreted as a vestige of the canal of the notochord.

Precondylar tubercles, unilateral or bilateral, just anterior to the anterior margin of the foramen magnum, were found in 10% of 1527 skulls.

The occipital condyles vary considerably in their position relative to the sides of the foramen magnum. The condyles articulate with the atlas; occasionally, a facet located on the anterior margin of the foramen magnum, the so-called *third occipital condyle*, articulates with the dens.

Various structures similar to parts of the *atlas* have been seen around the foramen magnum and have been described as *occipital vertebra* (termed *manifestation of occipital vertebra*). The atlas may be fused, in part or completely, with the occipital bone (termed *assimilation of atlas*). About 1% of skeletons show such a variation, which has been interpreted by some authors as a cranial shift in the regional grouping of vertebrae of the vertebral column. Signs believed associated with assimilated or occipital vertebrae around the foramen magnum include the following: (a) a massive paramastoid process; (b) an enlarged jugular process; (c) the anterior margin of the foramen magnum thickened and raised to form a bar of bone between the condyles; (d) the hypoglossal canal divided by a bony bridge; and (e) a tertiary condyle and a facet or other marking for the apex of the dens on the anterior margin of the foramen magnum.

If the atlas is "assimilated," the occipital condyles are replaced by the inferior facets of the atlas (atlantoaxial type). If there is duplication of the atlas, the condyles are of the occipitoatlantal type.

Elevation of an area between the supreme and superior nuchal lines is termed *torus occipitalis*. In these cases, the inion may be greatly enlarged.

Sutural bones are usually small, irregularly shaped ossicles, often found in the sutures of the cranium, especially in the *parietal bones*. When the lateral portions of the transverse occipital suture persist, the anomaly is termed *sutura mendosa*. As many as 100 sutural or *wormian* bones have been found in one skull. They are rarely found in the sutures of the face.

The hypoglossal (canal) foramen may be doubled. The jugular foramen may be divided into two parts by intrajugular processes. The condyloid canal is absent in 25% of skulls. In some cases it communicates with the hypoglossal canal. Accessory condyloid canals have been reported.

The grooving on the inner surface of the occipital bone is variable. In about 17% of cases the sagittal sulcus turns to join the left transverse sulcus. The sagittal sulcus may bifurcate, with the larger groove turning to join the right transverse sulcus and the smaller one joining the left transverse sulcus (about 15% of cases). In rare cases, the larger groove joins the left and the smaller the right. In very rare cases, the right and left groove appear equal in size. The crista occipitalis in some subjects is a sulcus (*occipital sulcus*) rather than a ridge. In a case reported by Le Double, the sagittal and occipital sulci were doubled or paired; both sulci joined the transverse sulci on their respective sides. The torcular fossa (housing the *torcular herophili* or confluens of the sinuses) is usually on the right side but may be in the median plane or on the left side. A shallow fossa of varying size may occasionally be found on the dorsal aspect of the foramen magnum, and has been named the *vermian fossa* or the *middle cerebellar fossa* of Verga. It is bounded by the limbs of the internal occipital crest or ridge, which diverge around the foramen magnum giving the fossa a somewhat triangular shape. It may be divided into upper and lower parts by a ridge of bone. This fossa houses part of the inferior vermis of the cerebellum.

References 34, 49, 90, 99, 138, 142, 147, 148, 163, 175, 199, 236, 262, 280, 286, 295, 309, 314, 337, 360, 367, 476, 490, 500, 506, 515, 528, 529, 536, 563, 569, 586, 588, 589, 592, 598, 607, 609, 618, 623, 627, 638, 643, 659, 666, 687, 688, 700, 706, 751, 762, 768, 774, 792, 807, 822, 838, 846, 881, 897, 905, 908, 949, 978, 979, 984, 1023, 1027, 1061, 1070, 1080, 1128–1130, 1135–1137, 1139–1141, 1146, 1148.

Parietal Bone

Occasionally, the interparietal region of the skull remains separate throughout life; the space-filling bone may be the *interparietal* or *Inca bone,* or a large sutural bone termed the *preinterparietal bone.*

Rarely, the parietal bone is composed of two pieces, one superior to the other, separated by an anteroposterior suture (*subsagittal suture*) which is almost parallel with the sagittal suture. In these cases the two primary centers of ossification are believed to have failed to fuse. The sagittal suture may be absent.

In very rare cases, vertical sutures divide the parietal bone into three or four parts. The parietal foramina are occasionally greatly enlarged. The foramen transmits a vein to the superior sagittal sinus and a small branch of the occipital artery to the dura mater. Parietal foramina occur in 60% of skulls; they may be absent, however, on one or both sides. The foramina are believed to be the remains of the fetal sagittal fontanelle. In the great fontanelle between the paired frontal and parietal bones, an accessory ossicle, *os bregmaticum,* may be found either free or fused with one of the frontals or parietals. In addition, at the anterior inferior angle (the so-called sphenoidal fontanelle) of the same (great) fontanelle, another ossicle may persist, the *os pterion.*

A large groove (*sinus of Breschet*) may occur on the intracranial face, parallel to the coronal suture and extending from the bregma to the lesser wing of the sphenoid.

References 5, 52, 53, 72, 75, 110, 141, 142, 175, 283, 332, 343, 365, 386, 387, 406, 414, 415, 420, 421, 442, 500, 509, 539, 541, 549, 607, 630, 640, 659, 670, 701, 754, 763, 766, 769, 792, 903, 945, 949, 957, 982, 993, 994, 998, 1020, 1024, 1042, 1093, 1114, 1128–1130, 1135–1141, 1146, 1148.

Frontal Bone

The *frontal sinuses* may be absent, small, or remarkably large, extending through orbital bones, zygomatic processes, and into the squamae.

The coronal suture may be absent. Persistence of the *interfrontal* or *metopic suture* has reported frequencies ranging from 1% to 12% of skulls. Metopism occurs, according to one study, in 8.7% of European Caucasians, 5.1% of Orientals, 1.2% of blacks, and only 1% of Australian Caucasians. Sutural or intercalated interfrontal ossicles may also be present.

Persistence of a complete or incomplete frontal (metopic) suture may be associated with a syndrome involving several anomalies of the visceral cranium and the phalanges. Failure of ossification in a broad and elongated area in the region of the frontal suture has been shown to be associated with a partial or complete absence of the clavicle (*cleidocranial dysostosis*).

A thickening of the internal lamina of the squama (*hyperostosis frontalis interna*) and orbital part of the frontal bone is found almost exclusively in women (about 98%). Hyperostosis frontalis interna may extend into the diploë but does not involve the external lamina or the posterior wall of the frontal sinus.

A frontal spine may be present, which can be mistaken for a perpendicular plate of the ethmoid.

References 4, 7, 37, 89, 98, 175, 190, 300–302, 481, 492, 493, 505, 551, 641, 663, 668, 713, 714, 743, 766, 800, 805, 809, 818, 827, 879, 938, 945, 995, 998, 1002, 1008, 1019, 1020, 1088, 1128–1130, 1136–1138, 1140, 1141, 1146, 1148.

Sphenoid Bone

The superior part of the dorsum sellae may be a separate bone or it may be joined to the petrous portion of the temporal bone.

Vesalius was the first to describe and illustrate the foramen that bears his name (*foramen vesalii*). It is located between the foramen rotundum and the foramen ovale on its medial side. It is traversed by a vein (vein of Vesalius), a small emissary, from the cavernous sinus to the pterygoid plexus. The presence of this foramen is rare.

Occasionally, foramina occur through bridges of bone between the posterior margin of the lateral pterygoid lamina and the spine medial to the foramen ovale; these transmit the nerves and vessels for the medial pterygoid muscle. This *pterygospinous foramen* occurs in about 5% of cases. A bridge of bone between the lateral surface and base of the lateral pterygoid plate and the greater wing of the sphenoid (lateral to the foramen ovale) provides passage for some or most of the motor fibers from the trigeminal nerve. This *pterygoalar foramen* (or *porus crotaphytico-buccinatorius,* Hyrtl) was reported in about 1% of skulls by Hyrtl and in 8% of specimens in another study. A bridge of bone between the anterior and posterior clinoid processes is reported to occur in 5% of individuals studied. The foramen spinosum may be absent, in which case the middle meningeal artery enters the foramen ovale. The foramina spinosum and ovale may be continuous. The *sphenoidal emissary* foramen may transmit the accessory meningeal artery (20% of cases). The usual diameter of the optic canal is 5.5 mm but varies from 3.5 to 6.5 mm. The foramen rotundum may be doubled. An extra bony canal may run from the medial angle of the superior orbital fissure through the body of the sphenoid, opening at the processus vaginalis. In 0.3% of adult skulls, a *craniopharyngeal* canal is present. An embryonic remnant, it opens in the hypophyseal fossa, traversing the sphenoid body.

The *middle clinoid* process may be elongated enough to form a bridge over the carotid sulcus, transforming it into a *caroticoclinoid* foramen. The optic canal may be divided into two parts by a bony partition. On occasion, the optic foramen is continuous with the orbital fissure. The foramen rotundum may also be continuous with the orbital fissure.

The angular spine of the greater wing of the sphenoid may participate with its lateral surface in forming the articular fossa for the mandible. The vaginal process may participate in creating a canal along its inferior aspect. The pterygoid fossa is sometimes subdivided by several partitions. The posterior aspect of the lateral pterygoid plate may be joined to the angular spine of the greater wing by a pterygopetrosal ligament that may calcify. At the base of the spina angularis, to the medial side of the foramen spinosum, there is occasionally a minute canal (*canaliculus innominatus*, Arnold) through which the lesser petrosal nerve passes. The medial aspect of the spina angularis is frequently grooved by the chorda tympani.

Sutural (*wormian*) bones may occupy the sphenoidal fontanelles between the second and fifteenth years, after which they may persist as separate ossicles or unite with the sphenoid, temporal, or frontal bone.

The *sphenoidal sinus* may be absent or small; it may extend into the basilar process of the occipital bone, the greater or lesser wings of the sphenoid, or the pterygoid processes. The bony plate separating the sphenoidal sinus from the optic nerve, the maxillary division of the trigeminal nerve, the nerve of the pterygoid canal, the cavernous sinus, the carotid artery, and the hypophysis may be very thin or absent. Hence these structures may become vulnerable in chronic sinus infections. It has been said that "the sphenoidal sinus is the most variable in form of any bilateral cavity or organ in the human body."[1]

The lesser wing of the sphenoid may project forward onto the roof of the orbit. The lateral part of the *superior orbital fissure* may be closed because of fusion of the greater and lesser wings. The ophthalmic artery in such cases passes through a separate foramen.

The groove for the meningeal branch of the lacrimal artery, located on the lateral inferior border of the superior orbital fissure, may be bridged by bone forming a foramen. Another canal (*canalis craniopharyngeus lateralis*) located at the junction of the basi-, ali-, and presphenoid is present in about 4% of cases.

The temporal aspect of the greater wing may be divided vertically by a fissure. An ossicle may be formed by a horizontal fissure separating off the apex of the greater wing of the sphenoid.

References 90, 156, 168, 173–175, 190, 193, 235, 330, 333, 339, 340, 430, 432, 466, 482, 530, 534, 567, 583, 584, 601, 608, 626, 628, 629, 756, 776, 781, 784, 802, 850, 884, 947, 948, 954, 974, 999, 1051, 1072, 1084, 1128–1130, 1136–1141, 1146, 1147, 1148.

Temporal Bone

Absence of the superior part of the squama temporalis has been reported. A *frontosquamous* suture frequently occurs, which is normally found in the skulls of the chimpanzee, gorilla, and gibbon.

The petrosquamous suture of the temporal bone sometimes persists. The mastoid cells vary in extent and may invade the squamous part of the bone; the air cells may not be symmetrical on the two sides. The squama in some cases is divided into two or four parts.

The styloid processes vary markedly in length, and the stylohyoid ligament may calcify to make a rigid connection with the hyoid bone. A 75-mm long styloid process has been reported. The styloid process may be composed of two to five osselets, articulating by synchondroses.

The greater horn of the hyoid may be joined to the body of the hyoid by an articulation. The zygomatic arch may be replaced by a fibrous band. In some cases the zygomatic process of the temporal bone is absent.

The mastoid process may possess a doubled apex, with the medial portion divided from the lateral by a fissure. Usually, the medial lip of the groove forms a distinct ridge and provides the point of attachment of the digastric muscle; an enlarged paramastoid process may result from the development of air cells within the medial portion of the process. Air cells, usually confined to the mastoid process, may invade the horizontal or vertical part of the squama, or even the pars petrosa.

The pars petrosa is very rarely variable. The *subarcuate fossa*, however, may persist, in part, as a bony canal. In cases of absence of the internal carotid artery, the carotid canal may also be absent. In two cases this was found on the left side; in a third case

---

[1] Congdon, E. D. The distribution and mode of origin of septa and walls of the sphenoid sinus. *Anat. Rec.* 18:97–23, 1920.

the side was not specified. The carotid canal may take one of several forms, including that of a sulcus. The squamous part of the temporal bone may exhibit a sizable, extra process just anterior to the external acoustic meatus, called the *postglenoid process*.

The suprameatal spine, which overlies the floor of the mastoid antrum, is absent in 20% of skulls. A bony process may project from the anterior edge of the squama that runs between the sphenoid and parietal bones to the lateral edge of the frontal bone. This process effectively intervenes between the parietal and sphenoid bones. The squama may also exhibit a *foramen jugulare spurium* by which intracranial venous blood may communicate with the external jugular system. The fetal suture between the squama and the mastoid process sometimes persists in the adult.

A foramen may be present in the petrobasilar fissure in front of the jugular foramen through which runs the inferior petrosal sinus. The jugular foramen, a passage between the petrous portion of the temporal bone and the jugular process of the occipital bone, may be divided into two parts by intrajugular processes.

The upper edge of the internal acoustic meatus may form a tubercle, the *crista tentori*, that roofs the superior petrosal sinus.

References 46, 79, 108, 126, 172, 175, 178, 194, 195, 201, 237, 257, 261, 272, 288, 293, 305, 332, 333, 388, 407, 419, 423, 424, 431, 437, 438, 531, 542, 569, 582, 605, 614, 630, 642, 648, 672, 784, 812, 847, 869, 944, 1011, 1061, 1120, 1128–1130, 1136–1138, 1140, 1141, 1146, 1148.

Ethmoid Bone

Ethmoidal variations include reduction in the size of the lamina orbitalis, which occurs when the frontal or maxillary bones or both participate in the formation of the medial wall of the orbit. The orbital plate may be divided into two, three, or four parts by vertical fissures. Occasionally, the anterosuperior or anteroinferior angles are separated from the orbital plate forming ossicles termed *ossiculum ethmolacrimale superius* and *ossiculum ethmolacrimale inferius*.

An increase in the number of nasal conchae is common. The ethmoid may exhibit two to four conchae (not including the inferior or maxilloturbinate). Three conchae are most frequent (80% of skulls). A bony process below the middle concha may descend over the aperture of the maxillary sinus. A fourth concha over the superior one is called *concha Santorini*.

The anterior ethmoidal foramen in the lamina cribrosa may lead into a bony canal through which an artery and nerve course. The crista galli may be pneumatized. The lamina papyracea that covers the labyrinth laterally may exist as two bones resulting from a perpendicular suture. The posterior ethmoidal canal is frequently absent, being replaced by a canal at the junction between the upper edge of the lamina and the sphenoid body.

References 81, 308, 408, 689, 777, 958, 1050, 1128–1130, 1136–1141, 1146, 1148.

Lacrimal Bone

The lacrimal bone may be divided into two or more parts or fused with neighboring bones. This bone may exhibit so many minute foramina that it takes the appearance of a bony net. It may be fused with or divided by the lamina papyracea. When absent, the lacrimal bone is replaced by processes from the frontal process of the maxilla and/or the lamina papyracea. A contribution from the frontal bone has been reported.

The pars lacrimalis may be deficient and this may be filled by the frontal process of the maxilla.

In some cases the crest is reduced in size or absent. The hamulus may be absent, represented by a fibrous band, or unusually long. An enlarged hamulus may articulate with the maxillary process of the zygomatic bone. The hamulus is sometimes replaced by an ossicle; additional perilacrimal ossicles have also been reported associated with the lacrimal bone.

The nasolacrimal duct may open into the middle meatus.

References 291, 308, 315, 320, 590, 1008, 1128–1130, 1135–1141, 1146, 1148.

Vomer

The inferior margin of the vomer has been found in the intermaxillary suture, participating with the palatine bones in the formation of the hard palate. The vomer may be separated from the perpendicular plate of the ethmoid by a strip of cartilage from the nasal septum. The vomer is sometimes fenestrated.

References 1128–1130, 1135–1138, 1140, 1141, 1146, 1148.

Nasal Bone

Nasal bones may be reduced in size with a concavity of the lateral margins. This is frequently accompanied by expansion of the frontal process of the maxillae. Nasal bones are rarely absent, but are frequently asymmetric. A process of the frontal bone may extend between the nasal bones and the frontal process of the maxilla. There may be only one nasal bone (unpaired) or there may be several. Internasal bones have been found at the edge of the nasal bones at the upper corner of the pyriform aperture, lying on the anterior tip of the perpendicular plate of the ethmoid. Transverse sutures have also been observed.

Rarely, the *internasal suture* is obliterated by the fusion of nasal bones.

References 40, 175, 177, 192, 239, 247, 333–335, 429, 809, 1047, 1065, 1068, 1128–1130, 1135–1141, 1146, 1148.

Maxilla

In the maxilla, the inferior wall of the infraorbital canal may be incomplete, and hence the infraorbital nerve is in contact with the mucosa of the maxillary sinus. In about 5% of skulls, the infraorbital groove is present as a canal. Two or more infraorbital foramina have been reported. The infraorbital foramen may lie anywhere from 3 to 7 mm below the infraorbital margin. Occasionally, a bony spine covers the foramen. The infraorbital canal (groove) may be shifted laterally as far as the suture between the maxilla and zygomatic bones. A supplementary infraorbital canal sometimes occurs medial or lateral to the normal canal.

Investigations have shown that enlargement of the maxillary sinus may be produced by (a) hollowing out of the alveolar process (alveolar recess); (b) excavation of the floor of the nasal fossa by extension of the alveolar recess between the plates of the hard palate (palatal recess); (c) encroachment of the sinus into the frontal process of the maxilla; (d) hollowing out of the zygomatic process of the zygomatic bone (malar recess); and (e) extension to and appropriation of an air cell within the orbital process of the palatine bone (palatal recess). Reduction of the maxillary sinus, on the other hand, may follow (a) imperfect absorption of cancellous bone on the floor of the sinus, or secondary thickening of its walls; or (b) encroachment due to approximation of the facial and nasal walls, unusual depression of the canine fossa, excessive bulging of the lateral nasal wall, or imperfectly erupted teeth. The crescentic projections that quite commonly protrude from the walls into the interior occasionally are replaced by septa that completely divide the sinus into two cavities, each having an independent opening into the nasal fossa, but having no communication with each other. These partitions vary in position and direction, in some cases subdividing the sinus into an anterior and a posterior compartment, and in others, into an upper and a lower chamber. In the latter case, the lower chamber may communicate with the inferior meatus.

Accessory ostia of the maxillary sinus are found in about 30% of skulls; as many as three in one skull have been reported.

The incisive part of the alveolar process may be an independent bone, the *os incisivum*. The incisive canals, which permit passage of the nasopalatine nerves, may open by several foramina.

Deficiency of the palatine processes result in *cleft palate*. Separate bones may occur on either side of the median palatine suture. The palatine processes of the maxilla sometimes project backward in the median plane, separating the horizontal plates from each other.

In the suture between the ethmoid and maxilla, beneath the anterior ethmoidal foramen, a 3-mm hiatus may occur through which the orbital and nasal cavities communicate.

References   153, 226, 228, 350, 439, 465, 503, 587, 602, 719, 888, 910, 1008, 1017, 1045, 1065, 1071, 1128–1130, 1136–1141, 1146, 1148.

Palatine Bone

The sphenopalatine notch of the palatine bone is frequently replaced by a foramen. This sphenopalatine foramen may be doubled. The size of the orbital process is quite variable. The horizontal plate of the palatine bone may contain a portion of the maxillary sinus.

The lesser palatine canal is frequently absent. It may also be located between the palatine and maxillary bones. The transverse palatine suture varies in its configuration. Wormian bones (ossicles) have been found in the palatomaxillary sutures.

The orbital process may extend upward behind the lamina papyracea of the ethmoid to articulate with the frontal bone, or a process of the frontal may descend in the same location to join the orbital process of the palatine bone.

References   107, 149, 1055, 1128–1130, 1136–1141, 1146, 1148.

Zygomatic Bone

The foramina of the zygomatic bone that transmit branches of the zygomatic nerve are often subject to variation in number and position. A bipartite zygomatic (*os japonicum* or *os ainoicum*) has frequently been observed. The incidence in Japanese is 7%, and in other ethnic groups 2.5–6.5%. A zygomatic bone may also be tripartite, possessing extra sutures and accessory bones.

An orbital tubercle, in the orbital surface of the zygomatic bone and inferior to the frontozygomatic suture, provides an attachment site for the lacertus of the lateral rectus muscle (about 90% of cases).

The zygomatic bone may or may not, with equal frequency, participate in forming the inferior orbital fissure. A *processus marginalis* is found arising from the posterior edge of the facies lateralis in about 50% of skulls, and may serve as the insertion to part of the temporal fascia. The openings of the zygomaticotemporal and zygomaticofacial canals may be increased in number.

Numerous cases have been reported in which the zygomatic bone does not contribute to the formation of the inferior orbital fissure as a result of the articulation of the greater wing of the sphenoid with the maxilla, or by the intervention of a wormian bone. The maxilla may by directly joined to the zygomatic process of the temporal bone by a dorsal extension of the zygomatic process of the maxilla. The maxillary process may articulate with the lacrimal hamulus.

References   39, 80, 202, 231, 398, 399, 405, 412, 422, 443, 544, 652, 671, 672, 680, 730–732, 740, 839, 840, 926, 1033, 1122, 1128–1130, 1136, 1137, 1139, 1140, 1141, 1146–1148.

Mandible

The mandible shows extensive variation in size and weight during an individual's lifetime. The chin may protrude or recede; there may be one rather than two mental tubercles.

Although the coronoid and condylar processes vary in size and form, they are rarely united by sutures with the ramus (*subcoronoid* and *subcondyloid sutures*). The condylar processes may be doubled. In some individuals, the two halves of the jaw fail to co-ossify. The *mental foramen* may be doubled or tripled. The location of the mental foramen has been found as far forward as the first premolar or as far back as the second premolar. In very rare cases a median mental foramen is present. The *mylohyoid groove* may be bridged over by a bony process and converted into a canal.

A process projecting from the posterior or inferior border (near the angle) of the mandible has been named *processus lemurinicus* by Sandifort (Albrecht). Anomalous teeth have been found widely distributed throughout the mandible.

References 111, 120, 246, 294, 394, 400, 433, 518, 816, 977, 989, 1128–1130, 1136, 1137, 1140, 1141, 1146, 1148.

Teeth

Medicolegal use of dentition has been established on the basis of width of the pulp chamber, stage of development, eruption of teeth, and resorption of roots of deciduous teeth. This use is believed to be valid in individuals up to 20 years of age.

It is common to find decreased numbers of teeth. Most often missing are the upper lateral incisors, then the second lower premolars, the wisdom teeth, and the medial incisors, in this order. Rarely the upper canines are missing, more rarely the upper premolars and second molars, and most rarely the first permanent molars.

The upper lateral incisors may be duplicated and the development of four molars or two supernumerary premolars may also occur. Rarely, the canines are duplicated.

Deciduous teeth are sometimes retained because of the absence of the bud of the permanent teeth in the jaw. The persistence occurs most often with the upper canine and the lower second molar, and less often with the upper lateral and lower mesial incisors.

The third lower molar has difficulty erupting; frequently, teeth are anomalously rotated causing difficult eruption. The roots of the lower third molar have been found surrounding the inferior alveolar nerve. A wisdom tooth has been reported facing the second molar, crown to crown. In another case, the crown faced downward and the root upward. Atypical arrangement, displacement, and retention of teeth is common. Supernumerary and retained teeth may show considerable variation in shape. Bizarre forms such as cases of "dens in dente," a tooth (with pulp) within the pulp chamber of another tooth, have been reported. A tooth has been found in the orbital cavity and others have been found throughout the mandible.

References 96, 153, 229, 473, 474, 546, 581, 591, 697, 698, 709, 715, 870, 922, 932, 987, 1036, 1043, 1081, 1128, 1130, 1136, 1137, 1139–1141, 1146, 1148.

Larynx

Thyroid Cartilage

In a study of 184 cartilages, 38% had a foramen in the lamina. In 90% of these the foramen transmitted blood vessels, and in 50% nerve fibers as well. The nerve may be a branch of the superior laryngeal (vagus) which may anastomose with the internal laryngeal (vagus).

A sesamoid cartilage has been reported in the superior (false) vocal fold.

References 46, 176, 349, 413, 543, 651, 1128, 1130, 1136–1141, 1146, 1148.

Ribs

The number of ribs may be increased by addition at either the cervical or lumbar end of the series. It is extremely rare to find an additional rib or pair of ribs in both the cervical and lumbar regions at the same time. Eight true ribs have been observed. The first rib may be shorter than usual, slender, and connected with the second costal cartilage. The number of ribs may also be decreased. The first rib is occasionally absent. A case of incomplete development of the third and fourth ribs, and the absence of ribs eight, nine, and ten has been reported. In another subject, ribs two, three, four, and five were absent.

The twelfth rib is variable in length; it may be markedly reduced in size, or as long as 14 cm. Differences in length are important considerations in the surgical approach to the kidney. The twelfth rib may be absent on one side and shortened on the other side. Asymmetric ribs are common.

A bicipital rib is seen in relation to the first thoracic rib. It appears to be the result of the fusion of two ribs, either of a cervical and first thoracic or of the first two thoracic ribs. Fusion of the first two ribs is common.

Unusual variations include the following. The costal cartilage and adjacent portions of the body of the rib are occasionally replaced by fibrous tissue. The first rib may be ossified at its extremities but fibrous between. Two adjacent ribs may be completely fused, or the bodies of two or more ribs may be joined by fusion due to small quadrilateral plates between the bones. Ribs bifurcated at their sternal ends are occasionally observed, with the two extremities joined to a bifid costal cartilage. Bifurcated ribs occurred in six of 14,000 cases in one study. The tenth rib may be a floating rib (a frequent observation by Japanese investigators). Two adjacent ribs, often five and six, and six and seven, sometimes articulate at their costal cartilages by a diarthrosis, synarthrosis, or syndesmosis.

Cervical ribs occur in about 0.5–1% of cadavers. They tend to be small but occasionally reach the sternum. The ventral extremity may (a) lie free above or between the scalene muscles, (b) be connected to the sternum by a ligamentous prolongation, (c) articulate with the superior surface of the first thoracic rib, or (d) form a complete rib that articulates with the sternum. Because of their location, cervical ribs may cause a disturbance to the vascular or nerve supply to the arm, forearm, and hand. Double cervical ribs have been reported. A cervical rib may pierce m. scalenus medius.

Lumbar ribs are usually quite small and seldom as complete as cervical ribs. Extra levator costae muscles may be associated with these ribs. Lumbar ribs have been reported to be more numerous than cervical ribs (8% of individuals). In another study, rib anomalies were found in 0.15% of 38,105 patients; of these, 35% were cervical ribs. Other costal anomalies found in the 38,105 chest films included lumbar ribs, dichotomies, forked ribs, and synostoses. In another study of 10,000 subjects, 75% of rib abnormalities were seen in the seventh cervical and first thoracic.

A rib rudiment was reported on the *coccyx*.

In a study of 6 million chest photofluorograms reviewed for costal anomalies the incidence was 0.31%. Females were involved more frequently than males, and the right side more commonly than the left.

References 8, 9, 21, 22, 55, 57, 74, 91, 100, 115, 152, 157, 159, 164, 166, 167, 186, 200, 225, 238, 240, 241, 252, 270, 292, 298, 322, 338, 347, 353, 368, 409, 410, 418, 435, 444, 447, 504, 522, 540, 561, 572, 597, 613, 637, 645, 647, 760, 761, 787, 797, 799, 801, 828, 886, 889, 890, 898, 911, 951, 952, 955, 970, 980, 1012–1014, 1025, 1035, 1075, 1082, 1083, 1086, 1096, 1128, 1130, 1136–1141, 1146, 1148.

Sternum

The two lateral halves of the sternum may fail to unite and thereby form a *cleft sternum* (so-called *fissura sterni*). In other cases, the union of the two halves occurs in the manubrial region but fails distally. The cranial and caudal may fuse but the intermediate part remains separate. The failure to fuse can be very slight or very extensive and is sometimes accompanied by ectopia cordis.

A more common variation in the sternum is asymmetry of the costal cartilages, which may articulate in an alternating pattern.

Cartilaginous nodules or suprasternal ossicles may occur singly or paired at the superior margin of the manubrium. The frequency of occurrence is reported to be 6–7% of cadavers studied. The joint between the manubrium and the body of the sternum may become ossified. The manubrium sometimes extends to the insertion of the third costal cartilages. This usually occurs in gibbons and occasionally in other anthropoid apes. The body of the sternum may be perforated, usually at its lower end.

The xiphoid process is typically variable in size and shape (it may be cleft) and may be absent. An abnormally long xiphoid, extending to the level of the umbilicus, has been reported. The xiphoid process may be perforated, more frequently than is the body of the sternum.

The body of the male sternum is relatively longer than that of the female; in males it is more than twice the length of the manubrium whereas in females it is usually less than twice the length.

Agenesis of the sternum in viable children is rare.

References 2, 36, 44, 73, 127, 132, 158, 169, 182, 188, 214, 249, 268, 270, 326, 366, 501, 502, 527, 576, 585, 622, 637, 644, 711, 717, 723, 738, 779, 803, 815, 872, 890, 973, 1028, 1037, 1128, 1130, 1136, 1137, 1140, 1141, 1146, 1148.

Upper Extremity

Clavicle

The clavicle is not infrequently (2–6% of cases) traversed by a small canal which provides passage for one of the intermediate supraclavicular nerves. This canal reportedly occurs more commonly on the left side.

The *costoclavicular ligament* may cause a deep impression in the clavicle, which may be mistakenly identified as a lesion when seen in x-ray films. The coracoclavicular syndesmosis may be a diarthrosis or a cartilaginous symphysis.

The roughened area on the superior surface of the clavicle, the so-called deltoid tubercle, may be elongated to form a *deltoid* process.

Bifurcate clavicle and duplication of the clavicle have been described.

The partial or complete absence of a clavicle was mentioned previously in relation to defects in the skull, and may be associated with a hereditary condition termed *cleidocranial dysostosis*.

References 284, 323, 390, 523, 530, 534, 583, 658, 695, 759, 818, 821, 853, 867, 916, 1003, 1066, 1116, 1128, 1130, 1136, 1138–1141, 1146, 1148.

Scapula

The scapular notch is frequently bridged by bone rather than a ligament (5% of cases studied). Accessory notches may be present; one frequently is found on the inferior angle. The acromion may fail to unite. In about 5% of individuals (more commonly males), the separate part (*os acromion*) is on the right side. Fascicles of the subclavius muscle may be inserted onto the coracoid process by passing through the clavipectoral fascia. The tendon of m. pectoralis minor, in part or entirely, sometimes passes over the coracoid process to insert elsewhere. As a result, it may produce a characteristic groove on the superior surface of the process. The coracoid process may exist as a separate bone. A plate of bone extending from the medial margin of the scapula to the vertebral column has been reported.

References 88, 289, 357, 358, 391, 393, 425, 469, 470, 519, 520, 654, 695, 727, 728, 746, 758, 813, 867, 981, 1004, 1005, 1128, 1130, 1136–1141, 1146, 1148.

Arm

The humeral head may be absent. The humerus may have an enlarged deltoid process. The olecranon fossa of the humerus is occasionally perforated to form a *supratrochlear foramen* or *septal aperture*. Septal apertures may occur in 4–13% of individuals, more frequently on the left side and in females. Racial variations of 4.1–58% have been reported.

A hook-like, bony spine of variable size, the *supracondylar process*, may project distally from the anteromedial surface of the humerus, about 5 cm proximal to the medial epicondyle. It is joined to the epicondyle by a fibrous band (ligament of Struthers) which may ossify. The process, band, and shaft of the humerus may form a ring or canal through which the median nerve and the brachial artery (or a branch of it) are transmitted. The nerve and/or artery may become compressed causing clinical symptoms. Occasionally, only the median nerve traverses the canal. The process and ligament sometimes provide origin for a part of the pronator teres muscle, as well as an insertion site for a portion of coracobrachialis. The supracondyloid process of the humerus has a reported frequency of 2.7% (Gruber). Adachi summarized the literature (up to 1928) and found that 12 authors (not including Gruber) had studied 9620 humeri and identified supracondylar processes in 78 or 0.8% of arms. The range of frequency in the 12 individual studies was 0.1–5.7%. The peoples studied were the Aino, Chinese, Koreans, Negroes, Melanesians, Australians, Japanese (including 142 from the stone age), Indians, Germans, French, Eskimos, Swedes, and Italians.

Gruber (1865) has remarked, "This knowledge cannot be entirely without interest to the surgeon."

References 11, 13, 14, 26, 45, 59, 62, 109, 137, 144, 255, 299, 310, 311, 372, 373, 468, 498, 517, 573, 574, 577, 634, 660, 664, 674, 707, 724, 745, 765, 785, 804, 806, 820, 832, 835, 863, 921, 931, 933, 937, 967, 969, 1046, 1057, 1094, 1124, 1128, 1130, 1136–1141, 1146, 1148.

Forearm

Congenital absence (partial or complete) of the *radius* with normal hand development or with absence of the thumb has been reported. A sesamoid bone may develop in the bicipital tendon over the radial tuberosity. The congenital bilateral absence of the radius in identical twins has been reported.

Congenital absence (partial or complete) of the *ulna* has also been reported. This bone may only be represented by an olecranon-like vestige. A sesamoid bone is occasionally contained within the tendon of the triceps muscle proximal to the olecranon. Another sesamoid (*os coronoides*) may also be found at the tip of the coronoid process. A separate ulnar styloid process (*os ulnostyloideum*) has been described.

Duplication of the radius and of the ulna have been reported and are rare.

References 133, 208, 209, 244, 299, 374, 377, 489, 552, 559, 594, 624, 634, 684, 685, 696, 710, 749, 750, 764, 862, 892, 935, 966, 1009, 1067, 1128, 1130, 1136–1141, 1146, 1148.

Wrist and Hand

The numbers of carpal bones are frequently and variably increased. Over 20 accessory ossicles have been described. These usually occur because of failure of fusion of ossification centers. Such is the case of the *os centrale,* an additional bone located on the dorsal aspect between the scaphoid, capitate, and trapezoid, or between the lunate, triquetrum, and hamate. It is formed when a small cartilaginous nodule fails to fuse with the scaphoid, and may be doubled. Additional elements appear in connection with the trapezium and hamate. Other accessory elements may occur in the angle between the hamate and the fifth metacarpal, between the second and third metacarpals, and between the trapezium and capitate bones. *Os vesalianum carpi* is a small bone at the lateral aspect of the carpus adjacent to the fifth metacarpal and hamate. First described by Vesalius in 1543, it was not reported again until 1870 by Gruber. It has a reported frequency of about 0.1%. *Os gruberi,* a rare palmar ossicle, is found, when present, between the capitate, hamate, and third and fourth metacarpal bones. Carpal fusion may also occur, most commonly the fusion of the lunate and triquetrum *(os lunatotriquetrum).* A case of bilateral absence of the scaphoid and lunate bones and a case of an enlarged scaphotrapezial ossicle *(os radiale externum)* have been reported.

A secondary scaphoid, lateral or medial to the normal scaphoid, has been reported. The scaphoid and the lunate may be bipartite or tripartite. An accessory triquetrum *(os triangulare),* when present, is located between the distal end of the radius, ulnar styloid process, lunate, and triquetrum. The accessory os centrale may fuse with the scaphoid. An appendage to the tubercle of the trapezium occasionally exists as a separate bone *(os trapezium secundarium).* The trapezoid may have an accessory bone *(os trapezoideum secundarium)* representing an unassimilated epiphysis. The styloid process of metacarpal III *(os styloideum)* is sometimes replaced by an accessory process of the trapezoid. The trapezoid may be split into two bones, a dorsal and a ventral. An independent subcapitate *(os subcapitatum)* may be found bilaterally.

The length and thickness of metacarpal bones is variable. The first metacarpal is the shortest and usually the thickest. The longest metacarpal is usually the second, and the most slender is usually the fourth. The styloid process of metacarpal III may exist as an independent, accessory carpal bone.

Macrodactyly has been reported. Phalanges may be increased or decreased in number and size. In one study, three-phalangeal thumbs were found twice in 75,000 consecutively examined hands. Duplication of the thumb is not uncommon, with or without a bony articulation. The number of digits may be increased or decreased. Septodactyly has been reported, as well as duplication of the index finger. The digits may be fused together or absent. *Ectrocheiry*, partial or complete absence of the hand, has been reported.

Sesamoid bones of the hand are variable but are found constantly at the metacarpophalangeal articulation of the thumb. They are found in the same location in the little finger in about 75% of cases, and in the index finger in about 47% of cases; they are found in the middle and ring finger in less than 1% of cases. Sesamoid bones are found at the interphalangeal joint of the thumb in about 73% of cases but are very rare in this location in the other digits. A dorsal metacarpophalangeal sesamoid has been observed in the thumb.

A critical and detailed review of carpal anomalies has been provided by O'Rahilly (reference 753).

References 3, 19, 24, 28, 31–33, 48, 60, 76, 77, 83, 84, 87, 93, 122, 123, 130, 135, 145, 154, 161, 171, 180, 181, 183, 184, 218, 219, 221, 222, 243, 256, 264, 265, 274, 278, 285, 287, 290, 299, 304, 319, 348, 353, 354, 370, 375, 376, 378–381, 383–385, 392, 397, 401, 402, 411, 426, 440, 445, 446, 448, 449, 451–453, 455–459, 461–464, 475, 478, 484, 486, 495, 499, 507, 514, 521, 545, 547, 550, 553, 557, 567, 568, 570, 579, 599, 600, 620, 621, 632, 634, 639, 662, 675, 677, 691, 694, 708, 735, 744, 748, 752, 753, 767, 770, 772, 773, 780, 782, 796, 808, 811, 824, 826, 831, 833, 843, 844, 851, 857, 875, 876, 909, 913, 923, 928–930, 941, 942, 960, 963–966, 968, 1001, 1006, 1039, 1048, 1090, 1092, 1095, 1103, 1128, 1130, 1131, 1136–1141, 1146, 1148.

Lower Extremity

Pelvis

The hip bone or *os coxae* is not subject to great variation; however, relatively few pelves are typically male or female. A canal may groove the iliac fossa, transmitting a vein that connects the external pelvic and internal pelvic veins. Conversion of the obturator sulcus (or groove) into a bony foramen has been reported. Defect or nonunion of the pubic and ischial rami is occasionally observed. There are reports of accessory articular surfaces on the ilium in the region of the posterior superior spine and tuberosity, which articulate with accessory facets on the sacrum at the level of the first and second posterior sacral foramina. They vary in diameter from 0.59 to 1.2 cm and are found in 20–50% of all pelves. The acetabular notch may be absent. A case of *os acetabuli persistens*, bilaterally, has been reported. Extra joint surfaces have also been reported to increase with age. The dimension of the greater sciatic notch is variable. An accessory ischial spine may project from the ilium/ischium synchondrosis, separating the greater sciatic foramen into two compartments. An accessory spine behind the iliopectineal eminence occasionally provides an attachment site for the psoas minor muscle. Sex differences in the size and form of the pelvis are related to its functions in parturition. Agenesis of the os pubis and duplication of the ischium have been reported.

References 16, 85, 140, 179, 185, 234, 260, 313, 321, 361, 362, 363, 467, 477, 625, 683, 716, 864, 871, 895, 915, 962, 1029, 1059, 1111, 1113, 1128, 1130, 1134–1141, 1146, 1148.

Thigh

The shaft of the *femur* has been reported congenitally absent bilaterally, and the unilateral absence of the entire femur has also been described.

Occasionally, the gluteal tuberosity is significantly enlarged and is termed the *third trochanter*. The gluteal tuberosity may also be absent and, instead, the area may be a hypotrochanteric fossa. In some cases the tuberosity and fossa coexist. The development of the linea aspera may also show considerable variation.

A bony process, about 3 cm in length, has been described arising at the origin of the short head of the biceps femoris. It is thought by some to be analogous to the supracondylar process of the humerus.

The *patella* is sometimes bipartite, probably due to failure of the ossification centers to unite, with the proximal lateral corner most frequently found as a separate piece. Congenital absence of the patella has been reported. In cases of congenital absence of the quadriceps muscle, the patella may be present but is usually dislocated.

References 6, 124, 125, 299, 327, 355, 359, 454, 498, 548, 558, 565, 604, 616, 667, 720, 741, 790, 793, 893, 906, 907, 937, 961, 971, 972, 976, 1015, 1026, 1104, 1106, 1107, 1128, 1130, 1136–1141, 1146, 1148.

Knee

In a study of 433 knee joints, the menisci were found to be "C"-shaped in 422, discoid in nine, and circular in two. The medial meniscus may not be "C"-shaped but rather a ring completed by a ligament.

In a study of 140 knees, a meniscofemoral ligament was present in 99 (71%). In 50 cases (36%) the ligament was the anterior and in the remaining 49 cases (35%) it was the posterior meniscofemoral ligament. Only eight knees (6%) had both anterior and posterior meniscofemoral ligaments. In 17 cases (12%) the anterior ligament was bilateral, in 12 cases (8%) the posterior was bilateral, and in two cadavers both ligaments were found in both knees. No sex differences were found.

The cruciate ligaments may have accessory slips or appendices. The anterior cruciate ligament may be absent.

The posterior horn of the medial meniscus occasionally supports a free-lying cartilage, which may be anchored posteriorly to the tibia. A lateral interarticular disk may be present.

The transverse ligament may be variably developed or absent.

References 65, 66, 170, 497, 554, 829, 845, 1053, 1054, 1105, 1108, 1128, 1130, 1134–1146, 1148.

Leg

The tibia may be absent. It may show side-to-side compression of the shaft (platycnemia) and retroversion of the proximal extremity of the bone. Additional articular facets on the anterior surface of the distal extremity of the tibia have been recorded. The distal part of the tibia occasionally has a bony canal through which may run a branch of the anterior tibial artery and a branch of the deep peroneal nerve.

It has been reported that the tibia presents more variation in length and form than any other long bone of human limbs.

Absence of the fibula has been frequently recorded. The primary variations are related to the form and relative size of the surfaces of the shaft. It has been reported that in 7% of cases studied, the posterior margin of the lateral malleolus was not grooved but was tranversely convex. A sesamoid bone may be found in the superior peroneal retinaculum, articulating with the lateral malleolus.

References 10, 15, 134, 139, 197, 213, 220, 227, 245, 279, 299, 472, 479, 705, 736, 778, 804, 823, 854, 887, 985, 1007, 1074, 1085, 1128, 1130, 1134, 1136–1141, 1146, 1148.

Foot

The *tarsal* bones are variable in size, form, and number. About 30 accessory tarsal ossicles have been described. The diversity has been related partially to their function in weight bearing. The tarsal bones are commonly reduced to six through the fusion of the calcaneus and the navicular. Additional elements to the normal seven may occur in the tarsus. The most frequent is the presence of a triangular bone (*os trigonum*), which may result from lack of fusion between the lateral tubercle of the posterior process of the talus and the primary ossification center of the bone. Its frequency is reported variously as 2.5–11.4%. The next most frequent variation involves the independent ossification of the plantar half and the dorsal half of the medial cuneiform bone (*os cuneiforme mediale bipartum*). Additional elements may be found, probably related to developmental problems of the calcaneus and navicular bones. Some of these additional elements are found (*a*) in relation to the medial cuneiform and the first and second metatarsals (*os intermetatarsum* is found in 10% of feet), (*b*) on the medial side of the cuneonavicular joint (*os paracuneiforme*), and (*c*) between the fifth metatarsal and cuboid (*os vesalianum*). The tuberosity of the navicular may be a separate bone (*os tibiale externum*). A secondary cuboid has been described arising from the navicular as a projection that separates from the lateral side of the plantar surface of the bone. At the posterior end of the sustentaculum there may be an anomalous bone called *os sustentaculi proprium*. The *os calcaneum secundarium* is a rare small bone present on the dorsum between the calcaneus, cuboid, navicular, and head of the talus. On the dorsum, at the proximal ends of cuneiforms I and II, a free intercuneiform bone (*os intercuneiforme*) may occasionally be found in the intercuneiform fossa.

The lateral portion of the navicular may be united with the cuboid, the lateral portion thus being separated from the remainder of the navicular. This results in the articulation of the cuboid with the talus. Congenital talonavicular synostosis has been reported. An additional facet has been observed on the lateral surface of the calcaneus anterior to the peroneal trochlea in about 7% of cases.

The metatarsals may be reduced in number to four elements.

Sesamoid bones occur at joints other than the first metatarsophalangeal joint, but these are more rare than those found in the hand. In a study of 246 feet, Pfitzner found two sesamoids in the fifth toe in about 6%, and one sesamoid in the second toe in about 2%. Sesamoids also were found at the interphalangeal joint of the great toe in 55% and in the same joint of the second toe in about 1%.

Phalanges may be increased or decreased in number and size. The number of digits may be increased or decreased. Octodactyly has been reported. The digits may be fused together or absent.

A critical and detailed review of tarsal anomalies has been provided by O'Rahilly (reference 753).

References 10, 28, 30, 43, 51, 58, 70, 84, 87, 97, 145, 155, 210, 212, 213, 215–217, 220, 251, 253, 258, 259, 273, 281, 282, 287, 299, 312, 356, 372, 382, 389, 403, 404, 427, 428, 436, 450, 460, 480, 491, 496, 510, 521, 567, 575, 595, 596, 600, 610–612, 615, 632, 649, 661, 676, 678, 690, 699, 712, 726, 739, 747, 753, 771, 772, 782, 796, 814, 855, 866, 873, 874, 894, 896, 917, 924, 925, 927, 939, 940, 963, 990, 1006, 1007, 1016, 1038, 1052, 1062, 1073, 1078, 1089, 1118, 1123, 1128, 1130, 1136, 1137, 1139, 1140, 1141, 1146, 1148.

References

* These journal articles are notable for their content, literature review, and/or illustrations.

+ The authors have not had access to these references. They are included because of their historic and/or scientific importance.

Journals

1. AASAR, Y.H. Three cases of fusion of axis with third cervical vertebra. *J. Anat.* 72:634–635, 1937–38.
2. ABBOTT, J. Absence de sternum chez une femme adulte; compatibilité de cette anomalie avec la vie et la santé. *Soc. Biol. Comptes Rendus des Séances et Mémoires* 4:163, 1852.
3. ABRAMOWITZ, I. Triphalangeal thumb in a Bantu family. *J. Bone Joint Surg. (Br.)* 41:766–771, 1959.
4. VAN ACKEN, F. Die Sutura frontalis im Röntgenbilde. *Fortschr. Röntgenstr.* 48:208–222, 1933.
5. ADAIR, F.L. and R.E. SCAMMON. Observations on the parietal fontanelle in the newborn and in young infants. *Am. J. Obstet. Gynecol.* 14:149–159, 1927.
6. ADAMS, J.D. and D.R. LEONARD. A developmental anomaly of the patella frequently diagnosed as fracture. *Surg. Gynecol. Obstet.* 41:601–604, 1925.
7. ADOLPHI. Über das Erscheinen der Spina frontalis in der Gesichtsfläche des menschlichen Schädels. *Anat. Anz.* 35:181–185, 1910.
8* ADSON, A.W. Surgical treatment for symptoms produced by cervical ribs and the scalenus anticus muscle. *Surg. Gynecol. Obstet.* 85:687–700, 1947.
9* AEBY, C. Seltene Rippenanomalie des Menschen. *Arch. Anat. Physiol. Wissen. Med.* 1868, pp. 68–71.
10. AITKEN, D.M. A note on the variations of the tibia and astragalus. *J. Anat. Physiol.* 39:489–491, 1905.
11. AKABORI, E. Septal apertures in the humerus in Japanese, Ainu and Koreans. *Am. J. Phys. Anthropol.* 18:395–500, 1934.
12. AKABORI, E. Additional notes on cranial variations. *Jpn. J. Med. Sci.* 4:317–318, 1933–34.
13. AKABORI, E. On the aperture in the fossa olecrani of some east Asiatics. *Folia Anat. Jpn.* 13:473–475, 1935.
14. ALBERS-SCHÖNBERG. Eine seltene, bisher nicht bekannte Strukturanomalie des Skelettes. Skelettanomalie von atavistischem Interesse, Proc. supracondyloideus oder entepicondyloideus. *Fortschr. Röntgenstr.* 23:174–177, 1915–16.
15* ALETTER, C. Über die angeborenen Defekte der Tibia. *Frankfurter Z. Pathol.* 43:196–221, 1932.
16. ALEXANDER, C. The aetiology of primary protrusio acetabuli. *Br. J. Radiol.* 38:567–580, 1965.
17. ALLEN, W. The varieties of the atlas in the human subject, and the homologies of its transverse processes. *J. Anat. Physiol.* 14:18–27, 1879.
18. ALLEN, W. Tertiary occipital condyle. *J. Anat. Physiol.* 15:60–68, 1881.
19. ALLENBACH, E. Ectrodactylie héréditaire et congénitale des quatre membres observée dans quatre générations. *Ann. Anat. Pathol.* 2:147–152, 1925.
20. ANCEL, P. and L. SENCERT. Variations numériques de la colonne vertébrale. *Assoc. Anatomistes Comptes Rendus* 3:158–165, 1901.
21. ANDERSON, E. Über Anomalien der Wirbelsäule und der Rippen. *Fortschr. Röntgenstr.* 34:491–499, 1926.
22. ANDERSON, R.J. Eight true ribs in man. *Anat. Anz.* 18:377–382, 1900.
23. ANDERSON, R.J. The presence of an astragala-scaphoid bone in man. *J. Anat. Physiol.* 14:452–455, 1879.
24. ANDERSON, R.J. Division of the scaphoid bone on the carpus, with notes on other varieties of the carpal bones. *J. Anat. Physiol.* 17:253–255, 1883.
25. ANDERSON, R.J. Observations on the diameters of human vertebrae in different regions. *J. Anat. Physiol.* 17:341–344, 1883.
26. ANDREASEN, A.T. Congenital absence of the humeral head. *J. Bone Joint Surg. (Br.)* 30:333–337, 1948.
27. ANDREW, J. Sacralization: An aetiological factor in lumbar intervertebral disk lesions, and a cause of misleading focal signs. *Br. J. Surg.* 42:304–311, 1954–55.
28* ANONYMOUS. Anatomie du Vivant. Septodactylie (manus) et Octodactylie (pedis). *La Presse Medicale* No. 69. 30:1447, 1922.
29. ANSON, B.J. and T.H. BAST. The surgical significance of stapedial and labyrinthine anatomy. *Q. Bull. Northwestern University Medical School* 32:307–325, 1958.
30. APPLETON, A.B. Note on a variable feature of the astragalus. *J. Anat. Physiol.* 67:123–142, 1913.
31. ARENS, W. Über die angeborene Synostose zwischen dem Os lunatum und dem Os triquetrum. *Fortschr. Röntgenstr.* 73:772–774, 1950.

32. ARENS, W. Über eine doppelseitige Rückbildung des Kleinfingerstrahles bei zwei Brüdern. *Fortschr. Röntgenstr.* 74:242–243, 1951.
33. ARENZ, J. Seltene Lokalisation von Nebenkernbildungen der Hand. *Fortschr. Röntgenstr.* 82:552–553, 1955.
34. ARMSTRONG, P.B. Considerations of a many-boned human skull. *Anat. Rec.* 38:97–128, 1928.
35. ARMSTRONG, J.R., CRISP, E.J., FREEDMAN, B., GILLESPIE, H.W., GOLDING, F.C., JACKSON, H. and K. LLOYD. Discussion on the significance of congenital abnormalities of the lumbosacral region. *Proc. R. Soc. Med.* 43:635–640, 1950.
36*. ASHLEY, G.T. The relationship between the pattern of ossification and the definitive shape of the mesosternum in man. *J. Anat.* 90:87–105, 1956.
37. AUGIER, M. Les os frontaux accessoires. *Assoc. Anatomistes Comptes Rendus* 14:22–29, 1912.
38. AUMANN, U. Apophyse am Margo vertebralis scapulae? *Fortschr. Röntgenstr.* 110:404, 1969.
39*. AYER, A.A. and S. RAMASWAMY. The marginal tubercle of the zygomatic bone. *J. Anat. Soc. India* 5:28–30, 1956.
40. AYMARD, J.L. Some new points in the anatomy of the nasal septum, and their surgical significance. *J. Anat.* 51:293–303, 1917.
41. BADGLEY, C.E. The articular facets in relation to low-back pain and sciatic radiation. *J. Bone Joint Surg.* 23:481–496, 1941.
42*. BALESTRA, A. and M. CHÉRIÈ-LIGNIÈRE. Sui derivati del secondo arco branchiale (cartilagine di Reichert) nell'uomo adulto. *Arch. Ital. Anat. Embriol.* 3:37–86, 1904.
43. BANCHI, A. Cunéiforme I biparti. Le cunéiforme comprend-t-il le tarsal distal du préallux. *Arch. Ital. Biol.* 45:280, 1906.
44. BÁNKI, Z. Über den Processus costalis sterni. *Fortschr. Röntgenstr.* 102:518–522, 1965.
45*. BÁNKI, Z. Die Apophyse mit Fortsatzbildung des Epicondylus medialis humeri. *Fortschr. Röntgenstr.* 107:815–816, 1967.
46. BANSAL, P.C. A case of abnormal os hyoid and thyroid cartilage. *J. Anat. Soc. India* 13:105–106, 1964.
47. BARBOSA SUEIRO, M.B. Note sur la fréquence de quelques variations du rachis humain. *Assoc. Anatomistes Comptes Rendus* 21:51–62, 1926.
48. BARBOSA SUEIRO, M.B. Un cas d'ankylose congénitale bilatérale de la phalangine avec la phalangette de l'annulaire et de clinodactylie externe bilatérale de l'auriculaire et de l'annulaire. *Assoc. Anatomistes Comptes Rendus* 28:28–29, 1933.
49. BARBOSA SUEIRO, M.B. Sur l'osselef odontoidien de l'axis humain. *Assoc. Anatomistes Comptes Rendus* 28:30–32, 1933.
50. BARBOSA SUEIRO, M.B. Démonstrations de quelques variations rachidiennes chez l'homme. *Assoc. Anatomistes Comptes Rendus* 28:709–710, 1933.
51. BARCLAY, M. A case of duplication of the internal cuneiform bone of the foot (Cuneiforme bipartitum). *J. Anat.* 67:175–177, 1932.
52. BARCLAY-SMITH, E. A rare condition of wormian ossification. *J. Anat. Physiol.* 43:277–278, 1909.
53. BARCLAY-SMITH, E. Two cases of wormian bones in the bregmatic fontanelle. *J. Anat. Physiol.* 44:312–314, 1910.
54. BARCLAY-SMITH, E. Multiple anomaly in a vertebral column. *J. Anat. Physiol.* 45:144–171, 1911.
55. BARDEEN, C.R. Costo-vertebral variation in man. *Anat. Anz.* 18:377–382, 1900.
56. BARDEEN, C.R. Numerical vertebral variation in the human adult and embryo. *Anat. Anz.* 25:497–519, 1904.
57. BARDELEBEN, K. VON. Über Verbindung zwischen dem 5. und 6. sowie zwischen dem 6. und 7. Rippenknorpel. *Anat. Anz.* 15:33–36, 1899.
58*. BARLOW, T.E. Os cuneiforme 1 bipartitum. *Am. J. Phys. Anthropol.* 29:95–111, 1942.
59*. BARNARD, L.B. and S.M. MCCOY. The supracondyloid process of the humerus. *J. Bone Joint Surg.* 28:845–850, 1946.
60. BARSKY, A.J. Congenital anomalies of the hand. *J. Bone Joint Surg. (Am.)* 33:35–64, 1951.
61. BÁRSONY, T. and K. WINKLER. Beiträge zur Röntgenologie der Wirbelsäule. *Röntgenpraxis* 9:601–608, 1937.
62*. BARTA, E. and L. PETROVITS. Das Ligamentum supracondyloideum. *Anat. Anz.* 63:177–184, 1927.
63*. BARTELS, P. Über Rassenunterschiede am Schädel. I. Untersuchungen und Material aus dem anatomischen Museum zu Berlin. *Int. Monatsschr. Anat. Physiol.* 21:137–194, 1904.
64. BARTON, P.N. The significance of anatomical defects of the lower spine. *Industr. Med.* 17:37–40, 1948.
65. BARUCHA, E. Unsere Erfahrungen über den Wert des Raubershen Röntgen-Zeichens bei der Meniscus-Diagnose. *Monatsschr. Unfallheilk.* 63:370–375, 1960.
66*. BASMAJIAN, J.V. A ring shaped medial semi-lunar cartilage. *J. Bone Joint Surg. (Br.)* 34:638–639, 1952.
67. BAUER, O. Totale Aplasie der Lendenwirbelsäule bei einem Neugeborenen. *Geburtsch. Frauenheilk.* 12:250–256, 1952.
68*. BELL, R.H., SWIGART, L.L. and B.J. ANSON. The relation of the vertebral artery to the cervical vertebrae. Based upon a study of 200 specimens. *Q. Bull. Northwestern University Medical School* 24:184–185, 1950.
69. BENIANS, T.H.C. and H.E. JONES. Congenital deficiency of a portion of the thoracic skeleton. *J. Anat. Physiol.* 44:67–72, 1910.
70. BERGGLAS, B. Über drei Fälle von Coalescentia calcaneonavicularis. *Anat. Anz.* 60:152–155, 1925–26.
71*. BERGMAN, P. On varieties of the atlas in man. *Folia Morphol.* 26:129–139, 1967.
72. BERGSTRAND, A. Foramina parietalia permagna. *Svenska Läkartidningen* 43:2388–2394, 1946. Cited in *Excerpta Medica* Sec. 1, Vol. 2, abstract 96, p. 55, 1948.
73. BERNARDO, G.F. A case of ectopia cordis c. fissura sterni. *J. Anat. Physiol.* 32:325–333, 1898.
74. BERNER, F. Über Rippenanomalien auf Grund von 6 Millionen Reihenbildern. *Fortschr. Röntgenstr.* 69:202–221, 1943.
75. BERRY, R.J.A. A case of os parietale bipartitum in an Australian aboriginal skull. *J. Anat. Physiol.* 44:73–82, 1910.
76. BERTACCHINI. Un cas de double pouce bilatéral chez l'homme et quelques considérations sur la valeur morphologique de l'hyperdactylie chez l'homme. *Arch. Ital. Biol.* 44:136, 1905.
77. BERTACCHINI, P. Un caso di doppio-pollice bilaterale nell'uomo e alcune considerazioni sul valore morfologico dell'iperdattilia nell'uomo. *Int. Monatsschr. Anat. Physiol.* 21:126–135, 1904.
78. BETTI, U. Sur un processus anormal en correspondance de l'empreinte deltoïdienne de l'humérus humain. *Arch. Ital. Biol.* 28:161, 1897.
79*. BEUTEL, A. Die Verknöcherung des Ligamentum stylohyoideum. *Röntgenpraxis* 7:665–670, 1935.
80. BHARGAVA, K.N. and T.C. Garg. Incidence of os japonicum (bipartite zygomatic) in Madhya Pradesh skulls. *J. Anat. Soc. India* 9:21–22, 1960.
81. BIANCHI, S. Sur la division de "l'os Planum" de l'Ethmoïde dans le crâne de l'homme et des antropoïdes et sur la non-existence de l'os lacrymal postérieur chez quelques mammifères. *Arch. Ital. Biol.* 24:310–311, 1985.
82. BIANCHI, S. Sur la fréquence des anomalies numériques vertebrales dans le squelette des individus normaux et des aliénés. *Arch. Ital. Biol.* 23:467, 1895.
83. BIENVENUE, F. Un cas de pouce supplémentaire a trois phalanges. *Rev. Orthop. (Paris)* 3:91–94, 1912.
84. BIRCHER, E. Neue Fälle von Varietäten der Handwurzel und des Fußgelenks. a) Os Trigonum traumaticum? b) Os sub tibiale. *Fortschr. Röntgenstr.* 26:85–88, 1918/19.
85. BIRKNER, R. Einseitiges Iliacalhorn. *Fortschr. Röntgenstr.* 105:910–911, 1966.
86. BIRKNER, R. and K. CONSENTIUS. Verknöcherungen in der Wand des Foramen obturatum. *Fortschr. Röntgenstr.* 127:72–74, 1977.
87*. BIZARRO, A.H. On sesamoid and supernumerary bones of the limbs. *J. Anat.* 55:256–268, 1921.
88. BLANCHARD, M. Anomalie vértébrale. *Soc. Biol. Comptes Rendus Hebdomadaires des Séances et Mémoires* 40:772–773, 1888.
89. BLANTON, P.L. and N.L. BIGGS. Eighteen hundred years of controversy: The paranasal sinuses. *Am. J. Anat.* 124:135–147, 1969.
90. BLASZCZYK, B., KASZUBA, A. and J. KOCHANOWSKI. Atypical foramina of base of the skull. *Folia Morphologica* 39:201–209, 1980.
91. BLOOMBERG, M.W. Bifurcate ribs – An unusual cause of deformity of the chest. *Can. Med. Assoc. J.* 16:807–808, 1926.
92*. BLUMEL, J., EVANS, E.B. and G.W.N. EGGERS. Partial and complete agenesis or malformation of the sacrum with associated anomalies. *J. Bone Joint Surg (Am.)* 41:497–518, 1959.
93*. BOGART, F.B. Variations of the bones of the wrist. *Am. J. Roentgenol.* 28:638–646, 1932.
94. BOHLIG, H. Ankylose des Atlanto-Epistropheal-Gelenkes. *Fortschr. Röntgenstr.* 81:216–218, 1954.
95. BOHLIG, H. and E. SIEMON. Ankylose von Atlas und Epistropheus (Bericht über 1 Fall). *Z. Orthop.* 85:231–236, 1954.
96*. BOHN, A. Dens in Dente. *Acta. Odont. Scand.* 8:53–80, 1948.
97. BÖKER, H. and W. MÜLLER. Das Os cuneiforme I bipartitum,

eine fortschreitende Umkonstruktion des Quergewölbes im menschlichen Fuß. *Anat. Anz.* 83:193–204, 1936–37.

98 BOLK, L. Os metopism. *Am. J. Anat.* 22:27–47, 1917.

99 BOLK, L. Über unvollständig assimilierte letzte Occipitalwirbel beim Menschen. *Anat. Anz.* 55:156–162, 1922.

100 BOONE, M.L., SWENSON, B.E. and B. FELSON. Ribnotching: Its many causes. *Am. J. Roentgenol.* 91:1075–1088, 1964.

101 BOOS, O. Zur Genese der Wirbelsäulendefekte. Ein weiterer klinischer Beitrag zur Kenntnis der Defektbildungen an den kaudalen Wirbelsäulenabschnitten. *Dtsch. Med. Wochenschr.* 78:485–487, 1953.

102* BORNSTEIN, P.E. and R.R. PETERSON. Numerical variation of the presacral vertebral column in three population groups in North America. *Am. J. Phys. Anthropol.* 25:139–146, 1966.

103 BOROVANSKY, L. Les soudures de l'axis et de la troisième vertèbrae cervicale chez l'homme. *Assoc. Anatomistes Comptes Rendus* 23:49–54, 1928.

104 BOROVANSKY, L. Observations craniologiques faites sur 1,000 cranes Tchèques. (Des ossuaires de la Bohême méridionale.) *Assoc. Anatomistes Comptes Rendus* 27:67–80, 1932.

105 BOSCO, L. Observations, dérivées de la statistique, sur les manifestations de vertèbres occipitales et sur l'assimilation de l'atlas à l'occipital. *Arch. Ital. Biol.* 75:160, 1925.

106 DU BOULAY, G. and T. BOSTICK. Linear tomography in congenital abnormalities of the ear. *Br. J. Radiol.* 42:161–183, 1969.

107 BOUVIER, M. and R. DIEULAFÉ. Variations des canaux palatins postérieurs accessoires. *Assoc. Anatomistes Comptes Rendus* 25:51–56, 1930.

108 BOVERO, A. Mancaza quasi completa della squama temporalis nel cranio umano associata ad altre anomalie. *Assoc. Anatomistes Comptes Rendus* 4:262–271, 1902.

109 BOVERO, A. Sur quelques modalités peu connues du processus supracondyloideus humeri internus. *Arch. Ital. Biol.* 53:150, 1910.

110* BOYD, G.I. The emissary foramina of the cranium in man and the anthropoids. *J. Anat.* 65:108–121, 1930.

111 BRADLEY, O.C. Two cases of dental anomaly. *J. Anat. Physiol.* 36:356–367, 1902.

112* BRAILSFORD, J.F. Deformities of the lumbosacral region of the spine. *Br. J. Surg.* 16:562–627, 1928–29.

113 BRANDT. Ein Fall von angeborenem Defect der Extremitäten. *Arch. Pathol. Anat. Physiol. Klin. Med.* 87:195–196, 1882.

114* BRANDT, K. Über die konstante Form der Lendenwirbelquerfortsätze 3 und 4. *Arch. Orthop. Unfall-Chir.* 34:445–450, 1933–34.

115* BRANNON, E.W. Cervical rib syndrome. An analysis of nineteen cases and twenty four operations. *J. Bone Joint Surg. (Am.)* 45:977–998, 1963.

116 BRASH, J.C. Vertebral column with six and a half cervical and thirteen true thoracic vertebrae with associated abnormalities of the cervical spinal cord and nerves. *J. Anat. Physiol.* 69:243–273, 1935.

117 BRAT, L. Klinische Fehldiagnose durch abnorm langen Griffelfortsatz des Felsenbeins. *Fortschr. Röntgenstr.* 75:758–759, 1951.

118 BRAUN, H. Seltene Anomalien der Wirbelsäule. *Fortschr. Röntgenstr.* 82:126–127, 1955.

119 BROCHER, J.E.W. Mehrfach angeborene Fehlbildungen der Wirbelsäule. *Fortschr. Röntgenstr.* 58:440–447, 1938.

120 BROCHER, J.E.W. Le processus paracondyleus s. apophyse paramastoide (dans un cas de dysostose mandibulo faciale unilatérale). *Radiol. Clin.* 16:393–396, 1947.

121 BROCHER, J.E.W. Atlas and Axis. *Fortschr. Röntgenstr.* 84:387–390, 1956.

122 BROEK, A.J.P. VAN DEN. Un cas d'absence bilatérale du pouce. *Assoc. Anatomistes Comptes Rendus* 21:107–112, 1926.

123 BROWN, J.B. and F. McDOWELL. Syndactylism with absence of the pectoralis major. *Surgery* 7:599–601, 1940.

124 BRÜCKE, H. VON. Über die Patella cubiti, eine seltene Abweichung des Ellenbogenskelettes. *Z. Orthop.* 73:158–160, 1942.

125 BRUNNER, G. Über Genese, congenitalen Mangel und rudimentäre Bildung der Patella. *Arch. Pathol. Anat. Physiol. Klin. Med.* 124:358–373, 1891.

126* BRZEZINSKI, D.K. VON. Zur Variabilität des Canalis caroticus. *Anat. Anz.* 113:164–170, 1963.

127 BUCHANAN, A.M. Abnormal sternum. *J. Anat. Physiol.* 28:313–314, 1894.

128 BUETTI, C. Zur Röntgendiagnostik seltener Fehlbildungen der Halswirbelsäule. *Radiol. Clin.* 22:141–161, 1953.

129* BULL, J. The normal variations in the position of the optic recess of the third ventricle. *Acta Radiol.* 46:72–80, 1956.

130 BURCKHARDT, E. Zur Klinik und zur pathologischen Histologie des Os styloideum carpi. *Röntgenpraxis* 16:108–115, 1944.

131 BURROWS, F.G.O. Some aspects of occult spinal dysraphism: A study of 90 cases. *Br. J. Radiol.* 41:496–507, 1968.

132* BURTON, J.F. Method of correction of ectopia cordis. *Arch. Surg.* 54:79–84, 1947.

133 BUSE, H. Beitrag zur Persistenz der Olecranonepiphyse. *Fortschr. Röntgenstr.* 104:867–869, 1966.

134 BUXTON, L.H.D. Platymeria and platycnemia. *J. Anat.* 73:31–36, 1938–39.

135 BUYSCH, K.H., DREWES, J. and D. GÜNTHER. Synostosen zwischen Multangulum minus und capitatum. *Fortschr. Röntgenstr.* 115:267–268, 1971.

136 BYSTROW, A. Assimilation des Atlas und Manifestation des Proatlas. *Z. Anat. Entwicklungsgesch.* 95:210–242, 1931.

137 CADY, L.D. The incidence of the supracondyloid process in the insane. *Am. J. Phys. Anthropol.* 5:35–49, 1921.

138* CAFFEY, J. On the accessory ossicles of the supraoccipital bone. Some newly recognized roentgen features of the normal infantile skull. *Am. J. Roentgenol.* 70:401–412, 1953.

139 CAIX, M. Etude comparée des différentes longueurs du tibia. *Assoc. Anatomistes Comptes Rendus* 45:227–230, 1958.

140* CALDWELL, W.E. and H.C. MOLOY. Anatomical variations in the female pelvis and their effect in labor with a suggested classification. *Am. J. Obstet. Gynecol.* 26:479–505, 1933.

141 CALORI, L. Intorno alle suture soprannumeraire del cranio umano e su quelle specialmente della ossa parietali. *Mem. R. Accad. Sci. Istituto di Bologna S.2*, 6:327–343, 1865.

142 CALORI, L. Die Wormiani occipitali ed interparietali posteriori del cranii nostrali e di quelli delle fontanelle laterali. *Mem. R. Accad. Sci. Istituto di Bologna S. 2* 7:303–321, 1867.

143* CALORI, L. Delle anomalie più importanti di ossa, vasi, nervi, e muscoli occorse nell'ultimo biennio facendo anatomia del corpo umano. *Mem. R. Accad. Sci. Istituto di Bologna S. 2* 8:417–482, 1868.

144* CALORI, L. Intorno al canale sopracondiloideo dell'omero nell'uomo. *Mem. R. Accad. Sci. Istituto di Bologna S. 4* 2:37–46, 1880.

145* CALORI, L. Sulla coesistenza di un'eccessiva divisione del fegato e di qualche dito soprannumerario nelle mani o nei piedi. *Memorie R. Accademia Scienze Istituto di Bologna S. 4* 2:335–343, 1880.

146 CALORI, L. Sur deux cas de variétés numérique des vertèbres, accompagnés de variétés numériques des cotes et d'autres anomalies. *Arch. Ital. Biol.* 11:425–426, 1889.

147 CALORI, L. Su varie particolarita osteologiche della base del cranio umano. *Mem. R. Accad. Sci. Istituto di Bologna S. 5* 2:85–109, 1891.

148 CALORI, L. Sur une union synostotique et une incorporation partielle de l'atlas avec l'os occipital, et sur deux autres anomalies correlatives. *Arch. Ital. Biol.* 25:152–153, 1896.

149 CAMERANO, L. Contribution à l'étude des os Wormiens palato-palatins et des os Wormiens médio-palatins de Calori chez les mammiferes. *Arch. Ital. Biol.* 64:280–281, 1915.

150 CAMERON, J. The naso-orbito-alveolar index. *Am. J. Phys. Anthropol.* 3:63–76, 1920.

151 CAMP, J.D. The roentgenologic manifestations of intracranial disease. *Radiology* 13:484–493, 1929.

152 CAMPBELL, J.A. Note of a case of abnormal union of several ribs. *J. Anat. Physiol.* 4:245–246, 1870.

153 CANNEYT, M.J.G. Dent dans la cavité orbitaire. *Bull. Soc. Belge Ophthalmol.* 82:123–127, 1945. Cited in *Excerpta Medica*, Sec. 1, Vol. 2, abstract 191, p. 97, 1948.

154 CAPEL, E.H. Macrodactyly, with a report of a case. *J. Anat.* 69:528–529, 1934–35.

155 CAREY, J.P. and M.C.W.L. DREXLER. Ein atypischer Knochen an der Planta pedis. *Fortschr. Röntgenstr.* 86:526–528, 1957.

156* CARSTENS, M. Die Selladiagnostik. *Fortschr. Röntgenstr.* 71:257–272, 1949.

157 CARTER, D.B. Arrest of development of the third rib in a foetus. *Lancet* 2:306, 1884.

158 CARWARDINE, T. The suprasternal bones in man. *J. Anat. Physiol.* 27:232–234, 1893.

159 CASCELLI, G. Über Rippenanomalien. *Röntgenpraxis* 12:375–385, 1940.

160 CASTALDI, L. Confronti encefalometrici. *Arch. Ital. Anat. Embriol.* 47:95–134, 1942.

161 CAVE, A.J.E. Fusion of carpal elements. *J. Anat.* 60:460–461, 1925.

162 CAVE, A.J.E. On fusion of the atlas and axis vertebrae. *J. Anat.* 64:337–343, 1930.
163 CAVE, A.J.E. Anatomical notes on the occipito-atlanto-axial articulations. *J. Anat.* 68:416–423, 1933–34.
164 CAVE, A.J.E. Cervical intercostal articulation. *J. Anat.* 68:521–524, 1933–34.
165 CAVE, A.J.E. The morphology of the last thoracic transverse process. *J. Anat.* 70:275–276, 1935–36.
166* CAVE, A.J.E. A further example of cervical intercostal articulation. *J. Anat.* 71:415–416, 1936–37.
167* CAVE, A.J.E. The earliest English example of bilateral cervical rib. *Br. J. Surg.* 29:47 51, 1941–42.
168 CAVE, A.J.E. The craniopharyngeal canal in Man and Anthropoids. *J. Anat.* 65:363–367, 1930–31.
169 CHANG, C.H. and W.C. DAVIS. Congenital bifid sternum with partial ectopia cordis. *American Journal Roentgenol.* 86:513–516, 1961.
170* CHARLES, C.M. On the menisci of the knee joint in American whites and negroes. *Anat. Rec.* 63:355–364, 1935.
171 CHARNALIA, V.M. Unilateral polydactyly of thumb. *J. Anat. Soc. India* 10:39–40, 1961.
172 DE CHAZAL, A. Elongated styloid process encountered during tonsillectomy. *Br. Med. J.* 2:199, 1946.
173 CHOUKE, K.S. On the incidence of the foramen of Civinini and the porus crotaphitico-buccinatorius in American Whites and Negroes. I. Observations on 1,544 skulls. *Am. J. Phys. Anthropol.* 4:203–225, 1946.
174 CHOUKE, K.S. On the incidence of the foramen of Civinini and the porus crotaphitico-buccinatorius in American Whites and Negroes. II. Observations on 2,745 additional skulls. *Am. J. Phys. Anthropol.* 5:79–86, 1947.
175* CHRISTENSEN, J.B., LACHMAN, E. and A.M. BRUES. A study of the roentgen appearance of cranial vault sutures: Correlation with their anatomy. *Am. J. Roentgenol.* 83:615–627, 1960.
176 CITELLI, S. Sur la présence de cartilages sésamoïdes dans la corde vocale supérieure de l'homme et sur leur signification morphologique. *Arch. Ital. Biol.* 47:497, 1907.
177 CIVALLERI, A. Observations sur les os nasaux. *Arch. Ital. Biol.* 47:478–479, 1907.
178 CLARKE, J.J. Some observations on the temporal bone, chiefly in childhood. *J. Anat. Physiol.* 27:411–414, 1893.
179 CLEAVES, E.N. Adolescent sacroiliac joints. Their normal development and their appearance in epiphysitis. *Am. J. Roentgenol.* 38:450–456, 1937.
180* CLERC, A. and BOBRIE. Malformations phalangiennes des doigts et des orteiles. *Soc. Biol. Comptes Rendus Hebdomadaires des Séances et Mémoires* 80:123–126, 1917.
181* CLERC, A. REGNARD, M. and J. BOBRIE. Un cas de polydactylie héréditaire. *Soc. Biol. Comptes Rendus Hebdomadaires des Séances et Mémoires* 81:259–262, 1918.
182 COBB, W.M. The ossa suprasternalia in whites and American negros and the form of the superior border of the manubrium sterni. *J. Anat.* 71:245–291, 1936–37.
183 COCKAYNE, E.A. An unusual form of brachyphalangy and syndactyly with double proximal phalanx in the middle fingers. *J. Anat.* 67:165–167, 1932.
184 COCKSHOTT, W.P. Carpal fusions. *Am. J. Roentgenol.* 89:1260–1271, 1963.
185 COHEN, A.S. The "normal" sacroiliac joint. *Am. J. Roentgenol.* 100:559–563, 1967.
186 COHN, B.N.E. Congenital absence of ribs. *Am. J. Roentgenol.* 52:495–499, 1944.
187 COLCHER, A.E. and A.M.W. HURSH. Pre-employment low-back x-ray survey. A review of 1,500 cases. *Industr. Med.* 21:319–321, 1952.
188 COLOSIMO, C. Xifosterno biforcuto e abnormente lungo. *Radiol. Med. (Torino)* 26:42–43, 1939.
189 CONDE ANDREU, J. Recherches sur quelques déviations vertébrales. *Assoc. Anatomistes Comptes Rendus* 22:59–70, 1927.
190 CONGDON, E.D. Simultaneous occurrence of very small sphenoid and frontal sinuses. *Anat. Rec.* 19:153–157, 1920.
191 CONSTANTIN, L. and R. PALEIRAC. Anomalies transitionnelles de la charnière dorso-lombaire et lombalgies. *J. Radiol. Electrol.* 33:167, 1952.
192 CONVERSE, J.M. The cartilaginous structures of the nose. *Ann. Otol. Rhinol. Laryngol.* 64:220–229, 1955.
193 COPE, V.Z. The internal structure of the sphenoidal sinus. *J. Anat.* 51:127–136, 1916.
194 CORSY, F. L'appareil hyoïdien n'est que la persistance de l'état foetal. *Soc. Biol. Comptes Rendus Hebdomadaires des Séances et Mémoires* 83:228–229, 1920.
195 CORSY, F. Appareil hyoïdien complet et appareil hyoïdien inverse essai d'interprétation. *Assoc. Anatomistes Comptes Rendus* 26:132–143, 1931.
196 COSTA, P. and F. PEROTTI. Sulla genesi e sul significato della frattura degli archi associata alla frattura del soma vertebrale. *Arch. Ortop. (Milan)* 64:34–45, 1951.
197 COVENTRY, M.B. and E.W. JOHNSON, JR. Congenital absence of the fibula. *J. Bone Joint Surg. (Am.)* 34:941–955, 1952.
198 CRAMER, K. Über Rückgratsverkrümmungen bei lumbosakralen Assimilationswirbeln. *Z. Orthop. Chir.* 22:68–89, 1908.
199 CSÁKÁNY, G. and T. DONÁTH. Vergleichende röntgenanatomische Untersuchung der beiderseitigen Foramine jugularia. *Fortschr. Röntgenstr.* 88:439–446, 1958.
200 CUMMING, J. Lumbar rib of unrecorded type. *Br. Med. J.* 1:55, 1926.
201 CUTORE, G. Fréquence et mode de se comporter des canaux perforants artériels dans la squama temporalis de l'homme. *Arch. Ital. Biol.* 45:277–278, 1906.
202 CUTORE, G. Sur un os malaire biparti. *Arch. Ital. Biol.* 48:477, 1907.
203 CYRIAX, E.F. Certain normal irregularities in the vertebral column in its lower dorsal area. *J. Anat. Physiol.* 56:147–148, 1921.
204 DANFORTH, C.H. Numerical variation and homologies in vertebrae. *Am. J. Phys. Anthropol.* 14:463–481, 1930.
205 DANFORTH, M.S. and P.D. WILSON. The anatomy of the lumbosacral region in relation to sciatic pain. *J. Bone Joint Surg.* 7:109–160, 1925.
206 DANKMEIJER, J. Congenital absence of the tibia. *Anat. Rec.* 62:179–194, 1935.
207 DASSEL, P.M. Agenesis of the sacrum and coccyx. *Am. J. Roentgenol.* 85:697–700, 1961.
208 DAVAINE. De l'absence congénitale du radius chez l'homme. *Soc. Biol. Comptes Rendus des Séances et Mémoires* 2:39–41, 1950.
209 DAVIDSON, A.J. and M.T. HORWITZ. Congenital club-hand deformity associated with absence of radius: Its surgical correction. *J. Bone Joint Surg.* 21:462–463, 1939.
210 DAVIS, G.G. Os vesalianum pedis. *Am. J. Roentgenol.* 17:551–553, 1927.
211 DAVIS, P.R. The thoraco-lumbar mortice joint. *J. Anat.* 89:370–377, 1955.
212 DECUVELAND, E. Die Apophyse des Metatarsale V und Os vesalianum. *Fortschr. Röntgenstr.* 82:251–257, 1955.
213 DECUVELAND, E. Über Beziehungen zwischen vorderer Außenknöchelapophyse und Os subfibulare mit differentialdiagnostischen Erwägungen. *Fortschr. Röntgenstr.* 83:213–221, 1955.
214 DECUVELAND, E. Über das sogenannte Os parasternale. *Fortschr. Röntgenstr.* 84:375–376, 1956.
215 DECUVELAND, E. Gibt es ein "Os accessorium supracalcaneum"? *Fortschr. Röntgenstr.* 85:58–59, 1956.
216 DECUVELAND, E. Beitrag zu den am medialen Fußrand vorkommenden inkonstanten Skelettelementen. *Arch. Orthop. Unfallchir.* 49:321–322, 1957.
217 DECUVELAND, E. Selten beobachtete Ossifikationen am menschlichen Tarsus. *Z. Orthop.* 89:268–270, 1957.
218 DECUVELAND, E. Als Epipyramis bzw. Epitriquetrum bezeichnete Skelettstücke und -abschnitte im und um den medialen (ulnaren) Interkarpalraum. *Fortschr. Röntgenstr.* 97:507–510, 1962.
219 DECUVELAND, E. Fehlbildung der Großzehengrundphalanx. *Fortschr. Röntgenstr.* 107:570, 1967.
220* DECUVELAND, E. and F. HEUCK. Osteochondropathie eines akzessorischen Knochenkernes am Malleolus tibiae (des sog. Os subtibiale). *Fortschr. Röntgenstr.* 79:728–732, 1953.
221 DEDERICH, R. Kongenitale Synostosen von Handwurzelknochen. *Monatsschrift Unfallheilkunde u. Versicherungsmedizin* 58:112–117, 1955.
222* DEGEN, S.T. Über das Auftreten der Knochenkerne am Handskelett von der Geburt bis zur Reife. Mit einem Hinweis auf die Sesambeine der Hand. *Med. Klin.* 46:1330–1332, 1951.
223 DEL BUONE, G. Studi per la determinazione della sede normale della pineale sul craniogramma normale. *Radiol. Med.* 37:997–1062, 1951.
224* DEL DUCA, V., DAVIS, E.V. and J.N. BARROWAY. Congenital absence of the sacrum and coccyx. *J. Bone Joint Surg. (Am.)* 33:248–253, 1951.
225 DERRY, D.E. Note on accessory articular facets between the sacrum and ilium, and their significance. *J. Anat. Physiol.* 45:202–210, 1911.

226 DERRY, D.E. Two skulls with absence of the premaxilla. *J. Anat.* 72:295–298, 1937–38.

227 DESFOSSES, P. Anatomie du vivant. Absence congénitale du péroné chez un jeune homme de 17 ans. *La Presse Medicale* No. 9. 30:175, 1922.

228 D'ESTE, L.S. Les incisures, les trous et les canaux sus-orbitares avec leur nerfs respectifs, et la résection du nerf sus-orbitaîre. *Arch. Ital. Biol.* 33:481, 1900.

229 D'ETIOLLES, R.L. Anomalie héréditaire des dents. *Soc. Biol. Comptes Rendus des Séances et Mémoires* 3:96, 1851.

230 DICKSON, L.M. Description of the skull of a girl of 17. *J. Anat.* 67:182–186, 1932.

231 DIDIO, L.J.A. The presence of the eminentia orbitalis in the os zygomaticum of Hindu skulls. *Anat. Rec.* 142:31–39, 1962.

232 DIDIO, L.J.A. Radiological study of the anatomical variations of the vertebra prominens in man. *Anat. Anzeiger* 120:210–218, 1967.

233 DIETHELM, L. Zur Kenntnis der Entwicklungsgeschichte der Wirbelsäule und der Wirbelkörperfehlbildungen. *Fortschr. Röntgenstr.* 69:143–150, 1943.

234 DIEULAFÉ and SAINT-MARTIN. Le type articulaire sacro-iliaque. *Assoc. Anatomistes Comptes Rendus* 14:95–109, 1912.

235 DIXON, F.W. A comparative study of the sphenoid sinus (a study of 1600 skulls). *Ann. Otol. Rhinol. Laryngol.* 46:687–698, 1937.

236 DORELLO, P. Sur plusieurs anomalies trouvées dan un occipital humain et spécialement sur le "troisième condyle occipital." *Arch. Ital. Biol.* 36:335, 1901.

237 DOUGLAS, T.E. JR. Facial pain from elongated styloid process. *Arch. Otolaryngol.* 56:635–638, 1952.

238 DOW, D.R. The anatomy of rudimentary first thoracic ribs, with special reference to the arrangement of the brachial plexus. *J. Anat.* 59:166–179, 1925.

239 DOWNIE, W. Unnamed supernumerary nasal cartilages. *J. Anat. Physiol.* 27:181–182, 1893.

240 DREXLER, C., STEWART, J.R. and O.K. KINCAID. Diagnostic implications of rib notching. *Am. J. Roentgenol.* 91:1064–1074, 1964.

241 DREYFUSS-LE-FOYER, A. and C. BRUNET. Anomalie par défaut de la primière côte. *Ann. Anat. Pathol.* 13:1028–1030, 1936.

242 VAN DRIESSCHE, R. Ein seltener Befund am Steißbein. *Fortschr. Röntgenstr.* 88:623, 1958.

243 DRINKWATER, H. Hereditary abnormal segmentation of the index and middle fingers. *J. Anat. Physiol.* 50:177–186, 1916.

244 DUBOST, E., PICARD, J.M., ECARLAT, B. and C. HERNANDEZ. Trois formes d'anomalies congénitales des membres. *J. Radiol. Electrol.* 41:579–583, 1960.

245 DUBOST-PERRET, T. Le traitement des absences congénitales du péroné et ses résultats. *Rev. Orthop.* 36:392–403, 1950.

246 DUBREUIL-CHAMBARDEL, L. Les trous de la symphyse du menton. *Assoc. Anatomistes Comptes Rendus* 8:14–21, 1906.

247 DUCKWORTH, W.L.H. On an unusual form of nasal bone in a human skull. *J. Anat. Physiol.* 36:257–259, 1902.

248 DUCKWORTH, W.L.H. Report on an abnormal first thoracic vertebra. *J. Anat. Physiol.* 45:65–68, 1911.

249 DUKES, L. and S.A. OWEN. Anomalies in the cervical and upper thoracic region, involving the cervical vertebrae, first rib, and brachial plexus. *J. Anat. Physiol.* 36:290–291, 1902.

250 DUNCAN, D. An anomaly of the knee joint: Lateral interarticular disc. *Anat. Rec.* 53:305–308, 1932.

251 DUROUX, BARRY and R. LATREILLE. Cunéiforme surnuméraire. *Assoc. Anatomistes Comptes Rendus* 39:155, 1952.

252* DWIGHT, T. Account of two spines with cervical ribs, one of which has a vertebra suppressed, and absence of the anterior arch of the atlas. *J. Anat. Physiol.* 21:539–550, 1887.

253* DWIGHT, T. Os intercuneiforme tarsi, Os paracuneiforme tarsi, Calcaneus secundarius. *Anat. Anz.* 20:465–472, 1902.

254 DWIGHT, T. A transverse foramen in the last lumbar vertebra. *Anat. Anz.* 20:571–572, 1902.

255* DWIGHT, T. A bony supracondyloid foramen in man with remarks about supracondyloid and other processes from the lower end of the humerus. *Am. J. Anat.* 3:221–228, 1904.

256 DWIGHT, T. A separate subcapitatum in both hands. *Anat. Anz.* 24:253–255, 1904.

257* DWIGHT, T. Stylo-hyoid ossification. *Ann. Surg.* 46:721–735, 1907.

258 DWIGHT, T. Description of a free cuboides secundarium, with remarks on that element, and on the calcaneous secundarius. *Anat. Anz.* 37:218–224, 1910.

259 DWIGHT, T. Free cuboides secundarium on both feet with some further remarks on Pfitzner's Theory. *Anat. Anz.* 39:410–414, 1911.

260 DYES, O. Os acetabuli persistens bilateralis. *Fortschr. Röntgenstr.* 39:658–659, 1929.

261 EAGLE, W.W. Elongated styloid process. Further observations and a new syndrome. *Arch. Otolaryngol.* 47:630–640, 1948.

262 EAST, C.F.T. A rare abnormality of the occipital bone. *J. Anat.* 60:416–417, 1926.

263 EGGELING, H. VON. Die Gabelung der Halswirbeldornen und ihre Ursache. *Anat. Anz.* 55:33–94, 1922.

264 EGGIMAN, P. Lunatum bipartitum. *Radiol. Clin. (Basel)* 18:203–205, 1949.

265 EGGIMAN, P. Zur Bipartition des Lunatum. *Radiol. Clin. (Basel)* 20:65–70, 1951.

266 EGLI, A. Beitrag zur Kenntnis der Fehlbildungen am Kreuzbein. *Z. Anat. Entwicklungsgesch.* 112:245–270, 1942.

267 EHRLICH, N. Untersuchungen über die congenitalen Defecte und Hemmungsbildungen der Extremitäten. *Arch. Pathol. Anat. Physiol. Klin. Med.* 100:107–138, 1885.

268 ELIAS, H. Ein Fall von Fissura sternalis congenita completa simplex. *Wien. Klin. Wochenschr.* 31:98–100, 1918.

269 ELLIOT SMITH, G. On a case of fusion of the atlas and axis. *Anat. Anz.* 31:166–168, 1907.

270 ETTER, L.E. Osseous abnormalities of the thoracic cage seen in forty thousand consecutive chest photoroentgenograms. *Am. J. Roentgenol.* 51:359–363, 1944.

271 EVANS, G.F., ALFARO, A. and S. ALFARO. An unusual anomaly of the superior extremities in a Tarascan Indian girl. *Anat. Rec.* 106:37–47, 1950.

272* EVANS, W.A. JR. The epihyal bone. A consideration of some small accessory bones of the neck. *Am. J. Roentgenol.* 44:714–715, 1940.

273 FABER, A. Über das Os intermetatarseum. *Z. Orthop. Chir.* 61:186–197, 1934.

274 FAWCETT, E. The sesamoid bones of the hand: A skiagraph confirmation of the work done by Pfitzner. *J. Anat. Physiol.* 31:157–161, 1897.

275 FAWCETT, E. On the complete ossification of the human sacrum. *Anat. Anz.* 30:414–421, 1907.

276 FAWCETT, E. The sexing of the human sacrum. *J. Anat.* 72:633, 1937–38.

277 FEDEROW, V. Zwei Fälle der seltenen Bildung von Querfortsätzen des ersten Brustwirbels. *Anat. Anz.* 36:556–560, 1910.

278 FERBER, C. Ein Beitrag zur Dreigliedrigkeit des "Daumens." *Z. Orthop.* 83:55–64, 1953.

279 FERGUSON, A.D. and R.B. SCOTT. Congenital absence of the tibia. *Am. J. Dis. Child.* 84:84–89, 1952.

280 FERRARI, G. Sur la soudure de l'os occipitis avec l'atlas. *Arch. Ital. Biol.* 59:472, 1913.

281 FIDDES, J. Case of supernumerary hallux and their etiology. *Anat. Anz.* 40:544–547, 1912.

282 FIEBELKORN, H.-J. Über ein wenig bekanntes inkonstantes Skelettelement der Fußwurzel. *Fortschr. Röntgenstr.* 77:624–625, 1952.

283 FISCHER, E. Zur Kenntnis der großen Foramina parietalia. *Fortschr. Röntgenstr.* 81:406–407, 1954.

284 FISCHER, E. Persistierende Klavikulaapophyse. *Fortschr. Röntgenstr.* 86:532, 1957.

285 FISCHER, E. Neues Skelettelement dorsal am Radio-Karpalgelenk. *Fortschr. Röntgenstr.* 91:530–531, 1959.

286 FISCHER, E. Akzessorische freie Knochenelemente in der Umgebung des Foramen magnum. *Fortschr. Röntgenstr.* 91:638–642, 1959.

287 FISCHER, H. Beitrag zur Kenntnis der Skelettvarietäten (überzählige Karpalia u. Tarsalia, Sesambeine, Kompaktainseln). *Fortschr. Röntgenstr.* 19:43–66, 1912/13.

288 FISCHER, H. Le sinus pétro-squameux chez l'homme: un cas de communication de la jugulaire externe avec le sinus latéral. *Assoc. Anatomistes Comptes Rendus* 21:210–211, 1926.

289 FISCHER, H. Quelques considérations sur la morphologie de l'omoplate. Echancrure coracoidienne transformée en un canal par un pont osseux (origine congénitale). *Assoc. Anatomistes Comptes Rendus* 22:95–98, 1927.

290 FISCHER, H. and PELLEGRINO. Un cas d'ectrokeirie. *Rev. Orthop.* 32:380–382, 1946.

291 FLECKER, H. Observations upon cases of absence of lacrimal bones and the existence of perilacrimal ossicles. *J. Anat. Physiol.* 48:52–72, 1914.

292 FLEISCHNER, F. Rippenanomalien als Quelle diagnostischer Irrtü-

mer und falscher therapeutischer Indikationsstellung. *Med. Klin.* 26:1366–1367, 1930.
293 FOURNIER, A.M. Un cas de persistance de l'appareil hyoïdien. *J. Radiol. Electrol.* 31:513, 1950.
294 FRANCESCHETTI, A. Un syndrome nouveau: la dysotose mandibulo-faciale. *Bull. Schweiz. Akad. Med. Wissen.* 1/2:60–66, 1944. Cited in *Excerpta Medica*, Sec. 1, Vol. 2, abstract 193, p. 98, 1948.
295 FRANCHI, M. Frequenza della principali variazoni ed anomalie dell'osso occipitale in 70 crani Senesi. *Studi Facoltà Med. Senese* 14:151–153, 1946. Cited in *Excerpta Medica*, Sec. 1, Vol. 2, abstract 1373, p. 484, 1948.
296 FRANCIS, C.C. Variations in the articular facets of the cervical vertebrae. *Anat. Rec.* 122:589–602, 1955.
297 FRANCIS, C.C. Dimensions of the cervical vertebrae. *Anat. Rec.* 122:603–609, 1955.
298 FRANK, J. Ein Fall von Halsrippe mit abnormem Nervenverlauf. *Anat. Anz.* 47:218–225, 1914–15.
299* FRANTZ, C.H. and R. O'RAHILLY. Congenital skeletal limb deficiencies. *J. Bone Joint Surg. (Am.)* 43:1202–1224, 1961.
300 FRASSETO, F. Sur une novelle soudure (soudure à tenon) dans les os du crâne d'un cerf, rencontrée dans les os du crâne de deux pirates chinois et d'un jeune indien. *Arch. Ital. Biol.* 32:453, 1899.
301 FRASSETTO, F. La variabilité du Crâne humain avec la methode quantitative statistique de Camerano et avec le méthode Sergi. *Arch. Ital. Biol.* 39:478, 1903.
302 HILTEMANN, H., Fronticulus metopicus und Sutura frontalis persistens mit Hypoplasie der Sinus frontales. *Fortschr. Röntgenstr.* 81:407–409, 1954.
303 FREEDMAN, B. Congenital absence of the sacrum and coccyx. Report of a case and review of the literature. *Br. J. Surg.* 37:299–303, 1949–50.
304 FRERE, J.M. A case having thumbs with three phalanges simulating fingers. *South Med. J.* 23:536–537, 1930.
305 FROMMER, J., MONROE, C.W. and B. SPECTOR. Joints between segments of the human stylohyoid chain. *Anat. Rec.* 142:305, 1962.
306 FULLENLOWE, T.M. Congenital absence of the odontoid process. Report of a case. *Radiology* 63:72–73, 1954.
307 FUNKE, E. Über einen Processus odontoideus atlantis hominis. *Anat. Anz.* 14:385–390, 1898.
308 FUSARI, R. Sur les différents modes de substitution de la partie postérieure de la lame papyraceé l'orbite de l'homme. *Arch. Ital. Biol.* 28:470–471, 1897.
309 GANFINI, C. Sur quelques facettes articulaires du basioccipital par rapport aux processus basilaires. *Arch. Ital. Biol.* 47:475–476, 1907.
310 GANTERT, F. and C. ALZHEIMER. Der Processus supracondylicus humeri als Ursache von Medianusschädigungen. *Nervenarzt* 27:349, 1956.
311 GANZ, E. Fraktur des Processus supracondylicus humeri. *Röntgenpraxis* 9:48, 1937.
312 GARDNER, E., GRAY, D.J. and R. O'RAHILLY. The prenatal development of the skeleton and joints of the human foot. *J. Bone Joint Surg. (Am.)* 41:848–873, 1959.
313* GARLAND, L.H. The shape of the female pelvis and its clinical significance. *Am. J. Roentgenol.* 40:359–370, 1938.
314 GASSMANN, W. Eine seltene Knochenvarietät am Os occipitale (Processus suboccipitalis). *Fortschr. Röntgenstr.* 85:633–635, 1956.
315 GEDDES, A.C. An abnormal nasal duct. *Anat. Anz.* 37:5–8, 1910.
316* GEIPEL, P. Zur Kenntnis der Spina bifida am Atlas. *Fortschr. Röntgenstr.* 42:583–589, 1930.
317* GEIPEL, P. Zur Kenntnis der Spaltbildung des Atlas und Epistropheus. *Fortschr. Röntgenstr.* 46:373–402, 1932.
318* GEIPEL, P. Doppelte Spaltbildung des Atlas. *Med. Klin.* 32:1302–1303, 1337, 1936.
319 GEORGY, H.U. and H. HILLGER. Beobachtung eines ungewöhnlich großen Os radiale externum. *Fortschr. Röntgenstr.* 111:715–716, 1969.
320 GÉRARD, G. Notion d'un éperon lacrymal antérieur. *Assoc. Anatomistes Comptes Rendus* 8:114–119, 1906.
321 GÉRARD, G. and H. FOURNET. Note statistique sur les variations de forme du bassinet humain. *Soc. Biol. Comptes Rendus Hebdomadaires des Séances et Mémoires* 84:893–894, 1921.
322 GERSHON-COHEN, J. and R.F. DELBRIDGE. Pseudarthrosis, synchondrosis and other anomalies of the first ribs. *Am. J. Roentgenol.* 53:49–54, 1945.
323 GIACCAI, L., SALAAM, M. and H. ZELLWEGER. Cleidocranial dysostosis with osteopetrosis. *Acta. Radiol.* 41:417–424, 1954.

324 GIACOMINI. De l'existence de l'os odontoïde chez l'homme. *Arch. Ital. Biol.* 8:40–48, 1887.
325 GIANNINI, M.J., BORRELLI, F.J. and W.B. GREENBERG. Agenesis of the vertebral bodies – A cause of dwarfism. *Am. J. Roentgenol.* 59:705–711, 1948.
326 GIBSON, G.A. and H. MALET. Presternal fissure uncovering the base of the heart. *J. Anat. Physiol.* 14:1–9, 1879.
327 GILES, R.G. A congenital anomaly of the patella. *Texas State J. Med.* 23:731–732, 1928.
328 GILLESPIE, H.W. The significance of congenital lumbo-sacral abnormalities. *Br. J. Radiol.* 22:270–275, 1949.
329* GILLMAN, E.L. Congenital absence of the odontoid process of the axis. *J. Bone Joint Surg. (Am.)* 41:345–348, 1959.
330 GILSE, P.H.G. VAN. The development of the sphenoidal sinus in man and its homology in mammals. *J. Anat.* 61:153–166, 1927.
331 GIRARD, P.M. Congenital absence of the sacrum. *J. Bone Joint Surg.* 17:1062–1064, 1935.
332 GIUFFRIDA-RUGGERI, V. Os fontanellaires et espaces suturaux dan la norma lateralis. *Arch. Ital. Biol.* 36:337, 1901.
333 GIUFFRIDA-RUGGERI, V. Division longitudinale de l'ala magna du sphénoïde. Extension de la squame du temporal en hauteur, comme caractère hiérarchique. Situation du squelette nasal. *Arch. Ital. Biol.* 36:337–338, 1901.
334 GIUFFRIDA-RUGGERI, V. Sur une très rare anomalie du squelette nasal. *Arch. Ital. Biol.* 36:338, 1901.
335 GIUFFRIDA-RUGGERI, V. Os nasal biparti, post-frontal et autres wormiens dan le squelette facial. *Arch. Ital. Biol.* 36:473, 1901.
336 GLADSTONE, R.J. Case of an additional presacral vertebra. *J. Anat. Physiol.* 31:530–538, 1897.
337 GLADSTONE, R.J. and W. ERICHSEN-POWELL. Manifestation of occipital vertebrae, and fusion of the atlas with the occipital bone. *J. Anat. Physiol.* 69:190–209, 1915.
338 GLADSTONE, R.J. and C.P.G. WAKELEY. Cervical ribs and rudimentary first thoracic ribs considered from clinical and etiological standpoints. *J. Anat.* 66:334–370, 1931–32.
339 GOALWIN, H.A. Die exakte radiographische Darstellung des Canalis opticus. *Fortschr. Röntgenstr.* 32:218–222, 1924.
340* GOALWIN, H.A. One thousand optic canals: Clinical, anatomic and roentgenologic study. *JAMA* 89:1745–1748, 1927.
341 GOCKEL, H.P. Über eine Form von Atlasfehlbildung. *Fortschr. Röntgenstr.* 88:485–487, 1958.
342 GOLDBERG, R.R. Congenital bilateral complete absence of the radius in identical twins. *J. Bone Joint Surg. (Am.)* 30:1001–1003, 1948.
343 GOLDHAMER, K. Welche Skelettvarietäten des Schädels können pathologische und traumatische Veränderungen vortäuschen? *Wien. Klin. Wochenschr.* 81:584–585, 1931.
344 GOLDSMITH, W.M. "The Catlin mark." The inheritance of an unusual opening in the parietal bones. *J. Hered.* 13:69–71, 1922.
345 GOLDSTEIN, R. Agenesis of sacrum and coccyx. *Acta Med. Orientalia* 6:202–204, 1947. Cited in *Excerpta Medica*, Sec. 1, Vol. 3, abstract 668, 1949.
346 GÓMEZ OLIVEROS, L. Observaciones anatómicas de tres casos de sacralización de la quinta vértebra lumbar. *Cirugía Aparato Locomotor* 5:29–35, 1948. Cited in *Excerpta Medica*, Sec. 1, Vol. 3, abstract 382, 1949.
347* GOODHART, S.P. Cervical rib and its relation to the neuropathies. *Am. J. Med. Sci.* 138:666–683, 1909.
348 GÓRSKI, K.A. Anatomy of hand in polydactylia. *Folia Morphol.* 36:141–147, 1977.
349* GOSSEREZ, M., MALRAISON, P. and TRÈHEUX. Anomalie thyrohyoidienne. *J. Radiol. Electrol.* 39:194–196, 1958.
350 GOZDZIEWSKI, S., NIZANKOWSKI, C. and R. KINDLIK. Die morphologische Analyse des Canalis infraorbitalis und des Foramen infraorbitale beim Menschen. *Anat. Anz.* 145:517–527, 1979.
351 GRÄFENBERG, E. Die Entwicklung der Knochen, Muskeln und Nerven der Hand und für die Bewegungen der Hand bestimmten Muskeln des Unterarms. *Anat. Hefte* 30:1–154, 1906.
352* GRAFFI BENASSI, E. Sulla coesistenza di variazioni ed anomalie in uno scheletro umano. *Arch. Ital. Anat. Embriol.* 44:282–301, 1940.
353 GRASHEY, R. Fehlerquellen, Varietäten Rippenanomalie: artikulierende Halsrippe. a) Hornförmiger Auswuchs am Os navic. manus. b) Abbruch der oberen Gelenkkante des Os navic. manus; Lücke im Karpus. *Röntgenpraxis* 7:209, 275, 1935.
354 GRASHEY, R. Hornförmiger Auswuchs am Os naviculare manus. *Röntgenpraxis* 7:275, 1935.
355 GRASHEY, R. Dorsale supracondyläre Femurexostose. *Röntgenpraxis* 9:278–280, 1937.

356 GRASHEY, R. Articulatio talo-calcanea (Os sustentaculi). *Röntgenpraxis* 14:139–142, 1942.
357 GRAVES, W.W. The types of scapulae. *Am. J. Phys. Anthropol.* 4:111–128, 1921.
358 GRAY, D.J. Variations in the human scapulae. *Am. J. Phys. Anthropol.* 29:57–72, 1942.
359 GREELEY, P.W. Multiple ossification centers of the patella. *Am. J. Surg.* 18:456–459, 1932.
360 GREEN, H.L.H.H. An unusual case of atlanto-occipital fusion. *J. Anat.* 65:140–144, 1930.
361* GREULICH, W.W. and H. THOMS. The dimensions of the pelvic inlet of 789 white females. *Anat. Rec.* 72:45–51, 1938.
362* GREULICH, W.W. and H. THOMS. An X-ray study of male pelvis. *Anat. Rec.* 75:289–299, 1939.
363* GREULICH, W.W. and H. THOMS. A study of pelvis type and its relationship to body build in white women. *JAMA* 112:485–493, 1939.
364 GREVE. Ein Fall von Schwanzbildung beim Menschen. *Arch. Pathol. Anat. Physiol. Klin. Med.* 72:129, 1878.
365 GREIG, D.M. Congenital and symmetrical perforation of both parietal bones. *J. Anat. Physiol.* 26:187–191, 1892.
366* GREIG, D.M. Cleft-sternum and ectopia cordis. *Edinb. Med. J.* 33:480–511, 1926.
367* GROB, M. Über die röntgenologischen Nahtverhältnisse der hinteren Schädelgrube beim Kinde mit spezieller Berücksichtigung der Sutura mendoza. *Fortschr. Röntgenstr.* 57:265–275, 1938.
368 GROSSE, U. Ein Fall von Mißbildung der ersten Rippe. *Anat. Anz.* 8:410–413, 1893.
369 GRUBER, W. Über den gesamten Apparat der Bänder zwischen dem Hinterhauptsbein und einem neuentdeckten Appendix superior des Ligamentum cruciatum insbesondere. *Arch. Anat. Physiol. Wissen. Med.* 1851, pp. 291–315.
370 GRUBER, W. Zergliederung oberer Extremitäten mit angeborenen Defecten an der Hand. *Arch. Anat. Physiol. Wissen. Med.* 1863, pp. 319–338.
371 GRUBER, W. Über einen am Malleolus externus articulirenden Knochen. *Arch. Pathol. Anat. Physiol. Klin. Med.* 27:205–206, 1863.
372 GRUBER, W. Ein Nachtrag zur Kenntnis des Processus supracondyloideus (internus) humeri des Menschen. *Arch. Anat. Physiol. Wissen. Med.* 1865, pp. 367–376.
373 GRUBER, W. Ein Nachtrag zur Kenntnis des Processus supracondyloideus (internus) humeri des Menschen. *Arch. Anat. Physiol. Wissen. Med.* 1865, p. 267.
374 GRUBER, W. Über congenitalen Radiusmangel. *Arch. Pathol. Anat. Physiol. Klin. Med.* 32:211–222, 1865.
375 GRUBER, W. Notiz über die Zergliederung einer rechten oberen Extremität eines Mannes mit Duplicität des Daumens an der Hand. *Arch. Pathol. Anat. Physiol. Klin. Med.* 32:223–228, 1865.
376 GRUBER, W. Über die secundären Handwurzelknochen des Menschen. *Arch. Anat. Physiol. Wissen. Med.* 1866, pp. 565–586.
377 GRUBER, W. Über congenitalen unvollständigen Radiusmangel. *Arch. Pathol. Anat. Physiol. Klin. Med.* 40:427–435, 1867.
378* GRUBER, W. Über ein dem Os intermedium s. centrale gewisser Säugethiere analoges neuntes Handwurzelknöchelchen beim Menschen. *Arch. Anat. Physiol. Wissen. Med.* 1869, pp. 331–341.
379* GRUBER, W. Über ein neuntes Handwurzelknöchelchen des Menschen mit der Bedeutung einer persistirenden Epiphyse des zum Ersatze des mangelnden Processus styloideus des Metacarpale III anomal vergrößerten Multangulum minus. *Arch. Anat. Physiol. Wissen. Med.* 1869, pp. 342–360.
380* GRUBER, W. Vorkommen des Processus styloideus des Metacarpale III als persistirende und ein neuntes Handwurzelknöchelchen repräsentirende Epiphyse. *Arch. Anat. Physiol. Wissen. Med.* 1869, pp. 361–366.
381 GRUBER, W. Über Mißbildungen der Finger an beiden Händen eines Lebenden. *Arch. Pathol. Anat. Physiol. Klin. Med.* 47:303–304, 1869.
382 GRUBER, W. Rechter Fuß mit 6 Fußwurzelknochen, 4 Mittelfußknochen und 4 Zehen. *Arch. Pathol. Anat. Physiol. Klin. Med.* 47:304–307, 1869.
383* GRUBER, W. Über das aus einer persistirenden und den Processus styloideus des Metacarpale III. repräsentirenden Epiphyse entwickelte, articulirende, neunte Handwurzelknöchelchen. *Arch. Anat. Physiol. Wissen. Med.* 1870, pp. 197–207.
384* GRUBER, W. Beiträge zu den secundären Handwurzelknochen des Menschen. *Arch. Anat. Physiol. Wissen. Med.* 1870, pp. 490–498.
385 GRUBER, W. Ungewöhnliches Ossiculum sesamoideum am Handrücken. *Arch. Anat. Physiol. Wissen. Med.* 1870, pp. 499–500.
386 GRUBER, W. Über Congenitale Theilung des Parietale durch eine quere oder schräge Sutur. *Arch. Pathol. Anat. Physiol. Klin. Med.* 50:113–124, 1870.
387 GRUBER, W. Über congenital abnorm weite Foramina parietalia. *Arch. Pathol. Anat. Physiol. Klin. Med.* 50:124–134, 1870.
388 GRUBER, W. Über enorm lange Processus styloides der Schläfebeine. *Arch. Pathol. Anat. Physiol. Klin. Med.* 50:232–234, 1870.
389 GRUBER, W. Über den Fortsatz des Höckers des Kahnbeines der Fußwurzel – Processus tuberositas navicularis – und dessen Auftreten als Epiphyse oder als besonderes articulirendes Knöchelchen. *Arch. Anat. Physiol. Wissen. Med.* 1871, pp. 281–285.
390 GRUBER, W. Über das Tuberculum deltoideum und den Processus deltoideus des Schlüsselbeines. *Arch. Anat. Physiol. Wissen. Med.* 1871, pp. 297–299.
391 GRUBER, W. Über ein congenitales Loch im unteren Schulterblattwinkel über dessen Epiphyse. *Arch. Anat. Physiol. Wissen. Med.* 1871, pp. 300–304.
392 GRUBER, W. Über einen Fall von Macrodactylie bei einem Lebenden. *Arch. Pathol. Anat. Physiol. Klin. Med.* 56:416–419, 1872.
393 GRUBER, W. Über einen fortsatzartigen, cylindrischen Höcker an der Vorderfläche des Angulus superior der Scapula. *Arch. Pathol. Anat. Physiol. Klin. Med.* 56:425–426, 1872.
394 GRUBER, W. Bemerkungen über das Foramen mentale. *Arch. Anat. Physiol. Wissen. Med.* 1872, pp. 738–745.
395 GRUBER, W. Zergliederung eines linken Armes mit Duplicität des Daumens. *Bull. Acad. Imp. Sci. St. Petersbourg* 17:24–31, 1872.
396 GRUBER, W. Sur les os du carpe surnumeraire chez l'homme. *Bull. Acad. Imp. Sci. St. Petersbourg* 17:408–413, 1872.
397* GRUBER, W. Über einige merkwürdige Oberkiefer-Abweichungen. *Arch. Anat. Physiol. Wissen. Med.* 1873, pp. 195–207.
398* GRUBER, W. Über den an der Schläfenfläche des Jochbeines gelagerten Kiefer-Schläfenbogen – Arcus maxillo-temporalis intra-jugalis – beim Menschen; nebst Nachträgen zum zweigetheilten Jochbeine – Os zygomaticum bipartitum – ohne oder mit Vorkommen des Kiefer-Schläfenbogens. *Arch. Anat. Physiol. Wissen. Med.* 1873, pp. 208–240.
399* GRUBER, W. Über supernumeräre Knochen im Jochbogen. *Arch. Anat. Physiol. Wissen. Med.* 1873, pp. 337–347.
400* GRUBER, W. Über das Semiinfundibulum inframaxillare, den Sulcus mylohyoideus und die beide deckenden knöchernen Brükken. *Arch. Anat. Physiol. Wissen. Med.* 1873, pp. 348–356.
401 GRUBER, W. Weitere neue Fälle des Vorkommens des Ossiculum supernumerarium in der unteren Handwurzelreihe als Vertreter des mangelnden Processus styloideus des Metacarpale III. *Arch. Anat. Physiol. Wissen. Med.* 1873, pp. 706–711.
402 GRUBER, W. Weitere Nachträge zum Vorkommen des Ossiculum intermedium carpi beim Menschen. *Arch. Anat. Physiol. Wissen. Med.* 1873, pp. 720.
403* GRUBER, W. Über den Fortsatz des Seitenhöckers – Processus tuberositas lateralis – des Metatarsale V. und sein Auftreten als Epiphyse. *Arch. Anat. Physiol. Wissen. Med.* 1875, pp. 48–58.
404 GRUBER, W. Über den Fortsatz des Höckers des großen vielwinkligen Beines – Processus tuberositas multanguli majoris – und dessen Auftreten als Epiphyse. *Arch. Anat. Physiol. Wissen. Med.* 1875, pp. 59–66.
405* GRUBER, W. Ein Nachtrag zum Vorkommen des zweigetheilten Jochbeins – Os zygomaticum bipartitum – beim Menschen. *Arch. Anat. Physiol. Wissen. Med.* 1875, pp. 194–201.
406 GRUBER, W. Nachtrag zu den Stirnfontanellknochen beim Menschen. *Arch. Pathol. Anat. Physiol. Klin. Med.* 63:92–93, 1875.
407 GRUBER, W. Enorm weiter Canalis mastoideus. *Arch. Pathol. Anat. Physiol. Klin. Med.* 65:9–10, 1875.
408 GRUBER, W. Die Crista galli des Os ethmoideum mit einer Höhle. *Arch. Pathol. Anat. Physiol. Klin. Med.* 65:11, 1875.
409 GRUBER, W. Gelenkige Verbindung der ersten mit der zweiten Rippe an deren Körpern. *Arch. Pathol. Anat. Physiol. Klin. Med.* 65:11–12, 1875.
410 GRUBER, W. Lateralwärts doppelästiger Rippenknorpel. *Arch. Pathol. Anat. Physiol. Klin. Med.* 65:12–13, 1875.
411* GRUBER, W. Über ein aus der Epiphyse eines durch einen fortsatzartigen Anhang vergrößerten Multangulum minus entwickeltes, articulirendes neuntes Ossiculum carpi. *Arch. Anat. Physiol. Wissen. Med.* 1876, pp. 221–229.
412* GRUBER, W. Ein Nachtrag zum Vorkommen des zweigetheilten Jochbeines – Os zygomaticum bipartitum –, bei Anwesenheit des

Kiefer-Schläfenbogens – Arcus maxillo-temporalis intrajugalis. *Arch. Anat. Physiol. Wissen. Med.* 1876, pp. 230–235.

413 GRUBER, W. Über das Foramen in den Laminae der Cartilago thyreoidea. *Arch. Pathol. Anat. Physiol. Klin. Med.* 66:455–457, 1876.

414* GRUBER, W. Zweiter Nachtrag zu den Stirnfontanellknochen beim Menschen. *Arch. Pathol. Anat. Physiol. Klin. Med.* 66:466–467, 1876.

415* GRUBER, W. Ein Nachtrag zum Vorkommen des Parietale partitum. *Arch. Pathol. Anat. Physiol. Klin. Med.* 66:468–470, 1876.

416 GRUBER, W. Über ein Gelenk zwischen den vorderen Querfortsätzen der rechten Seite des V. und VII. Halswirbels. *Arch. Pathol. Anat. Physiol. Klin. Med.* 67:327–330, 1876.

417 GRUBER, W. Erster Brustwirbel mit einem Foramen transversarium. *Arch. Pathol. Anat. Physiol. Klin. Med.* 67:341–343, 1876.

418* GRUBER, W. Über einen Fall von unvollkommen gebildeter erster Brustrippe. *Arch. Pathol. Anat. Physiol. Klin. Med.* 67:344–352, 1876.

419 GRUBER, W. Eine Bursa mucossa capituli ossis hyoidis. *Arch. Pathol. Anat. Physiol. Klin. Med.* 67:359–360, 1876.

420 GRUBER, W. Über einen neuen Fall eines congenital abnorm weiten Foramen parietale, mit Bemerkung über die zur Ermittlung der Ursache seines Auftretens aufgestellten Hypothesen und anderen Erörterungen. *Arch. Pathol. Anat. Physiol. Klin. Med.* 68:305–324, 1876.

421 GRUBER, W. Dritter Nachtrag zu den Stirnfontanellknochen. *Arch. Pathol. Anat. Physiol. Klin. Med.* 69:380–381, 1877.

422 GRUBER, W. Vierter Nachtrag zum Vorkommen des zweigetheilten Jochbeins – Os zygomaticum bipartitum – beim Menschen. *Arch. Pathol. Anat. Physiol. Klin. Med.* 69:381–383, 1877.

423 GRUBER, W. Über eine anomale, congenitale, von der Spina jugularis posterior des Temporale gebildete Knochenbrücke über dem Sulcus jugularis des Occipitale. *Arch. Pathol. Anat. Physiol. Klin. Med.* 69:283–285, 1877.

424 GRUBER, W. Eigenthümlicher stielförmiger Fortsatz der Schläfenbeinschuppe. *Arch. Pathol. Anat. Physiol. Klin. Med.* 69:386–387, 1877.

425 GRUBER, W. Zwei Scapulae mit je einem congenitalen Loche und eine Scapula mit einem congenitalen Fortsatze von zwei männlichen Skeletten. *Arch. Pathol. Anat. Physiol. Klin. Med.* 69:387–391, 1877.

426* GRUBER, W. Os naviculare carpi bipartitum. *Arch. Pathol. Anat. Physiol. Klin. Med.* 69:391–396, 1877.

427 GRUBER, W. Über den eine Thierbildung repräsentirenden normalen, und den exostotisch gewordenen Processus trochlearis calcanei. *Arch. Pathol. Anat. Physiol. Klin. Med.* 70:128–132, 1877.

428 GRUBER, W. Weitere Nachträge zum Vorkommen des Processus tuberositas navicularis und der navicularia secondaria tarsi. *Arch. Pathol. Anat. Physiol. Klin. Med.* 70:132–135, 1877.

429 GRUBER, W. Rechte Hälfte der Spina nasalis anterior von enormer Länge. *Arch. Pathol. Anat. Physiol. Klin. Med.* 70:135–136, 1877.

430 GRUBER, W. Die Fossa temporalis im Bereiche des aufsteigenden Theiles des Temporalflügels des Sphenoides als eine enorm tiefe Grube. *Arch. Pathol. Anat. Physiol. Klin. Med.* 70:136, 1877.

431 GRUBER, W. Processus styloides und Processus vaginalis des Felsenbeines von fast gleicher Länge. *Arch. Pathol. Anat. Physiol. Klin. Med.* 70:136–137, 1877.

432* GRUBER, W. Ungewöhnliche Grube von enormer Größe an der Wurzel des Processus pterygoides des Os sphenoides. *Arch. Pathol. Anat. Physiol. Klin. Med.* 70:137, 1877.

433 GRUBER, W. Persistirende unvollständige Theilung des Unterkiefers in zwei Hälften. *Arch. Pathol. Anat. Physiol. Klin. Med.* 70:137–139, 1877.

434 GRUBER, W. Congenital getheilter Processus anterior der rechten Seite eines V. Halswirbels. *Arch. Pathol. Anat. Physiol. Klin. Med.* 70:139, 1877.

435 GRUBER, W. Verwachsung zweier Rippen durch eine congenitale Knochenplatte zwischen deren Hälsen. *Arch. Pathol. Anat. Physiol. Klin. Med.* 70:139–140, 1877.

436 GRUBER, W. Über die beiden Arten des überzähligen Zwischenknöchelchens am Rücken des Metatarsus (Ossiculum intermetatarseum dorsale Gruber) und über den durch Ankylose eines dieser Knöchelchen entstandenen und eine Exostose am Os cuneiform I und Os metatarsale II vortäuschenden Fortsatz. *Arch. Pathol. Anat. Physiol. Klin. Med.* 71:440–452, 1877.

437* GRUBER, W. Zu den in der Sutura squamosa auftretenden Knochen. *Arch. Pathol. Anat. Physiol. Klin. Med.* 72:480–486, 1878.

438 GRUBER, W. Zweigetheilte Temporalschuppe – Squama temporalis bipartita. *Arch. Pathol. Anat. Physiol. Klin. Med.* 72:486–488, 1878.

439 GRUBER, W. Ein hakenförmiges Fortsätzchen über und vor dem Infraorbitalloche. *Arch. Pathol. Anat. Physiol. Klin. Med.* 72:494–496, 1878.

440 GRUBER, W. Über ein Multangulum minus mit einem, den mangelnden Processus styloides des Metacarpale III substituirenden, fortsatzartigen Anhange. *Arch. Pathol. Anat. Physiol. Klin. Med.* 73:337–338, 1878.

441 GRUBER, W. Über einen, den Eingang in die Orbita, unter dessen oberen Rande verlegenden knöchernen Bogen bei einem Knaben. *Arch. Pathol. Anat. Physiol. Klin. Med.* 77:110–112, 1879.

442 GRUBER, W. Vierter Nachtrag zu den Stirnfontanellknochen. *Arch. Pathol. Anat. Physiol. Klin. Med.* 77:112–113, 1879.

443 GRUBER, W. Fünfter Nachtrag zum Vorkommen des Os zygomaticum bipartitum, und Zurückweisung des Prädicates "Os japonicum" für dasselbe. *Arch. Pathol. Anat. Physiol. Klin. Med.* 77:113–123, 1879.

444 GRUBER, W. Ein Beispiel einer abnorm verbreiterten und tief gefurchten Rippe mit gabliger Spaltung am vorderen Ende und inselförmiger Spaltung am Körper. *Arch. Pathol. Anat. Physiol. Klin. Med.* 78:98–100, 1879.

445 GRUBER, W. Synostose des Os capitatum carpi und des Os metacarpale III an dem Processus styloideus des letzteren. *Arch. Pathol. Anat. Physiol. Klin. Med.* 78:100–101, 1879.

446 GRUBER, W. Zergliederung des linken Armes mit Doppeldaumen von einem Erwachsenen. *Arch. Pathol. Anat. Physiol. Klin. Med.* 78:101–105, 1879.

447 GRUBER, W. Congenitale Verwachsung der ersten mit der zweiten Brustrippe der rechten Seite – und congenitale Verkümmerung der ersten Brustrippe der linken Seite an ein und demselben Skelett. *Arch. Pathol. Anat. Physiol. Klin. Med.* 80:82–83, 1880.

448 GRUBER, W. Doppeldaumen besonderer Form. *Arch. Pathol. Anat. Physiol. Klin. Med.* 80:94–95, 1880.

449 GRUBER, W. Zergliederung des Doppeldaumens beider Hände einer Frau. *Arch. Pathol. Anat. Physiol. Klin. Med.* 86:495–505, 1881.

450 GRUBER, W. Zergliederung zweier Füße Erwachsener mit doppelter kleiner (und daran innerer supernumerärer) Zehe. *Arch. Pathol. Anat. Physiol. Klin. Med.* 86:505–510, 1881.

451 GRUBER, W. Navicularis carpi tripartita. *Arch. Pathol. Anat. Physiol. Klin. Med.* 94:343–348, 1883.

452 GRUBER, W. Os lunatum carpi mit einem Anhange am dorsalen Ende. Lunatum bipartitum 3. Art. *Arch. Pathol. Anat. Physiol. Klin. Med.* 94:349–353, 1883.

453 GRUBER, W. Ein im Centrum der Ulnarportion des Rückens des menschlichen Carpus zwischen dem Lunatum, Triquetrum und Hamatum gelagertes und articulirendes Ossiculum supernumerarium. Neue. (6). *Arch. Pathol. Anat. Physiol. Klin. Med.* 94:353–357, 1883.

454 GRUBER, W. In Bildungsanomalien mit Bildungshemmung begründete Bipartition beider Patellae eines jungen Subjectes. *Arch. Pathol. Anat. Physiol. Klin. Med.* 94:358–361, 1883.

455 GRUBER, W. Zergliederung des rechten Armes mit Duplicität des Daumens von einer Frau und von einem Jüngling. *Arch. Pathol. Anat. Physiol. Klin. Med.* 95:186–194, 1884.

456* GRUBER, W. Bericht über neue Befunde des Os centrale carpi radiale (mihi) beim Menschen. *Arch. Pathol. Anat. Physiol. Klin. Med.* 98:396–402, 1884.

457* GRUBER, W. Über das Os centrale carpi ulnare (mihi) bei dem Menschen und über das muthmaßlich homologe Carpalstück bei den Amphibien. *Arch. Pathol. Anat. Physiol. Klin. Med.* 98:402–408, 1884.

458 GRUBER, W. Drei neue Fälle von Os lunatum carpi bipartitum und ein Fall von Os lunatum tripartitum (vorher nicht gesehen) – Verhalten des Os lunatum secundarium dorsale wie ein "Os centrale carpi medium" in einem veröffentlichtem Falle und in den neuen Fällen. *Arch. Pathol. Anat. Physiol. Klin. Med.* 98:408–413, 1884.

459 GRUBER, W. Ossification an ungewöhnlichen Orten (vom Aussehen eines Processus am Triquetrum carpi und eines Ossiculum sesemoides in der Ursprungssehne des Musculus rectus femoris). *Arch. Pathol. Anat. Physiol. Klin. Med.* 98:413–416, 1884.

460 GRUBER, W. Auftreten der Tuberositas des Os metatarsale V. sowohl als persistirende Epiphyse, als auch mit einer an ihrem äußeren Umfange aufsitzenden Epiphyse. *Arch. Pathol. Anat. Physiol. Klin. Med.* 99:460–471, 1885.

461* GRUBER, W. Substitutio der Spitze des Processus styloides des

Metacarpale III durch ein articulierendes Ossiculum carpi supernumerarium. *Arch. Pathol. Anat. Physiol. Klin. Med.* 102:535–536, 1885.

462 GRUBER, W. Os centrale carpi ulnare (mihi). *Arch. Pathol. Anat. Physiol. Klin. Med.* 107:492–493, 1887.

463 GRUBER, W. Bipartition des Os multangulum minus in ein M. secundarium dorsale et volare. *Arch. Pathol. Anat. Physiol. Klin. Med.* 110:551–554, 1887.

464 GRUBER, W. Ein Corpusculum articulare mobile (Ossiculum carpi supernumerarium?) in der Carpo-Metacarpal-Amphiarthrose. *Arch. Pathol. Anat. Physiol. Klin. Med.* 110:549–551, 1887.

465 GRUBER, W. Über Fälle von Teilung des Sinus maxillaris durch ein Septum osseum perfuctum in zwei von einander völlig abgeschlossene Sinus maxillaris secundarii mit separaten Öffnungen in den Meatus narium medius. *Arch. Pathol. Anat. Physiol. Klin. Med.* 113:530–533, 1888.

466 GRUBER, W. Über Duplicität des Foramen rotundum ossis sphenoides. *Arch. Pathol. Anat. Physiol. Klinische Med.* 113:533, 1888.

467 GUILLEMINOT, J., NAOURI, G. and J.P. VINCENT. Un cas de malformation congénitale rare: L'agénésie du pubis. *Rev. Chir. Orthop.* 40:442–444, 1954.

468 GÜNSEL, E. Persistierende Apophyse des Epicondylus medialis humeri. *Fortschr. Röntgenstr.* 76:660–661, 1952.

469 GÜNSEL, E. Ein großer Processus styloideus an der Lendenwirbelsäule. *Fortschr. Röntgenstr.* 79:245–246, 1953.

470 GÜNSEL, E. Das Os coracoideum. *Fortschr. Röntgenstr.* 74:112–113, 1954.

471 GUPTA, S.C., GUPTA, C.D., ARORA, A.K. and B.B. MAHESHWARI. The retrotransverse groove (canal) in the Indian atlas vertebrae. *Anat. Anz.* 145:514–516, 1979.

472 GÜRKAN, K.I. Aplasie totale du tibia. *Rev. Orthop.* 35:389–390, 1949.

473 GUSTAFSON, G. Age determinations on teeth. *J. Am. Dent. Assoc.* 41:45–54, 1950.

474* GUSTAFSON, G. and S. SUNDBERG. Dens in dente. *Br. Dent. J.* 88:83–88, 1950.

475 HAAS, S.L. Three-phalangeal thumbs. *Am. J. Roentgenol.* 42:677–682, 1939.

476 HADLEY, L.A. Atlanto-occipital fusion, ossiculum terminale and occipital vertebra as related to basilar impression with neurological symptoms. *Am. J. Roentgenol.* 59:511–524, 1948.

477 HADLEY, L.A. Accessory sacroiliac articulations with arthritic changes. *Radiology* 55:403–409, 1950.

478 HADLICH, R. Eine vierfingerige rechte Hand als congenitale Mißbildung. *Arch. Pathol. Anat. Physiol. Klin. Med.* 174:392–401, 1903.

479 HAFFERL, A. Über einen abnormen Knochenkanal am unteren Ende der Tibia des Menschen. *Anat. Anz.* 46:271–272, 1914.

480 HAID, B. Beobachtung einer neuen Form des Os intermetatarseum. *Z. Orthop.* 80:298–303, 1950/51.

481* HALEY, J.C. A study of the extent, with variations, of the bony framework of the paranasal sinuses. *Med. Rec. Ann.* 42:693–696, 1948.

482 HAMMER, G. and C. RÅDBERG. The sphenoidal sinus. An anatomical and roentgenologic study with reference to transsphenoid hypophysectomy. *Acta Radiol.* 56:401–422, 1961.

483 HAMSA, W.R. Congenital absence of the sacrum. *Arch. Surg.* 30:657–666, 1935.

484* HANDFORTH, J.R. Polydactylism in the hand in southern Chinese. *Anat. Rec.* 106:119–125, 1950.

485 HANSON, R. Some anomalies, deformities and diseased conditions of the vertebrae during their different stages of development, elucidated by anatomical and radiological findings. *Acta Chir. Scand.* 60:309–368, 1926.

486 HARRIS, H. and J. JOSEPH. Variation in extension of the metacarpophalangeal and interphalangeal joints of the thumb. *J. Bone Joint Surg. (Br.)* 31:547–559, 1949.

487 HARRIS, H.A. Ossification in the lumbo-sacral region. *Br. J. Radiol.* 6:685–688, 1933.

488 HARRISON, R.G. On the occurrence of tails in man, with a description of the case reported by Dr. Watson. *Johns Hopkins Hosp. Bull.* 12:96–101, 1901.

489* HARRISON, R.G., PEARSON, M.A. and R. ROAF. Ulnar dimelia. *J. Bone Joint Surg. (Br.)* 42:549–555, 1960.

490* HARROWER, G. Variations in the region of the foramen magnum. *J. Anat.* 57:178–192, 1923.

491* HARROWER, G. A septidigitate foot in man. *J. Anat.* 60:106–109, 1925.

492* HASS, L. Über die sutura frontalis persistens. Eine röntgenanatomische Studie. *Fortschr. Röntgenstr.* 48:708–716, 1933.

493* HASS, L. Die supraorbitalen Pneumatisationen im Röntgenbilde. *Fortschr. Röntgenstr.* 50:71–78, 1934.

494 HAUCH, P.P. The fate of an accessory ossicle. *Br. J. Radiol.* 19:518–519, 1946.

495 HEFNER, R.A. Inherited abnormalities of the fingers. *J. Hered.* 15:323–329, 1924.

496 HEIMERZHEIM, A. Über einen seltsamen Knochenbefund am Calcaneus. *Dtsch. Z. Chir.* 187:281–283, 1924.

497* HELLER, L. and J. LANGMAN. The menisco-femoral ligaments of the human knee. *J. Bone Joint Surg. (Br.)* 46:307–313, 1964.

498* HENKEL, L. and H.G. WILLERT. Dysmelia. A classification and a pattern of malformation in a group of congenital defects of the limbs. *J. Bone Joint Surg. (Br.)* 51:399–414, 1969.

499 HENNECKE, U. Os radiale externum. *Fortschr. Röntgenstr.* 78:362, 1953.

500 HEPBURN, D. Anomalies in the supra-inial portion of the occipital bone, resulting from irregularities of its ossification, with constant variations in the interparietal bone. *J. Anat. Physiol.* 42:88–92, 1908.

501 HERDNER. Le sternum de l'enfant. Fréquence actuelle de troubles de dévelopment révélés par l'examen radiographique. *J. Radiol. Electrol.* 28:387–393, 1947.

502 HERDNER, M. Le sternum de l'enfant étude radiologique des anomalies de son développement. *Rev. Orthop.* 33:475–493, 1947.

503 HERPIN, A. Des anomalies maxillo-dentaires. *Assoc. Anatomistes Comptes Rendus* 21:270–272, 1926.

504 HERTSLET, L.E. and A. KEITH. Comparison of anomalous parts of two subjects, one with a cervical rib, the other with a rudimentary first rib. *J. Anat. Physiol.* 30:562–567, 1896.

505 HESS, L. The metopic suture and the metopic syndrome. *Hum. Biol.* 17:107–136, 1945.

506* HESS, L. Ossicula wormiana. *Hum. Biol.* 18:61–80, 1946.

507 HEYNOLD, H. Ein Fall von an allen vier Extremitäten gleichmäßig vorhandenen überzähligen Fingern und Zehen. *Arch. Pathol. Anat. Physiol. Klin. Med.* 72:502–503, 1878.

508 HEYNS, O.S. and J.E. KERRICH. The number of vertebrae in the fetal Bantu sacrum. *Am. J. Phys. Anthropol.* 5:67–78, 1947.

509 HILTEMANN, H. Fonticulus metopicus und Sutura frontalis persistens mit Hypoplasie der Sinus frontales. *Fortschr. Röntgenstr.* 81:407–409, 1954.

510 VON HOCHSTETTER, A. Ein Fall von "Articulus talotarsalis communis." *Anat. Anz.* 99:337–342, 1952–53.

511* HOFFMANN, L. Mißbildungen der oberen Extremität. *Fortschr. Röntgenstr.* 17:301–306, 1911.

512 HOHL, K. Das Os odontoideum (partielle Densaplasie). *Fortschr. Röntgenstr.* 91:518–521, 1959.

513 HOLL, M. Beitrag zu den Abnormitäten der Wirbelknochen. *Med. Jahrb. (Wien)* 1880, pp. 461–466.

514 HOLL, M. Über eine angeborene Coalition des Os lunatum und Os triquetrum carpi. *Med. Jahrb. (Wien)* 1882, pp. 499–502.

515* HORI, T. Über die Anomalien des Hinterhauptbeines. *Folia Anat. Jpn.* 3:291–312, 1925.

516* HOVELACQUE, A. Etude anatomo-pathologique de l'exstrophie complète de la vessie. *J. Urol. (Paris)* 1:205–234, 1912.

517 HRDLICKA, A. Incidence of the supracondyloid process in whites and other races. *Am. J. Phys. Anthropol.* 6:405–412, 1923.

518* HRDLICKA, A. Lower jaw: Double condyles. *Am. J. Phys. Anthropol.* 28:75–89, 1941.

519 HRDLICKA, A. The scapula: Visual observations. *Am. J. Phys. Anthropol.* 29:73–94, 1942.

520 HRDLICKA, A. The adult scapula. Additional observations and measurements. *Am. J. Phys. Anthropol.* 29:363–415, 1942.

521 HUBAY, C.A. Sesamoid bones of the hands and feet. *Am. J. Roentgenol.* 61:493–505, 1949.

522 HÜLSHOFF, T. Eine seltene Rippenanomalie. *Fortschr. Röntgenstr.* 103:231, 1965.

523 HULTKRANTZ, J. Über congenitalen Schlüsselbeindefect und damit verbundene Schädelanomalien. *Anat. Anz.* 15:237–241, 1899.

524 HUMPHRY, G.M. Loose bodies in joints. *J. Anat. Physiol.* 23:493–503, 1889.

525 HUNTER, R.H. An abnormal atlas. *J. Anat.* 58:140–141, 1924.

526 HUXLEY, T.H. On two widely contrasted forms of the human cranium. *J. Anat. Physiol.* 1:60–77, 1868.

527 HYMAN, C. Congenital defect of the sternum with congenital heart disease. *J. Med. Soc. NJ* 44:355–356, 1947.

528 HYRTL. Pneumatische Hinterhauptsknochen. *Wien. Med. Wochenschr.* 45:713–714, 1860.

529* INGELMARK, B.E. Über das craniovertebrale Grenzgebiet beim Menschen. *Acta Anat. [Suppl.]* 6:1–116, 1947.
530 INGHAM, F.L. Cranio-cleido-dysostosis. *Br. J. Radiol.* 20:332–334, 1947.
531 INUMARU, H. Über das Foramen mastoideum. *Folia Anat. Jpn.* 3:229–239, 1925.
532 IRNBERGER, T. Beobachtung eines Rippenrudimentes am menschlichen Steißbein. *Anat. Anz.* 86:369–378, 1938.
533 IVIE, J. McK. Congenital absence of the odontoid process. Report of a case. *Radiology* 46:268–269, 1946.
534 JACKSON, W.P.U. Osteo-dental dysplasia (cleido-cranial dysostosis). *Acta Med. Scand.* 139:292–307, 1951.
535 JACQUES, P. Anomalie rare du sinus latéral (S.L. sous-cutané). *Soc. Biol. Comptes Rendus Hebdomadaires des Séances et Mémoires* 83:360–361, 1920.
536 JAIN, S.P. A rare and unusual variety of atlanto-occipital fusion. *J. Anat. Soc. India* 5:72–73, 1956.
537 JANCU, F. and V. CHIRIAC. Foramina Transversaria im 1. Sakralwirbel. *Anat. Anz.* 115:403–404, 1964.
538 JANKER, R. Eine anatomische Variante am Querfortsatz des Atlas. *Röntgenpraxis* 7:399, 1935.
539 JANTZ, H. Fenestrae parietales symmetricae (permagnae). *Med. Monatsschr.* 2:241–243, 1948. Cited in *Excerpta Medica*, Sec. 1, Vol. 3, abstract 778, 1949.
540 JEFFERISS, F.B. A case of incomplete development of the third and fourth ribs. *Lancet* 1:1437–1438, 1900.
541 JEFFERSON, G. The morphology of the sulcus interparietalis (B.N.A.). *J. Anat. Physiol.* 47:365–380, 1913.
542 JELISIEJEW, T., SZMURLO, J. and I. KUDUK. Morphologic studies on the hyoid bone in man. *Folia Morphol.* 27:172–182, 1968.
543* JELISIEJEW, T. and J. SZMURLO. Structural variations of the thyroid cartilage. *Folia Morphol.* 31:117–122, 1972.
544 JEYASINGH, P., GUPTA, C.D., ARORA, A.K. and S.K. SAXENA. Study of os japonicum in Uttar Pradesh crania. *Anat. Anz.* 152:27–30, 1982.
545 JOACHIMSTHAL. Verdoppelung des linken Zeigefingers und Dreigliederung des rechten Daumens. *Berliner Klin. Wochenschr.* 37:835–838, 1900.
546 JOHNSON, W.T. Unusual variation in identical twins and its etiology. *Dental Rec.* 66:307–310, 1946.
547 JOHNSTON, H.M. Epilunar and hypolunar ossicles, division of the scaphoid, and other anomalies in the carpal region. (Illustrated by drawings from two specimens.) *J. Anat. Physiol.* 41:59–65, 1907.
548 JONES, H.C. and D.W. HEDRICK. Patellar anomalies, roentgenologic and clinical consideration. *Radiology* 38:30–34, 1942.
549 JONES, F.W. On the grooves upon the ossa parietalia commonly said to be caused by the arteria meningia media. *J. Anat. Physiol.* 46:228–238, 1912.
550* JOSEPH, J. The sesamoid bones of the hand and the time of fusion of the epiphyses of the thumb. *J. Anat.* 85:230–241, 1951.
551 JOVANOVIC, S. Supernumerary frontal sinuses on the roof of the orbit; their clinical significance. *Acta Anat.* 45:133–142, 1961.
552 KACZANDER, J. Über angeborenen Radiusmangel. *Arch. Pathol. Anat. Physiol. Klin. Med.* 71:409–413, 1877.
553 KANAVEL, A.B. Congenital malformations of the hands. *Arch. Surg.* 25:1–53, 283–320, 1932.
554 KAPLAN, E.B. Discoid lateral meniscus of the knee joint. *J. Bone Joint Surg. (Am.)* 39:77–87, 1957.
555 KARLEN, A. Congenital hypoplasia of the odontoid process. *J. Bone Joint Surg. (Am.)* 44:567–570, 1962.
556* KASPER, K.A. Nasofrontal connections. A study based on one hundred consecutive dissections. *Arch. Otolaryngol.* 23:322–343, 1936.
557 KASSATKIN, S. Die Sesambeine der Hand und des Fußes beim Menschen. *Z. Anat. Entwicklungsgesch.* 102:635–654, 1934.
558 KATE, B.R. A study of the regional variation of the Indian femur – the diameter of the head – its medico legal and surgical application. *J. Anat. Soc. India* 13:80–84, 1964.
559* KATO, K. Congenital absence of the radius. *J. Bone Joint Surg.* 6:589–626, 1924.
560 KATZ, J.F. Congenital absence of the sacrum and coccyx. *J. Bone Joint Surg. (Am.)* 35:398–402, 1953.
561 KEATING, D.R. and J.R. AMBERG. A source of potential error in the roentgen diagnosis of cervical ribs. *Radiology* 62:688–694, 1954.
562 KEATS, T.E. Four normal anatomic variations of importance to radiologists. *Am. J. Roentgenol.* 78:89–94, 1957.
563 KELLER, H.L. Eine seltene Form der Manifestation des Okzipitalwirbels. *Fortschr. Röntgenstr.* 93:370–372, 1960.
564 KELLER, H.L. Varianten der Arteria carotis interna, der Arteria meningea und der Arteria ophthalmica im Karotisangiogramm. *Fortschr. Röntgenstr.* 95:472–482, 1961.
565 KEMPSON, F.C. Emargination of the patella. *J. Anat. Physiol.* 36:419–420, 1902.
566 KENDRICK, G.S. and N.L. BIGGS. Incidence of the ponticulus posticus of the first cervical vertebra between the ages of six to seventeen. *Anat. Rec.* 145:449–453, 1963.
567* KERNER, D. Andreas Vesalius. 400 Jahre deskriptive Anatomie. *Fortschr. Röntgenstr.* 85:629–632, 1956.
568* KEWENTER, Y. Die Sesambeine des I. Metatarsophalangealgelenks des Menschen. Eine röntgenologische, klinische und pathologisch-histologische Studie. *Acta Orthop. Scand. [Suppl.]* 2:1–113, 1936.
569 KHOO, F.Y. Giant jugular fossa, with brief notes on the anatomical variations of the jugular fossa. *Am. J. Roentgenol.* 55:333–336, 1946.
570 KIDD, W. Abnormal phalanges in a human hand: A case of brachydactyly. *J. Anat. Physiol.* 44:64–66, 1910.
571 KIEFER, H. and J. EMMRICH. Abnorm großer Processus styloides der Lendenwirbelsäule mit reaktiver Pseudarthrosebildung. *Fortschr. Röntgenstr.* 100:280–281, 1964.
572 KIENBÖCK, R. Über angeborene Rippenanomalien. *Fortschr. Röntgenstr.* 13:269–298, 1908–09.
573 KIENBÖCK, R. Anomalie du Squelette de L'Articulation du Coude Ellenbogengelenks. Os sesamoideum de cubitus. *La Presse Médicale* 45:491–492, 1937.
574 KIENBÖCK, R. and G. DESENFANS. Über Anomalien am Ellenbogengelenk. Patella cubiti. *Brun's Beitr. Klin. Chir.* 165:524–529, 1937.
575 KIENBÖCK, R. and W. MÜLLER. Os tibiale externum und Verletzung des Fußes. *Z. Orthop. Chir.* 66:257–270, 1937.
576 KIPSHOVEN, H.J. Die röntgenologische Darstellung der Ossa suprasternalia. *Fortschr. Röntgenstr.* 74:320–323, 1951.
577 KIRCHMAIR, H. Abnorme Muskelapophyse am proximalen Humerusende. *Anat. Anz.* 99:332–336, 1952–53.
578 KIRCHHOFF, H.J. and H.H. ROHWEDDER. Über Mißbildungen der Wirbelsäule des Säuglings (ein klinischer Beitrag). *Arch. Kinderheilk.* 148:146–161, 1954.
579 KIRMISSION, E. Pouces a trois phalanges. *Rev. Orthop. (Paris)* 10:249–253, 1909.
580 KISTHINIOS, M. Quelques cas d'anomalies craniennes. *Ann. Anat. Pathol.* 5:582–584, 1928.
581 KITCHIN, C.P. Dens in dente. *Oral. Surg.* 2:1181–1193, 1949.
582 KLINEFELTER, E.W. The anomalous hyoid. Review of the literature and report of a case. *Radiology* 58:224–227, 1952.
583* KLINKE, K. and H. PAHLKE. Dysostosis cleidocranialis. Bericht über 2 Fälle. *Arch. Kinderheilk* 91:46–54, 1930.
584* KLÖPPNER, K. Die Sella turcica des Neugeborenen im Röntgenbild (Größe, Kontur, Form und Formvarianten der Sella turcica). *Fortschr. Röntgenstr.* 60:370–379, 1939.
585 KNIGHT, G.A.M. and G.H. MORLEY. Cleft sternum: Case report and brief commentary. *Br. J. Surg.* 24:60–64, 1936–37.
586 KOBLMÜLLER, L. Über einen Fall von Condylus tertius. *Anat. Anz.* 71:347–350, 1930–31.
587 KOERBIN. Anomale Nahtbildung am Oberkiefer-Stirnfortsatz. *Arch. Pathol. Anat. Physiol. Klin. Med.* 73:121–125, 1878.
588* KOLLMANN, J. Varianten am Os occipitale, besonders in der Umgebung des Foramen occipitale magnum. *Anat. Anz.* 30:545–563, 1907.
589 KOLTE, D.T. and V.R. MYSOREKAR. Cervico-occipital fusions. *J. Anat. Soc. India* 12:36–39, 1961.
590 KOPYLOW, M.B. Ein neues Verfahren zur röntgenologischen Darstellung des Canalis nasolacrimalis. *Röntgenpraxis* 2:686–692, 1930.
591* KOSKI, K. and S.M. GARN. Tooth eruption sequence in fossil and modern man. *Am. J. Phys. Anthropol.* 15:469–488, 1957.
592 KOSTANECKI, K. VON. Beiträge zur Kenntnis der Mißbildungen in der Kopf- und Halsgegend. *Arch. Pathol. Anat. Physiol. Klin. Med.* 123:401–428, 1891.
593* KRASTIN, L. Haemapophysis beim Menschen. *Anat. Anz.* 65:222–229, 1928.
594 KREMSER, K. Ein Beitrag zur Kasuistik der Skelettanomalie (Ellenbogenscheibe). *Röntgenpraxis* 6:371–374, 1934.
595 KREMSER, K. Os accessorium supracalcaneum. *Fortschr. Röntgenstr.* 82:279, 1955.
596 KREMSER, K. Os supratalare. *Fortschr. Röntgenstr.* 85:116–117, 1956.

597 KRISCH, H. Rudimentäre Halsrippen beiderseits und Otosklerose. *Dtsch. Med. Wochenschr.* 44:670, 1918.
598* KRMPOTIĆ-BOHACEK, J. Jugum Cerebellare intersemilunare. *J. Anat.* 87:358–361, 1953.
599* KUCZYNSKI, K. Development of the hand and some anatomical anomalies. *Hand* 4:1–9, 1972.
600* KUHNT. Eigenthümliche Doppelbildungen an Händen und Füßen. *Arch. Pathol. Anat. Physiol. Klin. Med.* 56:268–269, 1872.
601 KULLNIG, G. Persistierender offener Ductus craniopharyngicus. *Fortschr. Röntgenstr.* 79:127, 1953.
602* KURLEJ, W., GOŹDZIEWSKI, S. and J. MAREK. Morphology of the incisive fossa, canal and foramen in man. *Folia Morphol.* 42:129–138, 1983.
603 KURZ, E. Die Wirbelsäule einer 25jährigen Chinesin. *Anat. Anz.* 49:195–209, 1916–17.
604 KUTZ, E.R. Congenital absence of the patellae. *J. Pediatr.* 34:760–762, 1949.
605 KVASNICKA, I. Processus paramastoideus (P.p.) als Ursache einer schiefen Kopfhaltung. *Fortschr. Röntgenstr.* 88:744–746, 1958.
606 LABE, E. Cas de rein unique chez l'homme. *Soc. Biol. Comptes Rendus Hebdomadaires des Séances et Mémoires* 6:57–58, 1914.
607 LACHAPÈLE, A.P. and J. DE GROC. Suture supplémentaire de l'écaille occipitale: "os interpariétal," "os epactal." *J. Radiol. Electrol.* 27:461–463, 1946.
608 LACHI, P. Di una "fossetta del tubercolo pterigoideo" nel cranio umano. *Arch. Ital. Anat. Embriol.* 13:273–283, 1914–15.
609 LACHI, P. Sul significato dei canali basilari dell'osso occipitale dell'uomo. *Arch. Ital. Anat. Embriol.* 17:48–64, 1919–20.
610 LAIDLAW, P.P. The varieties of os calcis. *J. Anat. Physiol.* 38:133–143, 1904.
611 LAIDLAW, P.P. The os calcis. Part II. *J. Anat. Physiol.* 39:161–177, 1905.
612 LAMB, D. The ball and socket ankle joint – a congenital abnormality. *J. Bone Joint Surg. (Br.)* 40:240–243, 1958.
613 LANE, W.A. Supernumerary cervico-dorsal vertebra – bearing ribs, with vertebral asymmetry; abnormal articulation in a sternum. *J. Anat. Physiol.* 19:266–273, 1885.
614 LANGE, B. Ein menschlicher Schädel mit auffallendem Schläfenzwischenknochen. *Anat. Anz.* 59:197–199, 1924–25.
615 LANGE, K. Das Naviculare bipartitum. *Röntgenpraxis* 11:566–567, 1939.
616 LANGSTON, H.H. Congenital defect of the shaft of the femur. *Br. J. Surg.* 27:162–165, 1939–40.
617* LANIER, R.R. JR. The presacral vertebrae of American white and Negro males. *Am. J. Phys. Anthropol.* 25:341–420, 1939.
618 LANIER, R.R. JR. An anomalous cervico-occipital skeleton in man. *Anat. Rec.* 73:189–201, 1939.
619 LAPAYOWKER, M.S. An unusual variant of the cervical spine. *Am. J. Roentgenol.* 83:656–659, 1960.
620 LAPIDUS, P.W., GUIDOTTI, F.P. and C.J. COLETTI. Triphalangeal thumb. Report of six cases. *Surg. Gynecol. Obstet.* 77:178–186, 1943.
621 LAPIDUS, P.W. and G. GUIDOTTI. Triphalangeal bifid thumb. *Arch. Surg.* 49:228–234, 1944.
622 LARSEN, L.L. and H.F. IBACH. Complete congenital fissure of the sternum. *Am. J. Roentgenol.* 87:1062–1963, 1962.
623 LAURENT, Y. and F. CAELS. Processus paracondylicus mit Gelenkbildung. *Fortschr. Röntgenstr.* 93:136–137, 1960.
624 LAUSECKER, H. Der angeborene Defekt der Ulna. *Arch. Pathol. Anat. Physiol. Klin. Med.* 325:211–226, 1954.
625 LAZORTHES, G., POULHÈS, J. and G. BASTIDE. Etude sur 80 bassins de l'aire du détroit supérieur. *Assoc. Anatomistes Comptes Rendus* 39:295–302, 1952.
626 LAZORTHES, G., GAYRAL, L. and R. HURON. Etude anatomique de la fente sphénoidale. *Assoc. Anatomistes Comptes Rendus* 40:165–168, 1953.
627 LEDENYI, J. Ankylosis articuli atlantooccipitalis sinistra. *Anat. Anz.* 88:503–505, 1939.
628 LEDOUBLE. Sur quelques variations des trous optiques. *Assoc. Anatomistes Comptes Rendus* 4:209–212, 1902.
629 LEDOUBLE. A propos d'un cas de communication de la fente sphenoidale et du trou grand rond de l'alisphénoïde humain. *Assoc. Anatomistes Comptes Rendus* 4:207–208, 1902.
630 LEDOUBLE. Sillon temporo-pariétal externe. *Assoc. Anatomistes Comptes Rendus* 4:205–206, 1902.
631 LEHMANN-NITSCHE, R. Ein seltener Fall von angeborener medianer Spaltung der oberen Gesichtshälfte. *Arch. Pathol. Anat. Physiol. Klin. Med.* 163:126–134, 1901.
632 LEMAIRE, J. Malformations congénitales, familiales, héréditaires des mains et des pieds. *Rev. Orthop.* 10:267–273, 1909.
633* LETTERMAN, G.S. and M. TROTTER. Variations of the male sacrum. Their significance in caudal anesthesia. *Surg. Gynecol. Obstet.* 78:551–555, 1944.
634 LEVINGER, E. Mißbildung: Defekt des Humerus und Schultergürtels, Fehlen von Radius und Ulna, nur ein Finger vorhanden. *Anat. Anz.* 61:78–83, 1926.
635 LEWIN, T. Anatomical variations in lumbo-sacral synovial joints with particular references to subluxation. *Acta Anat.* 71:229–248, 1968.
636 LICHTOR, A. Sacral agenesis. *Arch. Surg.* 54:430–433, 1947.
637 LICKLEY, J.D. On the relations of the seventh and eighth ribs to the sternum. *Anat. Anz.* 24:326–331, 1904.
638 LILLIE, R.D. Variations of the canalis hypoglossi. *Anat. Rec.* 13:131–144, 1917.
639 VON DEN LINDEN and W.H.F. HELANDER. Synostosis lunatotriquetra. A case report of a rare congenital malformation. *Anat. Anz.* 151:101–103, 1982.
640 LISCHI, G. Foramina parietalia permagna (fenestrae parietales symmetricae) Rara malformazione ereditaria del cranio. *Radiol. Med.* 35:445–451, 1949.
641 LISCHI, G. Persistierender Fonticulus metopicus beim Erwachsenen. *Fortschr. Röntgenstr.* 92:460–461, 1960.
642 LOCKHART, R.D. Variations coincident with congenital absence of the zygoma (zygomatic process of the temporal bone). *J. Anat.* 58:233–236, 1929.
643* LOMBARDI, G. The occipital vertebra. *Am. J. Roentgenol.* 86:260–269, 1961.
644* LOSSEN, H. and R. HOFER. Ossa suprasternalia im Röntgenbild. *Röntgenpraxis* 3:34–36, 1931.
645 LOW, A. Description of a specimen in which there is a rudimentary first rib along with thirteen pairs of ribs and twenty-five presacral vertebrae. *J. Anat. Physiol.* 34:451–457, 1900.
646 LOWRANCE, E.W. and H.B. LATIMER. Weights and variability of components of the human vertebral column. *Anat. Rec.* 159:83–88, 1967.
647 LUCAS, M.F. (1) Two cases of cervical ribs. (2) An anomalous arrangement of the vagi. *J. Anat. Physiol.* 69:336–342, 1915.
648 LUCIEN. A propos du processus rétromastoideus chez l'homme. *Soc. Biol. Comptes Rendus Hebdomadaires des Séances et Mémoires* 84:803–805, 1921.
649 LUNGHETTI, B. Sur quelques cas d'Ossiculum intermetatarseum dorsale de Gruber. *Arch. Ital. Biol.* 47:480, 1907.
650 LUSCHKA, H. VON. Über angeborene Atresie der Choanen. *Arch. Pathol. Anat. Physiol. Klin. Med.* 18:168, 1860.
651 LUSCHKA, H. VON. Über den Mangel eines continuirlichen oberen Hornes der Cartilago thyroidea des Menschen. *Arch. Pathol. Anat. Physiol. Klin. Med.* 42:478–479, 1868.
652 LUSCHKA, H. VON. Der Processus marginalis des menschlichen Jochbeins. *Arch. Anat. Physiol. Wissen. Med.* 1869, pp. 326–330.
653 MACALISTER, A. Sesamoid bone in the tendon of the supinator brevis. *J. Anat. Physiol.* 3:108, 1869.
654* MACALISTER, A. Notes on the acromion. *J. Anat. Physiol.* 27:245–251, 1893.
655* MACALISTER, A. Notes on the development and variations of the atlas. *J. Anat. Physiol.* 27:519–542, 1893.
656 MACALISTER, A. The development and variations of the second cervical vertebrae. *J. Anat. Physiol.* 28:257–268, 1894.
657 MACAULAY, D. Digital markings in radiographs of the skull in children. *Br. J. Radiol.* 24:647–652, 1951.
658 MAGAUZZI-VALERI, R. Sur un cas d'articulation costo-claviculaire. *Arch. Ital. Biol.* 62:283, 1914.
659 MAGGI, L. Osselets fontellaires coronaux et lambdoïdiens dans le crâne de mammifères et de l'homme. *Arch. Ital. Biol.* 36:336–337, 1901.
660 MAHANTI, R. A case of acheiriea: Congenital absence of upper limbs. *J. Anat. Soc. India* 111:52, 1954.
661 MAIER, K. Über die Möglichkeit einer Verschmelzung des Os trigonum mit dem Kalkaneus. *Fortschr. Röntgenstr.* 92:715–717, 1960.
662 MAIER, K. Gibt es eine Navikulare-Aplasie an der Hand? *Fortschr. Röntgenstr.* 97:52–56, 1962.
663 MAIR, J. Zur Kenntnis der Fontanella metopica und der Sternnacht. *Anat. Anz.* 57:149–157, 1923.
664* MANDRUZZATO, F. Patologia e chirurgia del processo sopra epitrocleare. dell' omero. *Chir. Org. Movimento* 24:123–132, 1938.
665 MANNERS-SMITH, T. The variability of the last lumbar vertebra. *J. Anat. Physiol.* 43:146–160, 1909.

666 MANNU, A. Un cas indiscutable de manifestation de la vertèbre occipitale chez l'homme. *Arch. Ital. Biol.* 53:147–148, 1910.

667 MANOHAR, K. Congenital absence of the right femur. *Br. J. Surg.* 27:158–161, 1939–40.

668 MARCUS, G. Über einige Fälle von Pneumatisation des Orbitaldaches (Recessus supraorbitalis). *Anat. Anz.* 76:33–45, 1933.

669 MARRO, G. Sur quelques anomalies singulières dans des os humains. *Arch. Ital. Biol.* 41:469–470, 1904.

670 MARRO, G. Sur la division du pariétal. Trois observations originales dans des crâne d'idiots. *Arch. Ital. Biol.* 45:279, 1906.

671 MARRO, G. Sur la division de l'os malaire. *Arch. Ital. Biol.* 47:477, 1907.

672 MARRO, G. Contribution à la casuistique et la interprétation des anomalies de l'arcade zygomatique. *Arch. Ital. Biol.* 47:476–477, 1907.

673 MARRO, G. Sur la foveola coccygea. *Arch. Ital. Biol.* 50:156, 1908.

674 MARRO, G. Sur ce qu'on appelle la perforation olécrânienne et sur sa signification anatomique et anthropologique. *Arch. Ital. Biol.* 62:282–283, 1914.

675 MARTI, T. Über das Os centrale carpi. *Schweiz. Med. Wochenschr.* 80:280–281, 1950.

676 MARTI, T. Os cuneo-metatarsale I dorsale – fibulare. *Fortschr. Röntgenstr.* 81:830–831, 1954.

677 MARTI, T. Über das Os centrale carpi. *Fortschr. Röntgenstr.* 82:123–124, 1955.

678 MARTI, T. Über den Calcaneus secundarius. *Fortschr. Röntgenstr.* 82:124–125, 1955.

679 MASSLER, M. and I. SCHOUR. Growth of the child and the calcification pattern of the teeth. *Am. J. Orthodont. Oral Surg.* 32:495–517, 1946.

680 MATIEGKA, H. Über das "Os malare bipartitum." *Anat. Anz.* 16:546–557, 1899.

681 MATIEGKA, J. Sur les causes du developpement des os intercalaires. *Assoc. Anatomistes Comptes Rendus* 23:295–300, 1928.

682 MATIEGKA, J. Démonstration de quelques crânes présentant des anomalies de suture. *Assoc. Anatomistes Comptes Rendus* 23:555–557, 1928.

683 MAURER, H.J. Ungewöhnliche Form eines Tuberculum obturatorium anterius. *Fortschr. Röntgenstr.* 83:889, 1955.

684 MAURER, H.J. Zur Frage Os coronoides ulnae. *Fortschr. Röntgenstr.* 90:264–266, 1959.

685 MAURER, H.J. Os coronoides ulnae. *Fortschr. Röntgenstr.* 96:572–573, 1962.

686 MCDOUGALL, A. The os trigonum. *J. Bone Joint Surg. (Br.)* 37:257–265, 1955.

687 MCRAE, D.L. Bony abnormalities in the region of the foramen magnum; correlation of the anatomic and neurologic findings. *Acta Radiol.* 40:335–354, 1953.

688 MCRAE, D.L. and A.S. BARNUM. Occipitalization of atlas. *Am. J. Roentgenol.* 70:23–46, 1953.

689 MENZEL, K.M. Über das Vorkommen von überzähligen muschelartigen Gebilden im hinteren Teil des mittleren Nasenganges. *Anat. Anz.* 78:229–240, 1934.

690 MERCER, J. The secondary os calcis. *J. Anat.* 66:84–97, 1932.

691 MESCHEDE, F. Die Verrenkung der ersten Daumenphalanx nach der Volarseite. *Arch. Pathol. Anat. Physiol. Klin. Med.* 37:510–518, 1866.

692 MEX, W. Ungewöhnliche Form zweier Halsrippen. *Monatsschr. Unfallheilk. Versicherungsmed.* 62:307–310, 1959.

693 MEYER, A.W. Our inadequate terminology concerning the anterior palatine region. *Anat. Rec.* 49:19–30, 1931.

694 MIES. Ein Fall von angeborenem Mangel des 5. Fingers und Mittelhandknochens der rechten Hand. *Arch. Pathol. Anat. Physiol. Klin. Med.* 121:336–340, 1890.

695 MIESSEN, E. Ein Fall von doppelseitiger Gelenkbildung zwischen Clavicula und Processus coracoides. *Anat. Anz.* 83:392–394, 1936–37.

696 MILCH, H. Short radius. *Arch. Surg.* 59:856–869, 1370, 1949.

697* MILES, A.E.W. Malformations of the teeth. *Proc. R. Soc. Med.* 47:817–826, 1954.

698 MILES, A.E.W. The assessment of age from the dentition. *Proc. R. Soc. Med.* 51:1057–1060, 1958.

699 MILLIKAN, R.A. Os subcalcis. *American J. Surg.* 37:116–117, 1937.

700* MISRA, B.D. A case of atlanto-occipital fusion. *J. Anat. Soc. India* 3:45–46, 1954.

701 MISRA, B.D. Interparietal bone: A case report. *J. Anat. Soc. India* 9:39, 1960.

702 MITCHELL, G.A.G. The significance of lumbosacral transitional vertebrae. *Br. J. Surg.* 24:147–158, 1936–37.

703 MITRA, N.L. and A.N. DWIVEDI. A bilateral anomaly of the shoulder joint. *J. Anat. Soc. India* 8:72–73, 1959.

704 MIYAKAWA, G. Congenital absence of the odontoid process. A case report. *J. Bone Joint Surg. (Am.)* 34:676–677, 1952.

705 M'LAREN, J.S. A case of congenital absence of the tibia. *J. Anat. Physiol.* 23:598–605, 1889.

706 MONTEIRO, H. Fréquence de l'occipitalisation de l'atlas chez les Portugais. *Assoc. Anatomistes Comptes Rendus* 28:465–470, 1933.

707 DE MORAFS, F. Sur un cas d'apophyse sus-epitrochléenne avec manifestation clinique. *Rev. Orthop.* 36:30–34, 1950.

708 MORDEJA, J. Das Lunatum bipartitum und die multizentrische Kernanlage des Os pisiforme. *Z. Orthop.* 95:492–500, 1962.

709 MORGAN, G.A. and H.G. POYTON. Bilateral dens in dente. *Oral Surg.* 13:63–66, 1960.

710* MORI, A. Dello sprone dell'olecrano e del suo significato antropologico. *Arch. Ital. Anat. Embriol.* 33:222–241, 1934.

711 MORRIS, H. Suprasternal bones demonstrated in the living subject. *Br. J. Radiol.* 13:356, 1940.

712 MORRISON, A.B. The os paracuneiforme. *J. Bone Joint Surg. (Br.)* 35:254–255, 1953.

713 MOURET, J. Sinus frontaus supplementaires. *Assoc. Anatomistes Comptes Rendus* 4:25–27, 1902.

714* MÜKKE, G.C. and W. POPPE. Fenestrae parietales symmetricae unter phylogenetischem Aspekt. *Fortschr. Röntgenstr.* 111:300–303, 1969.

715 MÜLLER, O. Über die Röntgenologie in der Zahnheilkunde. *Radiol. Clin.* 14:10–18, 1945.

716 MURATORI, G. Contributi allo studio delle differenze sessuali della pelvi nell'uomo. *Arch. Ital. Anat. Embriol.* 55:33–50, 1950.

717 MURRAY, G. Surgical treatment of funnel sternum. *Am. J. Surg.* 82:144–148, 1951.

718 MUTEL. Les déhiscences normales lombo-sacres. *Assoc. Anatomistes Comptes Rendus* 21:422–429, 1926.

719* MYERSON, M.C. The natural orifice of the maxillary sinus. I. Anatomic studies. *Arch. Otolaryngol.* 15:80–91, 1932.

720 NADGIR, Y.G. The supracondyloid tubercles of the femur. *J. Anat.* 51:375, 1917.

721 NATHAN, H. and B. ARENSBURGH. An unusual variation in the fifth lumbar and sacral vertebrae. A possible cause of vertebral canal narrowing. *Anat. Anz.* 132:137–148, 1972.

722 NAYAK, U.V. Anatomical notes. A case of abnormal atlas and axis vertebra. *J. Anat.* 65:399–400, 1931.

723 NÈGRE, A. Ossicules suprasternaux. *J. Radiol. Electrol.* 33:537, 1952.

724 NEISS, A. Abrißfraktur eines Processus supracondylicus humeri. *Fortschr. Röntgenstr.* 83:120, 1955.

725* NEISS, A. Über wenig bekannte Skelettvariationen. *Fortschr. Röntgenstr.* 94:227–232, 1961.

726 NELSON, E.M. A report of a 7-toed foot. *Anat. Rec.* 147:1–3, 1963.

727 NEUMANN, W. Über das "Os acromiale." *Fortschr. Röntgenstr.* 25:180–191, 1917–1918.

728 DE NEUREITER, F. Contribution a l'étude de l'omoplate scaphoïde. *Soc. Biol. Comptes Rendus Hebdomadaires des Séances et Mémoires* 90:1123–1124, 1924.

729 NEWELL, J.L. A case of multiple congenital anomalies of the Mullerian and genito-urinary systems with absence of the coccyx. *N. Engl. J. Med.* 210:1217–1218, 1934.

730 NICOLA, B. Sur la suture zygomatico-frontalis. *Arch. Ital. Biol.* 40:313–314, 1903.

731 NICOLA, B. Sur la suture zygomatico-frontalis. *Arch. Ital. Biol.* 43:315–316, 1905.

732 NICOLA, B. Division verticale totale de l'os zygomaticum. *Arch. Ital. Biol.* 47:478, 1907.

733 NISHI, K. Beiträge zum Studium der Präsakralwirbel bei verschiedenen Menschenrassen. *Folia Anat. Jpn.* 6:1–101, 1928.

734 NITZSCHE, F. Über lokalisierte Doppelmißbildungen und ihre Genese. *Z. Orthop. Chir.* 55:601–617, 1931.

735 NOTICES, A. The os centrale in the human carpus. *J. Anat. Physiol.* 18:119–120, 1884.

736* NUTT, J.J. and E.E. SMITH. Total congenital absence of the tibia. *Am. J. Roentgenol.* 46:841–849, 1941.

737 NUTTER, J.A. Congenital anomalies of the fifth lumbar vertebra and the consequences. *J. Anat. Physiol.* 48:24–36, 1914.

738 OBERMEIER, A. Eine unvollständige Brustbeinspalte. *Arch. Pathol. Anat. Physiol. Klin. Med.* 46:209–217, 1869.

739 O'DONOGHUE, D.H. and L.S. SELL. Congenital talonavicular

synostosis. A case report of a rare anomaly. *J. Bone Joint Surg.* 25:925–927, 1943.

740* OERTEL, O. Zur Frage des Os interzygomaticum. *Anat. Anz.* 68:495–497, 1929–30.

741 OETTEKING, B. Anomalous patellae. *Anat. Rec.* 23:269–279, 1923.

742 OKAMOTO, K. Über das Foramen transversarium und die Eigentümlichkeit des Querfortsatzes am 7. Halswirbel. *Anat. Anz.* 58:401–430, 1924.

743 OKAMOTO, K. Über den Rest der Fontanella-metopica-Befunde bei Japanern. *Folia Anat. Jpn.* 5:225–242, 1927.

744* OLIVIER, G. Anatomie du pouce bifide. *Assoc. Anatomistes Comptes Rendus* 37:369–380, 1950.

745 OLIVIER, G. Signification de la perforation olecranienne. *Assoc. Anatomistes Comptes Rendus* 38:765–768, 1951.

746* OLIVIER, G. and R. RAOU. La facette sous-acromiale. *Assoc. Anatomistes Comptes Rendus* 39:747–750, 1952.

747 O'RAHILLY, R. Radiological investigations of a case of pedal hyperphalangism and supernumerary phalangeal epiphyses. *Br. J. Radiol.* 19:432–434, 1946.

748 O'RAHILLY, R. Supernumerary pollex. *Ir. J. Med. Sci.* 6:636, 1946. Cited in *Excerpta Medica*, Sec. 1, Vol. 2, abstract 1088, p. 408, 1948.

749 O'RAHILLY, R. An analysis of cases of radial hemimelia. *Arch. Pathol.* 44:28–33, 1948.

750* O'RAHILLY, R. Morphological patterns in limb deficiencies and duplications. *Am. J. Anat.* 89:135–193, 1951.

751* O'RAHILLY, R. Anomalous occipital apertures. *Arch. Pathol.* 53:509–519, 1952.

752* O'RAHILLY, R. Epitriquetrum, hypotriquetrum, and lunatotriquetrum. *Acta Radiol.* 39:401–411, 1953.

753* O'RAHILLY, R. A survey of carpal and tarsal anomalies. *J. Bone Joint Surg. (Am.)* 35:626–642, 1953.

754* O'RAHILLY, R.M. and M. TWOHIG. Foramina parietalia permagna. *Am. J. Roentgenol.* 67:551–561, 1952.

755 OTTIGER, J. Über eine extreme Minusform der Strahlenzahl der unteren Extremität beim Menschen. *Anat. Anz.* 104:122–141, 1957.

756 OTTOLENGHI, S. Sur un os surnumeraire de la paroi interne de l'orbite dans des crânes de dégénérés. *Arch. Ital. Biol.* 25:152, 1896.

757* OVERTON, L.M. and J.W. GROSSMAN. Anatomical variations in the articulation between the second and third cervical vertebrae. *J. Bone Joint Surg. (Am.)* 34:155–161, 1952.

758 OWEN, F. Bilateral glenoid hypoplasia. Report of five cases. *J. Bone Joint Surg. (Br.)* 35:262–267, 1953.

759* PAHL, R. Doppelter Nervenkanal der Klavikula als diagnostische Fehlerquelle. (Zugleich ein Beitrag über das Foramen nervi supraclavicularis.) *Fortschritte Röntgenstrahlen* 82:487–491, 1955.

760 PANCOAST, H.K. Cervical rib. *University Penn. Med. Bull.* 14:394–398, 1901–1902.

761* PANSCH, A. Über Anomalien am Thoraxskelette. *Arch. Anat. Physiol. Wissen. Med.* 1875, pp. 552–564.

762 PARAVICINI, G. Sur la fossette occipitale médiane. *Arch. Ital. Biol.* 40:310, 1903.

763 PARAVICINI, G. Sur quelques nouveaux osselets suturo-fontanellaires du crane humain jeune et adulte, appartenant à des aliénés et a des individus normaux. *Arch. Ital. Biol.* 39:483, 1903.

764 PARDINI, A.G. JR. Congenital absence of the ulna. *J. Iowa Med. Soc.* 57:1106–1112, 1967.

765 PARKINSON, C.E. The supracondyloid process. *Radiology* 62:556–558, 1954,

766 PARSONS, F.G. Notes on the coronal suture. *J. Anat. Physiol.* 40:242–243, 1906.

767 PARTRIDGE, E.J. Anomalous carpal bones. *J. Anat.* 57:378–379, 1923.

768 PATE, J.R. An unusual occipito-atloid articulation. *J. Anat.* 71:128–129, 1936–37.

769 PATERSON, A.M. and F.T. LOVEGROVE. Symmetrical perforations of the parietal bones: Including an account of a perforated and distorted cranium from the Liverpool Museum. *J. Anat. Physiol.* 34:228–237, 1900.

770 PATERSON, I. Congenital phalangeal fusion through many generations. *J. Bone Joint Surg. (Br.)* 34:509, 1952.

771 PATTEN, C. A case of six toes on the left foot. *Anat. Anz.* 34:478–480, 1909.

772 PATTERSON, R.F. Multiple sesamoids of the hands and the feet. *J. Bone Joint Surg.* 19:531–532, 1937.

773 PATTERSON, T.J.S. Congenital deformities of the hands. *Ann. R. Coll. Surg.* 25:306–330, 1959.

774* PAWLIK, H.J. Die Sutura mendosa. *Fortschr. Röntgenstr.* 84:698–702, 1956.

775* PECINA, M. and M. NOVOCELAE. Über einen Fall, in dem die Hypophysengrube die Nasenhöhle begrenzt. *Anat. Anz.* 127:32–37, 1970.

776* PEELE, J.C. Unusual anatomical variations of the sphenoid sinuses. *Laryngoscope* 67:208–237, 1957.

777 PEROVIC, D. Über eine ungewöhnliche Insertion der unteren Nasenmuschel. *Anat. Anz.* 88:133–139, 1939.

778 PETERSEN, F. Anomalie der distalen Fibula. *Fortschr. Röntgenstr.* 85:118, 1956.

779 PFEIFFER, K. Variationen und Anomalien des Brustbeines sowie Hinweise zu deren Entwicklung. *Fortschr. Röntgenstr.* 85:663–671, 1956.

780 PFEIFFER, K. Zweigeteiltes Os centrale carpi oder zweigeteiltes Os naviculare? *Radiol. Clin.* 28:43–49, 1959.

781 PFISTER, J. Über Form und Größe des Intervaginalraums des Sehnerven im Bereich des canalis opticus. *von Graefes' Arch. Ophthalmol.* 36:83–93, 1890.

782 PFITZNER, W. Morphol. Arbeiten (Schwalbe), Bd. 1 to 7, 1891–1897. Cited in *Quain's Anatomy*, 11th Ed., Vol. IV, Pt. I: Osteology and Arthrology, edited by E.A. Schaefer, J. Symington, and T.H. Bryce. Longmans, Green, and Co., London, 1915.

783 PHESANT, H.C. and P.C. SWENSON. The lumbosacral region. A correlation of the roentgenographic and anatomical observations. *J. Bone Joint Surg. (Am.)* 24:299–306, 1942.

784 PICOZZO, T. La soudure des sutures crâniennes en rapport avec le sexe et l'âge. *Arch. Ital. Biol.* 25:333–334, 1896.

785 PIEPER, I. On the incidence of the supracondyloid process. *Am. J. Phys. Anthropol.* 8:167–171, 1925.

786 PIGHINI, G. Un cas de microcéphalie pure. *Arch. Ital. Biol.* 53:156–157, 1910.

787* PIONNIER, R. and A. DEPRAZ. Les anomalies costales d'origine congénitale. (Etude statistique d'apres 10,000 radiographies.) *Radiol. Clin.* 25:170–186, 1956.

788* PIONTEK, J. Variation of the level of closure of the sacral canal in man. *Folia Morphol.* 30:460–464, 1971.

789 PIONTEK, J. and J. BUDZYŃSKA. Studies on the capacity of Polish crania. *Folia Morphol.* 29:435–437, 1970.

790 PIRES DE LIMA, J.A. On a bone found in the musculus vastus externus of the thigh. *J. Anat. Physiol.* 47:414–424, 1913.

791 PIRKEY, E.L. and J.H. PURCELL. Agenesis of lumbosacral vertebrae. A report of two cases in living infants. *Radiology* 69:726–729, 1957.

792 PITTARD, E. and A. RISENFELD. Os wormiens suturaux (sutures coronale et lamboïde) d'une série de 200 crânes brachycéphales du type Homo alpinus. *Actes Soc. Helvétiques Sci. Nat.* 125:186–190, 1945. Cited in *Excerpta Medica*, Sec. 1, Vol. 2, abstract 1067, p. 401, 1948.

793 PITTARD, E., MANUILA, L. and A. MANUILA. Troisième trochanter et fosse hypotrochantérienne chez 500 fémurs humains (250 de chaque sexe) provenant de la région de Genève. Caractères fonctionnels ou caractéres réversifs? *Actes Soc. Helvétiques Sci. Nat.*:175–176, 1946. Cited in *Excerpta Medica*, Sec. 1, Vol. 2, abstract 1372, p. 484, 1948.

794 PITZORNO, M. L'épistrophée. *Arch. Ital. Biol.* 31:179, 1899.

795 PITZORNO, M. Sur quelques novelles particularités da l'atlas et sur la coexistence de ses variétés. *Arch. Ital. Biol.* 31:179, 1899.

796 PLASTER, A. Angeborene Mißbildung an Händen und Füßen bei einem Chinesen. *Arch. Pathol. Anat. Physiol. Klin. Med.* 104:54–58, 1886.

797 POPOWSKY, J. Zur Frage der Rippenspaltung. *Anat. Anz.* 15:284–288, 1899.

798* PSENNER, L. Die anatomischen Varianten des Hirnschädels. *Fortschr. Röntgenstr.* 75:197–214, 1951.

799* PULAWSKI, A. Über eine Defectmißbildung einiger Rippen und Muskeln als Ursache einer seltenen Abnormität im Bau des Thorax (Fissura thoracis lateralis). *Arch. Pathol. Anat. Physiol. Klin. Med.* 121:598–600, 1890.

800 PUNTONI, L. Sur la présence de sinus frontaux multiples chez l'homme. *Arch. Ital. Biol.* 75:190, 1925.

801 PURVES, R.K. and P.H. WEDIN. Familial incidence of cervical ribs. *J. Thorac. Surg.* 19:952–956, 1950.

802 RADIOÉVITCH, S., JOVANOVIĆ, S. and N. LOTRIĆ. Les variations de la selle turcique, du corps du sphénoïde et des sinus sphénoïdaux en fonction des variations de l'angle sphénoïdal. *Acta Anat.* 44:354–362, 1961.

803 RAMOS, A. and A.S. TAVARES. Quelques aspects de la morphologie du sternum (a propos de deux observations radiologiques d'os supra-sternaux). *Assoc. Anatomistes Comptes Rendus* 43:717–735, 1956.

804 RANKE, H. Bemerkungen über ein fünf Monate altes Kind mit angeborenem Mangel aller Gliedmaßen. *Arch. Pathol. Anat. Physiol. Klin. Med.* 82:360–374, 1880.

805 RAU, R.K. Skull showing absence of coronal suture. *J. Anat.* 69:109–112, 1934–35.

806 RAU, R.K. and D. SIVASUBRAHMANIAN. Supracondylar process. *J. Anat.* 65:392–394, 1931.

807 RAU, R.K. and D. SIVASUBRAHMANIAN. Anomalous atlanto-occiput. *J. Anat* 67:622–623, 1933.

808* RAUBER, A. Zur Kenntnis des Os styloideum carpi ultimale. *Anat. Anz.* 22:210–214, 1903.

809* RAUBER, A. Zur Kenntnis des Os interfrontale und supranasale. *Anat. Anz.* 22:214–221, 1903.

810 RAUBER, A. Der Schädel von Kegel. Eine antropologische Studie. *Int. Monatsschr. Anat. Physiol.* 23:41–209, 1906.

811 RAVELLI, A. Zur Mondbeinossifikation. *Fortschr. Röntgenstr.* 76:265–266, 1952.

812 RAVELLI, A. Mehrfache Processus styloidei an der Lendenwirbelsäule. *Radiol. Clin.* 23:57–58, 1954.

813 RAVELLI, A. Persistierende Apophyse am Proc. coracoides. *Fortschr. Röntgenstr.* 84:500–502, 1956.

814 RAVELLI, A. Os accessorium supracalcaneum. *Fortschr. Röntgenstr.* 85:121–122, 1956.

815 RAVELLI, A. Das Ossiculum parasternale. *Fortschr. Röntgenstr.* 85:226–231, 1956.

816 REES, L.A. The structure and function of the mandibular joint. *Br. Dent. J.* 96:125–133, 1954.

817 REID, R.W. Deficiency of the left half of a dorsal vertebra. *J. Anat. Physiol.* 21:76–78, 1887.

818 REID, R.W. Cranio-cleido dysostosis and normal skull formation. *J. Anat.* 66:64–65, 1932.

819 REINHARDT, K. Agenesie und Dysgenesie des Kreuzbeines. *Fortschr. Röntgenstr.* 95:381–391, 1961.

820* REINHARDT, K. Der Processus supratrochlearis humeri. *Fortschr. Röntgenstr.* 105:125–127, 1966.

821 REINHARDT, K. Eine doppelseitige Anomalie am lateralen Klavikuladrittel, bestehend aus einer bogenförmigen Duplikatur des Knochens in Richtung auf das Coracoid und aus akzessorischen Knochenelementen. *Fortschritte Röntgenstrahlen* 113:527–530, 1970.

822* RENANDER, A. Anomalies roentgenologically observed of the craniovertebral region. *Acta Radiol.* 10:502–513, 1929.

823 RENDU, A. Absence congénitale du Péroné. *Rev. Orthop. (Paris)* 3:169–174, 1912.

824 RENON, C. Périarthrite due poignet ou os radial externe du carpe. *Rev. Orthop. (Paris)* 30:90–94, 1944.

825 RETTERER, E. Nouvelles observations sur l'evolution des sésamoides intratendineux. *Soc. Biol. Comptes Rendus Hebdomadaires des Séances et Mémoires* 81:237–240, 1918.

826 RETTERER, E. Des sésamoides de la main humaine. *Soc. Biol. Comptes Rendus Hebdomadaires des Séances et Mémoires* 81:829–832, 1918.

827 REVSKOI, Y.K. Variations in frontal sinus structure and their significance in selection of pilots. *Fed. Proc. [Translation Suppl.]* 24:T948–T950, 1965.

828 RICHES, E.W. The anatomy of cervical rib: With a case report. *Br. J. Surg.* 16:235–238, 1928–29.

829 RICHMOND, D.A. Two cases of discoid medial cartilage. *J. Bone Joint Surg. (Br.)* 40:268–269, 1958.

830 RINDFLEISCH, E. Die angeborene Spaltung der Wirbelkörper. *Arch. Pathol. Anat. Physiol. Klin. Med.* 27:137–145, 1863.

831 RIVA, G. Ein Fall von doppelseitigem Os triangulare carpi. *Radiol. Clin.* 18:78–82, 1949.

832 ROAF, R. Foramen in the humerus caused by the median nerve. *J. Bone Joint Surg. (Br.)* 39:748–749, 1957.

833 ROBBINS, R.H. The human pisiform. *J. Anat.* 51:150–152, 1917.

834 ROBERTS, S.M. Congenital absence of the odontoid process resulting in dislocation of the atlas on the axis. *J. Bone Joint Surg.* 15:988–999, 1933.

835 ROCA, C.A., IMAZ, J.I. and A.F. LAZARONI. Apofisis supraepitroclear dolorosa. *Rev. Sanidad Militar.* 45:435–446, 1946. Cited in Excerpta Medica, Sec. 1, Vol. 2, abstract 421, p. 210, 1948.

836 ROCHE, M.B. and G.G. ROWE. The incidence of separate neural arch and coincident bone variations. A summary. *J. Bone Joint Surg. (Am.)* 34:491–494, 1952.

837 ROLLER, G.T. and H.F.W. PRIBRAM. Lumbosacral intradural lipoma and sacral agenesis. *Radiology* 84:507–512, 1965.

838 ROMITI, G. La fossette pharyngienne dans l'os occipital de l'homme. *Arch. Ital. Biol.* 13:357, 1890.

839 ROMITI, G. Une observation d'arc maxillo-temporal sous-jugal et sur la genèse de la bipartition de l'os malaire chez l'homme. *Arch. Ital. Biol.* 11:420–421, 1889.

840 ROMITI, G. Sur l'inachèvement de l'arc zygomatique dan un crâne humain remarquable par d'autres variétés. *Arch. Ital. Biol.* 23:99, 1895.

841 ROMITI, G. Di una importante varietá vertebrale. *Arch. Ital. Anat. Embriol.* 8:1–3, 1909.

842* ROMITI, G. Sur une importante variété vertébrale. *Arch. Ital. Biol.* 53:147, 1910.

843 ROSENGARTEN, E. Occurrence of sesamoid bones in the hands in the Polish population. *Folia Morphol.* 26:110–117, 1967.

844 ROSS, E. Os capitatum bipartitum. *Fortschr. Röntgenstr.* 81:224–225, 1954.

845 ROSS, J.A., TOUGH, I.C.K. and T.A. ENGLISH. Congenital discoid cartilage. Report of a case of discoid medial cartilage, with an embryological note. *J. Bone Joint Surg. (Br.)* 40:262–267, 1958.

846 ROUSSEAUX, R., KISSEL, P., BEAU, A., MIDON, J. and G. ARNOULD. Synostose occipito-atloïdiene; rétrécissement du trou occipital et compression bulbo-médullaire par proéminences osseuses, vestiges des masses latérales de l'atlas. *Rev. Neurol.* 82:389–394, 1950.

847 ROUVIERE, H. and E. OLIVIER. Faisceau maxillaire du styloglosse et signification du ligament stylo-maxillaire. *Soc. Biol. Comptes Rendus Hebdomadaires des Séances et Mémoires* 87:337–339, 1922.

848 ROWE, G.G. Anomalous vertebrae from the lumbo-sacral column of man. *Anat. Rec.* 107:171–179, 1950.

849 RÜBE, W. Doppelanlage der letzten Steißbeinwirbel. *Fortschr. Röntgenstr.* 87:270–272, 1957.

850 RUPPRICHT, W. Über einige Fälle von Pneumatisation der kleinen Keilbeinflügel, besonders im Bereich des Processus clinoid ant. *Anat. Anz.* 58:480–496, 1924.

851 RUSHFORTH, A.F. A congenital abnormality of the trapezium and first metacarpal bone. *J. Bone Joint Surg. (Br.)* 31:543–546, 1949.

852* RUSSELL, H.E. and G.T. AITKEN. Congenital absence of the sacrum and lumbar vertebrae with prosthetic management. *J. Bone Joint Surg. (Am.)* 45:501–508, 1963.

853 RUTHERFORD, H. Bifurcate clavicle. *J. Anat.* 55:286–287, 1921.

854 SABATINI, G. Sur cas d'absence congénitale du pérone. *Arch. Ital. Biol.* 53:151, 1910.

855 SACK, G.M. Os trigonum und Shepheardsche Fraktur. *Röntgenpraxis* 4:1028–1035, 1932.

856* SACKETT, G.L. Errors in roentgen diagnosis. *Ohio State Med. J.* 36:137–144, 1940.

857 SALLAN, A.M. Triphalangeal thumbs. *Arch. Surg.* 71:257–259, 1955.

858 SAPPEY, C. Recherches sur le volume et la capacité du crâne, sur le volume et le poids de l'encéphale, comparés chez l'homme et chez l'femme. *Soc. Biol. Comptes Rendus des Séances et Mémoires* 13:109–120, 1861.

859 SCANNELL, R.C. Congenital absence of the odontoid process. A case report. *J. Bone Joint Surg.* 27:714–715, 1945.

860 SCHAEFFER, J.P. The genesis, development and adult anatomy of the nasofrontal region in man. *Am. J. Anat.* 20:125–145, 1916.

861* SCHERTLEIN, A. Über die häufigsten Anomalien an der Brustlendenwirbelsäulengrenze. *Fortschr. Röntgenstr.* 38:478–488, 1928.

862* SCHINZ, H.R. Der Abbruch des Processus styloideus ulnae. *Dtsch. Z. Chir.* 175:81–137, 1922.

863* SCHINZ, H.R. Das Foramen supratrochleare humeri. *Fortschr. Röntgenstr.* 29:193–200, 1922.

864 SCHINZ, H.R. Altes und Neues zur Beckenossifikation. Zugleich ein Beitrag zur Kenntnis des Os acetabuli. *Fortschr. Röntgenstr.* 30:66–81, 1923.

865* SCHINZ, H.R. Vergleichende Anatomie und Phylogenie des Skelettes in ihrer Bedeutung für den Röntgenologen. *Radiol. Clin.* 14:19–36, 1945.

866 SCHLÜTER, K. Der "calcaneus bifidus," eine Ossifikationsanomalie des Fersenbeins im Hackenplattfuß. *Fortschr. Röntgenstr.* 85:720–727, 1956.

867* SCHLYVITCH, B. Über den Articulus coracoclavicularis. *Anat. Anz.* 85:89–93, 1937–38.

868 SCHMIDT, H. and E. FISCHER. Über partielle einseitige Synostosen zwischen Atlas und Axis. *Fortschr. Röntgenstr.* 92:380–384, 1960.

869 SCHMIDT, O.V.M. Elongated styloid process which interfered with function of a singer's voice. Operation and recovery. *Arch. Otolaryngol.* 54:417–421, 1950.

870 SCHMITT, G.H. Dens in dente. *Fortschr. Röntgenstr.* 71:357, 1949.

871 SCHNEIDER, U. Das akzessorische Iliosakralgelenk (Articulus sacroilicus accessorius) im Röntgenbild. *Fortschr. Röntgenstr.* 85:426–432, 1956.

872* SCHOBERTH, H. Die Trichterbrust. *Ergeb. Chir. Orthop.* 43:122–202, 1961.

873 SCHOEN, H. Seltenere akzessorische Knochen am Fußrücken. *Röntgenpraxis* 7:775–776, 1935.

874 SCHOEN, H. Das Os vesalianum. *Fortschr. Röntgenstr.* 75:489, 1951.

875* SCHRÖDER, W. Das Os styloides und seine klinische Bedeutung. *Zentralbl. Chir.* 70:1371–1377, 1943.

876 SCHRÖDER, W. Über die überzähligen Handwurzelknochen, insbesondere das Os styloideum. *Röntgenpraxis* 14:190–196, 1940.

877 SCHULTE, E. Seltene Skelettanomalie am Schultergürtel. *Fortschr. Röntgenstr.* 92:226–227, 1960.

878 SCHULTE, K.J. Einige seltenere Knochenbefunde im Bereich des Schultergürtels. *Fortschr. Röntgenstr.* 86:231–235, 1957.

879 SCHULTZ, A.H. The fontanella metopica and its remnants in an adult skull. *Am. J. Anat.* 23:259–271, 1918.

880 SCHULTZ, A.H. Chevron bones in adult man. *Am. J. Phys. Anthropol.* 28:91–97, 1941.

881 SCHULTZ, A.H. The position of the occipital condyles and of the face relative to the skull base in primates. *Am. J. Phys. Anthropol.* 13:97–120, 1955.

882 SCHULTZ, E.H. JR., LEVY, R.W. and P.E. RUSSO. Agenesis of the odontoid process. *Radiology* 67:102–105, 1956.

883 SCHULZ, O. Beiträge zur Kenntnis der Fehlbildungen der Wirbelsäule. *Arch. Pathol. Anat. Physiol. Klin. Med.* 304:203–222, 1939.

884 SCHUMACHER, O. Weitgehende Asymmetrie des Sinus sphenoidalis. *Anat. Anz.* 68:419–423, 1929–30.

885* SCHWARTZ, C.W. Anomalies and variations in the normal skull from a roentgenological viewpoint. *Am. J. Roentgenol.* 42:367–373, 1939.

886 SCOTT, J.H. Note of a case of articulation between two ribs. *J. Anat. Physiol.* 13:577, 1879.

887* SCOTT, R.B. Congenital absence of the fibula. Its occurrence in a premature negro infant. *Am. J. Dis. Child.* 61:1037–1043, 1941.

888* SEDWICK, H.J. Form, size and position of the maxillary sinus at various ages studied by means of roentgenograms of the skull. *Am. J. Roentgenol.* 32:154–160, 1934.

889 SEIDEL, K. Ungewöhnliche symmetrische Rippenanomalie. *Fortschr. Röntgenstr.* 88:247–248, 1958.

890 SEITZ, C. Eine seltene Mißbildung des Thorax. *Arch. Pathol. Anat. Physiol. Klin. Med.* 98:335–337, 1884.

891* SELBY, S., GARN, S.M. and V. KANAREFF. The incidence and familial nature of a bony bridge on the first cervical vertebrae. *Am. J. Phys. Anthropol.* 13:129–141, 1955.

892 SENFTLEBEN, H. Notiz über eine angeborene Luxation des Radius mit Defect des mittleren Theils der Ulna. *Arch. Pathol. Anat. Physiol. Klin. Med.* 45:303–304, 1869.

893 SENTURIA, H.R. and B.D. SENTURIA. Congenital absence of the patellae associated with arthrodysplasia of the elbows and dystrophy of the nails. A hereditary syndrome. *Am. J. Roentgenol.* 51:352–358, 1944.

894 SEVER, J.W. Bifid os calcis. *Surg. Gynecol. Obstet.* 50:1012–1013, 1930.

895 SEYSS, R. Zu den Verknöcherungen im Bereich des Foramen obturatum. *Fortschr. Röntgenstr.* 91:525–527, 1959.

896 SHANDS, A.R. and I.W. WENTZ. Congenital anomalies, accessory bones, and osteochondritis in the feet of 850 children. *Surg. Clin. North Am.* 33:1643–1666, 1953.

897 SHEHATA, R. Occipitalisation of the atlas. *J. Anat. Soc. India* 13:97–99, 1964.

898 SHIMAGUCHI, S. Tenth rib is floating in Japanese. *Anat. Anz.* 135:72–82, 1974.

899 SHORE, L.R. Abnormalities of the vertebral column in a series of skeletons of Bantu natives of South Africa. *J. Anat.* 64:206–238, 1930.

900 SHORE, L.R. A report on the nature of certain bony spurs arising from the dorsal arches of the thoracic vertebrae. *J. Anat.* 65:379–387, 1931.

901* SHORE, L.R. A report on the spinous processes of the cervical vertebrae in the native races of South Africa. *J. Anat.* 65:483–505, 1931.

902 SHORE, L.R. A report on the spinous processes of the cervical vertebrae in a series of Egyptian skeletons. *J. Anat.* 67:422–442, 1933.

903 SHORE, L.R. A note on the interparietal groove in Egyptian skulls. *J. Anat.* 73:1–14, 1938–39.

904 SHUFELDT, R.W. Comparative data from 2000 Indian crania in the United States Army Medical Museum. *J. Anat. Physiol.* 22:191–214, 1888.

905 SIBATA, I. The supracondyloid process in Coreans. *Folia Anat. Jpn.* 20:357–369, 1941.

906 SIDDIQI, M.A.H. Variations in the lower end of the femur from Indians. *J. Anat.* 68:331–337, 1933–34.

907 SIDDIQI, M.A.H. Observations on the lower end of the femur from American Indians and modern Canadians. *J. Anat.* 70:410–412, 1935–36.

908 SIEBERT, P. Processus paracondylicus. *Fortschr. Röntgenstr.* 111:717, 1969.

909 SIEGERT, F. Das Problem des Metacarpale 1 und Metatarsale 1. *Z. Anat. Entwicklungsgesch.* 92:213–223, 1930.

910 SIMON, E. Anatomy of the opening of the maxillary sinus. *Arch. Otolaryngol.* 29:640–649, 1939.

911 SIMON, S. Zur Kenntnis der Rippenanomalien im Kindesalter. *Röntgenpraxis* 10:45–50, 1938.

912 SINCLAIR, J.G., DUREN, N. and J.C. RUDE. Congenital lumbosacral defect. *Arch. Surg.* 43:473–478, 1941.

913 SINGH, I. Variations in the metacarpal bones. *J. Anat.* 93:262–267, 1959.

914* SINGH, S. Variations of the superior articular facets of atlas vertebrae. *J. Anat.* 99:565–571, 1965.

915 SIRANG, H. Ein Canalis alae ossis ilii und seine Bedeutung. *Anat. Anz.* 133:225–238, 1973.

916* SKARBY, H.G. Das Foramen nervi clavicularis im Röntgenbild. *Acta. Radiol.* 17:397–402, 1936.

917 SKINNER, B.M. Note on the relative lengths of the first and second toes of the human foot. *J. Anat.* 66:123–124, 1931.

918 SMITH, E.B. Two rare vertebral anomalies. *J. Anat. Physiol.* 36:372–374, 1902.

919* SMITH, E.D. Congenital sacral anomalies in children. *Aust. N.Z. J. Surg.* 29:165–176, 1959.

920 SMITH, S. A contribution to the study of the modern Egyptian cranium. *J. Anat.* 60:121–129, 1925.

921 SMITH, S.A. Congenital absence of the humerus. *Br. J. Surg.* 15:385–386, 1927–28.

922 SMITH, W.R. Some rare abnormalities in teeth. *J. Anat. Physiol.* 41:216–220, 1907.

923 SMITHAM, J.H. Some observations on certain congenital abnormalities of the hand in African natives. *Br. J. Radiol.* 21:513–518, 1948.

924 SMOLA, L. Der Calcaneus bifidus. *Fortschr. Röntgenstr.* 85:120, 1956.

925 SMOOK, A.H. Doppelbildung eines menschlichen Fußes. *Anat. Anz.* 78:209–228, 1934.

926 SOENEN, R. Zwei bisher nicht veröffentlichte Fälle von "Os malare bipartitum" aus der Anthropologischen Sammlung des Anatomischen Instituts in Bonn. *Anat. Anz.* 70:320–324, 1930.

927* SOHIER, H.M.L. and A. GOUAZÉ. L'astragale de l'Africain d'A.O.F. *Association Anatomistes Comptes Rendus* 41:935–947, 1954.

928 SOKOLOWSKA-PITUCHOWA, J. and C. MIAŚKIEWICZ. Studies on the sesamoid bones of the hands in human males. *Folia Morphol.* 24:136–142, 1965.

929 SOKOLOWSKA-PITUCHOWA, J. and C. MIAŚKIEWICZ. Studies on the sesamoid bones in the hands of women. *Folia Morphol.* 26:24–31, 1967.

930* SOKOLOWSKA-PITUCHOWA, J. and C. MIAŚKIEWICZ. The sesamoid bones in Negroes. *Folia Morphol.* 28:179–183, 1969.

931* SOLIERI, S. Neuralgia del nervo mediano da processo soprapeitrocleare. *Chir. Org. Movimento* 14:171–180, 1929–30.

932 SONESSON, A. Odontogenic cysts and cystic tumors of the jaws. *Acta Radiol. [Suppl.]* 81:1–159, 1950.

933 SORGE, F. Über den atavistischen Processus supracondyloideus humeri. *Anat. Anz.* 60:467–474, 1925–26.

934* SOULE, A.B. JR., GROSS, S.W. and J.G. IRVING. Myelography by the use of pantopaque in the diagnosis of the intervertebral discs. *Am. J. Roentgenol.* 53:324–340, 1945.

935 SOUTHWOOD, A.R. Partial absence of the ulna and associated structures. *J. Anat.* 61:346–351, 1927.

936 SOUTHWORTH, J.D. and S.R. BERSACK. Anomalies of the lumbosacral vertebrae in five hundred and fifty individuals without

symptoms referable to the low back. *Am. J. Roentgenol.* 64:624–634, 1950.

937 DE SOUZA FONTES, J.R. Ein lebender Knabe mit angeborenem Mangel der Glieder. *Arch. Pathol. Anat. Physiol. Klin. Med.* 71:107–111, 1877.

938 SPERINO, G. and A. BOVERO. Sur la nature métopique basilaire ou frontale-basilaire dans le crâne humain. *Arch. Ital. Biol.* 28:159–161, 1897.

939 SPERINO, G. and R. BALLI. Os cuneiforme perfecte et imperfecte bipartitum. *Arch. Ital. Biol.* 59:473, 1913.

940 SPRONCK, C. Auftreten der ganzen Tuberositas (lateralis) des Os metatarsale V als ein für sich bestehendes am Metatarsale und Cuboides artikulierendes Skelett-Element. *Anat. Anz.* 2:734–739, 1887.

941 STACEY, R.S. A rare malformation of the hand. *J. Anat.* 72:456–457, 1937–38.

942 STADERINI, R. Un pouce avec trois phalanges et une main avec sept doigts chez l'homme. *Arch. Ital. Biol.* 23:84, 1895.

943 STADERINI, R. Recherches statistiques sur la fréquence des variétés numériques des vertèbres chez l'homme et considérations sur leur genèse. *Arch. Ital. Biol.* 22:149–150, 1895.

944* STAFNE, E.C. and W.H. HOLLINSHEAD. Roentgenographic observations on the stylohyoid chain. *Oral Surg.* 15:1195–1200, 1962.

945 STALLWORTHY, J.A. A case of enlarged parietal foramina associated with metopism and irregular synostosis of the coronal suture. *J. Anat.* 67:168–174, 1932.

946 STAŃCZYK, J.L. Canal of vertebral artery of atlas. *Folia Morphol.* 38:353–357, 1979.

947 STAURENGHI, C. Sur la topographie de l'"ostium sphenoidale" de l'homme. *Arch. Ital. Biol.* 25:154–155, 1896.

948 STAURENGHI, C. Rapports topographiques entre les petites ailes du présphénoïde et l'angle sphénoïden du pariétal humain (suture xiphopariétal, processus orbitaire de la grande aile du sphénoïde, suture fronto-grande-aile). Homologie des centres orbito-sphénoïde de l'homme. *Arch. Ital. Biol.* 36:333–334, 1901.

949 STAURENGHI, C. Variétés crâniennes trouvées dans le sépulcre de la rotonde du grand hôpital de Milan. *Arch. Ital. Biol.* 53:148, 1910.

950 STEHR, L. Variationen und Fehlbildungen im Bau des knöchernen Thorax. *Fortschr. Röntgenstr.* 62:67–107, 1940.

951* STEIN, M. Über einen Fall von beiderseitigen Halsrippen. *Anat. Anz.* 38:650–664, 1911.

952 STEINER, H.A. Roentgenologic manifestations and clinical symptoms of rib abnormalities. *Radiology* 40:175–178, 1943.

953 STEINHÄUSER, J. and G. MERHOF. Röntgenstudien an Handgelenken zur sogenannten Minusvariante der Elle (Hultén). *Z. Orthop.* 107:11–24, 1969.

954 STERNBERG, M. Ein bisher nicht beschriebener Kanal im Keilbein des Menschen. *Anat. Anz.* 3:784–786, 1888.

955 STEWART, T.D. Cervical intercostal articulation in a North American Indian. *J. Anat.* 69:124–126, 1934–35.

956 STEWART, T.D. The age incidence of neural arch defects in Alaskan natives, considered from the standpoint of etiology. *J. Bone Joint Surg. (Am.)* 35:937–950, 1953.

957 STIBBE, S.P. Anatomical note. Skull showing perforations of the parietal bone, or enlarged parietal foramina. *J. Anat.* 63:277, 1929.

958 STIEDA, L. Über den Sulcus ethmoidalis der Lamina cribrosa des Siebbeines. *Anat. Anz.* 6:232–237, 1891.

959* STIEFEL, D.M. Congenital absence of the odontoid process. Report of a case. *J. Bone Joint Surg. (Am.)* 32:946–947, 1950.

960 STIEVE, H. Über Hyperphalangie des Daumens. *Anat. Anz.* 48:565–581, 1915–16.

961 STOPFORD, J.S.B. The supracondyloid tubercles of the femur and the attachment of the gastrocnemius muscle to the femoral diaphysis. *J. Anat. Physiol.* 49:80–84, 1915.

962 STRAUS, W.L. JR. The human ilium: Sex and stock. *Am. J. Phys. Anthropol.* 11:1–28, 1927.

963* STRAUS, W.L. JR. and O. TEMKIN. Vesalius and the problem of variability. *Bull. Hist. Med.* 14:609–633, 1943.

964 STRECKER, C. Eine angeborene vierfingerige rechte Hand. *Arch. Pathol. Anat. Physiol. Klin. Med.* 127:181–187, 1892.

965 STRICKER, G. Großartiger Defect an beiden Vorderarmen und Händen eines Neugeborenen. *Arch. Pathol. Anat. Physiol. Klin. Med.* 72:144, 1878.

966 STRICKER, G. Doppelseitiger angeborener Defect des Radius und des Daumens. *Arch. Pathol. Anat. Physiol. Klin. Med.* 31:529–530, 1864.

967* STRUTHERS, J. On the supra-condyloid process of the human arm. *Lancet* 1:87–88, 1863.

968 STRUTHERS, J. Additional bone in the carpus. *J. Anat. Physiol.* 3:354–356, 1869.

969 STRUTHERS, J. On hereditary supra-condyloid process in man. *Lancet* 1: 231–232, 1873.

970 STRUTHERS, J. Variations of the vertebrae and ribs in man. *J. Anat. Physiol.* 9:17–96, 1875.

971* SUNDERLAND, S. The quadrate tubercle of the femur. *J. Anat.* 72:309–312, 1938.

972 SUSMAN, M.P. Rudimentary patellae, other skeletal defects, and dystrophy of the nails. *Med. J. Aust.* 1:685–686, 1934.

973* SUTHERLAND, I.D. Funnel chests. *J. Bone Joint Surg. (Br.)* 40:224–251, 1958.

974 SUTTON, J.B. The relation of the orbito-sphenoid to the region pterion in the side wall of the skull. *J. Anat. Physiol.* 18:219–222, 1884.

975 SUTTON, J.B. A critical study of cranial morphology. *J. Anat. Physiol.* 22:28–37, 1888.

976 SWATON, S. and Z. HUBER. Einseitige, angeborene, doppelte Kniescheibe (Patella duplex unilateralis congenita). *Zentralbl. Chir.* 85:2270–2274, 1960.

977 SWEET, A.P.S. Radiodontic study of the mental foramen. *Dent. Radiogr.* 32:28, 32–33, 1959.

978* SWJETSCHNIKOW. Über die Assimilation des Atlas und die Manifestation des Occipitalwirbels beim Menschen. *Arch. Anat. Physiol. Wissen. Med.* 1906, pp. 155–194.

979* SWJETSCHNIKOW. Über die Variationen des Occipitalwirbels. *Anat. Anz.* 32:50–61, 1908.

980* SYCAMORE, L.K. Common congenital anomalies of the bony thorax. *Am. J. Roentgenol.* 51:593–599, 1944.

981 SYMINGTON, J. Separate acromion process. *J. Anat. Physiol.* 34:287–294, 1899.

982 SYMMERS, W.ST.C. A skull with enormous parietal foramina. *J. Anat. Physiol.* 29:329–330, 1895.

983 SZAWLOWSKI, J. Über einige seltene Variationen an der Wirbelsäule beim Menschen. *Anat. Anz.* 20:305–320, 1902.

984 TAVARES, A. Sur les éminences basilaires de l'occipital (note préliminaire). *Assoc. Anatomistes Comptes Rendus* 21:534–541, 1926.

985 TAVERNIER, L. A propos du traitement des absences congénitales du péroné. *Lyon Chir.* 45:764–765, 1950.

986 TAYLOR, A.E. Note on a case of six lumbar vertebrae and abnormal middle sacral artery. *J. Anat. Physiol.* 32:687, 1898.

987 TAYLOR, A.E. Variations in the human tooth-form as met with in isolated teeth. *J. Anat. Physiol.* 33:268–272, 1899.

988* TAYLOR, R.G. Anomalies of the lumbosacral articulations. *JAMA* 113:463–465, 1939.

989 TEBO, H.G. and I.R. TELFORD. An analysis of the variations in position of the mental foramen. *Anat. Rec.* 107:61–66, 1950.

990 TEICHERT, G. Os calcaneocuboideum laterale. *Fortschr. Röntgenstr.* 84:647–648, 1956.

991 TENCHINI, L. Sur les variétés numériques vertébro-costales chez l'homme. *Arch. Ital. Biol.* 12:XLIII, 1889.

992 TENCHINI, L. Sur un noveau mode de compensation dans les anomalies numériques vertébrales de l'homme. *Arch. Ital. Biol.* 23:93, 1895.

993 TENCHINI, L. Sopra il canale infrasquamoso di Gruber nell'uomo. *Arch. Ital. Anat. Embriol.* 3:1–36, 1904.

994 TENCHINI, L. Sur un canal perforant artériel (infrapariétal) dans la voûte crânienne de l'homme adulte. *Arch. Ital. Biol.* 43:311, 1905.

995 TENCHINI, L. Sur la présence de canaux émissaires dans la squama frontalis de l'homme adulte. *Arch. Ital. Biol.* 43:312, 1905.

996 TENCHINI, L. Canali perforanti vascolari sagittali e parasagittali nel cranio dell'uomo adulto. *Arch. Ital. Anat. Embriol.* 4:116–152, 1905.

997 TENCHINI, L. Canaux perforants vasculaires sagittaux et parasagittaux dans le crâne de l'homme adulte. *Arch. Ital. Biol.* 45:276–277, 1906.

998 TENCHINI, L. and U. ZIMMERL. Sur les os Wormiens de la Fossa cranii anterior chez l'homme. *Arch. Ital. Biol.* 39:481, 1903.

999 TENCHINI, L. and U. ZIMMERL. Un nouveau processus anormal de l'Os sphenoidale de l'homme. *Arch. Ital. Biol.* 39:482, 1903.

1000 TEPLICK, J.G., STEINBERG, S. and B.P. ADELMAN. Congenital absence of the odontoid process. *Am. J. Roentgenol.* 83:653–655, 1960.

1001 TERLEP, H. Os radiale externum. *Fortschr. Röntgenstr.* 86:666, 1957.

1002 TERRAFRANCA, R.J. and A. ZELLIS. Congenital hereditary cranium bifidum occultum frontalis. *Radiology* 61:60–66, 1953.

1003 TERRY, R.J. Rudimentary clavicles and other abnormalities of the skeleton of a white woman. *J. Anat. Physiol.* 33:412–422, 1889.

1004 TERRY, R.J. A study of the supracondyloid process in the living. *Am. J. Phys. Anthropol.* 4:129–139, 1921.

1005 TERRY, R.J. New data on the incidence of the supracondyloid variation. *Am. J. Phys. Anthropol.* 9:265–270, 1926.

1006 THEWS, K. Fehldeutung und -behandlung auf Grund von Varietäten der Hand- und Fußwurzel im Röntgenbild. *Röntgenpraxis* 11:184–186, 1939.

1007 THOMSON, A. The influence of posture on the form of the articular surfaces of the tibia and astragalus in the different races of man and the higher apes. *J. Anat. Physiol.* 23:616–639, 1889.

1008 THOMSON, A. The orbito-maxillary frontal suture in man and apes, with notes on the varieties of the human lachrymal bone. *J. Anat. Physiol.* 24:349–357, 1890.

1009 TILLEY, A.R. Segmental duplication of ulnar elements of hand and forearm (mirror hand). *Med. Radiogr. Photogr.* 30:58, 1954.

1010 TITZE, A. Über einen akzessorischen Knochenschatten im Bereiche des jugendlichen Kniegelenkes. *Fortschr. Röntgenstr.* 87:668–669, 1957.

1011 TOBECK, A. Über den Verlauf des Facialkanals im Röntgenbild. *Arch. Ohr Nas Kehlk. Heilk* 144:276–289, 1938.

1012 TODD, T.W. "Cervical rib": Factors controlling its presence and its size. Its bearing on the morphology and development of the shoulder (with four cases). *J. Anat. Physiol.* 31:288–302, 1897.

1013 TODD, T.W. Costal anomalies of the thoracic inlet, their interpretation and significance. *Anat. Anz.* 41:257–271, 1912.

1014 TODD, T.W. "Cervical" rib: Factors controlling its presence and its size. Its bearing on the morphology and development of the shoulder. *J. Anat.* 46:244–288, 1912.

1015 TODD, T.W. and W.C. MCCALLY. Defects of the patellar border. *Ann. Surg.* 74:775–782, 1921.

1016 TOKMAKOFF, A.S. Zur Anatomie des "Os intermetatarseum Gruberi". *Anat. Anz.* 66:334–341, 1928–29.

1017 TÖNDURY, G. Zur Genese der Hasenscharte beim Menschen. *Rev. Suisse Zool.* 53:459–468, 1946. Cited in *Excerpta Medica*, Sec. 1, Vol. 2, abstract 299, p. 146, 1948.

1018 TORGERSEN, J. Anomalies of the spine in anomalies of viscera and constitution. A radiological contribution to the physiology of development of man and to the understanding of the bearing of the anomalies of the spine pathology. *Acta Radiol.* 29:311–320, 1948.

1019 TORGERSEN, J. A roentgenological study of the metopic suture. *Acta Radiol.* 33:1–11, 1950.

1020* TORGERSEN, J. Hereditary factors in the sutural pattern of the skull. *Acta Radiol.* 36:374–382, 1951.

1021 TORRIGIANI, C.A. Lo sviluppo delle cavità accessorie delle fosse nasali nell' uomo. *Arch. Ital. Anat. Embriol.* 12:35–109, 1913–14.

1022 TOYODA, J. Über den lumbosakralen Übergangswirbel bei den Japanern. *Folia Anat. Jpn.* 5:37–49, 1927.

1023* TRAMONTANO-GUERRITORE, G. Die Atlanto-occipital-Union. *Anat. Anz.* 64:173–184, 1927–28.

1024* TRAVERS, J.T. and L.C. WORMLEY. Enlarged parietal foramina. *Am. J. Röntgenol.* 40:571–579, 1938.

1025 TREDGOLD, A.F. Variations of ribs of primates, with especial reference to the number of sternal ribs in man. *J. Anat. Physiol.* 31:288–287, 1897.

1026 TREVES, F. A gluteal trochanter in the human subject. *J. Anat. Physiol.* 21:325–327, 1887.

1027 TROLARD, P. Note sur la présence d'un petit arc osseux dans l'épaisseur du ligament atloido-occipital postérieur. *Soc. Biol. Comptes Rendus Hebdomadaires des Séances et Mémoires* 44:226, 1892.

1028 TROTTER, M. Synostosis between manubrium and body of the sternum in whites and negroes. *Am. J. Phys. Anthropol.* 18:439–442, 1934.

1029 TROTTER, M. Accessory sacroiliac articulations. *Am. J. Phys. Anthropol.* 22:247–261, 1937.

1030 TROTTER, M. A common anatomical variation in the sacro-iliac region. *J. Bone Joint Surg.* 22:293–299, 1940.

1031* TROTTER, M. Variations of the sacral canal: Their significance in the administration of caudal analgesia. *Curr. Res. Anesth. Analgesia* 26:192–202, 1947.

1032* TROTTER, M. and G.S. LETTERMAN. Variations in the female sacrum. Their significance in continuous caudal anesthesia. *Surg. Gynecol. Obstet.* 78:419–424, 1944.

1033 TSUSAKI, T. Über das Jochbein der eingeborenen Formosaner (Seiban). *Folia Anat. Jpn.* 1:217–241, 1922.

1034* TSUSAKI, T. Some abnormalities in the skull of a Formosan aborigine. *Folia Anat. Jpn.* 1:247–250, 1923.

1035 TURNER, W. Supernumerary cervical ribs. *J. Anat. Physiol.* 4:130–139, 1870.

1036 TURNER, W. Case of supernumerary upper incisor tooth. *J. Anat. Physiol.* 21:142–143, 1878.

1037 TURNER, W. A description of cleft sternum. *J. Anat. Physiol.* 14:103–106, 1879.

1038 TURNER, W. A secondary astragalus in the human foot. *J. Anat. Physiol.* 17:82–83, 1883.

1039 TURNER, W. Some variations in the bones of the human carpus. *J. Anat. Physiol.* 17:244–249, 1883.

1040 TURNER, W. A first dorsal vertebra, with a foramina at the root of the transverse process. *J. Anat. Physiol.* 17:255–256, 1883.

1041 TURNER, W. The infra-orbital suture. *J. Anat. Physiol.* 19:218–220, 1885.

1042 TURNER, W. Double right parietal bone in an Australian skull. *J. Anat. Physiol.* 25:473–474, 1891.

1043 TURNER, W. An Australian skull with three supernumerary upper molar teeth. *J. Anat. Physiol.* 34:273–274, 1900.

1044 UHDE, C.W.F. Angeborene Verrenkung der Kniescheibe nach außen. *Arch. Pathol. Anat. Physiol. Klin. Med.* 44:412–417, 1868.

1045 UNDERWOOD, A.S. An inquiry into the anatomy and pathology of the maxillary sinus. *J. Anat. Physiol.* 44:354–369, 1910.

1046 VALENTI, G. Processus sus-condyloïdien de l'humérus chez deux criminels et chez une folle. *Arch. Ital. Biol.* 26:489, 1896.

1047 VALENTI, G. Variété de os nasaux chez un nègre du Soudan. *Arch. Ital. Biol.* 28:468–469, 1897.

1048 VALENTI, G. Pouces et gros orteils avec trois phalanges. *Arch. Ital. Biol.* 34:458, 1900.

1049 VALLOIS, H.V. La signification des apophyses mamillaires et accessoires des vertèbres lombaires. *Soc. Biol. Comptes Rendus Hebdomadaires des Séances et Mémoires* 83:113–115, 1920.

1050* VAN ALYEA, O.E. Ethmoid labyrinth. Anatomic study, with consideration of the clinical significance of its structural characteristics. *Arch. Otolaryngol.* 29:881–902, 1939.

1051 VAN ALYEA, O.E. Sphenoid sinus. Anatomic study, with consideration of the clinical significance of the structural characteristics of the sphenoid sinus. *Arch. Otolarnygol.* 34:225–253, 1941.

1052 VENERUSO, L.C. Unilateral congenital calcaneocuboid synostosis with complete absence of a metatarsal and toe. *J. Bone Joint Surg.* 27:718–719, 1945.

1053 VERMES, E. Ein Fall von hochgradiger Anomalie im inneren Bandapparat eines Kniegelenkes. *Anat. Anz.* 56:427–430, 1923.

1054 VERMES, E. Ein Beitrag zu den Varietäten der Bänder im Inneren des Kniegelenkes. *Anat. Anz.* 57:218–221, 1923–24.

1055* VIDIĆ, B. Incidence of torus palatinus in Yugoslav skulls. *J. Dent. Res.* 45:1511–1515, 1966.

1056 VIEHWEGER, G. Der Processus paratransversarius, eine Varietät am Atlas. *Fortschr. Röntgenstr.* 83:411–412, 1955.

1057 VIEHWEGER, G. Zum Problem der Deutung der knöchernen Gebilde distal des Epicondylus medialis humeri. *Fortschr. Röntgenstr.* 86:643–652, 1957.

1058* VIELBERG, H. Beobachtung seltener Wirbelkörperanomalien an der Lendenwirbelsäule eines Erwachsenen. Diskussionsbeitrag zur gestörten Chordasegmentierung. *Fortschr. Röntgenstr.* 113:60–67, 1970.

1059 VILLEMIN, F., MONTAGNÉ, M. and P. HUARD. Les variations due segment antérieur du bassin chez l'homme et chez la femme. *Soc. Biol. Comptes Rendus Hebdomadaires des Séances et Mémoires* 90:1035–1038, 1924.

1060 VIRCHOW, R. Schwanzbildung beim Menschen. *Arch. Pathol. Anat. Physiol. Klin. Med.* 83:560, 1881.

1061 VITALI, G. Osselets crâniens exoccipito-sus-occipitaux et pétro-exoccipito-sus-occipitaux chez l'homme. *Arch. Ital. Biol.* 44:115, 1905.

1062 VOLKMANN, J. Das Os subtibiale. *Fortschr. Röntgenstr.* 48:225–227, 1933.

1063 VOLKMANN, J. Über den Processus styloideus der Wirbel. *Zentralbl. Chir.* 61:1340–1342, 1934.

1064 VOLUTER, G. The V-test. *Radiol. Clin. [Suppl.]* 28:407, 1–32, 1959.

1065 VRAM, U. Sur un os interstitiel naso-maxillaire dans un crâne humain. *Arch. Ital. Biol.* 32:455, 1899.

1066 WAERN, A. Kongenitale Schlüsselbeinperforation. *Anat. Anz.* 77:221–229, 1933–34.
1067 WAGENSEIL, F. Über einen angeborenen doppelseitigen Ulnadefekt. *Anat. Anz.* 52:439–447, 1919–20.
1068 WAHBY, B. Abnormal nasal bones. *J. Anat. Physiol.* 38:49–51, 1904.
1069 WAKELEY, C.P.G. A case of congenital scoliosis due to suppression of half a vertebra. *J. Anat.* 57:147–148, 1923.
1070 WALDEYER, W. Bemerkungen über Gruben, Kanäle und einige andere Besonderheiten am Körper des Grundbeins (Os basilare). *Int. Monatsschr. Anat. Physiol.* 21:311–318, 1904.
1071 WALKER, C. Absence of pre-maxilla. *J. Anat.* 51:392–395, 1917.
1072* WARWICK, R. A juvenile skull exhibiting duplication of the optic canals and subdivision of the superior orbital fissure. *J. Anat.* 85:289–290, 1951.
1073 WASCHULEWSKI, H. Submalleoläre inkonstante Skelettelemente und malleoläre Knochenkerne. *Fortschr. Röntgenstr.* 86:492–499, 1957.
1074 WATT, J.C. Anatomy of a seven months' foetus exhibiting bilateral absence of the ulna accompanied by monodactyly (and also diaphragmatic hernia). *Am. J. Anat.* 22:385–437, 1917.
1075 WAYBURN, E. Solitary pulmonary tumor. Cyst-like tumors associated with anomalies of ribs. *Am. Rev. Tuberc.* 54:413–417, 1946.
1076 WEINOLD, H. Untersuchungen über das Wachstum des Schädels unter physiologischen und pathologischen Verhältnissen. *Beitr. Pathol. Anat.* 70:311–341, 1922.
1077 WEISS, O. Über die Anatomie der Übergangswirbel an der Grenze von Lendenwirbelsäule und Kreuzbein und ihre klinische Bedeutung. *Z. Anat. Entwicklungsgesch.* 90:533–550, 1930.
1078 WEITZNER, I. Congenital talonavicular synostosis associated with hereditary multiple ankylosing arthropathies. *Am. J. Roentgenol.* 56:185–188, 1946.
1079 WELLS, L.H. Congenital deficiency of the vertebral pedicle. *Anat. Rec.* 145:193–196, 1963.
1080 WENT, H. Zum klinischen Bild der Atlasassimilation. *Fortschr. Röntgenstr.* 89:213–219, 1958.
1081+ WETTSTEIN, W. Über heterotrope Zahnretention. Dissertation Zürich, 1933.
1082* WHITE, J.C., POPPEL, M.H. and R. ADAMS. Congenital malformation of the first thoracic rib. *Surg. Gynecol. Obstet.* 81:643–659, 1945.
1083 WHITE, J.D. Abnormalities of the bony thorax. *Br. J. Radiol.* 2:351–355, 1929.
1084 WHITE, L.E. An anatomic and x-ray study of the optic canal in cases of optic nerve involvement. *Ann. Otol. Rhinol. Laryngol.* 33:121–140, 1924.
1085 WICHTL, O. Tuberculum intercondylicum quartum tibiae. *Röntgenpraxis* 13:397–399, 1941.
1086 WILK, E. and T. HÜLSTER. Über eine seltene Rippenanomalie. *Fortschr. Röntgenstr.* 86:531–532, 1957.
1087* WILLIS, T.A. The lumbo-sacral vertebral column in man, its stability of form and function. *Am. J. Anat.* 32:95–123, 1923.
1088 WILLOCK, E.F. An os interfrontale. *J. Anat.* 59:439–441, 1925.
1089* WILNER, D. Diagnostic problems in fractures of the foot and ankle. *Am. J. Roentgenol.* 55:594–616, 1946.
1090 WILSON, G. Hereditary polydactylism. *J. Anat. Physiol.* 30:437–449, 1896.
1091 WINDLE, B. Three cases of malformations connected with the face. *Anat. Anz.* 4:219–223, 1889.
1092 WINDLE, B.C.A. Occurrence of an additional phalanx in the human pollex. *J. Anat. Physiol.* 26:100–116, 1892.
1093 WISSLER, H. Fenestrae parietales symmetricae. *Arch. Kinderheilk* 115:217–226, 1938.
1094* WITT, C.M. The supracondyloid process of the humerus. *J. Missouri Med. Assoc.* 47:445–446, 1950.
1095* WOOD, V.E. Duplication of the index finger. *J. Bone Joint Surg. (Am.)* 52:569–573, 1970.
1096 WOOD-JONES, F. Variations of the first rib, associated with changes in the constitution of the brachial plexus. *J. Anat. Physiol.* 45:249–255, 1911.
1097 WOOD-JONES, F. The non-metrical morphological characters of the skull as criteria for racial diagnosis. *J. Anat.* 65:179–195, 1931.
1098 WOOD-JONES, F. The non-metrical morphological characters of the skull as criteria for racial diagnosis. Part II. *J. Anat.* 65:368–378, 1931.
1099 WOOD-JONES, F. The non-metrical morphological characters of the skull as criteria for racial diagnosis. Part III. *J. Anat.* 65:438–445, 1931.
1100 WOOD-JONES, F. Contrasting types of Australian skulls. *J. Anat.* 68:323–330, 1933–34.
1101 WOOD-JONES, F. Skulls from the Purari Plateau, New Guinea. *J. Anat.* 70:405–409, 1935–36.
1102 WOOD-JONES, F. The cervical vertebrae of the Australian native. *J. Anat.* 72:411–415, 1937–38.
1103 WRIGHT, W. An Os centrale (Rosenberg) partially united to the scaphoid. *Anat. Anz.* 23:211–212, 1903.
1104 WRIGHT, W. Case of accessory patellae in the human subject, with remarks on emargination of the patella. *J. Anat. Physiol.* 38:65–67, 1904.
1105* WUTH, E.A. Über angeborenen Mangel, sowie Herkunft und Zweck der Kniescheibe. *Arch. Klin. Chir.* 58:900–917, 1899.
1106 WÜTSCHKE, J. Ein Fall von linksseitiger doppelter Patella. *Fortschr. Röntgenstr.* 78:218–220, 1953.
1107* WÜTSCHKE, J. Patella partita und Patella duplex (kritische Betrachtung zur Differentialdiagnose). *Fortschr. Röntgenstr.* 104:260–263, 1966.
1108 WUTZER, C.W. Angeborene Mißbildungen des Kniegelenks. *Arch. Anat. Physiol.* 385–390, 1835.
1109 YANO, K. Das Os sesamoideum muscli gastrocnemii lateralis bei den Japanern. *Folia Anat. Jpn.* 6:241–246, 1928.
1110 YOUNG, M. and J.G.H. INCE. Transmutation of vertebrae in the lumbosacral region of the human spine. *J. Anat.* 74:369–373, 1939–40.
1111* YOUNG, M. and J.G.H. INCE. A radiographic comparison of the male and female pelvis. *J. Anat.* 74:374–385, 1939–40.
1112* ZABOROWSKI, Z. The sagittal foramen of the atlas. *Folia Morphol.* 34:375–381, 1975.
1113* ZANDER, G. "Os acetabuli" and other bone nuclei, periarticular calcifications at the hip-joint. *Acta Radiol.* 24:317–327, 1943.
1114 ZARFL, M. Fenestrae parietales symmetricae. *Z. Kinderheilk* 57:54–66, 1936.
1115 ZELIGS, I.M. Congenital absence of the sacrum. *Arch. Surg.* 41:1220–1228, 1940.
1116* ZELLWEGER, H.K., THEILER, K. and F. LARCHER. Über die Dysostosis cleidocranialis. *Helv. Paediatr. Acta.* 5:264–278, 1950.
1117 ZIMMER, E.A. Die röntgenologische Untersuchung der Atlasspaltbildung. *Acta Radiol.* 18:842–850, 1937.
1118* ZIMMER, E.A. Krankheiten. Verletzungen und Varietäten des Os naviculare pedis. *Arch. Orthop. Unfall-Chir.* 38:396–411, 1938.
1119* ZIMMERN, A., LAURET and R. WEILL. Sacralisation vraie de la Ve lombaire et algies sciatiques. *La Presse Medicale* 30:698–701, 1922.
1120* ZIOLKOWSKI, M. and W. KURLEJ. Subarcuate fossa. *Folia Morphol.* 42:175–186, 1983.
1121 ZUCKERKANDL, E. Bericht des Wiener anatomischen Institutes über das Quinquennium 1874–1879. *Med. Jahrb. (Wien)* 1880, pp. 1–47.
1122 ZUCKERKANDL, E. Über die rudimentäre Bildung der Jochbeine und Jochbogen im Gesichtsskelete des Menschen. *Med. Jahrb. (Wien)* 1880, pp. 103–123.
1123 ZUCKERKANDL, E. Neue Mittheilungen über Coalition von Fußwurzelknochen. *Med. Jahrb. (Wien)* 1880, pp. 125–134.
1124 ZUKSCHWERDT, L. Processus supracondyloideus humeri und Foramen supracondyloideum. *Fortschr. Röntgenstr.* 40:79, 1929.

Books

1125 AASAR, Y.H. *Anatomical Anomalies*. Fouad I University Press, Cairo, 1947.
1126 ANSON, B.J. *Atlas of Human Anatomy*. W.B. Saunders Co., Philadelphia, 1950.
1127 ANSON, B.J. and W.G. MADDOCK. *Callander's Surgical Anatomy*. W.B. Saunders Co., Philadelphia and London, 1952.
1128 ANSON, B.J., Ed. *Morris' Human Anatomy*, 12th Ed. The Blakiston Division, McGraw-Hill Book Co., New York, 1966.
1129 BENNINGHOFF, A. *Lehrbuch der Anatomie des Menschen*. Urban & Schwarzenberg, Munich, 1952.
1130 BIRKNER, R. *Normal Radiologic Patterns and Variances of the Human Skeleton. An X-Ray Atlas of Adults and Children*. Urban & Schwarzenberg, Baltimore and Munich, 1978.
1131 BUNNELL, S. *Surgery of the Hand*, 2nd Ed. J.B. Lippincott Co., Philadelphia, 1948.

1132+ DWIGHT, T. *A Clinical Atlas. Variations of the Bones of the Hands and Feet*. J.B. Lippincott Co., Philadelphia, 1907.

1133+ EULER, H. *Die Anomalien, Fehlbildungen und Verstümmelungen der menschlichen Zähne*. Lehman, Munich, 1939.

1134 GARDNER, E., GRAY, D.J. and R. O'RAHILLY. *Anatomy*, 3rd Ed. W.B. Saunders Co., Philadelphia, 1969.

1135 GEGENBAUR, C. *Lehrbuch der Anatomie des Menschen*, 3rd Ed. V. Wilhelm Engelmann, Leipzig, 1888.

1136 HENLE, J. *Handbuch der Systematischen Anatomie des Menschen*, 3 volumes. Von Friedrich Vieweg und Sohn, Braunschweig, 1868.

1137 HUBER, G.C., Ed. *Piersol's Human Anatomy*, 9th Ed. J.B. Lippincott Co., Philadelphia, 1930.

1138 HYRTL, J. *Lehrbuch der Anatomie des Menschen*, 19th Ed. Wilhelm Braumüller, Wien, 1887.

1139 JACKSON, C.M., Ed. *Morris' Human Anatomy*, 9th Ed. P. Blakiston's Son & Co., Inc., Philadelphia, 1933.

1140 KOPSCH, F. *Rauber's Lehrbuch der Anatomie des Menschen*. Georg Thieme, Leipzig, 1908.

1141 LATARJET, A. *Testut's Traite d'Anatomie Humaine*, 9th Ed. G. Doin & Cie., Paris, 1948.

1142+ LE DOUBLE, A.F. *Traité des Variations des Os du Crâne de l'Homme*. Vigot Paris, 1903.

1143+ LE DOUBLE, A.F. *Traité des Variations des Os de la Face de l'Homme*, Vigot, Paris, 1906.

1144+ LE DOUBLE, A.F. *Traité des Variations de la Colonne Vertebrale*. Vigot, Paris, 1912.

1145+ MÜLLER, W. *Die angeborenen Fehlbildungen der menschlichen Hand*. Thieme, Leipzig, 1937.

1146 SCHAEFER, E.A., SYMINGTON, J. and T.H. BRYCE, Ed. *Quain's Anatomy*, 11th Ed. Longmans, Green, and Co., London, 1915.

1147 WHITNALL, S.E. *The Anatomy of the Human Orbit*, 2nd Ed. Oxford University Press, London, 1932.

1148 ZIMMER, E.A. *A. Köhler/E.A. Zimmer Borderlands of the Normal and Early Pathologic in Skeletal Roentgenology*, 11th Ed., translated and edited by S.P. Wilk. Grune & Stratton, New York, 1968.

Glossary of Terms

Their origin and definition related to the Skeleton.

The following abbreviations are used: G., Greek, and L., Latin.

Acromion (G. *akron*, tip, + *omos*, shoulder). The tip of the shoulder.
Alar (L. *ala*, wing). A wing-like structure.
Arm (L. *armus*, fore-quarter of an animal; G. *harmos*, a shoulder joint). The segment of the upper limb between the shoulder and elbow. The term *brachium* (L., G. *brachion*) is preferred usage.
Aspera (L. *asper*, rough). The roughened surface of a bone.
Astragalus (L., G. *astragalos*, ball of the ankle joint, dice). The talus.
Atlas (G. *Atlas*, from Greek mythology, a Titan who supported the world on his shoulders). First cervical vertebra.
Axis (L. *axis, axle*, axis; G. *axon*, axis). The odontoid or toothed vertebra. The second cervical vertebra about which the head rotates.
Bifid (L. *bifidus*, cleft into two parts). Referring to a bone that is more or less split or cleft.
Carpal (G. *carpus*, wrist). The eight bones of the wrist proximal to the hand.
Cervical (L. *cervix*, neck). Relating to the neck and to the "neck" of any structure (bone or soft tissue).
Clavicle (L. *clavicula*, a small key). The collar bone, os clavicula.
Cleido- (G. *kleis*, clavicle). Relating to the clavicle.
Clinoid (G. *kline*, bed, + *eidos*, form). Bed-shaped or resembling a bed.
Coccyx (G. *kokkyx*, a cuckoo, whose bill the bony structure is said to resemble). Os coccygis.
Conchae (L. *concha*, shell). Turbinate bones of the nose.
Condyle (L. *condylus*; G. *kondylos*, knuckle). Rounded articular surface of bone extremity.
Costo- (L. *costa*, rib). Referring to the ribs.
Cubitis (L. *cubitus*, elbow). Pertaining to the joint between arm and forearm.
Dens (L. *dens*, tooth). Referring to the tooth-like structure of the axis upon which the atlas rotates.
Dorsal (L. *dorsalis*, from *dorsum*, back). Term once used for notation of thoracic region, e.g., first dorsal vertebra.
Epistropheus (G. *epistropheus*, the pivot). Second cervical vertebra.
Ethmoid (G. *ethmos*, sieve). Relating to the os ethmoidale. A sieve-like structure, cribriform.
Femur (L. *femur*, thigh). The thigh bone.
Fibula (L. *fibula*, a clasp or buckle). A bone of the leg. The smaller and more lateral of the two bones of the leg.
Fissura (L. *findere*, to cleave). A fissure or deep cleft.
Foramen (L. *forare*, to pierce). An aperture.
Frontal (L. *frons*, front). Referring to the frontal bone.
Hemi- (G. *hemi*, half). A prefix indicating half of a structure.
Humerus (L. *humerus*, shoulder). Referring to the bone of the brachium or arm.
Hyoid (G. *hyoeides*, shaped like the Greek letter upsilon). A "U"- or "V"-shaped structure. The os hyoideum.
Lacrimal (L. *lacrima*, a tear). Relating to tears, their secretion, organ, and bone involved therewith. The os lacrimale.
Lumbar (L. *lumbus*, a loin). Referring to those vertebrae between the thorax and pelvis.
Lunato- (L. *lunatus*, semilunar). Pertaining to the os lunatum.
Mandible (L. *mandibula*, a jaw). The jaw bone.
Manubrium (L. *manubrium*, handle). Part of the sternum.

Mastoid (G. *mastos*, breast, + *eidos*, resemblance). Mastoid process of temporal bone.
Maxillary (L. *maxilla*, jaw bone). Relating to the upper jaw bone or cheek.
Nasal (L. *nasus*, nose). Os nasale.
Occipital (L. *occiput*, back of the head). Os occipitale.
Odontoideum (G. *odous*, a tooth, + *eidos*, form). See dens.
Olecranon (G. *olene*, ulna + *kranion*, head). Curved, pointed proximal end of the ulna.
Omo- (G. *omos*, shoulder). Related to the shoulder.
Orbital (L. *orbita*, a circle). Referring to the cavity formed from parts of seven bones.
Os (L. *os*, bone). Referring to a bone.
Palatine (L. *palatum*, palate). Os palatum.
Parietal (L. *pares*, wall). The wall of any cavity. Os parietale.
Patella (L. *patina*, a small shallow disk). The kneecap. The large sesamoid bone in the tendon of the extensors of the leg.
Pelvis (L., G. *pyelos*, an oblong trough, basin). Ring of bone at the lower end of the trunk.
Petrous (L. *petrosus*, rock-like; G. *petra*, stone). Petrous part of temporal bone.
Phalanx (L. from G. *phalanx*, line of soldiers). Bones of fingers and toes.
Pterygo- (G. *pteryx*, wing). A wing-like structure.
Radius (L. *radius*, spoke of a wheel, a rod). The shorter and more lateral of the two bones of the forearm.
Sacrum (L. *sacrum*, sacred bone). Os sacrum. So-named because it was believed to resist natural disintegration after death and to serve as the basis for the resurrected body.
Scapula (L. *scapula*, shoulder blade; G. *skaptein*, to dig). The shoulder blade.
Sinus (L. *sinus*, cavity, channel, hollow). A cavity in a bone, e.g., frontal sinus.
Sphenoid (G. *sphen*, wedge, + *eidos*, resemblance). Os sphenoidale.
Squamous (L. *squama*, scale or plate-like structure). Squama of temporal bone, for example.
Sternum (G. *sternon*, the chest). The breast bone consisting of the manubrium, body, and xiphoid process.
Styloid (G. *stylos*, pillar, post). Styloid process of temporal bone.
Supra- (L. *supra*, above). Prefix denoting a position above the named structure that follows it, e.g., suprasternal bones.
Suture (L. *sutura*, a seam, from *suere*, to sew). A specific type of joint between two bones.
Tarsal (G. *tarsos*, sole of foot). Any tarsal bone.
Temporal (L. *tempus*, time or temple). Temporal bone.
Thoracic (G. *thorax*, breastplate, the chest). Referring to the chest formed of 12 vertebrae, 12 ribs, and sternum.
Ulna (L., G. *olene*, elbow). The elbow bone. The larger and more medial of the two bones of the forearm.
Vertebra (L. *vertebra*, joint; *verto*, to turn). In humans, one of the 33 bony segments of the spinal column.
Volar (L. *vola*, a concave or hollow surface). Pertaining to the palm or sole. The flexor surface of the forearm, wrist, or hand.
Vomer (L. *vomer*, ploughshare). A flat bone forming part of the nasal septum.
Xiphoid (G. *xiphos*, sword + *eidos*, resemblance). Sword-shaped distal segment of the sternum; ensiform.
Zygomatic (G. *zygoma*, a bar or bolt). Os zygomaticum.

Terminology: Synonyms

Terminology for the skeleton from the older English, French, and German literature related to more modern usage.

Old Terminology	Modern Usage
Accessory calcaneus.	Os trochleare calcanei.
Accessory tarsal scaphoid s.* the sesamoid bone s. praehallux s. os naviculare secundarium s. os naviculare accessorium.	Os tibiale externum.
Ala descendens s. palatina.	Pterygoid process.
Ala lateralis.	Greater wing of sphenoid.
Ala magna.	Greater wing of sphenoid.
Ala orbitalis.	Lesser wing of sphenoid.
Ala parva.	Lesser wing of sphenoid.
Ala parva Ingrassiae.	Lesser wing of sphenoid.
Ala parva Ingrassiae.	Spine of sphenoid.
Ala temporalis.	Greater wing of sphenoid.
Os alare.	Os sphenoidale.
Antrum Highmori.	Maxillary sinus.
Antrum of Highmore.	Maxillary sinus.
Apertura pelvis abdominalis.	Greater pelvic aperture.
Apertura pelvis perinealis.	Inferior pelvic aperture.
Apertura sinus maxillaris.	Maxillary sinus.
Apertura spinalis.	Vertebral foramen.
Apertura spuria canalis Fallopiae.	Hiatus canalis n. petrosi minoris (inferior petrosal nerve).
Armspindel.	Radius.
Astragalus.	Talus.
Basis, os occipitis.	Pars basilar, occipital bone.
Bassinet.	Pelvis.
Os Bertin.	Sphenoidal conchae.
Os brachii.	Humerus.
Ossa bregmatis.	Ossa parietalia.
Os Breschet.	Os suprasternale.
Os calcis.	Calcaneus.
Canalis alveolaris inferioris.	Mandibular canal.
Canalis mandibularis.	Mandibular canal.
Canalis maxillaris superior.	Foramen rotundum.
Canalis recurrens.	Pterygoid canal.
Canalis rotundis.	Foramen rotundum.
Canalis tensoris tympani.	Semicanal for m. tensor tympani.
Canalis vidianus.	Pterygoid canal.
Canalis zygomatico-facialis.	Zygomaticofacial canal.
Canna major.	Tibia.
Canna minor.	Fibula.
Capitulum.	Head of radius.
Cartilage of Meckel.	Mandibular cartilage.
Cartilage of Morgagni.	Cuneiform cartilage.
Cartilage of Reichert.	Mesenchyme of second branchial arch from which the styloid processes, the stylohyoid ligaments, and the lesser cornu of the hyoid bone is formed.
Cartilage of Santorini.	Corniculate cartilage.
Cartilage of Wrisberg.	Cuneiform cartilage.
Cartilago ensiformis s. xiphoidea.	Xiphoid cartilage.
Cingulum extremitatis inferioris.	Pelvic girdle.

* s. = synonymus

Old Terminology	Modern Usage
Civinini's canal.	Iter chordae anterius.
Collum.	Neck of humerus.
Concha Morgagniana.	Superior concha.
Concha nasalis medialis.	Middle nasal concha.
Condylos extensorius.	Lateral epicondyle of humerus.
Condylus externus.	Lateral epicondyle of humerus.
Condylus internus.	Medial epicondyle of humerus.
Condylus flexorius.	Medial epicondyle of humerus.
Cornu sphenoidale.	Concha sphenoidalis.
Corpus, os occipitis.	Pars basilar, occipital bone.
Côte.	Rib.
Os cribriforme.	Os ethmoidale.
Crista s. pecten ossis pubis.	Pecten pubis.
Crista femoris.	Linea aspera.
Crista iliopectinea.	Pecten pubis.
Cubitis.	Ulna.
Os cuneiforme.	Os hamatum, manus.
Os cuneiforme.	Os sphenoidale.
Os cuneiforme.	Os triquetrum, manus.
Dens.	Odontoid process.
Dornfortsatz.	Spinous process.
Vv. dorsales.	Thoracic vertebrae.
Dorsum ephippii.	Dorsum sellae.
Dreiseitige Fontanelle.	Posterior fontanelle.
Ectrokeirie.	Partial or total absence of hand.
Elle.	Ulna.
Ellenbogenbein.	Ulna.
Eminentia capitata.	Capitulum of humerus.
Eminentia capitata.	Head of radius.
Eminentia trochlearis.	Capitulum of humerus.
Ensiform process.	Xiphoid process.
Epactalia.	Sutural bones.
Ephippium.	Sella turcica.
Epihyal bone.	An ossified stylohyoid ligament.
Episternal bone.	Os suprasternale.
Epipteric bone.	A wormian bone.
Epistropheus.	Axis, second cervical vertebra.
Erbsenbeine.	Os pisiforme.
Escrescentia ossis humeri insolita.	Supracondylar process of humerus.
Exitus pelvis.	Inferior pelvic aperture.
Exoccipital bone.	Pars lateralis ossis occipitalis.
Fenestra cochleae.	Cochlear window.
Fenestra ovalis.	Vestibular window.
Fenestra rotunda.	Cochlear window.
Fenestra semiovalis.	Vestibular window.
Fenestra vestibuli.	Vestibular window.
Fersenbein.	Calcaneus.
First cuneiform bone.	Os cuneiforme mediale.
Fissura Glaseri.	Petrotympanic fissure.
Fissura orbitalis superiorius.	Superior orbital fissure.
Fissura petrotympanica.	Petrotympanic fissure.
Fissura pterygo-palatina.	Pterygoid notch.
Fissura sphenoidalis.	Superior orbital fissure.
Flower's bone.	Epipteric or wormian bone.

Human Anatomic Variation

Flügelbein.	Os sphenoidale.	Grosse Fontanelle.	Anterior fontanelle.
Focile majus.	Tibia.		
Focile majus.	Ulna.	Hakenbein.	Os hamatum.
Focile minus.	Fibula.	Hamulus frontalis.	Pterygoid hamulus.
Focile minus.	Radius.	Heiligbein.	Os sacrum.
Fontanelle Casseri.	Mastoid fontanelle.	Hiatus canalis facialis.	Hiatus canalis n. petrosi minoris.
Fontanelle minor.	Posterior fontanelle.		
Fontanelle triangularis.	Posterior fontanelle.	Hiatus canalis Fallopiae.	Hiatus canalis n. petrosi minoris.
Fonticuli lateral anterior.	Sphenoidal fontanelle.		
Fonticulus anterior.	Anterior fontanellc.	Hiatus maxillaris.	Maxillary sinus.
Fonticulus major.	Anterior fontanelle.	Hinterhauptsbein.	Os occipitale.
Fonticulus quadrangularis.	Anterior fontanelle.	Hinterhauptsfontanelle.	Posterior fontanelle.
Foramen alveolare interioris.	Mandibular foramen.	Huguier's canal.	Iter chordae anterius.
Foramen condyloideum anterior.	Hypoglossal canal.	Os Incae.	Os interparietale.
Foramen condyloideum posterior.	Condyloid canal.	Incarial bone.	Os interparietale.
		Incisura frontalis.	Supraorbital foramen or notch.
Foramen lacerum posterius.	Jugular foramen.		
Foramen maxillare posterior.	Mandibular foramen.	Incisura iliaca major s. superior.	Greater sciatic notch.
Foramen maxillaris anterior.	Mental foramen.		
Foramen medullae spinalis.	Vertebral foramen.	Incisura ischiadica major.	Greater sciatic notch.
Foramen occipitale.	Foramen magnum.	Incisura ischiadica minor.	Lesser sciatic notch.
Foramen occipitale magnum.	Foramen magnum.	Incisura mandibulae.	Mandibular notch.
Foramen opticum.	Optic canal.	Incisura scapularis.	Scapular notch.
Foramen ovale.	Obturator foramen.	Incisura semilunaris.	Ulnar notch of radius.
Foramen palatinum anterius.	Incisive foramen or canal.	Incisura semilunaris.	Scapular notch.
Foramen of Stenson.	Incisive canal.	Incisura semilunaris s. sigmoidea.	Mandibular notch.
Foramen of Vesalius.	Foramen between rotundum and ovale (Vesalius' foramen).		
		Incisura supraorbitalis.	Supraorbital foramen or notch.
Foramen zygomatico anterior.	Foramen of zygomaticofacial canal.	Incisura suprascapularis.	Scapular notch.
		Incisura vertebralis inferioris.	Inferior vertebral notch.
Foramen zygomatico externus.	Foramen of zygomaticofacial canal.	Incisura vertebralis major.	Superior vertebra notch.
		Incisura vertebralis minor.	Inferior vertebral notch.
Foramen zygomatico-faciale.	Foramen of zygomaticofacial canal.	Incisura vertebralis superioris.	Superior vertebral notch.
Foramen zygomatico internus.	Foramen of zygomatico-temporal canal.	Os inferior maxillary.	Mandible.
		Os infranavicular s. os paracuneiforme I s. os naviculocuneiforme I dorsale.	Os infranaviculare.
Foramen zygomatico internus.	Foramen of zygomatico-orbital canal.		
Foramen zygomatico-orbitale.	Foramen of zygomatico-orbital canal.	Os innominatum.	Os coxae.
		Ossa intercalaria.	Sutural bones.
Foramen zygomatico posterior.	Foramen of zygomatico-temporal canal.	Intermaxillary bone.	Os incisivum.
		Os intermedium s. talus accessorius s. talus secundarius.	Trigonum, pedis.
Foramen zygomatico superior.	Foramen of zygomatico-orbital canal.		
Foramen zygomatico-temporalis.	Foramen of zygomatico-temporal canal.	Introitus pelvis.	Greater pelvic aperture.
		Iter chordae posterius.	Canaliculus chorda tympani.
Fossa anterior minus.	Radial fossa of humerus.		
Fossa articularis maxillae inferioris.	Mandibular fossa.	Jochbeine.	Ossa zygomatica.
		Jochbogen.	Zygomatic arch.
Fossa capituli radii.	Radial fossa of humerus.	Jochfortsatz.	Zygomatic process.
Fossa posterior.	Olecranon fossa of humerus.	Jugal bone.	Os zygomaticum.
Fossa processus coronoidei.	Coronoid fossa of humerus.	Ossa jugalia.	Ossa zygomatica.
Fossa suprasphenoidalis.	Hypophyseal fossa.	Jugum petrosum.	Eminentia arcuata, temporal bone.
Fossa supratrochlearis anterior.	Coronoid fossa of humerus.		
Fossa pituitaria.	Hypophyseal fossa.	Kahnbein.	Os naviculare, pedis.
Fossae costales.	Costal pit.	Kahnbein.	Os scaphoideum.
		Keilbein.	Os sphenoidale.
Gaumenbeine.	Ossa palatina.	Keilbeinfontanelle.	Sphenoidal fontanelle.
Gladiolus.	Body of sternum.	Kleine Becken.	True pelvis.
Gleider.	Limbs.	Kleine Fontanelle.	Posterior fontanelle.
Goethe's bone.	Sutural bone at the lambda or os incisivum.	Knöchelbein.	Talus.
		Kopfbein.	Os capitatum.
Gouttière de torsion.	Groove of the radial nerve.	Kreuzbein.	Os sacrum.
Grand os.	Os capitatum.	Kuckucksbein.	Os coccygis.
Grätenecke.	Acromion.		
Greater multangular bone.	Os trapezium.	Lendenwirbel.	Lumbar vertebrae.
Grosse Becken.	Greater pelvis.	Lentiform bone.	Os pisiforme.

Skeleton

Lesser multangular bone.	Os trapezoideum.	Pars horizontalis.	Horizontal plate of palatine bone.
Ligament, great sacrosciatic.	Sacrotuberous ligament.	Pars mamillaris.	Mastoid portion, temporal bone.
Ligament, lesser sacrosciatic.	Sacrospinous ligament.	Pars mastoideus.	Mastoid portion, temporal bone.
Ligamentum cruciatum s. ligamentum lambdoideum.	Inferior extensor retinaculum.	Pars nasalis.	Perpendicular plate of palatine bone.
Linea iliopectinae.	Pecten pubis.	Pars palatina.	Horizontal plate of palatine bone.
Linea intertrochanterica anterior.	Intertrochanteric crest.	Pars peronea metatarsalis I.	Os cuneometatarsale I plantare.
Linea obliqua femoris.	Intertrochanteric crest.	Pars perpendicularis.	Perpendicular plate of palatine bone.
Linea semicircularis ossis parietalis.	Parietal margin, temporal bone.	Os pelvis.	Os coxae.
Linea semicircularis superior ossis occipitis.	Superior nuchal line.	Periotic bone.	Petrous and mastoid parts of temporal bone.
Linea temporalis.	Parietal margin, temporal bone.	Os perone.	Fibula.
Lunula scapulae.	Scapular notch.	Os petroso-mastoideum.	Petrous portion, temporal bone.
Os magnum.	Os capitatum.	Pflugscharbein.	Vomer.
Os malar.	Os zygomaticum.	Os Pirie.	Dorsal talonavicular bone.
Ossa malaria.	Ossa zygomatica.	Plana orbitalia.	Pars orbitalis.
Margo coronalis.	Frontal margin, parietal bone.	Os planum.	Lamina papyracea.
Margo frontalis.	Frontal margin, parietal bone.	Planum popliteum.	Popliteal fossa.
Margo mastoideus.	Mastoid angle, parietal bone.	Pomum Adami.	Laryngeal prominence.
Margo parietalis.	Sagittal margin, parietal bone.	Poples.	Popliteal fossa.
Margo sphenoidalis.	Sphenoidal angle, parietal bone.	Porus acusticus externus.	External auditory (acoustic) meatus.
Margo superior.	Sagittal margin, parietal bone.	Porus acustica interior.	Internal auditory (acoustic) meatus.
Maxilla inferior.	Mandibula.	Porus auditorius externus.	External auditory meatus.
Maxilla superior.	Ossa maxillae.	Porus crotaphiticobuccinatorius.	Pterygoalar foramen.
Os maxillare inferior.	Mandibula.	Os precuneiform s. praecuneiforme s. praehallux s. os cuneometatarsale I tibiale s. os paracuneiforme.	Sesamum tibiale anterius.
Mesosternum.	Body of sternum.		
Metasternum.	Xiphoid process.		
Mondbein.	Os lunatum.		
Os multangulum majus.	Os trapezium.	Premaxillary bone.	Os incisivum.
Os multangulum minus.	Os trapezoideum.	Presternum.	Manubrium.
Muschelbeine.	Conchae inferiores.	Processus accessorius.	Accessory process, vertebral.
Nackenwirbel.	Cervical vertebrae.	Processus acromialis.	Acromion.
Nagelbeine.	Ossa lacrimalia.	Processus ad vomerem.	Processus vaginalis of sphenoid.
Nahtdoppler.	Sutural bones.	Processus alaris.	Pterygoid hamulus.
Ossa nasalia.	Ossa nasi.	Processus anconeus.	Olecranon.
Nasenbeine.	Ossa nasi.	Processus ensiformis.	Lesser wing of sphenoid.
Os naviculare (manus).	Os scaphoideum.	Processus ensiformis.	Processus xiphoideus.
Nebenhöcker.	Medial epicondyle of humerus.	Processus epicondyloideus ulnaris s. internus.	Supracondylar process.
Nodus extensorius.	Lateral epicondyle of humerus.	Processus epitrochlearis.	Supracondylar process.
Nodus externus.	Lateral epicondyle of humerus.	Processus lateralis.	Sustentaculum tali.
Nodus internus s. flexorius Arnold.	Medial epicondyle of humerus.	Processus malaris s. jugalis.	Zygomatic process.
		Processus mamillaris.	Mamillary process, vertebral.
Oberarmbein.	Humerus.	Processus mamillaris.	Mastoid process.
Obere Beckenöffnung.	Greater pelvic aperture.	Processus petrosus anterior.	Lingula sphenoidalis.
Oberkieferbeine.	Ossa maxillae.	Processus petrosus ossis sphenoidale posterior superior.	Posterior clinoid process.
Omoplata.	Scapula.	Processus petrosus posterior superior.	Posterior clinoid process.
Os orbiculare.	Os pisiforme.	Processus rostriformis.	Coracoid process.
Ossiculum Bertini.	Sphenoidal concha.	Prominentia laryngea.	Prominence of larynx.
Ossiculum talonaviculare dorsale s. intertaloscaphoid s. dorsal astragaloscaphoid ossicle.	Processus trochlearis ossis scaphoidis.	Pterygoalar foramen.	Foramen ovale.
		Os pyramidale.	Os trapezoidum.
		Os pyramidale.	Os triquetrum.
		Pyramidenbein.	Os triquetrum.
Palatum durum.	Hard palate.	Rabenschnabelfortsatz.	Coracoid process.
Pars ascendens.	Perpendicular plate of palatine bone.	Rami perpendiculares seu adscendentes.	Mandibular rami.
Pars basilaris os occipitis.	Pars basilar, occipital bone.	Ramus ascendens ossis ischii.	Ramus of ischium.
		Ramus descendens ossis ischii.	Body of ischium.

231

Human Anatomic Variation

Ramus descendens ossis pubis.	Inferior ramus of pubis.	Ossa suturarum.	Wormian (sutural) bones.
Ramus horizontalis.	Body of mandible.	Suture occipito-parietalis.	Lambdoid suture.
Ramus horizontalis ossis pubis.	Superior ramus of pubis.	Synchondrosis sacro-iliaca.	Sacroiliac articulation.
Raphogeminantia.	Sutural bones.	Tail bone.	Os coccygis.
Os rhomboides.	Os trapezium.	Os talus accessorius.	Os subtibiale.
Riechbein.	Os ethmoidale.	Thränenbeine.	Ossa lacrimale.
Riolan's bone.	A sutural bone (when present) in petro-occipital suture.	Three-cornered bone.	Os triquetrum.
		Tongue bone.	Os hyoideum.
Rotula.	Patella.	Trapezbein.	Os trapezium.
Rotula.	Trochlea of humerus.	Os trapezium minus.	Os trapezoideum.
Os rotundum.	Os pisiforme.	Trapezoidbein.	Os trapezoideum.
Rückgrat.	Vertebral column.	Os trapezoides.	Os trapezium.
Rückenmarksloch.	Vertebral foramen.	Os triangulare.	Os triquetrum.
Rückenwirbel.	Thoracic vertebrae.	Os tribasilare.	Os basilare.
		Trichterbrust.	Funnel chest.
		Triquetra.	Sutural bones.
Sacred bone.	Os sacrum.	Trochlear process of astragalus.	Os supranaviculare.
Sattelwulst.	Tuberculum sellae.	Tuberculum anterius.	Greater tubercle of humerus.
Os scaphoideum.	Os naviculare, pedis.	Tuberculum ephippii.	Tuberculum sellae.
Schaltknochen.	Sutural bones.	Tuberculum externum.	Greater tubercle of humerus.
Schambein.	Os pubis, superior ramus.	Tuberculum inferius.	Accessory process, vertebral.
Scheitelbeine.	Ossa parietalia.	Tuberculum internum.	Lesser tubercle of humerus.
Schienbein.	Tibia.	Tuberculum majus.	Greater tubercle of humerus.
Schiffbein.	Os naviculare, pedis.	Tuberculum minus.	Lesser tubercle of humerus.
Schiffbein.	Os scaphoideum, manus.	Tuberositas olecrani.	Olecranon process.
Schläfenbeine.	Os temporum or temporale.	Tuberculum posterius.	Greater tubercle of humerus.
Schlüsselbein.	Clavicle.	Tuberculum superius.	Mamillary process, vertebral.
Schulterblatt.	Scapula.	Tuberculum trochlearis.	Capitulum of humerus.
Schultergräte.	Scapular spine.	Tuberositas externus.	Lateral epicondyle of femur.
Schulterhöhe.	Acromion.	Tuberositas internus.	Medial epicondyle of femur.
Schulterschnabel.	Coracoid process.	Ossa turbinata.	Concha inferiores, turbinate bone.
Schwanzbein.	Os coccygis.		
Scroll bone.	Concha nasalis.	Os turbinatum s. spongiosum superior.	Superior concha.
Secondary astragalus s. supertalare.	Os supratalare.		
Secondary os calcis.	Os calcaneum secundarium.	Os unciforme.	Os hamatum.
Seitenwandbeine.	Ossa parietalia.	Unciformis.	Coracoid process.
Semicanalis humeri.	Intertubercular sulcus.	Uncinate process of third cuneiform bone s. processus uncinatus cuneiformis III.	Os unci.
Semicanalis tensor tympani.	Semicanal for m. tensor tympani.		
Os semilunare.	Os lunatum.	Ossa unguis.	Ossa lacrimale.
Septal bone.	Alveolar bone between tooth roots.	Untere Muscheln.	Conchae inferiores.
		Unterkieferbein.	Mandibula.
Siebbein.	Os ethmoidale.	Unterkinnlade.	Mandibula.
Sinus lunatus radii.	Ulnar notch of radius.		
Sinus maximus.	Fossa olecranon of humerus.	Vertebra prominens.	Spinous process, seventh cervical vertebra.
Speiche.	Radius.		
Spina angularis.	Spine of sphenoid.	Vertebrae abdomales.	Lumbar vertebrae.
Spina dorsi.	Vertebral column.	Vertebrae colli.	Cervical vertebrae.
Spina sphenoidalis.	Spine of sphenoid.	Vertebrae dorsalis.	Thoracic vertebrae.
Sprungbein.	Talus.	Ossa verticus.	Ossa parietalia.
Stachelfortsatz.	Spinous process.	Os vespiforme.	Os sphenoidale.
Steissbein.	Os coccygis.	Vierseitige Fontanellé.	Anterior fontanelle.
Stirnbein.	Os frontale.		
Stirnfontanelle.	Anterior fontanelle.	Wangenbeine.	Ossa zygomatica.
Os subcalcanea s. os tuberis calcanei.	Os subcalcis.	Warzenfontanelle.	Mastoid fontanelle.
		Wedge bone.	Os cuneiforme mediale.
Os subrotundum.	Os pisiforme.	Wedge bone (pedis).	Os cuneiforme laterale.
Sulcus bicipitalis.	Intertubercular sulcus.	Wespenbein.	Os sphenoidale.
Sulcus musculosus.	Semicanal for m. tensor tympani.	Wormiana.	Sutural or wormian bones.
Superficies articulares laterales.	Costal pit.	Yoke bone.	Os zygomaticum.
Suprainterparietal bone.	A sutural bone (when present) at the posterior part of the sagittal suture.	Zahnfortsatz.	Odontoid process.
		Zapfentheil.	Pars basilar, occipital bone.
		Zitzenfortsatz.	Mastoid process.
Sutura fronto-parietalis.	Coronal suture.	Zitzentheil.	Mastoid portion, temporal bone.
Sutura parietalis.	Sutura sagittalis, parietal bone.		
		Zwickelbeine.	Sutural bones.

Atlas

Table of Contents

The authors have retained wherever possible the style and content of the original figure legends.

Muscles

FIG. 1.
M. Omocervicalis (Levator Claviculae) · M. Infraspinatohumeralis · M. Sternocleidomastoideus (Anomalous) · M. Pectoralis Major (Anomalous) · M. Tensor Semivaginae Articulationis Humeroscapularis or M. Sternochondrocoracoideus 249

FIG. 2.
M. Deltoideus (Anomalous) · M. Infraspinatohumeralis · M. Sternocleidomastoideus (Anomalous) · M. Pectoralis Major (Anomalous) · M. Costocoracoideus 250

FIG. 3.
Congenital Absence of Muscles of Neck and Thorax 251

FIG. 4.
A Peculiar Digastric Muscle 251

FIG. 5.
Examples of Occipitohyoid and Cleido-occipital Muscles 252

FIG. 6.
M. Levator Glandulae Thyroideae 252

FIG. 7.
M. Thyroidei Marginales Inferior · M. Incisurae Mediae Obliquus · M. Incisurae Mediae Obliquus Bicaudatus · M. Cricothyroideus (Anomalous) · M. Keratoarytenoideus · M. Thyrotrachealis · M. Thyrotrachealis Biceps · M. Levator Glandulae Thyroideae 253

FIG. 8.
Anomalous Muscles of Larynx · M. Thyroarytenoideus Superior · M. Arymembranaceus . 254

FIG. 9.
M. Atlanticomastoideus · Supernumerary M. Rectus Capitis Minor · M. Cleidoepistrophicus · M. Transversalis Cervicis Anticus (Anomalous) 255

FIG. 10.
Diagram of Varieties of Muscles of Neck, Shoulder, and Chest 255

FIG. 11.
Varieties of Muscles of Neck, Shoulder, and Chest 256

FIG. 12.
M. Occipitoscapularis (Anomalous) 257

FIG. 13.
M. Omotrachelien (M. Levator Claviculae) . . 257

FIG. 14.
Quadricipital Type of Sternocleidomastoid, Showing Elemental Components of the Muscle (after Maubrac). 258

FIG. 15.
M. Sternalis 258

FIG. 16.
M. Sternalis 259

FIG. 17.
M. Axillary Arch · M. Pectoralis Quartus · M. Sternalis 260

FIG. 18.
Diagrammatic Sketch of Some Anomalous Muscles of the Pectoral Region 261

FIG. 19.
Subdivisions of M. Pectoralis Major · M. Sternalis 261

FIG. 20.
M. Tensor Semivaginae Beneath M. Pectoralis Major and Superficial to M. Pectoralis Minor 262

FIG. 21.
M. Tensor Semivaginae Deep to M. Pectoralis Major 262

FIG. 22.
M. Sternocleidomastoideus (Anomalous) · M. Sternoclavicularis · M. Sternoscapularis . . 263

Human Anatomic Variation

FIG. 23.
M. Sternochondrocoracoideus · M. Tensor
Semivaginae 264

FIG. 24.
M. Chondroepitrochlearis 264

FIG. 25.
M. Costoepitrochlearis. · M. Pectoralis Major
(Anomalous) 265

FIG. 26.
M. Chondroepitrochlearis · M. Dorsoepitrochlearis 265

FIG. 27.
M. Pectoralis Minor (Anomalous) · M. Chondrocoracoideus 266

FIG. 28.
M. Costocoracoideus 266

FIG. 29.
M. Supracostalis 267

FIG. 30.
M. Supracostalis (Anterior, Anomalous) ·
(Anomalous Transverse Cervical Artery) ·
(Anomalous Thyrocervical Trunk) 267

FIG. 31.
M. Supraclavicularis · M. Sternoclavicularis
Anticus s. Preclavicularis Medialis Singularis ·
M. Interclavicularis Anticus Digastricus · M.
Acromioclavicularis s. Preclavicularis
Lateralis 268

FIG. 32.
M. Trapezius (Anomalous) · M. Rhomboideus Minimus 269

FIG. 33.
M. Coracoclavicularis · (Suprascapular Artery
and Nerve Passing Beneath the Transverse
Scapular Ligament) 269

FIG. 34.
M. Deltoideus (Anomalous) · Deficient M.
Pronator Quadratus · M. Tensor Capsulae
Radiocubitalis 270

FIG. 35.
M. Coracoclavicularis 271

FIG. 36.
Axillary Arches 272

FIG. 37.
Unilateral Absence of M. Serratus Anterior
(Right Side) · Anomalous Fascicle of M.
Latissimus Dorsi 273

FIG. 38.
Unilateral Absence of M. Serratus Anterior
(Right Side) · Anomalous Fascicle of M.
Latissimus Dorsi 274

FIG. 39.
M. Infraspinatus Superficialis 274

FIG. 40.
M. Costodeltoideus (Albini) · M. Coracobrachialis (Anomalous) (Cruveilhier) · M.
Subscapularis (Anomalous) (Gruber) 275

FIG. 41.
Bilateral Accessory Subscapularis. Axillary
and Inferior Subscapular Nerves Bilaterally
Covered by Accessory Subscapularis Muscles. 276

FIG. 42.
Left Humerus with Supracondyloid Process
and Some Irregular Muscle Attachments . . . 277

FIG. 43.
M. Coracobrachialis Brevis (Superior) 277

FIG. 44.
M. Biceps Brachii (Anomalous) · Absence of
Long Head of M. Biceps Brachii 278

FIG. 45.
M. Biceps Brachii with Three Heads ·
Anomalous Subscapular Muscle (of Gruber) ·
Anomalous Fascicle of M. Teres Major ·
Anomalous Insertion of Long Head of M.
Triceps · (Musculocutaneous Nerve Not
Piercing M. Coracobrachialis) · (Branch of
Musculocutaneous Nerve Joining Median
Nerve) 279

FIG. 46.
M. Epitrochleoanconeus 280

FIG. 47.
Anomalous Supernumerary Head of M. Pronator Teres · (Supracondylar Canal of
Humerus) · (Anomalous High Division of Axillary Artery into Radial and Ulnar Artery
Branches) 281

FIG. 48.
M. Pronator Teres Accessorius 282

FIG. 49.
Anomalous Palmaris Longus Muscles ·
(Anomalous Ulnar Artery (Superficial
Course, High Division) Passing Through a
Split Tendon of M. Palmaris Longus) 282

FIG. 50.
Anomalous Palmaris Longus Muscles ·
Anomalous Double-Headed M. Flexor Carpi
Ulnaris · Supernumerary Head of M. Abductor Digiti Minimi · M. Palmaris Accessorius
Superficialis 283

Table of Contents

FIG. 51.
Anomalous Palmaris Longus Muscles · M. Palmaris Accessorius Profundus · Supernumerary Head of M. Abductor Digiti Minimi 284

FIG. 52.
Anomalous Palmaris Longus Muscles 285

FIG. 53.
Anomalous Palmaris Longus Muscles 286

FIG. 54.
Anomalous Palmaris Longus Muscles 287

FIG. 55.
Doubled M. Palmaris Longus, Joined by a Tendinous Slip, and an Accessorius Ad Flexorem Digiti Minimi Muscle 288

FIG. 56.
M. Flexor (Opponens) Digiti Minimi Accessorius · M. Palmaris Longus, Reversed Belly 288

FIG. 57.
M. Palmaris Longus (Anomalous) · M. Biceps Brachii (Anomalous) · M. Abductor Pollicis Longus Duplication 289

FIG. 58.
Absence of M. Palmaris Longus · Anomalous Head or Supernumerary M. Flexor Carpi Ulnaris · M. Palmaris Profundus (Variant of M. Palmaris Longus) · M. Flexor Pollicis Longus (Anomalous) 290

FIG. 59.
M. Tensor Capsulae · M. Flexor Carpi Radialis Accessorius · M. Pisiuncinatus · M. Subscapularis (Anomalous, Gruber) 291

FIG. 60.
Absence of M. Palmaris Longus · M. Pisipalmaris (Pisohamatus) · Duplication of M. Extensor Carpi Radialis Brevis 292

FIG. 61.
Absence of M. Palmaris Longus · Anomalous Muscle Fascicle from M. Brachioradialis Crossing Obliquely to the Medial Antibrachial Fascia. Radial Artery Passes Beneath the Anomalous Muscle Fascicle · (Variation in Superficial Palmar Arterial Arch) · (Presence of Median Artery Which Passes Beneath Transverse Carpal Ligament and Joins Radial Artery) · Variations in M. Flexor Digitorum Superficialis, Absence of Tendon to Fifth Digit 293

FIG. 62.
M. Extensor Digitorum Communis (Anomalous) · M. Extensor Digiti Minimi Proprius (Anomalous) 293

FIG. 63.
M. Extensor Pollicis Longus (Anomalous) · M. Extensor Digitorum Communis (Anomalous) 294

FIG. 64.
M. Extensor Pollicis Longus (Anomalous) . . 295

FIG. 65.
M. Tensor Ligamenti Annularis Radii Anterior · M. Tensor Ligamenti Annularis Radii Posterior 296

FIG. 66.
M. Radialis Internus Brevis Biceps · M. Cubitocarpeus · Supernumerary Head of M. Abductor Digiti Minimi 297

FIG. 67.
M. Cubitoradiocarpien 297

FIG. 68.
M. Pronator Quadratus (Anomalous) 298

FIG. 69.
Anomalous Muscles and Tendons of Forearm, Hand, and Foot 299

FIG. 70.
Anomalous Extensor of Middle Finger · Anomalous Short Extensor of Middle Finger · Anomalous Short Extensor of Middle and Third Fingers 299

FIG. 71.
Supernumerary Deep Radial Flexor Muscle of Hand · Supernumerary Muscle Extending Between Tendon of M. Brachioradialis and M. Flexor Pollicis Longus · Supernumerary Muscle Between M. Abductor Digiti Minimi and M. Opponens Digiti Minimi · Doubled M. Pronator Quadratus · Absence of Normal M. Extensor Indicis Proprius · Anomalous M. Indicis Brevis Manus 300

FIG. 72.
Accessory Slip Connecting Mm. Abductor Pollicis Longus and Brevis 301

FIG. 73.
Unusual Synovial Sheaths of Flexor Tendons 302

FIG. 74.
Unusual Insertions of M. Abductor Pollicis Longus 302

FIG. 75.
M. Extensor Digiti Brevis Manus (Anomalous) · M. Extensor Indicis Proprius (Anomalous) 303

FIG. 76.
Variations in Extensor Muscles 304

235

Human Anatomic Variation

FIG. 77.
Anomalous Dorsal Interosseous Muscle . . . 305

FIG. 78.
M. Pubotransversalis 305

FIG. 79.
M. Tensor Laminae Posterioris Vaginae Musculi Recti 306

FIG. 80.
Aberrant M. Psoas Major Fascicle 306

FIG. 81.
M. Gluteoperinealis 307

FIG. 82.
Congenital Absence of M. Quadriceps Femoris (Case Report) 308

FIG. 83.
M. Tensor Fasciae Suralis 309

FIG. 84.
M. Biceps Femoris Accessory Slip to M. Semitendinosus 309

FIG. 85.
Supernumerary M. Semimembranosus · Supernumerary M. Popliteus 310

FIG. 86.
M. Popliteus Biceps 311

FIG. 87.
Duplication of Medial and Lateral Heads of M. Gastrocnemius 312

FIG. 88.
Soleus Accessorius 312

FIG. 89.
M. Rectus Capitis Anticus (Anterior) Medius s. Minimus · M. Extensor Hallucis Longus Tricaudatus 313

FIG. 90.
Four Peronei Muscles in a Leg 313

FIG. 91.
M. Tibioastragalus Anticus · M. Accessorius Iliocostocervicalis 314

FIG. 92.
M. Extensor Hallucis Longus Minor Fibularis Anterior · M. Extensor Hallucis Longus Minor Fibularis Medialis · M. Extensor Hallucis Minor Tibialis · M. Extensor Hallucis Major · M. Extensor Hallucis Major Biceps . 315

FIG. 93.
M. Pisohamatus (Pisiuncinatus) · M. Extensor Hallucis Longus Bicaudatus · M. Extensor Digitorum Longus (Anomalous) · (Bipartite Zygomatic Bone) 316

FIG. 94.
M. Flexor Accessorius Longus or M. Accessorius ad Quadratum Plantae 317

FIG. 95.
Variations in Long Flexor Tendons of Toes · Various Forms of Union of Long Flexor Tendons of Toes 318

Cardiovascular System

FIG. 1.
Coronary Artery Arising from Pulmonary Artery 321

FIG. 2.
Anomalous Origin of Right Coronary Artery   321

FIG. 3.
Anomalous Coronary Arteries 322

FIG. 4.
Coronary Artery Preponderance 323

FIG. 5.
Distribution of Coronary Arteries on Back of Ventricles 323

FIG. 6.
Double Mitral Valve 324

FIG. 7.
Anomalous Origin of Left Pulmonary Artery · Retrotracheal Pulmonary Artery 325

FIG. 8.
Specimen of Right Aortic Arch 325

FIG. 9.
Vascular Rings Around Esophagus and Trachea 326

FIG. 10.
Right Aortic Arch and Sinuous Aorta 327

FIG. 11.
Double Aortic Arch 327

FIG. 12.
Nine Variations in Mode of Origin of Branches of Aortic Arch 328

FIG. 13.
Aortic Arch Anomalies 328

FIG. 14.
Aortic Arch Anomalies (continued) 330

FIG. 15.
Twenty-five Possible Combinations in Origin and Position of Arteries of Aortic Arch . . . 331

FIG. 16.
Aortic Arch Anomaly 332

FIG. 17.
Common Trunk for the Two Internal Carotids and Right Subclavian Arteries 333

FIG. 18.
Variations in Arrangement of the Bronchi and Pulmonary Arteries on the Right Side 333

FIG. 19.
Very Short Common Carotid Artery 334

FIG. 20.
Unusual Course of External Carotid Artery . . 335

FIG. 21.
Absence of Left Internal Carotid Artery . . . 336

FIG. 22.
Absence of Both Internal Carotid Arteries . . 336

FIG. 23.
Circle of Willis 337

FIG. 24.
Circle of Willis 338

FIG. 25.
Fourteen Anomalies of Circle of Willis and Related Vessels 339

FIG. 26.
Variations in Circle of Willis 340

FIG. 27.
Eighty-three Variations in Circle of Willis . . . 340

FIG. 28.
Eighty-three Variations in Circle of Willis (continued) 341

FIG. 29.
Eighty-three Variations in Circle of Willis (continued) 341

FIG. 30.
Eighty-three Variations in Circle of Willis (continued) 342

FIG. 31.
Eighty-three Variations in Circle of Willis (continued) 342

FIG. 32.
Eighty-three Variations in Circle of Willis (continued) 343

FIG. 33.
Eighty-three Variations in Circle of Willis (continued) 343

FIG. 34.
Eighty-three Variations in Circle of Willis (continued) 344

FIG. 35.
Variations in Anterior Communicating Artery 344

FIG. 36.
Anterior Cerebral Artery 345

FIG. 37.
Reduplication of Posterior Cerebral Artery . . 345

FIG. 38.
Types of Middle Meningeal Artery 346

FIG. 39.
Variation in Ophthalmic Artery · Lacrimal Artery Arising Chiefly from Middle Meningeal Artery 347

FIG. 40.
Variations in Appearance of Central Retinal Artery at the Optic Papilla 347

FIG. 41.
Vertebral Artery and Thoracic Duct 348

FIG. 42.
Rare Variation of Vertebral Artery 348

FIG. 43.
Thyroidea Ima Artery 349

FIG. 44.
Thyroidea Ima Artery (after Henle) 350

FIG. 45.
Origin of Right Vertebral Artery Between the Common Carotid and Right Subclavian with Inferior Thyroid Artery Crossing in Front of Its Course · Doubled Origin of Right Vertebral Artery · High Division of Popliteal Artery 351

FIG. 46.
Thyrocervical Trunk 352

FIG. 47.
Thyrocervical Trunk, Rarer Variations 352

FIG. 48.
Variation in Inferior and Superior Thyroid Arteries 353

FIG. 49.
Variations in Origin of Transverse Cervical Artery 354

FIG. 50.
Origin of Dorsal Scapular Artery (Three Major Sites) · Frequency of Position of Dorsal Scapular Artery as It Passes Through Brachial Plexus 355

FIG. 51.
Origin of Ascending Cervical Artery 356

FIG. 52.
Distribution of Superior and Inferior Thyroid Arteries to Thyroid Gland 356

FIG. 53.
Different Forms of Thyrocervical Trunk . . . 357

Human Anatomic Variation

FIG. 54. Subclavian Artery 358

FIG. 55. Vertebral Artery 359

FIG. 56. Thyrocervical Trunk 360

FIG. 57. Inferior Thyroid Artery 361

FIG. 58. Transverse Cervical Artery 362

FIG. 59. Suprascapular Artery 363

FIG. 60. Superficial Cervical Artery 364

FIG. 61. Ascending Cervical Artery 365

FIG. 62. Internal Thoracic Artery 366

FIG. 63. Costocervical Trunk 367

FIG. 64. Variations in Course and Vertebral Relations of Vertebral Artery 368

FIG. 65. Cervical Origins of Esophageal Arteries . . . 369

FIG. 66. Bronchial and Esophageal Arteries 370

FIG. 67. Lateral Internal Thoracic Artery 372

FIG. 68. Lateral Internal Thoracic Artery and Vein . . 373

FIG. 69. Branching Patterns of External Carotid Artery 374

FIG. 70. Composite Study of Axillary Artery 374

FIG. 71. Composite Study of Axillary Artery (continued) 375

FIG. 72. Composite Study of Axillary Artery (continued) 375

FIG. 73. Composite Study of Axillary Artery (continued) 376

FIG. 74. Anomalous Axillary Artery 377

FIG. 75. Partition of Left Axillary Artery into Two Trunks, One of Which Possessed All the Usual Branches of the Axillary While the Other Continued as the Brachial Artery · (Anomaly of Median Nerve) · (Anastomosis Between Radial and Ulnar Nerves). 378

FIG. 76. Spiral Course of Right Median Nerve Around Brachial Artery · (Anomaly of Median Nerve) · (Anomaly of Musculocutaneous Nerve) · Superficial Course of Ulnar Artery Through Lacertus Fibrosus of Biceps Brachii 379

FIG. 77. Absence of Ulnar Artery 380

FIG. 78. Some Variations in Arteries of Arm and Forearm (from the Literature) 381

FIG. 79. Some Variations in Arteries of Arm, Forearm, and Hand (from the Literature) 382

FIG. 80. Duplication of Ulnar Artery 383

FIG. 81. Variations in Brachial, Ulnar, and Radial Arteries · Persistent Median Artery 383

FIG. 82. Variations in Radial Artery · Persistent Median Artery 384

FIG. 83. Radial Artery with Three Roots · Rudimentary Radial Artery · Anomalous Branch of Median Artery · Anomalous Superior Branch of Anterior Interosseous Artery · Anomalous Inferior Branch of Anterior Interosseous Artery 385

FIG. 84. Superficial Palmar Arch 386

FIG. 85. Deep Palmar Arch 387

FIG. 86. Subcutaneous Course of Dorsal Branch of Radial Artery 388

FIG. 87. Variations in Unpaired Abdominal Arteries . 388

FIG. 88. Five Variations in Celiac Trunk 389

FIG. 89.
Bronchial Artery Derived from Celiac Trunk · Variation in Vertebral Artery and Circle of Willis · Absence of Confluens of Dural Sinuses · (Left-Sided Vermiform Appendix). ... 390

FIG. 90.
Origin of Cystic Artery 391

FIG. 91.
Variations in Origin of Common Hepatic Artery · Variations in Origin of Right Hepatic Artery 392

FIG. 92.
Variations in Origin of Left Hepatic Artery ... 393

FIG. 93.
Variations in Origin of Gastroduodenal Artery ... 394

FIG. 94.
Variations in Origin of Right Gastric Artery ... 395

FIG. 95.
Variations in Origin of Cystic Artery 396

FIG. 96.
Variations in Arterial Supply to the Colon from Superior Mesenteric Artery 397

FIG. 97.
Blood Supply of the Common Bile Duct ... 398

FIG. 98.
Variations in Blood Supply of the Common Bile Duct 399

FIG. 99.
Principal Patterns of Arterial Supply to the Liver 400

FIG. 100.
Hepatic Pedicle 401

FIG. 101.
Hepatic Pedicle (continued) 402

FIG. 102.
Hepatic Pedicle: Variations in Arteries and Bile Ducts 403

FIG. 103.
Hepatic Pedicle: Variations in Arteries and Bile Ducts (continued) 404

FIG. 104.
Hepatic Pedicle: Variations in Arteries and Bile Ducts (continued) 404

FIG. 105.
Hepatic Pedicle: Variations in Arteries and Bile Ducts (continued) 405

FIG. 106.
Hepatic Pedicle: Variations in Arteries and Bile Ducts (continued) 405

FIG. 107.
Major Variations of Appendiceal and Cecal Arteries (Based on 200 Specimens) 406

FIG. 108.
Variations in Inferior Mesenteric Artery ... 406

FIG. 109.
Double Inferior Mesenteric Artery 408

FIG. 110.
Renal Vascular Pedicle 408

FIG. 111.
Renal Vascular Pedicle (continued) 410

FIG. 112.
Renal Arterial Variations 412

FIG. 113.
Possible Sources of Accessory Renal Arteries 413

FIG. 114.
Variations in Renal Vessels 413

FIG. 115.
Variations in Renal Vessels (continued) 414

FIG. 116.
Variations in Renal Vessels (continued). ... 414

FIG. 117.
Variations in Renal Vessels (continued) 415

FIG. 118.
Variations in Ovarian Arterial Supply 416

FIG. 119.
Middle Sacral Artery 417

FIG. 120.
Common Iliac Artery Absent · Common Iliac Artery Short 417

FIG. 121.
Variation in Origin of the Parietal Branches of Internal Iliac Artery (Based on a Study of 169 Specimens (108 Males and 61 Females)). ... 418

FIG. 122.
Origins of Obturator Artery (Based on a Study of 169 Bodies (61 Females, 108 Males)). 419

FIG. 123.
Obliteration of Left Common, External, and Internal Iliac Arteries 419

FIG. 124.
Superior and Inferior Epigastric Arteries and Rectus Abdominis Muscle (Based on a Study of 115 Specimens). 420

FIG. 125.
Origin of Deep and Circumflex Femoral Group of Arteries (Based on a Study of 481 Arterial Patterns). 421

FIG. 126.
Femoral Artery and Branches 422

239

Human Anatomic Variation

FIG. 127.
Variations in Deep Femoral (Profunda Femoris) Artery 423

FIG. 128.
Femoral and Profunda Femoris Arteries . . . 424

FIG. 129.
Femoral Artery and Branches 424

FIG. 130.
Femoral Artery and Branches 425

FIG. 131.
Persistent Ischiadic (Sciatic) Artery 425

FIG. 132.
High Division and Reunion of Femoral Artery High Division of a Femoralis with Subsequent Reunion (from Quain's Plates). 426

FIG. 133.
Variations in Mode of Division of Popliteal Artery 426

FIG. 134.
Various Arterial Patterns on Dorsum of Foot (Meyer). Variations in the Plantar Arch (Dubreuil-Chambardel). 427

FIG. 135.
Persistent Left Superior Vena Cava 428

FIG. 136.
Closed Coronary Sinus (Obliterated Ostium to Right Atrium) 428

FIG. 137.
Absence of Coronary Sinus.
Great Cardiac (Cardiaca Magna) Vein Draining into Superior Vena Cava 429

FIG. 138.
Pulmonary Lobe of Azygos Vein 430

FIG. 139.
Venous Drainage of Hand and Forearm 431

FIG. 140.
Portal Vein and Its Origin 432

FIG. 141.
Gonadal Veins 433

FIG. 142.
Gonadal Veins 434

FIG. 143.
Gonadal Veins 435

FIG. 144.
Transposition of Inferior Vena Cava to Left Side Without Transposition of Viscera 436

FIG. 145.
Duplication of Postrenal Segment of Inferior Vena Cava 437

FIG. 146.
Double Inferior Vena Cava 438

FIG. 147.
Venous Drainage at Fossa Ovalis 439

FIG. 148.
Circumflex Veins of the Thigh: Modes of Termination (Based on a Study of 279 Cadavers, 541 Specimens). 440

FIG. 149.
Anomalies of Dural Sinuses · Absence of Confluens of Dural Sinuses · Bifurcation and Reunion of Superior Sagittal (Longitudinal) Sinus · Brief Doubling of Superior Sagittal (Longitudinal) Sinus 441

FIG. 150.
Various Types of Confluens of Sinuses (Posterior View or Convex Face of the Dura Mater). 442

FIG. 151.
Confluens of Sinuses: Types and Subtypes . . 442

FIG. 152.
Variations in Lateral and Sigmoid Sinuses . . . 443

FIG. 153.
Anomalous Thoracic Duct Emptying into Right Subclavian Vein · Retroesophageal Right Subclavian Artery 445

FIG. 154.
Some Types of Variation in Termination of Thoracic Duct 446

FIG. 155
Variations in Form of Thoracic Duct 446

Nervous System

FIG. 1.
Cavum Septi Pellucidi · Cavum Vergae 449

FIG. 2.
Absence of Corpus Callosum 450

FIG. 3.
Variation in Shape and Position of Ventricular Cavities (Lateral and Third Ventricles) in Agenesis of Corpus Callosum · Variations in Formation of Midline Extra Cavities 451

FIG. 4.
Bilateral Interruption of Fissure of Rolando . 451

FIG. 5.
Variation in Fissures of Cerebral Hemisphere Showing Doubled Rolandic Fissure 452

FIG. 6.
Variations in Gyri on the Medial Surface of
Frontal Lobe 452

FIG. 7.
Variations in Gyri and Sulci 453

FIG. 8.
Variations in Gyri and Sulci (continued) . . . 454

FIG. 9.
Variations in Gyri and Sulci (continued) . . . 455

FIG. 10.
Variations in Gyri and Sulci (continued) . . . 456

FIG. 11.
Variations in Gyri and Sulci (continued) . . . 457

FIG. 12.
Variations in Gyri and Sulci (continued) . . . 458

FIG. 13.
Variations in Gyri and Sulci (continued) . . . 459

FIG. 14.
Variations in Gyri and Sulci (continued) . . . 460

FIG. 15.
Spinal Cord Heterotopia 460

FIG. 16.
Cerebral Heterotopia (CT Scan) 461

FIG. 17.
Cerebral Heterotopia (MRI). 461

FIG. 18.
Incomplete Division of Cerebral Hemispheres
by Transverse Convolutions 462

FIG. 19.
Variations in Crossing Pattern of Pyramids in
Medulla and Subsequent Formation of Cor-
ticospinal Tracts in Spinal Cord 463

FIG. 20.
Absence of Optic Chiasma.
Bilateral Asymmetric Distribution of Nerve
Fibers in Optic Tracts 463

FIG. 21.
Variations in Position of Optic Chiasma in
Relation to Hypophyseal Stalk 464

FIG. 22.
Oculomotor Nerve 464

FIG. 23.
Abducens Nerve 466

FIG. 24.
Trochlear Nerve 467

FIG. 25.
Variations in Branching of Facial Nerve 468

FIG. 26.
Variations in Intraparotid Portion of Facial
Nerve 469

FIG. 27.
Variations in Vagus Nerve at the Lower Portion of Esophagus and Upper Stomach 469

FIG. 28.
Absence of Right Recurrent Laryngeal Nerve
in Retroesophageal Right Subclavian Artery   470

FIG. 29.
Anomalous Origin of Right Recurrent
Laryngeal Nerve Associated with Arteria
Lusoria, Retroesophageal Right Subclavian
Artery 471

FIG. 30.
Twenty-eight Variations in Relations of
Inferior Laryngeal Nerve to Inferior Thyroid
Artery 471

FIG. 31.
Four Variations in Ansa Hypoglossi 472

FIG. 32.
Variations in Relations Between Superficial
Origin of Posterior Roots of Spinal Nerves and
Spinous Processes of Vertebrae 472

FIG. 33.
Twenty-nine Variations in Form of Brachial
Plexus 474

FIG. 34.
Twenty-nine Variations in Form of Brachial
Plexus (continued) 474

FIG. 35.
Twenty-nine Variations in Form of Brachial
Plexus (continued) 475

FIG. 36.
Twenty-nine Variations in Form of Brachial
Plexus (continued) 475

FIG. 37.
Twenty-nine Variations in Form of Brachial
Plexus (continued) 476

FIG. 38.
Twenty-nine Variations in Form of Brachial
Plexus (continued) 476

FIG. 39.
Twenty-nine Variations in Form of Brachial
Plexus (continued) 477

FIG. 40.
Twenty-nine Variations in Form of Brachial
Plexus (continued) 477

Human Anatomic Variation

FIG. 41.
Twenty-nine Variations in Form of Brachial Plexus (continued) 478

FIG. 42.
Twenty-nine Variations in Form of Brachial Plexus (continucd) 478

FIG. 43.
Variation in Formation of Brachial Plexus with Union of Branches into a Single Cord (An Analysis of Variation 29 in Figure 42). 479

FIG. 44.
Thirty-eight Variations in Form of Brachial Plexus (Sketched Diagrammatically) 479

FIG. 45.
Thirty-eight Variations in Form of Brachial Plexus (Sketched Diagrammatically) (continued) 480

FIG. 46.
Thirty-eight Variations in Form of Brachial Plexus (Sketched Diagrammatically) (continued) 480

FIG. 47.
Thirty-eight Variations in Form of Brachial Plexus (Sketched Diagrammatically) (continued) 481

FIG. 48.
Thirty-eight Variations in Form of Brachial Plexus (Sketched Diagrammatically) (continued) 481

FIG. 49.
Thirty-eight Variations in Form of Brachial Plexus (Sketched Diagrammatically) (continued) 482

FIG. 50.
Prefixed Brachial Plexus 483

FIG. 51.
Rare Form of Prefixed Brachial Plexus 482

FIG. 52.
Postfixed Brachial Plexus 484

FIG. 53.
Spiral Course of Median Nerve Around Brachial Artery · Median Nerve and Brachial Artery Passing Behind Third Head of Biceps 485

FIG. 54.
Median-Ulnar Nerve Anastomoses in Forearm 486

FIG. 55.
Supracondylar Ligament Crossing Median Nerve and Brachial Artery 487

FIG. 56.
Axillary Artery and Vein Passing Through the Medial and Lateral Cords of Brachial Plexus . 487

FIG. 57.
Variations in Median and Musculocutaneous Nerves of the Arm 488

FIG. 58.
Variations in Course of Median Nerve at the Carpal Tunnel 489

FIG. 59.
Variations in Course of Median Nerve at the Carpal Tunnel (continued) 490

FIG. 60.
Variations in Cutaneous Innervation of Forearm and Hand (Based on Over 1,000 Gunshot Injuries to Peripheral Nerves) · Ulnar Nerve . 490

FIG. 61.
Variations in Cutaneous Innervation of Forearm and Hand (Based on Over 1,000 Gunshot Injuries to Peripheral Nerves) (continued) · Ulnar Nerve · Median Nerve 491

FIG. 62.
Variations in Cutaneous Innervation of Forearm and Hand (Based on Over 1,000 Gunshot Injuries to Peripheral Nerves) (continued) · Radial Nerve 491

FIG. 63.
Variations in Cutaneous Innervation of Forearm and Hand (Based on Over 1,000 Gunshot Injuries to Peripheral Nerves) (continued) · Radial Nerve · Musculocutaneous Nerve . . . 492

FIG. 64.
Variations in Cutaneous Innervation of Forearm and Hand (Based on Over 1,000 Gunshot Injuries to Peripheral Nerves) (continued) · Division of Median and Ulnar Nerves · Superficial Radial Nerve 493

FIG. 65.
Variations in Formation of Lumbosacral Plexus 494

FIG. 66.
Variations in Formation of Ilioinguinal Nerve 494

FIG. 67.
Variants of Origin of Genitofemoral Nerve from Lumbar Plexus 495

FIG. 68.
Variations in Connections Between Lumbar and Sacral Plexuses. 496

FIG. 69.
Variations in Sciatic Nerve · Variations in M. Piriformis) 497

FIG. 70.
Variations in Relation of Sciatic Nerve to
M. Piriformis 498

FIG. 71.
The Sural Nerve 498

FIG. 72.
Variations in Cutaneous Innervation of Dorsum of the Foot 499

FIG. 73.
Variations in Sympathetic Nerve Pathways to
the Heart 500

FIG. 74.
Variations in Sympathetic Nerve Pathways to
the Heart (continued) 501

FIG. 75.
Variations in Sympathetic Nerve Pathways to
the Heart (continued) 502

FIG. 76.
Variations in Sympathetic Nerve Pathways to
the Heart (continued) 503

FIG. 77.
Variations in Sympathetic Nerve Pathways to
the Heart (continued) 504

FIG. 78.
Variations in Sympathetic Nerve Pathways to
the Heart (continued) 505

FIG. 79.
Variations in Lumbar Sympathetic Trunks . . 506

FIG. 80.
Nerve Supply in Cases of Horseshoe Kidney . 507

FIG. 81.
Sacral Sympathetic Ganglia 508

Organs

FIG. 1.
Cutaneous Horn of the Scalp. 511

FIG. 2.
Cleft Nose 511

FIG. 3.
Situs Inversus (Transposition of Viscera) . . . 512

FIG. 4.
Positions of Vermiform Appendix 513

FIG. 5.
Doubled Spleen 513

FIG. 6.
Appendix Fibrosa Hepatis 514

FIG. 7.
Accessory Extrahepatic Ducts (Variations Encountered in 100 Cadavers) 515

FIG. 8.
Gallbladder Variations · Doubled Gallbladder · Bilobed Gallbladder · Abnormal Position
of Gallbladder 516

FIG. 9.
Duplication of Common Bile Duct 517

FIG. 10.
Insertion of Aberrant Bile Ducts 518

FIG. 11.
Pancreatic and Common Bile Ducts 519

FIG. 12.
Pancreatic Ducts 520

FIG. 13.
Variations in Peripheral Segmentation of
Right Lung · Variations in Fissures on the
Inferior Face of Right and Left Lungs 521

FIG. 14.
Congenital S-shaped Kidney · (Absence of
Sternocostal Heads of Pectoralis Major) ·
(Right-sided Heart) 522

FIG. 15.
Horseshoe Kidney 523

FIG. 16.
Horseshoe Kidney with Three Ureters 524

FIG. 17.
Supernumerary Third Kidney 524

FIG. 18.
Bilateral Duplication of the Ureter 525

FIG. 19.
Pelvic Kidney 526

FIG. 20.
Double Penis 526

FIG. 21.
Double Vagina 527

FIG. 22.
Double Vagina and Double Uterus 527

FIG. 23.
Uterine and Vaginal Variations 528

FIG. 24.
Supernumerary Mammae and Nipples in Men
and Women 529

FIG. 25.
Variations in Size, Number, Location, and
Structure of the Mammary Gland in Men and
Women 530

FIG. 26.
Variations in Form of Thyroid Gland 530

FIG. 27.
Thyroid Gland, Without an Isthmus (Glandula Thyroidea Partitia) 531

FIG. 28.
Thyroid Gland Variations · Accessory Thyroid Glands · Thyroglossal Duct 532

FIG. 29.
Accessory Thyroid Glands in Neck and Chest 533

FIG. 30.
Variations in Parathyroid Glands 534

FIG. 31.
Number and Location of Parathyroid Glands 535

FIG. 32.
Number and Location of Parathyroid Glands (continued) 536

FIG. 33.
Number and Location of Parathyroid Glands (continued) 537

Skeleton

FIG. 1.
Varieties of Different Vertebrae 541

FIG. 2.
Varieties of Atlas 542

FIG. 3.
Unusual Case of Absence of the Anterior Arch of Atlas 543

FIG. 4.
Varieties of Axis 543

FIG. 5.
Cervical Ribs 544

FIG. 6.
Supernumerary Cervical Ribs 545

FIG. 7.
Peculiar Thoracic Vertebrae 546

FIG. 8.
Congenital Absence of Dorsal Arches of the Fourth to Seventh Thoracic Vertebrae 547

FIG. 9.
Congenital Absence of Dorsal Arches of the Fourth to Seventh Thoracic Vertebrae 547

FIG. 10.
A Divided Thoracic Vertebra (after Turner) 548

FIG. 11.
Variations in Lumbar Ribs 548

FIG. 12.
Lumbar Sacralization · Unilateral Assimilation of Fifth Lumbar Vertebra 549

FIG. 13.
Lumbar Sacralization 549

FIG. 14.
Pubic Tubercles and Pubic Spines 550

FIG. 15.
Sacrum Composed of Five and Six Vertebrae · Coccyx Composed of Four Vertebrae 550

FIG. 16.
Accessory Iliosacral Articulations 552

FIG. 17.
Accessory Iliosacral Articulations 552

FIG. 18.
Absence of Sacral Dorsal Arches 553

FIG. 19.
Three Sacra 553

FIG. 20.
Persistence of the Tail 554

FIG. 21.
Sutural Bones · Sutures · Torus Occipitalis · Parietal Foramina 554

FIG. 22.
Absence of Coronal and Sagittal Sutures 555

FIG. 23.
Torus Occipitalis 555

FIG. 24.
Anomalies of Occipital Bone at Foramen Magnum (Occipital Condyles) 556

FIG. 25.
Vermian Fossa 556

FIG. 26.
Foramen Vesalii · Cranial Variations 557

FIG. 27.
Foramen Vesalii 558

FIG. 28.
Peculiar Spine Arising from the Sagittal Border of the Squama of Temporal Bone · Bipartite Scaphoid Bone · (M. Thyro-esophagus) 558

FIG. 29.
Bipartite Parietal Bone · Enlarged Parietal Foramina 559

FIG. 30.
Bilateral Parietal Foramina 560

FIG. 31.
Varieties of Occipital Wormian and Interparietal Bones 560

FIG. 32.
Supernumerary Parietal Bone Sutures and Wormian Bones 561

FIG. 33.
Bipartite Os Zygomaticum · (M. Pisohamatus) · (Anomalous M. Extensor Digitorum Longus Pedis) 562

FIG. 34.
Supernumerary Bones of the Zygomatic Arch   563

FIG. 35.
Metopic Suture 564

FIG. 36.
Metopic Suture 565

FIG. 37.
Variation in Incisive Canal · Variation in Greater and Lesser Palatine Foramina · Wormian Bones Between Palatine Process and Palatine Bones · Division of Zygomatic Bone into Two Horizontal Parts 565

FIG. 38.
Some Unusual Maxillary Bones 566

FIG. 39.
Variations in Mandibular Foramen, Canal, and Mylohyoid Groove 567

FIG. 40.
Unerupted Teeth 568

FIG. 41.
Various Shapes of Sella Turcica 568

FIG. 42.
Variation in Number of Meatuses in Nasal Fossae 569

FIG. 43.
Variations in Size of Frontal and Maxillary Sinuses 569

FIG. 44.
Variations in Sphenoidal Sinus 570

FIG. 45.
Styloid Processes 571

FIG. 46.
Variations in Sternum · Suprasternal Bones · Bifid Xiphoid Process · Apertures in Body of Sternum and in Xiphoid Process 572

FIG. 47.
Suprasternal Bones · Rib Anomalies. · Duplication of Xiphoid Process 573

FIG. 48.
Variations in Suprasternal Bones 574

FIG. 49.
Fused First and Second Ribs 574

FIG. 50.
Clavicular Canals for Supraclavicular Nerves   575

FIG. 51.
Radiographic Images of Some Rib Anomalies   576

FIG. 52.
Fissura Sterna 576

FIG. 53.
Various Types of Acromion Process 577

FIG. 54.
Ossification of Acromion Process 578

FIG. 55.
Os Acromiale 579

FIG. 56.
Supracondylar Process 579

FIG. 57.
Supracondylar Process · Supratrochlear Foramen 580

FIG. 58.
Supracondylar Process · (Supernumerary Belly of M. Biceps Brachii) · (Supernumerary Head of M. Pronator Teres) 581

FIG. 59.
Sesamoid Bones of the Hand 581

FIG. 60.
Os Centrale of Carpus · (M. Radialis Digiti I s. Pollicis) 582

FIG. 61.
Extensive Carpal Anomalies 583

FIG. 62.
Carpal Accessoria · Carpal Fusions 583

FIG. 63.
Metacarpostyloid Variations · Calcaneonavicular Variations 584

FIG. 64.
Supernumerary Thumb · (Supernumerary M. Flexor Pollicis Brevis) · (Supernumerary Median Nerve Branch) 585

FIG. 65.
Three-Phalangeal Thumb · Hexadactyly . . . 586

FIG. 66.
Septodactyly (Manus) · Octodactyly (Pedis) . 586

FIG. 67.
Medial Supracondylar Process of Femur . . . 587

FIG. 68.
Medial Supracondylar Process of Femur . . . 588

FIG. 69.
Variations in Shape of Patella Partita 589

FIG. 70.
Duplication of the Great Toe 589

FIG. 71.
Duplication of the Great Toe · Duplication of the Little Toe · Hexadactyly (Manus) · (Excessive Division of the Liver) 590

FIG. 72.
Os Vesalianum (Pedis) · Os Vesalianum (Manus) 591

FIG. 73.
Tarsal Accessoria · Tarsal Fusions 592

FIG. 74.
Supernumerary Tarsal Bones 593

Muscles

The authors have retained wherever possible the style and content of the original figure legends.

Muscles

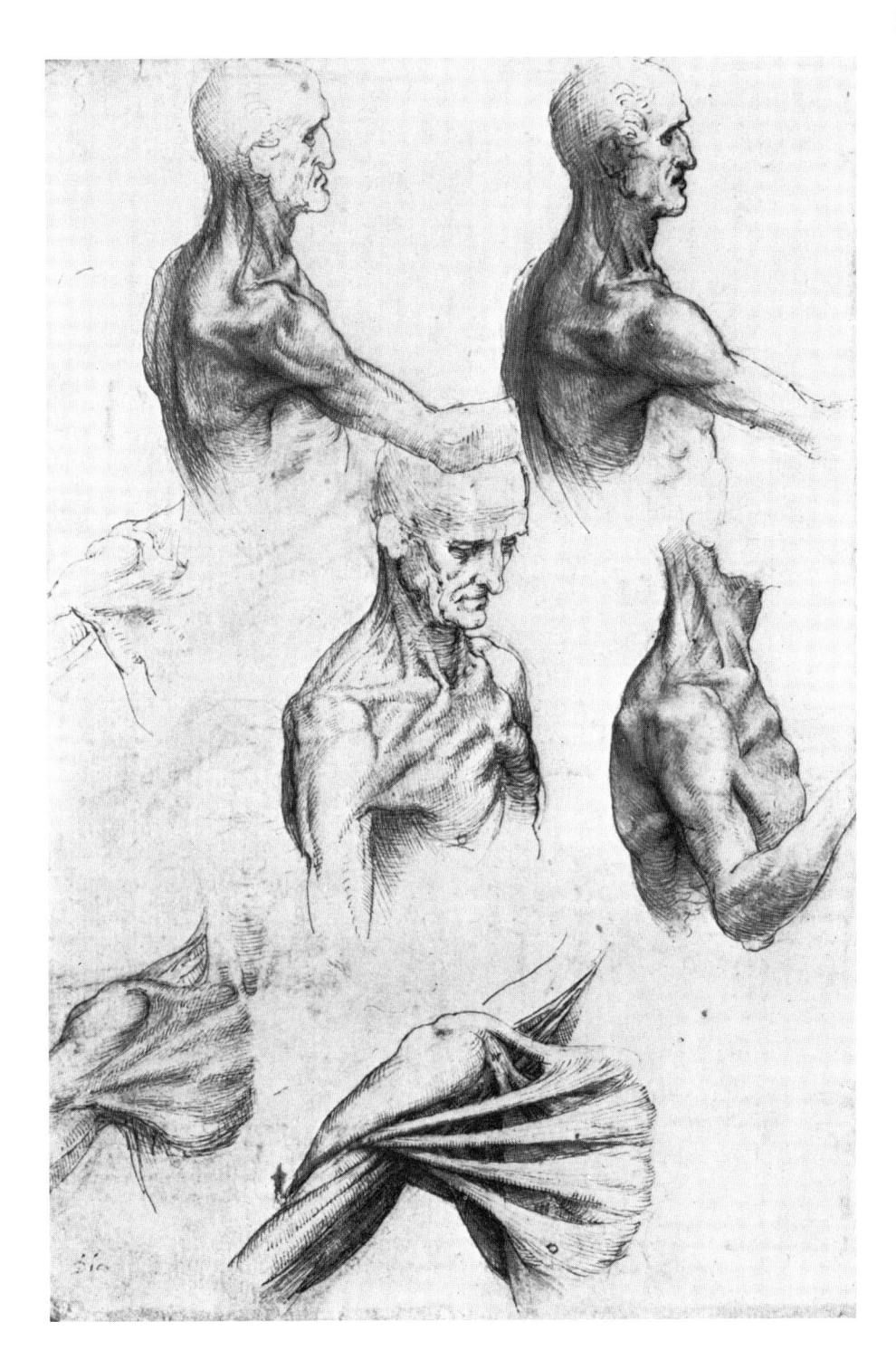

FIG. 1.
M. Omocervicalis (Levator Claviculae).
M. Infraspinatohumeralis.
M. Sternocleidomastoideus (Anomalous).
M. Pectoralis Major (Anomalous).
M. Tensor Semivaginae Articulationis Humeroscapularis or M. Sternochondrocoracoideus.

Leonardo's dissection drawings are particularly interesting because the subject appears to have several anatomic variations. In the upper two drawings, the slender omocervicalis (levator claviculae) muscle and an unnamed variation which could be termed infraspinatohumeralis may be identified. In the middle two drawings, note the distinct sternomastoid and cleidomastoid subdivision of the sternocleidomastoid muscle. In the lower two drawings, note the variation in the pectoralis major muscle. In addition, beneath pectoralis major there is the suggestion of the presence of either a tensor semivaginae articulationis humeroscapularis muscle or a sternochondrocoracoideus muscle. See listings under M. deltoideus, Mm. pectoralis major and minor, and M. sternocleidomastoideus.

Anatomic drawing by Leonardo da Vinci (1452–1519).
Reproduced with the gracious permission of Her Majesty Queen Elizabeth II.

Human Anatomic Variation

FIG. 2.
M. Deltoideus (Anomalous).
M. Infraspinatohumeralis.
M. Sternocleidomastoideus (Anomalous).
M. Pectoralis Major (Anomalous).
M. Costocoracoideus.

At the upper left is shown the deficient dorsal spinous part of m. deltoideus and the presence of m. infraspinatohumeralis. At the upper right is shown a variation in m. sternocleidomastoideus and m. pectoralis major. In the lower left drawing, positioned above m. pectoralis minor is m. costocoracoideus.

Anatomic drawing by Leonardo da Vinci. Reproduced with the gracious permission of Her Majesty Queen Elizabeth II.

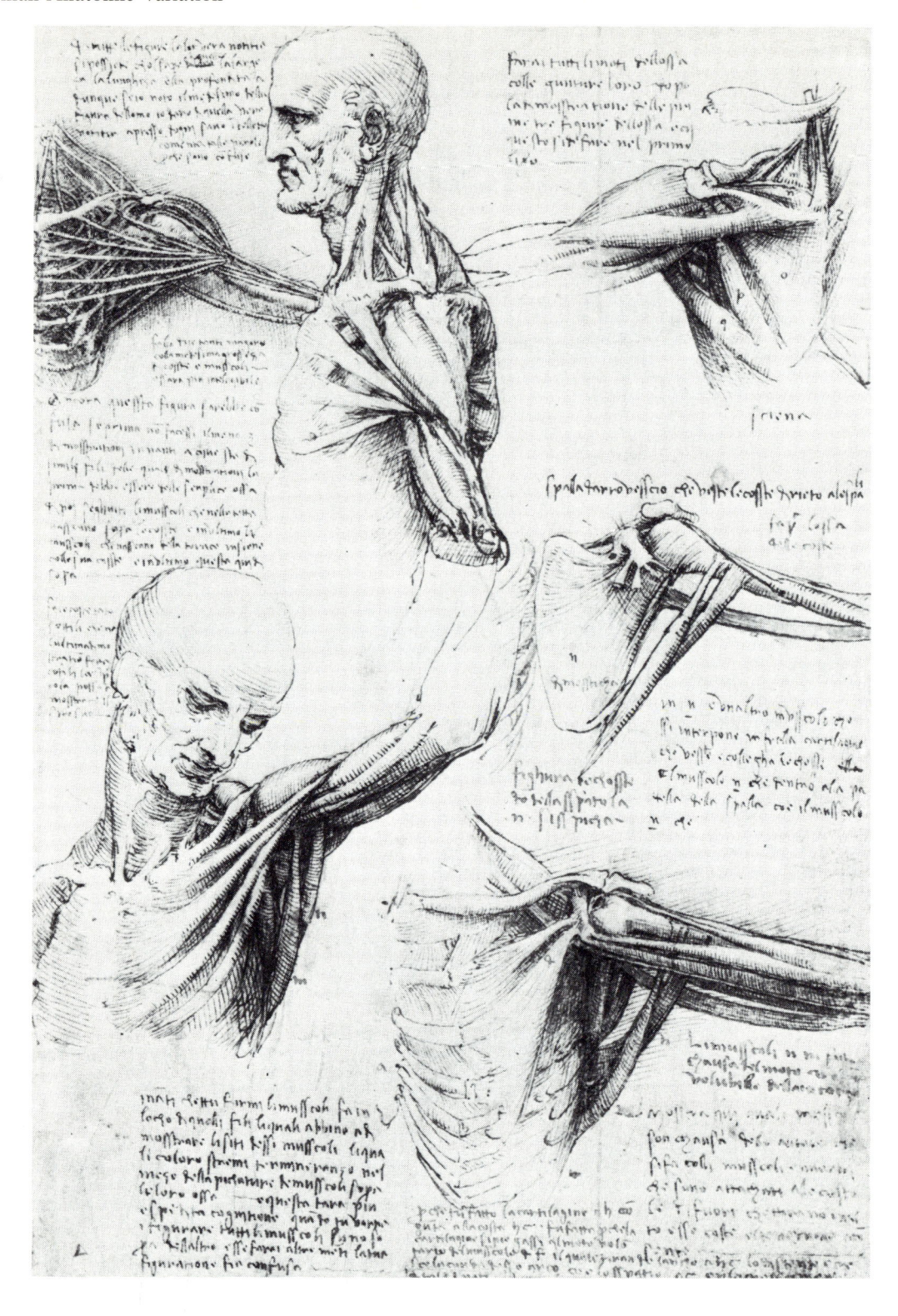

Muscles

FIG. 3.
Congenital Absence of Muscles of Neck and Thorax.
A: Congenital absence of the mm. pectoralis major and minor.
B: Torticollis due to absence of the left sternocleidomastoid muscle.
C: Bilateral absence of m. sternocleidomastoideus with hypertrophy of m. platysma.
Redrawn from La Presse Medicale. *Anatomie du Vivant*, No. 14, February 18, 1922, and No. 76, September 23, 1922.

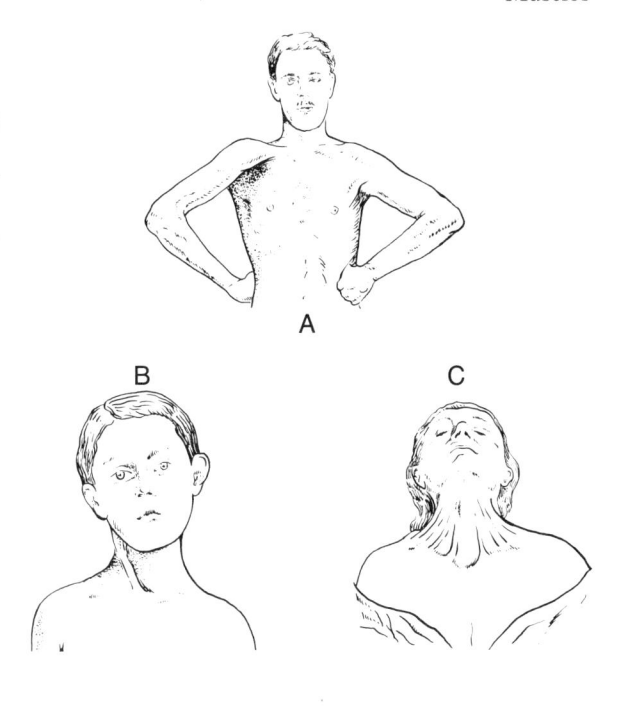

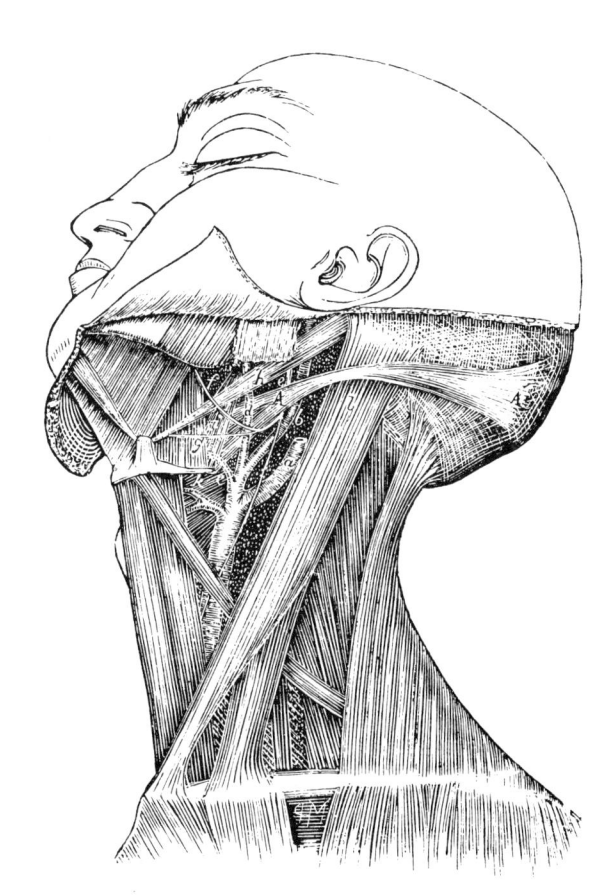

FIG. 4.
A Peculiar Digastric Muscle.
A,A', peculiar digastric muscle; *a*, internal carotid artery (cut); *b*, external carotid; *c*, lingual artery; *d*, facial; *e,f*, hypoglossal nerve; *g*, hyoglossus muscle; *h*, digastric and stylohyoid muscles (out of position); *i*, middle constrictor; *l*, sternocleidomastoid.

From Perrin, J.B. On a peculiar additional digastric muscle. *J. Anat. Physiol*. 5:251–256, 1871.

Human Anatomic Variation

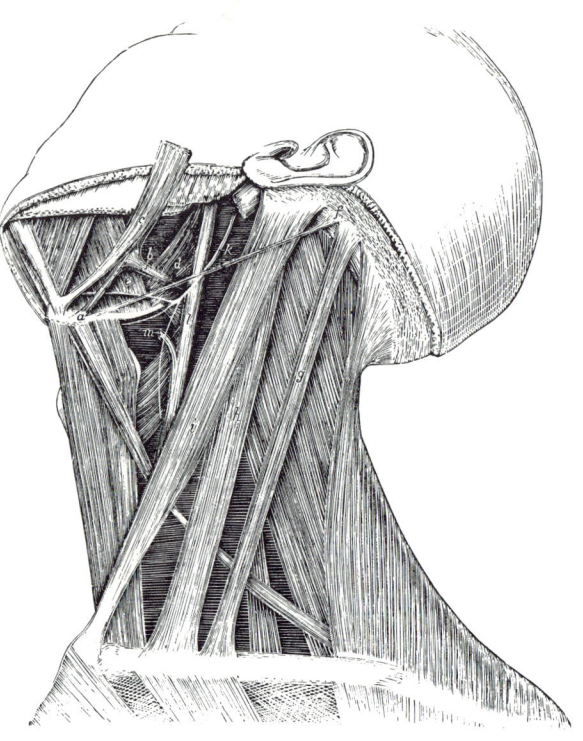

FIG. 5.

Examples of Occipitohyoid and Cleido-occipital Muscles.

Left side – female. *a*, Slip from posterior belly of digastric to hyoid bone; *b*, slip from middle constrictor (*d*); *c*, posterior belly of digastric cut through and turned up; *e*, hyoglossus muscle; *f*, occipitohyoid slip; *g*, cleido-occipital; *h,i*, sternocleidomastoid; *k*, 9th nerve; *l*, internal jugular vein; *m*, facial vein cut through.

From Perrin, J.B. On a peculiar additional digastric muscle. *J. Anat. Physiol.* 5:251–256, 1871.

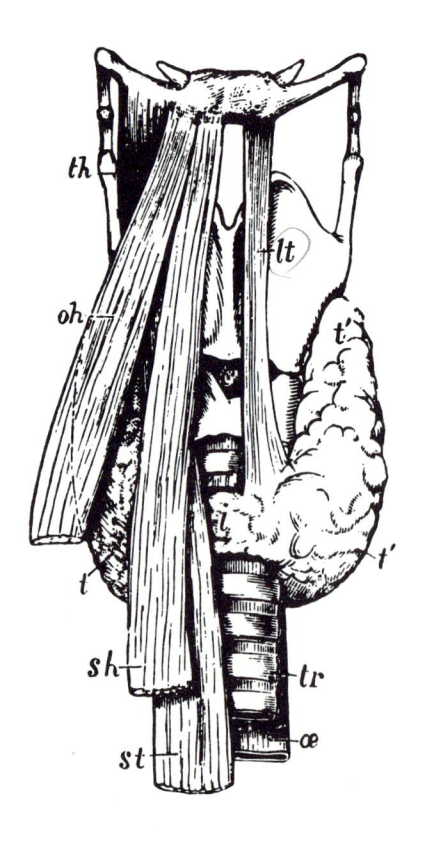

FIG. 6.

M. Levator Glandulae Thyroideae.

Sketch showing the form and position of the thyroid body, one-half natural size (Allen Thomson). The larynx and surrounding parts are viewed from before; on the right side, the muscles covering the thyroid body are retained, on the left side they are removed. *th*, right thyrohyoid muscle; *oh*, omohyoid; *sh*, sternohyoid; *st*, sternothyroid; *c*, cricothyroid membrane; *tr*, trachea; *oe*, esophagus; *t*, right lobe of the thyroid body; *t'*, the left lobe; *i*, isthmus; *lt*, the fibrous or muscular band termed *levator thyroideae*, which is more rarely found in the middle line or to the right side, and which existed in the case from which the figure was taken.

From Schafer, E.A., Symington, J. and T.H. Bryce, Eds. *Quain's Anatomy,* 11th Ed., Vol. II: Splanchnology. Longmans, Green, and Co., London, 1914.

Muscles

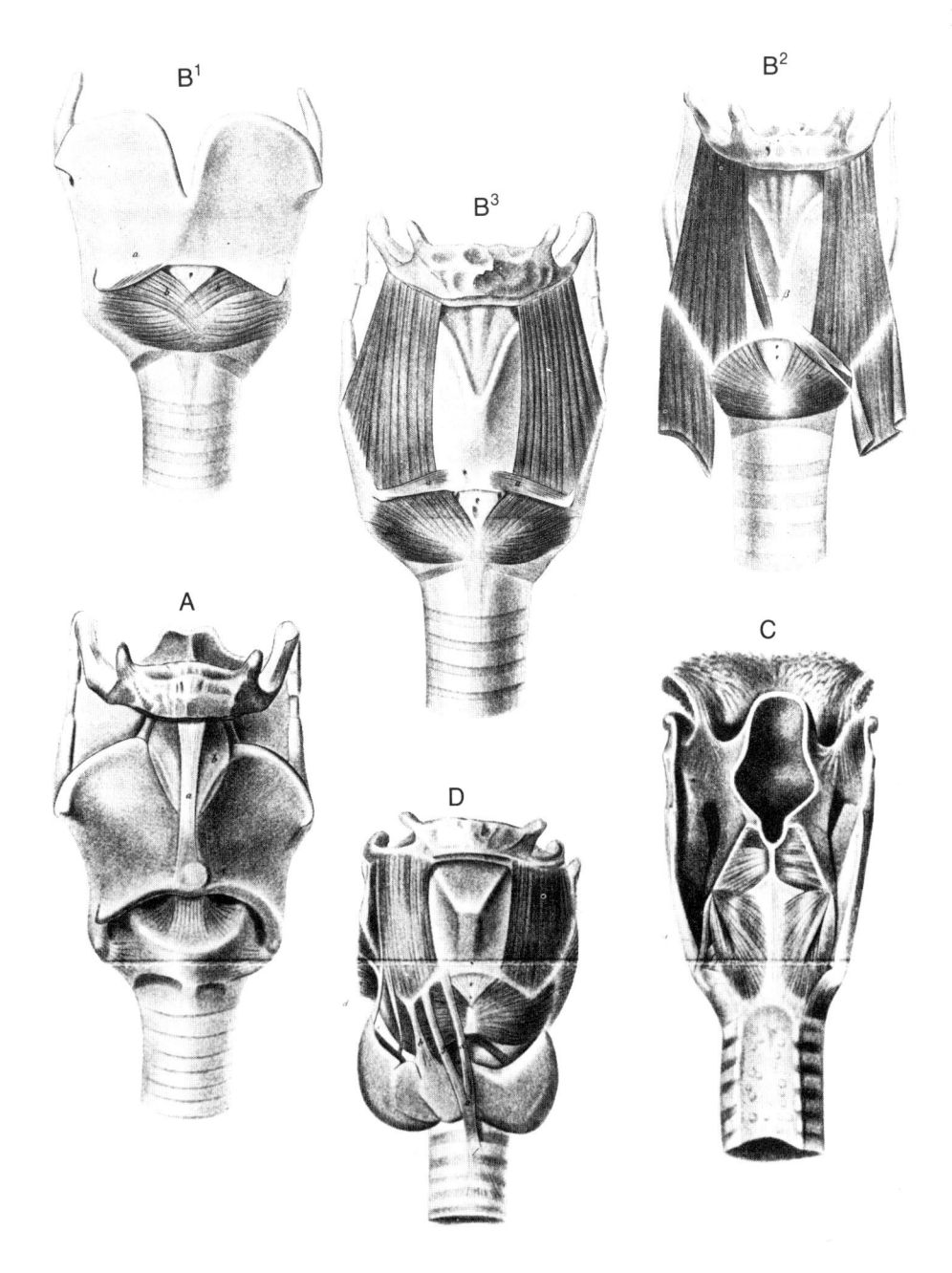

FIG. 7.
M. Thyroidei Marginales Inferior.
M. Incisurae Mediae Obliquus.
M. Incisurae Mediae Obliquus Bicaudatus.
M. Cricothyroideus (Anomalous).
M. Keratoarytenoideus.
M. Thyrotrachealis.
M. Thyrotrachealis Biceps.
M. Levator Glandulae Thyroideae.

A: *a*, Ligamentum hyothyroideum accessorium.
B1: *a*, Unpaired, right side, m. incisurae mediae obliquus; *b,b*, abnormal course of fibers of mm. cricothyroidei.
B2: *a*, Unpaired, left side, m. incisurae mediae obliquus bicaudatus; α, supernumerary inferior bundle of the same muscle; β, anomalous bundle of right thyrohyoid muscle.
B3: *a,a*, Paired m. incisurae mediae obliquus.

C: *a*, M. keratoarytenoideus.
D: *a*, M. thyrotrachealis biceps; *b, c, d*, mm. levator glandulae thyroideae.

A From Gruber, W. Über das Zungenbein-Schildknorpel-Hilfsband (Ligamentum hyothyreoideum accessorium). *Arch. Anat. Physiol. Wissen. Med.* 1868, pp. 633–634.

B1, B2, B3 From Gruber, W. Über die Muskeln des unteren Schildknorpelrandes — Musculi thyreoidei marginales inferiores. *Ibid.* 1868, pp. 635–639.

C From Gruber, W. Über den seltenen Schildknorpelhorn-Giessbeckenknorpelmuskel (Musculus kerato-arytenoideus). *Ibid.* 1868, pp. 640–641.

D From Gruber, W. Über eine neue Variante des Musculus thyreotrachealis und über den Musculus hyo-trachealis. *Ibid.* 1868, pp. 642–645.

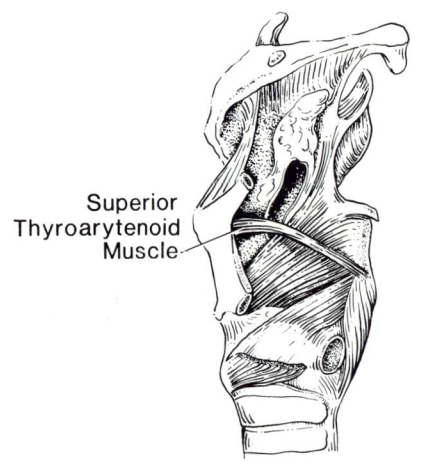

Human Anatomic Variation

FIG. 8.
Anomalous Muscles of Larynx.
M. Thyroarytenoideus Superior.
M. Arymembranaceus.
Two anomalous muscles of the larynx.

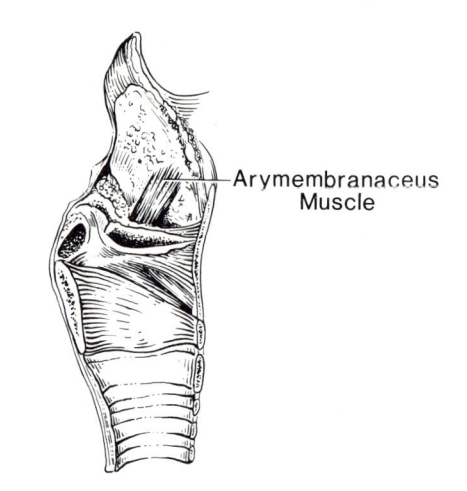

Redrawn from Toldt, C. *An Atlas of Human Anatomy for Students and Physicians.* The Macmillan Co., New York, 1926. Used by permission of the publisher.

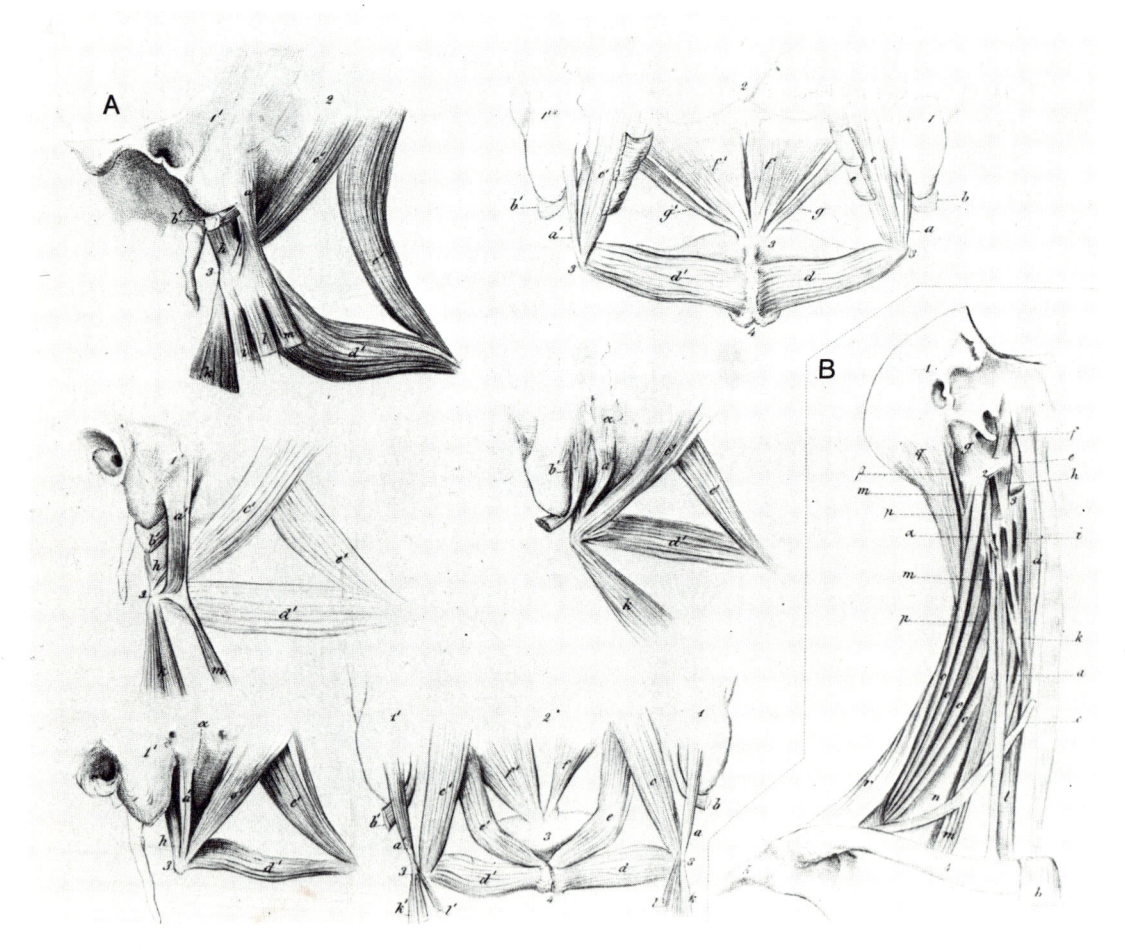

FIG. 9.
M. Atlanticomastoideus.
Supernumerary M. Rectus Capitis Minor.
M. Cleidoepistrophicus.
M. Transversalis Cervicis Anticus (Anomalous).

A, top, center, and bottom: *a,a'*, M. atlanticomastoideus; *g,g'*, supernumerary m. rectus capitis minor.

B: *a*, M. cleidoepistrophicus; *k*, anomalous transversalis cervicis anticus.

A From Gruber, W. Über den Musculus atlantico-mastoideus. *Arch. Anat. Physiol. Wissen. Med.* 1867, pp. 733–738.

B From Gruber, W. Ein Musculus cleido-epistrophicus bei Existenz des Musculus cleidomastoideus der Norm. *Ibid.* 1867, pp. 739–745.

254

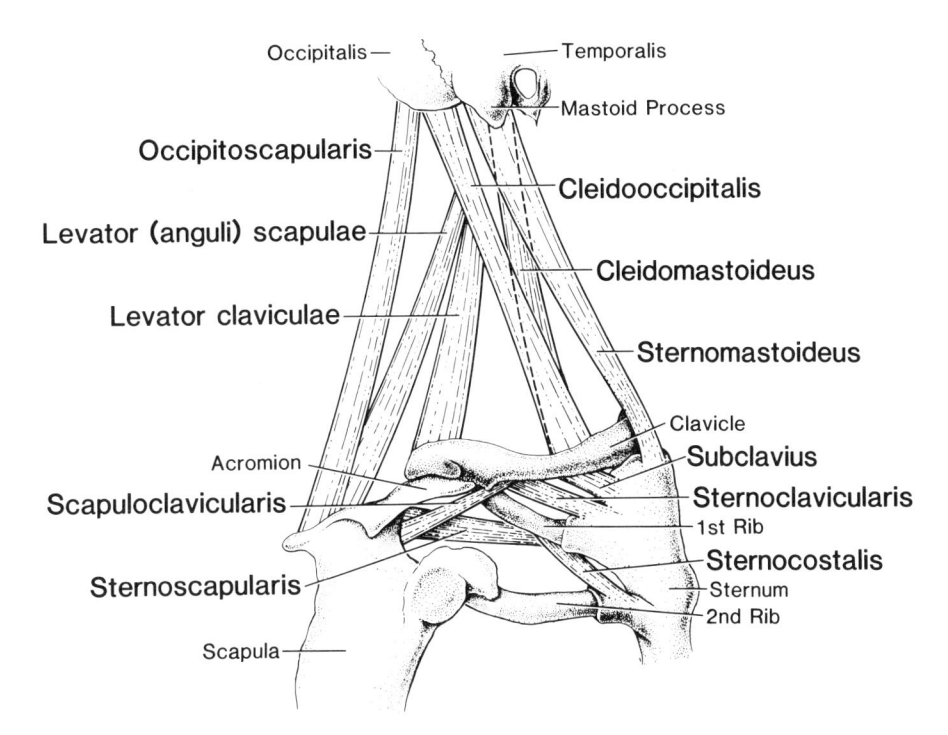

FIG. 10.
Diagram of Varieties of Muscles of Neck, Shoulder, and Chest. Modified and redrawn from an illustration in Wood, J. On a group of varieties of the muscle of the human neck, shoulder, and chest, and their transitional forms and homologies in the mammalia. *Phil. Trans. R. Soc. (London)* 160:83–116, 1870.

Human Anatomic Variation

FIG. 11.
Varieties of Muscles of Neck, Shoulder, and Chest.

A: Back view of the muscles of the human shoulder in a male subject, showing the occipitoscapular variety (d) on the right side.
B: Back view on both sides, showing varieties homologous with the occipitoscapular (d,d), and one other anomalous slip (θ) of rhomboideus major (R).
C: Back view on left side, showing homologous occipitoscapular (δ) and two slips (ε and η) from *levator anguli scapulae* (f) to serratus magnus (S) and serratus posticus superior (s).
D: Back view on left side, showing slips (ς and η) of levator anguli scapulae (f) to serratus posticus superior (s) and serratus magnus (S).
E: Back view on right side, showing homologous slips (ς and η) to the same muscles from the front surface of levator anguli scapulae (f).
F: Back view on right side, showing homologous slip (η) in front of levator anguli scapulae (f) to serratus magnus (S).
G: Back view on left side, showing slips (F) of levator anguli scapulae (f) to serratus magnus (S) and rhomboideus minor (r).
H: View of dissection of the left side of the neck and thorax of male subject, showing levator claviculae (e) and supracostalis (n) muscular varieties.
I: View of dissection of the left side of the neck and thorax of male subject, showing the cleido-occipital (c) and sternoclavicular (l) muscular varieties.
J: View of dissection of the right side of the neck and thorax of a female subject, showing the sternoscapular (i) and scapuloclavicular (k) muscular varieties.

m, Subclavius;
n, supracostalis or sternocostalis, the rectus thoracicus of Turner;
o, omohyoid;
P, pectoralis major;
p, pectoralis minor;
q, supraspinatus;
R, rhomboideus major;
r, rhomboideus minor;
Rt, rectus thoracicus, continuation of rectus abdominis;
S, serratus magnus;
s, serratus posticus superior;
T, trapezius;
t, subscapularis;
u, rectus capitis anticus major;
v, rectus capitis anticus minor;
w, rectus lateralis;
x, clavicle, or its rudimentary representative, or tendinous "inscription";
y, tuberosity of the humerus;
3, acromion process, or spine of scapula;
z, metacromion process of scapula;
l, transverse process of the atlas.

b,c, Levator humeri, lower part of cephalohumeral of some writers, when formed by the cleido-mastoid (b) and cleido-occipital (c) above, with the clavicular fibers of the deltoids (D) or pectoralis major (P) below.
b,e, The same compound muscle, when formed above by the cleidomastoid (b) and the acromiotrachélien or levator claviculae (e), and by the same muscles below.
θ, An abnormal anomalous muscular slip connected with the rhomboideus major, found only in one subject (male).

From Wood, J. On a group of varieties of the muscles of the human neck, shoulder, and chest, and their transitional forms and homologies in the mammalia. *Phil. Trans. R. Soc. (London)* 160:83–116, 1870.

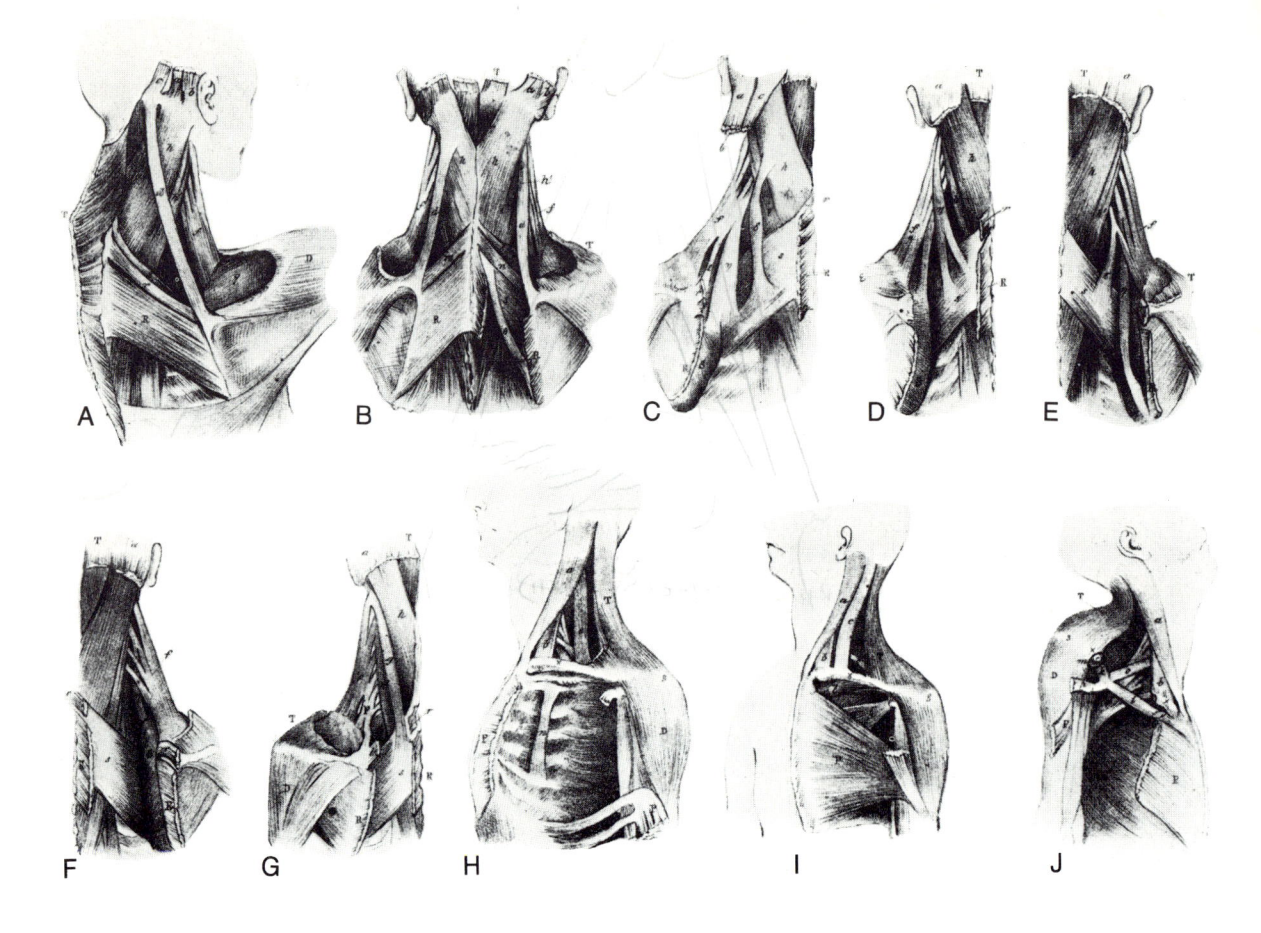

Muscles

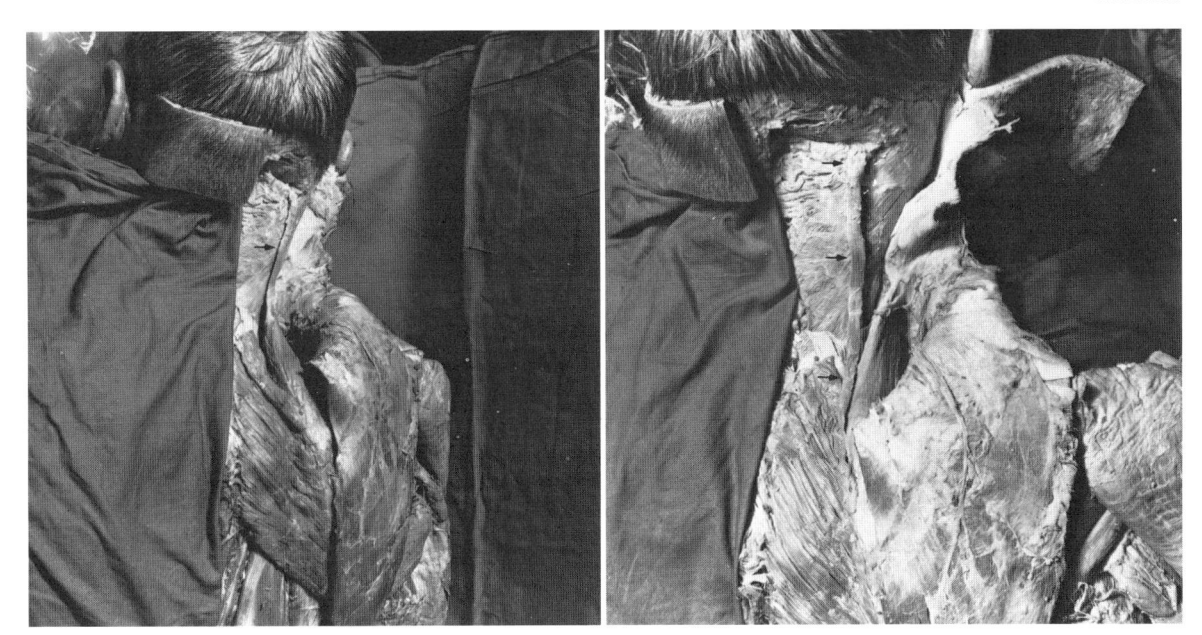

FIG. 12.
M. Occipitoscapularis (Anomalous).
Dissecting room specimen, The University of Iowa College of Medicine.

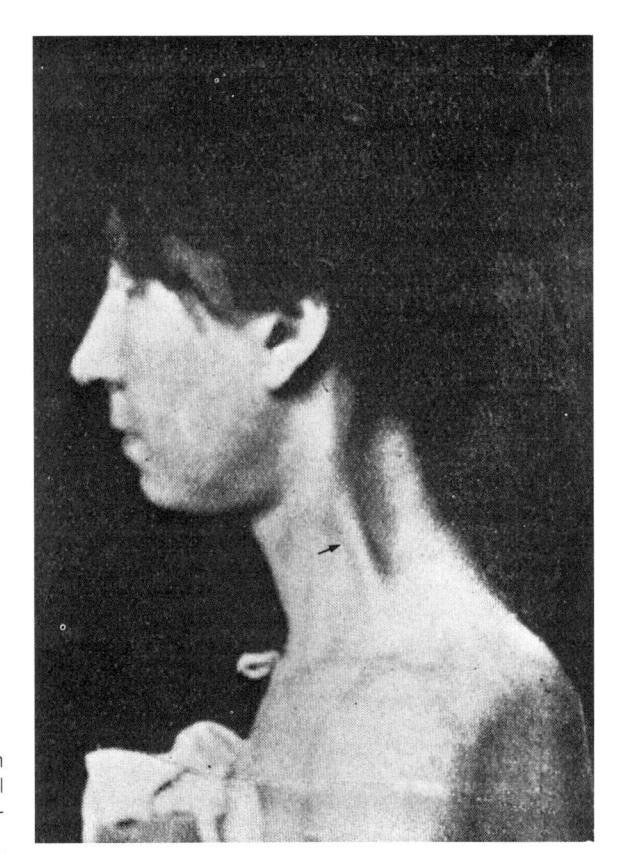

FIG. 13.
M. Omotrachelien (M. Levator Claviculae).
Arrow: Omotrachelian muscle.
From Corner, E.M. Communication of a case of omotrachelian muscle in a living subject. Proceedings of the Anatomical Society of Great Britain and Ireland, June 1900. *J. Anat. Physiol.* 35:iii, 1901.

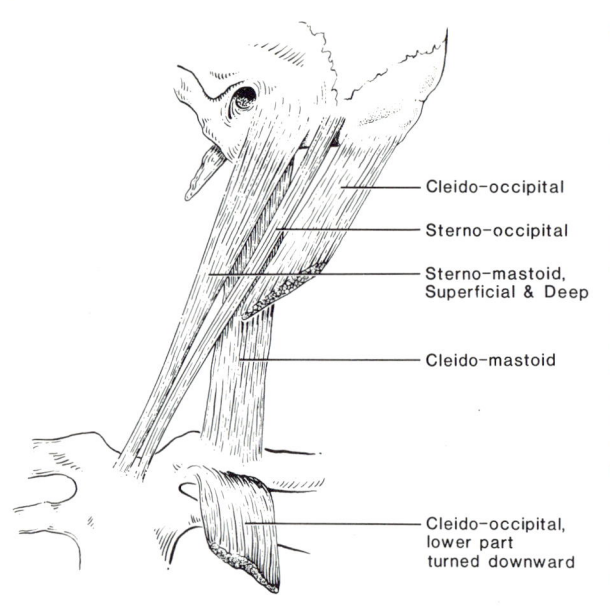

Human Anatomic Variation

FIG. 14.
Quadricipital Type of Sternocleidomastoid, Showing Elemental Components of the Muscle (after Maubrac).

Considerable variation exists in the amount of fusion of the two heads of sternocleidomastoid. Their complete distinctness being of so frequent occurrence as to be regarded as normal by some authors. But, in addition to these two portions, the muscle presents frequently a separation into other parts, and comparative anatomy reveals a primary constitution of the muscle from at least five distinct portions, any one or more of which may appear as distinct bundles. These portions are arranged in two layers, the superficial one consisting of a superficial sternomastoid, a sterno-occipital, and a cleido-occipital portion, while the deep one is formed by a deep sternomastoid and a cleidomastoid portion, the names applied indicating the attachments of the various bundles.

Occasionally the lower portion of the muscle is traversed by a tendinous intersection, a peculiarity of interest in connection with the formation of the muscle by the fusion of portions derived from different myotomes.

Redrawn from Piersol's *Human Anatomy*, G. C. Huber, J. B. Lippincott Company, Philadelphia, 1930.

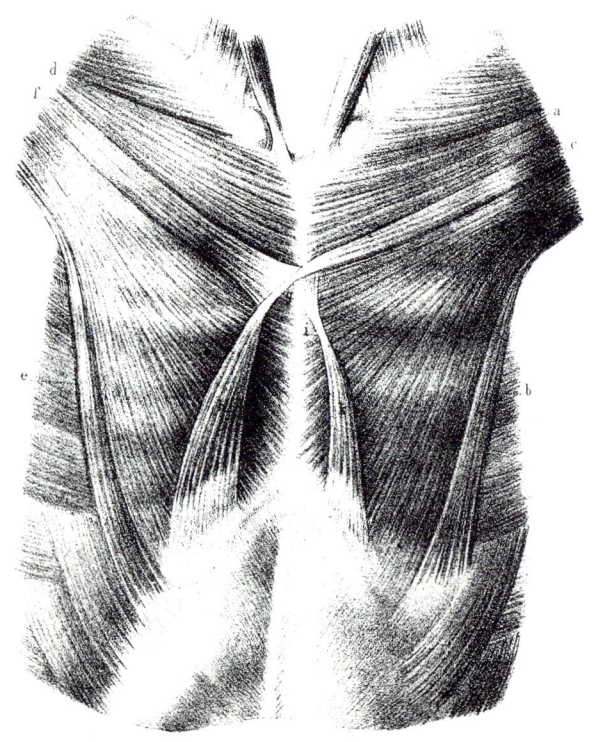

FIG. 15.
M. Sternalis.

a, Pectoralis major cut near the axilla; *b*, pectoralis major, abdominal origin; *c*, fascicle whose tendon of origin (*g*) serves as the origin of sternalis (*h*) and forms a muscle with two bellies; *d*, pectoralis major muscle cut as in *a*; *e*, same as *b*, right side; *f*, a large fascicle that joins the posterior fascia of the tendon (*g*); *h*, right side, sternalis muscle; *i*, left side, sternalis muscle with its tendon of origin (*f*).

From Calori, L. Sopra il muscolo episternale e le sue anatomiche interpretazioni. *Mem. Accad. Sci. Istituto di Bologna S. 4* 9:131–140, 1888.

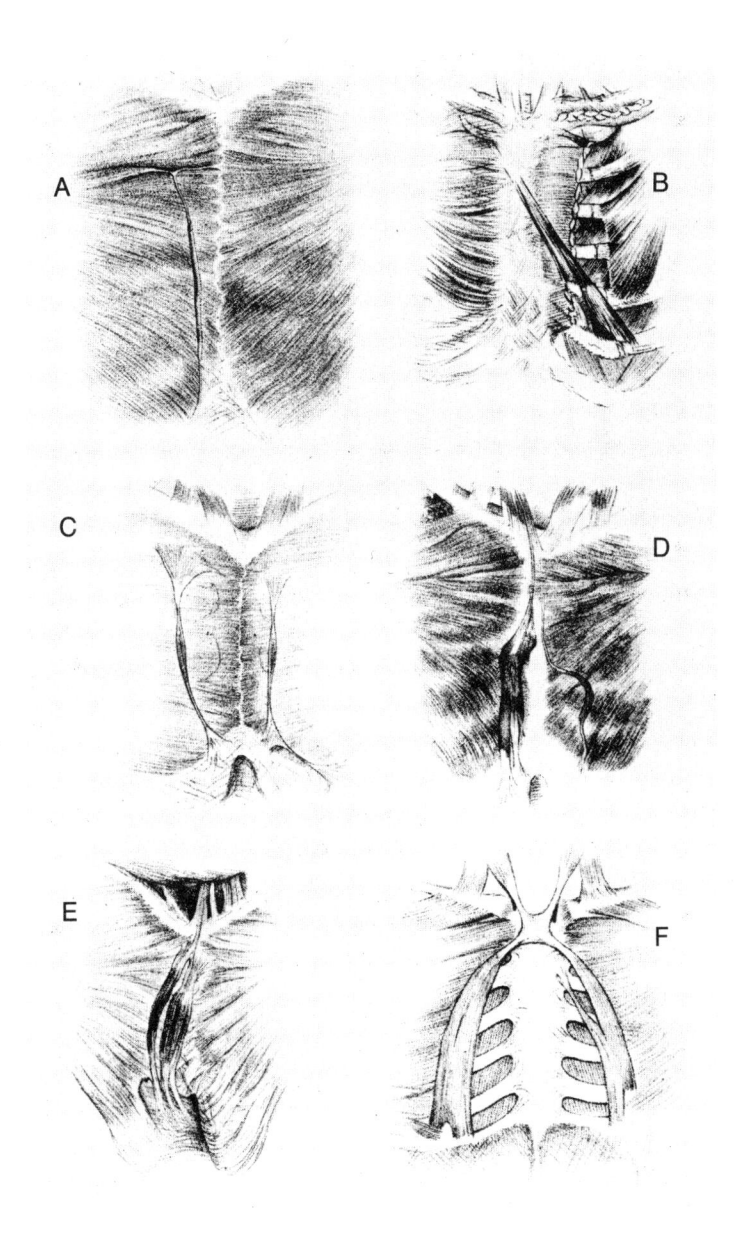

Muscles

FIG. 16.
M. Sternalis.

A: A slender right sternalis muscle arising from the decussating tendinous fibers in front of the lower end of the sternum, inserted into the front of the manubrium, and joining by a rounded slip the upper fibers of the pectoralis major.

B: A sternalis muscle arising from 5th and 6th left costal cartilages and 6th rib. It passed obliquely upwards and inwards across the front of the sternum to be inserted into the strong fibrous aponeurosis over the right pectoralis major, opposite the 2nd right costal cartilage.

C: Double sternalis muscle arising from 5th and 6th costal cartilages inserted into aponeurosis covering greater pectoral muscles.

D: Double sternalis, right much larger than left, extended from aponeurosis covering rectus to right sternomastoid, whilst left was inserted into aponeurosis covering manubrium sterni.

E: Right sternalis muscle arising from aponeurosis covering rectus, subdivided into various bundles, the external of which ascended to the sternomastoid, the internal into the manubrium close to sternal fibers of origin of left pectoralis major.

F: Double sternalis arising from cartilages of 5th and 6th ribs, inserted mostly into tendons of two sternomastoids; some fibers from the left muscle were attached to the manubrium sterni. Sternal origins of both greater pectoral muscles very defective.

Authors' note: Synonyms for musculus sternalis include episternalis, presternalis, rectus sternalis, sternalis brutorum, and thoracicus.

From Turner, W. On the musculus sternalis. *J. Anat. Physiol.* 1:246–253, 1867.

Human Anatomic Variation

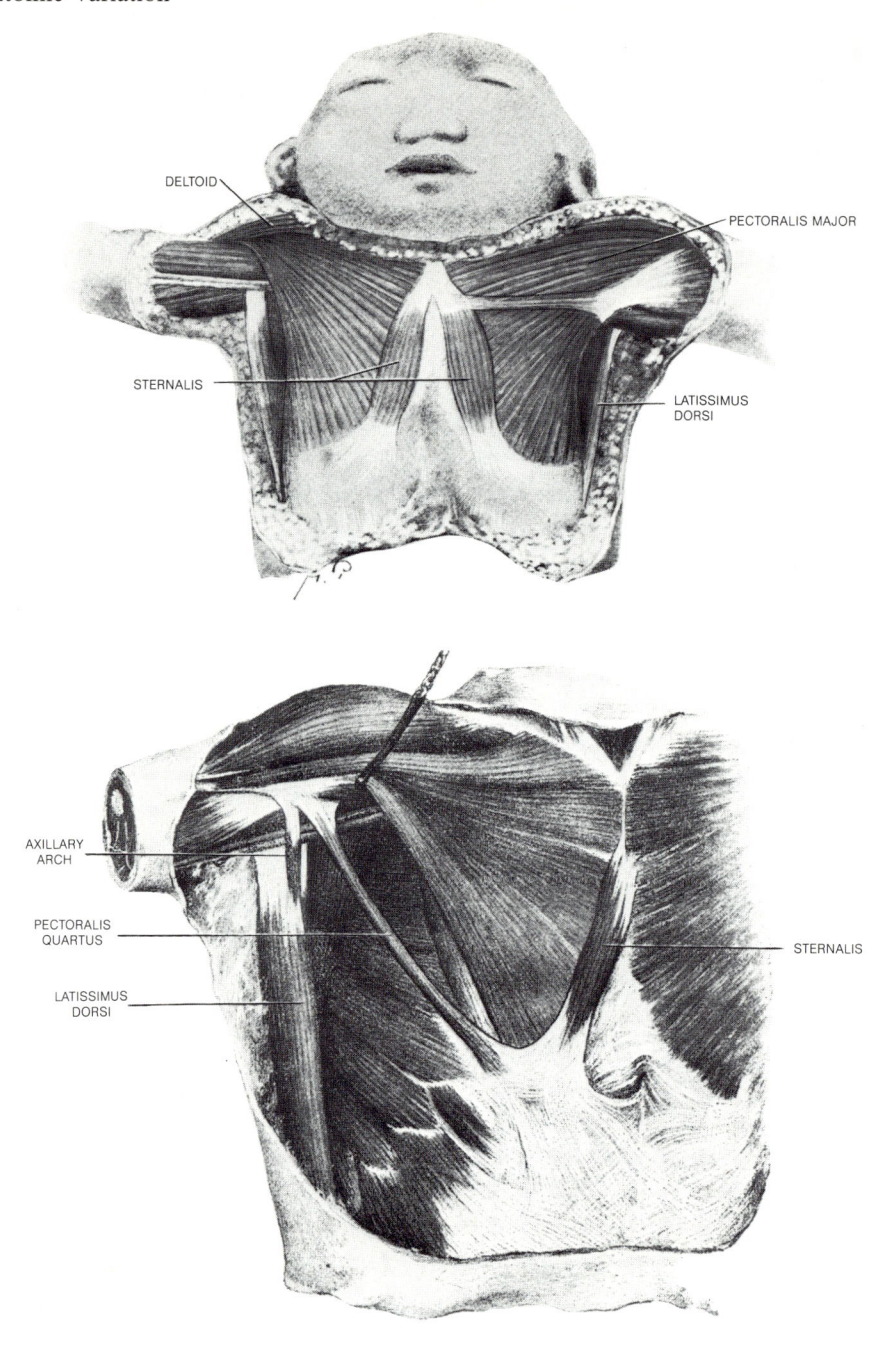

FIG. 17.
M. Axillary Arch.
M. Pectoralis Quartus.
M. Sternalis.

From Huntington, G.S. The derivation and significance of certain supernumerary muscles of the pectoral region. *J. Anat. Physiol.* 39:1–54, 1905.

Muscles

FIG. 18.
Diagrammatic Sketch of Some Anomalous Muscles of the Pectoral Region.

From Huntington, G.S. The derivation and significance of certain supernumerary muscles of the pectoral region. *J. Anat. Physiol.* 39:1–54, 1905.

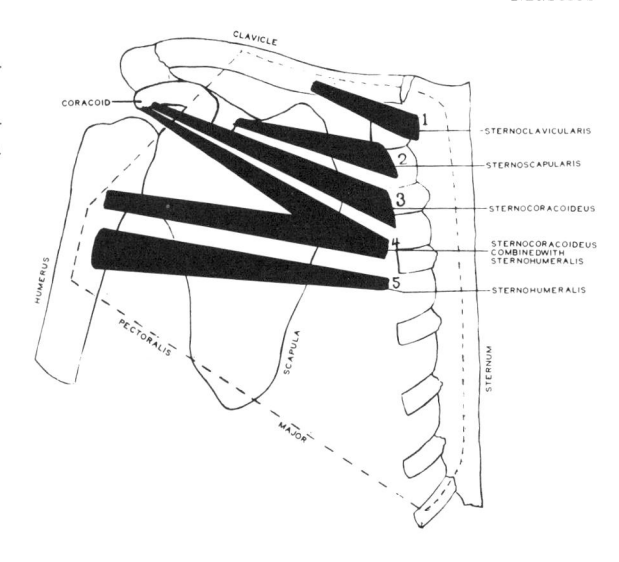

FIG. 19.
Subdivisions of M. Pectoralis Major.
M. Sternalis.

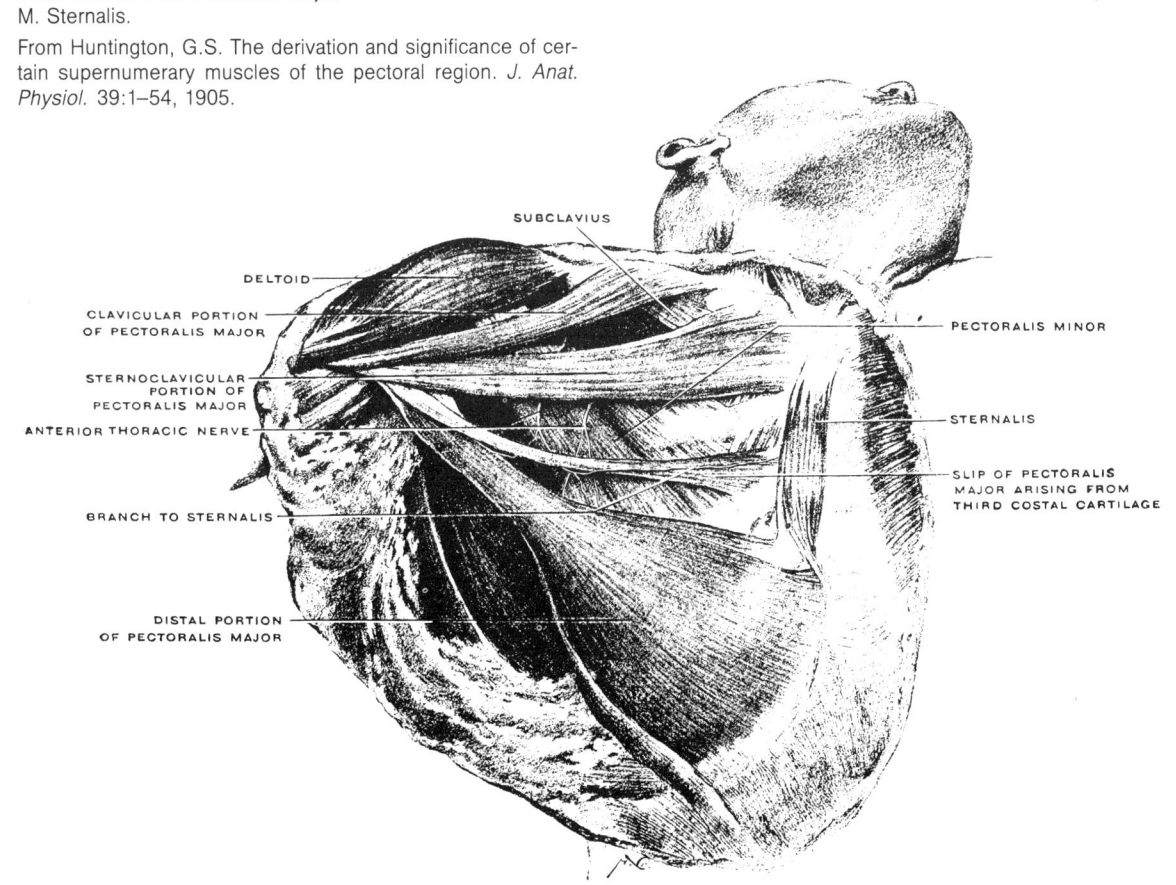

From Huntington, G.S. The derivation and significance of certain supernumerary muscles of the pectoral region. *J. Anat. Physiol.* 39:1–54, 1905.

Human Anatomic Variation

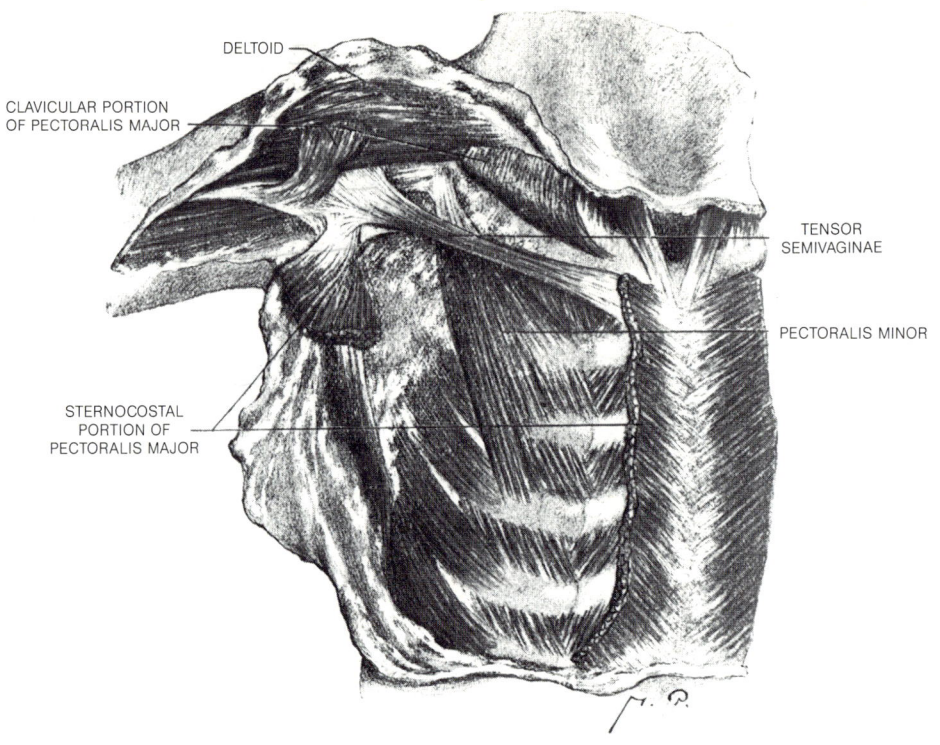

FIG. 20.
M. Tensor Semivaginae Beneath M. Pectoralis Major and Superficial to M. Pectoralis Minor.

From Huntington, G.S. The derivation and significance of certain supernumerary muscles of the pectoral region. *J. Anat. Physiol.* 39:1–54, 1905.

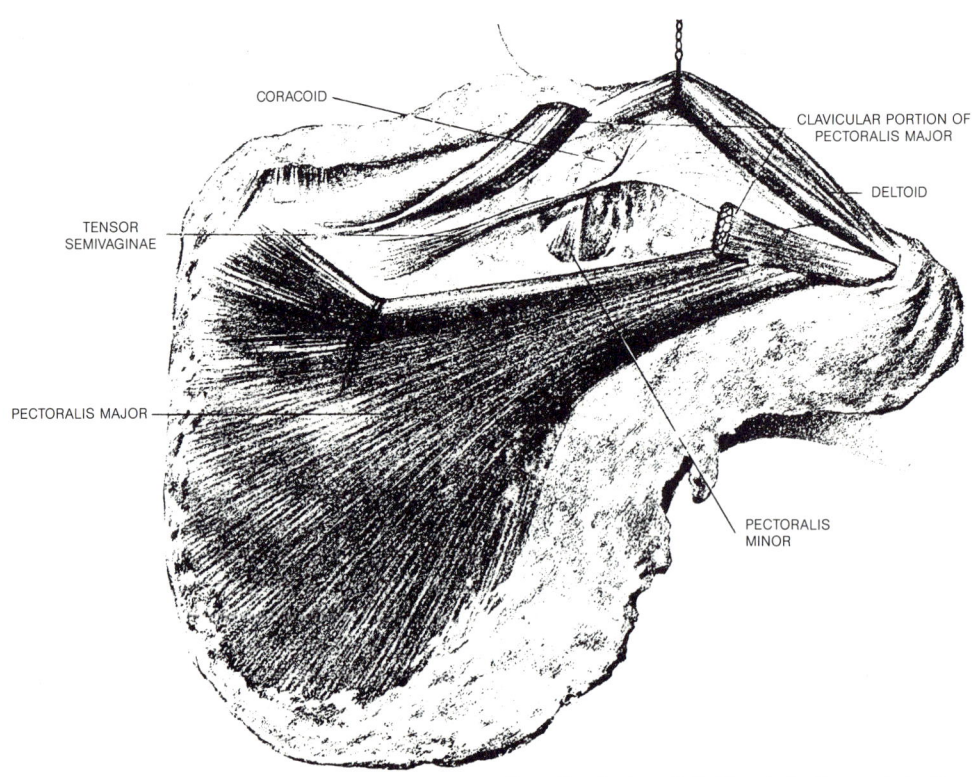

FIG. 21.
M. Tensor Semivaginae Deep to M. Pectoralis Major.

From Huntington, G.S. The derivation and significance of certain supernumerary muscles of the pectoral region. *J. Anat. Physiol.* 39:1–54, 1905.

FIG. 22.
M. Sternocleidomastoideus (Anomalous).
M. Sternoclavicularis.
M. Sternoscapularis.

M. sternocleidomastoideus having extensive origin of clavicular part.

From Huntington, G.S. The derivation and significance of certain supernumerary muscles of the pectoral region. *J. Anat. Physiol.* 39:1–54, 1905.

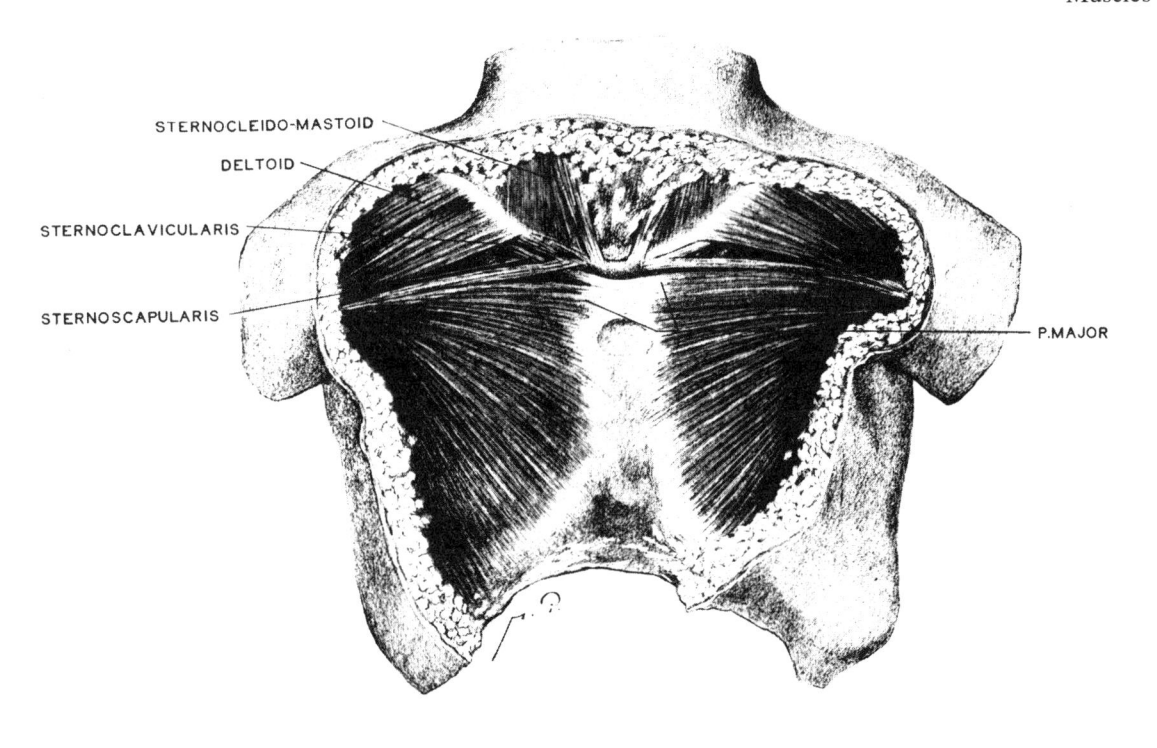

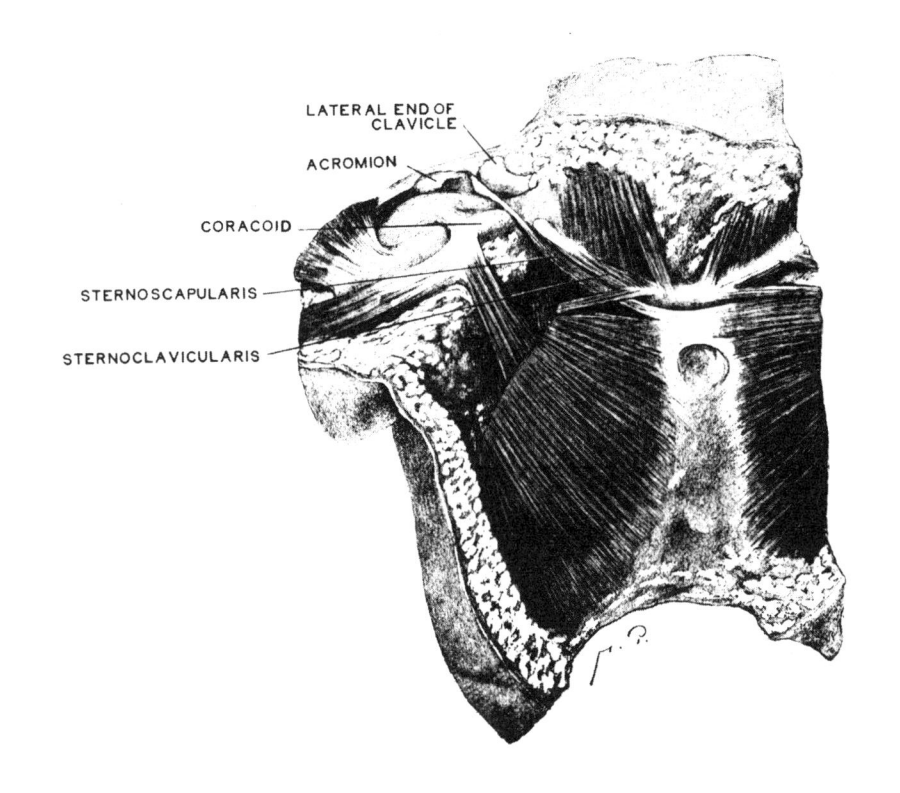

Human Anatomic Variation

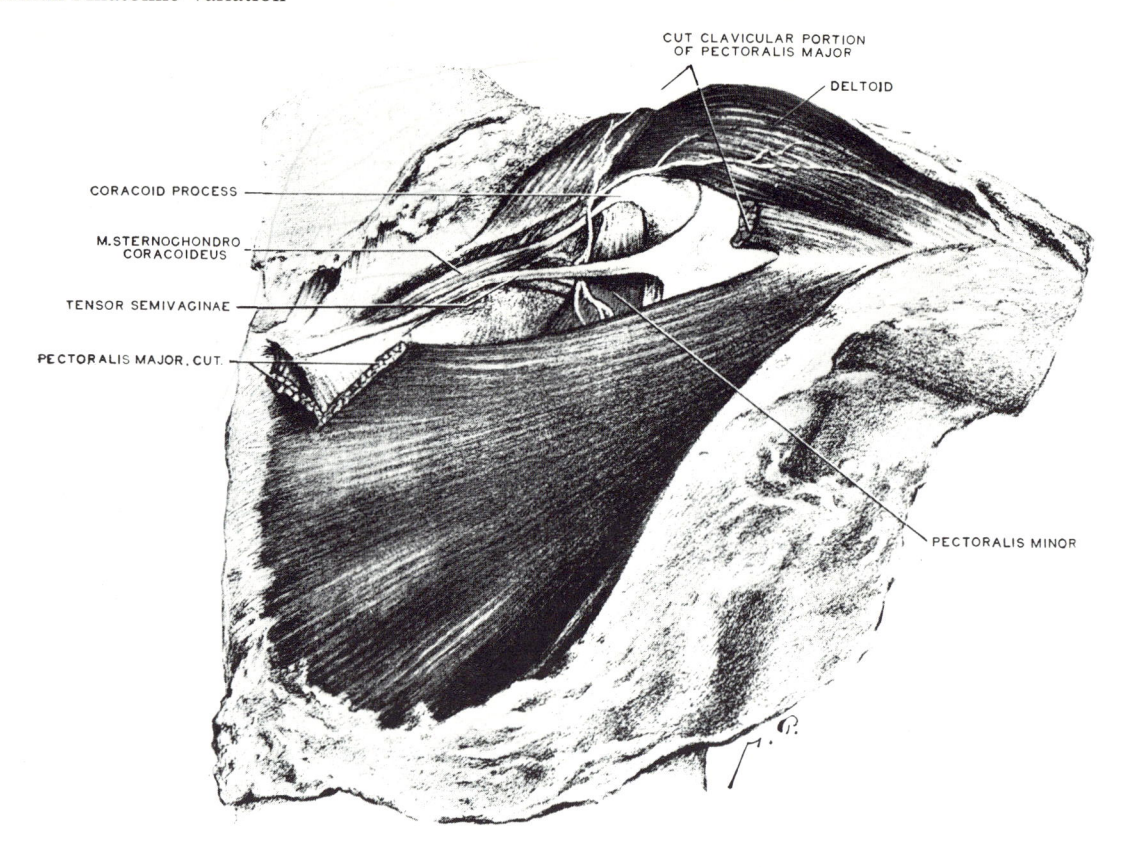

FIG. 23.
M. Sternochondrocoracoideus.
M. Tensor Semivaginae.

From Huntington, G.S. The derivation and significance of certain supernumerary muscles of the pectoral region. *J. Anat. Physiol.* 39:1–54, 1905.

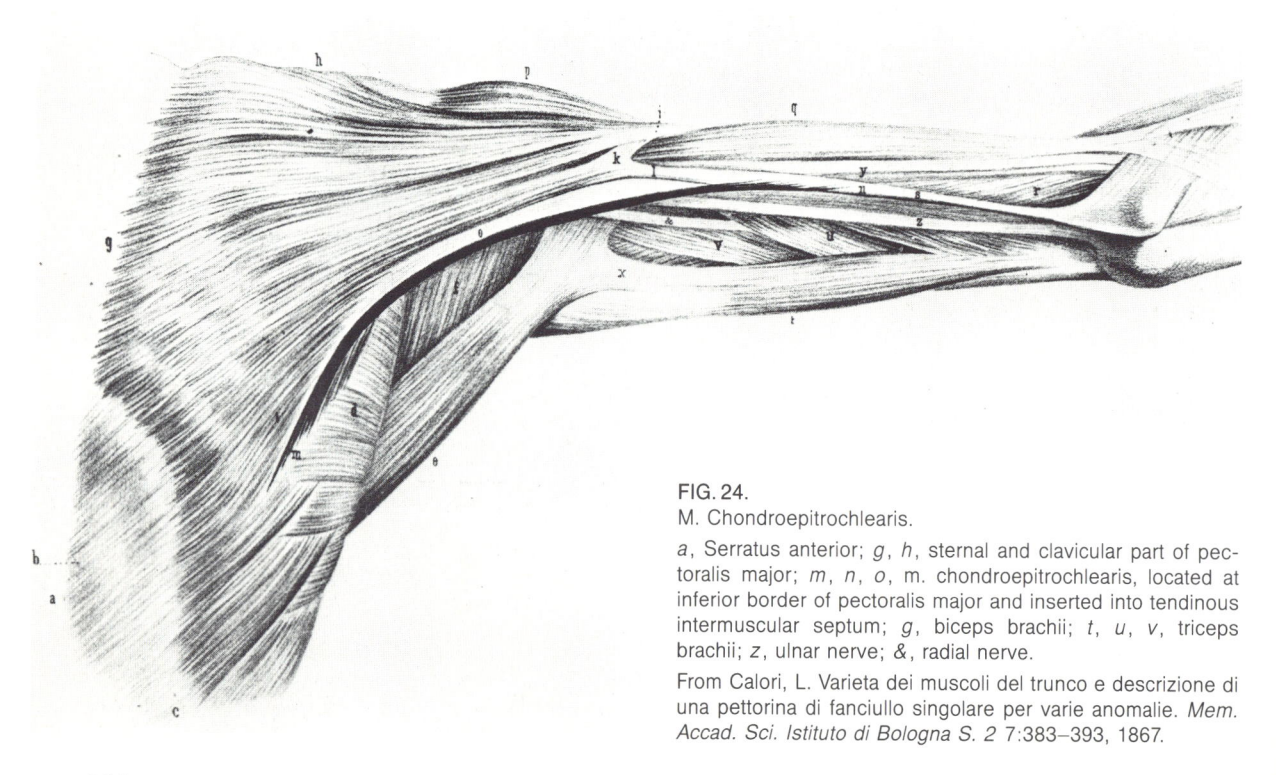

FIG. 24.
M. Chondroepitrochlearis.

a, Serratus anterior; *g*, *h*, sternal and clavicular part of pectoralis major; *m*, *n*, *o*, m. chondroepitrochlearis, located at inferior border of pectoralis major and inserted into tendinous intermuscular septum; *g*, biceps brachii; *t*, *u*, *v*, triceps brachii; *z*, ulnar nerve; &, radial nerve.

From Calori, L. Varieta dei muscoli del trunco e descrizione di una pettorina di fanciullo singolare per varie anomalie. *Mem. Accad. Sci. Istituto di Bologna S. 2* 7:383–393, 1867.

Muscles

FIG. 25.
M. Costoepitrochlearis.
M. Pectoralis Major (Anomalous).

M. costoepitrochlearis with absence of the sternal and costal heads of pectoralis major.

From Griffith, T.W. Note on a case of muscular abnormality observed during life. J. Anat. Physiol. 36:387–388, 1902.

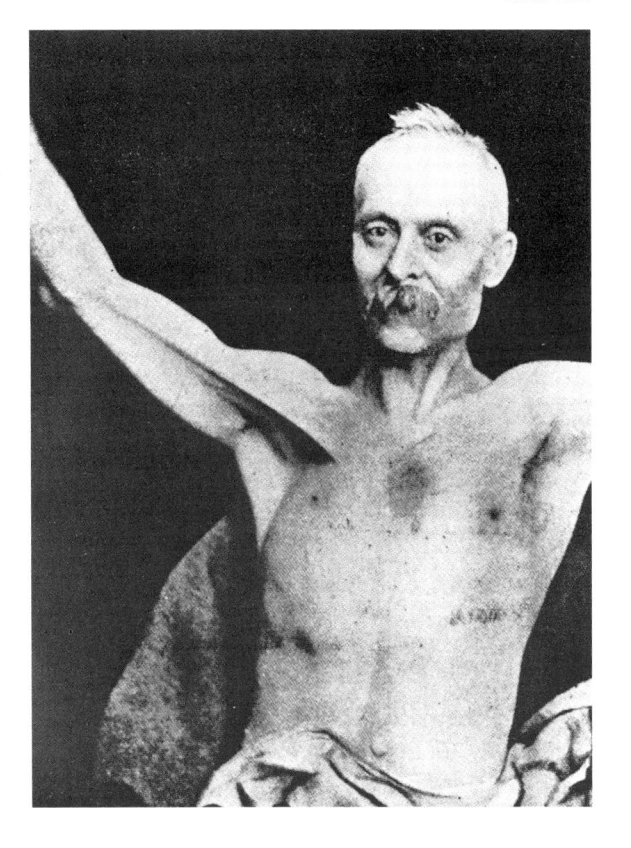

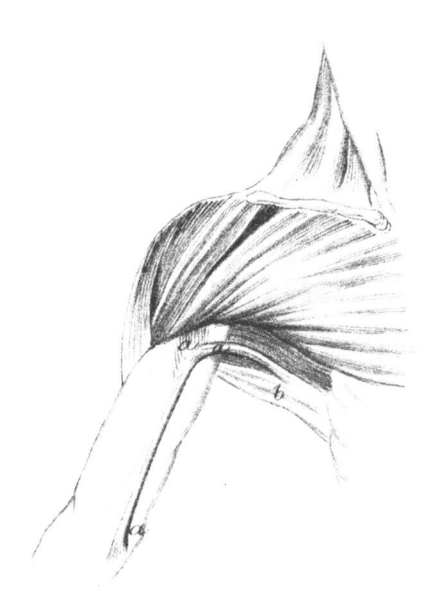

FIG. 26.
M. Chondroepitrochlearis.
M. Dorsoepitrochlearis.

Right drawing shows the left shoulder: a,a, chondroepitrochlearis; a', slip from coracobrachialis. Left drawing shows the right shoulder: a,a,a, dorsoepitrochlearis; b, latissimus dorsi.

From Perrin, J.B. Notes on some variations of the pectoralis major, with its associate muscles seen during Sessions 1868–69, 69–70, at King's College, London. J. Anat. Physiol. 5:233–240, 1871.

265

Human Anatomic Variation

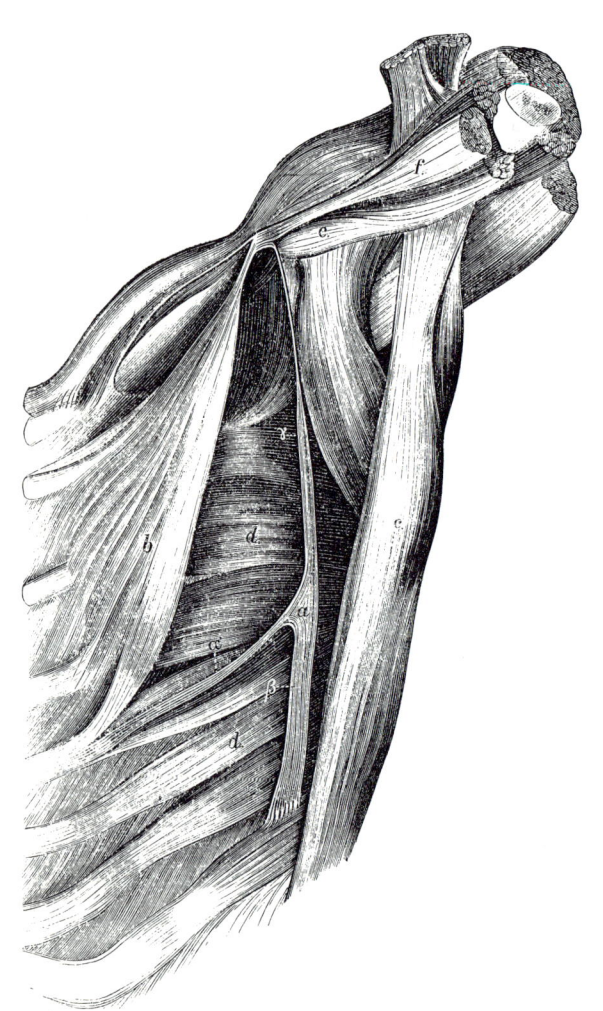

FIG. 27.
M. Pectoralis Minor (Anomalous).
M. Chondrocoracoideus.

e, M. serratus anterior; *f*, m. pectoralis minor; *g*, anomalous muscle fascicle arising from the abdominal aponeurosis and from the sixth and seventh ribs; *h*, belly of anomalous muscle; *i*, tendon of insertion continuous with those of the short head of m. biceps brachii and m. coracobrachialis.

From Calori, L. Varietá dei muscoli del tronco e descrizione di una pettorina di fanciullo singolare per varie anomalie. *Mem. R. Accad. Sci. Istituto di Bologna S. 2* 7:383–393, 1867.

FIG. 28.
M. Costocoracoideus.

a, M. costocoracoideus (supernumerarius): α, origin of an anterior limb; β, origin of a posterior limb; and γ, anterior limb; *b*, pectoralis minor; *c*, latissimus dorsi; *d*, serratus anterior; *e*, coracobrachialis; *f*, short head of biceps brachii.

From Gruber, W. Über einen Musculus costo-coracoideus supernumerarius beim Menschen. *Bull. l'Acad. Imp. Sci. St. Petersbourg* 17:408–414, 1872.

Muscles

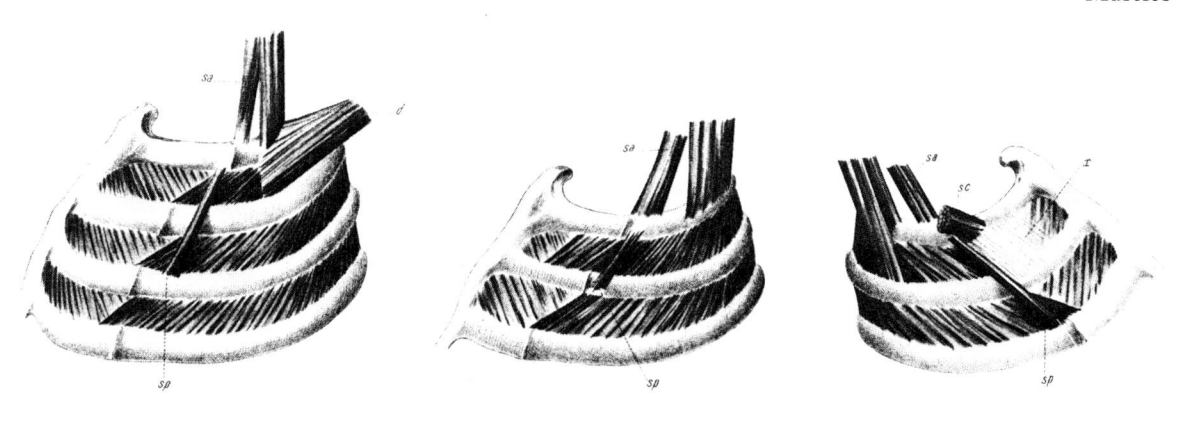

FIG. 29.
M. Supracostalis.

d, M. serratus anterior; sa, m. scalenus anterior; sc, m. subclavius; sp, m. supracostalis; x, cranial part of the tendon of rectus abdominis.

From Livini, F. Contribuzione alla morfologia del m. rectus abdominis e del m. supracostalis nell'uomo. Arch. Ital. Anat. Embriol. 4:81–115, 1905.

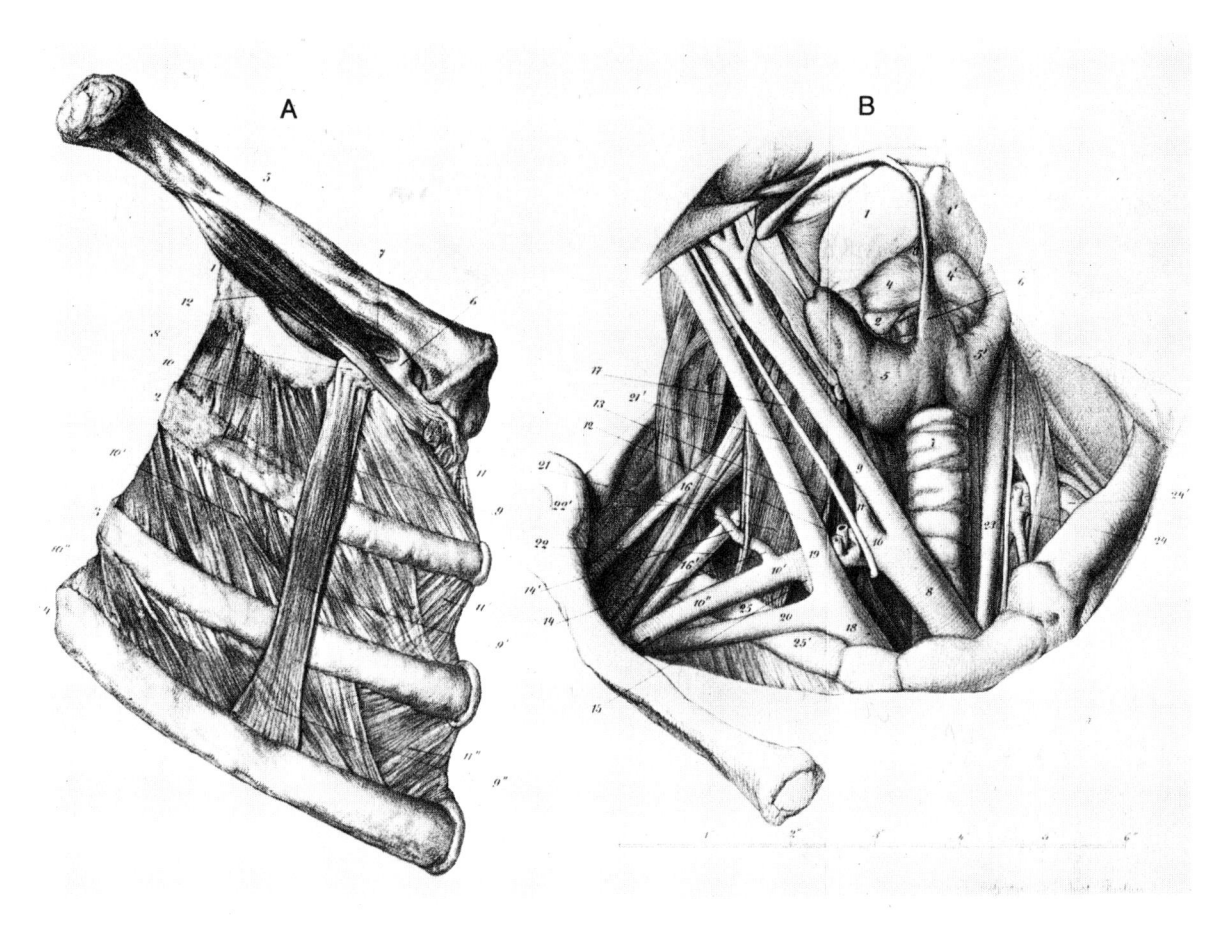

FIG. 30.
M. Supracostalis (Anterior, Anomalous).
(Anomalous Transverse Cervical Artery).
(Anomalous Thyrocervical Trunk).

A: 9,9',9", M. supracostalis; 10,10',10", intercostal muscles (external); 11,11',11", intercostal muscles (internal); 12, subclavius.

B: 1, Thyroid cartilage; 2, cricoid cartilage; 4, cricothyroid muscle; 5, thyroid gland; 6, pyramidal process of thyroid gland (see Organs); 12, thyrocervical trunk composed of two separate trunks; 13, ascending cervical artery; 14, transverse cervical artery, piercing m. scalenus anterior.

From Bochdalek, Jr. Ein anomaler Musculus supracostalis anterior. Arch. Path. Anat. Physiol. Klin. Med. 41:257–258, 1867.

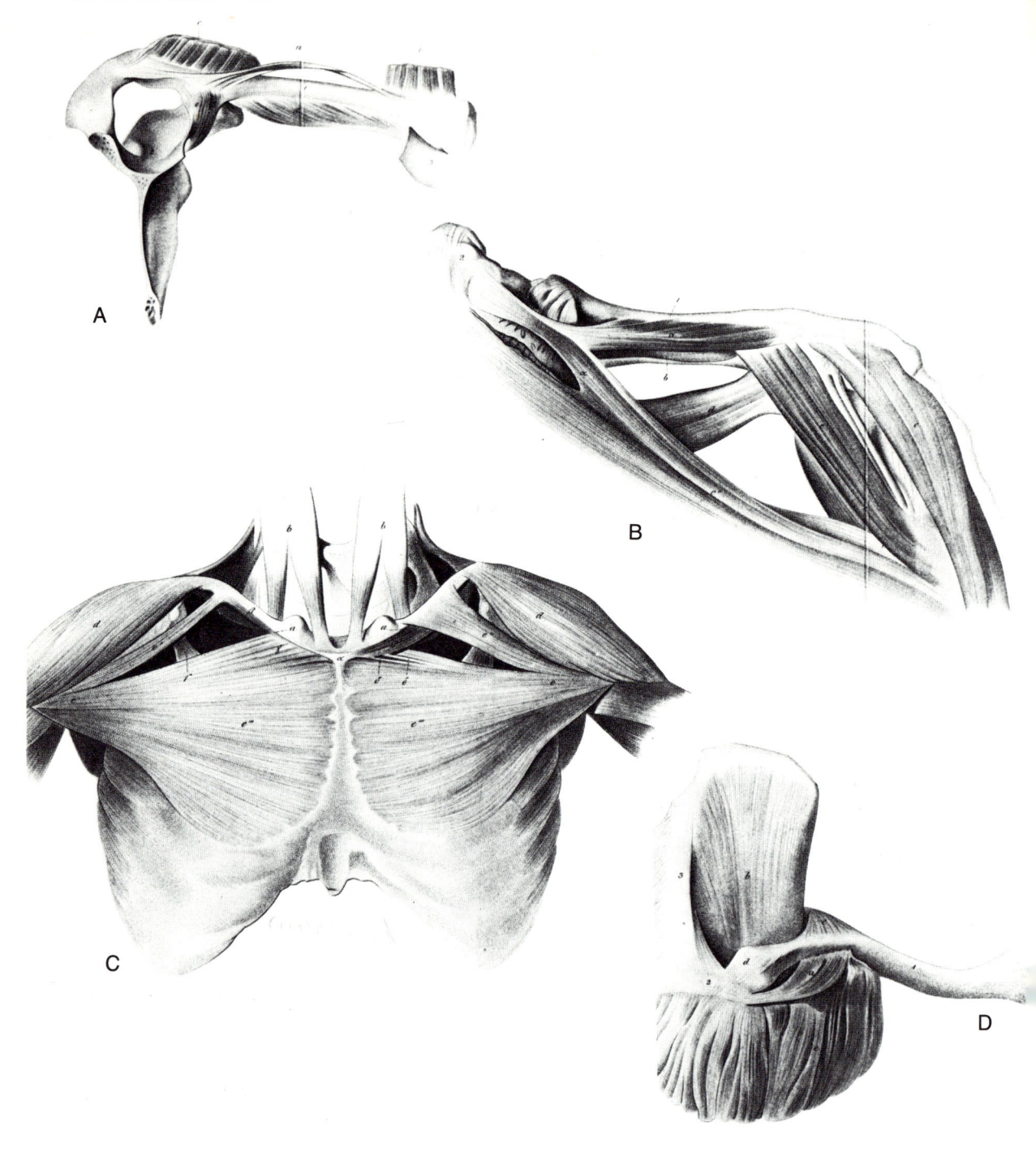

FIG. 31.
M. Supraclavicularis.
M. Sternoclavicularis Anticus s. Preclavicularis Medialis Singularis.
M. Interclavicularis Anticus Digastricus.
M. Acromioclavicularis s. Preclavicularis Lateralis.

A: Left shoulder girdle, seen from behind. *1*, Clavicle; *2*, shoulder blade; *a*, m. supraclavicularis (proprius); *b*, m. sternocleidomastoideus; *c*, clavicular part of trapezius; *d*, m. subclavius.
B: Left sternoinfraclavicular region. *1*, Clavicle; *2*, manubrium; *a*, m. sternoclavicularis anticus s. preclavicularis medialis singularis; *b*, subclavius; *c*, m. pectoralis major; *c'*, rudimentary clavicular portion of pectoralis major; *d*, m. pectoralis minor; *e*, clavicular portion of deltoideus; α, fascicle of pectoralis major attached to m. sternoclavicularis anticus medialis singularis.
C: Thorax, shoulder, etc. *a,a,* M. interclavicularis anticus digastricus: α, midline attachment; β, right side; and β', left side; *b,b*, m. sternocleidomastoideus; *c,c*, mm. subclavii; *d,d*, mm. deltoidei; *e,e*, mm. pectorales majores; *e',e"*, right and left clavicular parts of pectoralis major; *e"'*, sternocostal part of pectoralis major with tendinous insertions into m. interclavicularis anticus digastricus (δ,δ).
D: Right shoulder girdle. *1*, Clavicle; *2*, acromion; *a*, m. acromioclavicularis s. preclavicularis lateralis; *b*, m. supraspinatus; *c*, m. deltoideus; *d*, acromioclavicular capsule; *e*, coracoclavicular ligament.

From Gruber, W. Neue supernumeräre Schlüsselbeinmuskeln. *Arch. Anat. Physiol. Wissen. Med.* 1865, pp. 703–718.

Muscles

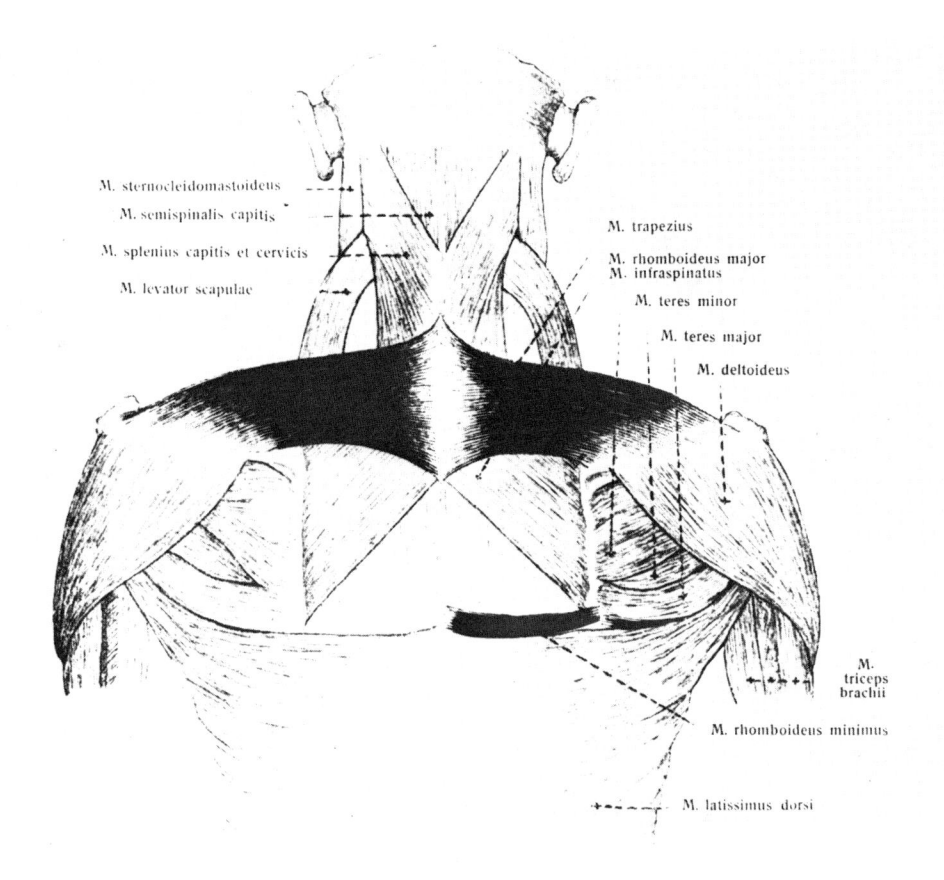

FIG. 32.
M. Trapezius (Anomalous).
M. Rhomboideus Minimus.
Bilateral deficiency of origin of m. trapezius.

From Haffner, H. von. Eine seltene doppelseitige Anomalie des Trapezius. *Int. Monatsschr. Anat. Physiol.* 20:313–318, 1903.

FIG. 33.
M. Coracoclavicularis.
(Suprascapular Artery and Nerve Passing Beneath the Transverse Scapular Ligament).

The left scapula is seen from the front, with the clavicle somewhat rotated on its own axis. *, M. coracoclavicularis; *p.m.*, m. pectoralis minor; *c.t.*, conoid and trapezoid ligaments; *s.s.a.*, *s.s.n.*, suprascapular artery and nerve: both passed under the ligament (suprascapular or transverse scapular ligament).

From Pye-Smith, P.H., Howse, H.G. and J.N.C. Davies-Colley. Notes of abnormalities observed in the dissecting rooms during the winter sessions of 1868–9 and 1869–70. *Guy's Hosp. Rep.* 16:147–164, 1871.

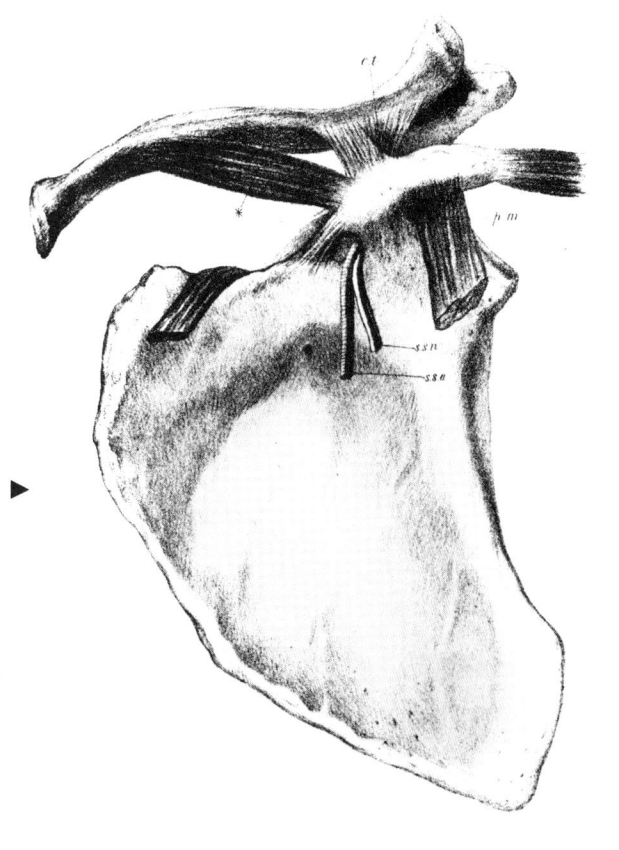

Human Anatomic Variation

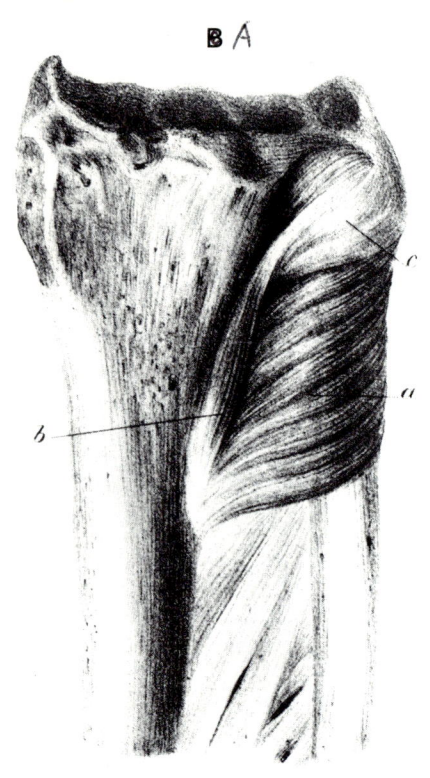

FIG. 34.
M. Deltoideus (Anomalous).
Deficient M. Pronator Quadratus.
M. Tensor Capsulae Radiocubitalis.

Absence of the middle part of deltoideus, deficient pronator quadratus, and m. tensor capsulae radiocubitalis.

A: Left forearm. *a*, Deficient pronator quadratus; *b*, m. tensor capsulae radiocubitalis inferioris; *c*, radiocubital (radioulnar) capsule.

B: Left shoulder. *1*, Clavicle; *2*, acromion; *3*, humerus; *a*, deltoideus: *a'*, clavicular part, and *a"*, spinous part; *b*, pectoralis major; *c*, pectoralis minor; α, coracoacromial ligament; β, greater tubercle of humerus; *,*, large gap in deltoideus (middle part).

From Gruber, W. Mangel der mittleren Portion des Musculus deltoides. *Arch. Path. Anat. Physiol. Klin. Med.* 54:184–185, 1872. Ein Spanner der unteren Radio-Cubitalkapsel – Tensor capsulae radio-cubitalis inferioris. *Arch. Pathol. Anat. Physiol. Klin. Med.* 54:186–187, 1872.

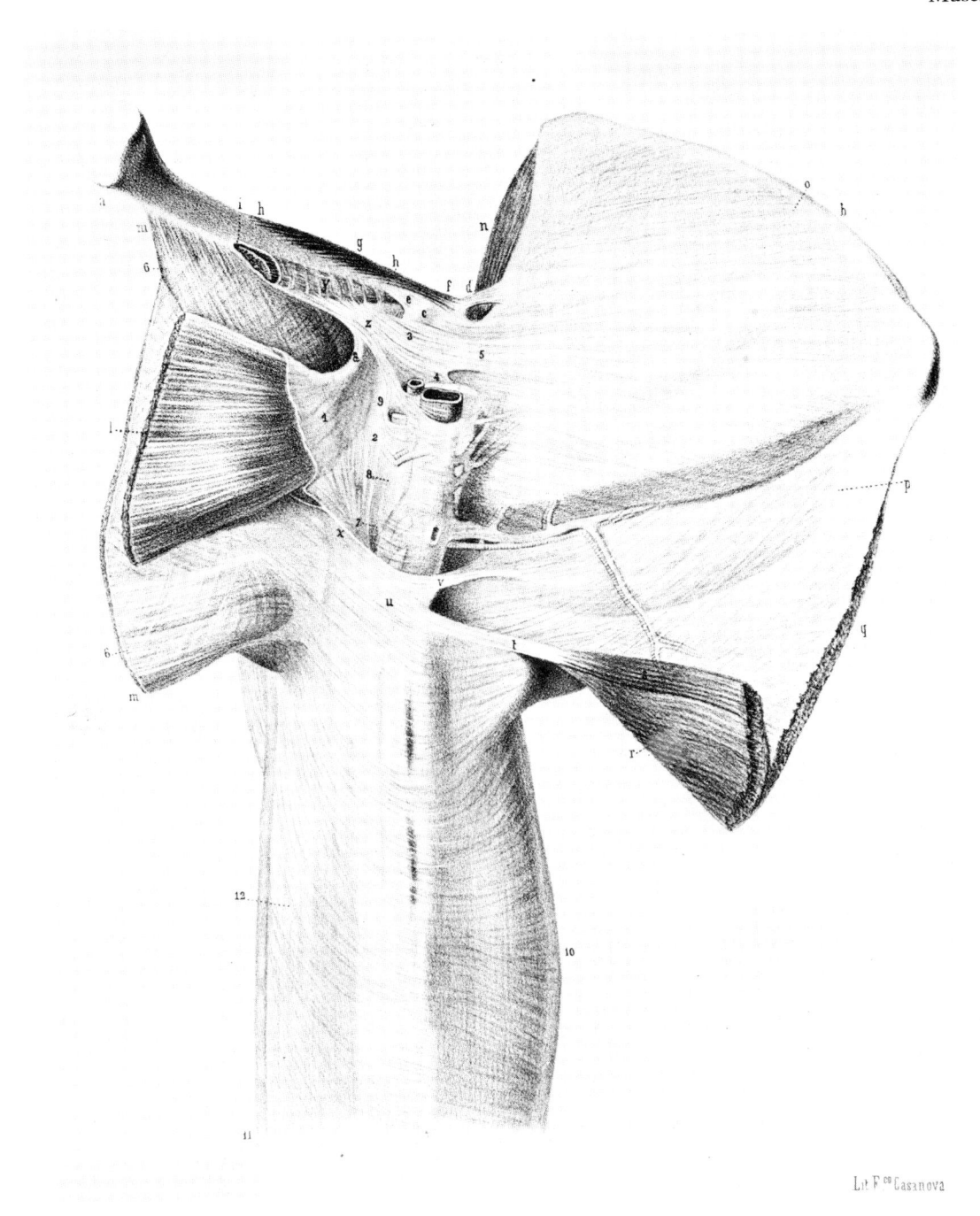

FIG. 35.
M. Coracoclavicularis.

a, Clavicle; *b*, scapula; *c*, coracoid process; *e*, part of the posterior coracoclavicular ligament; *f*, origin of m. coracoclavicularis; *g*, m. coracoclavicularis; *h,h*, insertion or termination of m. coracoclavicularis; *i*, m. subclavius (cut, covered with its fascia); *l*, m. pectoralis minor; *m,m*, m. pectoralis major (covered with its fascia); *n*, m. supraspinatus; *o*, m. subscapularis (covered with its fascia); *p*, m. teres major (covered with its fascia); *q*, m. latissimus dorsi; *y*, fascial sheath of m. subclavius; *7,8,9*, portion of the axillary fascia and neurovascular sheath; *10*, part of the arm; *11*, cephalic vein; *12*, brachial fascia.

From Calori, L. Intorno ad alcune varietà incontrate nella muscolatura degli arti superiori. *Mem. R. Accad. Sci. Istituto di Bologna* S. 2 6:157–174, 1866.

Human Anatomic Variation

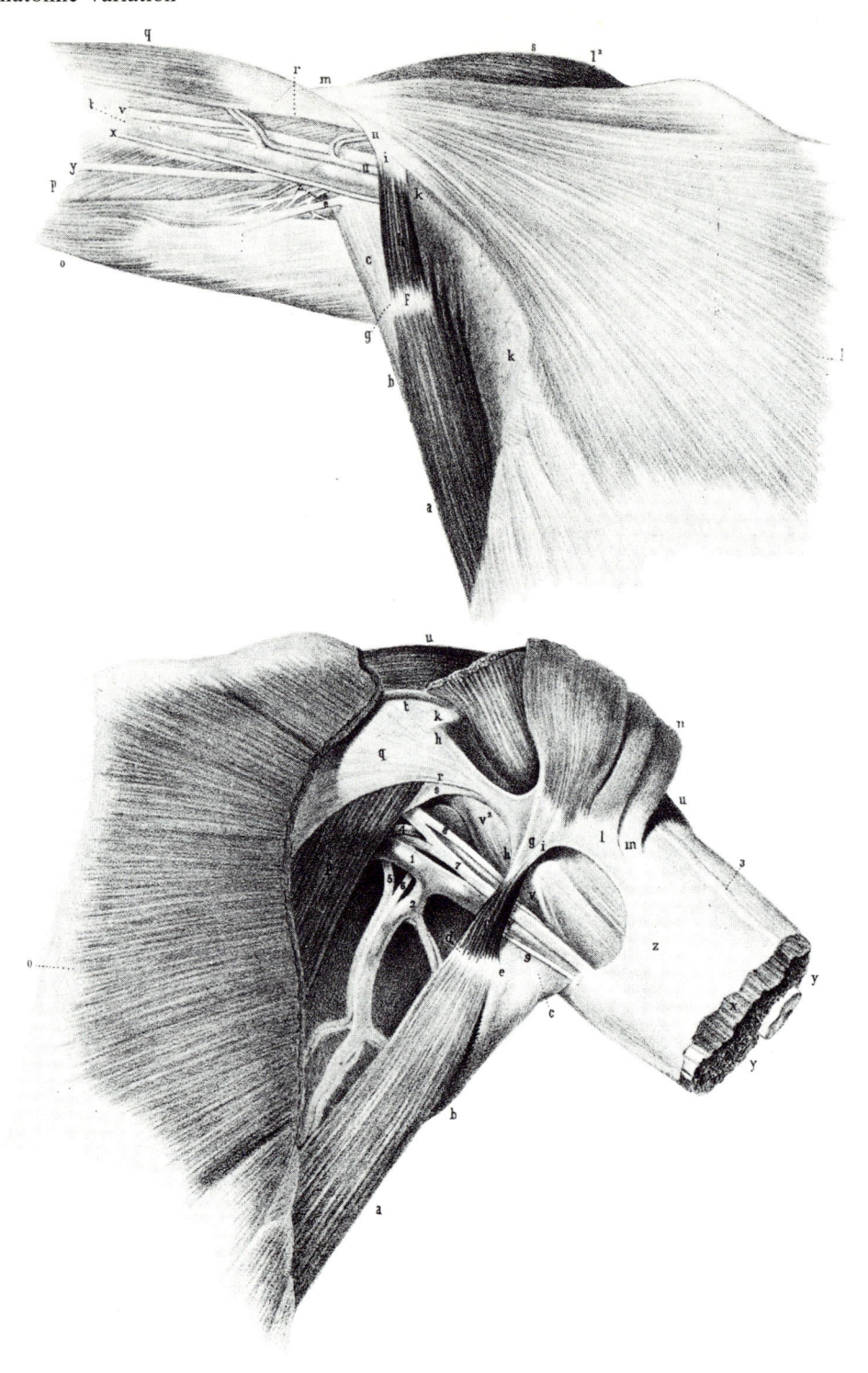

FIG. 36.
Axillary Arches.

A: *a,b*, M. latissimus dorsi; *c*, tendon of insertion of latissimus dorsi; *g,f*, tendinous origin of axillary arch muscle; *i,u*, axillary arch muscle and tendon crossing blood vessels and nerves of axilla; *k,k*, axillary fascia; *l*, pectoralis major.

B: *a*, M. latissimus dorsi; *d,e,g*, m. axillary arch, tendons of origin and insertion; *l,m,n,o*, m. pectoralis major, tendinous insertion of muscle; *p*, m. pectoralis minor; *u*, m. deltoideus; *x*, m. biceps brachii; *y*, m. triceps brachii; *1*, axillary vein; *2*, thoracodorsal vein; *3*, cephalic vein; *4*, axillary artery passing between the lateral and medial cords of the brachial plexus; *5*, thoracodorsal artery; *6*, thoracodorsal nerve; *7*, median nerve; *8*, musculocutaneous nerve; *9*, medial cutaneous nerve of forearm.

From Calori, L. Intorno ad alcune varietà incontrate nella muscolatura degli arti superiori. *Mem. Accad. Sci. Istituto di Bologna S. 2* 6:157–174, 1866.

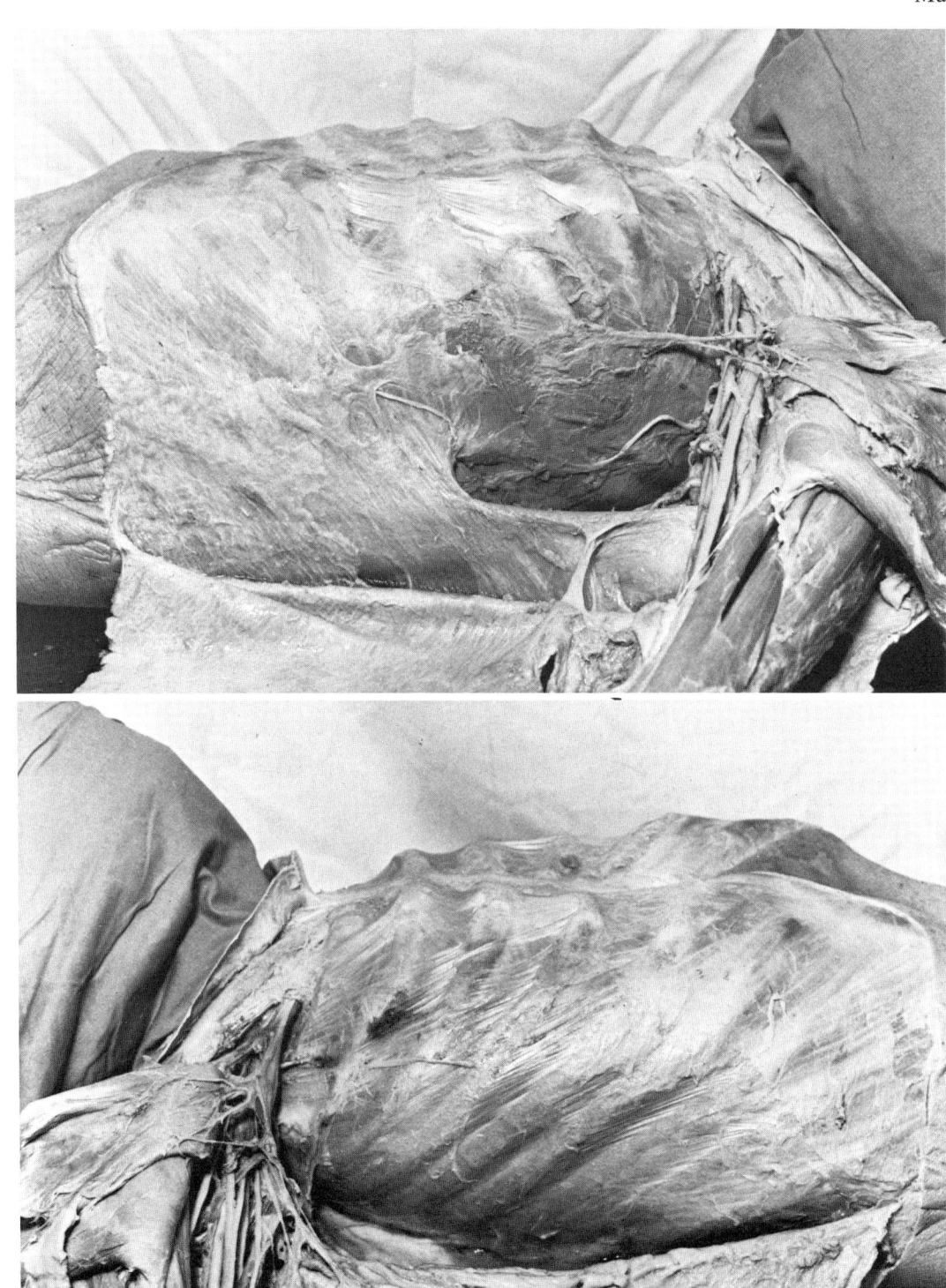

FIG. 37.
Unilateral Absence of M. Serratus Anterior (Right Side). Anomalous Fascicle of M. Latissimus Dorsi.

In addition to the absence of serratus anterior this specimen had an unusually high origin of latissimus dorsi (between the 5th and 6th thoracic vertebrae) which covered and was fused with the inferior angle of the scapula. An additional fascicle of latissimus arose from the medial border and (deep) fascia covering the inferior quarter of infraspinatus and joined the tendon of insertion of latissimus dorsi. The upper photograph shows the left side: serratus anterior present and normal in form; the lower shows the right side: serratus anterior absent.

Dissecting room specimen, The University of Iowa College of Medicine.

Human Anatomic Variation

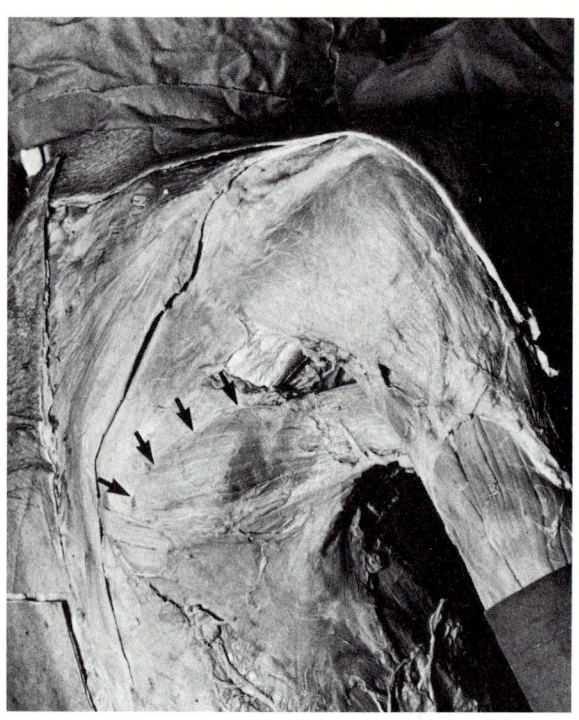

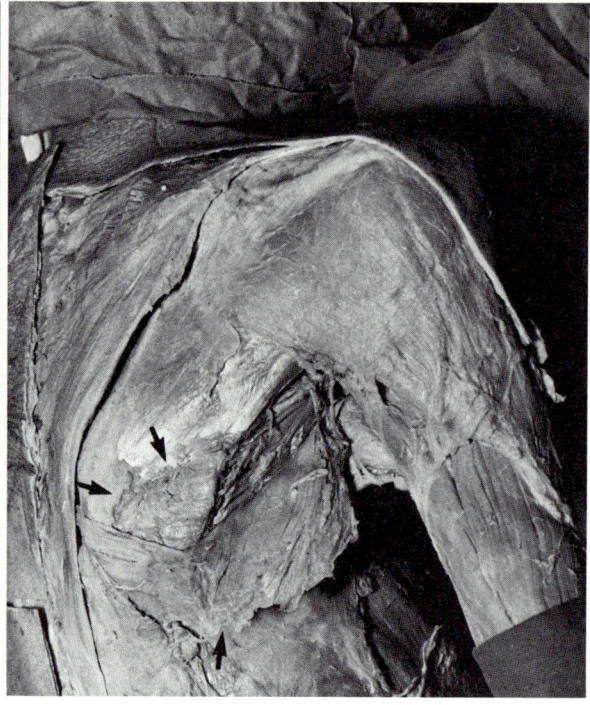

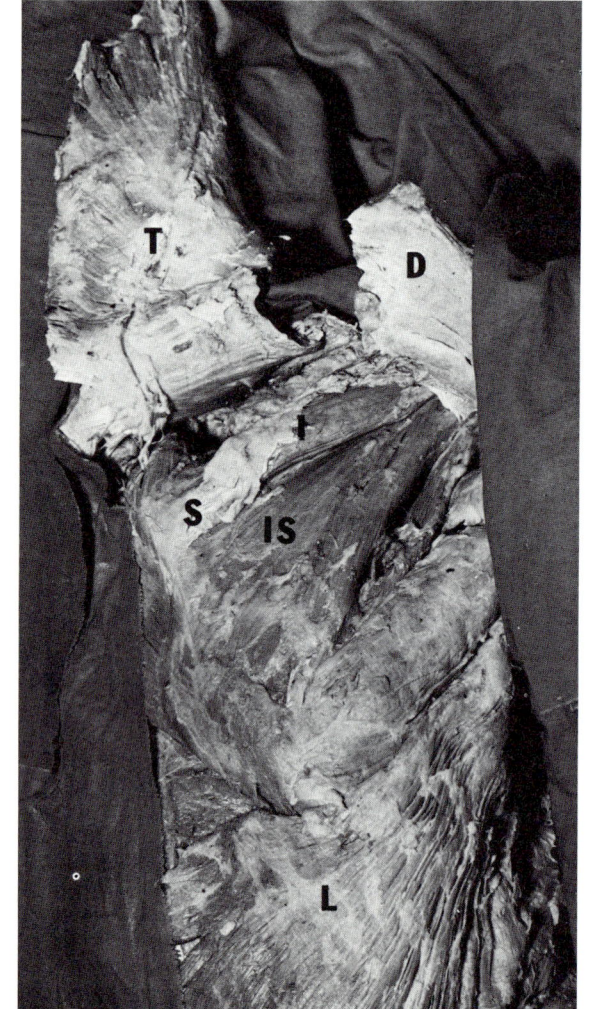

FIG. 38.
Unilateral Absence of M. Serratus Anterior (Right Side). Anomalous Fascicle of M. Latissimus Dorsi.

In addition to the absence of serratus anterior this specimen had an unusually high origin of latissimus dorsi (between the 5th and 6th thoracic vertebrae) which covered and was fused with the inferior angle of the scapula. An additional fascicle of latissimus arose from the medial border and (deep) fascia covering the inferior quarter of infraspinatus and joined the tendon of insertion of latissimus dorsi. The right photograph shows anomalous fascicle of latissimus dorsi (arrows). On the left, the anomalous fascicle is reflected to show origin from infraspinatus fascia (arrows).

Dissecting room specimen, The University of Iowa College of Medicine.

FIG. 39.
M. Infraspinatus Superficialis.
I, Infraspinatus superficialis (elevated upward); *D*, deltoid; *IS*, infraspinatus; *L*, latissimus dorsi; *S*, scapular spine; *T*, trapezius (reflected upward).

This common variation extends from the inferior spinous border of the scapula to the greater tubercle of the humerus.

Dissecting room specimen, The University of Iowa College of Medicine.

Muscles

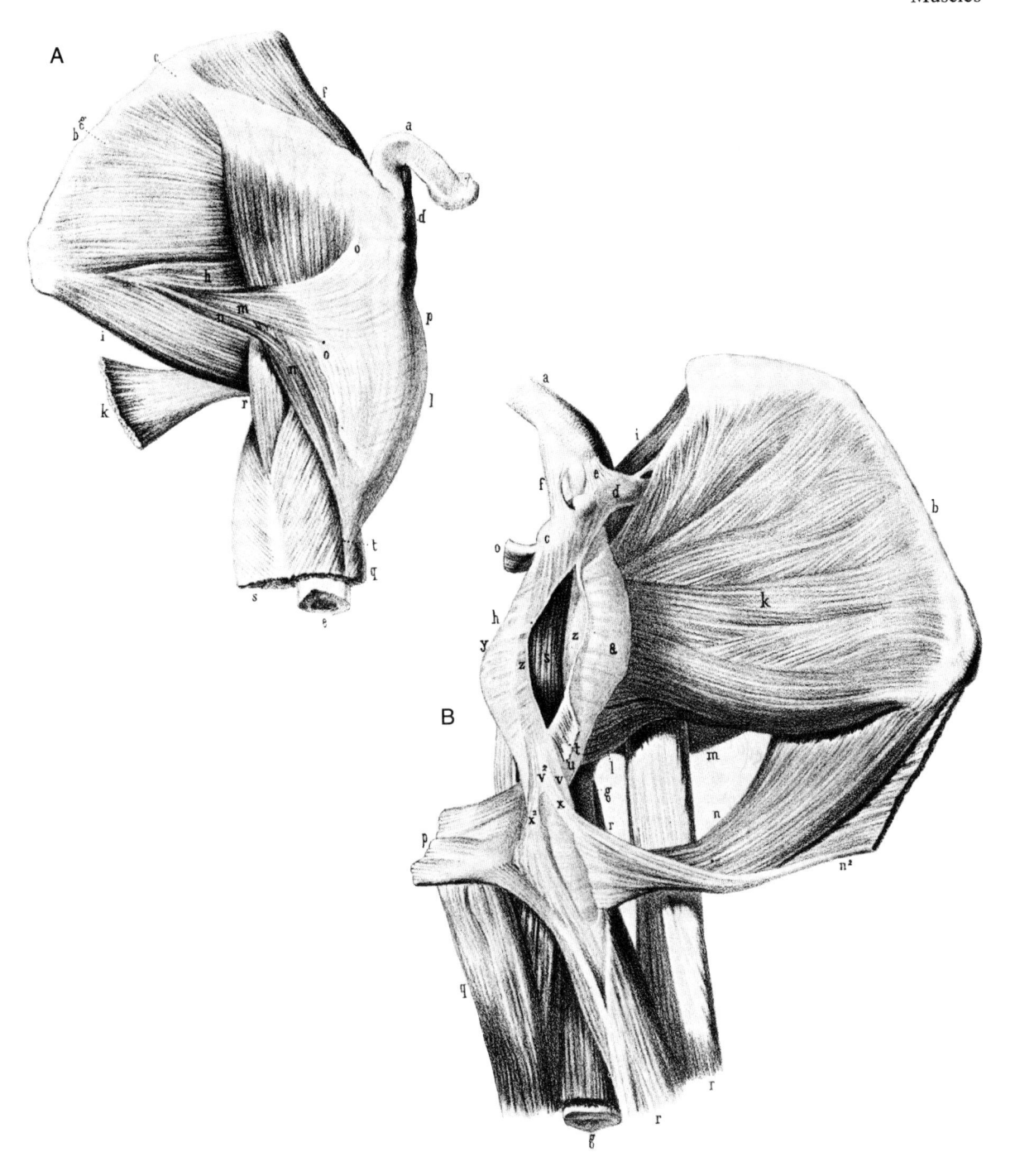

FIG. 40.

M. Costodeltoideus (Albini).

M. Coracobrachialis (Anomalous) (Cruveilhier).

M. Subscapularis (Anomalous) (Gruber).

A: a, Clavicle; b, scapula; c, scapular spine; d, acromion; e, humerus; f, m. supraspinatus; g, m. infraspinatus; h, m. teres minor; i, m. teres major; k, tendon of latissimus dorsi; l, m. deltoideus, invested in its fascia; m, major portion of m. costodeltoideus; n,n, minor portion of m. costodeltoideus; o,o, tendinous expansion of the major portion of m. costodeltoideus; r,s,t, m. triceps brachii.

B: a, Clavicle; b, scapula; c,d, coracoid process; e,f, coracoclavicular ligament; g, humerus; h, capsule of scapulohumeral articulation; i, m. supraspinatus; k, m. subscapularis; l, anomalous fascicle of m. subscapularis (Gruber); m, m. teres minor; n, m. teres major; n^2, m. latissimus dorsi; p, mm. coracobrachialis and short head of biceps brachii; q, m. biceps brachii; r,r,r, m. triceps brachii; s, m. coracobrachialis (anomalous) (Cruveilhier); v, tendon of insertion of anomalous coracobrachialis on the surgical neck of the humerus behind the tendon of insertion of m. teres major; v^2, separated portion of v which bifurcates (x,x^2): x joins the tendon of insertion of m. latissimus dorsi, x^2 continues to insert on the lesser tubercle of the humerus; z, fascial covering (partial) of anomalous coracobrachialis muscle.

From Calori, L. Intorno ad alcune varietà incontrate nella muscolatura degli arti superiori. *Mem. Accad. Sci. Istituto di Bologna S. 2* 6:157–174, 1866.

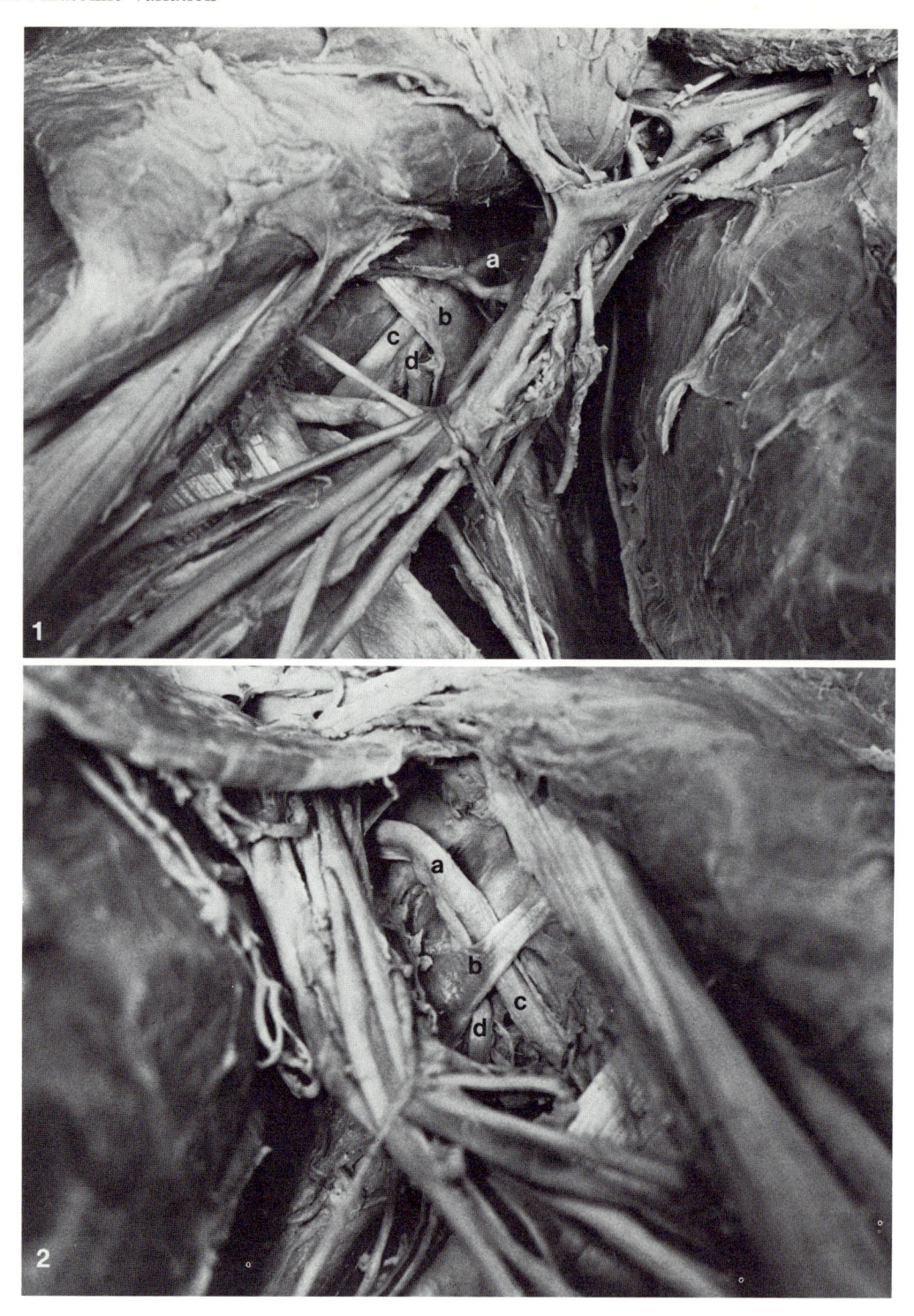

FIG. 41.
Bilateral Accessory Subscapularis.
Axillary and Inferior Subscapular Nerves
Bilaterally Covered by Accessory Subscapularis Muscles.

Fig. 1. Right Axilla.
a. Common axillary nerve; b. accessory subscapularis muscle.
Fig. 2. Left Axilla.
c. Axillary nerve; d. inferior subscapular nerve.

The relationship of the nerves to the variant subscapularis muscles has clinical significance with regard to entrapment neuropathy.

By permission from: Breisch, E.A. A rare human variation: The relationship of the axillary and inferior subscapular nerves to an accessory subscapularis muscle. *Anat. Rec.* 216:440–442, 1986.

FIG. 42.

Left Humerus with Supracondyloid Process and Some Irregular Muscle Attachments.

The supracondylar process projects downward from the medial border of the humerus about 5 cm (2 inches) above the medial epicondyle, to which it is joined by a ligament (Struthers). According to Terry it was found in 7 of 1000 subjects (0.7%) studied. Testut suggests that it has a frequency of about 1%, while Gruber found the process in 2.7% of cases. The ligament completes a ring through which the median nerve and brachial artery may pass; in some cases only the nerve is transmitted. The process and ring are clinically important because of their relationship to artery and nerve.

The pronator teres muscle may arise from the process, ligament, and supracondylar ridge. In addition, a persistent lower part of coracobrachialis may also insert on the process, as shown in the drawing. Occasionally, coracobrachialis is subdivided into an additional muscle, which is called coracobrachialis brevis (coracobrachialis superior or rotator humeri). Its attachment site on the humerus is shown.

Biceps brachii occasionally has more than two heads. The site of origin of a third and a fourth head is also given in this illustration.

Redrawn from Jackson, C.M., Ed. *Morris' Human Anatomy*, 7th Ed., Section III: Osteology, by R.J. Terry. P. Blackiston's Son & Co., Philadelphia, 1932.

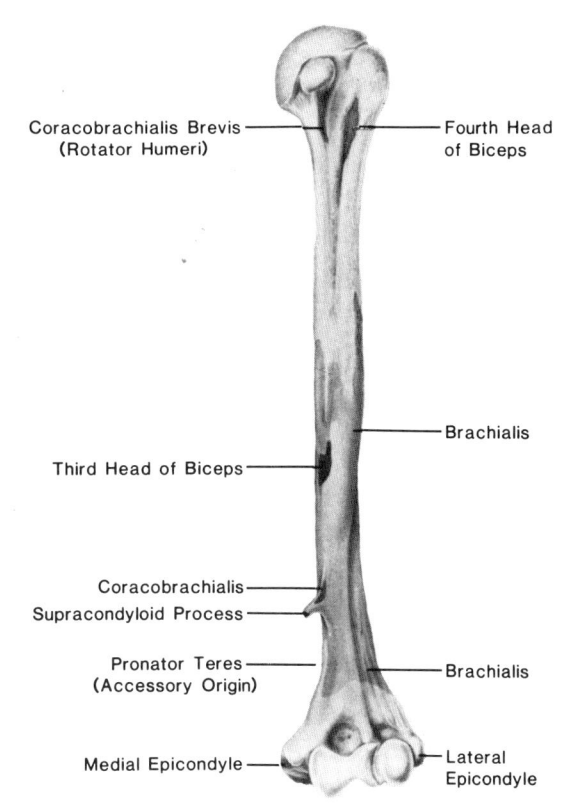

FIG. 43.

M. Coracobrachialis Brevis (Superior).

Occasionally a muscle described as coracobrachialis brevis (or superior) may arise from the coracoid process and insert onto the bicipital ridge of the humerus about 1 cm below the lesser tubercle.

Coracobrachialis is usually split by the musculocutaneous nerve to allow its passage. In some instances the cleavage is complete resulting in two separate muscles.

A more common variation, described by Wood in 1867, is designated as coracobrachialis longus and it extends from the coracoid process to the medial supracondylar ridge, or to the medial epicondyle, or to an anomalous supracondylar process if present.

Redrawn and modified from an illustration by Beattie. (Beattie, P.H. Description of bilateral coracobrachialis brevis muscle, with a note on its significance. *Anat. Rec.* 97:123–126, 1947).

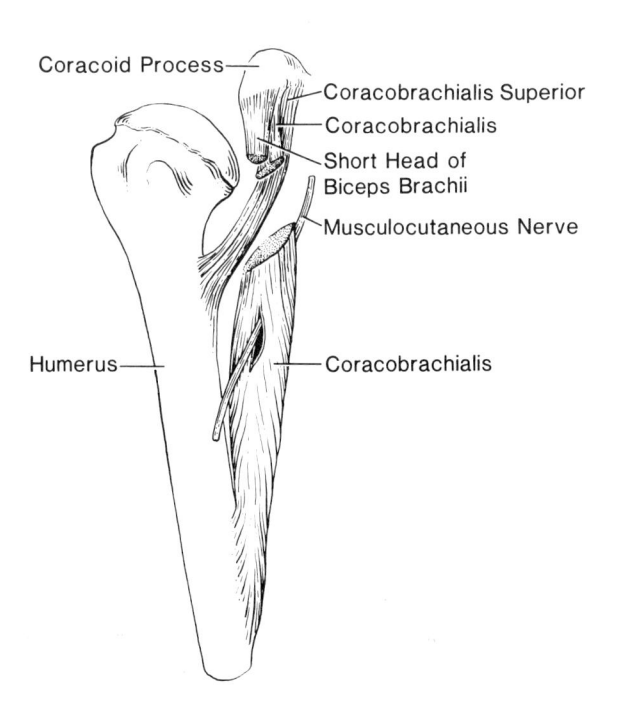

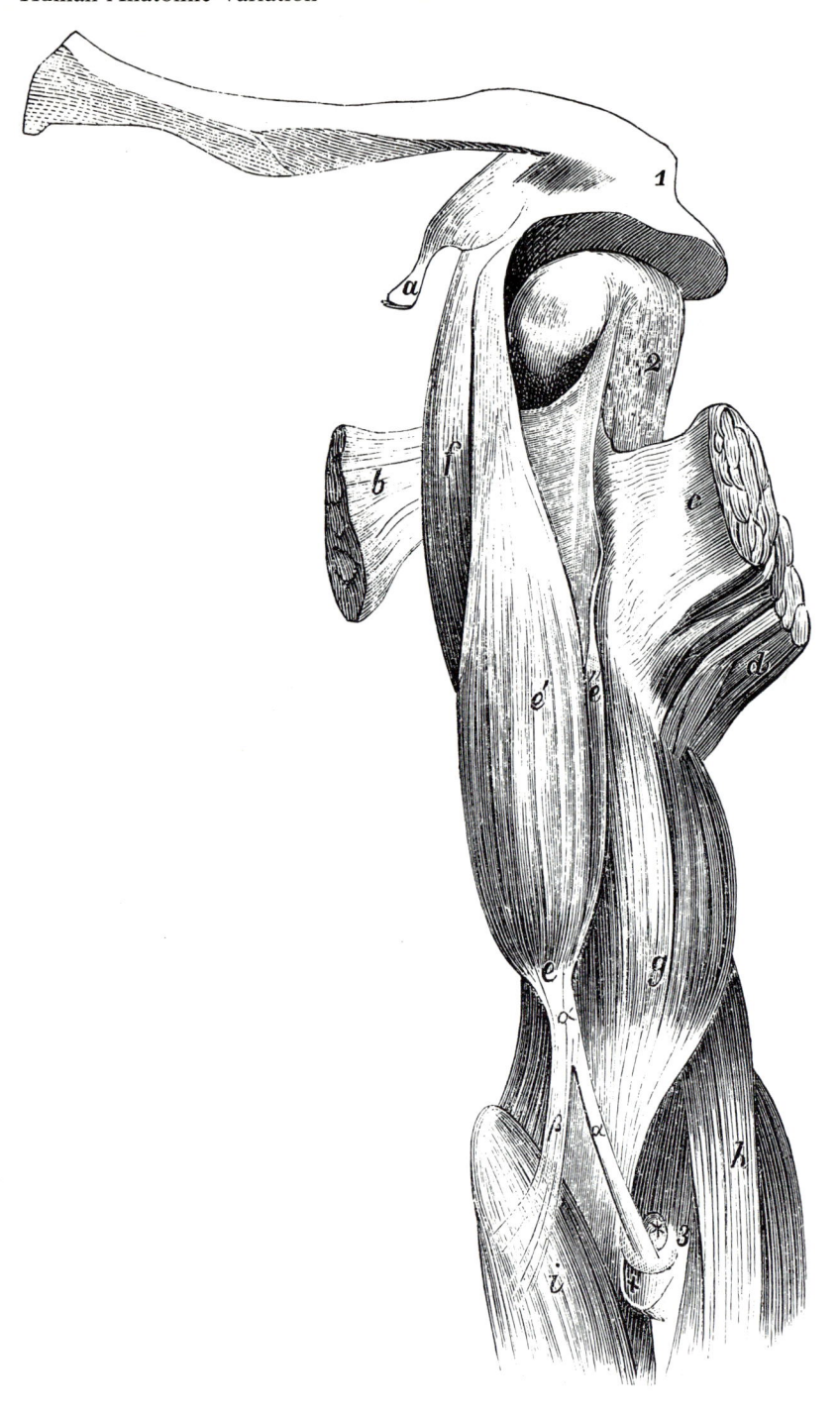

FIG. 44.
M. Biceps Brachii (Anomalous).
Absence of Long Head of M. Biceps Brachii.

1, Shoulder girdle; *2*, humerus; *3*, radial tuberosity; *a*, tendon of m. pectoralis minor; *b*, insertion of m. latissimus dorsi and m. teres major; *c*, insertion of m. pectoralis major; *d*, insertion of m. deltoideus; *e*, m. biceps brachii (anomalous): *e'*, short head of biceps, and *e"*, anomalous humeral head (frequency about 5% of limbs); α, insertion of m. biceps brachii (radial tuberosity); β, superficial insertion of m. biceps brachii; *f*, m. coracobrachialis; *g*, m. brachialis.

Gruber, W. Über einen Musculus biceps brachii mit einem Caput coracoideum und einem Caput humerale anomalum statt des mangelnden Caput glenoideum. *Bull. l'Acad. Imp. Sci. St. Petersbourg* 17:314–318, 1872.

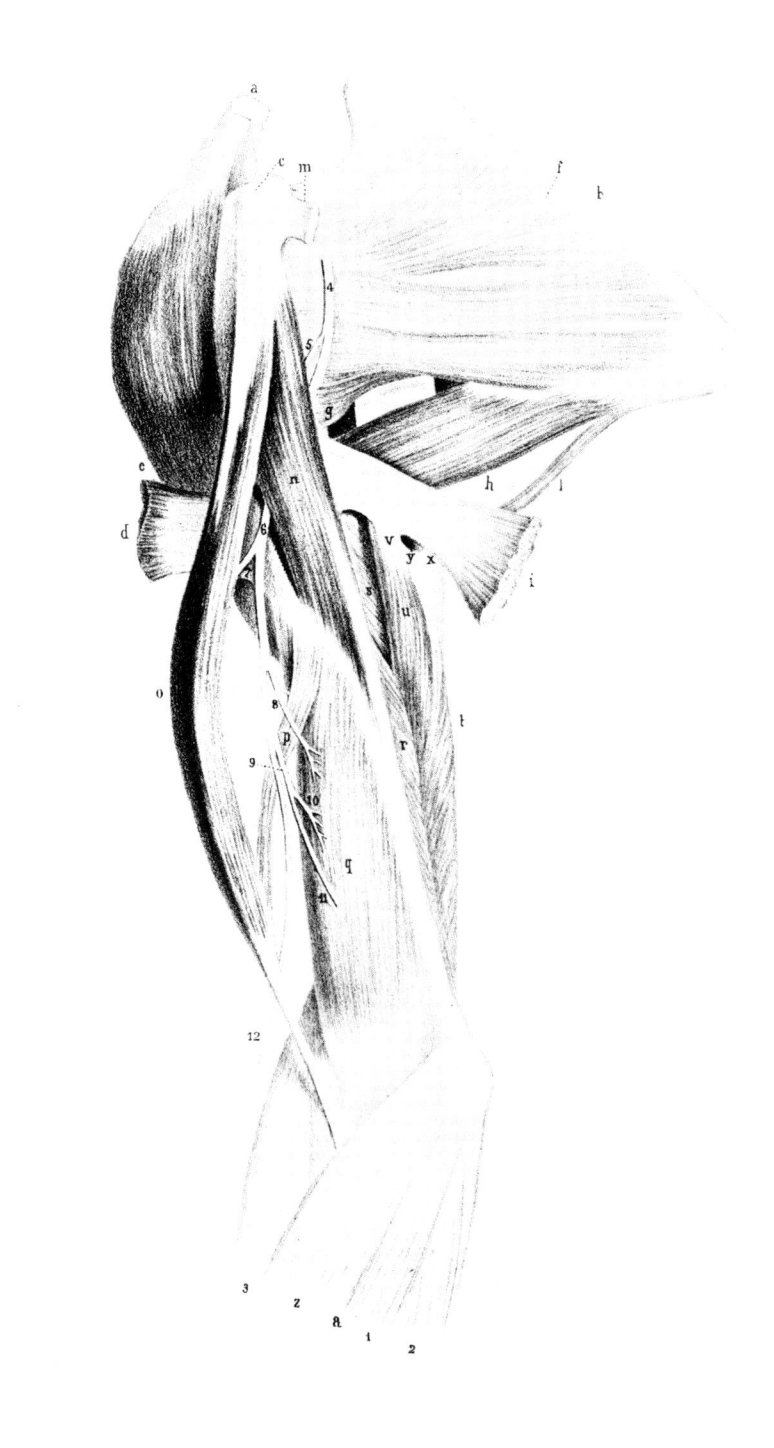

FIG. 45.
M. Biceps Brachii with Three Heads.
Anomalous Subscapular Muscle (of Gruber).
Anomalous Fascicle of M. Teres Major.
Anomalous Insertion of Long Head of M. Triceps.
(Musculocutaneous Nerve Not Piercing M. Coracobrachialis).
(Branch of Musculocutaneous Nerve Joining Median Nerve).

a, Clavicle (cut); *b*, scapula; *c*, coracoid process; *d*, tendon of m. pectoralis major; *e*, m. deltoideus; *f*, m. subscapularis; *g*, anomalous subscapular muscle (of Gruber); *h*, m. teres major; *i*, tendon of m. latissimus dorsi; *l*, anomalous fascicle of m. teres major joining the tendon of m. latissimus dorsi; *m*, tendon of origin of m. pectoralis minor; *n*, m. coracobrachialis; *o*, m. biceps brachii; *p*, third head of m. biceps brachii; *q*, m. brachialis; *r,s,t*, m. triceps brachii; *u,v*, anomalous insertion of the long head of triceps into tendon of latissimus dorsi; *6*, musculocutaneous nerve circling laterally around coracobrachialis; *7*, biceps (short head) nerve supply; *8*, third head nerve supply; *9,10*, m. brachialis nerve supply; *11*, anastomotic branch of musculocutaneous nerve to median nerve; *12*, cutaneous branch of musculocutaneous nerve.

From Calori, L. Delle corrispondenze del nervo muscolocutaneo con il capo sopranumerario del bicipite bracchiale e col bracchiale interno. *Mem. Accad. Sci. Istituto di Bologna S. 2* 6:149–155, 1866.

Human Anatomic Variation

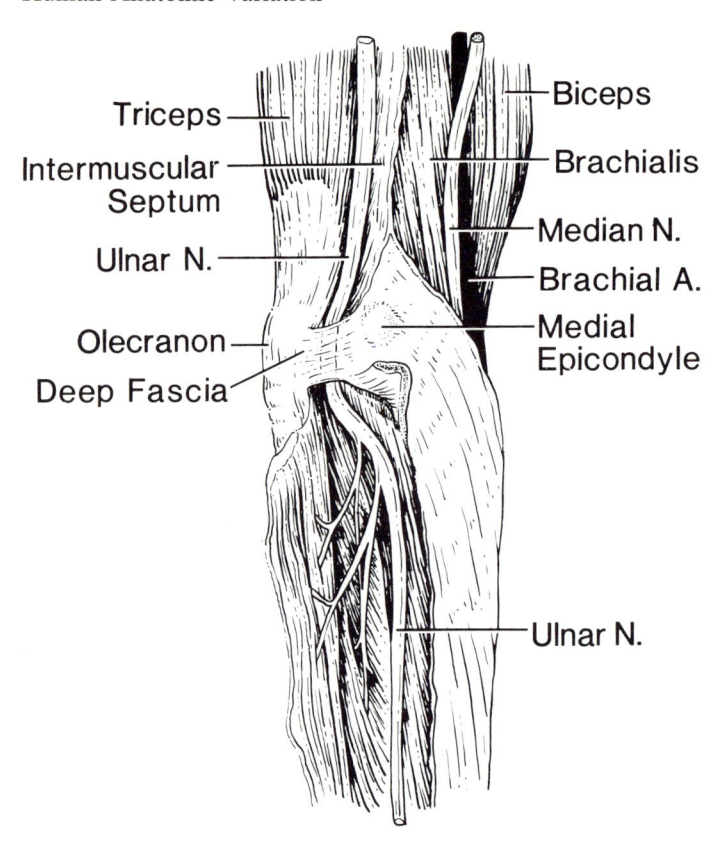

FIG. 46.
M. Epitrochleoanconeus.

This muscle is frequently found and arises from the posterior surface of the medial epicondyle of the humerus and inserts into the olecranon. It is superficial to the ulnar nerve which may become compressed and lead to impaired function of the nerve. When the muscle is absent it is replaced by a band of fascia.

Redrawn and modified from an illustration (Fig. 568) in Toldt, C. *An Atlas of Human Anatomy for Students and Physicians*, 2nd Ed., Vol. 1. The Macmillian Co., New York, 1928.

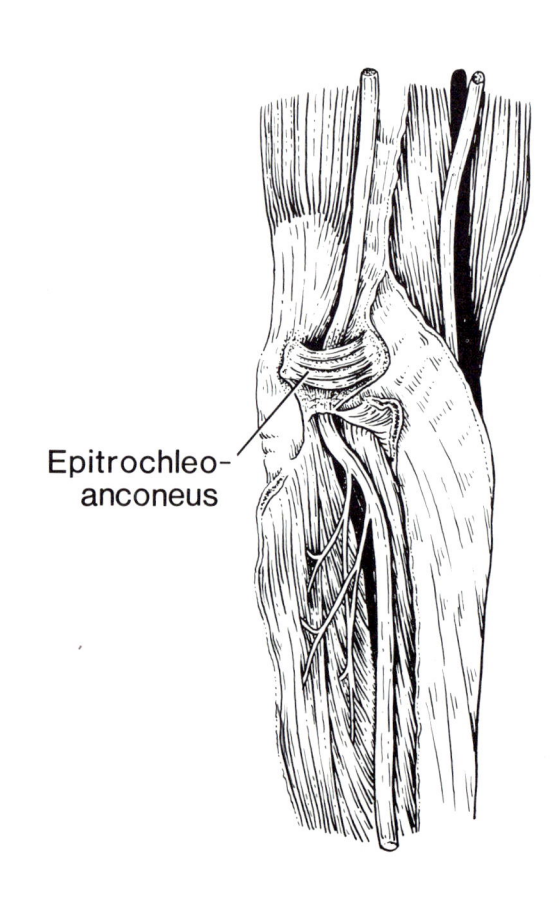

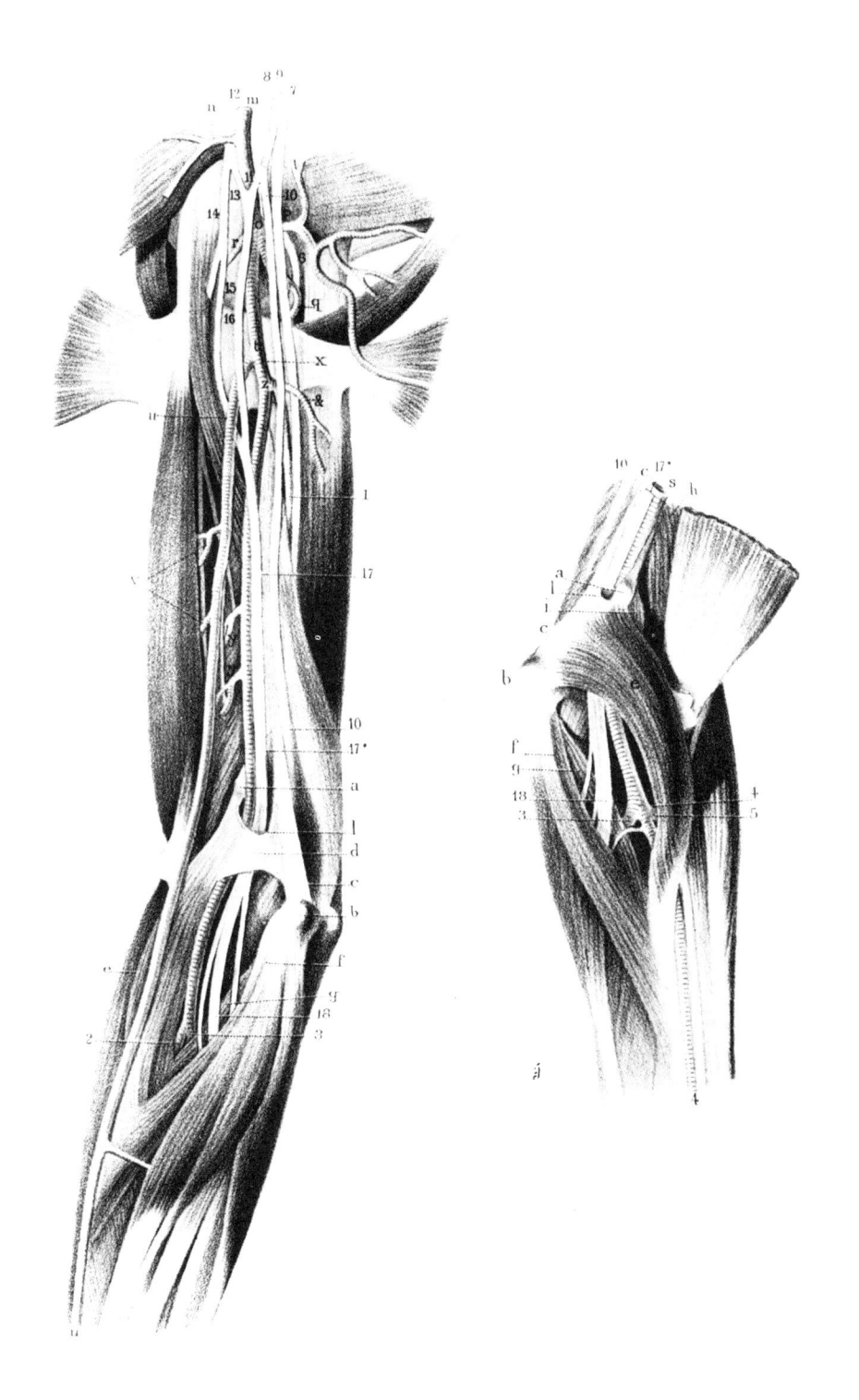

FIG. 47.
Anomalous Supernumerary Head of M. Pronator Teres.
(Supracondylar Canal of Humerus).
(Anomalous High Division of Axillary Artery into Radial and Ulnar Artery Branches).

a, Supracondylar process of humerus; b, medial epicondyle of humerus; c, medial intermuscular septum; d, aponeurotic sheet of the supracondylar process, origin of anomalous head of m. pronator teres, and anterior fascial covering of the supracondylar canal; e, supernumerary head of m. pronator teres; f, superior head of m. pronator teres; g, inferior head of m. pronator teres; h, inferior portion of brachialis muscle; i, muscle fascicle arising from ligament (Struthers) of supracondylar process; l, opening of supracondylar canal, passage of median nerve and brachial artery; m, axillary artery; s, brachial artery (right-hand drawing); t, axillary artery divides into radial and ulnar arteries; u, high division of radial artery; x, brachial artery continues as the ulnar artery; 10, ulnar nerve; 17, median nerve.

From Calori, L. Intorno al canale sopracondiloideo dell' omero nell' uomo. *Mem. R. Accad. Sci. Istituto di Bologna S. 4* 2:37–46, 1880.

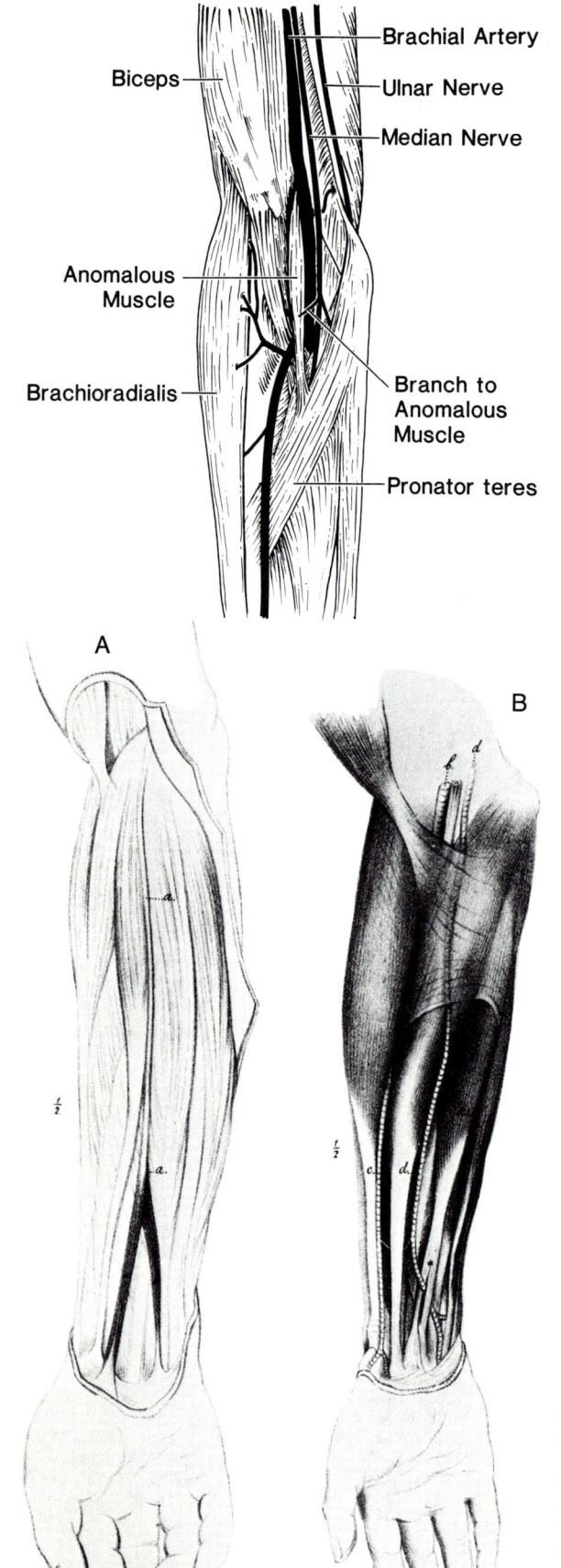

Human Anatomic Variation

FIG. 48.
M. Pronator Teres Accessorius.

An anomalous accessory pronator teres muscle bundle was found arising from the lateral edge of brachialis and joining pronator teres. It was innervated by the median nerve. Although this arm had a supracondylar process, the brachial artery and median nerve crossed anterior to the process and between the accessory fascicle and pronator teres.

Redrawn from Adachi, B. *Anatomie der Japaner. Das Arteriensystem der Japaner.* Verlag der Kaiserlich-Japanischen Universität zu Kyoto. In Kommission bei "Maruzen Co.", Kyoto und Tokyo. Gedruckt von "Kenkyusha" in Tokyo, 1928.

FIG. 49.
Anomalous Palmaris Longus Muscles.
(Anomalous Ulnar Artery (Superficial Course, High Division) Passing Through a Split Tendon of M. Palmaris Longus).

A: *a*, M. palmaris longus bicaudatus, reversed; α, lateral belly; β, medial belly.
B: *a*, Highly developed m. palmaris longus with a significant split in its tendon; *b,c*, radial artery; *d*, superficial ulnar artery, which arose anomalously from the axillary artery, with a superficial course in the forearm, passing through the split in the tendon of palmaris longus muscle.

From Gruber, W. Über die Varietäten des Musculus palmaris longus. *Mém. l'Acad. Imp. Sci. St. Petersbourg S. 7* 11:1–26, 1868.

FIG. 50.
Anomalous Palmaris Longus Muscles.
Anomalous Double-Headed M. Flexor Carpi Ulnaris.
Supernumerary Head of M. Abductor Digiti Minimi.
M. Palmaris Accessorius Superficialis.

A: *a*, M. palmaris longus; *b*, double-headed m. flexor carpi ulnaris: the tendons of both muscles are grown together and attach to the os pisiforme; α, superficial layer, on the ulnar side, of the transverse carpal ligament (flexor retinaculum) arises from the os pisiforme and the tendon of m. palmaris longus; β, ulnar artery; γ, ulnar nerve.

B: *a*, M. palmaris longus; *b*, supernumerary head of m. abductor digiti minimi. This bundle arises, in part, from the tendon of m. palmaris longus, partly from the fleshy belly of this muscle.
C: *a*, Normal palmaris longus; *a'*, m. palmaris accessorius superficialis whose tendon joins that of a double-headed m. flexor carpi ulnaris; *b*, double-headed m. flexor carpi ulnaris.

From Gruber, W. Über die Varietäten des Musculus Palmaris longus. *Mém. l'Acad. Imp. Sci. St. Petersbourg S. 7* 11:1–26, 1868.

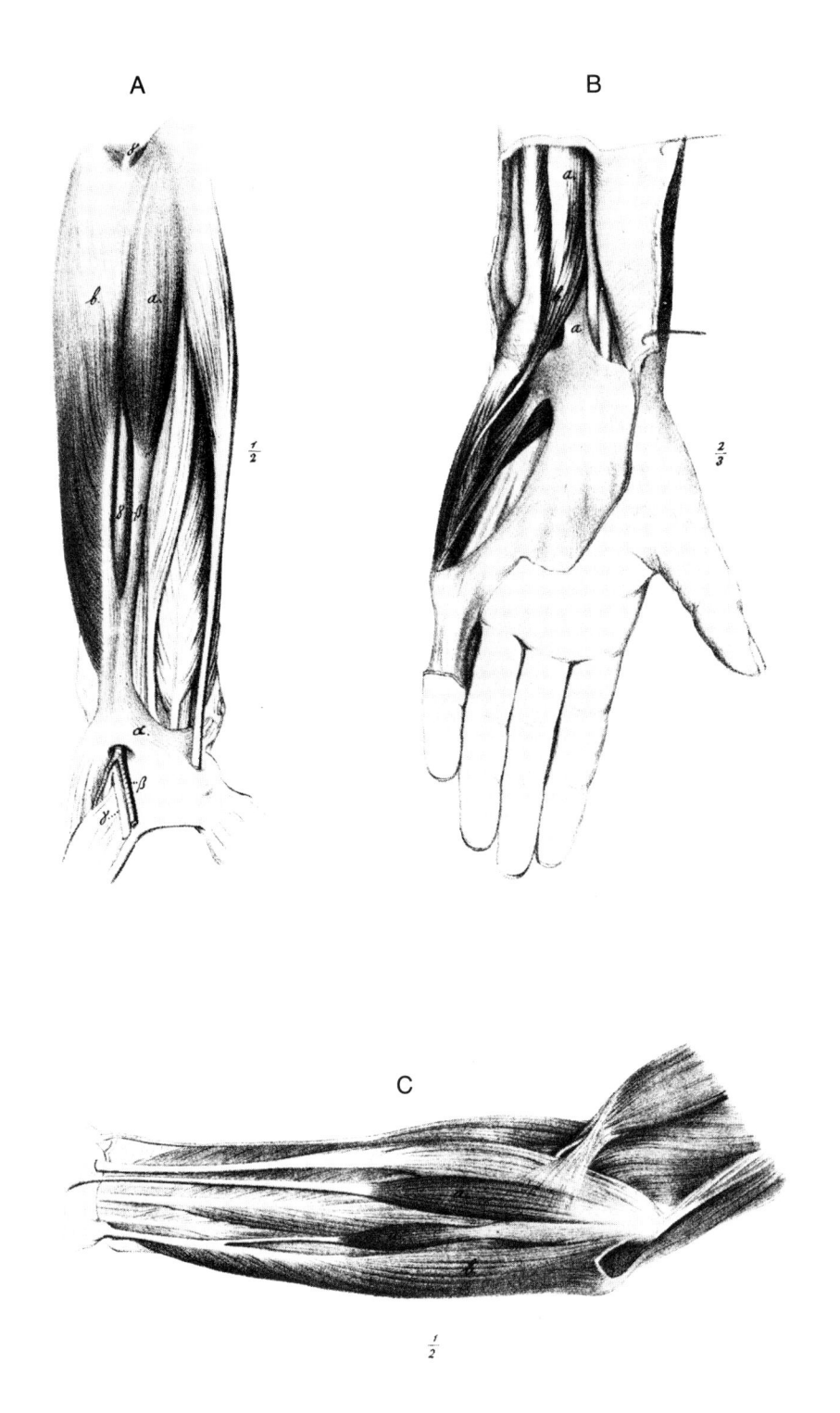

Human Anatomic Variation

FIG. 51.
Anomalous Palmaris Longus Muscles.
M. Palmaris Accessorius Profundus.
Supernumerary Head of M. Abductor Digiti Minimi.

A: *a*, M. palmaris longus normalis; *a'*, m. palmaris accessorius profundus, with a three-part origin from the ulna, radius, and pronator teres muscle.

B: *a'*, M. palmaris accessorius profundus arising from radius; *b*, radial head of m. flexor digitorum sublimis (superficialis) arising from the superior end of the tendon of m. palmaris accessorius profundus; *c*, supernumerary head of m. abductor digiti minimi.

C: *a*, M. palmaris longus normalis; *a'*, m. palmaris accessorius profundus arising from the radius; *b*, median nerve, with a cleft through which the tendon of m. palmaris accessorius profundus passes.

D: *a'*, M. palmaris accessorius profundus arising from radius; *b*, supinator muscle; *c*, m. flexor pollicis longus.

E: *a'*, M. palmaris accessorius arising from radius and attaching to the posterior side of the transverse carpal ligament (flexor retinaculum); *b*, insertion of part of m. pronator teres; *c*, origin of part of the radial head of m. flexor digitorum sublimis; *d*, head of m. flexor pollicis longus; *e*, m. pronator quadratus; *f*, tendon of m. flexor carpi radialis; α,α, cut edges of transverse carpal ligament.

From Gruber, W. Über die Varietäten des Musculus palmaris longus. *Mém. l'Acad. Imp. Sci. St. Petersbourg S. 7* 11:1–26, 1868.

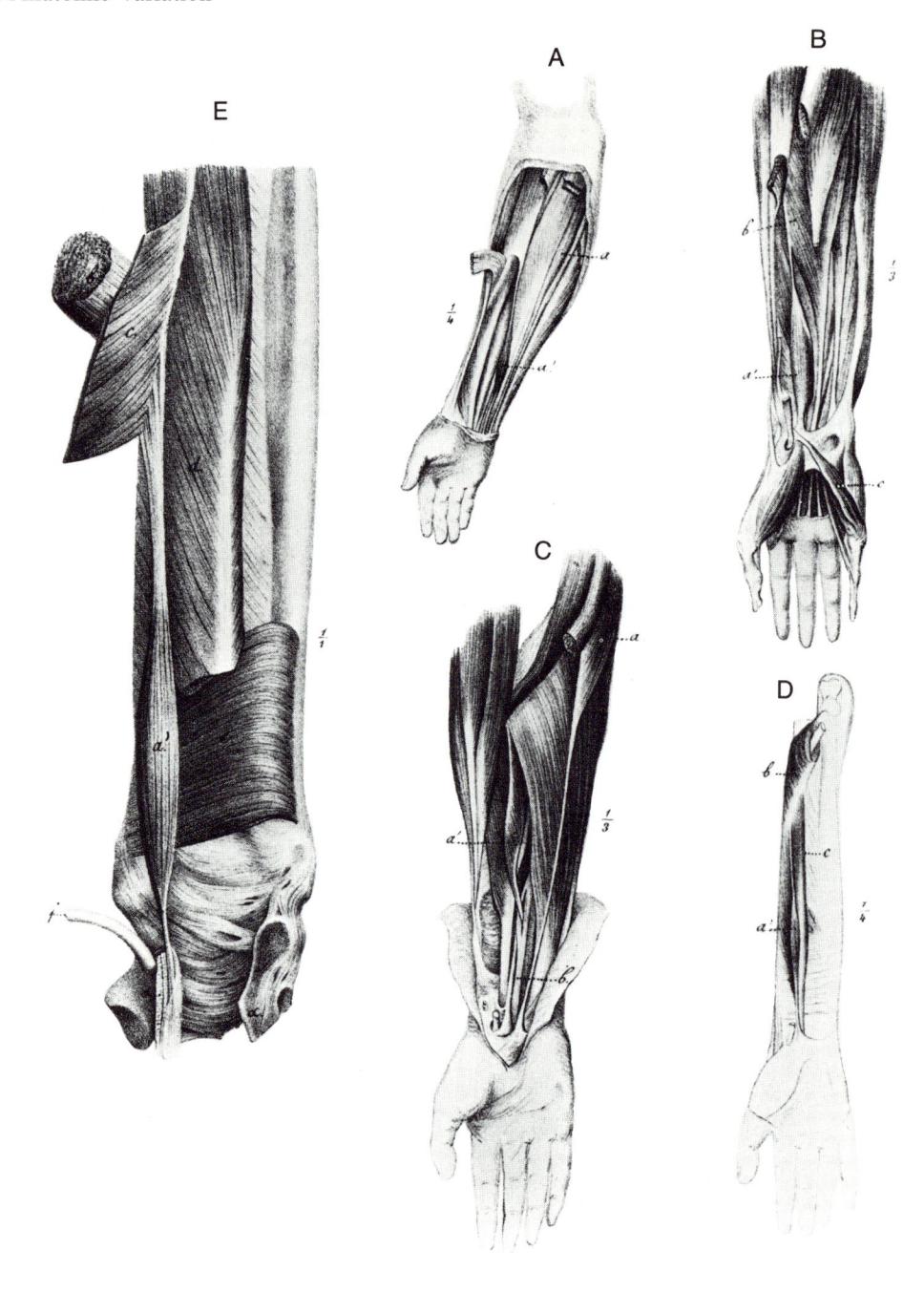

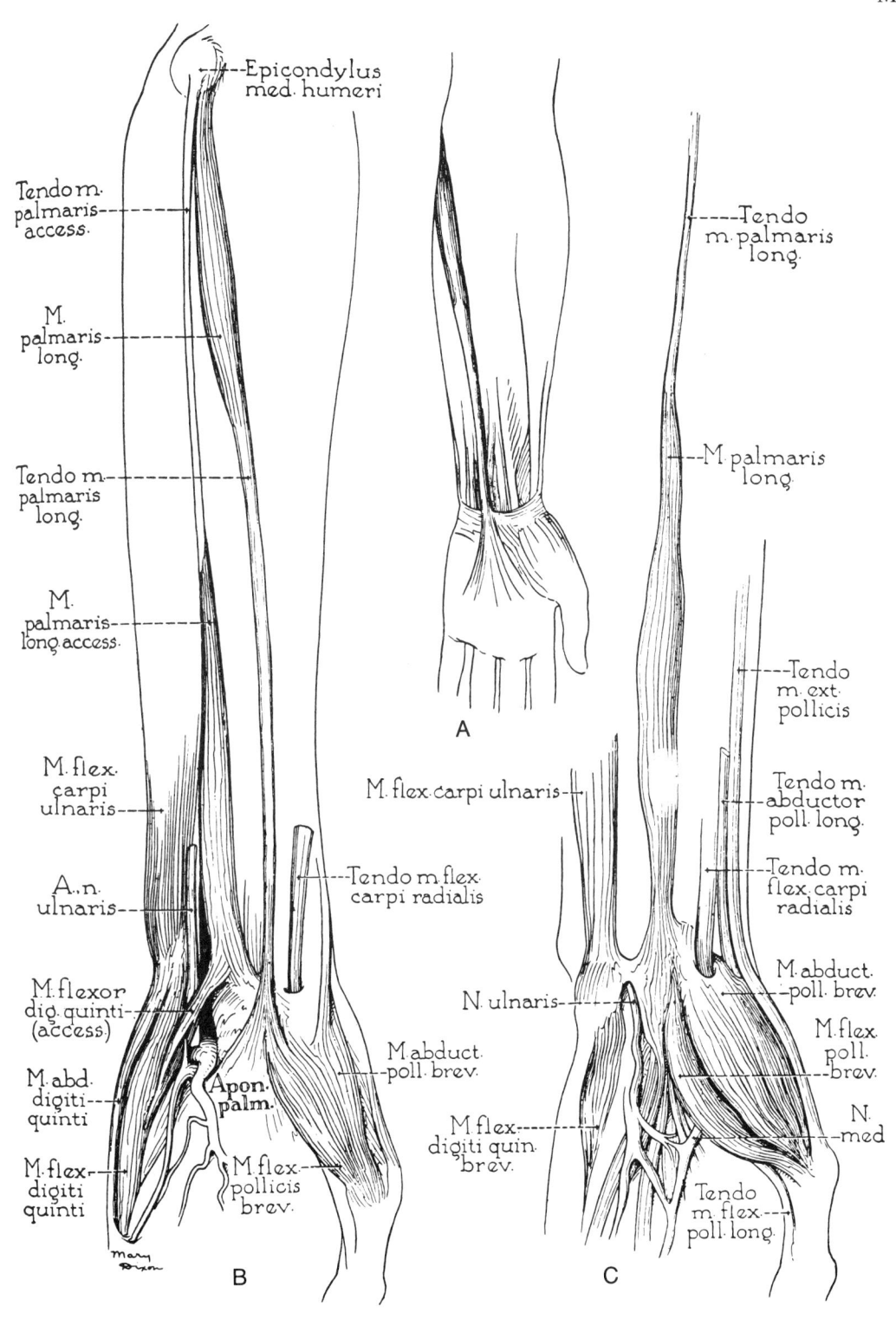

FIG. 52.
Anomalous Palmaris Longus Muscles.

A: Usual or regular palmaris longus muscle.
B: Duplication of muscle, the ulnar element possessing a distal muscular part.
C: M. palmaris with distally placed belly of broad insertion.

From Reimann, A.F., Daseler, E.H., Anson, B.J. and L.E. Beaton. The palmaris longus muscle and tendon. A study of 1600 extremities. *Anat. Rec.* 89:495–505, 1944.

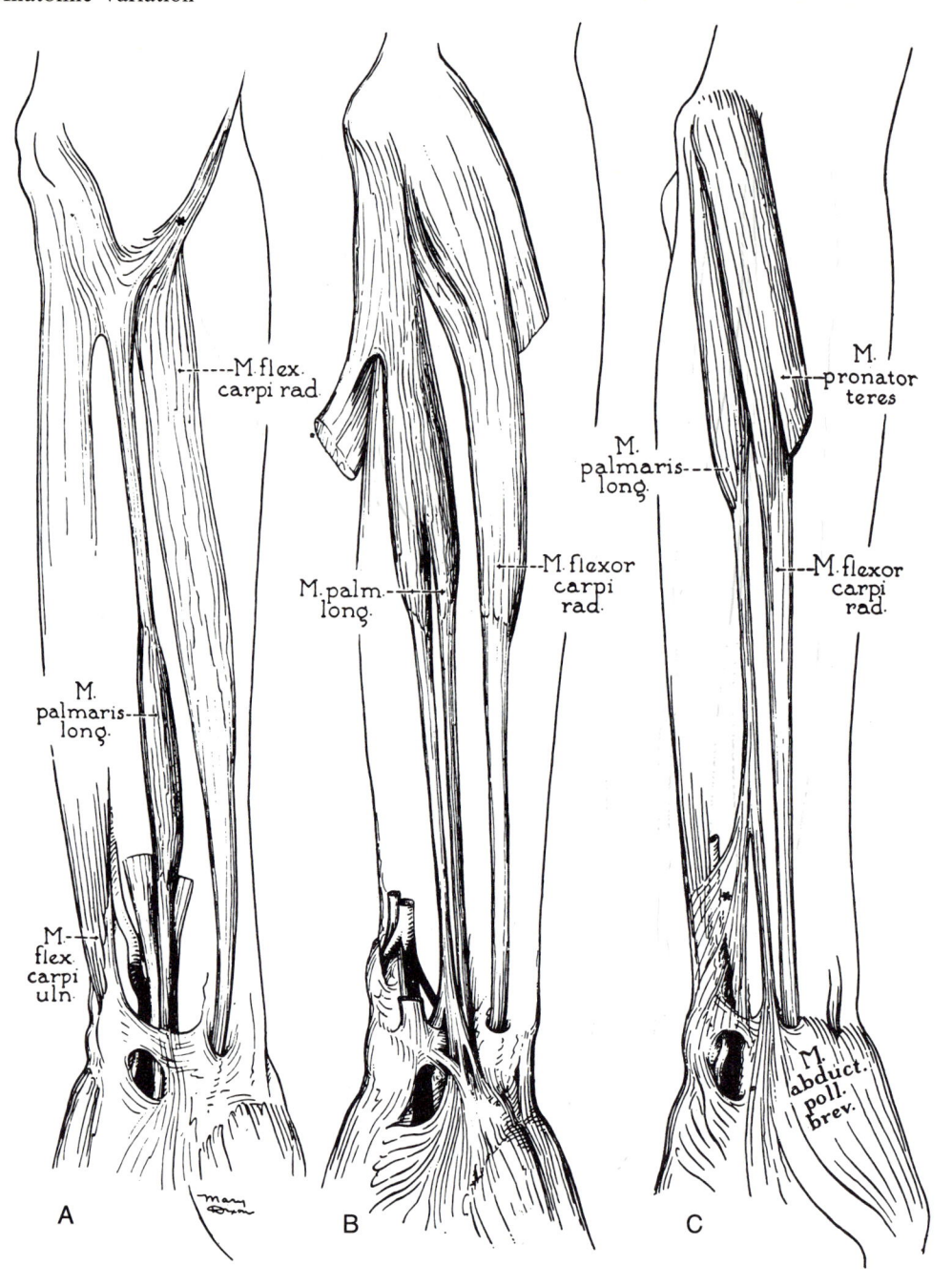

FIG. 53.
Anomalous Palmaris Longus Muscles.

A: Centrally placed muscular belly; proximally placed slip (*) joins lacertus fibrosus.

B: Bifid palmaris muscle, portions of equal size and with similar tendons.

C: Divided tendon, the ulnar slip (*) inserting into antibrachial fascia.

From Reimann, A.F., Daseler, E.H., Anson, B.J. and L.E. Beaton. The palmaris longus muscle and tendon. A study of 1600 extremities. *Anat. Rec.* 89:495–505, 1944.

FIG. 54.
Anomalous Palmaris Longus Muscles.

A: A short broad muscle replacing palmaris longus (tendon followed to deep aspect of aponeurosis).
B: Auxiliary muscle with long tendon of ulnar and radial origins (flexor digitorum longus, etc. removed).
C: Auxiliary muscle with robust, distally placed, fleshy part (flexor carpi radialis retracted).

Authors' note: The auxiliary muscles (shown in B and C) are identical to m. palmaris accessorius profundus described by Gruber in 1868.

From Reimann, A.F., Daseler, E.H., Anson, B.J. and L.E. Beaton. The palmaris longus muscle and tendon. A study of 1600 extremities. *Anat. Rec.* 89:495–505, 1944.

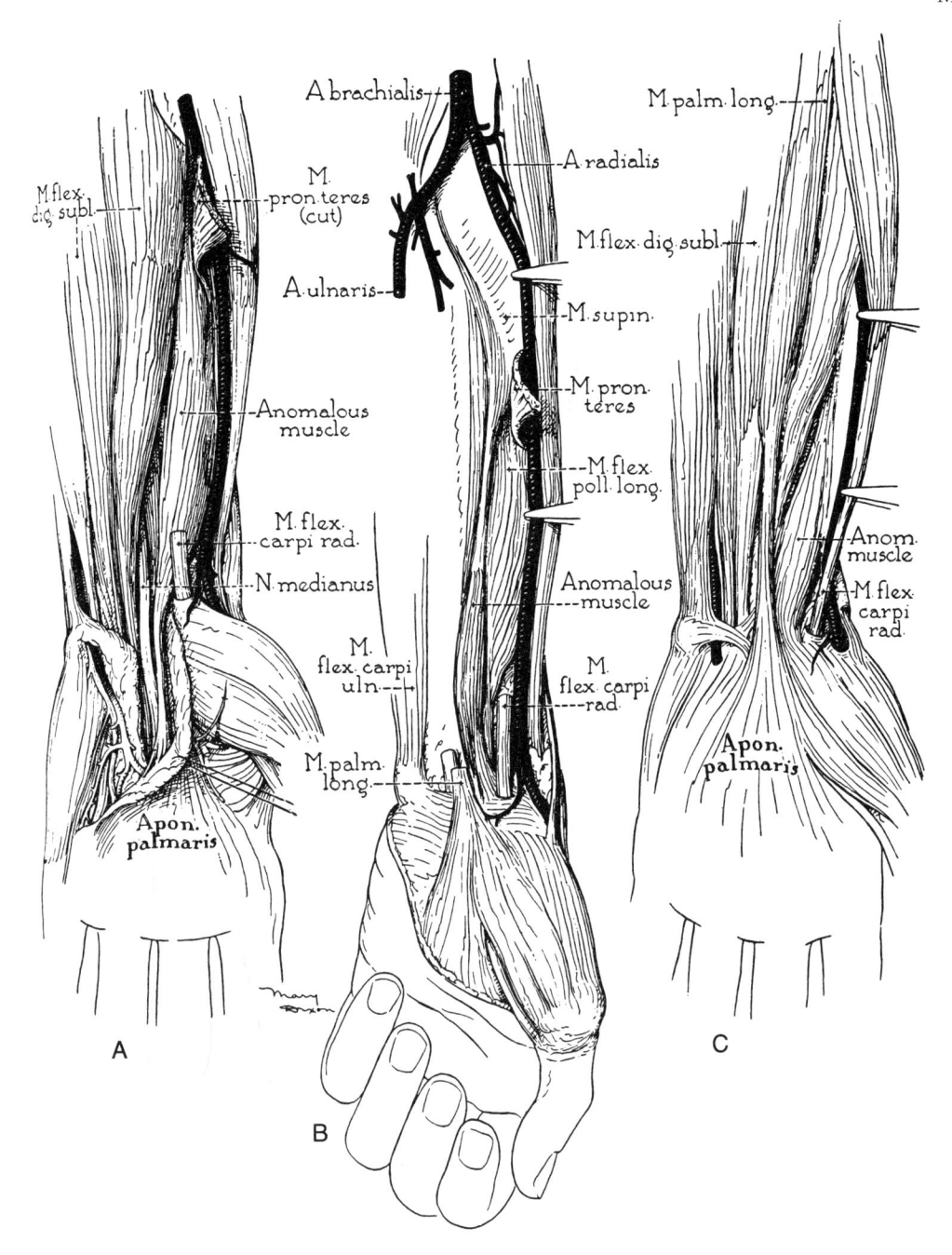

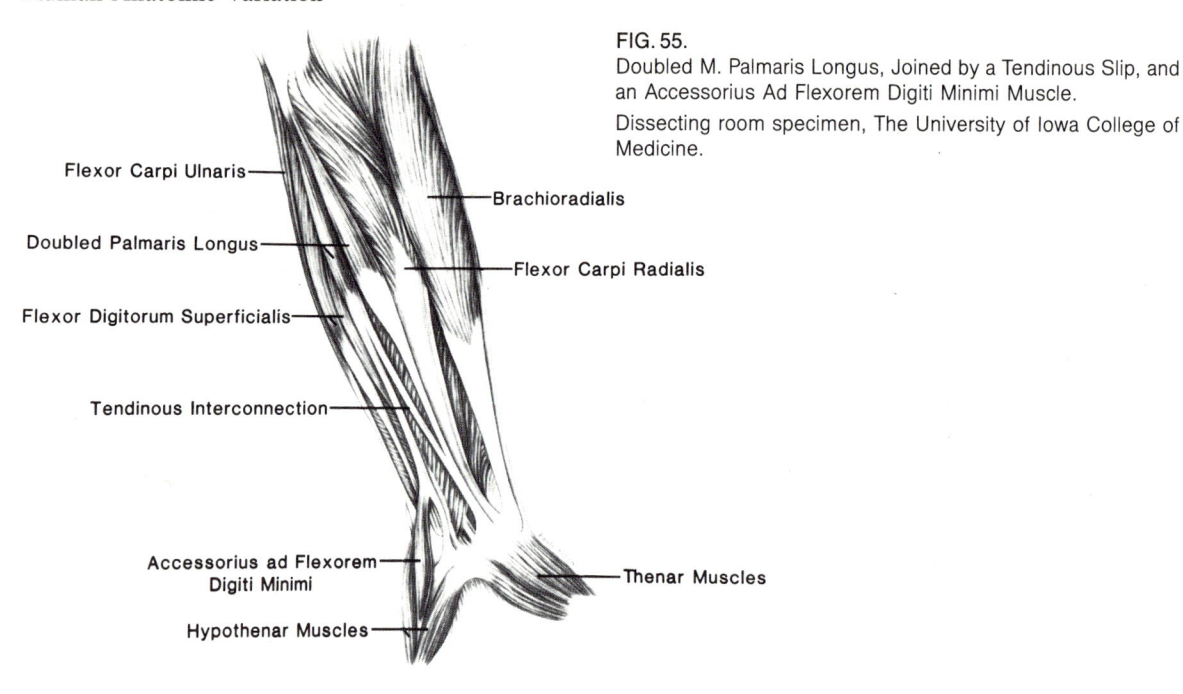

Human Anatomic Variation

FIG. 55.
Doubled M. Palmaris Longus, Joined by a Tendinous Slip, and an Accessorius Ad Flexorem Digiti Minimi Muscle.
Dissecting room specimen, The University of Iowa College of Medicine.

Labels: Flexor Carpi Ulnaris; Doubled Palmaris Longus; Flexor Digitorum Superficialis; Tendinous Interconnection; Accessorius ad Flexorem Digiti Minimi; Hypothenar Muscles; Brachioradialis; Flexor Carpi Radialis; Thenar Muscles.

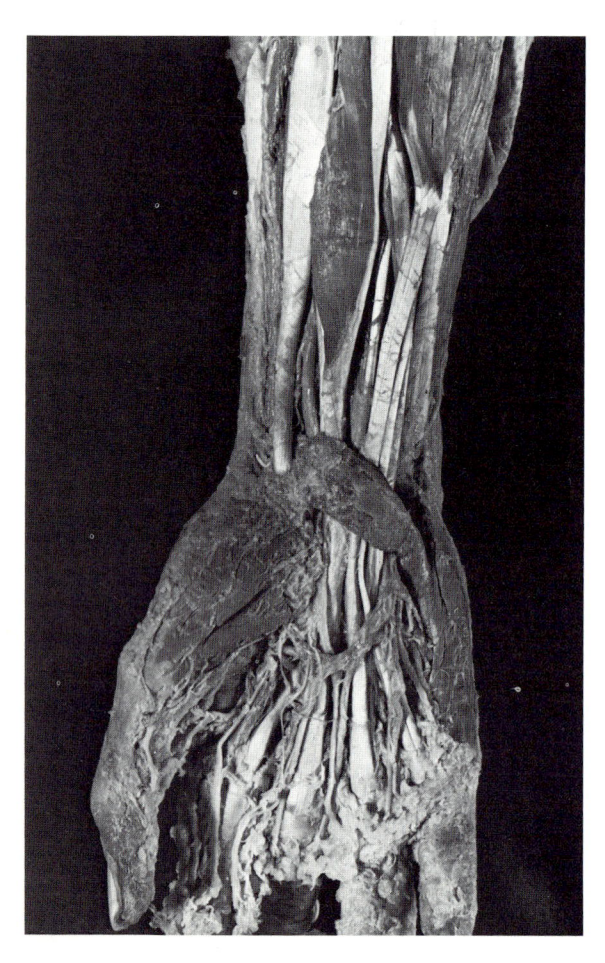

FIG. 56.
M. Flexor (Opponens) Digiti Minimi Accessorius.
M. Palmaris Longus, Reversed Belly.
Flexor digiti minimi accessorius muscle: origin, deep fascia of the radial side of the wrist; insertion, distal end of fifth metacarpal and base of proximal phalanx of the fifth digit; course, from the radial side across the transverse carpal ligament to the hypothenar mass.
Reversed m. palmaris longus inserts into the deep surface of the transverse carpal ligament.
Dissecting room specimen, The University of Iowa College of Medicine.

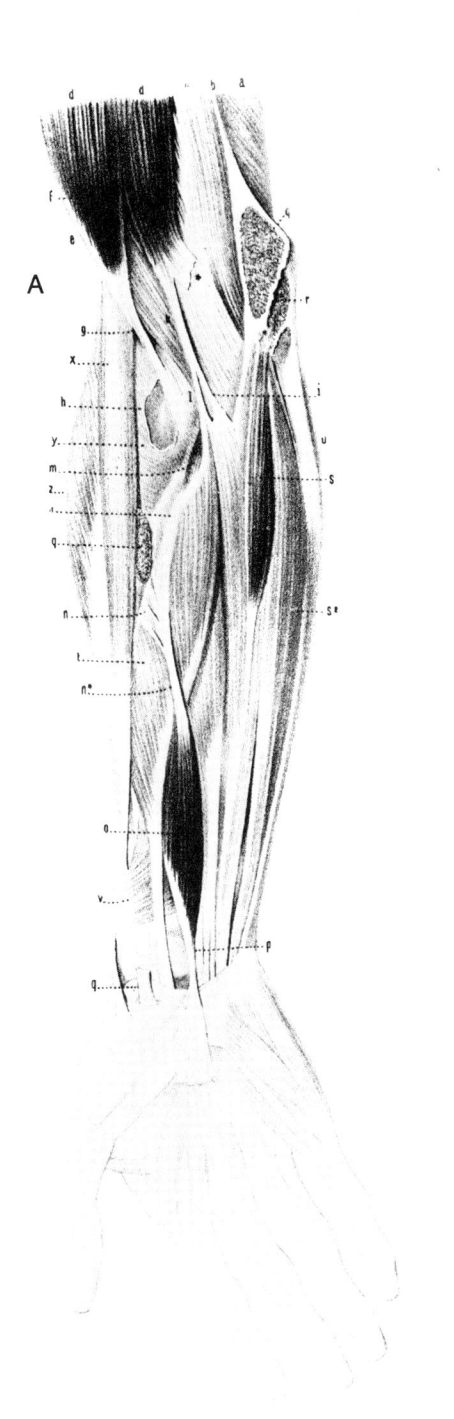

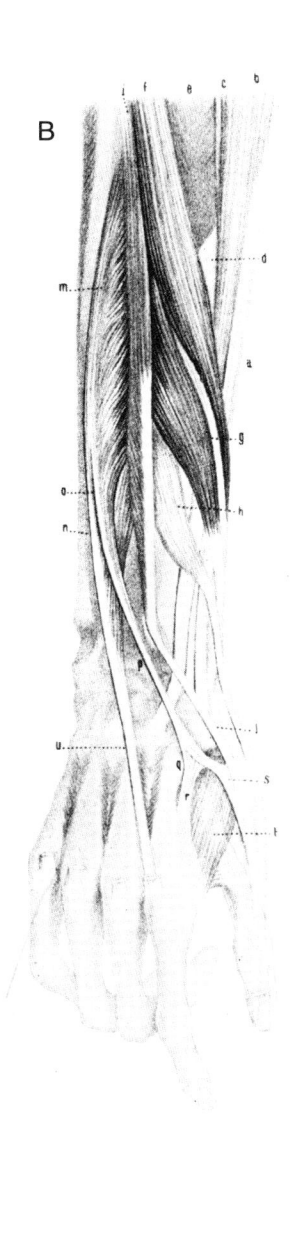

FIG. 57.
M. Palmaris Longus (Anomalous).
M. Biceps Brachii (Anomalous).
M. Abductor Pollicis Longus Duplication.

A: *a*, M. triceps brachii; *b*, medial intermuscular septum; *c*, m. brachialis; *d*, m. biceps brachii; *e,f*, division of the inferior belly and tendon of m. biceps brachii into two parts; *g*, tendon of portion *e*; *h*, bursa between the insertion of tendon *g*, the radial tuberosity, and m. supinator; *i*, major branch of tendon *k* inserting on the ulna distal to m. brachialis; *k*, tendon of medial part of the divided m. biceps brachii; *, fibrous expansion (bicipital aponeurosis) inserting into the antibrachial aponeurosis; *l*, tendon continuous with the tendon of origin of m. palmaris longus (*o*); *n,o,p*, anomalous m. palmaris longus; *q,r*, cut end of m. pronator teres, and origin of m. flexor carpi radialis and m. flexor digitorum superficialis from medial epicondyle; S,S^2, m. flexor digitorum profundus; *u*, m. flexor carpi ulnaris; *v*, m. pronator quadratus; *x*, m. brachioradialis; *z*, m. extensor carpi radialis longus.

B: *a*, Tendon of m. brachioradialis; *b,c*, m. extensor carpi radialis longus; *d*, part of the inferior tendon of m. pronator teres; *e*, m. supinator; *f*, m. abductor pollicis longus; *g*, second m. abductor pollicis longus; *h*, m. extensor pollicis brevis; *i*, m. extensor pollicis longus; *m*, belly of m. extensor indicis and division of its tendon into two parts: one major (*n*) and one minor (*o*); tendon *o* divides into two parts (*q*): *r* inserts on base of second metacarpal (index finger) and *s* joins the tendon of m. extensor pollicis longus (*l*); *u*, principal tendon of m. extensor indicis.

From Calori, L. Di alcune varietà muscolari dell'avambraccio e dell'eminenza ipothenar. *Mem. Accad. Sci. Istituto di Bologna S. 2* 7:359–381, 1867.

Human Anatomic Variation

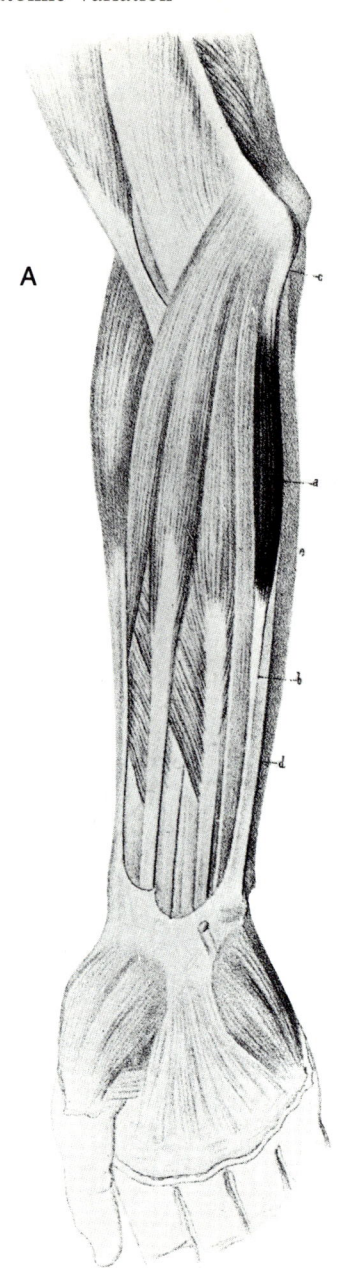

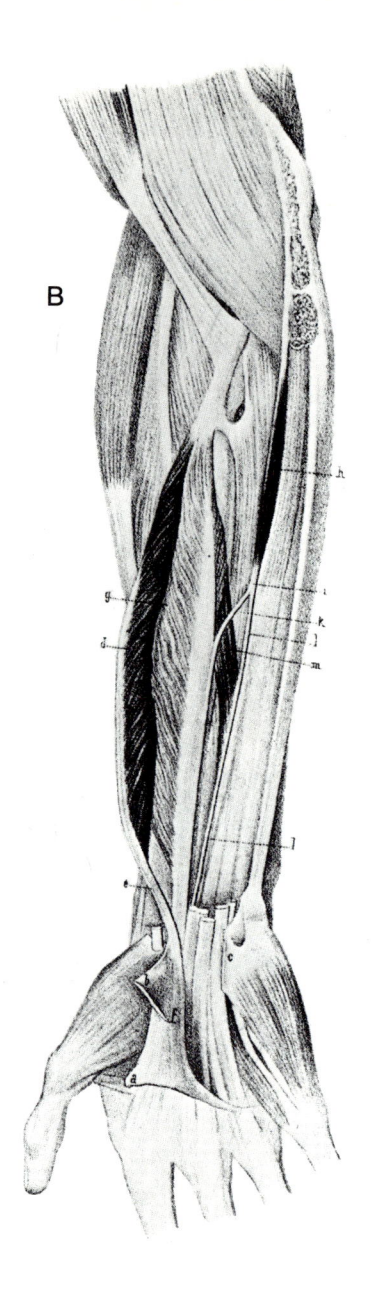

FIG. 58.
Absence of M. Palmaris Longus.
Anomalous Head or Supernumerary M. Flexor Carpi Ulnaris.
M. Palmaris Profundus (Variant of M. Palmaris Longus).
M. Flexor Pollicis Longus (Anomalous).

A: *a*, Anomalous muscle which resembles m. palmaris longus, but its tendon (*b*) unites (*d*) with the tendon of m. flexor carpi ulnaris (*e*) and may therefore be considered an anomalous head or supernumerary m. flexor carpi ulnaris; *c*, origin of the supernumerary muscle from the medial epicondyle of the humerus.

B: *a*, Palmar aponeurosis with the tendon of m. palmaris profundus attached to the deep (dorsal) side of the transverse carpal ligament (*b,c*); *d*, semipennate m. palmaris profundus; *e,f*, tendon of m. palmaris profundus passing beneath and attaching to the transverse carpal ligament and the palmar aponeurosis which are one continuous fascial sheet; *g*, m. flexor pollicis longus proprius; *h*, fusiform muscle fascicle or accessory head of m. flexor pollicis longus, which has a common origin with m. flexor digitorum superficialis and m. pronator teres; *i*, tendon of the anomalous slip which bifurcates with one joining the tendon (*k*) of flexor pollicis longus. The second (*l*) receives a large anomalous fascicle (*m*) from m. flexor pollicis longus, and terminates in the tendon of m. flexor digitorum profundus which supplies the index finger.

From Calori, L. Delle anomalie più importanti di ossa, vasi, nervi, e muscoli occorse nell'ultimo biennio facendo anatomia del corpo umano. *Mem. R. Accad. Sci. Istituto di Bologna S. 2* 8:417–482, 1868.

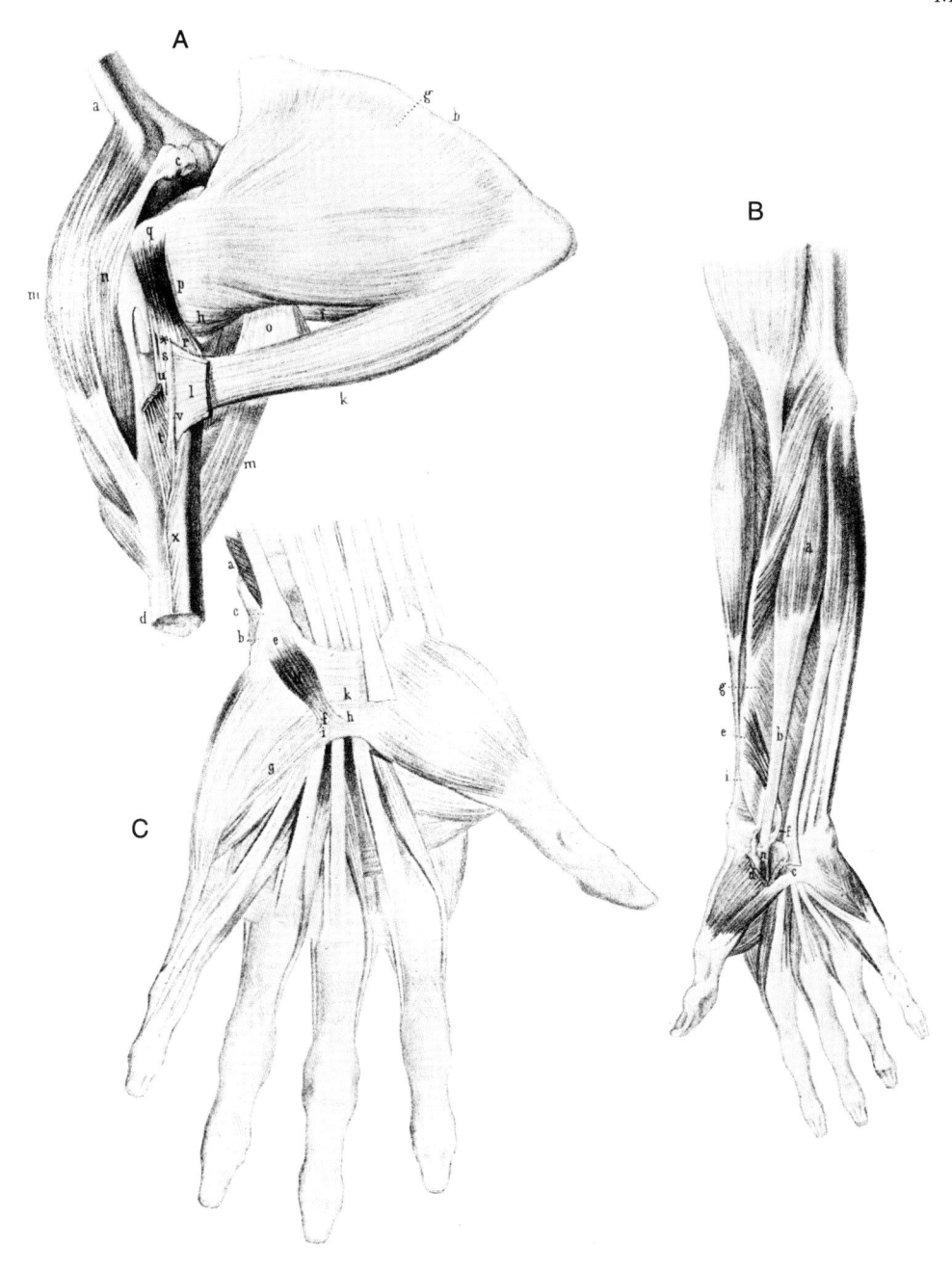

FIG. 59.
M. Tensor Capsulae.
M. Flexor Carpi Radialis Accessorius.
M. Pisiuncinatus.
M. Subscapularis (Anomalous, Gruber).

A: *a*, Clavicle, *b*, scapula; *c*, coracoid process; *d*, humerus; *g* m. subscapularis; *h*, anomalous m. subscapularis of Gruber; *i*, m. teres minor; *k*, m. teres major; *l*, tendon of m. latissimus dorsi; *m,m*, m. deltoideus; *o*, tendon of m. triceps brachii, long head; *p*, belly of m. tensor capsulae; *q*, origin of m. tensor capsulae; *r*, insertion of m. tensor capsulae onto the surgical neck of the humerus.

B: *a*, M. flexor carpi radialis; *b*, tendon of m. flexor carpi radialis; *c,d*, osseofibrous canal through which m. flexor carpi radialis passes; *e*, m. flexor carpi radialis accessorius; *f*, tendon of the accessory muscle; *g*, m. flexor digitorum superficialis; *i*, m. flexor pollicis longus.

C: *a*, M. flexor carpi ulnaris; *b*, pisiform bone onto which the flexor muscle inserts (*c*); *d*, m. pisiuncinatus; *e*, origin of m. pisiuncinatus from the pisiform bone; *f*, insertion of m. pisiuncinatus on the uncinate (hamulus of the hamate) bone; *i*, fibers of the termination of the insertion mingle with the origin of mm. opponens digiti minimi (*g*) and pollicis.

From Calori, L. Di alcuni nouvi muscoli soprannumerarii degli arti. *Mem. Accad. Sci. Istituto di Bologna S.2* 6:137–147, 1866.

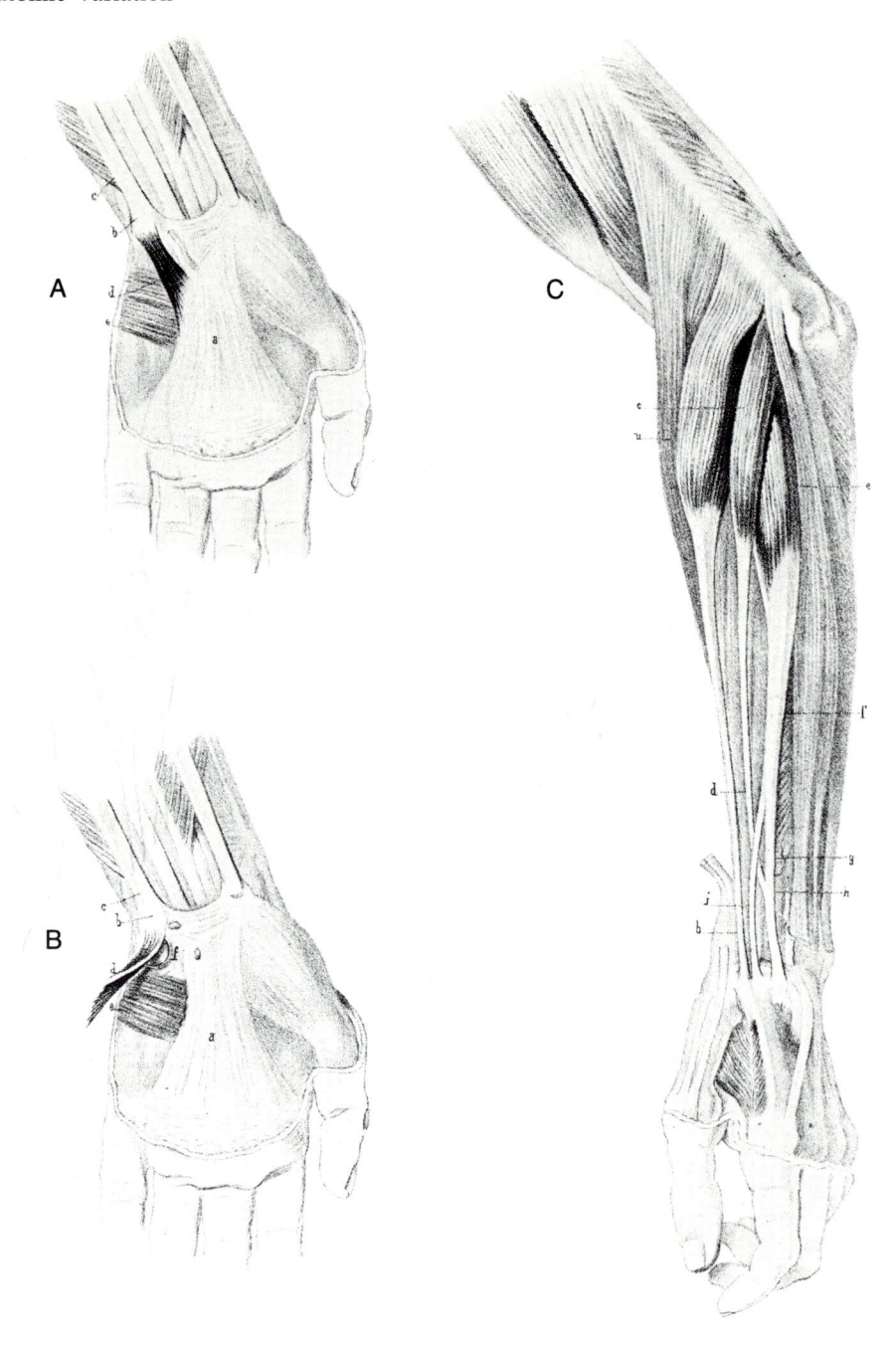

FIG. 60.
Absence of M. Palmaris Longus.
M. Pisipalmaris (Pisohamatus).
Duplication of M. Extensor Carpi Radialis Brevis.

A: *a*, Palmar aponeurosis: note absence of m. palmaris longus; *b*, pisiform bone with insertion of m. flexor carpi ulnaris (*c*), origin of m. pisipalmaris (*d*); *e*, m. palmaris brevis.

B: M. pisipalmaris detached from palmar aponeurosis. Beneath m. pisipalmaris is a small bursa (*f*).

C: *b*, Tendon of m. extensor carpi radialis longus; *c,d*, m. extensor carpi radialis brevis and its tendon, which joins the tendon of a second m. extensor carpi radialis brevis (anomalous); *e,f*, anomalous m. extensor carpi radialis brevis which divides to insert onto the base of the first and the second metacarpal bones.

From Calori, L. Delle anomalie più importanti di ossa, vasi, nervi e muscoli occorse nell'ultimo biennio facendo anatomia del corpo umano. *Mem. R. Accad. Sci. Istituto di Bologna S. 2* 8:417–482, 1868.

Muscles

FIG. 61.
Absence of M. Palmaris Longus.
Anomalous Muscle Fascicle from M. Brachioradialis Crossing Obliquely to the Medial Antibrachial Fascia. Radial Artery Passes Beneath the Anomalous Muscle Fascicle.
Variation in Superficial Palmar Arterial Arch.
Presence of Median Artery Which Passes Beneath Transverse Carpal Ligament and Joins Radial Artery.
Variations in M. Flexor Digitorum Superficialis, Absence of Tendon to Fifth Digit.

a, Anomalous muscle fascicle; *l,q*, median artery joining radial artery in palm of hand.

From Calori, L. Delle anomalie più importanti di ossa, vasi nervi, e muscoli occorse nell'ultimo biennio facendo anatomia del Corpo umano. *Mem. R. Accad. Sci. Istituto di Bologna S. 2* 8:417–482, 1868.

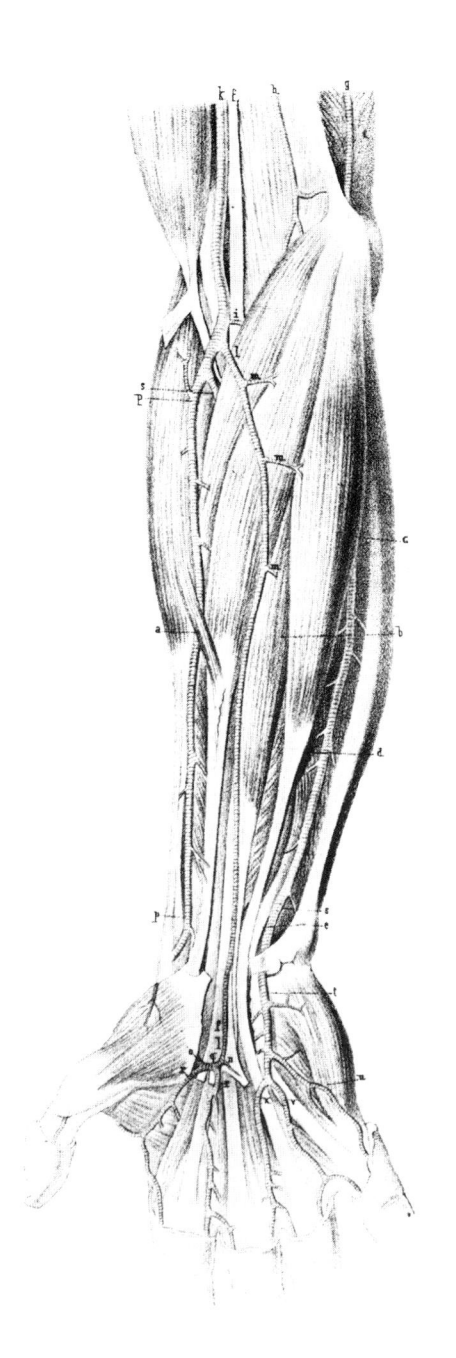

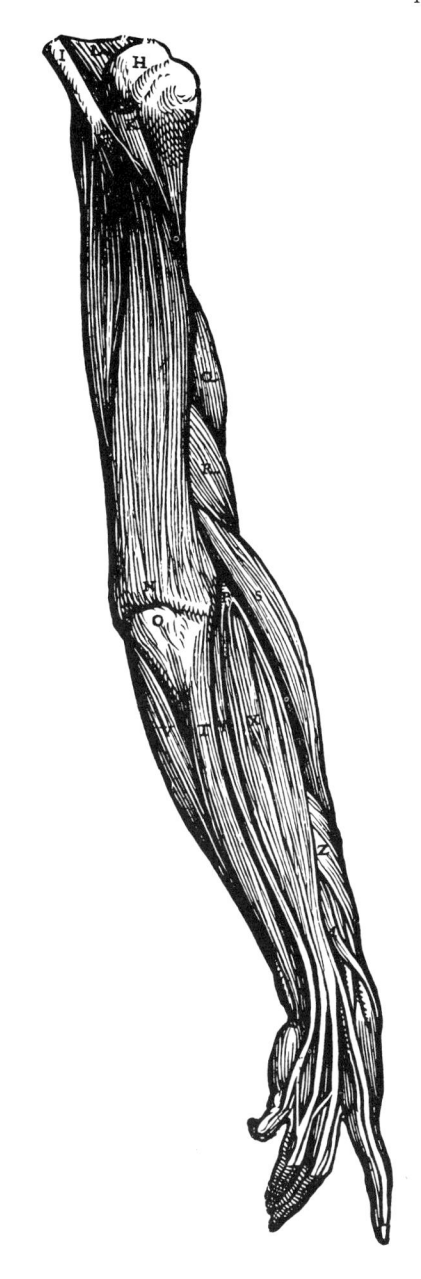

FIG. 62.
M. Extensor Digitorum Communis (Anomalous).
M. Extensor Digiti Minimi Proprius (Anomalous).

Posterior view of right arm. M. extensor digitorum communis (*x*) supplies tendons for the second, third, and fourth fingers. M. extensor digiti minimi proprius (*y*) supplies tendons for the fourth and fifth fingers.

Vesalius, Andreas. *Fabrica* (1543).

293

Human Anatomic Variation

FIG. 63.
M. Extensor Pollicis Longus (Anomalous).
M. Extensor Digitorum Communis (Anomalous).

In both hands, m. extensor pollicis longus (π) is distributed to both thumb and index finger. M. extensor digitorum communis (s) supplies tendons for fingers two, three, and four. In this subject, however, m. extensor digiti minimi proprius supplies only (as usual) the fifth digit.

Vesalius, Andreas. *Fabrica* (1543).

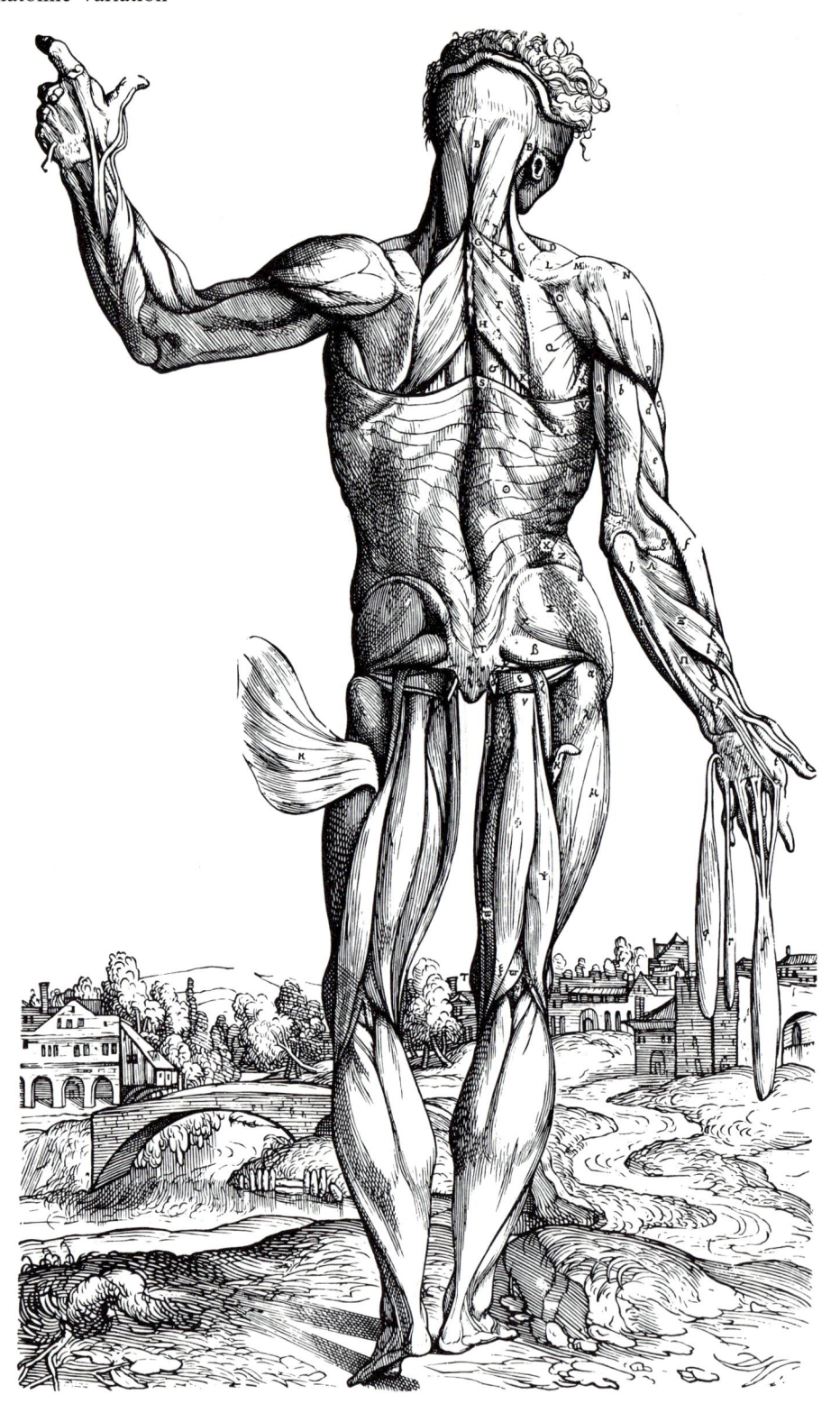

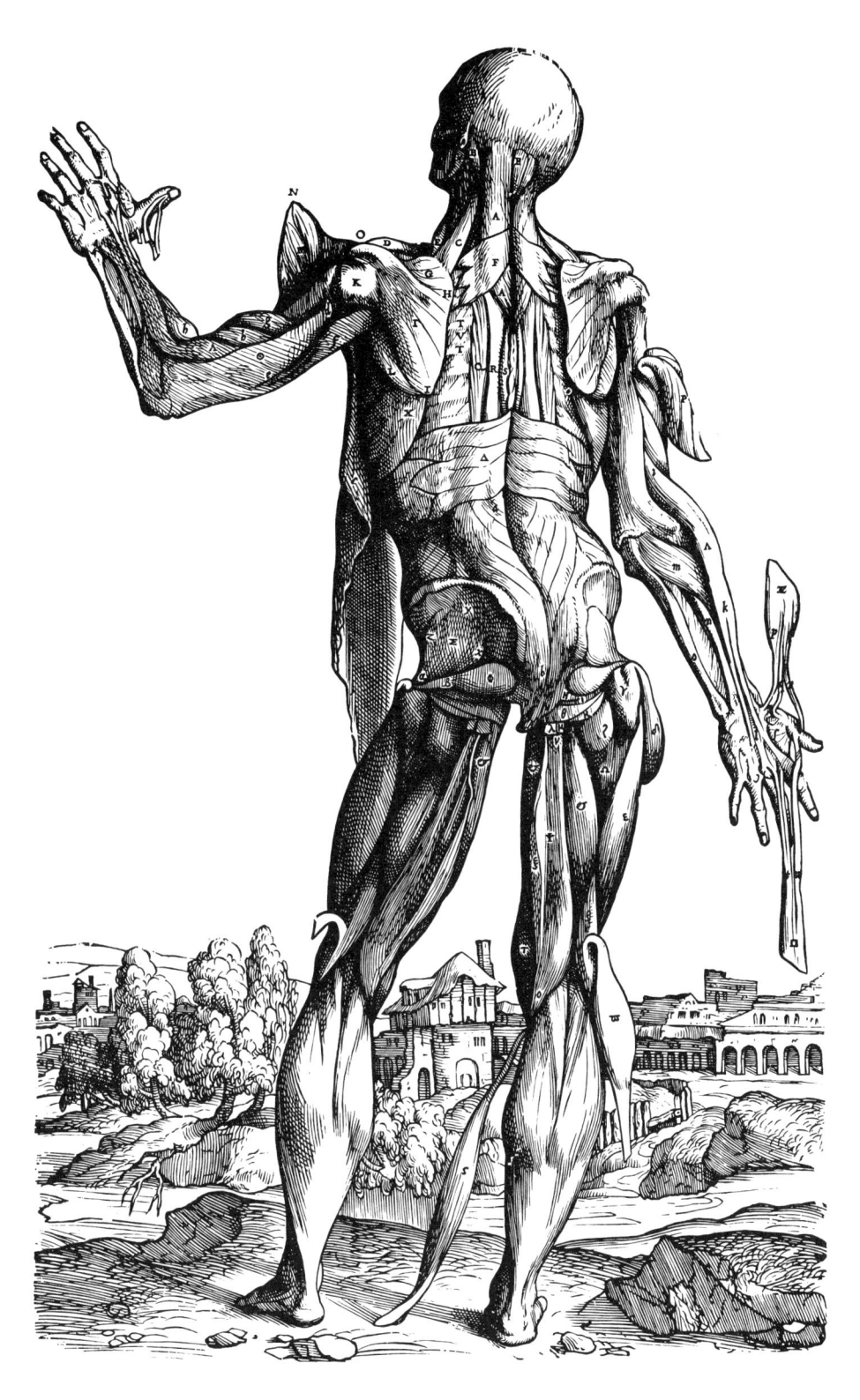

FIG. 64.
M. Extensor Pollicis Longus (Anomalous).
In the right hand, m. extensor pollicis longus (hanging down, π) supplies the thumb and both the index and middle fingers. This is an uncommon variant.
Vesalius, Andreas. *Fabrica* (1543).

Human Anatomic Variation

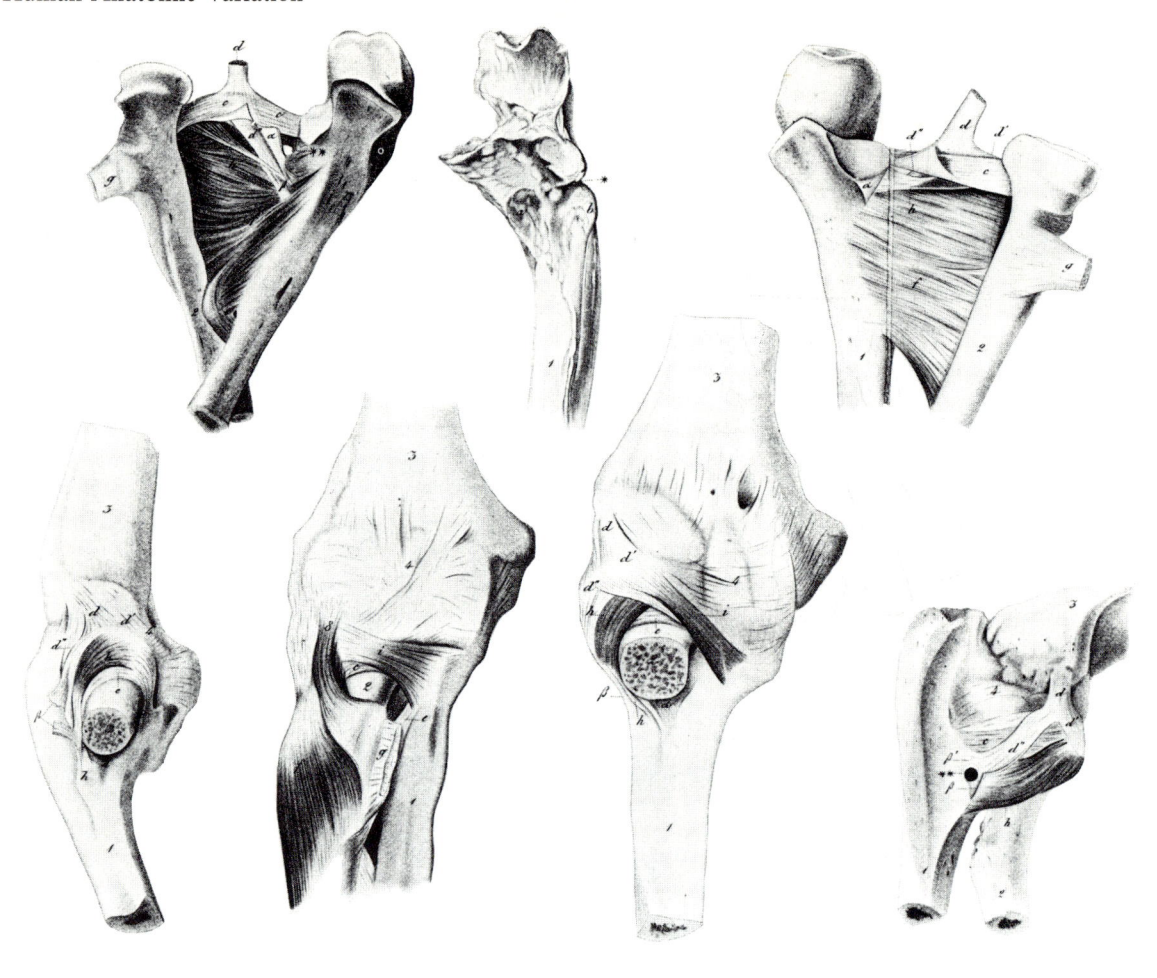

FIG. 65.
M. Tensor Ligamenti Annularis Radii Anterior.
M. Tensor Ligamenti Annularis Radii Posterior.

1, Ulna; *2*, radius; *3*, humerus; *4*, elbow joint capsule; *a*, ulnar tuberosity (insertion of brachialis muscle); *b*, ulnar tubercle (ulnar origin of m. flexor digitorum superficialis); *c*, annular ligament of radius; *d,d'*, radial collateral ligament; *e*, oblique cord; *f*, m. supinator; *g*, tendon of m. biceps brachii; *h*, m. tensor ligamenti annularis radii posterior; *i*, m. tensor ligamenti annularis radii anterior.

From Gruber, W. Die eigenen Spanner des Ringbandes des Radius – Musculi tensores proprii ligamenti annularis radii – bei dem Menschen. *Arch. Anat. Physiol. Wissen. Med.* 1865, pp. 377–391.

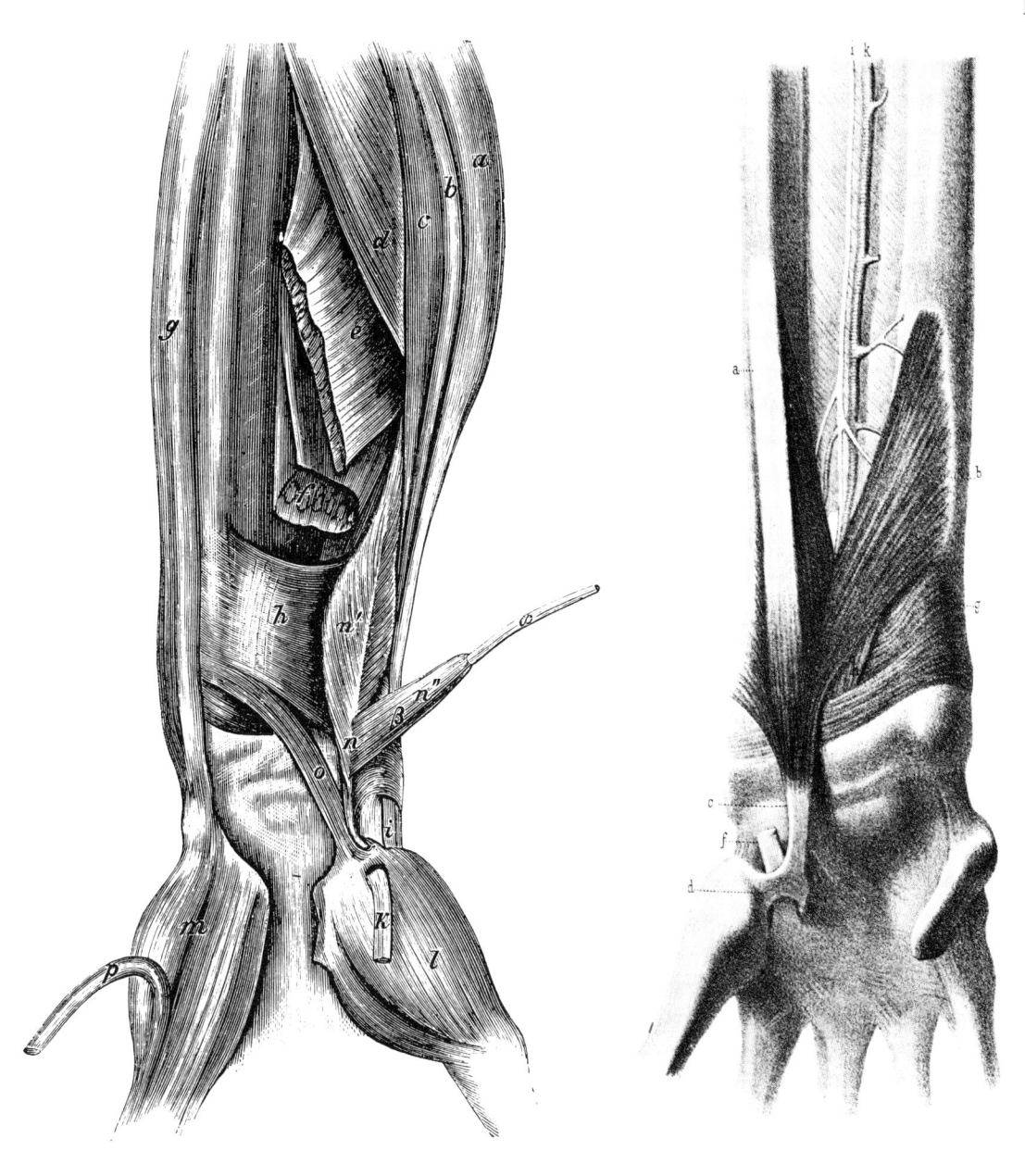

FIG. 66.
M. Radialis Internus Brevis Biceps.
M. Cubitocarpeus.
Supernumerary Head of M. Abductor Digit Minimi.

a, M. brachioradialis; b, m. extensor carpi radialis longus; c, m. extensor carpi radialis brevis; d, m. pronator teres; e, m. flexor digitorum sublimis (superficialis), radial origin; f, m. flexor pollicis longus (superior part); g, m. flexor carpi ulnaris; h, m. pronator quadratus; i, tendons of m. abductor pollicis longus and m. extensor pollicis brevis; k, tendon of flexor carpi radialis; l, thenar musculature; m, hypothenar musculature; n, m. radialis internus brevis biceps; n', larger head which arose from the radius; n", smaller supernumerary head which arose from forearm aponeurosis: α, tendon of origin, and β, belly of the small head; o, m. cubitocarpeus; p, supernumerary head of m. abductor digiti minimi.

From Gruber, W. Nachträge zu den Varietäten des Musculus radialis internus brevis. *Bull. l'Acad. Imp. Sci. St. Petersbourg* 17:380–388, 1872.

FIG. 67.
M. Cubitoradiocarpien.

a,b, The two heads of m. cubitoradiocarpien (a, radial head; b, ulnar head); c, tendon of this muscle, insertion in two parts; d, radial eminence of os trapezium, and e, palmar fascia of os trapezoidium; f, portion of tendon of m. flexor carpi radialis; g, m. pronator quadratus (h) augments the ulnar head (b) of the supernumerary muscle; i, interosseous nerve; k, anterior interosseous artery.

From Calori, L. Degli usi del muscolo pronatore quadrato e di un muscolo sopranumerario cubito-radio-carpeo nell'uomo. *Mem. R. Accad. Sci. Istituto di Bologna* S.2 10:647–657, 1870.

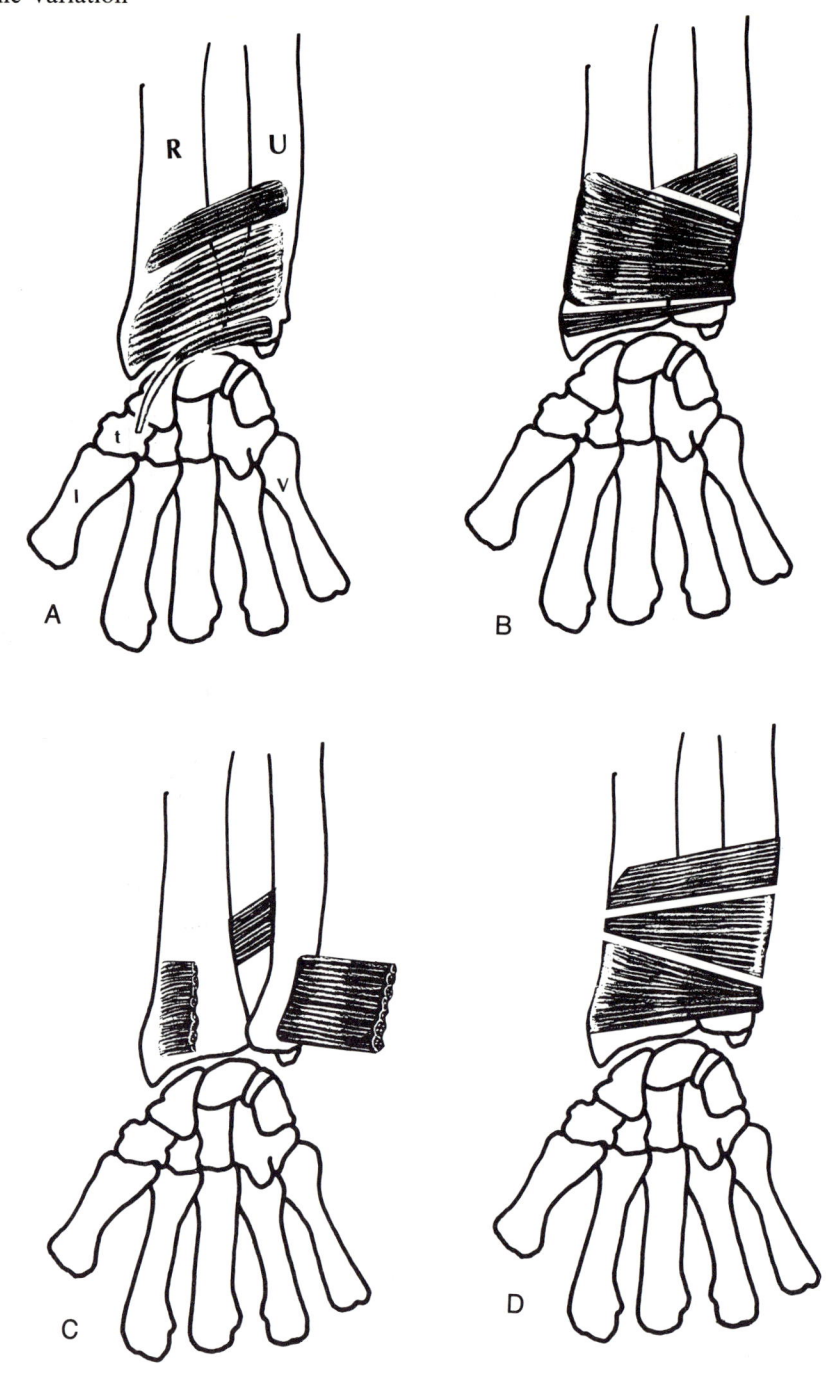

FIG. 68.
M. Pronator Quadratus (Anomalous).

A: Pronator quadratus, with a slip (cubitocarpeus of Gruber) to the trapezium (*t*); *R*, radius; *U*, ulna; *I*, first carpal; *V*, fifth metacarpal bone.
B: A trifid pronator quadratus.
C: A bilaminar pronator.
D: A complexly divided pronator quadratus.

Redrawn from Macalister, A. Observations on muscular anomalies in the human anatomy. (Third series with a catalogue of the principal variations hitherto unpublished.) *Trans. R. Irish Acad. Sci.* 25:1–130, 1875.

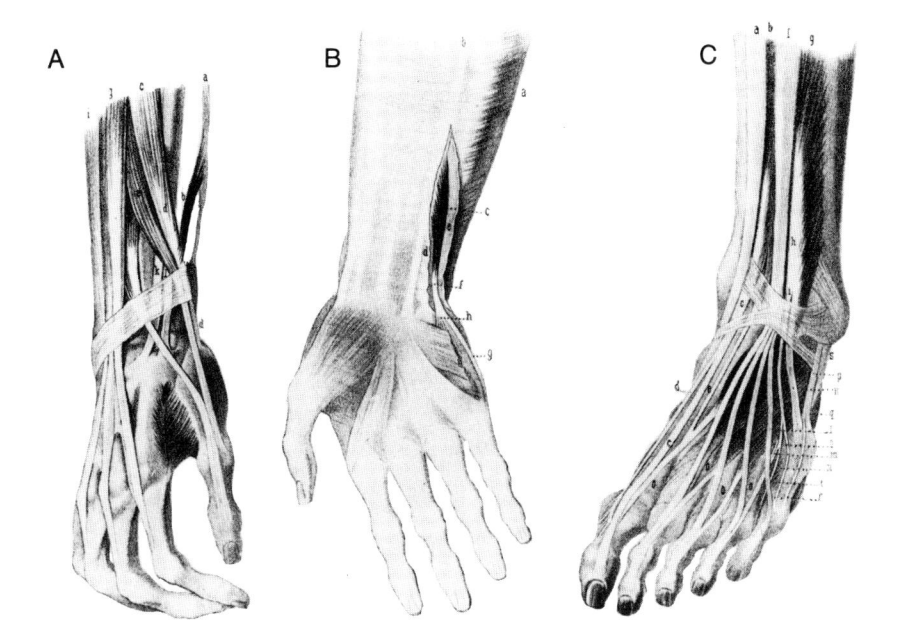

FIG. 69.
Anomalous Muscles and Tendons of Forearm, Hand, and Foot.

A: *a*, Tendon of m. brachioradialis; *b*, anomalous muscle extending between brachioradialis and the long abductor of the thumb; *c*, long abductor of the thumb; *d*, tendon of the long abductor; *e*, m. extensor pollicis brevis; *f*, m. extensor pollicis longus; *g*, m. extensor indicis; *i*, m. extensor carpi ulnaris; *k,l*, m. extensor radialis brevis and longus.
B: *a*, Antibrachial fascia; *b*, tendon of m. palmaris longus; *c*, belly of anomalous muscle (m. accessorius ad abductorem digiti minimi) arising from the fascia and tendon of m. palmaris longus; *f*, tendon of anomalous muscle; *g,h*, insertion into fifth metacarpal and lateral hypothenar fascia.
C: *a*, M. tibialis anterior; *b*, long extensor of great toe, *c,d,e*, multiple tendons of long extensor to the great toe and short extensor tendon of the great toe.

From Calori, L. Di alcune varieta muscolari dell'avambraccio e dell'eminenza ipothenar. *Mem. Accad. Sci. Istituto di Bologna S. 2* 7:359–381, 1867.

FIG. 70.
Anomalous Extensor of Middle Finger.
Anomalous Short Extensor of Middle Finger.
Anomalous Short Extensor of Middle and Third Fingers.

A: *a,b*, Two radial carpal muscles; *c*, m. abductor pollicis longus; *d*, m. extensor pollicis brevis; *e*, m. extensor pollicis proprius; *f*, m. extensor indicis; *f,g,i*, origin, belly, and tendon of anomalous extensor of the middle finger.
B: *b,i*, Anomalous extensor of middle finger; *l,n,p,q*, supernumerary short extensor with two heads, one to the middle and one to the third finger.

From Calori, L. Di alcune varieta muscolari dell'avambraccio e dell'eminenza ipothenar. *Mem. Accad. Scienze Istituto di Bologna S. 2* 7:383–393, 1867.

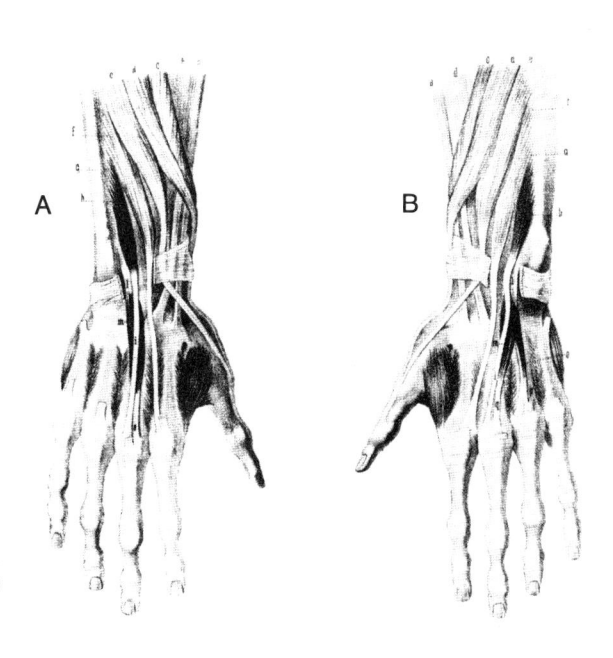

FIG. 71.
Supernumerary Deep Radial Flexor Muscle of Hand.
Supernumerary Muscle Extending Between Tendon of M. Brachioradialis and M. Flexor Pollicis Longus.
Supernumerary Muscle Between M. Abductor Digiti Minimi and M. Opponens Digiti Minimi.
Doubled M. Pronator Quadratus.
Absence of Normal M. Extensor Indicis Proprius.
Anomalous M. Indicis Brevis Manus.

A: e, M. pronator teres, twisted proximally to show origin of supernumerary deep radial flexor of the hand; f, m. flexor carpi radialis; g, tendon of m. palmaris longus; h, belly of m. palmaris longus (inversion of belly, distal to normal proximal position); i, insertion of m. palmaris longus (into tubercle of scaphoid); k, m. flexor digitorum superficialis; m, superior belly of doubled m. pronator quadratus; n, inferior belly of doubled m. pronator quadratus; o, m. flexor pollicis longus; p, tendon of m. flexor pollicis longus; q, origin of supernumerary deep radial flexor of hand; r, belly of anomalous muscle (q); s, tendon of insertion of anomalous muscle (q); t, m. brachioradialis; t^2, tendon of m. brachioradialis; u,v, the two radial carpal muscles; y, supernumerary muscle extending between tendon of m. brachioradialis and m. flexor pollicis longus; 1, flexor retinaculum; 2, short flexor of fifth digit.

B: a, M. extensor digiti minimi; b, m. extensor digitorum communis (drawn aside to show anomalous indicator muscle); c, m. extensor pollicis longus; d, m. extensor pollicis brevis; e, m. abductor longus; f,g, the two radial carpal muscles; h, origin of anomalous m. indicis brevis manus; i, tendon of insertion of anomalous muscle; m, fascicle of anomalous muscle reinforcing second dorsal interosseus muscle.

From Calori, L. Di alcune varieta muscolari dell'avambraccio e dell'eminenza ipothenar. *Mem. Accad. Sci. Istituto di Bologna S. 2* 7:383–393, 1867.

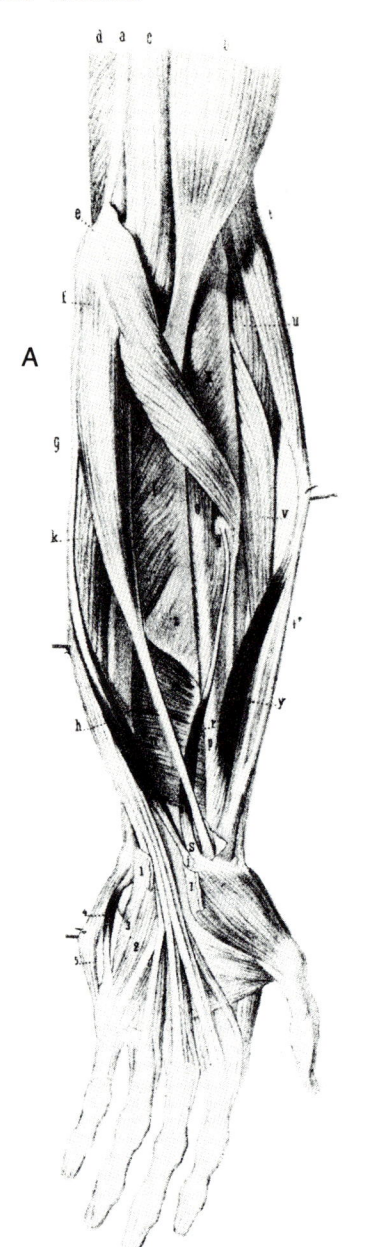

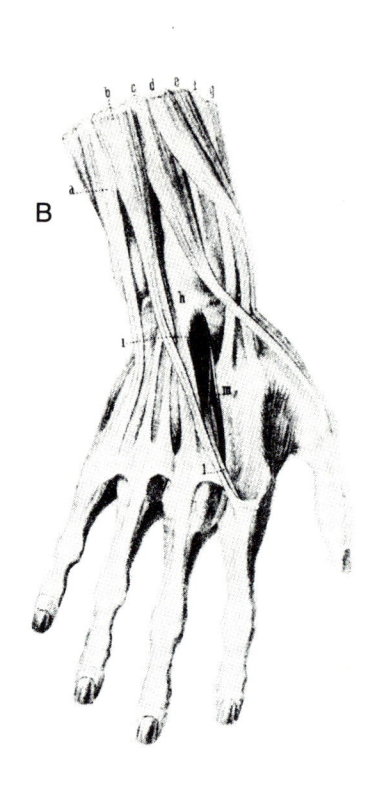

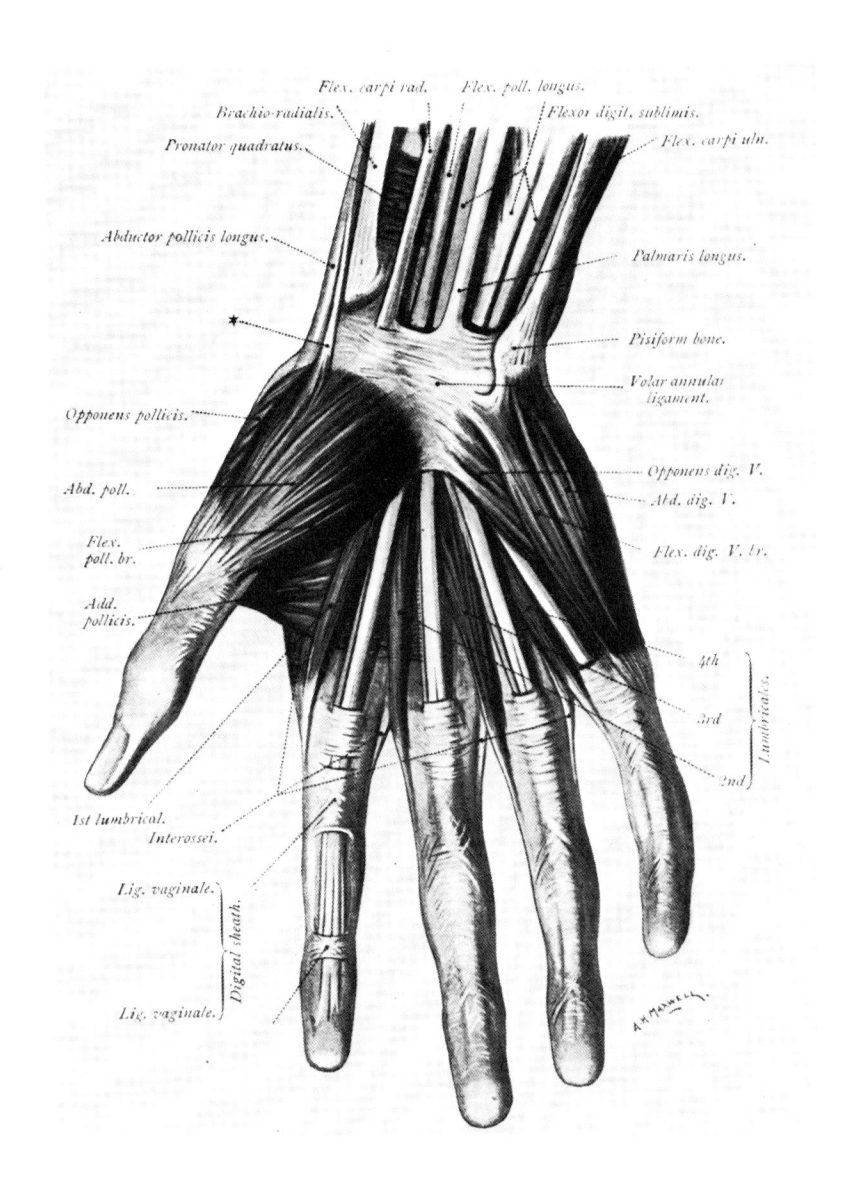

FIG. 72.
Accessory Slip Connecting Mm. Abductor Pollicis Longus and Brevis.

Superficial dissection of muscles of the palm of the hand. An accessory slip (*) is frequently present connecting the long and short abductors of the thumb.

From Schafer, E.A., Symington, J. and T.H. Bryce, Eds. *Quain's Anatomy*, 11th Ed., Vol. IV, Part II: Myology. Longmans, Green, and Co., London, 1923.

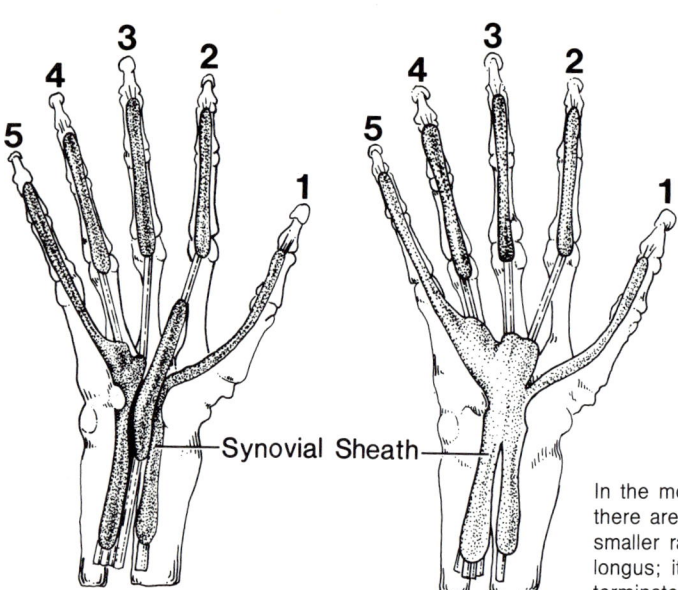

Human Anatomic Variation

FIG. 73.
Unusual Synovial Sheaths of Flexor Tendons.

In the hand on the left, the synovial sheath is divided into three parts: one sheath for the thumb; one for the index finger, and one for the middle, ring, and little finger. In the hand on the right, there is an incomplete division of the synovial sheath into the usual two parts, with all five tendons enclosed in a common sheath.

In the most common arrangement (75% of 500 dissections) there are two sheaths. On the ventral surface of the wrist a smaller radial sheath surrounds the tendon of flexor pollicis longus; it arises just proximal to the flexor retinaculum and terminates at the distal phalanx of the thumb. The larger ulnar sheath surrounds flexor tendons of digits two to five. It begins proximal to the flexor retinaculum and extends to about the mid-transverse line of the palm. The sheath of the fifth digit usually communicates with the ulnar sheath. In 17% of cases this communication is interrupted. (Scheldrup, E.W. Tendon sheath patterns in the hand. *Surg. Gynecol. Obstet.* 93:16–22, 1951.)

Redrawn and modified from Latarjet, A., Testut, L. and A. Latarjet. *Traité D'Anatomie Humaine*, 9th Ed. G. Doin & Cie, Paris, 1948.

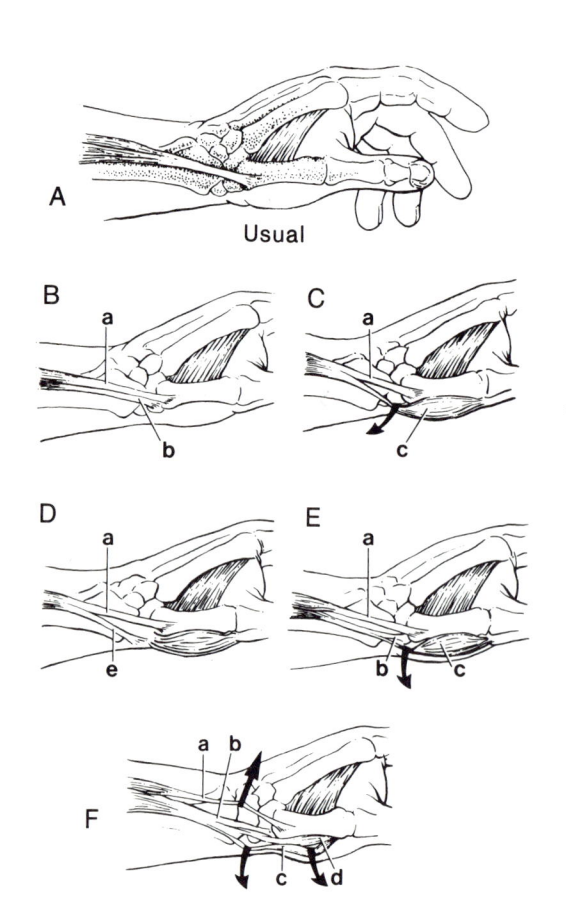

FIG. 74.
Unusual Insertions of M. Abductor Pollicis Longus.

A: Usual insertion of m. abductor pollicis longus, i.e., the radial side of the ventral aspect of the base of the first metacarpal.

B: *a*, M. abductor pollicis longus; *b*, slip to os trapezium (greater multangular).

C: *a*, M. abductor pollicis longus joins m. abductor pollicis brevis.

D: *a*, M. abductor pollicis longus; *e*, slip to volar carpal ligament.

E: *a*, M. abductor pollicis longus; *b*, slip to os trapezium (greater multangular); *c*, slip to m. abductor pollicis brevis.

F: *a*, M. abductor pollicis longus; *b*, slip to os trapezium (greater multangular); *c*, slip to m. abductor pollicis brevis; *d*, slip to M. opponens pollicis.

From Lacey, T. II, Goldstein, L.A. and C.E. Tobin. Anatomical and clinical study of the variations in the insertions of the abductor pollicis longus tendon, associated with stenosing tendovaginitis. *J. Bone Joint Surg.* 33A:347–350, 1951.

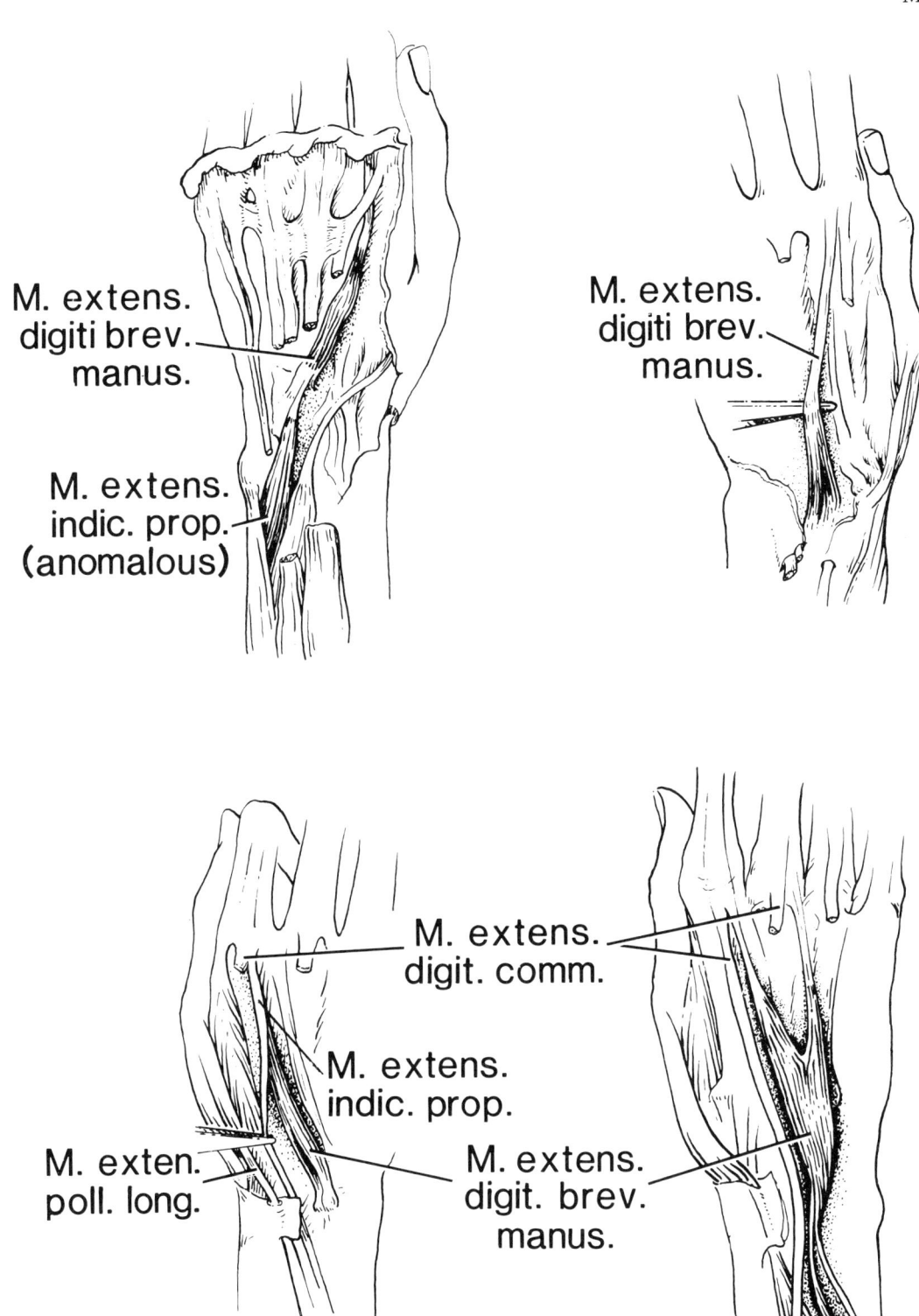

FIG. 75.
M. Extensor Digiti Brevis Manus (Anomalous).
M. Extensor Indicis Proprius (Anomalous).

Accessory muscle to ring finger is shown in upper left drawing.

The remaining drawings show extensor muscles of index finger.

Redrawn from an illustration in Anson, B.J. *An Atlas of Human Anatomy*. W.B. Saunders Company, Philadelphia, 1950. Used by permission.

303

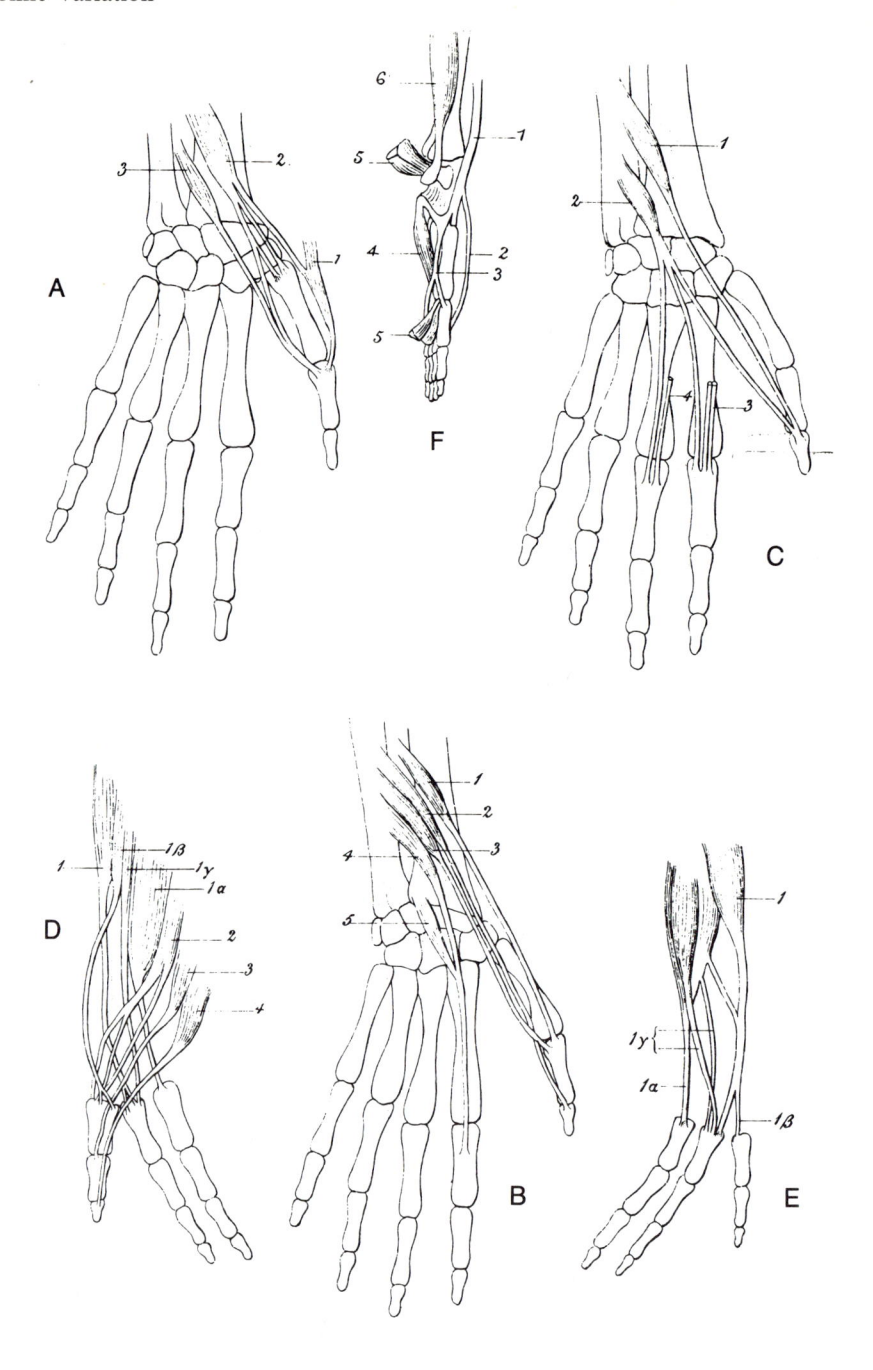

Human Anatomic Variation

FIG. 76.
Variations in Extensor Muscles.

A: *1*, Abductor pollicis brevis detached from origin on bones of carpus; *2*, extensor ossis metacarpi pollicis with quadruple tendon; *3*, extensor primi internodii pollicis.

B: *1*, Extensor primi internodii pollicis; *2*, extensor secundi internodii pollicis; *3*, additional extensor primi internodii pollicis; *4*, extensor indicis; *5*, extensor brevis digitorum manus joining tendon of extensor indicis.

C: *1*, Extensor secundi internodii pollicis; *2*, abnormal extensor indicis; *3*, double tendon to index from extensor communis digitorum; *4*, single tendon to middle digit from extensor communis digitorum.

D: *1*, Extensor carpi radialis longus; *1α*, extensor carpi radialis brevis; *1β*, extensor carpi radialis accessorius; *1γ*, extensor carpi radialis intermedius; *2*, extensor ossis metacarpi pollicis (triple tendon); *3*, extensor primi internodii pollicis; *4*, extensor secundi internodii pollicis.

E: *1*, *1α*, *1β*, *1γ*, as in drawing D.

F: *1*, Extensor carpi ulnaris; *2*, ulnaris digiti minimi; *3*, anterior slip from *1*; *4*, flexor brevis digiti minimi; *5*, abductor digiti minimi, cut; *6*, flexor carpi ulnaris.

Authors' note: M. extensor ossis metacarpi pollicis s. m. abductor pollicis longus; m. extensor primi internodii pollicis s. m. extensor pollicis brevis; m. extensor secundi internodii pollicis s. m. extensor pollicis longus.

From Curnow, J. Variations in the arrangement of the extensor muscles of the forearm. *J. Anat. Physiol.* 10:596–601, 1876.

Muscles

FIG. 77.
Anomalous Dorsal Interosseus Muscle.

id III, Third dorsal interosseus muscle; *id II*, second dorsal interosseus muscle; *id^x*, anomalous third head of the second dorsal interosseus muscle.

From Brunn, A. von. Varietät des Musc. interosseus dorsal. manus II. *Arch. Anat. Physiol. Wissen. Med.* 1873, p. 126.

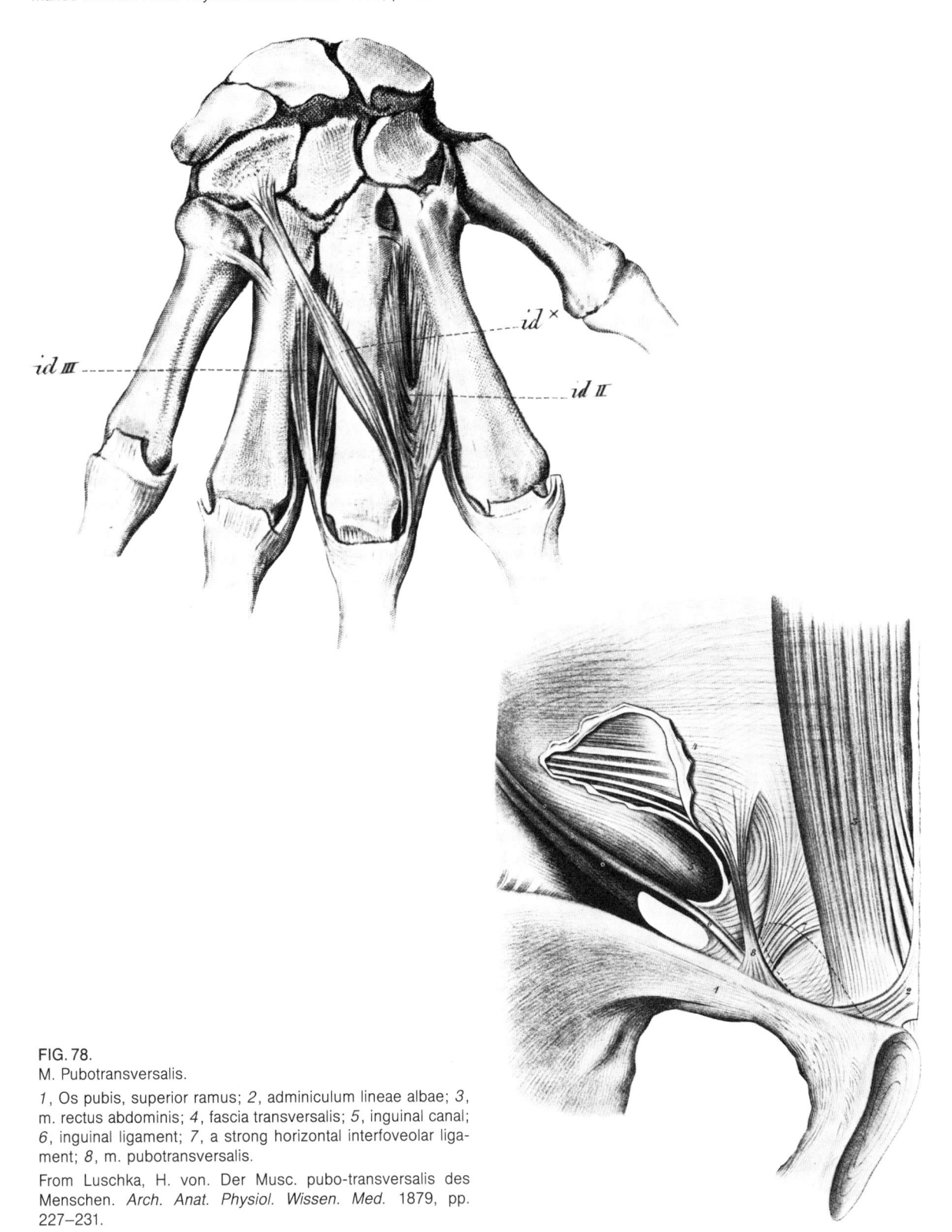

FIG. 78.
M. Pubotransversalis.

1, Os pubis, superior ramus; *2*, adminiculum lineae albae; *3*, m. rectus abdominis; *4*, fascia transversalis; *5*, inguinal canal; *6*, inguinal ligament; *7*, a strong horizontal interfoveolar ligament; *8*, m. pubotransversalis.

From Luschka, H. von. Der Musc. pubo-transversalis des Menschen. *Arch. Anat. Physiol. Wissen. Med.* 1879, pp. 227–231.

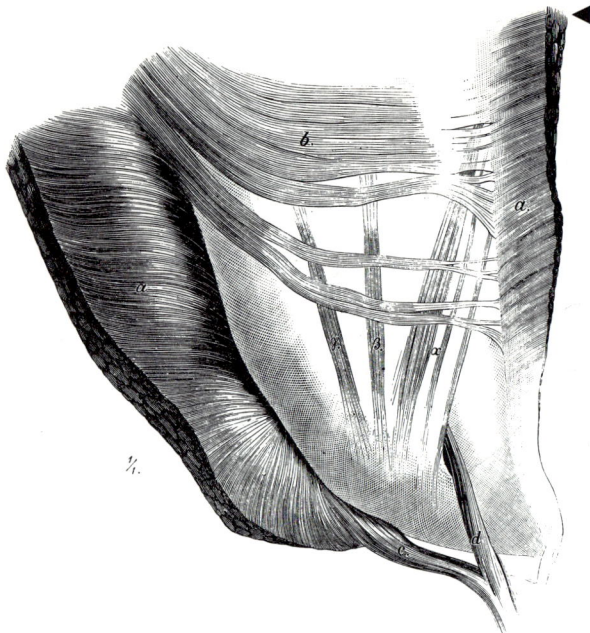

Human Anatomic Variation

◀ FIG. 79.
M. Tensor Laminae Posterioris Vaginae Musculi Recti.

α,β,γ, M. tensor laminae posterioris vaginae musculi recti; a, m. internal oblique; b, m. transversus abdominis; c, m. cremaster; d, spermatic cord.

From Gruber, W. Sur quelques muscles surnuméraires de l'abdomen, chez l'homme. *Bull. l'Acad. Imp. Sci. St. Petersbourg* 18:142–147, 1873.

FIG. 80.
Aberrant M. Psoas Major Fascicle.
Dissecting room specimen, The University of Iowa College of Medicine.
▼

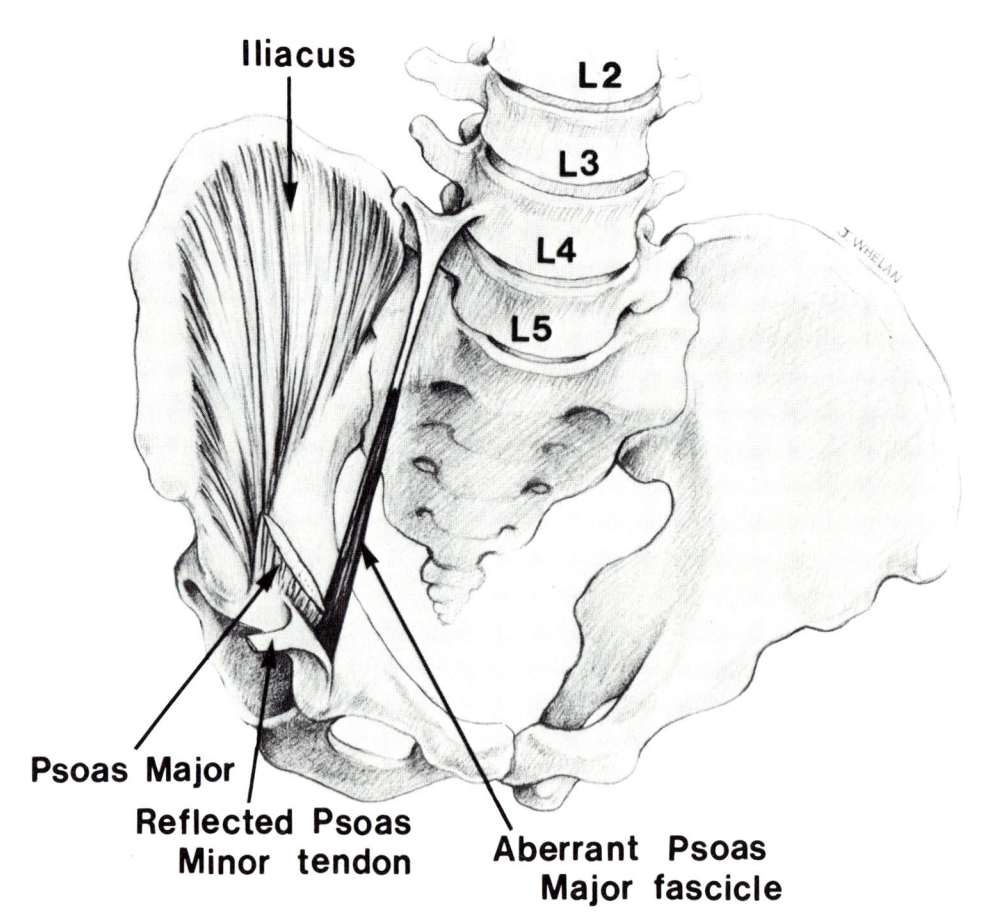

Iliacus

L2
L3
L4
L5

Psoas Major
Reflected Psoas Minor tendon
Aberrant Psoas Major fascicle

Muscles

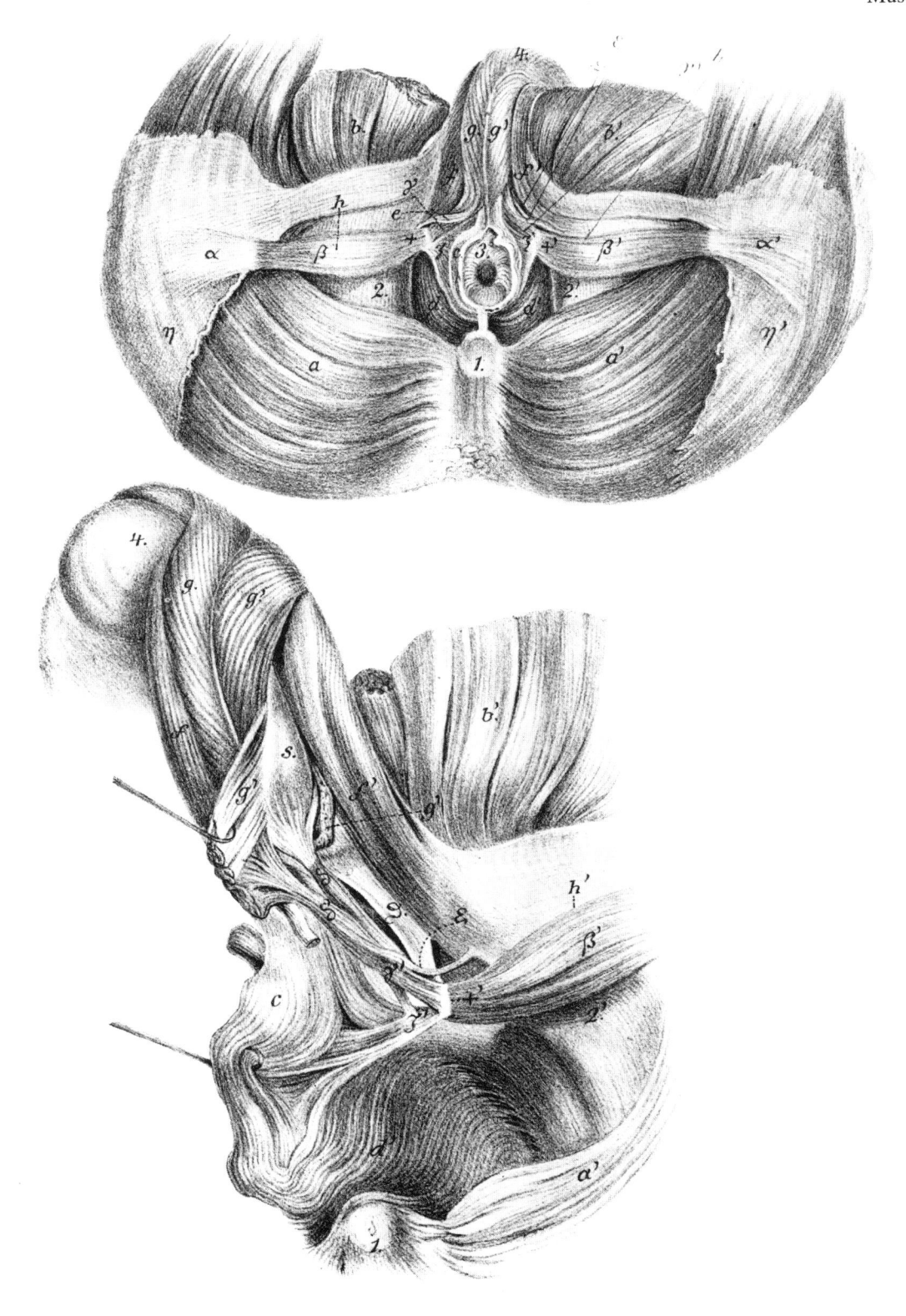

FIG. 81.
M. Gluteoperinealis.

a, Os coccygis; *2,2'*, ischial tuberosity; *3*, anus; *4*, root of penis; *a,a'*, m. gluteus maximus; *b,b'*, origin of thigh musculature from the ischial tuberosities; *c*, external anal sphincter; *d,d'*, m. levator ani; *e*, m. transversus perinei superficialis (right side); *f,f'*, m. ischiocavernosus; *g,g'*, m. bulbocavernosus; *h,h'*, m. gluteoperinealis; α,α', aponeurosis, origin of m. gluteoperinealis; β,β', belly of m. gluteoperinealis; γ,γ', insertion of m. gluteoperinealis; +,+', tendinous inscription in m. gluteoperinealis.

From Gruber, W. Über den Gesäß-Mittelfleischmuskel (Musculus gluteo-perinealis). *Arch. Pathol. Anat. Physiol. Klin. Med.* 74:456–460, 1878.

CASE OF CONGENITAL ABSENCE OF THE QUADRICEPS EXTENSOR CRURIS MUSCLE. Communicated by Prof. A. G. DRACHMANN, of Copenhagen[1].

MISS A. F., aged 28 years, consulted me in November of last year (1871), for an affection of the left knee, from which she stated she had suffered for a very long time—how long she could not remember, but it had increased of late years, and rendered walking more and more difficult, while she also usually felt pains in the knee-joint, especially when she at all exceeded her ordinary amount of walking. The knee-joint, too, became tender and swollen—a condition which, however, gradually disappeared if she remained at rest for any length of time. On examining the uncovered knee I was not a little surprised to find the knee-cap wanting; the outlines of both condyles of the femur very perceptibly exposed, covered only by the skin and some subcutaneous areolar tissue; the anterior inter-condyloid fossa similarly covered only by skin and filled with a rather soft, subcutaneous adipose tissue, which however was not present in such amount as to completely fill up the whole depression between the condyles, but left a visible sulcus behind. No trace of the ligamentum patellæ existed; but the tuberosity of the tibia, and the outlines of the condyles of this bone were seen and felt very plainly immediately beneath the skin. The whole anterior aspect of the thigh, from its upper third down to the knee, had lost its usual roundness and fulness. The thigh-bone itself throughout this space was felt immediately underneath the skin, without a trace of intervening muscular substance. A little above the external condyle of the femur, on the outer aspect of the thigh, the atrophied patella was found lying, without any attachment to any of the femoral muscles, freely moveable in all directions. The skin over the knee-joint was perfectly normal, no swelling of the tissues constituting or surrounding the articulation was present, no effusion into the *capsule of the knee*. On rather deep pressure being made over and a little above the external condyle, the patient felt some tenderness. The knee-joint freely admitted of passive motion; as regards active motion, the patient could easily bend the knee, but she could by no means extend it, nor was she able, while in the recumbent position, to lift the limb. The muscular structures on the posterior aspect of the thigh, and in the upper third of its inner aspect (the adductors) were strongly developed, while the triceps cruris (adductores longus, brevis, et magnus) was found more than ordinarily strong and largely developed. Measurement round the upper third of the thigh gave 18 inches, above the knee 10½ inches, round the leg 12½ inches. On bidding her strip the other, the right knee, of which she did not complain, I discovered to my surprise, that even in the most minute particulars it corresponded perfectly to the left one. On more closely interrogating the patient, I learned that she had not been aware of any defect in connection with her knees until her tenth year, although her gait had presented some irregularity. This had not however prevented her from playing and running about with her playfellows. She further stated that at this time she fell and hurt her knees, which a medical man had examined and had then declared that she suffered from a congenital and incurable defect. Since that time she has constantly, as the physician advised, worn a bandage on her knees, the result of which has been (as she positively affirmed) that the patella, which was before situated over the internal condyle of the femur, had by degrees glided over to the other, the opposite side, where it is now situated. It is only within the past two years, since she has been obliged to walk and stand a good deal (she is teacher in a large national school), that the inconveniences in walking and the troubles above mentioned have become more pronounced, and, as stated, have extended as yet only to the left lower extremity.

FIG. 82.
Congenital Absence of M. Quadriceps Femoris (Case Report). From Drachmann, A.G. Tilfalde af medfodt Mangel af musc. quadriceps. Nord. Med. Ark. (Stockholm) 4(1):6(II) SS7, 8, 1872. Translated by J.W. Moore, J. Anat. Physiol. 7:310–311, 1873.

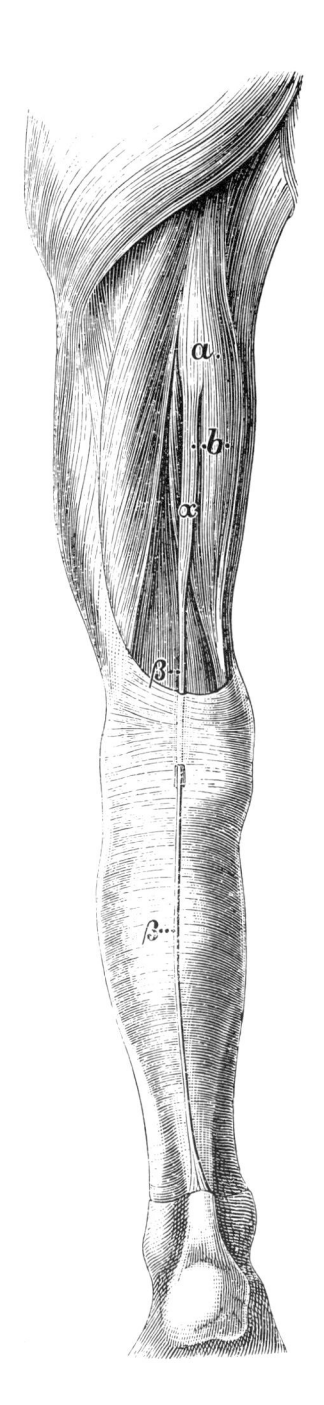

Muscles

◀ FIG. 83.
M. Tensor Fasciae Suralis.

a,b, M. semitendinosus; α, m. tensor fasciae suralis; β, tendon of tensor.

From Gruber, W. Sur une variante du muscle tenseur de l'aponeurose surale, partant du muscle demi-tendinex. *Bull. l'Acad. Imp. Sci. St. Petersbourg* 18:184–186, 1873.

FIG. 84.
M. Biceps Femoris Accessory Slip to M. Semitendinosus.

The muscle slip, 18 cm long and about 1 cm wide, extended from the midproximal part of biceps femoris (arrow) and terminated in the distal tendon of semitendinosus (curved arrow).

Dissecting room specimen, The University of Iowa College of Medicine.
▼

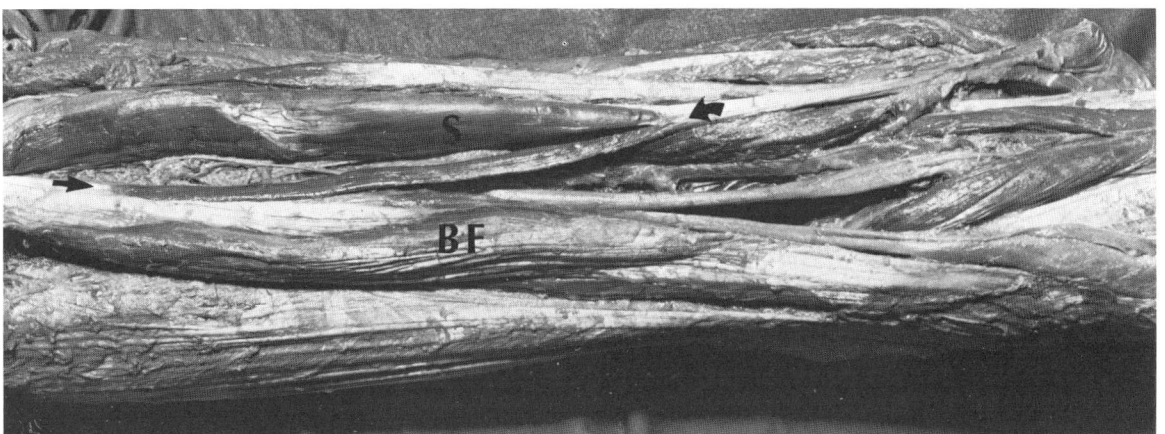

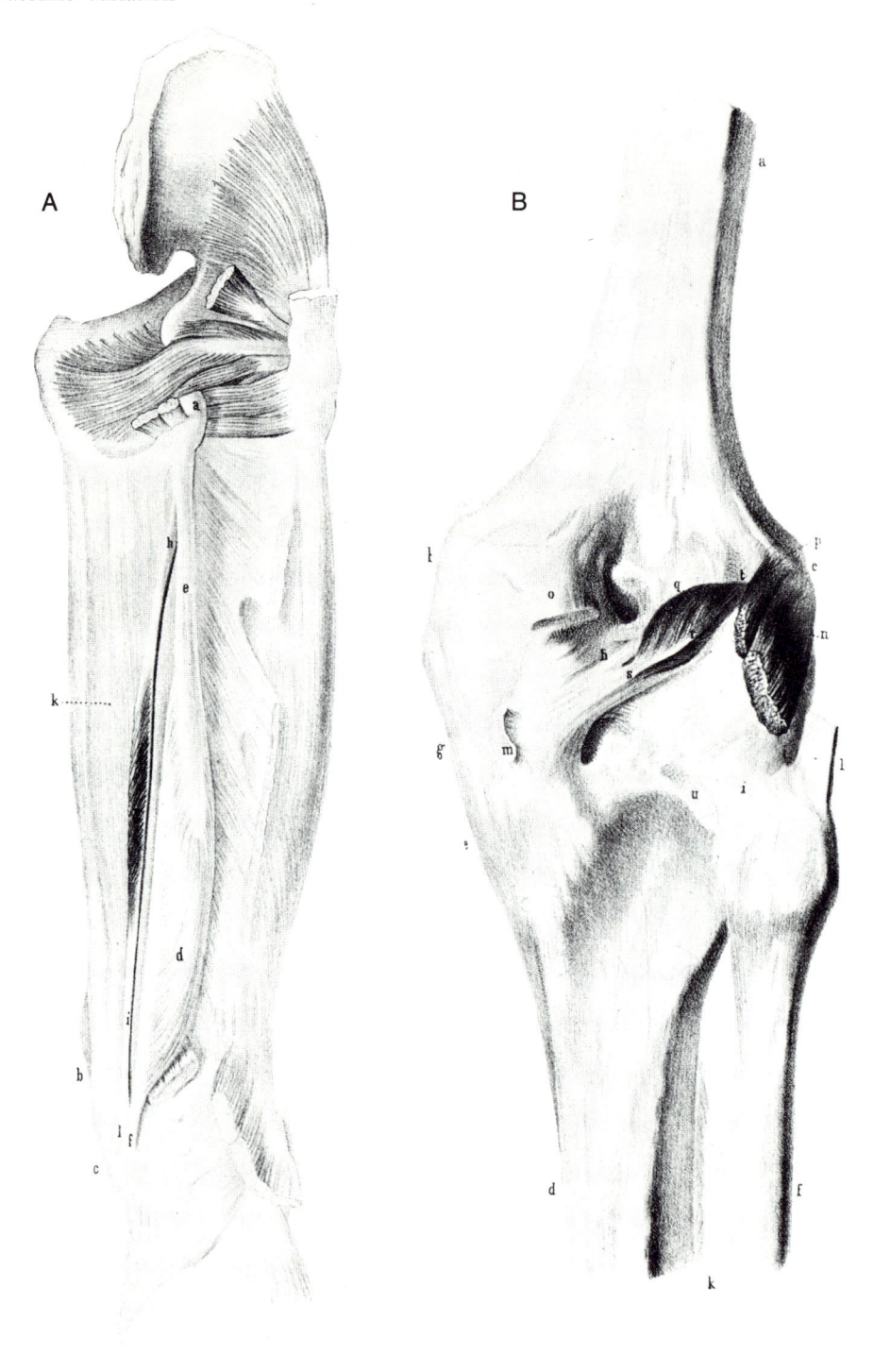

FIG. 85.
Supernumerary M. Semimembranosus.
Supernumerary M. Popliteus.

A: *a*, Ischial tuberosity; *b*, medial epicondyle of femur; *c*, medial tuberosity of tibia; *d*, belly of m. semimembranosus; *e*, its tendon of origin; *f*, its tendon of insertion; *g*, belly of supernumerary m. semimembranosus; *h*, its tendon of origin; *i*, its tendon of insertion; *k*, m. adductor magnus.

B: *a*, Femur; *b*, its medial epicondyle; *c*, its lateral epicondyle; *d*, tibia; *e*, medial tibial tuberosity; *f*, fibula; *k*, interosseus membrane; *l*, tendon of m. biceps femoris; *m*, tendon of insertion of m. semimembranosus; *n*, lateral head of m. gastrocnemius; *o*, origin of medial head of m. gastrocnemius; *p*, m. plantaris; *q*, supernumerary popliteus muscle; *r*, insertion of supernumerary m. popliteus; *s*, insertion of the same muscle into the medial tibial tuberosity; *t*, tendon of origin of supernumerary m. popliteus; *u*, tendon of origin of normal popliteus muscle.

From Calori, L. Di alcuni nuovi muscoli sopranumerarii degli arti. *Mem. R. Accad. Sci. Istituto di Bologna S. 2* 6:137–147, 1866.

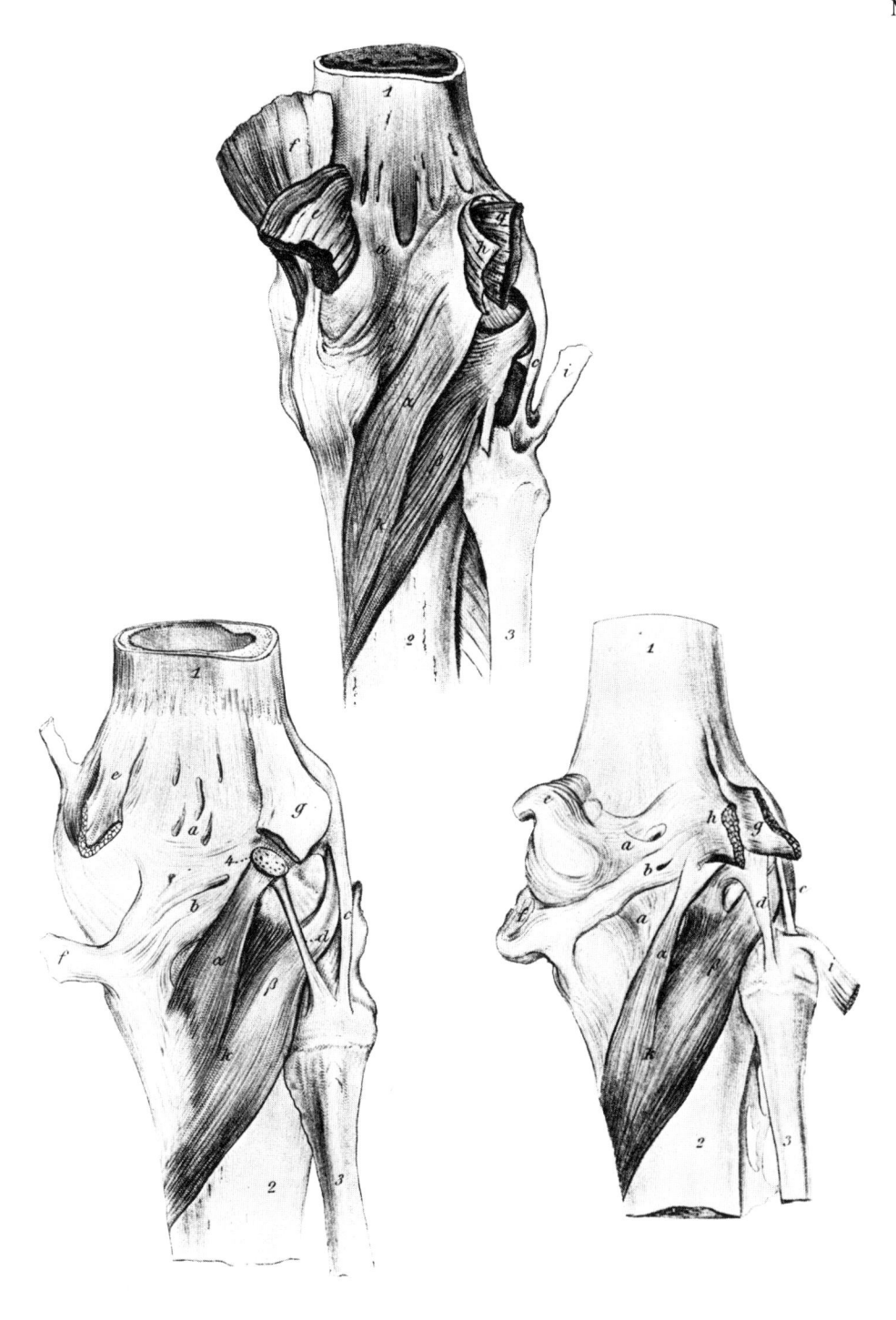

FIG. 86.
M. Popliteus Biceps.

1, Femur; *2*, tibia; *3*, fibula; *4*, ossiculum sesamoideum of m. gastrocnemius (lateral head); *a*, capsule of knee joint; *b*, ligamentum popliteum (oblique popliteal ligament); *c*, fibular collateral ligament (lateral); *d*, arcuate popliteal ligament; *e*, medial head, m. gastrocnemius; *f*, m. semimembranosus; *g*, lateral head, m. gastrocnemius; *h*, m. plantaris; *i*, m. biceps femoris; *k*, m. popliteus biceps: α, medial supernumerary head of m. popliteus biceps, and β, lateral normal (usual) head.

From Gruber, W. Über den Musculus popliteus biceps. *Arch. Anat. Physiol. Wissen. Med.* 1875, pp. 599–605.

Human Anatomic Variation

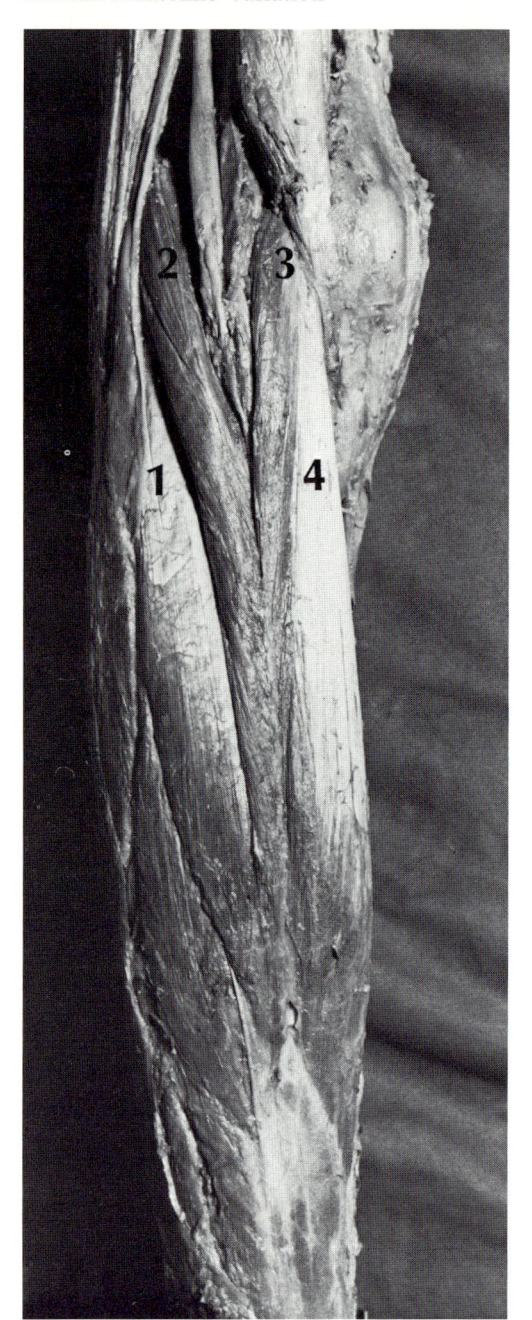

◀ FIG. 87.
Duplication of Medial and Lateral Heads of M. Gastrocnemius.

The four heads of gastrocnemius are numbered from left to right as follows:
1 and 2: Lateral heads
3 and 4: Medial heads

Dissecting room specimen, The University of Iowa College of Medicine.

FIG. 88.
Soleus Accessorius.

The accessory soleus muscle may arise from the soleal line with the soleus and may be detached from the muscle distally. It lies between the soleus and tendo Achillis (externally) and the flexor hallucis longus (internally). The muscle may end on tendo Achillis or it may have a separate attachment on the medial surface of the calcaneus or tibial collateral ligament of the ankle.

Authors' note: Additional references: Bardeen, C.R. Development and variation of nerves and musculature of the inferior extremity and neighboring regions of the trunk in man. *Am. J. Anat.* 6:259–390, 1906–1907; Ternowsky, W.N. and M. Sadikowa. Ein akzessorischer M. soleus. *Anat. Anz.* 61:280–284, 1926; Dunn, A.W. Anomalous muscles simulating soft-tissue tumors in the lower extremities. *J. Bone Joint Surg.* 47A:1397–1400, 1965; Gordon, S.L. and D.W. Matheson. The accessory soleus. *Clin. Orthop. Rel. Res.* 97:129–132, 1973; Ger, E. and E. Sedline. The accessory soleus muscle. *Clin. Orthop. Rel. Res.* 116:200–202, 1976; and Dokter, G. and L.A. Linclau. The accessory soleus muscle: Symptomatic soft-tissue tumour or accidental finding. *Netherland J. Surg.* 33:146–149, 1981.

▼

Extensoris hallucis longus — Tibialis anterior
Extensor digitorum longus — Great Saphenous Vein
— Tibia
Fibula — Tibialis posterior
Peroneus longus — Flexoris digitorum longus
Peroneus brevis — Flexor hallucis longus
Small Saphenous Vein — Tibial Nerve
— Plantaris tendon
Soleus Accessorius — Achilles tendon

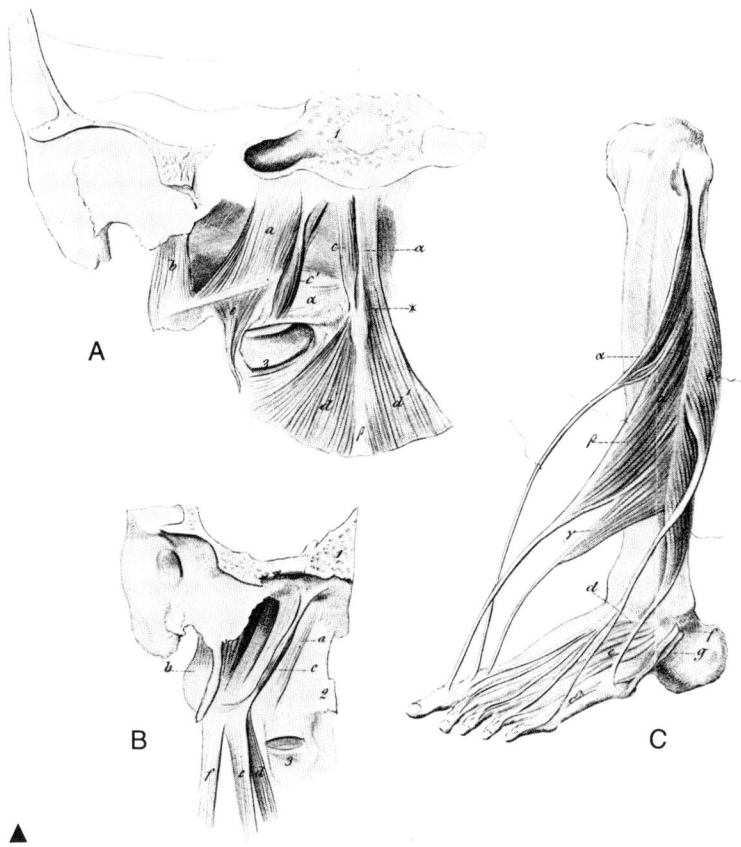

Muscles

FIG. 89.
M. Rectus Capitis Anticus (Anterior) Medius s. Minimus.
M. Extensor Hallucis Longus Tricaudatus.

A: *1*, Base of skull; *3*, epistropheus (axis); α, atlanto-occipital membrane (anterior); β, anterior longitudinal ligament; *a*, m. rectus capitis anterior; *b*, m. rectus capitis lateralis; *c*, m. rectus capitis anterior medius; *d,d'*, m. longus colli (on left side extending to base of skull and inserting onto the anterior atlanto-occipital ligament); *e*, m. anterior intertransversarius; *, anterior tubercle of atlas.

B: *1*, *3*, *a*, *b*, Same as in drawing A; *2.* atlas; *c*, anomalous point of origin of m. rectus capitis, anterior muscle from the transverse process of atlas; *d*, m. anterior intertransversarius; *e*, m. scalenus medius; *f*, m. levator scapulae.

C: *a*, M. extensor hallucis longus tricaudatus: α, superior belly; β, middle belly; and γ, inferior belly; *b*, m. extensor digitorum longus pedis; *c*, m. peroneus tertius; *d*, m. extensor hallucis brevis; *e*, m. extensor digitorum brevis pedis; *f*, tendon of m. peroneus brevis; *g*, tendon of m. peroneus longus; δ, tendon of m. peroneus brevis to the fifth toe.

From Gruber, W. Über den Musculus rectus capitis anticus medius s. minimus. *Arch. Anat. Physiol. Wissen. Med.* 1876, pp. 746–749. Ein neuer Fall von Musculus extensor hallucis longus tricaudatus. *Ibid.* 1876, pp. 750–752.

FIG. 90.
Four Peronei Muscles in a Leg.

The author illustrates two supernumerary peronei in one leg for the first time. One of the supernumerary muscles arises between the two common peronei and inserts onto the peroneal trochlea of the os calcaneus. The second muscle arises beneath the peroneus brevis muscle and its tendon inserts into the dorsal digital expansion of the fifth toe.

Illustration used by permission from Reimann, R. 4 Musculi peronei in einem menschlichen Unterschenkel. *Anat. Anz.* 145:205–207, 1979.

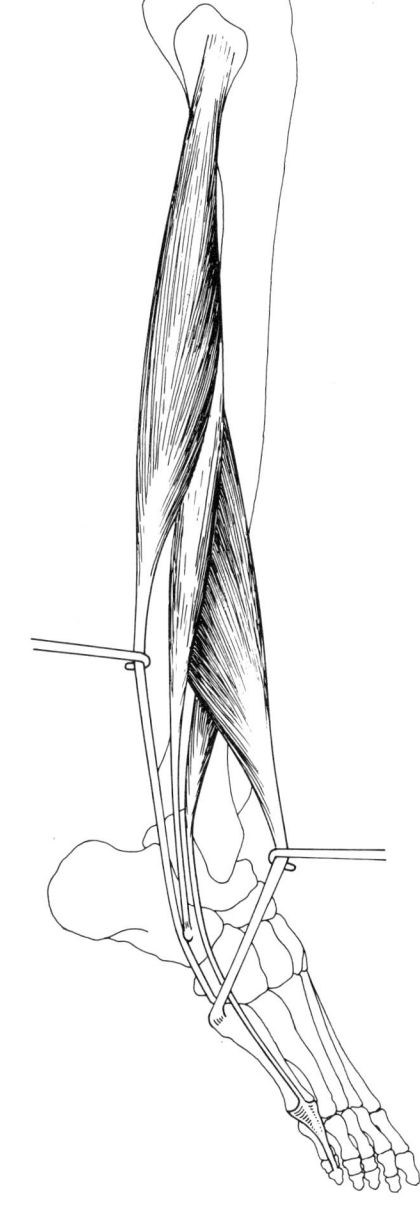

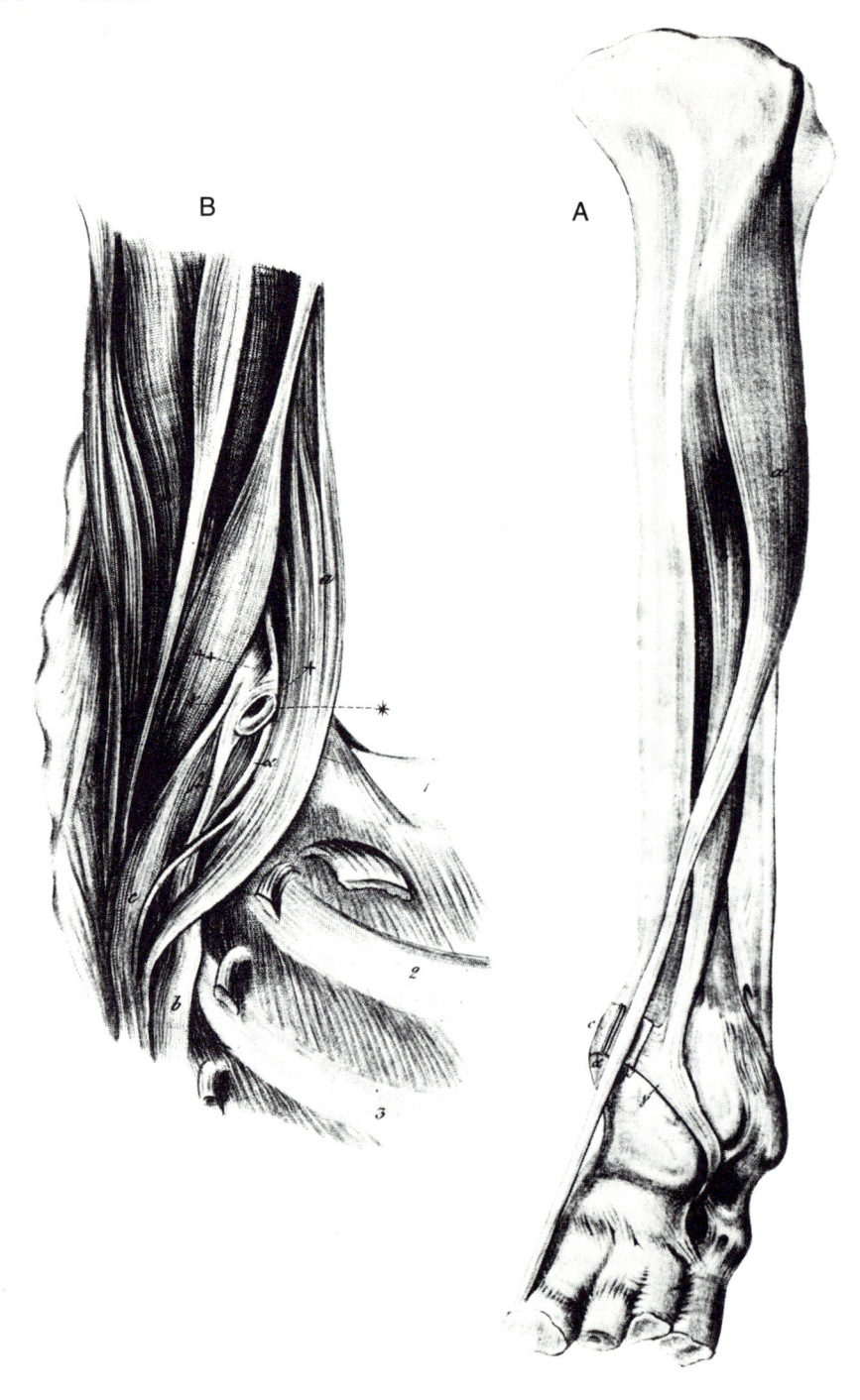

FIG. 91.
M. Tibioastragalus Anticus.
M. Accessorius Iliocostocervicalis.

A: *a*, M. tibialis anterior; *b*, m. tibioastragalus; *c*, inferior extensor retinaculum: α, superficial layer of sheath for tendon of m. tibialis anterior; β, deep layer of sheath for tendon of m. tibialis anterior; and γ, deepest layer of the retinaculum for the insertion of the tendon of m. tibioastragalus anticus.
B: *1*, First rib; *2*, second rib; *3*, third rib; *a*, m. iliocostalis cervicis; *b*, m. accessorius ad iliocostalem (cervicis): α, tendon to the transverse process of 7th cervical vertebra, and β, tendon to the tubercle of the 1st rib; *c*, *m*. longissimus thoracis: γ, tendon to the transverse process of 1st thoracic vertebra; +, tubercle of 1st rib, ++, tubercle of transverse process of 1st thoracic vertebra; *, bursa under the tendon of m. accessorius iliocostocervicalis.

From Gruber, W. Über einen Musculus tibio-astragaleus anticus des Menschen. *Arch. Anat. Physiol. Wissen. Med.* 1871, pp. 663–668. Über den Musculus und über die neue Bursa mucosa ilio-costocervicalis am Tuberculum der ersten Rippe, und über einige accidentelle Bursae mucosae am Rücken. *Ibid.* 1871, pp. 669–693.

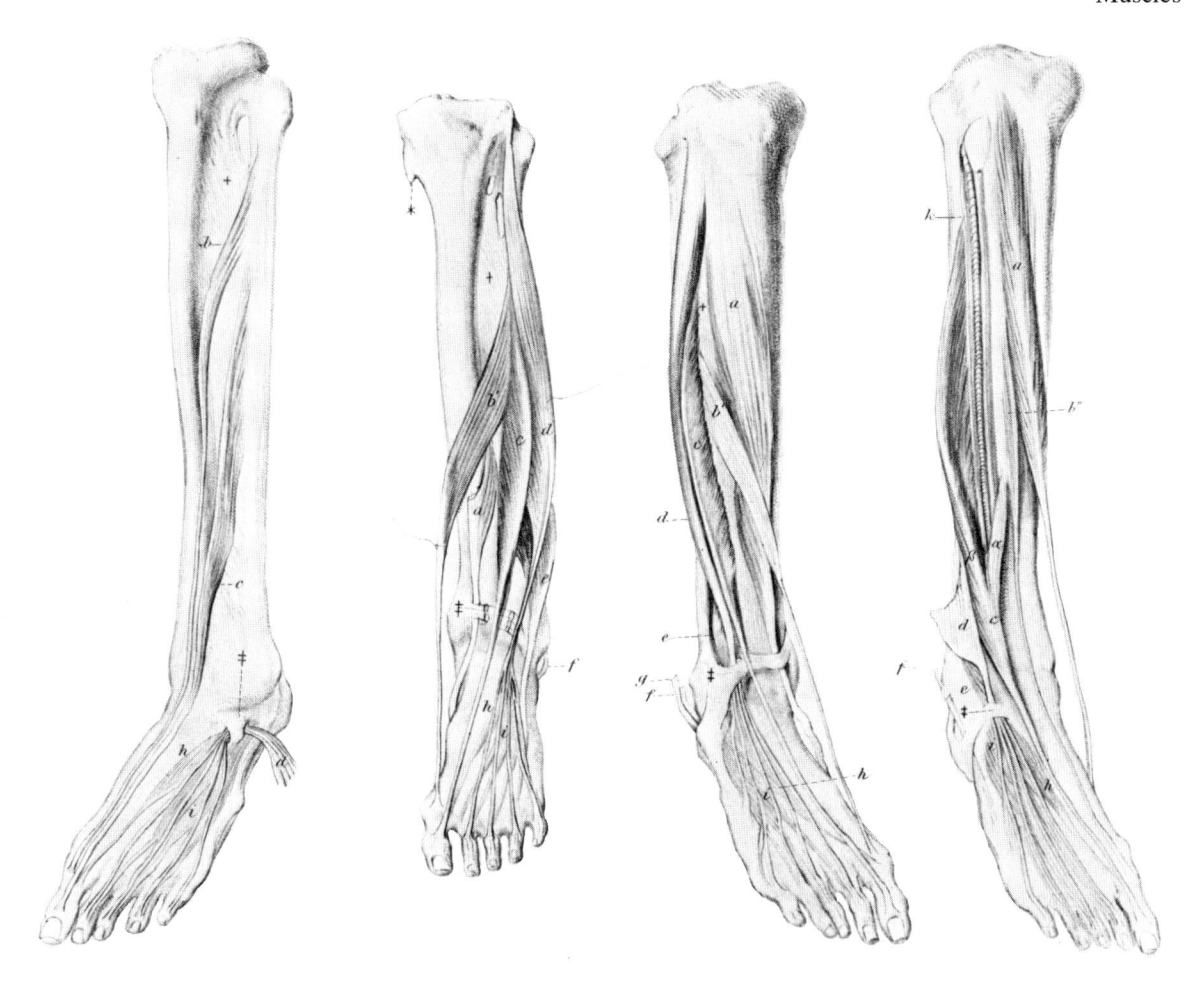

Muscles

FIG. 92.
M. Extensor Hallucis Longus Minor Fibularis Anterior.
M. Extensor Hallucis Longus Minor Fibularis Medialis.
M. Extensor Hallucis Minor Tibialis.
M. Extensor Hallucis Major.
M. Extensor Hallucis Major Biceps.

a, M. tibialis anterior; *b*, m. extensor hallucis longus minor fibularis anterior; *b'*, m. extensor hallucis longus minor fibularis medialis; *b"*, m. extensor hallucis longus minor tibialis; *c*, m. extensor hallucis longus major; *c'*, m. extensor hallucis longus major biceps: α, tibial head, and β, fibular head; *d*, m. extensor digitorum longus; *e*, m. peroneus tertius; *f*, m. peroneus brevis (tendon); *g*, m. peroneus longus (tendon), *h*, m. extensor hallucis brevis; *i*, m. extensor digitorum brevis; *k*, tibialis anterior artery and deep peroneal nerve; +, interosseous membrane; ‡, inferior extensor retinaculum (partial); *, trochlear process of tibia.

From Gruber, W. Über die Varietäten des Musculus extensor hallucis longus. *Arch. Anat. Physiol. Wissen. Med.* 1875, pp. 565–589.

315

Human Anatomic Variation

FIG. 93.
M. Pisohamatus (Pisiuncinatus).
M. Extensor Hallucis Longus Bicaudatus.
M. Extensor Digitorum Longus (Anomalous).
(Bipartite Zygomatic Bone).

B: *a*, Tendon of m. flexor carpi ulnaris; *b*, m. abductor digiti minimi; *c*, m. flexor digiti minimi; *d*, m. opponens digiti minimi; *e*, m. pisohamatus (pisiuncinatus); α, os pisiform; β, hamulus of hamate bone (hamulus ossis hamati); γ, ligamentum pisohamatum; *, triangular space of the hypothenar muscles.
C: *a*, M. tibialis anterior; *b*, m. extensor hallucis longus bicaudatus; *c*, m. extensor digitorum longus, anomalous, with 5 tendons (to all toes); *d*, m. peroneus tertius; *e*, m. peroneus longus; *f*, m. peroneus brevis; α, tendon of m. extensor hallucis longus to the proximal phalanx of the great toe; β, supernumerary tendon of m. extensor digitorum longus to the great toe.

From Gruber, W. Musculus piso-hamatus beim Menschen. *Arch. Anat. Physiol. Wissen. Med.* 1875, pp. 202–203. Über den Musculus extensor digitorum communis manus anomalous mit 5 Sehnen zu allen Fingern, und über den Musculus extensor digitorum longus pedis anomalous mit 5 Sehnen zu allen Zehen. *Ibid.* 1875; pp. 204–210.

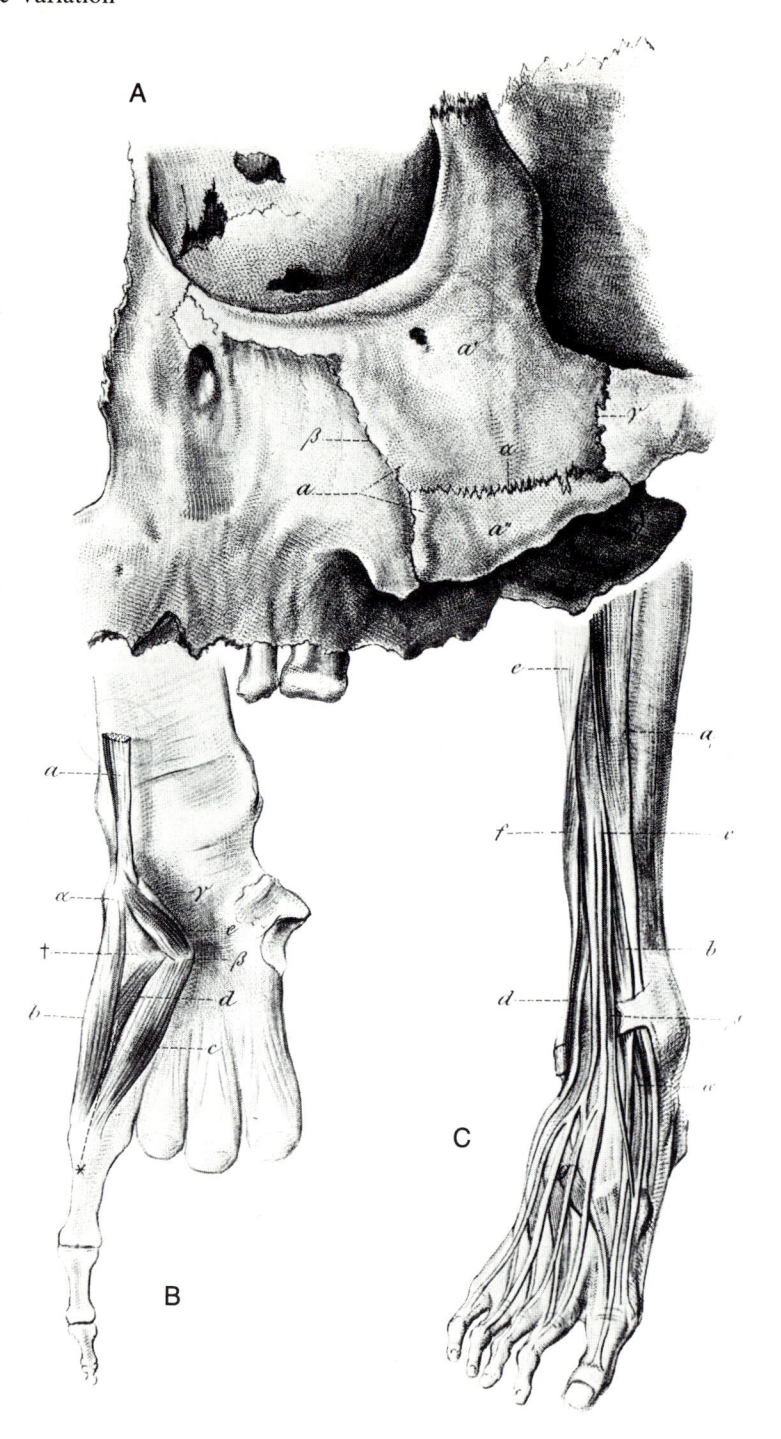

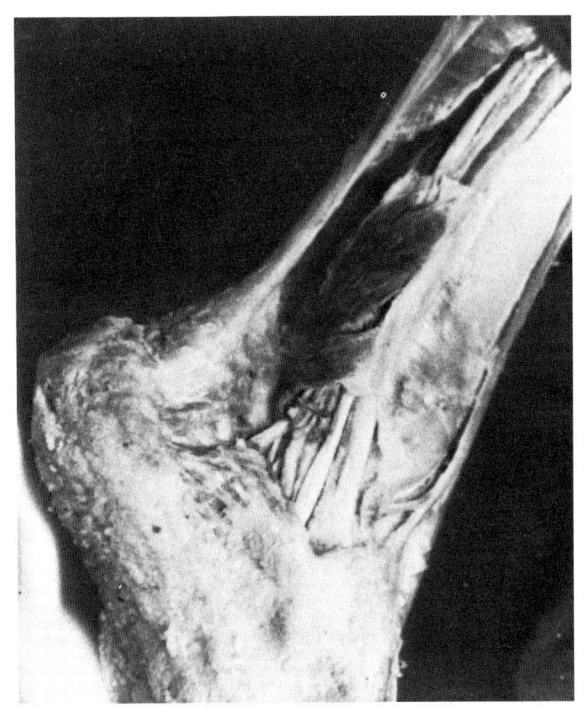

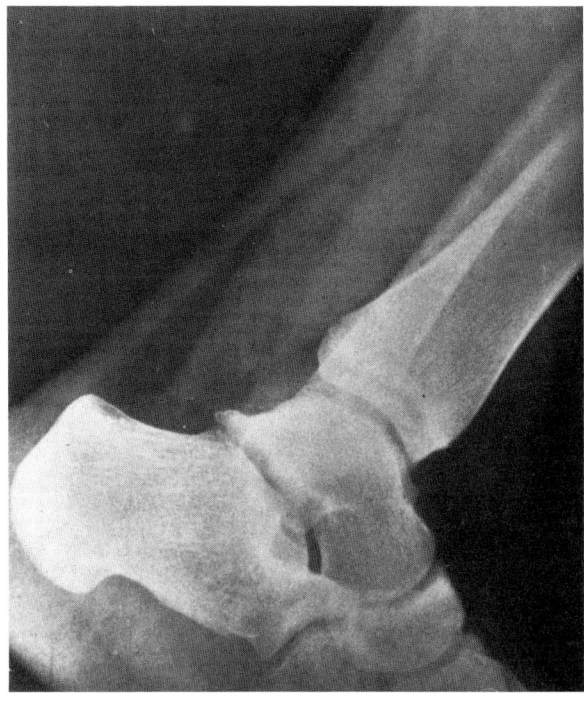

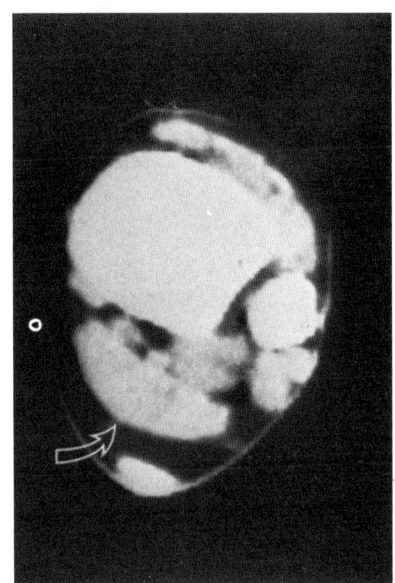

Muscles

FIG. 94.
M. Flexor Accessorius Longus or
M. Accessorius ad Quadratum Plantae

Left to Right: Dissection revealing the retrotibial accessory muscle. The Achilles tendon is seen to be normal, but a soft-tissue mass is visible in the pre-Achilles fat triangle. CT scan showing a well-defined soft-tissue mass behind the flexor tendons (arrow).

This muscle is found in about 1% of bodies. It is known by a variety of names including the following: m. accessorius ad quadratum plantae, m. accessorius ad flexorum accessorium, m. accessorius longus ad flexor digitorum longum, m. accessorius ad accessorium (Turner), m. peroneocalcaneus internus (Macalister), fibulocalcaneus medialis, and flexor accessorius digiti longus.

The authors have provided an elegant demonstration of this many-named but uncommon and very rarely portrayed muscle. In this case the muscle occupied the region of the pre-Achilles fat pad and covered the tibial neurovascular bundle. The muscle has been implicated in the tarsal tunnel syndrome (Bourrel, P. and M. Chickly. Anomalous muscles causing tunnel syndromes. *Anat. Clin.* 2:75–81, 1980).

Illustrations courtesy of Dr. A. Nidecker.
Nidecker, A.C., Hochstetter, A. von and H. Fredenhagen. Accessory muscles of the lower calf. *Radiology* 151:47–48, 1984.

Human Anatomic Variation

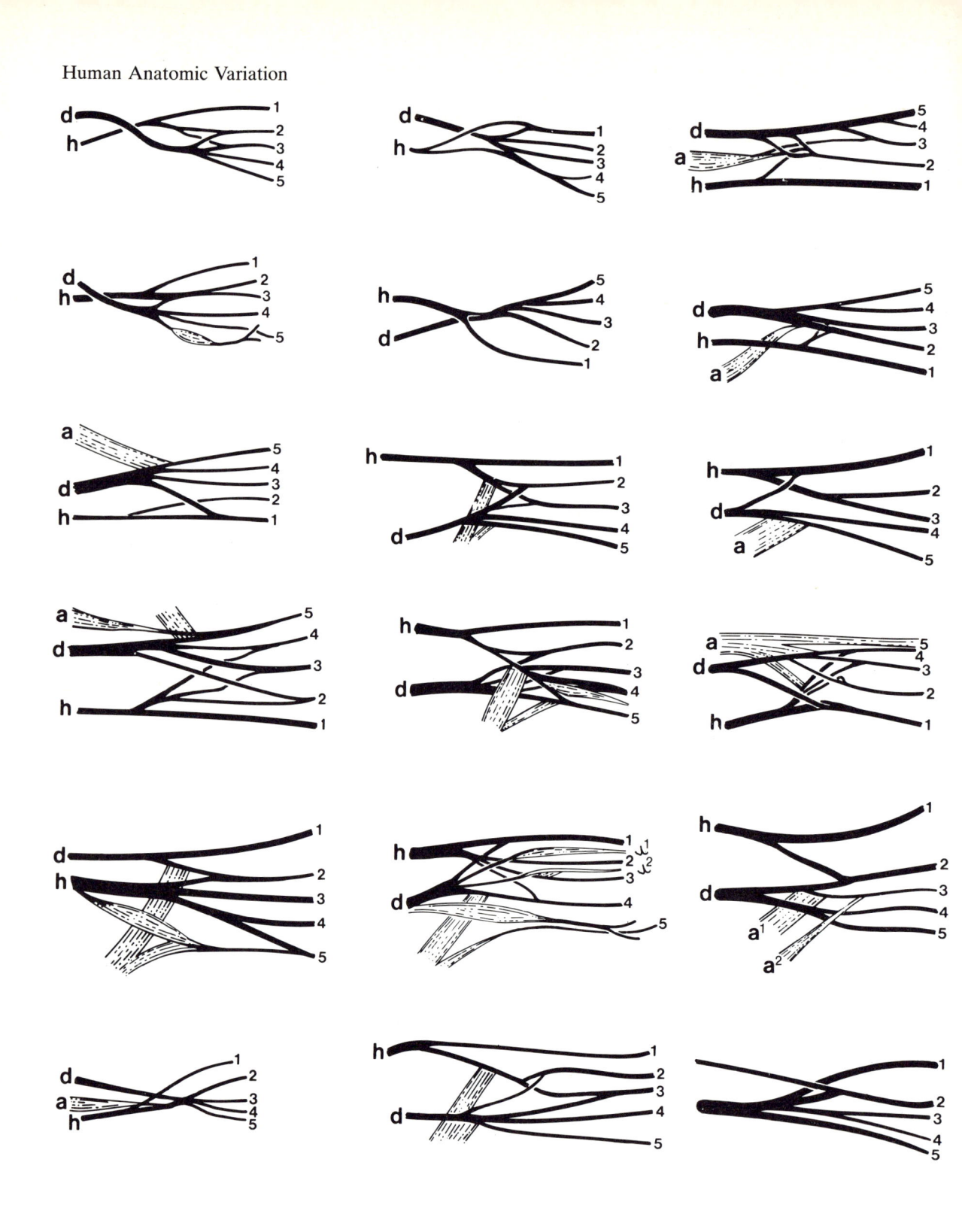

FIG. 95.
Variations in Long Flexor Tendons of Toes.
Various Forms of Union of Long Flexor Tendons of Toes.

d, M. flexor digitorum longus; *h*, m. flexor hallucis longus; *a*, m. accessorius (quadratus plantae); a^1, a^2, parts of the accessorius when slit; $1^1, 1^2$, mm. lumbricals; *1,2,3,4,5*, tendons for the supply of each of the respective toes.

Note: These sketches were copied by Macalister from his note book, but unfortunately they were not all in the same position; in some the under surface is shown, and in others the upper.

Redrawn from Macalister, A. Additional observations on muscular anomalies in human anatomy (third series), with a catalogue of the principle muscular variations hitherto published. *Trans. R. Irish Acad. Sci.* 25:1–130, 1875.

Cardiovascular System

FIG. 1.
Coronary Artery Arising from Pulmonary Artery.

A: Branch (*1*) from right coronary artery, about ⅓ inch from its origin and passing behind aorta and pulmonary artery (indicated by dotted line); reappears at (*2*), where it gives off three branches (*3*), and anastomoses with an anomalous coronary branch (*4*), which arises from right anterior sinus of Valsalva of the pulmonary artery. From this anastomosis two branches ascend in front of the bifurcation of the pulmonary artery and the transverse portion of arch of aorta and join anomalous branch (*5*) arising from subclavian artery near origin of vertebral.

B: The three branches (*3*) ascend, forming a cirsoid anastomosis between the pulmonary artery and the trachea, give off a branch (*6*) to right bronchus, and join anomalous branch (*7*) arising from aorta.

From Brooks, H.St.J. Two cases of an abnormal coronary artery of the heart arising from the pulmonary artery: With some remarks upon the effect of this anomaly in producing cirsoid dilatation of the vessels. *J. Anat. Physiol.* 20:26–29, 1886.

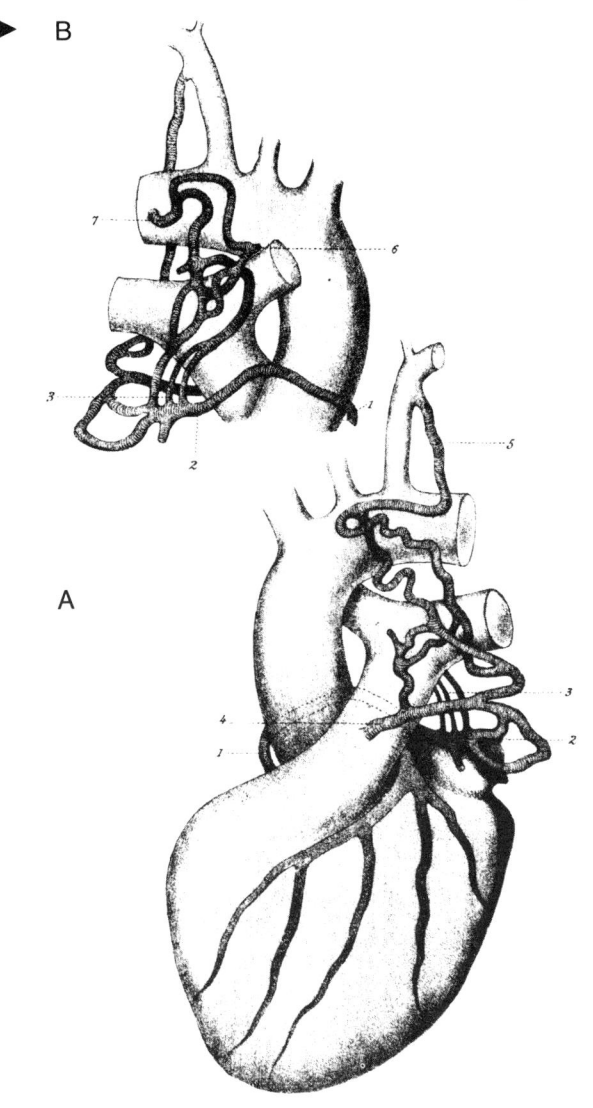

FIG. 2.
Anomalous Origin of Right Coronary Artery.

Anterior wall of the right ventricle was removed, and the interventricular septum (pars membranacea) and aorta (between the right semilunar and posterior valvulae as shown) were cut longitudinally.

In this subject, the ostium of right coronary artery had a very unusual origin high above the right sinus.

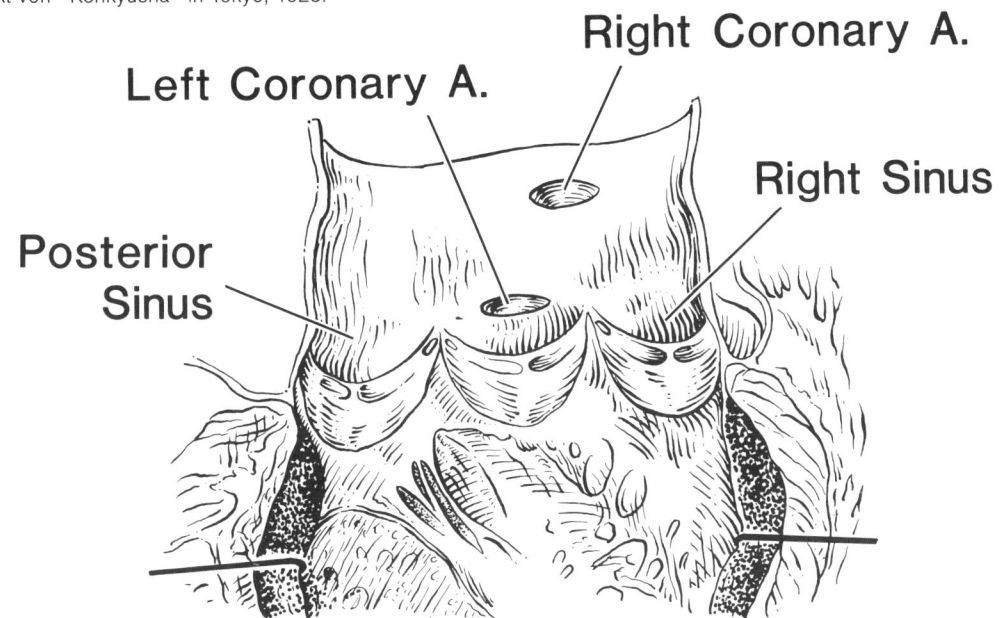

From Adachi, B. *Anatomie der Japaner. Das Arteriensystem der Japaner.* Verlag der Kaiserlich-Japanischen Universität zu Kyoto. In Kommission bei "Maruzen Co.", Kyoto und Tokyo. Gedruckt von "Kenkyusha" in Tokyo, 1928.

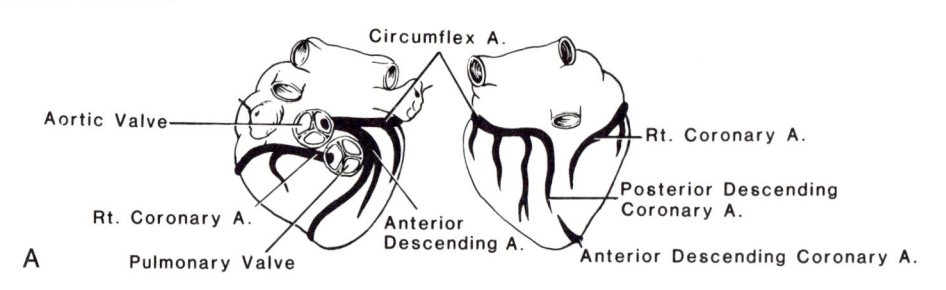

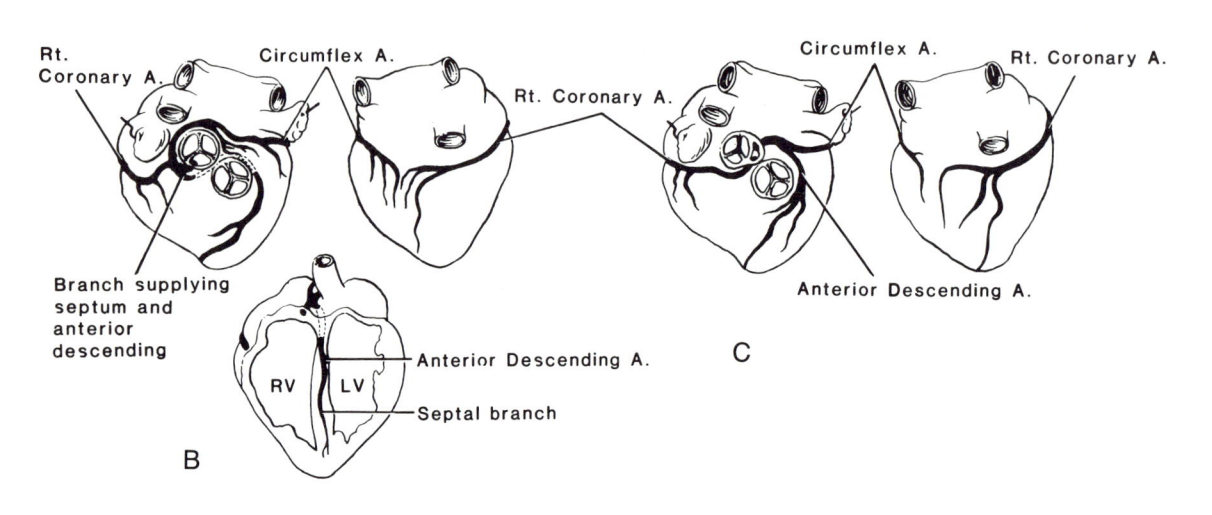

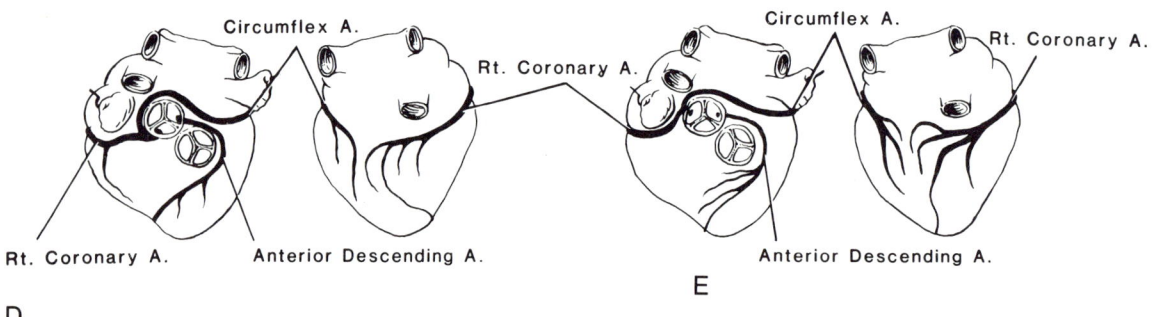

FIG. 3.
Anomalous Coronary Arteries.

A: The right coronary artery originated from the right sinus of the pulmonary trunk. In this case of a 74-year-old male, no disability was associated with the anomaly.
B: In this case, only one artery (the right coronary) arose from the aorta. The left circumflex and the anterior descending coronary arose from the right coronary sequentially. On the right, the posterior view of the exterior of the heart.
C: In this case, both coronary arteries arose from the left aortic sinus.
D: In this illustration of a case, the left circumflex coronary artery arose from the right coronary artery. The anterior descending coronary arose independently from the left aortic sinus.
E: Similar to D.

Redrawn from White, N.K. and J.E. Edwards. Anomalies of the coronary arteries: Report of four cases. *Arch. Pathol.* 45:766–771, 1948; and from Jordan, R.A., Dry, T.J. and J.E. Edwards. Anomalous origin of the right coronary artery from the pulmonary trunk. *Proc. Staff Meeting Mayo Clin.* 25:673–678, 1950.

Cardiovascular System

FIG. 4.
Coronary Artery Preponderance.

Three different coronary artery patterns are shown based on a study of 94 human hearts.
A: Balanced coronary artery distribution.
B: Left coronary artery preponderance.
C: Right coronary artery preponderance.

The heart is seen from the posterior aspect. The loop indicates the crux (the point where the posterior interventricular and atrioventricular sulci meet). In the study of 94 human hearts, the authors found left coronary artery preponderance to be more frequent in males (18.2%) than females (2.6%). Right coronary artery preponderance is more common in females (23.1%) than males (14.6%).

Redrawn from Ahmed, S.H., El-Rakhawy, M.T., Abdalla, A. and R.G. Harrison. A new conception of coronary preponderance. *Acta Anat.* 83:87–94, 1972.

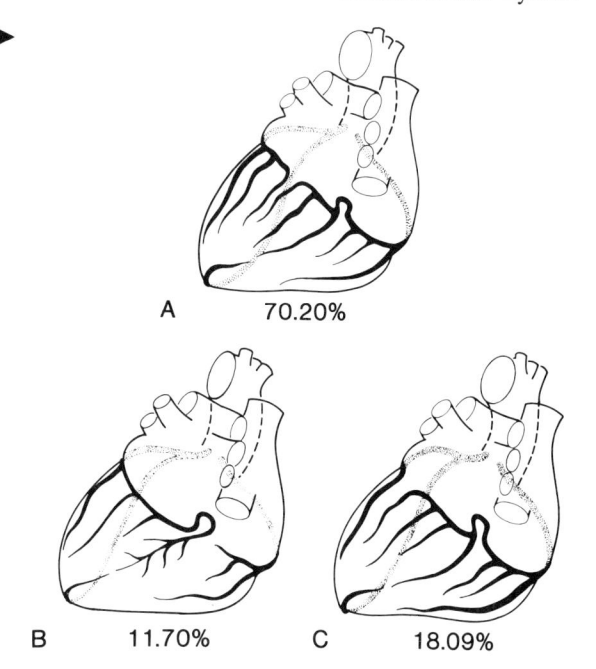

A 70.20%

B 11.70%   C 18.09%

FIG. 5.
Distribution of Coronary Arteries on Back of Ventricles

An analysis of the distribution of the coronary arteries on the back of the ventricles. (After Campbell.) *R* and *L*, right and left coronary arteries.

From Campbell, J. The coronary arteries of the heart. Thesis. Queen's University, Belfast, 1928.

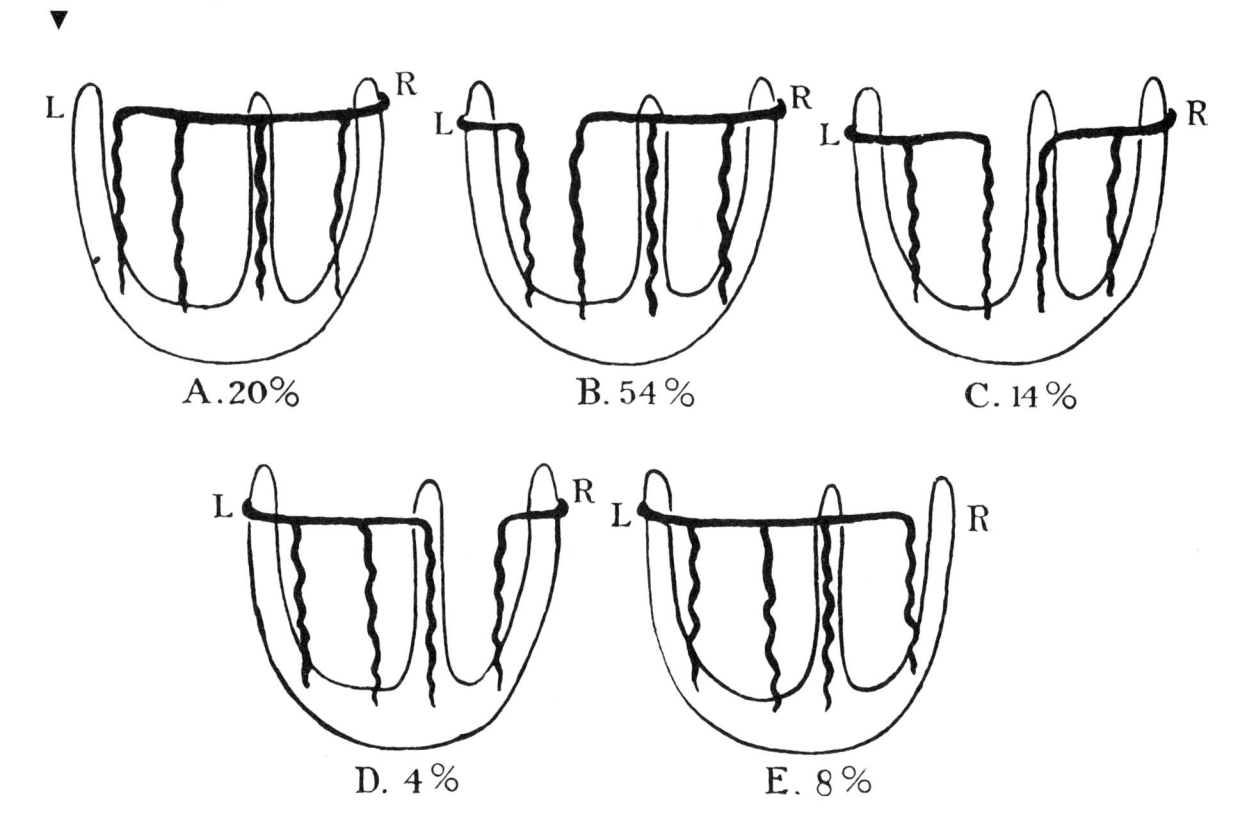

A. 20%   B. 54%   C. 14%

D. 4%   E. 8%

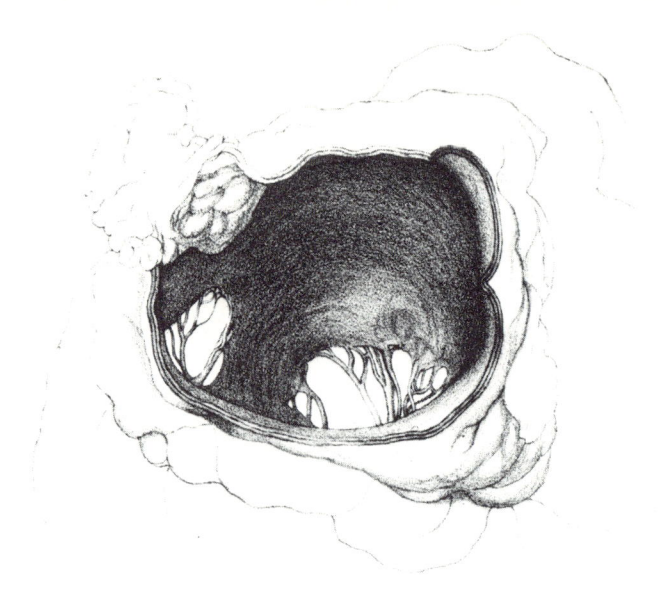

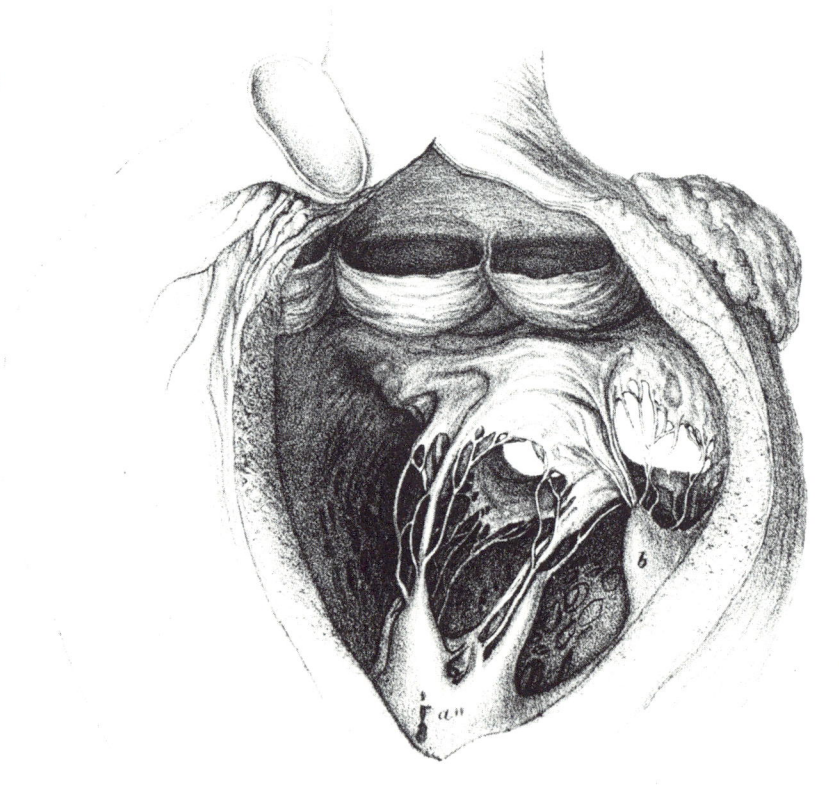

Human Anatomic Variation

FIG. 6.
Double Mitral Valve.
A: Seen from the auricle.
B: "This shows the cavity of the left ventricle laid open in front by an incision near the septum. The larger valve is seen receiving chordae tendineae from posterior musculi papillaris (*a*), the smaller from anterior (*b*). (The smaller orifice appears much larger than in the fresh condition, the valve having been necessarily stretched in the process of preserving the specimen.)"

From Greenfield, W.S. Double mitral valve. *Trans. Pathol. Soc. Lond.* 27:128–129, 1876.

Cardiovascular System

FIG. 7.
Anomalous Origin of Left Pulmonary Artery.
Retrotracheal Pulmonary Artery.

A: *ALPA*, Anomalous (retrotracheal) left pulmonary artery; *DA*, ductus arteriosus; *PT*, pulmonary trunk; *RPA*, right pulmonary artery.

B: Schematic diagram of the tracheobronchial tree and the pulmonary arterial system. In this case, there is no anomaly of tracheal bifurcation. The left pulmonary artery can be seen to arise from the right pulmonary artery. At the angle of the right main bronchus and the trachea, the anomalous left pulmonary curves posteriorly and toward the left side. In this position it can cause an obstruction of the airway.
DA, Ductus arteriosus; *EB*, eparterial bronchus; *HB*, hyparterial bronchus; *LB*, left bronchus; *LPA*, anomalous left pulmonary artery; *PT*, pulmonary trunk; *RPA*, right pulmonary artery; *T*, trachea.

C: In this case, the eparterial bronchus arises from the trachea. Below the anomalous eparterial bronchus, the trachea is hypoplastic. This hypoplastic trachea divides into the left main bronchus and the right hyparterial bronchus. The pulmonary artery does not give rise to the left pulmonary artery but, rather, continues to the right as the right pulmonary artery. Near the usual or normal origin of the left pulmonary artery is only the insertion of the ligamentum arteriosum. When the right pulmonary artery reaches the right side it gives rise to the anomalous left pulmonary artery and to branches which supply the right lung. The left pulmonary artery curves to the left, posteriorly around the hypoplastic trachea, causing compression of the airway between the eparterial bronchus and the bifurcation of the trachea. Labels and their designation are given above.

Redrawn from Jue, K.L., Raghib, G., Amplatz, K., Adams, P. Jr. and J.E. Edwards. Anomalous origin of the left pulmonary artery from the right pulmonary artery. Report of two cases and review of the literature. *Am. J. Roentgenol.* 95:598–610, 1965.

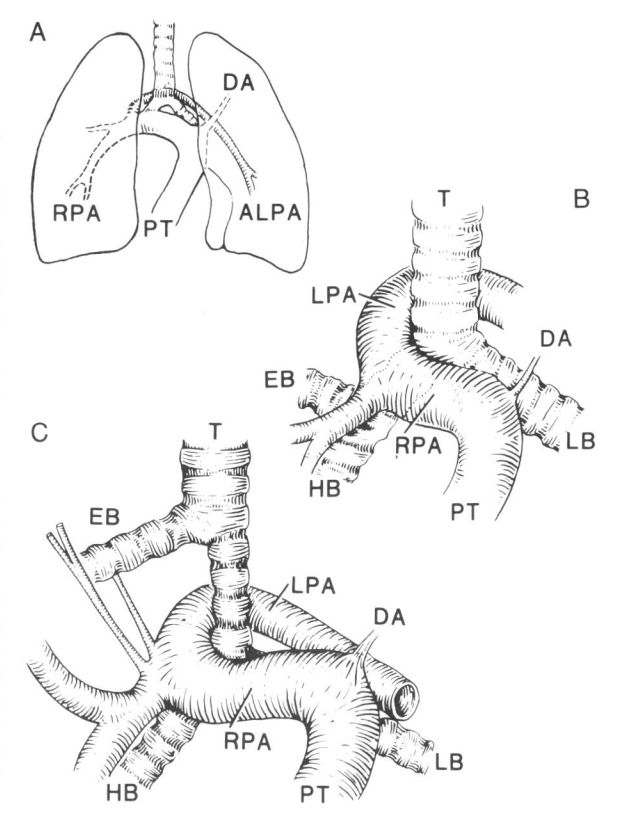

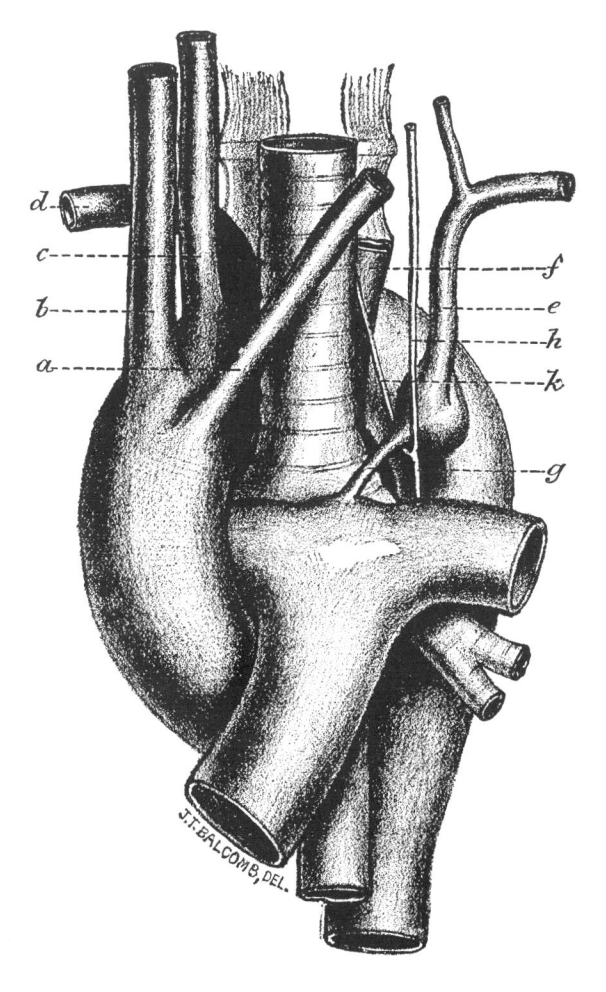

FIG. 8.
Specimen of Right Aortic Arch.

a, Left carotid; *b*, right carotid; *c*, right vertebral; *d*, right subclavian; *e*, left subclavian; *f*, esophagus; *g*, ductus arteriosus; *h*, vagus; *k*, recurrent laryngeal.

From Abbott, F.C. Specimen of right aortic arch. Proc. Anat. Soc. Great Britain and Ireland. *J. Anat. Physiol.* 26:581–582 (xiii–xiv), 1892.

Human Anatomic Variation

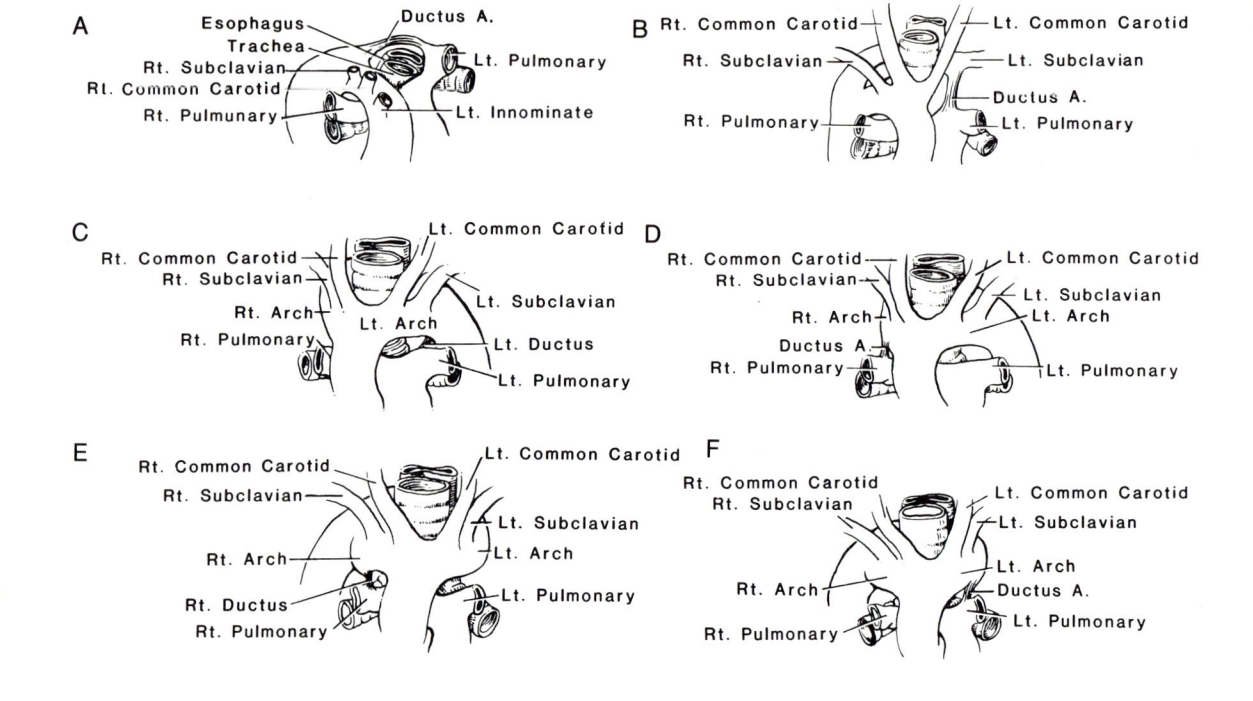

FIG. 9.
Vascular Rings Around Esophagus and Trachea.

A: Right-sided aortic arch and upper portion of the descending aorta, and left-sided ductus arteriosus in which the ductus arteriosus passed behind the esophagus to insert into an aortic diverticulum.
B: Right-sided aortic arch and upper portion of the descending aorta, and left-sided ductus arteriosus in which the ductus arteriosus inserted into a left subclavian artery which arose as the fourth branch from the aortic arch.
C: Left-sided descending aorta and left-sided ductus arteriosus.
D: Left-sided descending aorta and right-sided ductus arteriosus.
E: Right-sided upper portion of the descending aorta and right-sided ductus arteriosus.
F: Right-sided upper portion of the descending aorta and left-sided ductus arteriosus.

Kirklin, Clagett, and Edwards have provided the following classification of anomalies of aortic arch and a detailed discussion of these anomalies is given by Edwards, J.E. Anomalies of the derivatives of the aortic arch system. *Med. Clin. North Am.* 32:925–949, 1948.

Classification of Anomalies of Aortic Arch
I. Left-sided descending aorta.
   A. Functioning double aortic arch.
      1. Arches of approximately equal size.
      2. One arch narrowed (usually but not always the anterior or left one).
   B. Double aortic arch with partial atresia of one arch.
   C. Right-sided aortic arch with retroesophageal segment and left-sided descending aorta.
      1. Left subclavian artery originating from left-sided aortic diverticulum.
      2. Left subclavian artery originating from left innominate artery.
   D. Left-sided aortic arch.
      1. Right subclavian artery arising from the distal portion of the aortic arch from the descending aorta and passing to the right usually behind the esophagus.
      2. Normal arch and normal branches.
   E. Right-sided ductus arteriosus arising from right pulmonary artery. All possibilities occurring under A could occur here.
II. Right-sided upper portion of the descending aorta.
   A. Right-sided ductus arteriosus arising from right pulmonary artery.
      1. Functioning double aortic arch.
          a. Arches of approximately equal size.
          b. One arch narrowed.
      2. Double aortic arch with partial atresia of one arch.
      3. Left-sided aortic arch with retroesophageal segment and right-sided upper portion of the descending aorta.
          a. Right subclavian artery originating from right-sided aortic diverticulum.
          b. Right subclavian artery originating from innominate artery on right side.
      4. Right-sided aortic arch.
          a. Left subclavian artery arising from the distal portion of the aortic arch or from the descending aorta and passing to the left, usually behind the esophagus.
          b. Left subclavian artery originating from innominate artery on left side (mirror image of usual pattern).
   B. Left-sided ductus arteriosus arising from left pulmonary artery. All possibilities occurring under A could occur here.

Redrawn from Kirklin, J.W. and O.T. Clagett. Vascular "rings" producing respiratory obstruction in infants. *Proc. Staff Meeting Mayo Clin.* 25:360–367, 1950.

Cardiovascular System

FIG. 10.

Right Aortic Arch and Sinuous Aorta.

Diagram of right arch and its branching, showing tortuous condition of aorta, with various levels.

Authors' note: D4, D8, D12 s. T4, T8, T12 (thoracic vertebrae).

From Annan, J.L. Case of an abnormal sinuous aorta. *J. Anat. Physiol.* 44:241–243, 1910.

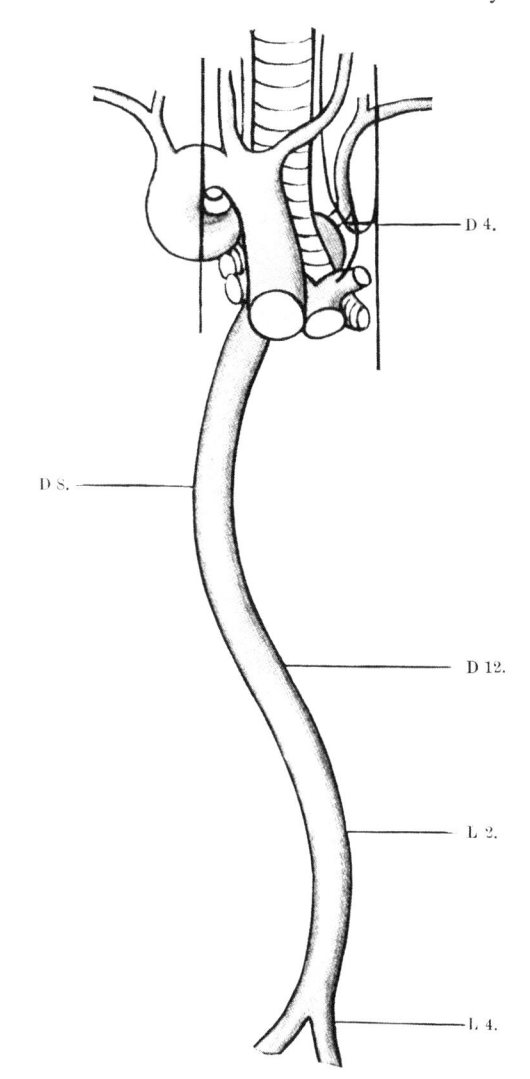

FIG. 11.

Double Aortic Arch.

r.c.a., *l.c.a.*, Right and left common carotid arteries; *r.v.a.*, *l.v.a.*, right and left vertebral arteries. The accessory vertebral artery of the left side is not figured. *r.s.a.*, *l.s.a.*, Right and left subclavian arteries; *r.i.v.*, *l.i.v.*, right and left innominate veins; *r.s.v.*, *l.s.v.*, right and left subclavian veins; *i.j.v.*, left internal jugular vein; *D.A.*, ductus arteriosus; *r.ph.n.*, right phrenic nerve; *r.p.n.*, *l.p.n.*, right and left pneumogastric nerves. The hook in the right-hand figure indicates the esophagus.

Authors' note: Pneumogastric nerve s. vagus nerve.

From Watson, M. Notes on a case of double aortic arch. *J. Anat. Physiol.* 11:229–234, 1877.

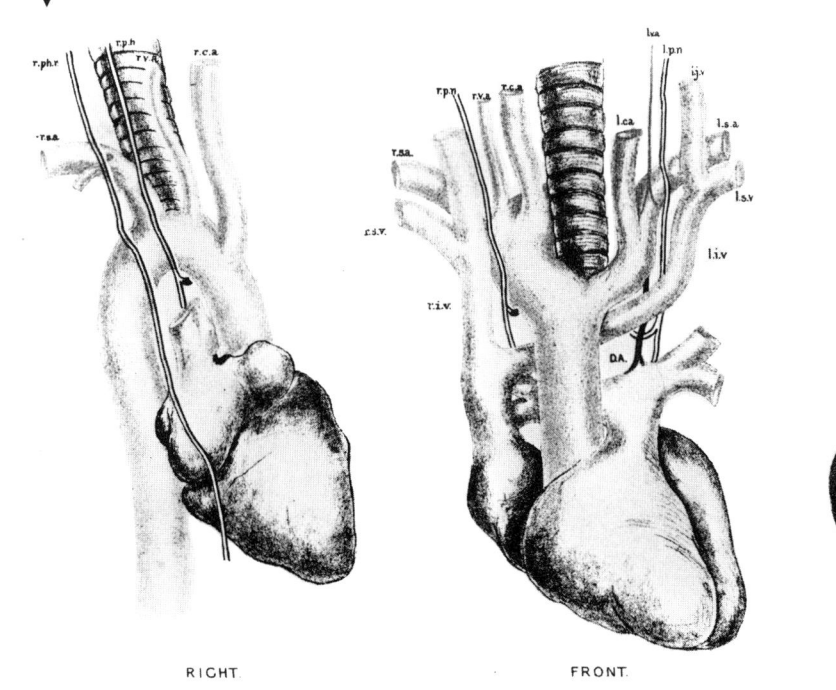

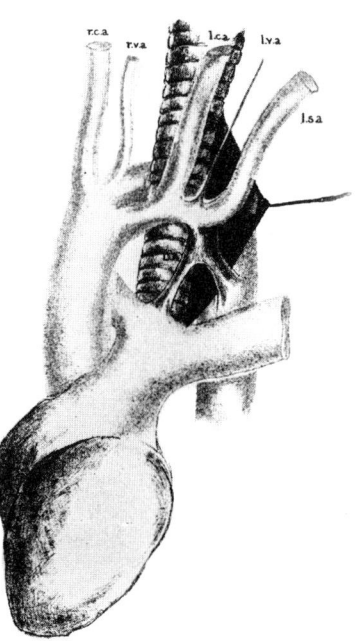

RIGHT.　　　　　FRONT.　　　　　LEFT.

Human Anatomic Variation

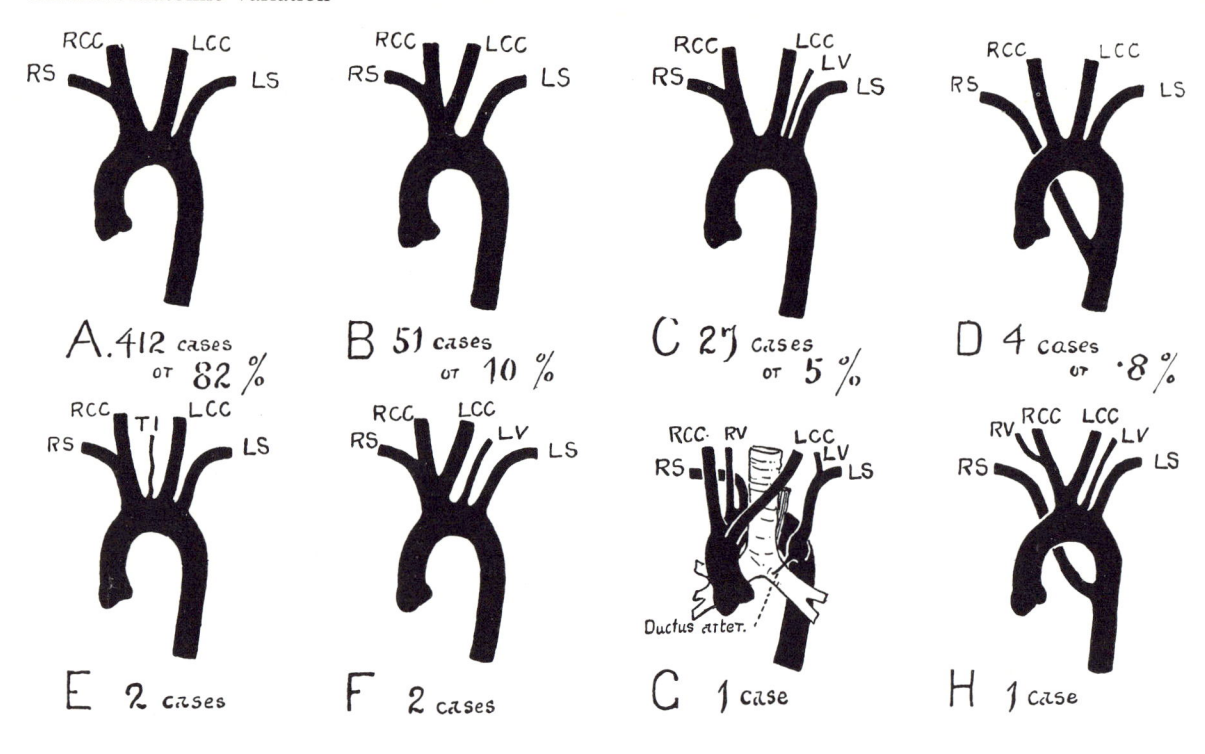

FIG. 12.
Nine Variations in Mode of Origin of Branches of Aortic Arch.

Variations in the mode of origin of the branches arising from the arch of the aorta. Total number examined, 500.
LCC, Left common carotid; *LS*, left subclavian; *LV*, left vertebral; *RCC*, right common carotid; *RS*, right subclavian; *RV*, right vertebral; *TI*, thyroidea ima.

From Thomson, A. Third annual report of the committee of collective investigation of the Anatomical Society of Great Britain and Ireland for the year 1891–92. *J. Anat. Physiol.* 27:183–194, 1893.

FIG. 13.
Aortic Arch Anomalies.

A: This figure, representing a case of true double aortic arch, was taken from Tiedemann (1822), Fig. 7, Plate IV, representing a case described by Malacarne (1788), Pt. 2, p. 119; it was copied by Quain (1844) as Fig. 8, Plate V, and by Krause (1876) as Fig. 108a.

B: This figure, representing a case in which the arch alone is double, was taken from Tiedemann (1822), Fig. 6, Plate IV, representing a case described by Hommel (1737). It is referred to in Haller Elementary Physiology, Tom. 2, p. 162, and copied by Quain (1844) as Fig. 7, Plate V, and by Krause (1876) as Fig. 108b.

C: This figure, representing a case of right aortic arch with the left subclavian artery as the last branch and the ductus arteriosus patent, was taken from Quain (1844), Fig. 2, Plate VII.

D: This figure, representing a case of right aortic arch with a left innominate trunk, was taken from Tiedemann (1822), Fig. 9, Plate IV.

E: This figure, representing a case of persistence of both the right and left ductus arteriosus, was taken from Breschet (1826), Fig. 9, Plate I.

F: This figure, representing the more or less common anomaly of left carotid artery springing from the innominate trunk, was copied from Tiedemann (1822), Fig. 5, Plate II.

G: This figure, representing a case of right aortic arch, with the left subclavian artery as the last branch and the right vertebral springing from the arch, was taken from Abbott (1892), Fig. I.

H: This figure, representing a case of atresia of the pulmonary artery, was taken from Keith (1909), Fig. 4; the pulmonary artery leaves the heart as a fibrous cord but rapidly enlarges to about normal, the ductus arteriosus is patent.

I: This figure, representing a case of thyroidea ima springing from the arch, was taken from Neubauer (1786), Fig. 2, Plate VII.

J: This figure, representing a case of bi-innominate trunks, was taken from Tiedemann (1822), Fig. 4, Plate II, and copied by Quain (1844), Fig. 9, Plate VI.

K: This figure was drawn from a specimen of left common carotid springing from the innominate, which is in the Warren Museum (Harvard University).

L: This figure, representing a single branch from the aortic arch, was taken from Tiedemann (1822), Fig. 3, Plate II, illustrating the case of Klinz (1793), p. 273. It was copied by Quain (1844), Fig. 6, Plate V.

Cardiovascular System

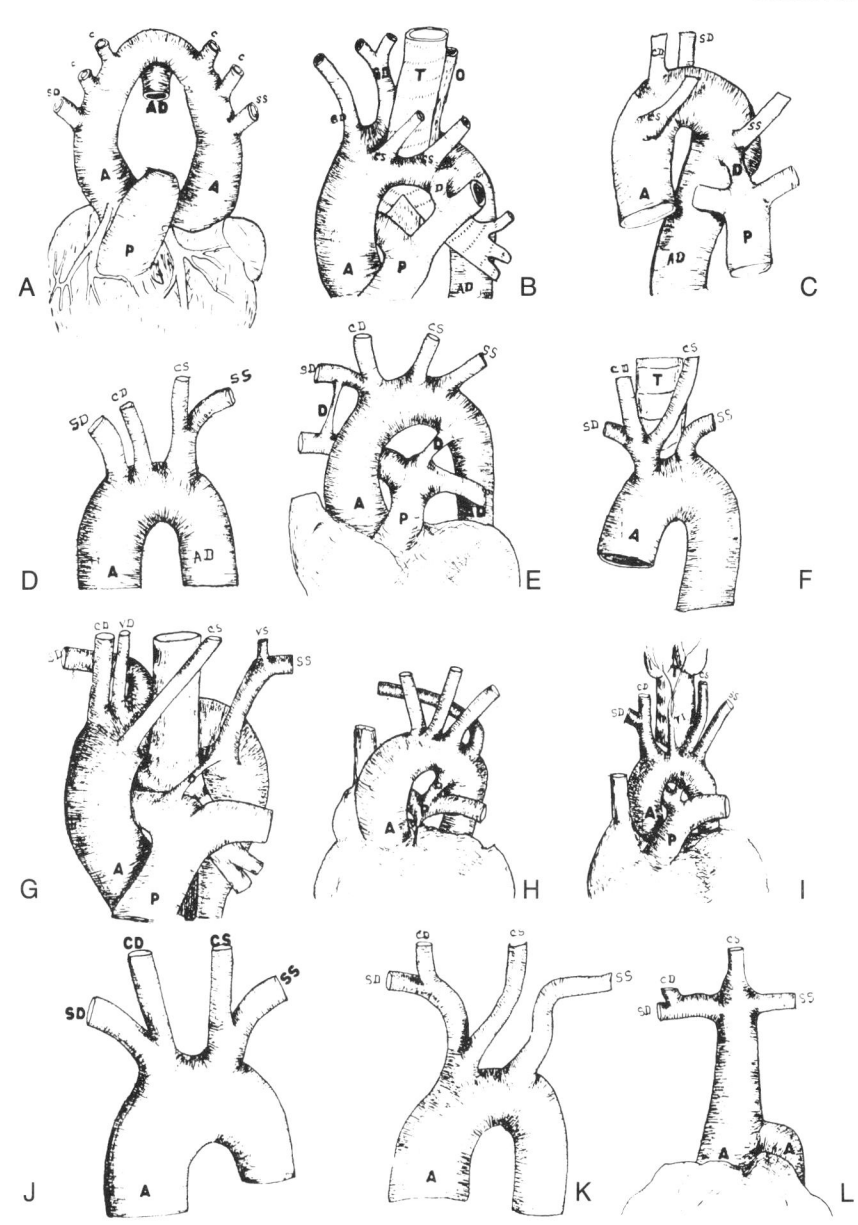

Abbreviations used in the figures: *A*, aorta; *AD*, aorta descendens (descending aorta); *C*, a. carotis (carotid); *CD*, a. carotis communis dextra (right common carotid); *CS*, a. carotis communis sinistra (left common carotid); *D*, ductus arteriosus (Botalli); *EC*, a. carotis externa (external carotid); *IC*, a. carotis interna (internal carotid); *O*, esophagus; *P*, a. pulmonalis (pulmonary); *SD*, a. subclavia dextra (right subclavian); *SS*, a. subclavia sinistra (left subclavian); *T*, trachea; *TD*, ductus thoracicus (thoracic duct); *VD*, a. vertebralis dextra (right vertebral); *VS*, a. vertebralis sinistra (left vertebral).

From Poynter, C.W.M. Arterial anomalies pertaining to the aortic arches and the branches arising from them. *The University Studies of the University of Nebraska* 16:229–345, 1916, Lincoln, Nebr.

Additional references from Poynter: Abbott, F.C. Specimen of right aortic arch. *J. Anat. Physiol.* 26:13, 1892.

Breschet, G. Memorie sur l'ectopie de l'appareils de la circulation et particulierement sur celle du coeur. *Repert. Genec. d'anat. et d. physiol. path.*, Paris. 2:14, 1826.

Haller, –. *Elementa physiologiae corporis humani.* Tom. 2. p. 162. Lausanne. 1760.

Hommel, –. *Commercium litterarium.* Norimbergae, p. 162. 1737.

Keith, A. The Hunterian Lectures on malformations of the heart. *Lancet* p. 20, 1909.

Klinz, –. Abhandl. d. med. chir. Josephs. Akadi. Wein. 1:273, 1787.

Krause, W. Varietäten des Aortensystems. In J. Henle. *Handb. der syst. Anat., Bd. 3, S. 1, Abt.: Gefässlehre.* Braunschweig 1876.

Malacarne, V. *Osservazioni sopra alcune arterie del corpo umano nello stato preternaturale.* Oserv. in Chir. p. 119. Torino. 1784.

Neubauer, J.E. *Descripto anatomica arteriae innominatae.* Jenae. 1722.

Quain, R. *The anatomy of the arteries of the human body with its application to pathology and operative surgery; in a series of plates and commentaries.* London, 1844.

Tiedemann, F. *Tabulae arteriarum corporis humani fabrica.* Karlsruhae. 1822.

Human Anatomic Variation

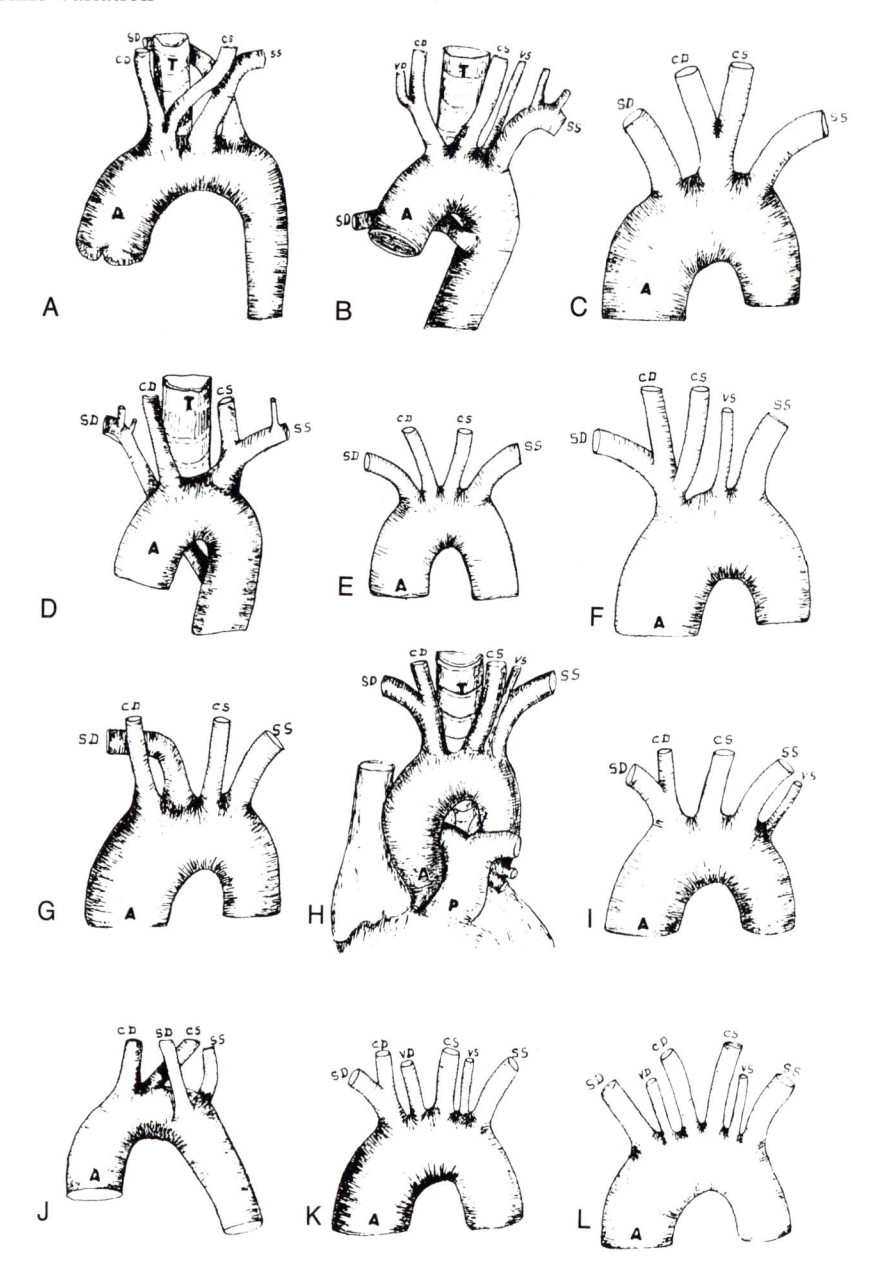

FIG. 14.

Aortic Arch Anomalies (continued).

A: This figure was drawn from a specimen of low origin of the right subclavian with both carotids springing from a common stem, which is in the Warren Museum (Harvard University).

B: This figure, taken from Krause (1876), Fig. 115 (Macartney), and Tiedemann, 1846, Fig. 6, Plate XXXIX, represents a case of low origin of the right subclavian artery with the right vertebral springing from the right common carotid and left vertebral from the aortic arch.

C: This figure, representing a case of common trunk for the carotids, was taken from Tiedemann (1822), Fig. 2, Plate III; copied by Quain (1844) as Fig. 4, Plate VII.

D: This figure, representing a case of low right subclavian artery, with a common stem for the left carotid and subclavian, was taken from Tiedemann (1822), Fig. 6, Plate II; copied by Quain (1844) as Fig. 8, Plate VII.

E: This figure, representing a case in which all four branches spring separately from the arch, was taken from Tiedemann (1822), Fig. 3, Plate III; copied by Quain (1844) as Fig. 10, Plate VI.

F: This figure, representing a case of the left common carotid arising from the innominate trunk and the left vertebral from the arch, was taken from Tiedemann (1822), Fig. 7, Plate II.

G: This figure, representing a case of the right subclavian arising as the second branch from the arch, was taken from Tiedemann (1822), Fig. 4, Plate III; copied by Quain (1844) as Fig. II, Plate VI. The original was by Huber (1777), Vol. 8, p. 75 and Fig. 3.

H: This figure is adapted from a case of left vertebral arising from the aortic arch which was found in our [Poynter's] dissecting rooms.

I: This figure, representing a case in which the left vertebral is the last branch from the arch, was taken from Tiedemann (1822), Fig. 10, Plate III; copied by Quain as Fig. 10, Plate VII.

J: This figure, representing the right subclavian as the third

Cardiovascular System

◄ branch from the arch, was reported by Walter (1785), p. 62, Fig. 5. Plate III; it was copied by both Tiedemann and Quain.
K: This figure, representing five branches from the arch, was reported by Penada (1801), p. 44, and illustrated by Tiedemann (1822) as Fig. 4, Plate IV.
L: This figure, representing six branches from the arch, was taken from Tiedemann (1822), Fig. 5, Plate IV; copied by Quain (1844) as Fig. 15, Plate VII.

Abbreviations used in the figures: *A*, aorta; *AD*, aorta descendens (descending aorta); *C*, a. carotis (carotid); *CD*, a. carotis communis dextra (right common carotid); *CS*, a. carotis communis sinistra (left common carotid); *D*, ductus arteriosus (Botalli); *EC*, a. carotis externa (external carotid); *IC*, a. carotis interna (internal carotid); *O*, esophagus; *P*, a. pulmonalis (pulmonary); *SD*, a. subclavia dextra (right subclavian); *SS*, a. subclavia sinistra (left subclavian); *T*, trachea; *TD*, ductus thoracicus (thoracic duct); *VD*, a. vertebralis dextra (right vertebral); *VS*, a. vertebralis sinistra (left vertebral).

From Poynter, C.W.M. Arterial anomalies pertaining to the aortic arches and the branches arising from them. *The University Studies of the University of Nebraska* 16:229–345, 1916, Lincoln, Nebr.

Additional references from Poynter: Huber, J.J. Observations. *Acta Helvet.* 8:85, and 100, 1777.

Krause, W. Varietäten des Aortensystems. In J. Henle. *Handb. der syst. Anat., Bd. 3, S. 1, Abt.: Gefässlehre.* Braunschweig 1876.

Penada, –. Sagg. terz. di osserv. e mem. path. anat. Padua. 1801.

Quain, R. *The anatomy of the arteries of the human body with its application to pathology and operative surgery: in a series of plates and commentaries.* London. 1844.

Tiedemann, F. *Tabulae arteriarum corporis humanic fabrica.* Karlsruhae. 1822.

Walter, M. Sur les maladies du coeur. *Nouveaux Memories de l'academie royal des sciences et belles-lettres.* Berlin, 1785:56–66.

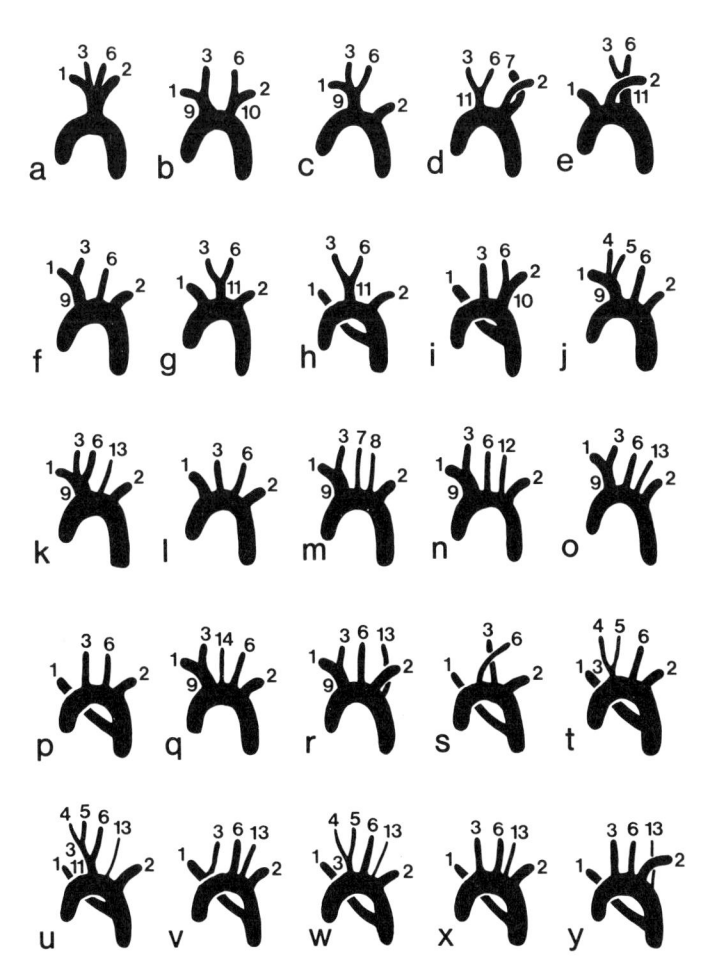

FIG. 15.
Twenty-five Possible Combinations in Origin and Position of Arteries of Aortic Arch.

Abnormalities in origin and number of arterial branches were studied in 453 bodies. On the basis of a literature review, the authors identified 25 variations. The authors draw attention to a very rare anomaly in which the brachiocephalic trunk arose on the left side below the left subclavian artery (e).

Possible combinations in origin and position of the arteries of the aortic arch: *1*, right subclavian artery; *2*, left subclavian artery; *3*, right common carotid artery; *4*, right internal carotid artery; *5*, external carotid artery; *6*, left common carotid artery; *7*, left external carotid artery; *8*, left internal carotid artery; *9*, right brachiocephalic trunk; *10*, left brachiocephalic trunk; *11*, common carotid trunk; *12*, inferior thyroid artery; *13*, left vertebral artery; *14*, lowest thyroid artery. These numbers correspond to those on the illustration and identify the branches.

From Nizankowski, C., Rajchel, Z. and M. Ziolkowski. Abnormal origin of arteries from the aortic arch in man. *Folia Morphol.* 34:109–116, 1975.

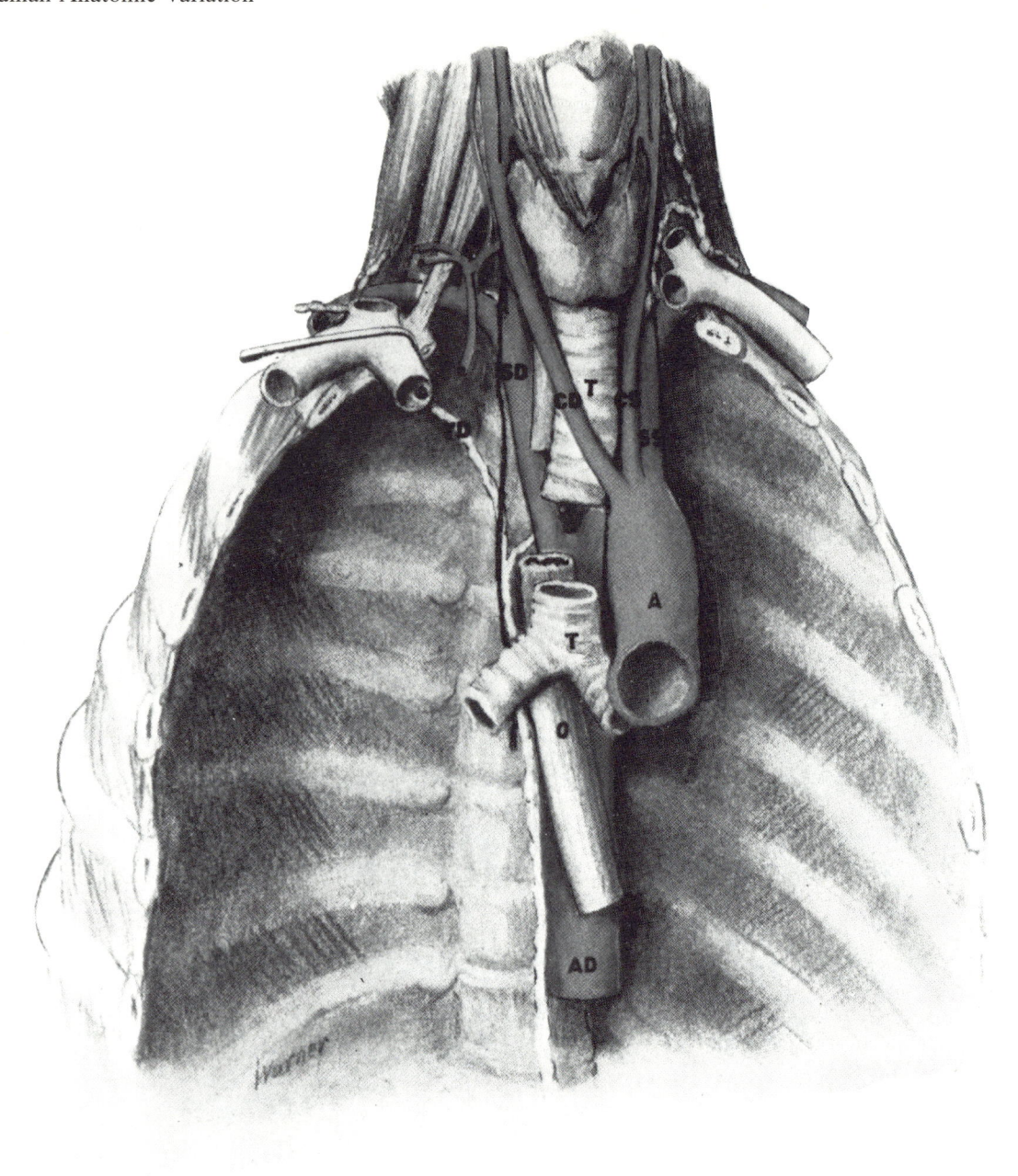

FIG. 16.
Aortic Arch Anomaly.

"This drawing was made from a dissection in our [Poynter's] laboratory of a case in which the right subclavian artery arose from the beginning of the dorsal aorta and passed behind the esophagus. The thoracic duct divided near the origin of the subclavian and opened on each side into the venous angle."

A, Aorta; *AD*, descending aorta; *CD*, right common carotid; *CS*, left common carotid; *O*, esophagus; *SD*, right subclavian artery; *SS*, left subclavian artery; *T*, trachea; *TD*, thoracic duct.

From Poynter, C.W.M. Arterial anomalies pertaining to the aortic arches and the branches arising from them. *The University Studies of the University of Nebraska* 16:229–345, 1916, Lincoln, Nebr.

Cardiovascular System

FIG. 17.
Common Trunk for the Two Internal Carotids and Right Subclavian Arteries.
From Vesalius, A. *Fabrica* (1543).

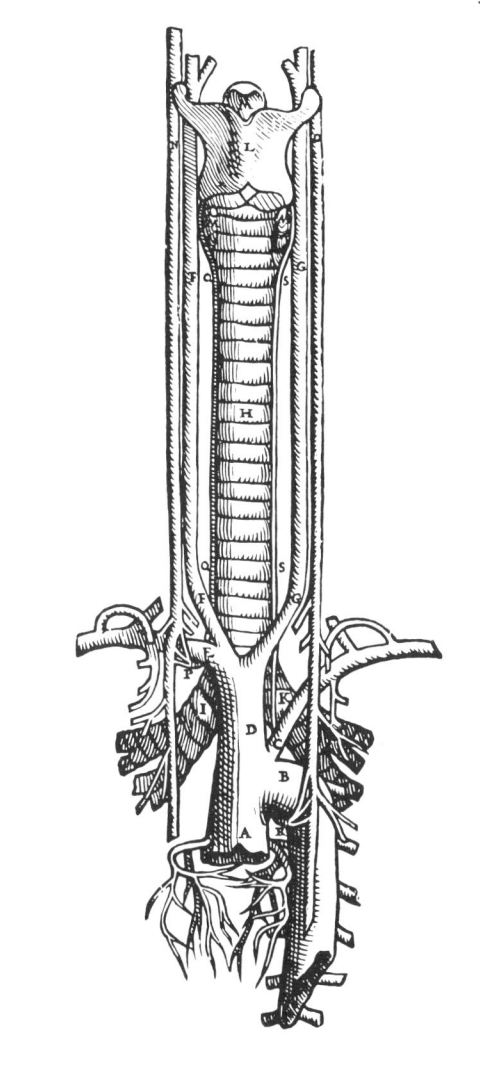

FIG. 18.
Variations in Arrangement of the Bronchi and Pulmonary Arteries on the Right Side.

Explanation of diagram (130 bodies examined):
A: Artery lower than bronchus.
B: Artery on same level with bronchus (to upper lobe).
C: Artery above level of bronchus (to upper lobe).
D: Right artery passes across trachea above its bifurcation and divides into two branches.

From Thomson, A. Fourth annual report of the committee of collective investigation of the Anatomical Society of Great Britain and Ireland for the year 1892–93. *J. Anat. Physiol.* 25:70–73, 1891.

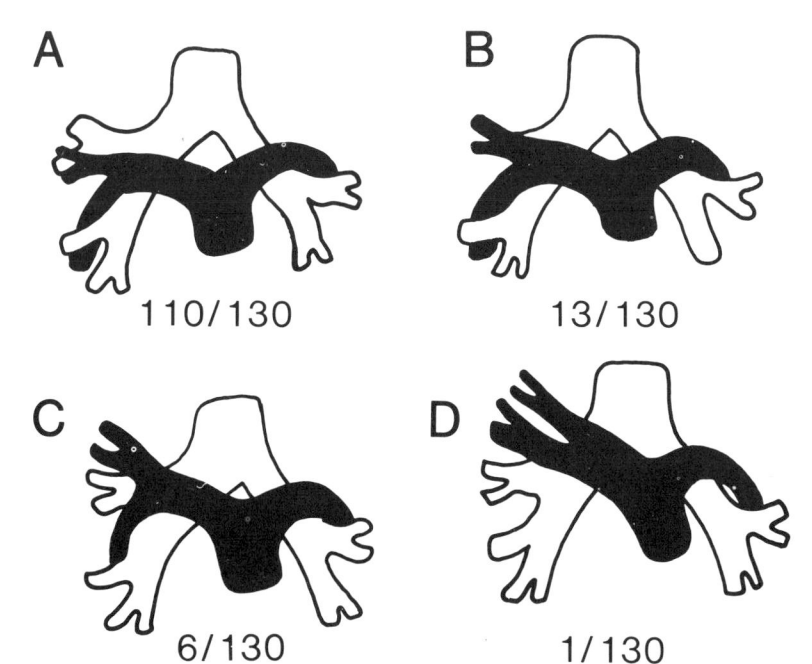

A 110/130

B 13/130

C 6/130

D 1/130

Human Anatomic Variation

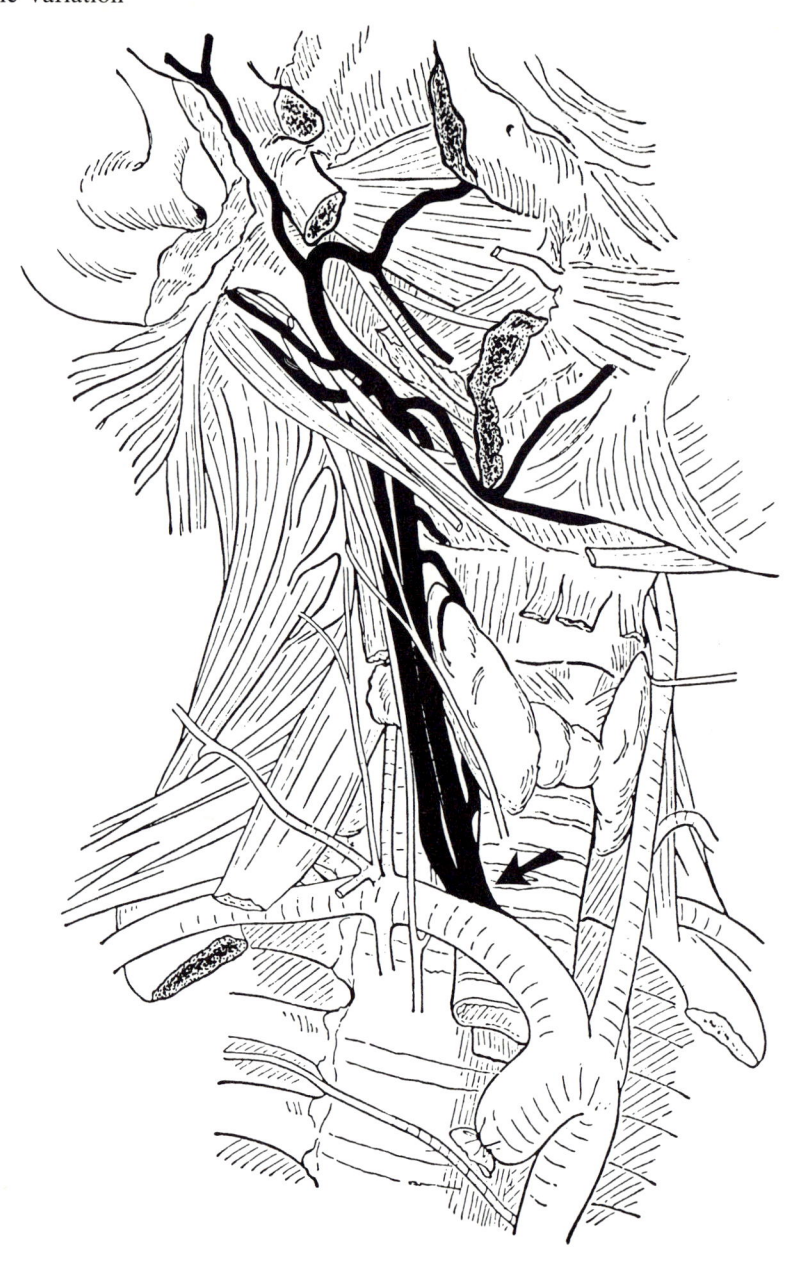

FIG. 19.
Very Short Common Carotid Artery.

From Adachi, B. *Anatomie der Japaner. Das Arteriensystem der Japaner.* Verlag der Kaiserlich-Japanischen Universität zu Kyoto. In Kommission bei "Maruzen Co.", Kyoto und Tokyo. Gedruckt von "Kenkyusha" in Tokyo, 1928.

FIG. 20.
Unusual Course of External Carotid Artery.

The external carotid artery courses between the digastric (D) and stylohyoid (S) muscles in this subject (*arrow*).

From Adachi, B. *Anatomie der Japaner. Das Arteriensystem der Japaner.* Verlag der Kaiserlich-Japanischen Universität zu Kyoto. In Kommission bei "Maruzen Co.", Kyoto und Tokyo. Gedruckt von "Kenkyusha" in Tokyo, 1928.

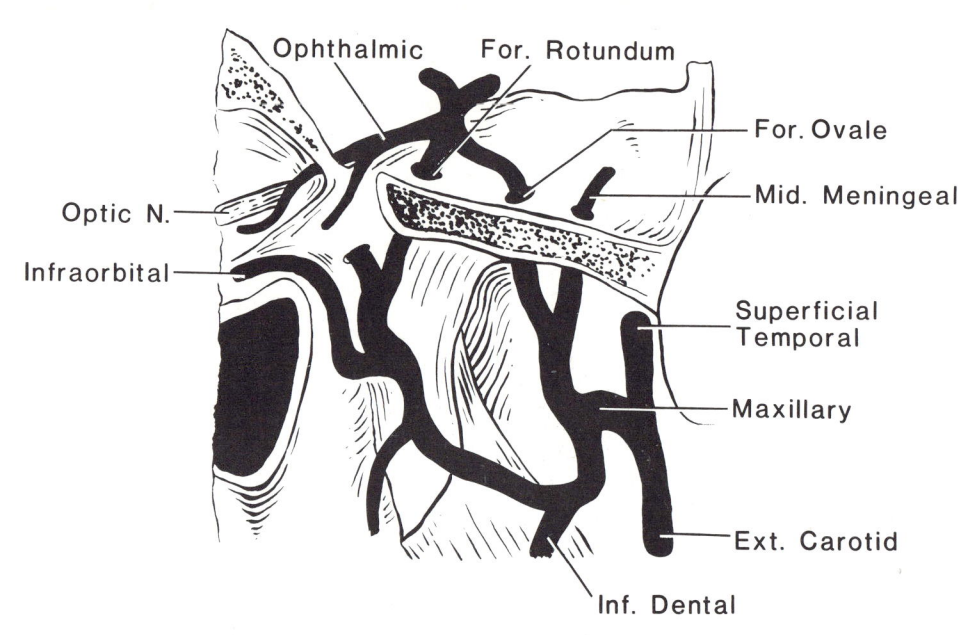

FIG. 21.
Absence of Left Internal Carotid Artery.

Regarding this illustration by Quain, Fisher quoted Quain on the absence of the left internal carotid as follows: "...the carotid artery furnishes the usual branches of the external carotid and divides into temporal and internal maxillary arteries. In the place of an internal carotid, of which there is no trace, two tortuous branches derived from the internal maxillary join within the skull to form a single vessel which lies to the inner side of the fifth nerve. The artery so constructed being, however, smaller than the usual size of the internal carotid, the deficiency is compensated by the artery of the opposite side, which is of large size."

From Fisher, A.G.T. A case of complete absence of both internal carotid arteries, with a preliminary note on the developmental history of the stapedial artery. *J. Anat. Physiol.* 48:37–46, 1914.

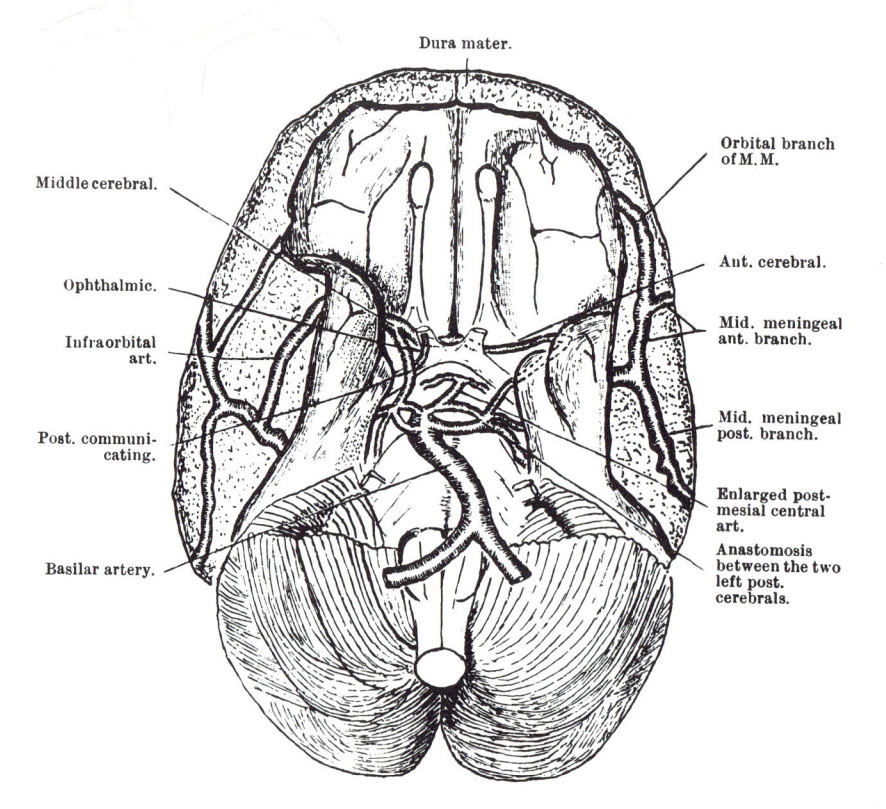

FIG. 22.
Absence of Both Internal Carotid Arteries.

Figure shows arteries at base of brain and in dura mater. (*m.m.* = middle meningeal artery.)

From Fisher, A.G.T. A case of complete absence of both internal carotid arteries, with a preliminary note on the developmental history of the stapedial artery. *J. Anat. Physiol.* 48:37–46, 1914.

Cardiovascular System

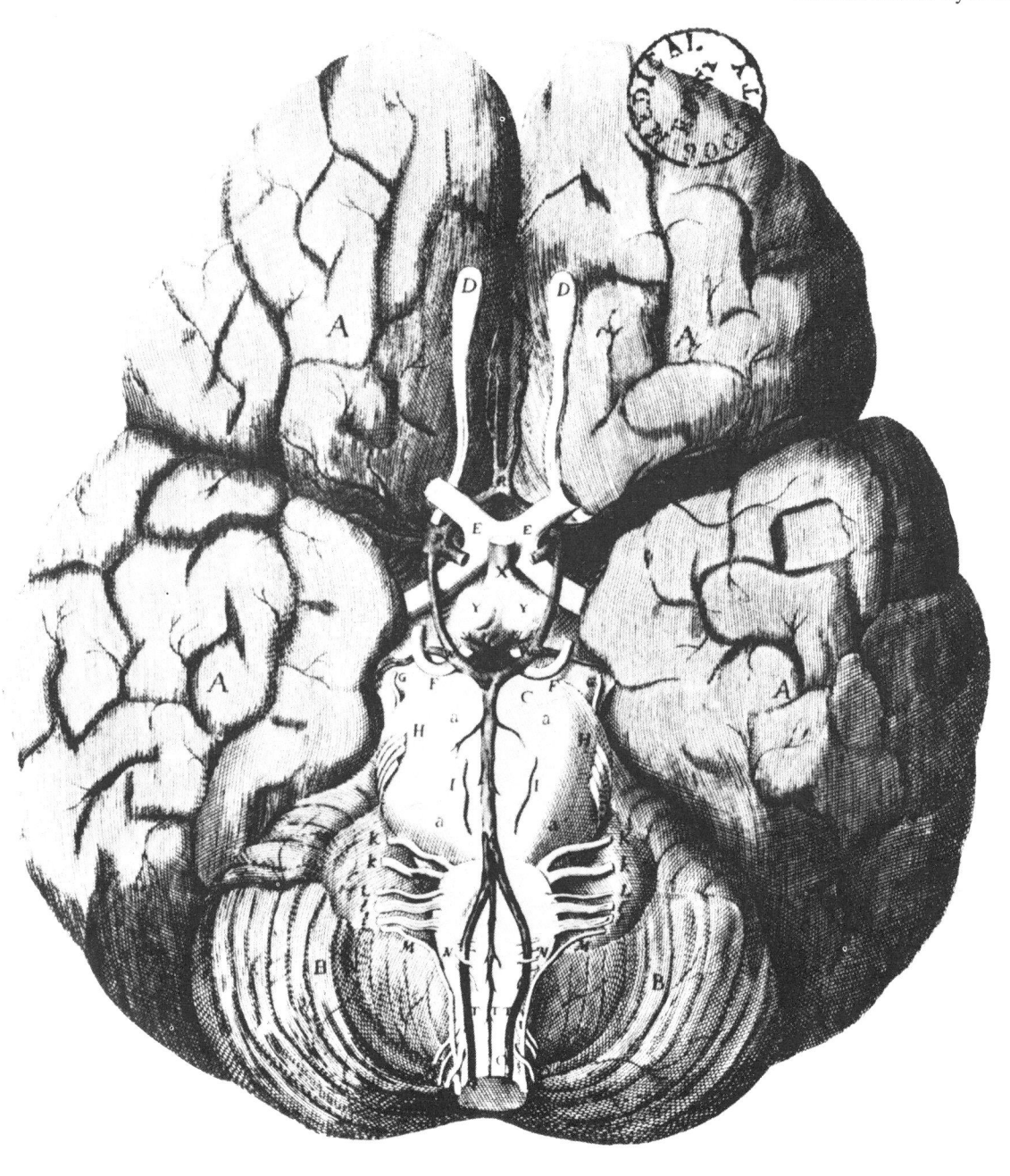

FIG. 23.
Circle of Willis.

The original illustration by Willis of the circle of anastomosing arteries at the base of the brain.

In *Heirs of Hippocrates. The Development of Medicine in a Catalogue of Historic Books in the Health Sciences Library.* Friends of The University of Iowa Libraries, The University of Iowa, Iowa City, 1980.

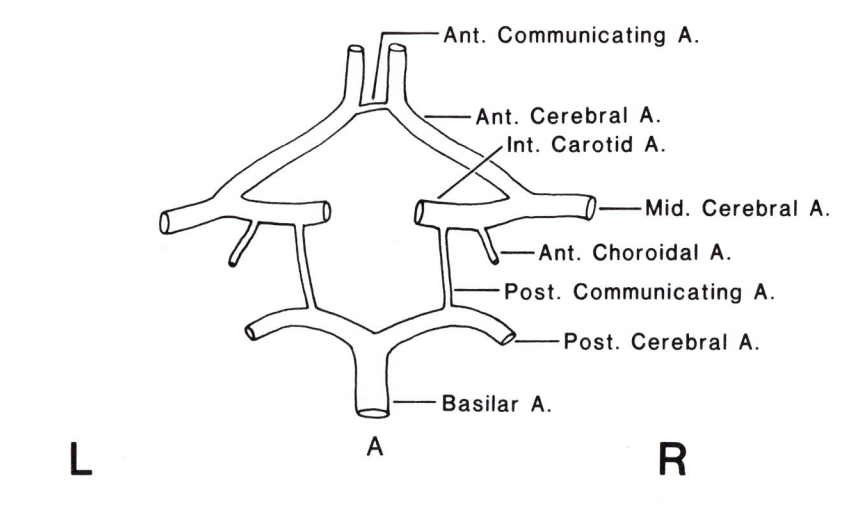

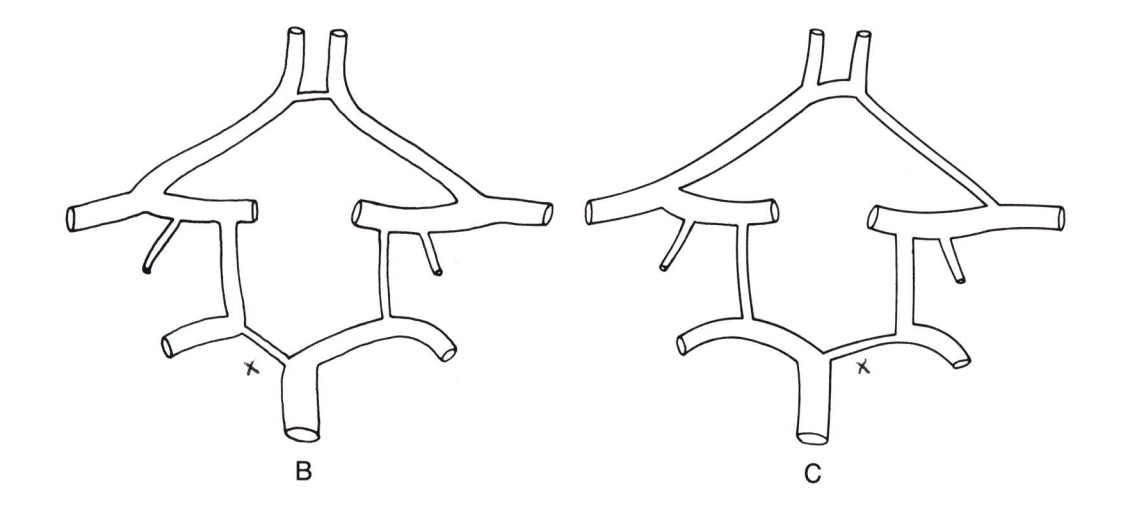

Human Anatomic Variation

FIG. 24.
Circle of Willis.

A: "Normal circle of Willis. It is complete, symmetrical and has thread-like communicating arteries.
B: Anomalous circle of Willis. It is complete but asymmetrical due to fetal trifurcation of the left internal carotid artery and normal bifurcation of the right internal carotid.
C: Anomalous circle of Willis. It is complete but asymmetrical due to anterior trifurcation of the left internal carotid artery and reverse bifurcation of the right internal carotid.

There are 2 relatively common, variations from the normal pattern of branching of internal carotid artery. One is the persistence of the fetal pattern of trifurcation in which all 3 of the ipsilateral major cerebral arteries arise from the internal carotid while the posterior cerebral segment of the circle continues, as in the early fetus, to be a small, thread-like, posterior communicating type of vessel [B]. In the other common variation the artery divides into the middle cerebral artery and an unusually large anterior cerebral branch. The latter in turn, divides at the anterior pole of the circle into both anterior cerebral arteries. This pattern of terminal branching of the internal carotid artery has been termed anterior trifurcation [C]. The proximal part of the branch to the opposite side is oriented in the manner of the usual anterior communicating artery but is several times larger. It usually is connected to the opposite internal carotid by a hair-like, communicating type of vessel. The opposite, anterior cerebral segment of the circle, however, may be missing, in which case the contralateral internal carotid simply continues as the middle cerebral artery or may bifurcate into the posterior cerebral artery and the middle cerebral artery [C].

Redrawn from Kirgis, H.D., Fisher, W.L., Llewellyn, R.C. and E. McC. Peebles. Aneurysms of the anterior communicating artery and gross anomalies of the circle of Willis. *J. Neurosurg.* 25:73–78, 1966.

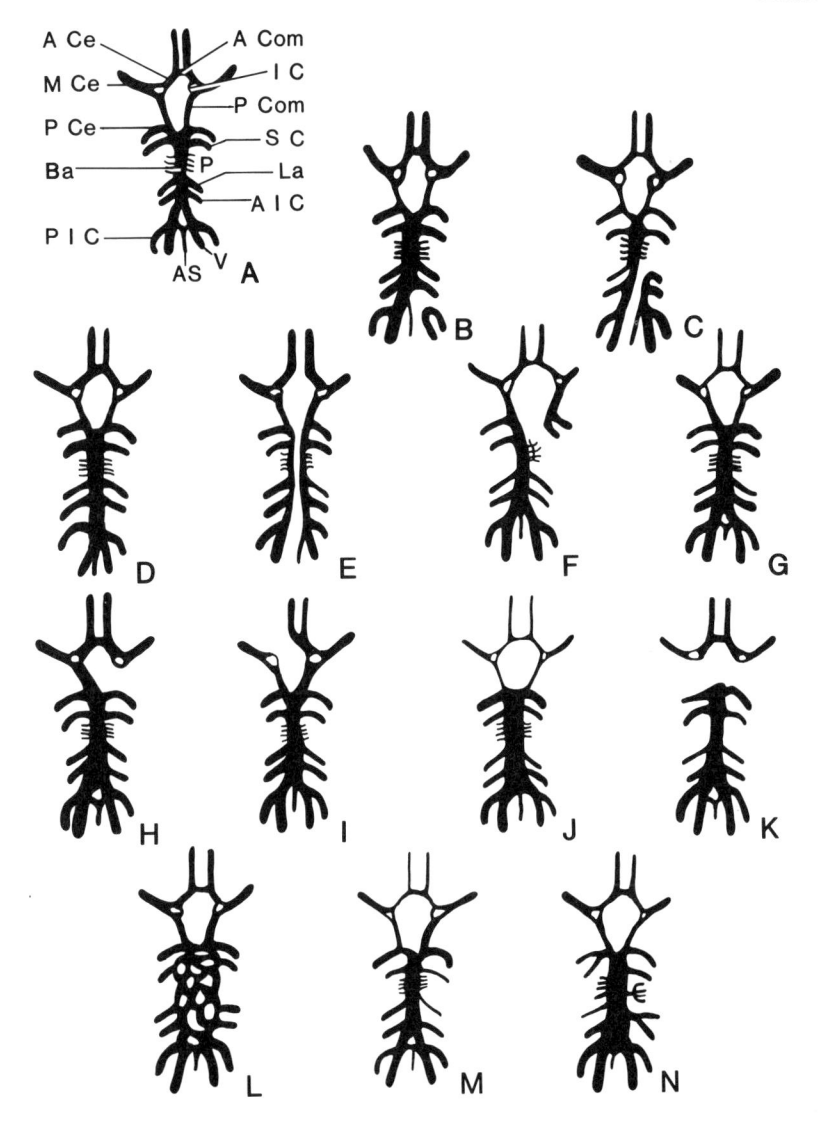

FIG. 25.
Fourteen Anomalies of Circle of Willis and Related Vessels.

Based on a study of 600 brains.

A: Normal cerebral arterial circle and associated vessels. *A Com*, anterior communicating; *A Ce*, anterior cerebral; *M Ce*, middle cerebral; *P Ce*, posterior cerebral; *IC*, internal carotid stump; *P Com*, posterior communicating; *SC*, superior cerebellar; *Ba*, basilar; *La*, labyrinthine; *AIC*, anterior inferior cerebellar; *PIC*, posterior inferior cerebellar; *V*, vertebral; *AS*, anterior spinal; *P*, pontine.

B, C: Failure of vertebral artery to unite with the basilar artery (extremely rare anomaly).

D: Inferior union of vertebral arteries to form basilar artery.

E: Double basilar artery with absence of anterior communicating artery (1st description of this anomaly).

F: Failure of left posterior communicating artery to unite with rostral end of basilar artery; all pontine branches arise from left side of basilar artery.

G: A specimen showing underdeveloped anterior cerebral, anterior communicating, and left posterior communicating arteries.

H: Absence of left posterior communicating artery.

I: Both anterior cerebral arteries are related only to the left internal carotid artery.

J: Specimen from a microcephalic idiot showing attenuation of internal carotid arteries and their branches, and a large saccular basilar artery.

K: Absence of both posterior communicating arteries.

L: A plexiform basilar artery (1st description of this anomaly).

M: Specimen showing difference in size of internal carotid, posterior communicating, middle, and anterior cerebral arteries, with right side underdeveloped.

N: A composite drawing showing several variations found in different specimens: (*a*) anterior inferior cerebellar and internal auditory arteries arise as branches from a common origin; (*b*) posterior cerebral and superior cerebellar arteries arise from a common origin; and (*c*) all pontine branches on one side arise from common origin.

Authors' note: Additional reference: Stehbens, W.E. Aneurysms and anatomical variation of cerebral arteries. *Arch. Pathol.* 75:45–64, 1963.

Redrawn from McCullough, A.W. (Rewritten posthumously for the author by Howard K. Suzuki). Some anomalies of the arterial circle (of Willis) and related vessels. *Anat. Rec.* 142:537–543, 1962.

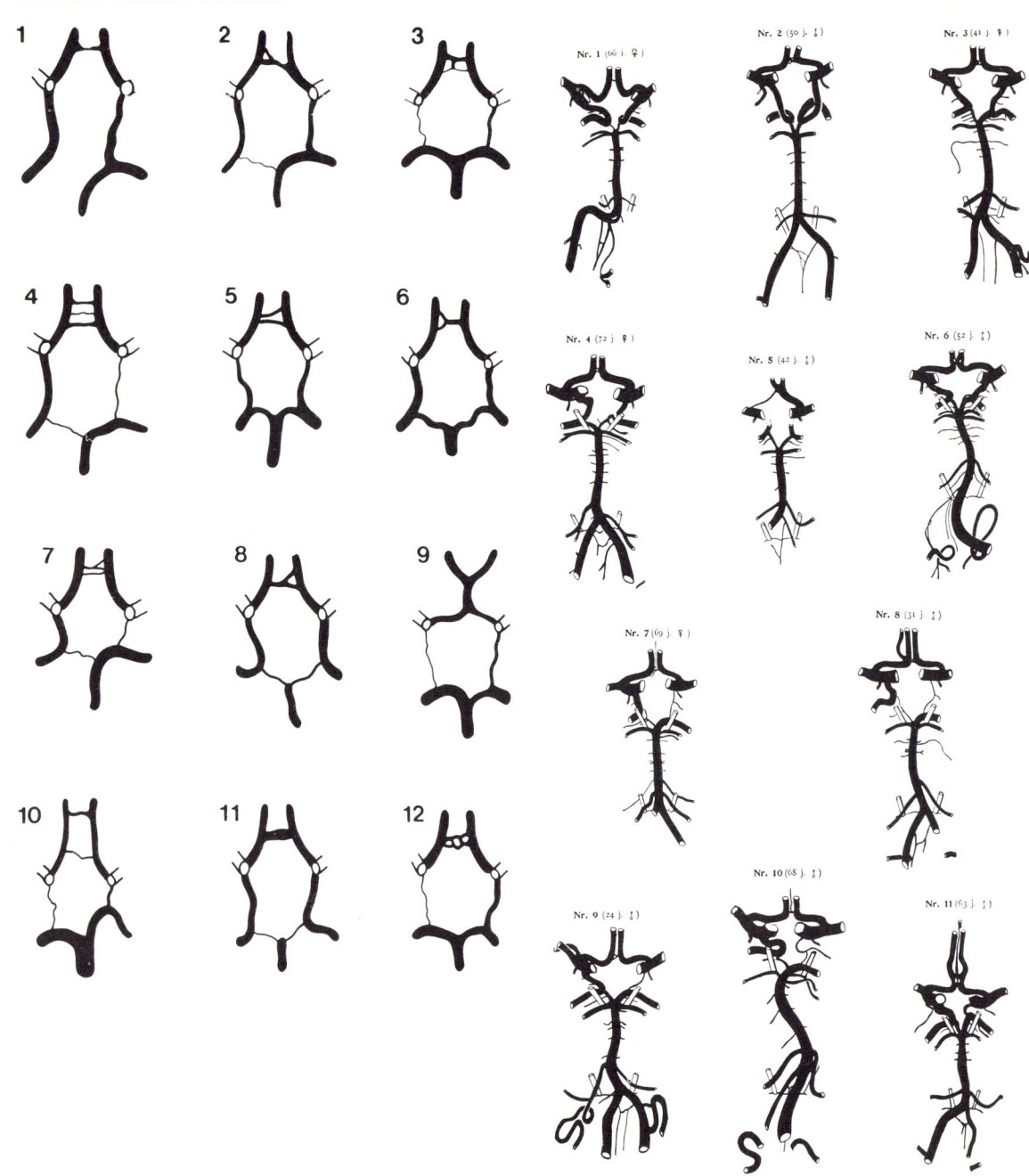

Human Anatomic Variation

FIG. 26.
Variations in Circle of Willis.

The author has found 12 different patterns in the arterial circle which may be incomplete or asymmetrical with duplication of components and absence or fusion of others.

Authors' note: Additional references: Stehbens, W.E. Aneurysms and anatomical variation of cerebral arteries. *Arch. Pathol.* 75:45–64, 1963; and Kirgis, J.D., Fisher, W.L., Llewellyn, R.C. and E. McC. Peebles. Aneurysms of anterior communicating artery and gross anomalies of the circle of Willis. *J. Neurosurg.* 25:73–78, 1966.

Redrawn from Kleiss, E. Die verschiedenen Formen des Circulus arteriosus cerebralis Willisi. *Anat. Anz.* 92:216–230, 1941–42.

FIG. 27.
Eighty-three Variations in Circle of Willis.

From Adachi, B. *Anatomie der Japaner. Das Arteriensystem der Japaner.* Verlag der Kaiserlich-Japanischen Universität zu Kyoto. In Kommission bei "Maruzen Co.", Kyoto und Tokyo. Gedruckt von "Kenkyusha" in Tokyo, 1928.

Cardiovascular System

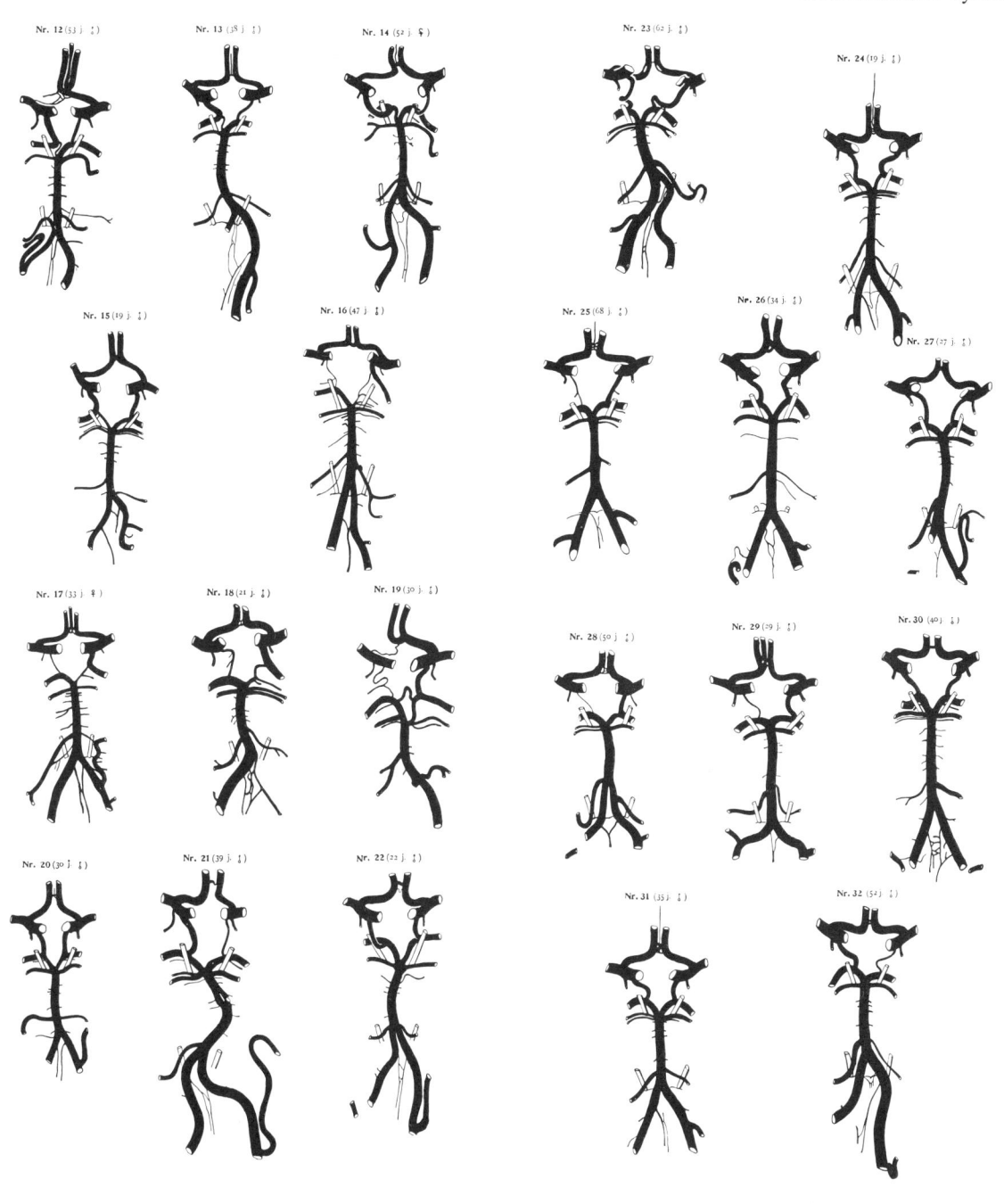

FIG. 28.
Eighty-three Variations in Circle of Willis (continued).

From Adachi, B. *Anatomie der Japaner. Das Arteriensystem der Japaner.* Verlag der Kaiserlich-Japanischen Universität zu Kyoto. In Kommission bei "Maruzen Co.", Kyoto und Tokyo. Gedruckt von "Kenkyusha" in Tokyo, 1928.

FIG. 29.
Eighty-three Variations in Circle of Willis (continued).

From Adachi, B. *Anatomie der Japaner. Das Arteriensystem der Japaner.* Verlag der Kaiserlich-Japanischen Universität zu Kyoto. In Kommission bei "Maruzen Co.", Kyoto und Tokyo. Gedruckt von "Kenkyusha" in Tokyo, 1928.

Human Anatomic Variation

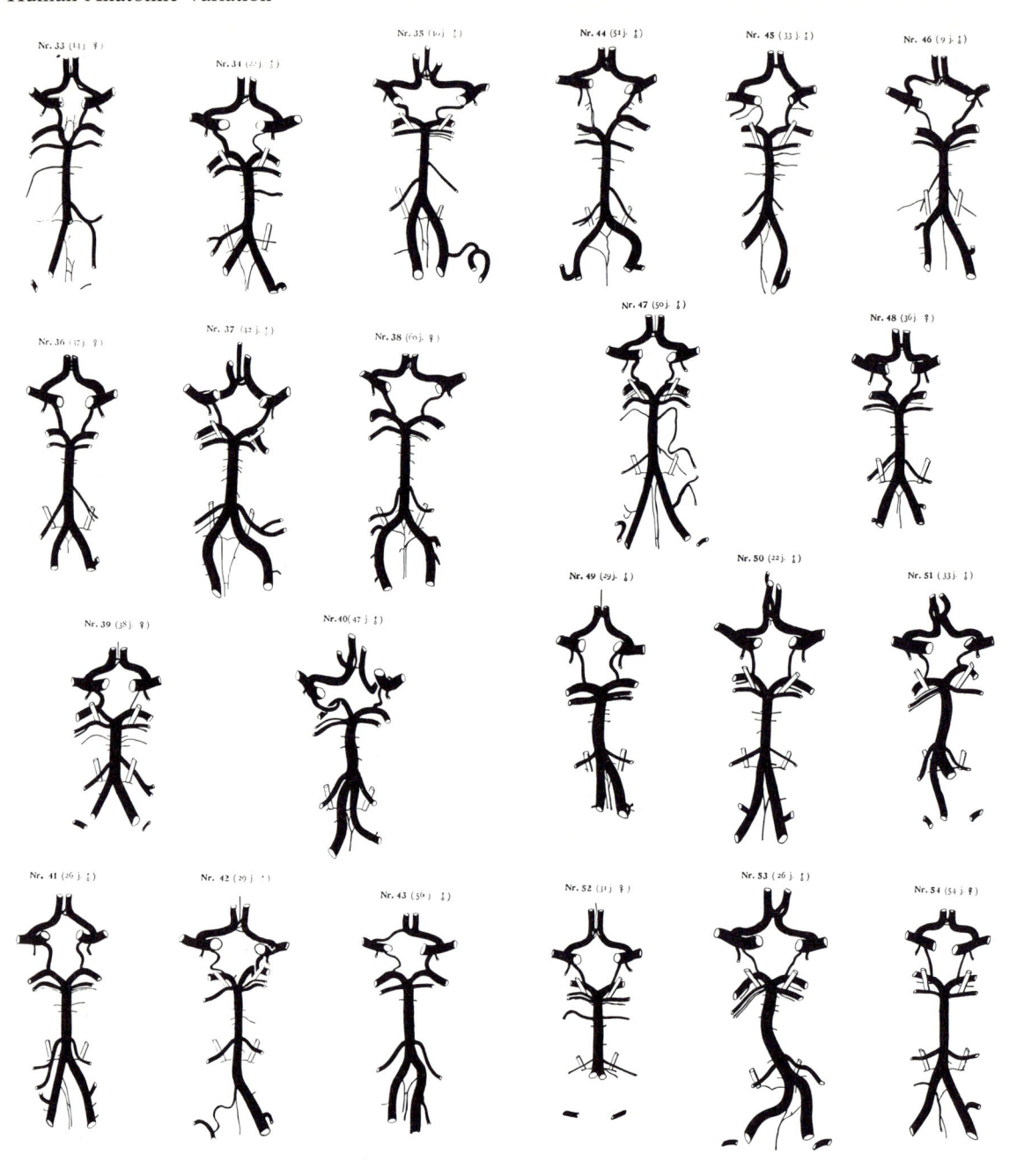

Fig. 30.
Eighty-three Variations in Circle of Willis (continued).

From Adachi, B. *Anatomie der Japaner. Das Arteriensystem der Japaner*. Verlag der Kaiserlich-Japanischen Universität zu Kyoto. In Kommission bei "Maruzen Co.", Kyoto und Tokyo. Gedruckt von "Kenkyusha" in Tokyo, 1928.

FIG. 31.
Eighty-three Variations in Circle of Willis (continued).

From Adachi, B. *Anatomie der Japaner. Das Arteriensystem der Japaner*. Verlag der Kaiserlich-Japanischen Universität zu Kyoto. In Kommission bei "Maruzen Co.", Kyoto und Tokyo. Gedruckt von "Kenkyusha" in Tokyo, 1928.

Cardiovascular System

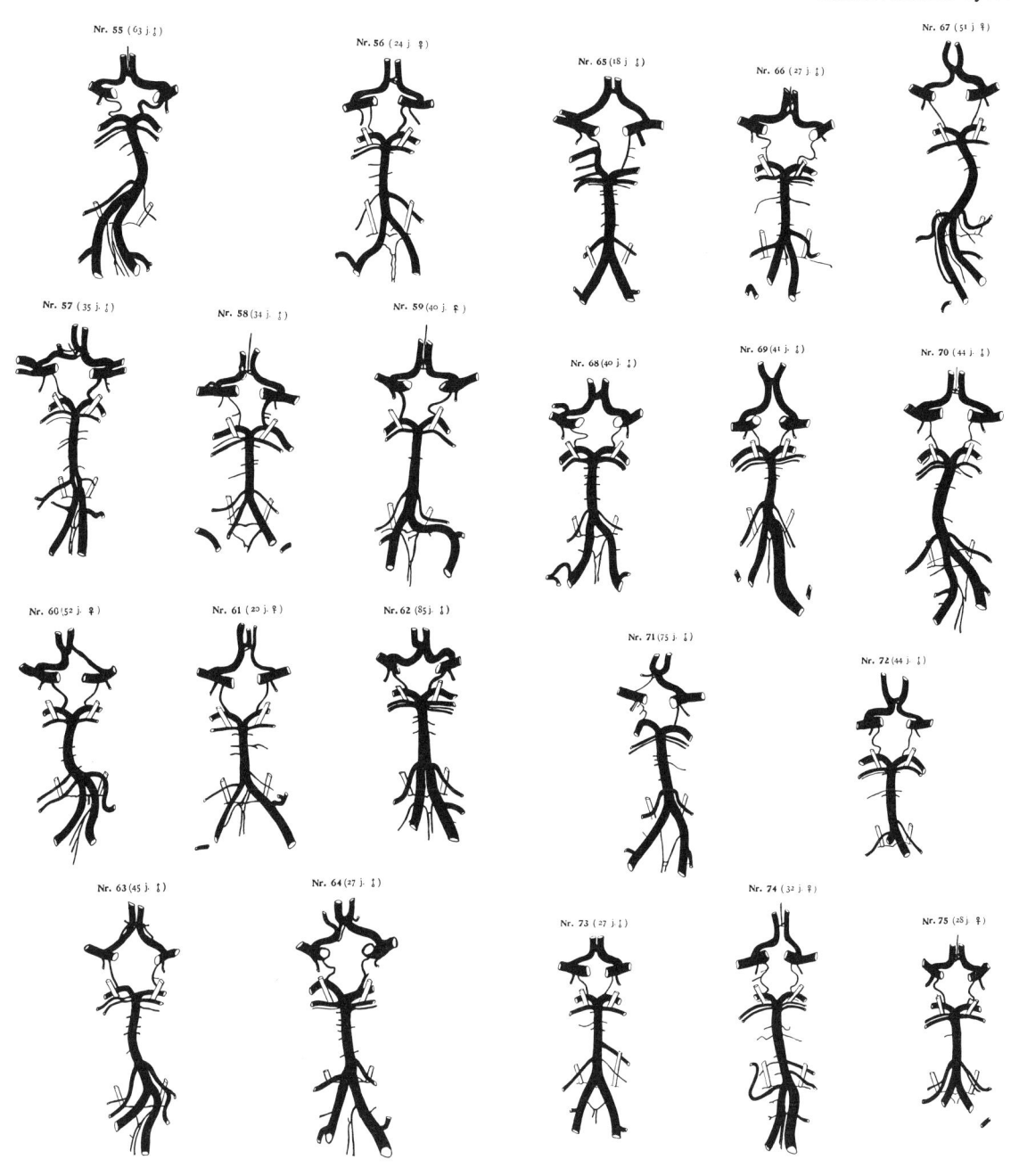

FIG. 32.
Eighty-three Variations in Circle of Willis (continued).

From Adachi, B. *Anatomie der Japaner. Das Arteriensystem der Japaner.* Verlag der Kaiserlich-Japanischen Universität zu Kyoto. In Kommission bei "Maruzen Co.", Kyoto und Tokyo. Gedruckt von "Kenkyusha" in Tokyo, 1928.

FIG. 33.
Eighty-three Variations in Circle of Willis (continued).

From Adachi, B. *Anatomie der Japaner. Das Arteriensystem der Japaner.* Verlag der Kaiserlich-Japanischen Universität zu Kyoto. In Kommission bei "Maruzen Co.", Kyoto und Tokyo. Gedruckt von "Kenkyusha" in Tokyo, 1928.

Human Anatomic Variation

FIG. 34.
Eighty-three Variations in Circle of Willis (continued).

From Adachi, B. *Anatomie der Japaner. Das Arteriensystem der Japaner.* Verlag der Kaiserlich-Japanischen Universität zu Kyoto. In Kommission bei "Maruzen Co.", Kyoto und Tokyo. Gedruckt von "Kenkyusha" in Tokyo, 1928.

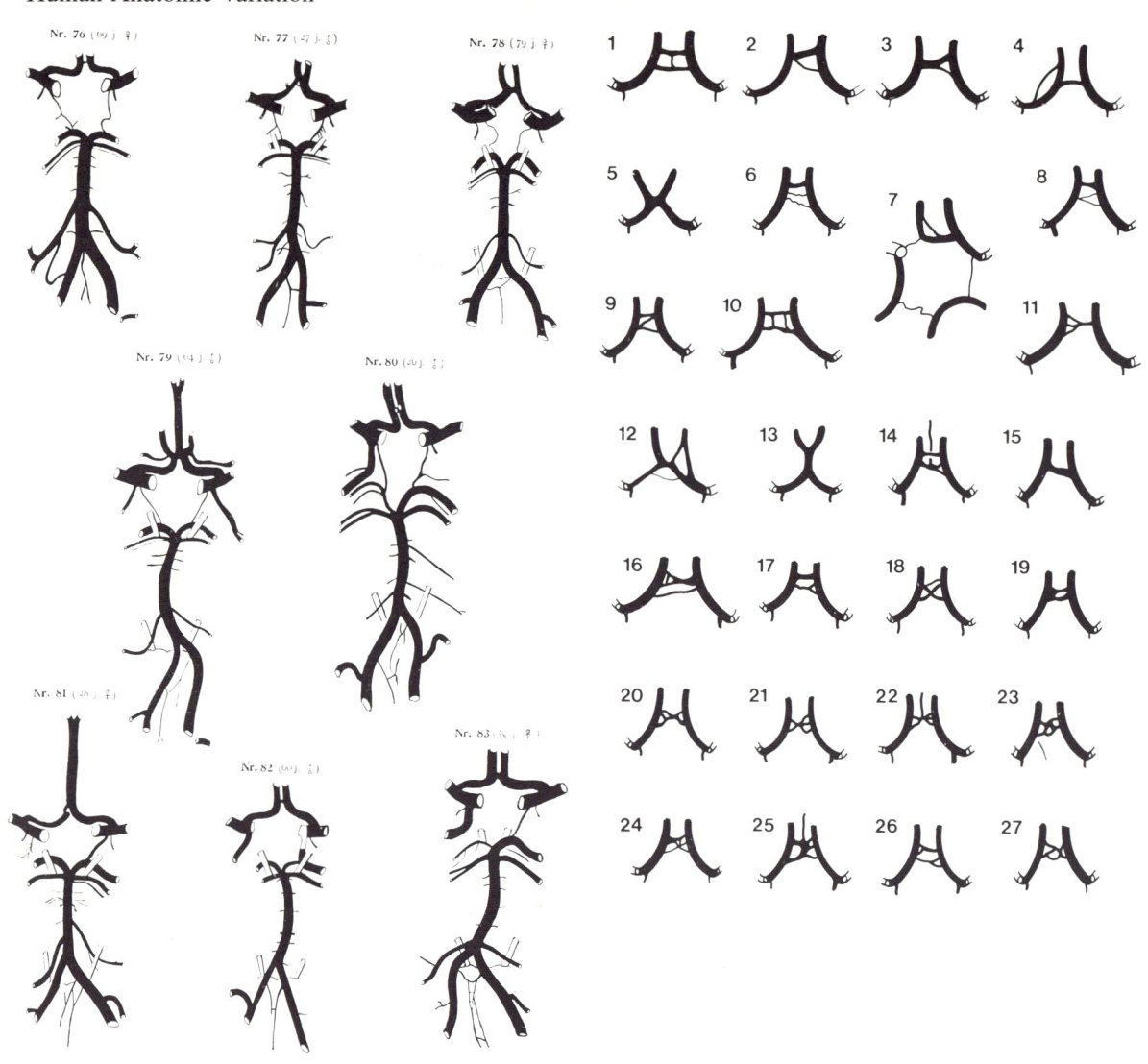

FIG. 35.
Variations in Anterior Communicating Artery.

The author has found 27 different patterns in the anterior communicating artery and its relationships to the anterior cerebral artery. This figure also shows an abnormal circle of Willis (7) and the absence of an anterior communicating artery due to fusion of the anterior cerebrals which subsequently branch (13).

Authors' note: Additional references: Stehbens, W.E. Aneurysms and anatomical variation of cerebral arteries. *Arch. Pathol.* 75:45–64, 1963; and Kirgis, J.D., Fisher, W.L., Llewellyn, R.C. and E. McC. Peebles. Aneurysms of anterior communicating artery and gross anomalies of the circle of Willis. *J. Neurosurg.* 25:73–78, 1966.

Redrawn from Kleiss, E. Die verschiedenen Formen des Circulus arteriosus cerebralis Willisi. *Anat. Anz.* 92:216–230, 1941–42.

Cardiovascular System

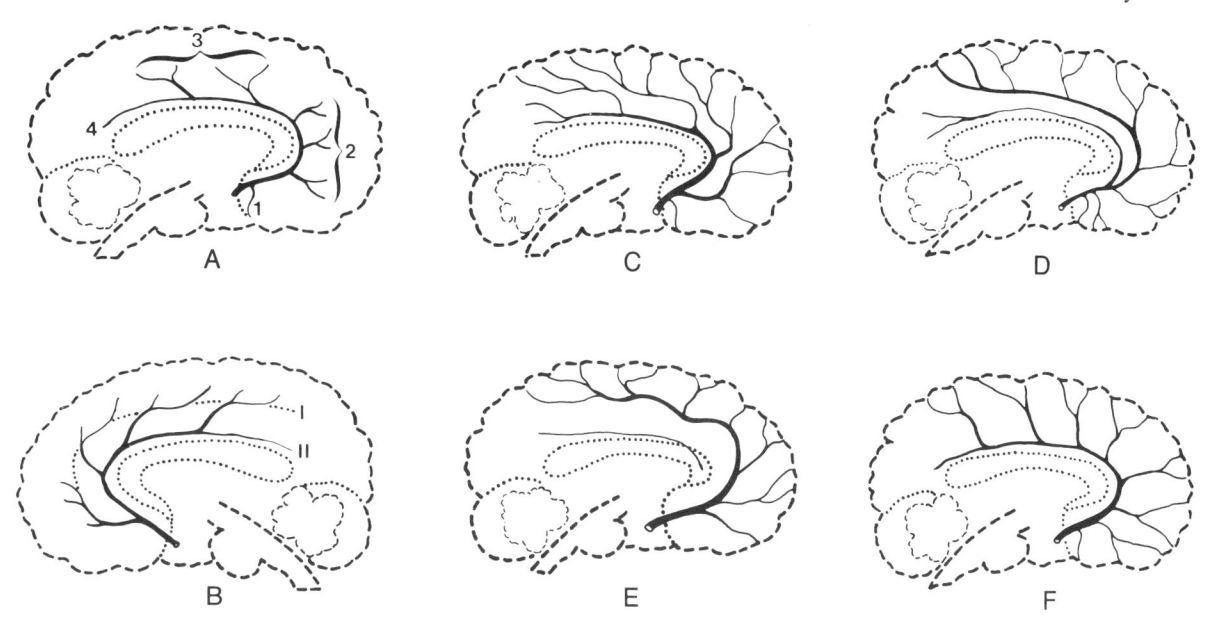

FIG. 36.
Anterior Cerebral Artery.

A, B: Branches of the anterior cerebral artery. *1*, orbital branch; *2*, precallosal branch; *3*, supracallosal branch; *4*, pericallosal branch; *I*, cingulate arch; *II*, pericallosal arch.
Variations in the distribution pattern of the anterior cerebral artery in 50 human brains are illustrated as four types.
C (Type I): A single large stem whose branching pattern runs over the corpus callosum forming the arcus pericallosus. 12 cases in 100 hemispheres studied.
D (Type II): Two separate arched vessels, one smaller inner and a larger branched outer vessel forming an arcus pericallosus and arcus cinguli. Found 9 times in 100 hemispheres studied.
E (Type III): A detached inner vessel forming an incomplete arcus pericallosus. Found in 10 cases of 100 hemispheres studied.
F (Type IV): Found in 60 cases, this type has an incomplete arcus cinguli.

Orbital artery variations: (*a*) absent, no branch, 14%; (*b*) 1 branch, 50%; (*c*) 2 branches, 30%; and (*d*) 3 branches, 6%. Precallosal artery variations: (*a*) absent, no branch, 1%; (*b*) 1 branch, 14%; (*c*) 2 branches, 49%; (*d*) 3 branches, 29%; and (*e*) 4 branches, 7%. Supracallosal artery variations: (*a*) absent, no branch, 1%; (*b*) 1 branch, 10%; (*c*) 2 branches, 28%; (*d*) 3 branches, 31%; (*e*) 4 branches, 20%; (*f*) 5 branches, 9%; and (*g*) 6 branches, 1%.

Redrawn from Kleiss, E. Die Arteria cerebralis anterior. *Anat. Anz.* 95:353–372, 1945.

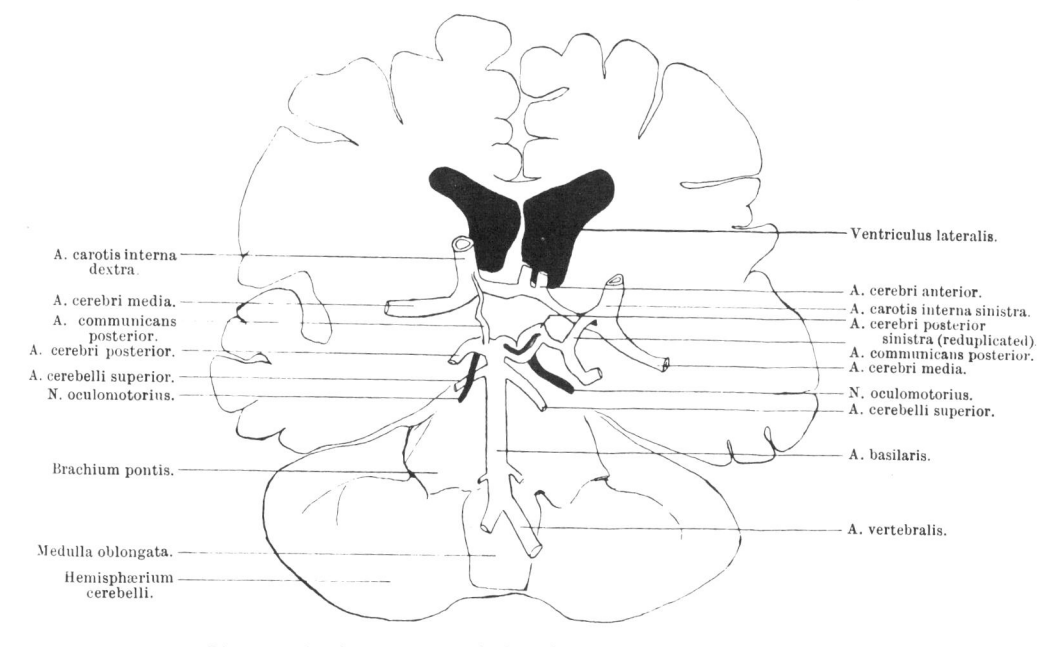

Dioptrographic drawing of the reduplicated a. cerebri posterior. (⅔ natural size.)

FIG. 37.
Reduplication of Posterior Cerebral Artery.
Dioptrographic drawing of the reduplicated a. cerebri posterior.

From Gordon-Shaw, C. Two cases of reduplication of the arteria cerebri posterior. *J. Anat. Physiol.* 44:244–248, 1910.

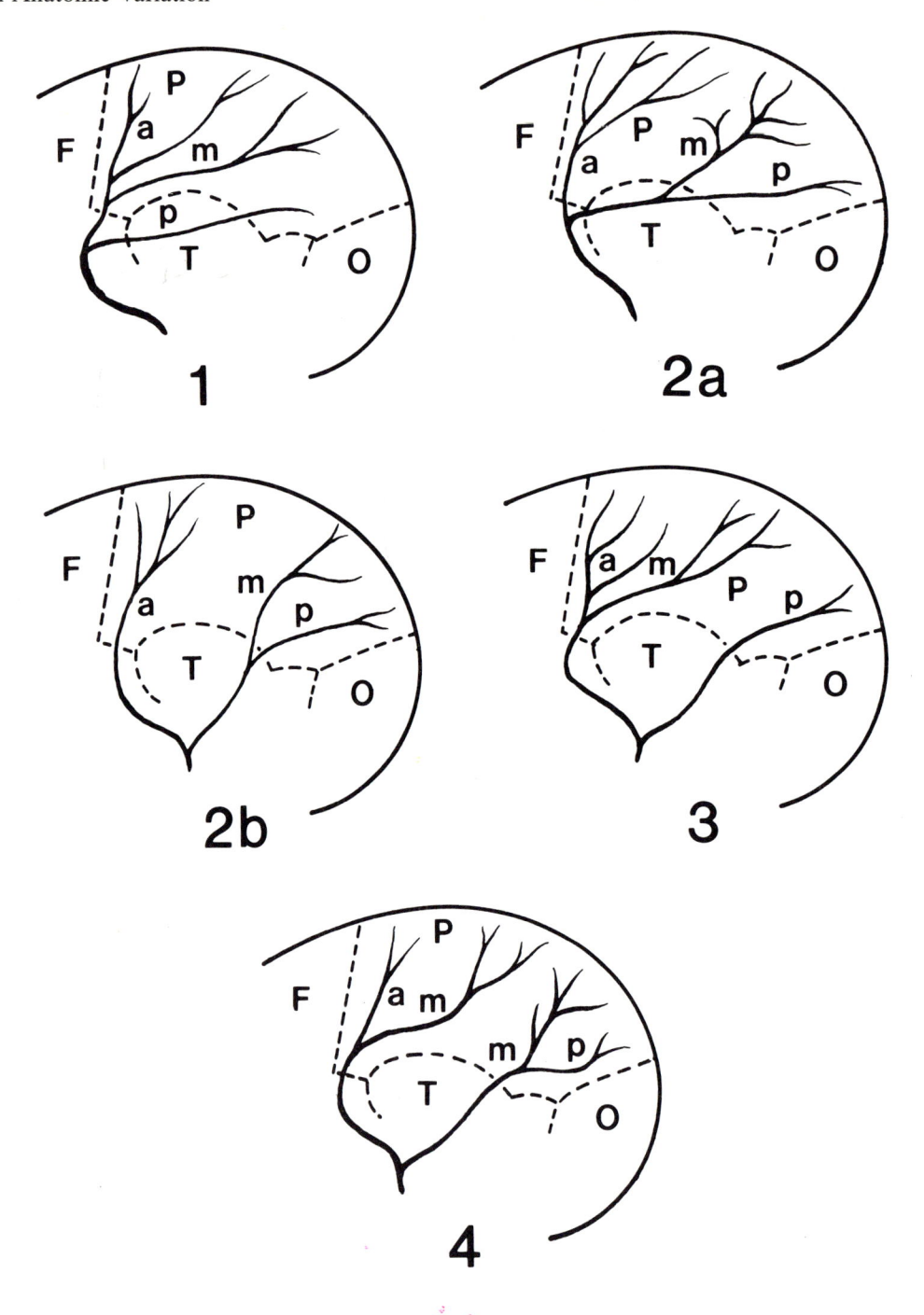

FIG. 38.
Types of Middle Meningeal Artery.

"Diagrams to illustrate the various patterns of the endocranial branches of the middle meningeal artery. F, P, T and O indicate the frontal, parietal, temporal and occipital bones, respectively; and a, m and p indicate anterior, middle and posterior branches, respectively. Type 1, pattern designated as type 1 in both Adachi and Giuffrida-Ruggeri's classifications. Type 2A, pattern designated as type 2 by Adachi and type 2A by Giuffrida-Ruggeri. Type 2B, pattern designated as type 2 by Adachi but as type 2B by Giuffrida-Ruggeri because of the low division. Type 3, pattern designated as type 1 by Adachi but designated as type 3 by Giuffrida-Ruggeri because of the low division of the artery into its terminal branches. Type 4, pattern designated as type 3 by Adachi and type 4 by Giuffrida-Ruggeri."

Authors' note: Additional references: Adachi, B. Das Arteriensystem der Japaner. *Bd.* I:93, 1928. (ref. 1348); Giuffrida-Ruggeri, V. Über die endokranischen Furchen der Arteria meningea media beim Menschen. *Zeitschr. f. Morph. u. Anthrop.* 15:401–413, 1913; Akiba, T. Über die endokranischen Furchen der Arteria meningea media bei Japanern. *Zeitschr. f. Morph. u. Anthropol.* 23:341–360, 1925; and Toida, N. Über die endokranischen Furchen der Arteria meningea media bei den Chinesen. *J. Oriental. Med.* 21:13–15, 1934.

Redrawn from Rothman, D. The endocranial course of the middle meningeal artery in American Whites and American Negroes. *Am. J. Phys. Anthropol.* 22:425–435, 1937.

Cardiovascular System

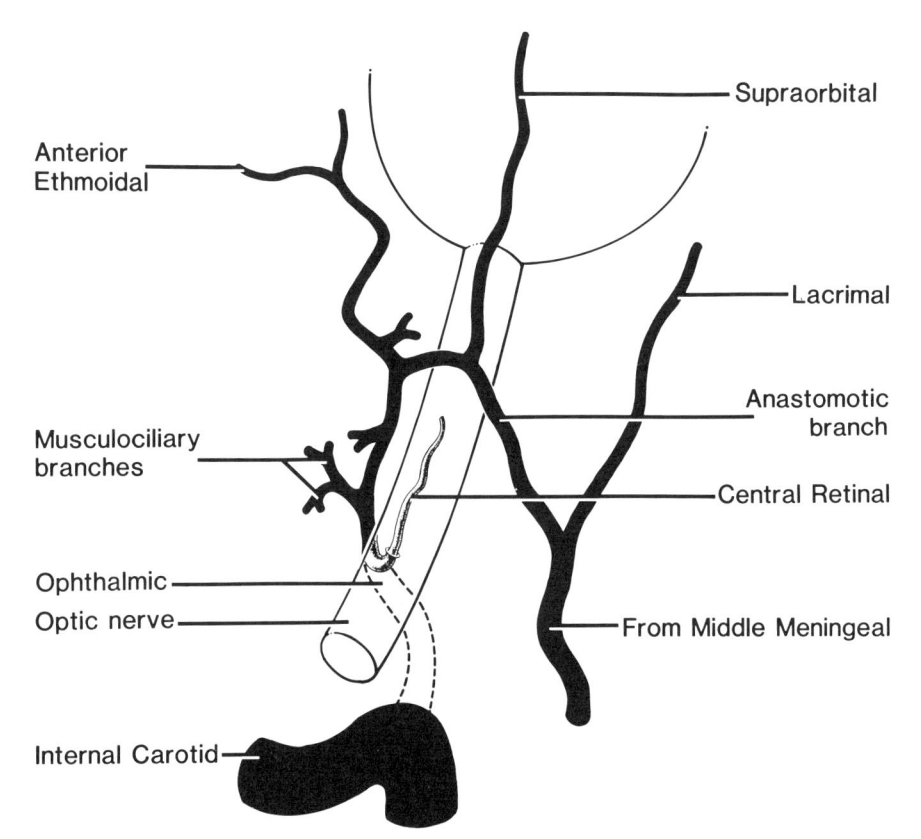

FIG. 39.
Variation in Ophthalmic Artery.
Lacrimal Artery Arising Chiefly from Middle Meningeal Artery.

"In addtition to the variations in the number and origins of its branches, the ophthalmic artery also presents variations in its course, in that, instead of passing to the inner wall of the orbit above the optic nerve, it sometimes passes below that structure. The most striking variation which it presents, however, is associated with the development of the branch of the lacrimal artery, which passes back through the sphenoidal fissure to anastomose with the middle meningeal. Occasionally this branch becomes exceptionally large and forms the main stem of the lacrimal artery, the connection of that vessel with the ophthalmic vanishing, so that it seems to be a branch of the middle meningeal. A further step in this process which sometimes occurs results in the origin of the entire ophthalmic system of vessels from the middle meningeal artery."

Redrawn from Huber, G.C. *Piersol's Human Anatomy Including Structure and Development and Practical Considerations*, 9th Ed., Vol. 1, p. 751. J.B. Lippincott Company, 1930.

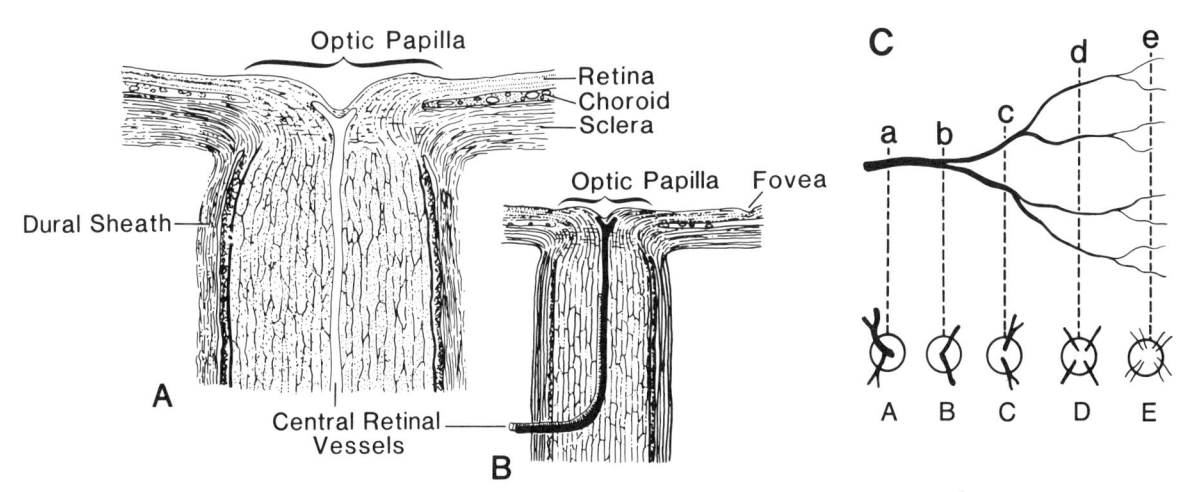

FIG. 40.
Variations in Appearance of Central Retinal Artery at the Optic Papilla.
A, B: Diagram of optic papilla and central retinal artery.
C: Diagram accounting for the varied appearance of the central retinal artery at the optic papilla as seen with an ophthalmoscope.

Redrawn and modified from Anson, B.J. *Morris' Human Anatomy*, 12th Ed. The Blakiston Division, McGraw-Hill Book Company, New York, 1966; and from Rollet, in Latarjet, A. Testut, L. et A. Latarjet. *Traité d'Anatomie Humaine*, 9th Ed. G. Doin & Cie., Paris, 1948.

Human Anatomic Variation

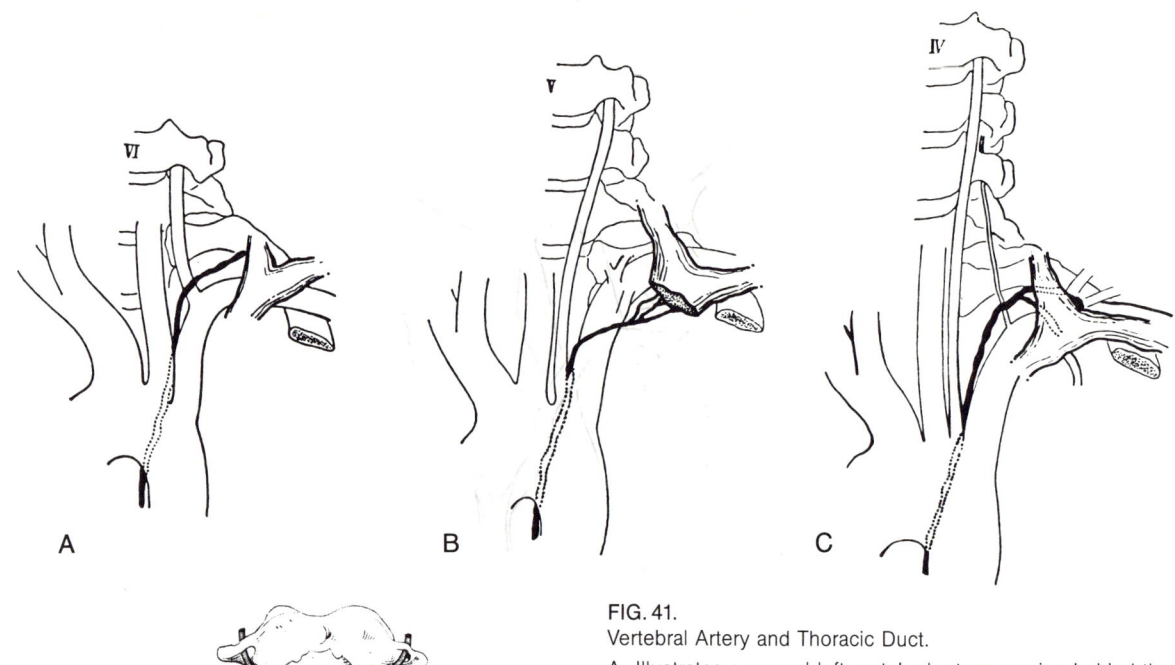

FIG. 41.
Vertebral Artery and Thoracic Duct.

A: Illustrates a normal left vertebral artery running behind the thoracic duct.
B: Illustrates a left vertebral artery with a very deep origin from the left subclavian and with an entrance into the fifth transverse foramen. The thoracic duct was located between the vertebral and left subclavian.
C: Illustrates a vertebral artery arising from the aortic arch and entering the fourth transverse foramen. The thoracic duct is located between the vertebral and left subclavian. A rudimentary second vertebral artery lies behind the duct and enters the sixth transverse foramen.

From Adachi, B. *Anatomie der Japaner. Das Arteriensystem der Japaner.* Verlag der Kaiserlich-Japanischen Universität zu Kyoto. In Kommission bei "Maruzen Co.", Kyoto und Tokyo. Gedruckt von "Kenkyusha" in Tokyo, 1928.

FIG. 42.
Rare Variation of Vertebral Artery.

The left vertebral artery arose from two trunks from the aortic arch: one from the arch which joined the vertebral, which was itself a branch of the thyrocervical trunk. The branching order was right subclavian, thyroidea ima, left common carotid, vertebral (medial trunk), thyrovertebral trunk, left common carotid. *Aod*, descending aorta; *Aa*, aortic arch; *Aoa*, ascending aorta; *sd*, a. right subclavian; *tim*, a. thyroidea ima; *Cs*, a. left common carotid; *Cd*, a. right common carotid; *mv*, medial vertebral branch; *tt*, thyrovertebral trunk; *ss*, a. left subclavian; *ttc* thyrocervical trunk; *vd*, a. right vertebral; *vs*, a. left vertebral; *lv*, lateral vertebral branch; *ti* a. left inferior thyroid; *Aan*, a. innominate.

From Kemmetmüller, H. Über eine seltene Varietät der Art. vertebralis. *Anat. Hefte* 44:305–362, 1911.

Cardiovascular System

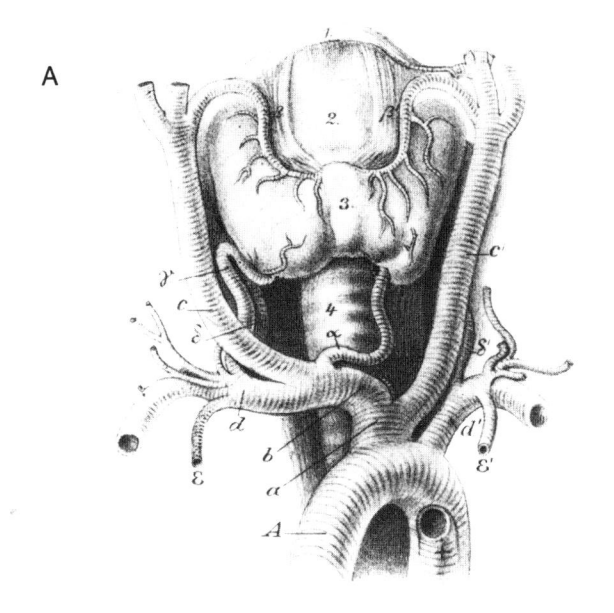

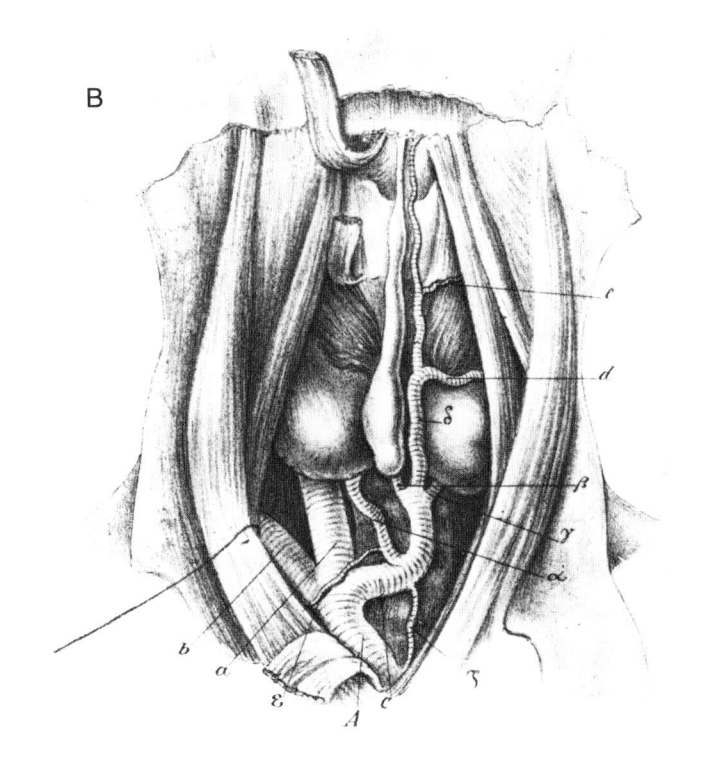

FIG. 43.
Thyroidea Ima Artery.

A: *1*, Os hyoideus; *2*, larynx; *3*, thyroid gland; *4*, trachea; *A*, aortic arch; *a*, common trunk from innominate (brachiocephalic) and left common carotid; *b*, brachiocephalic trunk; *c*, right common carotid; *d*, right subclavian; *d'*, left common carotid; α, thyroidea ima; β,β', superior thyroid; δ,δ', vertebral; ε,ε', internal thoracic; +, ductus arteriosus.

B: *A*, Brachiocephalic trunk; *a*, common carotid; *b*, subclavian; *c*, thyroidea ima; *d*, left superior thyroid; *e*, left cricothyroid, α, thyroidea ima branch to right lobe of thyroid gland; β, branch to isthmus; γ, branch to left lobe of thyroid; δ, end branch to root of tongue; ε, Muscular branches.

From Gruber, W. Über die Arteria thyreoidea ima. *Arch. Pathol. Anat. Physiol. Klin. med.* 54:445–485, 1872.

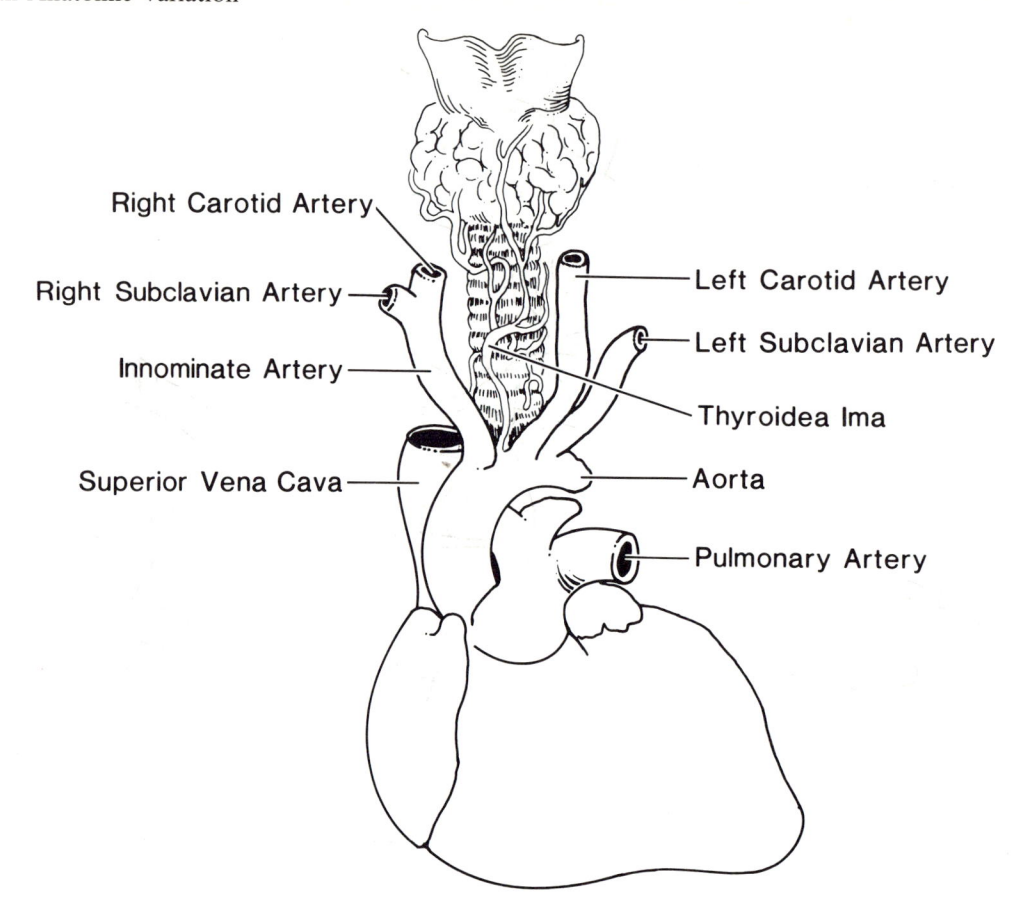

FIG. 44.
Thyroidea Ima Artery
(after Henle).
This artery may occur in 4 to 10% of individuals. In Henle's illustration, the artery is shown arising from the aortic arch. However, it may also arise from the internal thoracic, pericardiacophrenic, subclavian, thyrocervical trunk, inferior thyroid, or transverse scapular arteries.

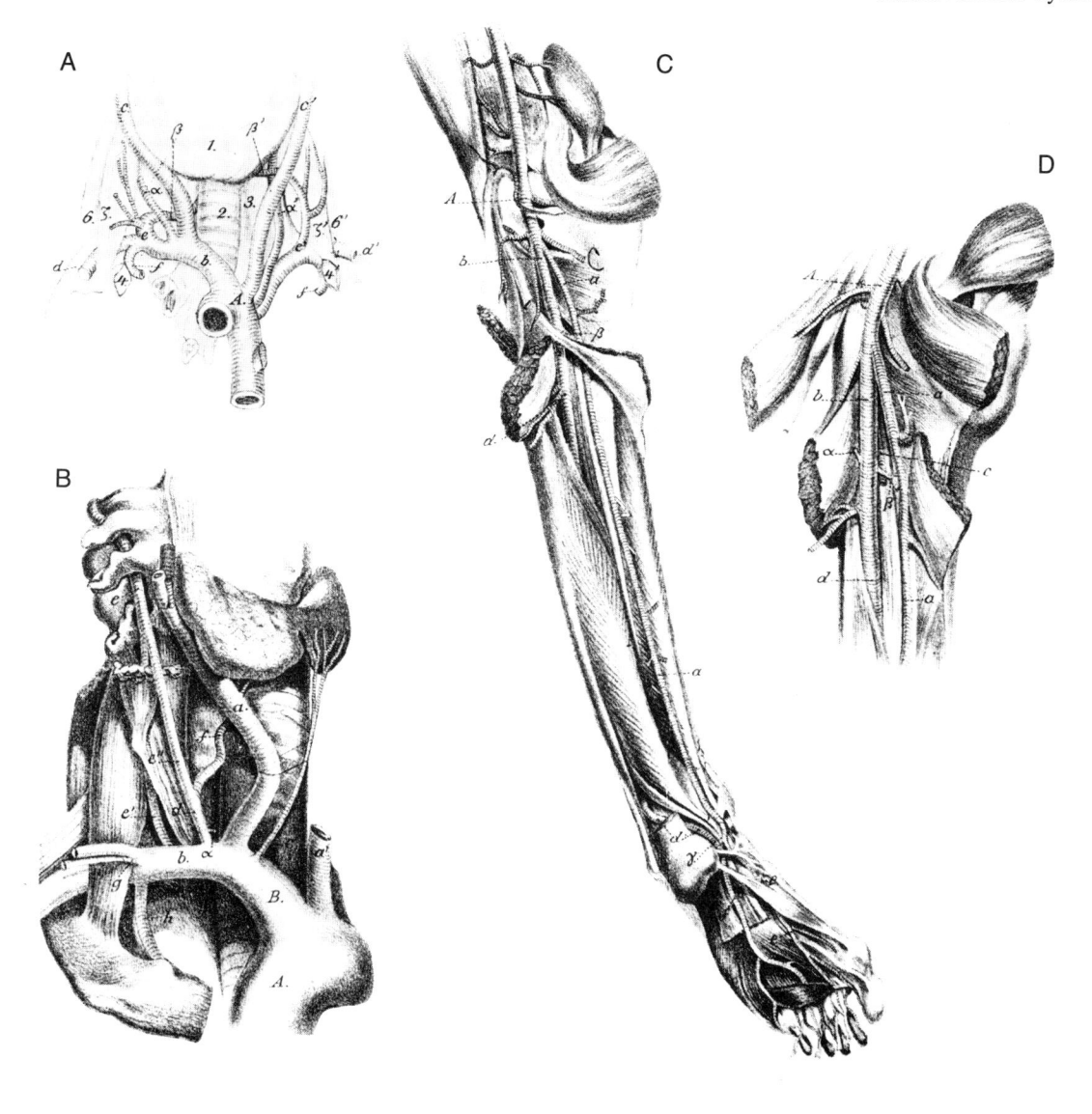

Cardiovascular System

FIG. 45.
Origin of Right Vertebral Artery Between the Common Carotid and Right Subclavian with Inferior Thyroid Artery Crossing in Front of Its Course.
Doubled Origin of Right Vertebral Artery.
High Division of Popliteal Artery.

A: *1*, Thyroid gland; *2*, trachea; *3*, esophagus; *4,4'*, first rib; *5,5'*, m. scalenus anterior; *6,6'*, m. scalenus medii; *A*, aortic arch; *b*, right brachiocephalic trunk; *c,c'*, common carotids; *d,d'*, subclavian arteries; *e,e'*, thyrocervical trunk. *f,f*, internal thoracic arteries; *α,α'*, vertebral arteries; *β,β'*, inferior thyroid arteries.
B: *A*, Aortic arch; *B*, right brachiocephalic trunk; *a*, right common carotid; *a'*, left common carotid; *b*, right subclavian; *c*, thyroidea ima; *d*, thyrovertebral trunk; *e*, doubled origin of right vertebral artery; *e'*, the usual branch of origin of the vertebral artery from a thyrovertebral trunk; *f*, right inferior thyroid artery; *g*, common trunk of origin for transverse cervical and suprascapular arteries; *h*, internal thoracic artery; *α*, mediastinal branch.

C includes the left knee, leg, and foot; D is the upper end of the same preparation. *A*, High division of the popliteal artery; *a*, tibialis posterior artery; *b*, common trunk for the peroneal and tibialis anterior arteries; *c*, tibialis anterior artery; *d*, peroneal artery; *e*, medial plantar artery; *f*, lateral plantar artery; *α*, superior fibular (peroneal) artery; *β*, nutrient fibular (peroneal) artery; *γ*, very short anastomotic branch at the end of a. tibialis posterior joining the end of the peroneal artery.

From Gruber, W. Ursprung der Arteria vertebralis dextra von der Subclavia knapp neben der Carotis, mit Kreuzung der Thyreoidea inferior von vorn während ihres Verlaufes. *Arch. Pathol. Anat. Physiol. Klin. Med.* 74:433–434, 1878; Zweiwurzlige Arteria vertebralis dextra bei Ursprung der accessorischen Wurzel von einem, vom Anfange der Subclavia entstandenen Truncus thyreo-vertebralis (und mit Vorkommen einer Arteria thyreoidea ima). (Neue Varietät.) *Ibid.* 74:435–438, 1878; and Hohe Theilung der Arteria poplitea in die A. tibialis postica und in den Truncus communis für die A. peronea und die A. tibialis antica, mit Endigung der A. tibialis postica als A. plantaris interna und der A. peronea als A. plantaris externa. *Ibid.* 74:438–443, 1878.

Human Anatomic Variation

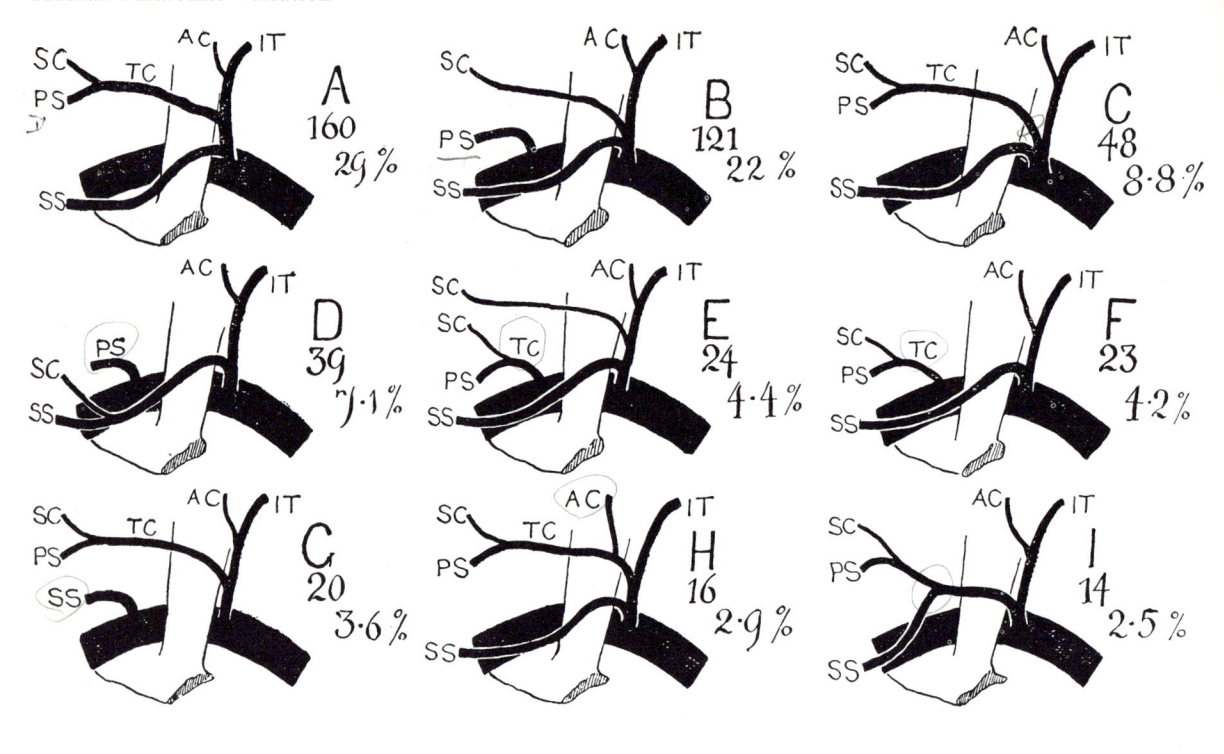

FIG. 46.
Thyrocervical Trunk.

Variations in the branches of the thyroid axis. Total number examined, 544. The various arteries are lettered as follows: inferior thyroid, *IT*; ascending cervical, *AC*; transverse cervical, *TC*; suprascapular, *SS*; superficial cervical, *SC*; posterior dorsal scapular, *PS*; internal (thoracic) mammary, *IM*; vertebral, *V*.

From Thomson, A. Second annual report of the committee of collective investigation of the Anatomical Society of Great Britain and Ireland for the year 1890–91. *J. Anat. Physiol.* 26:76–93, 1891.

FIG. 47.
Thyrocervical Trunk, Rarer Variations.

Variations in the branches of the thyroid axis. Total number examined, 544. The various arteries are lettered as follows: inferior thyroid, *IT*; ascending cervical, *AC*; transverse cervical, *TC*; suprascapular, *SS*; superficial cervical, *SC*; posterior dorsal scapular, *PS*; internal (thoracic) mammary, *IM*; vertebral, *V*.

From Thomson, A. Second annual report of the committee of collective investigation of the Anatomical Society of Great Britain and Ireland for the year 1890–91. *J. Anat. Physiol.* 26:76–93, 1891.

Cardiovascular System

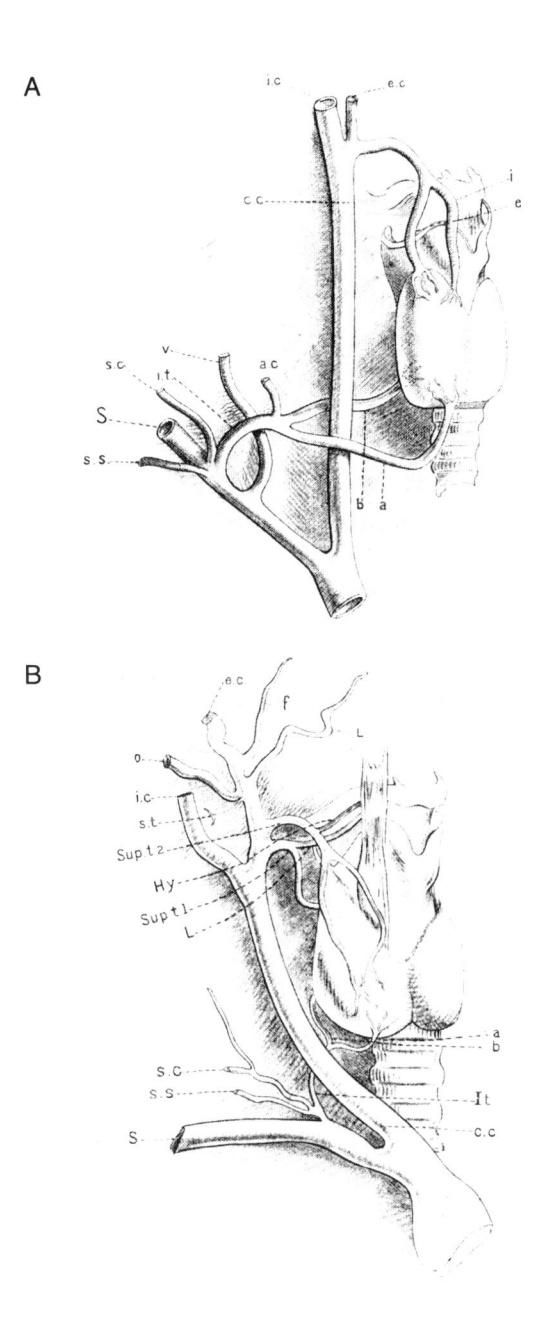

FIG. 48.
Variations in Inferior and Superior Thyroid Arteries.

A: "The right superior thyroid arose from the external carotid near its origin, and divided into medial and lateral branches. The inferior thyroid arose from the subclavian, gave rise to the ascending cervical, and immediately divided into superior and inferior branches. The former passed upwards and deep to the carotid sheath and beneath the right lobe of the thyroid gland, and distributed to the posterior, lower, and lateral part of that lobe. The latter passed in front of the carotid sheath at a more inferior level than the superior branch, and after a more prolonged course supplied the medial and inferior part of the right lobe of the gland, and the anterior surface."

B: "Two right superior thyroid arteries arose from the medial part of the external carotid. The upper superior thyroid, which was larger, passed downward and medially, and divided into two branches near the superior border of the thyroid cartilage; of these, the medial branch, passed beneath the sternohyoid and over the thyrohyoid, and then downward over the thyroid gland, and distributed to the medial part of the anterior surface and the isthmus, and gave rise to a cricothyroid branch at the lower border of the thyroid cartilage. The lateral branch passed beneath the sternothyroid, and crossed the right lobe obliquely from above downward and inward, and distributed to the anterior surface, lateral to the medial branch, reaching as far as the lower third of the gland."

Abbreviations for both illustrations: a.c, Ascending cervical artery; c.c, common carotid artery; e, superior thyroid artery, lateral branch; e.c, external carotid; f, facial artery; Hy, hyoid artery; i, superior thyroid artery, medial branch; i.c, internal carotid artery; i.t (l.t), inferior thyroid artery: a, anterior branch, and b, deep branch; L, lingual artery; o, occipital artery; S, subclavian artery, right; s.c, superficial cervical artery; s.s, suprascapular artery; Sup.t 1, superior thyroid artery, first; Sup.t 2, superior thyroid artery, second; V, vertebral artery.

Autors' note: Quain found three cases of double superior thyroid arteries in 292 cases. Tiedemann and Gruber also reported double superior thyroid arteries.

From Anderson, R.J. Abnormal arrangement of the thyroid arteries. *J.Anat. Physiol.* 14:353–356, 1879.

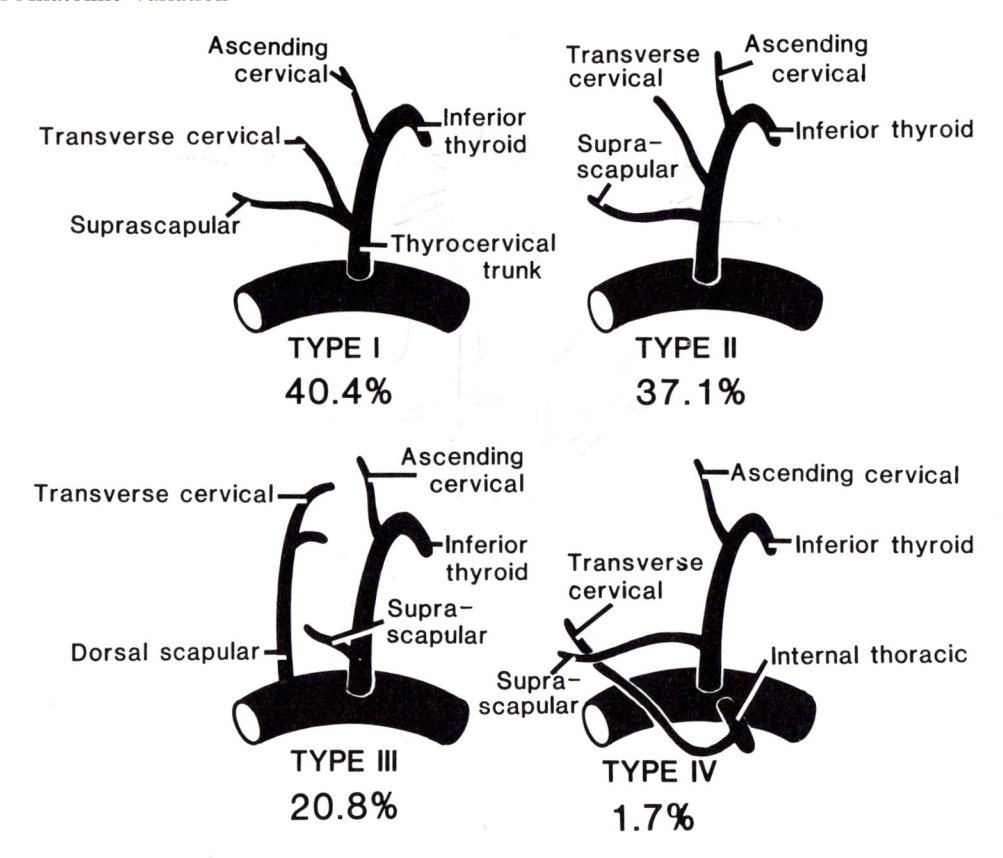

FIG. 49.
Variations in Origin of Transverse Cervical Artery.
Variations of the thyrocervical trunk and in the origin of the transverse cervical, dorsal scapular, and suprascapular arteries are shown, including the frequency of the four types reported by Huelke.

Redrawn from Huelke, D.F. A study of the transverse cervical and dorsal scapular arteries. *Anat. Rec.* 132:235–245, 1958.

Cardiovascular System

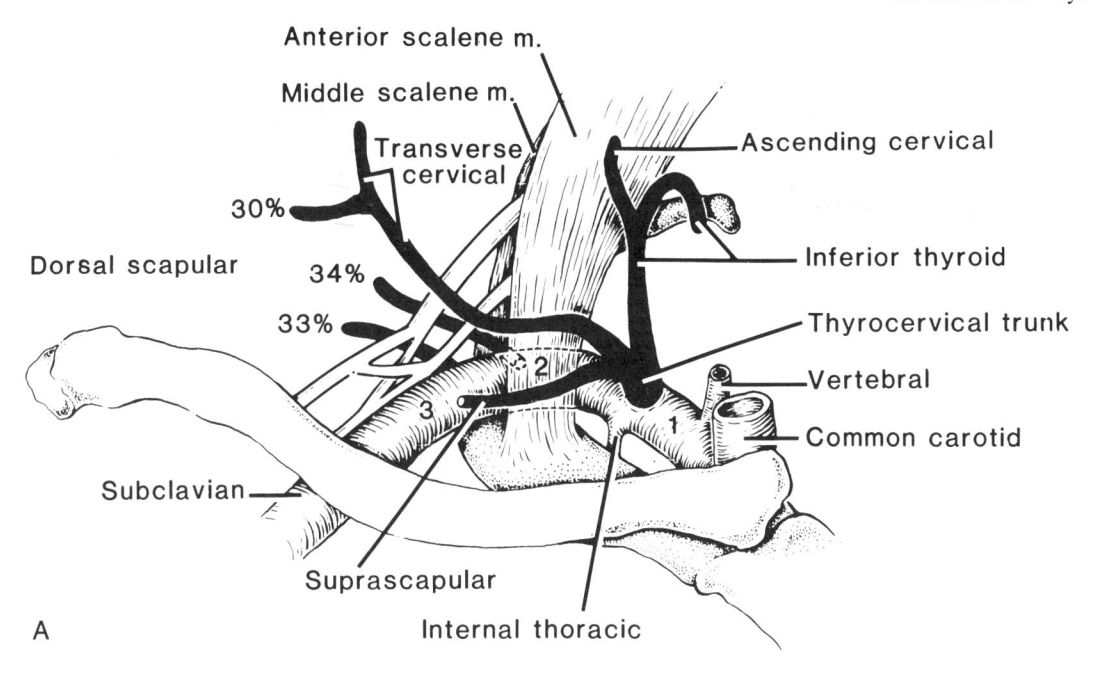

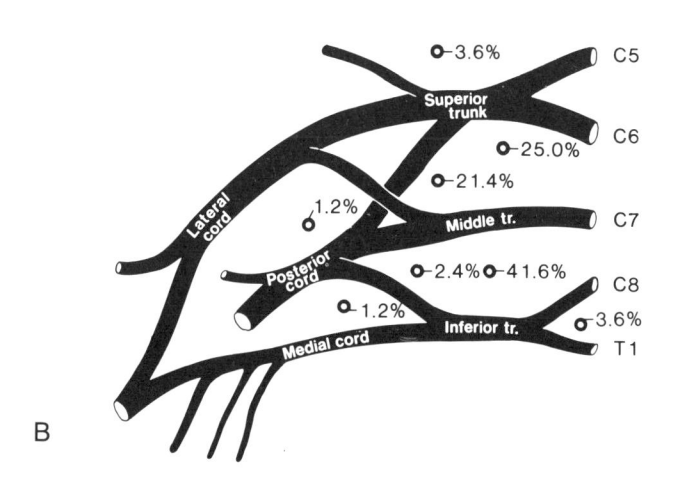

FIG. 50.
Origin of Dorsal Scapular Artery (Three Major Sites).
Frequency of Position of Dorsal Scapular Artery as It Passes Through Brachial Plexus.

A: The dorsal scapular artery arose from the thyrocervical trunk in 30% of cases, from the second part of the subclavian (behind the anterior scalene muscle) in 34%, and from the third part of the subclavian (lateral to the anterior scalene) in 33%. The following table (from Huelke) provides the number and frequency of occurrence of the various sites of origin of the dorsal scapular artery.

The subclavian artery is considered to be divided into three parts by the anterior scalene muscle. The segment of the artery medial to the muscle is subclavian I; the portion behind it is subclavian II, and the part of the artery lateral to the muscle is subclavian III.

B: This illustration shows the frequency of the various positions of the dorsal scapular artery as it passes through the branching of the nerves of the brachial plexus.

Origin	Number	Percent
A. From the transverse cervical artery	54	30.3
B. From subclavian II		
1. Directly	54	30.3
2. With the costocervical trunk	2	1.1
3. With the suprascapular artery	4	2.2
C. From subclavian III		
1. Directly	55	30.9
2. With the costocervical trunk	1	0.6
3. With the suprascapular artery	3	1.7
D. Directly from:		
1. Thyrocervical trunk	1	0.6
2. Subclavian I	1	0.6
3. Axillary I	2	1.2
4. Axillary II	1	0.6

Redrawn from Huelke, D.F. A study of the transverse cervical and dorsal scapular arteries. *Anat. Rec.* 132:235–245, 1958.

Human Anatomic Variation

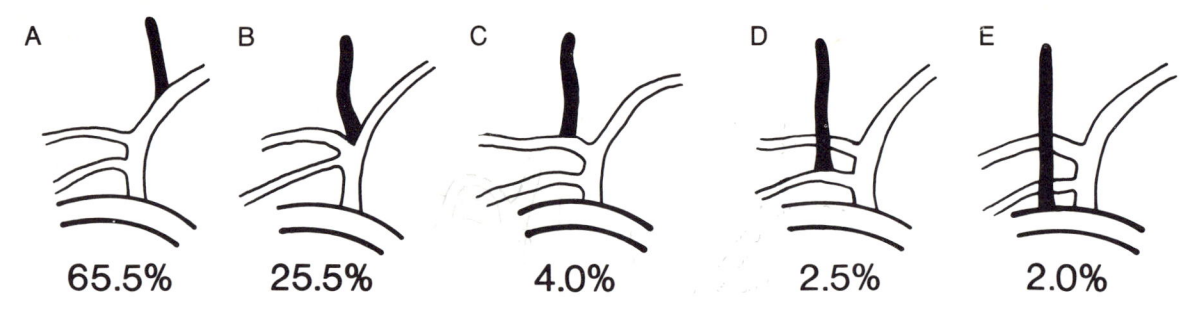

A	B	C	D	E
65.5%	25.5%	4.0%	2.5%	2.0%

FIG. 51.
Origin of Ascending Cervical Artery.

Yazuta found the origin and frequency of origin of the ascending cervical artery (solid black) to be as follows:
A: Inferior thyroid artery in 65.5%.
B: Thyrocervical trunk in 25.5%.
C: Transverse cervical in 4.0%.
D: Suprascapular in 2.5%.
E: Subclavian artery in 2.0%.

Redrawn from Yazuta, K. Zur Varietätenstatistik des Ursprunges und der Lage einiger Äste der A. subclavia. *Anat. Anz.* 63:139–143, 1927.

The following table (provided by Yazuta) is based on a study of 100 bodies (200 sides).

Origin of Ascending Cervical Artery	Right No. %	Left No. %	Total No. %	Dubrevil-Chambardel*
Inferior thyroid artery	64 64	67 67	131 65.5	72%
Thyrocervical trunk	25 25	26 26	51 25.5	–
Transverse cervical artery	4 4	4 4	8 4	11%
Suprascapular artery	3 3	2 2	5 2.5	–
Subclavian artery	3 3	1 1	4 2	8%
Absence of ascending cervical artery	1 1	– –	1 0.5	4%

* Dubrevil-Chambardel, L. *Variations des artères du membre superieur*, Masson, Paris, 1926.

Superior Thyroid Artery

Inferior Thyroid Artery

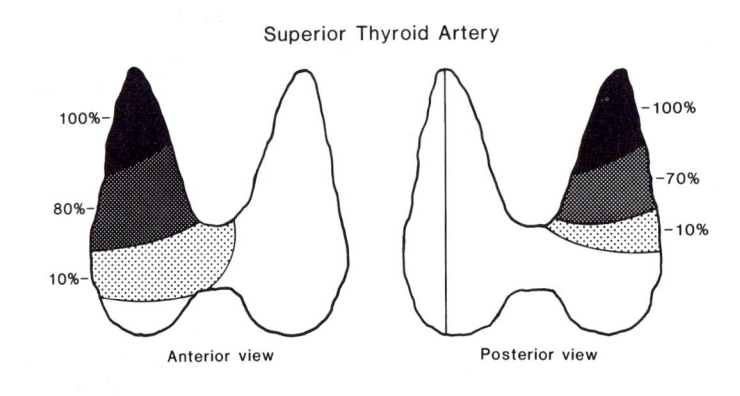

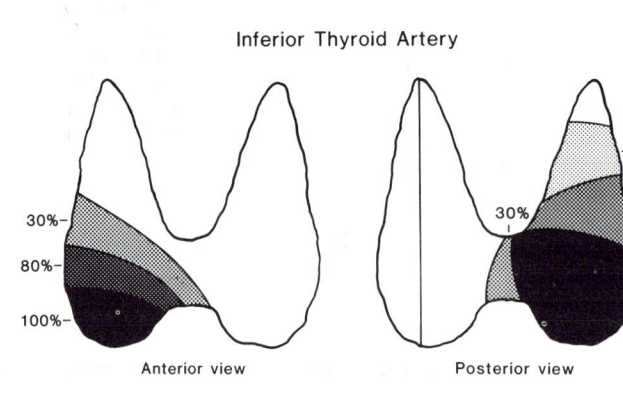

FIG. 52.
Distribution of Superior and Inferior Thyroid Arteries to Thyroid Gland.

The arterial supply to the thyroid gland is illustrated, showing the percent distrubution to various parts of the organ and the overlap between the primary sources. The authors also discuss another source, the thyroidea ima artery, which they encountered in four cases. This artery supplied the isthmus of the thyroid gland, which, in those four cases, was limited to an area left of the midline of the isthmus.

Redrawn from Poulhès, J., Hemous, G. and P. Metreau. La distribution des artères thyroidiennes. *Assoc. Anatomistes, Comptes Rendus* 41:478–488, 1954.

Cardiovascular System

FIG. 53.
Different Forms of Thyrocervical Trunk.

1, Inferior thyroid artery; *2*, ascending cervical artery; *3*, transverse cervical artery; *4*, suprascapular artery; *5*, internal thoracic artery; *6*, costocervical trunk; *7*, vertebral artery.

Inferior Thyroid Artery

A: The inferior thyroid arose from a common stem with the ascending cervical, transverse cervical, and suprascapular arteries. Right side, 85 cases (85.8%); left side, 63 cases (67.7%); total, 148 cases (77%); in men, 100 cases (80%); in women, 48 cases (71.6%).

Pellegrini (reference 979) found this type in Italians, 63 times in 104 cases; Casali (reference 204) found it in Italians, 58 times in 60 cases; and Adachi (reference 7) found it in Japanese, 78 times in 125 body-halves.

D: The inferior thyroid arose with the ascending cervical and suprascapular arteries from a common stem. Right side, 2 cases (2%); left side, 4 cases (4%); total, 6 cases (3.1%); in men, 5 cases (4%); in women, 1 case (1.4%).

Pellegrini found this type once in 104 cases.

I: The inferior thyroid arose with the ascending cervical and transverse cervical arteries from a common stem. Right side, 1 case (1%); left side, 3 cases (3.2%); total, 4 cases (2.0%); in men, 3 cases (2.4%); in women, 1 case (1.4%).

Pellegrini found this type in Italians, 9 times in 104 cases, and Adachi found it in Japanese, 10 times in 121 cases.

The inferior thyroid arose with the transverse cervical and suprascapular arteries from a common stem. Right side, 1 case (0.5%) in a woman.

Pellegrini reported this type once in 104 cases.

B: The inferior thyroid arose with the transverse cervical, ascending cervical, suprascapular, and internal thoracic arteries from a common stem. Right side, 1 case (1%); left side, 9 cases (9.6%); total, 10 cases (5.2%); in men, 4 cases (3.,2%); in women, 6 cases (8.9%).

Pellegrini found this type in Italians, 4 times in 104 cases, and Adachi found it in Japanese, 9 times in 121 cases (7.4%).

The inferior thyroid arose with the ascending cervical, transverse cervical, and internal thoracic from a common stem. Left side, 2 cases (2.1%) in men; total, 2 cases (1%).

Pellegrini found this type once in 104 cases.

E, H: The inferior thyroid arose with the ascending cervical from a common stem. Right side, 3 cases (3%); left side, 5 cases (5.3%); total, 8 cases (4.1%); in men, 3 cases (2.4%); in women, 5 cases (7.4%).

Pellegrini found this type in Italians, 6 times in 104 cases, and Adachi found it in Japanese, 2 times in 121 cases.

C: The inferior thyroid arose alone from the subclavian artery. Right side, 1 case (1%); left side, 4 cases (4.2%); total, 5 cases (2.6%); in men, 2 cases (2.4%); in women, 2 cases (2.9%).

This type was found in English in 1.2%, 4 times in 544 cases, (Committee of Collective Investigation); Germans in 1.7%, 27 times in 237 cases (Taguchi); English in 9.9%, 27 times in 274 cases (Quain); Italians in 14.2%, 15 times in 104 cases (Pellegrini); Americans in 33%, 25 times in 106 cases (Bean, reference 93); and Japanese in 4.5%, 13 times in 228 cases (Adachi).

The inferior thyroid arose from the vertebral artery. Left side only, 1 case (1%) in a man; total, ½ case (0.5).

Pellegrini found this one time in 104 cases, and Adachi found it twice in 286 cases.

F: The inferior thyroid artery was absent. Right side, 2 cases (2%); left side, 5 cases (5.3%); total, 7 cases (3.6%); in men, 4 cases (3.2%); in women, 3 cases (4.4%).

The absence of the inferior thyroid was reported by others as follows: in English in 0.2%, 1 in 544 cases (Committee of Collective Investigation); in Italians in 1.6%, 2 times in 273 cases (Pellegrini); in English in 2.2%, 6 times in 273 cases (Quain); in Americans in 2.3%, 3 times in 131 cases (Dwight); in Swiss in 3.6%, 4 times in 112 cases (Streckeisen); in Germans in 3.8%, 9 times in 237 cases (Taguchi); and in Japanese in 5.9%, 17 times in 287 cases (Adachi). Adachi found this to occur on the right side in 3.4%, left side in 8.5%, in men in 4%, and in women in 12.9%.

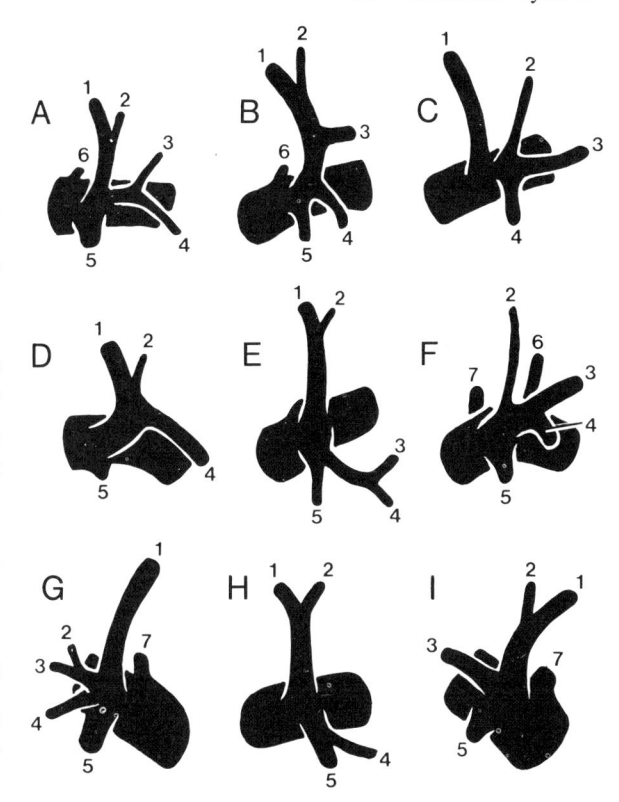

Ascending Cervical Artery

A, B, D, E, H, I: The ascending cervical arose from the inferior thyroid artery. Right side, 76 cases (79.1%); left side, 68 cases (74.7%); total, 144 cases (79.1%); in men, 95 cases (77.2%); in women, 53 cases (87.5%).

C, F: The ascending cervical arose with the transverse cervical artery from a common stem. Right side, 3 cases (3.1%); left side, 5 cases (5.7%); total, 8 cases (4.3%); in men, 6 cases (4.8%); in women, 2 cases (3.1%).

G: The ascending cervical arose with the transverse cervical artery from a common stem. Right side, 21 cases (21.8%); left side, 19 cases (20.8%); total, 40 cases (21.3%); in men, 19 cases (15.4%); in women, 11 cases (17.1%).

The ascending cervical arose with the suprascapular artery from a common trunk (with other vessels); one case, in a man (0.5%).

The ascending cervical artery was absent in only one case in Röhlich's series.

Transverse Cervical Artery

A: The transverse cervical arose in common with the suprascapular artery from the thyrocervical trunk. Right side, 41 cases (45.5%); left side, 33 cases (39.2%); total, 74 cases (42.6%); in men, 49 cases (44.5%); in women, 25 cases (39%).

The transverse cervical arose from the inferior thyroid which arose from a common stem with the suprascapular and the internal thoracic arteries. Right side, 2 cases (2.2%); left side, 6 cases (7.1%); total, 8 cases (4.5%); in men, 4 cases (3.6%); in women, 4 cases (6.2%).

B, I: The transverse cervical arose from the thyrocervical trunk. Right side, 41 cases (45.5%); left side, 36 cases (42.8%); total, 77 cases (44.2%); in men, 44 cases (41.8%); in women, 31 cases (48.4%).

Human Anatomic Variation

C, F: The transverse cervical arose from the inferior thyroid which arose from a common stem with the ascending cervical and suprascapular arteries. Right side, 3 cases (3.3%); left side, 5 cases (5.9%); total, 8 cases (4.5%); in men, 6 cases (4.5%); in women, 2 cases (3.1%).

The transverse cervical artery arose from the thyrocervical trunk from which the ascending cervical and internal thoracic arteries arose from a common stem. Left side, 1 case (1.1%); total, 1 case (0.5%); in men, left side, 1 case (1.9%); total, 1 case (0.9%).

D: The transverse cervical artery was absent. Right side, 4 cases (4.4%); left side, 5 cases (5.9%); total, 9 cases (5.1%); in men, 7 cases (6.3%); in women, 2 cases (3.1%).

Suprascapular Artery

B, D, G.: The suprascapular artery arose form the thyrocervical trunk. Right side, 41 cases (45.5%); left side, 34 cases (40.5%); total, 75 cases (43.1%); in men, 45 cases (40.9%); in women, 30 cases (46.8%).

The suprascapular artery arose from the thyrocervical trunk with the transverse cervical artery.

The suprascapular arose with the transverse cervical and the internal thoracic arteries from a common stem. The inferior thyroid artery was absent or it arose directly from the subclavian artery.

H: The suprascapular arose with the internal thoracic from a common stem. The inferior thyroid was absent or arose directly from the subclavian artery. Left side, 1 case (1.5%); total, 1 case (0.5%) in a man.

The suprascapular arose together with the ascending cervical artery. Left side, 1 case (1.1%); total, 1 case (0.5%) in a man.

I: The suprascapular artery was absent. Right side, 3 cases (3.3%); left side, 4 cases (4.7%); total, 7 cases (4%); in men, 4 cases (3.6%); in women, 3 cases (4.3%).

Redrwan from Röhlich, K. Über den Truncus thyreocervicalis des Menschen. Anat. Anz. 90:129–148, 1940–41.

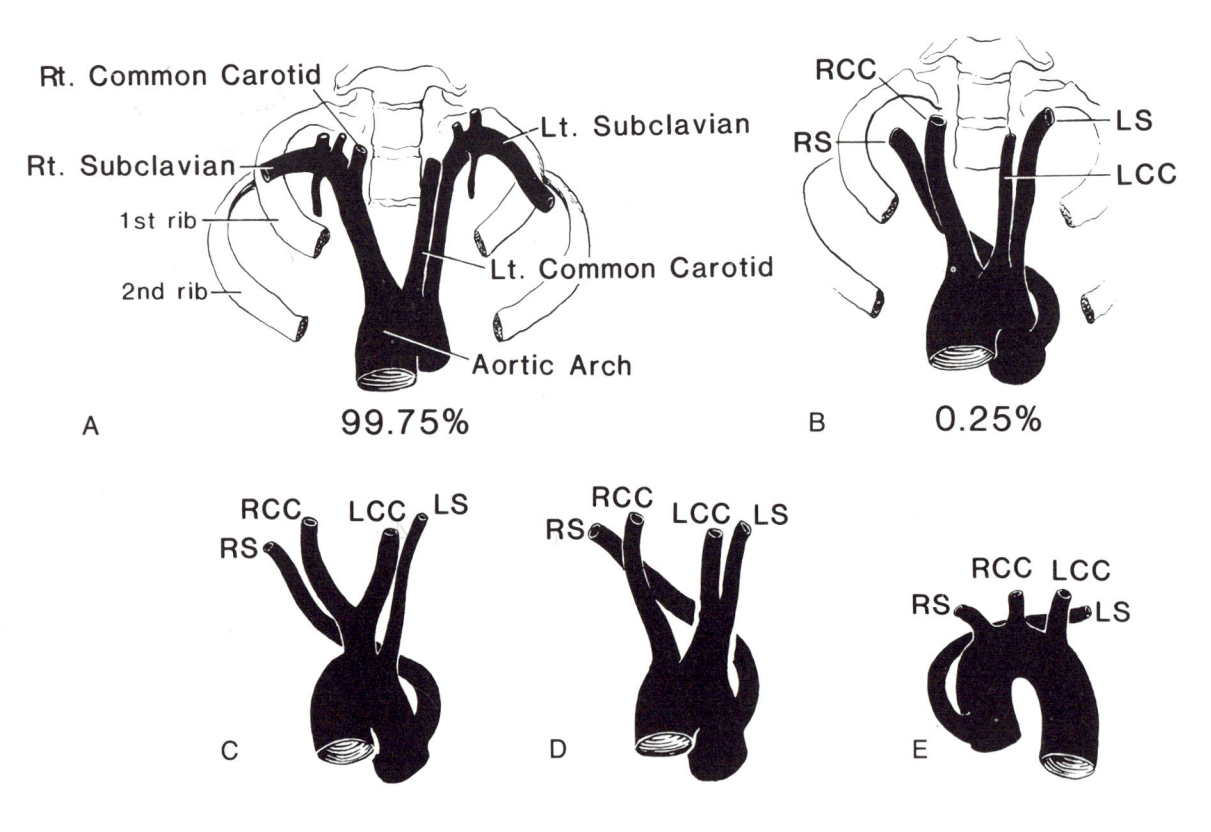

FIG. 54.
Subclavian Artery.

This study was based on 400 laboratory specimens.

A: The common vascular pattern encountered in 399 of the 400 bodies studied. On the right side, the common carotid [RCC] and subclavian [RS] arteries arise as branches of the innominate artery, while on the left [LCC, LS] they arise directly from the descending aortic arch.

B: An anomalous right subclavian artery arises from the descending thoracic aorta distal to the left subclavian artery.

C: In addition to the anomalous right subclavian artery, a bicarotid axis [occurs] from which both the right and the left common carotid arteries take origin.

D: An example of this same anomalous right subclavian artery arising from the descending aorta, but here associated with a left innominate artery.

E: A right aortic arch with the left subclavian artery arising as the last branch from the arch, and crossing behind the esophagus en route to the left upper extremity. In their series, Daseler and Anson found no examples of the variations depicted in the lower three illustrations.

Redrawn from Daseler, E.H. and B.J. Anson. Surgical anatomy of the subclavian artery and its branches. Surg. Gynecol. Obstet. 108:149–174, 1959.

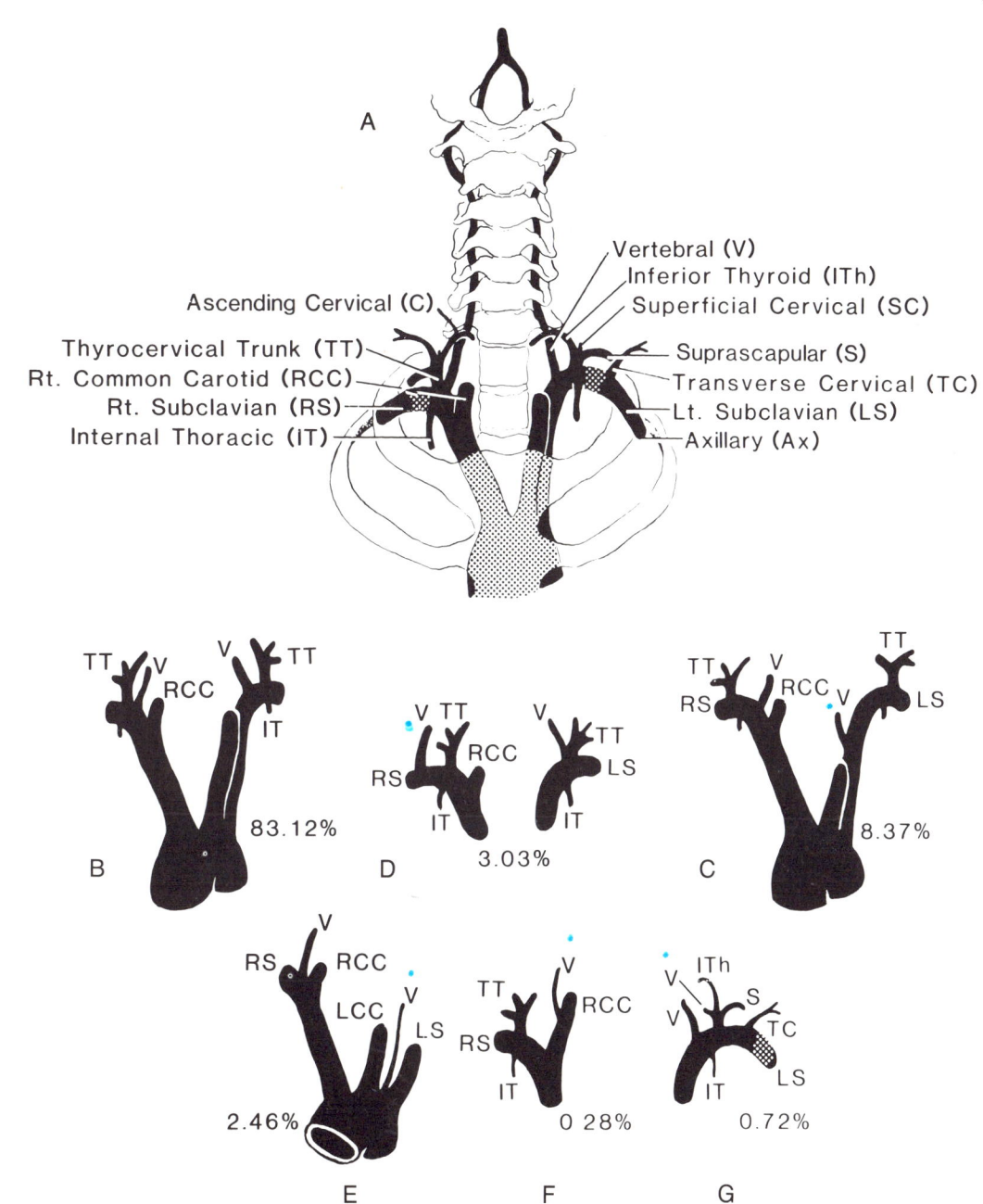

FIG. 55.
Vertebral Artery.

This study was based on 693 laboratory specimens.

A, B: Usual vascular pattern with vertebral artery arising from upper, posterior surface of subclavian artery ½ to 2 cm medial to thyrocervical trunk; 576 of 693 specimens, 83.12%.

C: Vertebral artery displaced more than 2 cm medial to the thyrocervical trunk but still derived from subclavian artery; 58 of 693 specimens, 8.37%; 25 right, 33 left.

D: Vertebral artery arises from subclavian artery lateral to or as common trunk with thyrocervical trunk; 21 of 693 specimens, 3.03%; 17 right, 4 left. In all of these cases, the vertebral artery arose from a common stem with the thyrocervical trunk on the left side. In only 3 of these specimens did the right vertebral artery arise as a branch of the second portion of the subclavian artery.

E: Left vertebral artery arising as direct branch of aortic arch between left common carotid and left subclavian arteries; 17 of 693 specimens, 2.46%. Right vertebral artery, however, arising at site of bifurcation of innominate artery into right subclavian and right common carotid arteries; 7 of 693 specimens, 1.11%.

F: Right vertebral artery arising as direct branch of right common carotid artery; 2 of 693 specimens, 0.28%.

G: Dual or accessory vertebral arteries were encounted in 5 of 693 specimens, 0.72%; all were left-sided. Three small accessory vertebral arteries arose as direct branches of the aortic arch, and 2 as a branch of the thyrocervical trunk. In all cases, the larger or main vertebral artery arose from the left subclavian artery.

Redrawn from Daseler, E.H. and B.J. Anson. Surgical anatomy of the subclavian artery and its branches. *Surg. Gynecol. Obstet.* 108:149–174, 1959.

Human Anatomic Variation

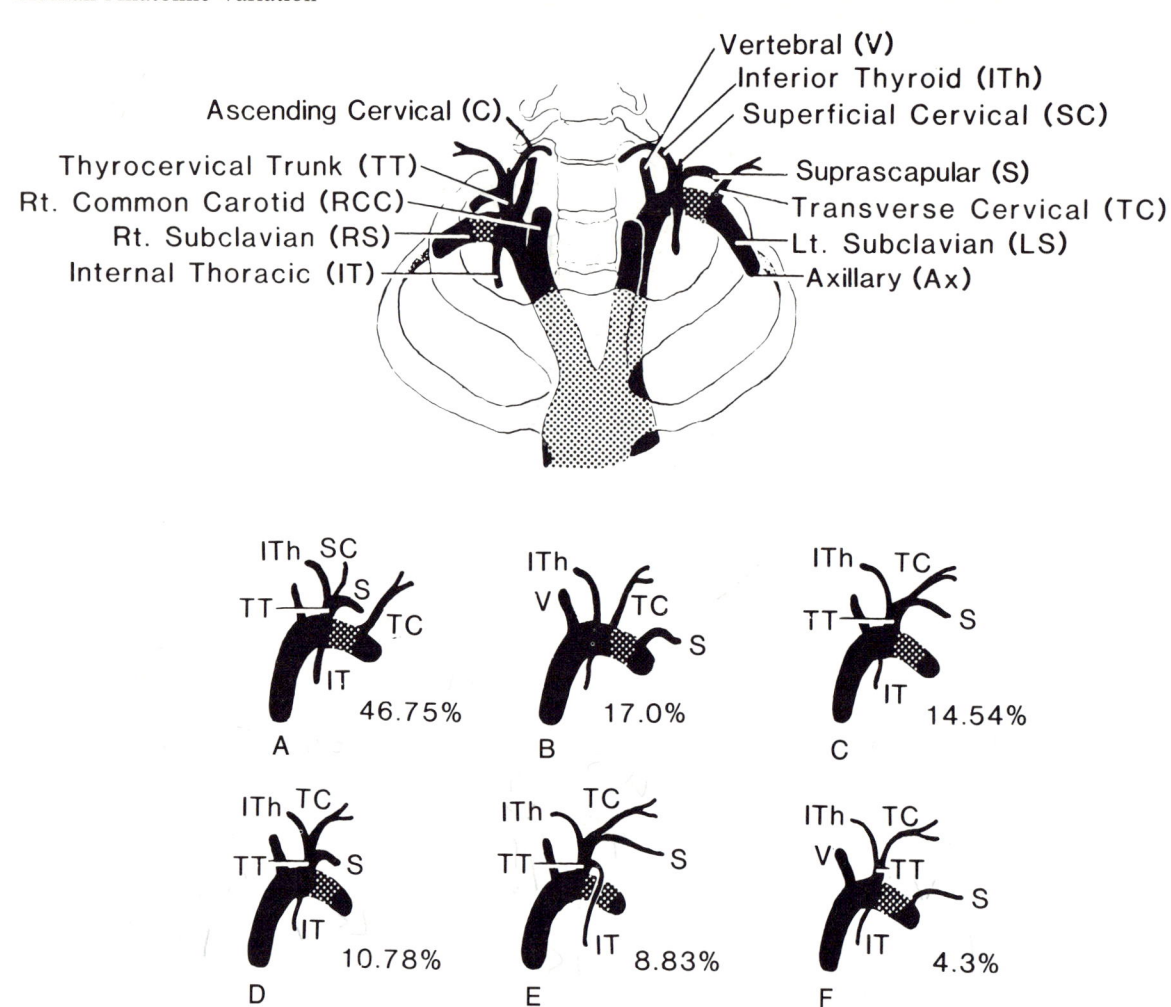

FIG. 56.
Thyrocervical Trunk.

This study was based on 770 laboratory specimens.
A: Transverse cervical directly from subclavian artery; 360 of 770, 46.75%; 180 right, 180 left.
B: Absence of thyrocervical trunk; 131 of 770, 17%; 74 right, 57 left.
C: Transverse scapular and cervical from common trunk off thyrocervical; 112 of 770, 14.54%; 54 right, 58 left.

D: Separate origins of branches from thyrocervical trunk; 83 of 770 specimens, 10.78%; 52 right, 31 left.
E: Internal thoracic from thyrocervical trunk; 68 of 770, 8.83%; 11 right, 57 left.
F: Transverse scapular arising separately from subclavian; 33 of 770, 4.3%; 12 right, 21 left.

Redrawn from Daseler, E.H. and B.J. Anson. Surgical anatomy of the subclavian artery and its branches. *Surg. Gynecol. Obstet.* 108:149–174, 1959.

Cardiovascular System

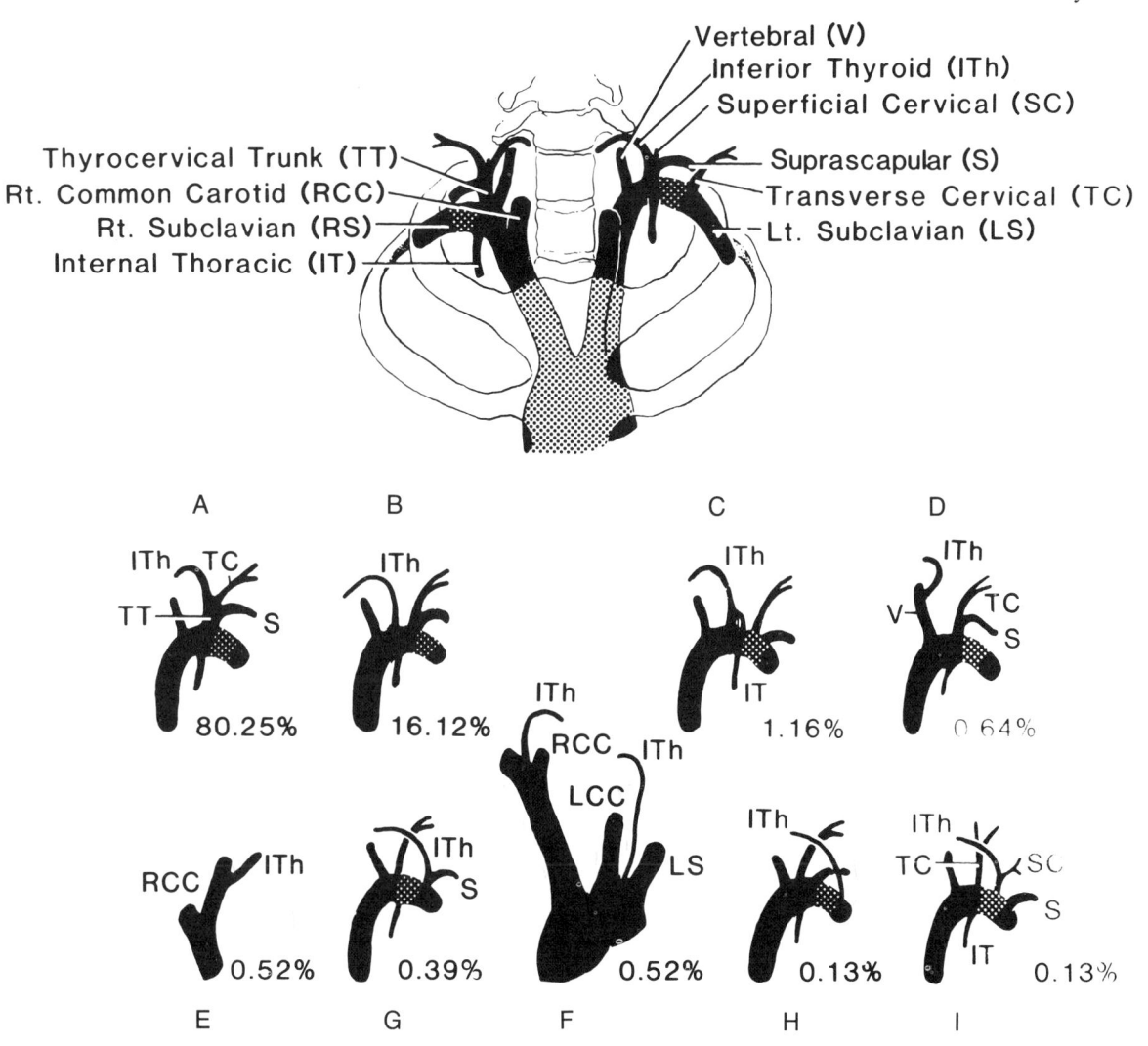

FIG. 57.
Inferior Thyroid Artery.
This study was based on 775 laboratory specimens.
A: Branch of the thyrocervical trunk derived from the first portion of the subclavian artery; 622 of 775 specimens, 80.25%; 301 right, 321 left.
B: Separate and distinct branch from first portion of subclavian artery; 125 of 775 specimens, 16.12%; 74 right, 51 left.
C: Common trunk with internal thoracic artery; 9 of 775 specimens, 1.16%; 5 right, 4 left.
D: Common trunk vertebral artery; 5 of 775 specimens, 0.64%; 4 right, 1 left.
E: Direct branch of common carotid artery; 4 of 775 specimens, 0.52%; all on right side.
F: Vessel of thyroidea ima type, arising at bifurcation of innominate artery on the right, and from the aortic arch between the left common carotid and subclavian arteries on the left; 4 of 775 specimens, 0.52%; 2 right, 2 left.
G: Common trunk with transverse scapular artery derived in turn from third part of subclavian artery; 3 of 775 specimens, 0.39%; 2 right, 1 left.
H: Separate and distinct branch from third portion of subclavian artery; 1 of 775 specimens, 0.13%; right side.
I: Common trunk with costocervical trunk derived in turn from second portion of subclavian artery; 1 of 775 specimens, 0.13%; right side.

Redrawn from Daseler, E.H. and B.J. Anson. Surgical anatomy of the subclavian artery and its branches. *Surg. Gynecol. Obstet.* 108:149–174, 1959.

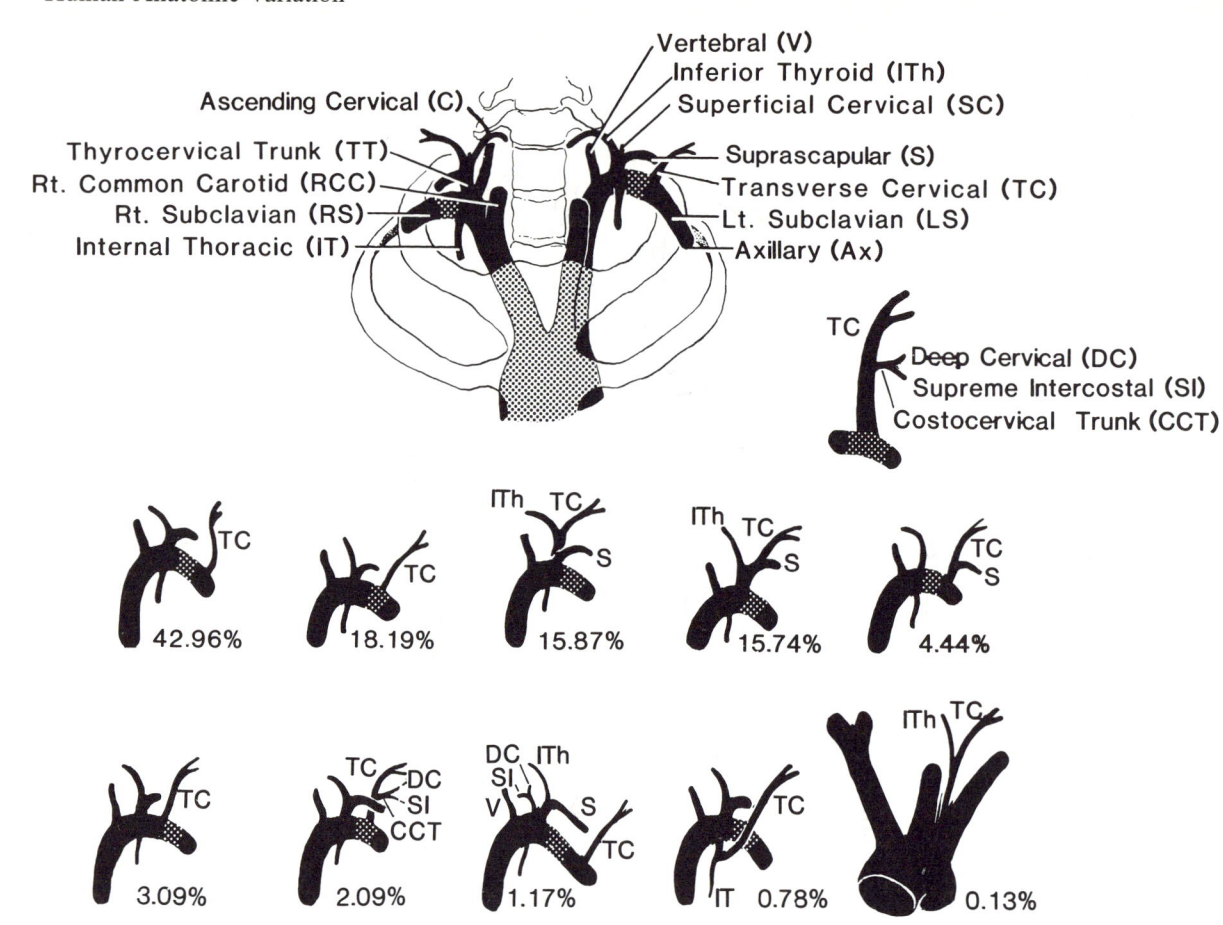

Human Anatomic Variation

FIG. 58.
Transverse Cervical Artery.

This study was based on 765 laboratory specimens.

A: Direct branch from third portion of subclavian artery; 333 of 765 extremities, 42.96%; 177 right, 156 left.
B: Direct branch from second portion of subclavian artery; 140 of 765 extremities, 18.19%; 70 right, 70 left.
C: Independent branch from thyrocervical trunk; 122 of 765 sides, 15.87%; 63 right, 59 left.
D: From fused common trunk with transverse scapular artery. Common branch in turn derived from thyrocervical trunk; 121 of 765 sides, 15.74%; 58 right, 63 left.
E: From fused common trunk with transverse scapular. Trunk in turn arising from subclavian artery; 34 of 765 sides. 4.44%; 20 right, 14 left. Twelve first portion, 2 second portion, 20 third portion of subclavian artery.
F: As direct branch from first portion of subclavian artery; 24 of 765 sides, 3.09%; 12 right, 12 left.
G: From a common vessel with costocervical trunk (or deep cervical artery); 16 of 765 extremities, 2.09%; 6 from first portion, 8 from second portion, and 2 from third portion of subclavian artery.
H: As a direct branch from the axillary artery; 9 of 765 sides, 1.17%; 4 right, 5 left.
I: Branch of internal thoracic artery; 6 of 765 sides, 0.78%; 2 right, 4 left.
J: Branch of a thyroidea ima artery derived in turn from aortic arch; 1 of 765 specimens, 0.13%; left side.

Redrawn from Daseler, E.H. and B.J. Anson. Surgical anatomy of the subclavian artery and its branches. *Surg. Gynecol. Obstet.* 108:149–174, 1959.

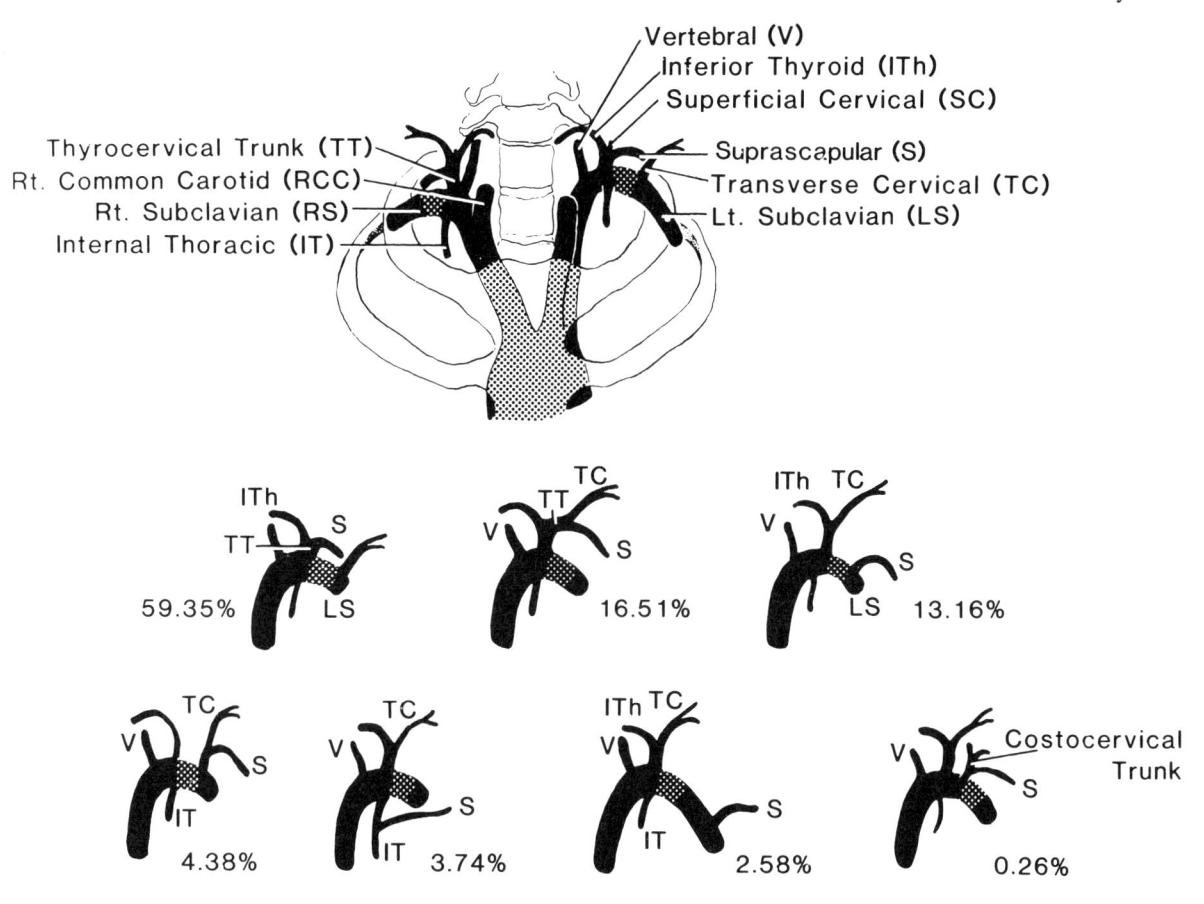

Cardiovascular System

FIG. 59.
Suprascapular Artery.

This study was based on 775 laboratory specimens.

A: Independent branch of thyrocervical trunk; 460 of 775 specimens, 59.35%; 236 right, 224 left.

B: Common trunk with transverse cervical artery derived in turn from the thyrocervical trunk; 128 of 775 specimens, 16.51%; 60 right, 68 left.

C: Direct branch of the subclavian artery; 102 of 775 sides, 13.16%; 53 right, 49 left. First portion subclavian artery, 51 sides; second portion, 8; third portion, 43.

D: Common trunk with transverse cervical artery derived from subclavian artery; 34 of 775 extremities, 4.38%; 20 right, 14 left. Branch of first portion subclavian, 12 specimens; second portion, 2, third portion, 20.

E: Common trunk with internal thoracic artery; 29 of 775 extremities, 3.74%; 9 right, 20 left.

F: Direct branch of the axillary artery; 20 of 775 sides, 2.58%; 8 right, 12 left.

G: From common trunk with costocervical trunk; 2 of 775 specimens, 0.26%; both left.

Redrawn from Daseler, E.H. and B.J. Anson. Surgical anatomy of the subclavian artery and its branches. *Surg. Gynecol. Obstet.* 108:149–174, 1959.

Human Anatomic Variation

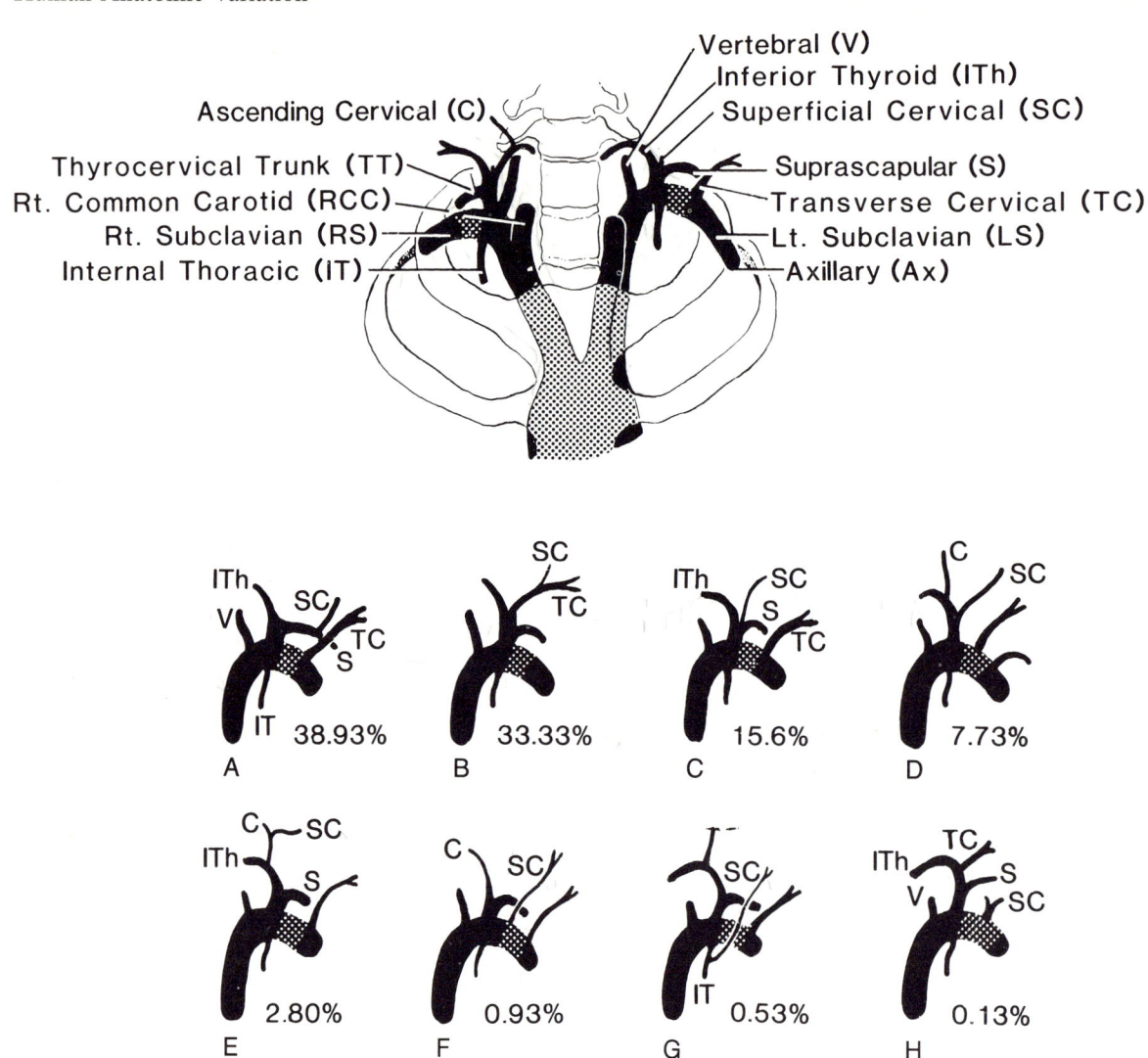

FIG. 60.
Superficial Cervical Artery.

This study was based on 750 laboratory specimens.

A: Branch of suprascapular artery; 292 of 750 sides, 38.93%; 139 right, 153 left.

B: Branch of transverse cervical artery; 250 of 750 sides, 33.33%; 125 right, 125 left.

C: Direct branch of thyrocervical trunk; 117 of 750 extremities, 15.6%; 68 right, 49 left.

D: Branch of the inferior thyroid artery; 58 of 750 extremities, 7.73%; 27 right, 31 left.

E: As common trunk with ascending cervical artery; 21 of 750 sides, 2.80%; 6 right, 15 left. In 31 additional cases, the small ascending cervical arose as a branch of the superficial cervical artery, derived in turn from some other source.

F: Direct branch from first portion of subclavian artery; 7 of 750 extremities, 0.93%; right side.

G: Branch of internal thoracic artery; 4 of 750 sides, 0.53%; 1 right, 3 left.

H: From common trunk with the deep cervical artery or costocervical trunk; 1 of 750 specimens, 0.13%; left side.

Redrawn from Daseler, E.H. and B.J. Anson. Surgical anatomy of the subclavian artery and its branches. *Surg. Gynecol. Obstet.* 108:149–174, 1959.

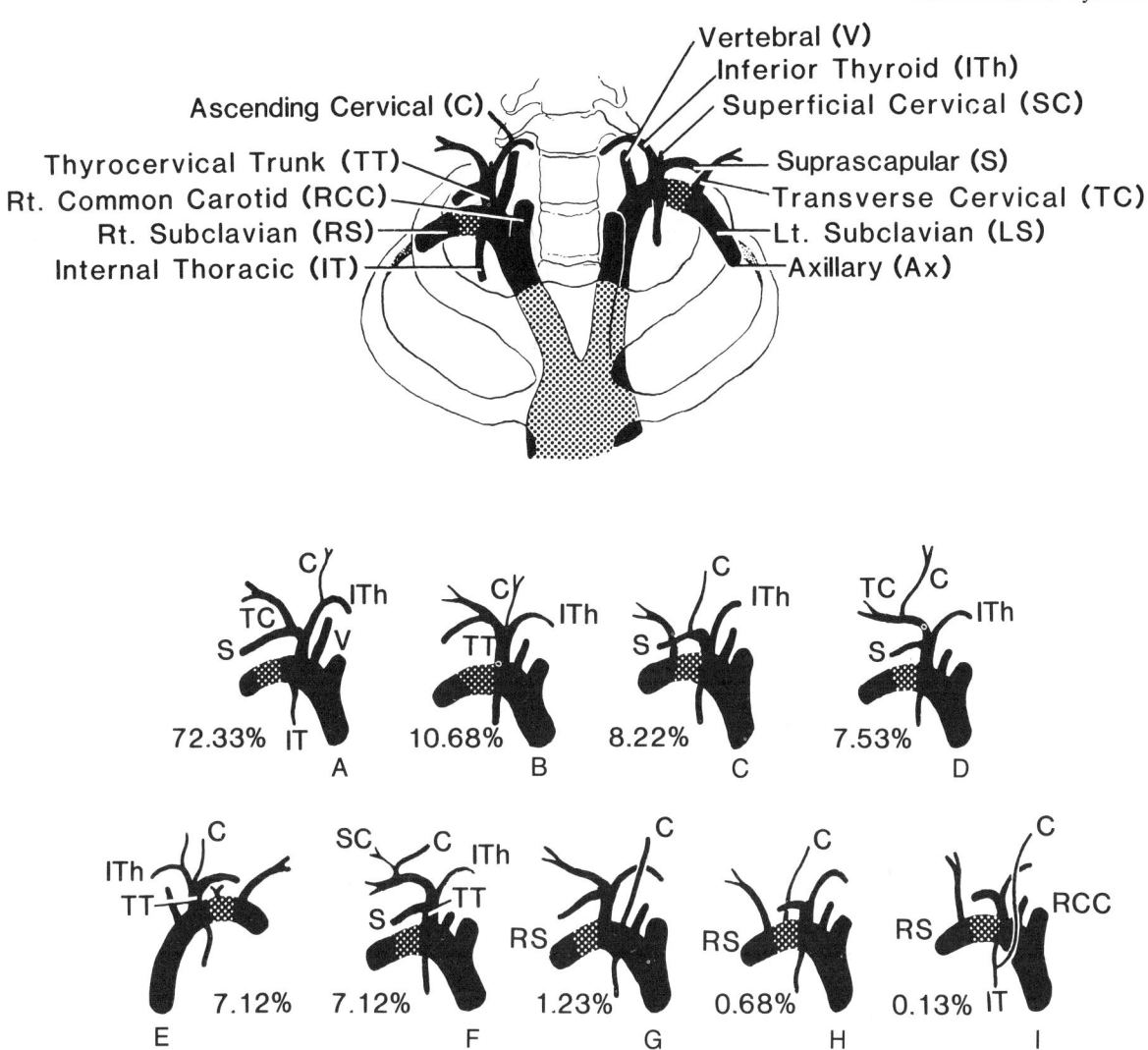

Cardiovascular System

FIG. 61.
Ascending Cervical Artery.

This study was based on 730 laboratory specimens.

A: From apex of arch of inferior thyroid artery; 528 of 730 sides, 72.33%; 266 right, 262 left.

B: As a direct branch of the thyrocervical trunk; 78 of 730 sides, 10.68%; 37 right, 41 left.

C: Branch of the suprascapular artery; 60 of 730 extremities, 8.22%; 29 right, 31 left.

D: Branch of the transverse cervical artery; 55 of 730 extremities, 7.53%; 32 right, 23 left.

E: Dual ascending cervical arteries; 52 of 730 sides, 7.12%; 29 right, 23 left. (In 16 instances, both vessels arose from the inferior thyroid artery; in 28 cases, 1 arose from the inferior thyroid, 1 from the thyrocervical trunk or one of its branches. In 8 instances, both vessels arose from the thyrocervical trunk, or one of its branches other than the inferior thyroid artery.)

F: Branch of the superficial cervical artery; 52 of 730 sides, 7.12%; 27 right, 25 left.

G, H: Direct branch of subclavian artery; 14 of 730 extremities, 1.91%; 11 right, 3 left. From the first portion of subclavian, 9 sides, 1.23%; from the second portion of the subclavian, 5 sides, 0.68%.

I: As a branch of the internal thoracic artery; 1 of 730 sides. 0.13%; 1 left side.

Redrawn from Daseler, E.H. and B.J. Anson. Surgical anatomy of the subclavian artery and its branches. *Surg. Gynecol. Obstet.* 108:149–174, 1959.

Human Anatomic Variation

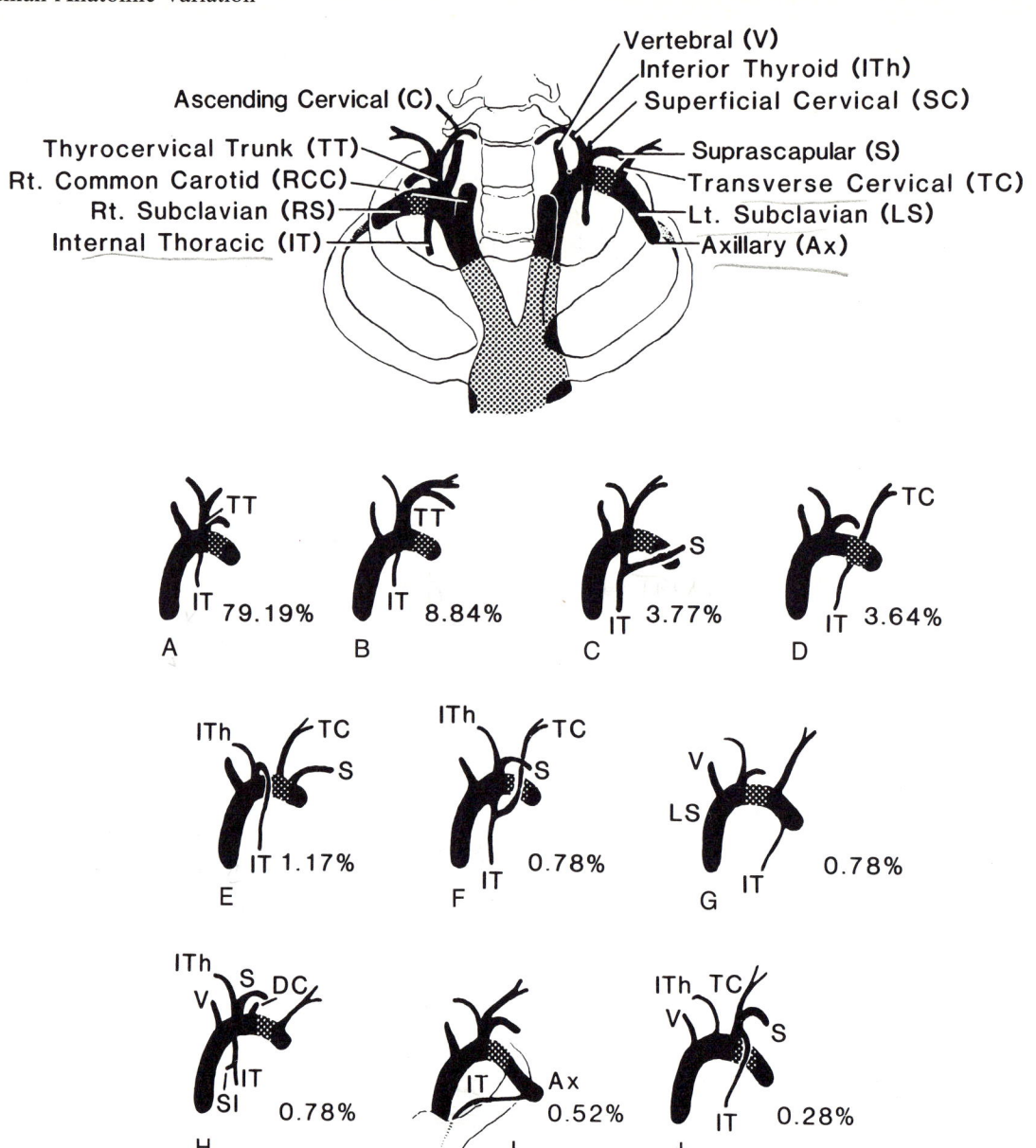

FIG. 62.
Internal Thoracic Artery.

Based on a study of 769 laboratory specimens.
A: Branch of first portion of subclavian artery; 609 of 769 sides, 79.19%; 324 right, 285 left.
B: From common stem with thyrocervical trunk; 68 of 769 extremities, 8.84%; 11 right, 57 left.
C: From common trunk with suprascapular artery; 29 of 769 extremities, 3.77%; 9 right, 20 left.
D: Direct branch of second portion of subclavian artery; 28 of 769 extremities, 3.64%; 6 right, 22 left.
E: From common trunk with inferior thyroid artery; 9 of 769 sides, 1.17%; 5 right, 4 left.
F: From common trunk with transverse cervical artery; 6 of 769 sides, 0.78%; 2 right, 4 left.
G: Direct branch from the third portion of the subclavian artery; 6 of 769 sides, 0.78%; 4 right, 2 left.
H: From common trunk with superior intercostal artery; 6 of 769 sides, 0.78%; 4 right, 2 left.
I: Direct branch from axillary artery; 4 of 769 extremities, 0.52%; 2 right, 2 left.
J: Internal thoracic, transverse cervical, and suprascapular arteries, all from common trunk; 2 of 769 extremities, 0.28%; 1 right, 1 left.

Redrawn from Daseler, E.H. and B.J. Anson. Surgical anatomy of the subclavian artery and its branches. *Surg. Gynecol. Obstet.* 108:149–174, 1959.

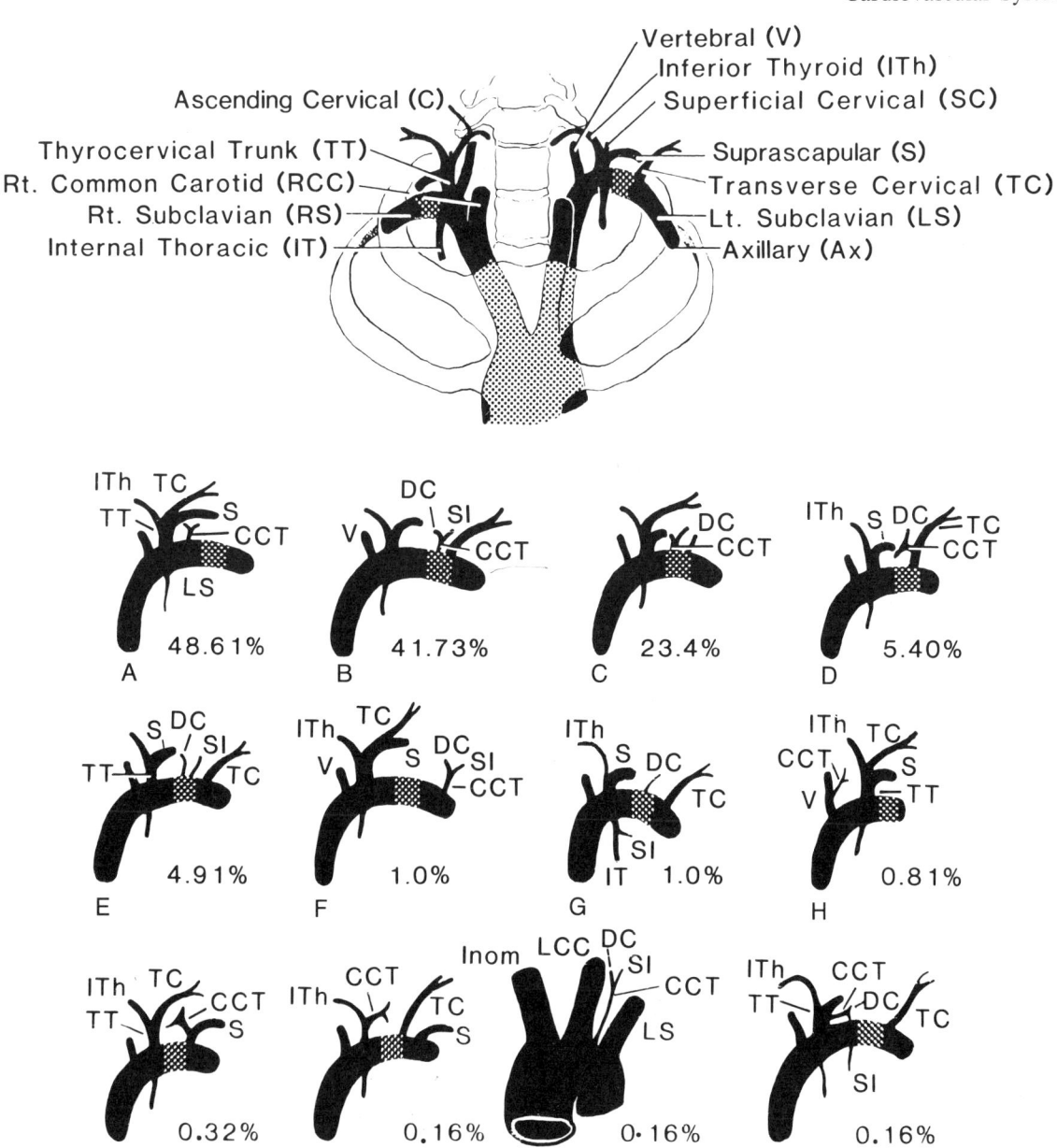

FIG. 63.

Costocervical Trunk.

This study was based on 611 laboratory specimens.

A: Trunk arises from first portion of subclavian artery; 297 of 611 sides, 48.61%; 93 right, 204 left.

B: Trunk arises from second portion of subclavian artery; 255 of 611 sides, 41.73%; 187 right, 68 left.

C: Accessory deep cervical arteries were present in 144 of 611 sides, 23.4%; 55 right, 89 left. Single accessory deep cervical, 133 sides. Dual accessory deep cervicals occurred in 9 sides, triple accessory deep cervicals, 1 side. These vessels may occur more frequently than our [Daseler and Anson] present study might indicate, owing to loss in dissection.

D: Costocervical trunk proper, or its deep cervical branch, arose as a branch from the transverse cervical artery; 33 of 611 extremities, 5.40%; 11 right, 22 left. Costocervical trunk proper, 13 of 611 sides, 2.12%; 5 right, 8 left.

E: Absent costocervical trunk; 30 of 611 extremities, 4.91%; 18 right, 12 left. Both branches from subclavian artery, 24 sides. Deep cervical from subclavian, superior intercostal from internal thoracic, 6 sides.

F: Trunk arises from third portion of subclavian artery; 6 of 611 extremities; 1.0%; 4 right, 2 left.

G: Superior intercostal artery arises as direct branch of internal thoracic artery; 6 of 611 sides, 1.0%; 4 right, 2 left.

H: Trunk arises as branch of vertebral artery; 5 of 611 sides, 0.81%; 2 right, 3 left.

I: Trunk arises as branch of suprascapular artery; 2 of 611 sides, 0.32%; both left.

J: Trunk arises as branch of inferior thyroid artery; 1 of 611 sides, 0.16%; right side.

K: Costocervical trunk arising as direct branch of aortic arch between left common carotid and subclavian arteries; 1 of 611 sides, 0.16%; left side.

L: Trunk arises as branch of thyrocervical trunk; 1 of 611 sides, 0.16%; left side.

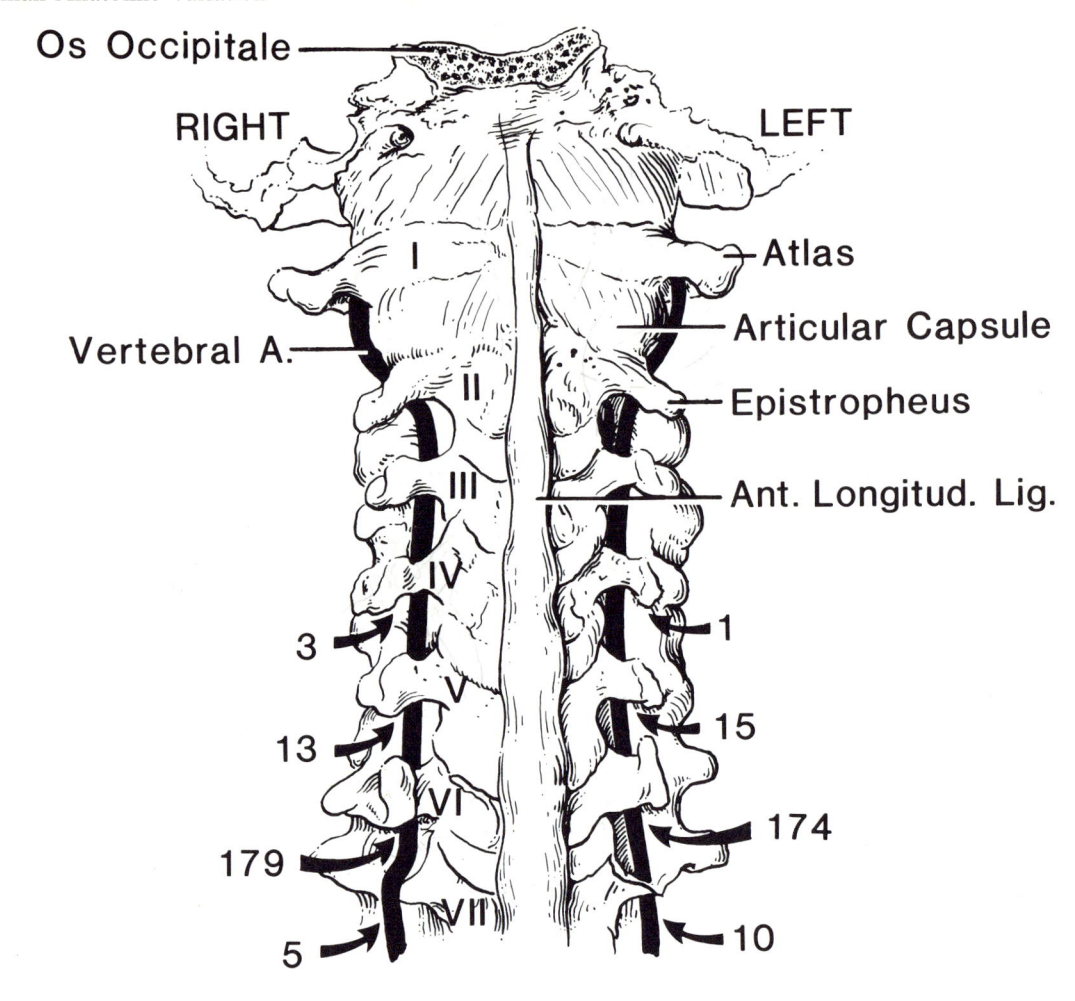

FIG. 64.

Variations in Course and Vertebral Relations of Vertebral Artery.

The point of entry into cervical transverse foramina is variable; in 200 specimens the most frequent point of entrance is the sixth cervical vertebra on both sides; entrance at the level of the fourth or seventh vertebra is rare. None were found to enter cranial to that of the fourth cervical vertebra.

Daseler and Anson (reference 282) also studied 379 bodies in which "detailed studies were made of the site of entrance of the vertebral arteries into the foramina of the transverse processes of the cervical vertebrae. In 41 specimens, 5.41 percent, 16 on the right and 25 on the left side, the vertebral artery entered the foramen in the transverse process of the seventh cervical vertebra. In 663 of the 758 sides examined, 87.46 percent, 337 on the right and 326 on the left, the vertebral artery entered the foramen at the level of the sixth cervical vertebra. In 50 of the sides, 6.6 percent, 23 on the right and 27 on the left, the vertebral artery ascended to the level of the fifth cervical vertebra before entering the foramen of its transverse process. In 5 specimens, 3 right and 2 left, the vertebral artery rose to the level of the fourth cervical vertebra before entering the foramen.

Yazuta (reference 1321) has pointed out that in approximately 50 percent of cases in which the vertebral artery ascends to a level above the sixth cervical vertebra before entering its foramen, the inferior thyroid artery will cross beneath or dorsal to this anomalous vertebral trunk. This is particulary true when the origin of the vertebral artery is also anomalous, as from the aorta, innominate, or common carotid arteries. A surgeon unaware of this relatively common variation might well encounter serious or even fatal hemorrhage and do untold damage to such important adjacent structures as the brachial plexus, thoracic duct, common carotid artery, or jugular vein if this vessel were accidentally torn or severed while attempting to mobilize or ligate the inferior thyroid artery as it crosses beneath the carotid sheath in doing a thyroidectomy, or attempting to excise a pharyngeal diverticulum."

Redrawn from Anson, B.J. *An Atlas of Human Anatomy.* W. B. Saunders Co., Philadelphia, 1950; and Bell, R.H., Swigart, L.L. and B.J. Anson. The relation of the vertebral artery to the cervical vertebrae. Based on a study of 200 specimens. *Q. Bull. Northwestern Universitiy Medical School* 24:184–185, 1950.

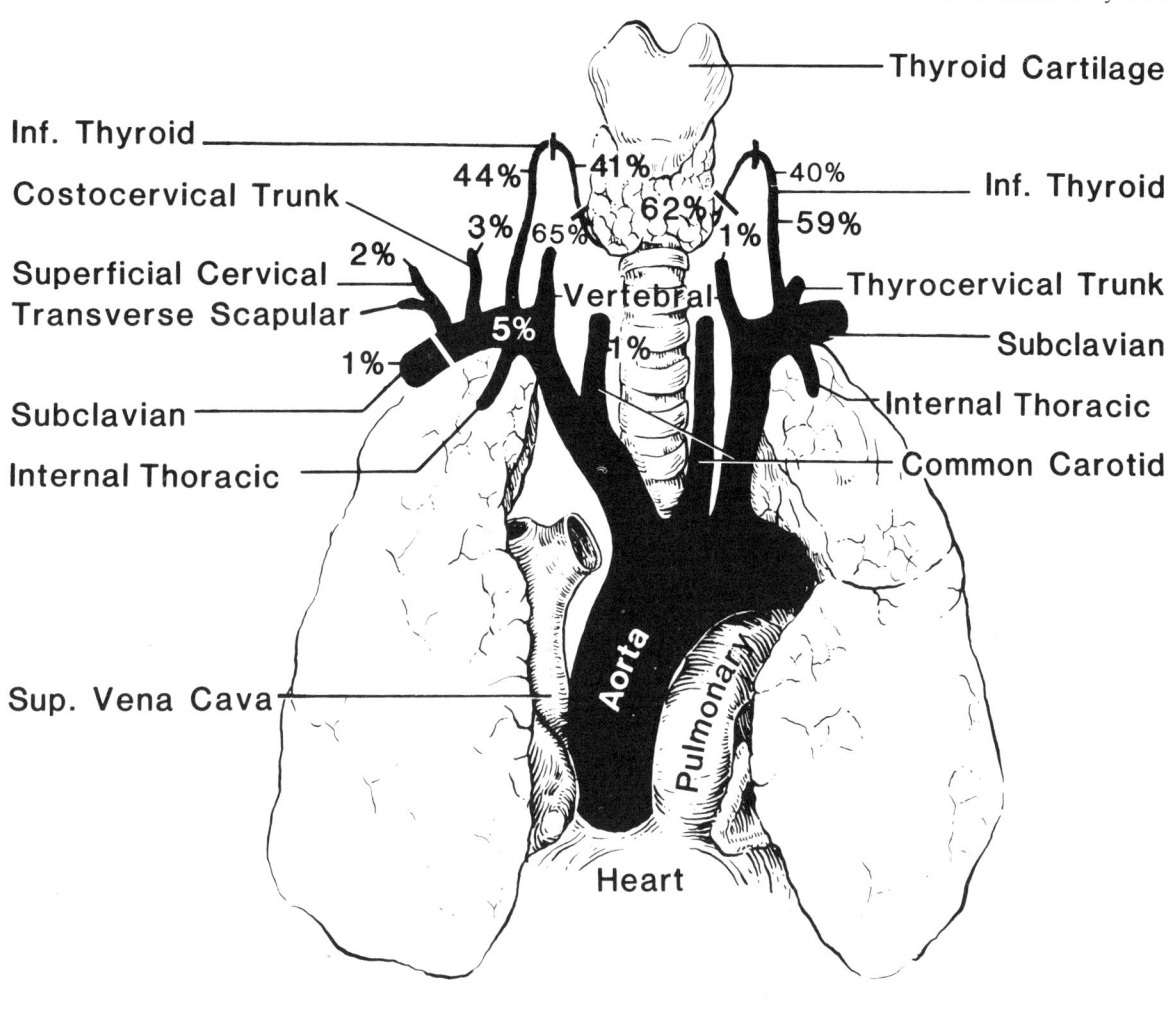

FIG. 65.
Cervical Origins of Esophageal Arteries.

The perpendicular lines that transect the inferior thyroid arteries divide the vessel into ascending, descending, and terminal segments. The number in percent either near or on a vessel indicates the instances in which esophageal rami originate from a part of a vessel or from a particular vessel.

Daseler and Anson (reference 282) also reported the results of detailed dissections made in 62 bodies (115 body-halves), concerning the arterial supply to the cervical and upper thoracic portions of the esophagus and trachea.

"The pharynx, esophagus, and trachea are usually supplied by several very fine arterial twigs from the terminal branches of the inferior thyroid arteries. Although these arterial branches are quite small in size, they are usually multiple in number, ranging from 1 to 4 on either side. These branches may pass directly to the esophagus and trachea, or may pursue a long and devious course downward into the superior mediastinum, before breaking up into their terminal branches. As a rule, this terminal branching occurs in the region of the groove formed between the trachea and esophagus, with the terminal twigs supplying the adjacent surfaces of both structures.

Single arterial branches were present 31 times on the right side and 27 times on the left. Dual arterial branches were found 18 times on the right and 21 times on the left. Triple arterial twigs were encountered on 15 sides, 8 on the right and 7 on the left, while quadruple branches were present on 3 left sides. Since many of the dissections were performed by our freshman medical students, it is possible that multiple branches occur much more frequently than is indicated by this study.

The arterial branches to the esophagus and trachea were derived from the distal half of the inferior thyroid artery 63 times on the right and 65 times on the left side. The esophageal and tracheal arterial branches originated from the proximal half of the inferior thyroid artery 20 times on the right and 23 times on the left side. These arterial branches took origin from the thyrocervical trunk proper, before its division into terminal branches, in 10 cases, 4 on the right and 6 on the left side. In 2 cases, 1 on each side, these arterial branches were derived from the costocervical trunk, whereas in 2 additional cases, 1 on either side, the arterial branches took origin from the first part of the subclavian artery. In 1 specimen, on the right, the arterial twig to the esophagus and trachea arose as a branch of the innominate artery, while in another specimen, also on the right side, this vessel arose as a direct branch of the common carotid artery. On the left side of this specimen, this twig arose as a branch of the vertebral artery, the inferior thyroid artery being absent. In another specimen, in which the inferior thyroid artery was also absent, a large, short trunk dividing into 4 separate twigs, arose as a branch from the common trunk of a vessel from which the superficial and ascending cervical arteries and suprascapular artery took origin."

Redrawn from Anson, B.J. *An Atlas of Human Anatomy.* W.B. Saunders Co., Philadelphia, 1950.

FIG. 66.
Bronchial and Esophageal Arteries.

The most commonly encountered pattern of bronchial arterial supply, Type I (A), occurred in 47.67% of specimens (143 of 300 cases); the pattern is that described as "normal" in standard textbooks. In these specimens, 2 left and 1 right bronchial artery are present. Next in the order of frequency is the pattern, Type II (B), in which a bronchial artery passes to each lung; this schema of bronchial arterial origin occurred in 24.33% of cases (73 of 300). The third major group, Type III (C), is characterized by bilateral duplication of the arteries; it was encountered in 14.33% of cases (43 of 300). These three types comprised 86.33% of the arterial patterns in all cadavers studied (that is, together they represent approximately 7 of 8 specimens). In 7.67% of cases (23 of 300), Type IV (D), the pattern was found to be the reverse of Type I – with 1 artery to the left lung, and 2 to the right. In Type V (E), 3 arteries were sent to the left lung, a single vessel to the right; this schema of bronchial supply occurred in 3% of specimens (9 of 300 cases). In Type VI (F), 3 bronchial arteries were present on the left side, 2 on the right. In 2 cases, Type VII (G), (2 in 300), 0.67%, the reverse of this pattern was encountered (3 on the right, 2 on the left). In a single specimen, Type VIII (H), 0.33% (1 in 300), 4 bronchial arteries were present on the left side; in another, Type IX (I), 0.33%, the pattern was the reverse of that representing Type VIII.

Most frequently there were 3 bronchial arteries, 2 left and 1 right. In some specimens, however, there were 2 right bronchial arteries; in such cases the second right bronchial artery coursed either ventral or dorsal to the esophagus. The bronchial artery of the ventral surface of the esophagus usually passed anterior to the bifurcation of the trachea, ultimately reaching the inferior border of the right main bronchus.

In 72% (108 specimens) of 150 cadavers examined, the inferior left bronchial artery gave rise to esophageal vessels. The latter vessels were of slightly smaller caliber than the bronchial arteries; they descended a variable distance on the anterior surface of the esophagus. Anastomoses between these esophageal vessels and the ascending branch of esophageal arteries of thoracic aortic origin occurred in many instances. Esophageal arteries from the superior left bronchial artery and superior right bronchial artery, 34.67% and 66.67% (52 and 100 cases, respectively, of 150 specimens), ascended on the anterolateral or posterolateral surface of the esophagus. When a second right bronchial artery was present (as an inferior right bronchial artery), esophageal branches derived therefrom supplied the esophagus, as the bronchial artery passed either ventral or dorsal to the visceral tube. In 14% (21 cases), the inferior right bronchial artery gave rise to a vessel of appreciable size, which coursed a short distance inferiorly on the right anterior surface of the esophagus. Esophageal arteries of this latter origin rarely formed a gross anastomosis with any other esophageal vessel. In fact, the portion of the thoracic esophagus appeared to be less well supplied by esophageal vessels than any other portion of the tube.

In 2.67% of cases (4 specimens) there were 3 left bronchial arteries; in each instance esophageal arteries arose from the middle left bronchial artery.

From Intercostal Arteries. Right intercostal arteries were the source of origin of esophageal arteries in 20% of the cases studied (30 of 150 specimens). In a single specimen an esophageal artery arose from a left intercostal artery. Usually an intercostal artery gave origin to a single esophageal artery; however, in 5 specimens 2 branches arose from the same intercostal artery, and in 3 specimens, 3 esophageal rami were derived from the same intercostal.

From Other Sources. Exceptional origins of esophageal arteries occurred on the right side in 3 specimens. The internal thoracic, the costocervical trunk, and the subclavian artery were, each, the source of origin for an esophageal artery. In each instance the vessel coursed downward on the anterior surface of the esophagus.

Arteries to Abdominal Portion of Esophagus

From Left Gastric Artery. The arterial supply to the abdominal portion of the esophagus was examined in 150 specimens. In 94.67% of cases (142 specimens), the left gastric artery was a source of origin for esophageal arteries.

The esophageal arteries originated from the left gastric artery, either just proximal to the point where the vessel reverses its direction to supply the lesser curvature of the stomach, or from the area of the bend itself. These esophageal arteries followed the longitudinal axis of the esophagus, supplying predominately the right anterior and posterior surfaces of the esophagus. Throughout the course of the vessels, under the visceral peritoneum of the esophagus, small branches penetrated the muscular tunic of the viscus. No constant relationship of the esophageal arteries to the vagus nerves was noted. Of the 142 specimens in which the left gastric artery was a source of esophageal supply, 78% of the cases possessed 1 to 3 esophageal arteries which arose from the left gastric. In but 1.33% (2 specimens) there were as many as 6 esophageal arteries arising from the left gastric.

From Inferior Phrenic Artery. The left inferior phrenic artery was a source of origin of esophageal vessels in 56% of cases (84 of 150 specimens). In all but 8% (12 specimens of the 84), a single esophageal branch was given off. The course of these esophageal vessels was similar to that of vessels arising from the left gastric artery, that is, parallel with the longitudinal axis of the esophagus and through the esophageal hiatus of the diaphragm.

Of the 150 cases, esophageal supply originated from the right inferior phrenic artery in 3.33% (5 specimens). In 1 case there were 2 branches; in each of the remaining 4 cases there was a single branch, distributed to the right posterior surface of the esophagus.

From Accessory Left Hepatic Artery. In 10% of the cases (15 of 150 specimens), esophageal arteries originated from an accessory left hepatic artery. With the exception of 3 specimens, only 1 esophageal artery in each instance was derived from the above vessel; however, the main branch divided into 2 or 3 twigs when the esophagus was reached.

In 5.56% of the above 15 specimens (8 cases), no esophageal artery arose from the left gastric artery. In the 8 specimens, the supply to the right aspect of the esophagus was derived from the accessory left hepatic artery. The artery divided in such a way as to send rami twigs to both anterior and posterior surfaces.

From Splenic Artery. In 1.33% of cases (2 specimens), 1 to 2 esophageal arteries originated from the proximal one-third of the splenic artery.

From Celiac Artery. In a single specimen (0.67%), 1 esophageal artery originated from the celiac axis.

Arteries to Cervical Portion of Esophagus

The arteries sent to this portion of the esophagus usually originated from the inferior thyroid artery; however, other origins were not infrequent.

From Inferior Thyroid Artery. In 125 specimens, esophageal arteries originated from the inferior thyroid artery at three separate sites: the ascending portion, the descending portion, and the artery's terminal rami.

The esophageal arteries originated from the terminal branches of the inferior thyroid artery more frequently than from the ascending or descending portions of the vessel. However, the total number of branches from the latter two segments of the inferior thyroid artery exceeded the number derived from the terminal divisions of the parent vessel. Usually a single esophageal branch originated from a segment of the inferior thyroid artery; however, two, but rarely more, esophageal branches arose from the same or different segments of the

Cardiovascular System

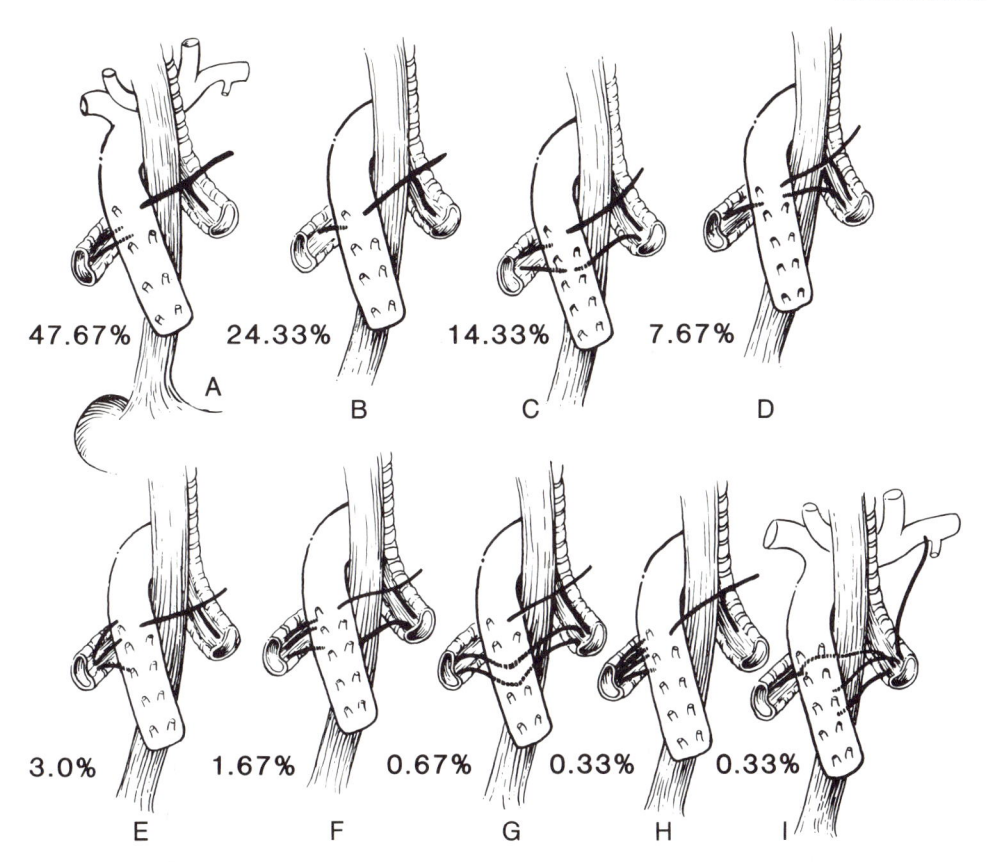

47.67% 24.33% 14.33% 7.67%
A   B   C   D

3.0% 1.67% 0.67% 0.33% 0.33%
E   F   G   H   I

thyroid artery. Esophageal vessels which originated from the ascending segment of the thyroid artery were not only a larger size, but also descended a greater distance on the anterolateral surface of the esophagus than did similar vessels derived from the terminal branches and the descending portion of the inferior thyroid artery. In the majority of cases, there occurred a tracheoesophageal vessel which arose from the ascending portion of the inferior thyroid artery; it coursed downward in company with the recurrent laryngeal nerve into the superior mediastinum. This vessel supplied the trachea and the hilar lymph nodes; additionally, in several instances, the artery established a gross anastomosis with the right or left superior bronchial artery. Near the level of the suprasternal notch, 1 to 3 branches arose from the tracheoesophageal artery to supply the posterolateral aspect of the esophagus.

With the exception of a single instance, the arteries, after reaching the cervical segments of the esophagus, gave off no ascending branches similar to those seen in the thoracic region. In the exceptional specimen, the esophageal artery, after arising from the superior terminal branch of the inferior thyroid artery, ascended on the anterior aspect of the esophagus to the level of the cricoid cartilage. In another unusual case, in which the left inferior thyroid artery was absent, the left half of the thyroid gland was supplied by the superior thyroid artery. In this same case the esophagus received its only arterial supply by means of a single branch derived from one of the terminal divisions of the right inferior thyroid artery. There was no demonstrable arterial supply to the left side of the cervical esophagus from any source.

In still another specimen, there was no esophageal supply from either the right or the left inferior thyroid artery; the lower cervical and upper thoracic portions of the esophagus were supplied by rami derived from the bronchial arteries.

Esophageal arteries were found to arise from the ascending segment of the inferior thyroid artery in 36.8% (46 cases) on the right, 51.11% (64 cases) on the left side. As has already been stated, esophageal arteries which arise from the descending portion of the inferior thyroid artery or its terminal branches do not descend for a considerable distance inferiorly on the esophagus. It would appear that unless an artery supplying the esophagus near the level of the jugular notch be demonstrable, it would be well to assume that the esophageal supply is derived either from the descending portion of the inferior thyroid artery, or from one of the latter's terminal branches.

Important, too, is the fact that, were the inferior thyroid artery ligated near its terminal branches (as in a previous subtotal thyroidectomy), an adequate arterial supply for the esophagus (near the jugular notch of the sternum) would remain, in the majority of specimens studied – being derived from the ascending and descending portions of the inferior thyroid.

From Other Sources. Although the majority of esophageal arteries took origin from the inferior thyroid artery, there were 13 cases in which the esophageal arteries arose from other sources representing 100 dissections. The addition of 25 dissections, bringing the total of cervical dissections to 124, revealed two cases, on the right side, in which esophageal arteries arose from the ascending pharyngeal artery and the common carotid artery. One case, on the left side, revealed an esophageal artery arising from the left common carotid artery. Thus, in 12.8% of cases (16 of 125 cervical dissections), esophageal arteries originated from sources other than the inferior thyroid arteries.

Authors' note: Original text has been slightly edited.

Redrawn from Cauldwell, E.W., Siekert, R.G., Lininger, R.E. and B.J. Anson. The bronchial arteries. An anatomic study of 150 human cadavers. *Surg. Gynecol. Obstet.* 86:395–412, 1948; and Swigart, LaV.L., Siekert, R.G., Hambley, W.C. and B.J. Anson. The esophageal arteries. An anatomic study of 150 specimens. *Ibid.* 90:234–243, 1950.

Human Anatomic Variation

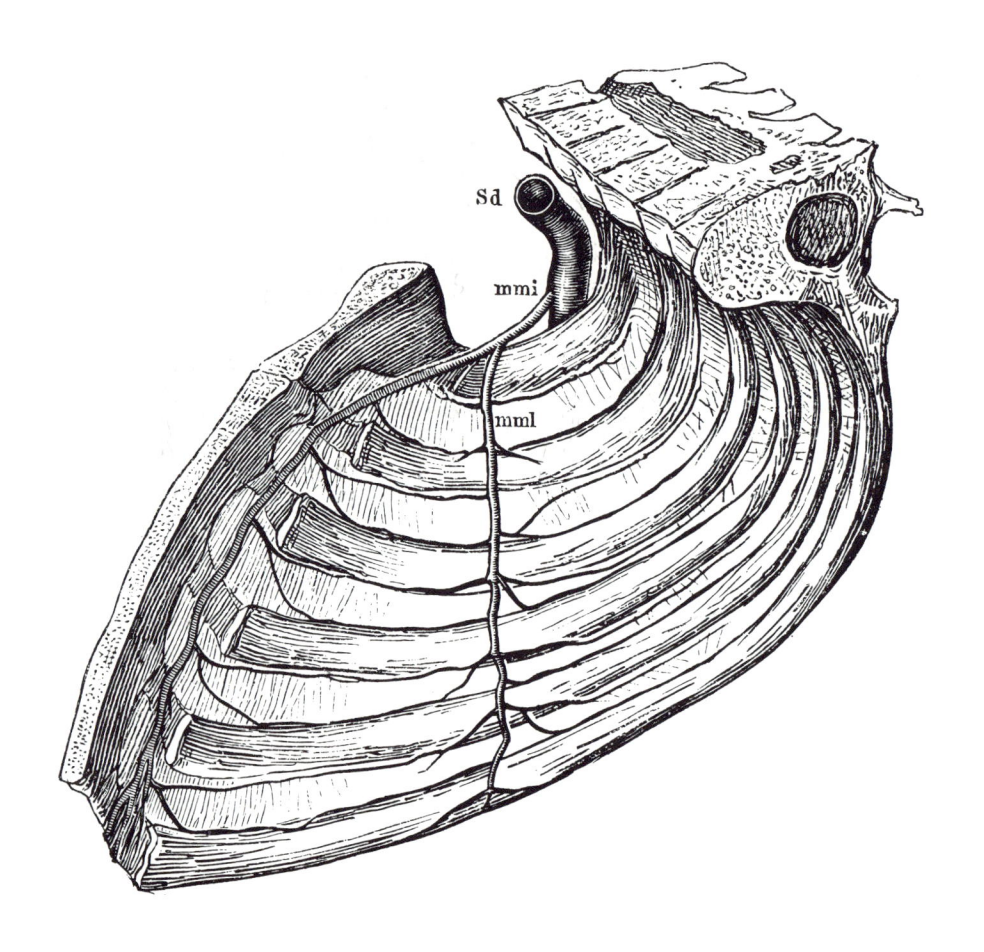

FIG. 67.
Lateral Internal Thoracic Artery.

Sd, Right subclavian artery; *mmi*, internal thoracic artery; *mml*, lateral thoracic artery.

The internal thoracic artery may give rise to a lateral internal thoracic artery as it enters the thorax. The lateral internal thoracic runs about midway between the spine and sternum or somewhat forward; it spans up to six intercostal spaces and anastomoses with intercostal arteries of those spaces. A lateral internal thoracic artery may also arise from the costocervical trunk, thyrocervical trunk, or ascending cervical. The possibility of the existence of this vessel must be remembered when paracentesis is performed.

Henle, J. *Handbuch der Systematischen Anatomie des Menschen*, 3rd Ed. V. Wilhelm Engelmann, Leipzig 1888.

Cardiovascular System

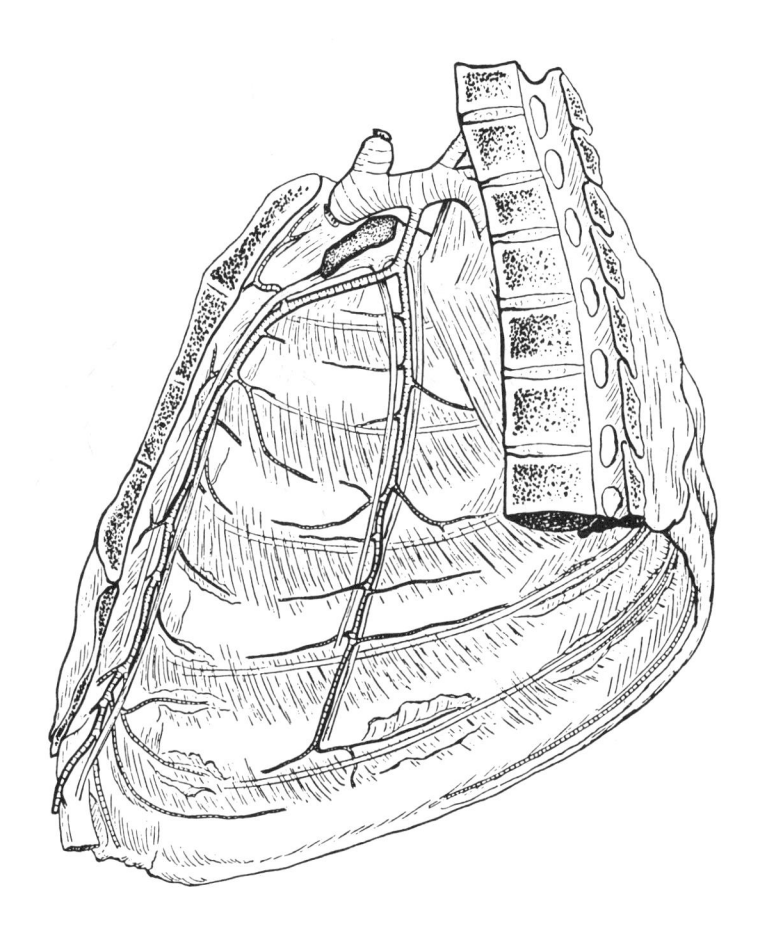

FIG. 68.
Lateral Internal Thoracic Artery and Vein.
From Adachi, B. *Anatomie der Japaner. Das Arteriensystem der Japaner*. Verlag der Kaiserlich-Japanischen Universität zu Kyoto. In Kommission bei "Maruzen Co.", Kyoto und Tokyo. Gedruck von "Kenkyusha" in Tokyo, 1928.

Human Anatomic Variation

Branching Patterns of External Carotid

Truncus Faciolingualis

Truncus Thyrolingualis **Truncus Thyrolinguofacialis** **Truncus Thyrofacialis** **Truncus Faciomaxillaris**

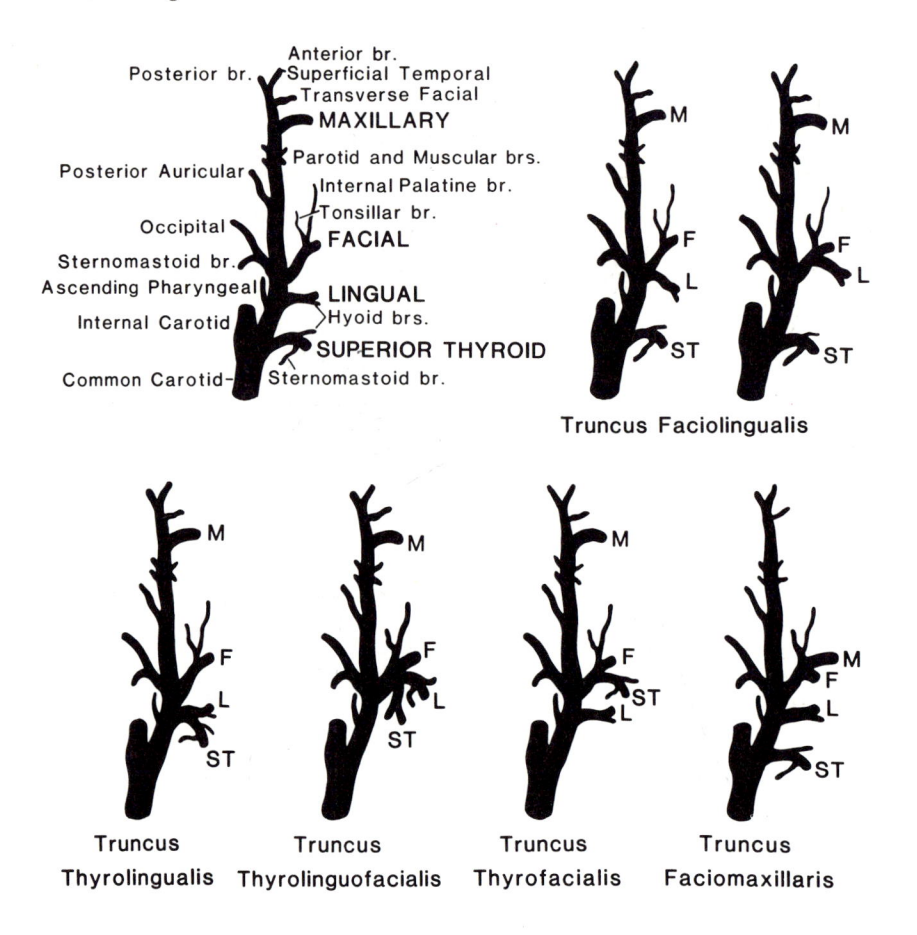

FIG. 69.
Branching Patterns of External Carotid Artery.
This composite drawing illustrates a common branching pattern and combinations of four major arteries into trunks from the external carotid artery: faciolingual trunk, faciomaxillary trunk, thyrofacial trunk, thyrolingual trunk, and thyrolinguofacial trunk.

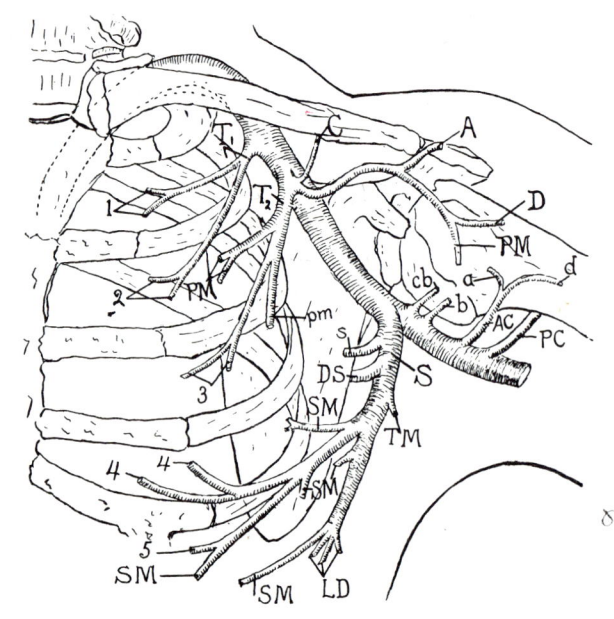

FIG. 70.
Composite Study of Axillary Artery.
Type I of the axillary artery. Present 20 times in 47 cases.
A, Ramus acromialis; a, ramus ascendens; AC, a. circumflexa humeri anterior; b, m. biceps; C, r. clavicularis; cb, m. coracobrachialis; D, m. deltoideus; d, ramus anastomoticus; DS, a. circumflexa scapulae (dorsal scapular); LD, m. latissimus dorsi; P, r. pectoralis; PC, a. circumflexa humeri posterioris; PM, m. pectoralis major; pm, m. pectoralis minor; S, a. subscapularis; s, m. subscapularis; SM, m. serratus anterior (magnus); SP, a. profunda brachii (superior profunda); T_1, a. thoracalis suprema; T_2, a. thoracoacromialis; T_3, a. thoracalis lateralis; TM, m. teres major; 1, 2, 3, 4, 5, 1st, 2d, 3d, 4th, and 5th intercostal spaces.

From Hitzrot, J.M. A composite study of the axillary artery in man. *Johns Hopkins Hosp. Bull.* 12:136–145, 1901.

Cardiovascular System

FIG. 71.
Composite Study of Axillary Artery (continued).
A: Type II of the axillary artery. Present 9 times in 47 cases.
B: Type III of the axillary artery. Present 7 times in 47 cases.
A, Ramus acromialis; *a*, ramus ascendens; *AC*, a. circumflexa humeri anterior; *C*, r. clavicularis; *D*, m. deltoideus; *d*, ramus anastomoticus; *DS*, a. circumflexa scapulae (dorsal scapular); *LD*, m. latissimus dorsi; *PC*, a. circumflexa humeri posterioris; *PM*, m. pectoralis major; *pm*, m. pectoralis minor; *S*, a. subscapularis; *s*, m. subscapularis; *SM*, m. serratus anterior (magnus); T_1, a. thoracalis suprema; T_2, a. thoracoacromialis; T_3, a. thoracalis lateralis; *TM*, m. teres major; *1, 2, 3, 4, 5*, 1st, 2d, 3d, 4th, and 5th intercostal spaces.

From Hitzrot, J.M. A composite study of the axillary artery in man. *Johns Hopkins Hosp. Bull.* 12:136–145, 1901.

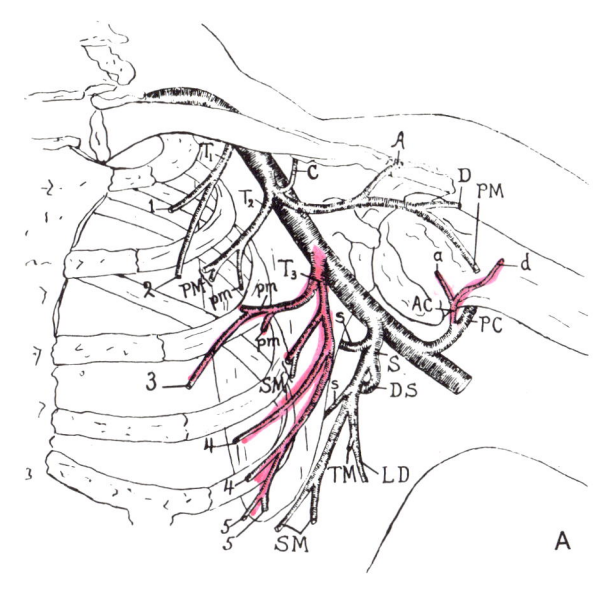

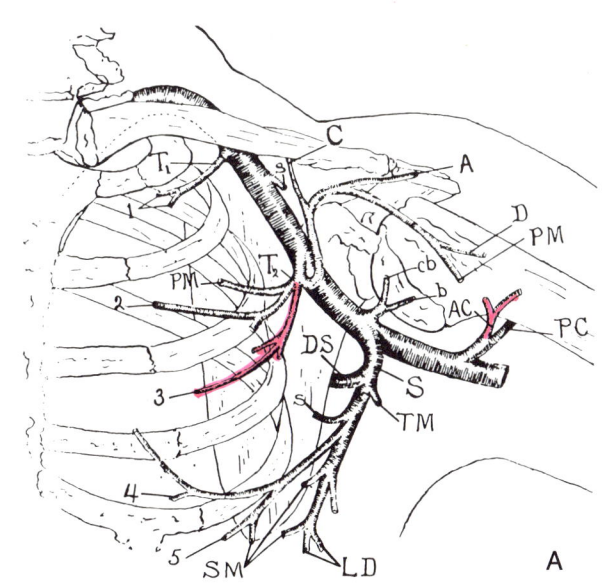

FIG. 72.
Composite Study of Axillary Artery (continued).
A: Type IV of the axillary artery. Present 4 times in 47 cases.
B: Type V of the axillary artery. Present 3 times in 47 cases.
A, Ramus acromialis; *AC*, a. circumflexa humeri anterior; *b*, m. biceps; *C*, r. clavicularis; *cb*, m. coracobrachialis; *D*, m. deltoideus; *DS*, a circumflexa scapulae (dorsal scapular); *LD*, m. latissimus dorsi; *PC*, circumflexa humeri posterioris; *PM*, m. pectoralis major; *pm*, m. pectoralis minor; *S*, a. supscapularis; *s*, m. subscapularis; *SM*, m. serratus anterior (magnus); T_1, a. thoracalis suprema; T_2, a thoracoacromialis; *TM*, m. teres major; *1, 2, 3, 4, 5*, 1st, 2d, 3d, 4th, and 5th intercostal spaces.

From Hitzrot, J.M. A composite study of the axillary artery in man. *Johns Hopkins Hosp. Bull.* 12:136–145, 1901.

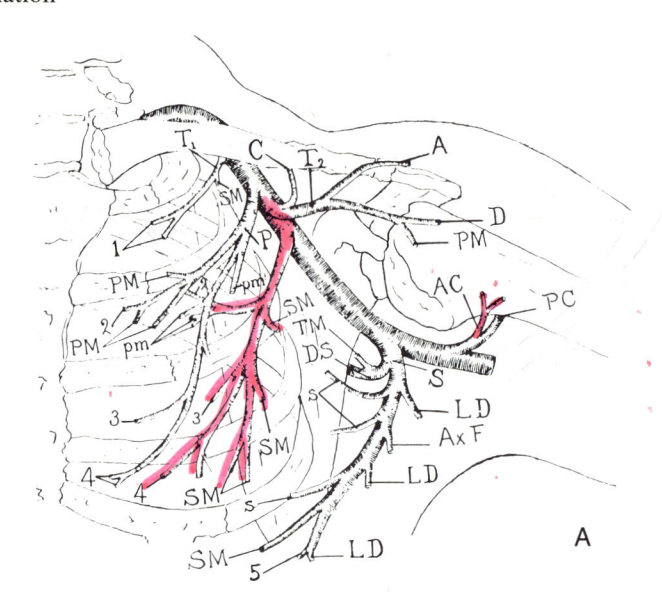

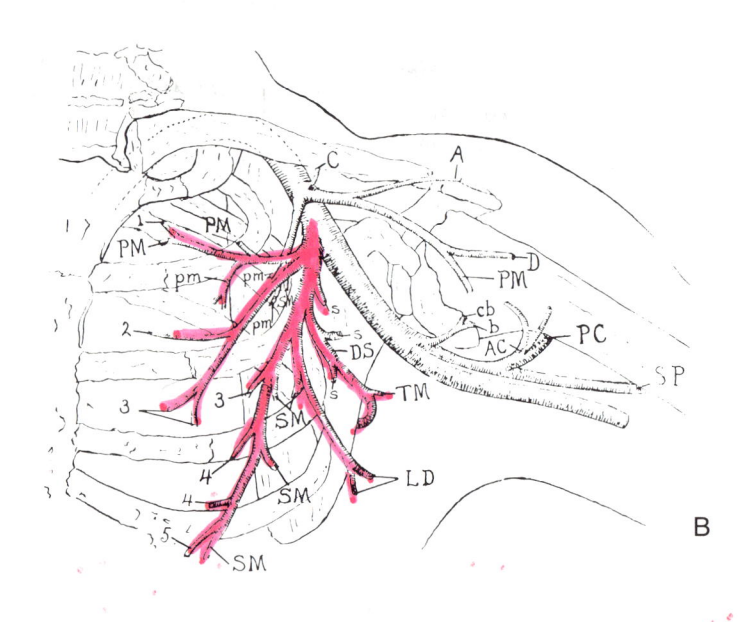

FIG. 73.
Composite Study of Axillary Artery (continued).

A: Type VI of the axillary artery. Present 2 times in 47 cases.
B: Type VII of the axillary artery. Present 2 times in 47 cases.
A, Ramus acromialis; *AC*, a circumflexa humeri anterior; *b*, m. biceps; *C*, r. clavicularis; *cb*, m. coracobrachialis; *D*, m. deltoideus; *DS*, a. circumflexa scapulae (dorsal scapular); *LD*, m. latissimus dorsi; *P*, r. pectoralis; *PC*, a. circumflexa humeri posterioris; *PM*, m. pectoralis major; *pm*, m. pectoralis minor; *S*. a. subscapularis; *s*, m. subscapularis; *SM*, m. serratus anterior (magnus); *SP*, a. profunda brachii (superior profunda); T_1, a. thoracalis suprema; T_2, a. thoracoacromialis; *TM*, m. teres major; *1, 2, 3, 4, 5,* 1st, 2d, 3d, 4th, and 5th intercostal spaces.

From Hitzrot, J.M. A composite study of the axillary artery in man. *Johns Hopkins Hosp. Bull.* 12:136–145, 1901.

Cardiovascular System

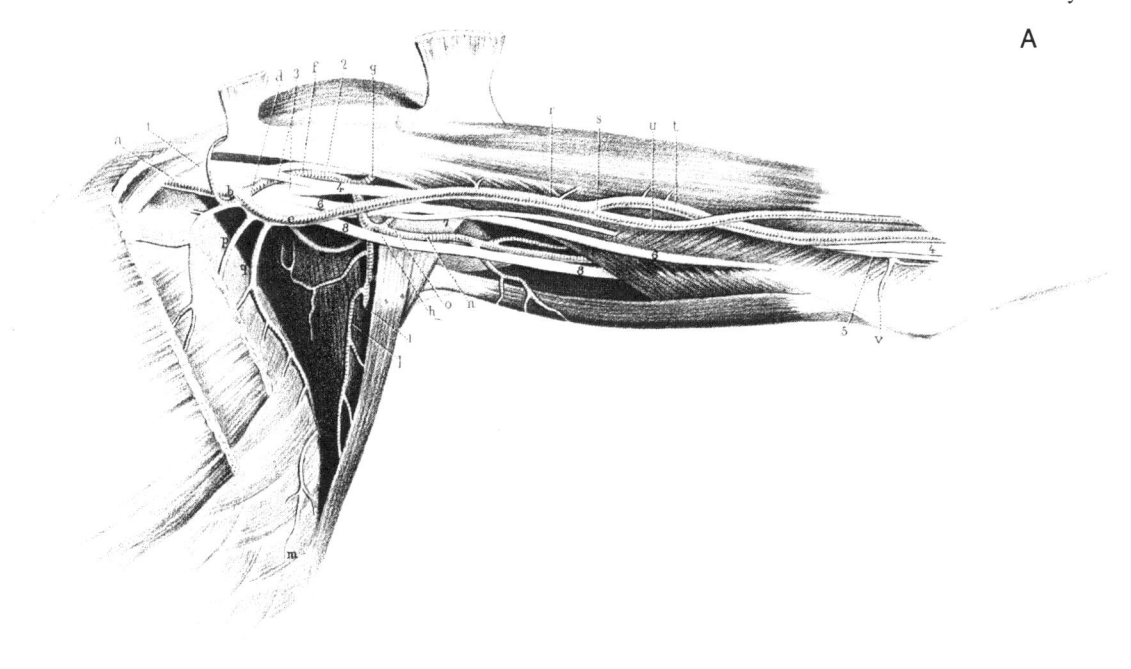

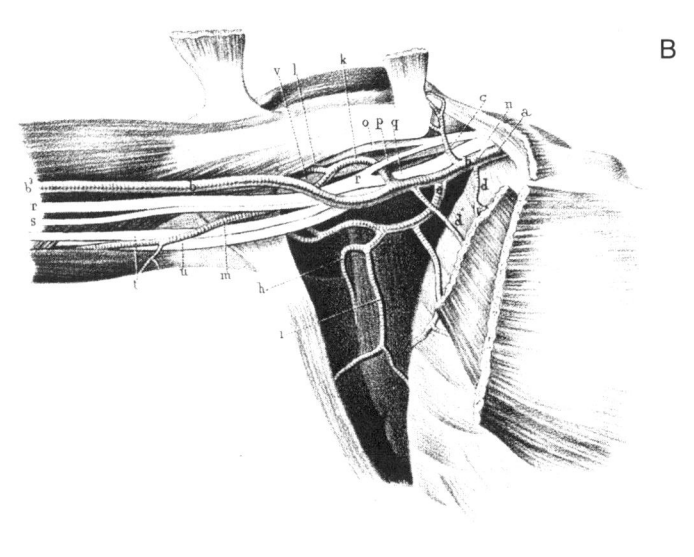

FIG. 74.
Anomalous Axillary Artery.

A: *a*, Subclavian artery, third part; *b*, axillary artery; *d*, division of axillary artery; *e*, medial branch (continuation of axillary artery); *f*, lateral branch, forming subscapular, circumflex humeral, profunda brachii, and circumflex scapular arteries; *g*, anterior circumflex humeral artery; *h*, division of lateral branch into subscapular and into posterior circumflex humeral and profunda brachii; *i*, circumflex scapular; *l*, thoracodorsal; *p,q*, a second and third lateral thoracic artery; *r*, brachial artery; *s, t, u*, division into ulnar (continuation of brachial) and radial arteries in arm; *v*, inferior ulnar collateral artery; *1*, subclavicular part of brachial plexus; *2*, musculocutaneous nerve; *3*, two roots of median nerve separated by anomalous branch of axillary artery; *4*, median nerve separated from ulnar nerve by anomalous branch of axillary artery; *5*, median nerve branch to pronator teres; *6*, ulnar nerve; *7*, medial cutaneous nerve perforating brachial fascia; *8*, radial nerve; *9*, axillary nerve.

B: *a*, Third part of subclavian; *b*, axillary artery; *c*, acromial artery; *d*, lateral thoracic artery, superior; *e*, anomalous trunk giving rise to long thoracic, subscapular, and posterior circumflex humeral branches; *g*, subscapular artery; g^2, posterior circumflex humeral artery; *h*, circumflex scapular artery; *i*, thoracodorsal artery; *k*, anomalous branch of axillary artery passing through lateral and medial cords of brachial plexus (*p,q*); *l*, anterior circumflex humeral artery; *m*, profunda brachii artery; *n*, subclavicular part of brachial plexus; *o*, musculocutaneous nerve; *r*, median nerve passing behind axillary artery; *s*, medial cutaneous nerve of forearm; *t*, ulnar nerve; *u*, radial nerve; *v*, axillary nerve.

From Calori, L. Sul variato rapporto vascolare delle due radici del nervo mediano e sulle condizioni atte a produrlo. *Mem. Accad. Sci. Istituto di Bologna S. 3* 8:443–456, 1877.

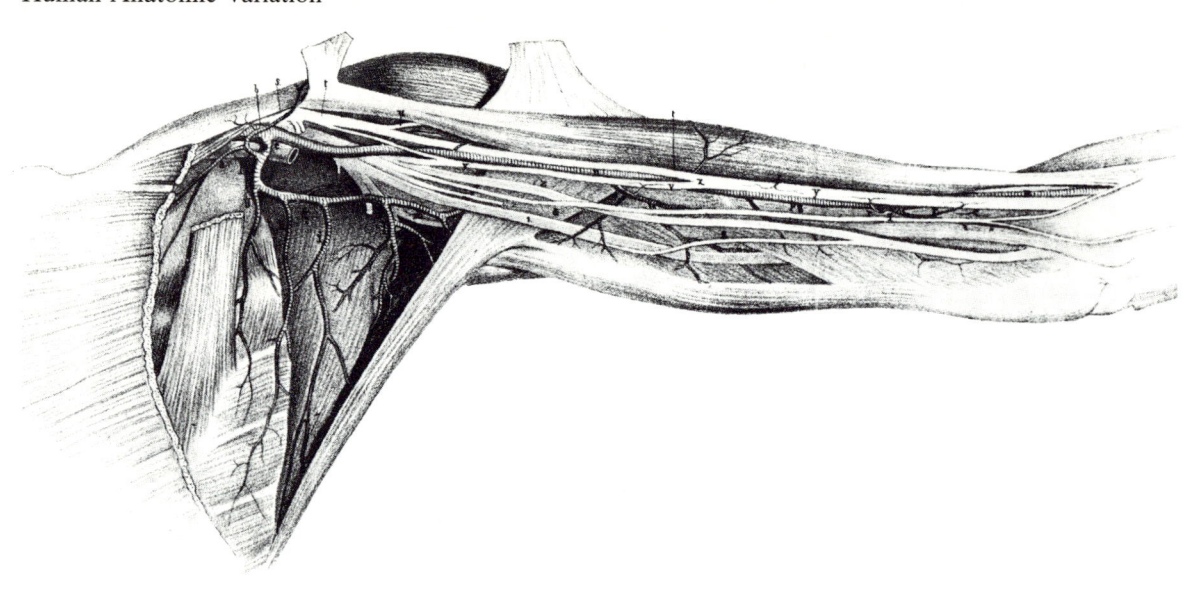

FIG. 75.
Partition of Left Axillary Artery into Two Trunks, One of Which Possessed All the Usual Branches of the Axillary While the Other Continued as the Brachial Artery.
(Anomaly of Median Nerve).
(Anatomosis Between Radial and Ulnar Nerves).
a, Portion of the axillary vein; *b*, axillary artery; *c*, partition of the axillary artery; *d*, "axillary trunk"; *e*, continuation of the vessel as the brachial artery; *f*, thoracoacromial artery; *g*, lateral thoracic artery; *h*, accessory lateral thoracic or lateral mammary branches (lateral external thoracic); *i*, thoracodorsal branch; *l*, circumflex scapular; *m*, posterior circumflex humeral artery; *n,n*, brachial artery; *o*, profunda brachii artery; *p,q*, ulnar collateral arteries; *r,r*, arteries of the median nerve; *t*, lateral origin of the median nerve from which the musculocutaneous nerve (*u*) arises, its prolongation ends at the middle of the arm running close together with the lateral side of the brachial artery; *v,v*, other origin of the median nerve running close to the medial side of the brachial artery; the union of the two origins of the median nerve (*x*) forms the trunk (*y*) of the median nerve; *&*, medial cutaneous nerve of the forearm; *1*, radial nerve; *2*, anastomosis between the radial and ulnar nerves.

From Calori, L. Delle anomalie più importanti di ossa, vasi, nervi, e muscoli occorse nell' ultimo beinnio fascendo anatomia dell corpo umano. *Mem. R. Accad. Sci. Istituto di Bologna S. 2* 8:417–482, 1868.

FIG. 76.
Spiral Course of Right Median Nerve Around Brachial Artery. (Anomaly of Median Nerve).
(Anomaly of Musculocutaneous Nerve).
Superficial Course of Ulnar Artery Through Lacertus Fibrosus of Biceps Brachii.

a, Axillary artery; b, small subscapular artery; c, thoracoacromial artery; d, lateral thoracic artery; e, large subscapular artery; f, anterior circumflex humeral artery; g, posterior circumflex humeral artery; h,h, brachial artery; i, profunda brachii artery; k, superior ulnar collateral artery; l, inferior ulnar collateral with anastomosis to k; m, division of the brachial into an ulnar artery (n) and a trunk (o) which provides the radial artery (p) and interosseous (q) arteries; s,t, origin of the median nerve; u, superior portion of the trunk of the median nerve which descends at first to the medial side, then above, and finally to the lateral side of the artery and connects with (v) the musculocutaneous nerve (y,z).

x, Portion of the trunk of the median nerve situated to the lateral side of the brachial artery and giving rise to the lateral cutaneous nerve of the forearm (1); the same part (x) passes beneath (2) the brachial artery and finally courses in its usual position, that is to the medial side of the brachial artery as seen at 3; 4, aponeurotic tape-like structure of the tendon of pectoralis major muscle which forms a small canal for the tendon (9) of the long head of the biceps.

From Calori, L. Delle anomalie più importanti di ossa, vasi, nervi, e muscoli occorse nell' ultimo biennio fascendo anatomia del corpo umano. *Mem. Accad. Sci. Istituto di Bologna S. 2* 8:417–482, 1868.

Human Anatomic Variation

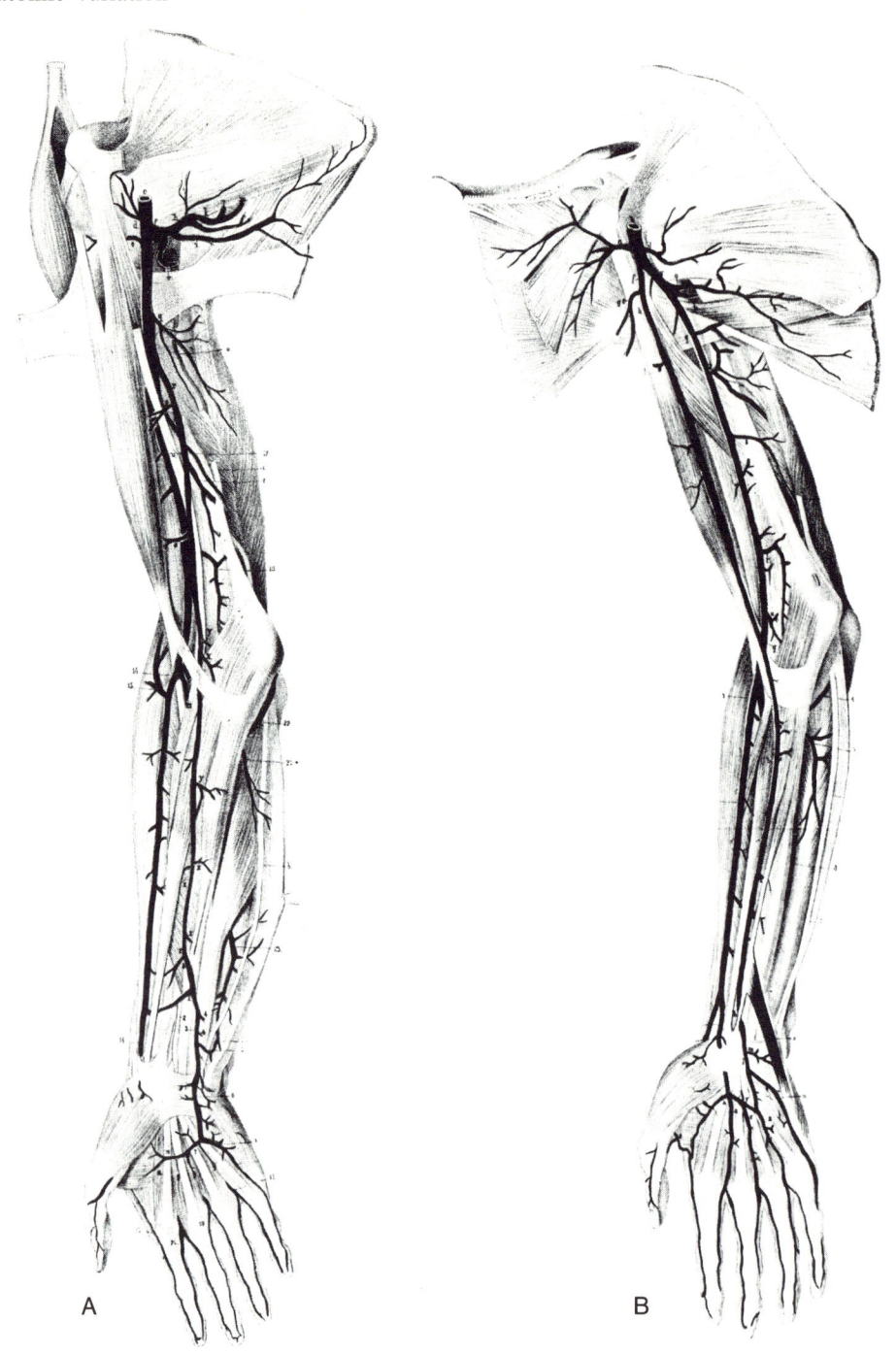

FIG. 77.
Absence of Ulnar Artery.

A: Partition of the axillary artery into two trunks of equal size, one of which becomes the brachial artery and the second a common trunk for the subscapular, posterior circumflex humeral, and profunda brachii. Persistence of a superficial median artery (arteria mediana superficialis communis artuum superiorum) which forms the superficial palmar arch. The median artery has several anastomotic branches to the radial artery.
B: Division of the axillary artery into two trunks resulting in the formation of a brachial artery which passes into the forearm to continue as the radial artery. The radial artery provides interosseous branches and pierces the lateral side of the transverse carpal ligament to participate significantly in the formation of the superficial palmar arch. The second trunk provides the thoracoacromial and lateral thoracic arteries, and after having a superficial course in the forearm, continues as a persistent median artery which passes superficially over the transverse carpal ligament to participate in the superficial palmar arch. The tendon of palmaris longus muscle gives rise to the anomalous accessorius ad flexorem digiti minimi muscle.

From Calori, L. Delle anomali più importanti di ossa, vasi, nervi, e muscoli occorse nell' ultimo bienno fascendo anatomia cell corpo umano. *Mem. R. Accad. Sci. Istituto di Bologna S. 2* 8:417–482, 1868.

Cardiovascular System

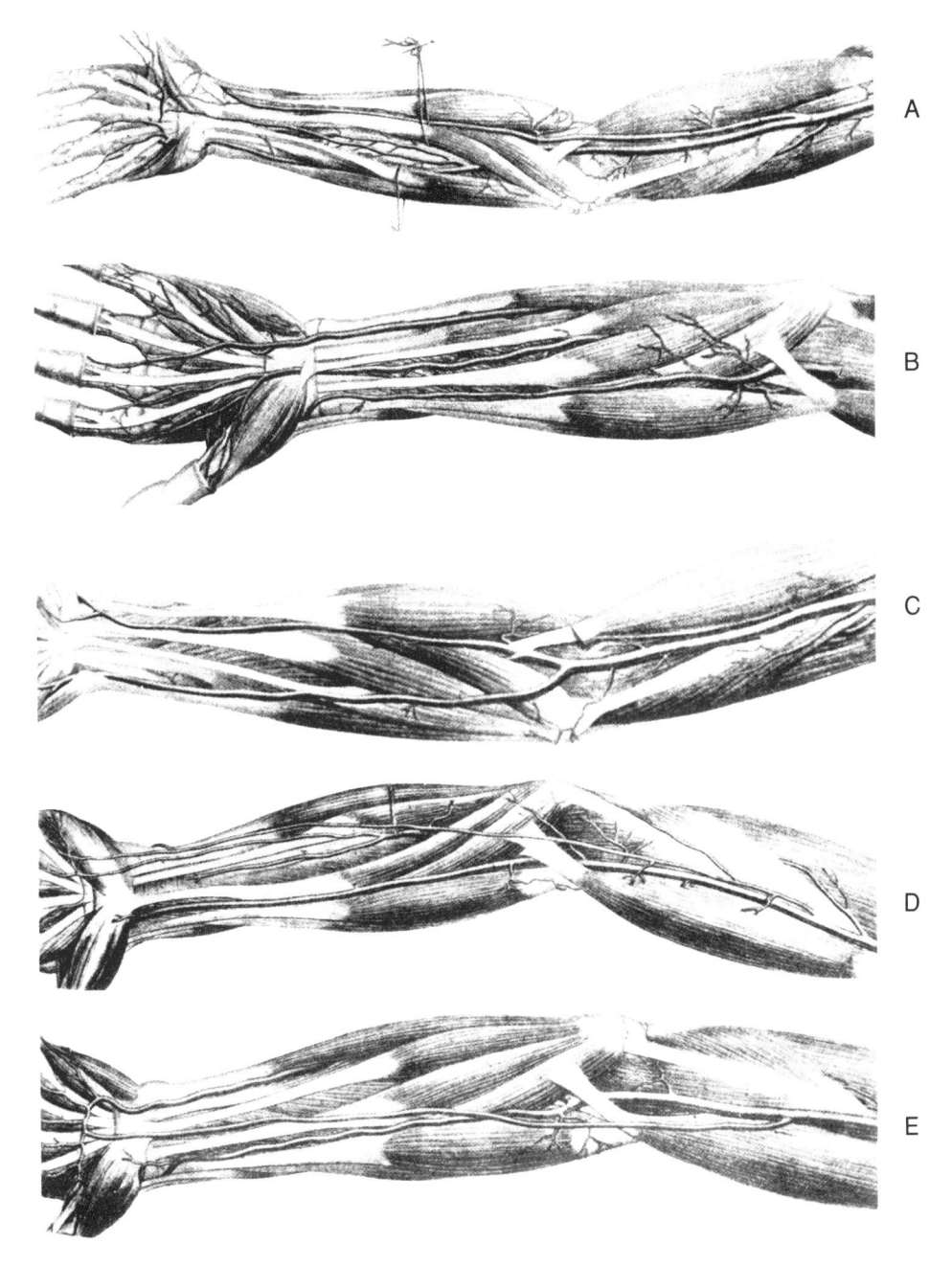

FIG. 78.
Some Variations in Arteries of Arm and Forearm (from the Literature).

A: Tiedemann (1822, Tab. XVI, Fig. 1). A superficial brachial artery continues as the radial. A distal superficial median artery crosses in front of the transverse carpal ligament and has a deep antibrachial origin (thus not arising from a superficial brachial).

B: Idem (1822, Tab. XVI, Fig. 2). Radial and ulnar arteries are normal. A distal superficial median artery crosses in front of the transverse carpal ligament and has a deep antibrachial origin (thus not arising from a superficial brachial).

C: Idem (1822, Tab. XVII, Fig. 7). A superficial ulnar artery arises from the brachial (proper).

D: Idem (1846, Tab. XLVI, Fig. 1). A superficial brachial artery continues as the superficial ulnar. A second superficial ulnar (or median) artery crosses in front of the transverse carpal ligament and over muscles of the hypothenar eminence, and has a deep antibrachial origin (thus not arising from a superficial brachial).

E: Idem (1846, Tab. XLVI, Fig. 3). A superficial brachial artery continues as the complete superficial median which crosses in front of the transverse carpal ligament.

From De Garis, C.F. Aortic axillary collaterals and the pattern of arm arteries in anomalous right subclavian artery. *Am. J. Anat.* 51:189–213, 1932.

Additional references: Tiedemann, F. *Tabulae Arteriarum Corporis Humani Fabrica.* Karlsruhae. 1822. *Supplementa ad Tabulas Arteriarum Corporis Humani.* Heidelbergae. 1846.

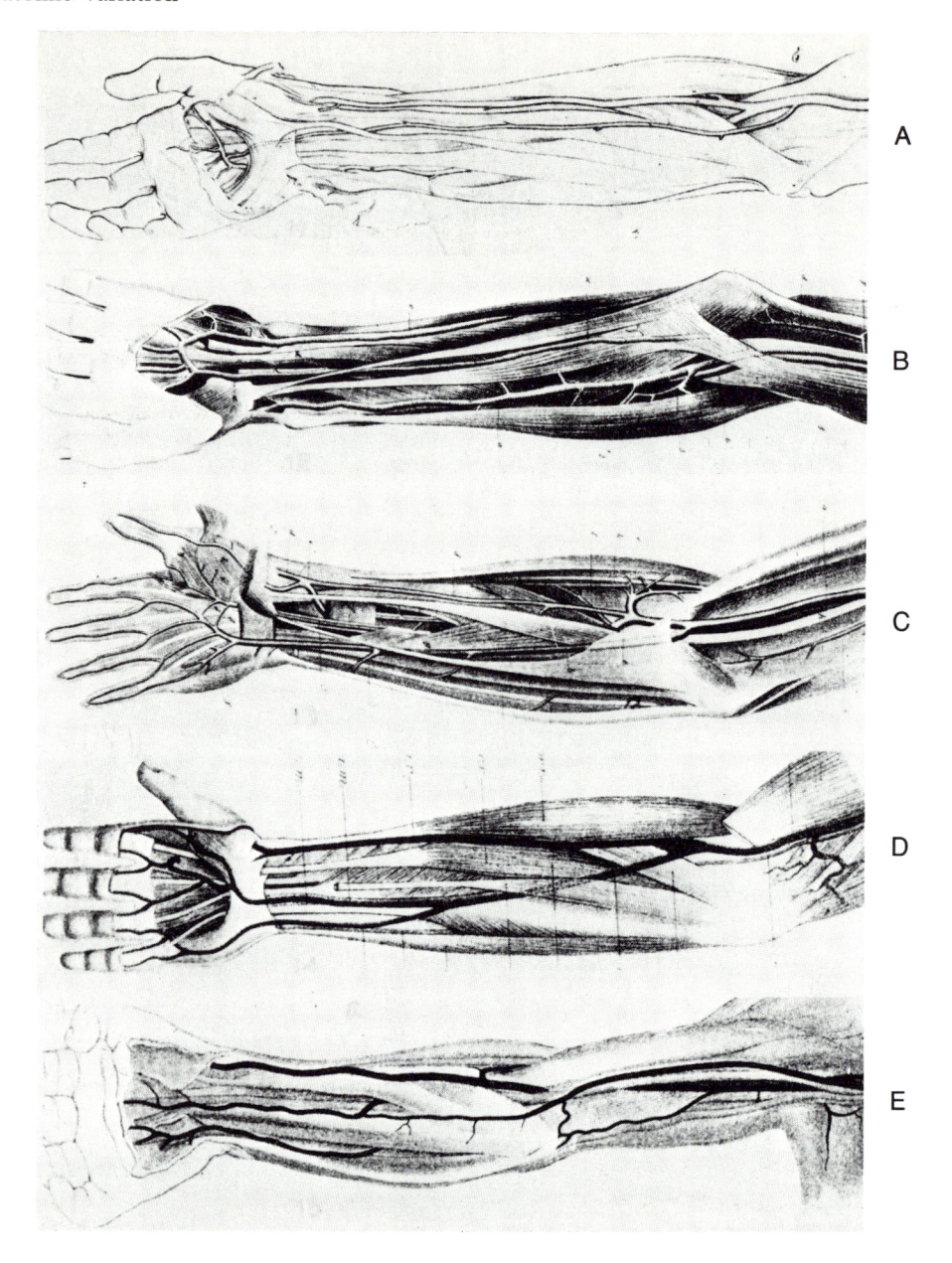

Human Anatomic Variation

FIG. 79.
Some Variations in Arteries of Arm, Forearm, and Hand (from the Literature).

A: Quain (1844, Plate 45, Fig. 1). Radial and ulnar arteries are normal. The proximal superficial median artery arises from the brachial (proper) and crosses behind the transverse carpal ligament.

B: Gruber (1867, Fig. 1). The proximal superficial median artery continues from a superficial brachial and crosses behind the transverse carpal ligament.

C: Idem (Fig. 2). Duplication of the ulnar artery, i.e., the superficial ulnar artery continuing from a superficial brachial and the deep ulnar artery continuing from the common interosseous.

D: Schwalbe (1898, Taf. I, Fig. 2). A superficial volar antibrachial artery of the medianoulnar type arises from the brachial (proper), both median and ulnar components passing in front of the transverse carpal ligament.

E: Müller (1903, Taf. XIX–XX, Fig. 6). The proximal superficial median artery continues from a superficial brachial and joins the deep median artery beyond the muscle mass of the superficial pronator-flexor group.

From De Garis, C.F. Aortic axillary collaterals and the pattern of arm arteries in anomalous right subclavian artery. *Am. J. Anat.* 51:189–213, 1932.

Additional references: Gruber, W. Über die Arteria mediana antibrachii superficialis, Arteria ulnaris antibrachii superficialis and Duplicität der Arteria ulnaris. *Arch. Anat. Physiol. Wissen. Med.* 1867:668–687.

Müller, E. Beiträge zur Morphologie des Gefäßsystems. I. Die Armarterien des Menschen. *Anat. Hefte* 22:379–574, 1903.

Quain, R. *Anatomy of the arteries of the human body with its application to pathology and operative surgery: In a series of plates and commentaries.* London, 1844.

Schwalbe, E. Beitrag zur Kenntnis der Arterienvarietäten des Menschlichen. *Armes. Morph. Arbeiten.* 8:1–42, 1898.

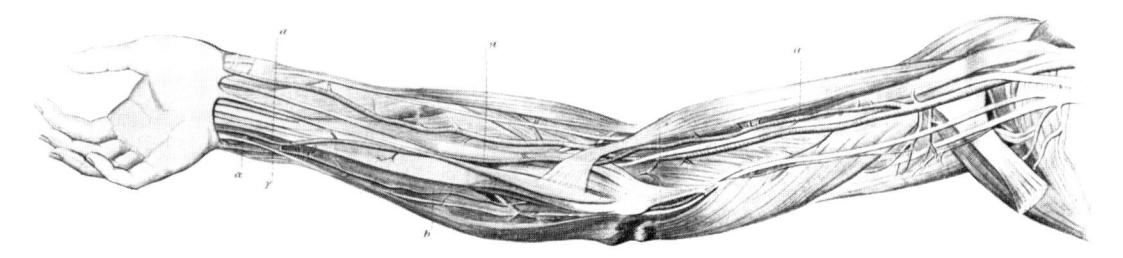

FIG. 80.
Duplication of Ulnar Artery.

This illustration shows the right arm.
a, Superficial ulnar artery: α, small dorsal ramus (I); b, ulnar artery (deep): β, large anastomotic ramus; and γ, smaller dorsal ramus (II).

From Gruber, W. Duplicität der Arteria ulnaris – neuer Fall – (nebst Berichtigungen in gelegentlichen Bemerkungen). *Arch. Anat. Pathol. Wissen. Med.* 1871, pp. 286–296.

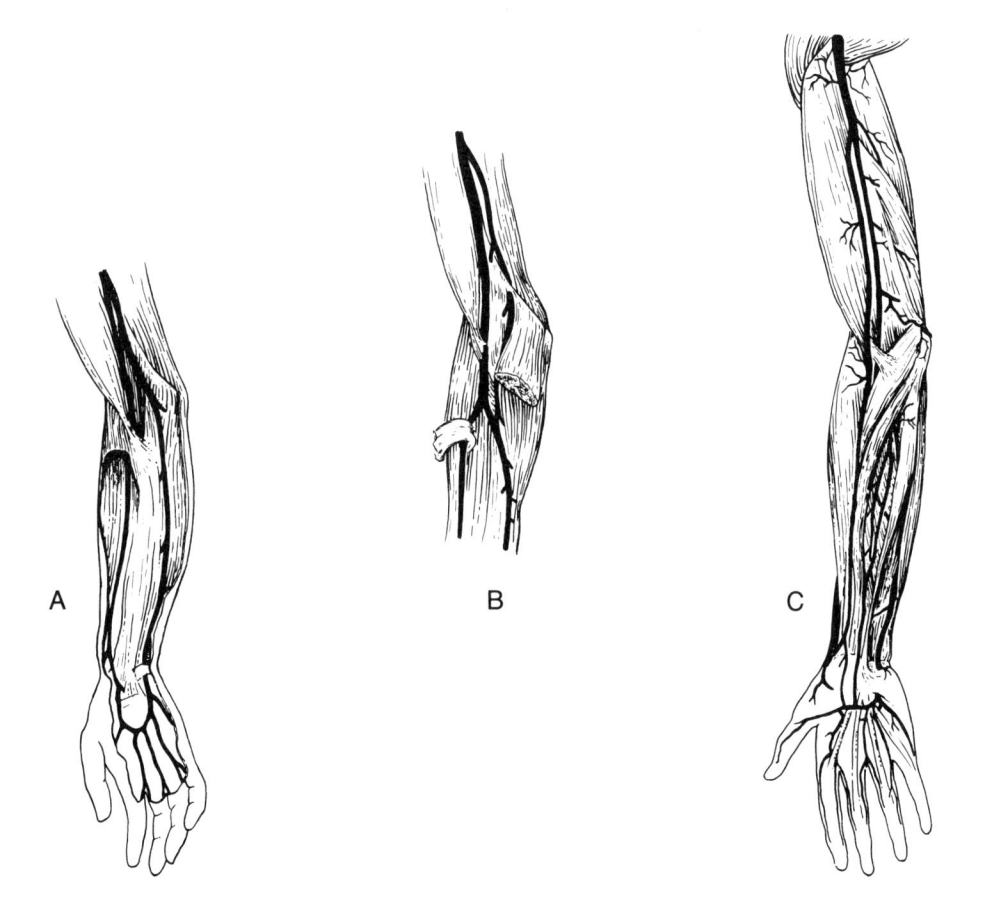

FIG. 81.
Variations in Brachial, Ulnar, and Radial Arteries.
Persistent Median Artery.

A: Ulnar artery, high origin from the brachial artery, superficial course.
B: Accessory brachial artery supplying ulnar and radial arteries. Brachial artery passes behind supracondylar process of the humerus, and joins the ulnar.
C: Radial artery, high origin from the brachial artery. Superficial palmar branch passes over transverse carpal ligament to join superficial palmar arch. Median artery in forearm with superficial course over the transverse carpal ligament to participate in superficial palmar arch.

Human Anatomic Variation

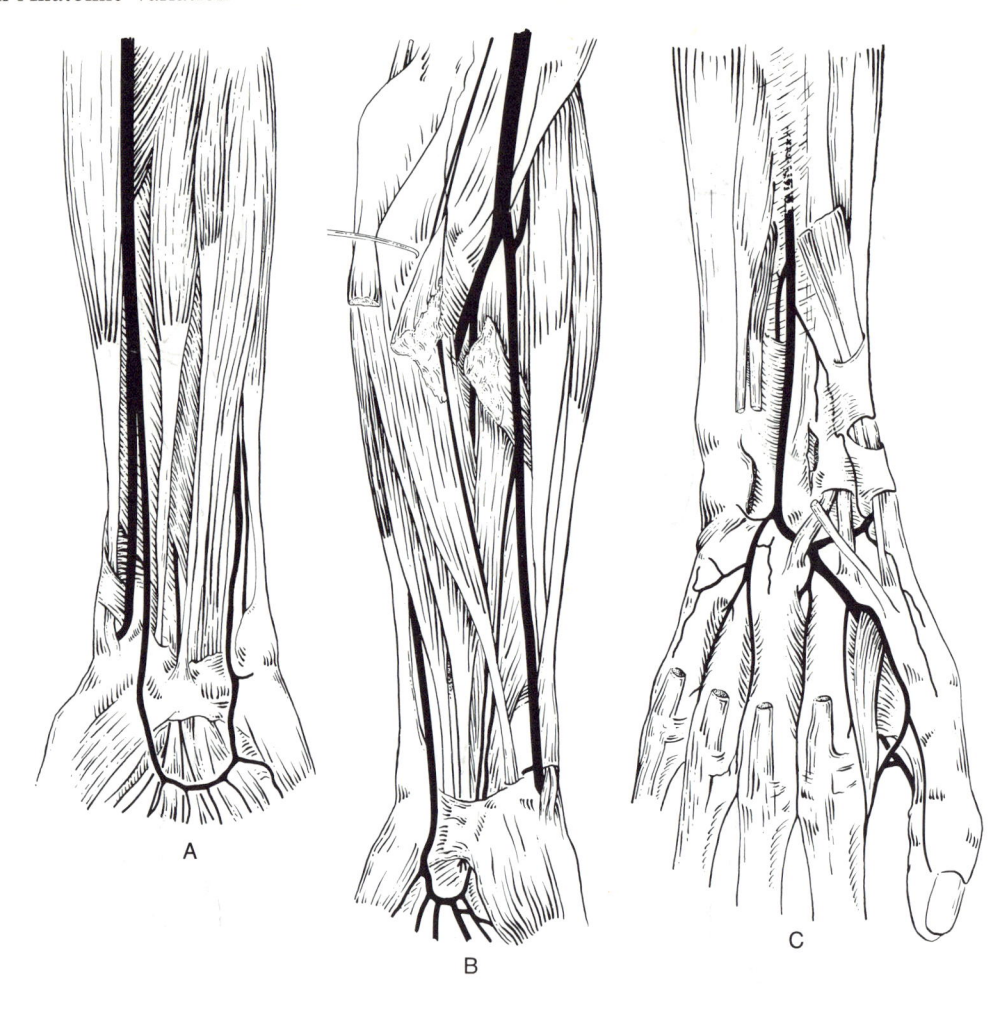

FIG. 82.
Variations in Radial Artery.
Persistent Median Artery.
A: High origin of superficial palmar branch from the radial artery.
B: A large median artery arising from the radial artery traverses the carpal canal and enters into the formation of the superficial palmar arch with the ulnar artery.

C: Extensive branching of the anterior (volar) interosseous artery to the back of the hand in a case of a small (inadequate) radial artery.

From Adachi, B. *Anatomie der Japaner. Das Arteriensystem der Japaner.* Verlag der Kaiserlich-Japanischen Universität zu Kyoto. In Kommission bei "Maruzen Co.", Kyoto und Tokyo. Gedruckt von "Kenkyusha" in Tokyo, 1928.

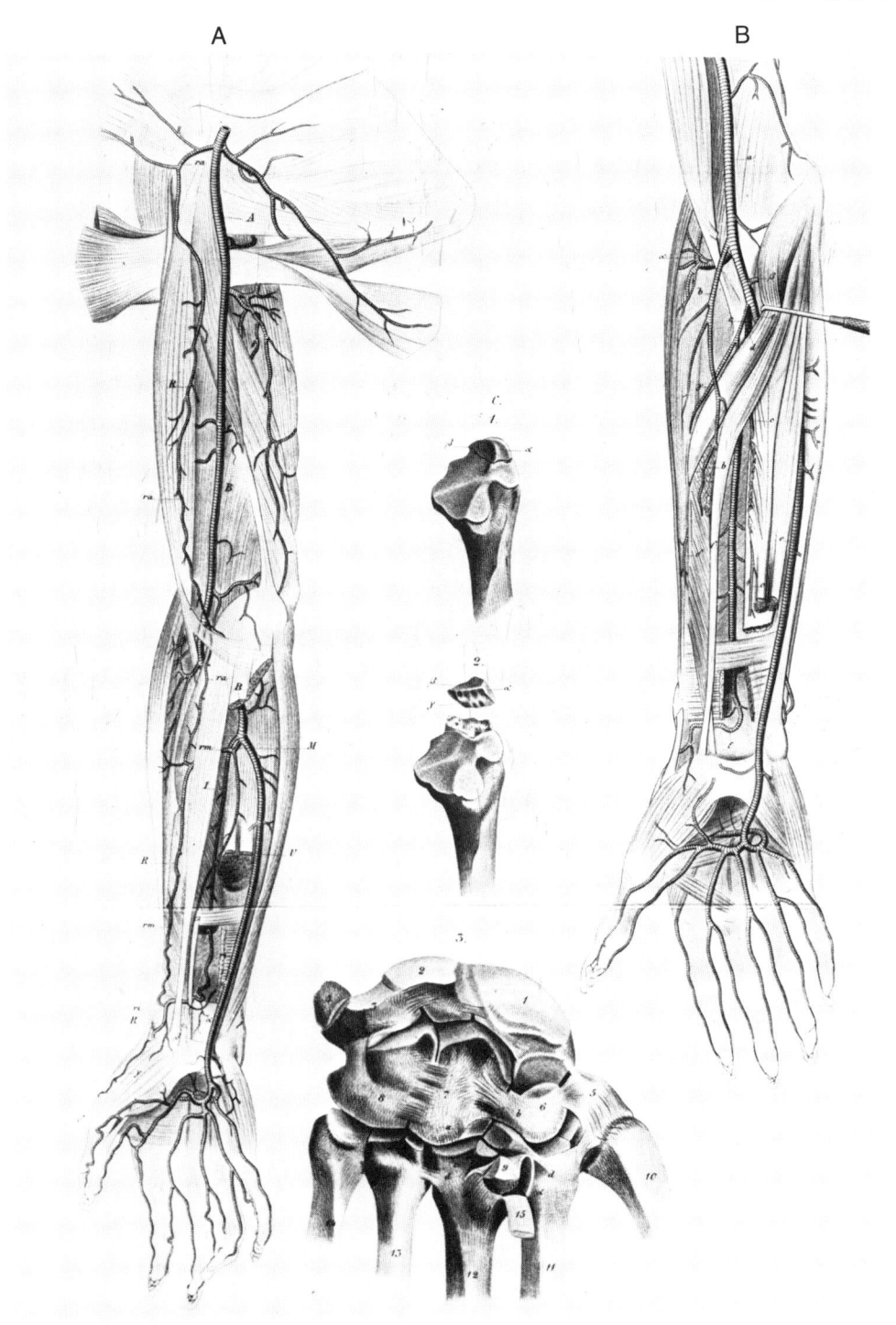

FIG. 83.
Radial Artery with Three Roots.
Rudimentary Radial Artery.
Anomalous Branch of Median Artery.
Anomalous Superior Branch of Anterior Interosseus Artery.
Anomalous Inferior Branch of Anterior Interosseus Artery.

A: *A*, Axillary artery; *B*, brachial artery; *J*, anterior interosseous artery; *M*, median (antibrachial, deep) artery; *R*, three roots of the radial artery: *ra*, superior root of radial artery from the axillary artery, *rm*, middle root of radial artery from the median artery (deep), and *ri*, inferior root of radial artery from the anterior interosseous artery; *U*, ulnar artery; α, carpal volar artery; β, superficial palmar artery; γ, dorsal radial branch of thumb; δ, volar radial branch of thumb. α,β,γ, and δ form the lower root and its continuations.

B: *a*, Brachial artery, *b*, rudimentary radial artery; *c*, median (antibrachial, deep) artery; *d*, common interosseous artery; *e*, ulnar artery; *f*, anterior interosseous artery; α, radial recurrent artery; β, ulnar recurrent artery; γ, anomalous branch of median artery; δ, anomalous superior branch of anterior interosseous artery; ε, anomalous inferior branch of anterior interosseous artery. γ,δ, and ε replace the missing part of the rudimentary radial artery.

From Gruber, W. Dreiwurzelige Arteria radialis. *Arch. Anat. Physiol. Wissen. Med.* 1870, pp. 180–188; Rudimentäre Arteria radialis. *Ibid.* 1870, pp. 189–196.

Human Anatomic Variation

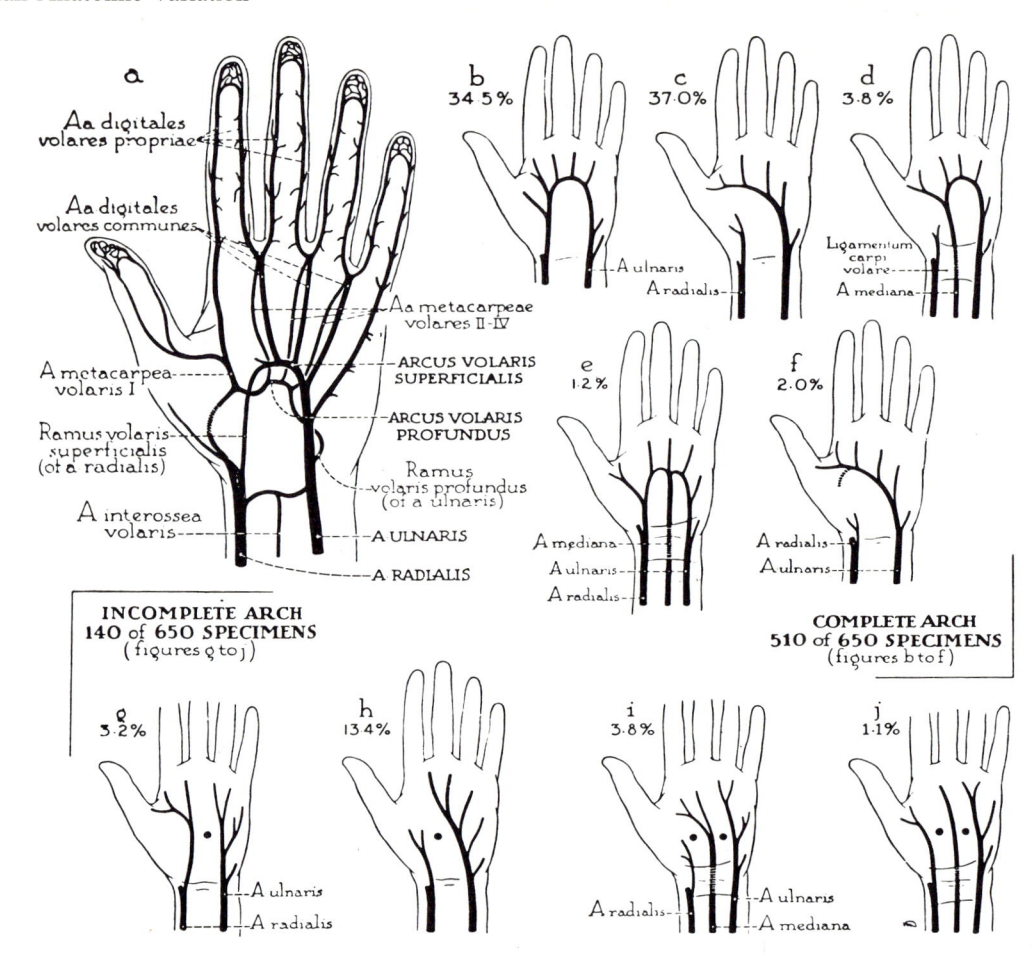

FIG. 84.
Superficial Palmar Arch.

"Types of superficial volar arterial arch (*b* to *j*) encountered in 650 dissections of the hand, together with the "textbook normal" (*a*). Shown schematically, with the percentage occurrence of each of the nine types. Variations (*b* to *f*) encountered in 510 specimens in which the arch was complete; patterns (*g* to *j*) found in the 140 hands in which the arch was incomplete – in the areas indicated by marker.

a: The standard, and regularly pictured, pattern of superficial arterial arch with its branches at metacarpal and phalangeal levels – the common and the proper volar digital arteries, respectively.

b: The schematized form of the complete radioulnar communication present in slightly more than one-third of the total number of extremities studied.

c: A transpalmar arciform continuation of the ulnar artery with a full complement of common volar digital branches. This type occurs more frequently, 37% of cases, than the traditional or "standard" type.

d: An arciform arrangement to which the contributors are ulnar and median arteries – the median artery replacing the radial of the type shown in *a* and *b*. The ulnomedian type initiates the series of patterns whose frequencies of occurrence show a sharp decline from the most common varieties; namely, those depicted in *b* and *c*.

e: An arch to which the three arteries contribute. Here a median communication is sent to the center of the arch formed by anastomosis of the radial and ulnar arteries.

f: In this type a transpalmar continuation of the ulnar artery (compare with *c*) receives a midpalmar contribution from the deep palmar arterial arch, not from the radial artery itself. In the remaining specimens, 140 out of a total of 650 (21.5%), the superficial arterial arch is incomplete, the area of interruption indicated in each diagram by a black dot.

g: In specimens of this type, the proper volar arteries are derived equally from the radial and ulnar arteries, without communication across the middle line of the hand.

h: Here the ulnar artery is the chief contributor to the set of digital vessels, supplying three and one half digits; that is, toward the thumb, to include the ulnar aspect of the index finger.

i: In specimens of this type the median artery reaches the hand to furnish digital arteries (compare with *d* and *e*) but without anastomosing the radial and ulnar arteries. Inclining toward the radial side of the hand, the median artery gives off a branch to the thumb.

j: In this variety of palmar arterial supply, the deviation of the three source-vessels toward the radial side of the hand is of lesser degree than in the preceding type; the branches of the common digital artery derived from the median artery pass only to the facing aspects of the index and middle fingers, the radial and ulnar arteries caring for the areas marginal thereto."

From Coleman, S.S. and B.J. Anson. Arterial patterns in the hand based upon a study of 650 specimens. *Surg. Gynecol. Obstet.* 113:409–424, 1961.

FIG. 85.
Deep Palmar Arch.

"Variations in constituency of the deep volar arch, in comparison with the typical patterns of radioulnar communication.
a: The "standard" pattern of deep volar arch.
b,c,d: The arrangements dependent upon differences in ulnar supply.
e,f,g,h: The subgroups of patterns in which the deep palmar arterial arch is complete.
i,j: Derivative of the incomplete arch. Four types of complete deep arch were found (a to h) and two varieties of incomplete arch (i and j)."

From Coleman, S.S. and B.J. Anson. Arterial patterns in the hand based upon a study of 650 specimens. *Surg. Gynecol. Obstet.* 113:409–424, 1961.

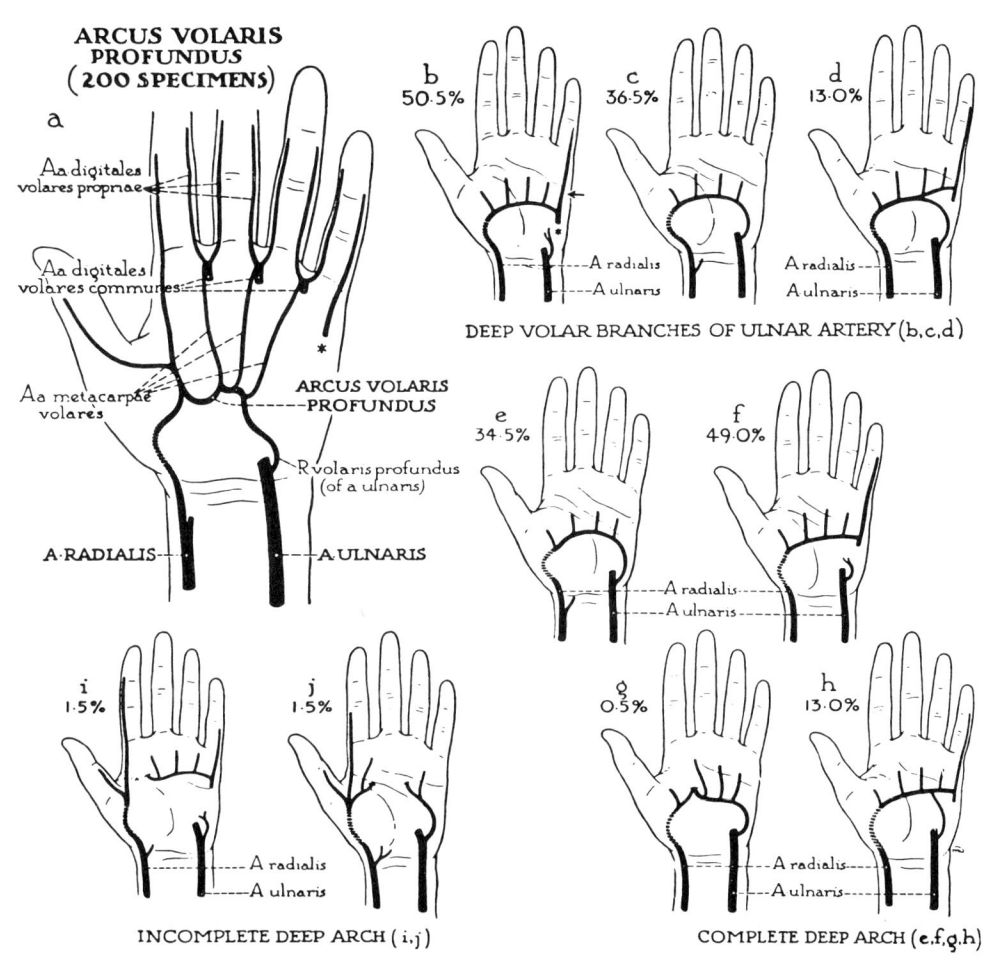

Human Anatomic Variation

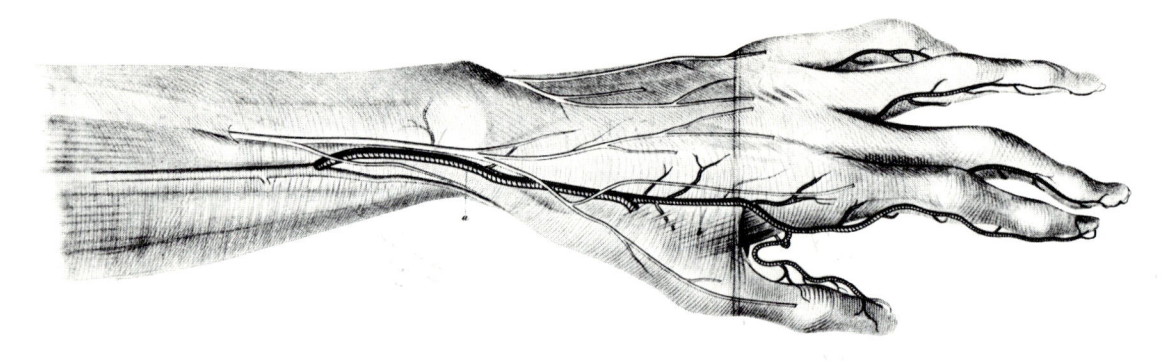

FIG. 86.
Subcutaneous Course of Dorsal Branch of Radial Artery.
a, Subcutaneous course of the dorsal branch of the radial artery; b, volar digital artery communis.

From Gruber, W. Subcutaner Verlauf des Ramus dorsalis der Arteria radialis am Unterarm- und Handwurzelrücken. (Zweiter eigener Fall.) *Arch. Anat. Physiol. Wissen. Med.* 1870, pp. 484–489.

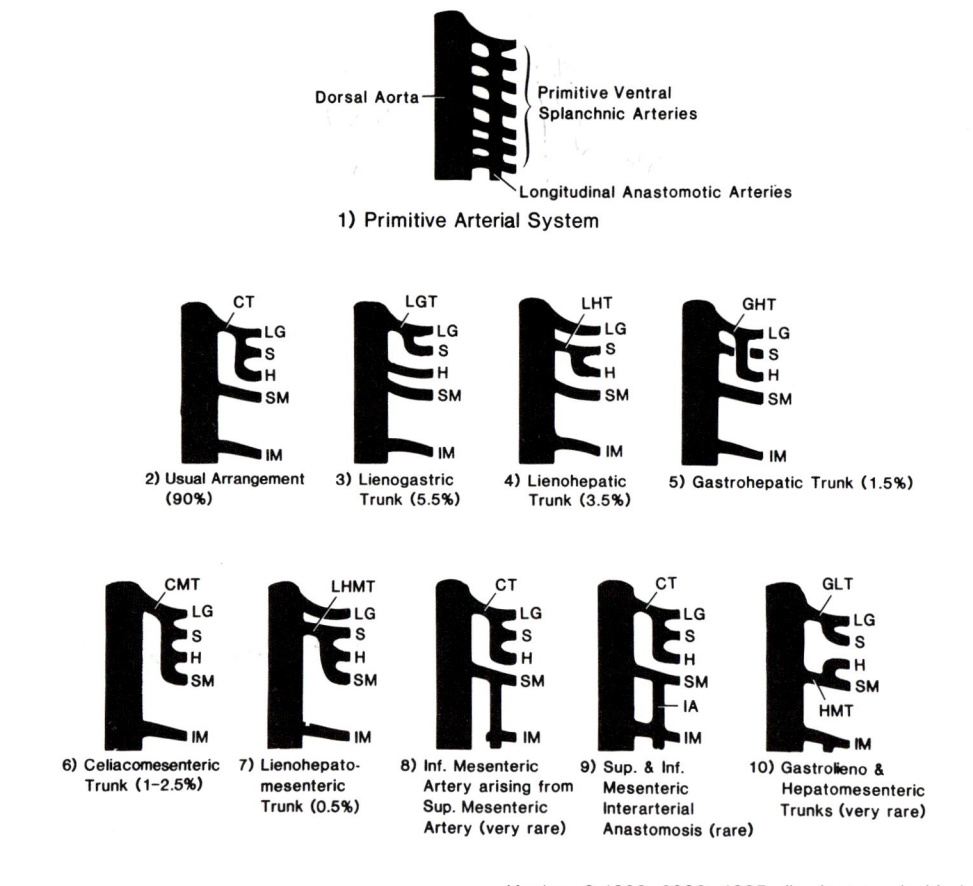

1) Primitive Arterial System

2) Usual Arrangement (90%)
3) Lienogastric Trunk (5.5%)
4) Lienohepatic Trunk (3.5%)
5) Gastrohepatic Trunk (1.5%)
6) Celiacomesenteric Trunk (1–2.5%)
7) Lienohepatomesenteric Trunk (0.5%)
8) Inf. Mesenteric Artery arising from Sup. Mesenteric Artery (very rare)
9) Sup. & Inf. Mesenteric Interarterial Anastomosis (rare)
10) Gastrolieno & Hepatomesenteric Trunks (very rare)

FIG. 87.
Variations in Unpaired Abdominal Arteries.

The variations shown above can be related to the embryologic development of segmental splanchnic arteries.

Authors' note: The concept derived is based on the following studies including that of Kitamura et al.: Tandler, J. Zur Entwicklungsgeschichte der menschlichen Darmarterien. *Anat. Hefte* 23:187–210, 1903; Tandler, J. Über die Varietäten der Arteria coeliaca und deren Entwicklung. *Anat. Hefte* 25:473–500, 1904; Morita, S. Über 3 Fälle von seltenen Variationen der A. coeliaca und A. mesenterica superior. *Igaku Kenkyu* 9:1993–2006, 1935. (in Japanese); Mori, Y., Ito, I., Hatashita, S. and K. Yoshikawa. An anomalous case of the mesenteric arteries, absence of A. mesenteric superior. *J. Osaka Med. Coll.* 20:77–79, 1960. (in Japanese); and Gwyn, D.G. and J.S. Skilton. A rare variation of the inferior mesenteric artery in man. *Anat. Rec.* 156:235–238, 1966.

Redrawn and modified from Kitamura, S., Nishiguchi, T., Sakai, A. and K. Kumamoto. Rare case of the inferior mesenteric artery arising from the superior mesenteric artery. *Anat. Rec.* 217:99–102, 1987.

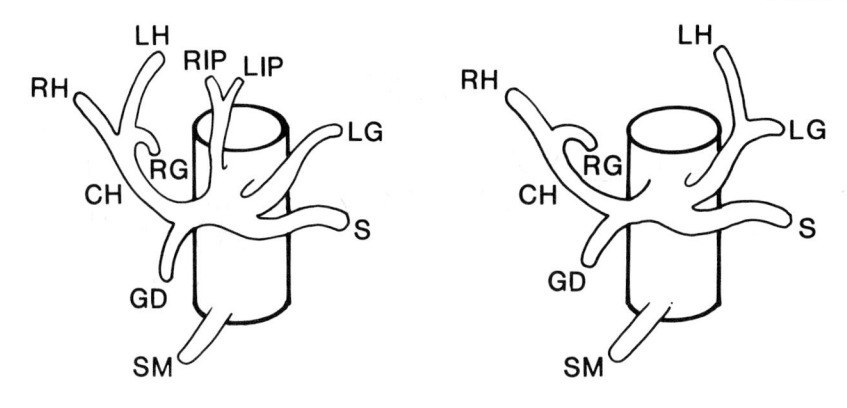

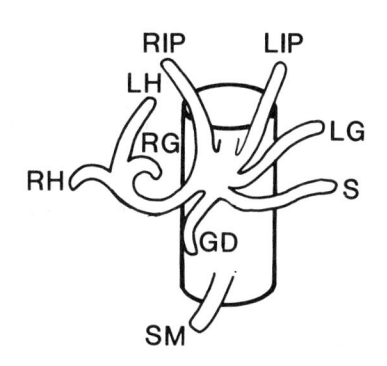

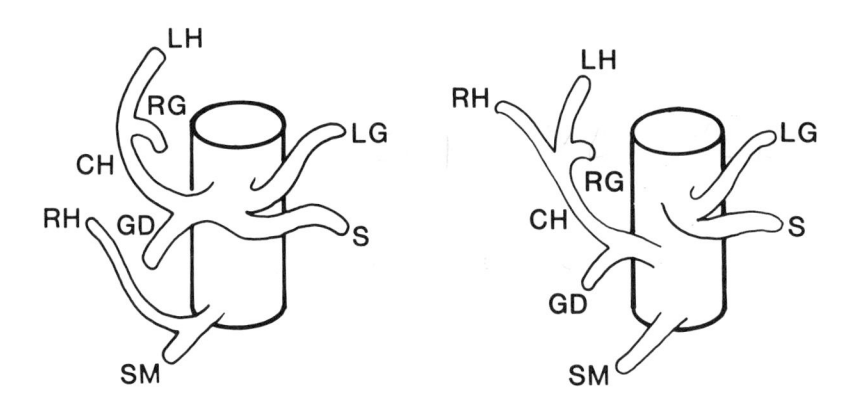

FIG. 88.
Five Variations in Celiac Trunk.

CH, Common hepatic artery; GD, gastroduodenal artery; LG, left gastric artery; LH, left hepatic artery; LIP, left inferior phrenic artery; RG, right gastric artery; RH, right hepatic artery; RIP, right inferior phrenic artery; S, splenic artery; SM, superior mesenteric artery.

Human Anatomic Variation

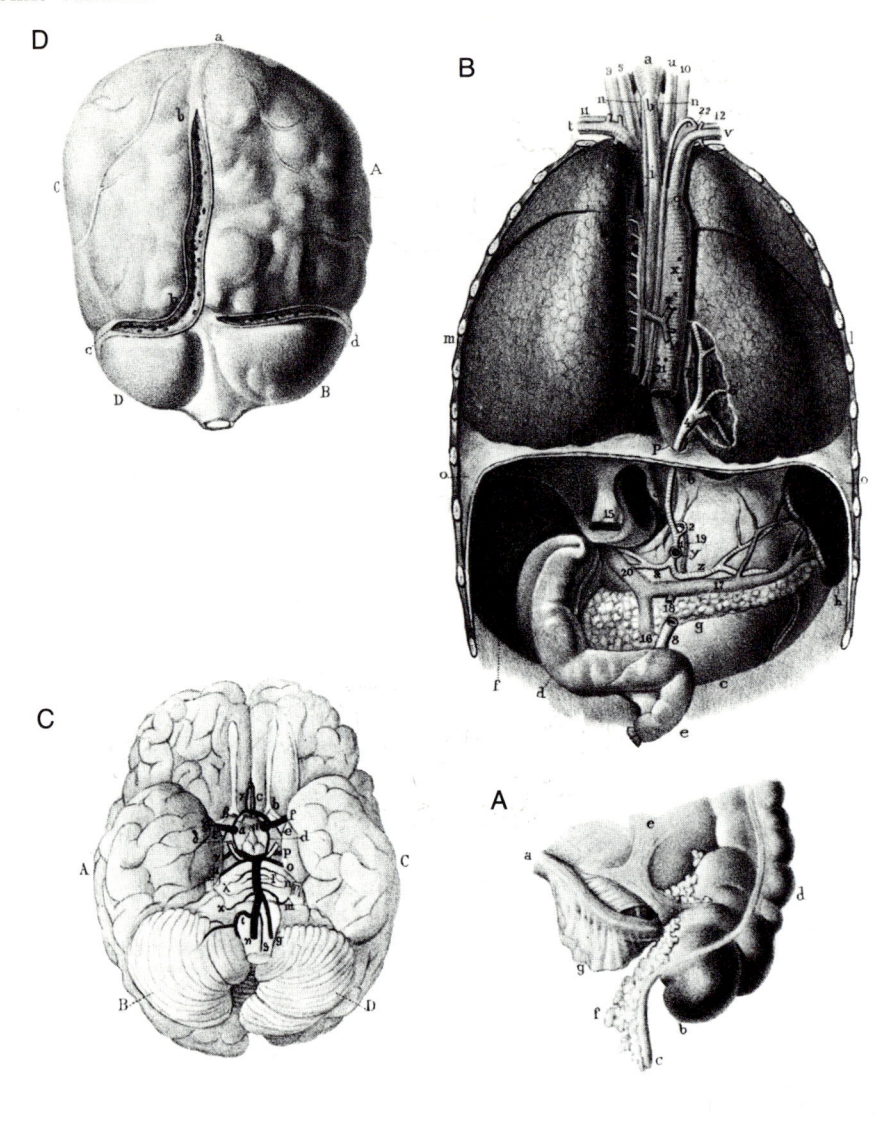

FIG. 89.
Bronchial Artery Derived from Celiac Trunk.
Variation in Vertebral Artery and Circle of Willis.
Absence of Confluens of Dural Sinuses.
(Left-Sided Vermiform Appendix).

A: *a*, Ileum; *b*, cecum; *c*, vermiform appendix (left side); *d*, ascending colon; *e*, part of the mesentery; *f*, mesentery of the appendix.

B: *a*, inferior extremity of the pharynx; *b,b*, esophagus; *c*, stomach; *d*, duodenum; *e*, fasting, or empty, intestine; *f*, liver; *g*, pancreas; *h*, spleen; *i*, trachea; *k*, bronchus on the left side; *l*, lung, bilobed on the right side; *m*, lung, trilobed on the left side; *n*, thyroid gland; *o*, diaphragm; *p*, diaphragmatic esophageal hiatus; *r*, brachiocephalic artery; *s*, common carotid, left side; *t*, subclavian artery on the left side; *u*, common carotid, right side; *v*, right subclavian artery; *x*, descending thoracic aorta; *y*, celiac trunk; *z*, splenic artery; *&*, hepatic artery; *1*, celiac trunk; *2*, left gastric artery; *3*, branch of the celiac artery, passing through the esophageal hiatus, supplying the inferior medial part of the right lung; *4*, varicosity of *3*; *5,6,7*, branches to the right lung; *8*, superior mesenteric artery; *9,10*, internal jugular vein; *11,12*, subclavian vein; *13,14*, azygos and hemiazygos veins; *15*, inferior vena cava; *16*, superior mesenteric vein; *17*, splenic vein; *18*, inferior mesenteric vein; *19*, left gastric vein; *20*, portal vein; *21*, thoracic duct ending on the right side into the right subclavian vein.

C: Variations of the arteries at the base of the brain, asymmetry. Variations in size of arteries forming an asymmetric circle of Willis. Imbalance in size of the vertebral arteries (*g* and *η*). Anterior spinal artery arising from right vertebral.

D: Absence of confluens of sinuses. The superior sagittal sinus drains only into the left transverse sinus.

From Calori, L. Di una inversione splancnica generale nell'uomo accompagnata da alcuni notabili del capo con esso lei convenienti e da estranee anomalie. *Mem. R. Accad. Sci. Istituto di Bologna S. 4* 2:597–622, 1880.

Cardiovascular System

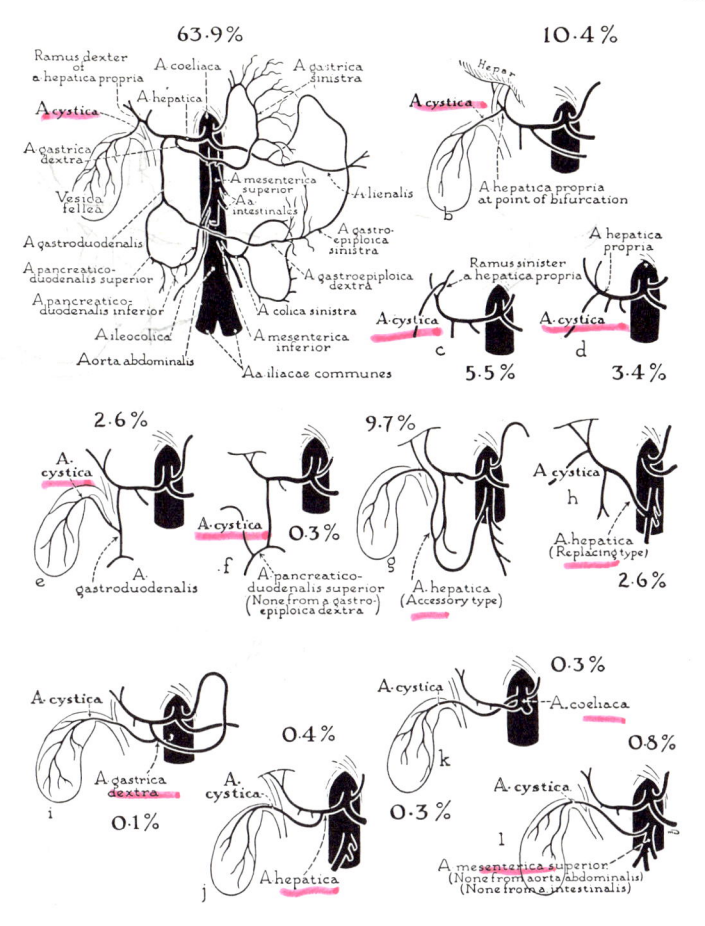

FIG. 90.
Origin of Cystic Artery.

"The (primary) source of the cystic arteries is still of the classical variety, that is to say, they arise from the closest possible source (the right ramus of the hepatic artery proper).

As might reasonably be expected, the aforementioned type of origin of arteries to the gallbladder is the commonest of all. Other origins, of course, occur – to make a total of twelve types encountered in a study of 800 specimens carried out in the author's laboratory. The twelve types fall conveniently into four categories, which now will be reviewed.

GROUP I. In those specimens which belong in a first (and most numerous) category, the cystic artery arises from a ramus of the proper hepatic artery, or from the latter at the point of division, or from the same vessel proximal to the point of bifurcation (figs. a to d). In succession, and in the order of decreasing frequency, these are: from the right ramus of the hepatic proper in 63.9%; from the latter vessel at the point of division in 10.4%; from the left ramus in 5.5%; and from the hepatic artery proper, proximal to the point of bifurcation, in 3.4%. Totally, 83% for the four near source-vessels.

GROUP II. To a second category may be assigned those specimens in which the cystic artery takes origin from the downward-directed vessel, the gastroduodenal artery, or the latter's branch, the superior pancreaticoduodenal (figs. e and f). Origin from the more distant source is far less common (0.3%) than the nearer (2.6%).

GROUP III. In a third general group belong those specimens in which the immediate source of the cystic artery matches the regular pattern which exists in the cases illustrated in the first category, but differs from them in the derivation of the parent vessel (figs. g and h). The parent stem may be the superior mesenteric in each, but the hepatic artery itself may be either an accessory artery or a replacing one. Together, these instances (9.7% and 2.6%) make up a total of 12.3%. From the point of view of nearness of source to area of supply, these cases could be regarded as belonging to Group I (to make a combined total of 95.5%).

GROUP IV. To the last, or fourth, category may be assigned those specimens in which the cystic artery is "moved," so to speak, toward the aorta. These have been observed in four types: from the right gastric in 1 case (fig. i); from the hepatic artery near the coeliac source in 3 instances (fig. j); from the coeliac itself in 2 cases (fig. k) and from the superior mesenteric in 6 instances (fig. 1). Together these cases make a total of only 12."

Conclusions

"Presented in this segregated way, the variations hereinbefore described tend to appear somewhat less critical than they actually are, since, in the structure of any individual patient, unexpected arrangements in the biliary duct-system may accompany equally unpredictable patterns of vascularity – even to the occurrence of a portal vein in a position anterior to the hepatic artery and the common bile duct, together with a cystic artery derived from a distant source and a supernumerary hepatic artery of large caliber.

This means, obviously, that there can be no substitute for full exposure of the extrahepatic element of the so-called "triad" and of the accessory structure – which may be arteries, veins, or ducts, or these in combination."

Authors' note: Additional reference: McVay, C.B. Anson and McVay *Surgical Anatomy*, 6th Ed, Vol. 2. W.B. Saunders Co., Philadelphia, 1984.

From Anson, B.J. Anatomical considerations in surgery of the gallbladder. *Q. Bull. Northwestern University Medical School* 30:250–259, 1956.

Human Anatomic Variation

Common Hepatic Artery

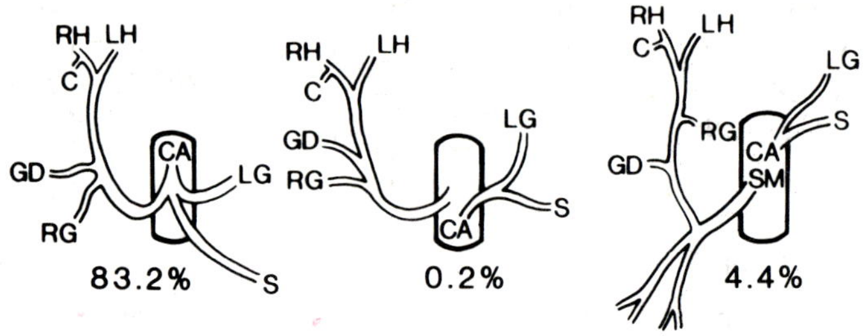

Right Hepatic Artery

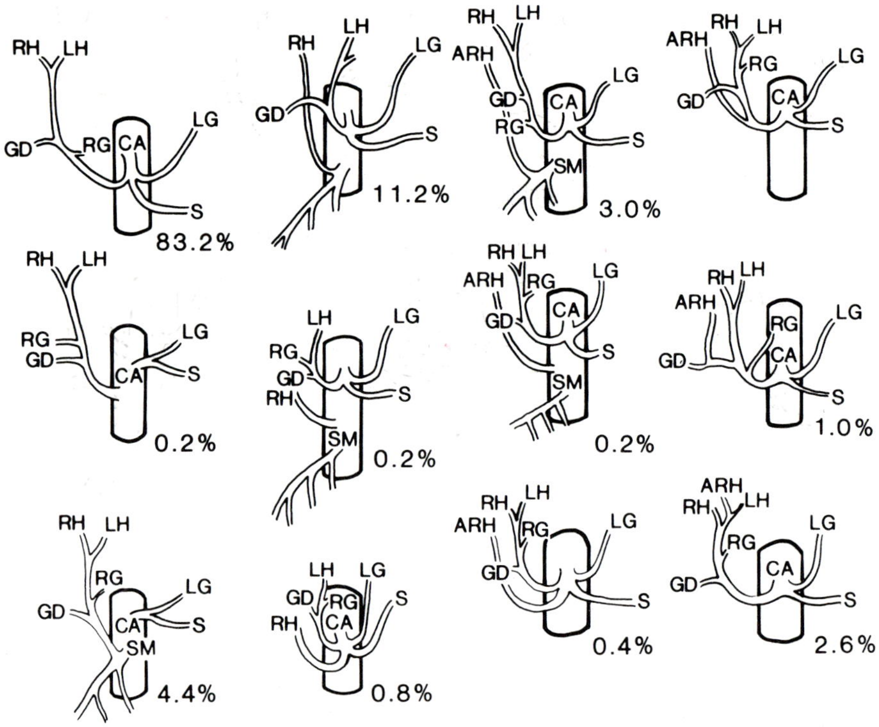

FIG. 91.
Variations in Origin of Common Hepatic Artery.
Variations in Origin of Right Hepatic Artery.

ALH, accessory left hepatic artery; *ARH*, accessory right hepatic; *C*, cystic; *CA*, celiac (axis); *GD*, gastroduodenal; *LG*, left gastric; *LH*, left hepatic; *RG*, right gastric; *RH*, right hepatic; *RRH*, replacing right hepatic; *S*, splenic; *SM*, superior mesenteric.

Daseler et al. provide the following summary:
Origin, of common hepatic (or hepatic) artery; 500 cases: celiac, 416; superior mesenteric, 22; aorta, 1; absent, 61.
Location of hepatic (common hepatic) artery; 439 cases: left of common bile duct, 431; posterior to same, 7; anterior to same, 1.
Origin of right hepatic artery (right ramus of hepatic proper); 500 cases: common hepatic, 415; superior mesenteric, 56; replacing common hepatic, 23; celiac, 4; aorta, 1.
Origin of accessory right hepatic artery; 36 cases: superior mesenteric, 15; left hepatic, 13; gastroduodenal, 5; celiac, 2; aorta, 1.
Location of right hepatic artery; 500 cases: posterior to common hepatic duct, 325; anterior to common hepatic duct, 58; posterior to common bile duct, 58; right of duct system, 22; posterior to right and left hepatic ducts, 18; left of duct system, 8; anterior to common bile duct, 7; anterior to right and left hepatic ducts, 4.

Redrawn from Daseler, E.H., Anson, B.J., Hambley, W.C. and A.F. Reimann. The cystic artery and constituents of the hepatic pedicle. A study of 500 specimens. *Surg. Gynecol. Obstet.* 85:47–63, 1947.

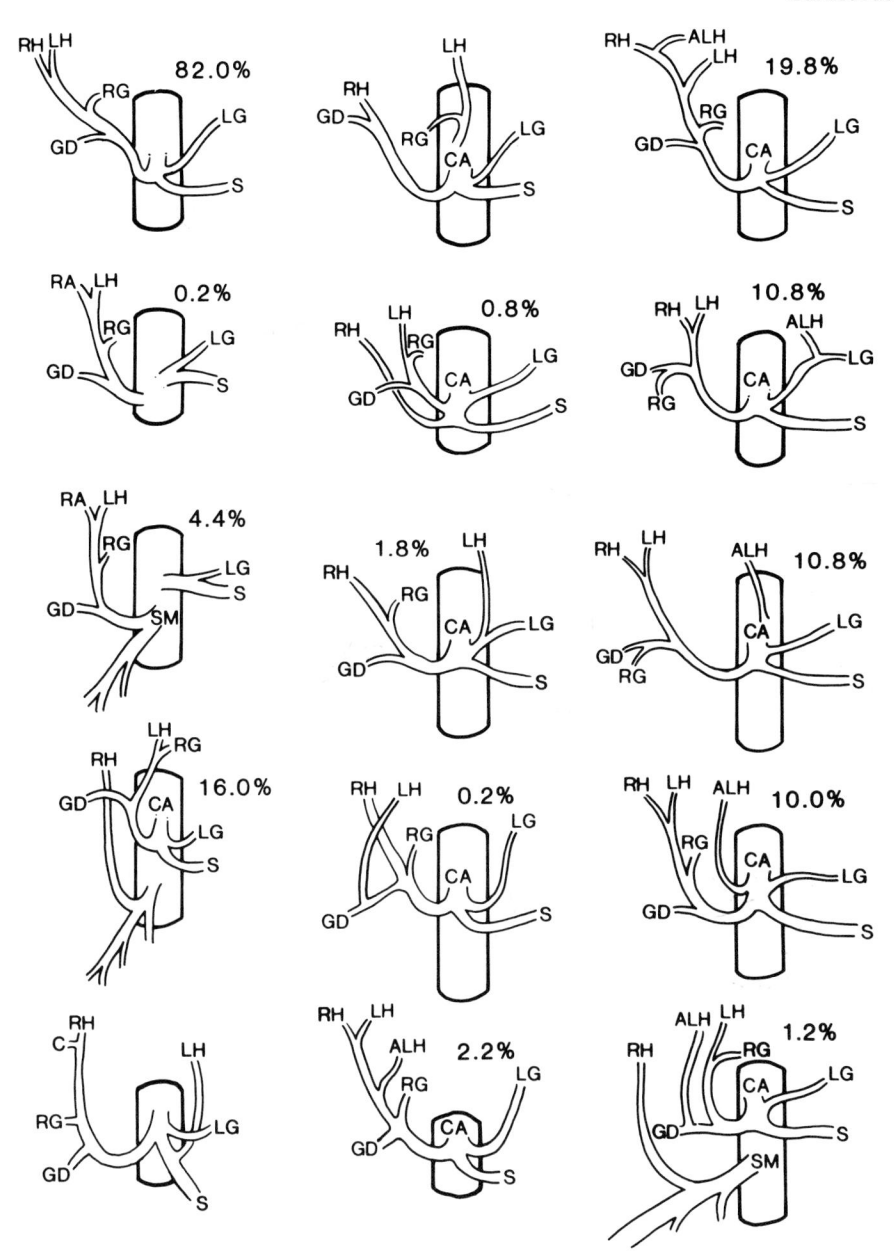

FIG. 92.
Variations in Origin of Left Hepatic Artery.

ALH, accessory left hepatic artery; *ARH*, accessory right hepatic; *C*, cystic; *CA*, celiac (axis); *GD*, gastroduodenal; *LG*, left gastric; *LH*, left hepatic; *RG*, right gastric; *RH*, right hepatic; *RRH*, replacing right hepatic; *S*, splenic; *SM*, superior mesenteric.

Daseler et al. provide the following summary:
Origin of left hepatic artery, 500 cases: common hepatic, 410; celiac, 57; replacing common hepatic, 23; left gastric, 9; gastroduodenal, 1.

Origin of accessory left hepatic artery; 175 cases: right hepatic, 99; left gastric, 54; common hepatic, 11; gastroduodenal, 6; celiac, 5.

Redrawn from Daseler, E.H., Anson, B.J., Hambley, W.C. and A.F. Reimann. The cystic artery and constituents of the hepatic pedicle. A study of 500 specimens. *Surg. Gynecol. Obstet.* 85:47–63, 1947.

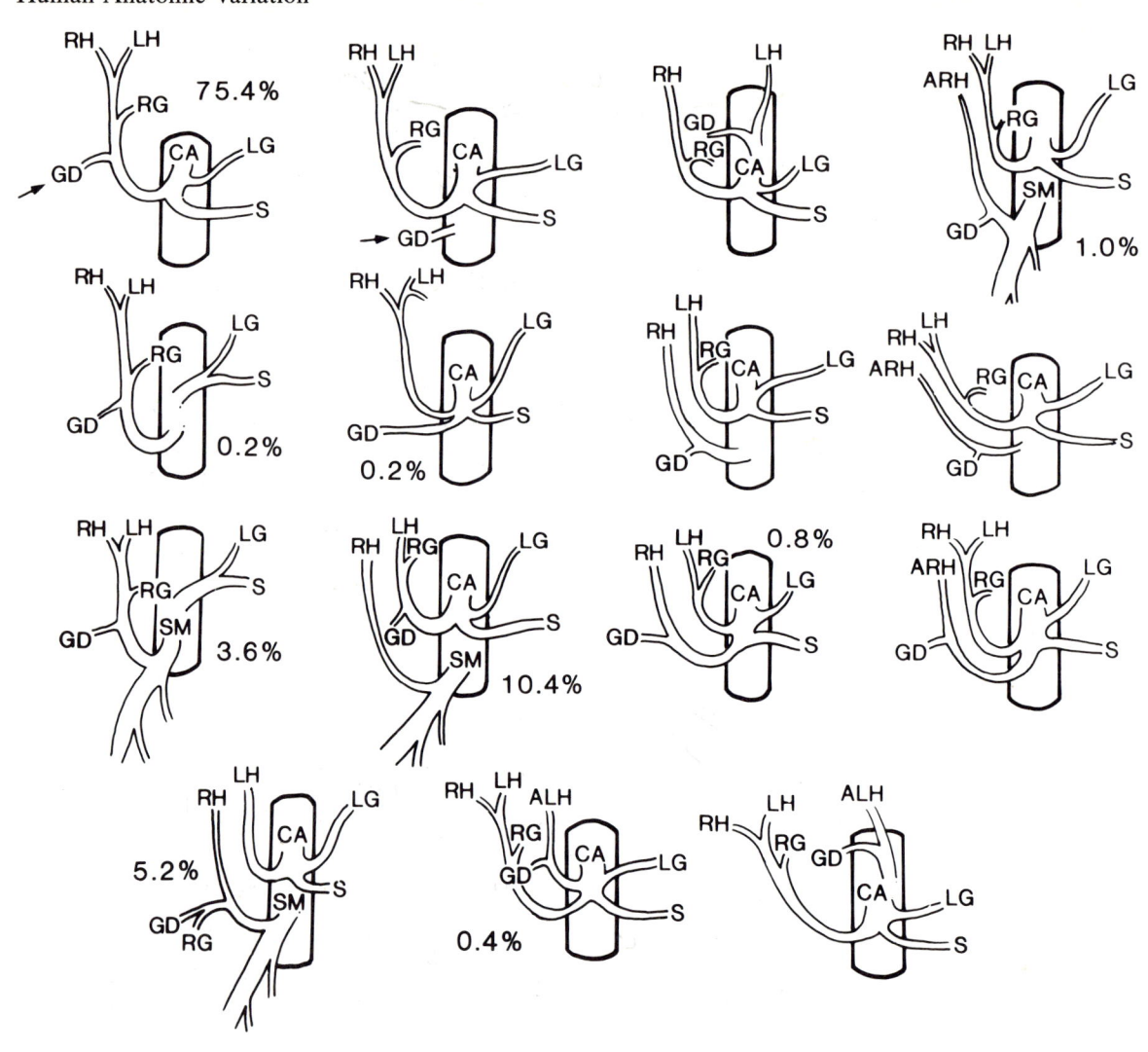

Human Anatomic Variation

FIG. 93.
Variations in Origin of Gastroduodenal Artery.

ALH, accessory left hepatic artery; *ARH*, accessory right hepatic; *C*, cystic; *CA*, celiac (axis); *GD*, gastroduodenal; *LG*, left gastric; *LH*, left hepatic; *RG*, right gastric; *RH*, right hepatic; *RRH*, replacing right hepatic; *S*, splenic; *SM*, superior mesenteric.

Daseler et al. provide the following summary:
Origin of gastroduodenal artery; 500 cases: common hepatic, 377; replacing left hepatic, 52; right hepatic, 35; replacing common hepatic, 19; accessory left hepatic, 2; celiac, 1; absent, 14.

Redrawn from Daseler, E.H., Anson, B.J., Hambley, W.C. and A.F. Reimann. The cystic artery and constituents of the hepatic pedicle. A study of 500 specimens. *Surg. Gynecol. Obstet.* 85:47–63, 1947.

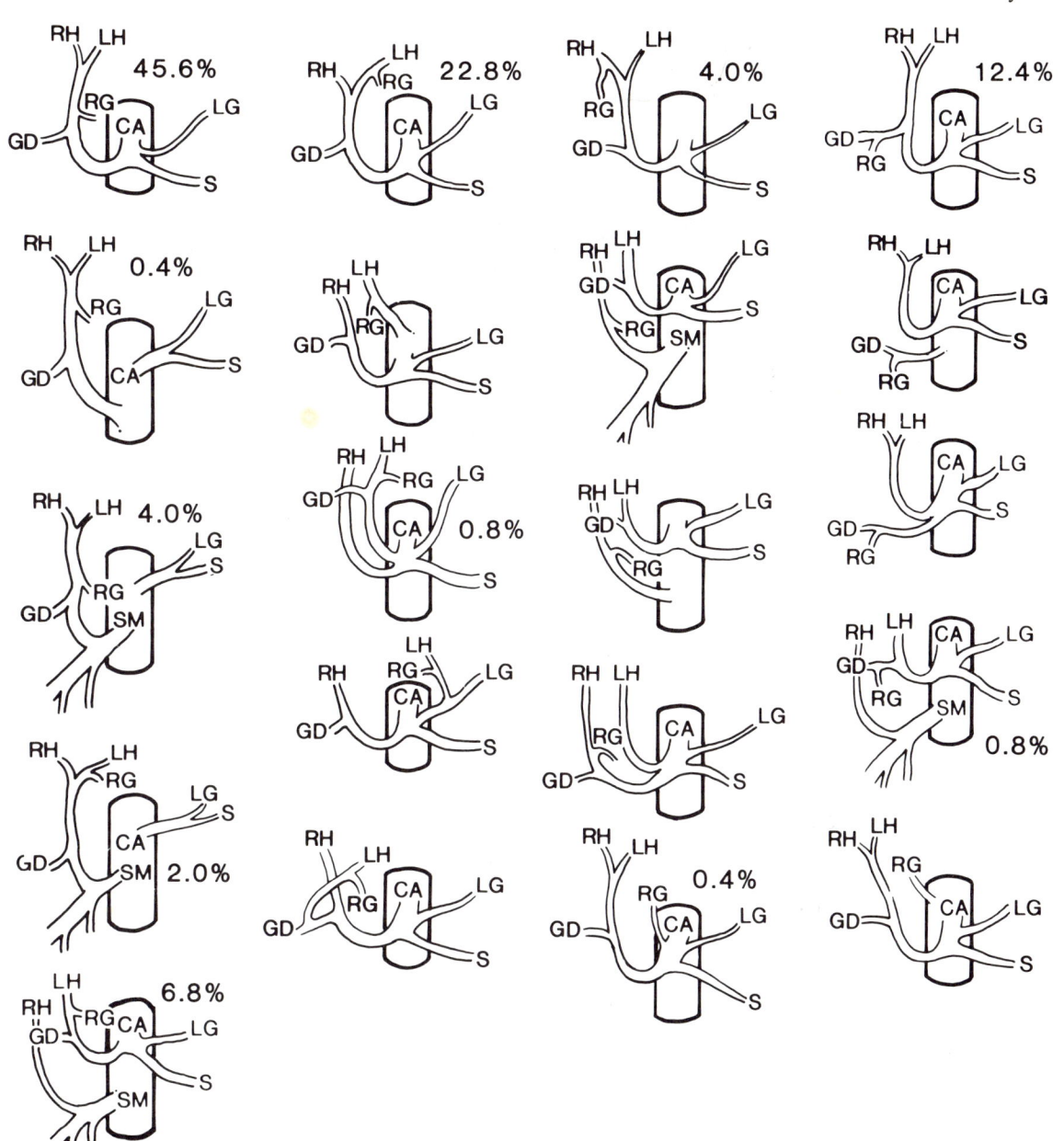

Cardiovascular System

FIG. 94.
Variations in Origin of Right Gastric Artery.

ALH, accessory left hepatic artery; *ARH*, accessory right hepatic; *C*, cystic; *CA*, celiac (axis); *GD*, gastroduodenal; *LG*, left gastric; *LH*, left hepatic; *RG*, right gastric; *RH*, right hepatic; *RRH*, replacing right hepatic; *S*, splenic; *SM*, superior mesenteric.

Daseler et al. provide the following summary:
Origin of right gastric artery; 250 cases: common hepatic, 125; left hepatic, 81; gastroduodenal, 33; right hepatic, 10; celiac, 1.
Redrawn from Daseler, E.H., Anson, B.J., Hambley, W.C. and A.F. Reimann. The cystic artery and constituents of the hepatic pedicle. A study of 500 specimens. *Surg. Gynecol. Obstet.* 85:47–63, 1947.

Human Anatomic Variation

FIG. 95.
Variations in Origin of Cystic Artery.

ALH, accessory left hepatic artery; *ARH*, accessory right hepatic; *C*, cystic; *CA*, celiac (axis); *GD*, gastroduodenal; *LG*, left gastric; *LH*, left hepatic; *RG*, right gastric; *RH*, right hepatic; *RRH*, replacing right hepatic; *S*, splenic; *SM*, superior mesenteric.

In addition to the origins shown here, the authors found that the cystic artery also arose in three instances from an accessory or a replacing type of right hepatic artery which arose from the celiac trunk. (Number of specimens, 500.)

Daseler et al. provide the following summary:

Origin of cystic artery; 580 cases: right division of regular hepatic, 416; right hepatic from superior mesenteric, 90; left hepatic, 36; common hepatic, 16; gastroduodenal, 15; replacing or accessory right hepatic, 3; celiac, 2; superior mesenteric, 1; superior pancreaticoduodenal, 1.

Origin of accessory cystic artery; 65 cases: right hepatic, 50; gastroduodenal, 6; common hepatic, 4; left hepatic, 3; accessory right hepatic from superior mesenteric, 2.

Origin of single replacing cystic artery; 120 cases: replacing right hepatic from superior mesenteric, 36; left hepatic, 27; right hepatic from replacing common hepatic, 21; accessory right hepatic from superior mesenteric, 12; common hepatic, 11; gastroduodenal, 6; accessory right hepatic from celiac, 3; celiac, 2; superior mesenteric, 1; superior pancreaticoduodenal, 1.

Origin of dual replacing cystic arteries, 15 cases: both from right hepatic branch of superior mesenteric, 7; replacing right hepatic and left hepatic, 3; both from left hepatic, 1; gastroduodenal and accessory right hepatic, 1; gastroduodenal and right hepatic branch from celiac, 1; gastroduodenal and left hepatic, 1; gastroduodenal and right hepatic branch of superior mesenteric, 1.

Location of cystic artery; 580 vessels: *Chalot's triangle, 405; anterior to common hepatic duct, 123; anterior to common bile duct, 17; posterior to common hepatic duct, 12; between hepatic ducts, 6; anterior to cystic duct and right hepatic duct, 6; right of duct system, 5; posterior to common bile duct, 3; left of duct system, 2; posterior to cystic duct and right hepatic duct, 1.

Authors' note: *Chalot's triangle is formed by the common hepatic and cystic ducts.

Redrawn from Daseler, E.H., Anson, B.J., Hambley, W.C. and A.F. Reimann. The cystic artery and constituents of the hepatic pedicle. A study of 500 specimens. *Surg. Gynecol. Obstet.* 85:47–63, 1947.

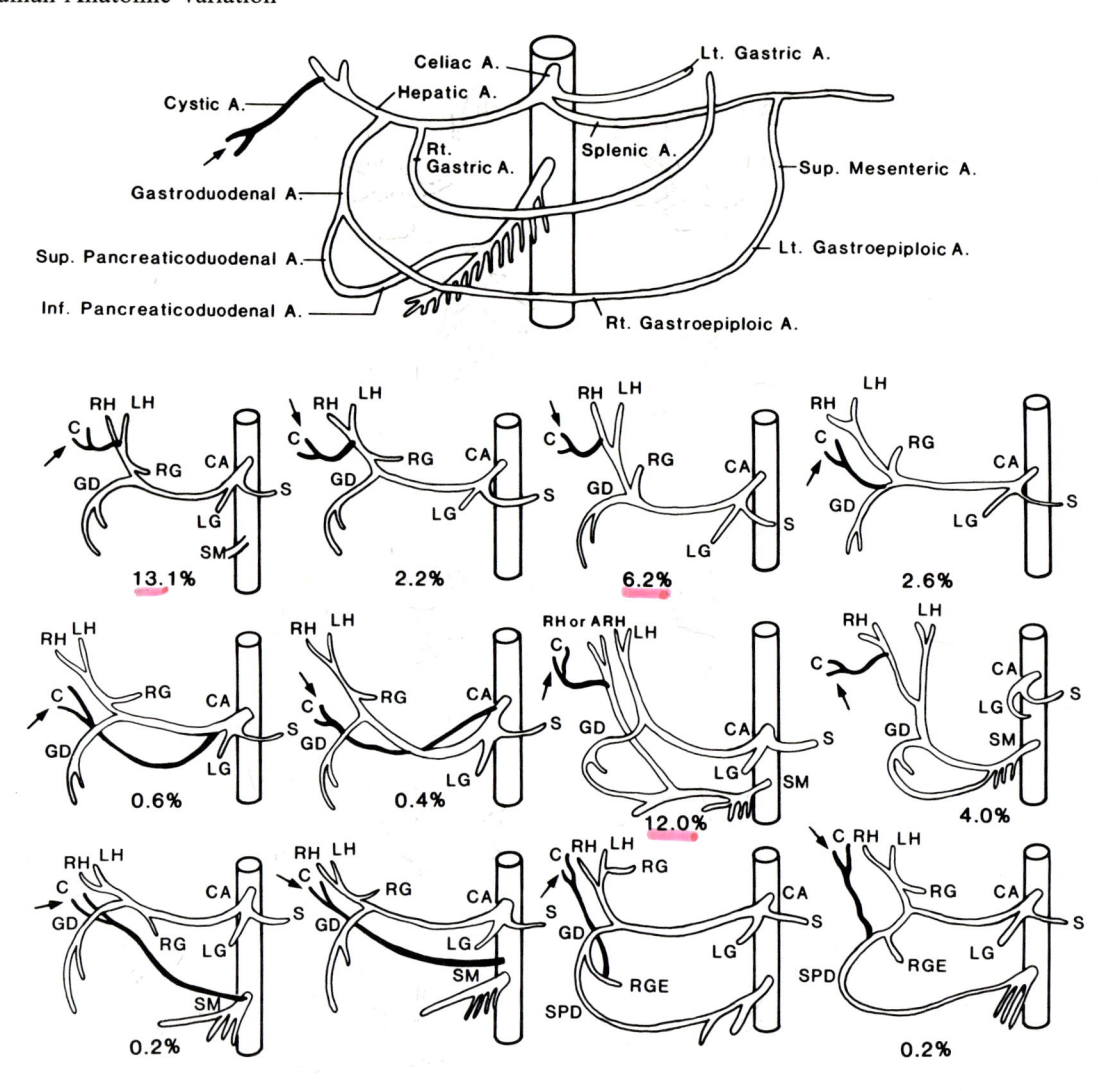

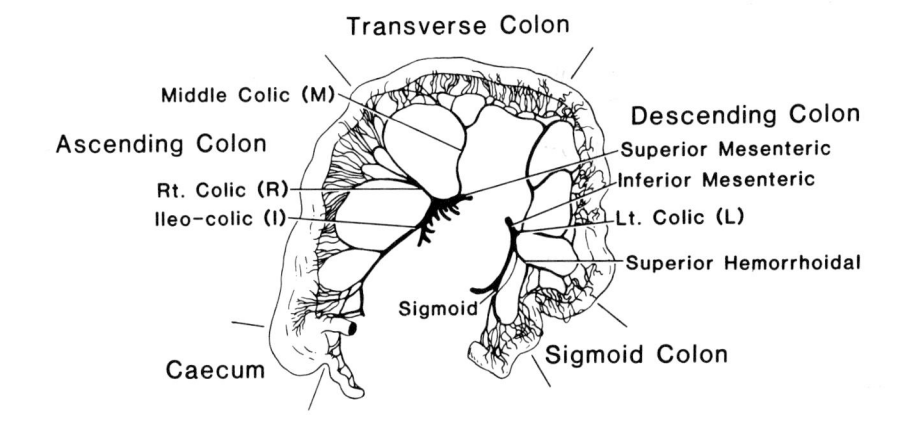

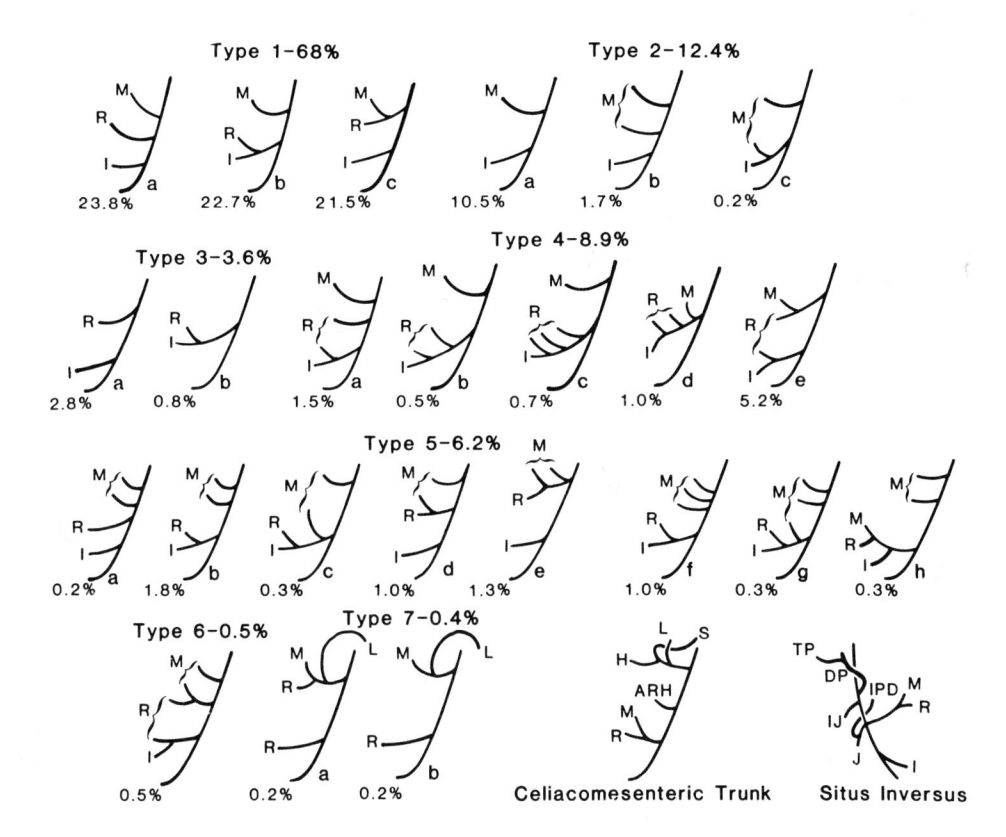

FIG. 96.

Variations in Arterial Supply to the Colon from Superior Mesenteric Artery.

Types of colic branching of the superior mesenteric artery and varieties within each of 7 categories (types 1 to 7). Schematized from records on 600 specimens. Abbreviations for arteries: *I*, ileocolic; *L*, left colic; *M*, middle colic; *R*, right colic.

In type 1 (representing 68% of all specimens studied) the 3 typical branches, of approximately equal caliber, arise either independently from the superior mesenteric, and thus as 3 vessels, or from 3 roots (with the right colic derived from ileocolic or middle colic). In type 2 (12.4%) a right colic artery is wanting and the middle colic is either single or double. The distinguishing feature in type 3 (3.6%) is the absence of a middle colic in the presence of a right colic of either direct or indirect mesenteric source. In the specimens catalogued as belonging to type 4 (8.9%) 2 or 3 right colic arteries occur, accompanied in each variety of subtype by both ileocolic and middle colic branches. Type 5 (6.2%) is characterized by the occurrence of 2 or 3 middle colic arteries in the presence of a right colic (in contrast, type 2) and of an ileocolic artery. In type 6 (0.5%) both middle and right colic arteries are doubled. In type 7 (0.4%) a left colic artery is added to the typical set of 3 branches.

Authors' note: Two additional drawings unrelated to the above study were added to this illustration (*lower right*) of arterial branches from a celiacomesenteric trunk (1–2.5%) and from a case of situs inversus. Abbreviations for these arteries: *ARH*, accessory right hepatic; *DP*, dorsal pancreatic; *H*, hepatic, *I*, ileocolic; *IJ*, ileojejunal; *IPD*, inferior pancreatoduodenal; *J*, jejunal; *L*, left colic; *M*, middle colic; *R*, right colic; *S*, splenic, and *TP*, transverse pancreatic.

This article, drawing on the experience of these authors and their colleagues with 600 cases, provides a statistical basis for predicting the probability of variation of colic arteries. Thus,

Human Anatomic Variation

these authors "demonstrate statistically that the classical pattern of colic arteries is present in but 23.8 per cent of their 600 cases. It will be established that variations, of the classical pattern are the rule, not the exception. As we know during surgery, variants for the individual patient are not merely interesting anatomic facts, but may be a matter of life and death. In writing about surgery of the colon, Mayo emphasized this point: 'One cannot operate with implicit faith equipped only with a textbook picture. Many, and sometimes tragic, are the lessons learned from experience.' [reference 831]. The term 'anticipatory variants' aptly describes the central thought of this article. Study and knowledge of 'anticipatory variants' of the arterial supply to the colon, we believe, should be of material aid to the operating surgeon."

Redrawn from Sonneland, J., Anson, B.J. and L.E. Beaton. Surgical anatomy of the arterial supply to the colon from the superior mesenteric artery based upon a study of 600 specimens. *Surg. Gynecol. Obstet.* 106:385–398, 1958.

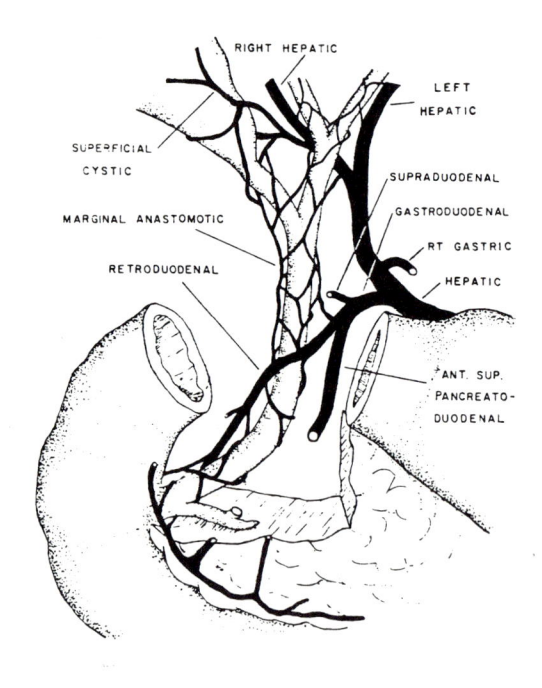

FIG. 97.
Blood Supply of the Common Bile Duct.
Schematic drawing of most frequent type of arterial pattern of common bile duct. Relationship of retroduodenal or posterior-superior pancreatoduodenal artery to duct makes it the primary source of vascularization to both its supraduodenal and infraduodenal segments. Note continuity of ensheathing and anastomotic epicholedochal plexus on cystic and hepatic ducts and contributions made by hepatic and cystic arteries (based upon 41 fetal and neonatal specimens).

Parke, W.W., Michels, N.A. and G.M. Ghash. Blood supply of the common bile duct. *Surg. Gynecol. Obstet.* 117:47–55, 1963.

FIG. 98.
Variations in Blood Supply of the Common Bile Duct.

a: The gastroduodenal artery (GD) is the main contributor to epicholedochal plexus. Later it ascends the cystic duct to unite with superficial cystic artery (SC). Anterior-superior pancreatoduodenal artery (ASP) supplies upper infraduodenal part of plexus. Marginal artery (A) from retroduodenal artery.

b: Specimen lacks marginal anastomotic artery shown in a (A). Substitute contributions are made to epicholedochal plexus by branches from proper hepatic artery (H).

c: Specimen shows right hepatic artery (RH) of low origin functioning as a substitute for marginal anastomotic artery by providing direct branches to an epicholedochal plexus.

d: Low origin of cystic artery (C) from hepatic (H) enables it to supply branches to epicholedochal plexus as functional substitute for missing anastomotic loops.

e: Only specimen encountered in this investigation which shows proper epicholedochal artery (EA). Derived from retroduodenal (RD), it supplies the common, cystic, and hepatic ducts.

f: One of two cases lacking true epicholedochal plexus. Small arteries from relatively distant paracholedochal vessels derived from retroduodenal (RD) supply common duct.

From Parke, W.W., Michels, N.A. and G.M. Ghash. Blood supply of the common bile duct. *Surg. Gynecol. Obstet.* 117:47–55, 1963.

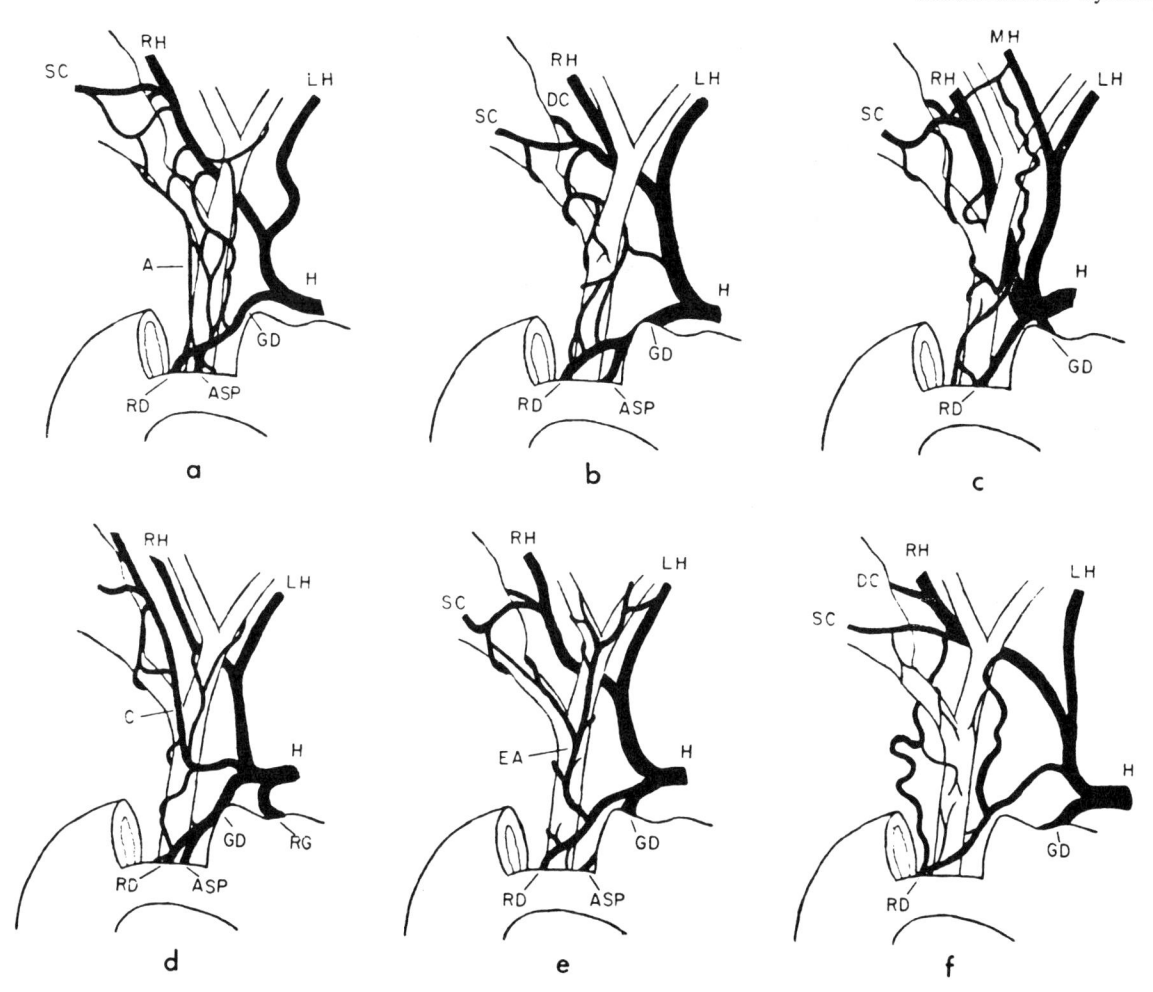

Human Anatomic Variation

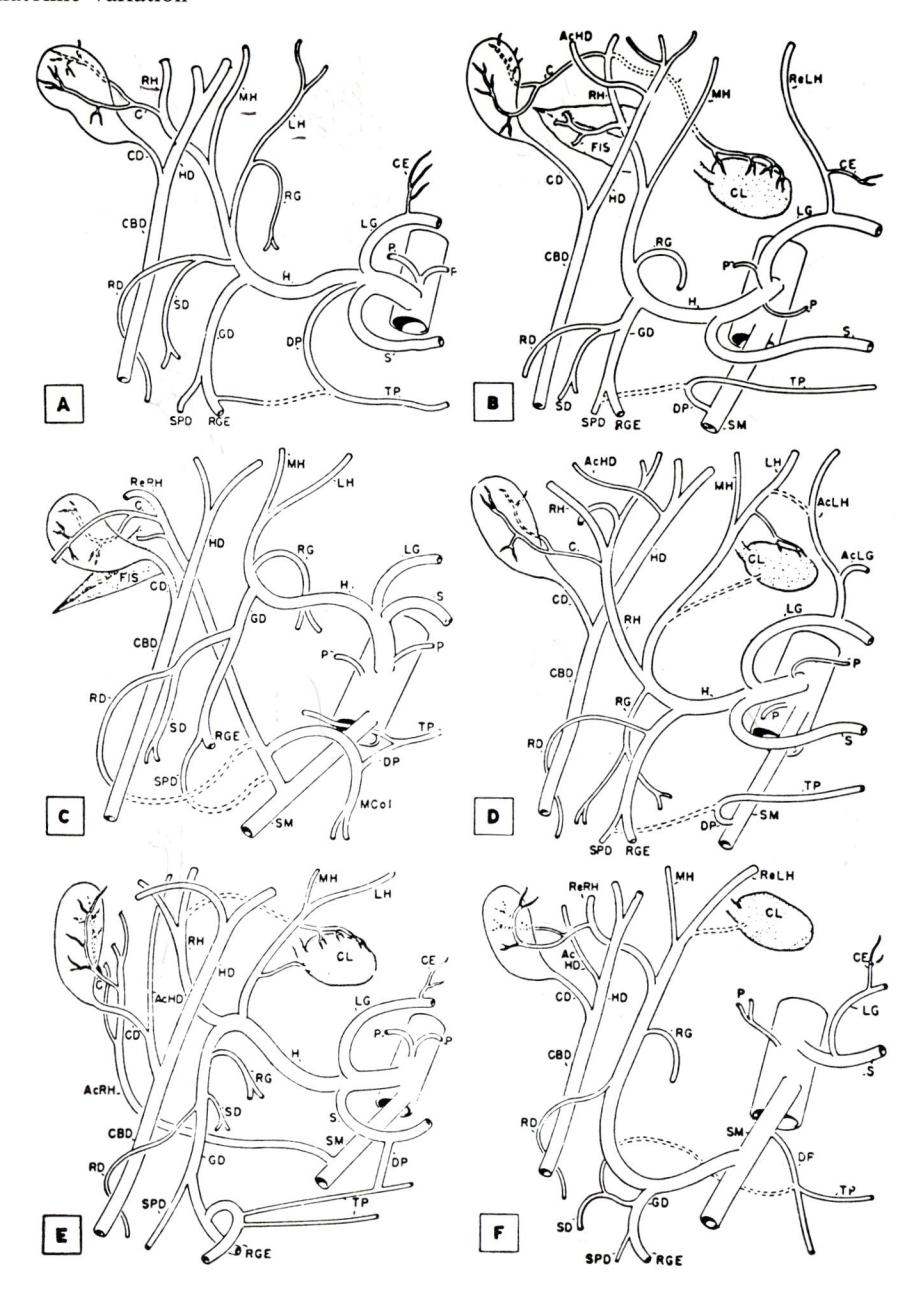

FIG. 99.
Principal Patterns of Arterial Supply to the Liver. Selected from Dissections of 200 Cadavers).

A: Typical textbook pattern of the hepatic artery with its three hepatic branches, right, left, and middle.
B: Hepatic gives off right hepatic and middle hepatic; left hepatic is replaced from left gastric.
C: Hepatic gives off left hepatic and middle hepatic; right hepatic is replaced from superior mesenteric.
D: Hepatic gives off right, middle, and left hepatic; an accessory left hepatic from left gastric.
E: Hepatic gives off right, middle, and left hepatic; an accessory right hepatic from superior mesenteric.
F: Entire hepatic trunk arises from superior mesenteric, there being no celiac hepatic artery.

AcHD, Accessory hepatic duct; *AcLG*, accessory left gastric; *AcLH*, accessory left hepatic; *AcRH*, accessory right hepatic; *C*, cystic artery; *CBD*, common bile duct; *CD*, cystic duct; *CE*, cardioesophageal branches; *CL*, caudate lobe; *DP*, dorsal pancreatic; *Fis*, fissure under gallbladder; *GD*, gastroduodenal; *H*, hepatic; *HD*, hepatic duct; *HL*, left hepatic; *MCol*, middle colic; *MH*, middle hepatic; *P*, inferior phrenic; *ReLH*, replaced left hepatic; *ReRH*, replaced right hepatic; *RD*, retroduodenal; *RG*, right gastric; *RGE*, right gastroepiploic; *RH*, right hepatic; *S*, splenic; *SD*, supraduodenal; *SPD*, superior pancreatoduodenal; *SM*, superior mesenteric; *TP*, transverse pancreatic.

From Michels, N.A. Collateral arterial pathways to the liver after ligation of the hepatic artery and removal of the celiac axis. *Cancer* 6:708–724, 1953.

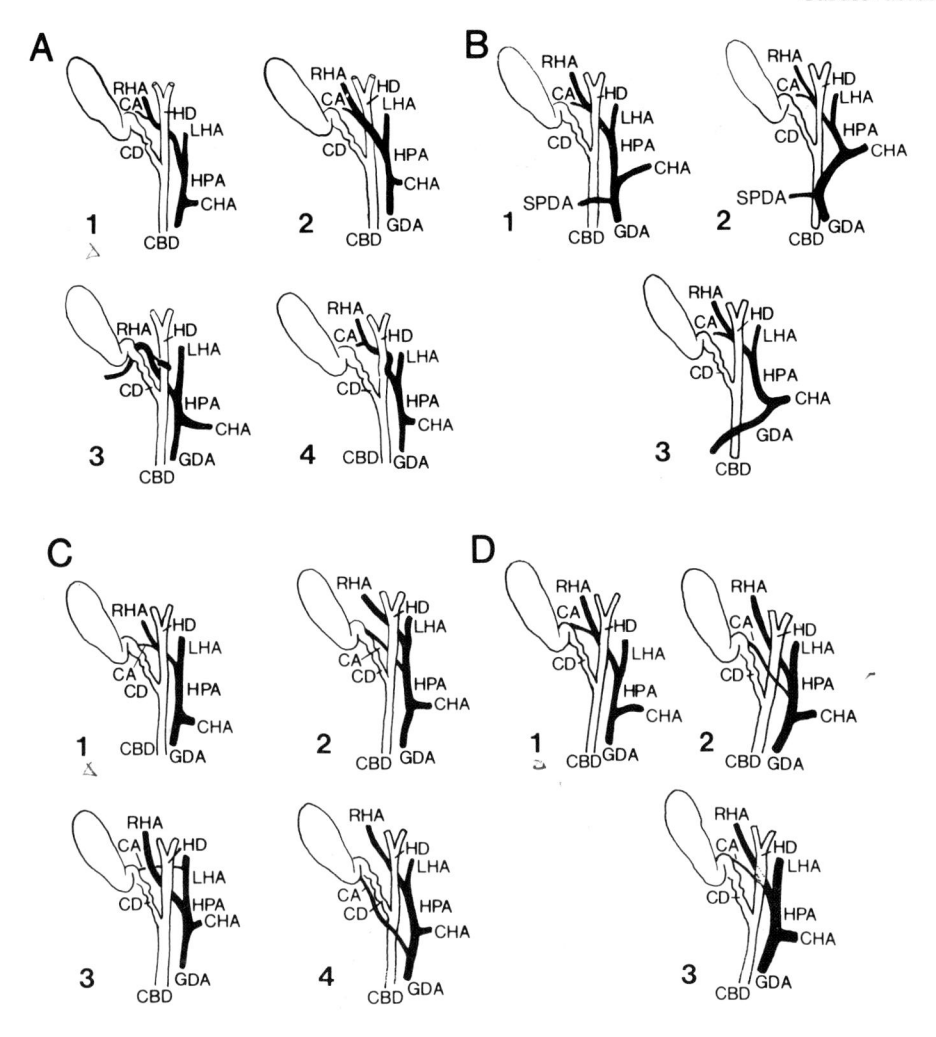

FIG. 100.
Hepatic Pedicle.

RHA, Right hepatic artery; *LHA*, left hepatic artery; *HPA*, main hepatic artery; *CHA*, common hepatic trunk; *GDA*, gastroduodenal artery; *CA*, cystic artery; *HD*, main hepatic duct; *CD*, cystic duct; *CBD*, common bile duct; *SPDA*; superior pancreatoduodenal artery.

A: Relations of the right hepatic artery to the main hepatic duct.
   1: In 70% the right hepatic artery runs behind the main hepatic duct and the cystic artery arises just beyond right edge of duct.
   2: In 12% the right hepatic artery passes in front of the main hepatic duct before entering right lobe of the liver.
   3: In 10% the right hepatic artery runs parallel to the cystic duct throughout its course and lies in close proximity to the neck of the gallbladder before entering right lobe of the liver. Artery could be easily injured during cholecystectomy.
   4: In 8% the right hepatic artery passes across right edge of the main hepatic duct and then enters liver or forms a ring around the hepatic duct.

B: Variations of the gastroduodenal artery in relation to the common duct.
   1: In 42% a branch of the gastroduodenal artery crosses the common duct.
   2: In 38% the gastroduodenal artery arches across the left border of the common duct.
   3: In 20% the gastroduodenal artery itself crosses in front of the common duct.

C: Abnormalities in the origin of the single cystic artery.
   1: There is only one cystic artery in 88% of individuals. In 82% of these it arises from the right hepatic artery.
   2: In 3% of the 88% the cystic artery arises from the main hepatic.
   3: In 2% of the 88% the cystic artery arises from the left hepatic.
   4: In 1% of the 88% the cystic artery arises from the gastroduodenal artery. This is a very important anomaly if the cystic artery is only looked for in its usual place, i.e., arising from the right hepatic.

D: Relation of the single cystic artery to the main hepatic duct.
   1: In 71% the cystic artery arises to the right of the main hepatic duct.
   2: In 27% the cystic artery arises on the left side of the hepatic duct or common duct and must cross one of these to reach neck of gallbladder.
   3: In 2% the cystic artery arises behind the main hepatic duct, and if it should be retracted, it is easy to include duct in grasp of forceps.

Redrawn from Eisendrath, D.N. The clinical importance of anatomical anomalies in biliary surgery. *Boston Med. Surg.* 182:573–578, 1920.

Human Anatomic Variation

FIG. 101.
Hepatic Pedicle (continued).

RHA, Right hepatic artery; *LHA*, left hepatic artery; *HPA*, main hepatic artery; *CHA*, common hepatic trunk; *GDA*, gastroduodenal artery; *CA*, cystic artery; *HD*, main hepatic duct; *CD*, cystic duct; *CBD*, common bile duct; *SPDA*, superior pancreatoduodenal artery.

E: Various modes of origin of double cystic arteries.
  1: In 8% of the cases having two cystic arteries, both vessels arise from the right hepatic artery.
  2: In 2% one artery arises from the right hepatic artery.
  3: In 1% one artery arises from the right hepatic and the other from the main hepatic (hepatica propria) artery.
  4: In 1% both vessels arise from the right hepatic artery.
F: Variations in the mode of union of the cystic and main hepatic ducts.
  1: Normal (75%) unite at acute angle. Terminal 2 cm paralleled and held together firmly by fibrous tissue.
  2: Short parallel type. Parallel for 5 cm or more as far as upper border of pancreas.
  3: Long parallel type. Parallel almost throughout course, i.e., to within ½–1 cm from ampulla of Vater.
  2,3: Together occurred in 17%.
  4,5: Anterior and posterior spiral types occurred in 8%. Note how cystic duct winds around anterior (or posterior) surface of the hepatic duct to enter its left border.
G: Anomalies in the right hepatic duct.
  1: Right hepatic duct empties into cystic duct.
  2: Cystic duct empties into the right hepatic duct.
H: Possible locations of calculi in cases with anomalies of the mode of union of the cystic and hepatic ducts.
  1: Calculus in cystic duct of short parallel type. Can compress hepatic duct and cause same symptom as calculus in that duct.
  2: Calculi in long parallel type of ducts. Could cause great technical difficulty in removal, with possible injury of ducts.
  3: Calculi in spiral cystic duct. Very puzzling clinical picture if one (*single arrow*) compressed hepatic duct and other obstructed a cystic duct (*double arrow*) emptying into hepatic duct on its left side.
  4: Similar possibilities from clinical and operative standpoint as in 3.

Redrawn from Eisendrath, D.N. The clinical importance of anatomical anomalies in biliary surgery. *Boston Med. Surg. J.* 182:573–578, 1920.

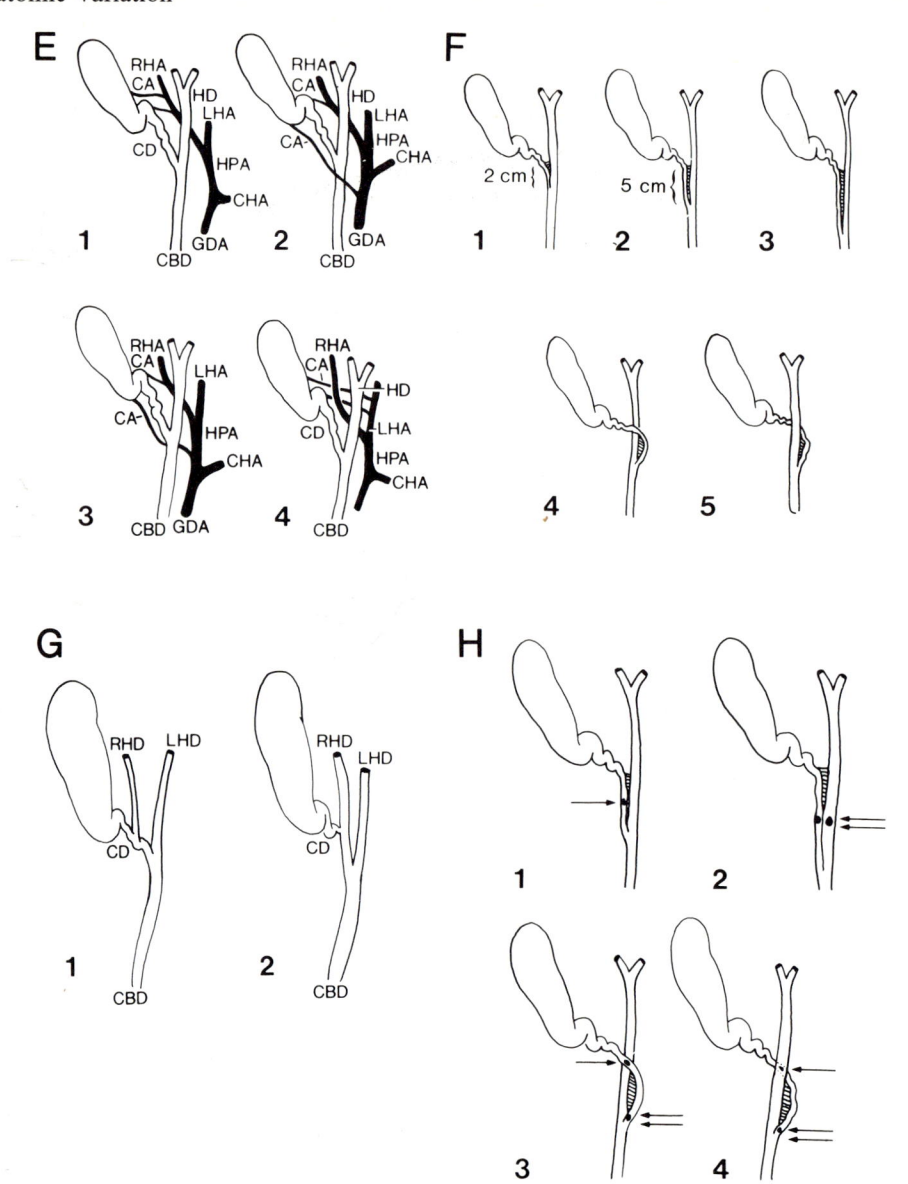

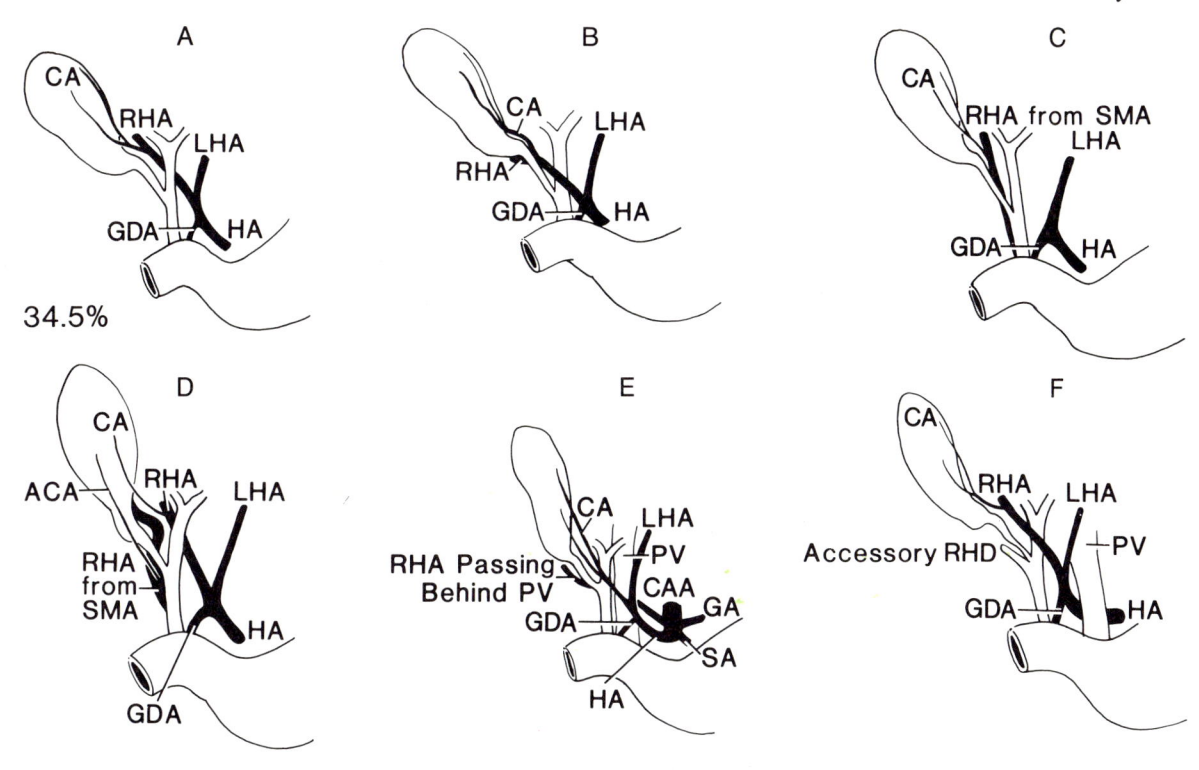

FIG. 102.

Hepatic Pedicle: Variations in Arteries and Bile Ducts.

The arrangement of the vessels and ducts given as normal constitute only 69 cases of the 200 studied (34.5%) by Flint (A). "So frequent are variations that it is impossible to regard any one type as normal; the arrangement found in the 69 cases can only be described as the most usual one."

ACA, Accessory cystic artery; CA, cystic artery; CAA, celiac trunk; CD, cystic duct; GA, gastric artery; GDA, gastroduodenal artery; HA, hepatic artery; LHA, left hepatic artery; PV, portal vein; RHA, right hepatic artery; RHD, right hepatic duct; SA, splenic artery; SMA, superior mesenteric artery; SPDA, superior pancreatoduodenal artery.

Right Hepatic Artery

This artery arose from the common hepatic in 158 cases, and reached the liver by passing behind the common hepatic duct in 136 cases (A), and in front of the duct in 22 cases (B). In 42 cases the right hepatic artery arose from the superior mesenteric (C), and always passed behind the common duct. In 7 cases there were two right hepatic arteries, one from the hepatic trunk and one from the superior mesenteric artery (D). In 2 cases there were two right hepatic arteries both from the common hepatic, one passed in front of, and the other behind, the common hepatic duct.

In 4 cases, in addition to passing behind the ducts, the common hepatic or the right hepatic artery also passed behind the portal vein (E and W).

Authors' note: The right hepatic artery may also arise from the aorta, right renal, gastric, or inferior mesenteric artery. No such cases were found in Flint's series.

Cystic Artery

The cystic artery arose from the right hepatic artery in 196 of the 200 cases; in 3 cases from the left hepatic (I and Y), and in 1 case from the gastroduodenal artery (H).

In 32 cases it passed in front of the common hepatic duct (E and G), and in 168 it arose from the right side of the common hepatic duct (A) or behind it (A'), the former being the more common.

Accessory Cystic Artery

In 31 cases there was an accessory cystic artery and in 169 cases a single cystic. The accessory cystic artery arose from the right hepatic in 16 (D, J, and K), from the left hepatic in 3 (L and Y), from the gastroduodenal in 11 (M), and from the superior pancreatoduodenal in 1 case (N) out of the 200.

Left Hepatic Artery (100 cases studied)

In 32 cases there were two left hepatic arteries with one coming from the common hepatic and the other from the gastric, and in 1 case the left lobe of the liver received its arterial supply from the gastric only.

Bile Ducts

According to Flint, the most common point at which union actually occurs is within 1 cm of the upper border of the duodenum (H). In 28 cases there were no supraduodenal common bile ducts at all (T), the union taking place at a point anywhere from behind the upper border of the duodenum to the part embedded in the pancreas, and in 3 cases the only representative of the common duct was the part that lies in the wall of the duodenum (U).

Accessory Bile Ducts

Flint reported 29 examples of accessory right hepatic ducts (F, X, Y, A', and C').

Redrawn from Flint, E.F. Abnormalities of the right hepatic, cystic, and gastroduodenal arteries, and of the bile ducts. *Br. J. Surg.* 10:509–519, 1923–24.

Human Anatomic Variation

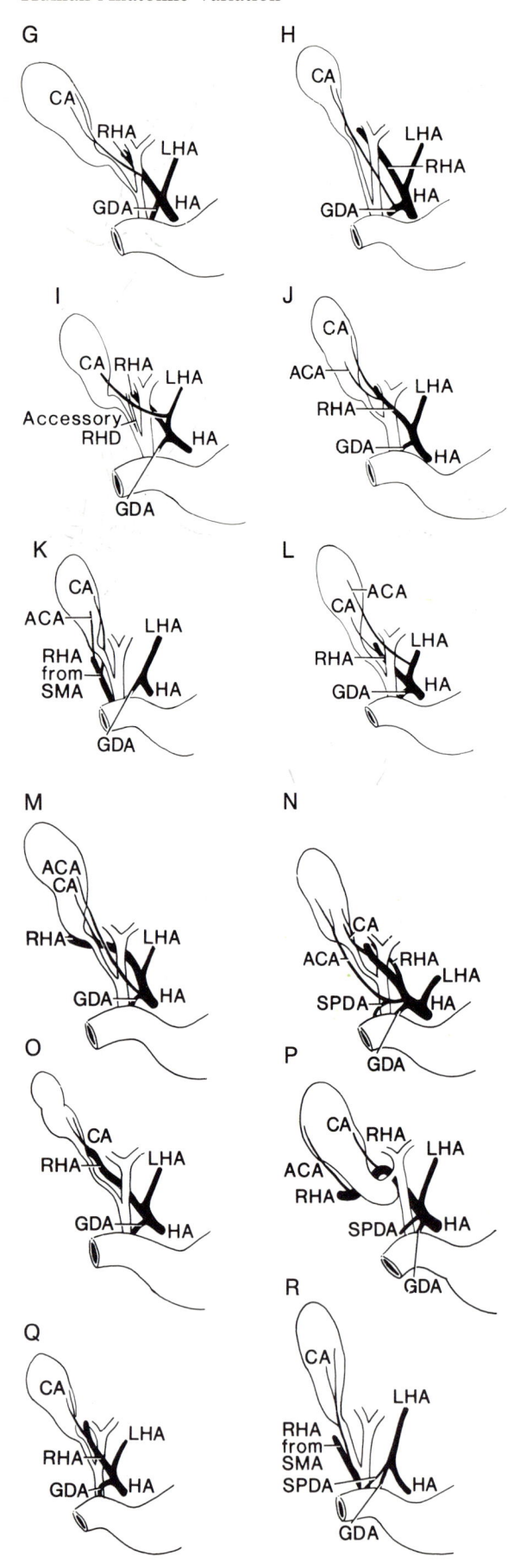

FIG. 103.

Hepatic Pedicle: Variations in Arteries and Bile Ducts (continued).

ACA, Accessory cystic artery; *CA*, cystic artery; *CAA*, celiac trunk; *CD*, cystic duct; *GA*, gastric artery; *GDA*, gastroduodenal artery; *HA*, hepatic artery; *LHA*, left hepatic artery; *PV*, portal vein; *RHA*, right hepatic artery; *RHD*, right hepatic duct; *SA*, splenic artery; *SMA*, superior mesenteric artery; *SPDA*, superior pancreatoduodenal artery.

See Figure 102. for further description.

Redrawn from Flint, E.F. Abnormalities of the right hepatic, cystic, and gastroduodenal arteries, and of the bile ducts. *Br. J. Surg.* 10:509–519, 1923–24.

FIG. 104.

Hepatic Pedicle: Variations in Arteries and Bile Ducts (continued).

ACA, Accessory cystic artery; *CA*, cystic artery; *CAA*, celiac trunk; *CD*, cystic duct; *GA*, gastric artery; *GDA*, gastroduodenal artery; *HA*, hepatic artery; *LHA*, left hepatic artery; *PV*, portal vein; *RHA*, right hepatic artery; *RHD*, right hepatic duct; *SA*, splenic artery; *SMA*, superior mesenteric artery; *SPDA*, superior pancreatoduodenal artery.

See Figure 102. for further description.

Redrawn from Flint, E.F. Abnormalities of the right hepatic, cystic, and gastroduodenal arteries, and of the bile ducts. *Br. J. Surg.* 10:509–519, 1923–24.

Cardiovascular System

FIG. 105.
Hepatic Pedicle: Variations in Arteries and Bile Ducts (continued).

ACA, Accessory cystic artery; *CA*, cystic artery; *CAA*, celiac trunk; *CD*, cystic duct; *GA*, gastric artery; *GDA*, gastroduodenal artery; *HA*, hepatic artery; *LHA*, left hepatic artery; *PV*, portal vein; *RHA*, right hepatic artery; *RHD*, right hepatic duct; *SA*, splenic artery; *SMA*, superior mesenteric artery; *SPDA*, superior pancreatoduodenal artery.

See Figure 102. for further description.

Redrawn from Flint, E.F. Abnormalities of the right hepatic, cystic, and gastroduodenal arteries, and of the bile ducts. *Br. J. Surg.* 10:509–519, 1923–24.

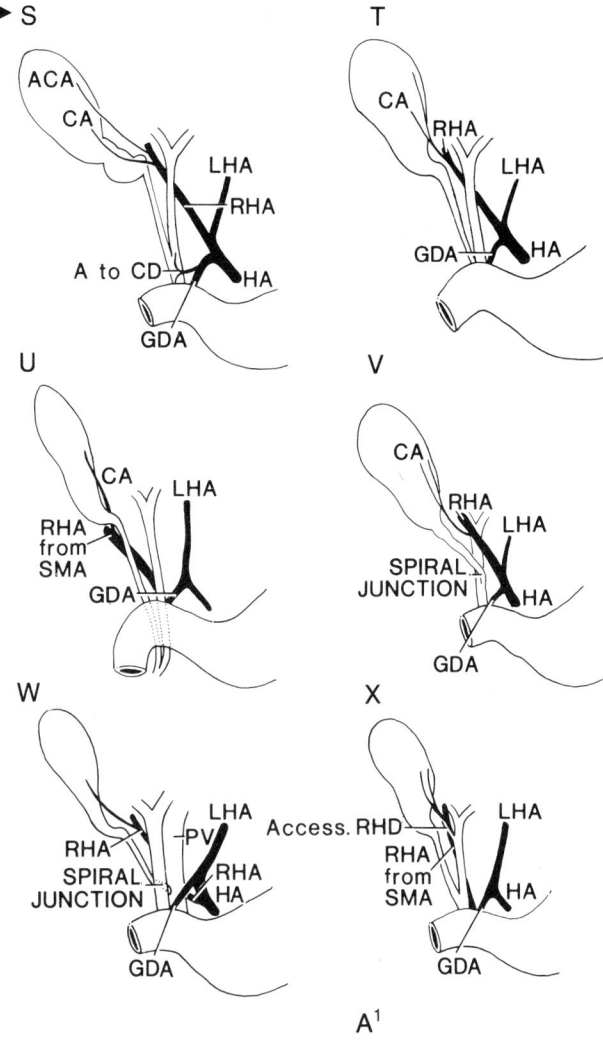

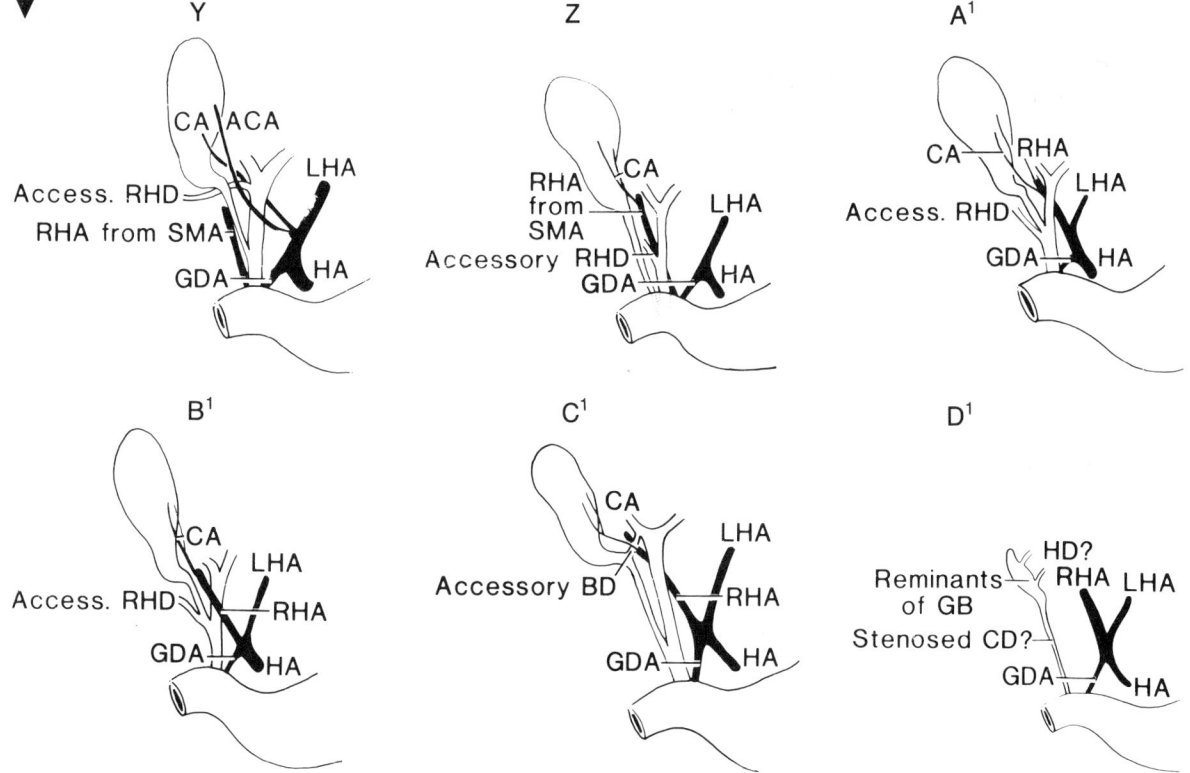

FIG. 106.
Hepatic Pedicle: Variations in Arteries and Bile Ducts (continued).

ACA, Accessory cystic artery; *CA*, cystic artery; *CAA*, celiac trunk; *CD*, cystic duct; *GA*, gastric artery; *GDA*, gastroduodenal artery; *HA*, hepatic artery; *LHA*, left hepatic artery; *PV*, portal vein; *RHA*, right hepatic artery; *RHD*, right hepatic duct; *SA*, splenic artery; *SMA*, superior mesenteric artery; *SPDA*, superior pancreatoduodenal artery.

See Figure 101. for further description.

Redrawn from Flint, E.F. Abnormalities of the right hepatic, cystic, and gastroduodenal arteries, and of the bile ducts. *Br. J. Surg.* 10:509–519, 1923–24.

405

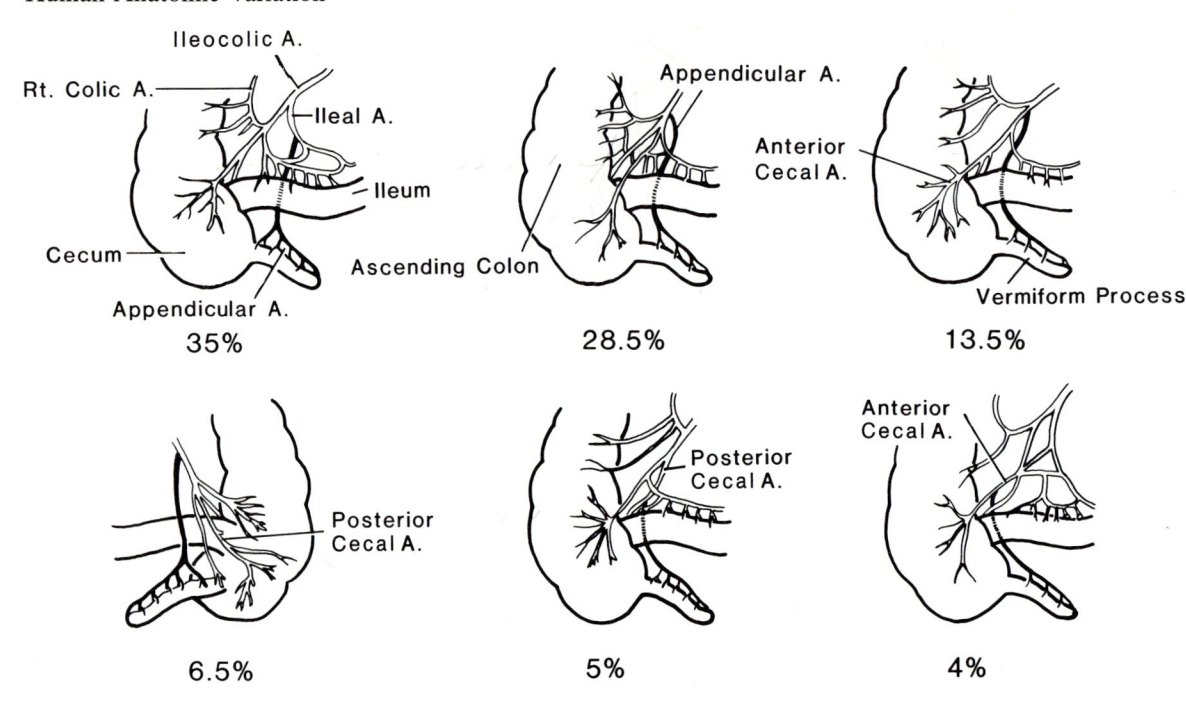

Human Anatomic Variation

[Diagrams of cecal/appendicular artery variations with percentages: 35%, 28.5%, 13.5%, 6.5%, 5%, 4%]

FIG. 107.
Major Variations of Appendiceal and Cecal Arteries (Based on 200 Specimens).
(Unpublished report of Beaton, Anson, Swigart, and Jamieson).
See also Anson, B.J. *Atlas of Human Anatomy*. W.B. Saunders Co., Philadelphia, 1950. Three additional minor variations are reported amounting to about 4.6% of the 200 specimens studied.
Redrawn from Anson, B.J. and W.G. Maddock. *Callender's Surgical Anatomy*, 3rd Ed. W.B. Saunders Co., Philadelphia, 1952.

FIG. 108.
Variations in Inferior Mesenteric Artery.

In this study based on 115 inferior mesenteric arteries, the authors provide information regarding the point of origin of the main trunk of the artery, its course and length, and the course and anastomoses of its main branches (i.e., left colic, sigmoid, and superior rectal arteries). The authors reported the following observations.

"In all cases the inferior mesenteric artery was present, arising from the anterior wall of the aorta in 112 cases. In 2 cases it arose from the left side, and in one case from the right side of the aorta.

Usually, the origin of the artery from the aorta was at the level of L3 (in 77 cases), in a few cases at the level of the intervertebral disc between L3–L4 (23 cases), or at the level of L4 (9 cases). In a smaller number of cases, origin of the artery at the level of the intervertebral disc between L2–L3 (in 4 cases) or L4–L5 (2 cases) was observed.

The distance from the point of origin of the artery to its first branch, called the length of the trunk, ranged between 1.0–7.0 cm. Arteries 6–7 cm long predominated in both sexes; arteries with intermediate length of the trunk, 4–5 cm, were least frequent.

In the studied material, various forms of the inferior mesenteric artery, and within each form subgroups characterized by different manner of branching were distinguished.

In addition, various forms of anastomosis between the branches of the inferior mesenteric artery and with the branches of neighboring arteries (middle mesenteric, middle rectal, and inferior rectal arteries) were noted.

Form I. After leaving the aorta and a long obliquely downward course, the inferior mesenteric artery divided into the left colic artery and common rectal-sigmoid trunk. This form had two varieties of branching.

Variety A [A]. Two sigmoid arteries arising from the common rectal-sigmoid trunk independently, divided dichotomously into secondary branches running to the sigmoid colon.
A few anastomoses were present between the branches of the sigmoid arteries and descending branch of the left colic artery, and numerous anastomoses between the branches of the left colic and middle colic arteries.

Variety B [B]. Two sigmoid arteries arose by a common sigmoid trunk and then divided dichotomously into secondary branches running to the walls of the sigmoid colon. Secondary divisions and numerous anastomoses between these branches and with the descending branch of the colic artery and branches of the superior rectal artery were present.

Form II. The inferior mesenteric artery divided into the superior rectal and colosigmoid trunk. Two varieties of this form were described.

Variety A [C]. The trunk of the colosigmoid artery ran to the left, giving off a common sigmoid trunk in the downward direction and continued as a separate left colic artery, which divided near the wall of the descending colon into an ascending and a descending branch. The common sigmoid trunk divided dichotomously into two sigmoid arteries, which formed numerous anastomoses between themselves and weak anastomoses between the sigmoid arteries and the superior rectal artery.

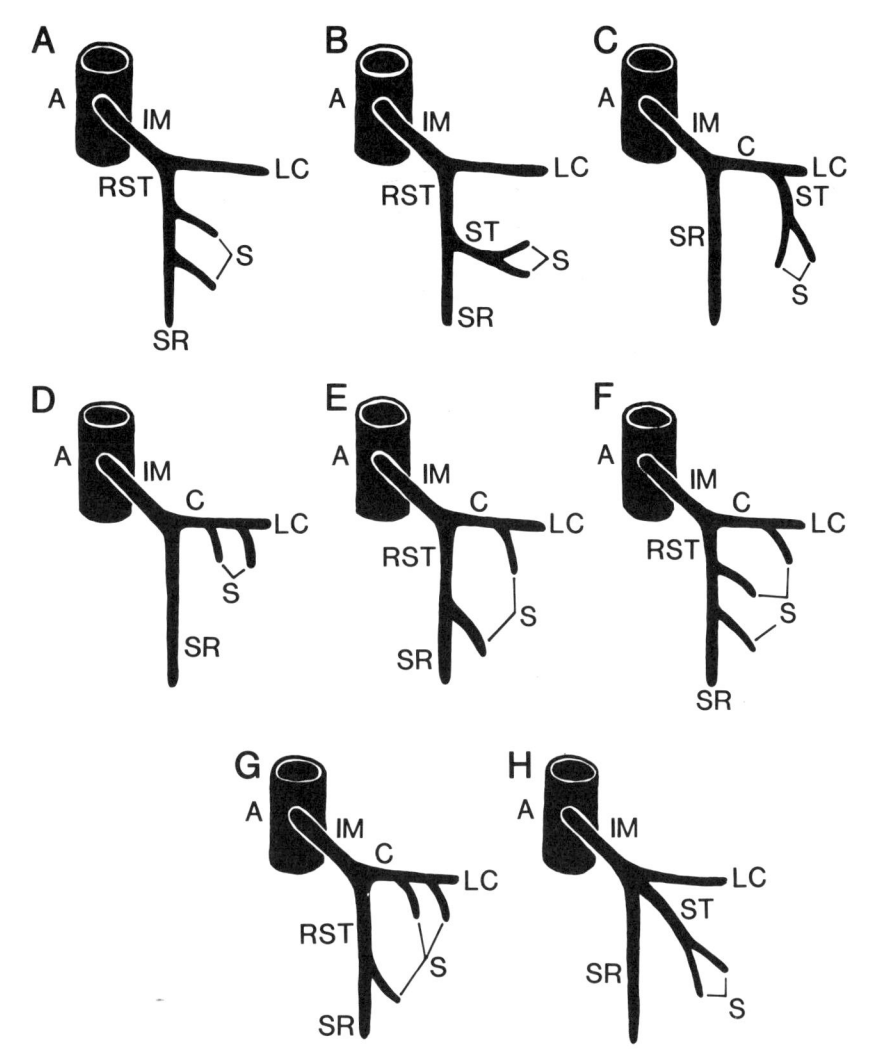

Variety B [D]. The thick trunk of the inferior mesenteric artery divided into the superior rectal artery and colosigmoid trunk, which continued to the left and gave rise to two sigmoid arteries. These broke up into smaller branches forming various anastomoses among themselves and between the sigmoid arteries and descending branch of the left colic artery and branches of the superior rectal artery. In this group, numerous internal connections were observed not only among the branches of the inferior mesenteric artery, but also with the left branch of the middle colic and middle and inferior rectal arteries.

Form III. This was the most numerous group of arteries and was subdivided into three varieties.

Variety A [E]. The inferior mesenteric artery divided into two large arterial trunks: rectal-sigmoid and colosigmoid, each of which gave off one large sigmoid artery. The arteries ran to the walls of the sigmoid colon where they divided into a number of branches, usually 5–6, producing typical arcades.

Variety B [F]. In this variety, the colosigmoid trunk gave off one large sigmoid artery, and the rectal-sigmoid trunk two large sigmoid arteries anastomosing similarly to the preceding form.

Variety C [G]. In this variety, the common colosigmoid trunk gave off two sigmoid arteries, and the rectal-sigmoid trunk one sigmoid artery, which anastomosed as in varieties A and B, above [E,F].

Form IV [H]. In this group the inferior mesenteric artery differed fundamentally from the other forms, its main trunk constituting a common colosigmoid-rectal trunk. The trunk of the inferior mesenteric artery was short, in the form of a characteristic tripod, dividing simultaneously into the left colic and superior rectal arteries and the common sigmoid trunk. This trunk after a short course, divided into two sigmoid arteries, which near the wall of the sigmoid colon broke up into a large number of secondary sigmoid branches running to the intestinal walls. Anastomoses with the left colic and with the middle rectal artery, of some interest to the surgeon, were present."

Redrawn from Zebrowski, W., Augustyniak, E. and S. Zajac. Variation of origin and branches of the inferior mesenteric artery and its anastomoses. *Folia Morphol.* 30:510–517, 1971.

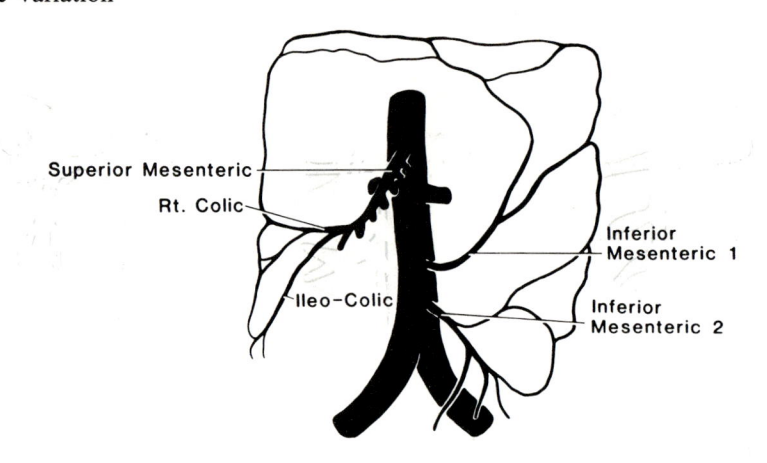

FIG. 109.
Double Inferior Mesenteric Artery.

"In place of a single inferior mesenteric trunk, two arteries originated from the aorta 2.6 cm apart, the inferior one located 3 cm superior to the bifurcation of the aorta. The superior artery was directed superiorly and gave origin to two branches: one to the superior portion of the descending colon and another which supplied the entirety of the transverse colon. The inferior artery also gave rise to two branches: one to the inferior portion of the descending colon and the other to the sigmoid colon and rectum.

The superior mesenteric artery gave rise to one major trunk to the colon, which divided into ileocolic and right colic branches only.

The marginal artery of the colon was well developed and patent throughout its length.

Search of the literature, to date, has revealed no description of this variation. Two references to a variation in which the inferior mesenteric artery gave rise to a branch corresponding to the middle colic artery have been found. There is a paucity of literature regarding variations of the inferior mesenteric artery. Consequently it is uncertain whether this is due to lack of observation, small amount of variation, or both."

Authors' note: In their literature review, the authors report that the inferior mesenteric artery has been reported absent by Fleishmann in 1815 (cited by Henle, reference 1356) and by Adachi in 1930 (reference 8). The absence of the left colic branch has been reported by Dubreuil, 1847 (reference 1352) and Sonneland et al., 1958 (reference 1152). Two additional references are given: Okinczyc, J. Anatomie chirurgicale des colons. In Hartmann, H. *Travaux de chirurgie anatomo-clinique, 3 serie, Chirurgie de l'intestine.* G. Steinheil, Paris, 1907; and Andrejew, J.D. Ueber den Recessus arcus Riolani, den Recessus A. Colicae sinistrae und die Beziehungen der Bauchfelltaschen des Mesogastrium zu den Blutgefäßen. *Anat. Anz.* 71:135–155, 1931.

Petsche, 1736 (cited in reference 1358) reported the inferior mesenteric arising from the left common iliac artery.

Redrawn from Benton, R.S. and W.B. Cotter. A hitherto undocumented variation of the inferior mesenteric artery in man. *Anat. Rec.* 145:171–173, 1963.

FIG. 110.
Renal Vascular Pedicle.

AA, Abdominal aorta; *ARV*, accessory renal vein; *CA*, celiac artery; *CIA*, common iliac artery; *CIV*, common iliac vein; *GA*, gonadal artery, *GVv*, gonadal veins; *HV*, hemiazygous vein; *IMA*, inferior mesenteric artery; *IPA*, inferior phrenic artery; *IPV*, inferior phrenic vein; *ISA*, inferior suprarenal artery; *ISAa*, inferior suprarenal arteries; *IVC*, inferior vena cava; *K*, kidney; *MSV*, middle sacral vein; *RA*, renal artery; *SSA*, superior suprarenal artery; *SV*, suprarenal vein; *U*, ureter; *VL*, venous Loops.

A: *Right side, arteries*: A supernumerary renal artery, coursing transversely as a constituent of the pedicle, gives off 3 suprarenal twigs and, in midcourse, an internal spermatic artery. The main artery sends a suprarenal branch cranialward, and accessory polar ramus lateralward; from the latter, in turn, a suprarenal twig arises. The right phrenic gives off three small suprarenal twigs.

B: *Right, arteries*: Three renal arteries; from the lowermost, in addition to a hilar branch, 1 branch enters the inferior extremity. Branches to the superior extremity arise from the intermediate one of the 3 renal arteries and from the inferior phrenic.

C: *Right, arteries*: Four renal arteries of approximately equal caliber. The superior 2 arise near the origin of the superior mesenteric, pass behind the vena cava; the next artery is precaval, enters the lower part of the hilus. The inferior member arises at the level of the inferior mesenteric, is also precaval, reaches the lower extremity just beyond the true hilus. *Veins*: The superior 1 of 4 renal veins is small, passes from the superior extremity; the next appears from behind the most cranial of the right renal arteries; the third passes along the caudal margin of the latter artery; the fourth vein, placed between the preaortic renal arteries, enters the vena cava by a short channel which it shares with the internal spermatic vein.

Left, arteries: A single renal artery, concealed by the renal vein. *Veins*: A single renal vein, of preaortic course, crosses behind the superior mesenteric artery; it receives suprarenal, internal spermatic, and parietal vein.

D: *Left side, arteries*: Four renal arteries, 3 of them small. The first supplies the superior extremity; the second, the extremity and hilus; the third, the inferior aspect of the hilus; the fourth, the inferior extremity. *Veins*: A single large vessel.

E: *Left, arteries*: Two segmentally arranged renal arteries. *Veins*: A single large renal vein which receives a single supra-

Cardiovascular System

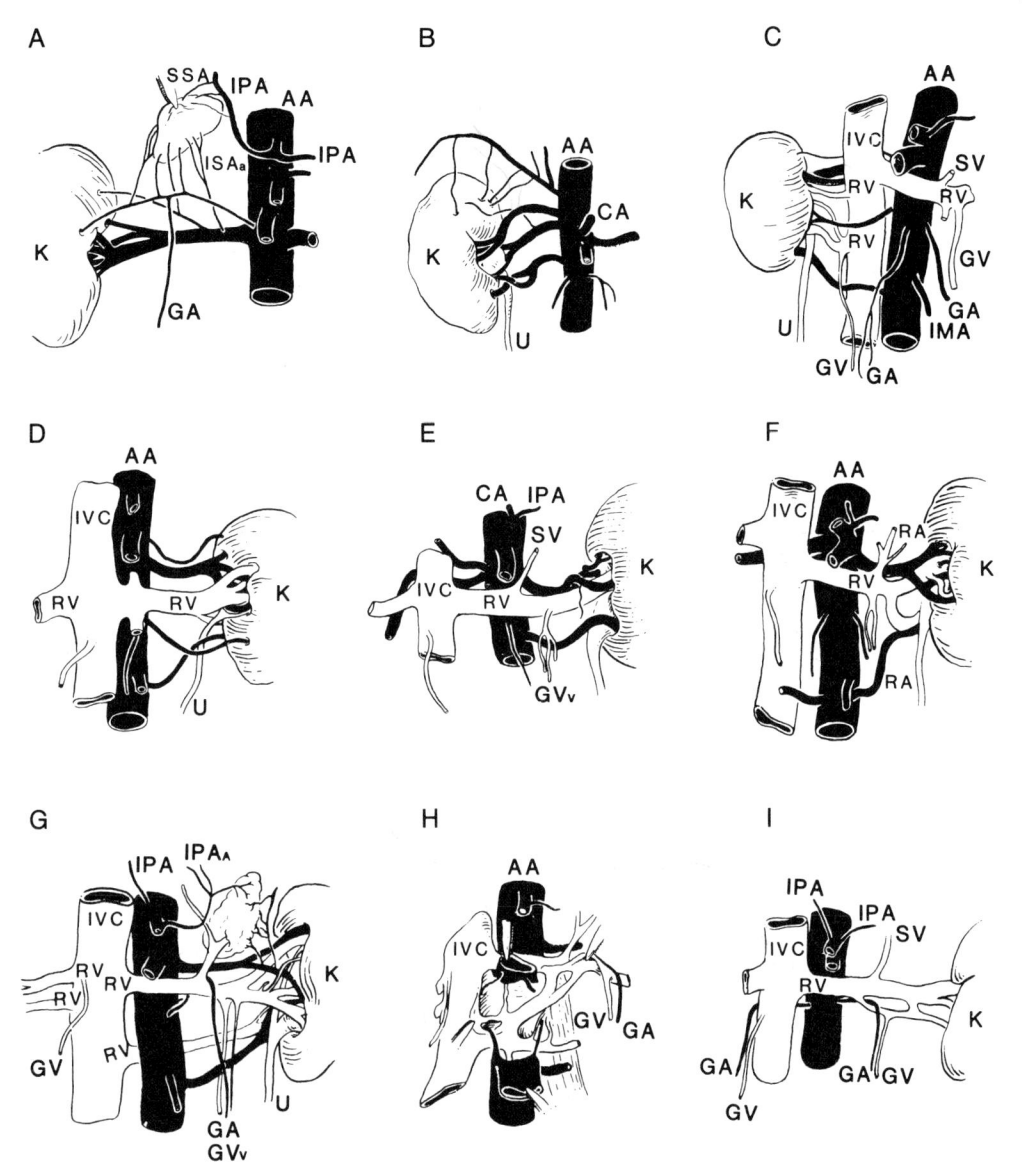

renal vein above, a parietal tributary behind (out of view), an internal spermatic below.

F: *Left, arteries*: Two renal arteries of large size arise, 1 at the level of the superior, the other at that of the inferior mesenteric arteries; they correspond in position to those of the opposite side. *Veins*: A single renal vein, the confluence of 2 large hilar tributaries.

Right, arteries: Three aortic renal arteries, the superior and inferior large, the intermediate small (concealed); the superior and intermediate are retrocaval, the inferior is precaval. The ureter passes anterior to the lowermost of the three arteries.

G: *Left, arteries*: Two renal arteries, 1 to each end of the hilus. The upper arises near the origin of the superior mesenteric, the lower near that of the inferior mesenteric, artery. *Veins*: The upper, or preaortic segment, is formed by the conjunction of 3 hilar tributaries; proximal to junction the lowest tributary receives 2 spermatic veins, and distal thereto the single suprarenal vein; around the latter the spermatic artery hooks. The suprarenal vein receives the phrenic. The lower member of the circumaortic ring is retroaortic in position.

H: Deeper dissection of the same specimen: The retroaortic venous connections form part of a circumaortic ring. The upper vein is preaortic, a lower one is retroaortic. The latter vein receives from above a small vein from the retroaortic connective tissue and a communicating stem from the inferior vena cava. The large retroaortic channel joins the lower 1 of the 2 renal veins, the latter receiving lumbar tributaries.

I: *Left, arteries*: A single renal artery occurs (concealed). *Veins*: A major renal vein and a lesser one join to form a common channel before entering the inferior vena cava; between the 2 passes an anastomotic vessel which forms the lateral boundary of a venous hiatus; the hiatus transmits the internal spermatic artery; into the lower constituent of the loop drains the spermatic vein.

Authors' note: See Figure 111 for the continuation of this study and conclusions.

Redrawn from Pick, J.W. and B.J. Anson. The renal vascular pedicle. An anatomical study of 430 body-halves. *J. Urol.* 44:411–434, 1940; and Davis, R.A., Milloy, F.J. Jr. and B.J. Anson. Lumbar, renal, and associated parietal and visceral veins based upon a study of 100 specimens. *Surg. Gynecol. Obstet.* 107:1–22, 1958.

FIG. 111.
Renal Vascular Pedicle (continued).
Abbreviations: see Fig. 110.

J: *Right, arteries*: A single renal artery whose hilar branches attain an anterior position. *Veins*: A single renal vein in which a triangular hiatus occurs; through this orifice passes the right internal spermatic artery; from the latter arise suprarenal arteries, before and after the contributing artery passes through the hiatus.

K: *Right, arteries*: Two renal arteries, transverse in course; their branches reach the extreme ends of the renal hilus. The superior one is retrocaval, supplies the superior extremity; the inferior one is antecaval, gives off 3 hilar branches. The internal spermatic artery arises from the superior one of the 2 renals, passes through a cleft at the caval end of the internal spermatic vein. A ureteric artery leaves the lowermost hilar branch. From the superior renal artery; near its aortic origin, arises a short stem which divides into inferior phrenic and supernumerary renal artery, the latter piercing the parenchyma of the superior extremity. From both renal and phrenic subdivisions come suprarenal branches. *Veins*: A single large renal vein with a terminal hiatus transmitting the spermatic artery.

L: *Right, arteries*: A single renal artery from which arises a common stem for an inferior phrenic (with suprarenal twigs) and a suprarenal artery. From one of the hilar branches is derived an inferior suprarenal artery. The internal spermatic artery, of high aortic origin, passes obliquely downward and lateralward, at first in retrocaval course; emerging through an hiatus formed by the subdivision of the renal vein as the latter reaches the vena cava, it descends into the pelvis.

M: *Veins*: The double inferior vena cava is divided to a position superior to the point of entrance of the renal veins. Into the channel of the right side the renal vein is received; into the left, the left renal veins are received.

N: *Left, arteries*: A major renal artery and a supernumerary, the latter descending to perforate the lower extremity of the kidney (passing around the suprarenal vein and crossing the renal vein anteriorly); from the point of perforation an accessory vein emerges, the 2 vessels thus producing a minute supernumerary "hilus," the vein joins the internal spermatic before entering the renal. *Veins*: A single renal vein, retroaortic in course. Into this is received, on its cranial aspect, a suprarenal vein; the latter receives in turn a phrenic vein. Into the renal vein is received a common channel formed by the internal spermatic and the polar tributaries.

O: *Right, arteries*: Three renal arteries which enter the hilus and kidney substance in the order of their origin (i.e., do not cross). The upper and middle area postcaval, the lower, antecaval, in position. A spermatic branch arises from the lowermost of 3, and is accompanied by the corresponding vein (which hooks over the lower renal artery). *Veins*: A single renal vein occurs, the superior tributary of which lies anterior to the upper artery, the lower and middle tributaries posterior to the hilar subdivisions of the intermediate artery. *Ureter*: Double throughout its course to the wall of the bladder.

P: *Left, arteries*: Double renal arteries from the inferior one of which the ovarian artery arises. Lumbar arteries are accompanied by communicating veins from the psoas major muscle, etc. *Veins*: One large renal vein which superiorly receives a single suprarenal vein, inferiorly, the 2 divisions of an ovarian vein. A second, or supernumerary, renal vein occurs; leaving the anterior surface of the inferior extremity, this vessel is one of 4 veins tributary to a short venous trunk which enters the inferior vena cava; in transverse course, it rests upon the psoas major muscle and the body of the fourth lumbar vertebra; it passes posterior to the ureter, ovarian vessels, abdominal aorta, and anterior to the sympathetic trunk and fourth lumbar artery. The lowermost of the 4 tributaries is a confluence of an anastomotic connection with the common iliac vein, and ureteric vein. The 2 intermediate tributaries of the short venous trunk are lumbar veins (here shown by removing a portion of the psoas major muscle.)

Three examples (Q, R, S) of persistent left caval vein were encountered in the course of 100 dissections by Davis et al. They are arranged in the order of decreasing caliber of the persistent, "anomalous" vessel. In each instance the caval vein connected the iliac with the renal.

Q: The large left inferior vena cava communicated with the fifth, fourth, and third lumbar veins in its course to the renal, and also received the internal spermatic vein.

R: The most caudally situated vein (fifth) of the lumbar series terminated in the left iliac near the point of origin of the caval vein; the third crossed to the "normal" inferior vena cava of the right side. The internal spermatic vein ended in the renal extremity of the left caval vein, the latter then being continued cranialward as the intermediate hemiazygos root.

S: The small, left inferior vena cava communicated with a regular caval vein of the right side through a third lumbar vein; as a paravertebral channel (represented in the thorax by the hemiazygos root), it was interrupted by connection with the caudal limb of a bifid renal vein. This point of renal connection was one of convergence of the lumbar, suprarenal, and thoracic (hemiazygos) vessels.

Conclusions

From a study of the renal vascular pedicles of 200 consecutive cadavers (400 sides), the following conclusions were provided by Pick and Anson:

There is a striking difference in frequency of occurrence, depending upon whether the figures are calculated on the basis of specimens or on that of sides.

Multiple renal veins are rare on the left side (occurring in only 2 of 200 cadavers; 1 per cent) while they are common on the right (54 of 194 sides; 27.8 per cent).

Variety in form of the renal pedicles, based on the number of constituent vessels, is greater on the right than on the left; this difference is owing to the fact that multiple renal veins are a rarity on the left, this variable being important on the right side only.

Based only on the number of vessels (and considering both sides) the renal pedicle in 39.0 per cent of the specimens (78 of 200) is composed bilaterally of 2 vessels. This type, therefore, represents merely a plurality of the specimens.

Classifying the renal pedicles on the basis of the number of constituent vessels, there are 6 different forms. The majority, 59.75 per cent (239 of 400 sides), are confined to 1 group, in which there are but 2 vessels in the pedicle. The highest number of vessels in any single pedicle is 8 (single instance, right side).

Bilateral equality in the total number of vessels in the pedicles occurs in 49.5 per cent of the cadavers (99 of 200). However, 39 per cent (78 of 200) of the specimens included in this group belong to the simplest type of pedicle (2 right and 2 left renal vessels); actually then, bilateral equality in the number of renal vessels is rare in the presence of supernumerary renal arteries or veins (10.5 per cent, 21 of 200 cadavers).

In 64.25 per cent (257 of 400) of the sides an equal number of arteries and veins occur in their renal pedicles. This feature is not, however, as important as it might seem, since symmetry of this kind encountered in 59.5 per cent (238 of 400) are sides with single artery and vein. In the presence of supernumerary renal arteries or veins equality is encountered in only 4.75 per cent (19 of 400 sides).

Single renal arteries and veins on both sides in the same specimen occur in 38.5 per cent of cadavers (77 of 200), whereas a single renal artery and vein make up the pedicle in 59.5 per cent of the sides (239 of 400).

Classifying specimens on the basis of the number of arteries and veins in the renal pedicles of the 2 sides, 27 different varieties of arrangement are encountered, none of which represent over 38.5 per cent of the total (77 of 200). Thus there is no majority group of specimens, the most common type merely representing a plurality.

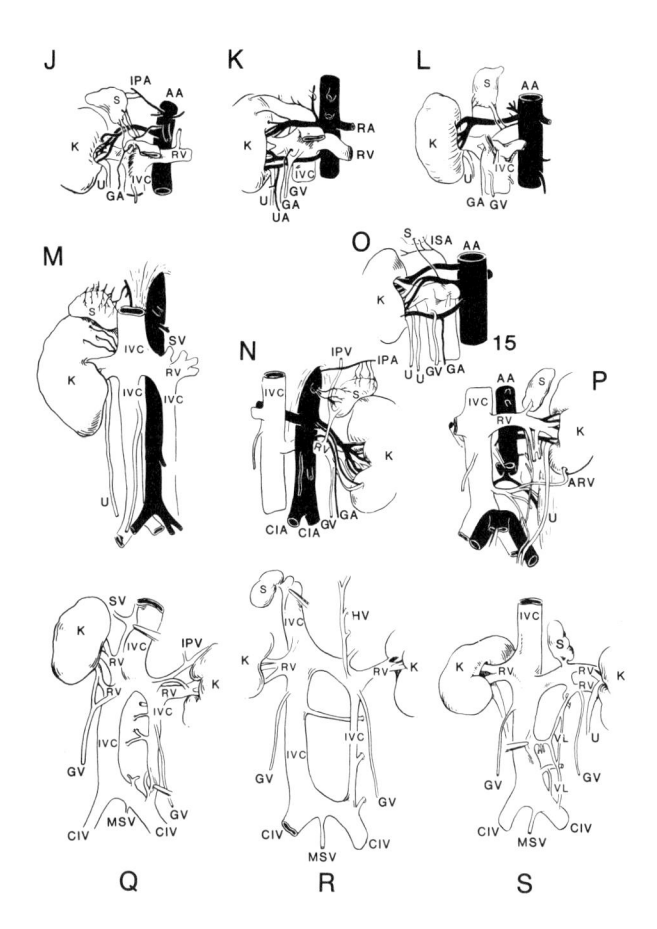

Classifying sides, however, on the basis of the number of arteries and veins in the pedicles, there are 12 different types, but the most common is represented by the majority, 59.5 per cent (238 of 400).

Frequently, the so-called "normal" form represents a plurality rather than a majority of the cases, depending on whether the incidence is based upon a consideration of side or of specimen. Failure to distinguish between these methods of calculating frequency of occurrence of types accounts for the discrepant reporting, in the literature, of results obtained by other investigators; it also accounts for failure to detect significant differences between sides.

Differences between the renal pedicles, depending on side, are accounted for chiefly by adult asymmetry in the renal venous drainage – the basis for which is the developmental transformation of a symmetrical into an asymmetrical caval plan.

It is regularly stated that the right is longer than the left renal artery, the left renal vein longer than the right. While this is anatomically true, actually the surgical lengths of the pedicles (permitting free mobilization) are about equal; on the right side the lateral border of the vena cava offers a fixed point for the renal arteries, while on the left, the lateral border of the aorta acts in a similar manner. From the urological standpoint the difference in the length of the pedicles is more apparent than real.

While it is true that ligation of the large collaterals of the dorsum of the left renal vein, just lateral to the aorta, would free the vein more completely, the left renal arteries are short; therefore, such a procedure would be of little surgical benefit. On the right it would be necessary to liberate the renal arteries from the dorsum of the vena cava – a dangerous method of mobilization. Even if such a procedure were successful, the right renal veins, being short, would nullify the effect of securing greater freedom for the right renal arteries.

In some instances the presence of supernumerary vessels does not occasion a pronounced spreading of the renal pedicle; in others, vessels affect the entire abdominal portion of the aorta and the vena cava.

The pedicle is frequently complicated by the occurrence of arteries sent from the renal stems to the suprarenal gland or from a suprarenal arterial source to the kidney.

Additional renal veins, which occur less frequently than supernumerary arteries, often communicate, on the left side, with other visceral veins (suprarenal and spermatic or ovarian), with a retroaortic venous plexus, and with the left lumbar and hemiazygos veins. Not infrequently the regular vein passes in front of the aorta, the supernumerary vein behind, forming (between left kidney and vena cava) a circumaortic venous ring.

Capacious hiatuses occur in the renal veins, and more commonly on the right than on the left; since the hiatus usually transmits the internal spermatic or the ovarian artery, the gonadal vessel, in such instances, lies definitely in the field of urological surgery. Of surgical interest, also, are those specimens in which the gonadal vessel merely crosses the renal pedicle, and those in which the vessel follows the renal artery in transverse course before descending – on the left side, hooking around the suprarenal tributary of the renal vein. Spermatic or ovarian arteries may arise directly from the renal, thus being, proximally, a constituent of the pedicle.

Redrawn from Pick, J.W. and B.J. Anson. The renal vascular pedicle. An anatomical study of 430 body-halves. *J. Urol.* 44:411–434, 1940; and Davis, R.A., Milloy, F.J. Jr. and B.J. Anson. Lumbar, renal, and associated parietal and visceral veins based upon a study of 100 specimens. *Surg. Gynecol. Obstet.* 107:1–22, 1958.

Human Anatomic Variation

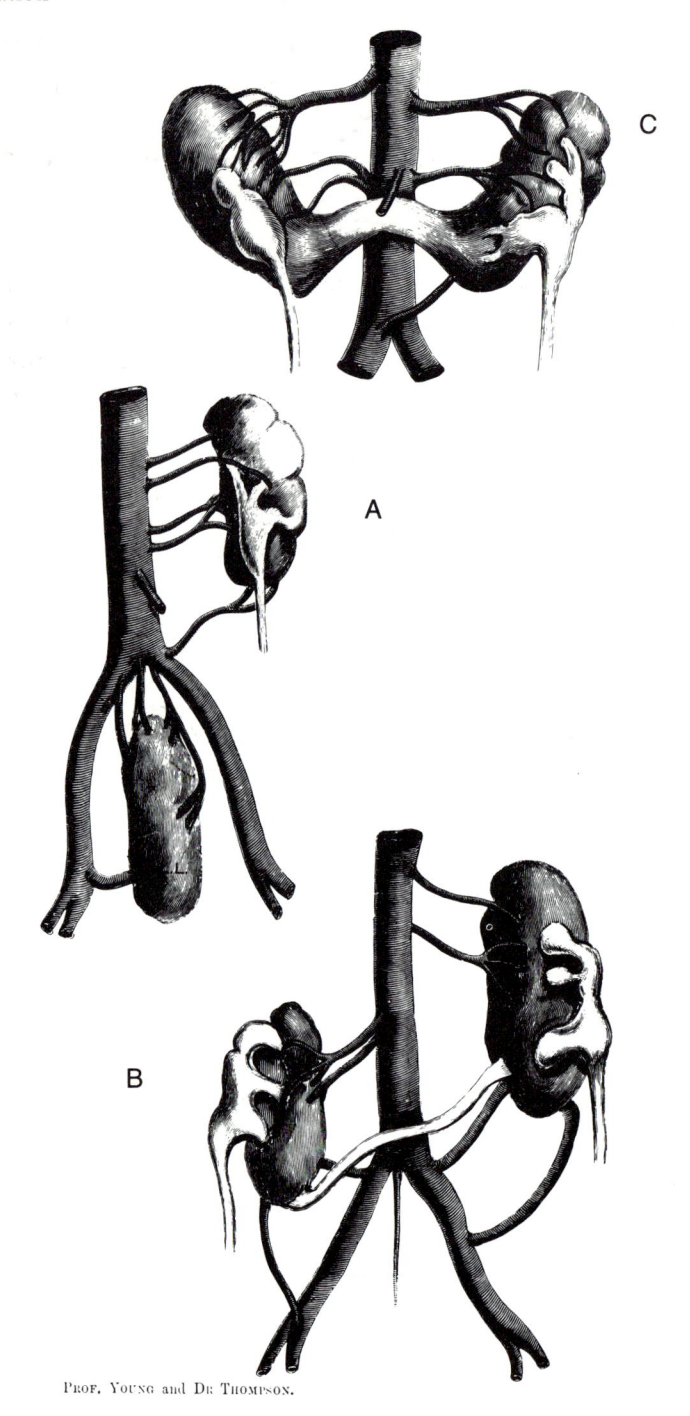

FIG. 112.
Renal Arterial Variations.

A: Showing multiple renal arteries, malformation of both kidneys, and malposition of the right kidney. The pelvis of the ureter on the left side is ventral in position.
B: Showing multiple renal arteries, with malposition and malformation of both kidneys. On each side the pelvis of the ureter lies on the ventral aspect of the kidney. The two kidneys are united near their lower extremities by a fibrous band, which passes in front of the aorta immediately above its bifurcation.
C: Showing multiple renal arteries, and union of the two kidneys to form a well-marked horseshoe-shaped kidney. The pelvis of each ureter is situated on the ventral surface of the corresponding kidney.

From Young, A.H. and P. Thompson. Abnormalities of the renal arteries, with remarks on their development and morphology. *J. Anat. Physiol.* 38:1–14, 1904.

Cardiovascular System

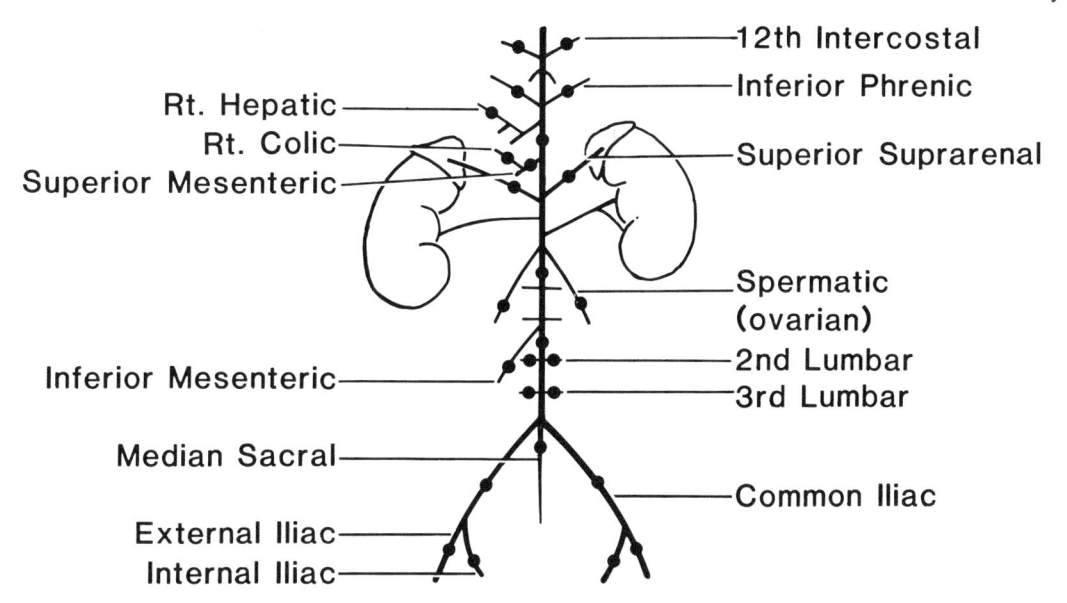

FIG. 113.
Possible Sources of Accessory Renal Arteries.
The black circles indicate the site of origin of accessory arteries as reported by Poisel and Spängler.

Redrawn from Poisel, S. and H.P. Spängler. Über aberrante und akzessorische Nierenarterien bei Nieren in typischer Lage. *Anat. Anz.* 124:244–259, 1969.

FIG. 114.
Variations in Renal Vessels.

Based on a study of 102 bodies.

A: Diagram showing origin of upper polar (accessory) artery from the aorta (*UPA*).
B: Diagram of mode of origin of lower polar (accessory) artery from the aorta (*LPA*).
C: Lower polar artery arising from the iliac artery.
D: *Upper*, most important feature is origin of accessory lower polar artery from main renal on left side. *Lower*, posterior view of right kidney of upper figure showing superior, middle, and inferior types of retropelvic arteries.
E: *Upper*, chief feature is presence of an inferior polar artery on right side. The lower of two main renal arteries crosses the posterior aspect of the pelvis. *Lower*, posterior view of the upper figure, showing two retropelvic veins on each side crossing middle of back of pelvis. Note accessory artery crossing posterior aspect of pelvis.
F: *Upper*, anterior view of right kidney showing accessory superior polar artery arising from renal and also division of main renal artery into equal-sized pre- and retropelvic branches. *Lower*, posterior view of the upper figure showing fan-like distribution of retropelvic artery as in E.
G: Front view showing on both sides a superior polar artery arising from the main renal artery. On the left side is seen a typical example of bifid pelvis.
H: Retropelvic artery arising from renal and crossing, as in E, the middle of posterior aspect of pelvis. Note lower polar accessory artery directly from aorta.

See Figures 115–117 for continuation of this study and conclusions.

Redrawn from Eisendrath, D.N. The relations of variations in the renal vessels to pyelotomy and nephrectomy. *Ann. Surg.* 71:726–743, 1920.

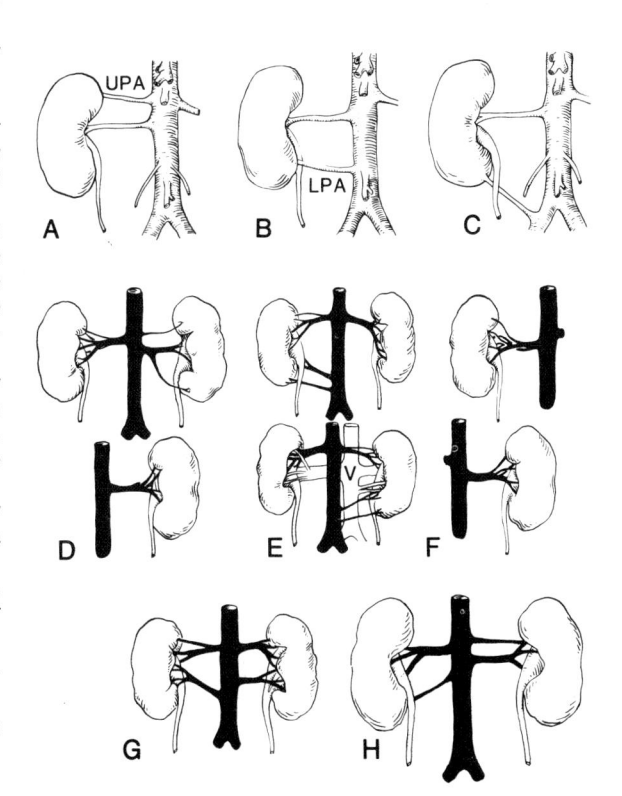

Human Anatomic Variation

FIG. 115.
Variations in Renal Vessels (continued).

I: *Upper left*, superior polar artery arising from one or two main renals. *Upper right*, accessory artery to lower pole crossing renal vein obliquely. This specimen shows how difficult it would be to secure the bleeding trunk if such an artery were overlooked. *Lower left*, upper of two main renal arteries gives off a superior polar artery, and a retropelvic artery. *Lower right*, main renal artery divides into pre- and retropelvic branches of equal size.

J: On the right side is seen a typical division of the main renal into pre- and retropelvic branches of equal size. Note lower polar accessory artery to right kidney lying behind ureter. There is also an accessory superior polar on the left side.

K: Anterior view showing retropelvic artery on right side arising from renal, close to aorta. On the left side note the superior polar and the retropelvic artery, both arising from one of the two main renals. Note inferior polar (accessory) arteries arising from aorta on both sides.

L: *Upper*, right kidney. Typical example on right side of division of main renal artery into equal-sized pre- and retropelvic branches. *Lower*, example of fan-like division of retropelvic artery over posterior aspect of pelvis.

M: *Upper*, anterior view. Right kidney shows a single accessory artery to lower pole. Left kidney shows two accessory arteries to lower pole, all from aorta. *Lower*, posterior view of right kidney showing typical fan-like distribution of retropelvic artery.

Redrawn from Eisendrath, D.N. The relations of variations in the renal vessels to pyelotomy and nephrectomy. *Ann. Surg.* 71:726–743, 1920.

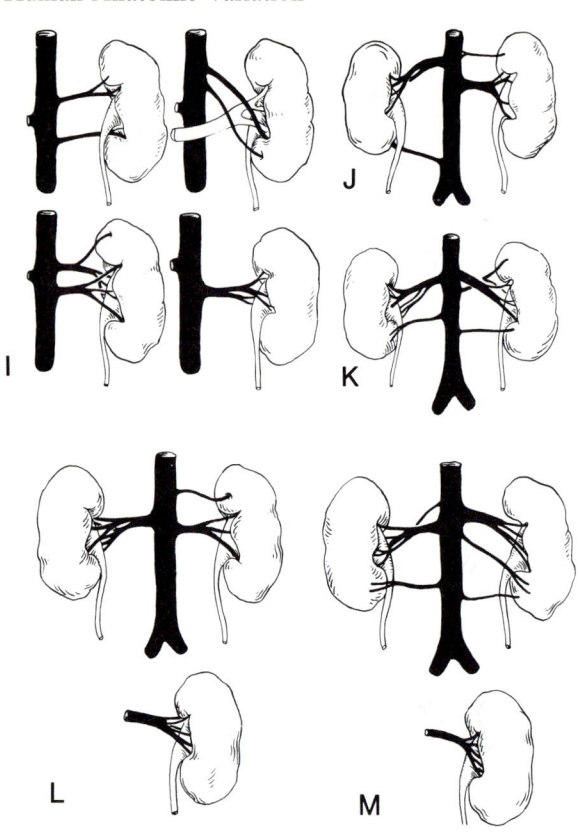

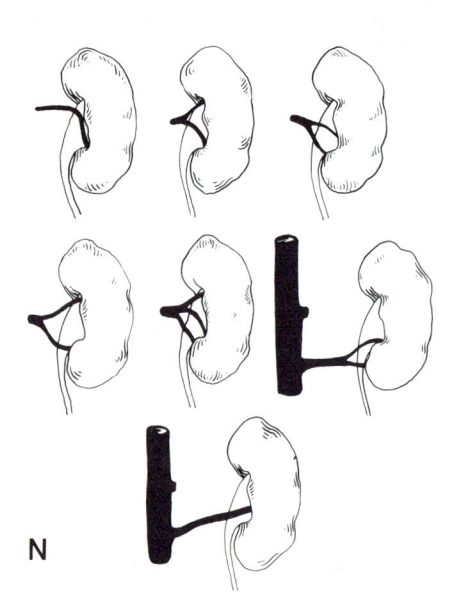

FIG. 116.
Variations in Renal Vessels (continued).

N: Diagrammatic representation of various forms of retropelvic arteries. *Upper left*, typical single retropelvic artery forming an arch at uppermost portion of renal pelvis. *Upper middle*, upper and middle retropelvic arteries. *Upper right*, lower and middle retropelvic arteries. *Middle left*, lower and upper retropelvic arteries. *Middle center*, fan-like retropelvic artery, there being a middle, upper, and lower branch (see also D and L). *Middle right*, retropelvic artery arising from a lower polar accessory artery. *Lower*, retropelvic artery directly from aorta (E).

Redrawn from Eisendrath, D.N. The relations of variations in the renal vessels to pyelotomy and nephrectomy. *Ann. Surg.* 71:726–743, 1920.

Cardiovascular System

FIG. 117.
Variations in Renal Vessels (continued).

O: *Upper*, anterior view. *Lower*, posterior view of right kidney showing main renal vein crossing posterior aspect of pelvis.
P: Right kidney. Upper, Accessory upper polar artery arising close to aorta from one of two main renal arteries. Note also accessory lower polar artery arising from aorta close to common iliac. Similar condition in left kidney but superior polar artery arises from one of two main renals close to kidney. *Lower*, posterior view of upper figure, showing a large renal vein and artery entering kidney back of pelvis on left side, and a smaller retropelvic vein directly from the vena cava.
Q: *Upper*, front view showing superior polar artery arising from main renal on right side and from aorta on left side. *Lower*, posterior view of upper figure, showing a fairly large renal vein crossing middle of back of pelvis.
R: Diagrammatic representation of various types of retropelvic veins: *Upper left*, division of the main renal vein into pre- and retropelvic branches of equal size; *upper right*, main renal vein crosses posterior aspect of pelvis to reach kidney; *lower left*, accessory renal vein to lower pole of kidney; *lower right*, main renal vein divides into two branches, the larger of which is retropelvic. There is also an accessory renal vein directly from aorta.

Eisendrath provided the following practical conclusions from all published statistics.

"It is of importance for the surgeon to remember that the examination of 1237 kidneys by various investigators reveals the fact that upper polars from the renals occurred in 68, or 16 percent, of 518 kidneys. Upper polars from the aorta were found in 68, or .5 percent, and lower polars from the aorta in 71, or nearly .6 percent, of 1237 kidneys. Lower polars from the iliacs were found in only 6, or .04 percent, of the 1237 kidneys."

"According to my Eisendrath's present series of dissections, one can expect to find upper polar arteries arising from the main renals in about one kidney out of five. Upper polars arising from the aorta were found in one out of about 17 kidneys and lower polars (from the aorta and iliacs) in one kidney out of about 7 kidneys.

Adding together the observations of all previous investigators and our own we find that (*a*) upper polars arising from the main renals occur in about one out of 200 kidneys; (*b*) upper polars arising from the aorta in about one out of about 190 kidneys; and (*c*) lower polars from the main renal, the aorta, or common iliacs in one out of about 185 kidneys. Although accessory polar vessels did not occur as frequently as stated by Quain, i.e., 20 percent, they are found often enough to be constantly borne in mind during operation."

The Retropelvic Vessels

"The tradition still exists that one needs only to guard against injury of a retropelvic artery which pursues a more or less typical course in the sinuses formed at the point where the kidney tissue slightly overlaps the renal pelvis. That there may be (*a*) variations from this arch-like distribution of the artery and (*b*) that one or more large veins, even the main renal, may cross our field of operations are two anatomical facts which deserve more widespread knowledge in order to avoid injury to these anomalous vessels during pyelotomy.

In Albarran's book, published in 1910, reference is made to retropelvic artery and vein to the effect that the main renal artery may divide into pre- and retropelvic trunks of equal size, and that the retropelvic artery on its way to the sinus gives off branches similar to those arising from the prepelvic artery. His only statement in regard to the retropelvic vein is that it is not constant – was found in 5 of 29 cases by Hauch – and finally that it may prove a source of trouble during pyelotomy.

In view of the results of my present dissections, I believe we must abandon the view that the posterior aspect of the renal pelvis is the avascular field which we have generally believed it to be. The distribution of the prepelvic vessels seldom, if ever, enters into consideration in the operation of pyelotomy, because the route of election is through the less vascular field.

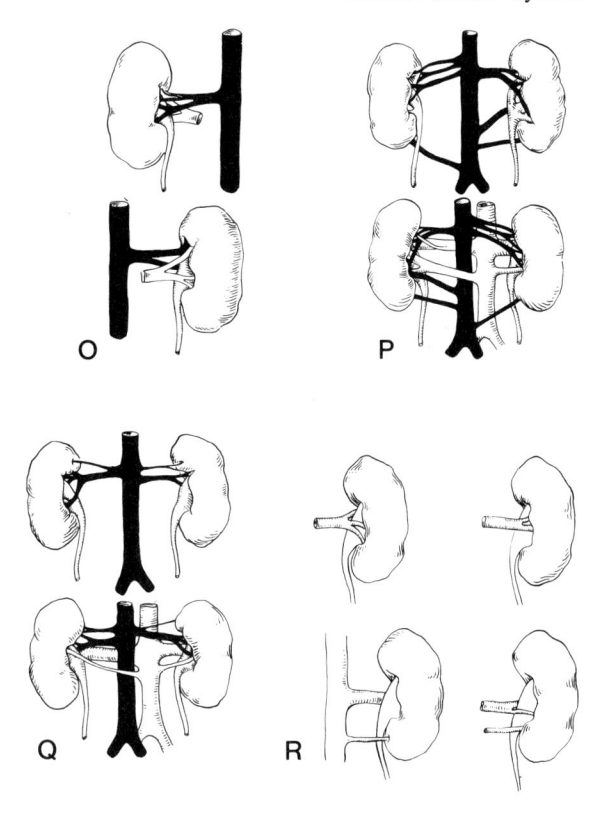

In a total of 218 kidneys, the following observations of variations of the retropelvic vessels of surgical importance were made:

A. Anomalies of the Retropelvic Artery Alone.
  1. Division of the single main renal artery into equal-sized branches was found very frequently.
  2. When there were two main renal arteries, one of these frequently became the retropelvic, i.e., the latter arose directly from the aorta instead of the main renal artery. This was found in 5 kidneys out of 124.
  3. The main renal or one of two renal arteries was found to be retropelvic in 2 kidneys out of 124.
  4. The retropelvic artery had its origin from an accessory lower polar artery in 2 kidneys out of 124.
  5. A retropelvic artery directly from the aorta was found in 2 kidneys out of 124.
  6. The main retropelvic artery does not cling in an arch-like manner to the renal sinus. This generally accepted course of the vessel may be described as the high type to distinguish it from various combinations which were found in both the Illinois and Harvard dissections. These extra vessels which may give rise to troublesome bleeding during pyelotomy are (*a*) a high middle and low or fan-like distribution found twice in 124 kidneys; (*b*) a high and middle type of branching found twice in 124 kidneys; (*c*) a high and low type found twice in 124 kidneys; (*d*) a single artery crossing the middle of the pelvis, found four times in 124 kidneys; and finally (*e*) a middle and low type found seven times in 124 kidneys.

B. Anomalies of the Retropelvic Veins.
  1. One large vein arising from the vena cava was found five times in 218 kidneys, passing directly across the back of the renal pelvis.

Human Anatomic Variation

2. The main renal vein divided into equal-sized pre- and retropelvic branches in 3 of 218 kidneys. The retropelvic branch passed directly across the pelvis and, as in the case of the preceding anomaly, could be easily injured during pyelotomy.
3. The most important anomaly, so far as the veins were concerned, was that the main renal vein, instead of being prepelvic, was retropelvic in 9 out of 218 kidneys.

C. Anomalies Involving Both Retropelvic Vein and Artery.
1. One large vein directly from the vena cava and one artery directly from the aorta crossed the back of the pelvis in one of 94 kidneys.
2. Two large veins directly from the vena cava and one artery from the aorta crossed the back of the pelvis in one of 94 kidneys."

Types of Pelvis

Observations made as to the relative frequency of the various types of renal pelvis revealed the following:
1. The single or ampullary pelvis was found in 84 (89 percent) out of 94 kidneys.
2. The divided or bifid type was found in 7 (8 percent) out of 94 kidneys. In 4 of these it was present on both sides.
3. The trifid type was found in 3 percent of 94 kidneys.

Authors' note: Eisendrath's references to a book by Albarran (1910) and to the work of Hauch were not found in his bibliography.

Redrawn from Eisendrath, D.N. The relations of variations in the renal vessels to pyelotomy and nephrectomy. *Ann. Surg.* 71:726–743, 1920.

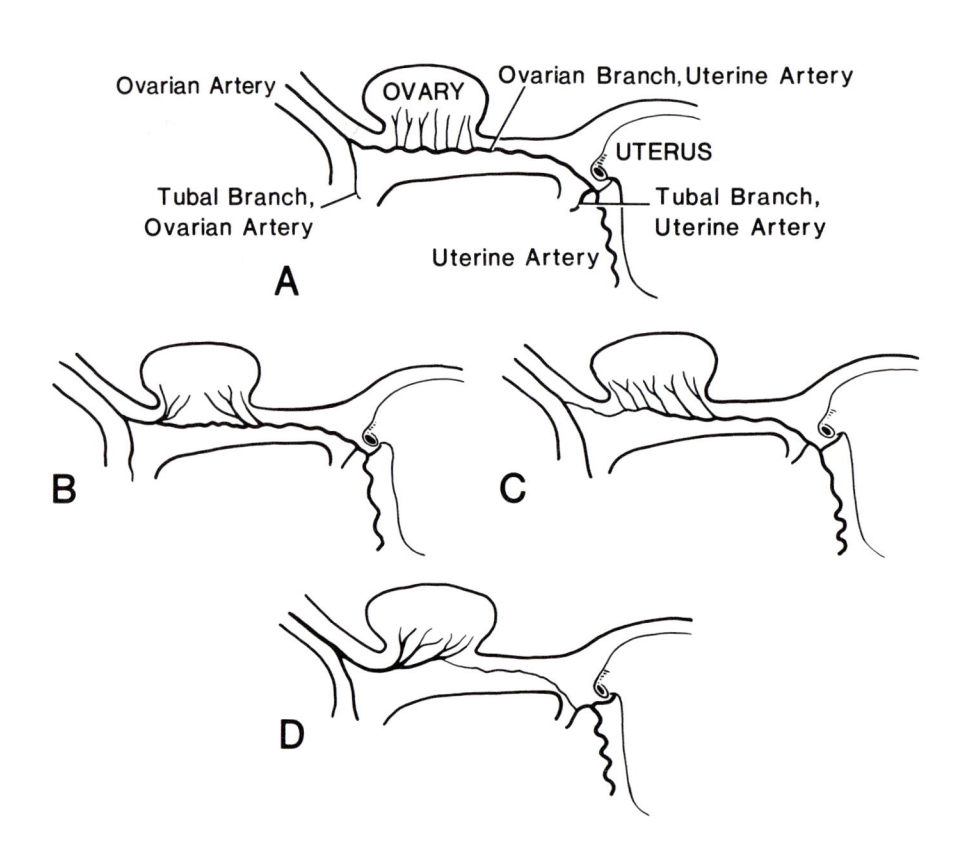

FIG. 118.
Variations in Ovarian Arterial Supply.

The authors found four types of arterial blood supply, which are shown in the illustration.

Type I (A): Anastomosis between the ovarian and uterine arteries giving rise to arterial vessels, equally, to supply the ovary.

Type II (B): The ovarian and uterine arteries each directly supply the ovary and also anastomose with each other.

Type III (C): The primary source (sole source) of blood to the ovary is the uterine artery which has a small anastomotic branch with the ovarian artery.

Type IV (D): The ovarian artery is the sole source of the blood supply to the ovary. It also supplies a small anastomotic branch which joins the uterine artery or the tubal branch of the uterine artery.

Types I and II were considered to occur with equal frequency and commonly, whereas types III and IV were considered to be of rare occurrence.

Redrawn from Mocquot, P. and C. Rouvillois. La vascularisation artérielle de l'ovaire étudiée en vue de la chirurgie conservatrice. *J. Chir.* (Paris) 51:161–176, 1938.

FIG. 119.
Middle Sacral Artery.

Variations in the mode and place of origin of the middle sacral artery. Total number of cases examined, 400.

LCI, Left common iliac; *LL*, left lumbar; *MS*, middle sacral; *RCI*, right common iliac; *RL*, right lumbar.

From Thomson, A. Third annual report of the committee of collective investigation of the Anatomical Society of Great Britain and Ireland for the year 1891–92. *J. Anat. Physiol.* 27:183–194, 1893.

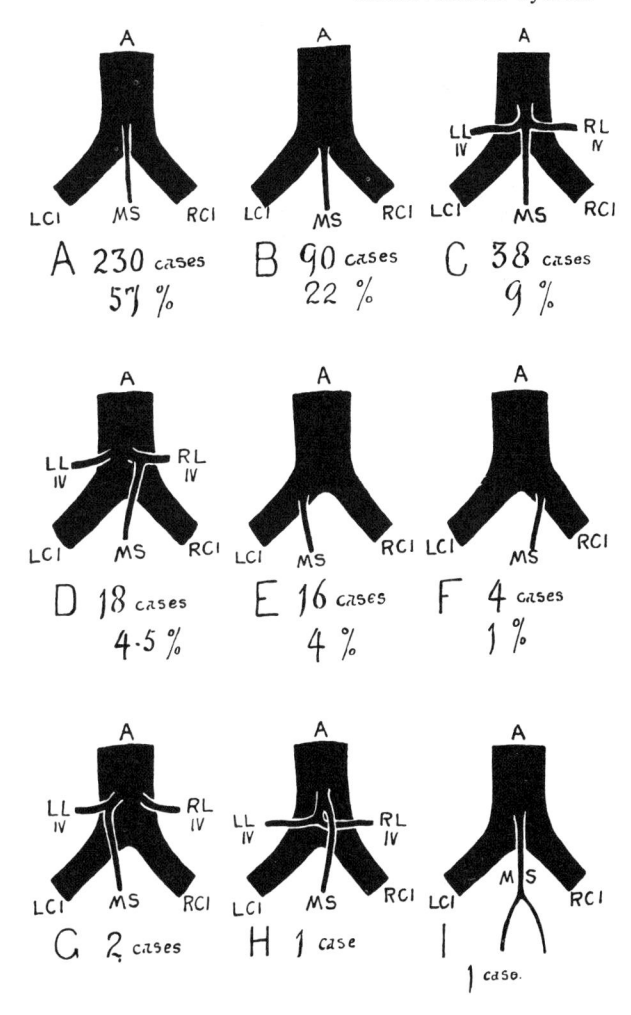

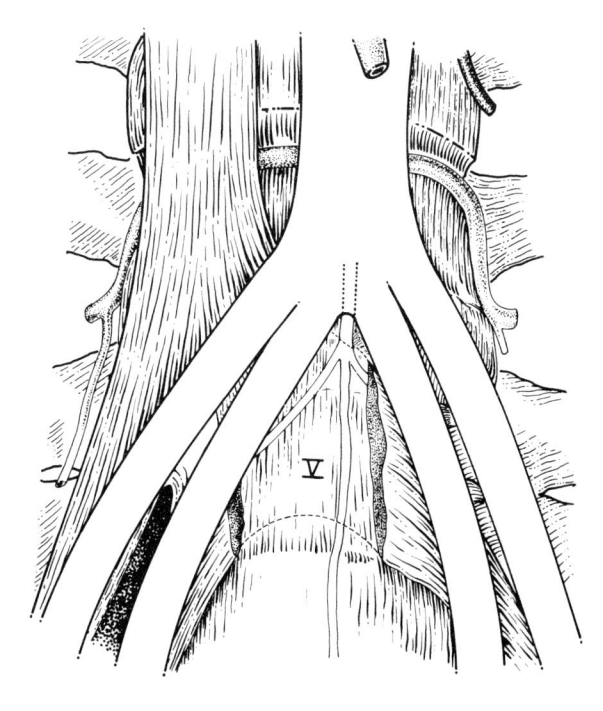

A 230 cases 57%
B 90 cases 22%
C 38 cases 9%
D 18 cases 4.5%
E 16 cases 4%
F 4 cases 1%
G 2 cases
H 1 case
I 1 case

FIG. 120.
Common Iliac Artery Absent.
Common Iliac Artery Short.

In this subject, Adachi found a very short right common iliac artery. The left common iliac was absent.

From Adachi, B. *Anatomie der Japaner. Das Arteriensystem der Japaner.* Verlag der Kaiserlich-Japanischen Universität zu Kyoto. In Kommission bei "Maruzen Co.", Kyoto und Tokyo. Gedruckt von "Kenkyusha" in Tokyo, 1928.

Human Anatomic Variation

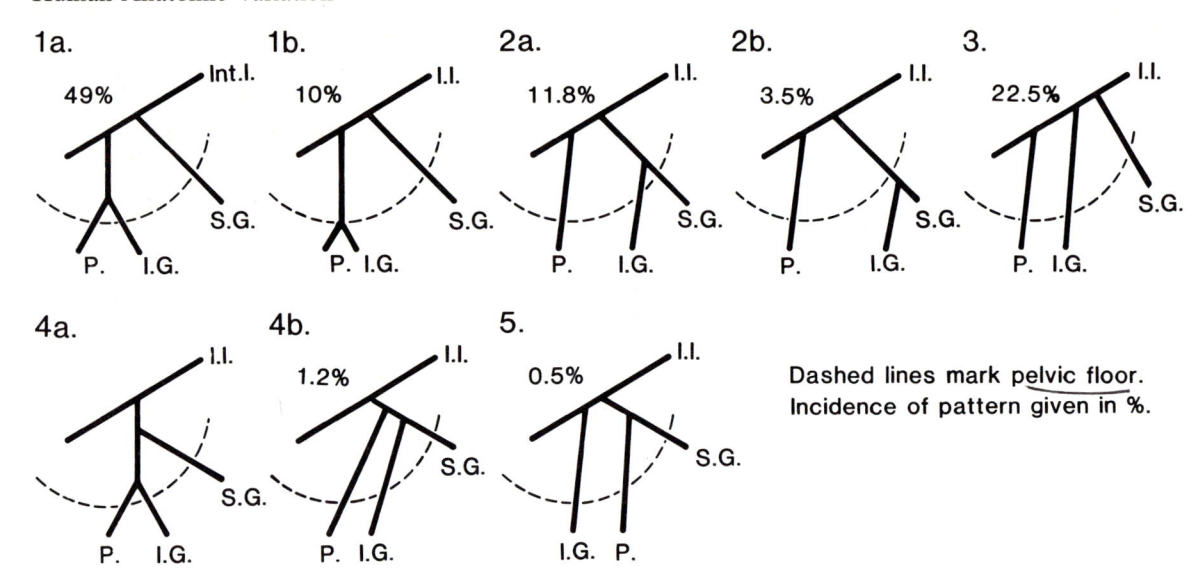

FIG. 121.
Variation in Origin of the Parietal Branches of Internal Iliac Artery (Based on a Study of 169 Specimens (108 Males and 61 Females.))

Int. I (*I.I.*), Internal iliac artery; *I.G.*, inferior gluteal artery; *P.*, pudendal artery; *S.G.*, superior gluteal artery.

Type 1: The superior gluteal artery arises separately from the internal iliac artery, and the inferior gluteal and internal pudendal vessels are given off by a common trunk. If the latter divides within the pelvis it is considered to be Type 1a, whereas if the bifurcation occurs below the pelvic floor it is classified as Type 1b.

Type 2: The superior and inferior gluteal arteries arise by a common trunk and the internal pudendal vessel separately. In this category, as in the previous one, two subtypes are described. Type 2a includes those specimens in which the trunk common to the two gluteal arteries divides within the pelvis, and Type 2b those in which the division occurs outside the pelvis.

Type 3: The three branches arise separately from the internal iliac artery.

Type 4: The three arteries arise by a common trunk. The subtyping in this group is based on the sites of origin of the superior gluteal and the internal pudendal arteries from the parent stem. In Type 4a the trunk first gives rise to the superior gluteal artery before bifurcating into the other two branches; in Type 4b the internal pudendal is the first vessel to spring from the common trunk, which then divides into superior and inferior gluteal arteries.

Type 5: The internal pudendal and the superior gluteal arteries arise from a common trunk, and the inferior gluteal has a separate origin.

Braithwaite provides the following summary:

1. The sites of origin of the superior gluteal, inferior gluteal, internal pudendal, and obturator arteries have been investigated in 169 specimens (108 males and 61 females).

2. The superior gluteal, inferior gluteal, and internal pudendal arteries are comparatively constant in their origins, conforming to a Type [1] arrangement on the Adachi scale in 58.5% of cases, Type [3] pattern being found in 22.5% and Type [2] in 15.3%. Type [4] is less frequent and occurs in 3.6% of specimens. In 52.7% of instances, a similar origin of vessels is noted on both sides.

3. The obturator artery is more variable and arises as a direct branch from the anterior division of the internal iliac artery in 41.4% of instances, from the inferior epigastric artery in 19.5%, from the superior gluteal artery in 10%, from the inferior gluteal-internal pudendal trunk in 10%, and by a double origin in 6.4%. In only 23% of instances is a similar origin noted on both sides.

4. The present findings show a lower incidence from the inferior epigastric artery and a higher rate of occurrence of a vessel with a double origin than previous series. Reasons are suggested to explain these differences.

Redrawn from Braithwaite, J.L. Variations in origin of the parietal branches of the internal iliac artery. *J. Anat.* 86:423–430, 1952.

Cardiovascular System

FIG. 122.

Origins of Obturator Artery.

(Based on a Study of 169 Bodies (61 Females, 108 Males)).

C.I., Common iliac artery; *E.I.*, external iliac artery; *I.E.*, inferior epigastric artery; *I.G.*, inferior gluteal artery; *I.I.*, internal iliac artery; *O.*, obturator artery; *O.U.A.*, obliterated umbilical artery; *P.*, pudendal artery; *S.G.*, superior gluteal artery. *1–7*, corresponds to following listing: *1*, from anterior division of internal iliac artery, 41.4%; *2*, from inferior epigastric artery, 19.5%; *3*, from superior gluteal artery; 10%; *4*, from inferior gluteal-internal pudendal trunk, 10%; *5*, from inferior gluteal artery, 4.7%; *6*, from internal pudendal artery, 3.8%; *7*, from external iliac artery; 1.1%.

The following tabular summary is provided by Braithwaite:

Origins of Obturator Artery

Origin	Female No.	Female %	Male No.	Male %	Total No.	Total %
A. From internal iliac artery:						
Direct branch from anterior division	24	39.3	46	42.6	70	41.4
Inferior gluteal-internal pudendal trunk	5	8.2	12	11.1	17	10.0
Inferior gluteal artery	2	3.2	6	5.5	8	4.7
Internal pudendal artery	2	3.2	4	3.7	6	3.8
Superior gluteal artery	10	16.4	7	6.4	17	10.0
Iliolumbar artery	3	4.9	2	1.8	5	3.5
B. From external iliac artery:						
Direct branch	1	1.6	1	0.9	2	1.1
Inferior epigastric artery	10	16.4	23	21.3	33	19.5
C. From internal and external iliac arteries:						
By double origin	4	6.4	7	6.4	11	6.5
Totals	61	99.6	108	99.7	169	99.5

From Braithwaite, J.L. Variations in origin of the parietal branches of the internal iliac artery. *J. Anat.* 86:423–430, 1952.

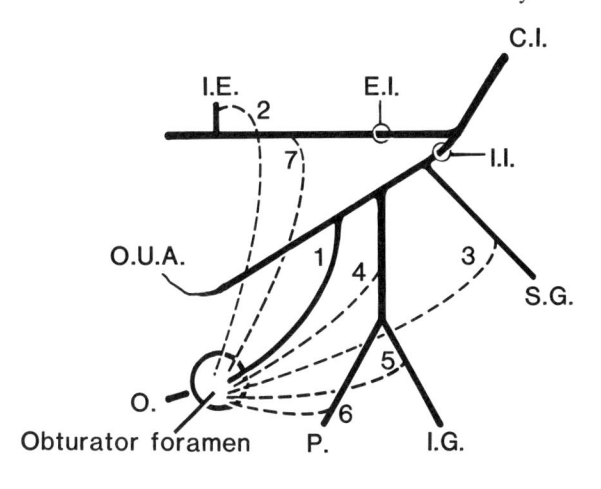

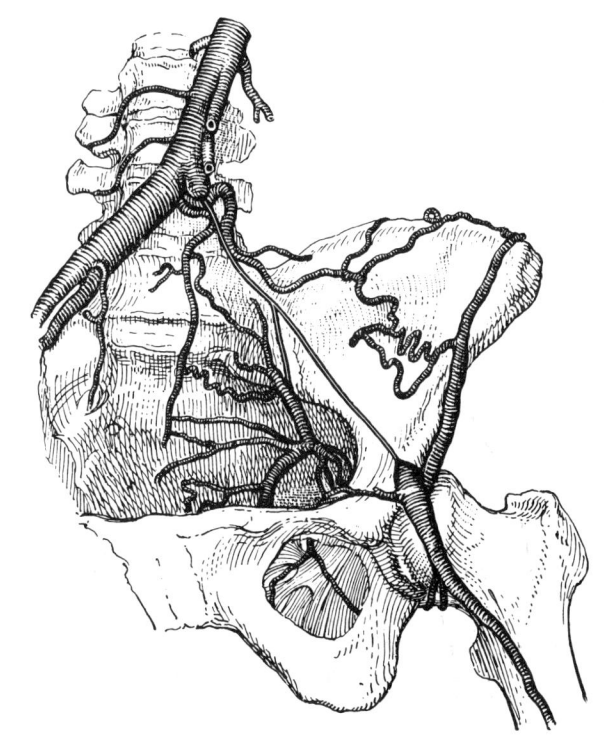

FIG. 123.

Obliteration of Left Common, External, and Internal Iliac Arteries.

From Paterson, A.M. Obliteration of the left common, external, and internal iliac arteries. *J. Anat. Physiol.* 44:56, 1910.

Human Anatomic Variation

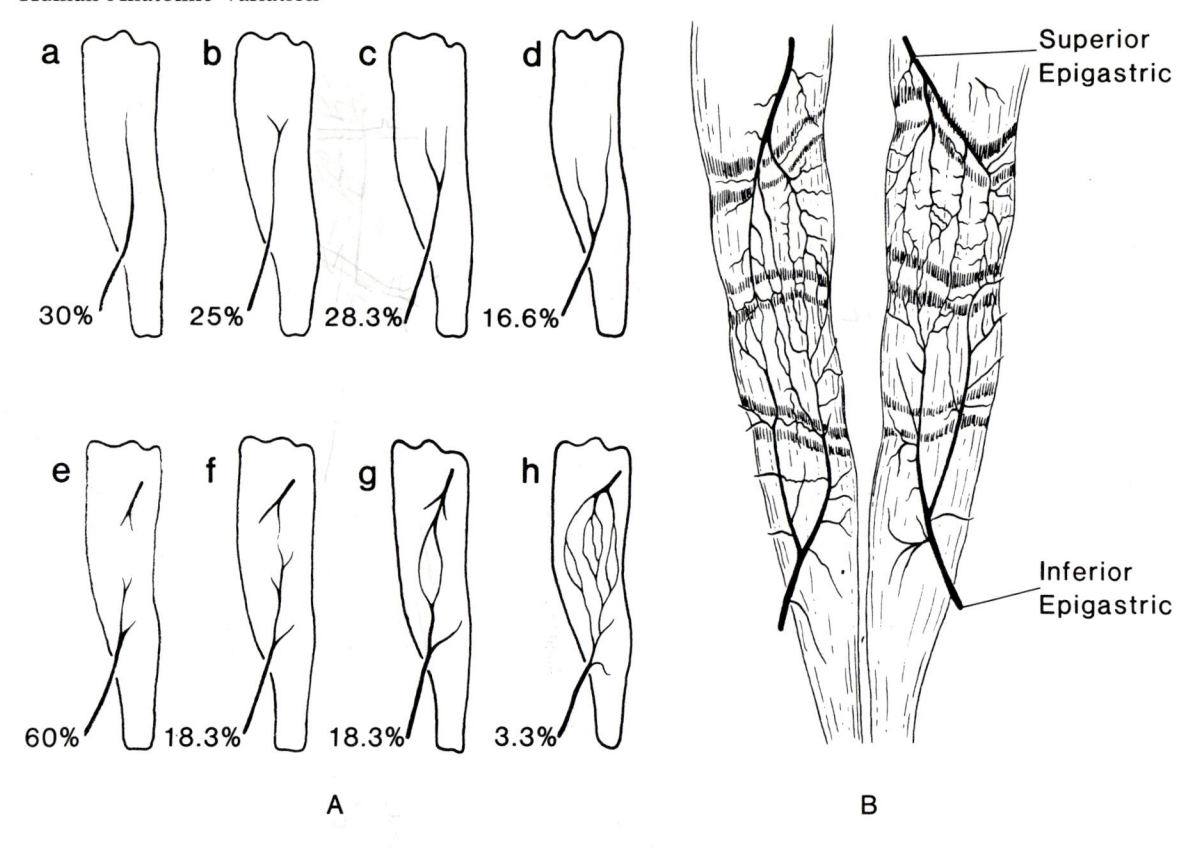

FIG. 124.
Superior and Inferior Epigastric Arteries and Rectus Abdominis Muscle (Based on a Study of 115 Specimens).

A: The frequency of various configurations of the inferior epigastric artery after dissection are shown in *a*, *b*, *c*, and *d*. The lower set of figures shows the anastomoses or absence thereof as revealed by dissection: *e*, no anastomosis; *f*, 1 anastomosis; *g*, 2 anastomoses; and *h*, multiple anastomoses.
B: The rectus muscles with the epigastric arteries dissected. A special dissection of a specimen with many anastomoses between superior and inferior epigastric arteries. This dissection also illustrates the manner in which arteries tend to follow the tendinous inscriptions.

The authors provide the following conclusions from their study: The rectus muscles are variable in length and width, tending to parallel body build in their relative dimensions. In the majority of cases the two recti are of equal length, and measure between 34 and 38 centimeters.

The two rectus muscles in a given individual usually have the same number of inscriptions, and in half of the bodies in this series the number was 3 on each side. No muscle had more than 4 inscriptions.

The deep inferior epigastric artery was present in all bodies bilaterally, most frequently as a single-stem vessel lying on the posterior aspect of the rectus, and entering the substance of the muscle in its middle third.

The superior epigastric artery is occasionally not seen on the dorsal surface of the rectus, being either absent or buried in muscle tissue. When present, it is usually a single stem that enters the medial portion of the muscle after a short course on the surface.

In only 1 instance, in this series of 162 muscles examined, did the superior and inferior epigastric vessels anastomose on the surface of the rectus muscle in the pattern frequently depicted in anatomy texts. Furthermore, when muscle fibers had been dissected away from the arteries, only 40 per cent of the superior and inferior epigastric arteries were found to have any gross anastomoses.

Three or 4 small branches of the inferior epigastric artery commonly pass laterally through the linea semilunaris into the plane between the transversus abdominis and internal oblique muscles of the abdomen. These vessels usually leave the middle two-fourths of the muscle from its posterior aspect, and they most frequently accompany branches of the ninth, tenth, or eleventh thoracic nerves. These vessels were dissected out to their termination, and it was found that only 15 per cent anastomosed with intercostal or circumflex iliac arteries. The remaining vessels terminated most commonly in the transversus abdominis in the upper abdomen and in the internal oblique in the lower abdomen.

Finally, the data on arterial anastomoses were re-examined to discover whether there was an increased number of anastomoses between epigastric and intercostal vessels in those bodies with grossly demonstrable anastomoses between superior and inferior epigastric arteries. It was found that there was no increased tendency for such anastomoses to occur between epigastric and intercostal vessels when grossly demonstrable connections between the superior and the inferior epigastric arteries were present.

Redrawn from Milloy, F.J., Anson, B.J. and D.K. McAfee. The rectus abdominis muscle and the epigastric arteries. *Surg. Gynecol. Obstet.* 110:293–302, 1960.

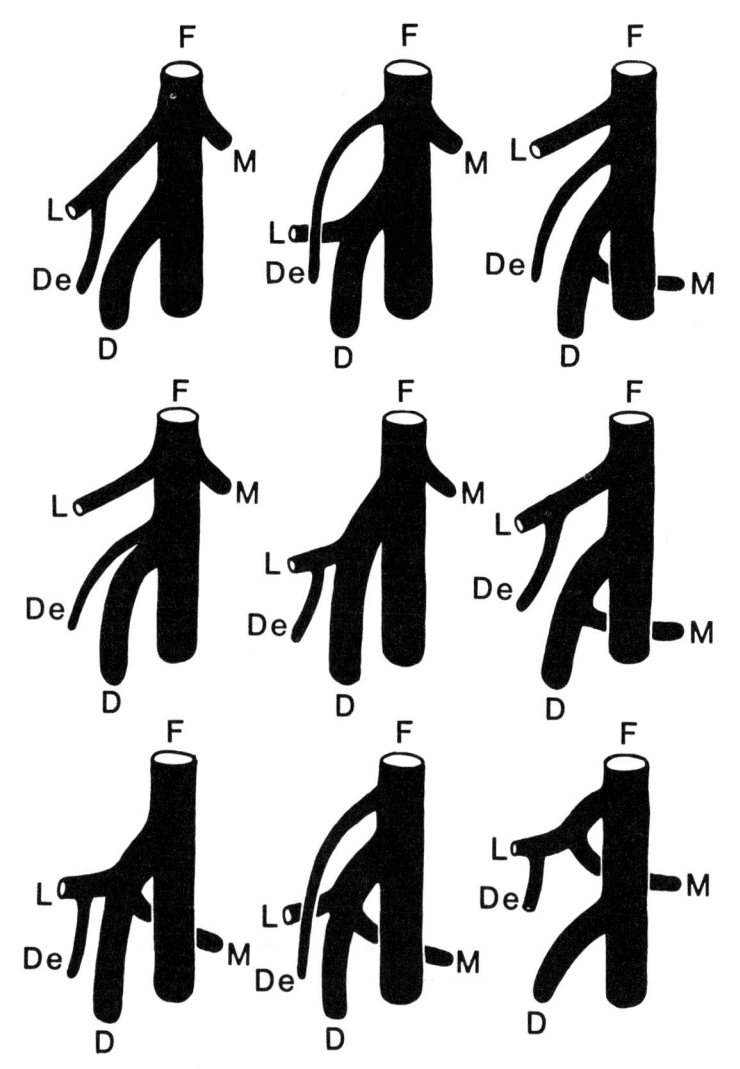

FIG. 125.
Origin of Deep and Circumflex Femoral Group of Arteries (Based on a Study of 481 Arterial Patterns).

D, Deep femoral artery; *De*, descending circumflex femoral artery; *F*, femoral artery; *L*, lateral circumflex femoral artery; *M*, medial circumflex femoral artery.

Types 1, 1a, 1b (upper row, left to right); types 1c, 2, 3 (middle row, left to right); and types 4, 5, 6 (lower row, left to right).

Type 4 is the variety of origin commonly described in the textbooks. There is evidently a tendency for this form of origin to appear more often in the left thigh than in the right in the male but not the female.

Adachi's statistical tables for right and left sides of the bodies of Japanese are based upon a population of about three-quarters males and one-quarter females. Our [Williams et al] findings agree with his: in that type 4 (Adachi's truncus profundo-circumflexus perfectus) is oftener seen on the left than on the right; in that type 3 (Adachi's truncus profundo-circumflexus medialis) is more frequently a right-sided than a left-sided phenomenon; there may be a difference as to sidedness in the occurrence of type 2 (Adachi's truncus profundo-circumflexus lateralis).

Redrawn from Williams, G.D., Sindelar, R.J., Peart, J.C., Martin, N.A., McIntire, L.R. and C.H. Martin. Origin of the deep and circumflex femoral group of arteries. *Anat. Rec.* 46:273–279, 1930; and Williams, G.D., Martin, C.H. and L.R. McIntire. Origin of the deep and circumflex femoral group of arteries. *Anat. Rec.* 60:189–196, 1934.

Percentage Distribution of Types of Origin of the Deep Circumflex Femoral Group of Arteries

Comparison of sides, regardless of race

Type	Male Right (197)	Male Left (198)	Difference	Female Right (43)	Female Left (43)	Difference
4	51.3	60.6	9.3 ± 5.0	58.1	53.5	4.6 ± 10.7
2	20.3	20.7	0.4	16.3	25.6	9.3 ± 8.8
3	16.8	11.6	5.2 ± 3.5	16.3	7.0	9.3 ± 6.9
5	3.0	2.5	0.5	0.0	0.0	0.0
1	4.1	4.0	0.1	7.0	7.0	0.0
6	1.0	0.0	1.0	2.3	2.3	0.0
1a	2.0	0.5	1.5	0.0	2.3	2.3
1b	1.0	0.0	1.0	0.0	2.3	2.3
1c	0.5	0.0	0.5	0.0	0.0	0.0

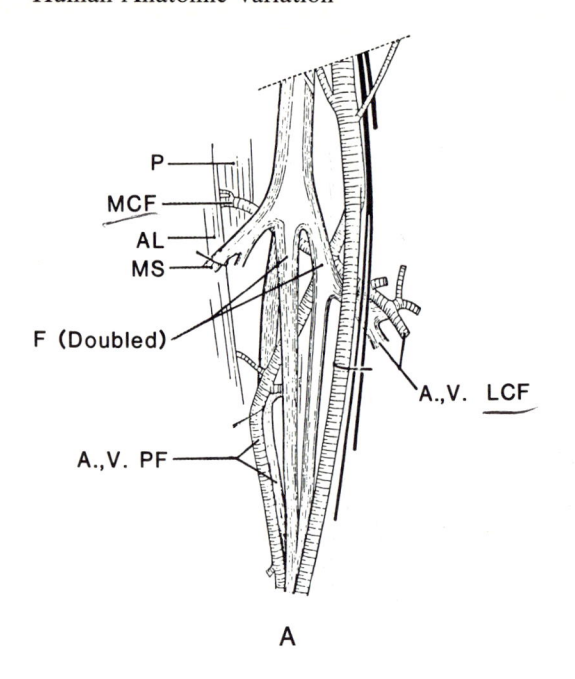

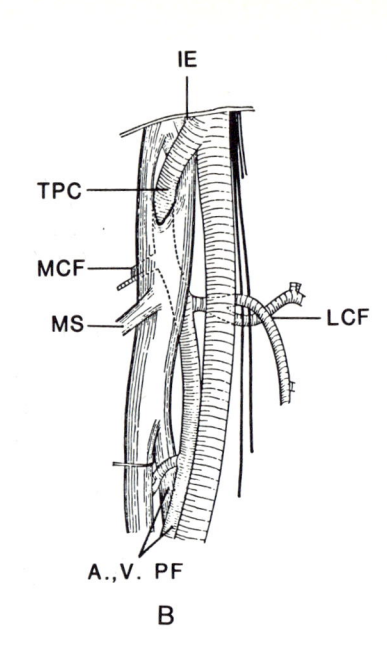

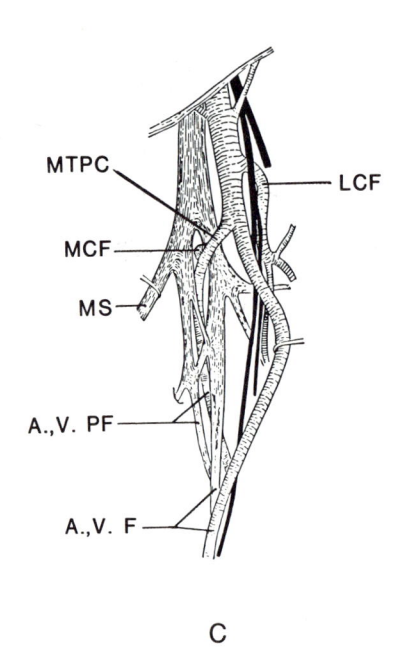

FIG. 126.
Femoral Artery and Branches.
AL, Adductor longus muscle; A.,V.F, femoral artery and vein; A.,V.LCF, lateral femoral circumflex artery and vein; A.,V.PF, profunda femoris artery and vein; F, femoral vein; IE, inferior epigastric artery; LCF, lateral femoral circumflex artery; MCF, medial femoral circumflex artery; MS, magna saphena vein; MTPC, medial circumflex-profunda trunk; TPC, profunda femoris-medial circumflex (perfectus) trunk.
A: The femoral vein is doubled. No large arterial vessel passes in front of the femoral vein.
B: The profunda femoris-circumflex trunk is seen passing through an aperture formed in the femoral vein.
C: The profunda femoris-medial circumflex trunk is seen passing through an aperture formed in the femoral vein.

From Adachi, B. *Anatomie der Japaner. Das Arteriensystem der Japaner.* Verlag der Kaiserlich-Japanischen Universität zu Kyoto. In Kommission bei "Maruzen Co.", Kyoto und Tokyo. Gedruckt von "Kenkyusha" in Tokyo, 1928.

Cardiovascular System

Origin of Profunda Femoris, 100 Legs

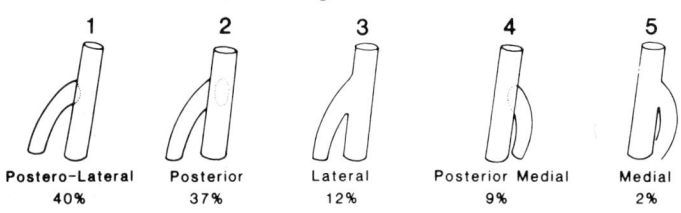

1	2	3	4	5
Postero-Lateral	Posterior	Lateral	Posterior Medial	Medial
40%	37%	12%	9%	2%

Origin of Lateral Circumflex Artery, 100 Legs

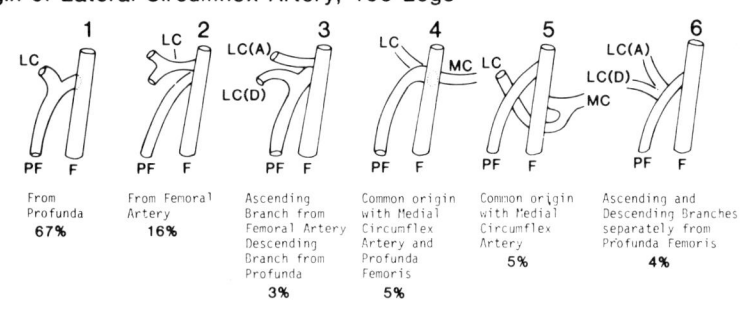

1	2	3	4	5	6
From Profunda	From Femoral Artery	Ascending Branch from Femoral Artery Descending Branch from Profunda	Common origin with Medial Circumflex Artery and Profunda Femoris	Common origin with Medial Circumflex Artery	Ascending and Descending Branches separately from Profunda Femoris
67%	16%	3%	5%	5%	4%

Origin of Medial Circumflex Artery, 100 Legs

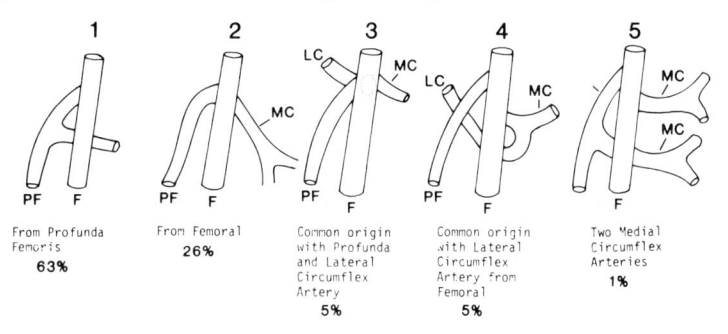

1	2	3	4	5
From Profunda Femoris	From Femoral	Common origin with Profunda and Lateral Circumflex Artery	Common origin with Lateral Circumflex Artery from Femoral	Two Medial Circumflex Arteries
63%	26%	5%	5%	1%

Median Measurements of Femoral, Profunda Femoris and Perforating Arteries

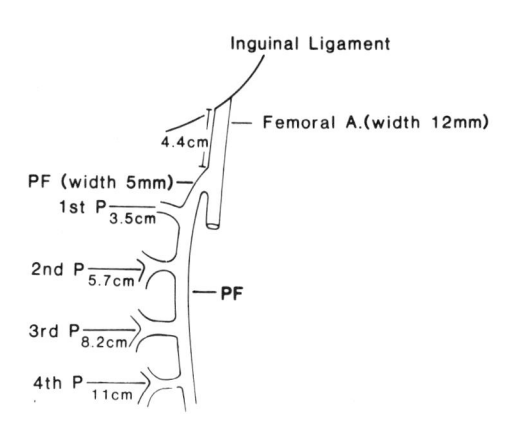

Inguinal Ligament
Femoral A. (width 12mm)
4.4 cm
PF (width 5mm)
1st P — 3.5 cm
2nd P — 5.7 cm
3rd P — 8.2 cm
4th P — 11 cm
PF

FIG. 127.
Variations in Deep Femoral (Profunda Femoris) Artery.
PF, Profunda femoris; LC, lateral circumflex; F, femoral; MC, medial circumflex; P, perforating artery.

Redrawn from Siddharth, P., Smith, N.L., Mason, R.A. and F. Giron. Variational anatomy of the deep femoral artery. *Anat. Rec.* 212:206–209, 1985.

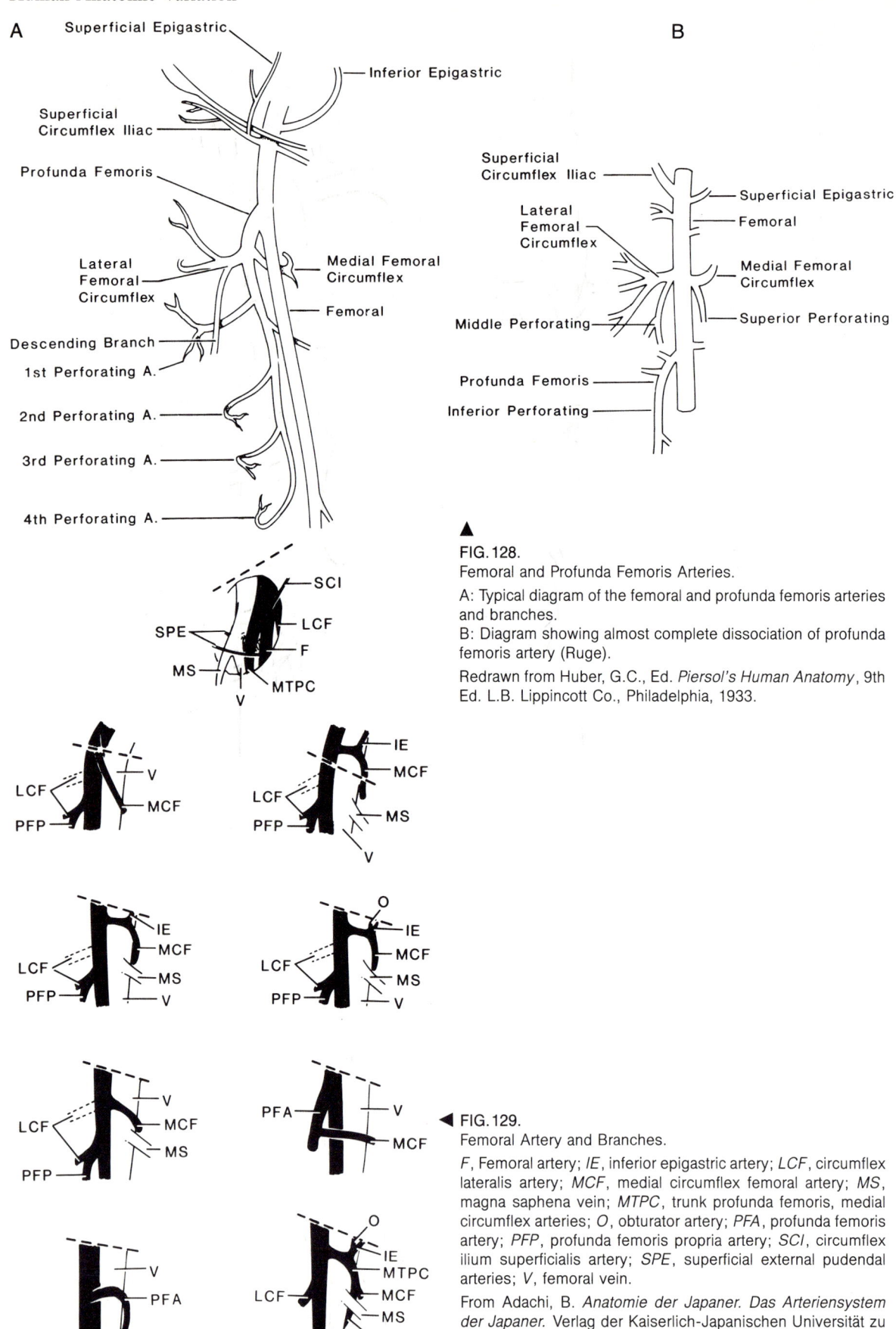

Human Anatomic Variation

FIG. 128.
Femoral and Profunda Femoris Arteries.

A: Typical diagram of the femoral and profunda femoris arteries and branches.

B: Diagram showing almost complete dissociation of profunda femoris artery (Ruge).

Redrawn from Huber, G.C., Ed. *Piersol's Human Anatomy*, 9th Ed. L.B. Lippincott Co., Philadelphia, 1933.

FIG. 129.
Femoral Artery and Branches.

F, Femoral artery; *IE*, inferior epigastric artery; *LCF*, circumflex lateralis artery; *MCF*, medial circumflex femoral artery; *MS*, magna saphena vein; *MTPC*, trunk profunda femoris, medial circumflex arteries; *O*, obturator artery; *PFA*, profunda femoris artery; *PFP*, profunda femoris propria artery; *SCI*, circumflex ilium superficialis artery; *SPE*, superficial external pudendal arteries; *V*, femoral vein.

From Adachi, B. *Anatomie der Japaner. Das Arteriensystem der Japaner*. Verlag der Kaiserlich-Japanischen Universität zu Kyoto. In Kommission bei "Maruzen Co.", Kyoto und Tokyo. Gedruckt von "Kenkyusha" in Tokyo, 1928.

Cardiovascular System

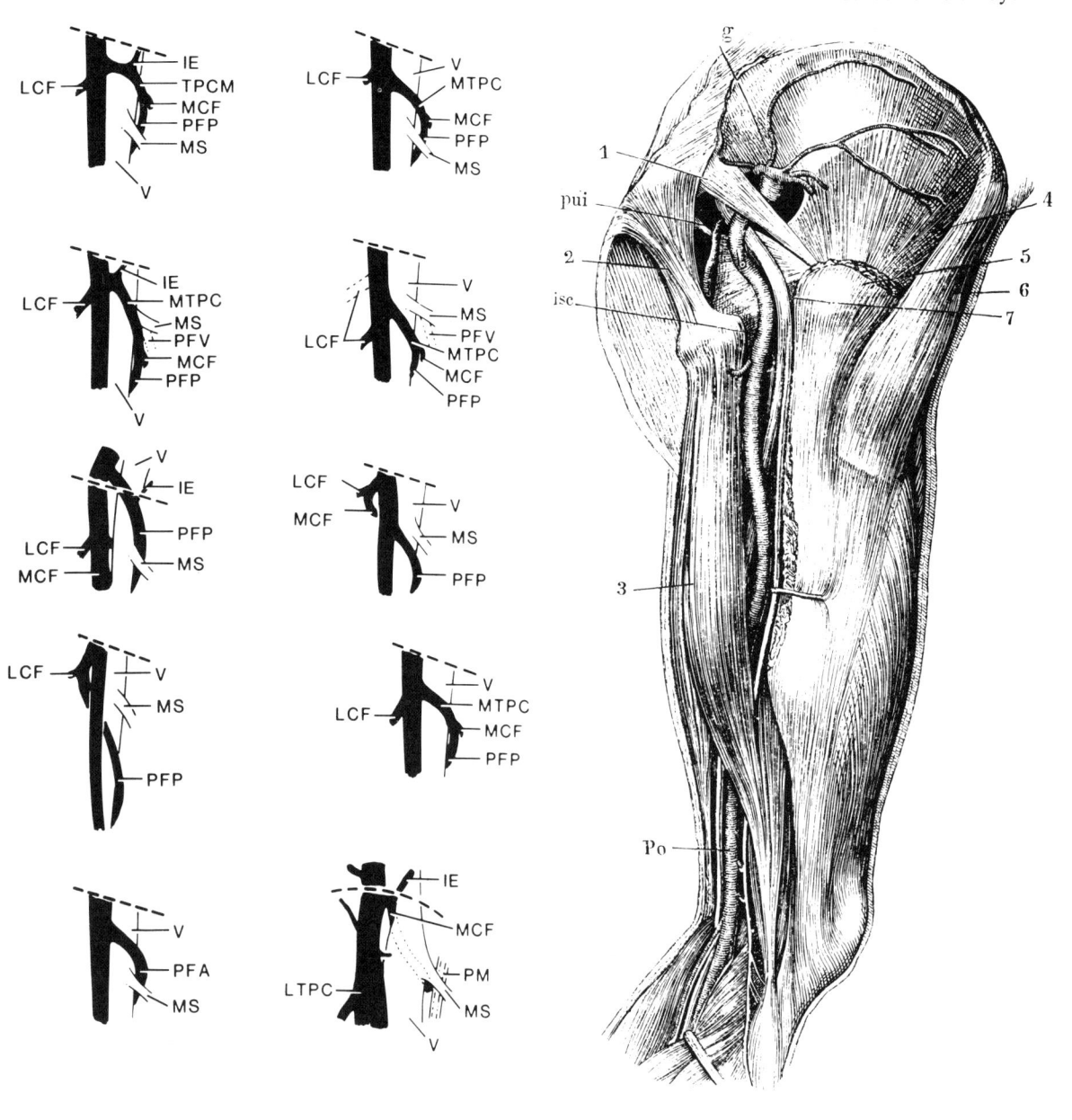

FIG. 130.
Femoral Artery and Branches.

IE, Inferior epigastric artery; *LCF*, circumflex lateralis artery; *LTPC*, trunk profunda femoris, lateral circumflex arteries; *MCF*, medial circumflex femoral artery; *MS*, magna saphena vein; *MTPC*, trunk profunda femoris, medial circumflex arteries; *PFA*, profunda femoris artery; *PFP*, profunda femoris propria artery; *PFV*, profunda femoris vein; *PM*, musculus pectineus; *V*, femoral vein.

From Adachi, B. *Anatomie der Japaner. Das Arteriensystem der Japaner.* Verlag der Kaiserlich-Japanischen Universität zu Kyoto. In Kommission bei "Maruzen Co.", Kyoto und Tokyo. Gedruckt von "Kenkyusha" in Tokyo, 1928.

FIG. 131.
Persistent Ischiadic (Sciatic) Artery.

g, Gluteal artery; *isc*, ischiadic artery; *pui*, internal pudendal artery; *po*, popliteal artery; *1*, piriformis muscle; *2*, sacrotuberous ligament; *3*, biceps femoris muscle; *4*, gluteus minimus muscle; *5*, insertion of gluteus medius muscle; *6*, tensor fasciae latae muscle; *7*, sciatic nerve (n. ischiadicus).

The ischiadic artery (a. comitans n. ischiadici) is usually a long, thin vessel which is normally the arterial supply to the sciatic nerve. It arises from the inferior gluteal artery and enters the nerve immediately below the piriformis muscle. The ischiadic artery was developmentally the important axial artery of the lower limb arising originally from the umbilical artery. The artery may persist in the adult as a large vessel continuous with the popliteal artery.

From Henle, J. *Handbuch der Systematischen Anatomie des Menschen*, in three volumes. Von Friedrich Vieweg und Sohn, Braunschweig 1868. After Dubreuil, J.M. *Des Anomalies Arterielles Considerees dans Leur Rapports avec l'Pathologie et les Operations Chirurgicales*. J.B. Bailliere, Paris, 1847.

Human Anatomic Variation

FIG. 132.
High Division and Reunion of Femoral Artery.

High Division of a. Femoralis with Subsequent Reunion (from Quain's Plates).

From Manners-Smith, T. The limb arteries of primates. *J. Anat. Physiol.* 44:95–172, 1912.

FIG. 133.
Variations in Mode of Division of Popliteal Artery.

AT, Anterior tibial artery; *DG*, descending genicular artery; *P*, peroneal artery; *PT*, posterior tibial artery; *TAP*, anterior tibioperoneal trunk; *TP*, tibioperoneal trunk.

A: Usual branching pattern.
B: Anterior tibioperoneal trunk.
C: The three arteries of the leg separating from a common source.
D: Anastomosis between the anterior tibial artery and the tibioperoneal trunk.
E: Exaggeration of length of the tibioperoneal trunk.

After Dubreuil-Chambardel, L. in Latarjet, A. *Testut, L. Traité d'Anatomie Humaine*, 8th Ed. G. Doin & Cie., Paris, 1929.

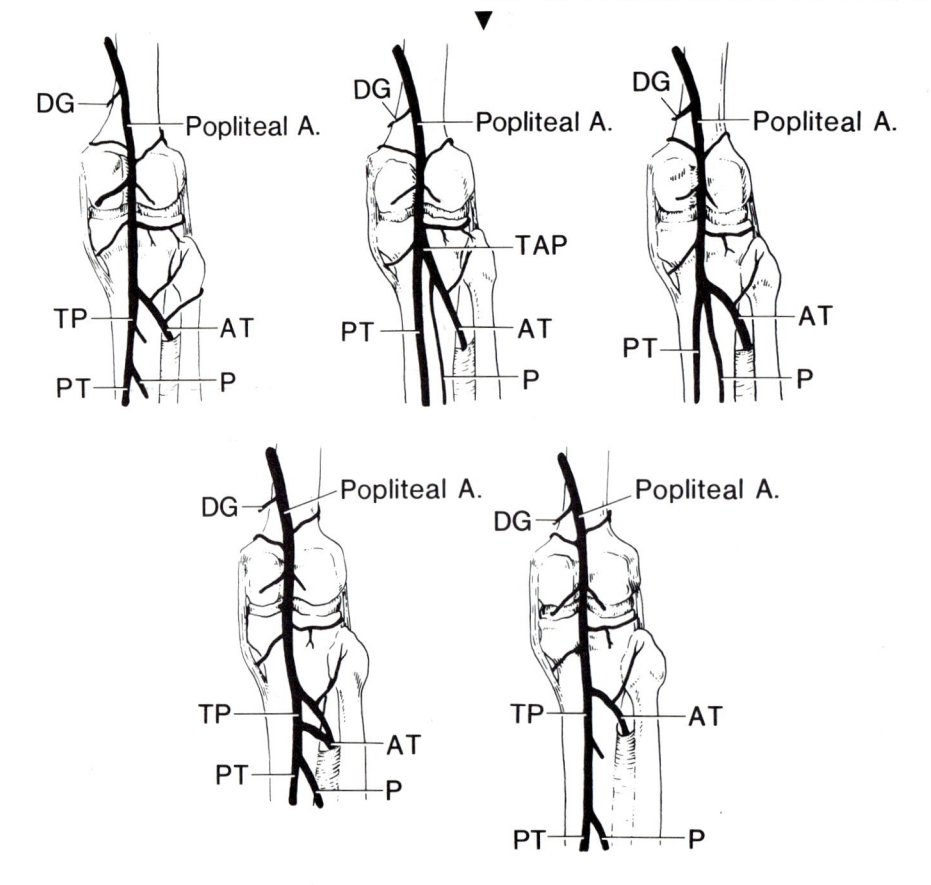

FIG. 134.
Various Arterial Patterns on Dorsum of Foot (Meyer). Variations in the Plantar Arch (Dubreuil-Chambardel).

A–E (upper row): Variations in the branching pattern of the anterior tibial artery on the dorsum of the foot.

A–F (middle and lower rows): Variations in the plantar arch. Identification of numbers: *1*, posterior tibial artery; *2*, medial plantar artery; *3*, lateral plantar artery; *4*, deep plantar arch arising from *3*; *5*, deep plantar artery.

After Meyer, H. in Kopsch, F. *Rauber's Lehrbuch und Atlas der Anatomie des Menschen*. Georg Thieme, Leipzig 1908; and after Dubreuil-Chambardel, L. in Latarjet, A. *Testut, L. Traité d'Anatomie Humaine*. 8th Ed. G. Doin & Cie., Paris, 1929.

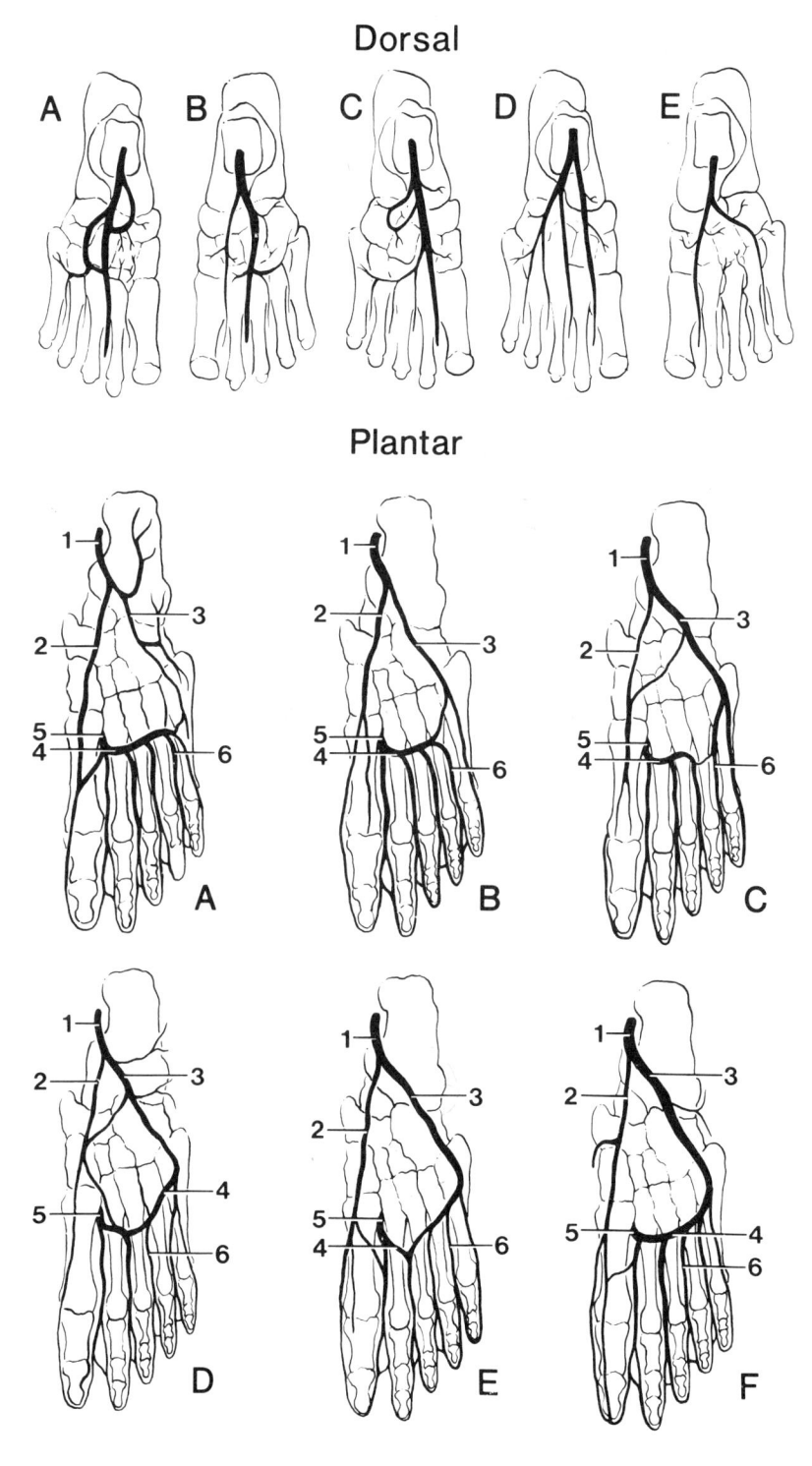

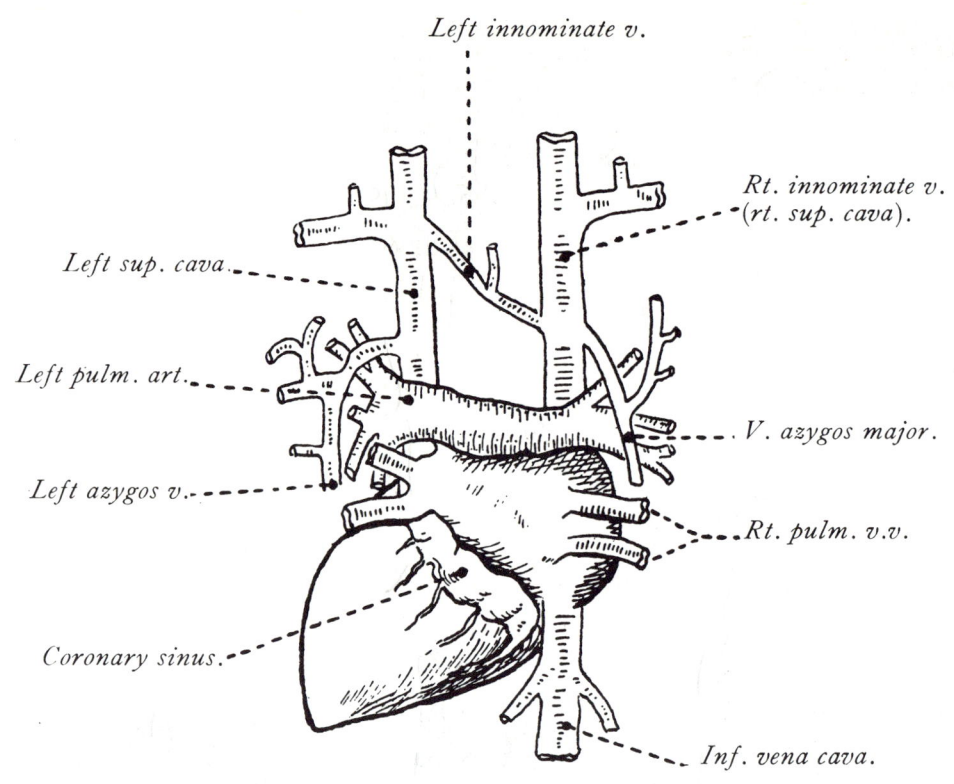

FIG. 135.
Persistent Left Superior Vena Cava.
Diagram of the heart and great veins, from behind, showing the arrangement in a case of persistence of the left superior cava.

From Sharpey-Schafer, E., Symington, J. and T.H. Bryce, Eds. *Quain's Anatomy*, 11th Ed., Vol. IV, Pt. III. The Heart. Longmans, Green, and Co., London, 1919.

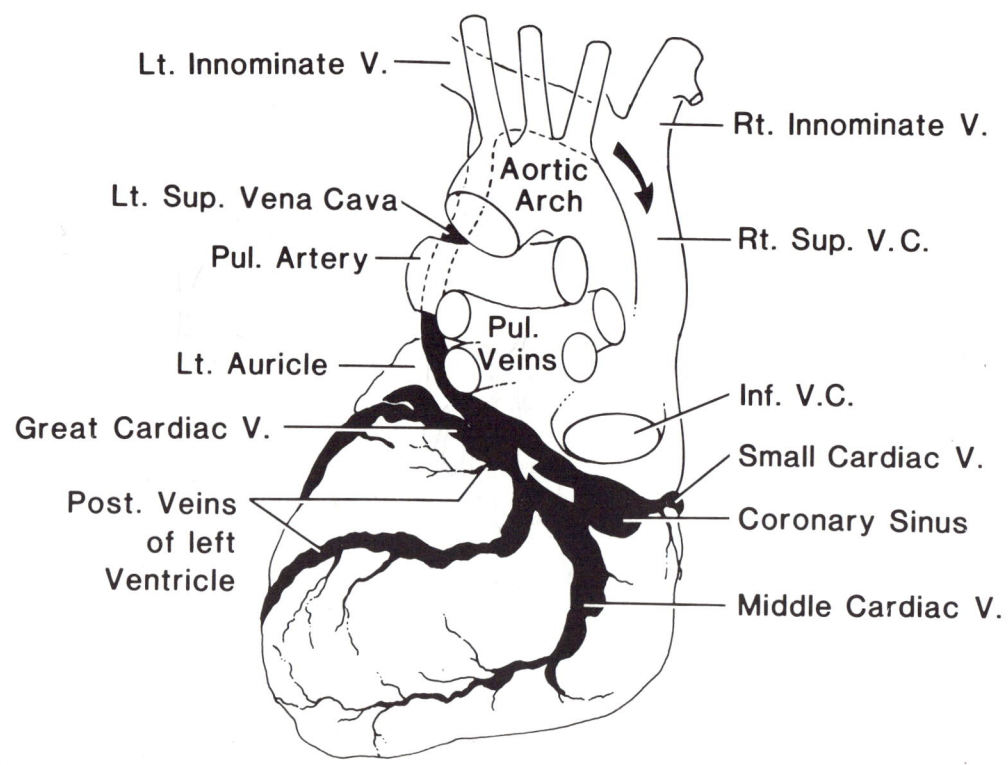

FIG. 136.
Closed Coronary Sinus (Obliterated Ostium to Right Atrium).
In this case of Prows, the cardiac veins drained into the right atrium via the coronary sinus, oblique vein, left superior vena cava, left innominate vein, and right superior vena cava.

Redrawn from Prows, M.S. Two cases of bilateral superior venae cavae, one draining a closed coronary sinus. *Anat. Rec.* 87:99–106, 1942.

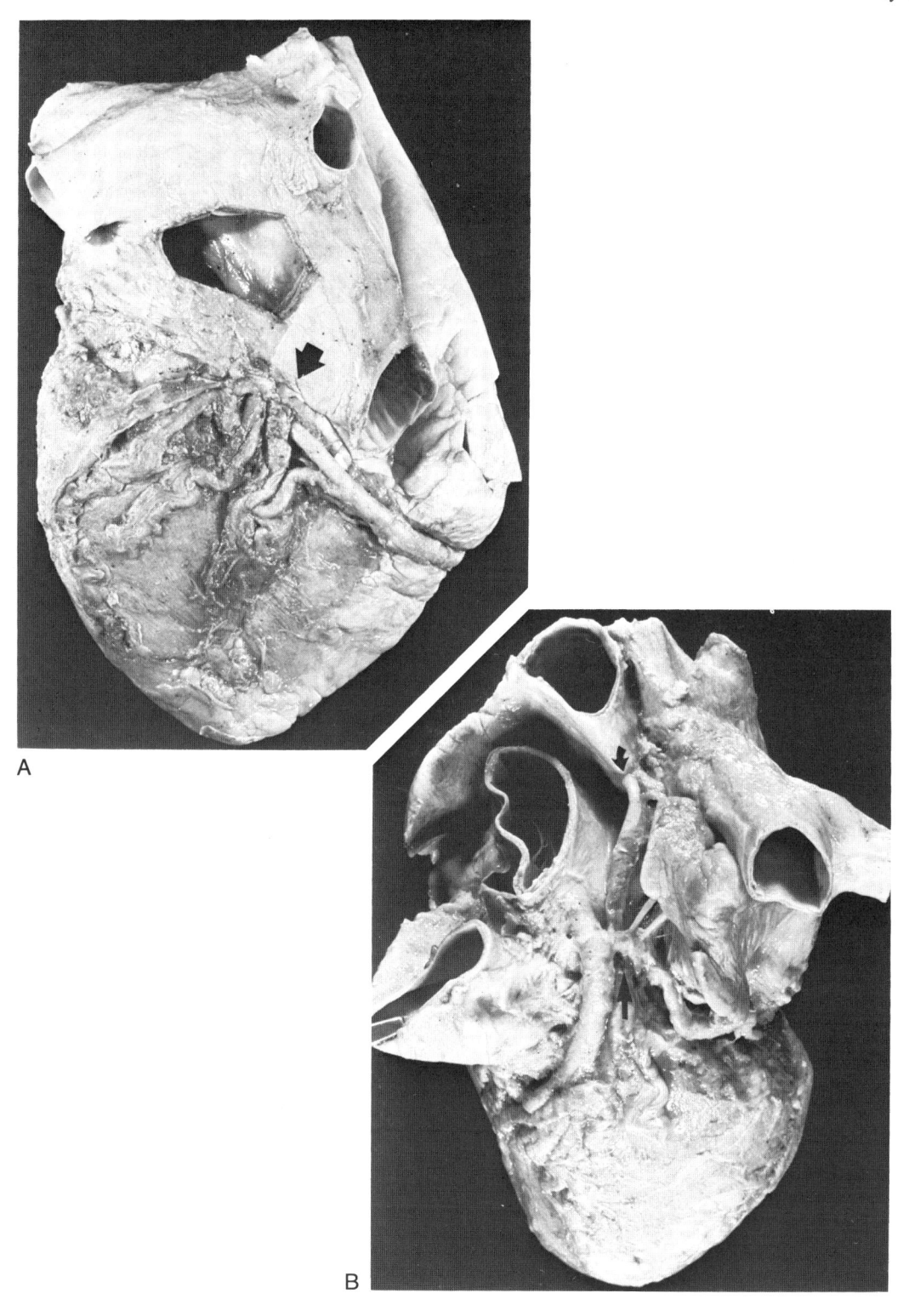

FIG. 137.
Absence of Coronary Sinus.
Great Cardiac (Cardiaca Magna) Vein Draining into Superior Vena Cava.

These rare variations were found in a specimen obtained from a 78-year-old male subject.

Huber reported (in Piersol's *Human Anatomy*, 9th Ed. J.B. Lippincott Co., Philadelphia, 1930) that the great cardiac vein may, in the absence of the coronary sinus, terminate in the left brachiocephalic vein. In this specimen the great cardiac vein terminated in the superior vena cava.

A: Veins converging on ostium to right atrium (*arrow*).
B: Great cardiac vein draining into superior vena cava (*curved arrow*) after passing beneath a branch of the left coronary artery (*straight arrow*).

Dissecting room specimen. The University of Iowa College of Medicine.

Human Anatomic Variation

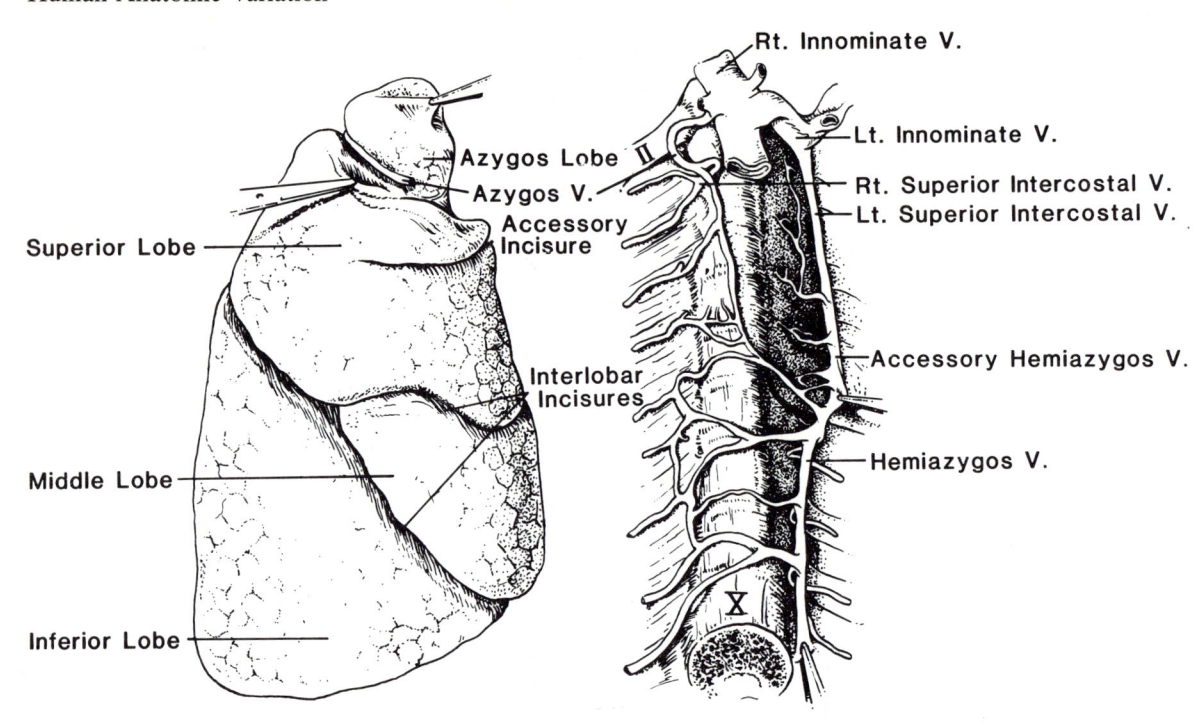

FIG. 138.
Pulmonary Lobe of Azygos Vein.

The vena azygos may be displaced outward, so that, instead of curving over the roof of the lung, it may make a deep fissure in the upper part of the right lung, which is then called the azygos lobe. It can be identified radiographically because it has the shape of an inverted "teardrop" or comma.

Authors' note: Data from 323,641 roentgen examinations indicate an incidence of about 0.57%.

Redrawn from Anson, B.J., Siekert, R.G., Richmond, T.E. and W.E. Bishop. The accessory pulmonary lobe of the azygos vein. *Q. Bull. Northwestern University Medical School* 24:285–290, 1940.

Cardiovascular System

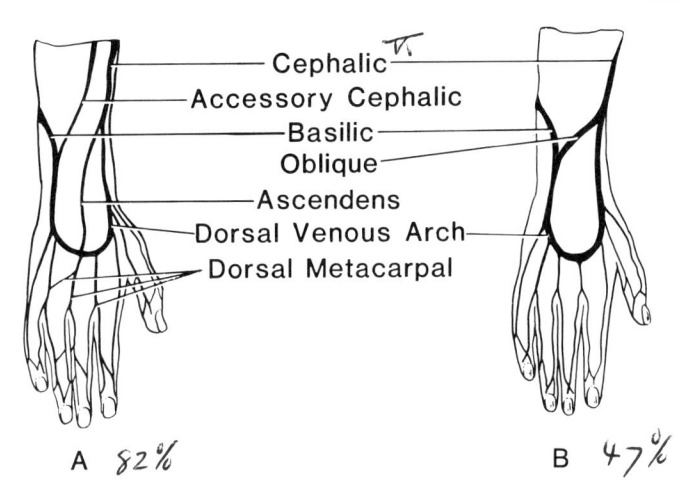

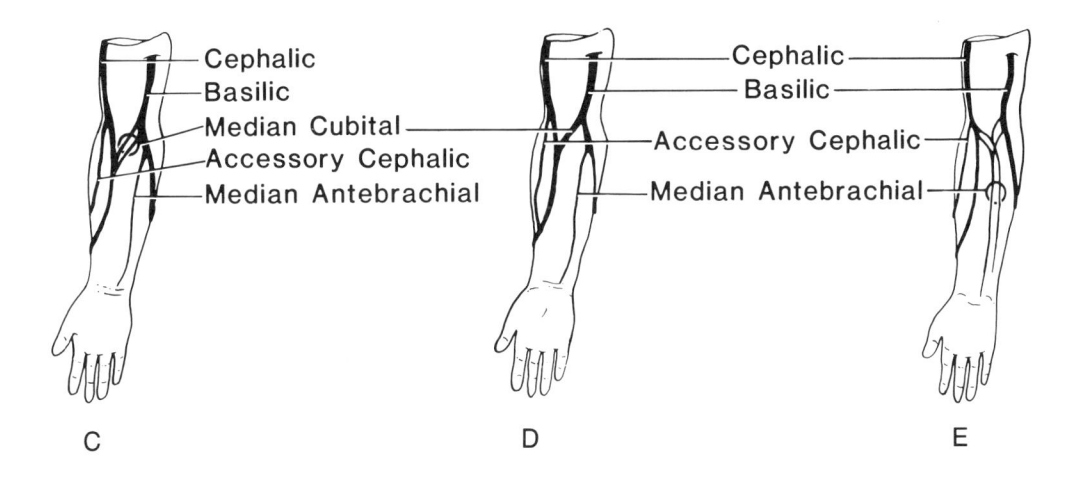

FIG. 139.
Venous Drainage of Hand and Forearm.

A: Dorsal venous arch. In 82% of 300 individuals a large vein passed proximally from the center of the concavity of the dorsal venous arch to terminate in 65% in the cephalic vein, and in the remaining 17% in the basilic vein.

B: The oblique vein. In 47% of individuals the authors found a large well-marked vein issuing from the ulnar end of the dorsal venous arch, and passing obliquely proximally and radially across the dorsal surface of the wrist to terminate in the cephalic vein.

C,D,E: An accessory cephalic vein was found to arise in the ulnar end of the dorsal venous arch in about 28% (A). The accessory cephalic vein arises from variable sources but was present in 82% of the 300 cases: in 39% as some small venous radicles in the distal part of the dorsal surface of the forearm; in 28% arising from the dorsal venous arch; in 16% arising from the cephalic vein at a point where the vein turns round the radial border of the forearm (C,D,E).

The median cubital vein was present in 84%, and in 43% it received, as a tributary, the median antebrachial vein. In 4% it arose from the cephalic vein usually at about the middle of the forearm.

The authors found the median cubital vein duplicated or doubled in 4% of their 300 cases (C). In addition, they reported duplication of the median antebrachial vein (E).

Redrawn from Berry, R.J.A. and H.A.S. Newton. A study of the superficial veins of the superior extremity in 300 living subjects. *Anat. Anz.* 33:591–602, 1908.

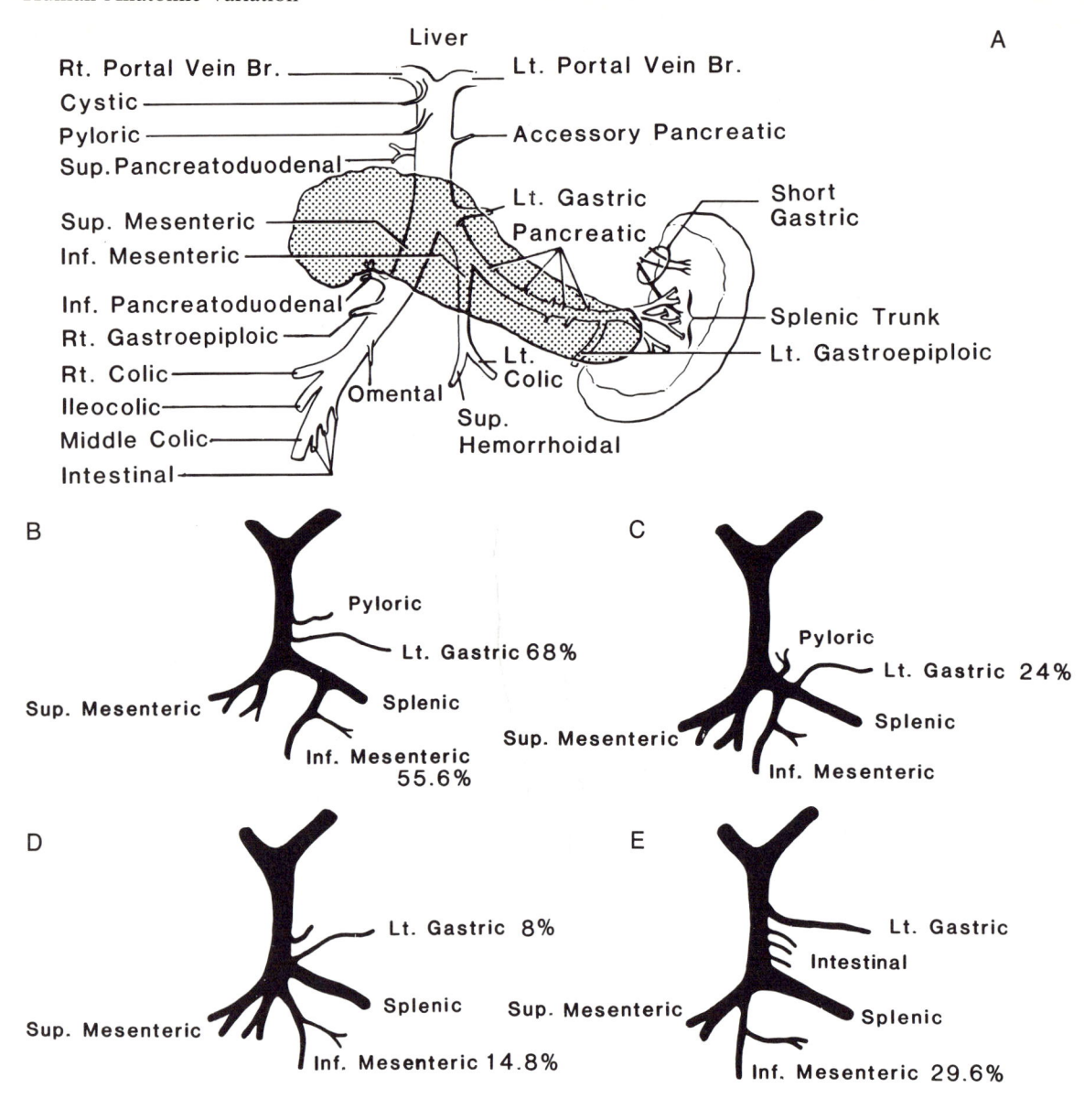

Human Anatomic Variation

FIG. 140.
Portal Vein and Its Origin.

A (from Douglass et al): The termination of each vein is shown as it was encountered most frequently in 92 dissections.
B–D (from Gilfillan): Variations in the morphology of the portal vein found in 59 dissections.

Redrawn from Gilfillan, R.S. Anatomic study of the portal vein and its main branches. *Arch. Surg.* 61:449–461, 1950; and from Douglass, B.E., Baggenstoss, A.H. and W.H. Hollinshead. The anatomy of the portal vein and its origin. *Surg. Gynecol. Obstet.* 91:562–575, 1950.

Cardiovascular System

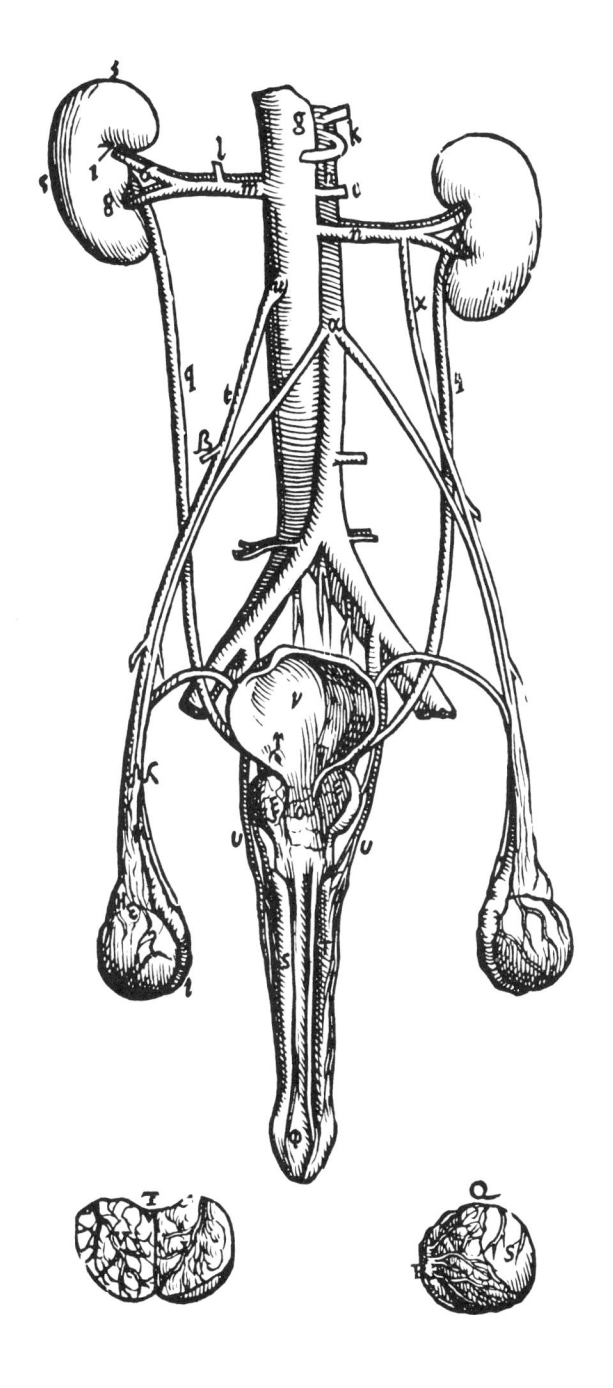

FIG. 141.
Gonadal Veins.
The usual arrangement of the internal spermatic veins. The left spermatic vein (x) empties into the left renal vein (n), whereas the right spermatic vein (t) drains into the inferior vena cava (g).
From Vesalius, A. *Fabrica.* 1543, p. 374.

Human Anatomic Variation

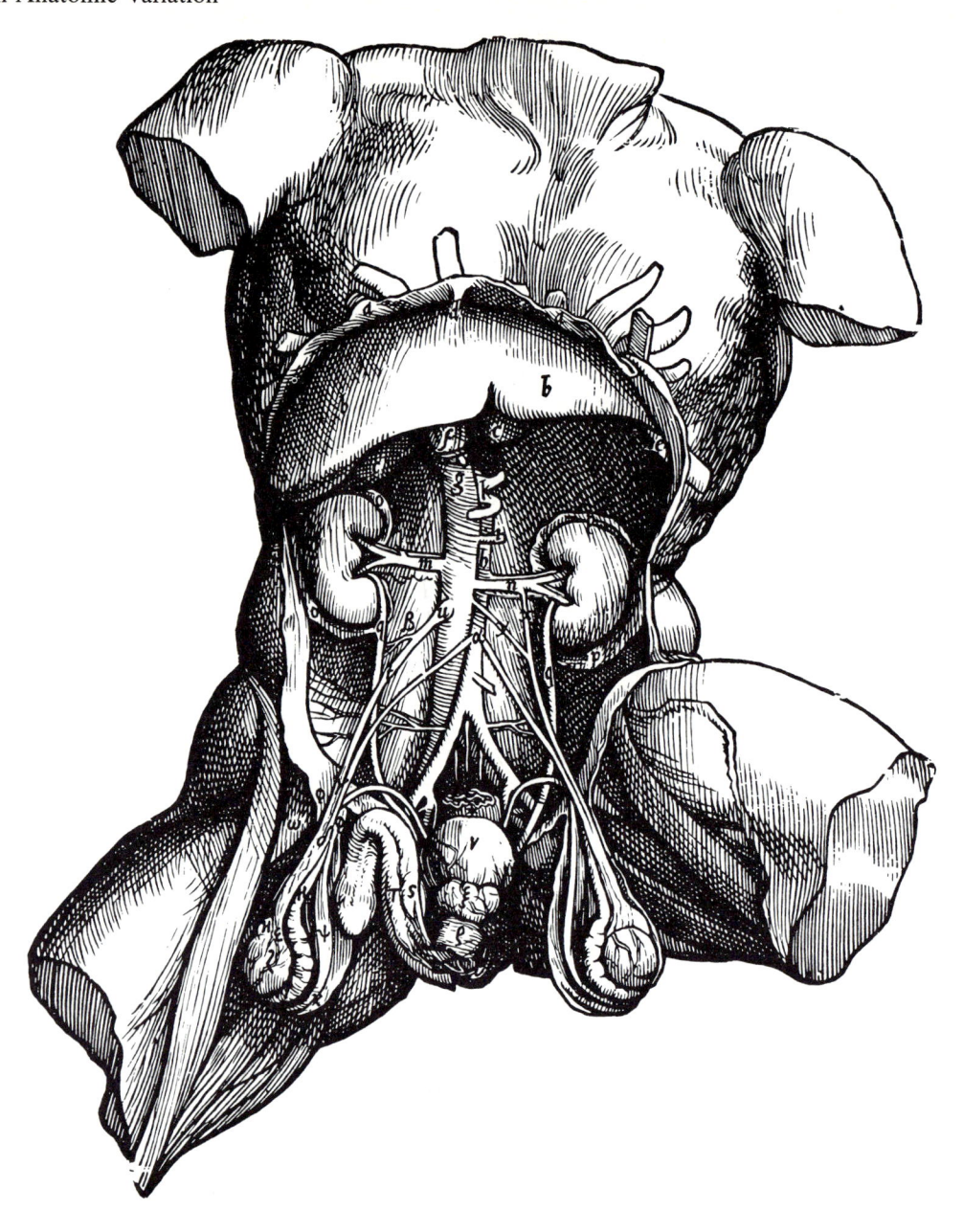

FIG. 142.
Gonadal Veins.

The right internal spermatic vein (*t*) is as usually found. The left internal spermatic vein, however, empties by two branches (*x,y*), into the left renal vein and the inferior vena cava, respectively.

From Vesalius, A. *Fabrica*, 1543, p. 372.

Cardiovascular System

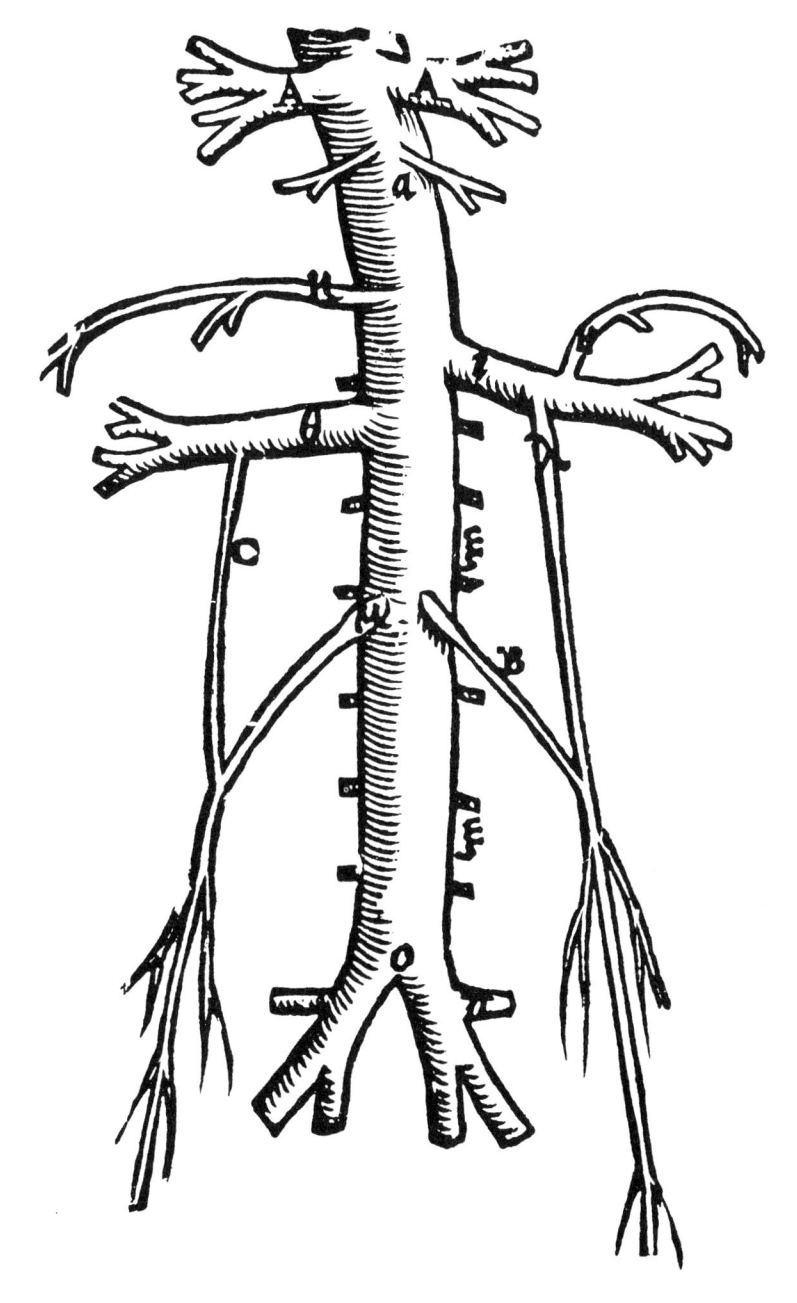

FIG. 143.
Gonadal Veins.
Inferior vena cava, showing the two internal spermatic veins each of which ends as two branches entering the inferior vena cava and renal vein, respectively (β and λ on left side, μ and C on right side).
From Vesalius, A. *Fabrica*, 1543, p. 289.

435

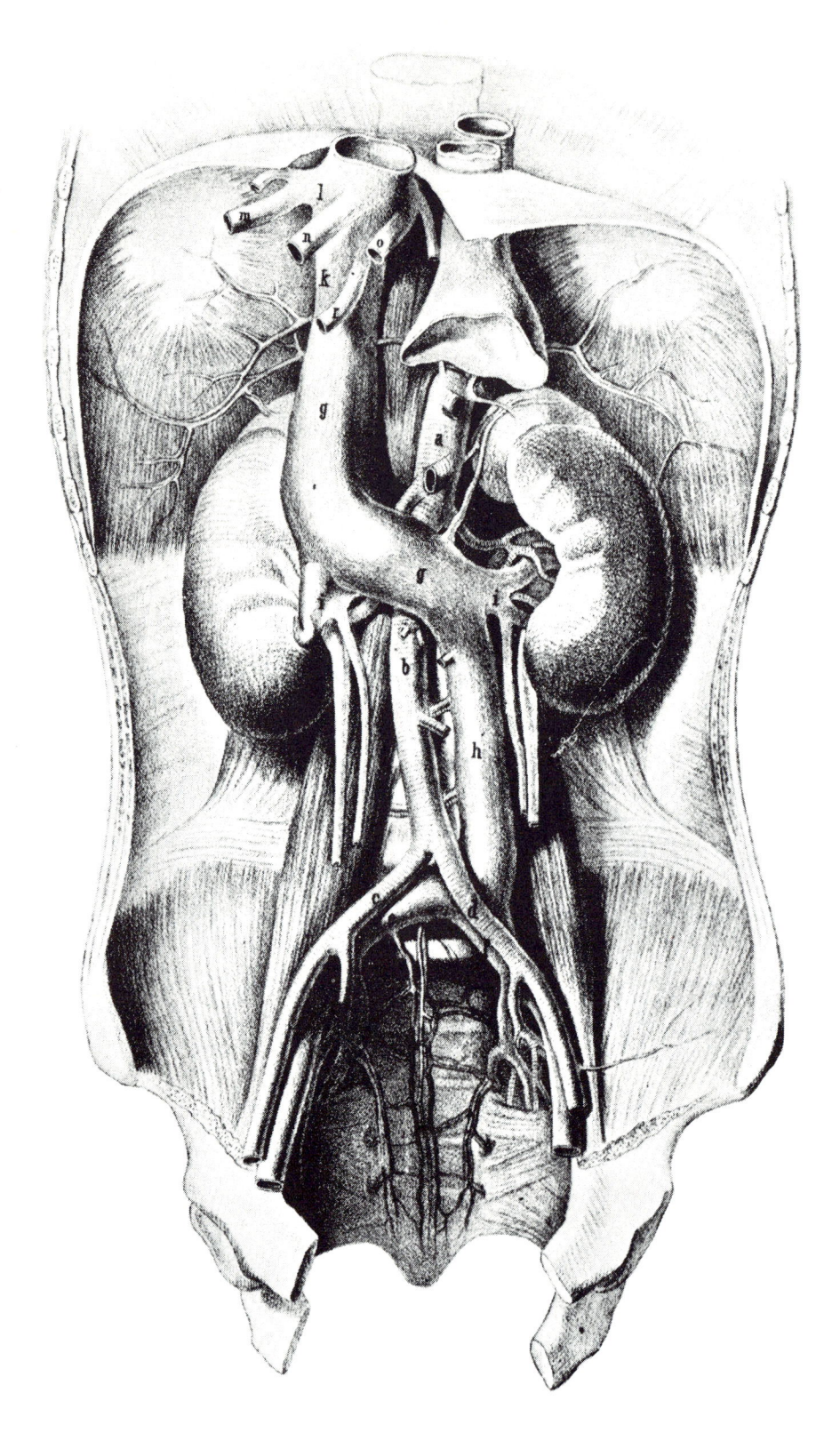

Human Anatomic Variation

FIG. 144.
Transposition of Inferior Vena Cava to Left Side Without Transposition of Viscera.

a, Abdominal aorta in normal position; *b*, portion of aorta shifted to left; *c*, common iliac artery (right); *d*, common iliac artery (left); *e*, right common iliac vein (very long); *f*, left common iliac vein (very short); *g*, dilated part of inferior vena cava; *h*, ascending inferior vena cava on left side of vertebral column; *i*, renal veins; *k*, ascending vena cava, hepatic portion; *l,m,n,o,r*, termination of hepatic veins.

From Calori, L. Delle anomalie più importanti di ossa, vasi, nervi, e muscoli occorse nell'ultimo biennio facendo anatomia del corpo umano. *Mem. Accad. Sci. Istituto di Bologna S. 2* 8:417–482, 1868.

FIG. 145.

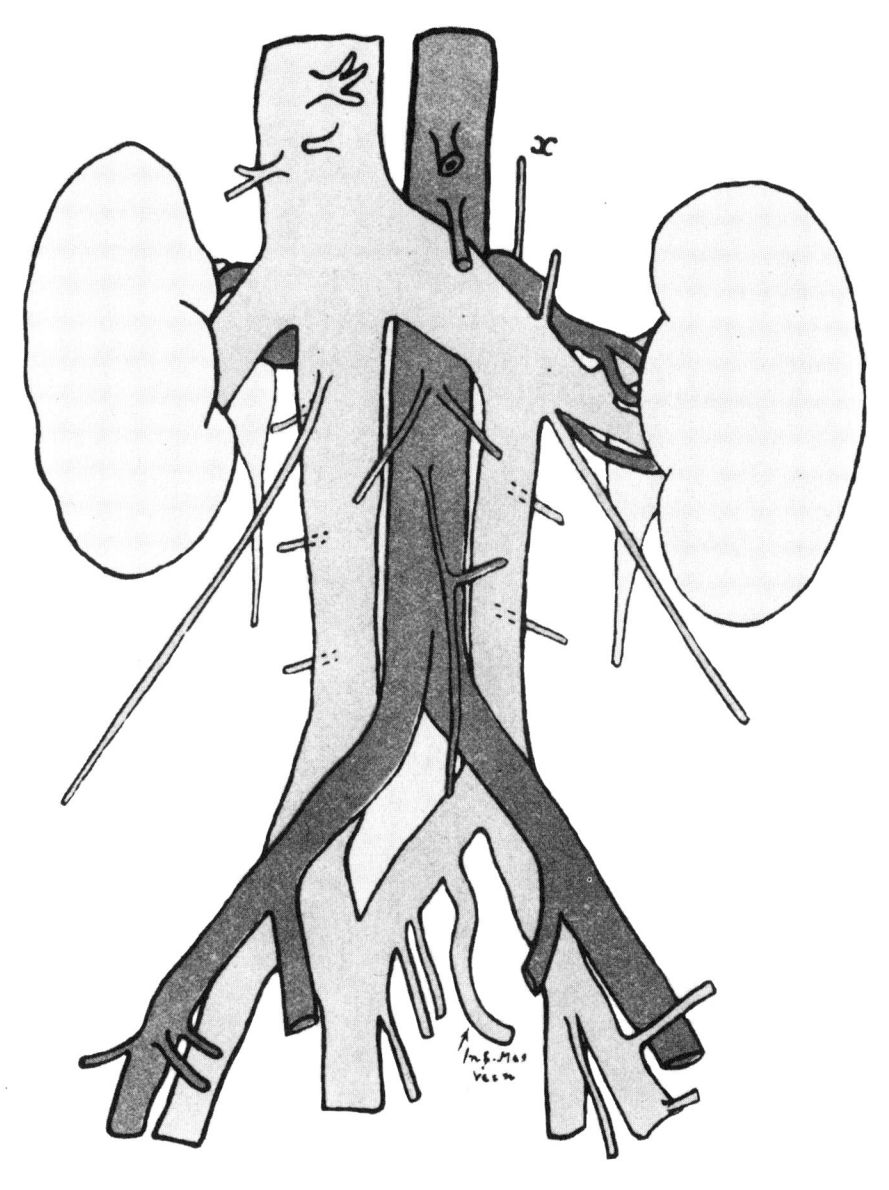

Duplication of Postrenal Segment of Inferior Vena Cava.

From Rischbieth, H. Anomaly of the inferior vena cava: Duplication of the post-renal segment. *J. Anat. Physiol.* 48:287–292, 1914.

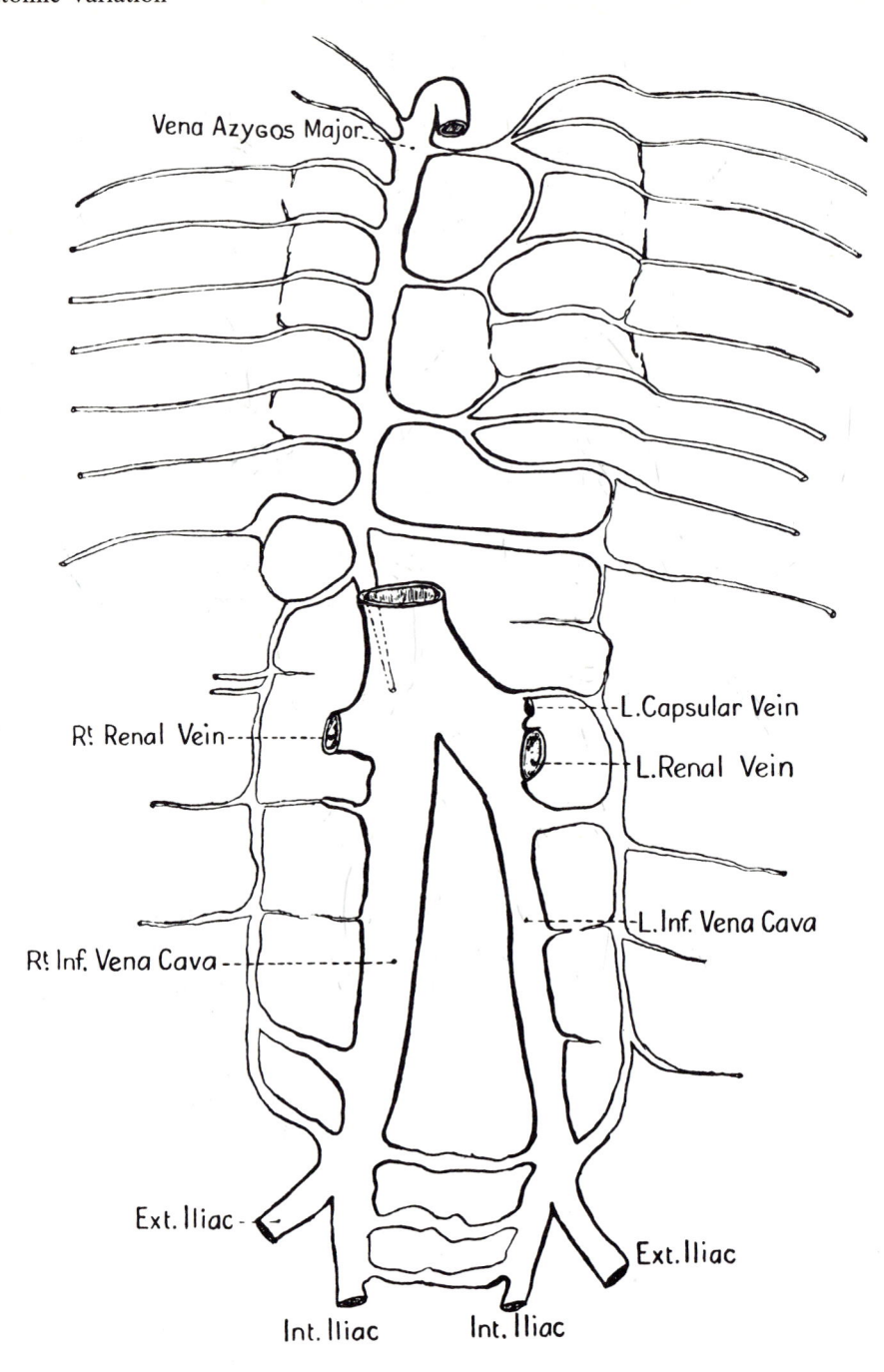

FIG. 146.
Double Inferior Vena Cava.

From Lucas, M.F. A case of double inferior vena cava. *J. Anat. Physiol.* 51:69–70, 1917.

Cardiovascular System

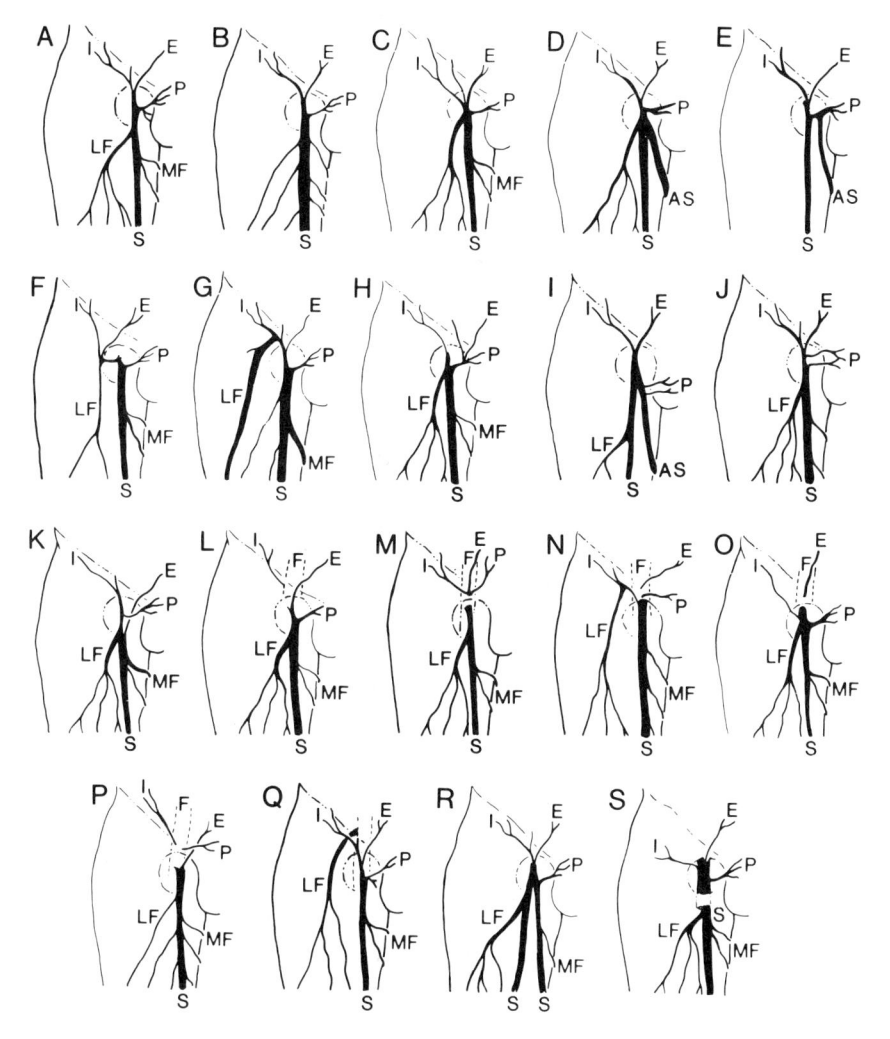

FIG. 147.
Venous Drainage at Fossa Ovalis.

I, Superficial circumflex iliac vein; *E*, superficial epigastric vein; *P*, superficial external pudendal vein; *LF*, lateral superficial femoral vein; *MF*, medial and superficial femoral vein; *AS*, accessory saphenous vein; *SS*, double saphenous vein; *S*, vena saphena magna; *F*, femoral vein.

A: Average "textbook" diagram of venous drainage at fossa ovalis. Incidence, 37%.

B: Multiple divisions of the medial and lateral femoral veins of small caliber. Incidence, 6%.

C: The large lateral superficial femoral vein drains into the fossa ovalis. The inconstant thoracoepigastric vein drains into the vena saphena magna instead of into the femoral vein. Incidence, 2%.

D: The lateral superficial femoral and the accessory saphenous vein drain into the fossa ovalis. Incidence, 2%.

E: The accessory saphenous vein forms a common stem with the superficial external pudendal vein before joining the vena saphena magna. Incidence, 6%.

F: A common trunk formed by the lateral superficial femoral, superficial circumflex iliac, and superficial epigastric veins drains into the fossa ovalis. Incidence, 9%.

G: A common trunk formed by the lateral superficial femoral and the superficial circumflex iliac vein drains into the fossa ovalis. Incidence, 9%.

H: The superficial epigastric and the superficial external pudendal vein form a common trunk. A large lateral superficial femoral vein is present. Incidence, 2%.

I: An accessory saphenous vein is present. Note the drainage of double superficial external pudendal veins. Incidence, 1%.

J: Double superficial external pudendal veins drain into the fossa ovalis. Incidence, 3%.

K: The superficial epigastric vein drains into the vena saphena magna below the fossa ovalis. Incidence, 3%.

L: The superficial circumflex iliac vein drains into the femoral vein. Incidence, 1%.

M: All high collateral veins drain directly into the femoral vein. Incidence, 6%.

N: The lateral femoral and the superficial circumflex iliac vein form a common trunk. The other high collateral veins drain directly into the femoral vein. Incidence, 1%.

O: The lateral femoral vein drains into the fossa ovalis. The superficial epigastric vein drains directly into the femoral vein. Incidence, 6%.

P: Note the small caliber multiple medial and lateral superficial femoral veins. The superficial circumflex iliac and the superficial external pudendal veins drain directly into the femoral vein. Incidence, 1%.

Q: The lateral superficial femoral vein drains directly into the femoral vein. Incidence, 1%.

R: A double vena saphena magna with joining at the fossa ovalis. Incidence, 3%.

S: The saphena magna pierces the deep fascia to enter the femoral vein about 1 inch below the fossa ovalis. Incidence, 1%.

Redrawn from Glasser, S.T. An anatomic study of venous variations at the fossa ovalis. *Arch. Surg.* 46:289–295, 1943.

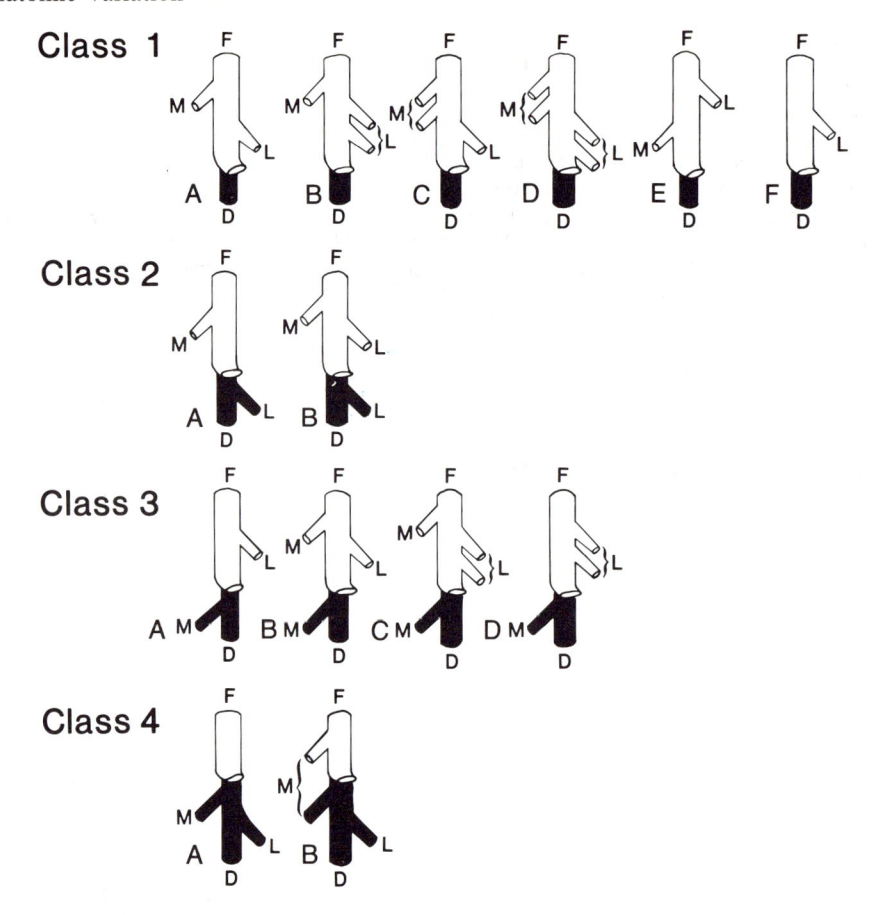

Human Anatomic Variation

Class 1

A B C D E F

Class 2

A B

Class 3

A B C D

Class 4

A B

FIG. 148.
Circumflex Veins of the Thigh: Modes of Termination
(Based on a Study of 279 Cadavers, 541 Specimens).

D, Deep femoral vein (shaded black); F, femoral vein; L, lateral femoral circumflex vein; M, medial femoral circumflex vein.
The deep femoral vein empties into the femoral vein about 8 cm distal to the inguinal ligament.
The authors make the following observations based on their study (the classes are here referred to as I,II,III, and IV):

Observations

"It was found by Charles et al. [reference 226] that the modes of termination might be placed in one of four classes, and the same method of classification has been followed in this paper although the number of types has increased with the increase in the number of observations.
Classes. Class I consists of those cases in which the circumflex veins terminate directly in the femoral vein [see figure]. In class II the lateral circumflex vein terminates in the deep femoral, while the medial circumflex vein ends directly in the femoral. The reverse is true of class III, the medial circumflex terminating in the deep femoral, and the lateral circumflex vein emptying into the femoral itself. In class IV both the medial and lateral circumflex veins terminate in the deep femoral vein.
Types. The several types composing each class permit classification as to the relative position, duplicity or absence of one or other of the circumflex veins.

Discussion

It was noted in an accompanying table 1 [not provided here] that class I, in which the circumflex veins terminate directly in the femoral, includes 86.14 per cent of the 541 observations. This is in contradiction to the statements made in textbooks which describe the circumflex veins as terminating in the deep femoral (our class IV). In the present series, only 1.85 per cent of the veins ended in that manner.
Of the six types making up class I, type IA constitutes 64.87 per cent of all observations. As is shown in the diagram, there are in this type single medial and lateral circumflex veins, both terminating in the femoral, the medial circumflex vein being the more proximal.
Next in frequency is type IB (12.57 per cent), which is quite similar to IA with the exception that there are two lateral circumflex veins distal to a single medial circumflex vein, all of which empty directly into the femoral. Type IC (3.70 per cent) has two medial circumflex veins proximal to a single lateral circumflex vein.
In all three of the above types (IA, IB, and IC) the termination of the medial circumflex vein or veins is proximal to that of the lateral circumflex vein. Since these three types together make up 81.14 per cent of all the cases observed, it may be concluded that in a large majority of the cases the termination of the lateral circumflex is distal to that of the medial circumflex vein. Type ID is an interesting example of this tendency. In this type there are two medial circumflex veins and two lateral circumflex veins and although the two medial circumflex veins are not both proximal to the two lateral circumflex veins, each medial circumflex is proximal to its corresponding lateral circumflex. In addition, in classes other than class I, where the circumflex veins do not characteristically end in the femoral, there is still a tendency for the medial circumflex to end above the lateral circumflex.
In type IE (2.96 per cent) we find for the first time a change in this characteristic relationship of the medial and lateral circumflex veins, the lateral circumflex being proximal to the medial circumflex. In type IF (0.92 per cent) the medial circumflex vein is absent.

Classes II and III constitute together but 11.46 per cent of the observations, class III predominating slightly (6.28 per cent as compared with 5.18 per cent).

Class II is made up of two types, the first and predominating (IIA, 4.07 per cent) having a single medial circumflex emptying into the femoral, and a single lateral circumflex emptying into the deep femoral. Type IIB (1.11 per cent) differs from IIA only in having an accessory lateral circumflex ending in the femoral, just distal to the medial circumflex. Here again in both these types the position of the medial circumflex is characteristically proximal to that of the lateral circumflex vein.

There are four types making up class III, type IIIA being found most frequently (3.51 per cent). This type has a single medial circumflex terminating in the deep femoral, while the lateral circumflex terminates superiorly in the femoral. Type IIIB differs from IIIA only in having an accessory medial circumflex emptying into the femoral itself, immediately proximal to the lateral circumflex vein. Again there is but a small variation that distinguishes IIIB from IIIC, namely, the presence of two adjacent lateral circumflex veins rather than a single vein. Type IIID has duplicate lateral circumflex veins emptying into the femoral and a single medial circumflex emptying into the deep femoral. These two last-named types are only rarely found, representing but 0.74 per cent and 0.37 per cent of the present series, respectively.

As mentioned above, in only 1.85 per cent of the cases do both the lateral and medial circumflex veins end in the deep femoral (class IV). The most common type in this class is type IVA which has a single medial circumflex and a single lateral circumflex, both emptying into the deep femoral. The other type IVB, has an accessory medial circumflex ending in the femoral, and was found in only one case."

Redrawn from Baird, R.D. and J.S. Cope. On the termination of the circumflex veins of the thigh and their relations to the origins of the circumflex arteries. *Anat. Rec.* 57:325–337, 1933.

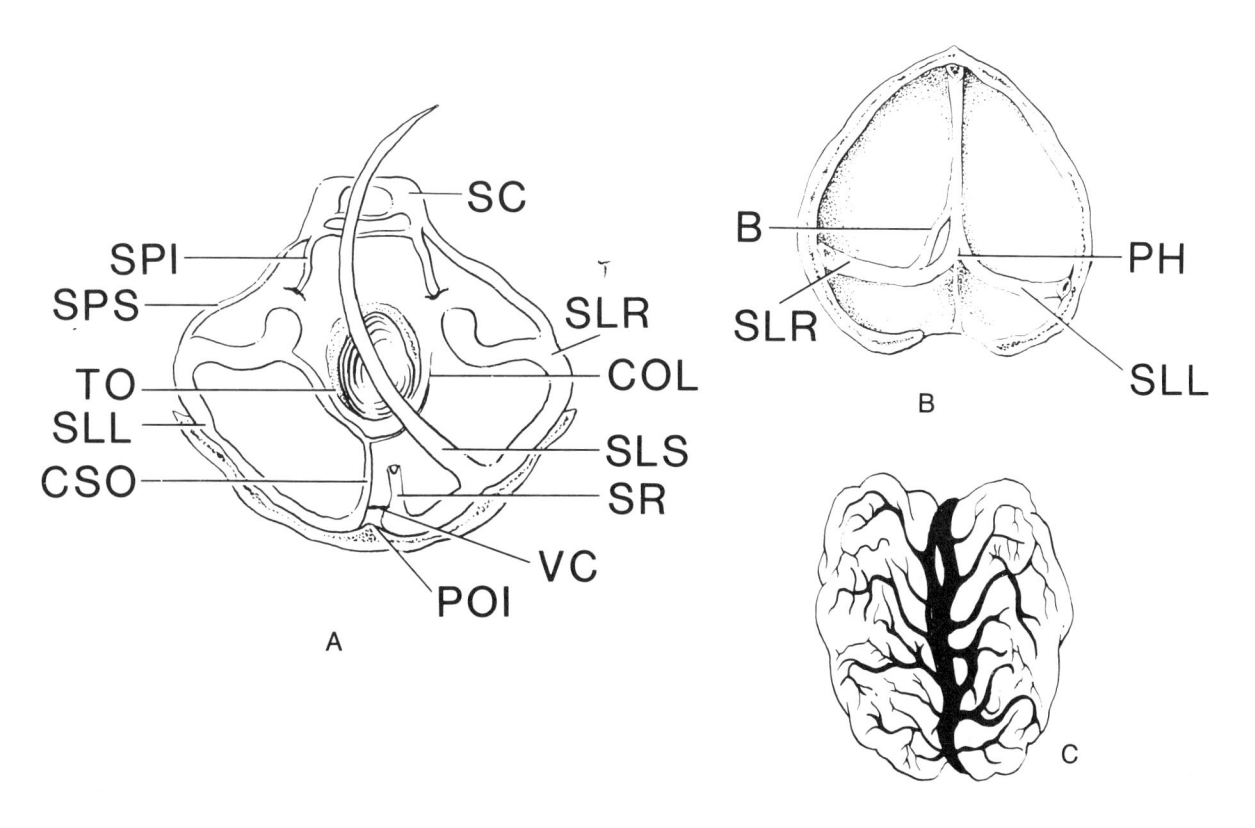

FIG. 149.
Anomalies of Dural Sinuses.
Absence of Confluens of Dural Sinuses.
Bifurcation and Reunion of Superior Sagittal (Longitudinal) Sinus.
Brief Doubling of Superior Sagittal (Longitudinal) Sinus.

A: Absence of confluens (after Hallett). *COL*, Communication between the right occipital sinus and the sigmoid sinus (on the same side) is absent; *CSO*, common canal of the occipital sinuses; *POI*, internal occipital protuberance; *SC*, cavernous sinus; *SLL*, left transverse sinus; *SLR*, right transverse sinus; *SLS*, superior sagittal sinus; *SPI*, inferior petrosal sinus; *SPS*, superior petrosal sinus; *SR*, sinus rectus (straight); *TO*, foramen magnum; *VC*, vein (occipital emissary) communicating to region of confluens (rare observation).

B: Bifurcation of superior sagittal sinus. *B*, Bifurcating branch of superior sagittal sinus; *PH*, confluens of sinuses (torcular herophili); *SLL*, left transverse sinus; *SLR*, right transverse sinus.

C: Bifurcation of superior sagittal sinus (after Vicq d'Azyr).

Redrawn from Labbé, C. Anomalies des sinus de la dure-mere. Dévelopment de ces sinus. Considérations sur la suppléance réciproque de ces canaux veineux dans les cas d'absence de l'un d'eux description de quelques sinus peu connus. *Arch. Physiol. S. 3* 1:1–27, 1883.

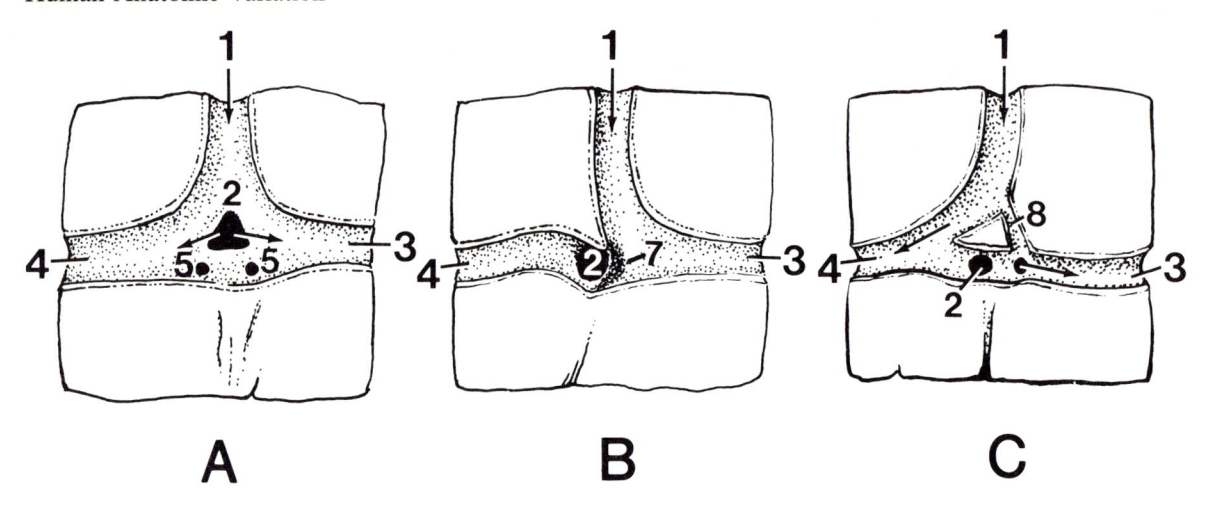

Human Anatomic Variation

FIG. 150.
Various Types of Confluens of Sinuses
(Posterior View or Convex Face of the Dura Mater).

1, Superior sagittal sinus; *2*, right straight sinus; *3*, right transverse sinus; *4*, left transverse sinus; *5*, occipital sinus; *7*, fibrous partition; *8*, small anastomotic branch from superior sagittal sinus which empties into the right transverse sinus.
A: Confluens of the dural venous sinuses.
B: The superior sagittal ends in the right transverse sinus. The straight sinus empties into the left transverse sinus.
C: The superior sagittal sinus continues in large measure into the left transverse sinus. A small branch from the superior sagittal sinus empties into the right transverse sinus as does some of the drainage from the straight sinus.
Redrawn after Dumont, in Latarjet, A. Testut, L. *Traité d'Anatomie Humaine*, G. Doin & Cie., Paris, 1929.

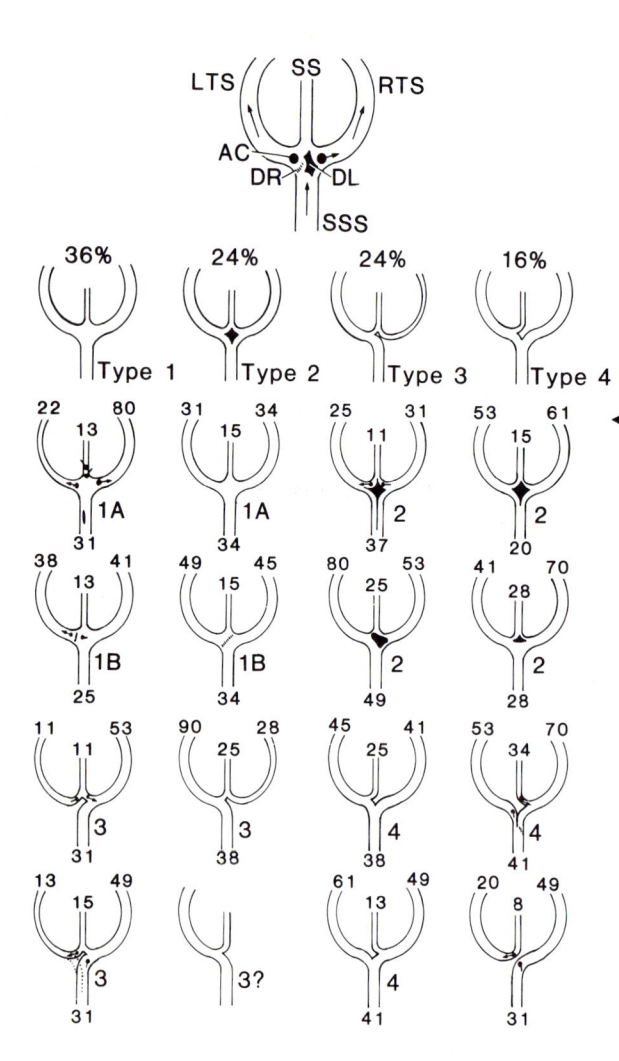

FIG. 151.
Confluens of Sinuses: Types and Subtypes

AC, Anastomotic channel; *DL*, apposition of dural layers; *DR*, projecting dural ridges; *LTS*, left transverse sinus; *RTS*, right transverse sinus; *SS*, Straight sinus; *SSS*, superior sagittal sinus.

Conventionalized diagrams of the 4 major types of sinus system. Explanation of the conventional representation for all figures is given in the upper center hypothetical diagram, where the arrows indicate the direction of blood flow in life and of fluid flow under experimental conditions.

Type 1 (1A,1B): Conventionalized diagrams of variations in sinus systems of subtypes 1A and 1B. Figures at the cut extremities of the sinuses are cross-sectional areas in square millimeters.

Type 2 (2): Conventionalized diagrams of variations in sinus systems of type 2.

Type 3 (3): Conventionalized diagrams of variations in sinus systems of type 3. Dotted lines in lower left figure represent a partially obliterated channel.

Type 4 (4): Conventionalized diagrams of variations in sinus systems of type 4.

Redrawn from Browning, H. The confluence of dural venous sinuses. *Am. J. Anat.* 92:307–329, 1953.

Cardiovascular System

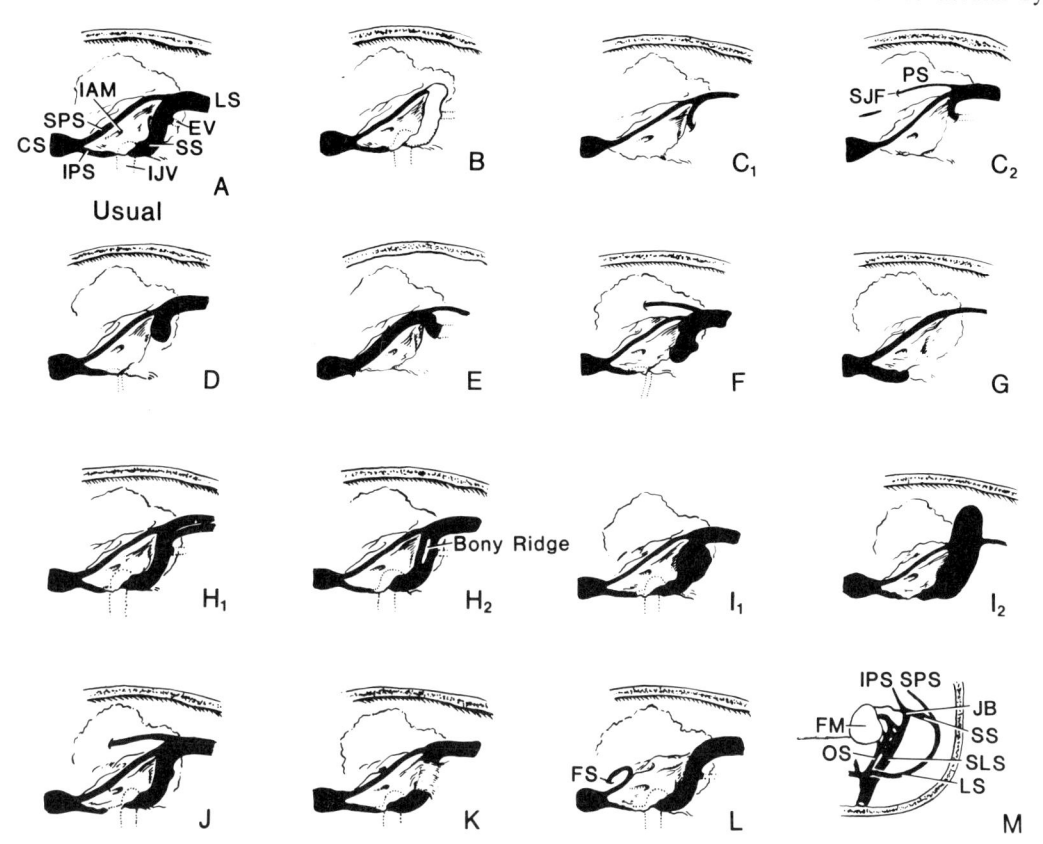

FIG. 152.
Variations in Lateral and Sigmoid Sinuses.

"Anatomic variations of the sinuses of the dura mater, however infrequent, may present puzzling diagnostic and operative problems in the presence of thrombophlebitis. Some of the variations are extremely rare. In this paper a hitherto unknown anomaly of the sigmoid sinus will be reported and a short description of most of the published observations which are important for the otologist will be given. In view of the existing differences in the nomenclature, in this paper the horizontal portion of the transverse sinus will be called the lateral sinus and the vertical portion will be called the sigmoid sinus."

Abbreviations: CS, Cavernous sinus; EV, mastoid emissary vein; FM, foramen magnum; FS, foramen spinosum; IAM, internal auditory meatus; IJV, internal jugular vein; IPS, inferior petrosal sinus; JB, jugular bulb; LS, transverse sinus; OS, ophthalmopetrosal sinus; PS, petrosquamosal sinus; SJF, spurious jugular foramen; SLS, superior sagittal sinus; SPS, superior petrosal sinus; SS, sigmoid sinus.

A, Diagrammatic representation of the normal sinus region (intracranial view). Other drawings, from the same view, show anatomic variations of venous sinuses: B, lateral sinus absent; C₁, small transverse sinus leaving endocranium through mastoid foramen; C₂, persistent petrosquamous sinus present; D, petrous bone infantile; E, superior petrosal sinus passing through mastoid foramen; F, sigmoid sinus ending in a blind pouch; G, complete absence of sigmoid sinus.

Diagrammatic representations of the other anomalies of the region of the sinuses (intracranial view). H₁, H₂, duplications of lateral and sigmoid sinuses; I, hernia-like bulging of outer wall: (1) in upper part of region of knee and (2) with jugular bulb missing; J, persistent petrosquamous sinus draining lateral sinus; K, cortical layer thinned to paper-like sheet by enlarged sigmoid sinus; L, superior petrosal sinus turning downward into foramen spinosum of middle fossa; M, superior longitudinal sinus continuing directly with jugular bulb.

Waltner suggests "a short outline of the development of the sinuses will help to explain the genesis of the variations in question. The sinuses develop from three venous plexuses, which in the primary type of circulation are drained by the 'head vein' (Streeter[1]) [references listed numerically at end]. The 'head vein' runs along side the brain tube. In early embryonic life the primitive anterior and middle venous plexuses of the brain drain mainly into the vena capitis lateralis, which leaves the endocranium through the spurious jugular foramen and drains into the system of the external jugular vein. The spurious jugular foramen is located above the jaw joint and is usually obliterated in fully developed skulls. The posterior venous plexuses drain into the internal jugular vein, which gives origin to the sigmoid sinus. Later the sigmoid sinus becomes connected with the anterior and middle plexuses by means of a horizontally running channel which represents the lateral sinus. The vena capitis lateralis is obliterated (Labbé[2]).

In most mammals, both before and after birth, the venous blood of the brain is drained almost entirely through the spurious jugular foramen into the external jugular vein; this venous structure seems to be phylogenetically the older one. The internal jugular vein plays only a secondary role in the drainage. In man and in some anthropoid monkeys the main venous channel of the endocranium is the internal jugular vein.

An incomplete or missing connection between these two types of venous plexuses or persistence of the primitive venous channels explains most of the anomalies. A large number of persistent, patent spurious jugular foramina were reported by different authors: first by Krause,[3] in 1842, and then by Hyrtl,[4] Knott[5] (in 26 cases), Luschka,[6] Cheatle[7] (in 23 of 2,585 skulls examined) and others. This fact confirms the embryologic data and suggests a more frequent occurrence of the primitive, persistent petrosquamous sinus which passes through that foramen than is commonly assumed. Cheatle found 3 spurious jugular foramina located in the glenoid fossa of the temporomandibular joint, 3 in the zygomatic process, 6 at the base of the malar bone, and 11 close to the glaserian fissure. It may be mentioned here that studies of the skulls alone may lead to faulty conclusions about the venous sinuses, because frequently a small sinus is located in a large groove of the skull or a large sinus in a small groove.

I found few or no data about anatomic variations of the venous sinuses in textbooks of anatomy. In the following paragraph many of these variations important to the otologist will be discussed:

1. A contracted jugular foramen with a small internal jugular vein was found by Linser[8] in 30 of 1,022 skulls examined. The contralateral jugular vein sometimes was four times as large in such cases. A contracted jugular foramen was five times more frequent on the left side. Swift[9] reported a case in which both jugular foramina of a newborn child were rudimentary; the infant died three days after birth.

2. Absence of the lateral sinus is a rare anomaly and usually occurs on the left side. A few cases were reported by Lieutaud,[10] Hallett,[11] and Furstenberg.[12] The superior petrosal sinus was present in these cases, and continued directly into the sigmoid sinus. This anomaly occurs if the internal jugular vein is not connected or is only insufficiently united with the anterior and middle venous plexuses of the brain. The sigmoid sinus, which develops from the internal jugular vein, is present, while the lateral sinus is either absent or is filiform (fig. B).

3. A small transverse sinus leaving the endocranium through the mastoid foramen was observed by Knott[5] in 2 cases; the sinus was 1.5 mm. in diameter (fig. C). In Hoople's[13] case the lateral sinus measured 3 mm. and continued with the mastoid emissary vein just below the superior knee of the sigmoid sinus. A persistent petrosquamous sinus was present (fig. C_2).

4. A normal lateral sinus leaving the skull through an enlarged mastoid foramen was reported by Malacarne[14]; it was a bilateral anomaly and the emissary foramen measured 15 by 10 mm. In Laff's[15] case the lateral sinus abruptly curved downward into a large mastoid foramen; the jugular foramen was small, the jugular bulb was absent and the inferior petrosal sinus emptied into a small jugular vein. The whole petrous bone was infantile and underdeveloped, although the patient was 35 years old (fig. D). In Barkow's[16] case the sigmoid sinus was missing between the large mastoid foramen and the jugular fossa.

5. The sigmoid sinus absent, the lateral sinus threadlike, and the large superior petrosal sinus passing through the mastoid foramen were reported in a case of Williams.[17] There is no information about the jugular bulb and inferior petrosal sinus because the anomalies just mentioned were discovered at operation. Ten years later Williams and Hallberg[18] reported a case in which the operation revealed no venous sinuses on the right side (fig. E).

6. A sigmoid sinus ending in a blind pouch and draining through a large mastoid foramen and a persistent petrosquamous sinus were reported by Furstenberg[12] (fig. F).

7. Complete absence of the sigmoid sinus with a large inferior petrosal and a narrow lateral sinus has already been described (fig. G).

8. Duplication of the lateral sinus is a more frequent anomaly. Hahn[19] and Streit[20] collected 12 cases of duplicated lateral sinus. Streit and Brown[21] each reported 1 case of duplicated sigmoid sinus. The two channels sometimes are separated by a bony ridge; at other times a fibrous septum divides the lumen of the sinus for a shorter or a longer distance (figs. H_1 and $[H]_2$). The change in position of the dural veins in the fetus is accomplished by the formation of a collateral vein which persists while the substituted primary one is obliterated. Duplication is due to the persistence of both these veins.

9. Hernia-like bulging of the outer wall of the sinus, especially in the region of the upper part of the knee, occurs with relative frequency (fig. I_1). Ruttin[22] described the following variation: A huge sigmoid sinus ended in a blind sac in the squamous portion of the temporal bone. In addition, the jugular bulb was missing and a small vein connected the enlarged sigmoid sinus with the internal jugular vein. The superior petrosal sinus and the thin lateral sinus drained into the sigmoid sac at the usual location (fig. I_2).

10. A persistent petrosquamous sinus draining the lateral sinus is a more frequent anomaly. Its course shows great variations. Usually it crosses the ridge of the petrous bone, follows the course of the petrosquamous fissure, and leaves the middle cranial fossa either through the spurious jugular foramen (fig. J) or with the middle meningeal vein through the foramen spinosum. Knott[5] found 7 bilateral and 19 unilateral persistent petrosquamous sinuses in 44 heads examined. Zuckerkandl[23] reported a 5 mm wide petrosquamous sinus. Vernieuwe[24] found a more or less developed petrosquamous sinus in 12 of 60 fresh heads examined. Laff[15] reported 3 and Streit[20] reported 2 such anomalies, most of them with normal lateral and sigmoid sinuses. However, the petrosquamous sinus may be combined with an anomaly of the sigmoid sinus, as in figure F. Veins of the middle ear may drain directly into the petrosquamous sinus through the petrosquamous fissure. Cheatle[7] reported 2 cases in which the infection spread from the middle ear to the lateral sinus by this short route of the petrosquamous sinus, which lies just above the roof of the middle ear and the antrum. This venous communication may give the explanation of some of the early septic symptoms in acute otitis media, if no pathologic condition is found in the mastoid bone.

11. The sinus may lie directly under the periosteum, or the cortical layer may be thinned to a paper-like sheet by an enlarged sigmoid sinus (Bezold[25]) (fig. K). Dehiscences and perforations of the cortical layer are not uncommon along the course of the sinus. Furstenberg[12] described a variation in which the lateral sinus was located lateral to the skull in the soft tissues of the scalp.

The wide variations in size and location of the sigmoid sinus in the mastoid process are well known to the otologist. It may be, however, of a certain interest to mention those anatomic variations of the outer venous sinuses with which one frequently deals during operations.

1. Absence of the superior petrosal sinus was reported in a few cases (Knott,[5] 2; Hyrtl,[4] 1); in Streit's[20] case the superior petrosal sinus turned downward into the foramen spinosum of the middle fossa (fig. L).

2. The superior longitudinal sinus continued directly with the jugular bulb in a case of Streit (fig. F).

3. The inferior petrosal sinus drained into the superior thyroid vein in a case reported by Theile.[26] Major variations of the inferior petrosal sinus are extremely rare.

4. Absence of the mastoid foramen and several foramina have been frequently reported. According to Coudert,[27] the mastoid foramen is relatively better developed in children than it is in adults. Occasionally the mastoid emissary vein crosses the mastoid cells in a long bony canal, and an isolated thrombophlebitis of the emissary vein may occur. In many cases the emissary vein was found lying on the dura of the posterior fossa for variable distances. Both the internal and external openings of the mastoid foramen show the widest variations in location. A direct connection between the emissary vein and the petrosquamous sinus was described by Otto.[28]

5. The ophthalmopetrosal sinus connects the ophthalmic vein with the superior petrosal or the lateral sinus in 8 to 9 per cent of the cases, according to Knott[5] and Hyrtl.[4]

Comment

Variations of the lateral and sigmoid sinuses are independent of each other, because they are developed from separate anlages; a normal sigmoid sinus or a normal lateral sinus may each be present with the other absent.

The sigmoid sinus shows greater constancy and fewer variations than does the lateral sinus. This could be explained by the fact that the lateral sinus has to adapt itself to the increasing size and changing form of the surrounding structures, for example, the brain and the otic capsule. Therefore, the lateral sinus is more likely to be interfered with in its development than is the sigmoid sinus. The latter is located close to the base of the brain from the very beginning of its development.

A knowledge of the anatomic variations is of importance in cases of thrombophlebitis, not only for determining the surgical management but for understanding unusual symptoms and signs."

Authors' note: References for Waltner's study:

1. Streeter, G.L. Contrib. Embryol. 8:5, 1918.
2. Labbé, C. Arch. Physiol. Norm. Pathol. 1:1, 1883.
3. Krause, C.F.T. Handbuch der menschlichen Anatomie, 2nd Ed. Hahn, Hannover, 1842, p. 914.
4. Hyrtl. Wien. Med. Wochenschr. 12:289, 1862.
5. Knott, J.F. J. Anat. Physiol. 16:27, 1881.
6. Luschka. Z. Rat. Med. 7:72, 1859.
7. Cheatle, A. Trans. Int. Congr. Oto., Sect. 6, p. 160, 1899.
8. Linser. Beitr. Klin. Chir. 28:642, 1900.
9. Swift, G.W. Transverse sinus and its relation to choked disk. Arch. Ophthalmol. 3:47, 1930.
10. Lieutaud, J. Essais anatomiques. P.M. Huart, Paris, 1742, p. 332.
11. Hallett. Med. Times, London 17:72, 1847.
12. Furstenberg, A.C. Trans. Am. Acad. Ophthalmol. 42:424, 1937.
13. Hoople, G.D. Ann. Otol. Rhinol. Laryngol. 45:1019, 1936.
14. Malacarne, M.V. Encefalotomia nuova universale. G. Briolo, Torino, 1780; cited by Labbé.[2]
15. Laff, H.J. Unilateral absence of sigmoid sinus. Arch. Otolaryngol. 11:151, 1930.
16. Barkow, J.C.L. Anatomische Abhandlungen. F. Hirt, Breslau, 1851, p. 1.
17. Williams, H.L., Jr. Apparent unilateral absence of sigmoid sinus noted at operation. Arch. Otolaryngol. 12:339, 1930.
18. Williams, H.L. and O.E. Hallberg. Congenital absence of cranial venous sinuses on the right. Arch. Otolaryngol. 33:78, 1941.
19. Hahn, R. Arch. Ital. Otol. 25:203, 1914.
20. Streit, H. Arch. Ohrenh. 58:85, 161, 1903.
21. Brown, J.M. Trans. Am. Laryngol. Rhinol. Otol. Soc. 27:302, 1921.
22. Ruttin, E. Acta Oto-laryngol. 9:217, 1926.
23. Zuckerkandl. Monatsschr. Ohrenh. 7:102, 1873.
24. Vernieuwe. Rev. Laryngol. 52:207, 1921.
25. Bezold. Monatsschr. Ohrenh. 7:130, 1873.
26. Theile, cited by Streit.[20]
27. Coudert, R. Anatomie et pathologie de la veine émissaire mastoidienne. Thesis, no. 492, Paris, 1900.
28. Otto, cited by Streit.[20]

Redrawn from Waltner, J.G. Anatomic variations of the lateral and sigmoid sinuses. Arch. Otolaryngol. 39:307–312, 1944.

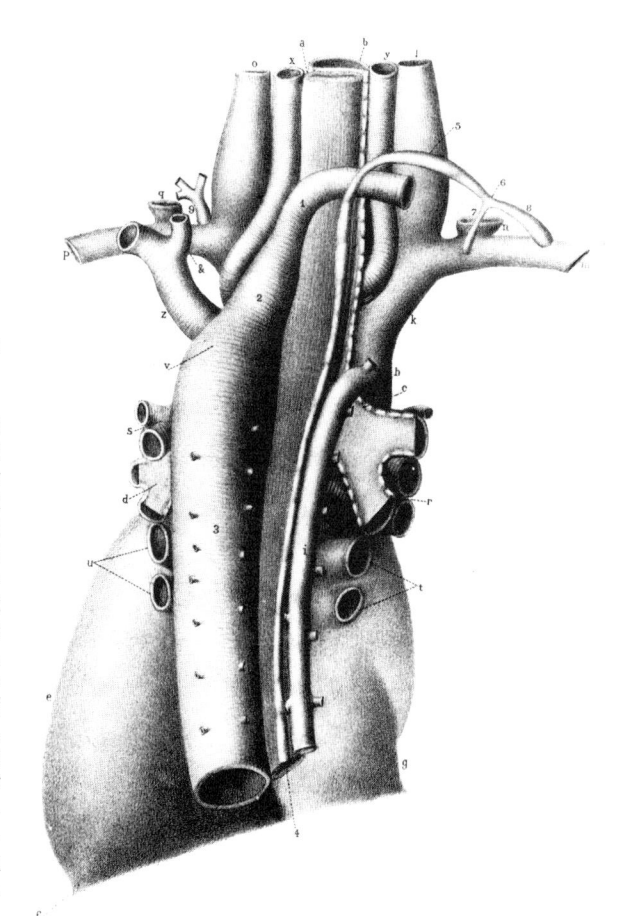

FIG. 153.
Anomalous Thoracic Duct Emptying into Right Subclavian Vein.
Retroesophageal Right Subclavian Artery.

a, Esophagus; b, trachea; c, right primary bronchus; d, left primary bronchus; e, pericardial sac; f, line indicating site of adherence of the pericardium to the diaphragm; g, inferior vena cava; h, superior vena cava; i, azygos vein; k, right brachiocephalic vein; l, right internal jugular vein; m, right subclavian vein; n, right external jugular vein; o, left internal jugular vein; p, left vein; q, left external jugular vein; r, branches of the right pulmonary artery; s, branches of the left pulmonary artery; t,u, right and left pulmonary vein; v, aortic arch at its posterior extremity; x,y, common carotid arteries; z, left subclavian artery; &, vertebral artery branch of left subclavian; 1, right subclavian artery; 2, dilated conus at origin of right subclavian artery; 3, descending (thoracic) aorta; 4, thoracic duct; 5, arch of thoracic duct; 6, division of duct into two branches; 7, minor branch of thoracic duct; 8, major branch of thoracic duct; 9, small thoracic duct on the left side.

From Calori, L. Sopra um caso d'inversione dei condotti toracici accompagnato da inverse origine dell'arteria succlavia destra e sulla gnesi delle due anomalie. Mem. R. Accad. Sci. Istituto di Bologna S. 5 1:189–196, 1890.

Human Anatomic Variation

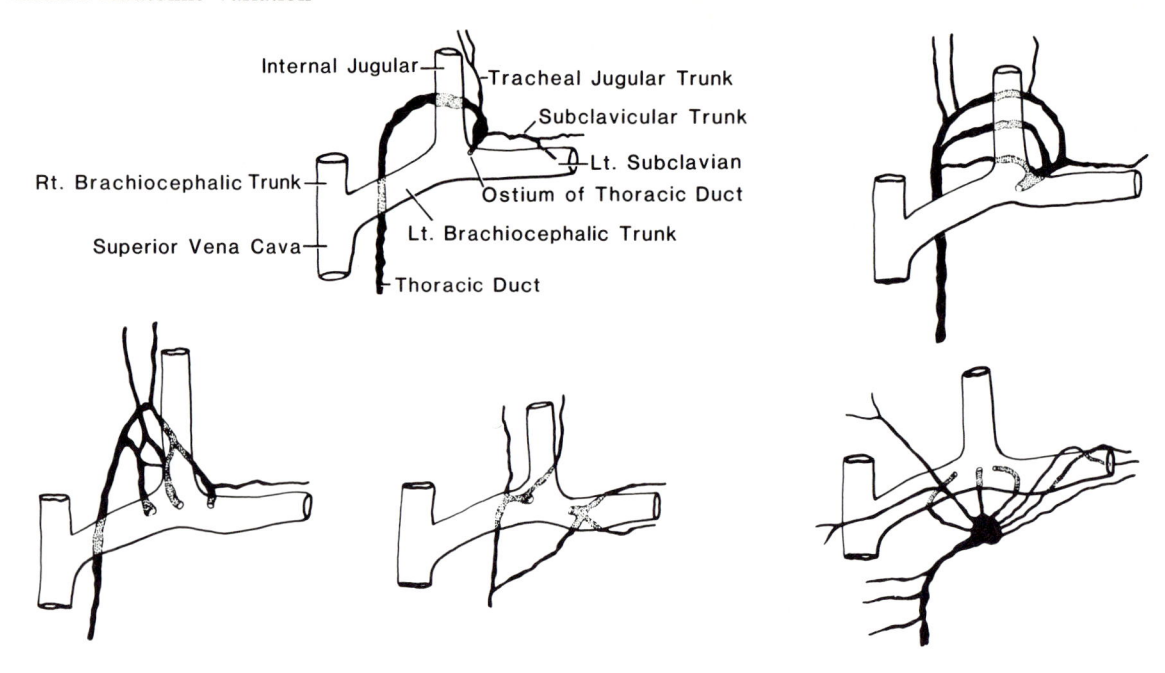

FIG. 154.
Some Types of Variation in Termination of Thoracic Duct. Redrawn from various sources.

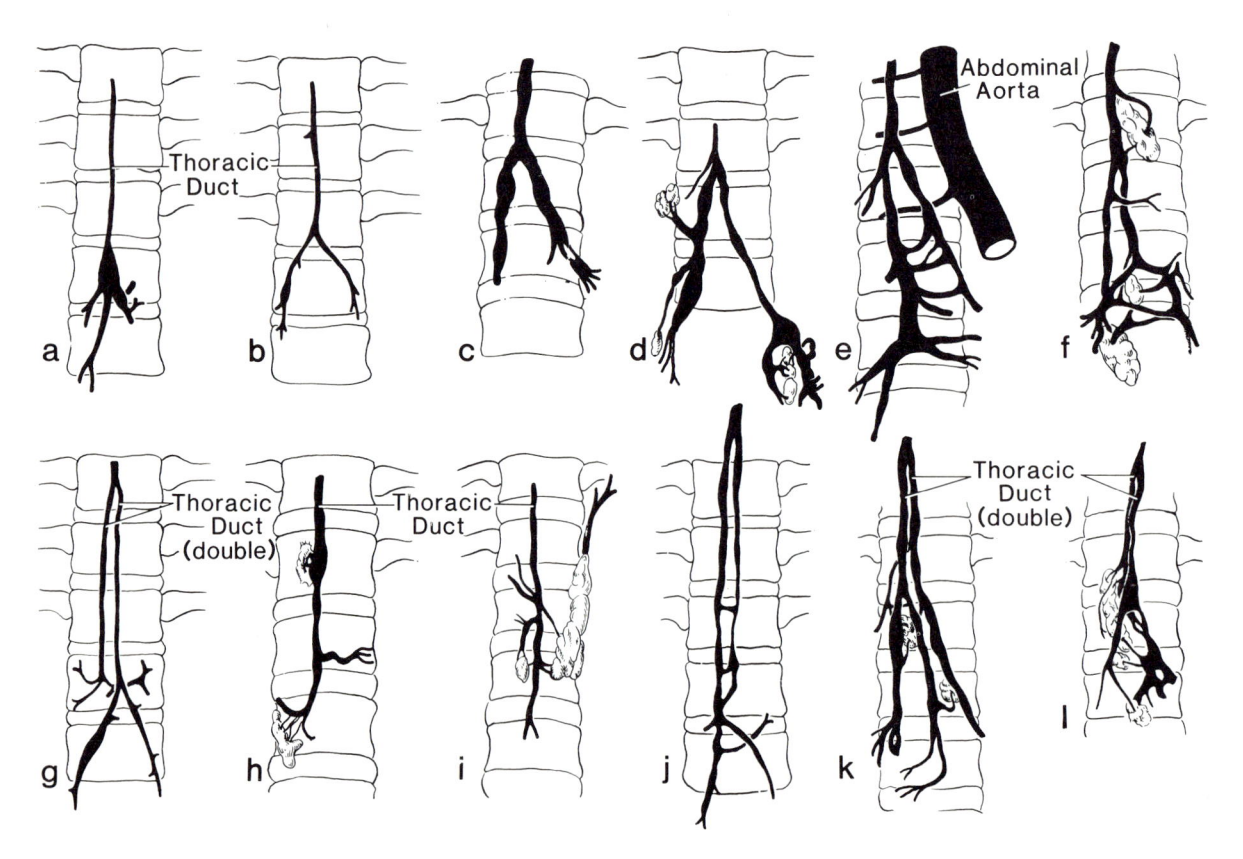

FIG. 155.
Variations in Form of Thoracic Duct.

Redrawn from Anson, B.J. *Atlas of Human Anatomy.* W.B. Saunders Co., Philadelphia, 1950.

Nervous System

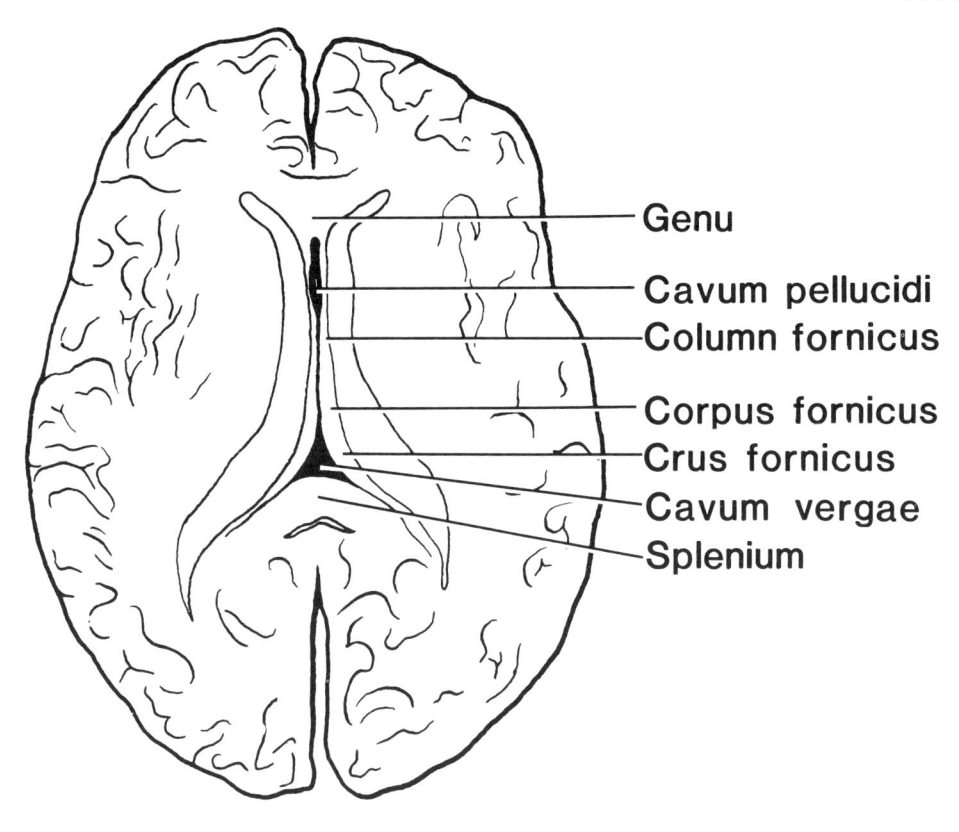

FIG. 1.
Cavum Septi Pellucidi.
Cavum Vergae.

In many older textbooks of anatomy there is a brief but accurate description of these two cavities. Dandy describes these as follows:

"The nomenclature, however, is not uniform. For example, the cavum septi pellucidi is perhaps better known as the fifth ventricle, and the cavum vergae is called Verga's ventricle, the sixth ventricle, the ventricle of Strambio, ventriculus fornicis, ventriculus triangularis, and the canal aqueduct.

Position and Boundaries of the Cavities

The corpus callosum defines the anterior, superior and posterior limits of the two cavities which, when not continuous, are separated from each other by the anterior limit of the fornix as it courses obliquely backward and upward from the anterior commissure to the body of the corpus callosum. Being of congenital origin, the two cavities are doubtless dependent on the development of the corpus callosum and the fornix. They may coexist and be isolated from each other when the fornix is intact; they may coexist and be in communication through a defect in the fornix, or they may form a single large cavity when the fornix is not attached to the corpus callosum. The cavum septi pellucidi is frequently present when the cavum vergae is absent, and Verga's cavity may be present when the cavum septi pellucidi is absent. Kauffmann stated that the cavum vergae may exist on one side of the midline and be absent on the other.

The cavum septi pellucidi has the following boundaries: anteriorly, the genu of the corpus callosum; superiorly, the body of the corpus callosum; posteriorly, the anterior limb and pillars of the fornix; inferiorly, the rostrum of the corpus callosum and the anterior commissure; laterally, the layers of the septum pellucidum. Viewed laterally, the cavum septi pellucidi is roughly triangular with the base at the corpus callosum. Viewed in cross-section the cavity is also triangular with the base at the corpus callosum.

The cavum vergae has the following boundaries: anteriorly, the anterior limb of the fornix; superiorly, the body of the corpus callosum; posteriorly, the splenium of the corpus callosum; inferiorly, the psalterium (lyra davidis) and hippocampal commissure, the fibers of which bridge the space between the diverging posterior pillars of the fornix. This cavity is also triangular when viewed from the side. The cavum vergae flares out laterally on both sides with the curve of the fornix and pushes under the lateral ventricles at its extreme lateral extensions.

In most adult brains both spaces are absent or are at most potential, but in every 100 necropsies actual cavities of varying size will be seen. Neither cavity can be regarded as part of the great ventricular system in which cerebrospinal fluid forms and through which it circulates."

From Dandy, W.E. Congenital cerebral cysts of the cavum septi pellucidi (fifth ventricle) and cavum vergae (sixth ventricle). Diagnosis and treatment. *Arch. Neurol. Psychiatry* 25:44–66, 1931.

Human Anatomic Variation

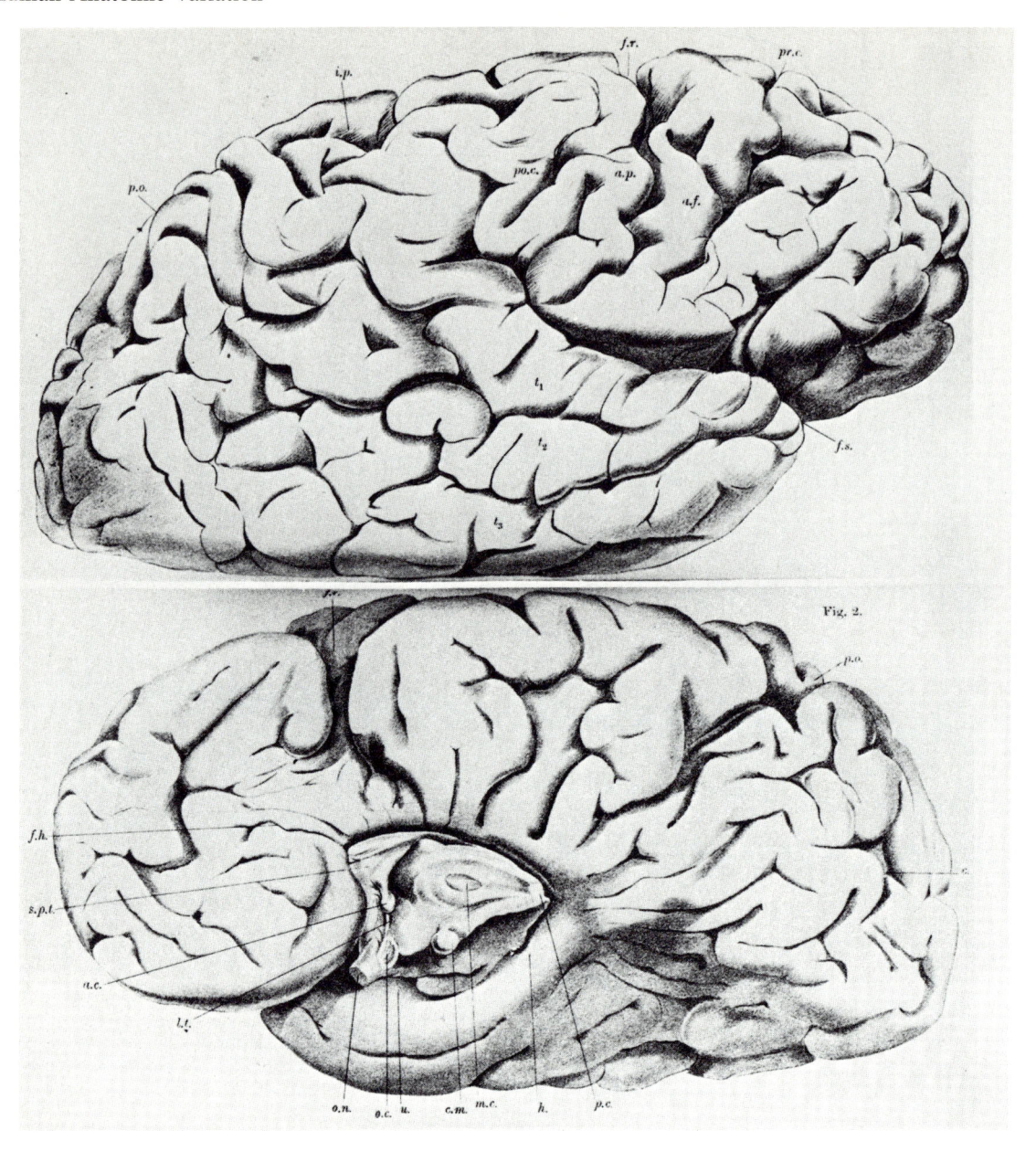

FIG. 2.
Absence of Corpus Callosum.
Schematic drawing of the lateral and medial surfaces of the brain showing agenesis of the corpus callosum in the medial surface view.

From Bruce, A. On the absence of the corpus callosum in the human brain, with the description of a new case. *Brain* 12:171–190, 1890.

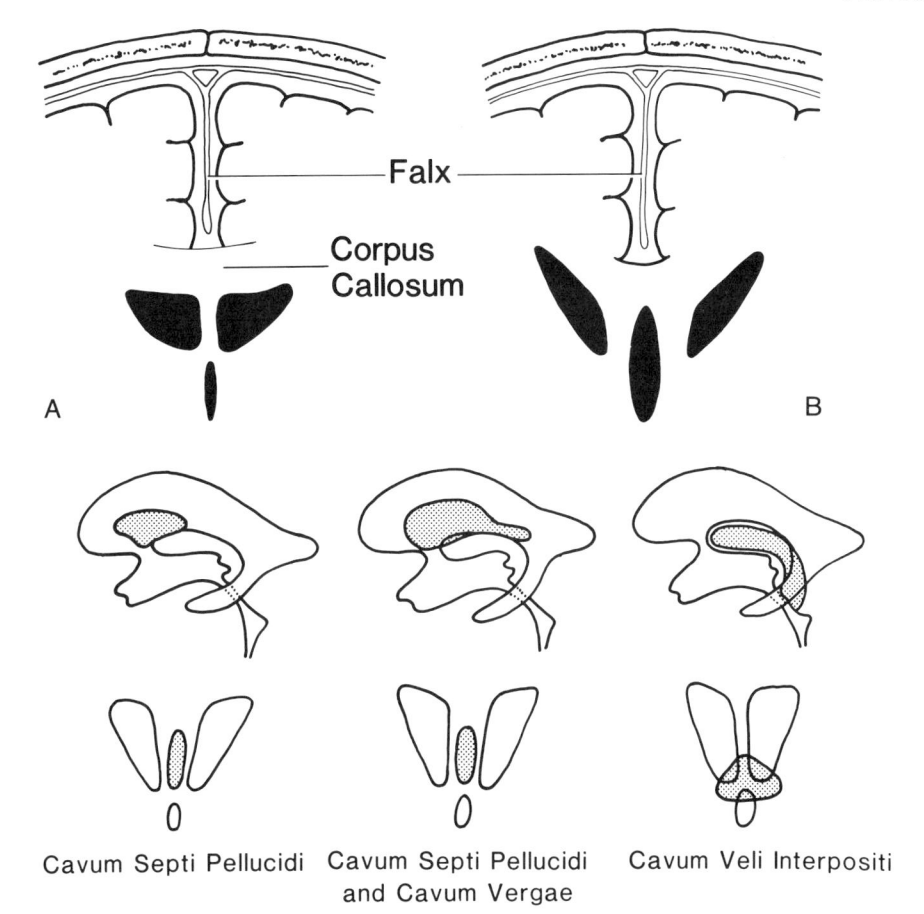

Nervous System

FIG. 3.
Variation in Shape and Position of Ventricular Cavities (Lateral and Third Ventricles) in Agenesis of Corpus Callosum. Variations in Formation of Midline Extra Cavities.

A: Normal appearance of lateral and third ventricles.
B: Appearance of lateral and third ventricles in cases of absence of corpus callosum.

C: Variations in formation of midline extra cavities related to the ventricular system: their radiographic appearance.

From Mori, K. *Anomalies of the Central Nervous System. Neuroradiology and Neurosurgery.* Thieme-Stratton Inc., New York, 1985.

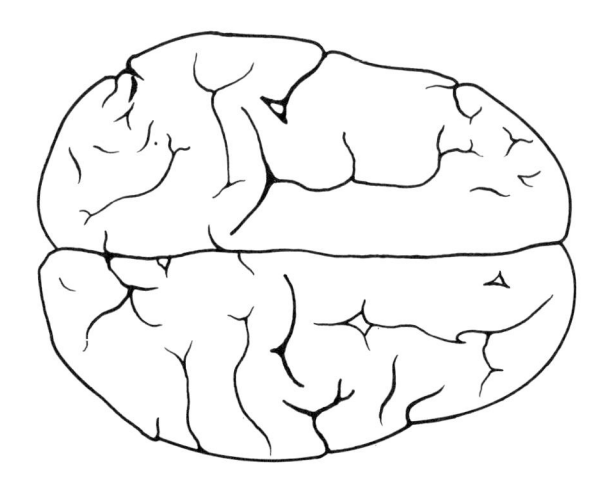

FIG. 4.
Bilateral Interruption of Fissure of Rolando
The author demonstrated the rarely found interruption of the rolandic fissure by a bridging gyrus.

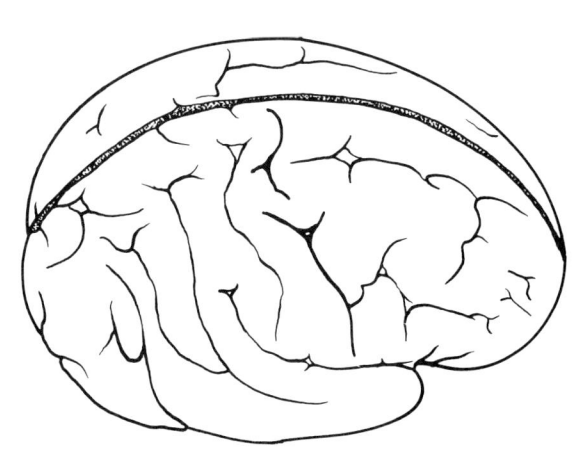

Redrawn from Waterston, D. Complete bilateral interruption of the fissure of Rolando. *J. Anat. Physiol.* XLI:143–146, 1907.

Human Anatomic Variation

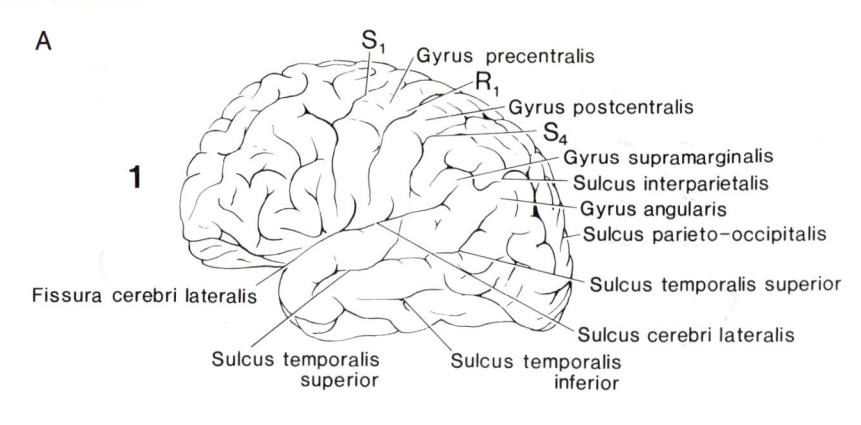

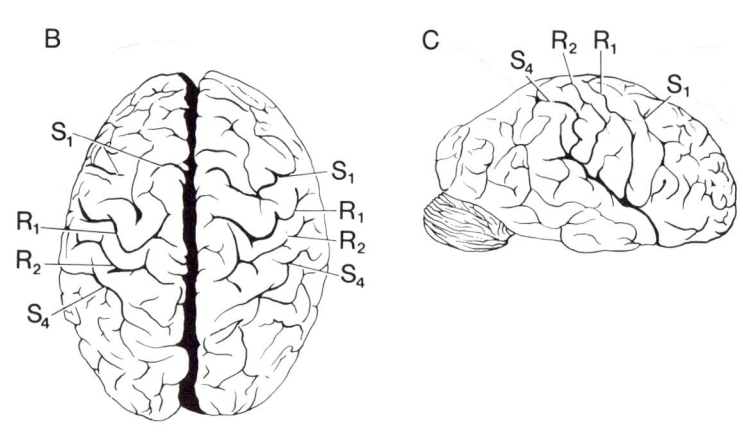

FIG. 5.
Variation in Fissures of Cerebral Hemisphere Showing Doubled Rolandic Fissure.

A: Normal appearance of brain, lateral view.
B: Bilateral doubled rolandic fissures (R_1, R_2). The gyrus between R_1 and R_2 is termed the rolandic gyrus.
C: Same as B, lateral view.

Redrawn and slightly modified from Gerlach, E.J. and H. Weber. Über ein menschliches Gehirn mit beiderseitiger Verdoppelung der Zentralfurche. *Anat. Anz.* 67:440–452, 1929.

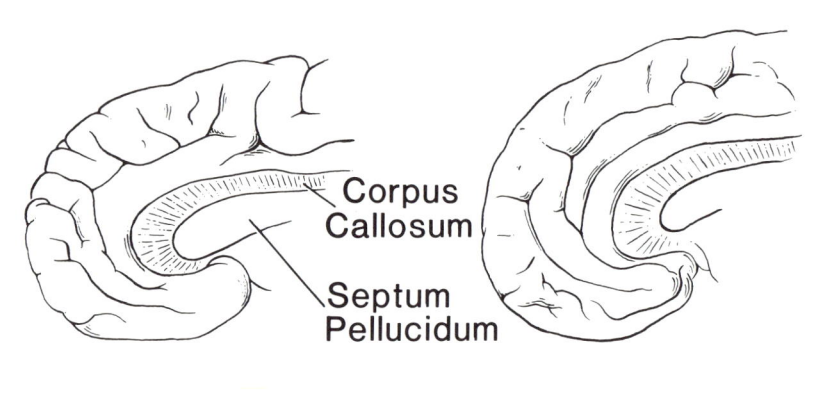

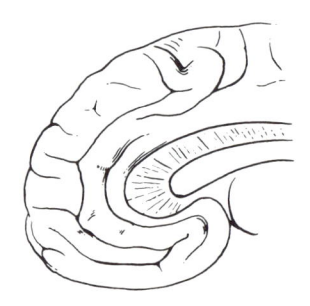

FIG. 6.
Variations in Gyri on the Medial Surface of Frontal Lobe.

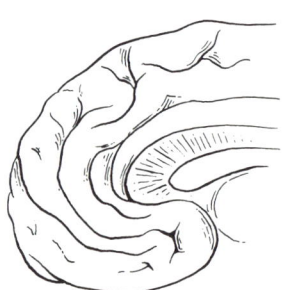

Redrawn from Henle, J. *Handbuch der Systematischen Anatomie des Menschen.* Friedrich Vieweg und Sohn, Braunschweig 1868.

Nervous System

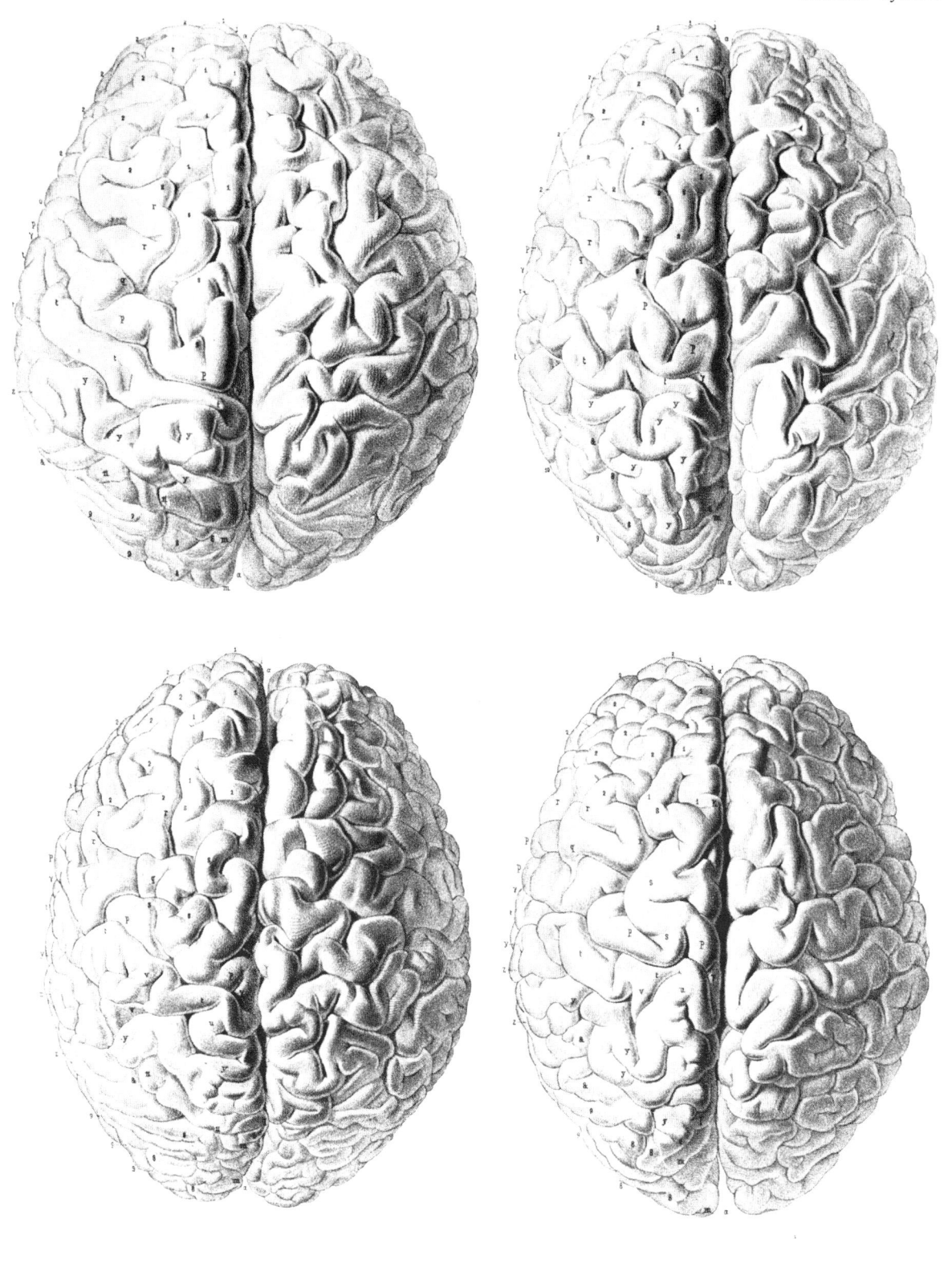

FIG. 7.
Variations in Gyri and Sulci.

From Calori, L. Del cervello nei due tipi brachicefalo e dolicocefalo italiani. *Mem. R. Accad. Sci. Istituto di Bologna* S.2 10:35–156, 1870.

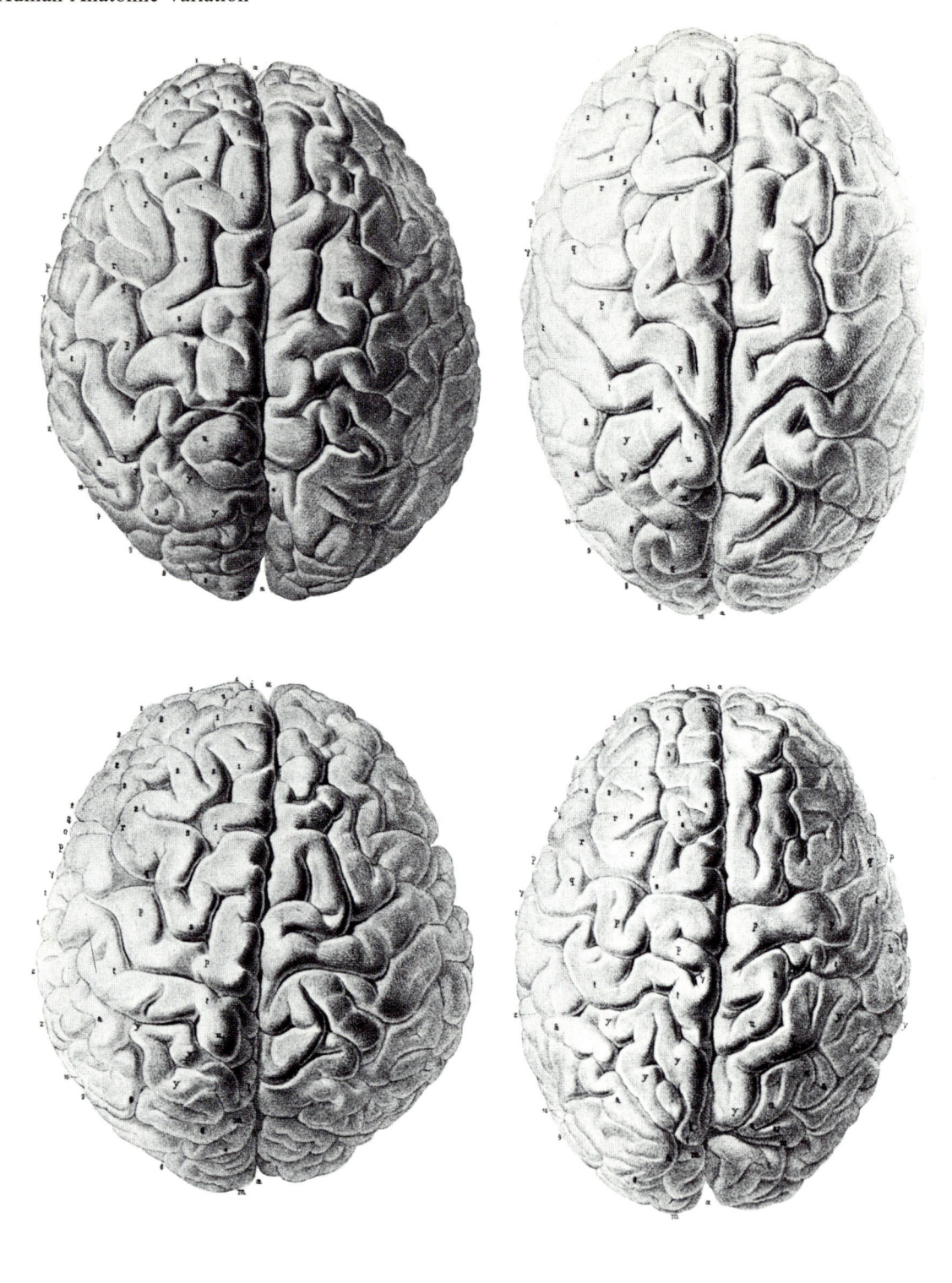

FIG. 8.
Variations in Gyri and Sulci (continued).

From Calori, L. Del cervello nei due tipi brachicefalo e dolicocefalo italiani. *Mem. R. Accad. Sci. Istituto di Bologna S. 2* 10:35–156, 1870.

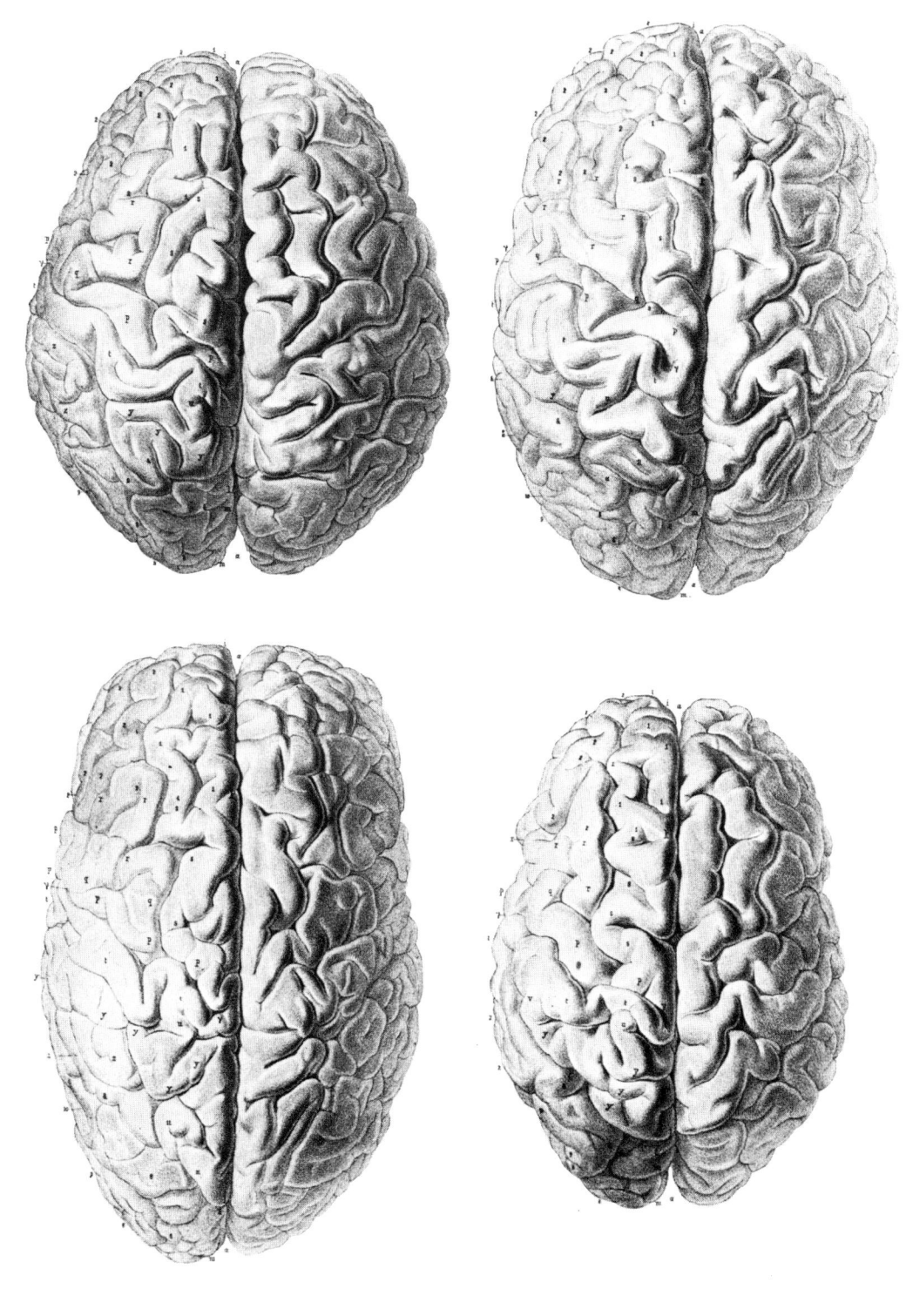

FIG. 9.
Variations in Gyri and Sulci (continued).

From Calori, L. Del cervello nei due tipi brachicefalo e dolicocefalo italiani. *Mem. R. Accad. Sci. Istituto di Bologna* S.2 10:35–156, 1870.

Human Anatomic Variation

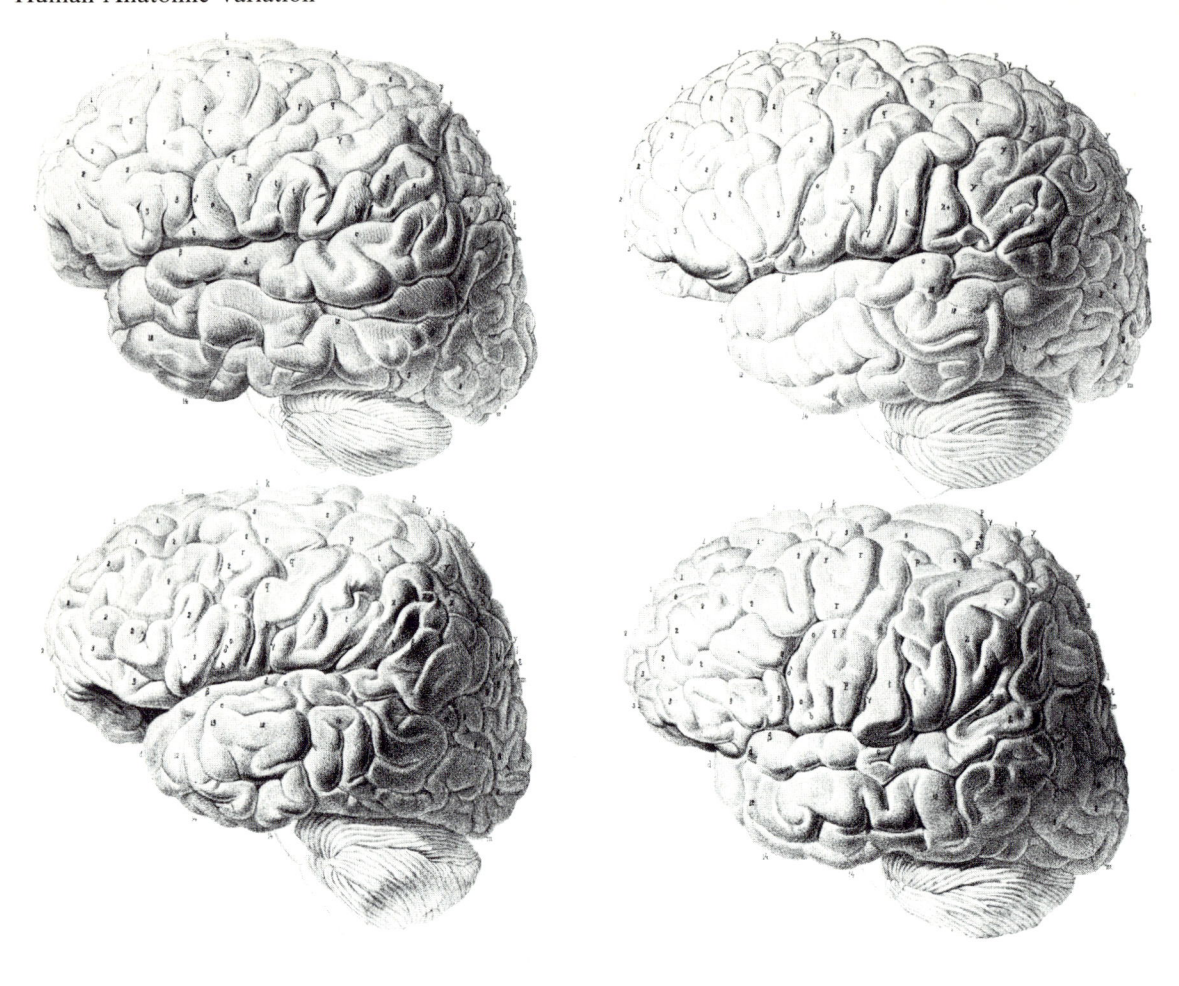

FIG. 10.
Variations in Gyri and Sulci (continued).

From Calori, L. Del cervello nei due tipi brachicefalo e dolicocefalo italiani. *Mem. R. Accad. Sci. Istituto di Bologna* S. 2 10:35–156, 1870.

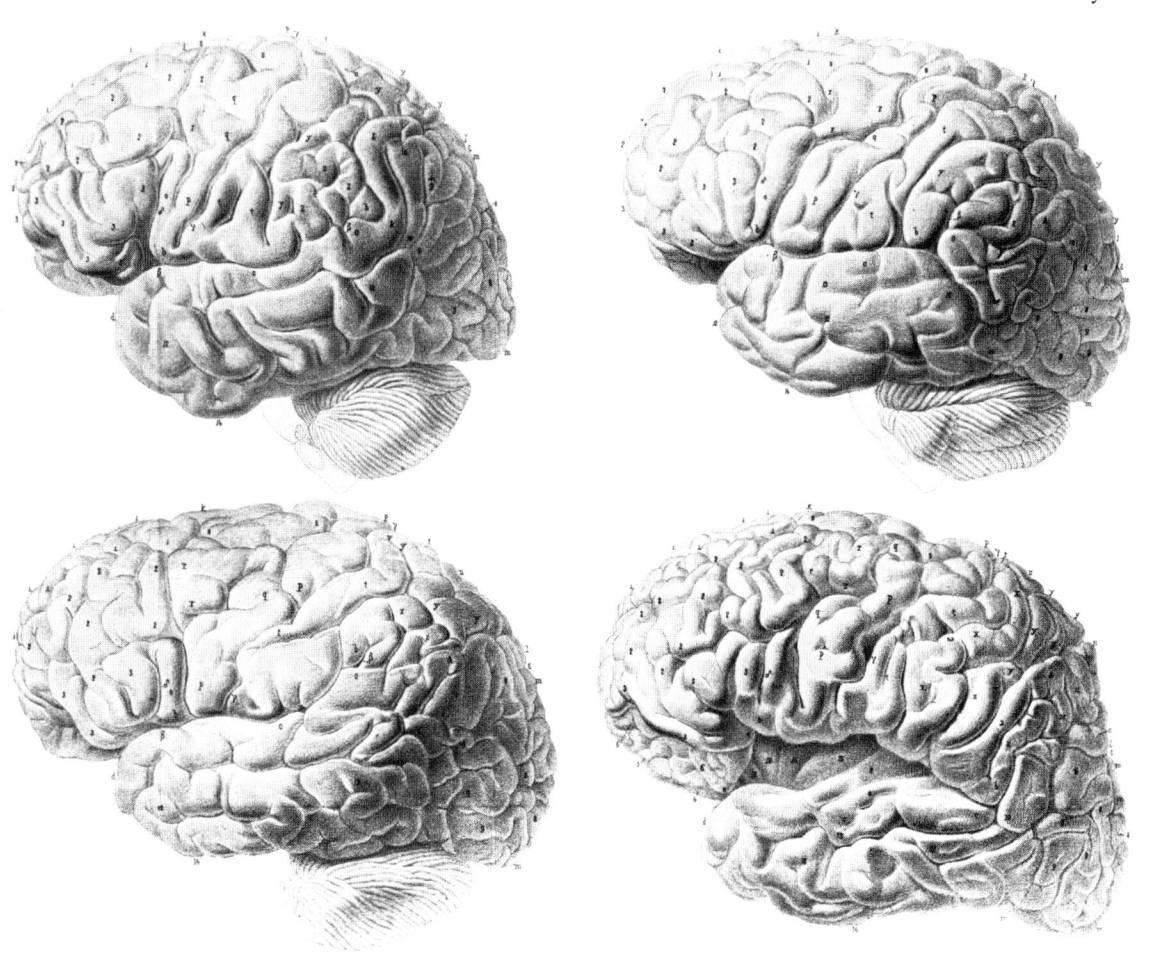

FIG. 11.
Variations in Gyri and Sulci (continued).

From Calori, L. Del cervello nei due tipi brachicefalo e dolicocefalo italiani. *Mem. R. Accad. Sci. Istituto di Bologna* S.2 10:35–156, 1870.

Human Anatomic Variation

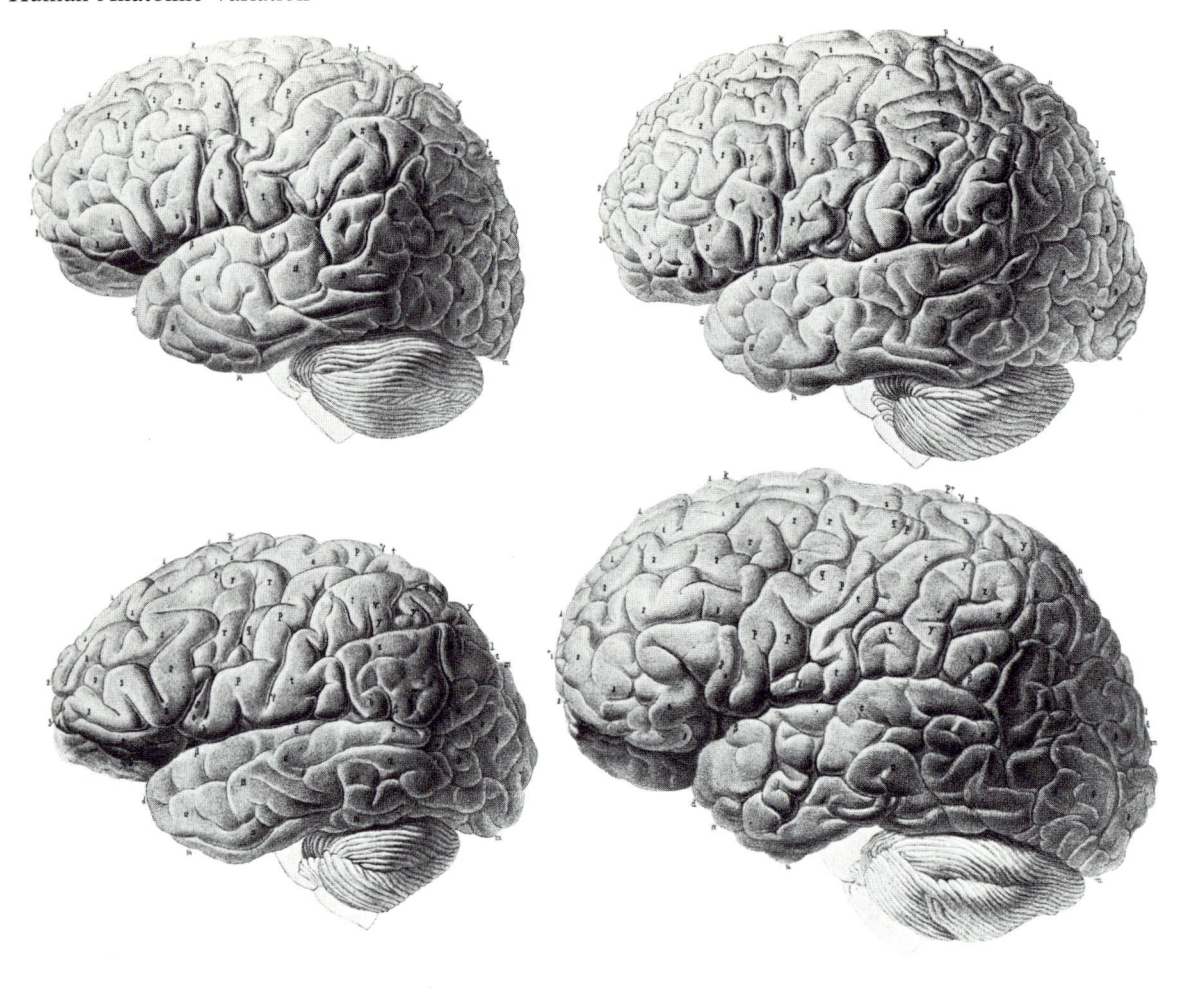

FIG. 12.
Variations in Gyri and Sulci (continued).

From Calori, L. Del cervello nei due tipi brachicefalo e dolicocefalo italiani. *Mem. R. Accad. Sci. Istituto di Bologna* S. 2 10:35–156, 1870.

Nervous System

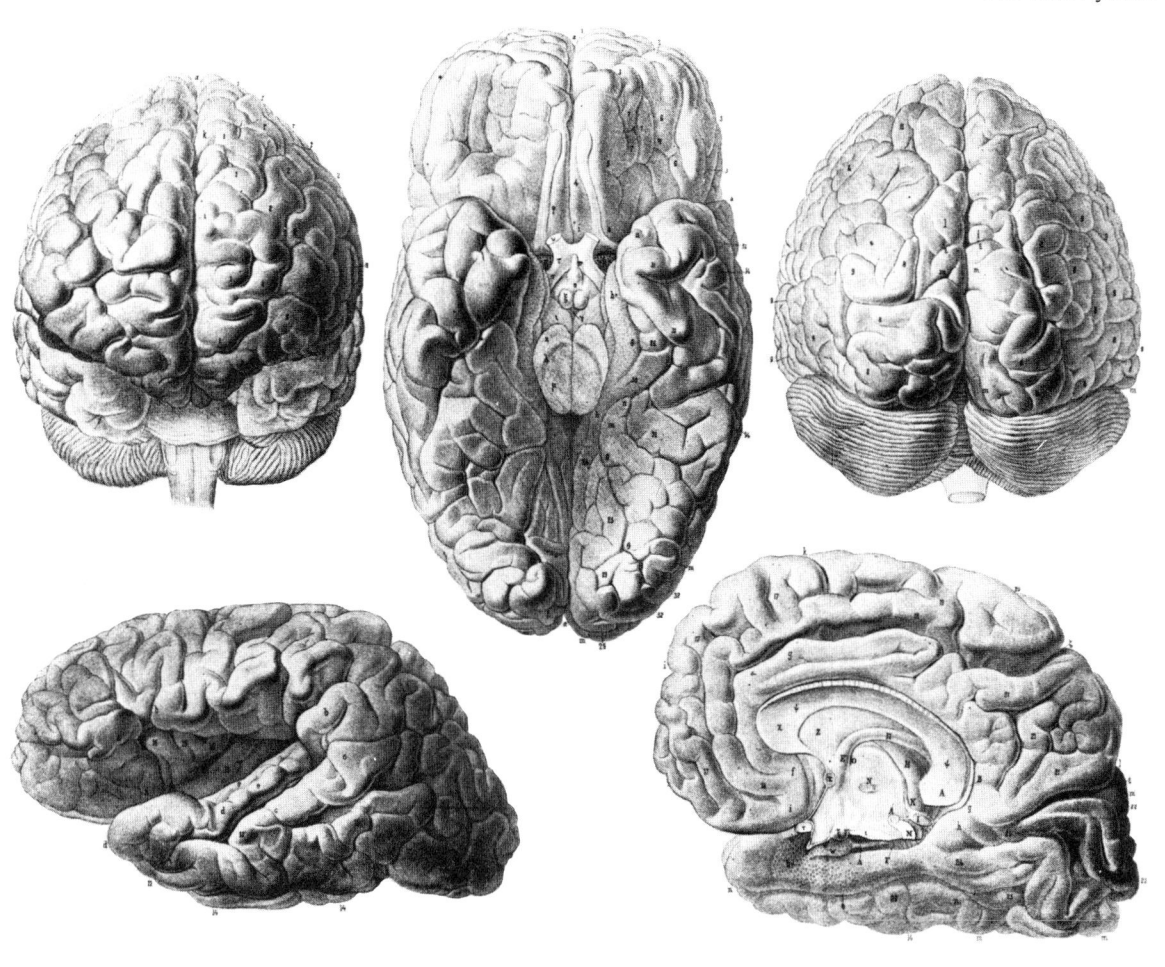

FIG. 13.
Variations in Gyri and Sulci (continued).

From Calori, L. Del cervello nei due tipi brachicefalo e dolicocefalo italiani. *Mem. R. Accad. Sci. Istituto di Bologna* S. 2 10:35–156, 1870.

Human Anatomic Variation

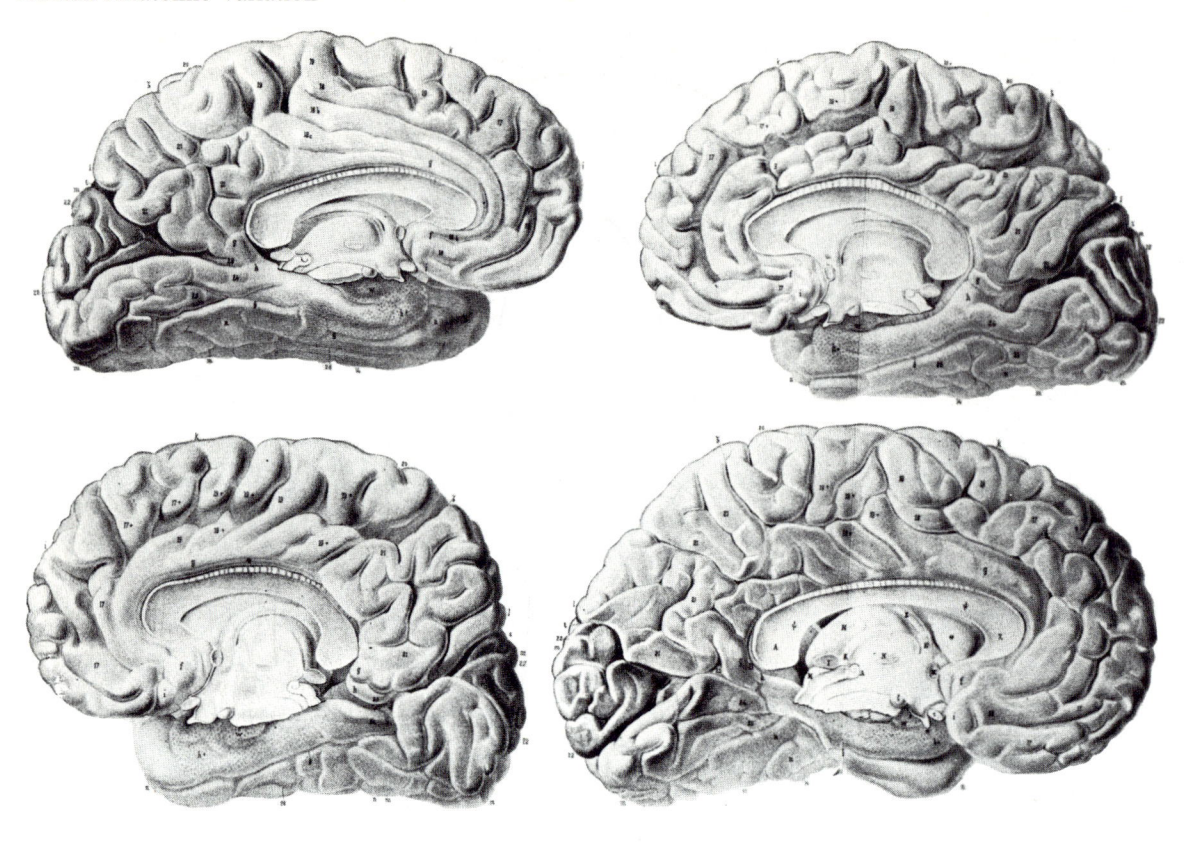

FIG. 14.
Variations in Gyri and Sulci (continued).

From Calori, L. Del cervello nei due tipi brachicefalo e dolicocefalo italiani. *Mem. R. Accad. Sci. Istituto di Bologna* S. 2 10:35–156, 1870.

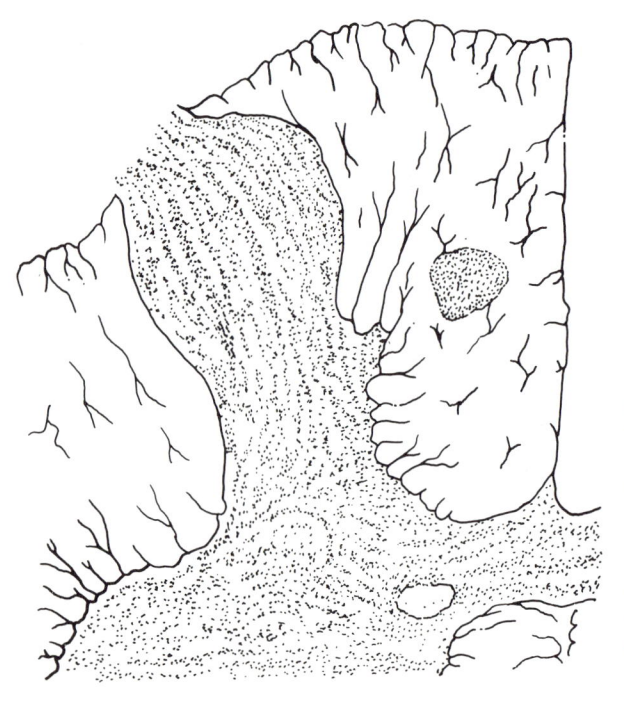

FIG. 15.
Spinal Cord Heterotopia.

Heterotopia of the spinal cord was first described by Arnold Pick in 1878. (Pick, A. Beiträge zur normal. und path. Anat. des centralen Nervensystems – Heterotopie grauer Substanz im menschlichen Rückenmark. Archiv. f. Psychiatrie, 7, 1878.) The drawing shows Pick's case of heterotopia of the spinal cord. The shaded circular area in the posterior column indicates the heterotopic fragment of gray matter.

From Van Gieson, I. A study of the artefacts of the nervous system. *N.Y. Med. J.* 56:337–346, 422–437, 1892.

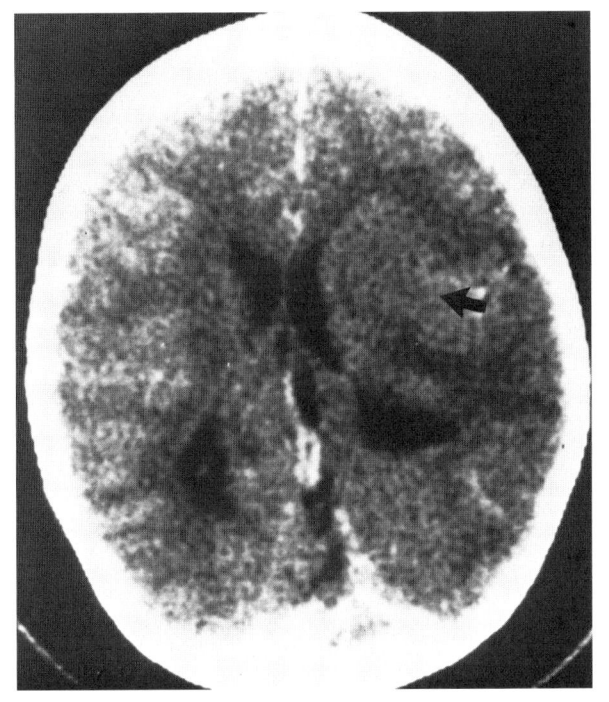

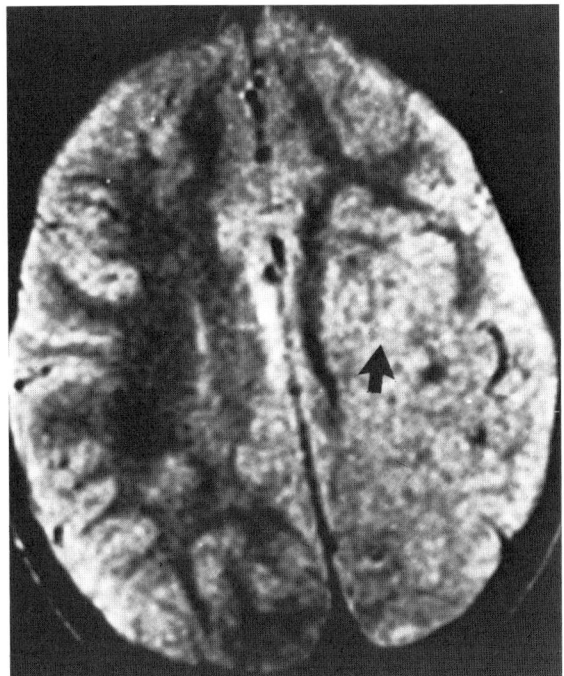

Nervous System

FIG. 16.
Cerebral Heterotopia (CT Scan).

Computed tomography (CT) scan showing heterotopic gray matter (*arrow*) within white matter of the cerebral hemispheres. This variation results from a defect in neuronal migration in early fetal life. Heterotopias may be associated with seizures, microcephaly, or mental retardation.

Courtesy of Dr. Charles Jacoby and Dr. William Bell.

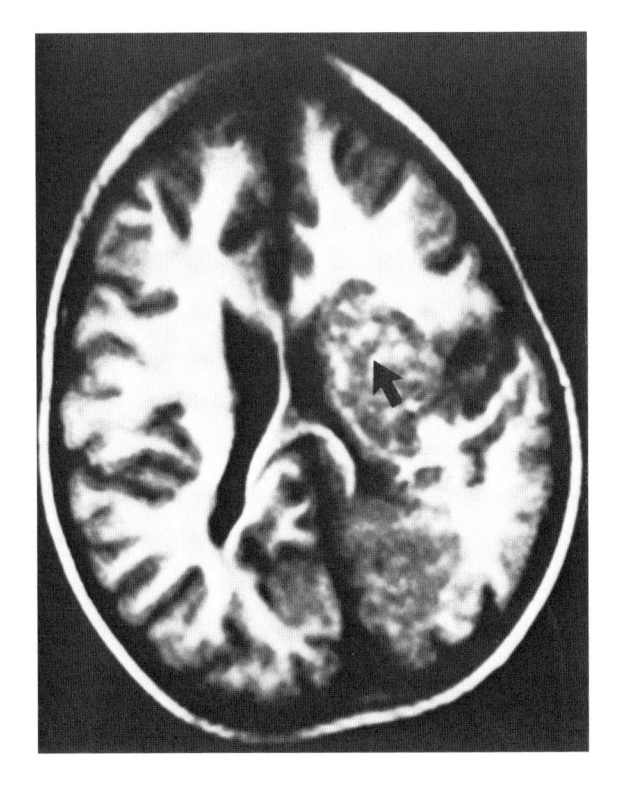

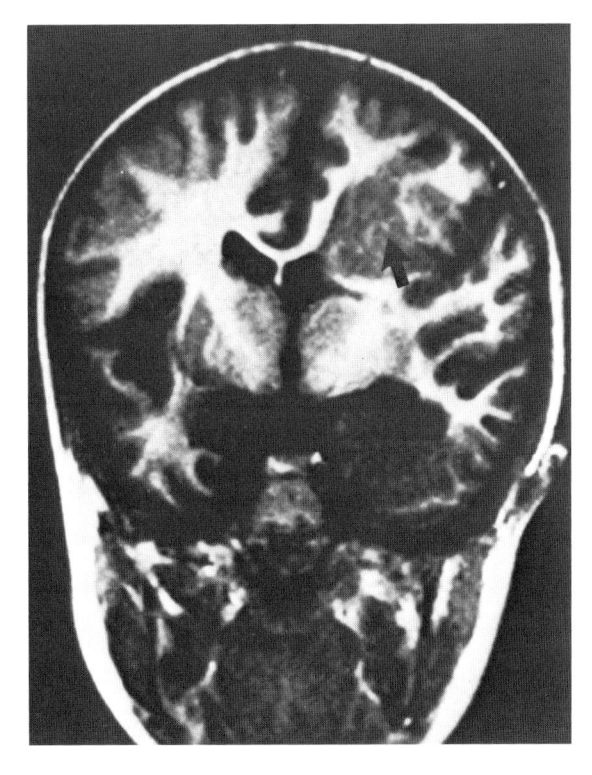

FIG. 17.
Cerebral Heterotopia (MRI).

Magnetic resonance images (MRI) of the brain of the same patient shown in Figure 16, showing heterotopic gray matter (*arrows*) within white matter of the cerebral hemispheres.

Courtesy of Dr. Charles Jacoby and Dr. William Bell.

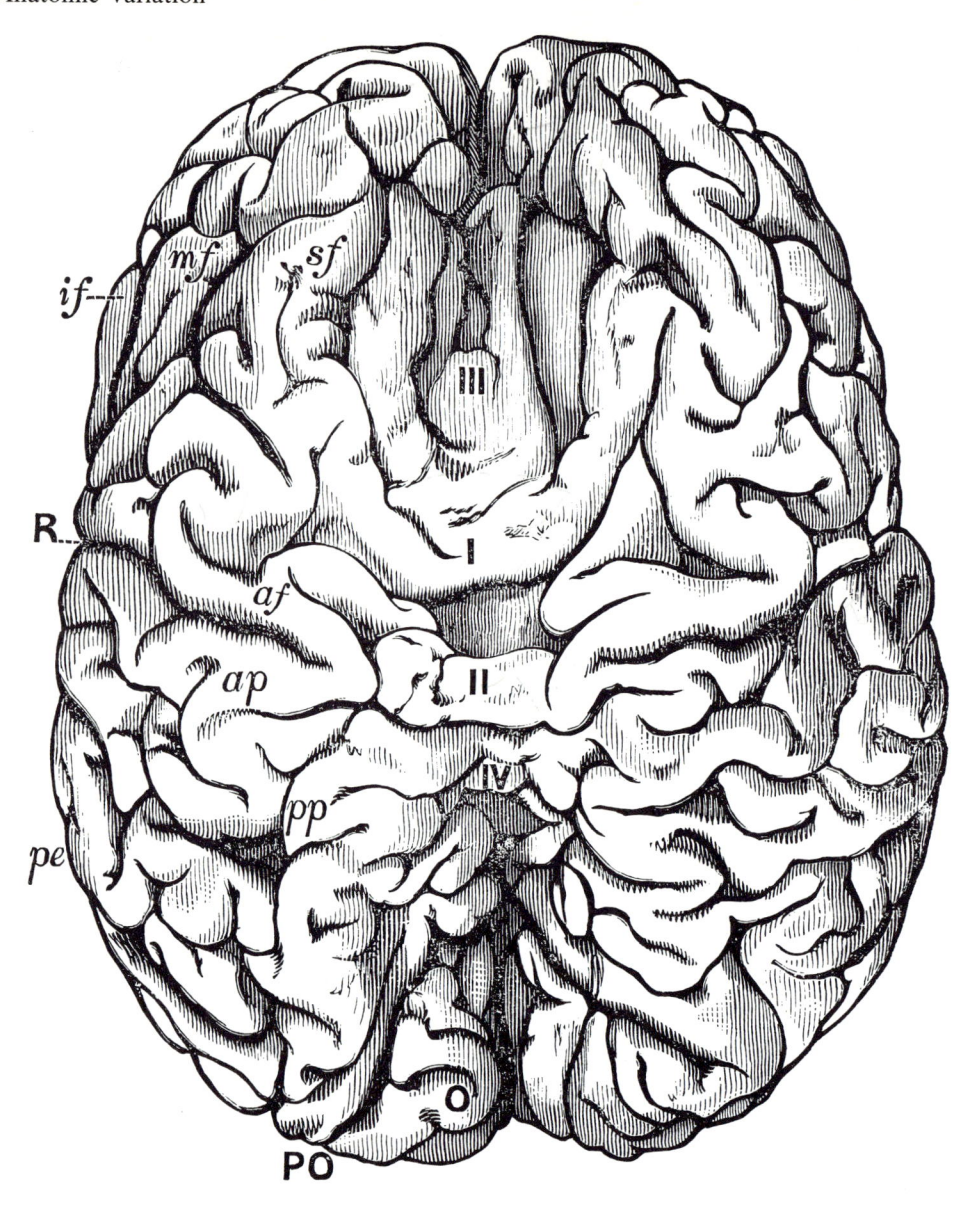

FIG. 18.

Incomplete Division of Cerebral Hemispheres by Transverse Convolutions.

Vertex view of cerebrum. *R*, fissure of Rolando; *PO*, parieto-occipital fissure; *O*, occipital lobe; *pe*, convolution of parietal eminence; *pp*, posteroparietal convolution; *ap*, ascending parietal convolution; *af*, ascending frontal convolution; *sf*, *mf*, *if*, superior, middle, and inferior frontal convolutions; *I*, anterior, and *II*, posterior transverse convolutions; *III*, gray matter in front of anterior transverse convolutions; *IV*, mesial convolutions behind posterior transverse convolution.

From Turner, W. A human cerebrum imperfectly divided into two hemispheres. *J. Anat. Physiol.* 12:241–253, 1878.

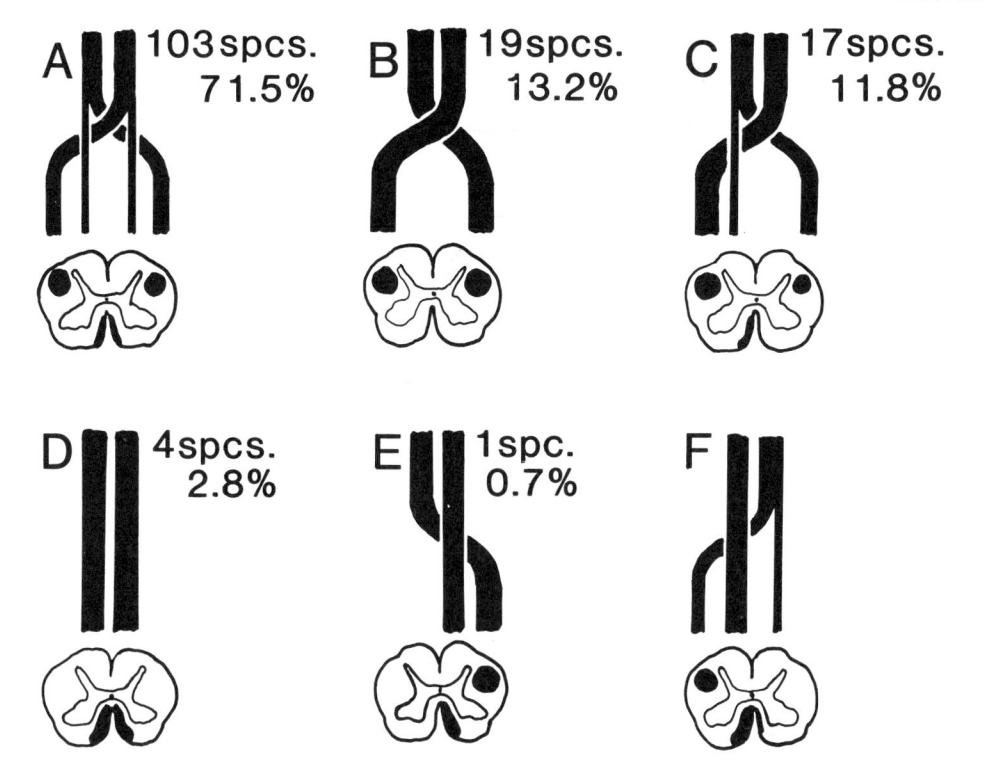

FIG. 19.
Variations in Crossing Pattern of Pyramids in Medulla and Subsequent Formation of Corticospinal Tracts in Spinal Cord.

"Pyramidal tracts in spinal cords of 144 human fetal or neonatal specimens. Although the commonest pattern (A) is shown in the diagram as bilaterally symmetrical, the pyramidal tracts were 'patently unequal' in 81 of these 103 specimens." The illustration F was not encountered in the study of 144 brains and may not be associated with viable individuals. SPCS, specimens.

Redrawn from Sweet, W.H. Recent observations pertinent to improving anterolateral cordotomy. *Clin. Neurosurg.* 23:80–95, 1976.

FIG. 20.
Absence of Optic Chiasma.
Bilateral Asymmetric Distribution of Nerve Fibers in Optic Tracts.

A: A diagram of the optic nerve and tract from a case in which Vesalius states that he was unable to find the optic chiasma.
B: Variation in nerve and optic tract. Unequal distribution of nerve fibers in the optic tracts beyond the optic chiasma.

Redrawn from Vesalius, A. *De Corporis Humani Fabrica*, Edit. II, lib. IV, cap. IV, S. 518. Basil, 1555; and from Weinberg, R. Ungekreuzte Sehnervenfasern. *Z. Anat. Entwicklungsgesch.* 79:433–446, 1926.

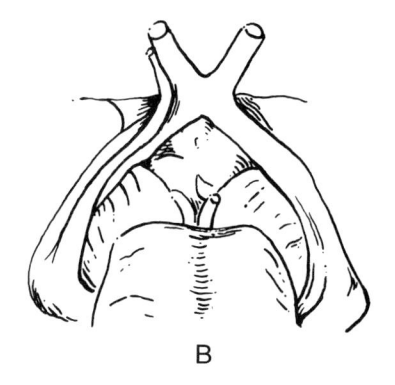

Human Anatomic Variation

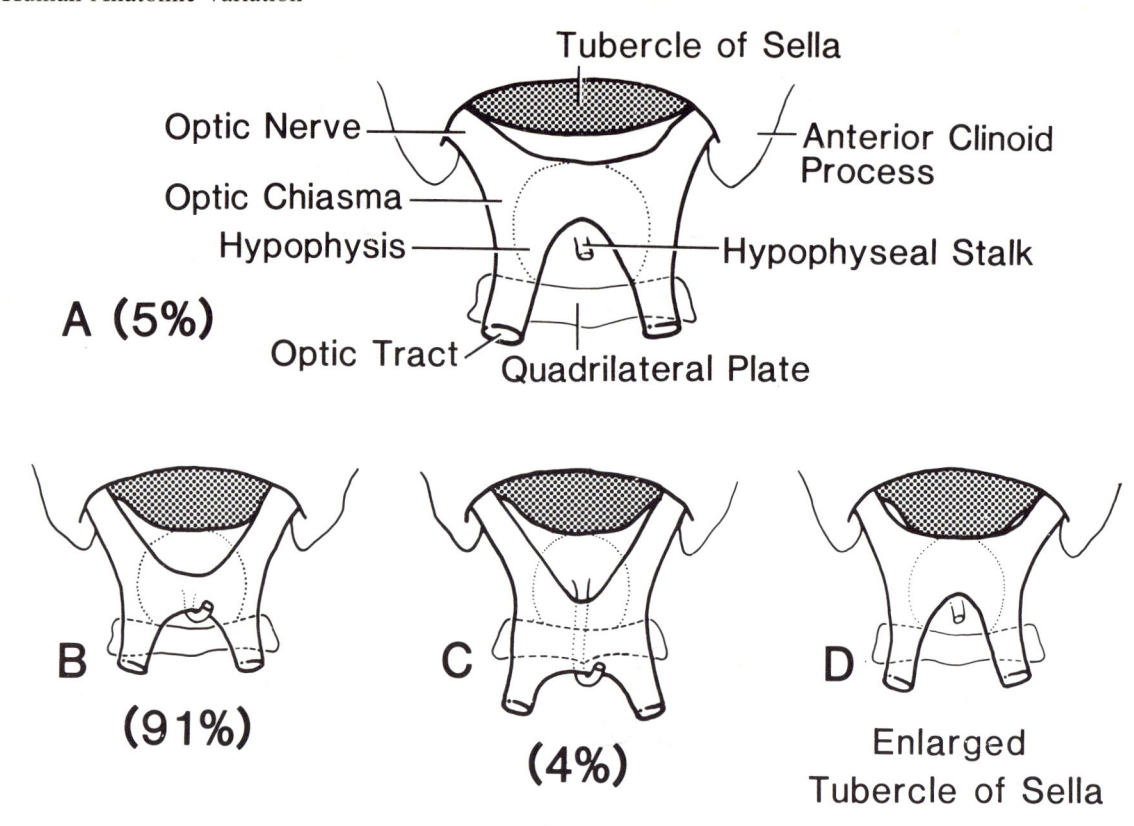

A (5%)
- Tubercle of Sella
- Optic Nerve
- Optic Chiasma
- Hypophysis
- Optic Tract
- Quadrilateral Plate
- Anterior Clinoid Process
- Hypophyseal Stalk

B (91%)

C (4%)

D Enlarged Tubercle of Sella

FIG. 21.
Variations in Position of Optic Chiasma in Relation to Hypophyseal Stalk.

After Schaeffer, redrawn and modified from Testut, L. and A. Latarjet. *Traité d'anatomie humaine*. G. Doin & Cie., Paris, 1948.

FIG. 22.
Oculomotor Nerve.

Twenty formaldehyde-fixed brains were examined under the stereoscopic microscope. In 12 brains (60%), the oculomotor nerves were penetrated by the circumflex mesencephalic artery or by a branch of the perforating vessels of the posterior cerebral artery, either on one side (40%) or on both (20%). In one brain (5%), a particular relationship was noticed between the trochlear nerve and the superior cerebellar artery. The abducens nerves were penetrated by the corresponding pontine veins in three brains (15%). These anatomic findings might have important clinical implications.

A: Ventral view of right oculomotor nerve (*1*) penetrated by long circumflex mesencephalic artery (*arrows*). Terminal part of basilar (*2*) and initial parts of right (*3*) and left (*3'*) posterior cerebral arteries are elevated. Note also interpeduncular fossa (*4*) with numerous perforating branches of posterior cerebral arteries, left crus cerebri (*5*), and right optic nerve (*6*).

B: Ventral view of short circumflex mesencephalic artery (*large arrows*) running through most peripheral part of right oculomotor nerve (*1*). Note tiny arterioles (*small arrows*) supplying nerve. Note also small vein (*2*) running around fiber bundle of oculomotor nerve, terminal portion of basilar artery (*3*), superior cerebellar artery drawn caudally (*4*), initial portion of right posterior cerebral artery drawn rostrally (*5*), and large posterior communicating artery (*6*).

C: Ventral view of left long circumflex mesencephalic artery (*1*) dividing (*white arrow*) into two terminal stems (*2* and *2'*). One stem (*2'*) runs through middle of oculomotor nerve (*3*). Terminal division of other stem (*2*) encircles small fiber bundle (*black arrow*) of nerve. Note initial portion of posterior cerebral artery (*4*) and small posterior communicating artery (*5*). Oculomotor nerve is drawn caudally.

From Milisavljević, M., Marinković, S., Lolić-Draganić, V. and M. Kovacević. Oculomotor, trochlear, and abducens nerves penetrated by cerebral vessels. Microanatomy and possible clinical significance. *Arch. Neurol.* 43:58–61, 1986. Copyright 1986, American Medical Association.

Nervous System

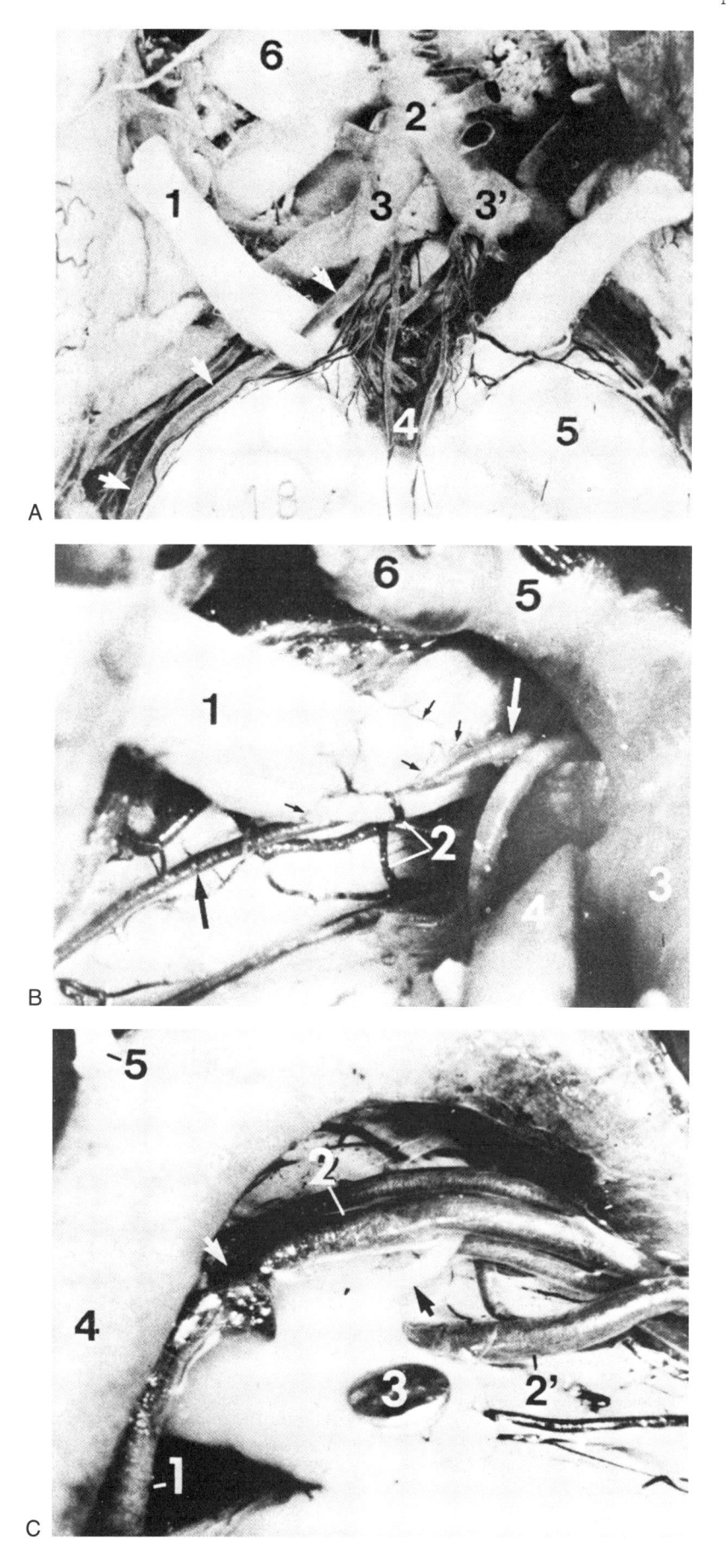

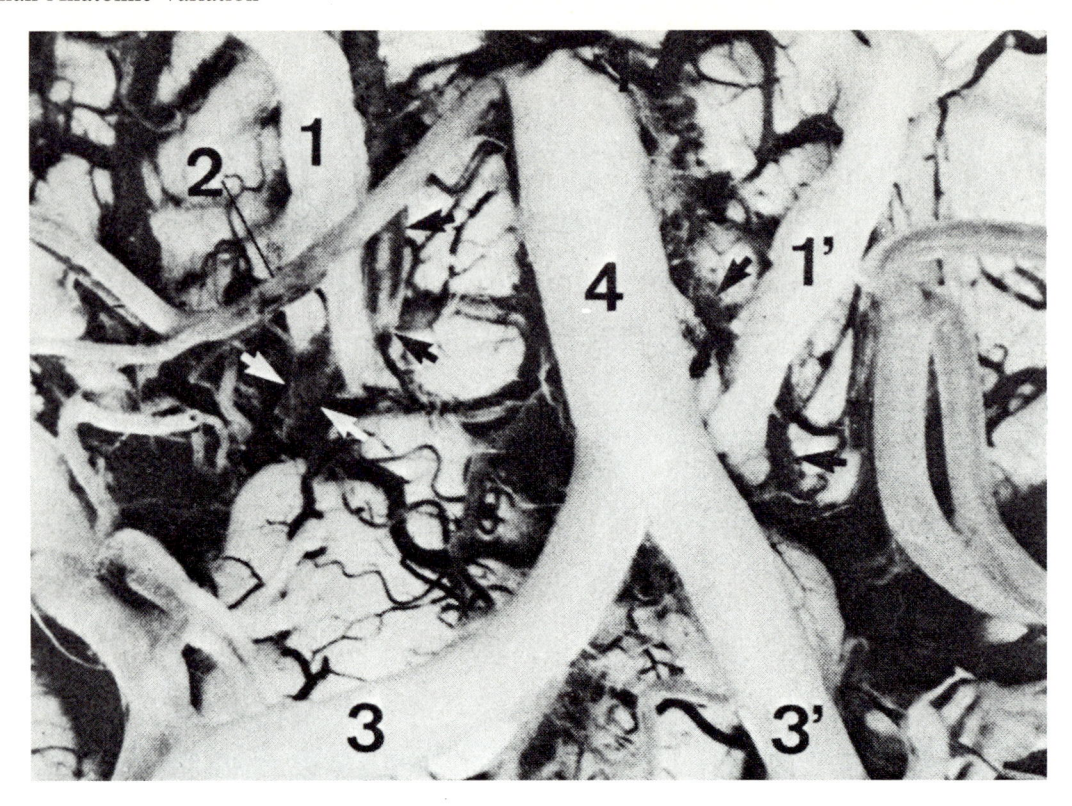

FIG. 23.
Abducens Nerve.

Twenty formaldehyde-fixed brains were examined under the stereoscopic microscope. In 12 brains (60%), the oculomotor nerves were penetrated by the circumflex mesencephalic artery or by a branch of the perforating vessels of the posterior cerebral artery, either on one side (40%) or on both (20%). In one brain (5%), a particular relationship was noticed between the trochlear nerve and the superior cerebellar artery. The abducens nerves were penetrated by the corresponding pontine veins in three brains (15%). These anatomic findings might have important clinical implications.

Ventral view of right abducens nerve (1) penetrated by anterolateral pontine vein (arrows). Through left nerve (1') runs tributary of anteromedian pontine vein (arrows). In addition, anterior inferior cerebellar artery (2) traverses right nerve. Note right (3) and left (3') vertebral arteries and basilar artery (4).

From Milisavljević, M., Marinković, S., Lolić-Draganić, V. and M. Kovacević. Oculomotor, trochlear, and abducens nerves penetrated by cerebral vessels. Microanatomy and possible clinical significance. Arch. Neurol. 43:58–61, 1986. Copyright 1986, American Medical Association.

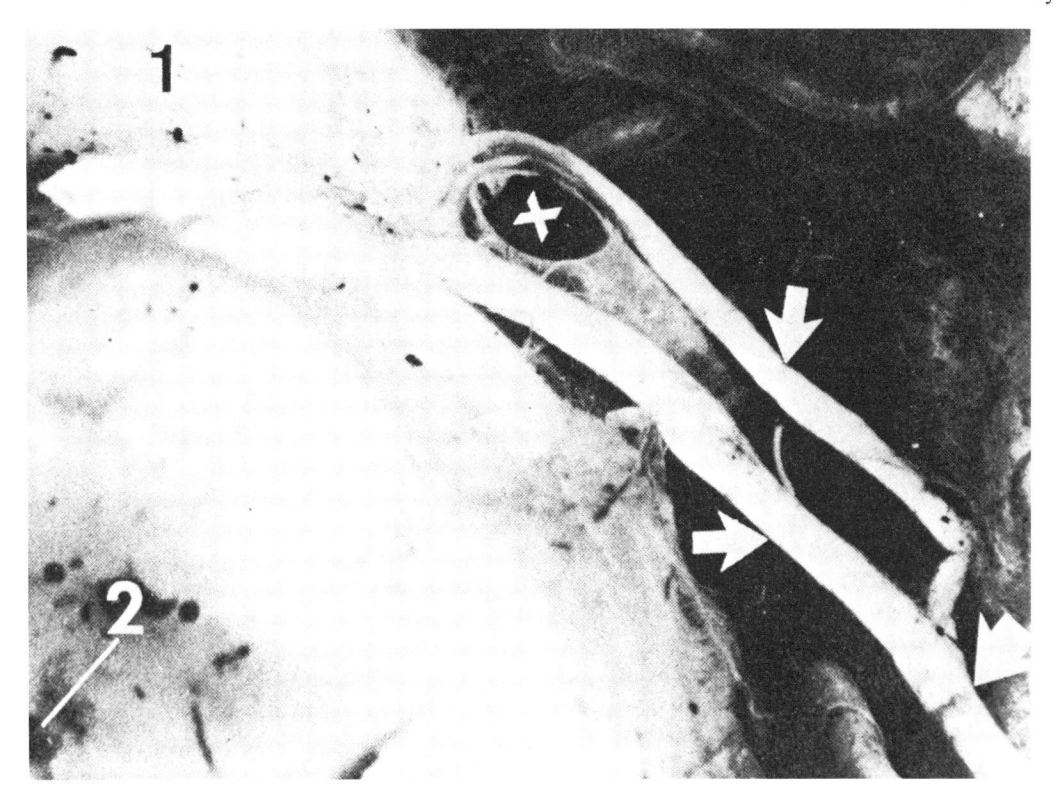

FIG. 24.
Trochlear Nerve.

Twenty formaldehyde-fixed brains were examined under the stereoscopic microscope. In 12 brains (60%), the oculomotor nerves were penetrated by the circumflex mesencephalic artery or by a branch of the perforating vessels of the posterior cerebral artery, either on one side (40%) or on both (20%). In one brain (5%), a particular relationship was noticed between the trochlear nerve and the superior cerebellar artery. The abducens nerves were penetrated by the corresponding pontine veins in three brains (15%). These anatomic findings might have important clinical implications.

Caudal view of right trochlear nerve (*large arrow*) arising from two bundles (*smaller arrows*). Between them runs superior cerebellar artery (cut) (*x*). Note right inferior colliculus (*1*) and most rostral part of fourth ventricle (*2*).

From Milisavljević, M., Marinković, S., Lolić-Draganić, V. and M. Kovacević. Oculomotor, trochlear, and abducens nerves penetrated by cerebral vessels. Microanatomy and possible clinical significance. *Arch. Neurol.* 43:58–61, 1986. Copyright 1986, American Medical Association.

Human Anatomic Variation

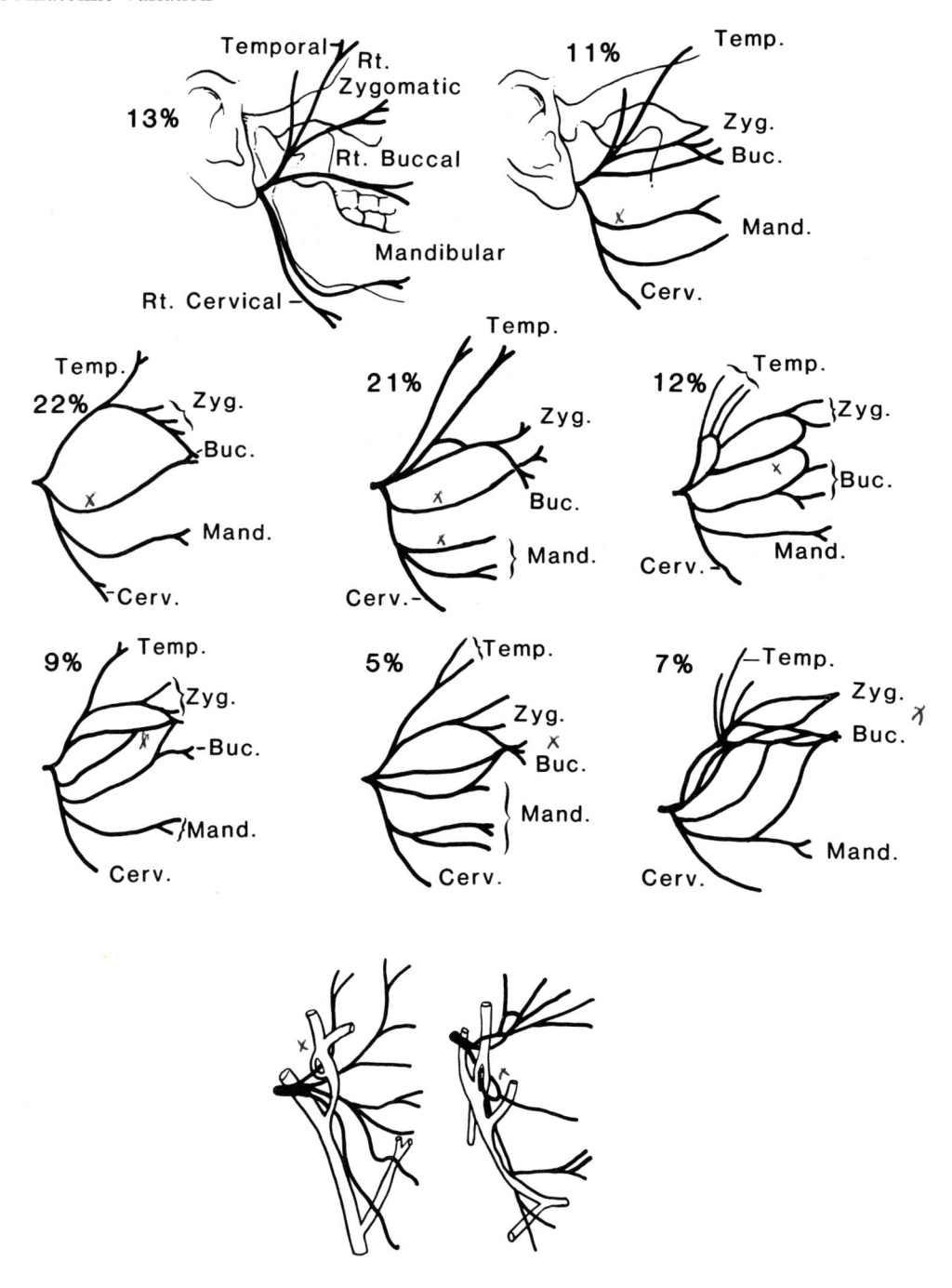

FIG. 25.
Variations in Branching of Facial Nerve.

Major types and percentages of facial nerve branching and anastomoses. In 13%, major divisions (temporal and facial) are independent; in 11%, anastomoses occur between rami of the temporal division; in 22%, connections occur between adjacent rami from the major divisions; in 21%, anastomoses representing a composite of those in the 11% and 22% categories occur; in 12%, proximal anastomoses occur within the temporal component, as well as distal interconnection between the latter and the cervical component; in 9%, two anastomotic rami connect the buccal division of the cervical to the zygomatic part of the temporal; in 5%, a transverse ramus, from the trunk of the nerve, contributes to the buccal ramus formed by anastomosis between the two major divisions; in 7% richly plexiform communications occur, especially within the temporal portion of the nerve.
Variations of the facial nerve include branches that pass through clefts in superficial veins or the formation of nerve loops through which pass superficial veins.

Redrawn from McCormack, L.J., Cauldwell, E.W. and B.J. Anson. The surgical anatomy of the facial nerve with special reference to the parotid gland. *Surg. Gynecol. Obstet.* 80:620–630, 1945; and from Dargent, M. and P.E. Duroux. Données anatomiques concernant la morphologie et certains rapports du facial intra-parotidien. *La Presse Medicale* No. 37, pp. 523–524, 1946.

Nervous System

FIG. 26.

Variations in Intraparotid Portion of Facial Nerve.

These authors dissected 68 facial nerves from within the substance of the parotid gland. They recognized two major classes and five "types" of facial nerve branching from 59 of the 68 dissections. Class 1 (35 cases): Facial nerves without anastomoses between branches after their initial branching from the nerve trunk. Class 2 (24 cases): Facial nerves with anastomoses between the cervicotemporal branches, which form intraglandular plexuses.

Class 1. A (Type 1): Classical variety (15 cases).

        B (Type 2): Variety with precocious branching (11 cases).

        C (Type 3): Rare variety with ladder-like branching (9 cases). It was considered a difficult type to dissect.

Class 2. D (Type 4): Variety with long anastomotic loops (13 cases).

        E (Type 5): Variety with short anastomotic loops between the cervical and temporal branches and all arising from a common stem (11 cases).

Redrawn and modified from Dargent, M. and P.E. Duroux. Données anatomiques concernant la morphologie et certains rapports du facial intra-parotidien. *La Presse Medicale* No. 37, pp. 523–524, 1946.

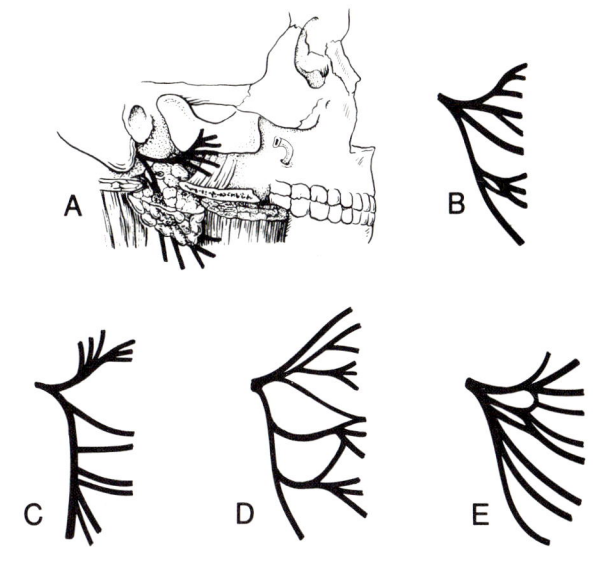

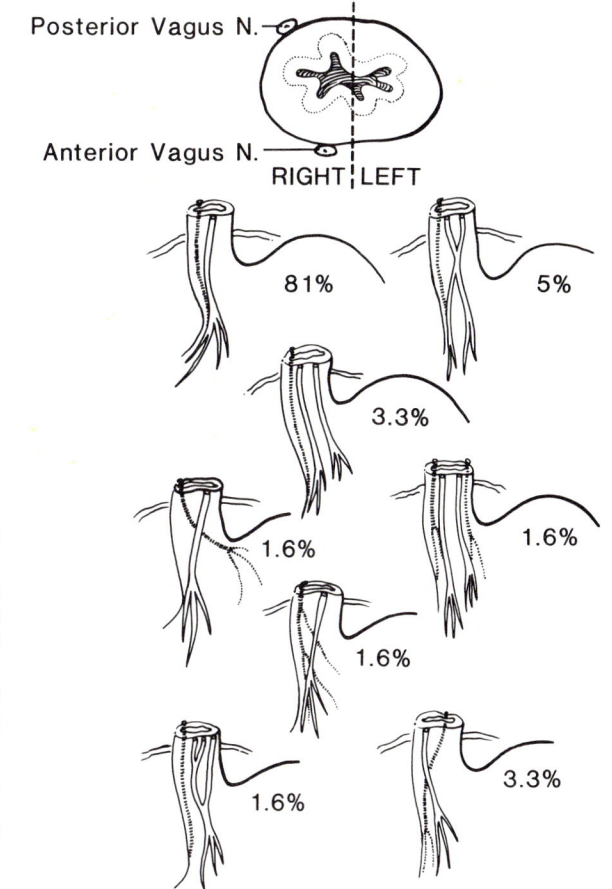

FIG. 27.

Variations in Vagus Nerve at the Lower Portion of Esophagus and Upper Stomach.

In 81% of cases the distribution of the right and left vagus nerves, located just below the diaphragm, lies on the right side of the esophagus. In less than 20% of cases, variation in branching patterns and side (position) was found.

Authors' note: Additional reference: Jackson, R.G. Anatomy of the vagus nerve in the region of the lower esophagus and the stomach. *Anat. Rec.* 103:1–18, 1949.

Redrawn from Dragstedt, L.R., Fournier, H.J., Woodward, E.R., Tovee, E.B. and P.V. Harper, Jr. Transabdominal gastric vagotomy. A study of the anatomy and surgery of the vagus nerves at the lower portion of the esophagus. *Surg. Gynecol. Obstet.* 85:461–466, 1947.

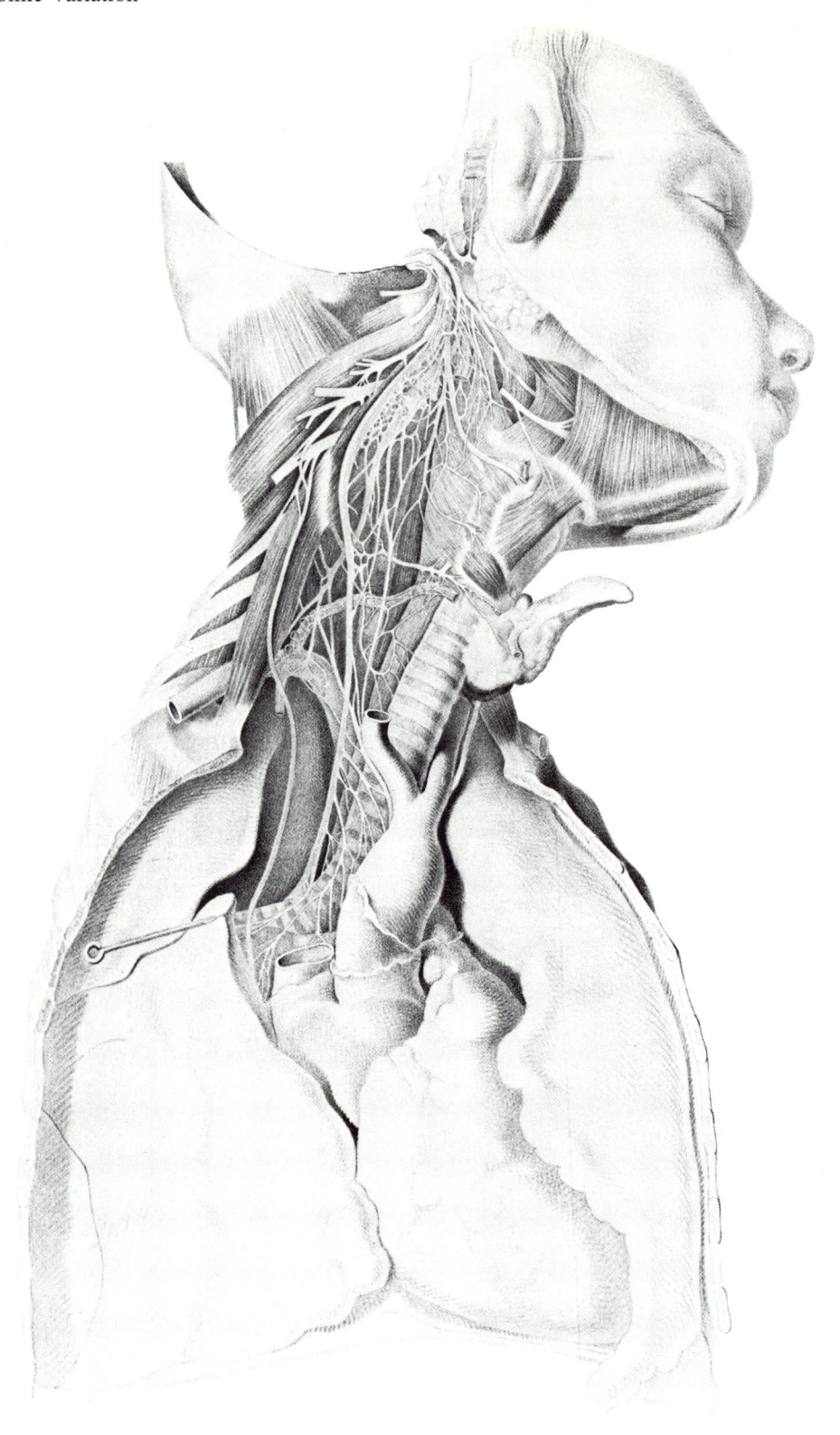

FIG. 28.
Absence of Right Recurrent Laryngeal Nerve in Retroesophageal Right Subclavian Artery.

Anomalous origin and retroesophageal course of right subclavian artery, with consequent absence of right recurrent loop of vagus.

From Calori, L. Delle anomalie più importanti di ossa, vasi, nervi, e muscoli occorse nell'ultimo biennio facendo anatomia del corpo umano. *Mem. Accad. Sci. Istituto di Bologna S. 2* 8:417–482, 1868.

Nervous System

FIG. 29.
Anomalous Origin of Right Recurrent Laryngeal Nerve Associated with Arteria Lusoria, Retroesophageal Right Suclavian Artery.

Redrawn from Pemberton, J. de J. and J.M. Miller. Anatomy of the right inferior laryngeal nerve. *Arch. Surg.* 42:712–718, 1941.

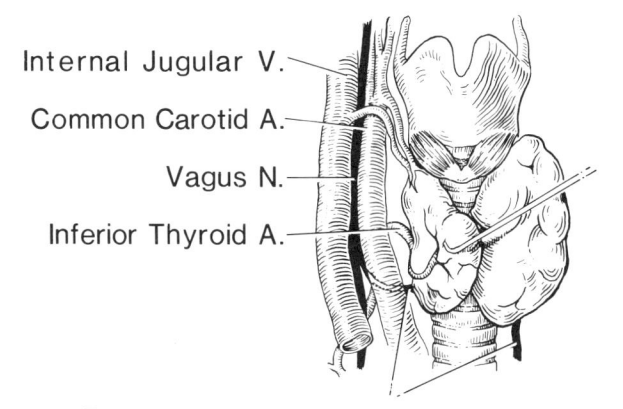

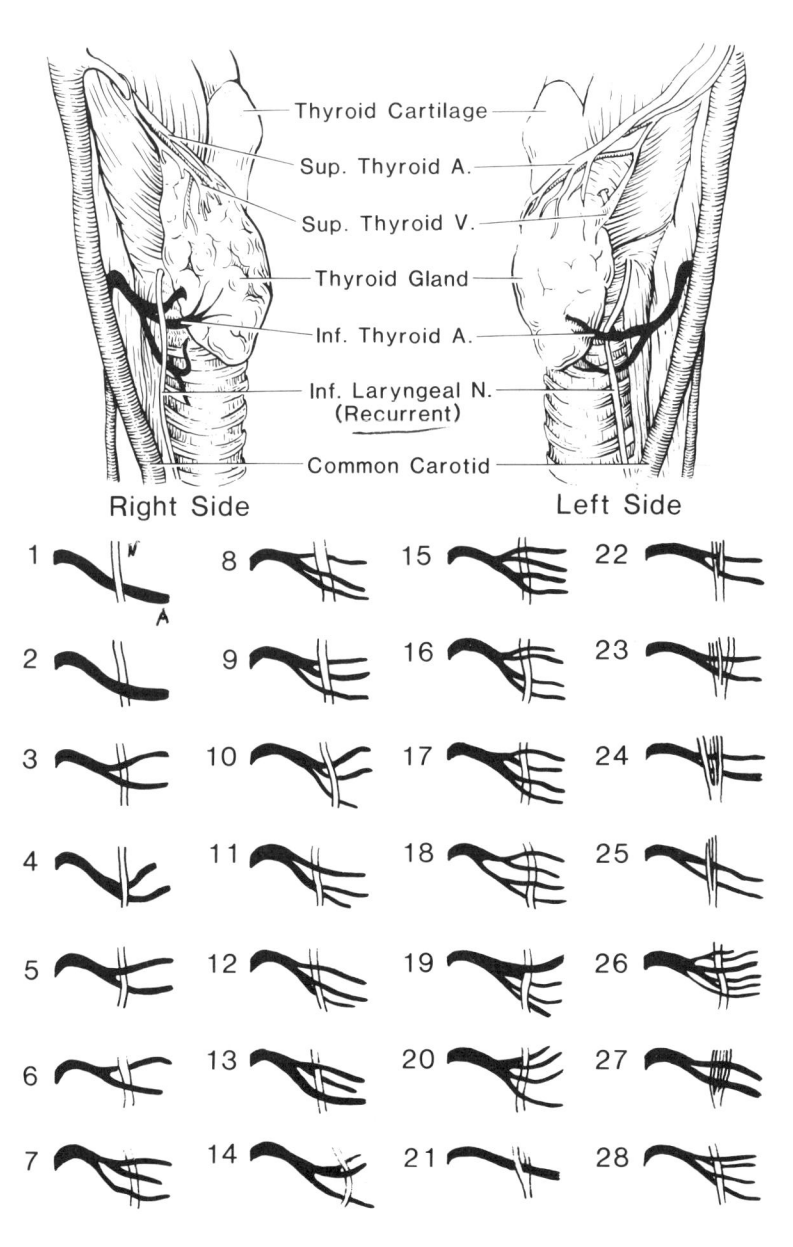

FIG. 30.
Twenty-eight Variations in the Relations of Inferior Laryngeal Nerve to Inferior Thyroid Artery.

Redrawn from Reed, A.F. The relations of the inferior laryngeal nerve to the inferior thyroid artery. *Anat. Rec.* 85:17–23, 1943.

Human Anatomic Variation

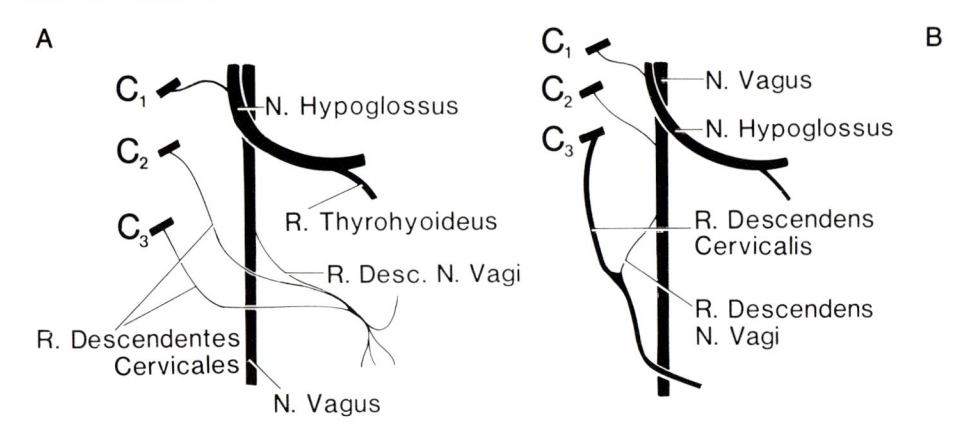

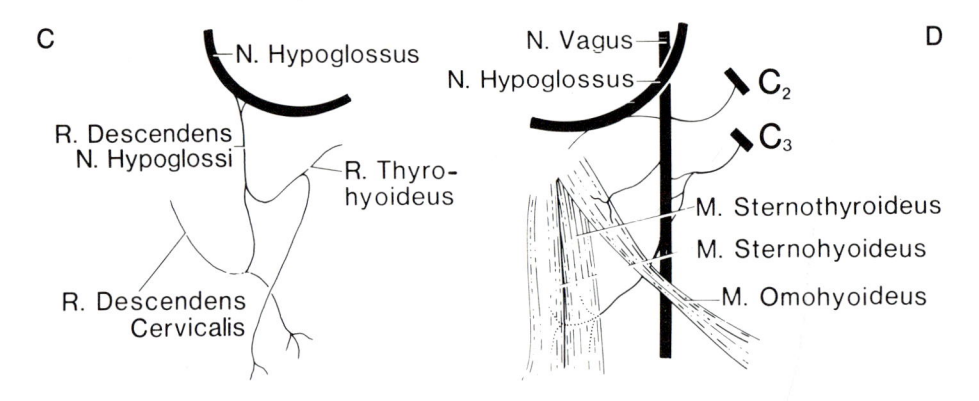

FIG. 31.
Four Variations of Ansa Hypoglossi.

A: Absence of descendens hypoglossi. A branch of C₁ joins the hypoglossal nerve. Branches of C₂, C₃ join anomalous descendens vagi.

B: Absence of descendens hypoglossi. A branch from C₁ joins the hypoglossal nerve. A branch from C₂ joins the vagus. C₃ provides the major part of descendens cervicalis along with a branch from the vagus, descendens vagi, to form ansi vagi.

C: The branch to m. thyrohyoideus departs from the ansa hypoglossi rather than being carried by and leaving from the hypoglossal nerve.

D: Absence of descendens hypoglossi. Branches from C₂ and C₃ join the vagus, which then provides the branches to infrahyoid muscles.

Redrawn and modified from Wimmer, A. Einige, darunter seltene Nervenvariationen des Halsgebietes. Anat. Anz. 105:391–396, 1958.

FIG. 32.
Variations in Relations Between Superficial Origin of Posterior Roots of Spinal Nerves and Spinous Processes of Vertebrae.

A: Drawing of a dissection made by R.W. Reid showing the relation between the superficial origins of the posterior roots of the spinal nerves and the spinous processes of the vertebrae. C, cervical; D (dorsal), thoracic; L, lumbar; S, sacral vertebrae.

B: Chart showing the limits of the variations in the topographical relation between the superficial origins of the nerve roots from the spinal cord and the spinous processes of the vertebrae. Abbreviations as in drawing A.

The following summary may therefore be made of the limits within which Reid found the posterior and anterior nerve roots take their superficial origin from the cord, in relation to the posterior ends of the spinous processes in the six specimens examined. (A signifies the highest point of origin; B, the lowest point of origin.)

Nerves

2nd Cervical: A, A little above posterior arch of atlas; B, midway between posterior arch of atlas and spine of axis.
3rd Cervical: A, A little below posterior arch of atlas; B, junction of upper ⅔ and lower ⅓ of spine of axis.
4th Cervical: A, Just below upper border of spine of axis; B, middle of spine of C₃.
5th Cervical: A, Just below lower border of spine of axis; B, just below lower border of spine of C₄.
6th Cervical: A, Lower border of spine of C₃; B, lower border of spine of C₅.
7th Cervical: A, Just below upper border of spine of C₄; B, just above lower border of spine of C₆.
8th Cervical: A, Upper border of spine of C₅; B, upper border of spine of C₇.
1st Dorsal: A, Midway between spines of C₅ and C₆; B, junction of upper ⅔ and lower ⅓ of interval between C₇ and D₁.
2nd Dorsal: A, Lower border of spine of C₆; B, just above lower border of spine of D₁.

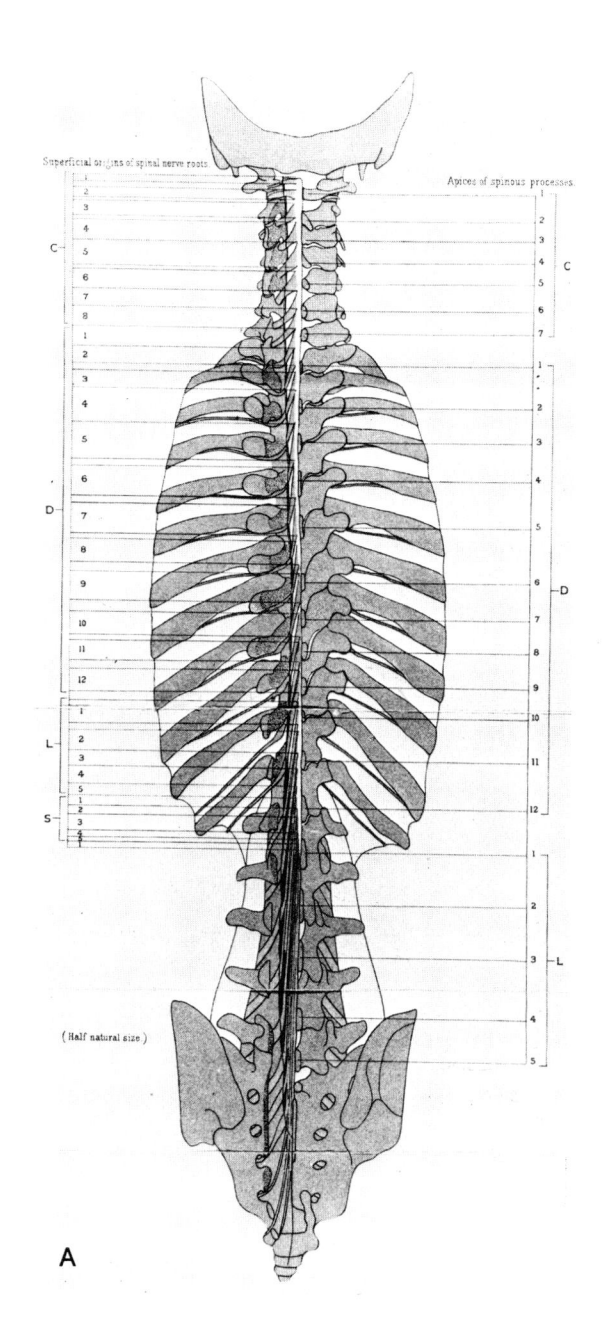

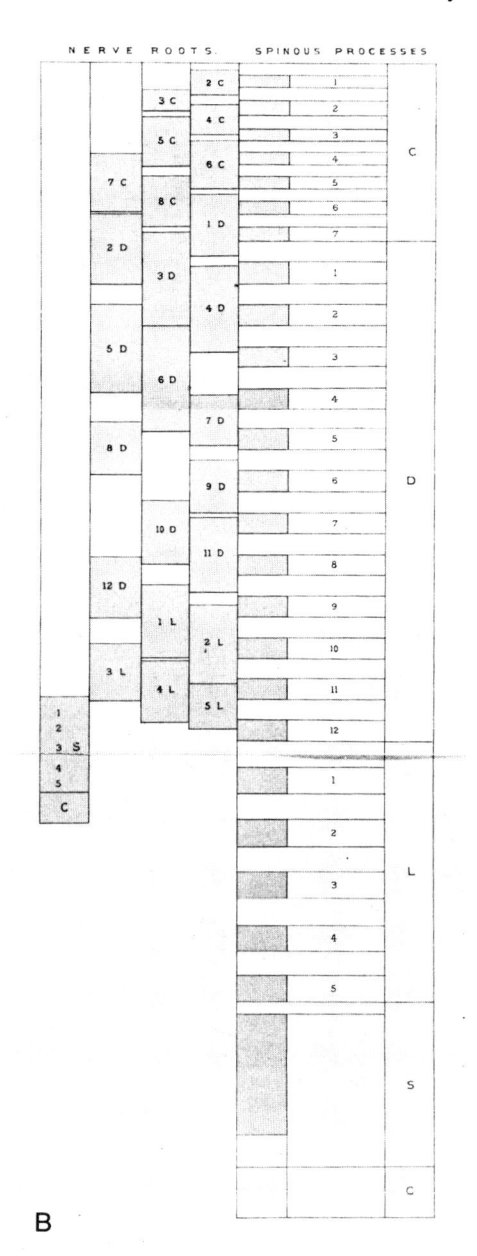

3rd Dorsal: A, Just above middle of spine of C₇; B, lower border of spine of D₂.
4th Dorsal: A, Just below upper border of spine of D₁; B, junction of upper ⅓ and lower ⅔ of spine of D₃.
5th Dorsal: A, Upper border of spine of D₂; B, junction of upper ¼ and lower ¾ of spine of D₄.
6th Dorsal: A, Lower border of spine of D₂; B, just below upper border of spine of D₅.
7th Dorsal: A, Junction of upper ⅓ and lower ⅔ of spine of D₄; B, just above lower border of spine of D₅.
8th Dorsal: A, Junction of upper ⅔ and lower ⅓ of interval between spines of D₄ and D₅; B, junction of upper ¼ and lower ¾ of spine of D₆.
9th Dorsal: A, Midway between spines of D₅ and D₆; B, upper border of spine of D₇.
10th Dorsal: A, Midway between spines of D₆ and D₇; B, middle of spine of D₈.
11th Dorsal: A, Junction of upper ¼ and lower ¾ of spine of D₇; B, just above spine of D₉.

12th Dorsal: A, Junction of upper ¼ and lower ¾ of spine of D₈; B, just below spine of D₉.
1st Lumbar: A, Midway between spines of D₈ and D₉; B, lower border of spine of D₁₀.
2nd Lumbar: A, Middle of spine of D₉; B, junction of upper ⅓ and lower ⅔ of spine of D₁₁.
3rd Lumbar: A, Middle of spine of D₁₀; B, just below spine of D₁₁.
4th Lumbar: A, Just below spine of D₁₀; B, junction of upper ¼ and lower ¾ of spine of D₁₂.
5th Lumbar: A, Junction of upper ⅓ and lower ⅔ of spine of D₁₁; B, middle of spine of D₁₂.
1st Sacral: A, Just above lower border of spine of D₁₁.
5th Sacral: B, Lower border of spine of L₁.
Coccygeal: A, Lower border of spine of L₁; B, just below upper border of spine of L₂.
From Reid, R.W. The relations between the superficial origins of the spinal nerves from the spinal cord and the spinous processes of the vertebrae. J. A. Physiol. 23:341–353, 1889.

Human Anatomic Variation

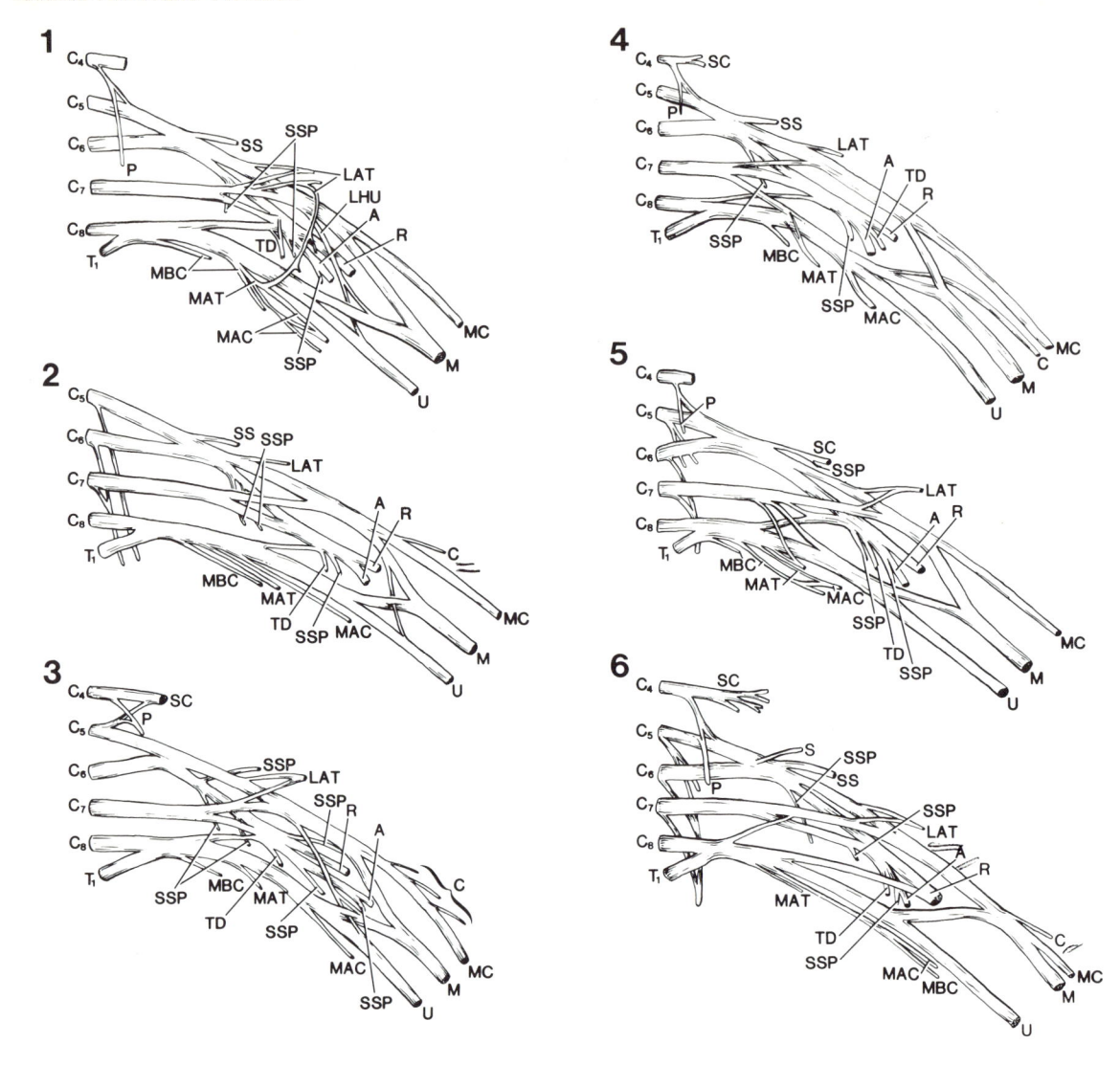

FIG. 33.
Twenty-nine Variations in Form of Brachial Plexus.

A, Axillary nerve; *C*, nerve to coracobrachialis; C_4 through C_8, cervical spinal nerves, 4 through 8; *IC*, intercostal nerves; *LAT*, lateral thoracic nerve; *MAC*, medial antibrachial cutaneous nerve; *MAT*, medial anterior thoracic nerve; *MBC*, medial brachial cutaneous nerve; *LHU*, lateral head of ulnar nerve; *LP*, nerve to pectoralis major; *LT*, long thoracic nerve; *M*, median nerve; *MC*, musculocutaneous nerve; *MP*, nerve to pectoralis minor; *NR*, nerve to rhomboids; *P*, phrenic nerve; *R*, radial nerve; *S*, nerve to subclavius; *SA*, spinal accessory nerve; *SC*, supraclavicular nerve; *SS*, suprascapular nerves; *SSP*, subscapular nerve; T_1, first thoracic spinal nerve; *TD*, thoracodorsal nerve; *U*, ulnar nerve.
Variations are continued in Figures 33 through 41.

Redrawn from Kerr, A.T. The brachial plexus of nerves in man. *Am. J. Anat.* 23:285–395, 1918.

FIG. 34.
Twenty-nine Variations in Form of Brachial Plexus (continued).
Abbreviations as in Figure 33.

Redrawn from Kerr, A.T. The brachial plexus of nerves in man. *Am. J. Anat.* 23:285–395, 1918.

Nervous System

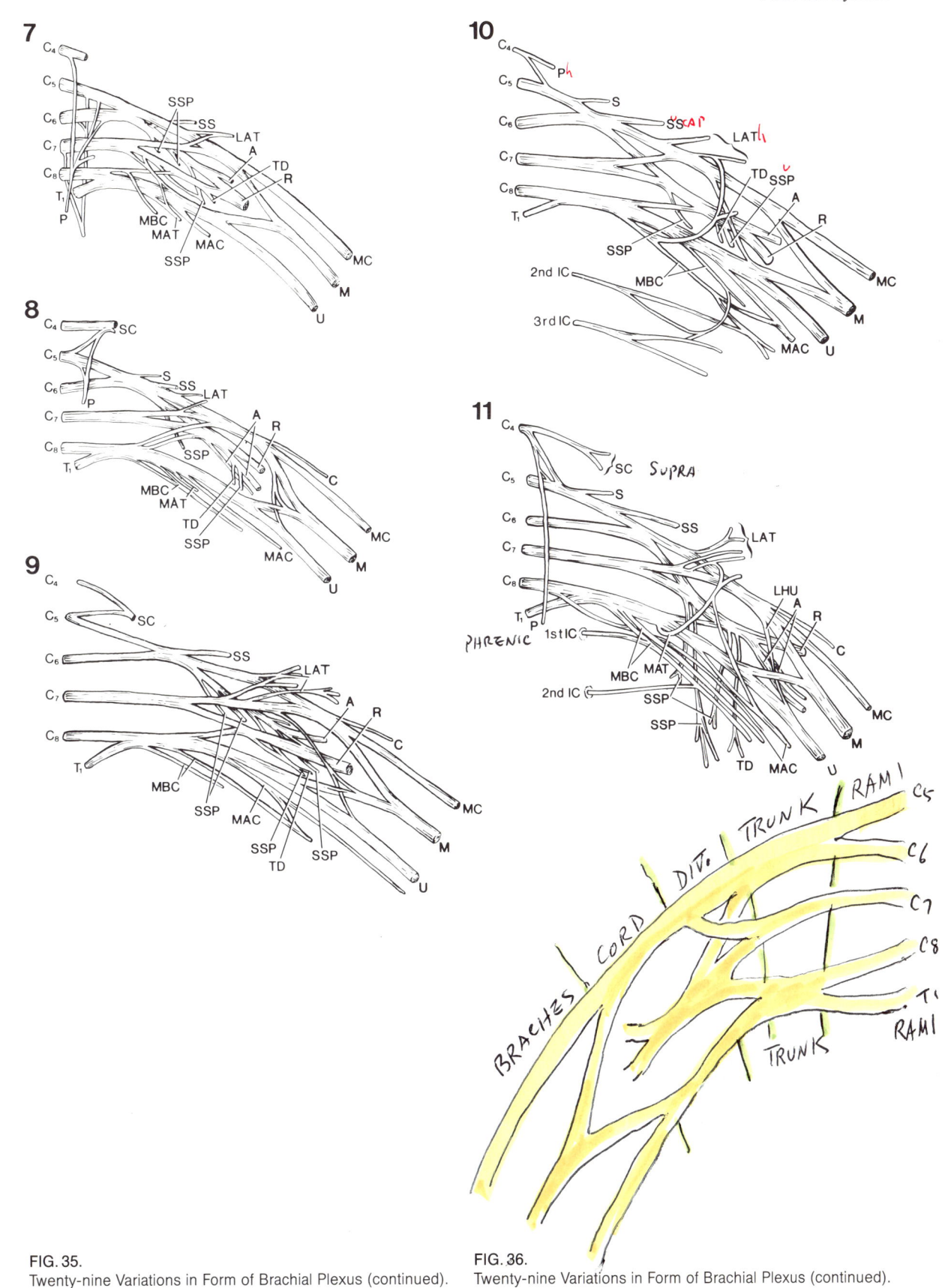

FIG. 35.
Twenty-nine Variations in Form of Brachial Plexus (continued).
Abbreviations as in Figure 33.

Redrawn from Kerr, A.T. The brachial plexus of nerves in man. *Am. J. Anat.* 23:285–395, 1918.

FIG. 36.
Twenty-nine Variations in Form of Brachial Plexus (continued).
Abbreviations as in Figure 33.

Redrawn from Kerr, A.T. The brachial plexus of nerves in man. *Am. J. Anat.* 23:285–395, 1918.

475

Human Anatomic Variation

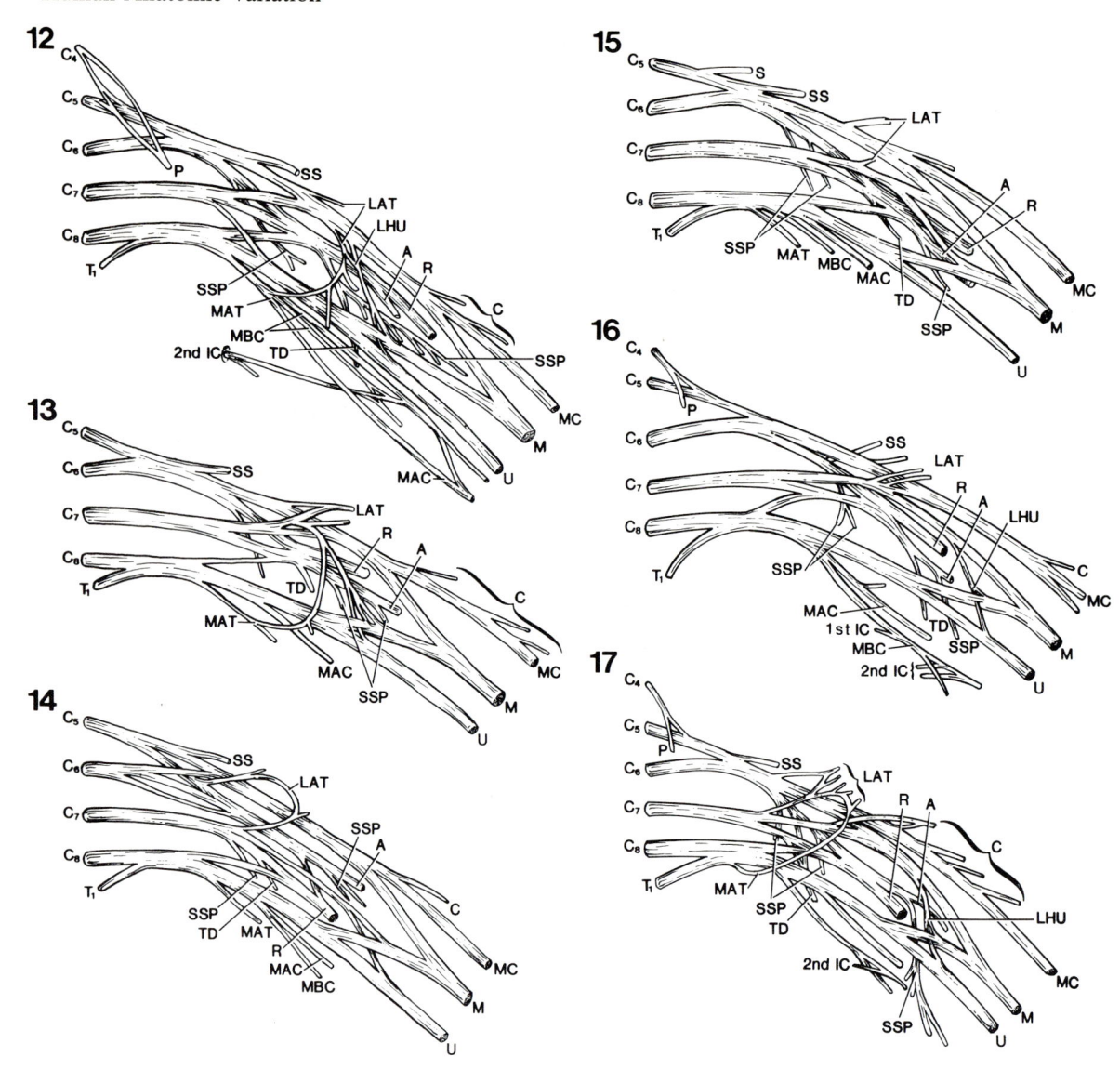

FIG. 37.
Twenty-nine Variations in Form of Brachial Plexus (continued). Abbreviations as in Figure 33.

Redrawn from Kerr, A.T. The brachial plexus of nerves in man. *Am. J. Anat.* 23:285–395, 1918.

FIG. 38.
Twenty-nine Variations in Form of Brachial Plexus (continued). Abbreviations as in Figure 33.

Redrawn from Kerr, A.T. The brachial plexus of nerves in man. *Am. J. Anat.* 23:285–395, 1918.

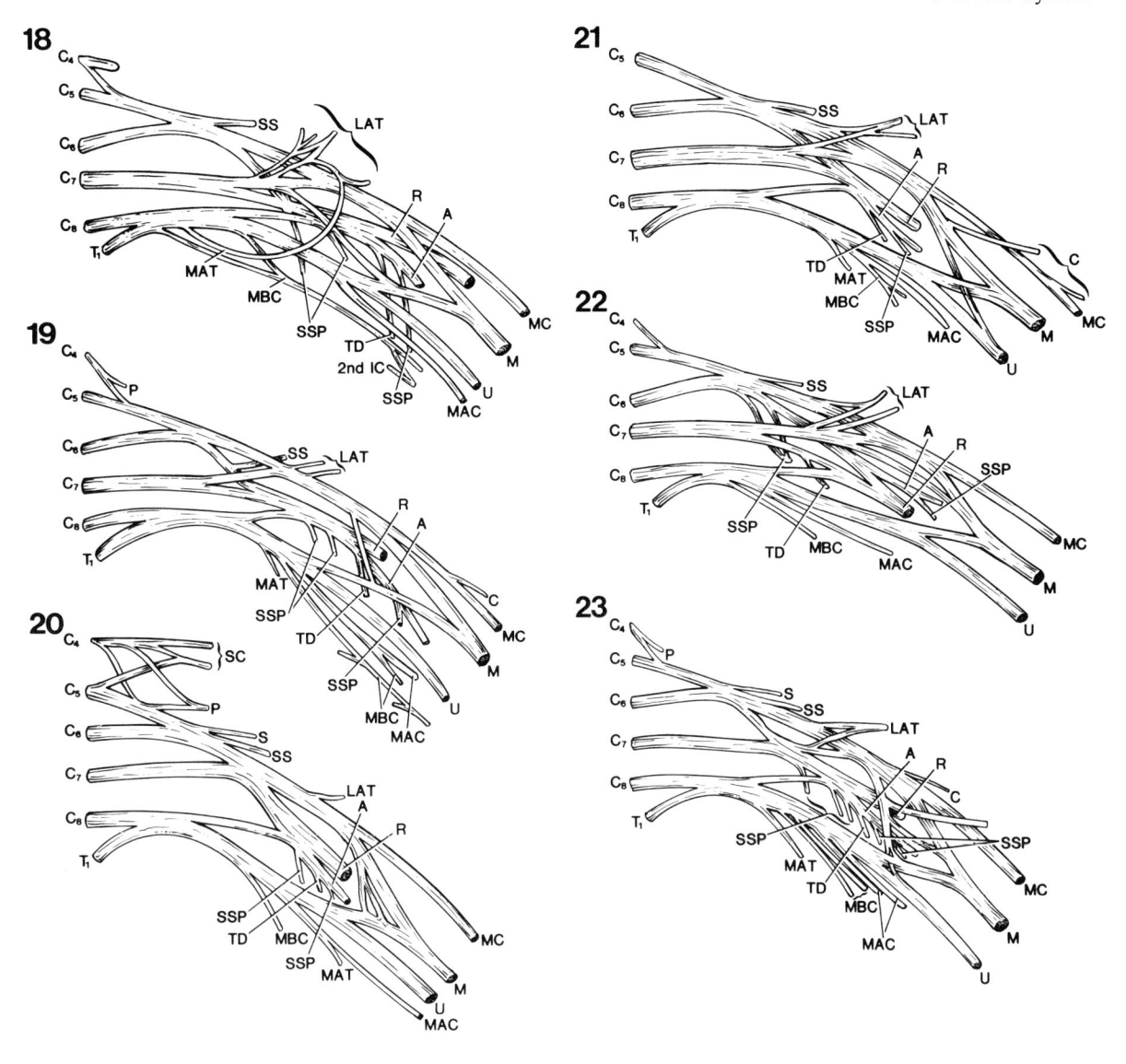

FIG. 39.
Twenty-nine Variations in Form of Brachial Plexus (continued).
Abbreviations as in Figure 33.
Redrawn from Kerr, A.T. The brachial plexus of nerves in man. *Am. J. Anat.* 23:285–395, 1918.

FIG. 40.
Twenty-nine Variations in Form of Brachial Plexus (continued).
Abbreviations as in Figure 33.
Redrawn from Kerr, A.T. The brachial plexus of nerves in man. *Am. J. Anat.* 23:285–395, 1918.

Human Anatomic Variation

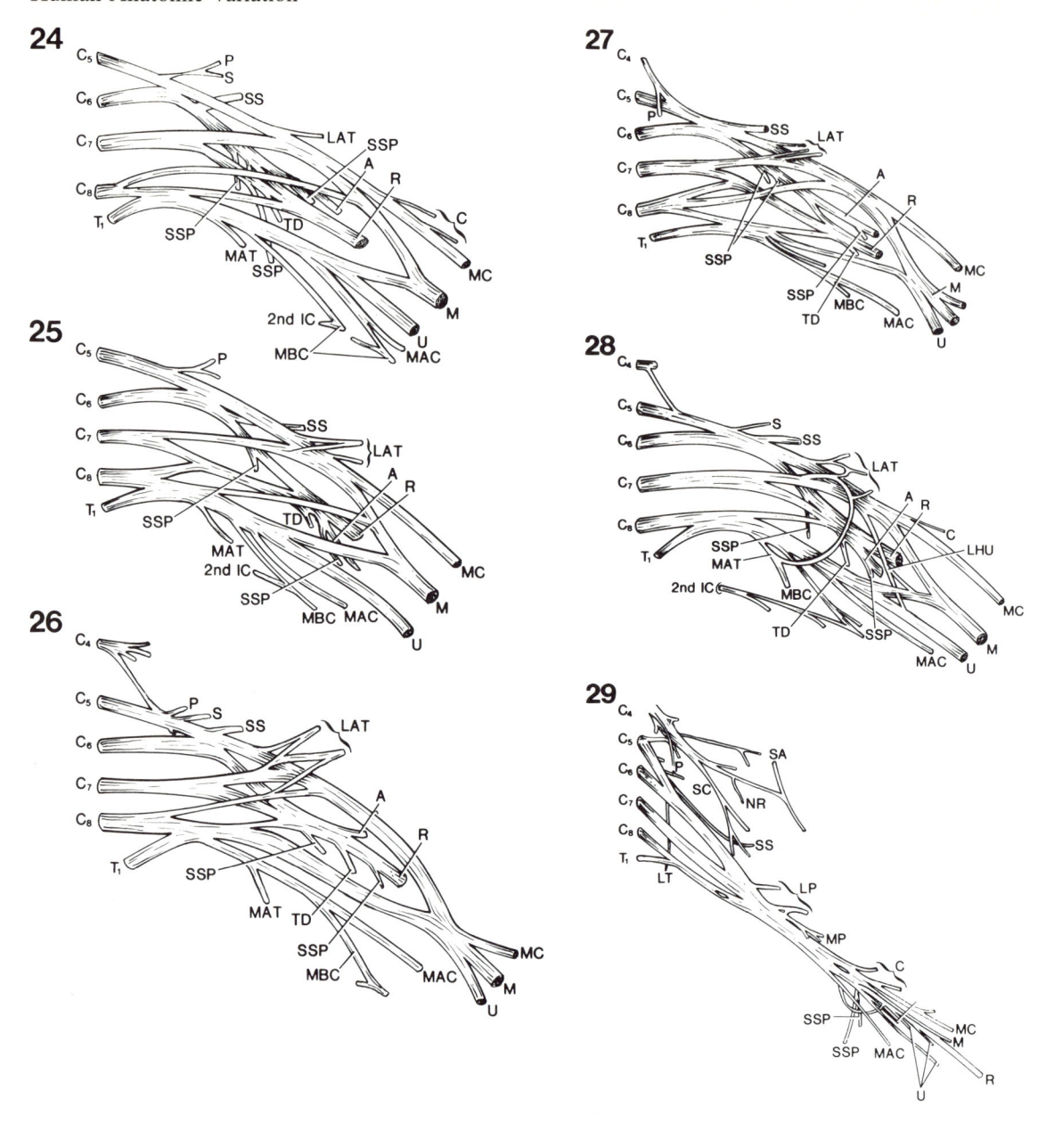

FIG. 41.
Twenty-nine Variations in Form of Brachial Plexus (continued). Abbreviations as in Figure 33.

Redrawn from Kerr, A.T. The brachial plexus of nerves in man. *Am. J. Anat.* 23:285–395, 1918.

FIG. 42.
Twenty-nine Variations in Form of Brachial Plexus (continued). Abbreviations as in Figure 33.

Redrawn (variations 27, 28) from Kerr, A.T. The brachial plexus of nerves in man. *Am. J. Anat.* 23:285–395, 1918; and (variation 29) from Singer, E. Human brachial plexus united into a single cord. Description and interpretation. *Anat. Rec.* 54:411–419, 1932.

Nervous System

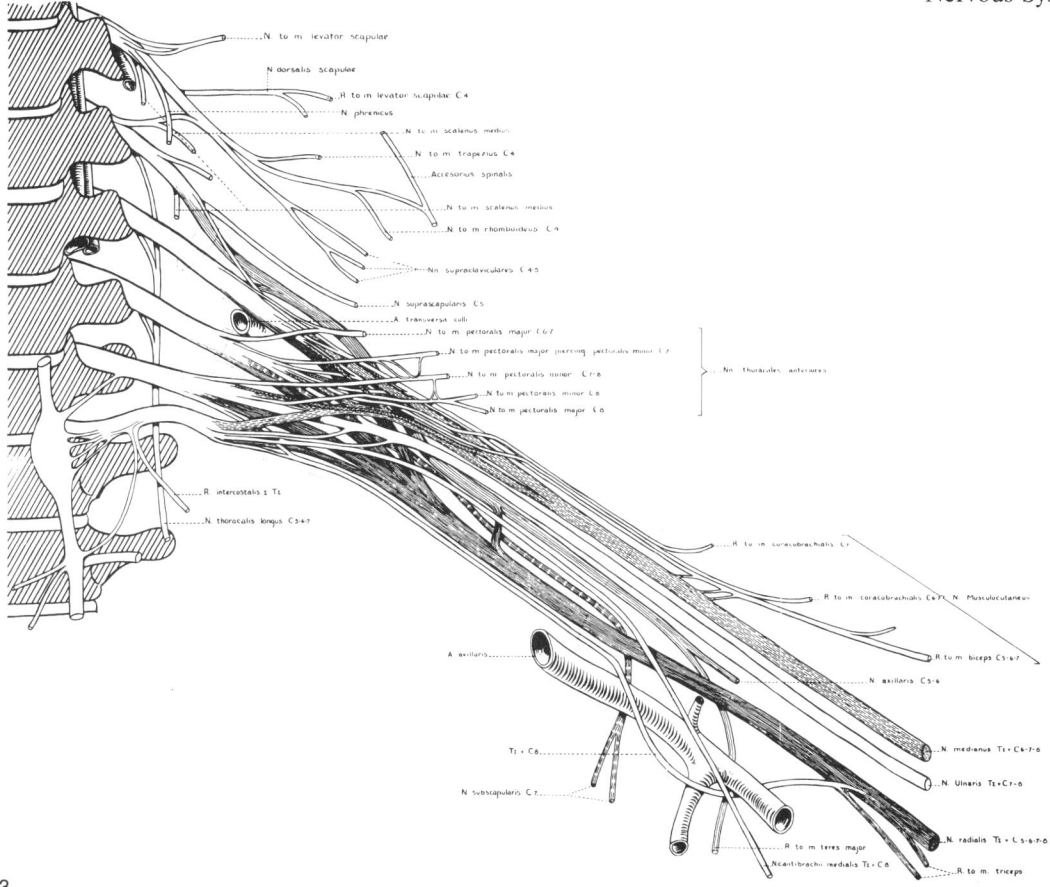

FIG. 43.
Variation in Formation of Brachial Plexus with Union of Branches into a Single Cord (An Analysis of Variation 29 in Figure 42.)

From Singer, E. Human brachial plexus united into a single cord. Description and interpretation. Anat. Rec. 54:411–419, 1932.

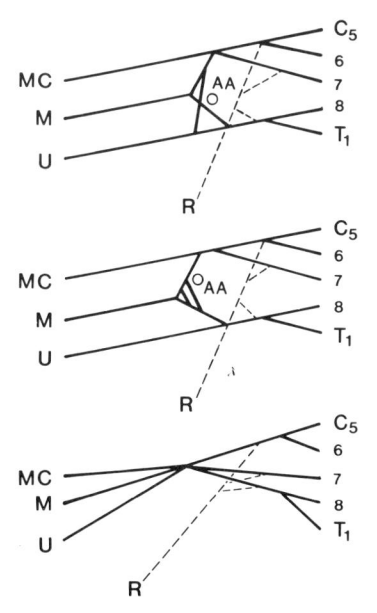

FIG. 44.
Thirty-eight Variations in Form of Brachial Plexus (Sketched Diagrammatically).

A, Axillary nerve; AA, axillary artery (position); M, median nerve; MAC, medial antebrachial cutaneous nerve; MBC, medial brachial cutaneous nerve; MC, musculocutaneous nerve; R, radial nerve; U, ulnar nerve.

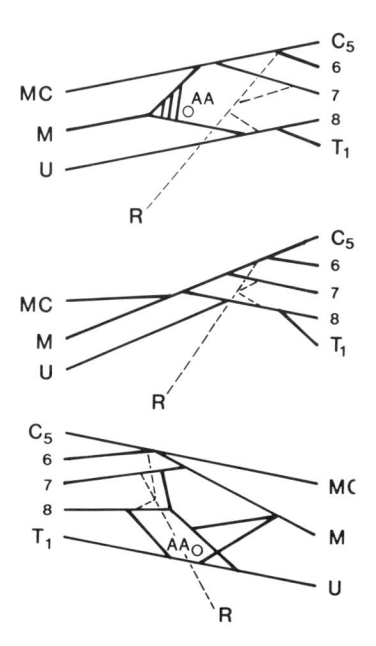

Variations are continued in Figures 45–49.

From Hirasawa, K. Arbeiten aus der dritten Abteilung des Anatomischen Institutes der Kaiserlichen Universität Kyoto. Series A: Untersuchungen über das periphere Nervensystem. Book 2: Plexus brachialis und die Nerven des peripheren Nervensystems. Kyoto 1931.

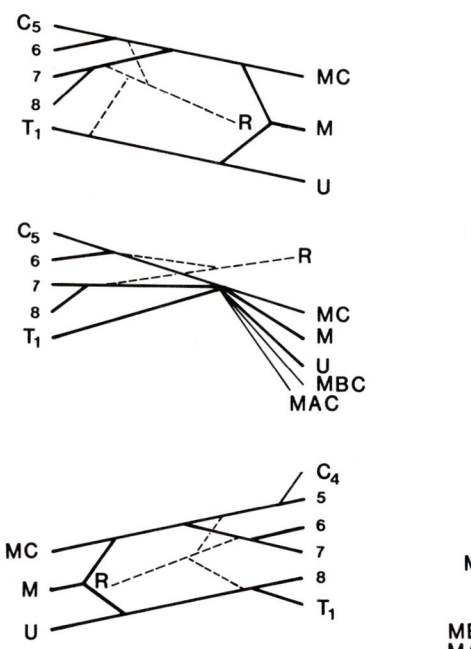

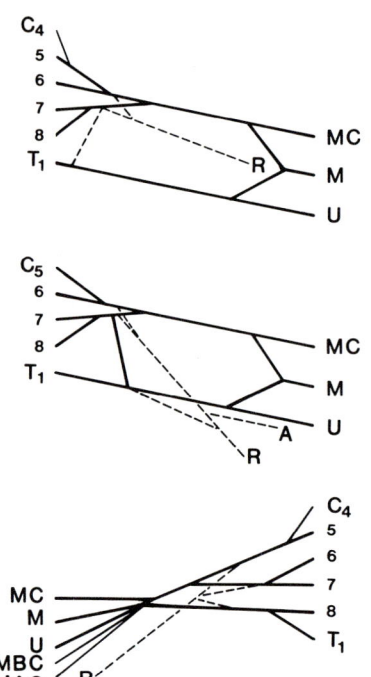

Human Anatomic Variation

FIG. 45.
Thirty-eight Variations in Form of Brachial Plexus (Sketched Diagrammatically).
Abbreviations as in Figure 44.

From Hirasawa, K. *Arbeiten aus der dritten Abteilung des Anatomischen Institutes der Kaiserlichen Universität Kyoto.* Series A: Untersuchungen über das periphere Nervensystem. Book 2: Plexus brachialis und die Nerven des peripheren Nervensystems. Kyoto 1931.

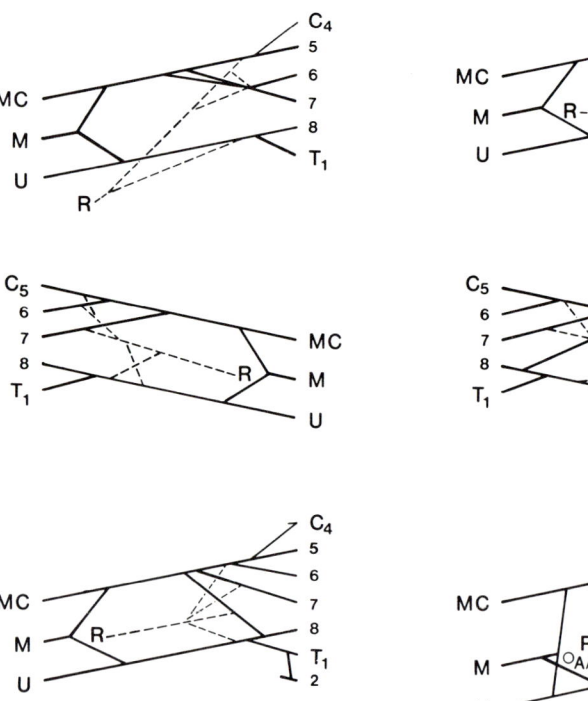

FIG. 46.
Thirty-eight Variations in Form of Brachial Plexus (Sketched Diagrammatically) (continued).
Abbreviations as in Figure 44.

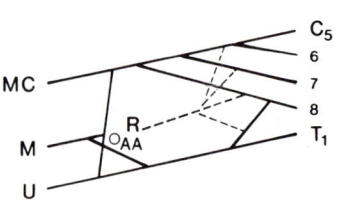

From Hirasawa, K. *Arbeiten aus der dritten Abteilung des Anatomischen Institutes der Kaiserlichen Universität Kyoto.* Series A: Untersuchungen über das periphere Nervensystem. Book 2: Plexus brachialis und die Nerven des peripheren Nervensystems. Kyoto 1931.

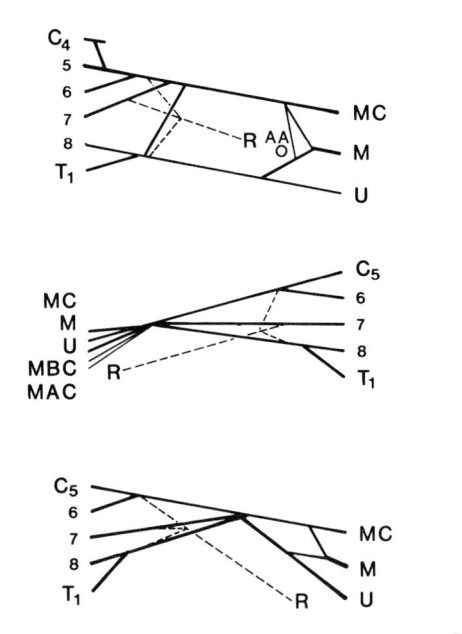

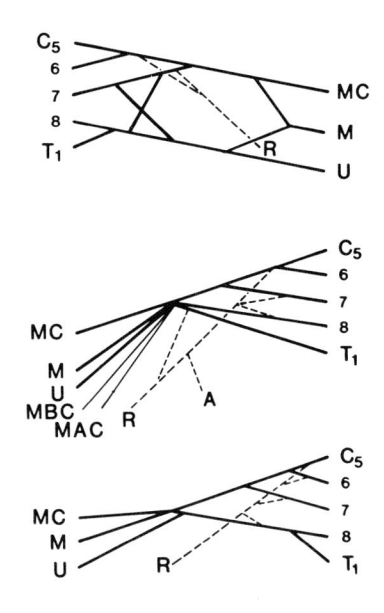

FIG. 47.
Thirty-eight Variations in Form of Brachial Plexus (Sketched Diagrammatically) (continued).
Abbreviations as in Figure 44.

From Hirasawa, K. *Arbeiten aus der dritten Abteilung des Anatomischen Institutes der Kaiserlichen Universität Kyoto*. Series A: Untersuchungen über das periphere Nervensystem. Book 2: Plexus brachialis und die Nerven des peripheren Nervensystems. Kyoto 1931.

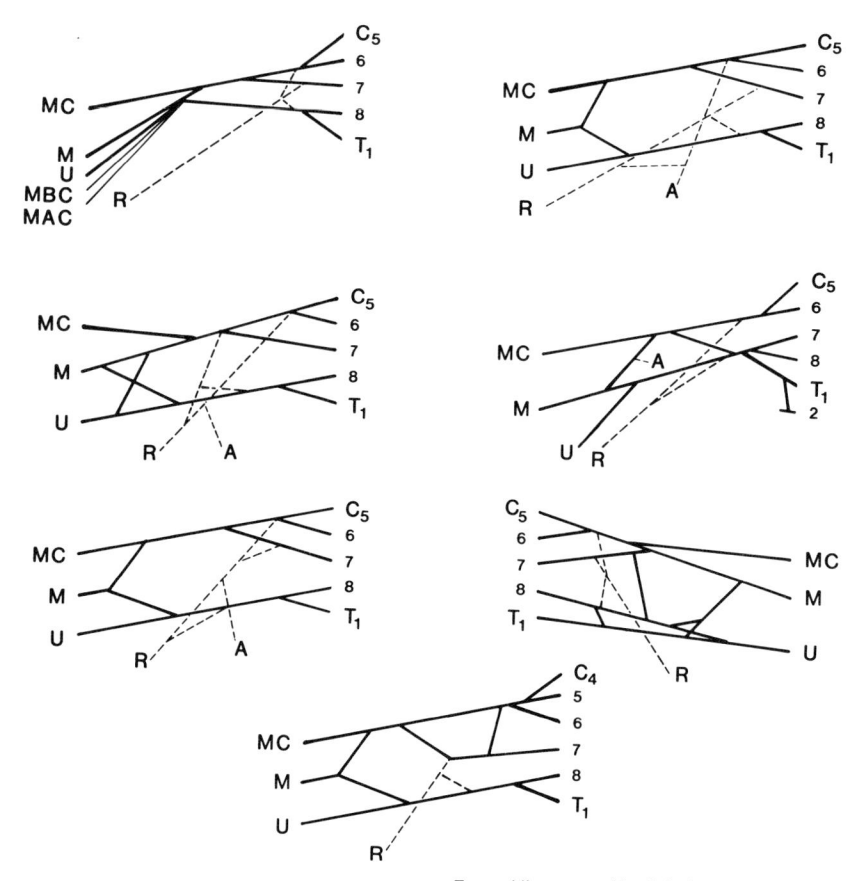

FIG. 48.
Thirty-eight Variations in Form of Brachial Plexus (Sketched Diagrammatically) (continued).
Abbreviations as in Figure 44.

From Hirasawa, K. *Arbeiten aus der dritten Abteilung des Anatomischen Institutes der Kaiserlichen Universität Kyoto*. Series A: Untersuchungen über das periphere Nervensystem. Book 2: Plexus brachialis und die Nerven des peripheren Nervensystems. Kyoto 1931.

Human Anatomic Variation

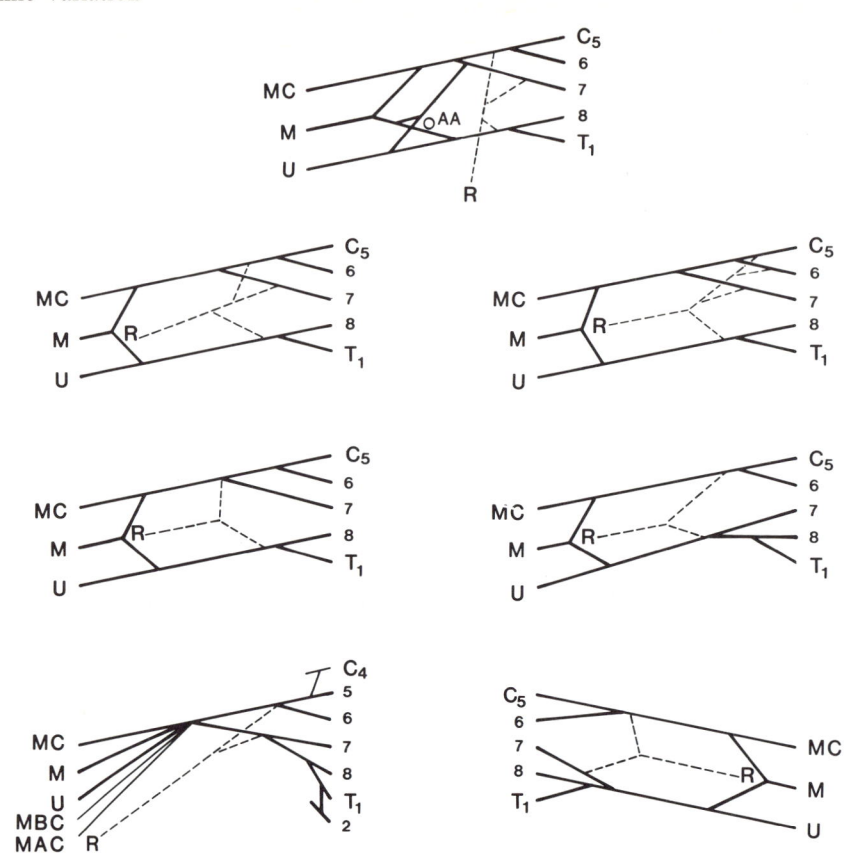

FIG. 49.
Thirty-eight Variations in Form of Brachial Plexus (Sketched Diagrammatically) (continued).
Abbreviations as in Figure 44.

From Hirasawa, K. *Arbeiten aus der dritten Abteilung des Anatomischen Institutes der Kaiserlichen Universität Kyoto*. Series A: Untersuchungen über das periphere Nervensystem. Book 2: Plexus brachialis und die Nerven des peripheren Nervensystems. Kyoto 1931.

FIG. 51.
Rare Form of Prefixed Brachial Plexus.

"In this case the nerve sheaths have not been dissected away at all, and only the chief branches of the plexus are shown. There is no outer cord as usually formed, but the musculo-cutaneous is seen arising from the ventral branch of 5–6 C., which also gives a descending branch, which is joined by a ventral branch from 7 C., to form the outer head of the median. Note the large lower ventral branch of 7 C. which passed behind the subclavian artery to reach the inner cord; this gave a large bundle to the ulnar. This arrangement of branches to the outer cord is very unusual, with the musculo-cutaneous being given off from 5–6 before the ventral branches of 5–6–7 have joined to form the outer cord. I have only seen it amongst animals in some of the Marsupials, and in one of the Edentata; in a large Wallaroo (*Macropus robustus*), and in *Dasyurus viverrinus*, a quadrupedal carnivorous marsupial; and also in the Armadillo (*Dasypus villosus*)."

From Harris, W. The true form of the brachial plexus, and its motor distribution. *J. Anat. Physiol.* 38:399–422, 1904.

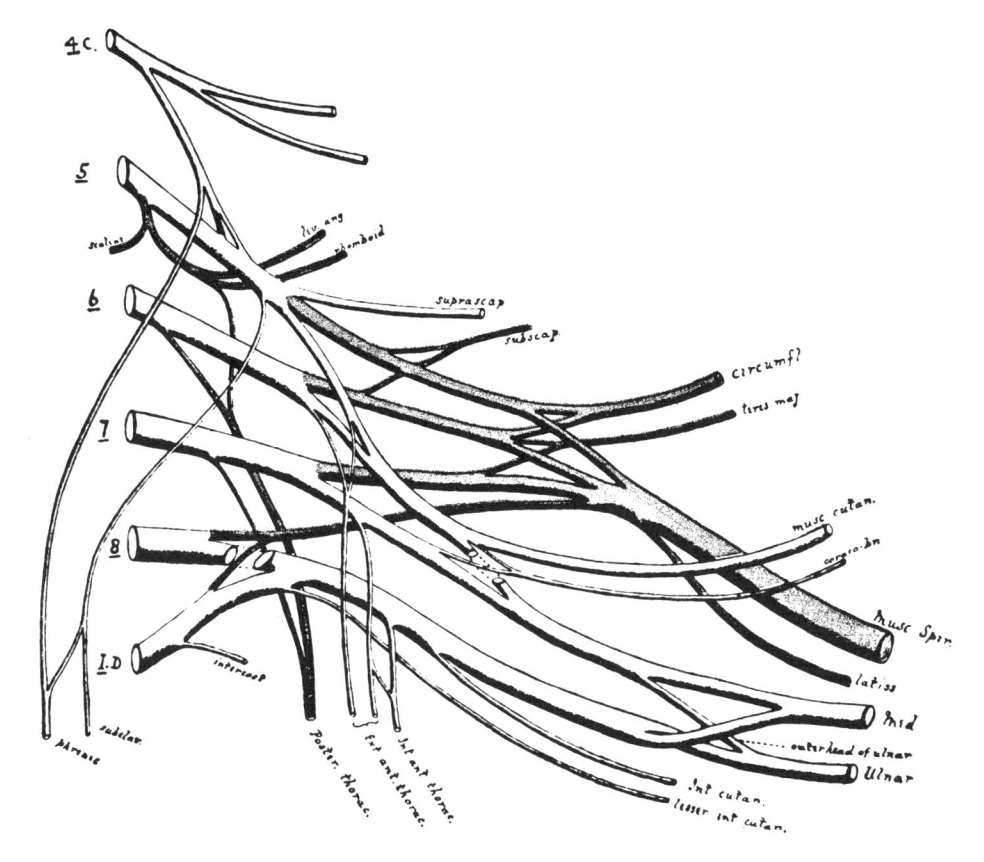

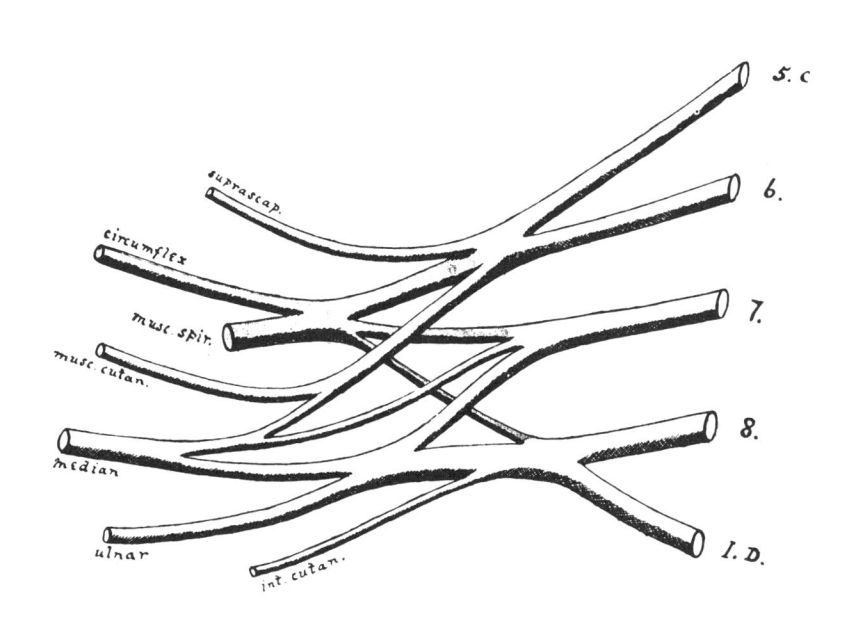

FIG. 50.
Prefixed Brachial Plexus.

"The connective tissue sheaths of the nerves are all dissected off, and thus the posterior and outer cords are seen resolved into their component parts. The branch to the outer cord from 5–6 C. has been divided in order to show the origin of the nerve to the coraco-brachialis; the branch from 8 C. to the inner cord has also been similarly divided to show the posterior branch of 1 D. Note the large size of the outer head of the ulnar, piercing directly through the inner head of the median. Note also that the inner head of the median is given off from the lower and anterior portion of the inner cord, and passes in front of the ulnar, which is given off from the posterior bundles of the inner cord."

From Harris, W. The true form of the brachial plexus, and its motor distribution. *J. Anat. Physiol.* 38:399–422, 1904.

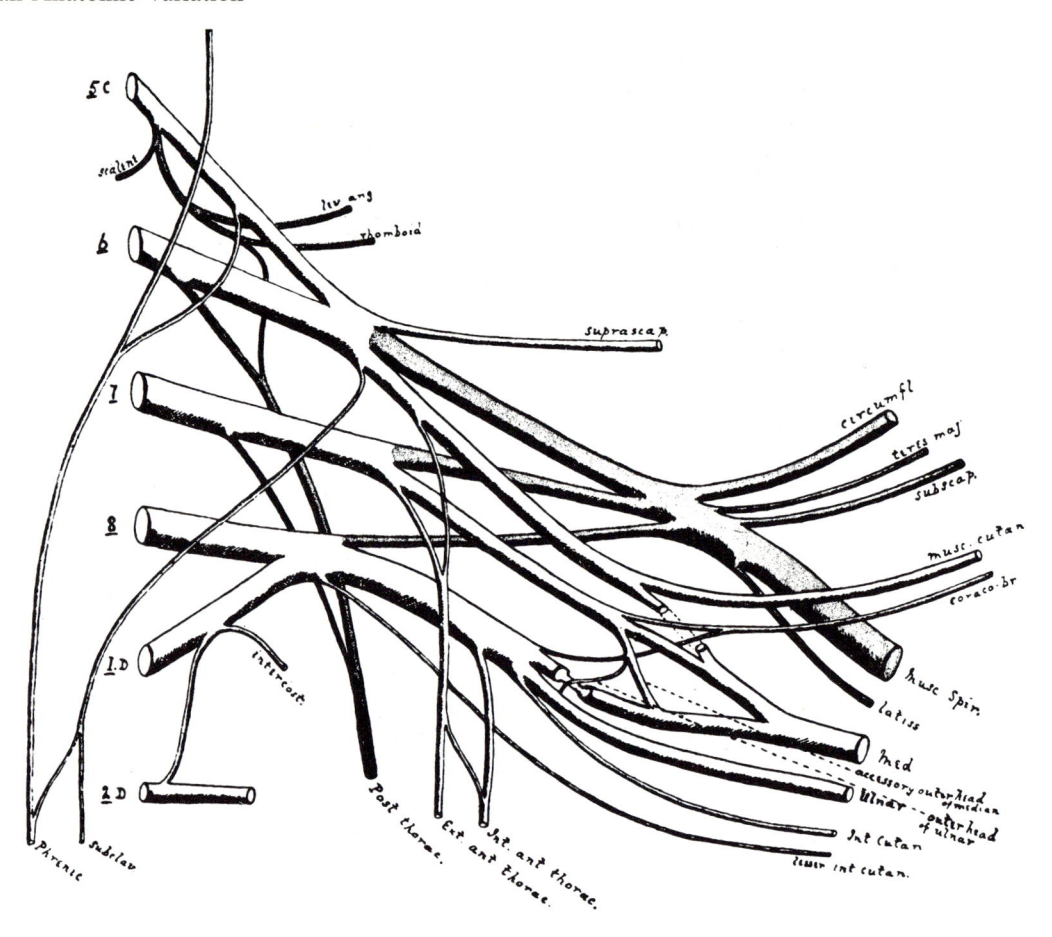

FIG. 52.
Postfixed Brachial Plexus.

"In this case the nerve sheaths have not been dissected off the posterior cord, but the outer cord has been dissected to show the origin of the nerve to the coraco-brachialis from 7–8 C. Note the small size of 5 C., and also the fine recurrent branch which leaves the accessory outer head of the median, and perforates the inner head of the median to form the outer head of the ulnar."

From Harris, W. The true form of the brachial plexus, and its motor distribution. *J. Anat. Physiol*. 38:399–422, 1904.

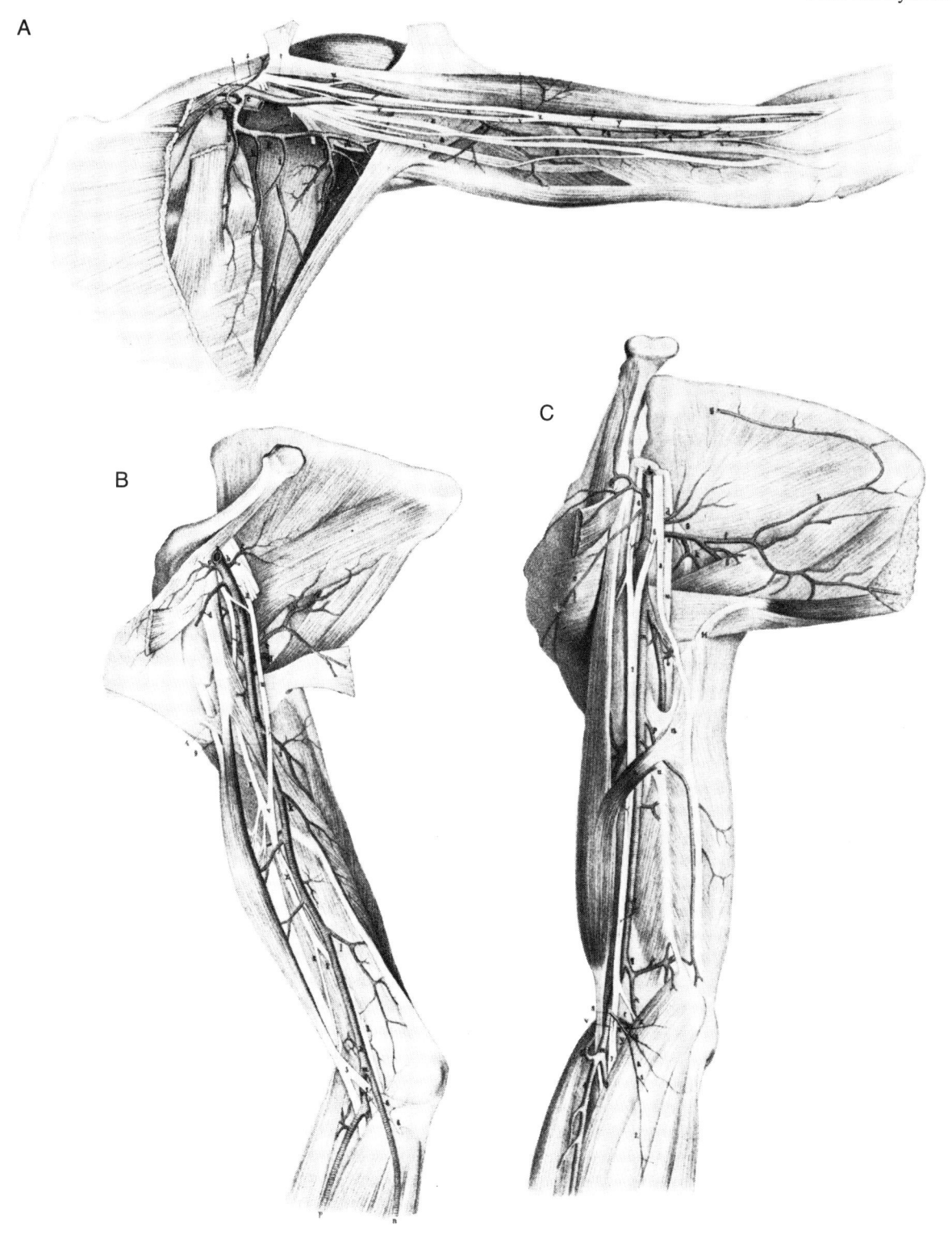

FIG. 53.
Spiral Course of Median Nerve Around Brachial Artery.
Median Nerve and Brachial Artery Passing Behind Third Head of Biceps.

A: Subscapular artery arising from the thoracoacromial artery. Variation in the form of brachial plexus.

B: Spiral course of median nerve around brachial artery. Aponeurotic expansion of fascia passing between short head of biceps brachii and humerus and between cords of brachial plexus.

C: Third head of biceps joining brachial fascia, passing over brachial artery and median nerve and between median and ulnar nerves. Tendinous slip from latissimus dorsi muscle to brachial fascia.

From Calori, L. Delle anomalie piu importanti di ossa, vasi, nervi, e muscoli occorse nell'ultimo biennio facendo anatomia del corpo umano. *Mem. Accad. Sci. Istituto di Bologna S. 2* 8:417–482, 1868.

Human Anatomic Variation

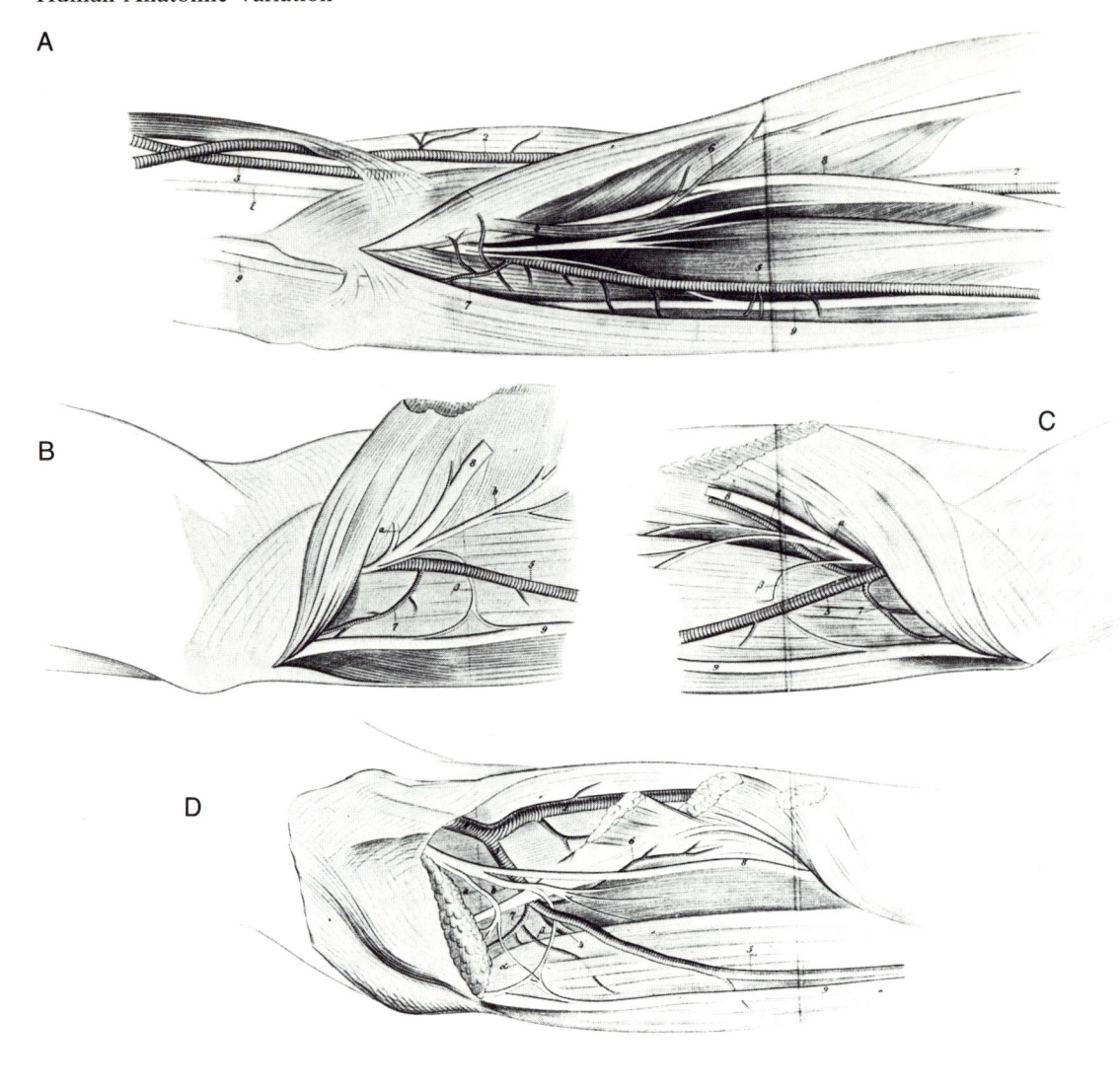

FIG. 54.
Median-Ulnar Nerve Anastomoses in Forearm.

1, Brachial artery; *2*, radial artery; *3*, anterior interosseous artery; *4*, common interosseous artery; *5*, ulnar artery; *6*, deep median artery; *7*, ulnar recurrent artery; *8*, median nerve; *9*, ulnar nerve; *a*, superficial branch of the median nerve; *b*, deep branch of the median nerve; α, superior anastomotic branch between the median nerve and the ulnar nerve (in D, as a root to the ulnar nerve from the median nerve); β, inferior anastomosis between the two nerves (in A, as a root to the ulnar nerve from the median nerve; in B and D, as a root of the ulnar with recurrent loops between the two nerves; in C, as a single loop.)

A: Upper elbow and forearm region of the left side. This preparation also had a high origin of the radial artery from the brachial artery.
B: Upper elbow region of the left side.
C: Upper elbow region of the right side.
D: Upper elbow and forearm region of the left side.

From Gruber, W. Über die Verbindung des Nervus medianus mit dem Nervus ulnaris am Unterarme des Menschen und der Säugethiere. *Arch. Anat. Physiol. Wissen. Med*. 1870, pp. 501–522.

Nervous System

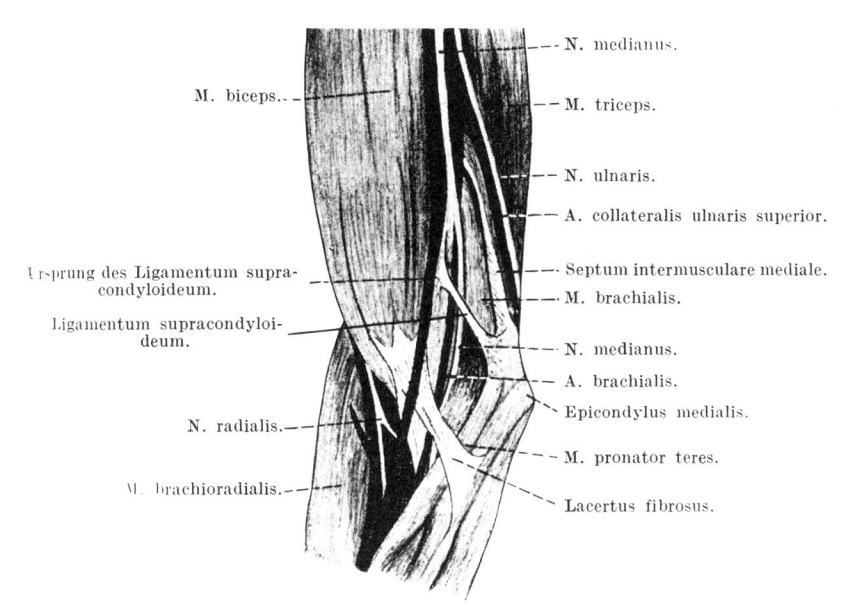

FIG. 55.
Supracondylar Ligament Crossing Median Nerve and Brachial Artery.

Supracondylar ligament (of Struthers) without a supracondylar process forming a canal through which the median nerve and brachial artery pass. Supracondylar canals are potential entrapment sites for median nerves and brachial arteries that may require surgical treatment.

From Barta, E. and L. Petrovits. Das Ligamentum supracondyloideum. *Anat. Anz.* 63:177–184, 1927.

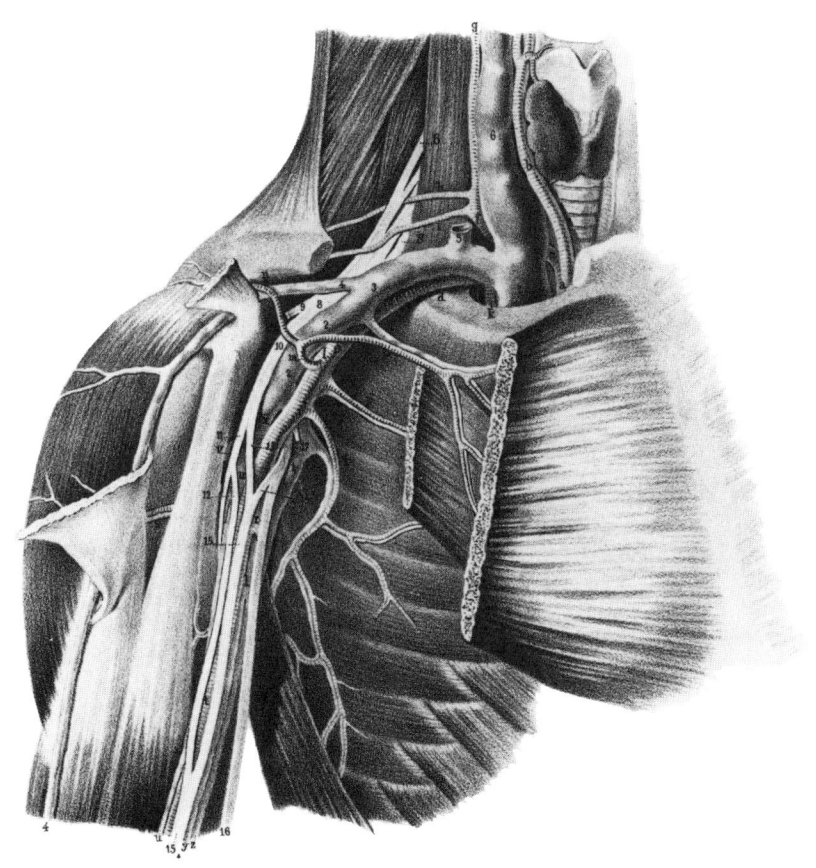

FIG. 56.
Axillary Artery and Vein Passing Through the Medial and Lateral Cords of Brachial Plexus.

The axillary artery and vein are found passing between the medial and lateral cords of brachial plexus which form the median nerve. The median nerve is formed by two branches of the lateral cord and one branch (normal) from the medial cord. Transverse cervical artery passes between ventral rami of C_6 and C_7.

From Calori, L. Sul variato rapporto vascolare delle due radici del nervo mediano e sulle condizioni atte a produrlo. *Mem. Accad. Sci. Istituto di Bologna S. 3* 8:443–456, 1877.

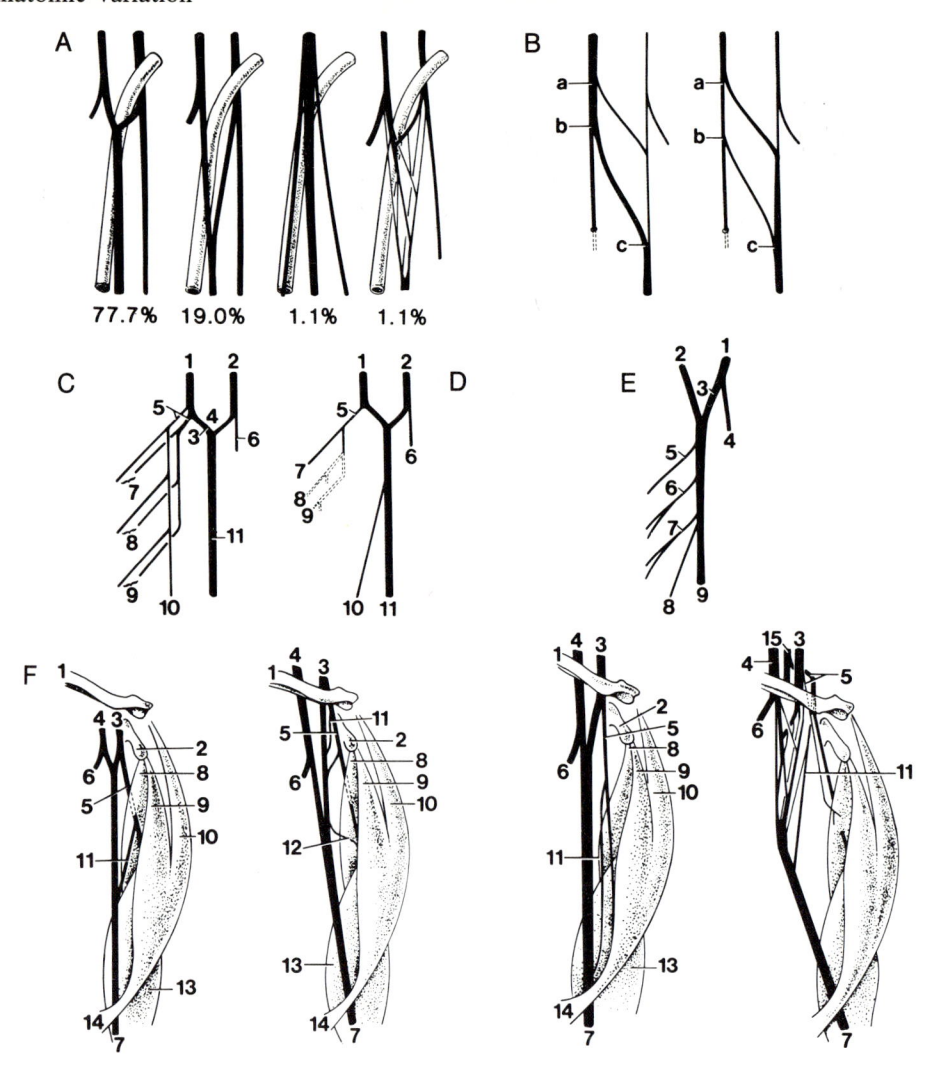

FIG. 57.

Variations in Median and Musculocutaneous Nerves of the Arm.

A: Formation of the median nerve. Relationship of axillary artery to the median nerve.

B: Variation in branching between the musculocutaneous nerve and the median nerve.

C: Doubled musculocutaneous nerve. 1, Ventral-lateral fasciculus (lateral cord); 2, ventro-medial fasciculus (medial cord); 3, branch from lateral cord to form median nerve; 4, branch from medial cord to form median nerve; 5, musculocutaneous nerve; 6, ulnar nerve; 7, branches for m. coracobrachialis; 8, branches for m. biceps brachii; 9, branches for m. brachialis; 10, cutaneous branch of musculocutaneous nerve (lateral cutaneous nerve of the forearm); 11, median nerve.

D: Short musculocutaneous nerve. In this case the lateral cutaneous nerve of the forearm (10) arose from the median nerve.

E: Absence of the musculocutaneous nerve. 1, Medial cord; 2, lateral cord; 3, branch from the medial cord to form the median nerve; 4, ulnar nerve; 5, branch for m. coracobrachialis; 6, branch for m. biceps brachii; 7, branch for m. brachialis; 8, lateral cutaneous nerve of forearm of musculocutaneous nerve; 9, median nerve.

F: Variations in the musculocutaneous nerve. 1, Clavicle; 2, coracoid process; 3, lateral cord; 4, medial cord; 5, musculocutaneous nerve; 6, ulnar nerve; 7, median nerve; 8, m. coracobrachialis; 9, m. biceps brachii, short head; 10, m. biceps brachii, long head; 11, lateral-medial anastomosis between the musculocutaneous and median nerves; 12, medial-lateral anastomosis between the median and musculocutaneous nerves; 13, m. brachialis; 14, lacertus fibrosus of m. biceps brachii; 15, posterior cord.

From Buch-Hansen, K. Über Varietäten des Nervus medianus und des Nervus musculocutaneus und deren Beziehungen. *Anat. Anz.* 102:187–203, 1955–56.

Nervous System

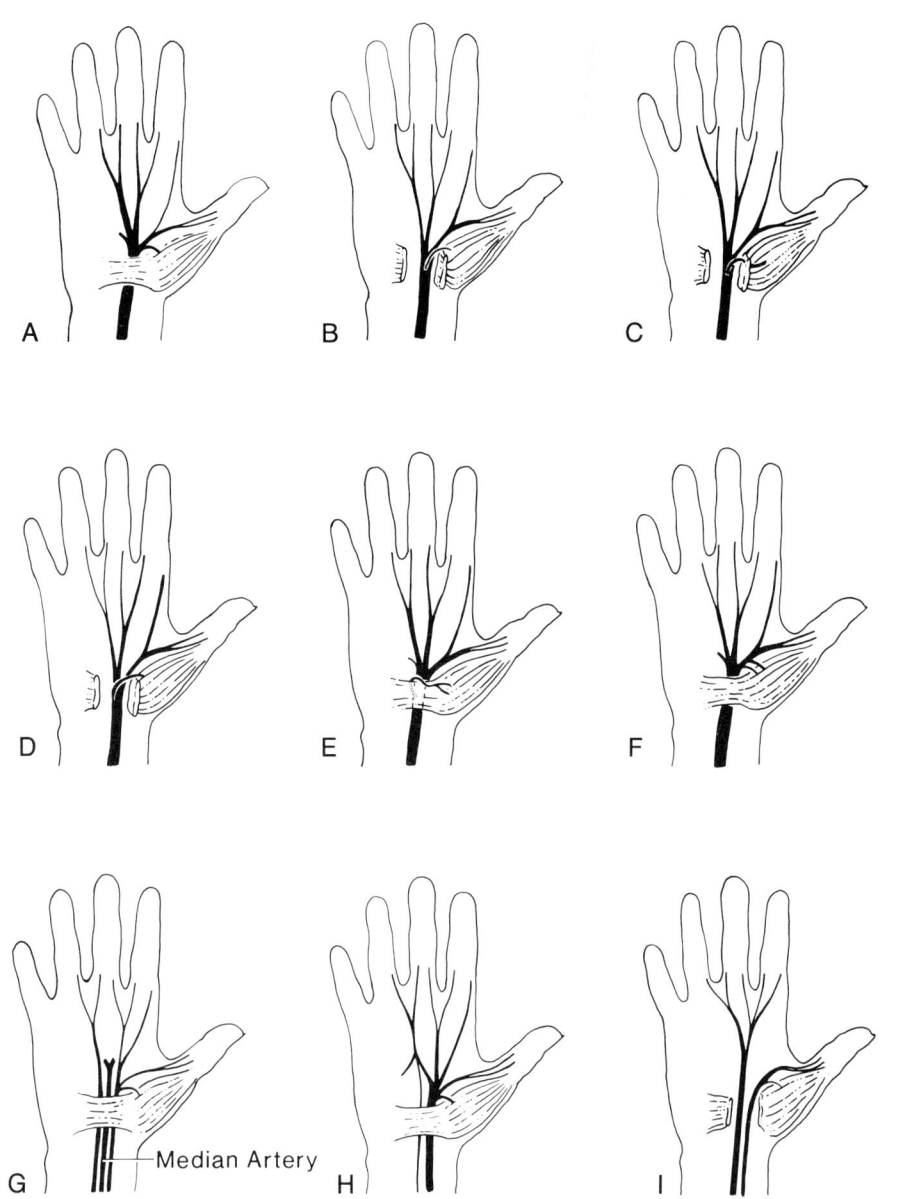

FIG. 58.
Variations in Course of Median Nerve at the Carpal Tunnel.

A: Usual branching of the median nerve.
B: Thenar branch leaving the median nerve within the carpal tunnel (subligamentous).
C: Transligamentous course of thenar branch.
D: Thenar branch leaving median nerve at its ulnar aspect.
E: Thenar branch crosses over the top of the transverse carpal ligament.
F: Doubled thenar motor branch.
G: High division of the median nerve with a persistent median artery between the two parts.
H: High division of the median nerve with a thinner ulnar part.
I: High division of the median nerve with a thicker ulnar part.

Redrawn from Lanz, U. Anatomical variations of the median nerve in the carpal tunnel. *J. Hand Surg.* 2:44–53, 1977.

Human Anatomic Variation

FIG. 59.
Variations in Course of Median Nerve at the Carpal Tunnel (continued).

A: High division of the median nerve with an accessory lumbrical muscle between the divided nerve.
B: "Accessory" median nerve branch proximal to the carpal tunnel perforating the transverse carpal ligament.
C: "Accessory" median nerve branch proximal to the carpal tunnel.
D: "Accessory" median nerve branch arising from the ulnar side of the median nerve proximal to the carpal tunnel.
E: "Accessory" median nerve branch proximal to the carpal tunnel resulting in two nerves running through the transverse carpal ligament directly into the thenar musculature.
F: Incidence of the extraligamentous, subligamentous, and transligamentous course of the thenar (recurrent) branch of the median nerve.

Redrawn from Lanz, U. Anatomical variations of the median nerve in the carpal tunnel. *J. Hand Surg.* 2:44–53, 1977.

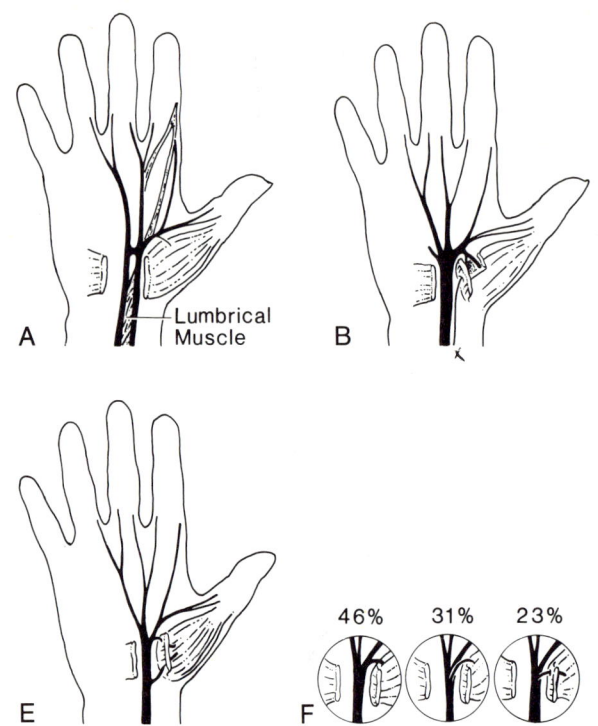

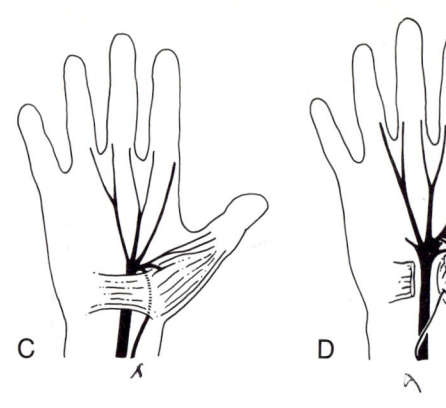

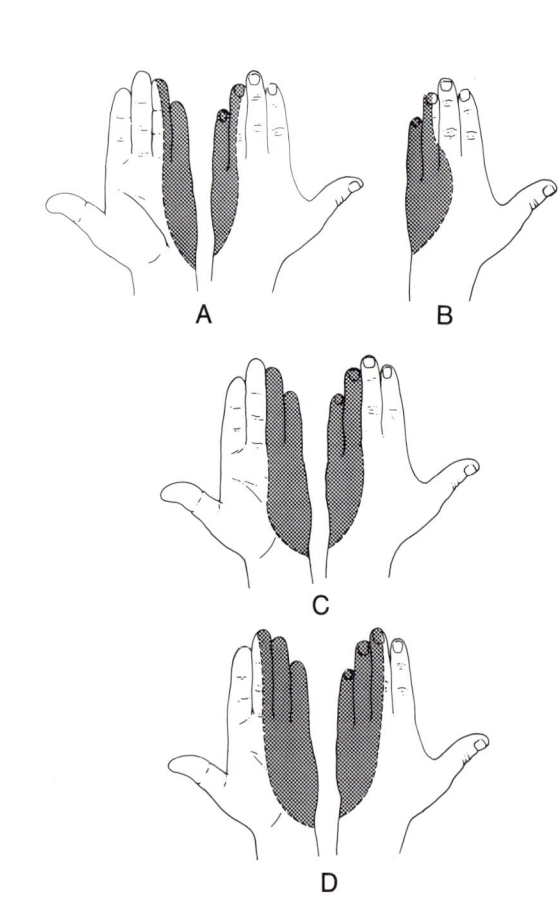

FIG. 60.
Variations in Cutaneous Innervation of Forearm and Hand (Based on Over 1,000 Gunshot Injuries to Peripheral Nerves). Ulnar Nerve.

A: In 79% of 102 cases, the ulnar nerve supplied the little and ulnar half of the ring finger, on both dorsal and palmar aspects of the hand.
B: In 30% of the 79% of 102 cases, the ulnar nerve supply in the dorsum of the hand extended as far laterally as a line drawn through the axis of the middle finger.
C: In 14%, the whole of the palmar and dorsal surfaces of the ring and little fingers were supplied by the ulnar nerve.
D: In 3%, there was an additional distribution on the ulnar half of the middle finger.

See Figures 61–64 for continuation of this study and conclusion.

Redrawn from Stopford, J.S.B. The variation in distribution of the cutaneous nerves of the hand and digits. *J. Anat.* 53:14–25, 1919.

Nervous System

FIG. 61.
Variations in Cutaneous Innervation of Forearm and Hand (Based on Over 1,000 Gunshot Injuries to Peripheral Nerves) (continued).
Ulnar Nerve.
Median Nerve.

A: In one case, the ulnar was shown to supply the skin of the whole of the middle, ring, and little finger. In another case, however, the nerve only supplied digital branches to the little finger.

B: The usual description of the cutaneous supply of the median nerve occurred in Stopford's study in 72% of 51 cases. The distribution of the median nerve on the dorsum of the fingers was very constant and was always found to extend proximally as far as the level of the first interphalangeal joint, and in more than half of cases studied, as far as the middle of the proximal phalanx.

C: In 4%, the median nerve extended medially as far as a line drawn through the center of the middle finger (midaxial line).

Redrawn from Stopford, J.S.B. The variation in distribution of the cutaneous nerves of the hand and digits. *J. Anat.* 53:14–25, 1919.

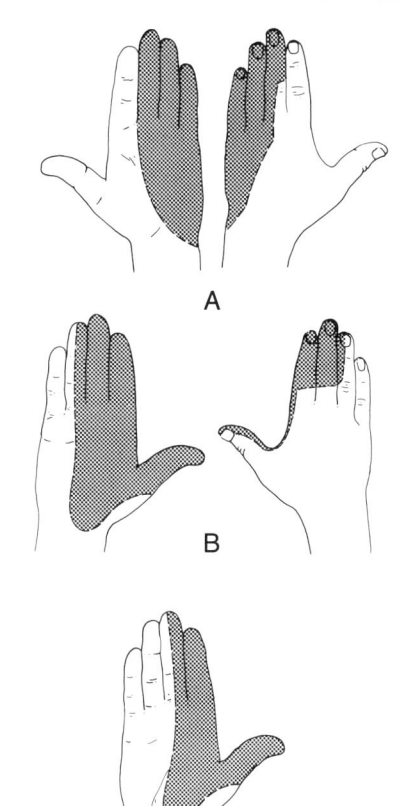

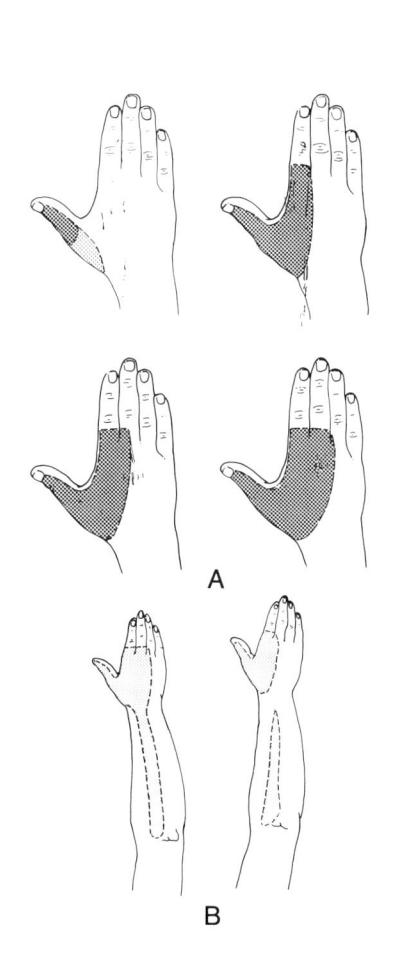

FIG. 62.
Variations in Cutaneous Innervation of Forearm and Hand Based on Over 1,000 Gunshot Injuries to Peripheral Nerves) (continued).
Radial Nerve.

The sensory distribution on the radial side of the dorsum of the hand is subject to considerable variation. Stopford studied this problem by examining the distribution of sensory loss after division of I. the musculospiral (radial) nerve above and below the point of origin of its lower external cutaneous branch (lower lateral cutaneous branch); II. the radial at various levels in the forearm; and III. the musculocutaneous nerve.

A: *Upper left*, the anesthesia is confined to the dorsum of the thumb in most cases, including the dorsum of the first metacarpal bone, but in a few, is restricted to the phalanges.
When the division of the radial occurred above the origin of the lower lateral cutaneous nerve, the anesthesia conformed to the following distribution in 33 cases examined. *Upper left*, in 13%; *upper right*, in 13%; *lower left*, in 16%; *lower right*, in 58%.
When the radial was severed distal to the origin of the lower lateral cutaneous nerve, the anesthesia conformed to the following distribution in 20 cases. *Upper left*, in 20%; *upper right*, in 30%; *lower left*, in 20%, *lower right*, in 30%.
When the radial was severed in the forearm, the anesthesia conformed to the following distribution in 7 of 8 cases. *Upper left*, in 33%; *upper right*, in 33%; *lower left*, in 17%; *lower right*, in 17%.

B: This figure shows the most common pattern of sensory loss in the forearm and hand following division of the radial in the upper third of the arm. *Left*, area supplied by both the lower lateral cutaneous nerve and the superficial radial nerve; *right*, where the sensory loss on the hand is more restricted there is a distinct separation of the two nerves.

Redrawn from Stopford, J.S.B. The variation in distribution of the cutaneous nerves of the hand and digits. *J. Anat.* 53:14–25, 1919.

Human Anatomic Variation

FIG. 63.
Variations in Cutaneous Innervation of Forearm and Hand Based on Over 1,000 Gunshot Injuries to Peripheral Nerves) (continued).
Radial Nerve.
Musculocutaneous Nerve.
A: Two cases of division of the radial high in the arm demonstrate the distribution of the lower lateral cutaneous nerve, i.e., over the dorsum of the thumb and a strip over the dorsum of the third metacarpal which is continuous with the area of the skin supplied in the forearm. Stopford suggested that the above description holds in 50% of hands.
B: Usual area of supply of the musculocutaneous nerve in the forearm, i.e., a strip of skin on both aspects of the radial border of the forearm as far distally as the lower end of the radius.
C: In a few cases, the musculocutaneous nerve (lateral antebrachial cutaneous nerve) was found supplying the lateral part of the thenar eminence and the dorsal surface of the first metacarpal and the first interosseous space.
D: In two cases the radial (lower lateral cutaneous nerve) had been severed; in another case there was division of the superficial radial nerve as well as the musculocutaneous, and the resultant anesthesia was similar to that shown in this figure.

Redrawn from Stopford, J.S.B The variation in distribution of the cutaneous nerves of the hands and digits. J. Anat. 53:14–25, 1919.

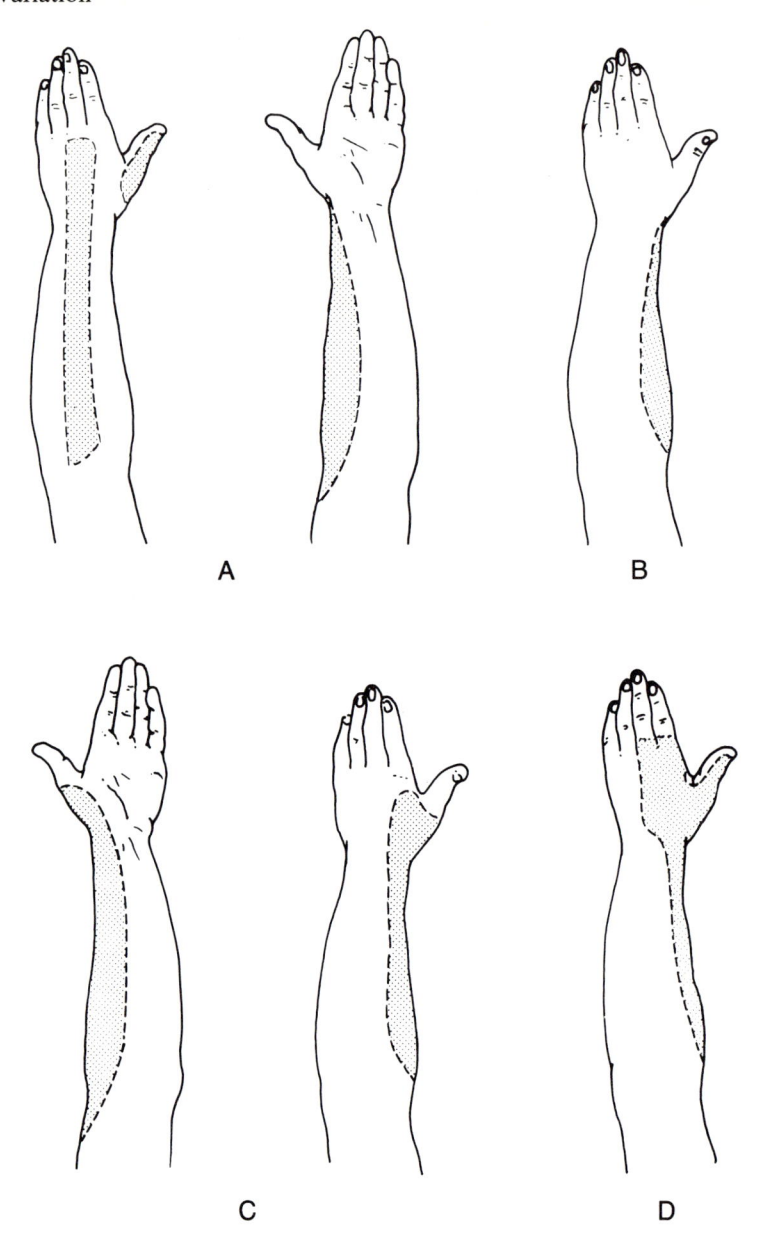

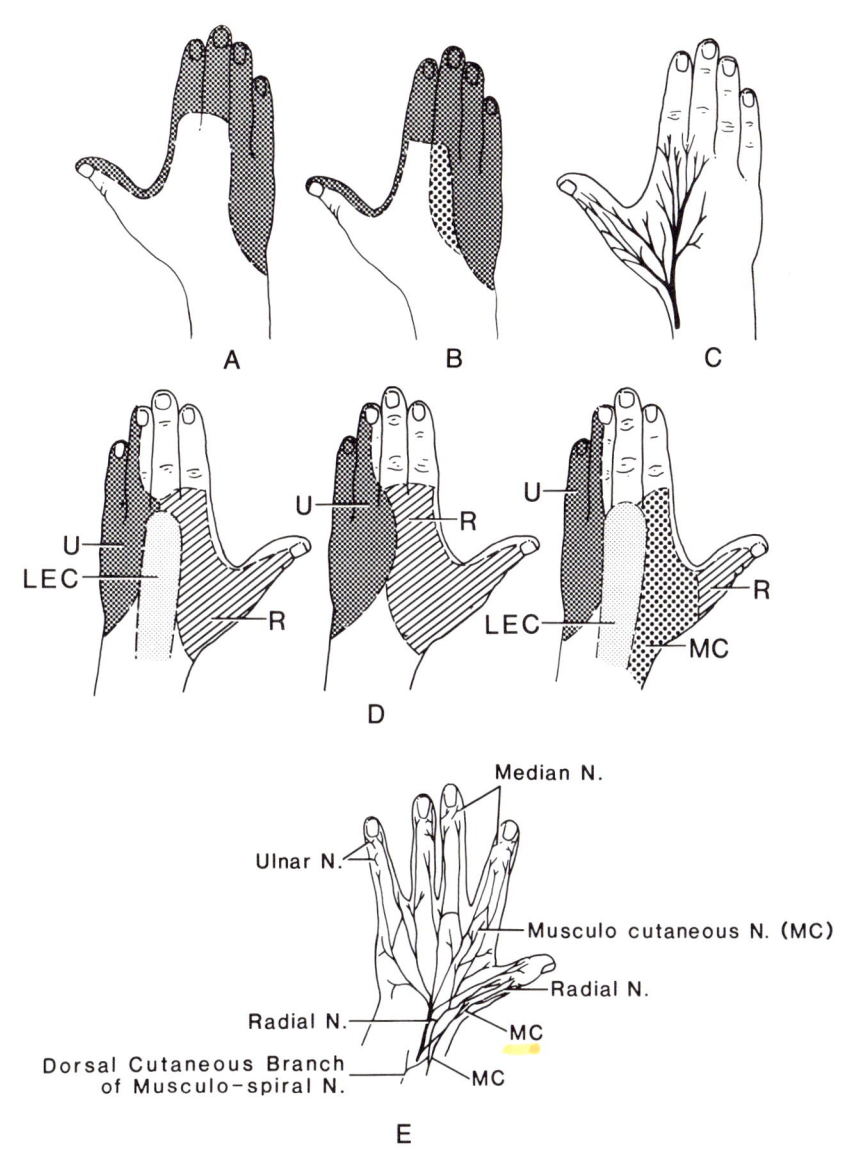

FIG. 64.
Variations in Cutaneous Innervation of Forearm and Hand (Based on Over 1,000 Gunshot Injuries to Peripheral Nerves) (continued).
Division of Median and Ulnar Nerves.
Superficial Radial Nerve.

A: In 14 of 20 patients, after the division of both nerves, the residual area of normal skin extended medially as far as the third interosseous space.
B: In 6 of the same 20 patients, it was less extensive and only reached, on the ulnar side, the second interspace or the middle of the third metacarpal bone.
C: Dissection specimen of the superficial radial nerve (corresponds to Figure 62, A, *lower right*, which resulted from severance of the same nerve).
D: In conclusion (Figures 60–64), the skin of the hand may be supplied with cutaneous branches from four sources (excluding the distribution of the median nerve), and the most frequent combinations are represented in this figure.

LEC, Lower lateral cutaneous nerve from the radial nerve; *MC*, musculocutaneous nerve (lateral antebrachial cutaneous nerve); *R*, superficial radial nerve; *U*, ulnar nerve.
E: In this specimen the radial (superficial radial nerve) supplied almost the entire dorsum of the hand. The ulnar nerve supplied one-half of the distal phalanx of the fifth digit, the median nerve supplied the medial half of the distal phalanx of the ring finger and the distal phalanx of the middle finger and the medial side of the distal phalanx of the index finger. On the lateral side of the hand, the lateral cutaneous nerve of the musculocutaneous nerve and the superficial radial nerve made a minor contribution.

Redrawn (A–D) from Stopford, J.S.B. The variation in distribution of the cutaneous nerves of the hand and digits. *J. Anat.* 53:14–25, 1919; and (E) from Learmonth, J.R. A variation in the distribution of the radial branch of the musculospiral nerve. *J. Anat.* 53:371–372, 1919.

Human Anatomic Variation

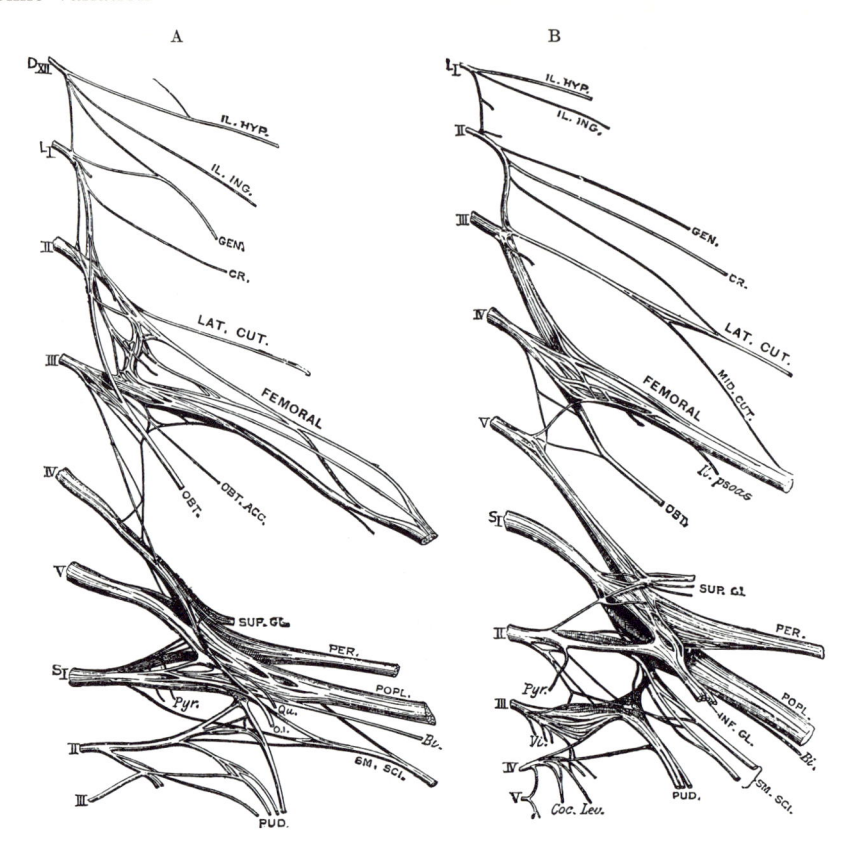

FIG. 65.
Variations in Formation of Lumbosacral Plexus.

A: "High form of plexus: ventral aspect. From a subject in which the fifth lumbar vertebra was assimilated to the sacrum, and articulated on both sides with the ilium.
B: Low form of plexus: dorsal aspect. From a subject in which there were thirteen thoracic and five lumbar vertebrae."

From Eisler, P. Der Plexus lumbosacralis des Menschen. Halle 1892, in *Quain's Anatomy*, 11th Ed., Vol. III, Pt. II: The Peripheral Nerves and Sense Organs, edited by E.A. Schafer, J. Symington and T.H. Bryce. Longmans, Green, and Co., London, 1915.

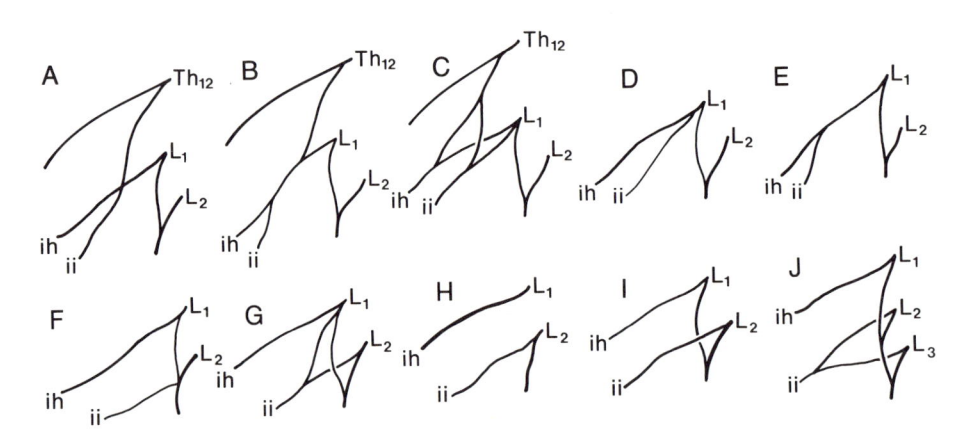

FIG. 66.
Variations in Formation of Ilioinguinal Nerve.

The ilioinguinal nerve was present in 97.5% of 200 bodies. It originated from the lumbar plexus independently in 72.5% of cases, together with the iliohypogastric nerve in 25%, and was absent in 2.5% of cases.
Ten types of origin of the nerve were distinguished depending on its origin from the lumbar plexus and formation.
The ilioinguinal nerve was formed by one root in 92.5% of cases (A, B, D, E, F, H, I), and by two roots in 5% (C, G, J). In 86.5% of cases, the ilioinguinal nerve arose from one spinal root (primarily L_1, rarely L_2) and in 11% of cases from two spinal nerves (T_{12}, L_1; L_1, L_2; or L_2, L_3).
Uniradicular nerves arose by union of the ventral branches of nerves T_{12}, L_1 (A, B) and L_1, L_2 (F) or directly from ventral branches L_1 (D, E) and L_2 (H, I). In biradicular nerves, the upper root arose from ventral branches of T_{12} (C), L_1 (G), or L_2 (J); and the lower root from L_1 (C), L_2 (G), or L_3 (J).
Th_{12}, Twelfth thoracic spinal nerve; L_1–L_3, respective lumbar spinal nerves; *ih*, iliohypogastric nerve; *ii*, ilioinguinal nerve.

Redrawn from Zaluska, S. External structure of the ilioinguinal nerve in postfetal life in man. *Folia Morphol.* 34:419–424, 1975.

FIG. 67.
Variants of Origin of Genitofemoral Nerve from Lumbar Plexus.
In this study of 200 bodies (400 nerves), 80.2% presented as a single trunk and 19.75% as two rami: genital and femoral. The trunk or rami were usually formed by nerve fibers from L_1, L_2 from the lumbar plexus. The level of emergence of the nerve or nerves from the psoas major muscle and the level of division into terminal branches were individually highly variable. These characteristics were unrelated to each other, sex, age, height or weight or side of the body. The genitofemoral nerve was present in all cases. Two types were recognized. In type I, the trunk of the nerve divided into two main rami: genital and femoral. In type II, the rami arose separately from the lumbar plexus. Type I was present bilaterally in 70.5%, and type II in 10% of cases.

The manner in which the genitofemoral nerve arose from the lumbar plexus varied markedly. Nerves of type I arose from a single root in 75.5%, and from two roots in 4.5% of cases. Single root nerves arose by union of the ventral branches of spinal nerves L_1, L_2 (A) in 75%, from L_2 (B) in 0.5%, and by union of L_2, L_3 (C) in 0.25% of cases. In double-rooted nerves (C–G) the upper root arose from L_1 in 3.75%, and from L_1, L_2 in 0.75% of cases; and the lower root from L_1, L_2 in 1.75%, from L_2 in 2.5%, and from L_3 in 0.25%. Both roots, after a short course, joined within the substance of the psoas major muscle. In type II, three variations of origin of the nerve from the lumbar plexus were observed. In the first variant, present in 18.25% of cases, both the genital and femoral branch arose from the plexus by one root. Both branches were derived from L_1, L_2 in 17.25% of cases (H); the genital branch from L_1 and the femoral branch from L_1, L_2 (I) in 0.5%; and the genital branch from L_1, L_2 and the femoral branch from L_1 (J) in 0.5% of cases. In the second variant, present in 1.0% of cases, the genital branch arose by one root, and the femoral branch by two (K–N). In three cases the genital branch arose from L_1, L_2, and in one case from L_1. The upper root of the femoral branch in three cases arose from L_1, in one case from L_1, L_2; and the lower root in three cases arose from L_1, L_2, and in one case from L_2. After a short course, the two roots of the femoral branch united within the psoas major muscle. In the third variant, represented in 0.5% of cases, the genital branch had two roots, and the femoral branch had one (O, P). The upper root of the genital branch arose from L_1 or L_1, L_2, and the lower root from L_1, L_2 or L_3. Both roots united just after leaving the plexus. The femoral branch in this variant was formed from L_1, L_2. In type II of the genitofemoral nerve, the genital branch arose from the plexus above the femoral branch in 18.5%, and from the femoral branch in 1.25% of cases.

The origin of the genitofemoral nerve from the lumbar plexus was symmetric in 64% of cases in type I, and in 7.5% in type II. Spinal nerve L_1 was present in nearly all cases, and L_2 in all cases sent fibers to the genitofemoral nerve. Fibers from L_3 were present in only 0.75% of cases.

L_1 to L_3, respective lumbar nerves; *gf*, genitofemoral nerve; *rf*, femoral branch; *rg*, genital branch.

Redrawn from Urbanowicz, Z. External structure of the genitofemoral nerve in postfetal life in man. *Folia Morphol.* 34:425–435, 1975.

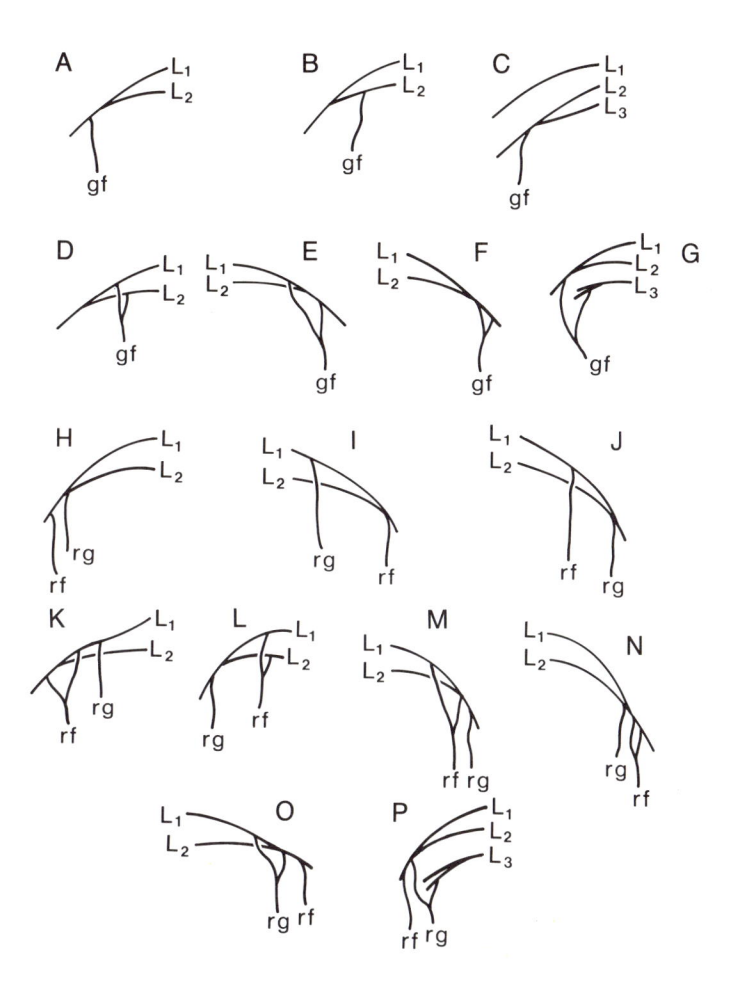

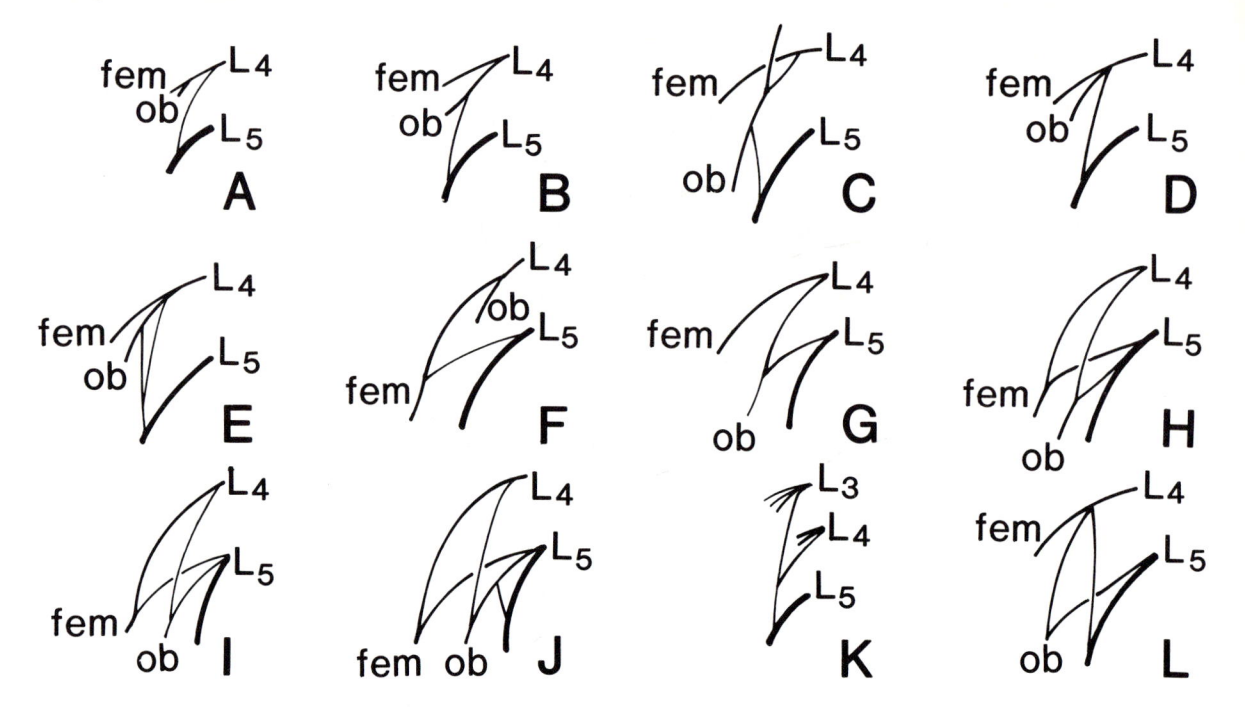

FIG. 68.

Variations in Connections Between Lumbar and Sacral Plexuses.

The connections between the lumbar and the sacral plexus were examined bilaterally in 122 subjects. In 91.8% of subjects the plexuses were joined by a single nervus furcalis and in 0.8% by a doubled nervus furcalis. The single furcal nerve was formed by the abdominal nerve L_4 in 80% and L_5 in 7.7% of cases. The doubled furcal nerve arose from L_3 and L_4 in 0.4% of cases and from L_4 and L_5 in 0.4%. Usually, the major part of the furcal nerve arising from L_4 went to the lumbar plexus, and from L_5 to the sacral plexus. In 7.4% of cases no connection between the plexus could be found.

Authors' note: Urbanowicz considers illustrations A through E to be type I variations, F through J to be type II, K to be type III, and L to be type IV.

L_3 to L_5, respective lumbar nerves; *fem*, femoral nerve; *ob*, obturator nerve.

Redrawn from Urbanowicz, Z. Connections between the lumbar and the sacral plexus in man. *Folia Morphol.* 40:271–279, 1981.

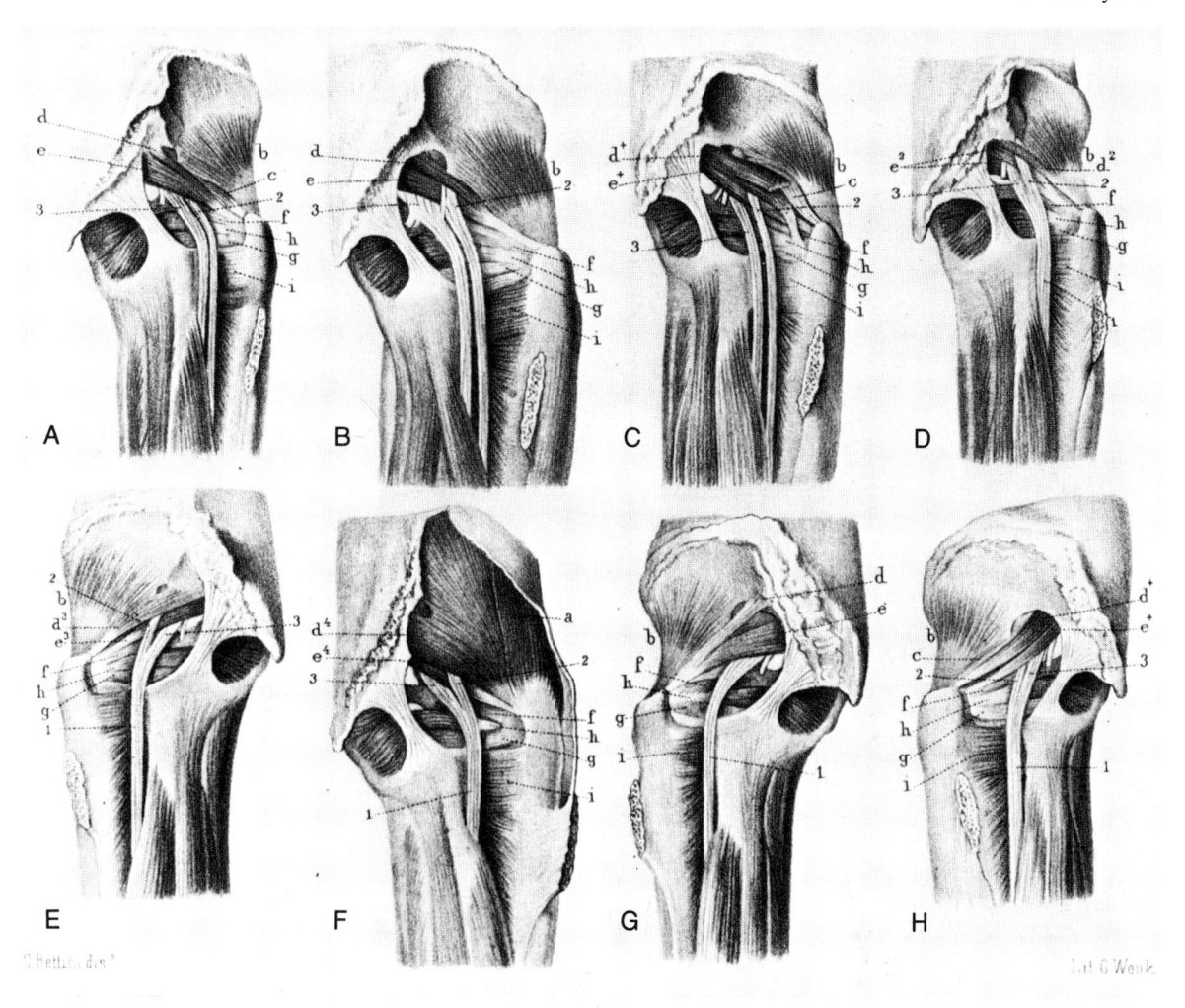

FIG. 69.
Variations in Sciatic Nerve.
(Variations in M. Piriformis).

A: The sciatic nerve issues from the pelvis and remains divided into peroneal and tibial divisions passing beneath m. piriformis which in this case has two bellies.

B: The sciatic nerve issues from the pelvis divided into peroneal and tibial divisions. The peroneal division passes through the division of m. piriformis.

C: Similar to preceding, m. piriformis is divided allowing passage of the peroneal division of the sciatic nerve.

D: The sciatic nerve issues from the pelvis divided into peroneal and tibial divisions, which unite after passing through the divided m. piriformis. The superior part of piriformis is joined with m. gluteus minimus and the inferior part has two bellies. In this case the tibial division passes through the division and the peroneal division passes above m. piriformis.

E: Similar to D, except the union of the division of the sciatic nerve takes place at a lower level, at the tendon of m. obturator internus.

F: Similar to D and E except that the larger superior division of m. piriformis is intimately united with m. gluteus medius.

G: The sciatic nerve issues from the pelvis as usual beneath piriformis, but m. piriformis is divided into a smaller superior part and a larger inferior part.

H: The sciatic nerve issues from the pelvis divided, but rejoins immediately below m. piriformis which is completely divided into two parts.

a, M. gluteus medius; b, m. gluteus minimus, which has a part (c) separated by a deep sulcus (A, C, H); d,e, m. piriformis with two bellies: the upper belly is small (d) and the inferior belly (e) is large (A–C, G); d^+, e^+, m. piriformis duplicated: the upper muscle is larger (d^+) and the lower muscle is smaller (e^+); d^2, e^2, m. piriformis duplicated, united by a single tendon (D); d^3, e^3, m. piriformis duplicated and intimately united with m. gluteus medius (E); d^4, e^4, m. piriformis intimately united with m. gluteus medius which has a distinct, partial, division (c); f, g, mm. gemelli; h, m. obturator internus; i, m. quadratus femoris.

From Calori, L. Sull'alta divisione del nervo grande ischiatico considerata come differenza nazionale e sulle varietà del muscolo piriforme. *Mem. R. Accad. Sci. Istituto di Bologna S. 4* 2:623–633, 1880.

Human Anatomic Variation

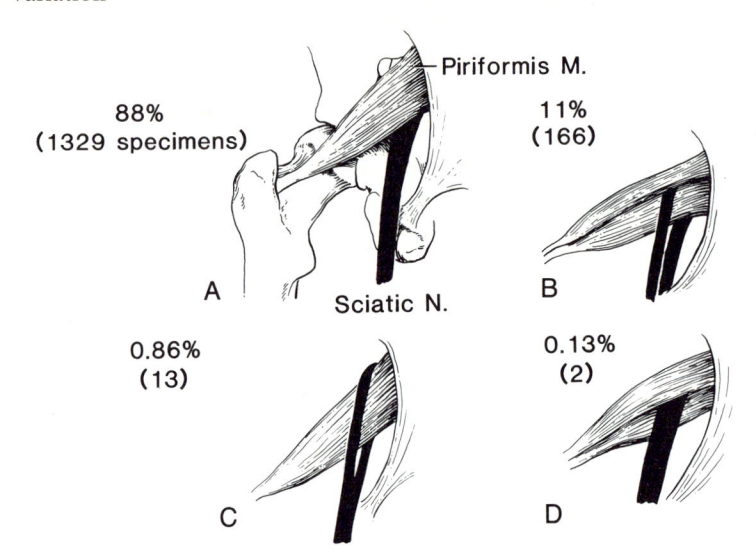

FIG. 70.
Variations in Relation of Sciatic Nerve to M. Piriformis.

Number of extremities studied, 1510.
A: Usual relationships with the sciatic nerve passing from the pelvis beneath m. piriformis.
B: M. piriformis divided into two parts with the peroneal division of the sciatic nerve passing between the two parts of piriformis.
C: The peroneal division of the sciatic nerve passes over m. piriformis and the tibial division passes beneath the undivided muscle.
D: In these cases the entire nerve passes through the divided m. piriformis.

From Beaton, L.E. and B.J. Anson. The relation of the sciatic nerve and its subdivisions to the piriformis muscle. *Anat. Rec.* 70:1–5, 1938.

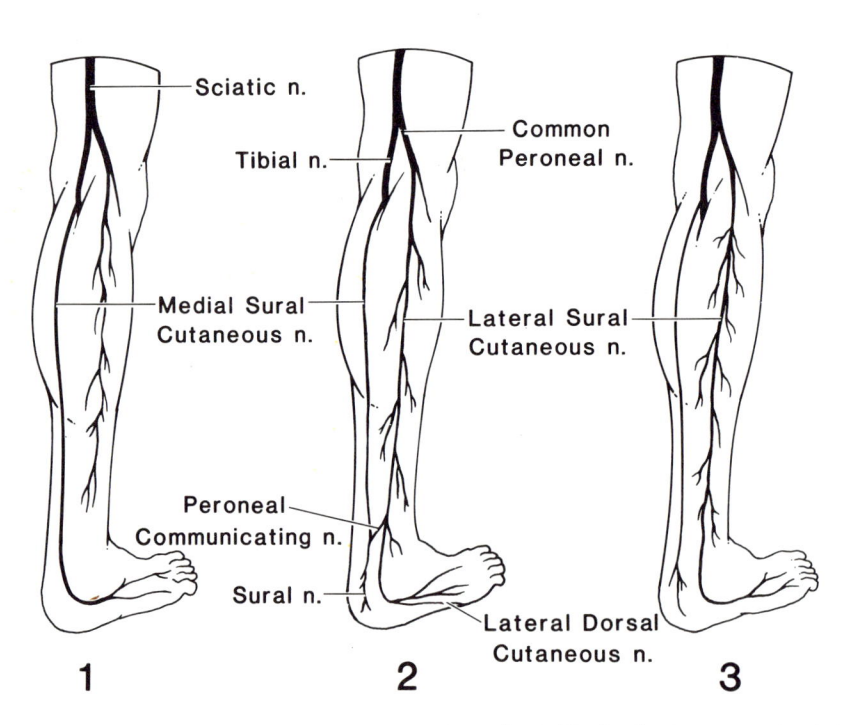

FIG. 71.
The Sural Nerve

Description of Illustration (after Huelke).
1. The medial sural cutaneous nerve passes onto the foot as the lateral dorsal cutaneous nerve. There is no sural nerve.
2. The sural nerve terminates as branches to the heel, with the lateral sural cutaneous nerve passing onto the foot.
3. The medial sural cutaneous nerve terminates as cutaneous branches to the heel with the lateral sural cutaneous nerve passing onto the foot. There is no sural nerve.

According to Huelke, the frequency of variation 1 was 19.1% of 550 limbs, of 2 about 80.5%, and of 3 about 0.4%. A sural nerve is present in 80.5% of bodies.
The peroneal communicating nerve has a variable origin but the same type of origin occurred on both sides in only 58 of 99 cadavers (58.6%).

Redrawn from Huelke, D.F. The origin of the peroneal communicating nerve in adult man. *Anat. Rec.* 132:81–92, 1958.

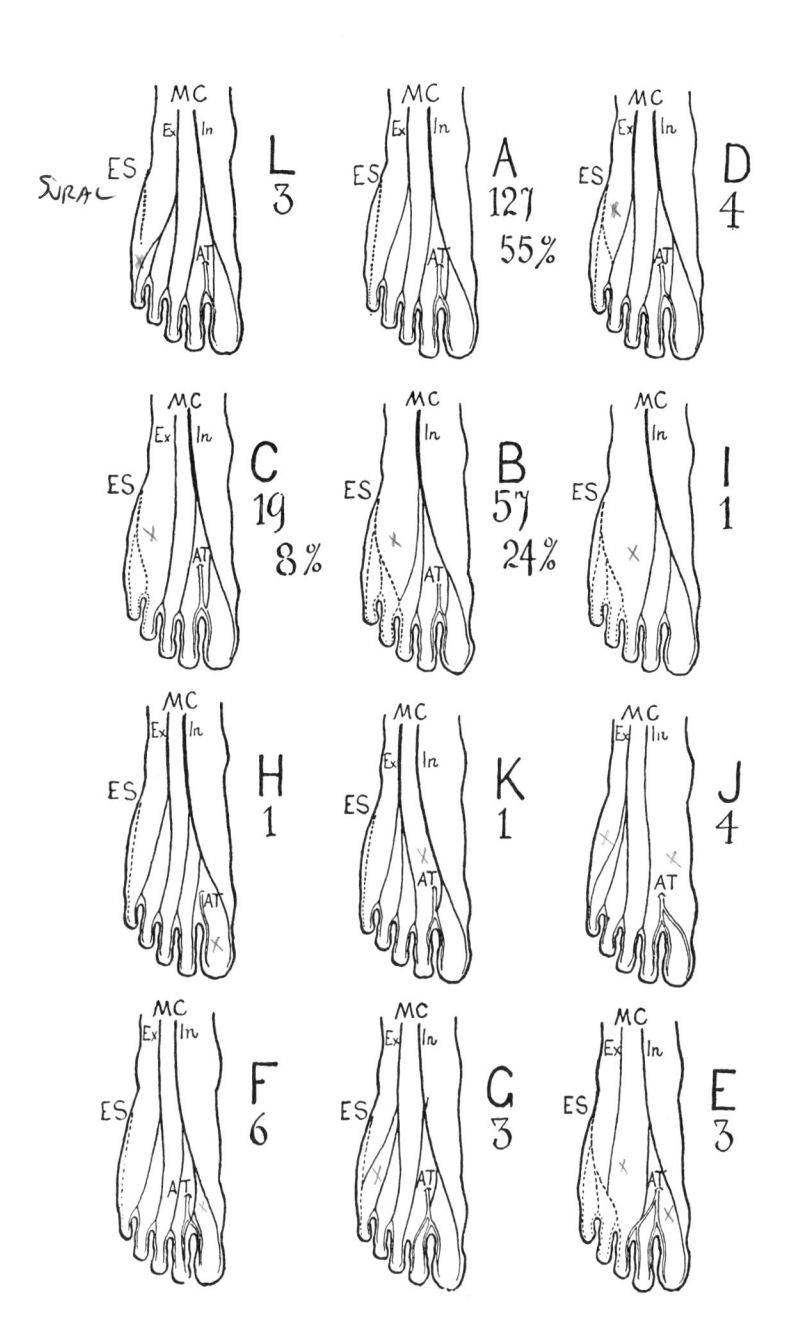

FIG. 72.
Variations in Cutaneous Innervation of Dorsum of the Foot.

"A total of 229 feet have been examined, and in tabulating the results the graphic method has been adopted in preference to the descriptive. Twelve types are figured, and lettered A, B, C, & c., in order of their frequency. The first six are arranged in such a way as to show how the external saphenous replaces the external division of the musculo-cutaneous nerve, or vice versa. The last six (H, K, I, F, G, E) are arranged to display the progressive variations in the cutaneous distribution of the anterior tibial nerve.

In these types connecting loops between the nerves have been in some cases omitted, as the information necessary was not sufficiently complete.

The figures attached to the types indicate the numbers of such specimens out of the total of 229 examined. In some cases the percentage is also given.

The musculo-cutaneous is represented by a solid black line; the external saphenous by a dotted line. The anterior tibial by a double outline.

The so-called normal arrangement represented by type A occurs in 55 percent of cases. Next in order is type B, where the external saphenous supplies the two and a half outer toes. This distribution of the nerves was met with in 24 percent of the feet examined.

In one case, the internal saphenous supplied the inner side of the great toe; in another, it reached the inner side of the head of the metatarsal bone."

AT, Anterior tibial (deep peroneal) nerve; *ES*, external saphenous (sural) nerve; *MC*, musculocutaneous (superficial peroneal) nerve: *Ex*, external (lateral) branch; and *In*, internal (medial) branch.

From Thomson, A. Second annual report of the committee of collective investigation of the Anatomical Society of Great Britain and Ireland for the year 1890–91. *J. Anat. Physiol.* 26:76–93, 1891.

Human Anatomic Variation

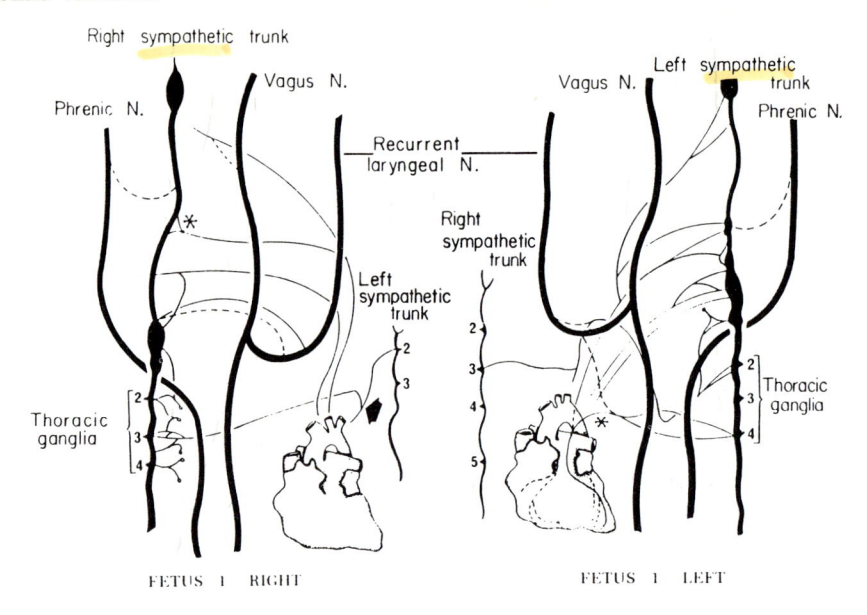

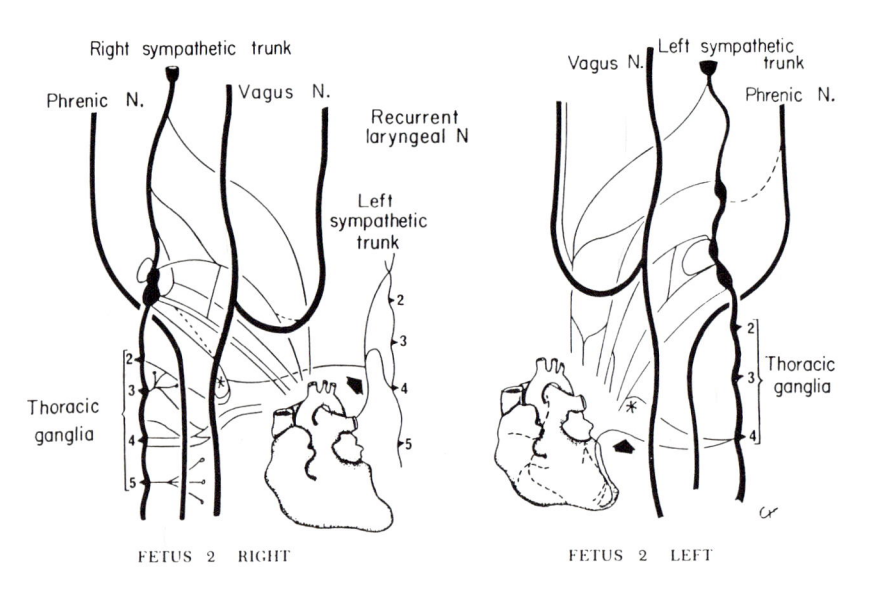

FIG. 73.

Variations in Sympathetic Nerve Pathways to the Heart.

"Description of individual dissections. For the convenience of the reader, the findings are presented in a schematic fashion, using the following conventions: (1) Sympathetic cardiac branches are shown as continuous dark lines. (2) Connections made by the sympathetic trunks or their cardiac rami with other nerves are indicated by interrupted lines. (3) Sympathetic rami to structures other than the heart are shown as continuous lines ending in small blobs. Many rami contain, in addition to cardiac elements, nerve fibers destined for other structures and this explains why, in some diagrams, a branch from a cardiac nerve is shown ending in a blob. (4) Points of special interest are denoted by asterisks, stars and arrows. Explanations of these symbols are given in the relevant descriptions.

Fetus 1. Both sympathetic trunks lack vertebral ganglia (i.e., ganglia in close relationship to the vertebral artery). On the right side, a distal intermediate ganglion (*) lies on the middle cervical cardiac ramus; the middle cervical ganglion being absent.

In addition to connections between cervical sympathetic cardiac rami and other nerves, thoracic branches from both trunks join together (*arrow*) before running to the heart. On the left side a nerve (*) has been traced to a plexus on the descending branch of the left coronary artery.

Fetus 2. There are no ganglia on the right sympathetic trunk between the superior cervical and cervicothoracic.

As in fetus 1, thoracic cardiac rami from the two sides join (*arrow*) and, in addition, a strand was traced from the left thoracic trunk to a plexus on the descending branch of the left coronary artery."

Variations continued in Figures 74–78.

From Ellison, J.P. and T.H. Williams. Sympathetic nerve pathways to the human heart, and their variations. *Am. J. Anat.* 124:149–162, 1969.

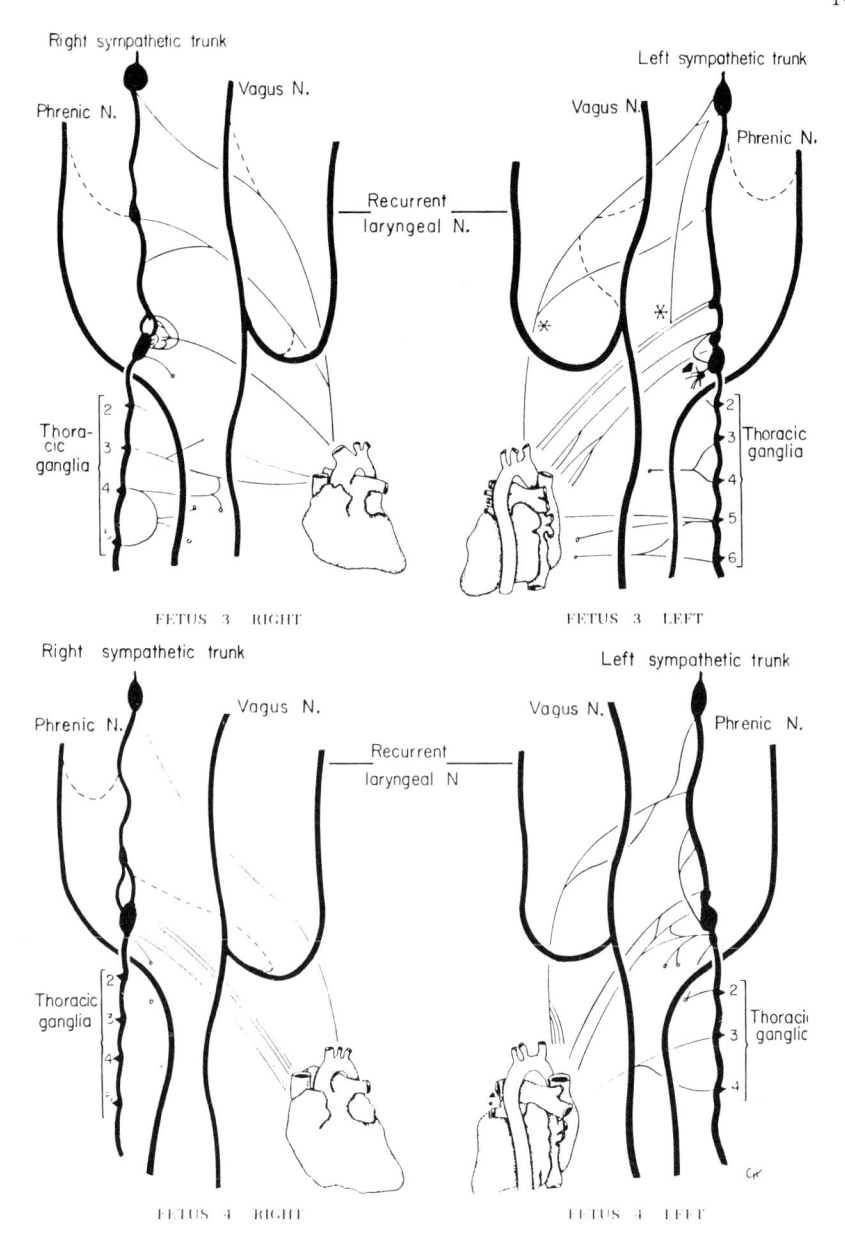

FIG. 74.
Variations in Sympathetic Nerve Pathways to the Heart (continued).

"Fetus 3. On the left side, a distal intermediate ganglion, giving vertebral and aortic rami only, lies on a loop between the cervicothoracic and the second thoracic ganglion (arrow). The superior and middle cervical cardiac rami that are marked with asterisks receive contributions from different levels of the trunk.

Fetus 4. On the right side, all but one of the sympathetic cardiac rami arise from the cervicothoracic ganglion. The left side shows an additional thoracic cardiac ramus."

From Ellison, J.P. and T.H. Williams. Sympathetic nerve pathways to the human heart, and their variations. *Am. J. Anat.* 124:149–162, 1969.

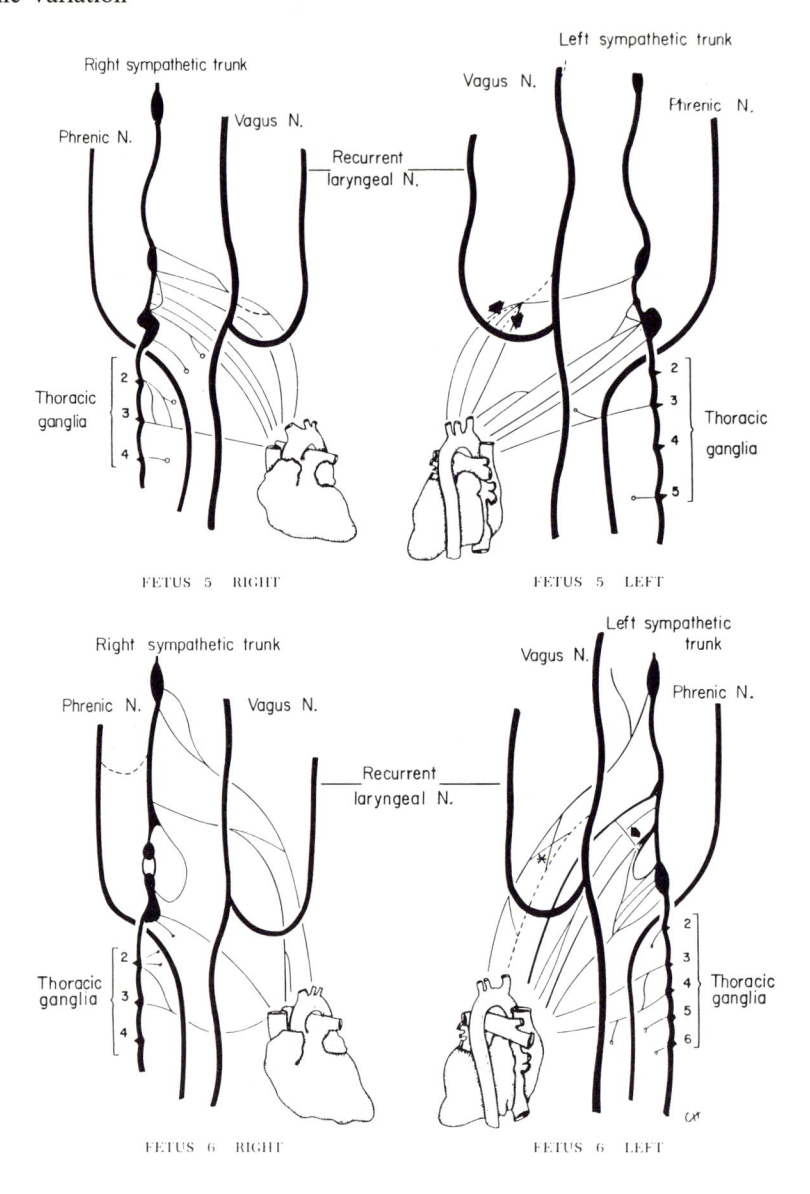

FIG. 75.

Variations in Sympathetic Nerve Pathways to the Heart (continued).

"Fetus 5. This fetus presents certain features of bilateral symmetry, and in particular the sites of origin of the sympathetic cardiac rami are similar. Superior cervical contributions are absent on both sides. There are no phrenic connections and the vagal connection is bilateral. However, links with the recurrent laryngeal nerve (*arrows*) are present on the left side only. Fetus 6. On the left side sympathetic and parasympathetic cardiac rami (*) exchange nerve fibers as they proceed towards the heart. A cross-link (*arrow*) occurs between a middle cervical cardiac ramus and the ansa subclavia."

From Ellison, J.P. and T.H. Williams. Sympathetic nerve pathways to the human heart, and their variations. *Am. J. Anat.* 124:149–162, 1969.

Nervous System

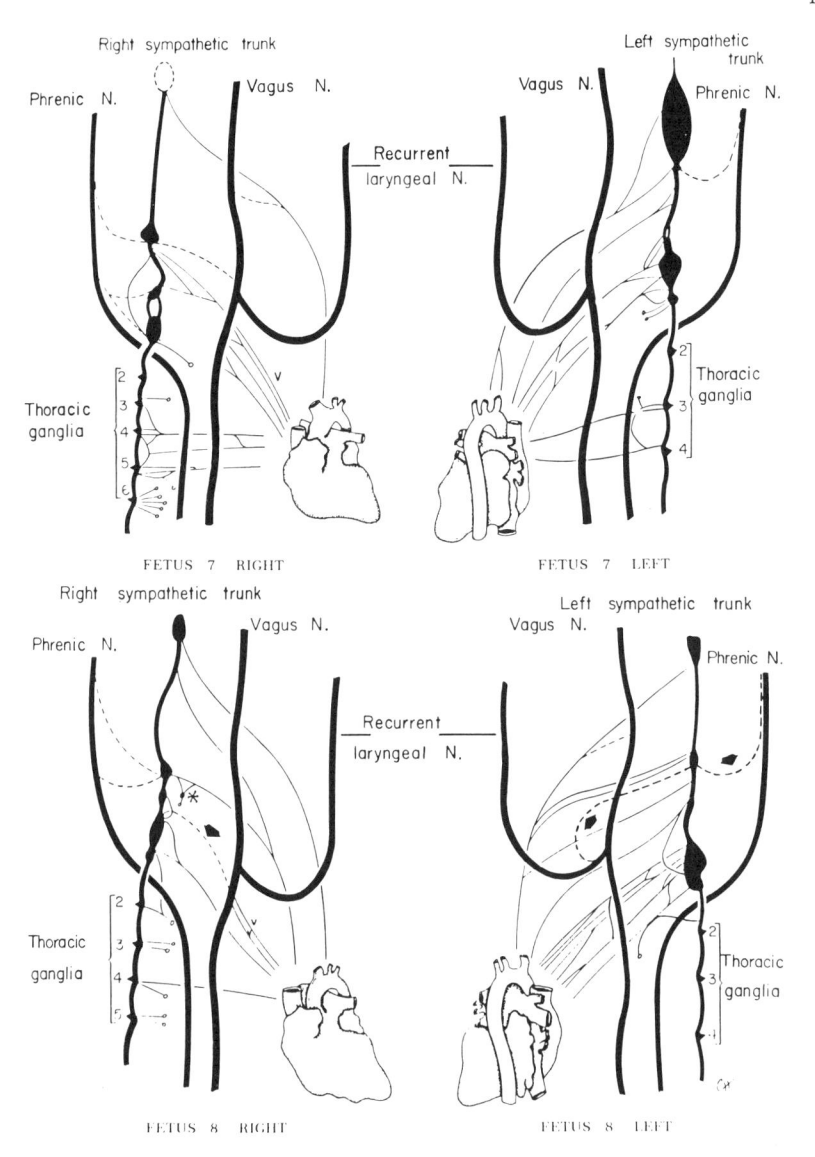

FIG. 76.

Variations in Sympathetic Nerve Pathways to the Heart (continued).

"Fetus 7. The upper end of the right superior cervical sympathetic ganglion was inadvertently cut off during preparation, but its cardiac branch was preserved. This dissection provided the only example of a link between the cervicothoracic ganglion and the phrenic nerve. A group of vagal cardiac rami (v) are shown.

A single link with the phrenic nerve was the only connection that was evident on the left sympathetic trunk.

Fetus 8. A distal intermediate ganglion (*) is attached by two roots to the middle right cervical cardiac nerve. The right vagus receives a thick sympathetic connection (arrow), and just below this point the vagus gives off two cardiac rami (vv).
On the left side the sympathetic trunk connects at the same level with the vagus and phrenic nerves (arrows)."

From Ellison, J.P. and T.H. Williams. Sympathetic nerve pathways to the human heart, and their variations. Am. J. Anat. 124:149–162, 1969.

Human Anatomic Variation

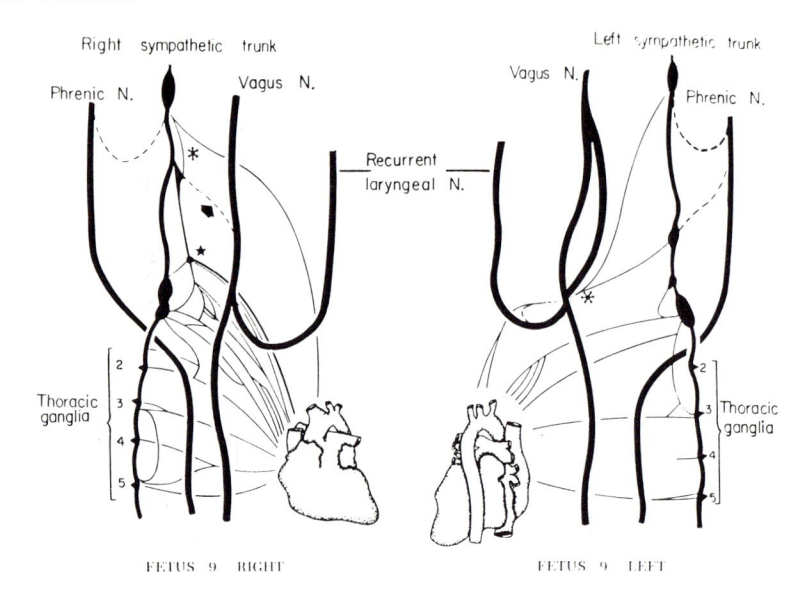

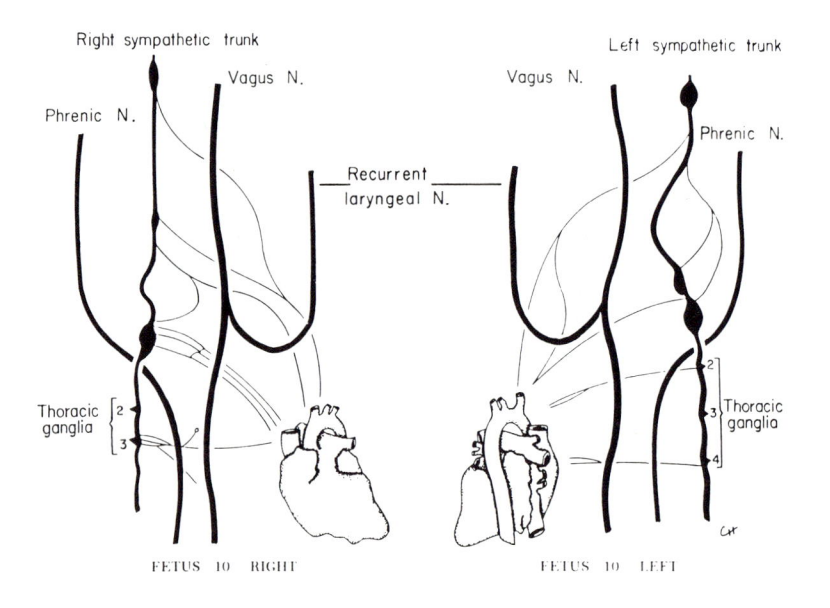

FETUS 9 RIGHT

FETUS 9 LEFT

FETUS 10 RIGHT

FETUS 10 LEFT

FIG. 77.
Variations in Sympathetic Nerve Pathways to the Heart (continued).

"Fetus 9. On the right side, a small triangular ganglion is attached by its upper pole to the sympathetic trunk, and one of its other corners links with the vagus (*arrow*). Issuing from a distal intermediate ganglion (*star*), and from the cervicothoracic, are strong contributions to the cardiac plexus. The left superior and middle sympathetic cardiac rami join (*) before communicating with the recurrent laryngeal nerve.
Fetus 10. The most noteworthy feature of this specimen is the absence of communications between the sympathetic trunks and neighboring nerves. This feature is seen on both sides."

From Ellison, J.P. and T.H. Williams. Sympathetic nerve pathways to the human heart, and their variations. *Am. J. Anat.* 124:149–162, 1969.

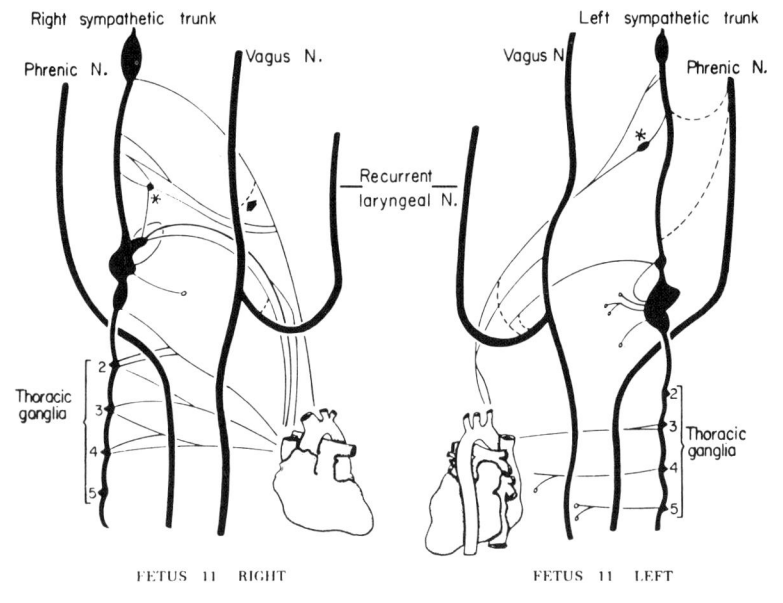

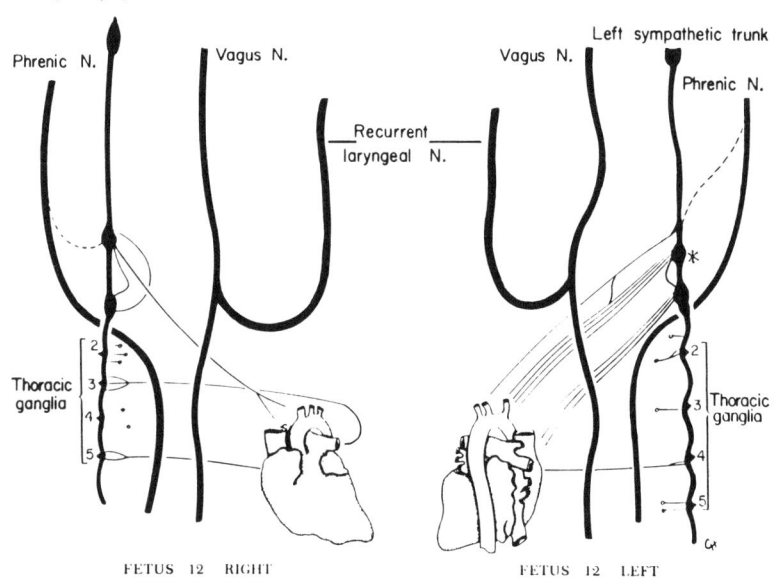

FIG. 78.
Variations in Sympathetic Nerve Pathways to the Heart (continued).

"Fetus 11. The right superior cervical sympathetic cardiac nerve communicates with the vagus (*arrow*) and joins the middle cervical sympathetic cardiac nerve.
On both sides there is a distal intermediate ganglion on a middle cervical cardiac ramus (*, *), whilst the middle cervical sympathetic ganglion is lacking.

Fetus 12. The superior cervical sympathetic cardiac nerve is bilaterally absent. The right middle cervical ganglion lies at the upper limit of the ansa subclavia and provides the only cervical sympathetic cardiac ramus.
On the left side, the vertebral ganglion (*) is large and globular."

From Ellison, J.P. and T.H. Williams. Sympathetic nerve pathways to the human heart, and their variations. *Am. J. Anat.* 124:149–162, 1969.

Human Anatomic Variation

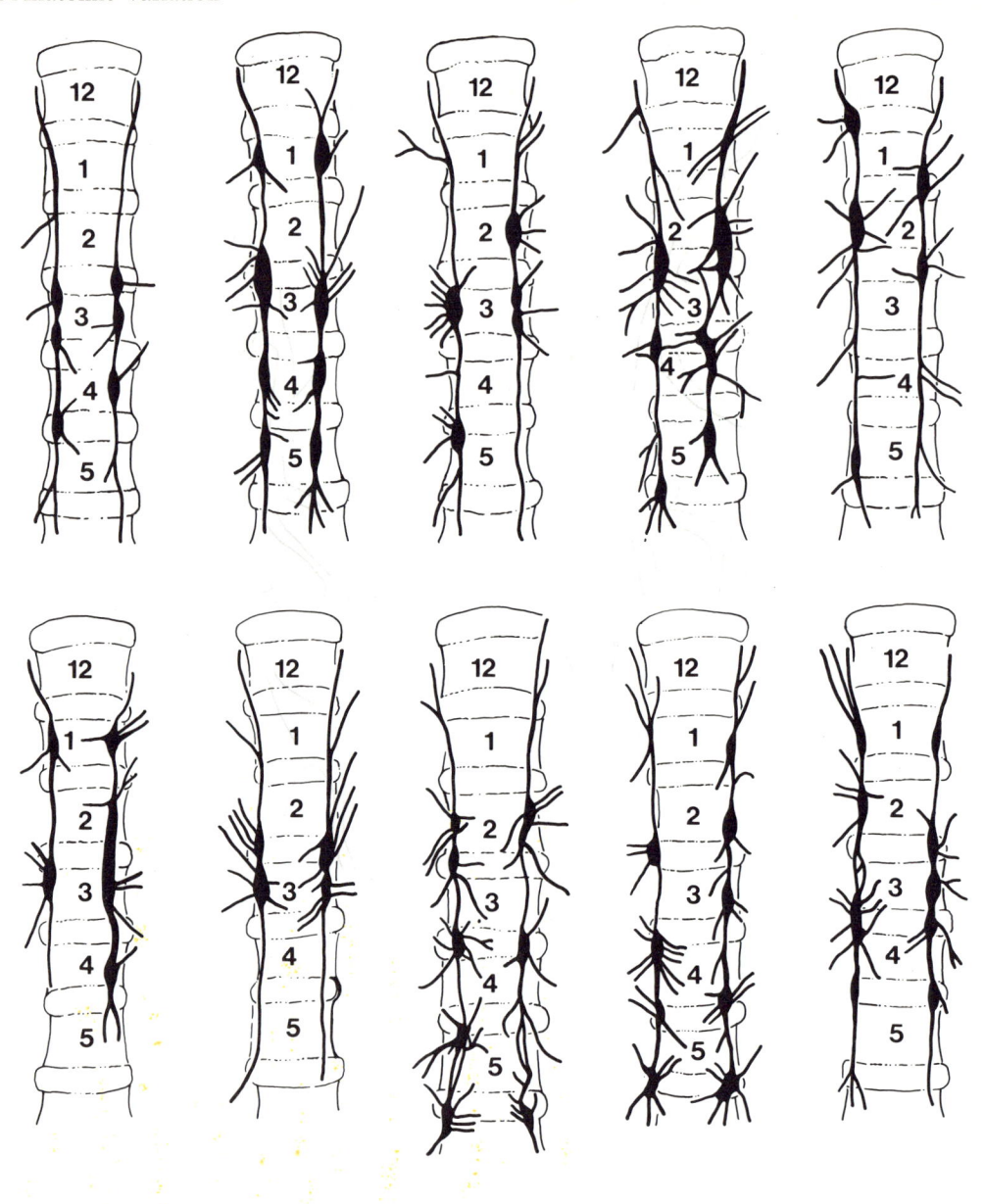

FIG. 79.
Variations in Lumbar Sympathetic Trunks.
The ganglia on the two sides of the body are rarely symmetric in their appearance, branches, number, and location.

Redrawn from Yeager, G.H. and R.A. Cowley. Anatomical observations on the lumbar sympathetics with evaluation of sympathectomies in organic peripheral vascular disease. *Ann. Surg.* 127:953–967, 1948.

Nervous System

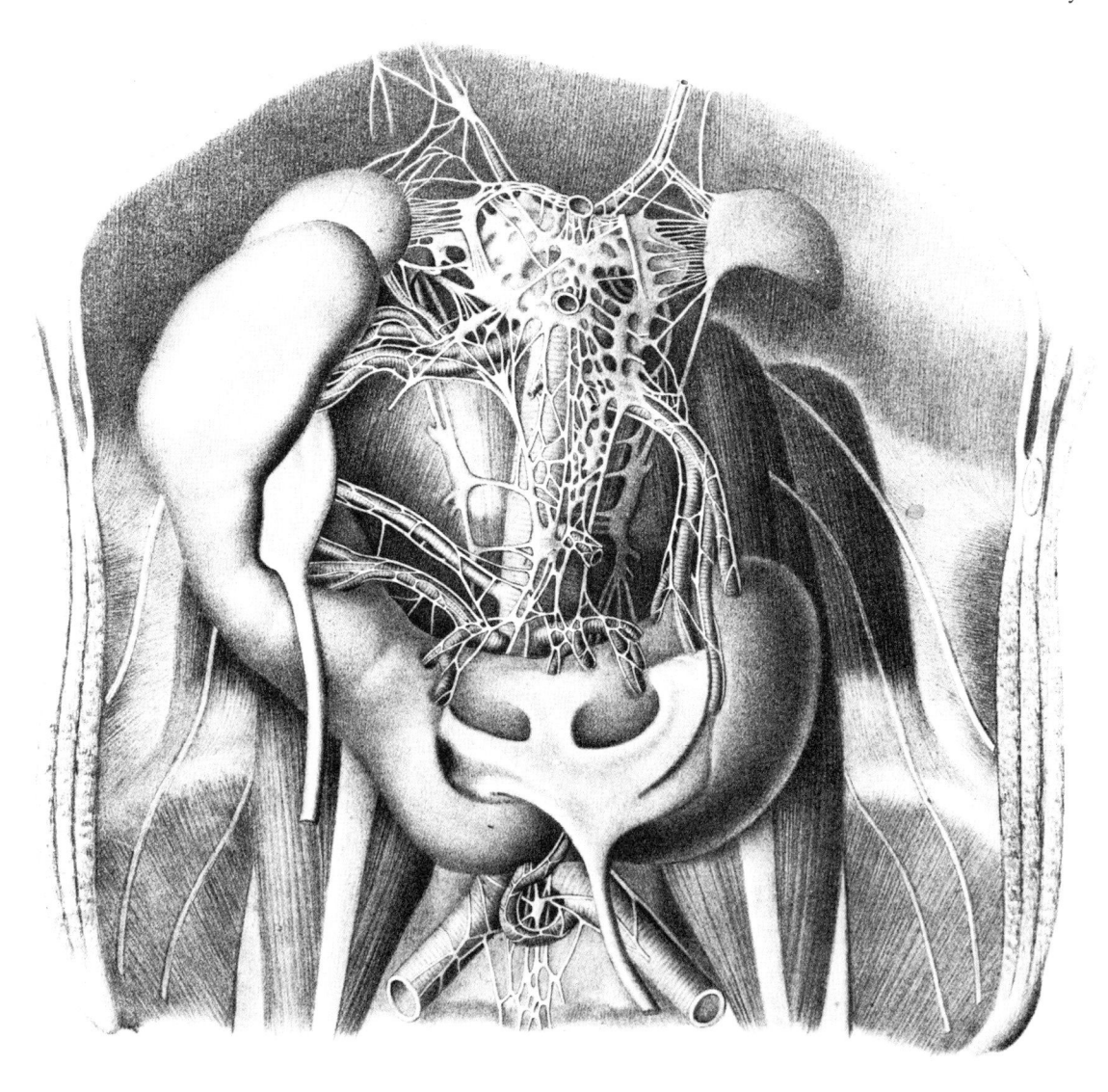

FIG. 80.
Nerve Supply in Cases of Horseshoe Kidney.
This rare variation (0.19%) demonstrates multiple "accessory" renal arteries all of which carry sympathetic nerve fibers into the substance of the organ.

From Calori, L. Sui nervi di un rene a ferro di cavallo con ectopia del rene sinistro. *Mem. R. Accad. Sci. Istituto di Bologna S.4* 9:451–460, 1888.

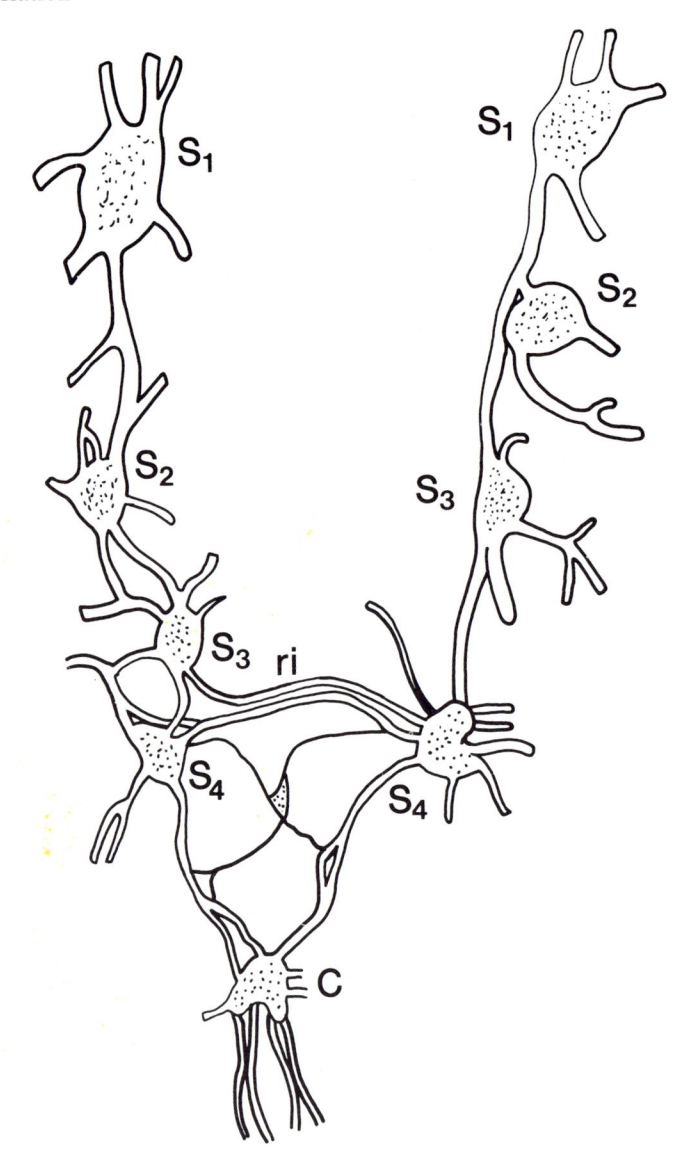

FIG. 81.
Sacral Sympathetic Ganglia.
Sacral portion of the sympathetic cords of an infant, slightly enlarged (Rauber.).
S_1–S_4, Sacral ganglia; C, coccygeal ganglion, showing indications of its origin by the fusion of two ganglia, and sending downwards branches which run with the middle sacral artery; ri, interfunicular branches uniting the ganglia of the two sides; below these is seen a small interfunicular ganglion.

From Schafer, E.A. and J. Symington. *Quain's Elements of Anatomy*, 11th Ed., Vol. III, Pt. II: Neurology, edited by E.A. Schafer, J. Symington, and T.H. Bryce. Longmans, Green and Co., London, 1909.

Organs

"For there are eunuchs who have been so from birth, and there are eunuchs who have been made eunuchs by men, and there are eunuchs who have made themselves eunuchs for the sake of the kingdom of Heaven."

Matthew 19:12
~ 80 AD

Organs

FIG. 1.
Cutaneous Horn of the Scalp.
The cutaneous horn in this case attained a length of 20 to 30 cm.
From Anonymous. Anatomie du vivant. Corne cutanée du cuir chevelu. *Le Presse Médicale* No. 104. 30:2183, 1922.

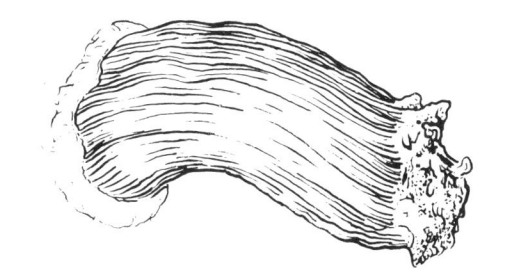

FIG. 2.
Cleft Nose.
From Taruffi, C. Nuovo caso di meso-rino-schisi nell'uomo. *Mem. R. Accad. Sci. Istituto di Bologna S.5* 1:227–233, 1890.

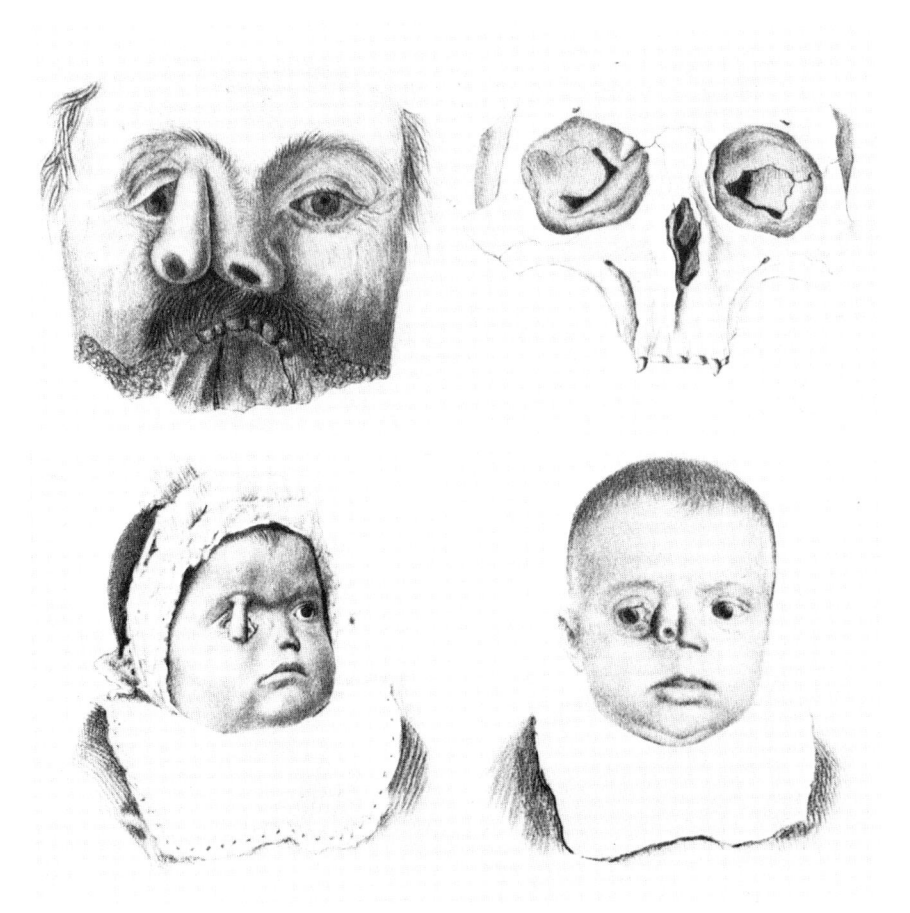

Human Anatomic Variation

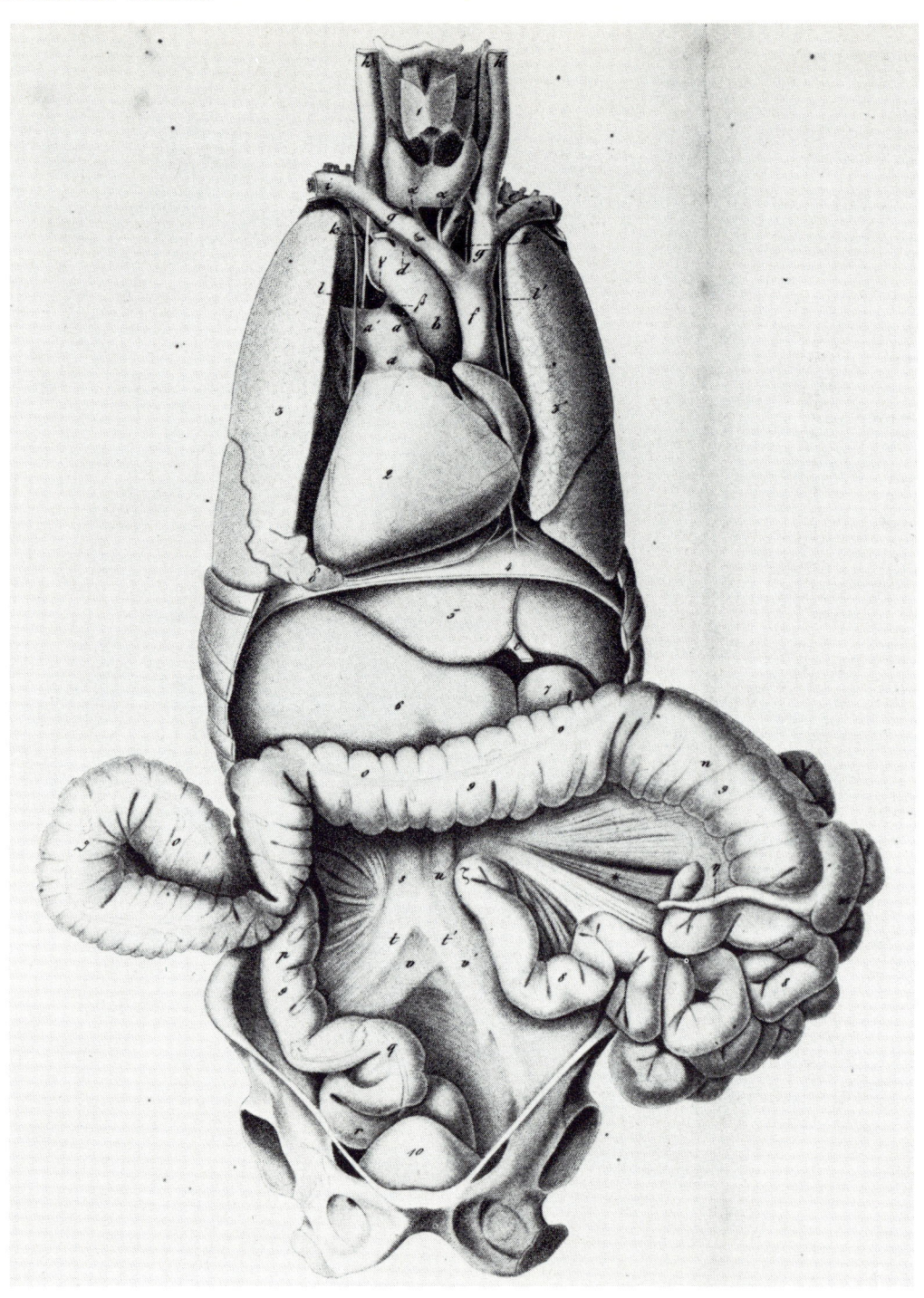

FIG. 3.
Situs Inversus (Transposition of Viscera).

1, Thyroid cartilage; *2*, heart; *3*, lungs; *4*, diaphragm; *5*, liver; *6*, stomach; *7*, duodenum; *8*, jejunum; *9*, ascending colon, transverse colon, and descending colon; *10*, bladder; *a*, pulmonary trunk; *a'*, right pulmonary trunk; *a''*, left pulmonary trunk; *b*, aortic arch; *c*, innominate artery; *d,d'*, right and left carotid artery; *e,e'*, right and left subclavian artery; *f*, superior vena cava; *g,g'*, right and left innominate vein; *h,h'*, right and left internal jugular vein; *i,i'*, right and left subclavian vein; *k,k'*, right and left vagus nerve; *l,l'*, right and left phrenic nerve; *m*, cecum; *n*, ascending colon; *o*, transverse colon; *p*, descending colon; *q*, sigmoid flexure; *r*, rectum; *s*, bifurcation of aorta; *t,t'*, right and left common iliac artery; *u*, inferior vena cava; *v,v'*, right and left common iliac vein; *d,d'*, inferior thyroid vein; β, ligamentum arteriosum; γ, recurrent laryngeal of right vagus; δ, lingula of superior lobe of right lung; ε, ligamentum teres; η, vermiform appendix.

From Gruber, W. Über das Vorkommen eines Mesenterium commune für das Jejuno-Ileum und die größere Anfangshälfte des Dickdarmes bei seitlicher Transposition der Viscera aller Rumpfhöhlen. Resultate aus den bis jetzt gemachten Beobachtungen seitlicher Transposition der Viscera aller Rumpfhöhlen zugleich, oder jener der Bauch- und Beckenhöhle. *Arch. Anat. Physiol. Wissen. Med.* 1865, pp. 558–569.

Organs

FIG. 4.
Positions of Vermiform Appendix.

The diagram illustrates the positions the appendix may occupy in relation to the cecum and ileum, with frequencies of occurrence (%).

Redrawn from Wakeley, C.P.G. The position of the vermiform appendix as ascertained by an analysis of 10,000 cases. *J. Anat. Physiol.* 67:277–283, 1933.

FIG. 5.
Doubled Spleen.

A,B, The two spleens; *a*, part of the inferior portion of the esophagus which traverses the diaphragm (*b*); *c,c'*, stomach seen from its superior and posterior face; *d*, portion of the duodenum; *e,e'*, portion of the greater omentum; *f*, celiac artery; *g*, hepatic branch of celiac artery; *h*, splenic branch of celiac artery; *i*, splenic artery branching into a superior or anterior vessel, which is the smaller of the two arterial branches; *l*, inferior or posterior branch of the splenic artery, which is the larger of the two arterial branches; *m*, splenic vein, superior or anterior, which is the smaller of the two veins; *n*, splenic vein, inferior or posterior, which is the larger of the two veins; *o*, common splenic vein; *p,q*, mesenteric veins; *r*, trunk of part of the portal vein; *s*, celiac ganglion; *t*, common splenic plexus; *u*, superior or anterior splenic nerve plexus; *v*, inferior or posterior splenic nerve plexus.

From Calori, L. Sulla duplicità congenita della milza. *Mem. R. Accad. Sci. Istituto di Bologna S.2* 2:343–360, 1862.

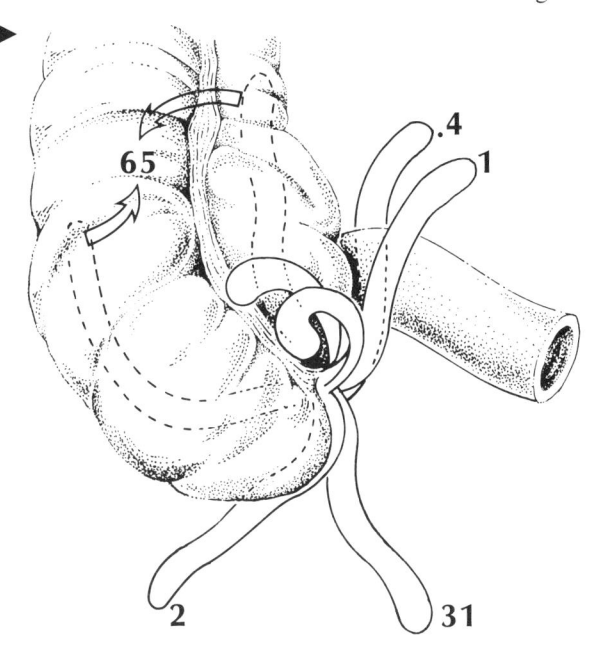

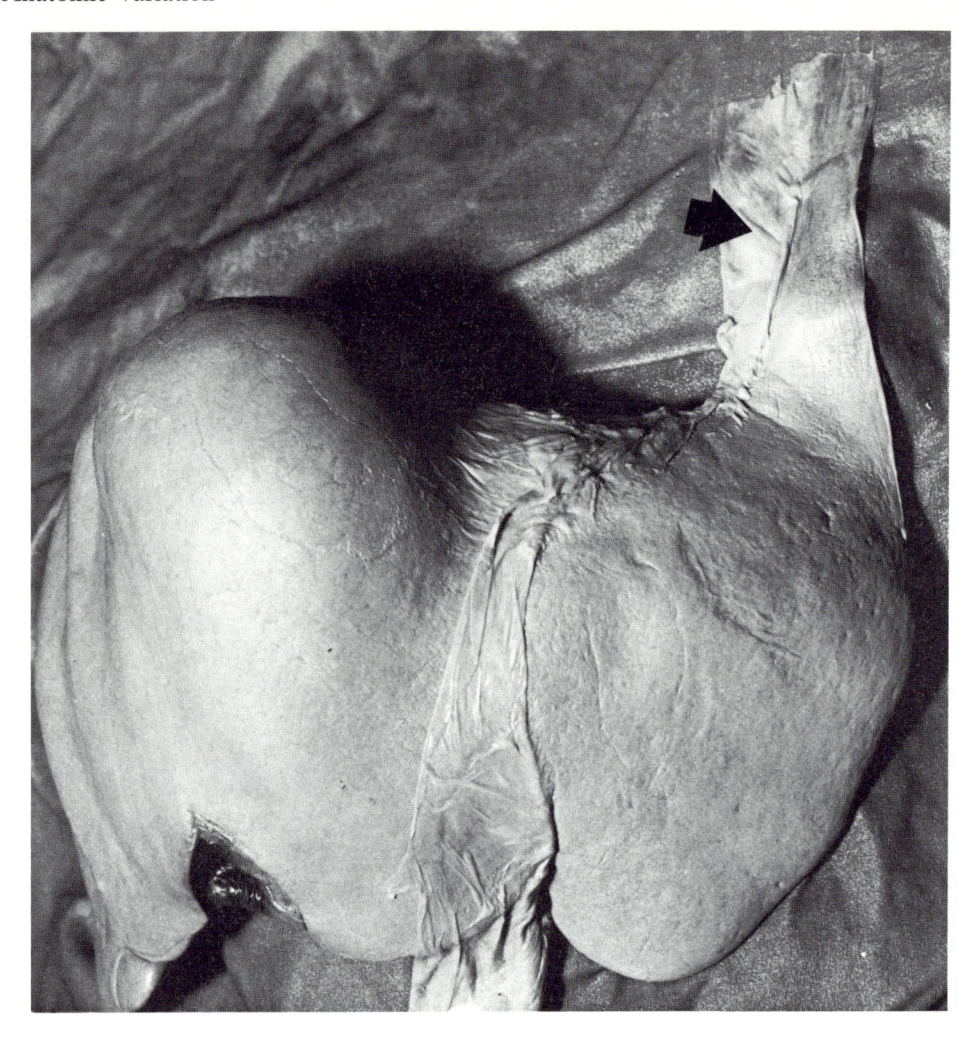

FIG. 6.
Appendix Fibrosa Hepatis.

An extension of the extremity of the left lobe of the liver, in an 81-year-old specimen, was found to curve over the spleen and attach to the posterior abdominal wall and diaphragm. Termed appendix fibrosa hepatis, it is a remnant of a larger organ (relatively) in fetal life. In the fibrous portion of the appendix, the bile ducts of the atrophied liver tissue are said to persist as vasa aberrantia hepatis. The appendix is variable in composition as well as in length. Since this specimen contained viable liver tissue the organ is named a "beaver-tail" liver.

Dissecting room specimen, The University of Iowa College of Medicine.

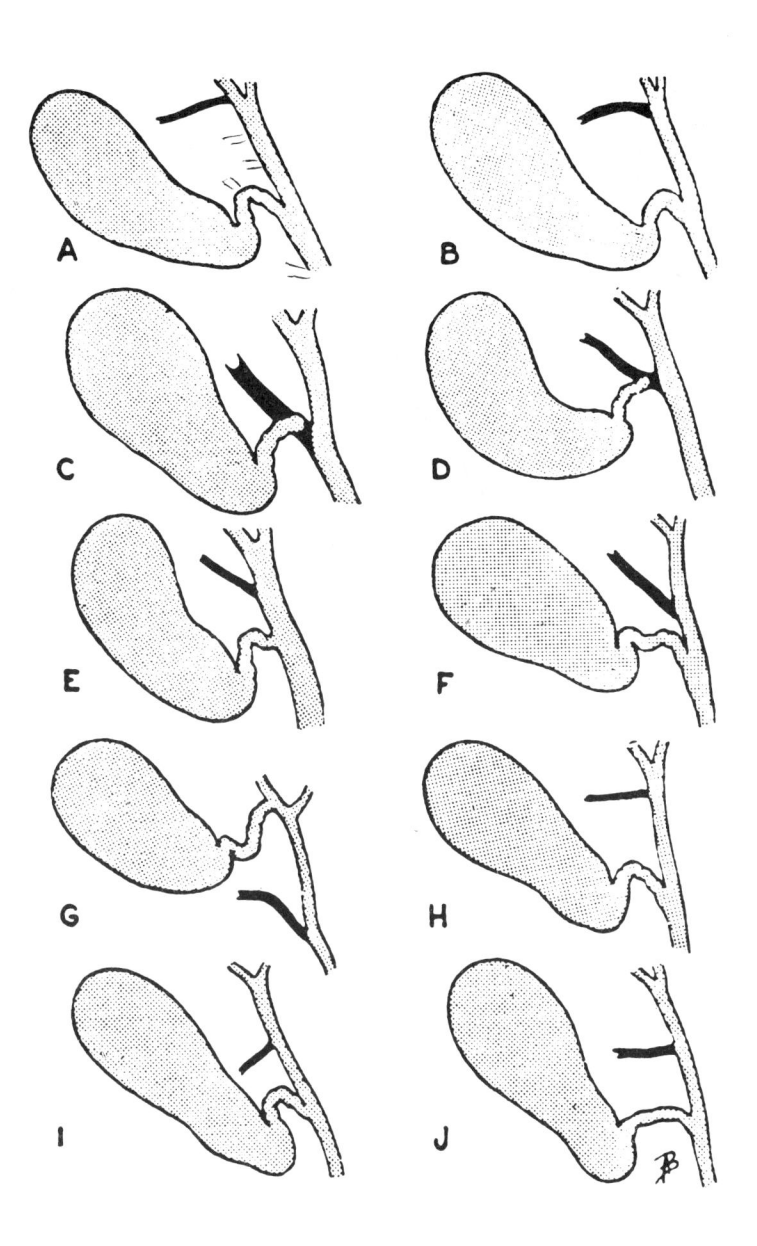

FIG. 7.
Accessory Extrahepatic Ducts
(Variations Encountered in 100 Cadavers.)

A: Accessory duct empties into the right hepatic duct in transverse fissure (porta hepatis). Distance between the accessory duct and junction of the cystic duct with the common hepatic duct is 3.5 cm.

B: Essentially as in A. The distance between the cystic duct and accessory duct is 2.1 cm.

C: Cystic duct empties into the accessory duct close to its junction with the common hepatic duct. Accessory duct approximates the diameter of the common hepatic duct. Distance between accessory duct junction with common hepatic duct and junction of right and left hepatic ducts is 2.7 cm.

D: Characteristics essentially the same as C. Distance between junction of right and left hepatic ducts and junction of accessory duct with common hepatic duct is 1.2 cm.

E, I, J: These are essentially the same in that the accessory duct empties into the common hepatic duct midway between junction of right and left hepatic ducts and junction of the cystic duct with the common hepatic duct. Length of the common hepatic duct in E and I is 3 cm, and in J is 2.2 cm.

F: Accessory hepatic duct empties into the common hepatic duct simultaneously with the cystic duct. Distance of accessory duct junction from beginning of the common hepatic duct is 3.5 cm. Diameters of the accessory and common hepatic ducts are the same.

G: Cystic duct empties into the right hepatic duct 1 cm from hilum of liver. Junction of right and left hepatic ducts is approximately 2 cm from hilum of liver. Accessory hepatic duct empties into the common bile duct 2.5 cm from junction of right and left hepatic ducts and 3.5 cm from the cystic duct junction. Diameter of accessory duct is the same as that of the common bile duct.

H: Accessory duct empties into the common hepatic duct. Distance between the accessory duct and cystic duct junction with the common hepatic ducts is 2.2 cm. Accessory duct lies under the surface of the liver.

From Lichtenstein, M.E. and A.J. Nicosia. The clinical significance of accessory hepato-biliary ducts. *Ann. Surg.* 141:120–124, 1955.

Human Anatomic Variation

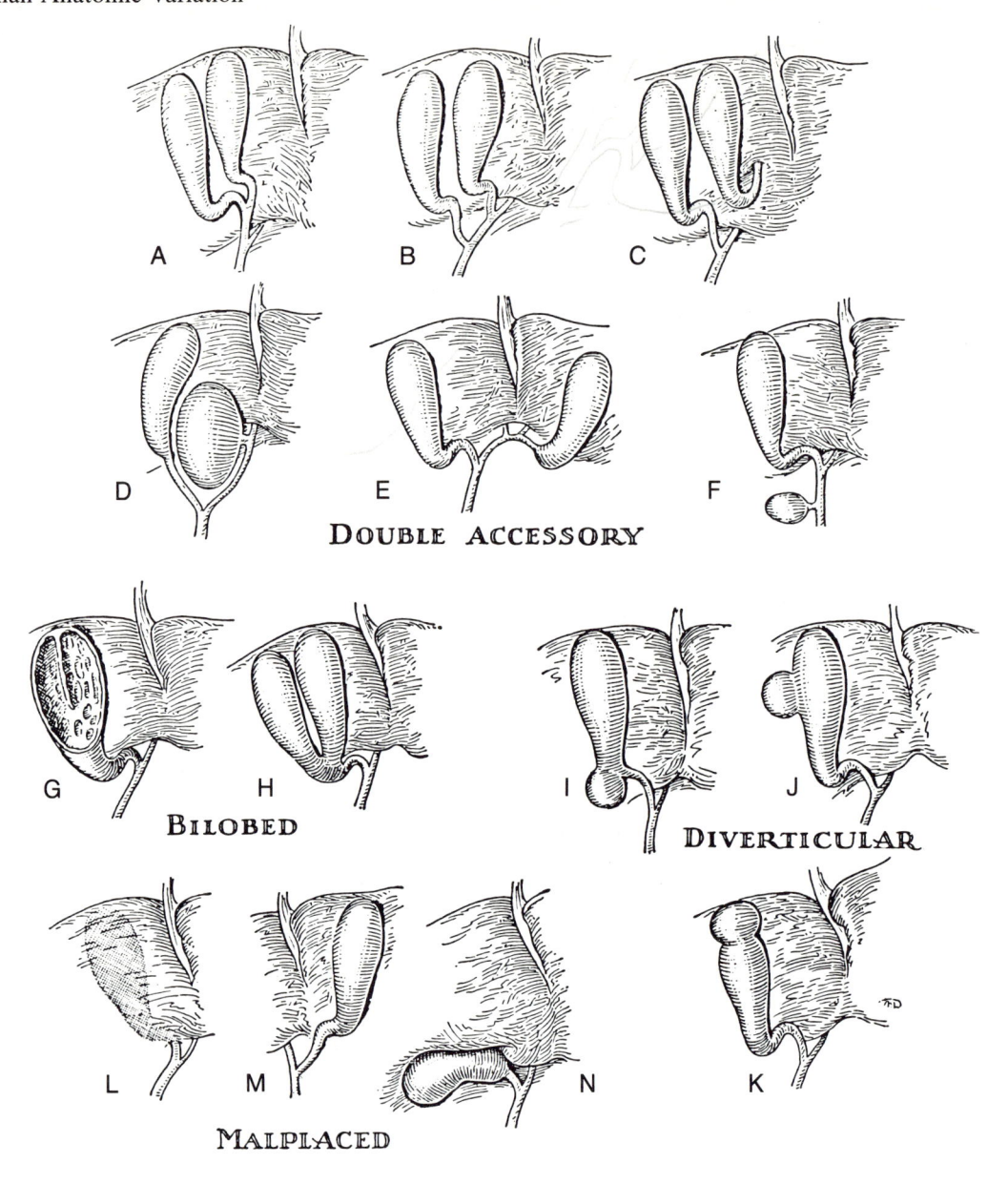

FIG. 8.
Gallbladder Variations.
Doubled Gallbladder.
Bilobed Gallbladder.
Abnormal Position of Gallbladder.

Types of Gallbladder
A: Regular and accessory gallbladders lodged in a fossa of normal position and possessing a common cystic duct.
B: A pattern differing from the preceding in the occurrence of separate cystic ducts.
C: Termination of one of the cystic ducts in the liver.
D: An accessory gallbladder, with a cystic duct terminating in the common hepatic duct.
E: An accessory vesica situated beneath the left lobe of the liver.
F: An instance of doubling, in which the accessory gallbladder, situated in the lesser omentum, emptied into the common bile duct.

Forms of Bilobed Gallbladder
G: Partially separated by an internal septum.
H: Paired through the fundus and body, but joined at the neck to form a single cystic duct.
Types of Congenital Diverticula
I: Diverticulum at the neck of the gallbladder.
J: Cul-de-sac from the body of the organ.
K: Similar expansion at the fundus.
Abnormality in Position of Gallbladder
L: Lodgment within the substance of the liver.
M: Placement under the left hepatic lobe.
N: Posterior situation under the right lobe.

Redrawn from Gross, R.E. Congenital anomalies of the gallbladder. A review of 148 cases with report of double gallbladder. *Arch. Surgery* 32:131–162, 1936. In McVay, C.B. *Anson and McVay Surgical Anatomy*, 6th Ed, Vol. 1. W.B. Saunders Co., Philadelphia, 1984.

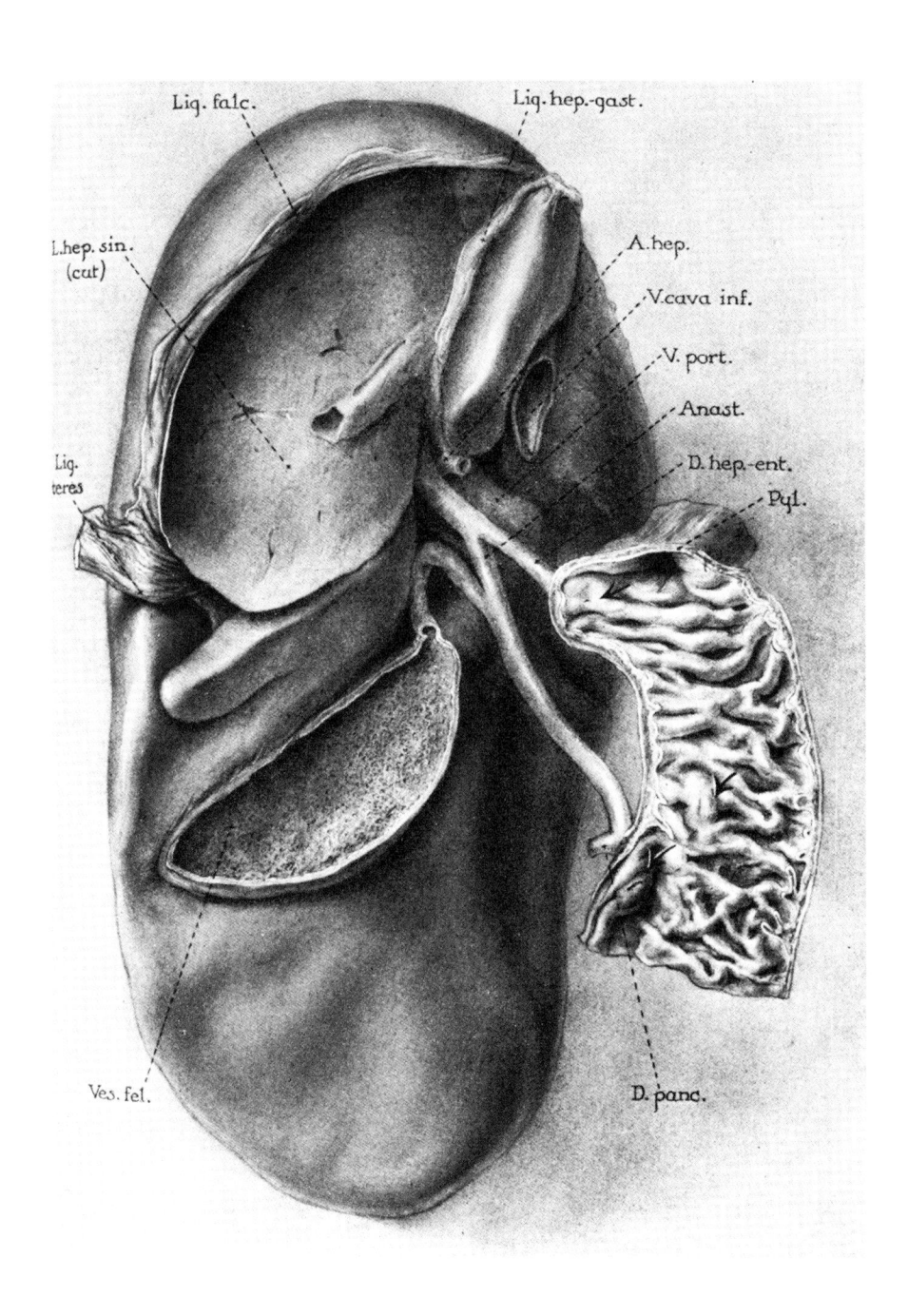

FIG. 9.
Duplication of Common Bile Duct.

Boyden explains the figure as follows:

"Liver and duodenum of white male, aged 48, exhibiting duplication of common-bile duct. Left lobe of liver (*L.hep.sin.*) severed and removed at level of falciform ligament (*Lig.falc.*). Same orientation as in figure 140. Cunningham's Manual of Anatomy, vol. 2, 8th ed., '27, except that left wall of duodenum has been removed and the remaining half rotated to observer's right. Common hepatic duct, 4 cm. long; hepato-enteric duct (*D.hep.ent.*), 1.8 cm. long; common-bile duct, 6 cm. long; anastomotic segment (*Anast.*) between common hepatic duct and cystic duct, 1 cm. long; cystic duct, 4 cm. long. Hepato-enteric duct, virtually occluded at point of insertion on pars superior of duodenum; site marked by small pore, 1.5 cm. from pylorus (see uppermost of 3 arrows). Common-bile duct empties low down, on pars descendens, in common with ductus pancreaticus (*D.panc.*), 8 cm. below pylorus (see lowest of 3 arrows). Accessory pancreatic duct opens on duodenal papilla (middle arrow), 2 cm. above ampulla of Vater. Edge of duodenum to observer's right is anterior wall; posterior edge, to which ducts are attached, shows post mortem (?) contracture."

From Boyden, E.A. The problem of the double ductus choledocus (an interpretation of an accessory bile duct attached to the pars superior of the duodenum). *Anat. Rec.* 55:71–93, 1932.

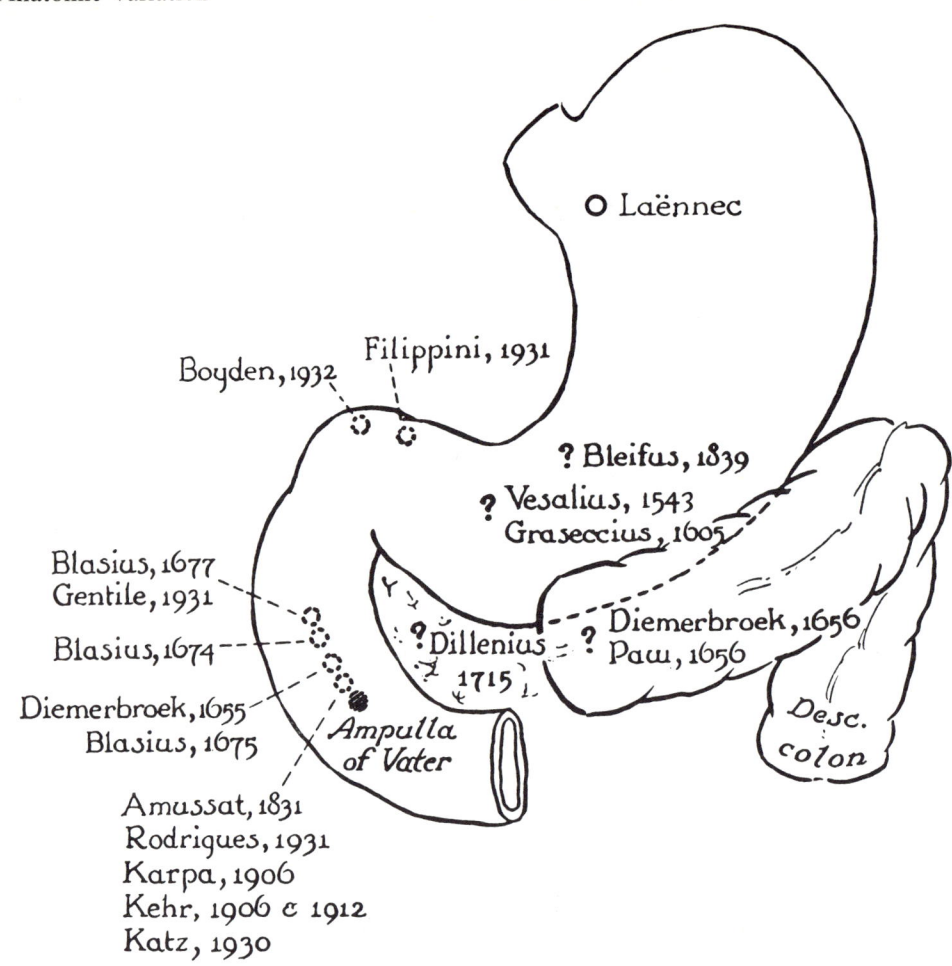

FIG. 10.
Insertion of Aberrant Bile Ducts.

The sites of aberrant insertion of bile ducts is shown in the illustration. Inspection of the illustration "reveals the fact that the ducts tend to fall into two general groups – those bordering upon the pylorus or extending further into the stomach (group A) and those (group B) situated on the duodenum within 3 cm. of the usual of the ampulla of Vater."

From Boyden, E.A. The problem of the double ductus choledocus (an interpretation of an accessory bile duct attached to the pars superior of the duodenum). *Anat. Rec.* 55:71–93, 1932.

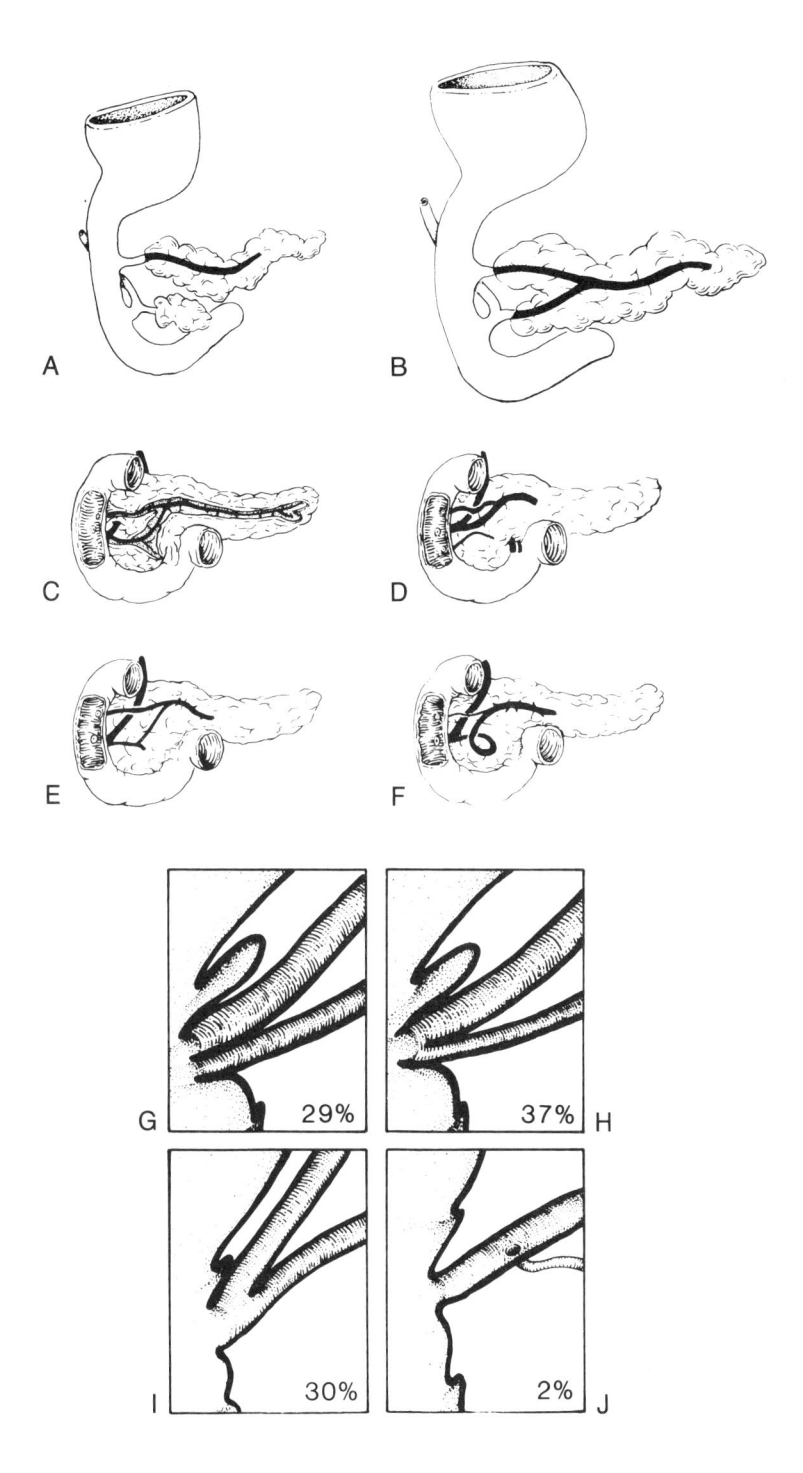

Organs

FIG. 11
Pancreatic and Common Bile Ducts.

A: Pancreas of a human embryo, fifth week.
B: Pancreas of a human embryo, seventh week.
C: Most common arrangement of pancreatic ducts.
D: Specimen with three papillae.
E: Adult pancreas showing an embryonic type of duct system in which the accessory duct carries most of the pancreatic secretion.
F: Adult pancreas showing an unusual loop configuration of the main pancreatic duct.
G: In 73 of 250 specimens, about 29%, no connection between the pancreatic and bile ducts was found, each entering the duodenum contiguously or at separate points.
H: In 92 specimens, 37%, the two ducts were contiguous with the dividing septum terminating about 2 mm from the apex.
I: In 81 instances, 30%, the duct emptied into a common ampulla which extended from 3 to 14 mm from the apex of the duodenal orifice. In four specimens the length of the diverticulum exceeded 10 mm and was as shown.
J: In four instances the pancreatic duct was reduced to a fibrous cord (not shown), and the accessory duct drained the entire gland.

Redrawn from Rienhoff, W.F. Jr. and K.L. Pickrell. Pancreatitis. An anatomic study of the pancreatic and extrahepatic biliary systems. *Arch. Surg.* 51:205–219, 1945.

Human Anatomic Variation

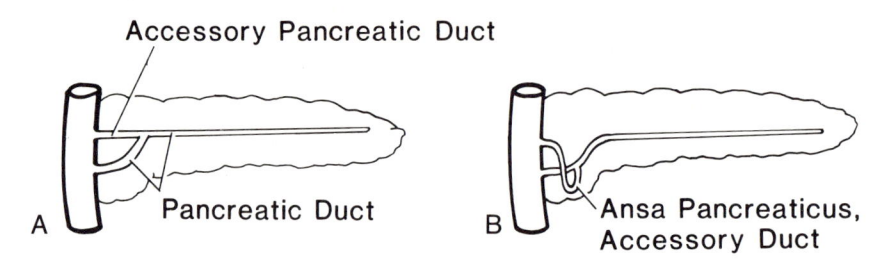

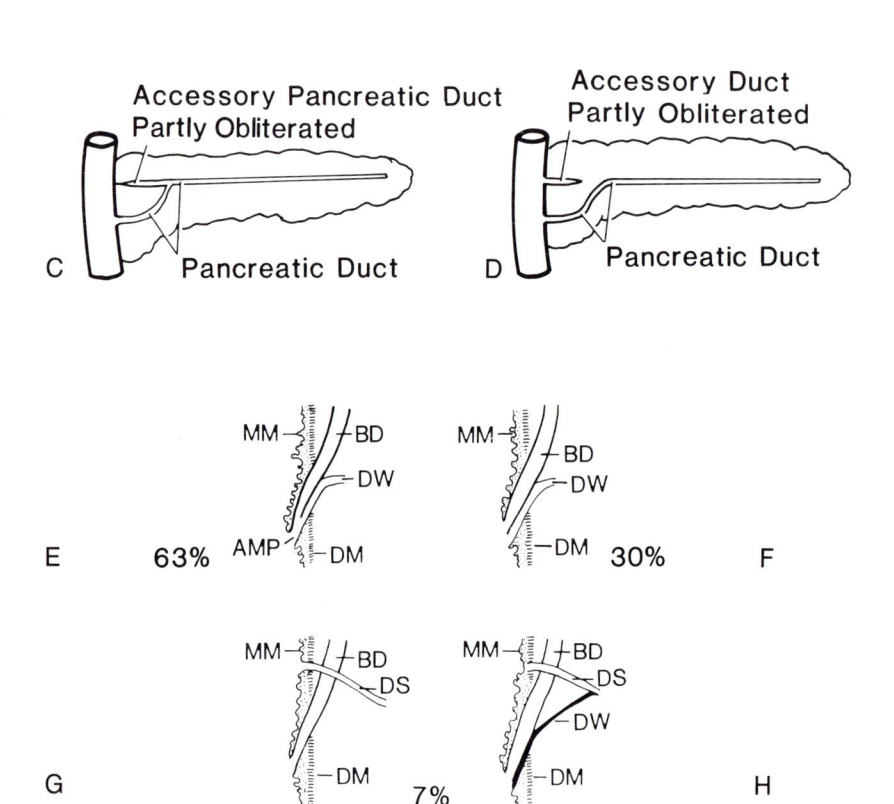

FIG. 12.
Pancreatic Ducts.

A: Pancreatic duct system in which dorsal and ventral pancreatic ducts have fused to form the combined pancreatic duct (Wirsung) and the accessory pancreatic duct (Santorini) entering the duodenum at the minor papilla.
B: Pancreatic duct system showing the "ansa pancreatica" formed by the proximal part of the accessory duct, the inferior branch of the dorsal duct, and the inferior branch of the ventral duct.
C: Pancreatic duct system showing obliteration of the accessory duct at its entrance into the duodenum.
D: Pancreatic duct system showing obliteration of the accessory duct at its entrance into Wirsung's duct.
E: Type I. The bile duct and the duct of Wirsung join together in the duodenal wall to form the hepatopancreatic duct and open through it into the duodenum.
F: Type II. The two ducts are separated throughout their extent by a septum and open separately at the summit of the major duodenal papilla.
G, H: Type III. The whole pancreas is drained by the duct of Santorini. The duct of Wirsung is either reduced to a fibrous cord or is absent.
In the present series of 100 specimens, 63 belonged to type I, 30 to type II, and 7 to type III. In one case classified as type II the duct of Santorini was larger than the duct of Wirsung. In the 63 cases belonging to type I, the hepatopancreatic duct varied in length from 1 mm to 14.5 mm.
AMP, Hepato-pancreatic duct; BD, bile duct; DM, duodenal muscle; DS, duct of Santorini; DW, duct of Wirsung; MM, mucous membrane.

Redrawn (A–D) from Dawson, W. and J. Langman. An anatomical-radiological study on the pancreatic duct pattern in man. Anat. Rec. 139:59–68, 1961; and (E–H) from Singh, I. Observations on the mode of termination of the bile and pancreatic ducts: Anatomical factors in pancreatitis. J. Anat. Soc. India. 5:54–60, 1956.

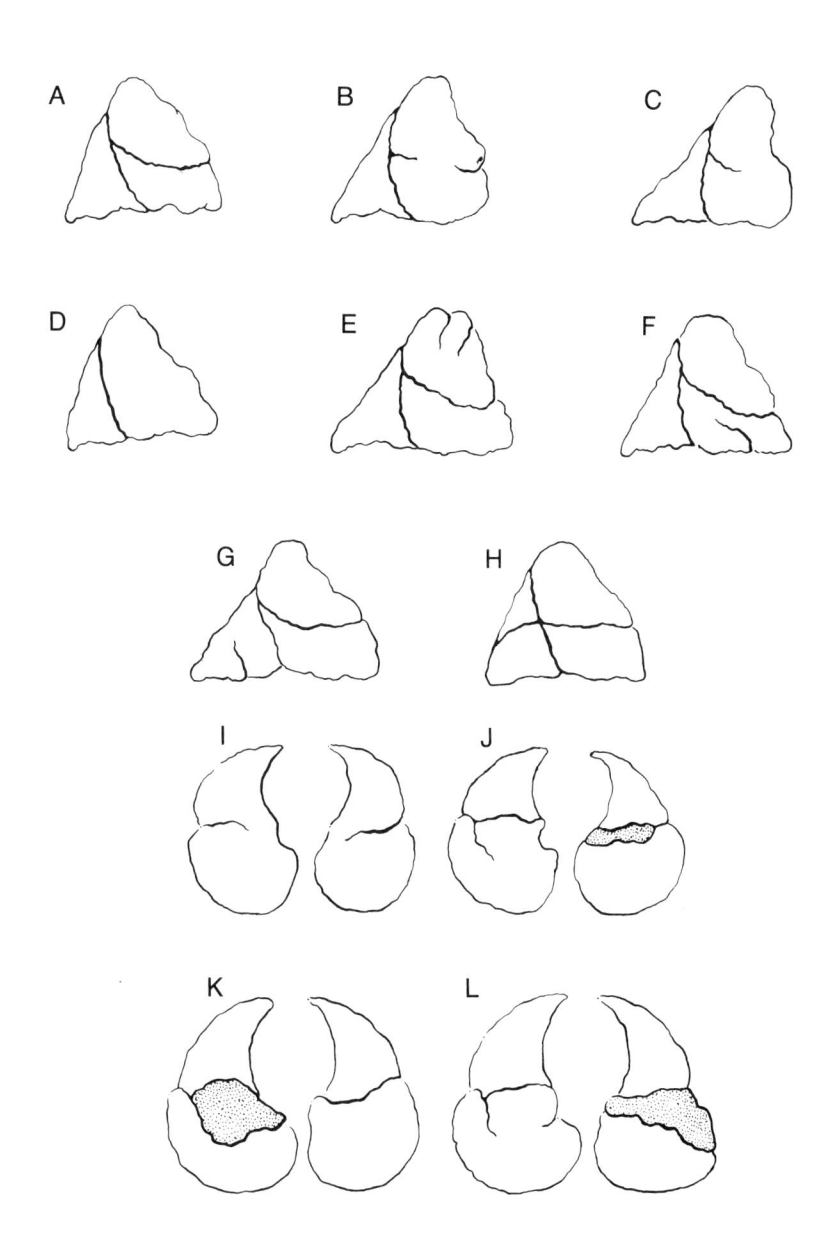

FIG. 13.
Variations in Peripheral Segmentation of Right Lung.
Variations in Fissures on the Inferior Face of Right and Left Lungs.

A: Normal right lung.
B: Incomplete horizontal fissure.
C: Another type of incomplete horizontal fissure.
D: Absence of horizontal fissure.
E: Subdivision of the superior lobe.
F: Subdivision of the middle lobe.
G: Subdivision of the inferior lobe.
H: More complete subdivision of the inferior lobe.
I: Incomplete oblique fissure.
J: Right infracardiac lobe.
K: Left infracardiac lobe.
L: Left infracardiac lobe with a second ventral lobe.

Redrawn from Lucien, in Testut, L. and A. Latarjet. *Traité d'Anatomie Humaine*. G. Doin & Cie., Paris, 1949.

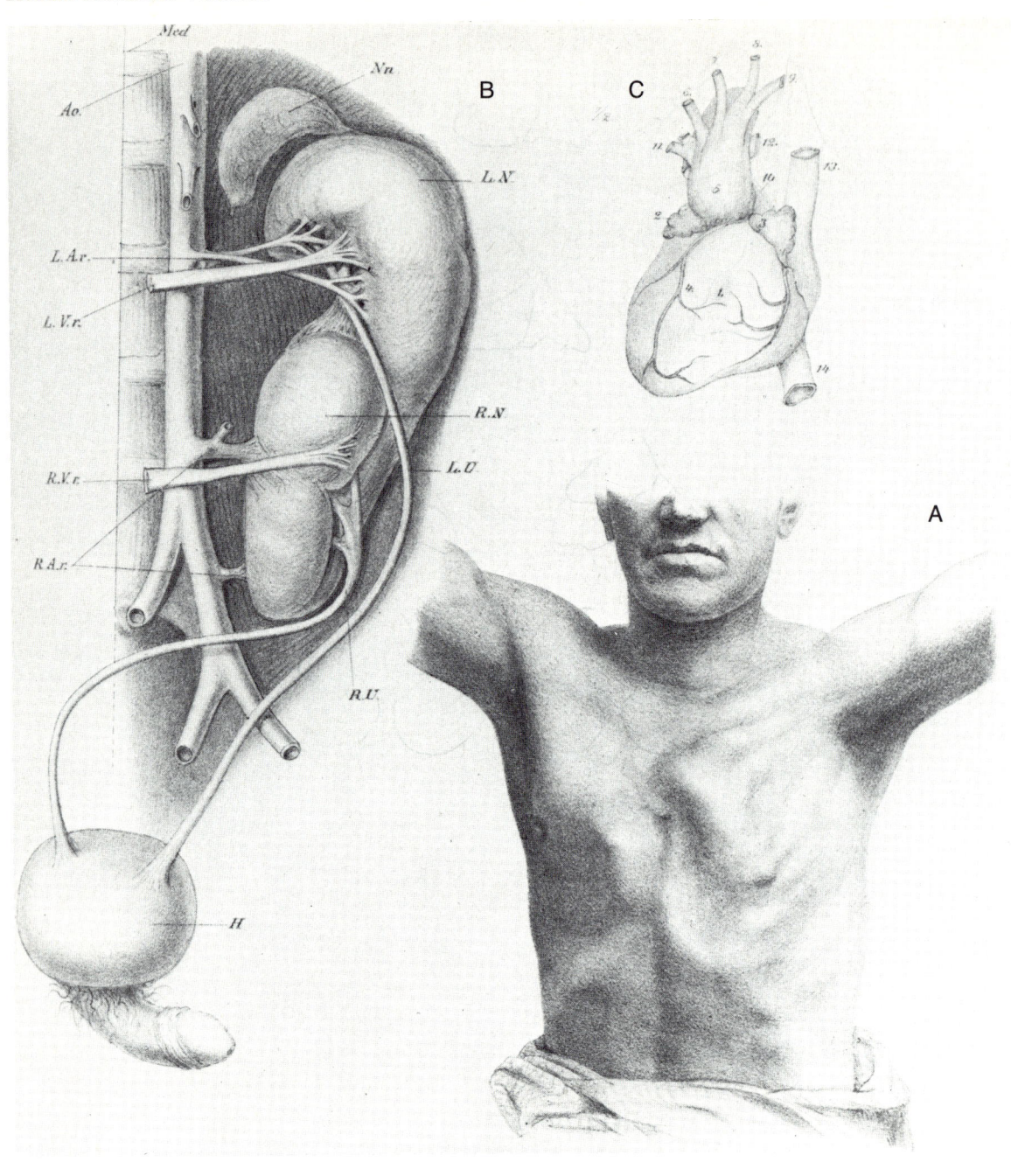

FIG. 14.
Congenital S-shaped Kidney.
(Absence of Sternocostal Heads of Pectoralis Major).
(Right-sided Heart).

A: Absence of the sternocostal heads of pectoralis major in a living subject. Rib anomalies were also present.
B: Congenital S-shaped kidney. *Med*, Midsagittal line; *Ao.*, aorta; *Nn.*, suprarenal gland; *L.N.*, left kidney; *R.N.*, right kidney; *L.U.*, left ureter; *R.U.*, right ureter; *L.A.r.*, left renal artery; *L.V.r.*, left renal vein; *R.A.r.*, right renal artery; *R.V.r.*, right renal vein; *H.*, urinary bladder.

C: Right-sided heart (dextrocardia) in situs transversus (situs inversus viscerum). *1.*, Left ventricle; *2.*, right atrium; *3.*, left atrium; *4.*, anterior longitudinal sulcus; *5.*, aorta; *6.*, right subclavian artery; *7.*, right common carotid; *8.*, left common carotid; *9.*, left subclavian artery; *10.*, pulmonary artery; *11.*, pulmonary artery, right branch; *12.*, pulmonary artery, left branch; *13.*, superior vena cava; *14.*, inferior vena cava.

From (A) Seitz, C. Eine seltene Mißbildung des Thorax. *Arch. Pathol. Anat. Physiol. Klin. Med.* 98:335–337, 1884; (B) Broesike, G. Ein Fall von congenitaler S-förmiger Verwachsung beider Nieren. *Ibid.* 98:338–341, 1884; and (C) Wehn. Zur Frage des Situs transversus. *Ibid.* 98:354–363, 1884.

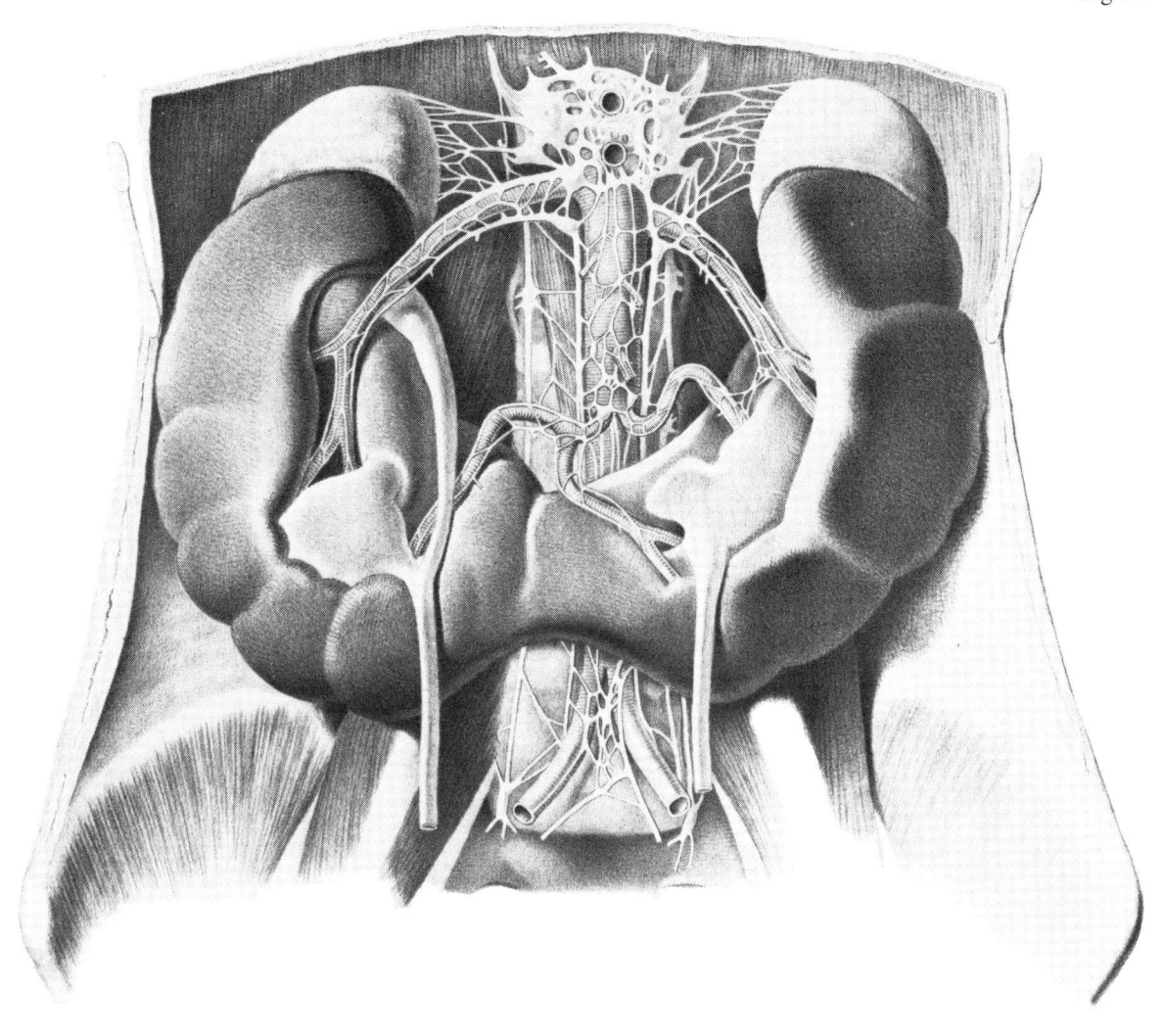

Organs

FIG. 15.
Horseshoe Kidney.

From Calori, L. Caso di un rene a ferro di cavallo. *Mem. Accad. Sci. Istituto di Bologna S. 4* 6:73–83, 1884.

Human Anatomic Variation

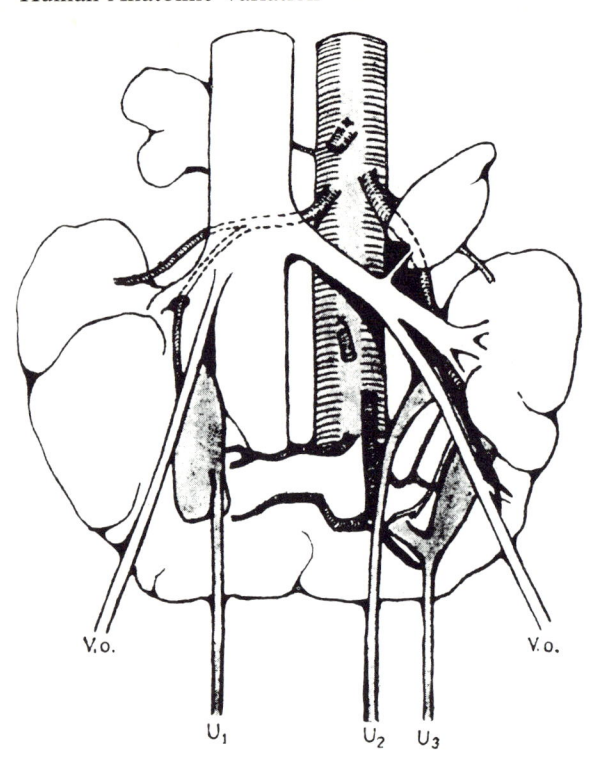

FIG. 16.
Horseshoe Kidney with Three Ureters.

Veins are white, unshaded; arteries are shaded gray. +, origin of superior mesenteric artery; x, origin of inferior mesenteric artery; U_1, U_2, U_3, three ureters from the horseshoe kidney; *V.o.*, ovarian veins.

From Serbeny, A. Ein Fall von Hufeisenniere mit drei getrennt in die Blase mündenden Ureteren. *Anat. Anz.* 63:131–138, 1927.

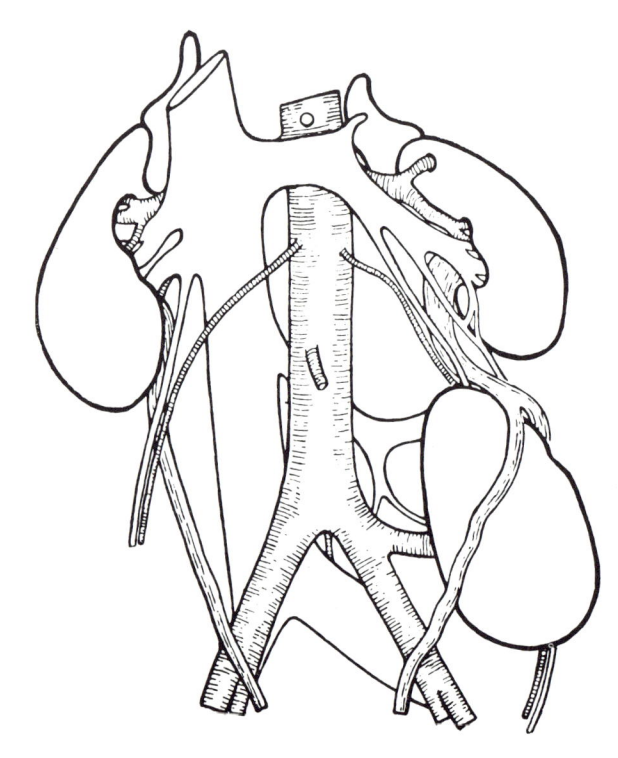

FIG. 17.
Supernumerary Third Kidney.

Camera lucida drawing made from an adult male subject which possessed three kidneys.

From Dixon, A.F. Supernumerary kidney: The occurrence of three kidneys in an adult male subject. *J. Anat. Physiol.* 45:117–121, 1911.

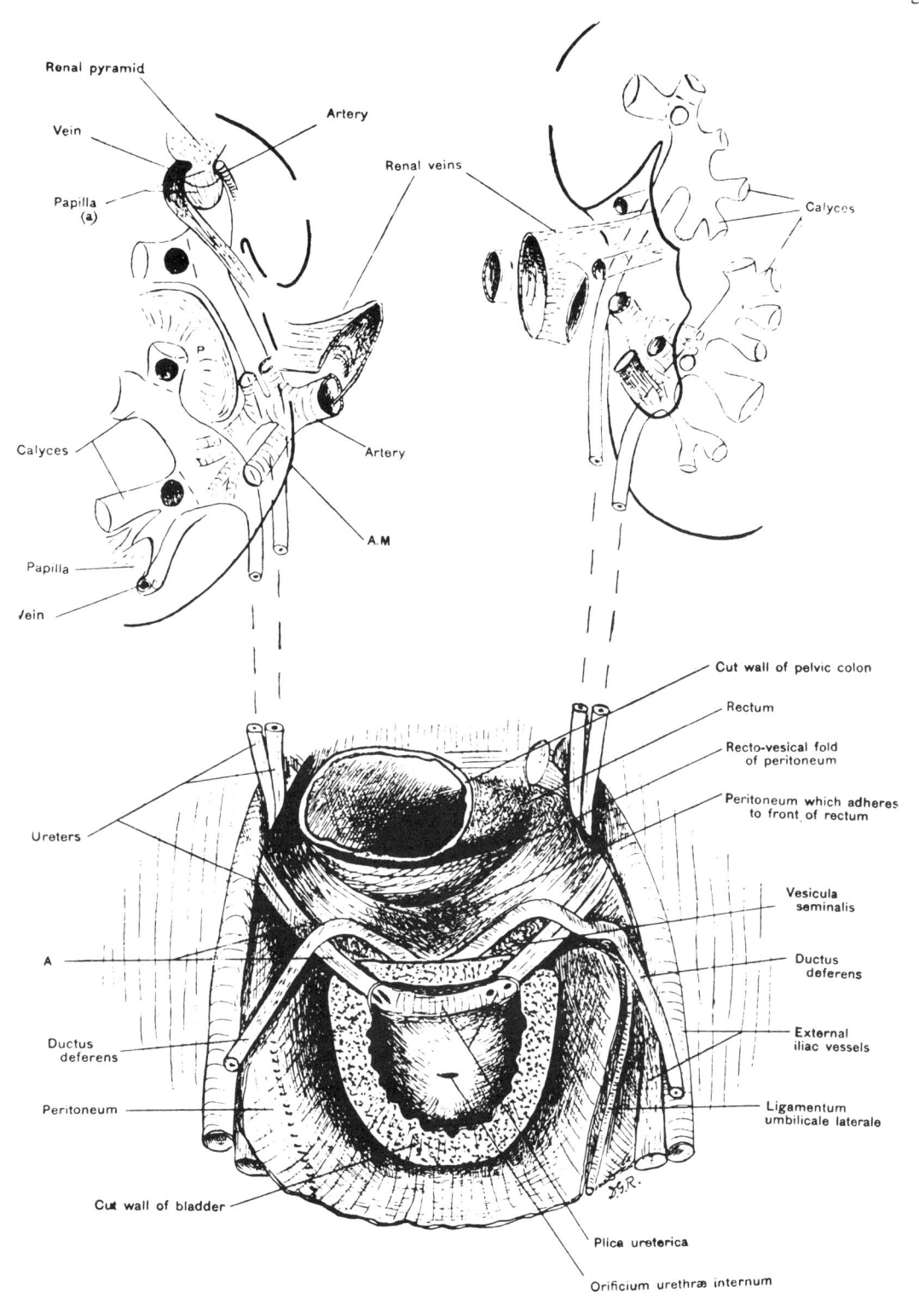

FIG. 18.
Bilateral Duplication of the Ureter.

"The kidneys are represented as lying lower than was the case. The bladder is viewed from above and slightly from in front. A indicates the apparent single portion of the right ducts. The whole of the convex border of the recto-vesical fold of peritoneum adheres to the rectum, so that the fold is purely rectal."

From Reid, D.G. Bilateral duplication of the ureter. *J. Anat.* 51:30–35, 1917.

Human Anatomic Variation

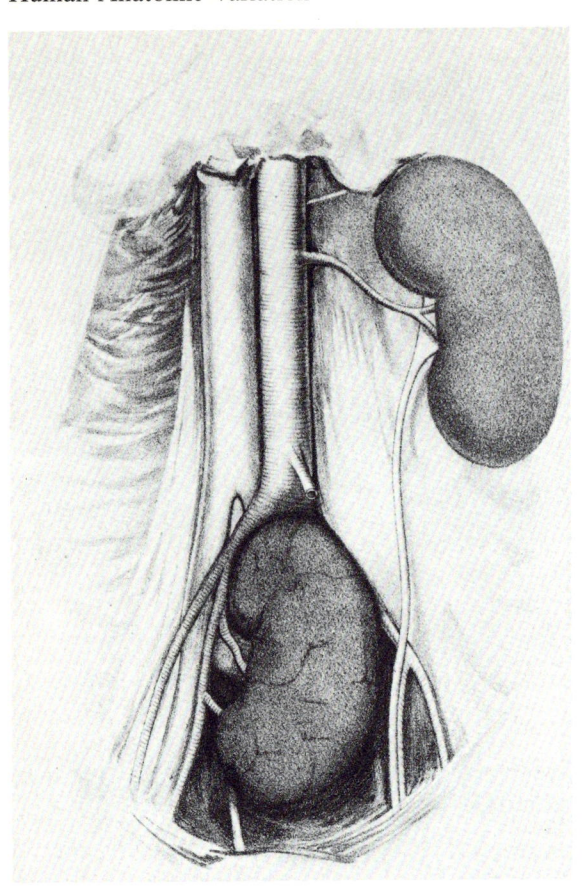

FIG. 19.
Pelvic Kidney.
From Brown, M. Variations in the position and development of the kidneys. *J. Anat. Physiol.* 28:193–194, 1894.

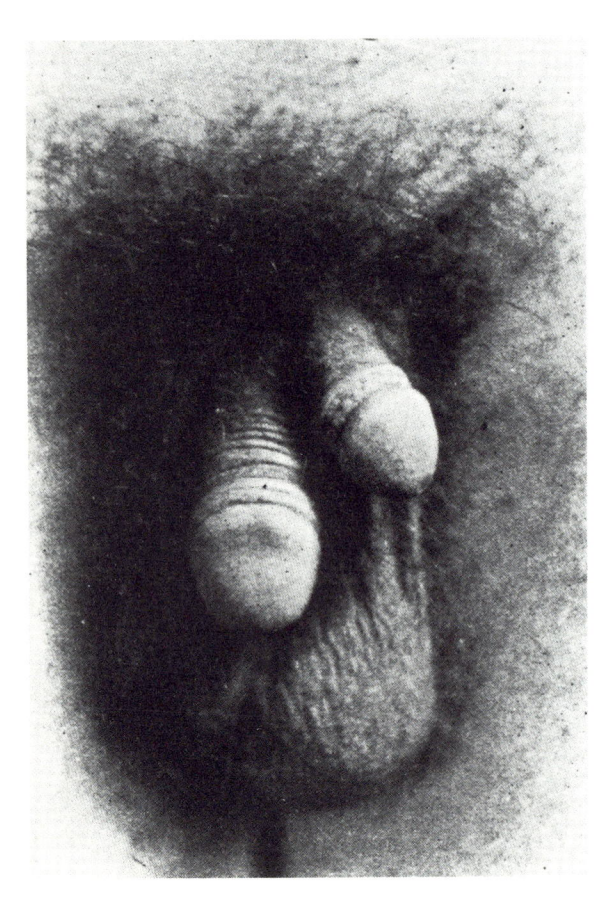

FIG. 20.
Double Penis.
From Seth, R.E. and A.H. Peacock. Double penis. *Urol. Cutan. Rev.* 36:590–592, 1932.

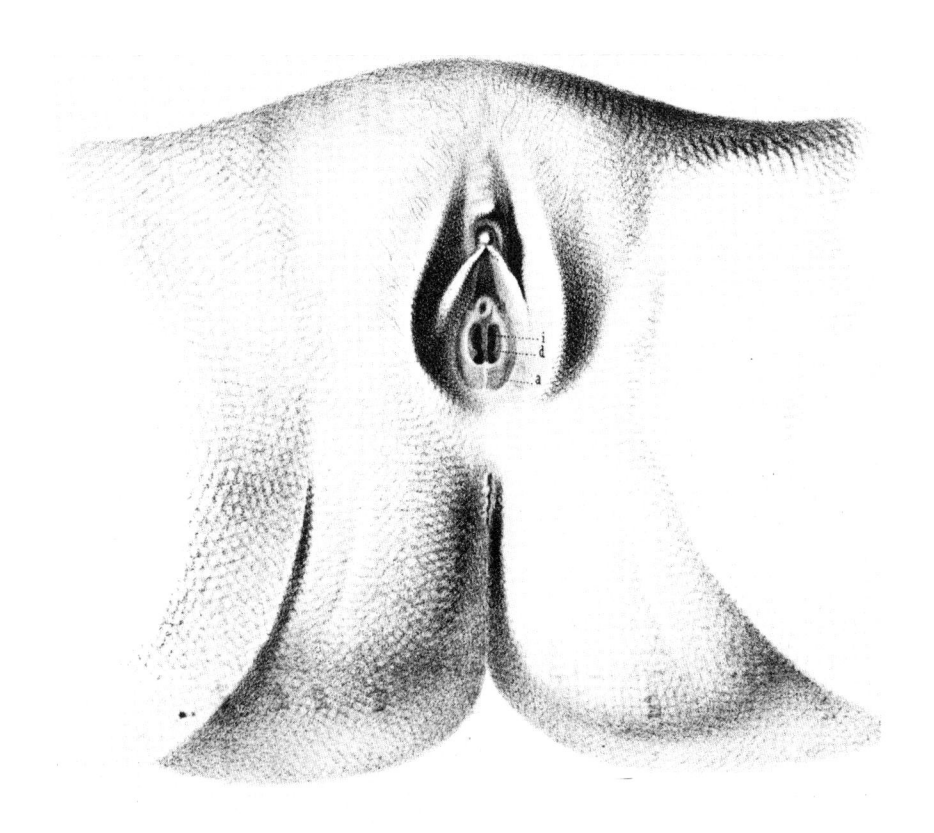

FIG. 21.
Double Vagina.

a, Longitudinal raphe; *d*, lateral caruncula; *i*, medial perpendicular septum.

From Calori, L. Di alcuni particolari intorno le parti genitali; muliebri. *Mem. Accad. Sci. Istituto di Bologna S.2* 1:171–201, 1862.

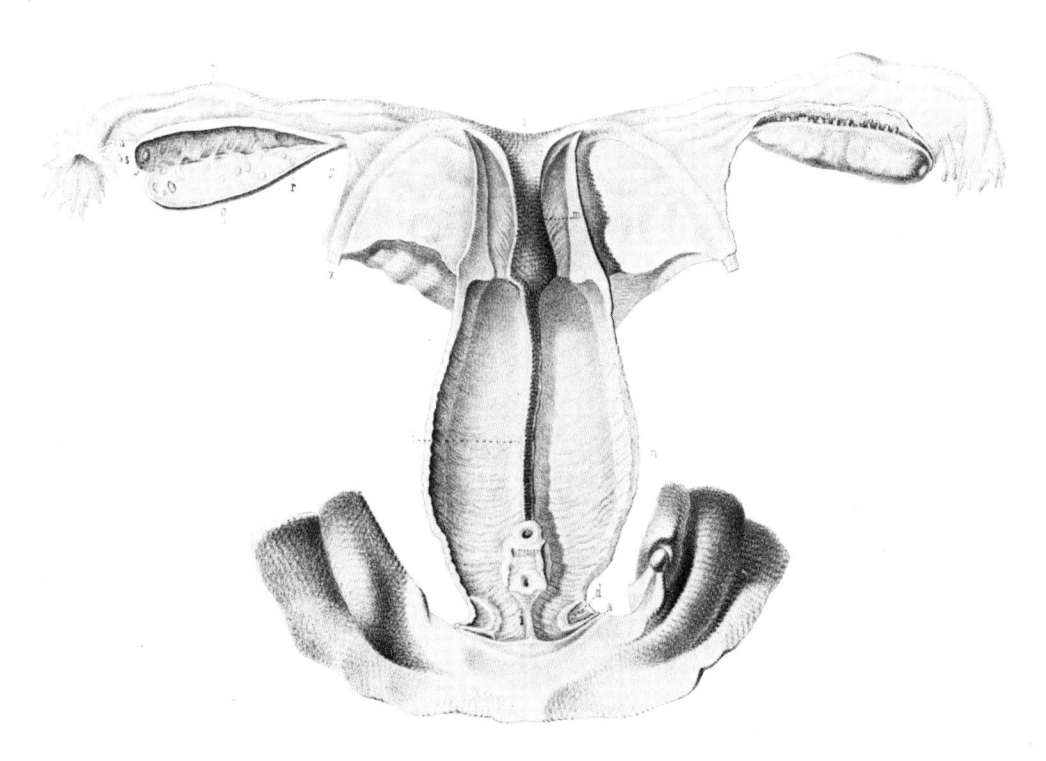

FIG. 22.
Double Vagina and Double Uterus.

a, External fascia; *d*, lateral caruncula; *i*, medial perpendicular septum of vagina; *m*, perpendicular septum of uteri; *p*, fallopian tube; *q*, ovary; *r*, graafian follicles; *s*, appendices; *x*, round ligament of uterus; *y*, broad ligament.

From Calori, L. Di alcuni particolari intorno le parti genitali muliebri. *Mem. Accad. Sci. Istituto di Bologna S.2* 1:171–201, 1862.

Human Anatomic Variation

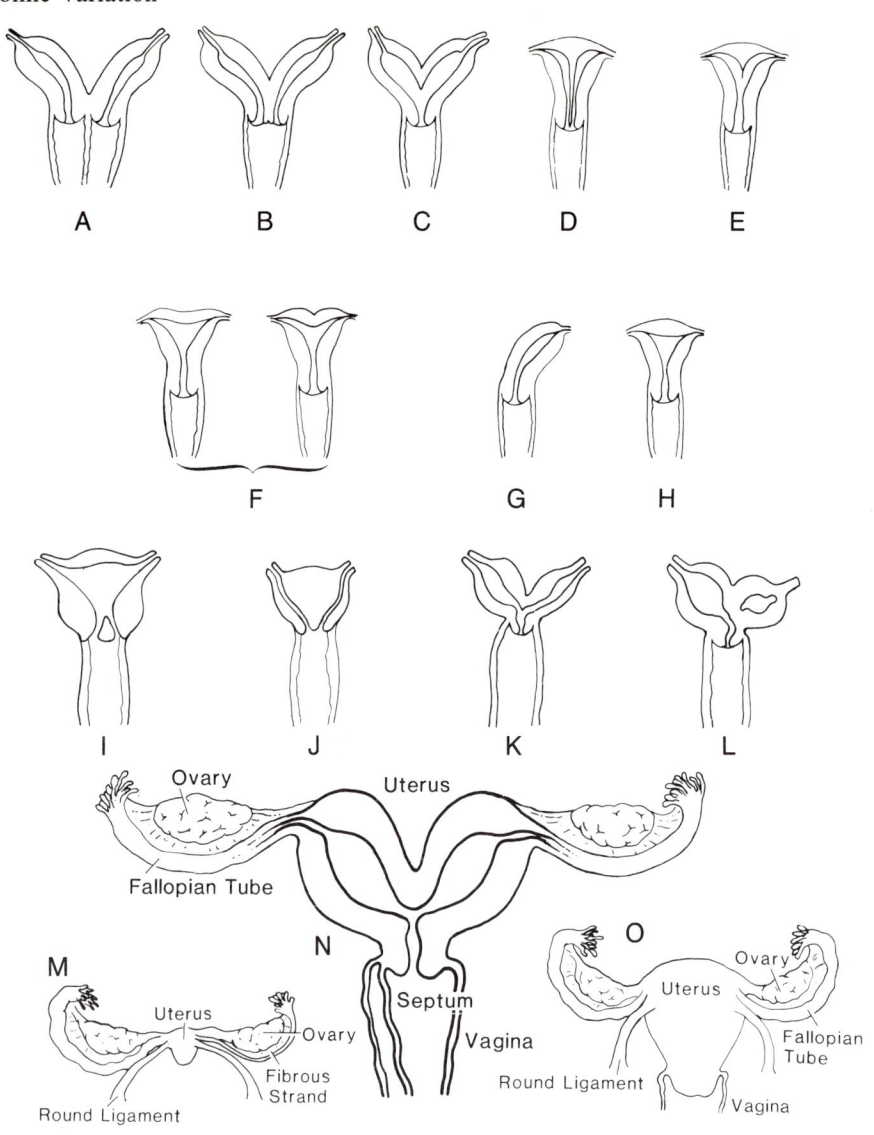

FIG. 23.
Uterine and Vaginal Variations.

A: Uterus didelphys.
B: Uterus duplex bicornis bicollis, vagina simplex.
C: Uterus bicornis unicollis, vagina simplex.
D: Uterus septus.
E: Uterus subseptus.
F: Uterus arcuatus or uterus cordiformis.
G: Uterus unicornis.
H: Uterus simplex (the usual form) (also O).
I: Uterus subseptus unicorporeus with septate cervix.
J: Uterus septus duplex (uterus bilocularis).
K: Uterus bicornis, unilateral rudimentarius.
L: Uterus with closed accessory horn.
M: Case: Uterus, small as a thumbnail. Right tube normal. Left tube replaced by a strand of fibrous tissue. Both ovaries appeared grossly normal and were actually twice the size of the uterus. There was no cervix. The vagina was one inch long (2.54 cm) and about three-eighths of an inch wide (0.95 cm).
N: Case: Illustration demonstrating uterus duplex (bicornis) unicollis with a vaginal septum dividing the vagina into a blind pouch and a left normal vagina. Each member had a single normal tube and an associated normal ovary.
O: Normal uterus.

Redrawn from Weintrob, M. Abnormalities of the female genitalia. *J. Int. Coll. Surgeons* 7:381–396, 1944; and from Jarcho, J. Malformations of the uterus. Review of the subject, including embryology, comparative anatomy, diagnosis and report of cases. *Am. J. Surg.* 71:106–166, 1946.

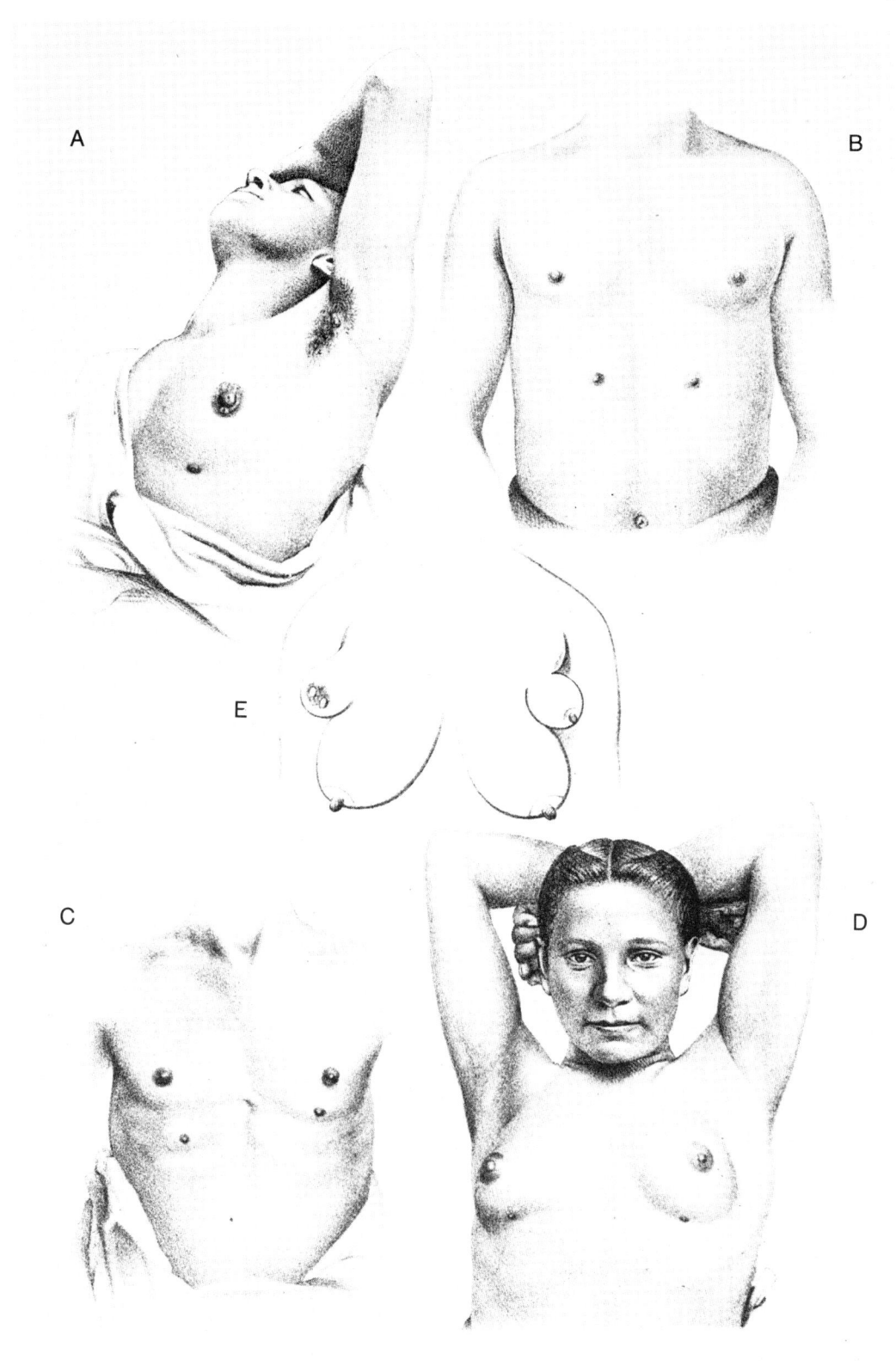

FIG. 24.
Supernumerary Mammae and Nipples in Men and Women. From Leichtenstern. Über das Vorkommen und die Bedeutung supernumerärer (accessorischer) Brüste und Brustwarzen. Auf Grund 13 eigener und 19 aus der Literatur gesammelter Beobachtungen. *Arch. Pathol. Anat. Physiol. Klin. Med.* 73:222–256, 1878.

Human Anatomic Variation

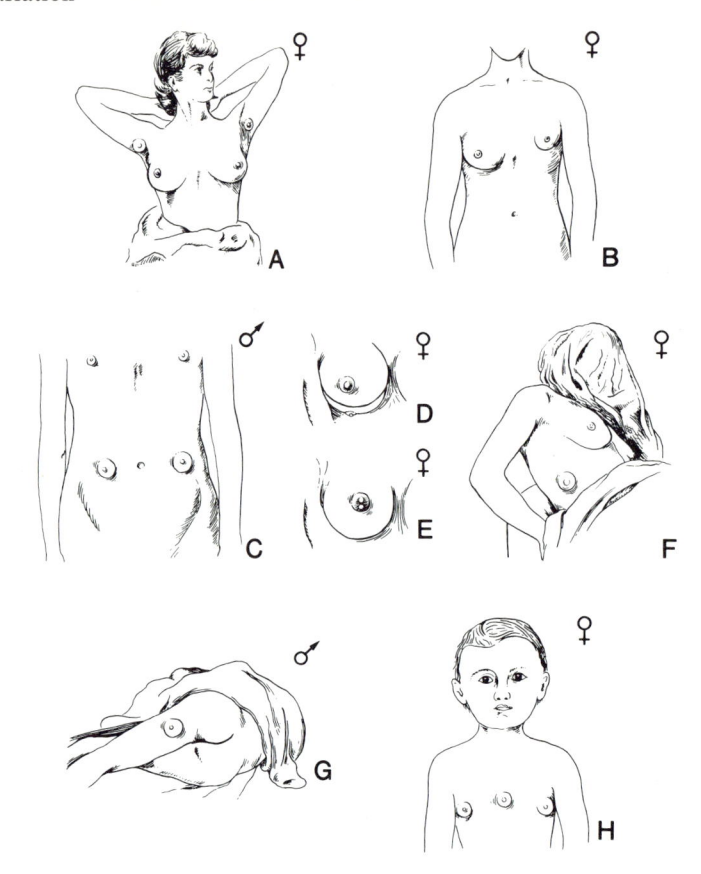

FIG. 25.
Variations in Size, Number, Location, and Structure of the Mammary Gland in Men and Women.

Symbols: ♂, male; ♀, female.
A: Axillary mammary glands.
B: Marked variation in size (left side rudimentary).
C: Paired, accessory glands at level of navel.
D: Doubled mammary gland (second gland beneath normal gland).
E: Doubled nipple.
F: Unusual location, inferior lateral chest wall between 7th and 9th ribs.
G: Supernumerary gland in the unusual location of the posterior position of the left thigh, just below the gluteal fold.
H: Unusual supernumerary gland between those in the normal position, in a child.

Redrawn from McGillicuddy, T.J. The mammae and their anomalies. *Med. Rec.* 40:446–449, 1891.

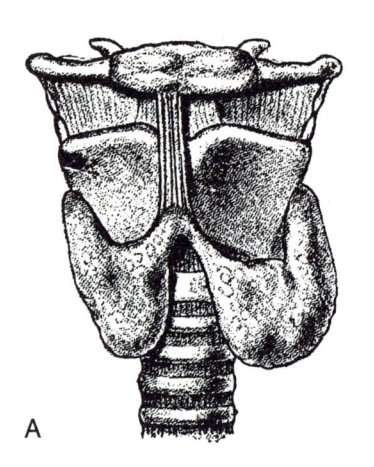

FIG. 26.
Variations in Form of Thyroid Gland.

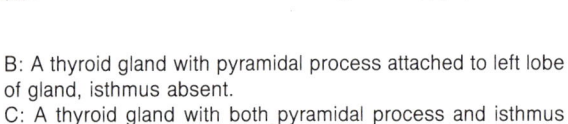

A: A thyroid gland showing bifurcation of the lower end of pyramidal process, one part going to each lateral lobe.
B: A thyroid gland with pyramidal process attached to left lobe of gland, isthmus absent.
C: A thyroid gland with both pyramidal process and isthmus absent.

From Marshall, C.F. Variations in the form of the thyroid gland in man. *J. Anat. Physiol.* 29:234–239, 1895.

Organs

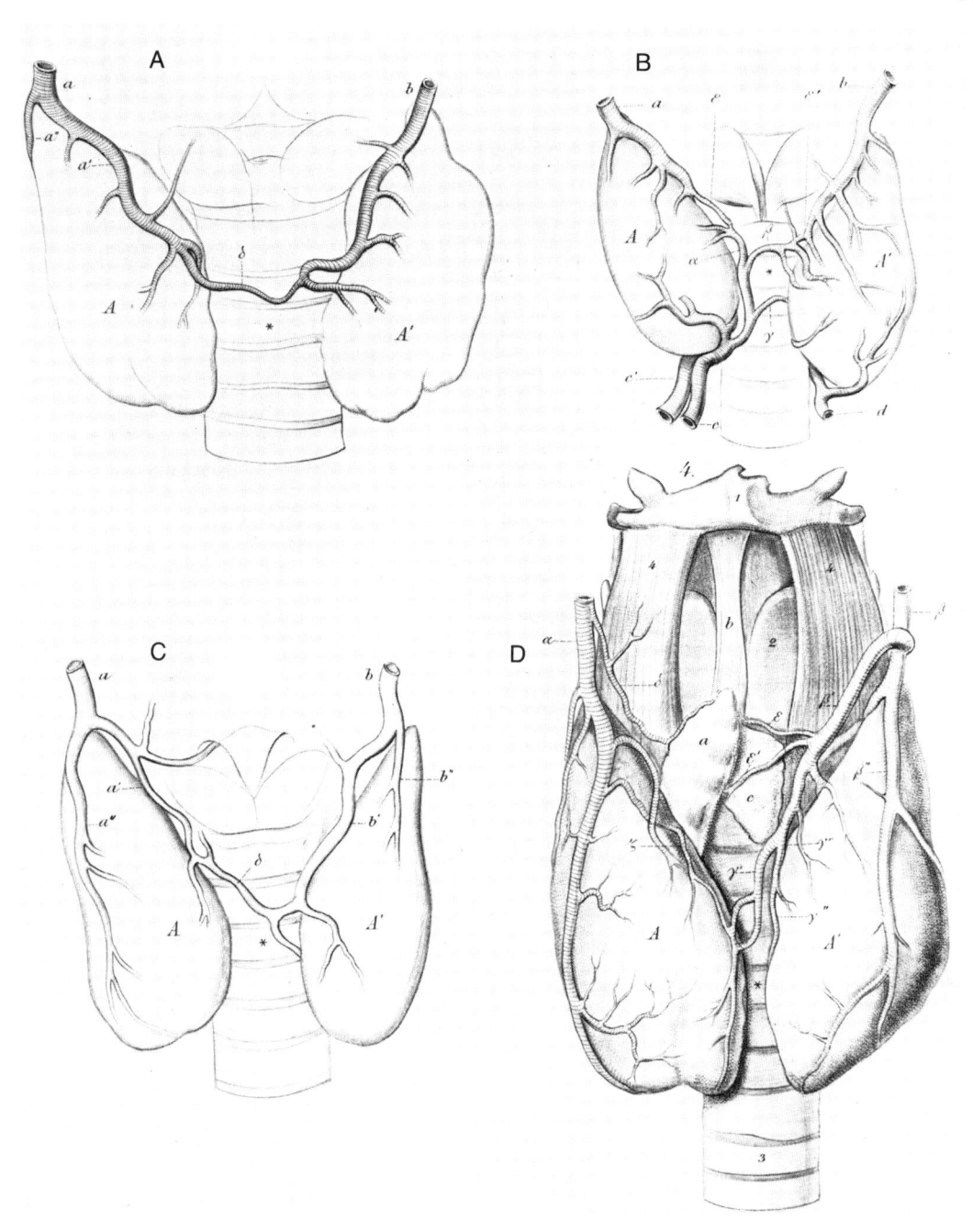

FIG. 27.

Thyroid Glands Without an Isthmus (Glandula Thyroidea Partitia).

A–C: *A,A'*, Lobes of divided thyroid; *, pretracheal gap; *a*, right superior thyroid artery: *a'*, medial branch, and *a''*, lateral branch; *b*, left superior thyroid artery: *b'*, medial branch, and *b''*, lateral branch; *c*, medial branch, and *c'*, lateral branch of right inferior thyroid artery; *d*, left inferior thyroid artery; α, right secondary branch, and β, left secondary branch of medial branches of right inferior thyroid artery; δ, communication of superior thyroid arteries in pretracheal gap; ε, communicating branch between right superior and right secondary branch, and ε', between left superior and left secondary branch of medial branches of the right inferior thyroid artery.

D: *1*, Os hyoideus; *2*, larynx; *3*, trachea; *4*, thyrohyoid muscle; *A,A'*, lobes of divided thyroid; *a*, medial lobe of right lobe of thyroid; *b*, suspensory ligament of medial lobe; *c*, accessory thyroid, left superior; *, pretracheal gap; α, right superior thyroid artery; β, left superior thyroid artery: β', medial branch, as anomalous cricothyroid artery, and β'', lateral branch; γ, terminal to branches of left superior thyroid artery; γ'', branches to left lobe and across pretracheal gap communicating with right superior thyroid artery; ε,ε', branches of medial branches of left superior thyroid artery, which represent the cricothyroid branches; ζ, long anterior branch of right superior thyroid, which communicates with branches from left superior thyroid artery.

From Gruber, W. Über die Glandula thyreoidea ohne Isthmus beim Menschen. *Arch. Anat. Physiol. Wissen. Med.* 1876, pp. 208–220.

Human Anatomic Variation

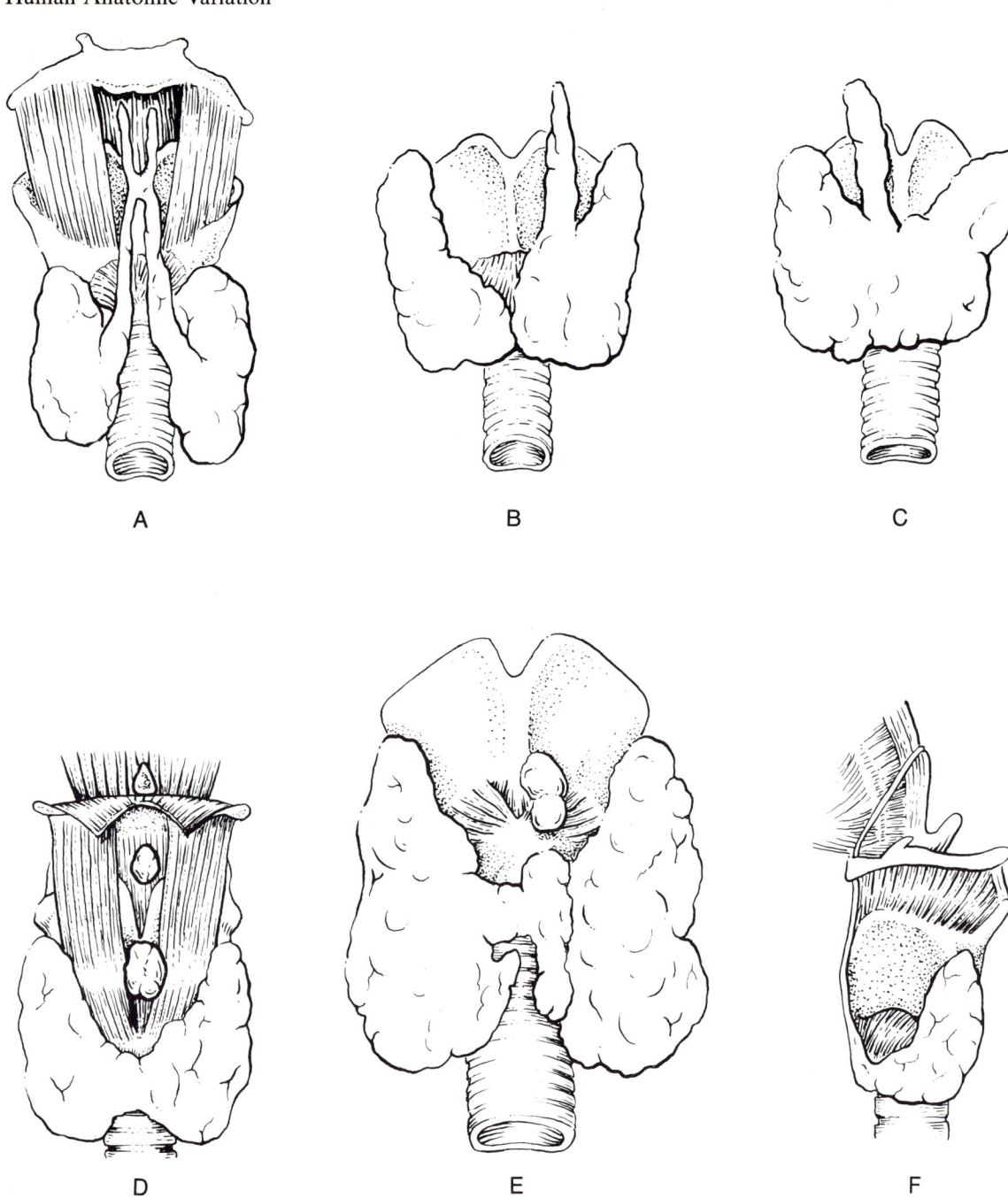

FIG. 28.
Thyroid Gland Variations.
Accessory Thyroid Glands.
Thyroglossal Duct.

A: Absence of thyroid isthmus. Lateral lobes each have pyramidal lobes.
B: Pyramidal lobe arising from the union of the left lobe at the isthmus.
C: Pyramidal lobe arising from the isthmus of the gland.
D: Accessory thyroids may be located on the trachea, thyroid cartilage, thyrohyoid muscle, geniohyoid muscle, and hyoid bone, under and above the hyoid bone.
E: Accessory thyroid gland on cricothyroid muscle. Pyramidal lobe reduced to left inferior part of the isthmus.
F: Persistent thyroglossal duct in an adult, originating at the foramen cecum of the tongue.

Redrawn from various sources.

Organs

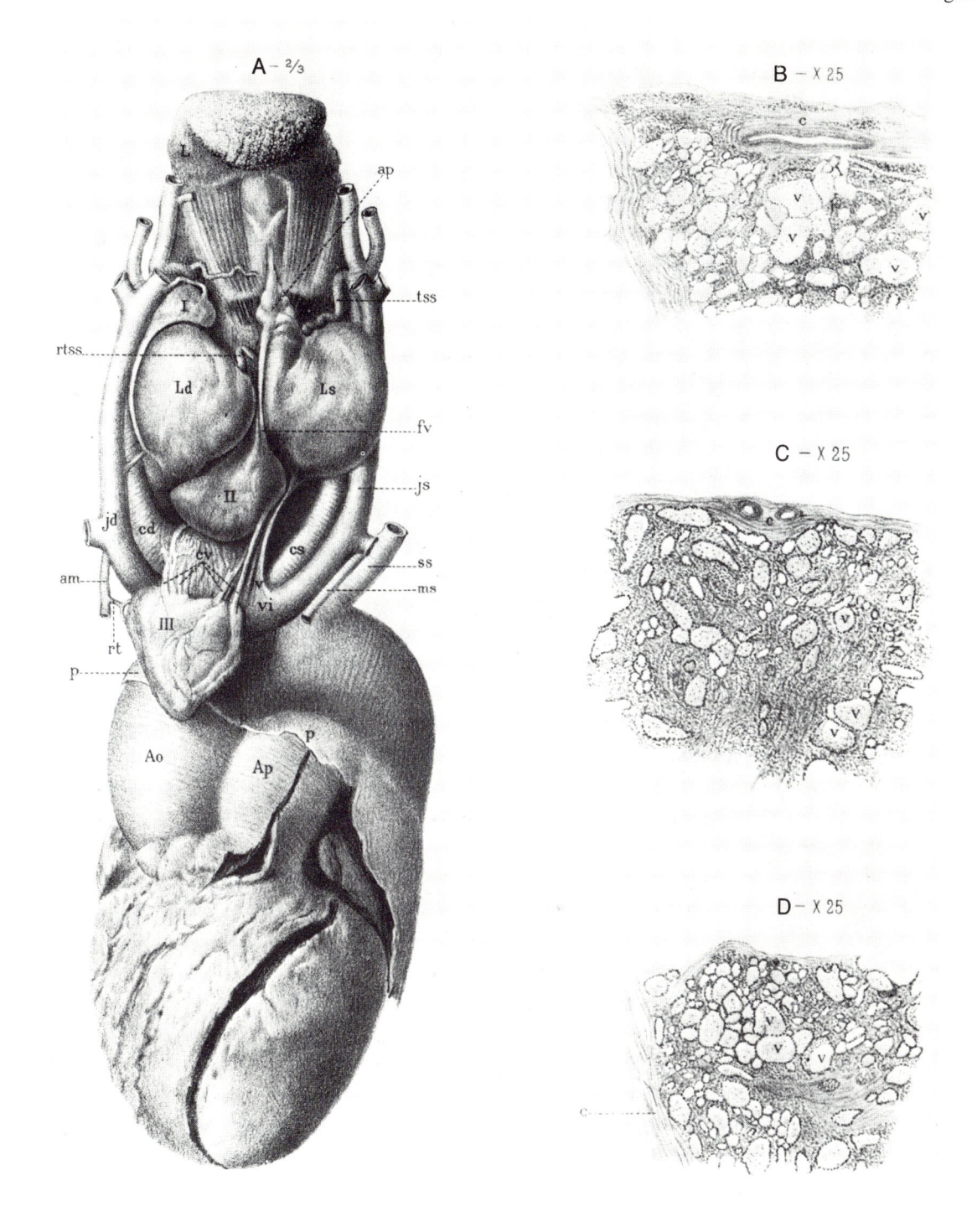

FIG. 29.
Accessory Thyroid Glands in Neck and Chest.

A: *L*, Tongue; *Ld*, right lobe of thyroid; *Ls*, left lobe of thyroid; *I*, superior accessory thyroid; *II*, middle accessory thyroid; *III*, inferior accessory thyroid; *ap*, pyramidal lobe of thyroid; *Ao*, aorta; *Ap*, pulmonary artery; *p,p*, pericardium; *cd*, right common carotid artery; *cs*, left common carotid artery; *jd*, right internal jugular vein; *js*, left internal jugular vein; *am*, right internal thoracic artery; *tss*, left superior thyroid artery; *rtss*, branch of right superior thyroid artery supplying right lobe of thyroid and middle accessory thyroid; *fv*, vascular bundle; *v*, vein of left lobe of thyroid, which opens into left innominate vein; *cv*, connective tissue and vessels, which unite accessory thyroids II and III.

B: Section of superior accessory thyroid gland.
C: Section of middle accessory thyroid gland.
D: Section of inferior or mediastinal accessory thyroid gland. *c*, Fibrous capsule; *v*, glandular vesicles containing colloid.

From D'Ajutolo, G. Delle strume tiroidea accessorie ed in particolare di una mediastinica e di due cervicali nello stesso individuo. *Mem. R. Accad. Sci. Istituto di Bologna S.4* 10:773–782, 1889.

Human Anatomic Variation

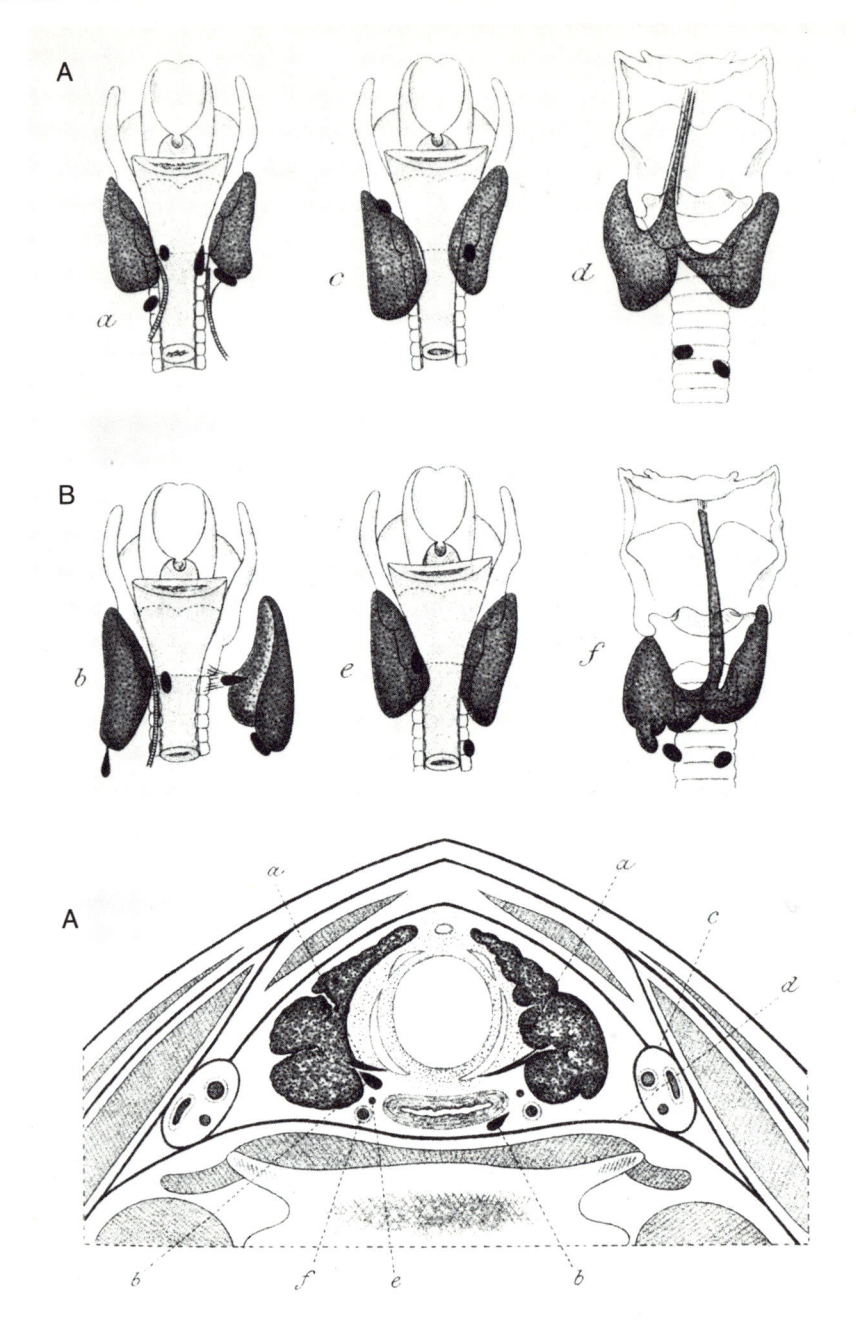

FIG. 30.
Variations in Parathyroid Glands.

A: "Diagrammatic representation of a transverse section through the front of the neck at the level of the lower border of the cricoid cartilage, to illustrate the relations of the posterior superior parathyroids on the right to the prevertebral fascia, etc., on the left, to the lateral thyroid ligament.
The lateral thyroid lobes, the trachea, oesophagus, and part of a vertebral body, are easily identified; other structures are indicated by letters."

a, Lateral thyroid ligament; *b*, posterior superior parathyroid; *c*, carotid sheath; *d*, prevertebral fascia; *e*, recurrent laryngeal nerve; *f*, inferior thyroid artery.

B: "Outline sketches, reduced from actual dissections, to illustrate possible positions of the parathyroids. *a* and *b* represent posterior views of two different cases; *c* and *d* represent the posterior and anterior views, respectively, in a third case; *e* and *f* represent corresponding views in a fourth case."

Welsh, D.A. Concerning the parathyroid glands: A critical, anatomical, and experimental study. Part II. Anatomy. *J. Anat. Physiol.* 32:380–402, 1898.

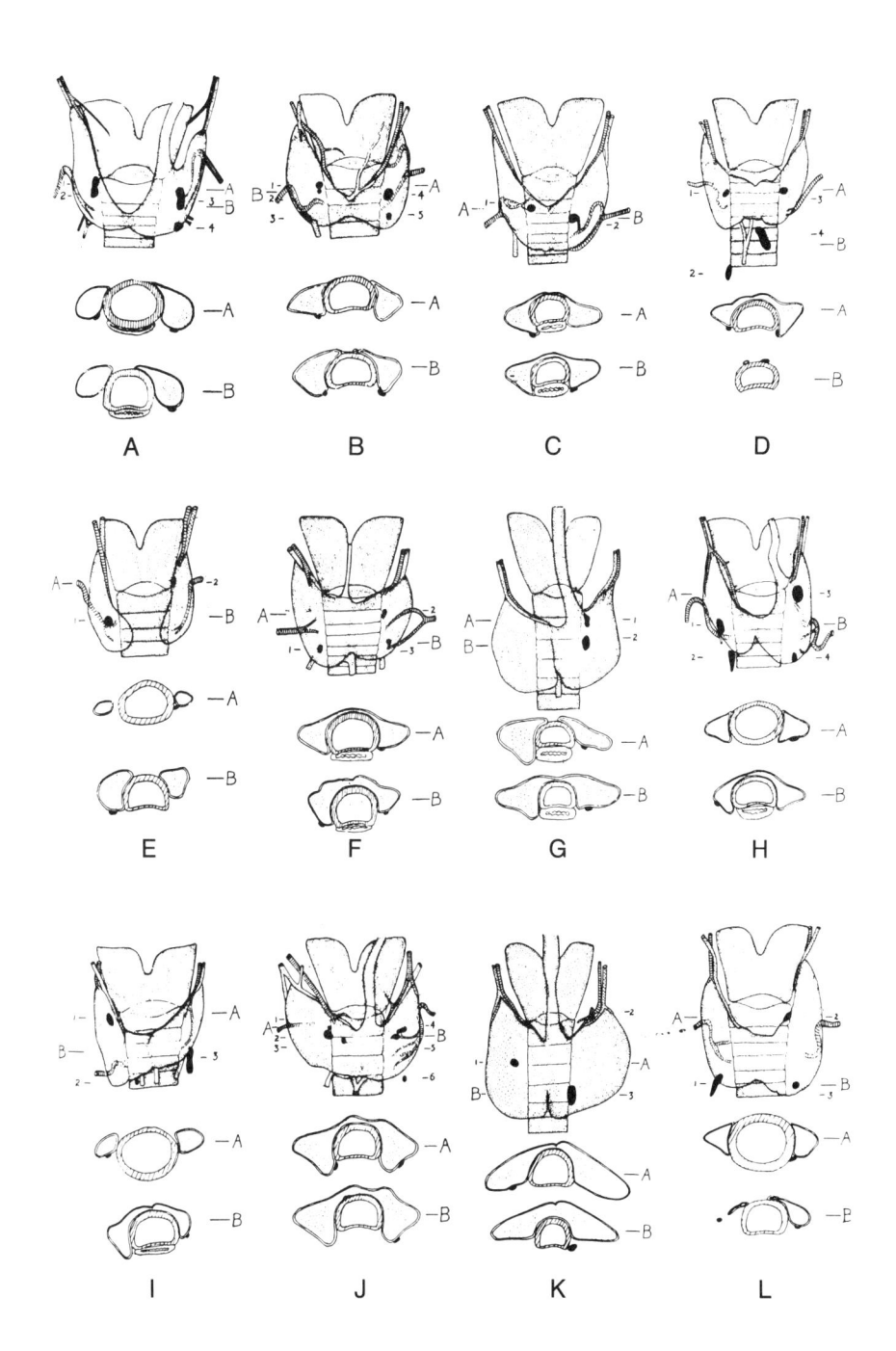

FIG. 31.
Number and Location of Parathyroid Glands.

The illustrations in Figures 31 through 33 are "reproductions of the thyroid and the parathyroid glands and neighboring structures."

A–L: These, and Y in Figure 33, are "from dissecting room material; the locations of the glands are shown in cross section in addition to the ventral view."

"An attempt has been made to indicate the parathyroids which might be called 'superior parathyroids' in outline and 'inferior parathyroids' in black. Those glands which could not be classified either by location or by being one of a pair, are crosshatched."

From Heinbach, W.F. Jr. A study of the number and location of the parathyroid glands in man. Anat. Rec. 56:251–261, 1933.

Human Anatomic Variation

FIG. 32.
Number and Location of Parathyroid Glands (continued).

M–X: These are "from autopsy material; both the ventral and the dorsal views are given. In the ventral views the parathyroids, which are dorsal to the thyroid, are drawn as if the latter were transparent. The various parathyroid glands in each case are numbered from one to six."

From Heinbach, W.F. Jr. A study of the number and location of the parathyroid glands in man. *Anat. Rec.* 56:251–261, 1933.

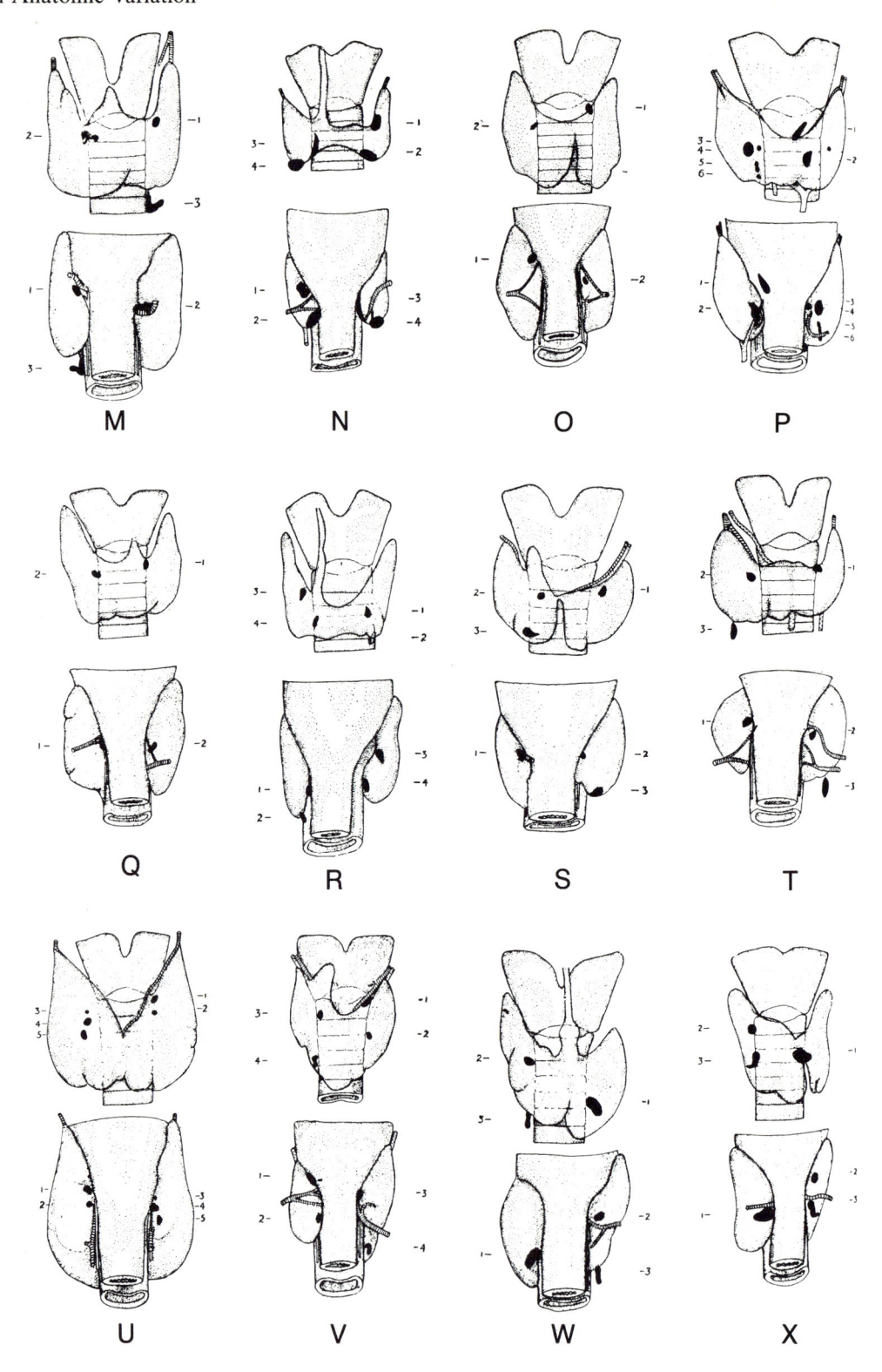

FIG. 33.
Number and Location of Parathyroid Glands (continued).

Y: This is from dissecting room material (see also Figure 31, A–L).

Z: "This diagram represents a composite of all the parathyroids found in the twenty-five cases [Figures 31 through 33, A–Y], plotted on a thyroid gland of average size and shape. The upper brackets indicate approximately the upper, middle, and lower thirds of the thyroid; the lower bracket the space below the thyroid gland; the numbers giving the counts of the parathyroids within each bracket. The parathyroids are represented as if the thyroid were transparent."

From Heinbach, W.F. Jr. A study of the number and location of the parathyroid glands in man. *Anat. Rec.* 56:251–261, 1933.

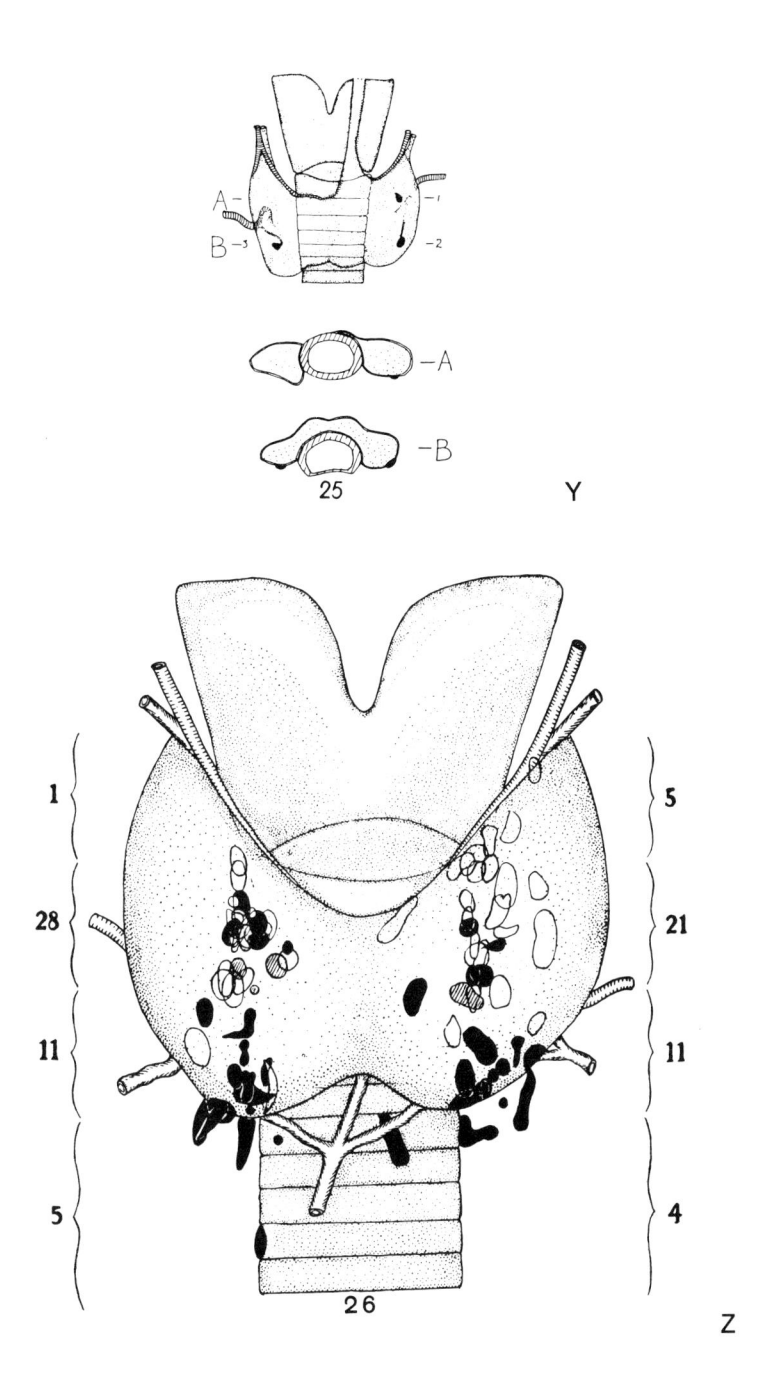

Skeleton

"And there was again war at Gath, where there was a man of great stature, who had six fingers on each hand, and six toes on each foot, twenty-four in number; and he also was descended from the giants."
2 Samuel 21:20

"From an anatomical standpoint, instances of cervical rib are not exceedingly uncommon, and indeed reports of this anomaly are to be found in very early medical literature.

Sir Thomas Brown in his *Pseudodoxia Epidemica* or *Commentaries on Vulgar Errors* (Second Edition, London, 1650, p. 292) speaks of the heated discussion over the biblical story, which arose from the description by Columbus of a female skeleton which possessed a supernumerary rib.

'That a Man hath one Rib lesse than a Woman, is a common conceit derived from the history of Genesis, wherein it stands delivered, that Eve was framed out of a Rib of Adam; whence 'tis concluded the sex of Man still wants that rib our Father lost in Eve. And this is not only passant with the many, but was urged against Columbus in an Anatomy of his at Pisa; where having prepared the sceleton of a Woman that chanced to have thirteen ribs on one side, there arose a party that cried him down, and even unto oathes affirmed, this was the rib wherein a woman exceeded. Were this true, it would autoptically silence that dispute out of which side Eve was framed; it would determine the opinion of Oleaster, that she was made out of the ribs of both sides; or such as from the expression of the Text maintain there was a pleurality required, and might indeed decry the parabolicall exposition of Origen, Cajetan, and such as fearing to concede a monstrosity, or mutilate the integrity of Adam; preventively conceive the creation of thirteen ribs'."

Harvey Cushing
Bulletin of the Johns Hopkins Hospital 14:318–31, 1903

Skeleton

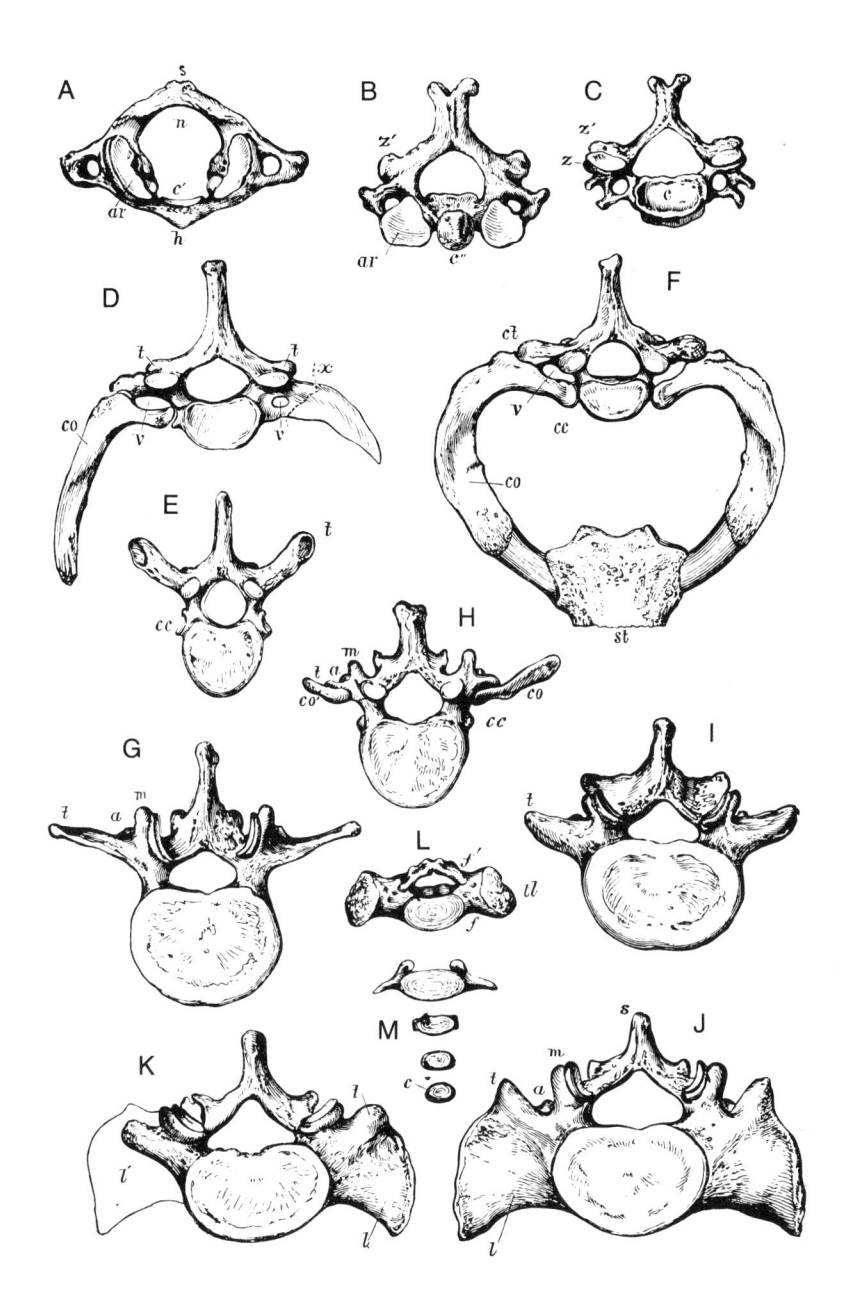

FIG. 1.

Varieties of Different Vertebrae.

Views of different vertebrae from above to illustrate their homologies and some of their varieties. (Allen Thomson.)

A: Atlas. (s, spine; n, neural arch; c', the space occupied by the dens, or displaced centrum; h, anterior arch; ar, superior articular service.)

B: Axis. (c'', dens; ar, superior articular surface; z', inferior articular process.)

C: Fifth cervical. (c, corpus; z,z', superior and inferior articular processes.)

D: Seventh cervical with supernumerary ribs. (t, transverse process; v, vertebrarterial foramen; co, movable right supernumerary or cervical rib; x, with a dotted line marking the place where an ankylosed rib on the left side may be considered to be superadded to the transverse process of the vertebra.)

E: Middle thoracic. (t, transverse process with costal fovea; cc, fovea costalis.)

D and F: First thoracic, with costal arch and sternum attached. (v, vascular interval; CT, costo-transverse, cc, costo-central articulations; co, first rib.)

G: Third lumbar. (m, processus mamillaris; a, processus accessorius; t, processus transversus.)

H: First lumbar with supernumerary ribs. (co,co', supernumerary ribs.)

I: Fifth lumbar.

J: First sacral. (l, pars lateralis.)

K: Lumbosacral vertebra. (e', place of pars lateralis remaining undeveloped in this instance.)

L: Fourth sacral vertebra in a young subject. (tl, transverse process and pars lateralis, which unite with the corresponding parts on the preceding vertebrae; f,f', anterior and posterior sacral foramina thus formed.)

M: Four coccygeal vertebrae. (c, corpus which alone remains in the lower coccygeal vertebrae.)

From Schafer, E.A., Symington, J. and T.H. Bryce, Eds. *Quain's Anatomy*, 11th Ed., Vol. IV, Pt. I: Osteology and Arthrology. Longman, Green, and Co., London, 1915.

Human Anatomic Variation

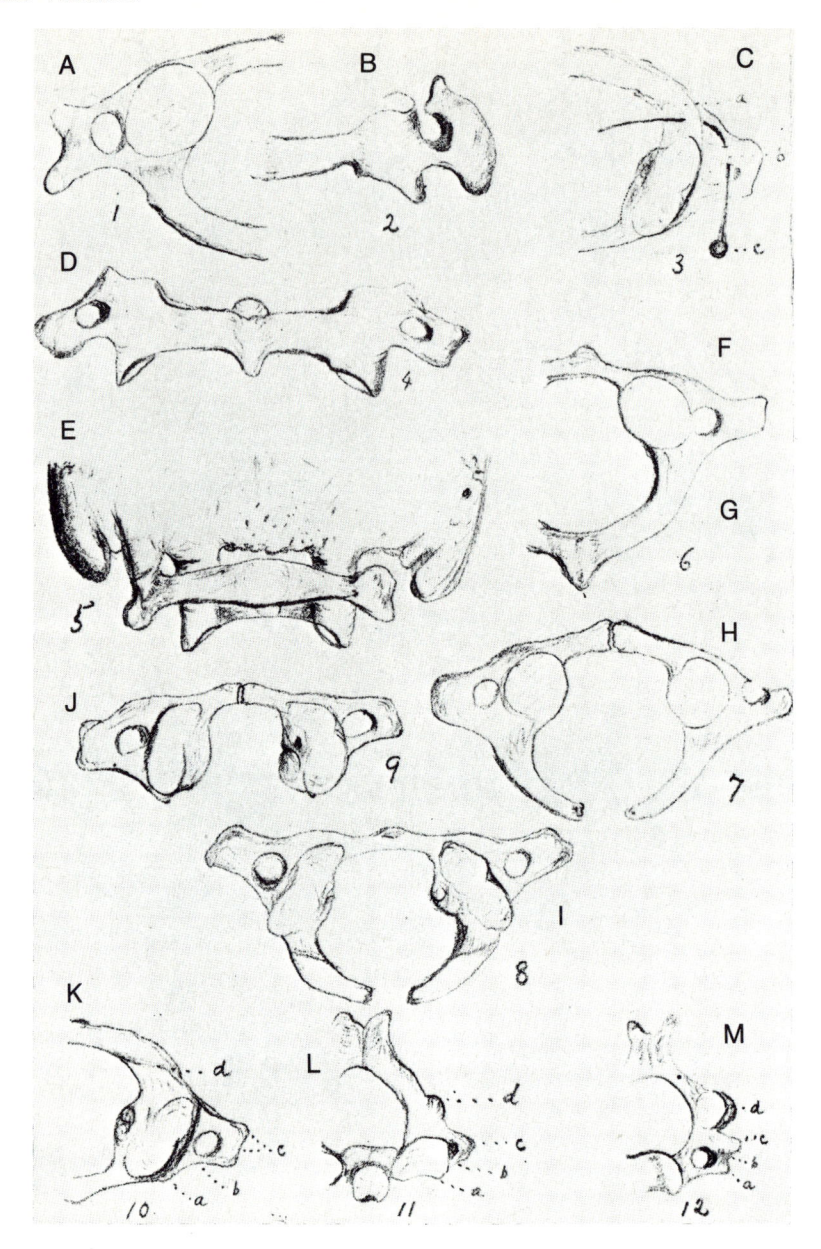

FIG. 2.
Varieties of Atlas.

A: "Undersurface of atlas with bifid transverse process.
B: Left half of atlas seen from the front, showing a growth of bone springing from the upper surface of the transverse process and passing toward the jugular process of the occipital bone.
C: Atlas from above showing: *a*, spicule of bone springing from posterior arch at the upper surface of the point of junction of lamina and pedicle and passing over the position of nerve and artery to the back part of the superior articular surface; *b*, a second spicule of bone passing from upper surface of transverse process to the outer edge of the superior articular surface and crossing the place of exit of the anterior division of the first spinal nerve. A probe, *c*, is represented passing through the abnormal foramina.
D: The same atlas seen from the front, showing the foramen which looks forward and gives passage to the anterior division of the first spinal nerve.
E: Occiput and atlas fused together, posterior view; the abnormal pillar described in [Allen's] text is shown; also the atlas and base of skull are seen to be further separated on that side.
F: Shows a large anterior and posterior tubercle.
G: Shows an imperfect posterior arch, also a suture in the middle of the anterior arch, and on the left side a deficiency of the anterior root of the transverse process.
H: Atlas with posterior arch open and transverse process thrown forward, probably from the cause suggested in [Allen's] text.
I: Semidiagrammatic view of the abnormal atlas described, in which the posterior arch is absent and the anterior is in two parts.
J–L: These figures represent the left halves of a normal atlas, axis, and typical cervical vertebrae; the same letters in each point to the homologous parts: *a* represents the position of the point in each which corresponds to the tip of a rib, *b* to the rib tubercle, and *c* to the tip of the transverse process proper.

From Allen, W. The varieties of the atlas in the human subject, and the homologies of its transverse processes. *J. Anat. Physiol.* 14:18–27, 1879.

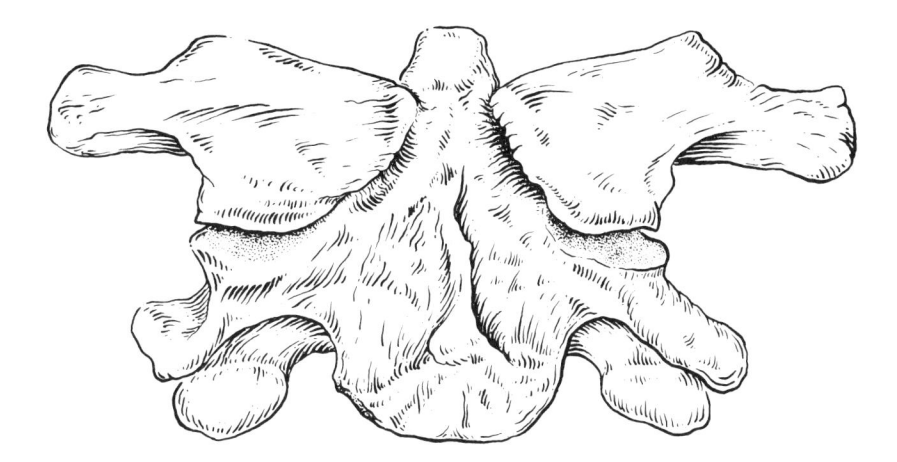

FIG. 3.
Unusual Case of Absence of the Anterior Arch of Atlas.

In this case, the two lateral masses were joined to the dens of the axis.

Redrawn from Huber, G.C. *Piersol's Human Anatomy*, 9th Ed. J.B. Lippincott Co., Philadelphia, 1930.

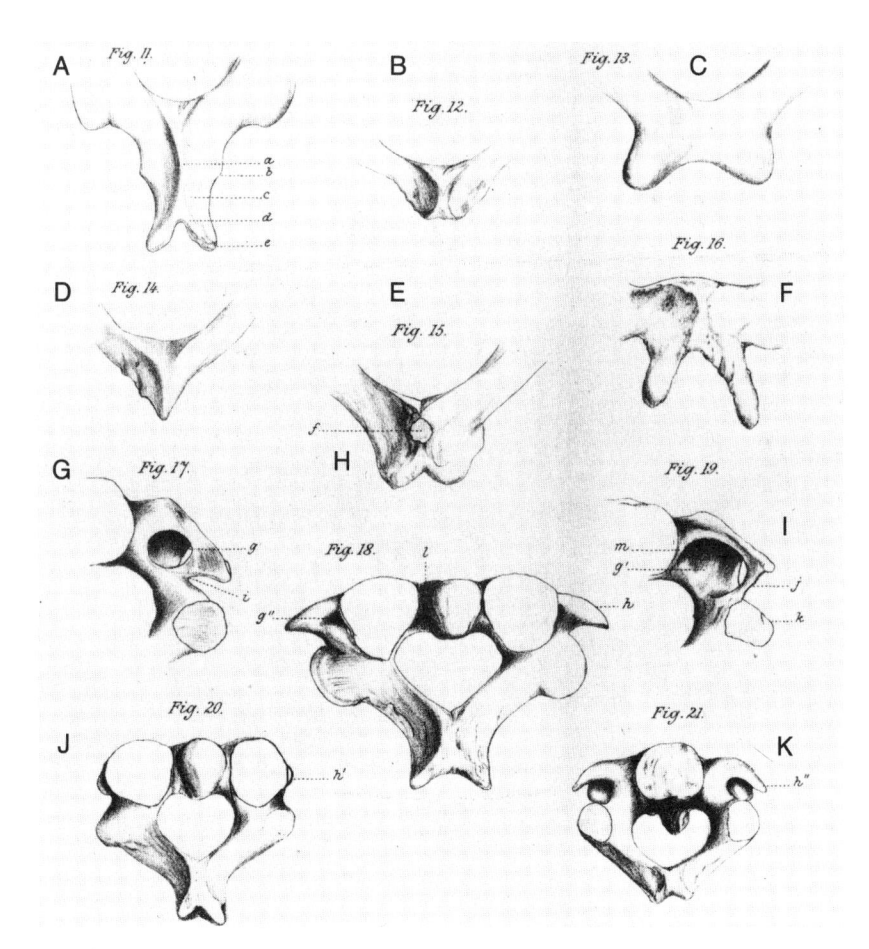

FIG. 4.
Varieties of Axis.

A: Elongated spine of axis, natural size. *a*, Median superior ridge; *b*, inferior tubercle; *c*, lateral oblique surface; *d*, posterior surface; *e*, lateral tubercle.
B: Rudimental spine in the axis of ancient Egyptian.
C: Wide bituberculate spine in axis of Saxon, from secondary interment on Bowl's barrow, Wiltshire.
D: Axis with acuminate spine.
E: Axis with friction facet on superior ridge.
F: Spine of axis with long lateral tubercles.
G: Arterial foramen, round variety.
H: Axis with exserted lateral processes.
I: Arterial foramen, ovate variety.
J: Axis with lateral processes under cover of upper articular surface.
K: Axis with deficiency of outer wall of arterial foramen.

From Macalister, A. The development and varieties of the second cervical vertebrae. *J. Anat. Physiol.* 28:257–268, 1894.

Human Anatomic Variation

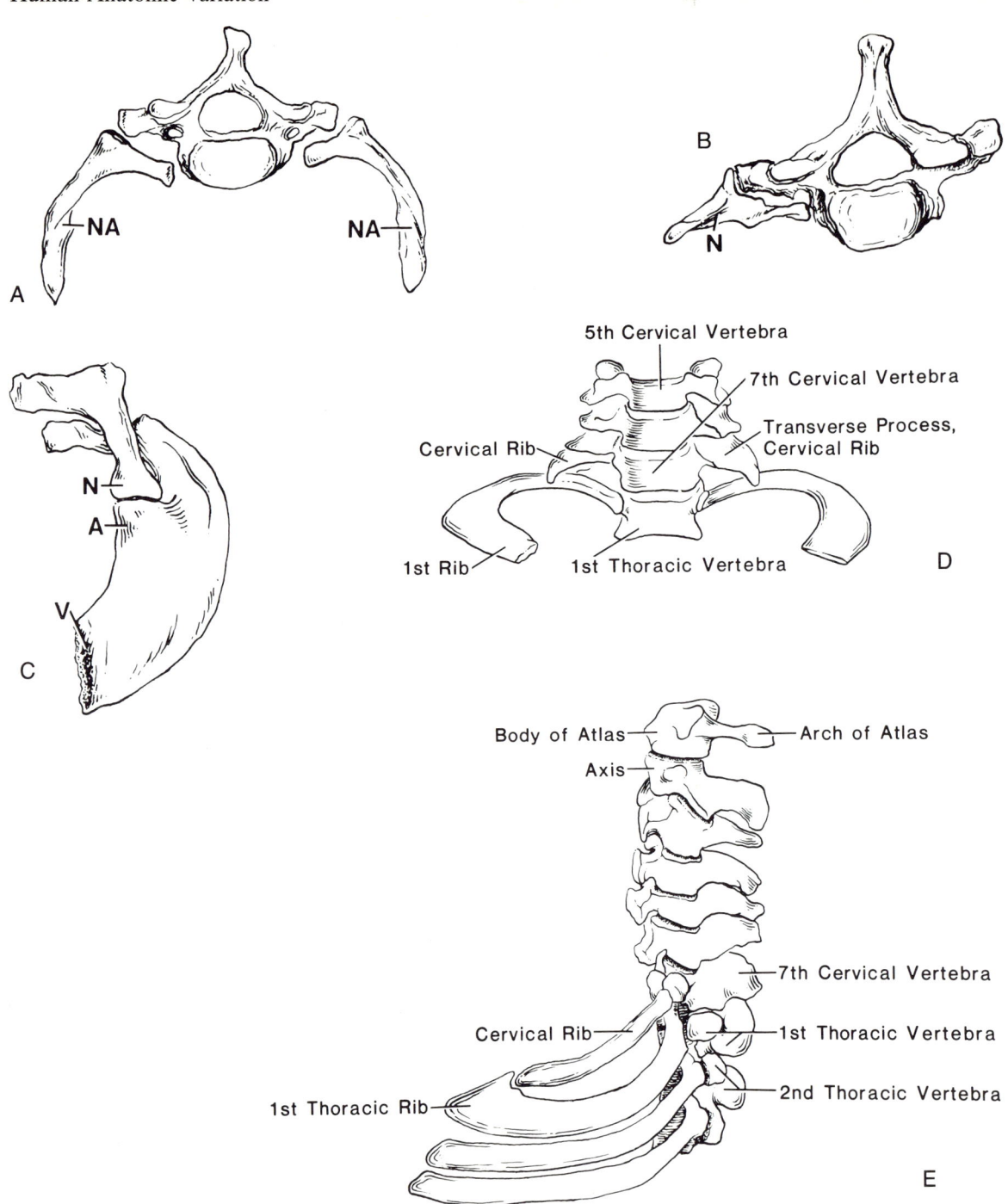

FIG. 5.
Cervical Ribs.

Grooves are indicated as follows: *A*, artery (subclavian); *N*, nerve; *NA*, nerve and artery; *V*, vein (subclavian).

A: Seventh cervical vertebra and ribs; adult male. This shows cervical ribs whose ossification extends as far forward as the crossing of the subclavian vein. A common groove is shown for the nerve and artery. Note the sigmoid appearance of this groove.

B: Seventh cervical vertebra and rib, right side; adult male. This illustrates the type of rib that is ossified only as far as the crossing of the lowest brachial nerve trunk. In this case, no ligamentous connection with the sternum was present.

C: Seventh cervical rib, left side, showing articulation of the first thoracic rib; adult male. This illustrates the type of cervical rib whose independent ossification proceeds only as far the crossing of the subclavian artery. The groove for the lowest brachial trunk is shown at *N*. This figure shows the variety of this type that has an intermediary cartilage between the two ribs.

D: Cervical rib, anterior view.

E: Cervical rib, lateral view showing cervical rib articulating with the first thoracic rib.

Authors' note: Cervical ribs are normally present in reptiles and birds.

Redrawn from Braus, H. *Anatomie des Menschen. Ein Lehrbuch für Studierende und Ärzte.* Erster Band. Julius Springer, Berlin 1921.

Skeleton

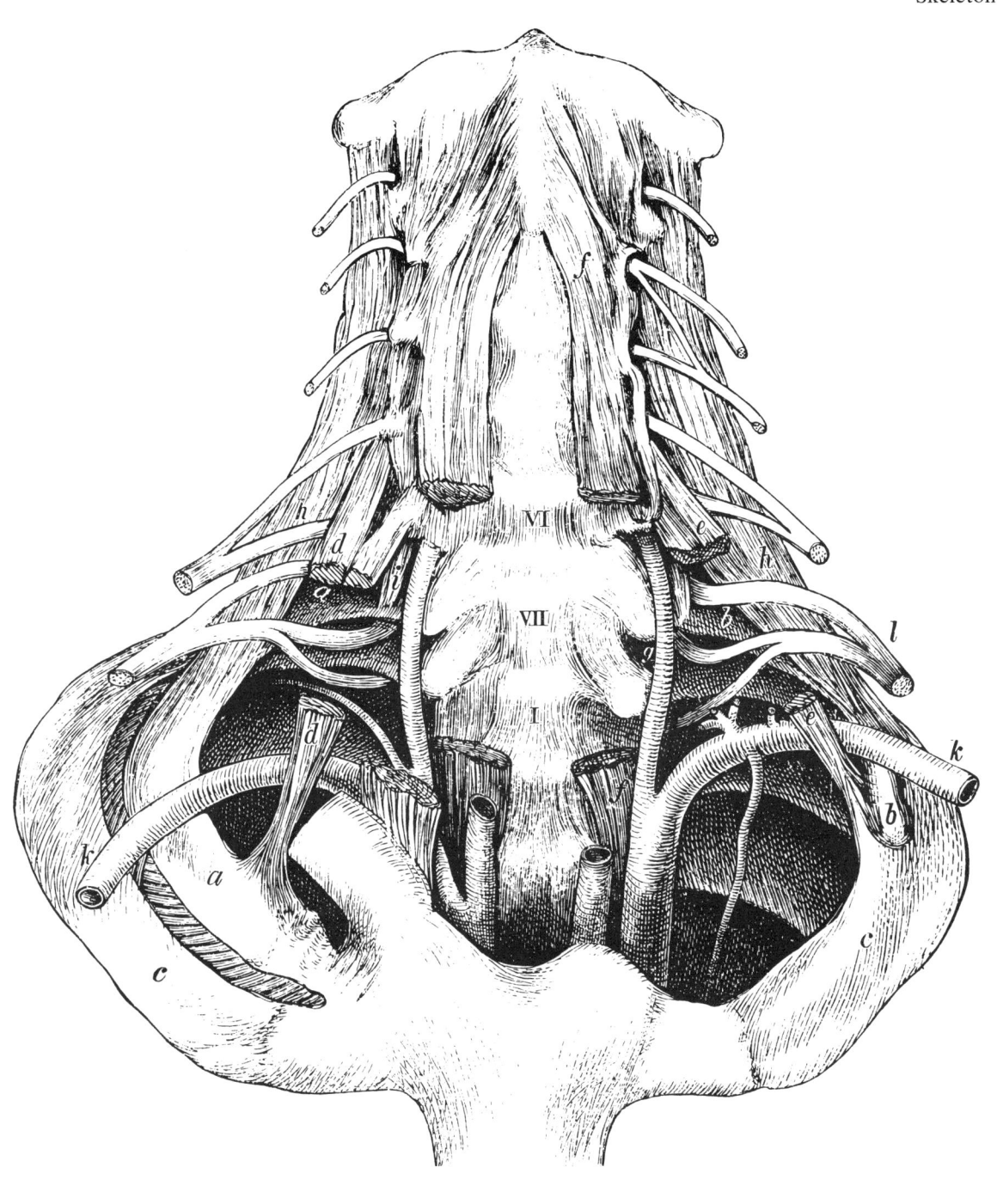

FIG. 6.
Supernumerary Cervical Ribs.

Sixth and seventh cervical vertebrae. *I*, First dorsal vertebra; *a,a*, right cervical rib; *b,b*, left cervical rib; *c,c*, first pair of thoracic ribs; *d,d*, right, and *e,e*, left scalenus anticus: these muscles are represented with pieces cut out to show the neck of the supernumerary and 1st thoracic ribs, and the 7th and 8th cervical and 1st dorsal nerves; *f,f*, longus colli, divided to show the heads of the supernumerary and 1st thoracic ribs with their stellate ligaments; *g*, vertebral artery; *h,h*, right and left middle scalene muscles; *i*, inter-transverse muscle; *k,k*, right and left subclavian arteries; *l*, lower cord of brachial plexus. From a careful drawing of the specimen by Turner's pupil, Mr. Millen Coughtrey.

From Turner, W. Supernumerary cervical ribs. *J. Anat. Physiol.* 4:130–139, 1870.

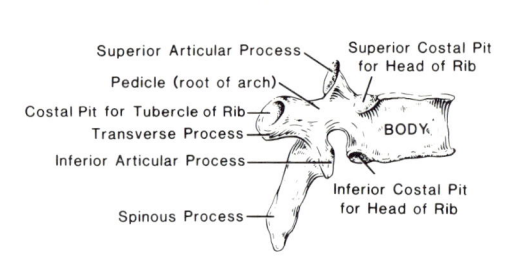

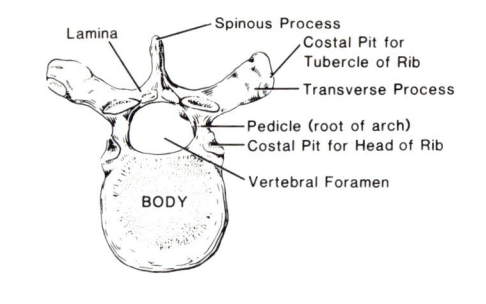

Human Anatomic Variation

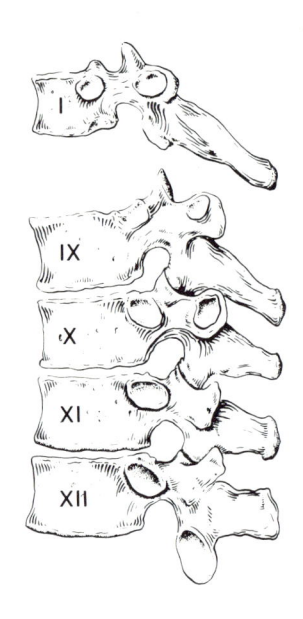

FIG. 7.
Peculiar Thoracic Vertebrae.

"The first thoracic vertebra is a transitional vertebra. The body in its general conformation approaches very closely the seventh cervical, in that the greatest diameter is transverse, and its upper surface is concave from side to side. On each side is an entire pit, close to the upper border, for the head of the first rib, and a very small pit (inferior costal pit) below for the head of the second rib. The spinous process is thick, strong, almost horizontal and usually more prominent than that of the seventh cervical, an important point to remember when counting the spines in the living subject.

The ninth has superior costal pits, and usually no inferior; when the inferior pits are present, this vertebra is not exceptional.

The tenth usually has an entire costal pit at its upper margin, on each side, but occasionally only a superior costal pit. It has no lower pits and the pits on the transverse process are usually small.

The eleventh has a large body resembling that of a lumbar vertebra. The pits are on the pedicles and they are complete and of large size. The transverse processes are short, show evidence of subdivision into three parts, and have no pits for the tubercles of the eleventh pair of ribs.

In many mammals, the spines of the anterior vertebrae are directed backward, and those of the posterior directed forward, whilst in the center of the column there is usually one spine vertical. The latter is called the anticlinal vertebra, and indicates the point at which the thoracic begin to assume the characters of lumbar vertebrae. In man the eleventh thoracic is the anticlinal vertebra.

The twelfth resembles in general characters the eleventh, but may be distinguished from it by the articular surfaces on the inferior articular processes being convex and turned laterally as in the lumbar vertebrae. The transverse process is rudimentary and tripartite, presenting for examination three tubercles, superior, inferior, and lateral, which correspond respectively to the mammillary, accessory, and transverse processes of a lumbar vertebra. There is one complete pit on the root (pedicle) for the head of the twelfth rib.

The twelfth thoracic, in the absence of the twelfth pair of ribs, commonly conforms to the type of a lumbar vertebra. The transverse process of the tenth occasionally lacks the facet for costal articulation. The lumbar form of transverse process may be present in the eleventh thoracic vertebra. A peculiarity, more frequent in the thoracic and lumbar than in the cervical and sacral regions of the column, is the existence of a half-vertebra. Such specimens have a wedge-shaped half-centrum, to which are attached a lamina, a transverse, superior, and inferior articular, and half a spinous process. As a rule, a half-vertebra is ankylosed to the vertebrae above and below."

Redrawn from Terry, R.J. in Jackson, C.M., Ed. *Morris' Human Anatomy*, 9th Ed. P. Blakiston's Son & Co., Philadelphia, 1933.

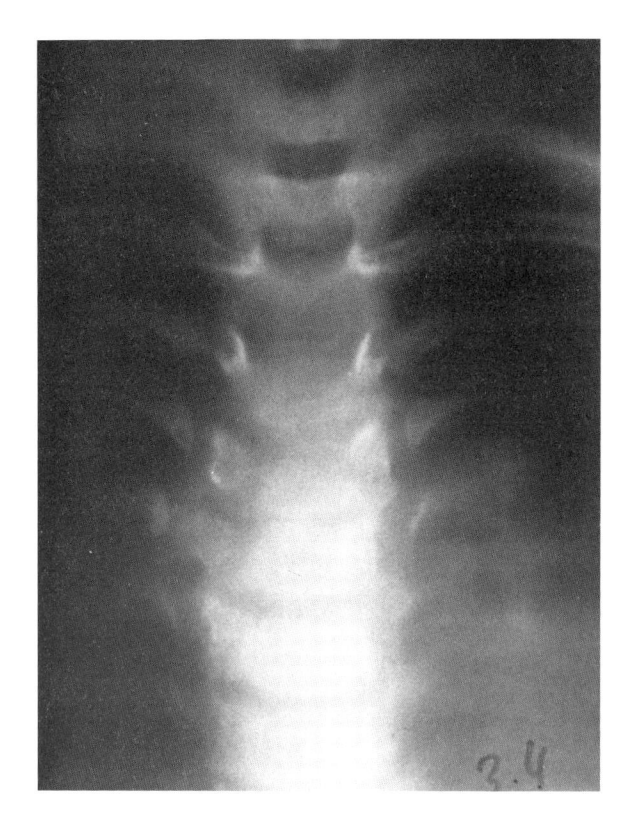

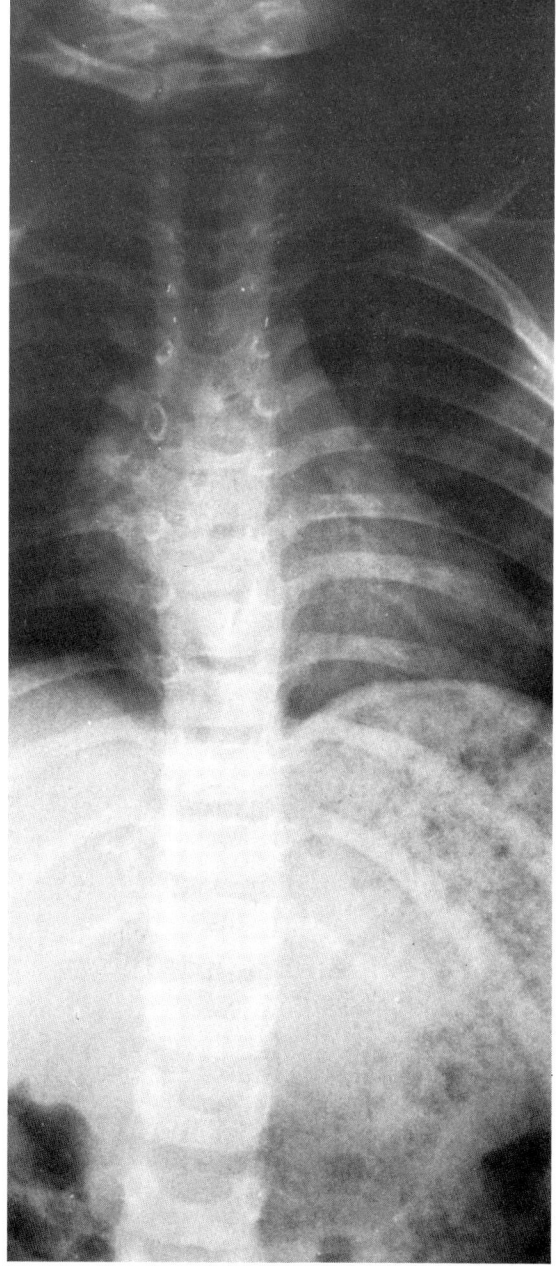

FIG. 8, FIG. 9.
Congenital Absence of Dorsal Arches of the Fourth to Seventh Thoracic Vertebrae.

Radiograph of the dorsal spine of a 3-year-old female, showing absence of the posterior arches of D4–D7 [T4–T7], with marked widening of the spinal canal and multiple hemivertebrae involving the bodies of D6, 7, and 8 [T6–T8].

Courtesy of Dr. Rafic Melhem, Department of Radiology, American University of Beirut, Beirut, Lebanon. May, 1980.

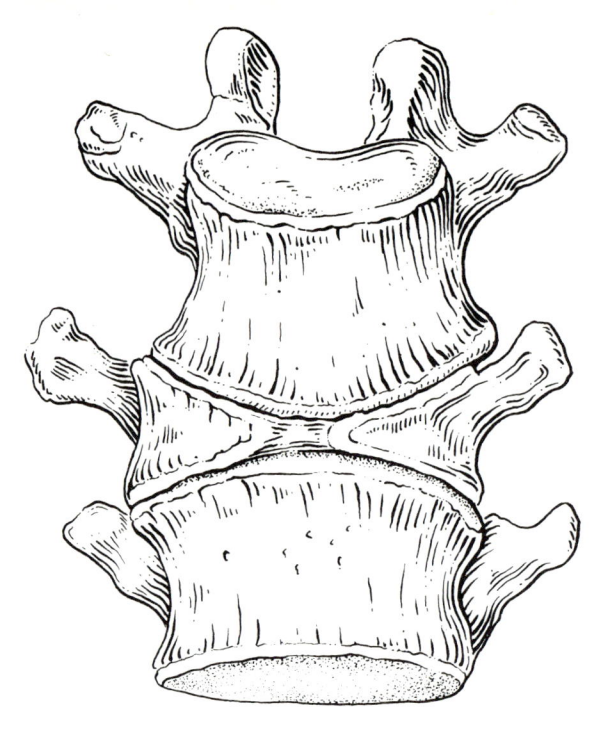

Human Anatomic Variation

FIG. 10.
A Divided Thoracic Vertebra (after Turner).

The developmental nucleus for the vertebral body may (very rarely) be doubled and the two parts of the body may remain separate throughout life, as seen in this illustration.

Redrawn from Jackson, C.M., Ed. *Morris' Human Anatomy*, Section III: Osteology, by R.J. Terry. P. Blakiston's Son & Co., Philadelphia, 1923.

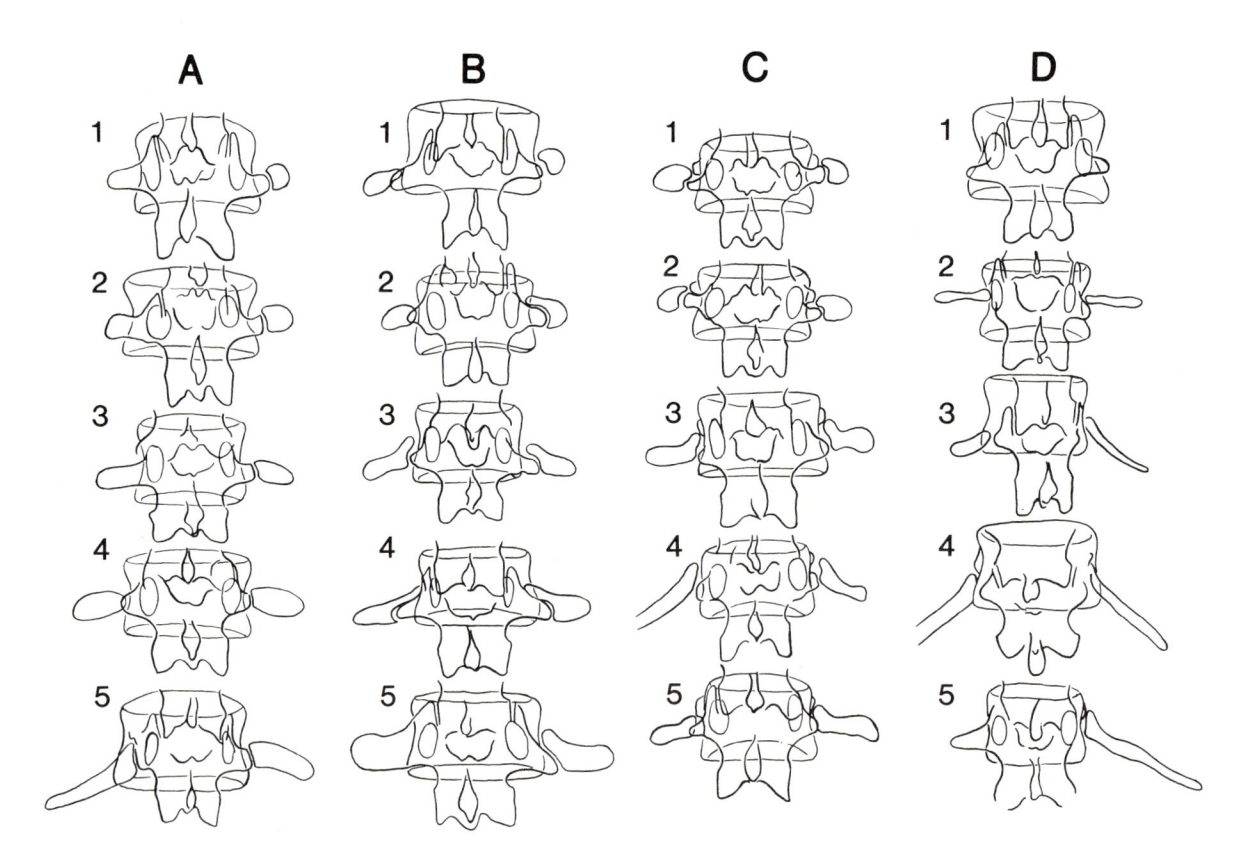

FIG. 11.
Variations in Lumbar Ribs.

The borders between regions of the spine (e.g., cervical and thoracic, cervical rib) may involve many variations. In this drawing from radiographs, Schertlein illustrated 20 varieties of lumbar ribs that he found on the twentieth vertebra.

Redrawn from Schertlein A. Über die häufigsten Anomalien an der Brustlendenwirbelsäulengrenze. *Fortschr. Röntgenstr.* 38:478–488, 1928.

Skeleton

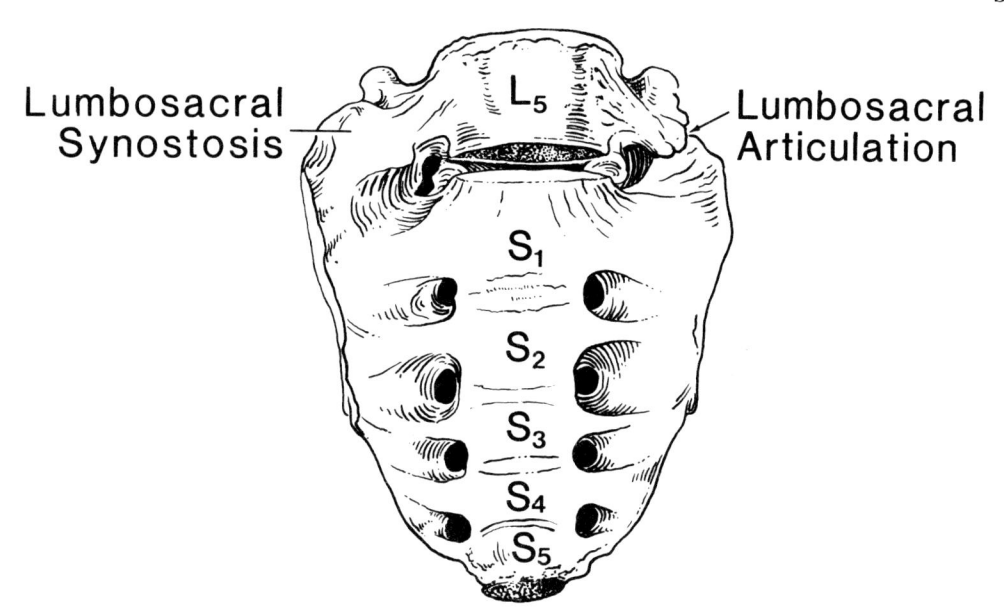

FIG. 12.
Lumbar Sacralization.
Unilateral Assimilation of Fifth Lumbar Vertebra.

Anterior aspect of the lumbosacral joint. The joint is fused on the right side, while on the left side the joint is represented by an articulation.

Redrawn from Toldt, C. *An Atlas of Human Anatomy for Students and Physicians*, 2nd Ed., Vol. 1. The Macmillan Co., New York, 1928.

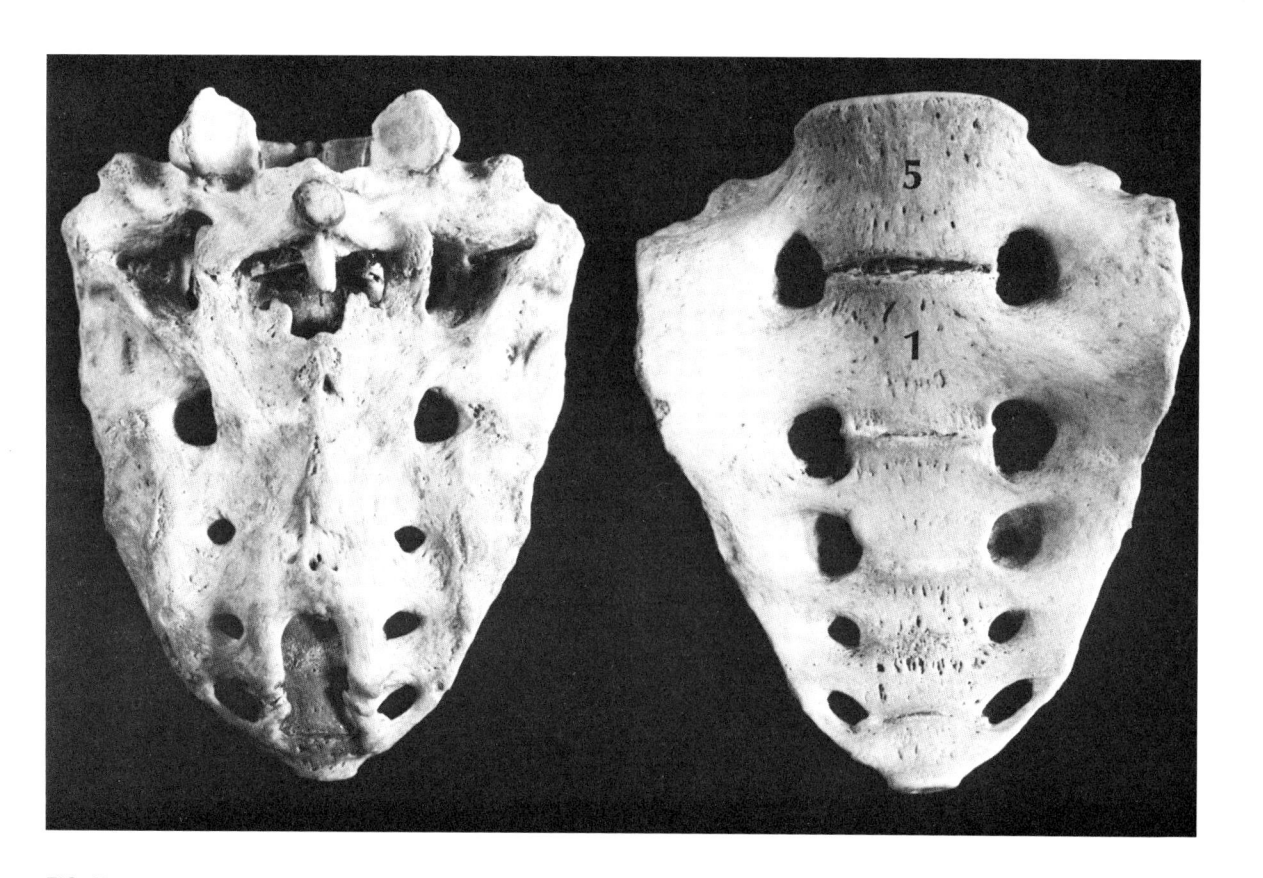

FIG. 13.
Lumbar Sacralization.
The fifth lumbar vertebra is completely fused with the first sacral vertebra.

Dissecting room specimen, The University of Iowa College of Medicine.

Human Anatomic Variation

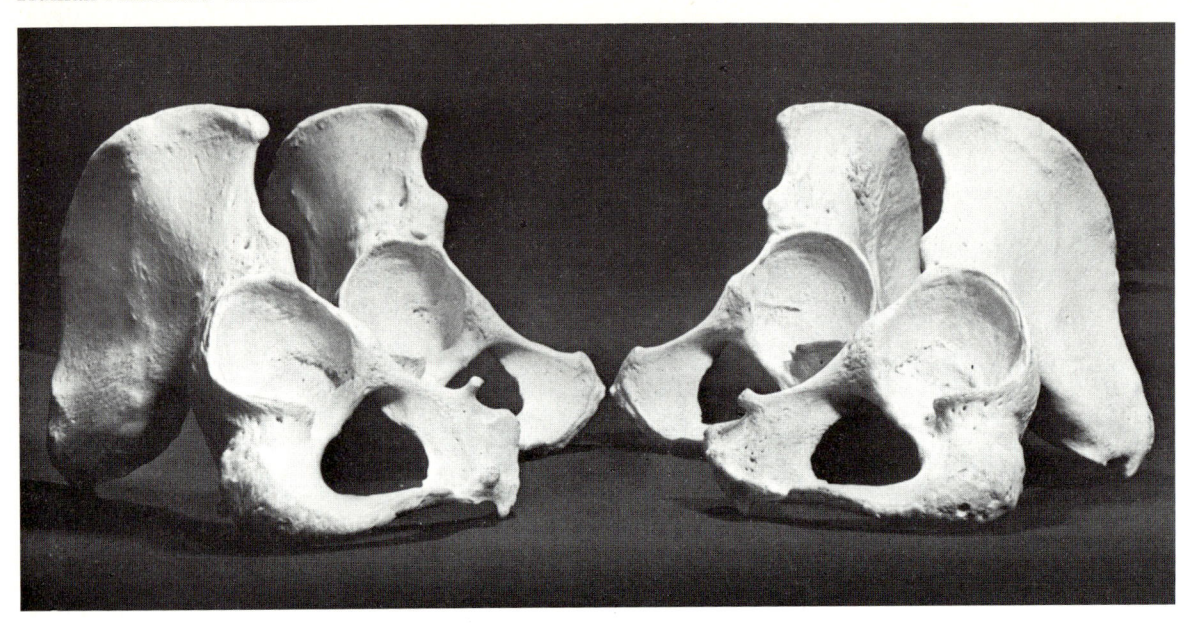

FIG. 14.
Pubic Tubercles and Pubic Spines.

The illustration demonstrates variation in the pubic tubercle which may become, as can be seen here, an enlarged spinous process. It may be recalled that the pubic tubercle is the attachment site of the inguinal ligament.

Dissecting room specimen, The University of Iowa College of Medicine.

FIG. 15.
Sacrum Composed of Five and Six Vertebrae.
Coccyx Composed of Four Vertebrae.

The upper illustration shows a sacrum composed of five vertebrae and coccyx composed of four vertebrae (*Tabulae Sex*, 1538). The lower drawing shows a sacrum composed of six vertebrae and coccyx composed of four vertebrae (*Fabrica*, 1543).

From Vesalius, A. *Tabulae Sex*, 1538, and *Fabrica*, 1543. In Straus, W.L. Jr. and O. Temkin. Vesalius and the problem of variability. *Bull. Hist. Med.* 14:609–633, 1943.

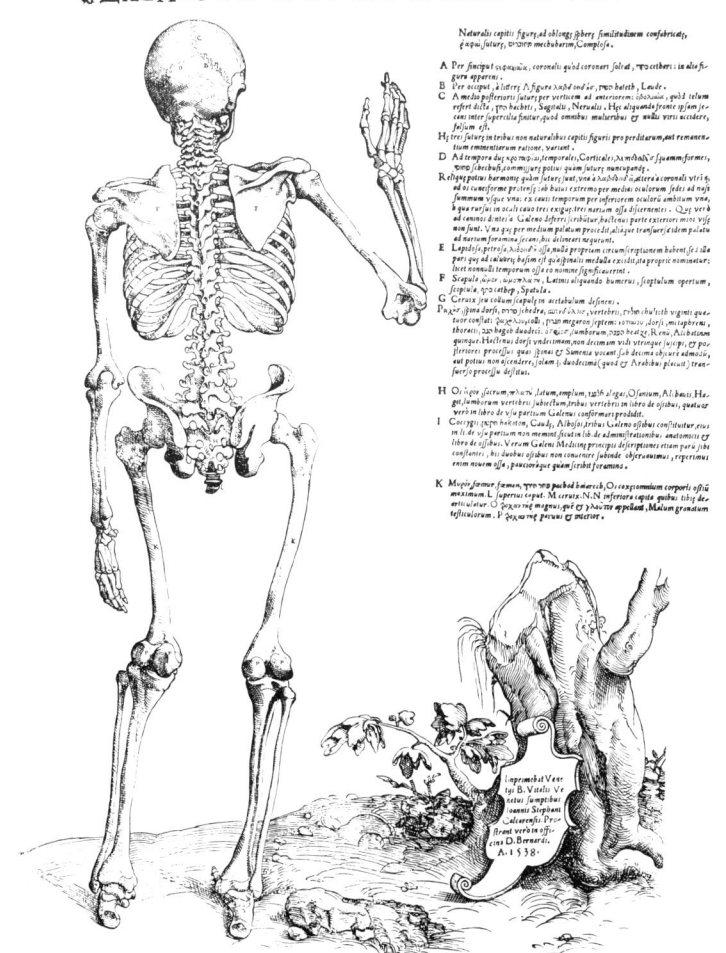

Skeleton

ΣΚΕΛΕΤΟΝ A TERGO DELINEATVM.

CAVTVM EST DECRETO PAV. III. PONT. MAX. ET SACRATISSIMAE CAESAREAE MAISTATIS

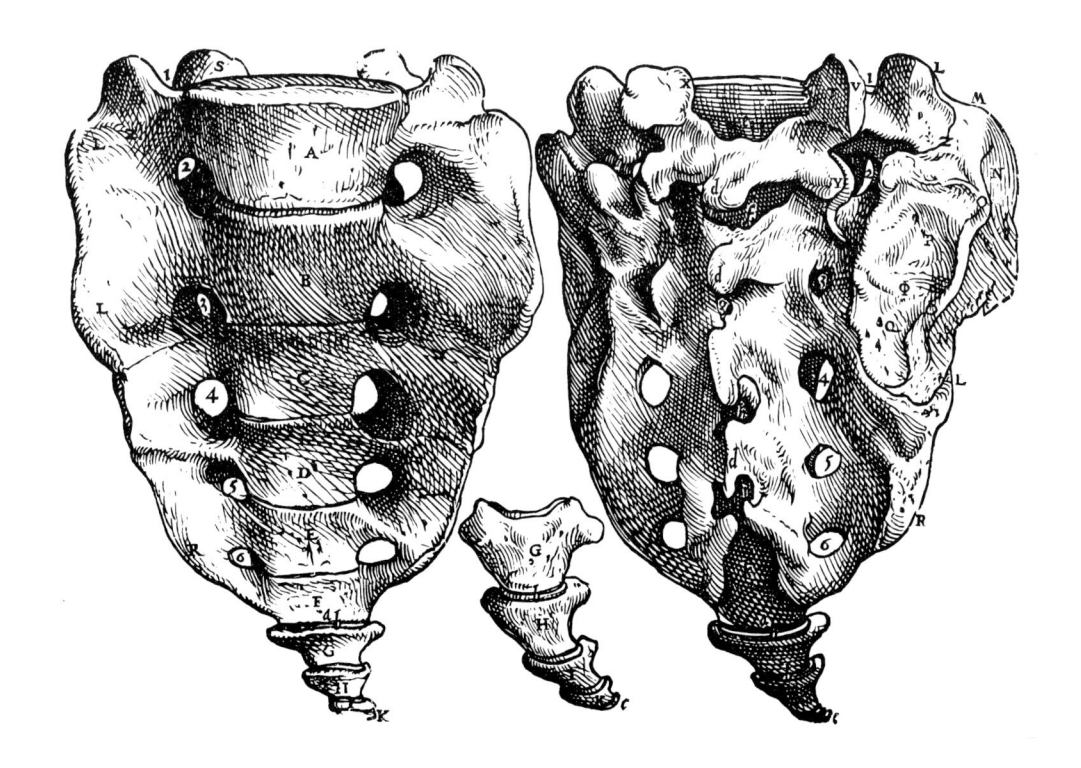

Human Anatomic Variation

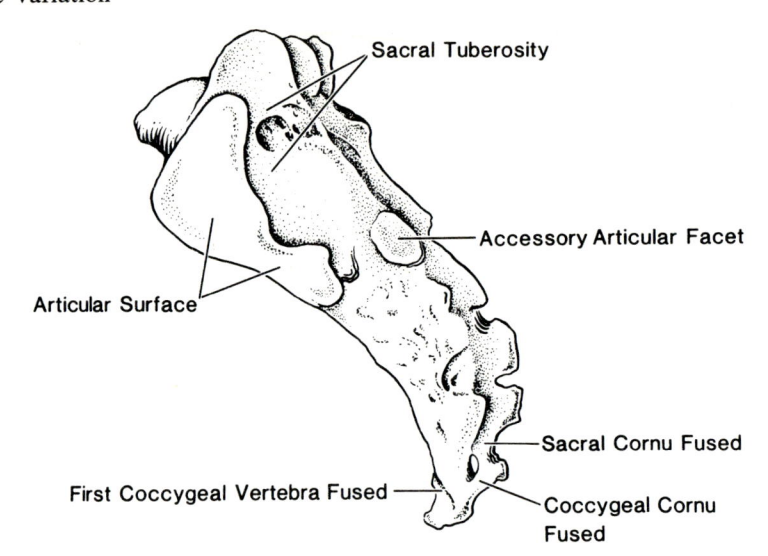

FIG. 16.
Accessory Iliosacral Articulations.

Accessory articulations vary in number, according to author, from 16% to 50% of all specimens. They vary in diameter from 0.59 to 1.2 cm and are usually located near and between the first and second posterior sacral foramina.
The drawing shows the left lateral view of the sacrum.

In addition to accessory articulations, the sacrum may possess ribs. In one reported case the rib was found to be partially embedded on the undersurface of the gluteus maximus. The distal end of the rib was attached to the greater trochanter of the femur by a fibrous band.

Halloran, W. Report of sacral ribs. *Q. Bull. Northwestern Univ. Med. School.* 34:304–305, 1960.

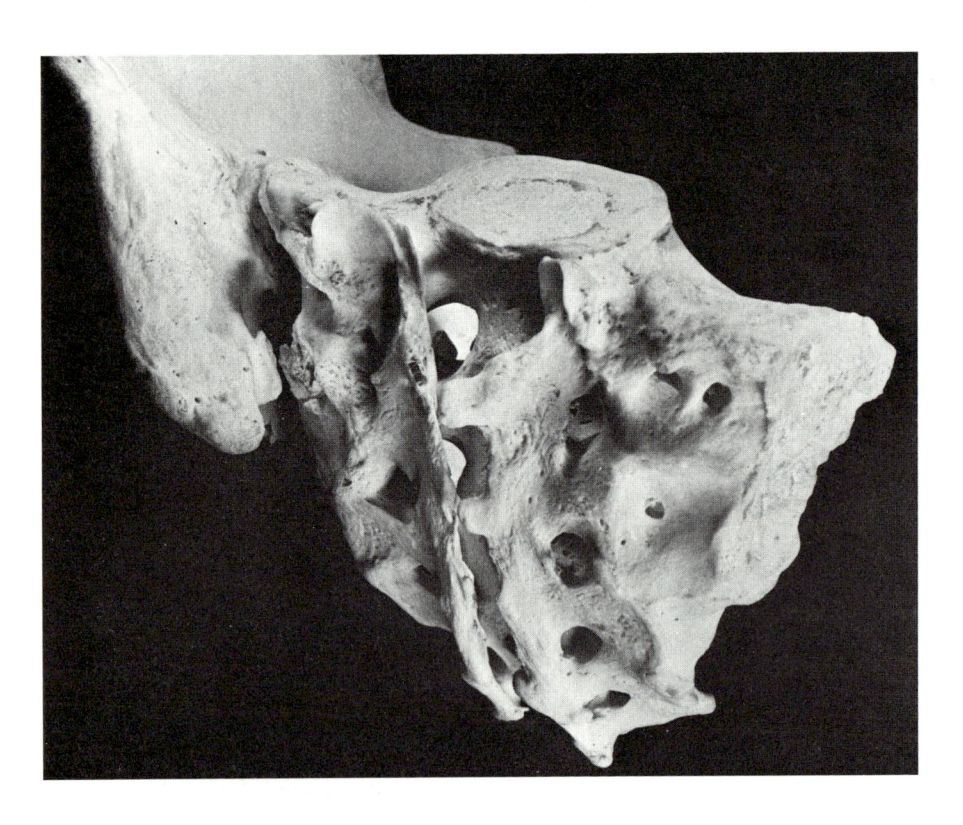

FIG. 17.
Accessory Iliosacral Articulations.

Accessory articulations are commonly found and usually occur at the level of the first or second dorsal sacral foramen; they may be paired or unpaired as seen in this illustration.

In females the coalescence of the coccyx and sacrum takes place less often and later in life than in males.

Dissecting room specimen, The University of Iowa College of Medicine.

Skeleton

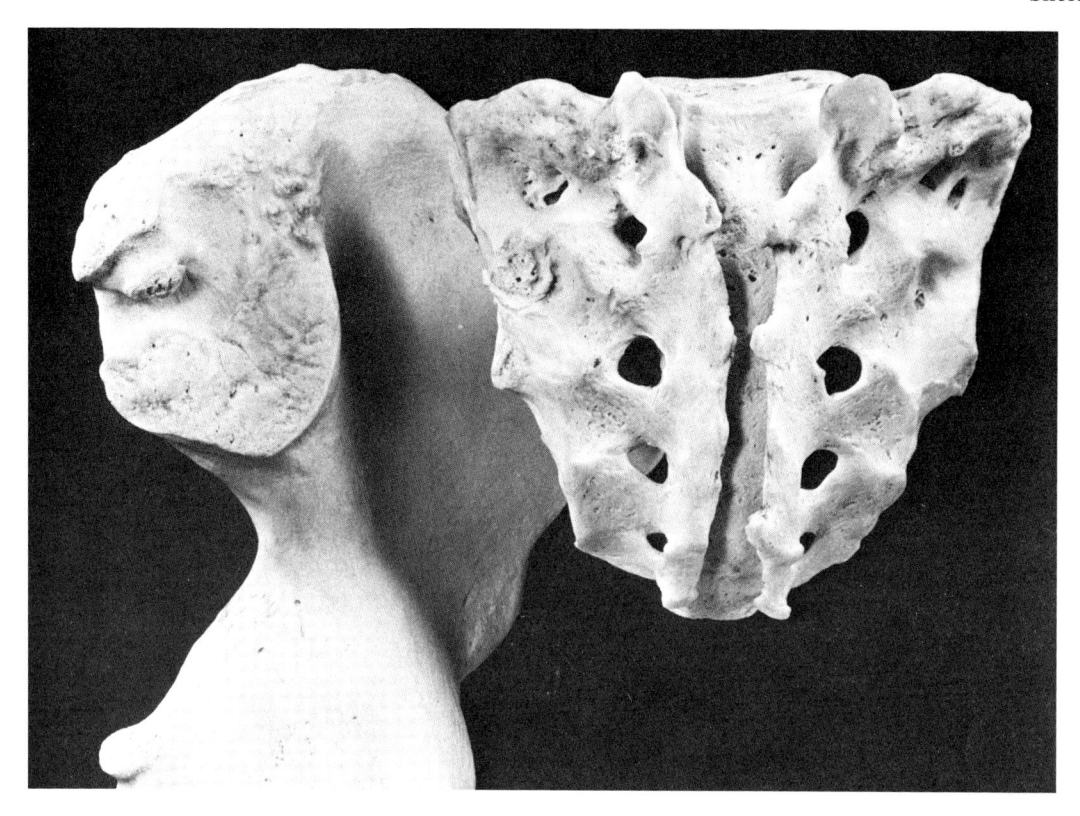

FIG. 18.
Absence of Sacral Dorsal Arches.

In 1% of a series of 1,275 sacra, all of the laminae failed to meet in the dorsal midline. As a result, the sacral canal is completely open. This anomaly occurred almost exclusively in males and twice as frequently in whites as in blacks (Anson, reference 1128).

Dissecting room specimen, The University of Iowa College of Medicine.

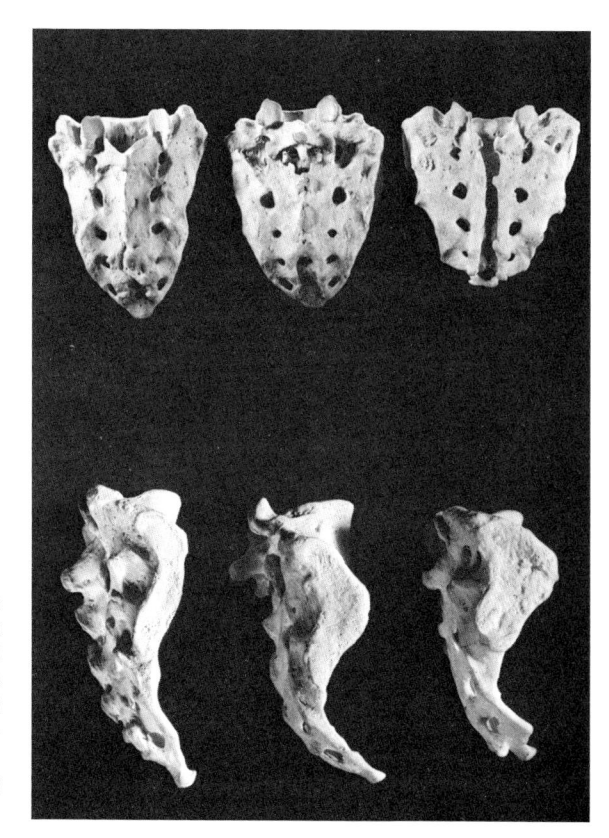

FIG. 19.
Three Sacra.

These three sacra are interesting because they show variable absence of the dorsal arches. In addition, it has been noted that in the female the sacrum is broader and less curved compared to the males whose sacrum is usually narrower and more curved. Paradoxically here, the broadest sacrum is the most curved and the narrowest least curved. The sex of the individuals from which these specimens were obtained is unknown.

Dissecting room specimen, The University of Iowa College of Medicine.

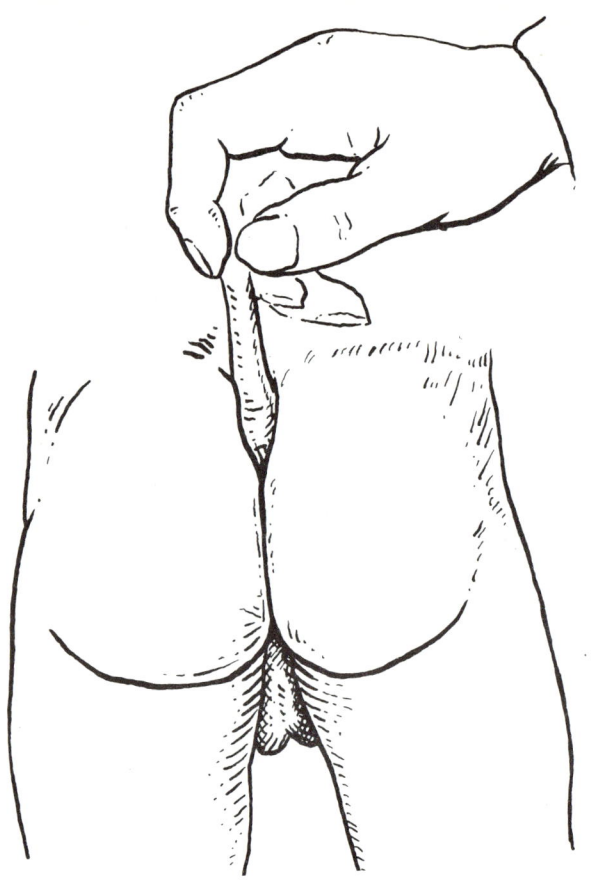

Human Anatomic Variation

FIG. 20.
Persistence of the Tail.

A well-developed tail is characteristic of the human embryo in the second month. Usually during the third month the tail regresses and disappears as an anatomic external feature. Occasionally the tail persists and grows with the rest of the body. Tails as long as 23 cm have been reported.

Redrawn from Harrison, R.G. Occurrence of tails in man, with a description of a case reported by Dr. Watson. *Johns Hopkins Hosp. Bull.* 12:96–101, 1901.

Additional references: Keith, A. The disappearance and reappearance of the human tail. *Nature* (Lond.) 106:845, 1921. Wood Jones, F. Human and other tails. *Nature* (Lond.) 107:487, 1921.

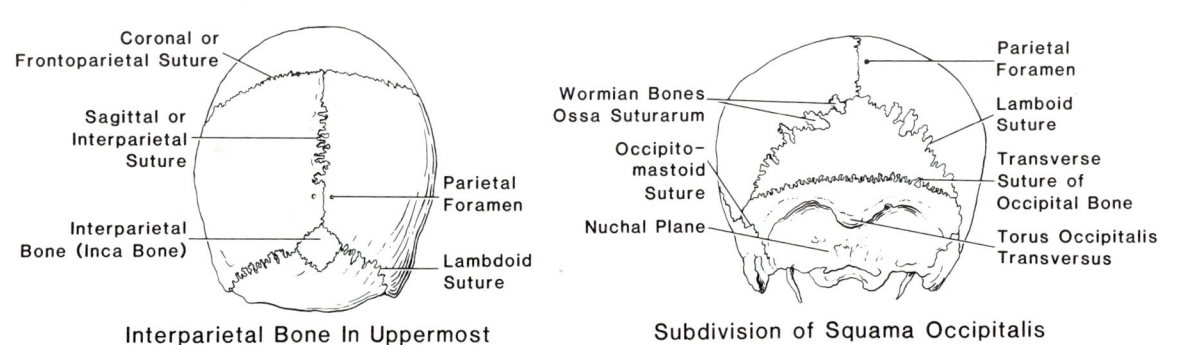

Interparietal Bone In Uppermost Part of Lambdoid Suture

Subdivision of Squama Occipitalis by Transverse Suture

Lamboid suture with Inca bone and mendosal suture.

Apical and sutural bones.

Lamboid suture with apical bones.

FIG. 21.
Sutural Bones.
Sutures.
Torus Occipitalis.
Parietal Foramina.

Redrawn from Toldt, C. *An Atlas of Human Anatomy for Students and Physicians*, 2nd Ed., Vol. 1. The Macmillan Co., New York, 1928.

Skeleton

FIG. 22.
Absence of Coronal and Sagittal Sutures.
Dissecting room specimen, The University of Iowa College of Medicine.

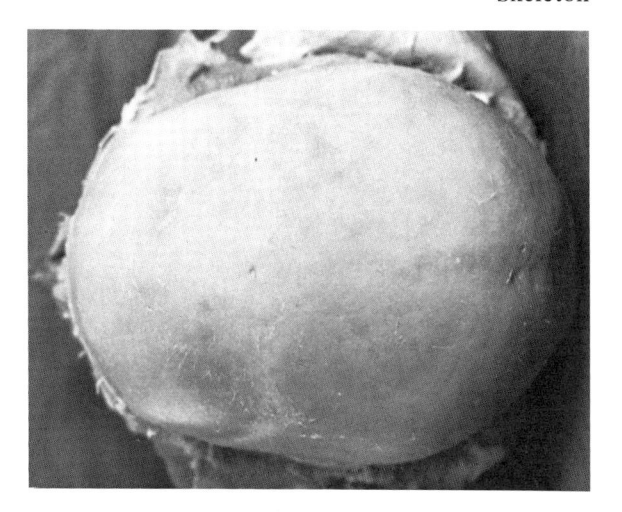

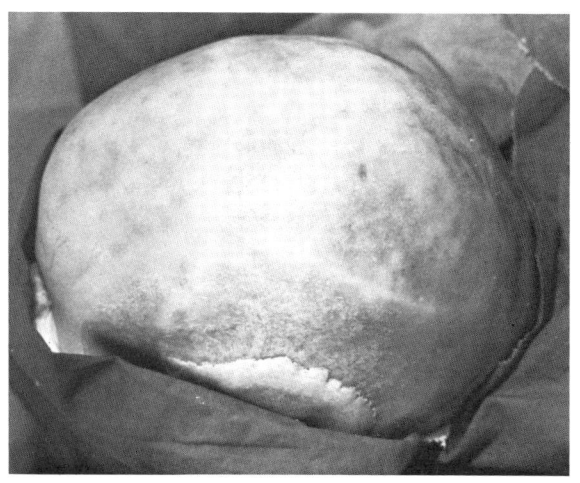

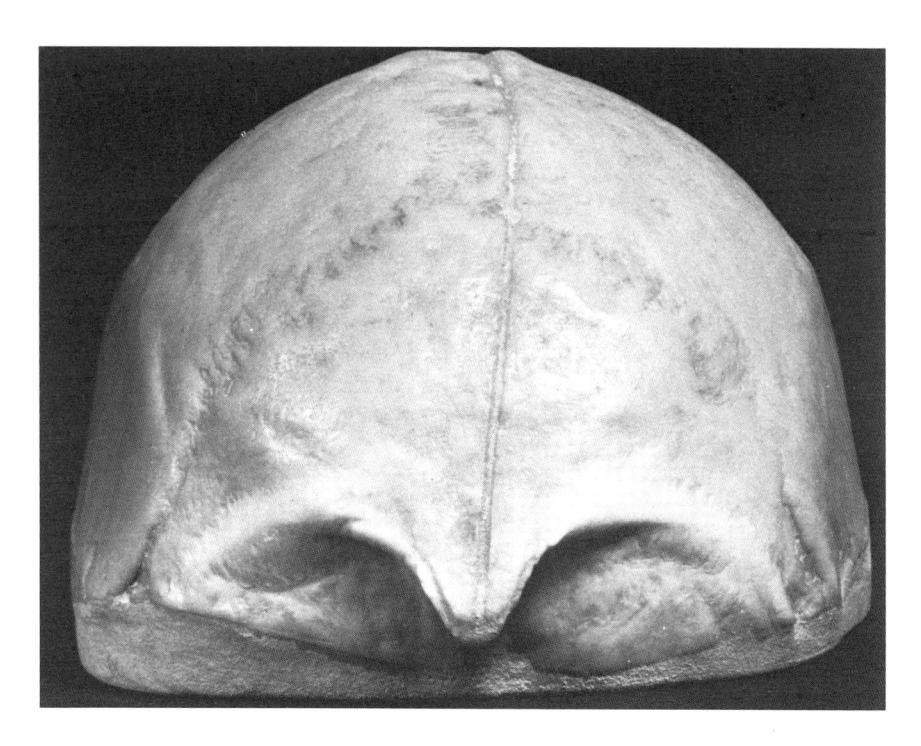

FIG. 23.
Torus Occipitalis.

Dissecting room specimen, The University of Iowa College of Medicine.

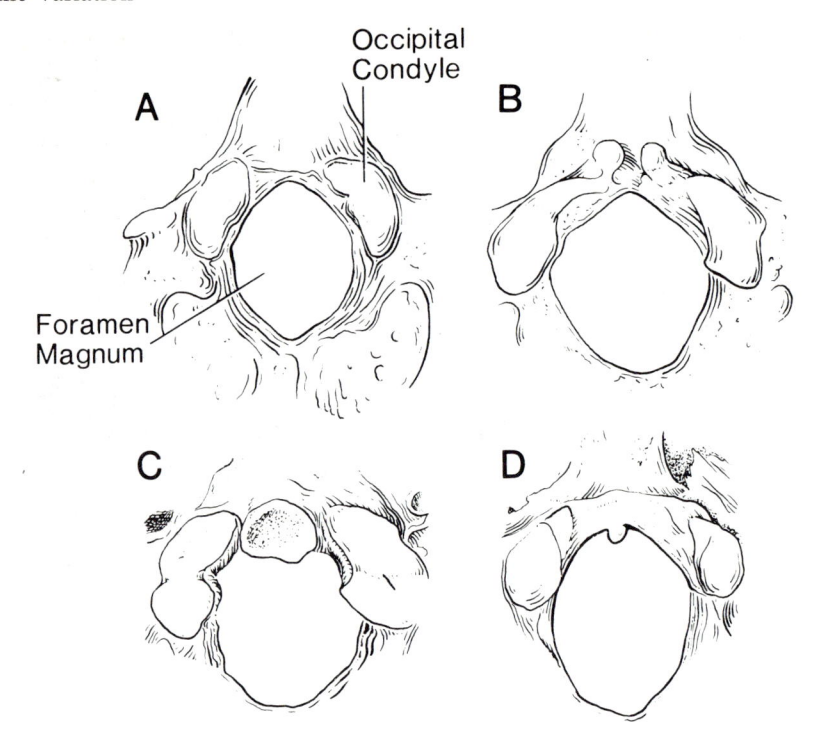

Human Anatomic Variation

FIG. 24.
Anomalies of Occipital Bone at Foramen Magnum (Occipital Condyles).
A: The usual appearance of occipital bone at the foramen magnum.
B: Precondylar tubercles.
C: Third occipital condyle.
D: Ossification of the ligament of the odontoid process of the axis.

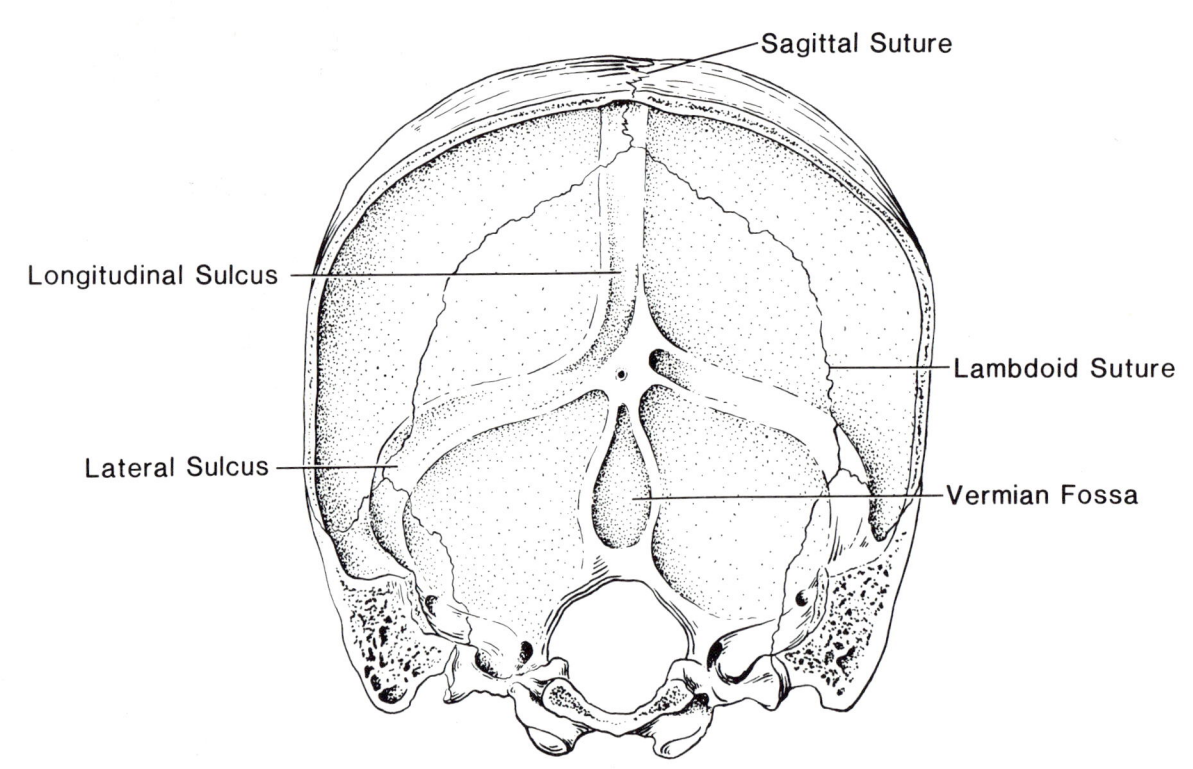

FIG. 25.
Vermian Fossa.

View of the posterior portion of the cranial cavity. The vermian fossa is a depression, varying in size, near the lower part of the internal occipital crest which lodges part of the inferior vermis of the cerebellum.

From Toldt, C. *An Atlas of Human Anatomy for Students and Physicians*, 2nd Ed., Vol. 1. The Macmillan Co., New York, 1928.

Skeleton

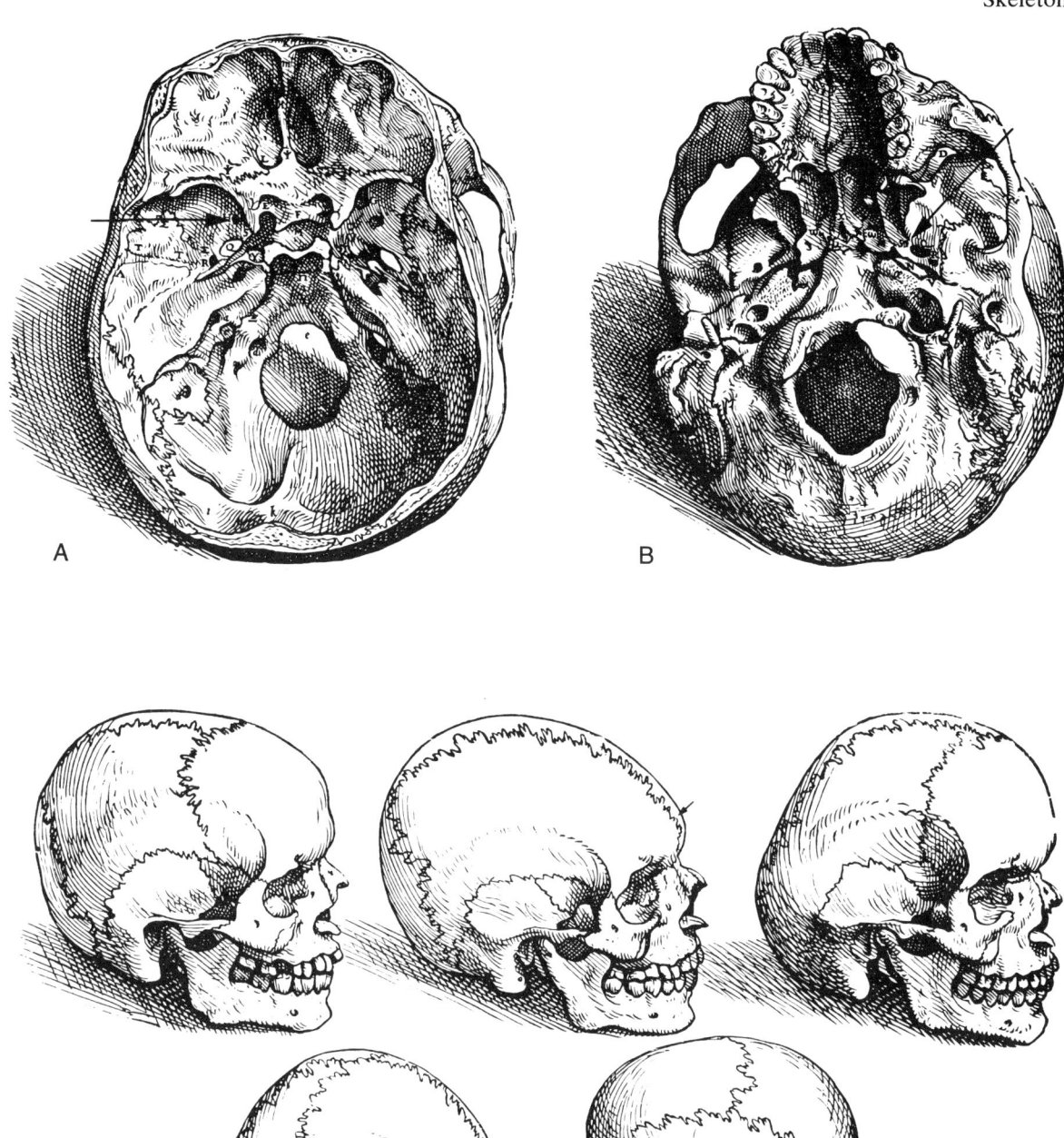

FIG. 26.
Foramen Vesalii.
Cranial Variations.

A: Base of skull seen from above. *Arrow* indicates the Vesalian foramen.
B: Base of skull seen from below. *Arrow* indicates the Vesalian foramen.
C: Five types of human skull. Of these, only the one in the upper left was considered by Vesalius to be normal ("natural").

Note absence of the coronal suture and presence of a metopic suture (*arrow*) in the upper middle figure. Note the metopic suture also in the lower left illustration (*arrow*). Note the horizontal suture of the zygomatic bone (*long arrow*) in the lower right figure.

Authors' note: Arrows have been added for use in this Atlas.
From Vesalius, A. *Fabrica*, 1543. In Straus, W.L. Jr. and O. Temkin. Vesalius and the problem of variability. *Bull. Hist. Med.* 14:609–633, 1943.

Human Anatomic Variation

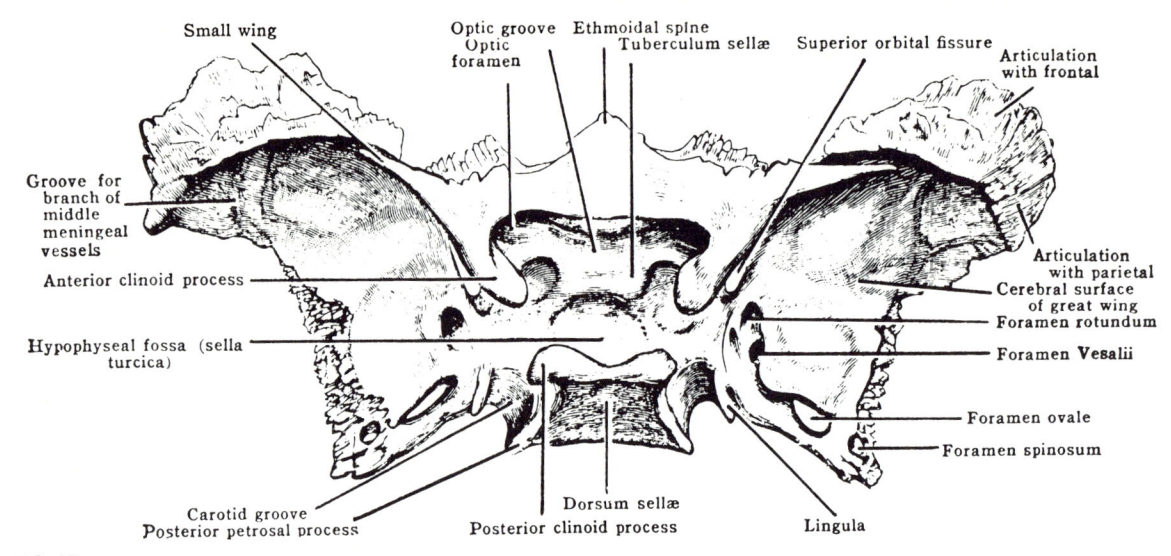

FIG. 27.
Foramen Vesalii.

First described by Andreas Vesalius, this rare foramen in the sphenoid bone transmits a vein (vein of Vesalius) from the cavernous sinus to the pterygoid plexus.

From Jackson, C.M., Ed. *Morris' Human Anatomy,* Section III: Osteology, by R.J. Terry, P. Blakiston's Son & Co., Philadelphia, 1923.

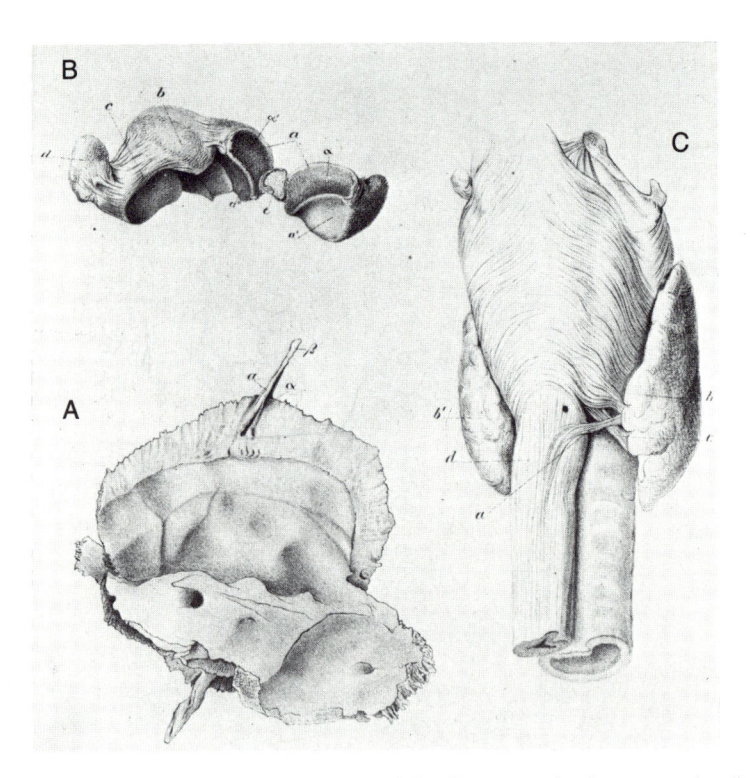

FIG. 28.
Peculiar Spine Arising from the Sagittal Border of the Squama of Temporal Bone.
Bipartite Scaphoid Bone.
M. Thyro-esophagus.

A: Right temporal bone of a man. *a*, Peculiar spine arising from temporal bone; α, root; β, spine.
B: Superior row of bones in the left carpus of a male corpse. *a*, Os scaphoid, bipartite; *b*, os lunatum; *c*, os triquetrum; *d*, os pisiforme; *e*, bone fragment from dorsal end of the divided part of the scaphoid bone.
Authors' note: O'Rahilly (reference 753) reported finding a bipartite scaphoid only once in 743 roentgenograms (0.13%).

O'Rahilly summarized an extensive literature by saying that most cases of divided scaphoid are traumatic in origin, with most workers also conceding that a few reported instances are congenital.
C: Musculus thyro-esophagus. Pharynx with adnexae. *a*, M. thyro-esophagus; *b,b'*, lobes of the thyroid gland; *c*, suspensory ligament of the thyroid gland; *d*, esophagus.

From Gruber, W. Eigenthümlicher stielförmiger Fortsatz der Schläfenbeinschuppe. *Arch. Pathol. Anat. Physiol. Klin. Med.* 69:386–387, 1877; Os naviculare carpi bipartitum. *Ibid.* 69:391–396, 1877; and Über das neue Anheftungsbündel des Oesophagus an die Glandula thyreoidea – Musculus thyreo-oesophagus. *Ibid.* 69:396–398, 1877.

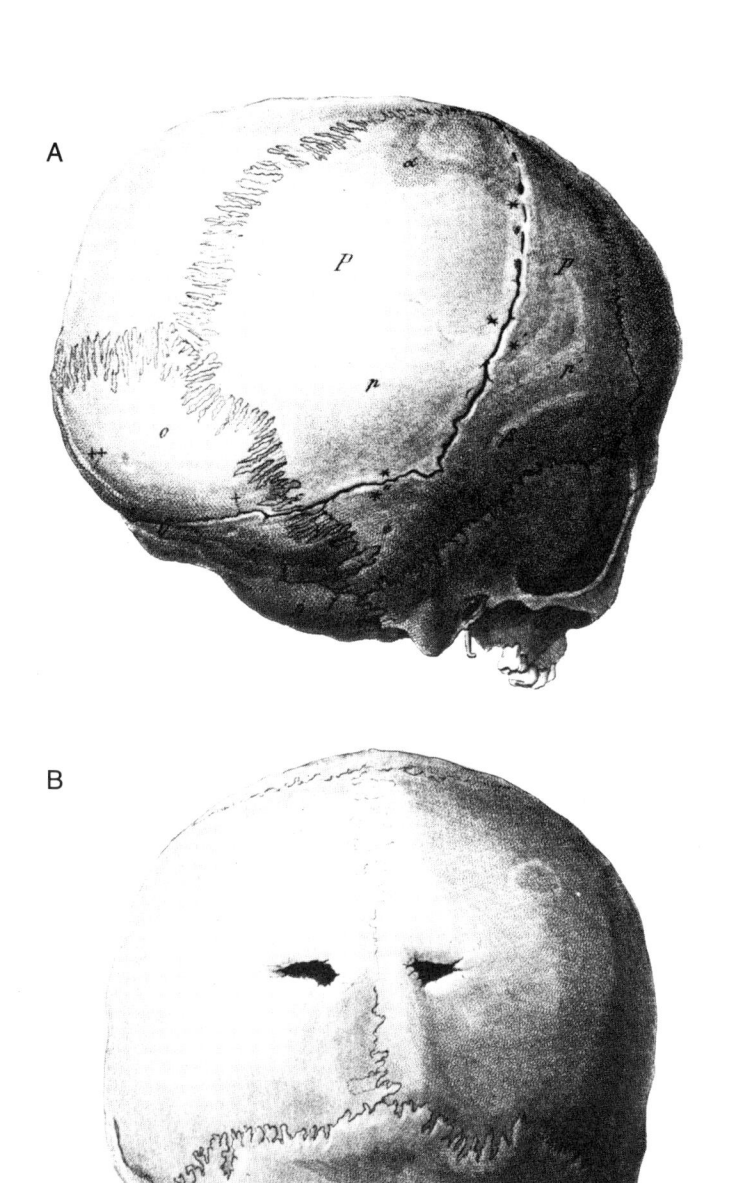

FIG. 29.
Bipartite Parietal Bone.
Enlarged Parietal Foramina.

A: Skull of a 15- to 18-year-old youth. *P,P*, Right bipartite parietal bone; *, congenital suture.

B: Bilateral parietal foramina.

From Gruber W. Über congenitale Theilung des Parietal durch eine quere oder schräge Sutur. *Arch. Pathol. Anat. Physiol. Klin. Med*. 50:113–124, 1870; and Über congenital abnorm weite Foramina parietalia. *Ibid*. 50:124–134, 1870.

Human Anatomic Variation

FIG. 30.
Bilateral Parietal Foramina.

Parietal foramina transmit veins that drain into the superior sagittal sinus and an artery, a small branch of the occipital artery, that supplies the dura mater. Although the foramina have a reported frequency of 60%, they may be absent on one or both sides.

Dissecting room specimen, The University of Iowa College of Medicine.

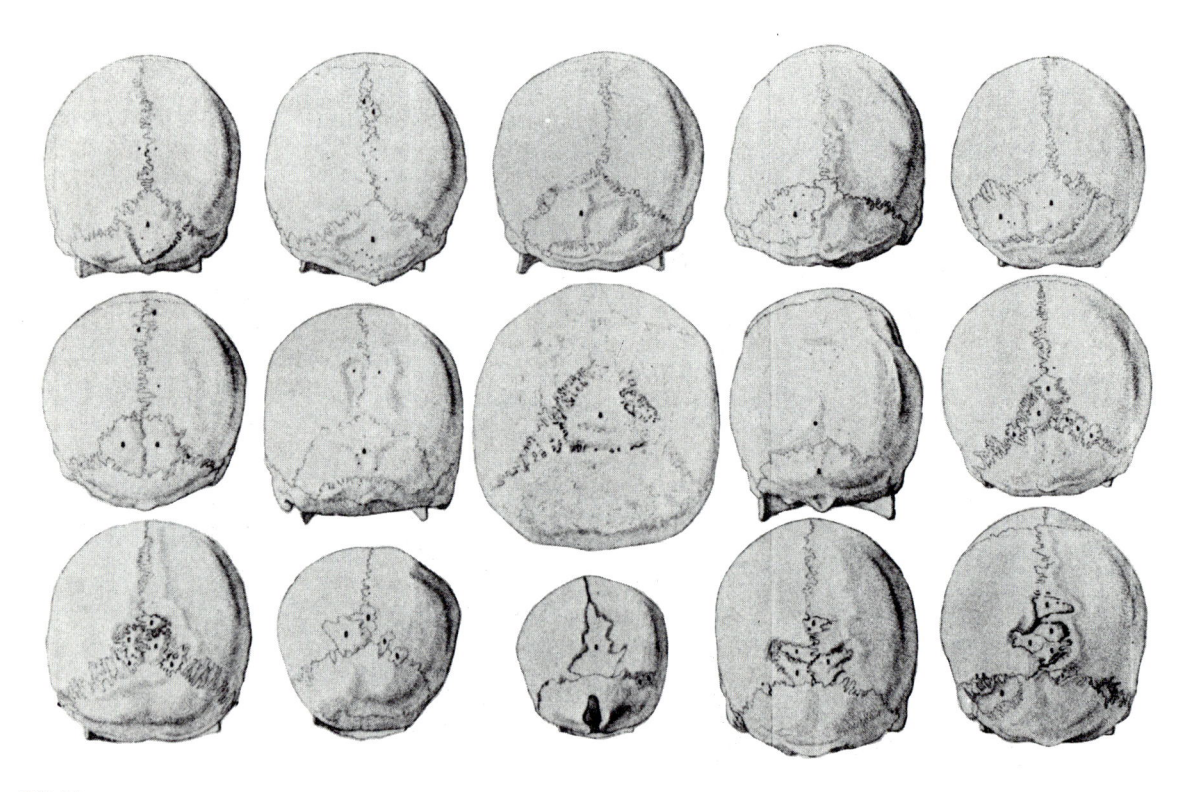

FIG. 31.
Varieties of Occipital Wormian and Interparietal Bones.

From Calori, L. De' wormiani occipitali ed interparietali posteriori dei cranii nostrali e di quelli delle fontanelle laterali ne' cranii di negro. Lettera responsira del Prof. Luigi Calori al celebre craniologo Prof. Giustniano Nicolucci. *Mem. R. Accad. Sci. Istituto di Bologna S. 2* 7:303–321, 1867.

Skeleton

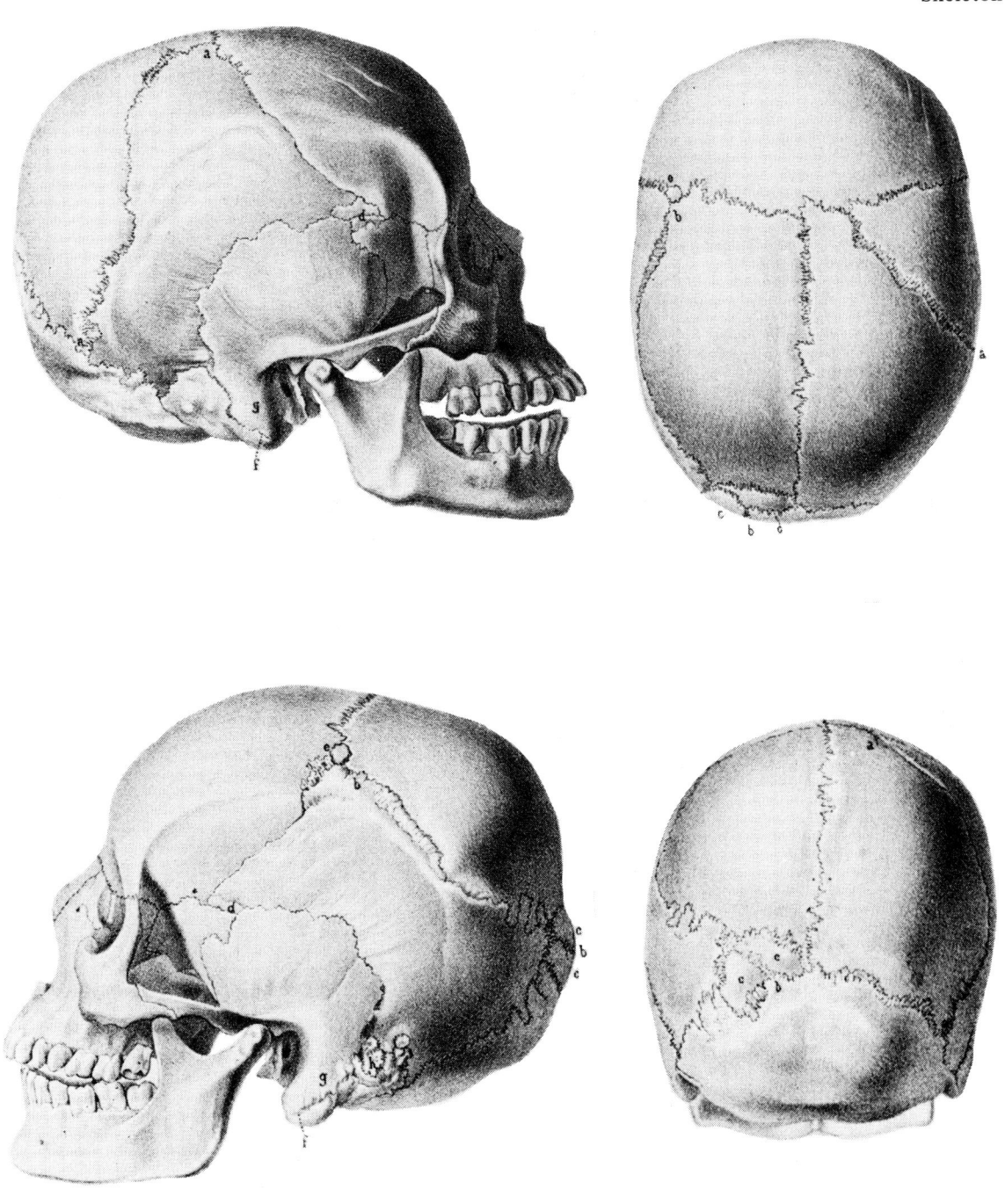

FIG. 32.
Supernumerary Parietal Bone Sutures and Wormian Bones.

From Calori, L. Intorno alle suture soprannumeraire del cranio umano e su quelle specialmente della ossa parietali. *Mem. Accad. Sci. Istituto di Bologna S.2* 6:327–343, 1866.

Human Anatomic Variation

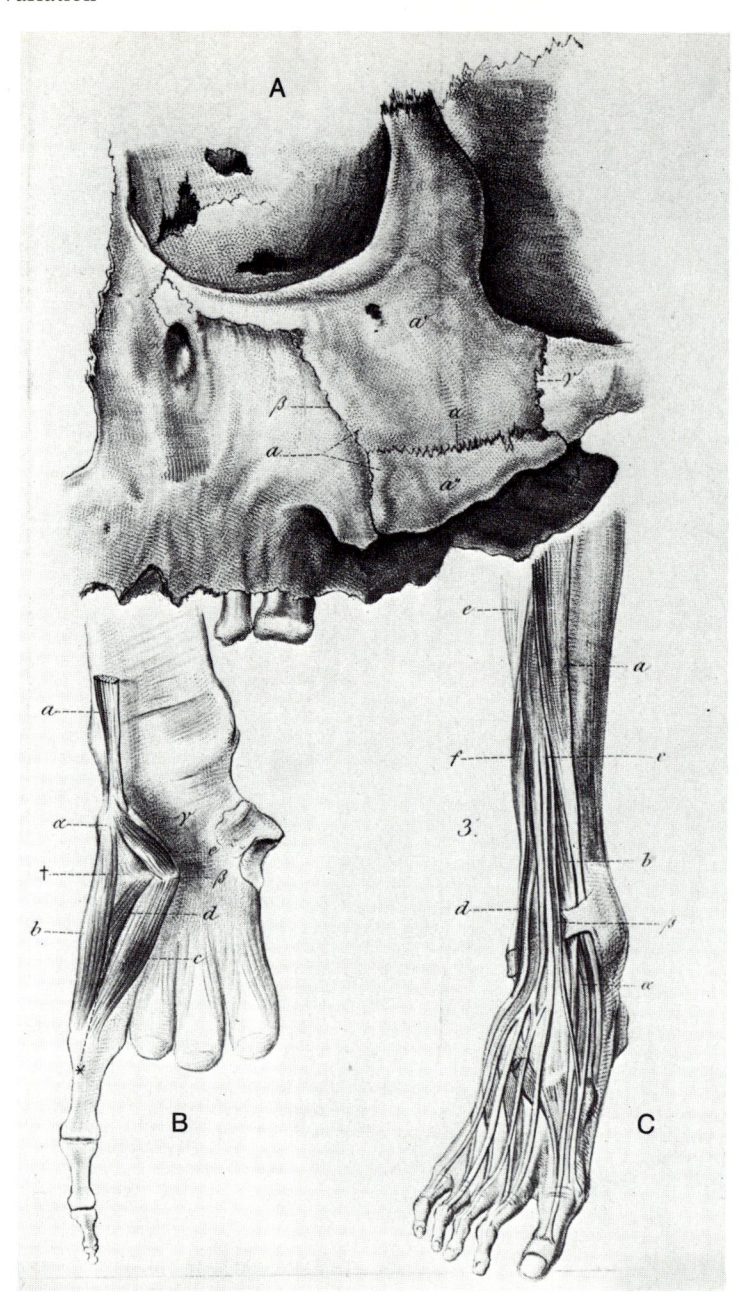

FIG. 33.
Bipartite Os Zygomaticum.
(M. Pisohamatus).
(Anomalous M. Extensor Digitorum Longus Pedis).

A: a, Bipartite zygomatic bone; a', orbital portion of bipartite zygomatic bone; a'', maxillotemporal portion of bipartite zygomatic bone; α, zygomatic suture; β, zygomaticomaxillary suture; γ, zygomaticotemporalis suture.
B: a, Tendon of flexor carpi ulnaris muscle; b, abductor digiti minimi manus; c, flexor digiti minimi brevis manus; d, opponens digiti minimi manus; e, m. pisohamatus; α, pisiforme bone; β, hamulus of the hamate bone; γ, pisohamate ligament; *, triangular hypothenar space; †, gap between anomalous muscle and passageway for vessels and nerves.

C: a, M. tibialis anterior; b, m. extensor hallucis longus bicaudatus; c, anomalous m. extensor digitorum longus with 5 tendons, one for each toe; d, m. peroneus tertius; e, m. peroneus longus; f, m. peroneus brevis; α, tail of m. extensor hallucis longus to proximal phalanx of great toe; β, supernumerary tendon of m. extensor digitorum longus to great toe.

From Gruber, W. Ein Nachtrag zum Vorkommen des zweigetheilten Jochbeins – Os zygomaticum bipartitum – beim Menschen. Arch. Anat. Physiol. Wissen. Med. 1875, pp. 194–201; Ein Musculus piso-humatus beim Menschen. Ibid. 1875, pp. 202–203; and Über den Musculus extensor digitorum communis manus anomalus mit 5 Sehnen zu allen Fingern, and über den Musculus extensor digitorum longus pedis anomalus mit 5 Sehnen zu allen Zehen. Ibid. 1875, pp. 204–210.

FIG. 34.
Supernumerary Bones of the Zygomatic Arch.

a, Zygomatic bone; *b*, zygomatic process of temporal bone; *c*, supernumerary bones of zygomatic arch at the place of masseteric muscle tubercle; *d*, supernumerary bones located at the end of zygomatic arch of temporal bone. α, masseteric tubercle of zygomatic arch; β, masseteric spine of zygomatic arch.

From Gruber, W. Über supernumeräre Knochen im Jochbogen. *Arch. Anat. Physiol. Wissen. Med.* 1875, pp. 337–347.

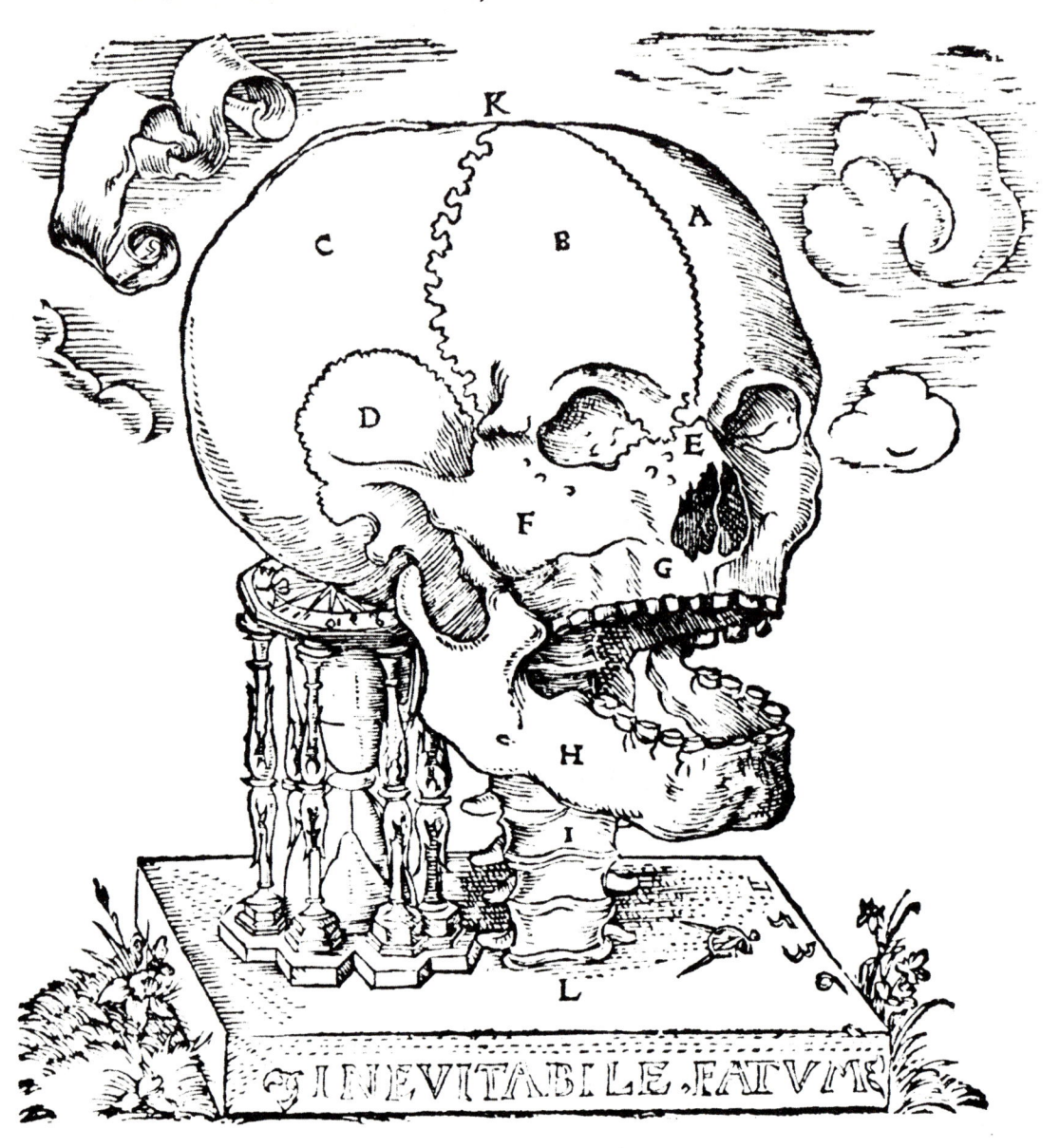

FIG. 35.
Metopic Suture.

A, Metopic suture; B, frontal bone; C, parietal bone; D, temporal bone; E, nasal bone; F, zygomatic bone; G, maxilla; H, mandible; I, cervical vertebra; K, coronal suture.

"JOHANNES DRYANDER (ca. 1500-1560). *Anatomiae*. Marburg: Apud Eucharium Ceruicornum, 1537. [72] pp., illus. (woodcuts), fold. table. 20 cm. Full antique calf.
Dryander (also known as Eichmann), professor of surgery at Marburg, was a friend of Vesalius and was "among the first anatomists who made illustrations after their own dissections."
... His *Anatomiae* appeared six years before Vesalius' great work.
This was the first significant book on the anatomy of the head and contains twenty full-page woodcuts made from Dryander's own dissections. Sixteen of the plates are of the head and brain and were done to show successive stages of dissection. The first eleven plates appeared earlier in his *Anatomia Capitis Humani* (1536) and the remaining four plates of the chest and lungs were added as an appendix, possibly as models for a second book which was never published. A folding chart containing an explanation of the parts of the head, included here in fascimile, is not present in most copies.
This work also contains a brief tract on the anatomy of the pig traditionally attributed to Copho (ca. 1110), an early teacher at Salerno, and a short essay by Gabriele de Zerbis (1445-1505), professor at Padua, on the anatomy of the fetus."

From Dryander, J. *Anatomiae*. Marburg, 1537. In *Heirs of Hippocrates. The Development of Medicine in a Catalogue of Historic Books in the Health Sciences Library*. Friends of The University of Iowa Libraries, The University of Iowa, Iowa City, 1980.

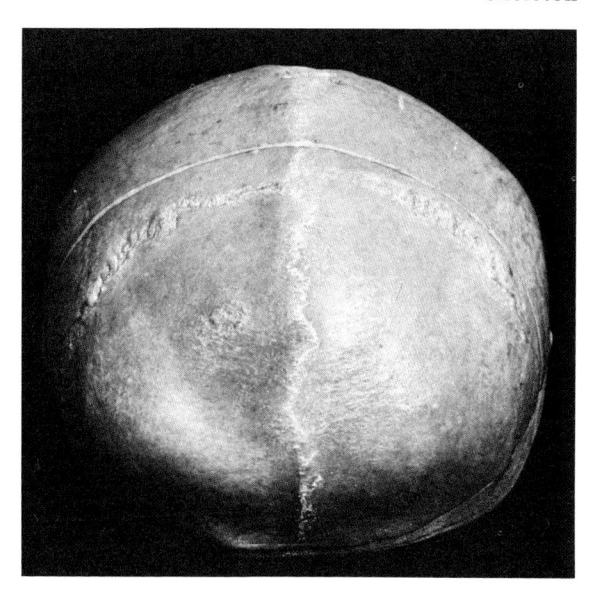

Skeleton

FIG. 36.
Metopic Suture.
Dissecting room specimen, The University of Iowa College of Medicine.

FIG. 37.
Variation in Incisive Canal.
Variation in Greater and Lesser Palatine Foramina.
Wormian Bones Between Palatine Process and Palatine Bones.
Division of Zygomatic Bone into Two Horizontal Parts.

A–C: Variation in size of the incisive canal and in the formation of greater and lesser palatine foramina. Wormian bones between the maxilla (palatine process) and palatine bones.
D: Division of zygomatic bone into two horizontal parts (*m*, *z*).

From Calori, L. Delle anomalie piu importanti di ossa, vasi, nervi, e muscoli occorse nell'ultimo biennio facendo anatomia del corpo umano. *Mem. Accad. Sci. Istituto di Bologna S.2* 8:417–482, 1868.
▼

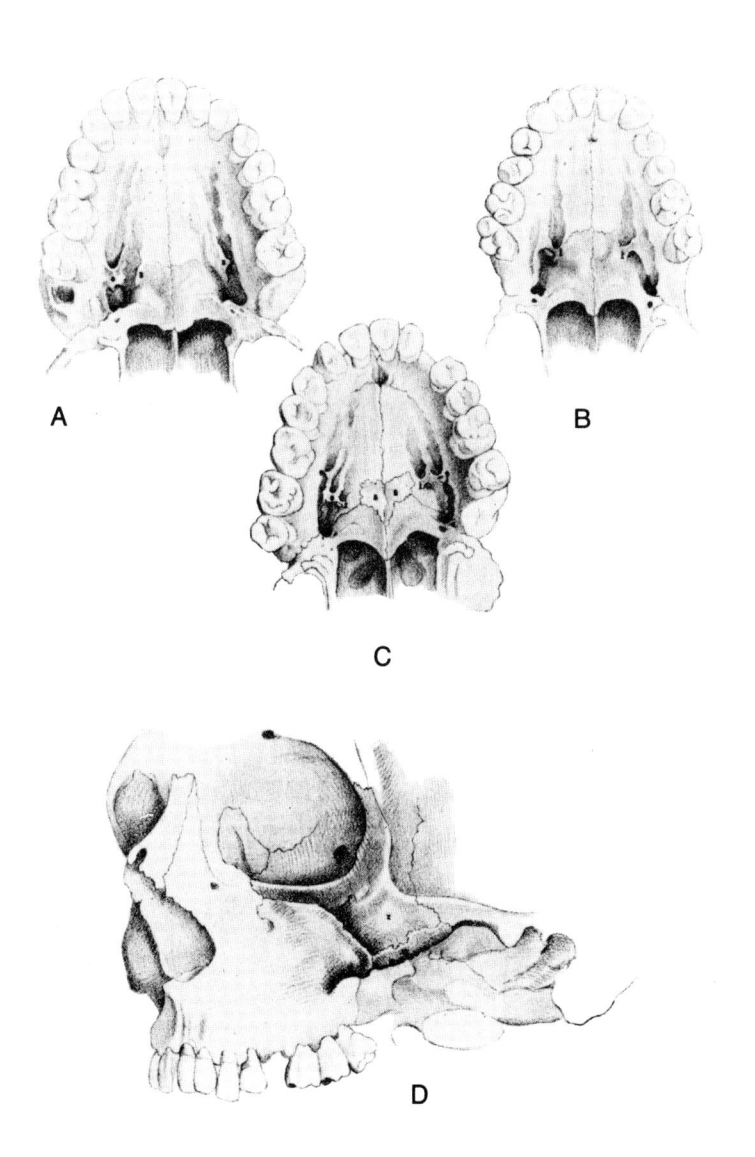

Human Anatomic Variation

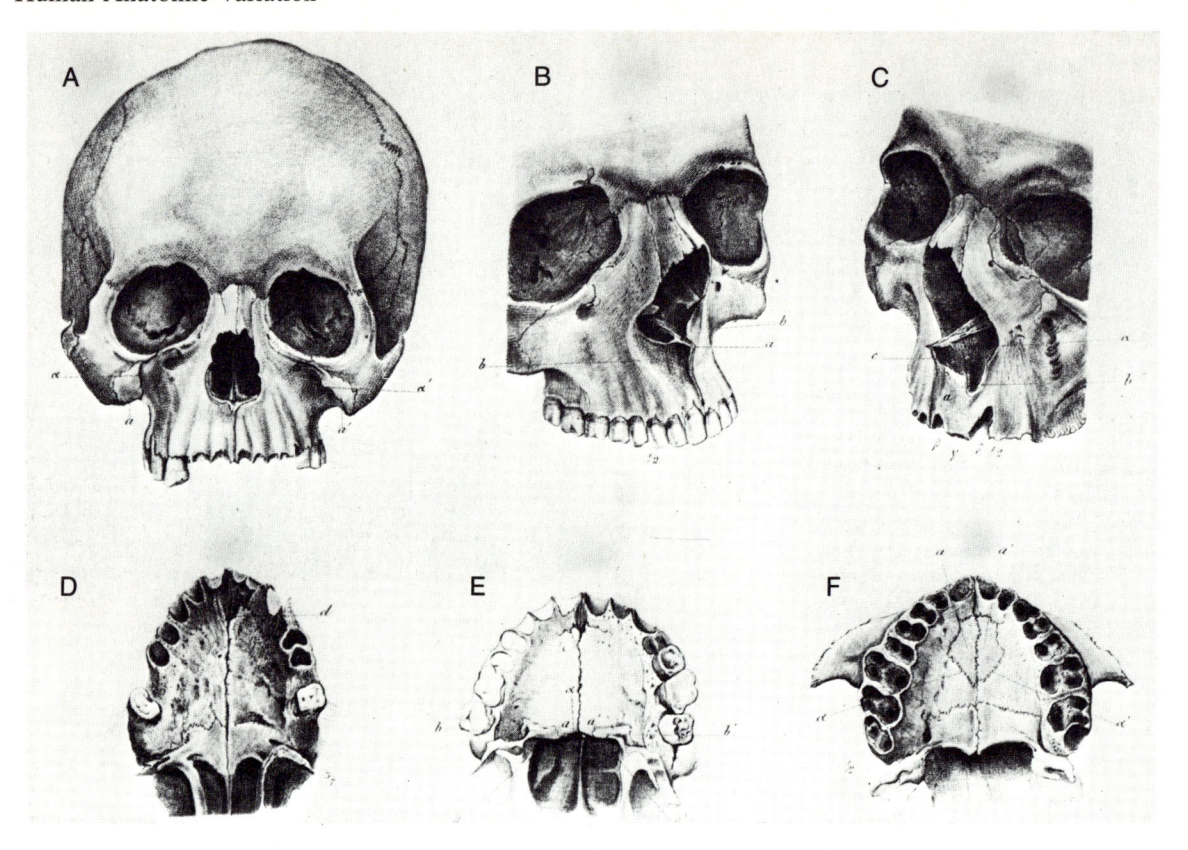

FIG. 38.
Some Unusual Maxillary Bones.

A: *a*, Division of the zygomatic process of the right maxillary bone; α,α', supernumerary suture and process on the lower bone.
B: *a*, Anterior nasal spine; *b,b'*, pit (fossa) on the alveolar process, pars incisiva, of the maxilla.
C: *b*, Unusual groove of the nasal maxilla.
D–F: Supernumerary bones and sutures of the palatine process of the maxillary bone.

From Gruber, W. Über einige merkwürdige Oberkiefer-Abweichungen. *Arch. Anat. Physiol. Wissen. Med.* 1873, pp. 195–207.

FIG. 39.
Variations in Mandibular Foramen, Canal, and Mylohyoid Groove.

a, a', Mandibular foramen (infundibulum); b, mylohyoid groove; b', mylohyoid groove doubled by a bony septum; c, supernumerary mylohyoid canal; α, anterior infundibular line; α', crest of the anterior infundibular line; α'', lingula mandibulae; β, posterior infundibular line; β', crest of posterior infundibular line; γ, anterior mylohyoid sulcal line; γ', crest of anterior mylohyoid sulcal line; δ, posterior mylohyoid sulcal line.

From Gruber, W. Über das Semiinfundibulum inframixillare, den Sulcus mylohyoideus und die beide deckenden knöchernen Brücken. *Arch. Anat. Physiol. Wissen. Med.* 1873, pp. 348–356.

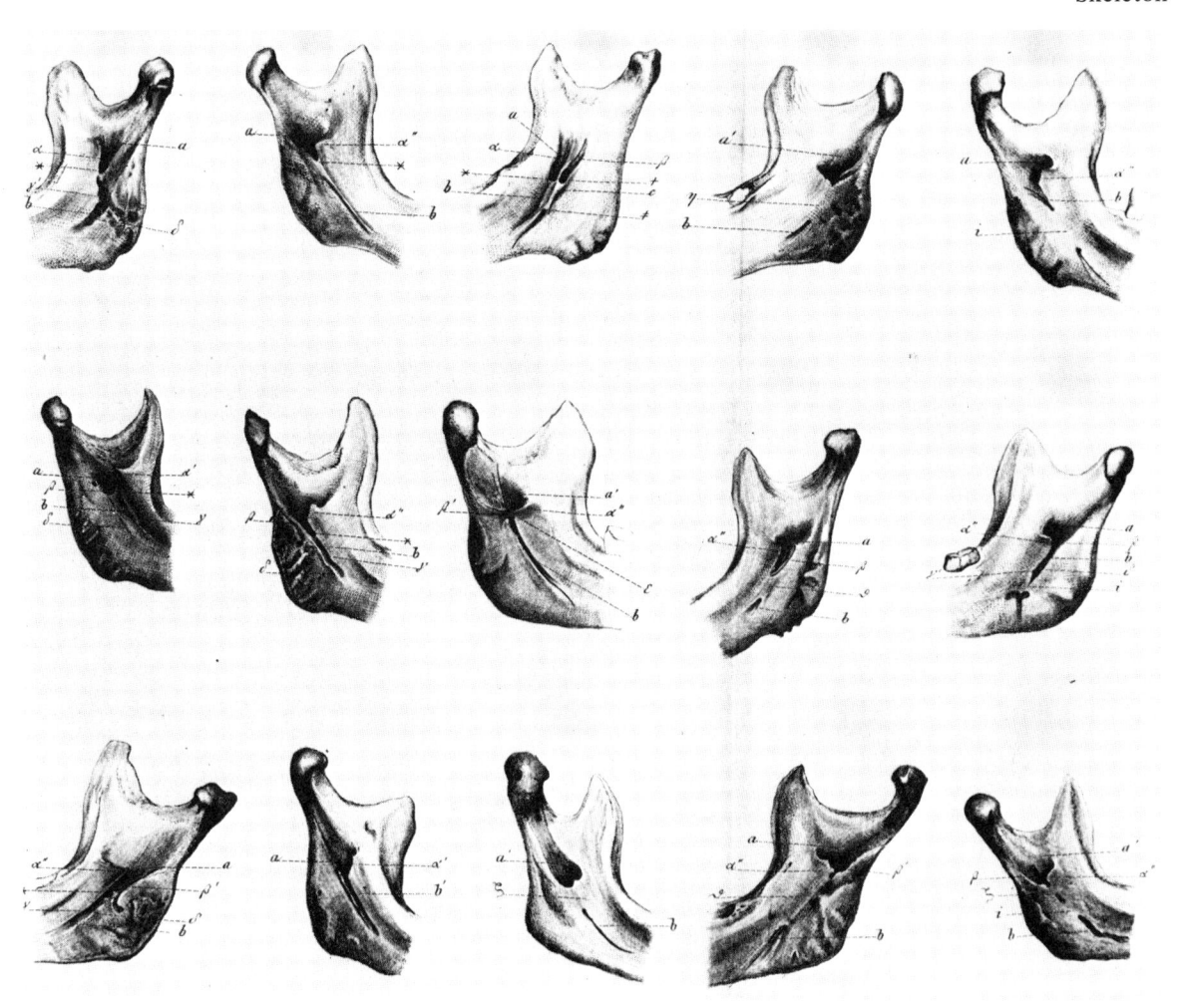

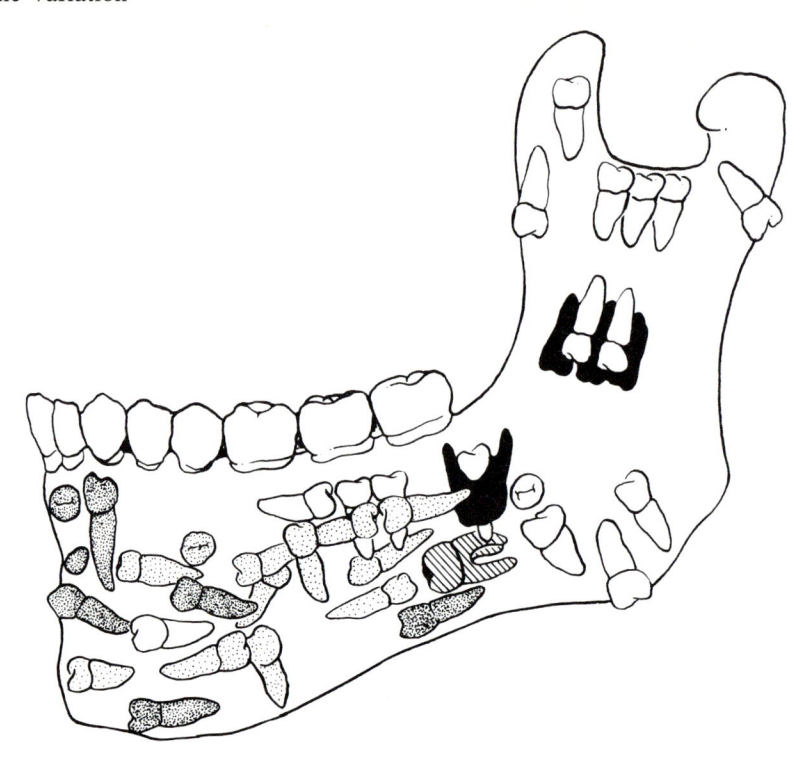

Human Anatomic Variation

FIG. 40.
Unerupted Teeth.
Incisors and canines, heavily dotted; third molars, white; premolars, lightly dotted; and supernumerary molars, black.

Redrawn from Wettstein. In Zimmer, E.A. *A. Köhler and E.A. Zimmer Borderlands of the Normal and Early Pathologic in Skeletal Roentgenology*, 3rd American Ed., translated and edited by S.P. Wilk. Grune & Stratton, New York, 1968.

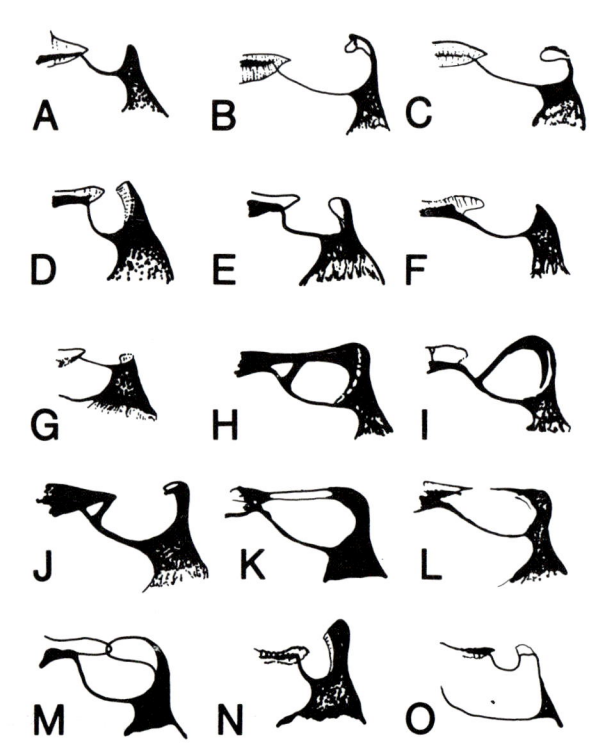

FIG. 41.
Various Shapes of Sella Turcica.

A–F: Principal variations in sellar shape according to H.O. Martin: A, round; B, oval; C, flat (shallow); D, deep; E, quadrangular; F, distorted, contracted.
G: "Sellar monstrosity" of Carstens. It lacks any real fossa; correspondingly, the hypophysis is inadequate.
H–L: Various types of bridging.
M: Schneider's "pseudo-bridging."
N: Raab's variant; tall, clumsy dorsum.
O: Fossa rounded and small, low wide dorsum, and posterior clinoid processes may be present or lacking.

Redrawn from Zimmer, E.A. *A. Köhler and E.A. Zimmer Borderlands of the Normal and Early Pathologic in Skeletal Roentgenology*, 3rd American Ed., translated and edited by S.P. Wilk, Grune & Stratton, New York, 1968.

Skeleton

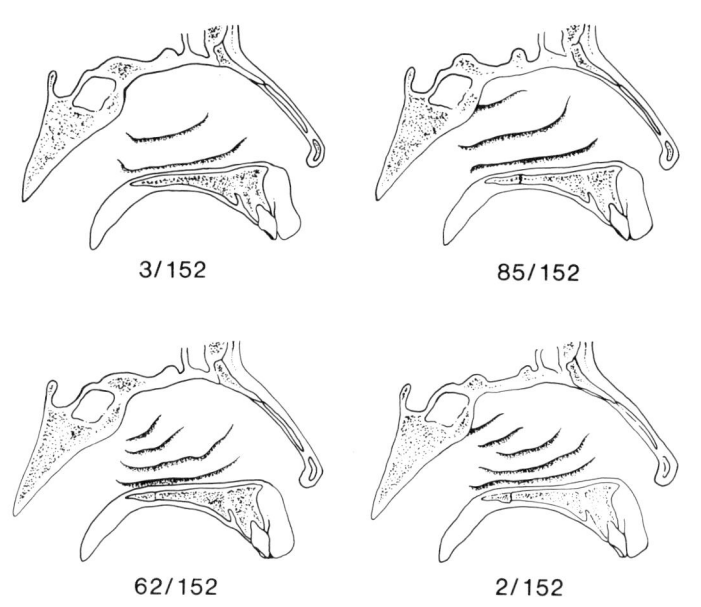

FIG. 42.
Variation in Number of Meatuses in Nasal Fossae.

Nasal fossae characterized by two, three, four, or five meatuses. The number observed out of a total of 152 cases is shown below each figure.

From Thomson, A. Report of Committee of Collective Investigation of the Anatomical Society of Great Britain and Ireland for the year 1890–91. *J. Anat. Physiol.* 26:74–93, 1892.

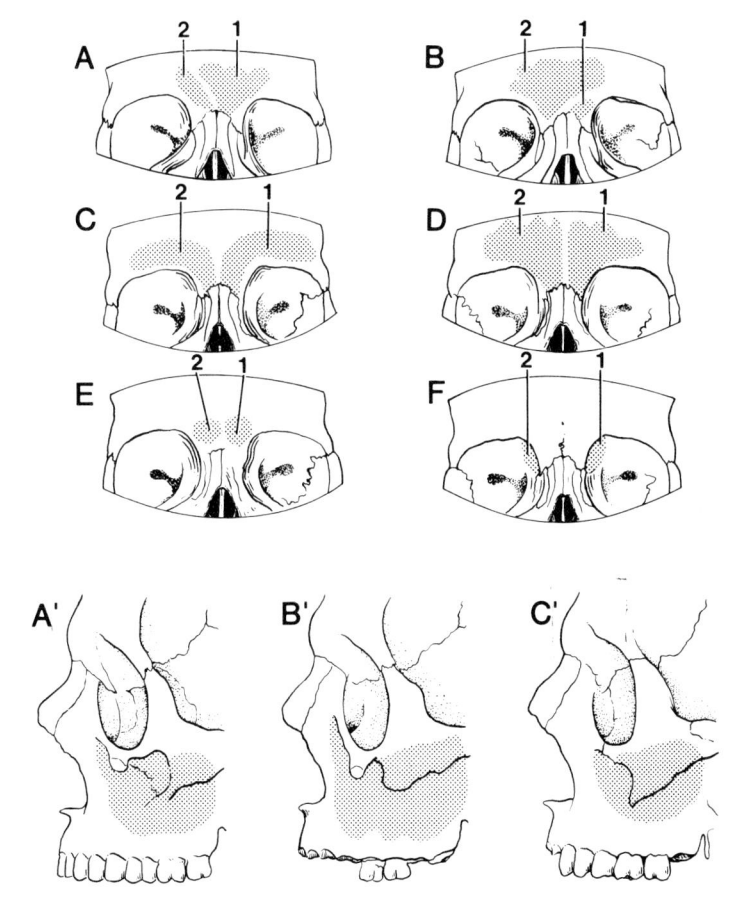

FIG. 43.
Variations in Size of Frontal and Maxillary Sinuses.

A–F: *1*, Left frontal sinus; *2*, right frontal sinus.

A'–C': Three types of maxillary sinus: A', usual sinus; B', enlarged sinus; C', small sinus.

Redrawn from Testut, L. *L. Testut et A. Latarjet Traité d'anatomie humaine*. G. Doin & Cie, Paris, 1948.

Human Anatomic Variation

FIG. 44.
Variations in Sphenoidal Sinus.

A: *1*, Choanae; *2*, maxillary nerve; *3*, ophthalmic vein; *4*, abducens nerve; *5*, ophthalmic nerve; *6*, oculomotor nerve; *7*, trochlear nerve; *8*, ophthalmic artery; *9*, optic nerve; *10*, intersinus septum; *11*, optic canal; *12*, sphenoidal foramen; *13*, extension of sphenoidal sinus into anterior clinoid process; *14*, foramen rotundum; *15*, vidian canal; *16*, middle concha; *17*, inferior concha.

B: Sphenoidal sinus anterior to the hypophysis. *1*, Optic nerve; *2*, internal carotid artery; *3*, ridge formed by optic nerve; *4*, ridge formed by carotid artery; *5*, arrow indicates orifice of sinus; *6*, pituitary (sella turcica) fossa; *7*, basilar part of occipital bone.

C: Large sphenoidal sinus anterior to and below the hypophysis. *1*, Anterior clinoid process; *2*, optic foramen; *3*, pituitary fossa; *4*, nasal cavity; *5*, basilar part of occipital bone; *6*, ridge for optic nerve; *7*, infraorbital diverticulum; *8*, ridge for carotid artery; *9*, pterygoidal diverticulum running anteriorly in contact with the foramen rotundum, and extended posteriorly by an alar diverticulum coming almost up to the foramen ovale.

Redrawn from Testut, L. *L. Testut et A. Latarjet Traité d'Anatomie Humaine.* G. Doin & Cie, Paris, 1948.

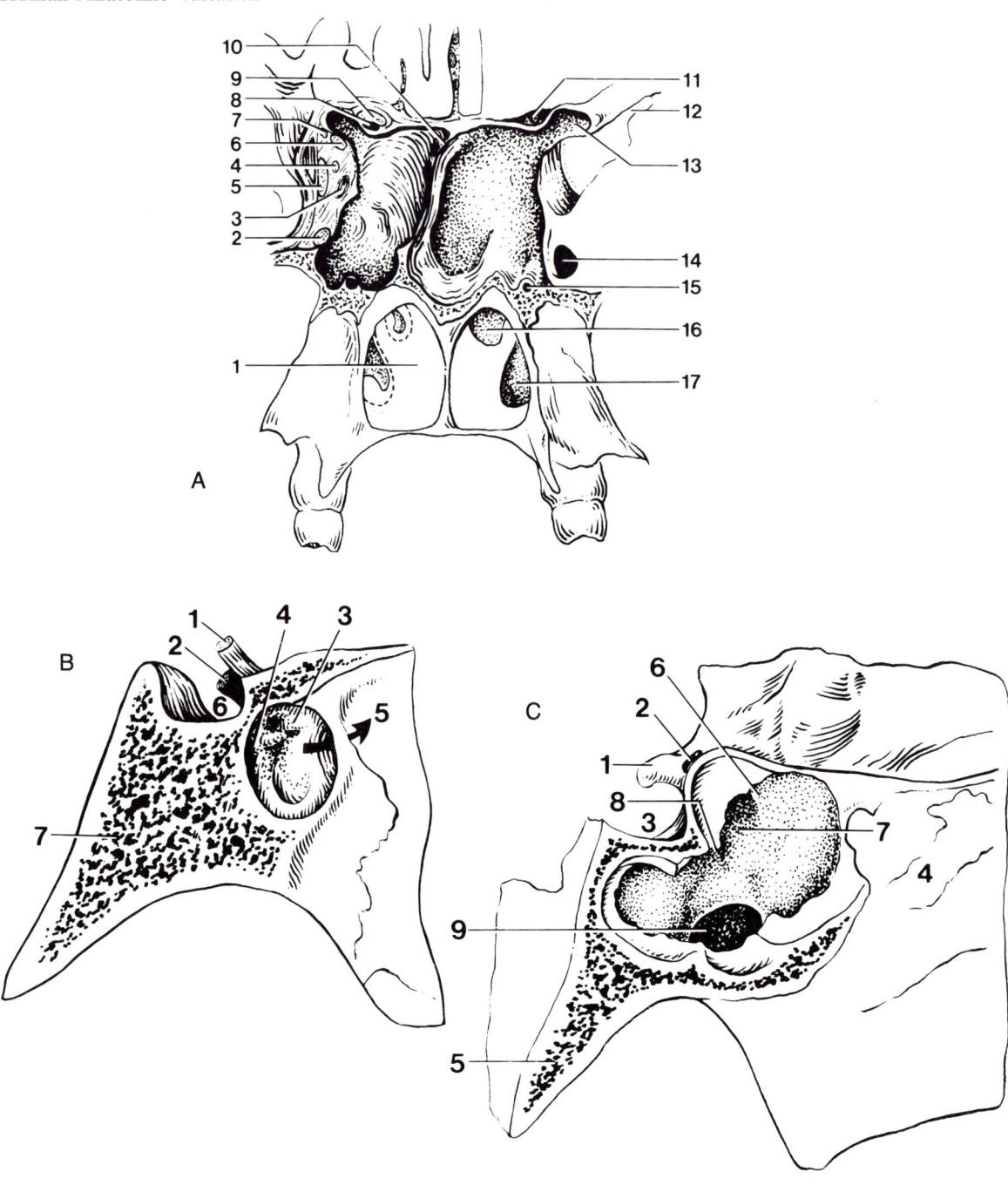

FIG. 45.
Styloid Processes.

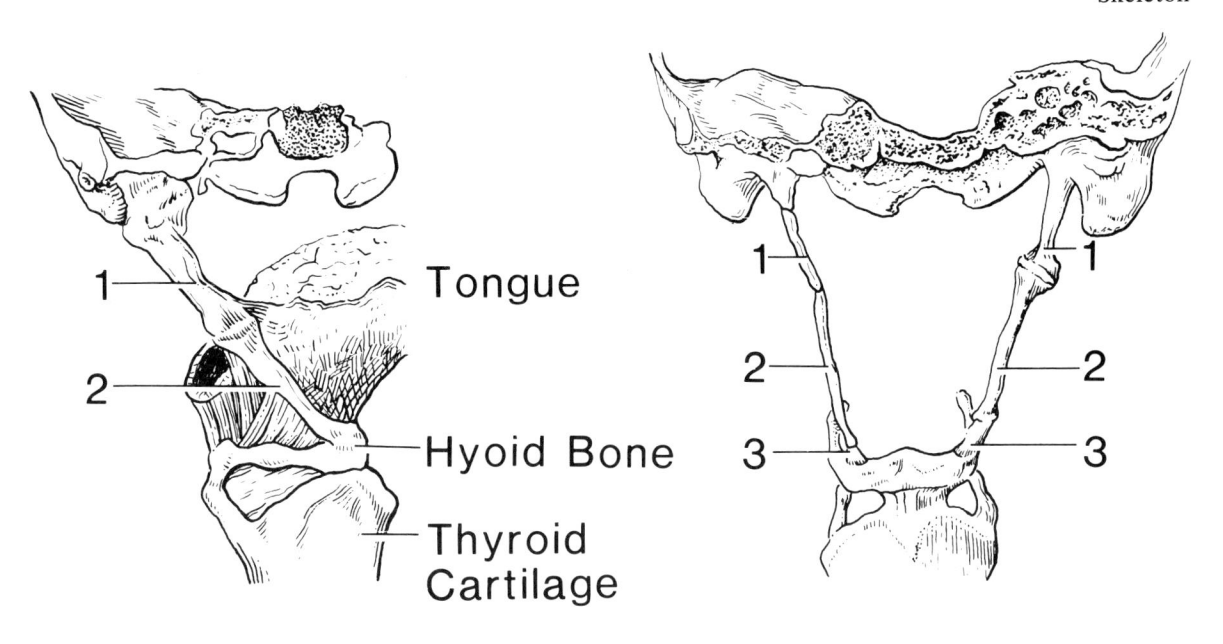

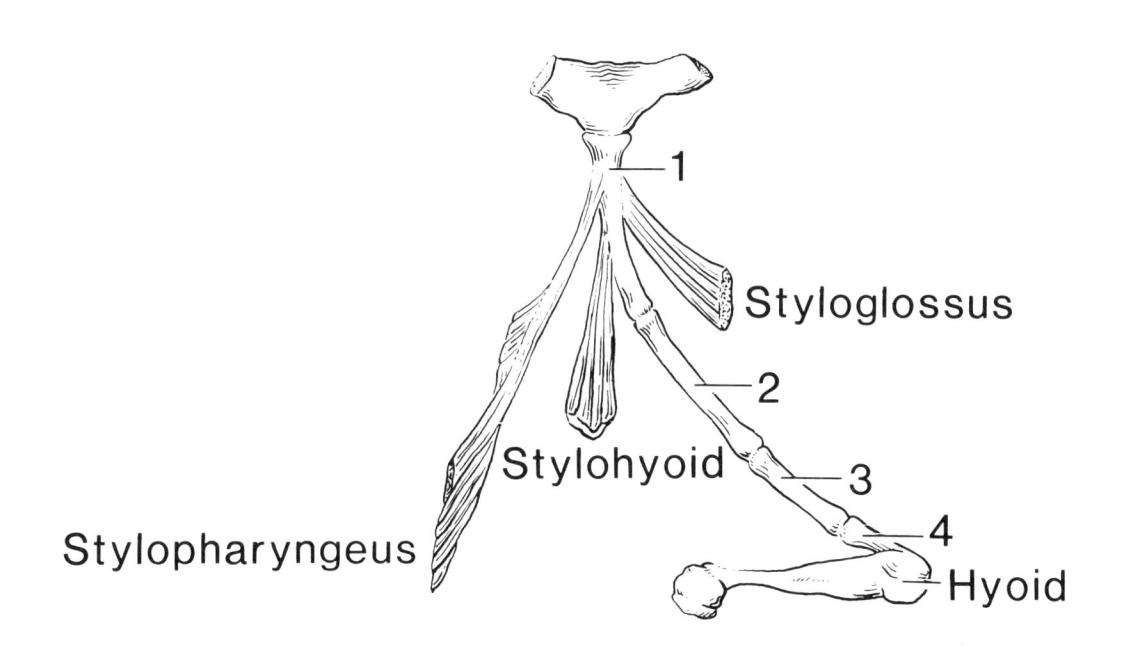

Styloid processes composed of 1, 2, 3, and 4 bones. See A.M. Fournier, reference 293.

Redrawn from various sources.

Human Anatomic Variation

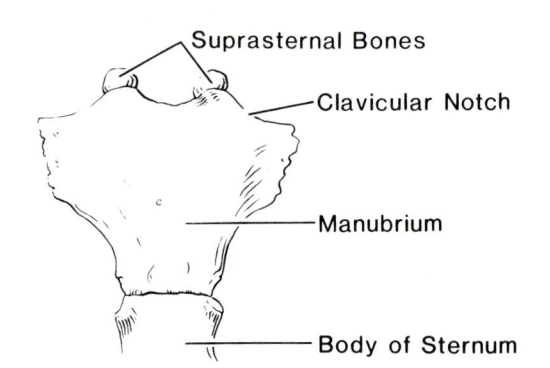

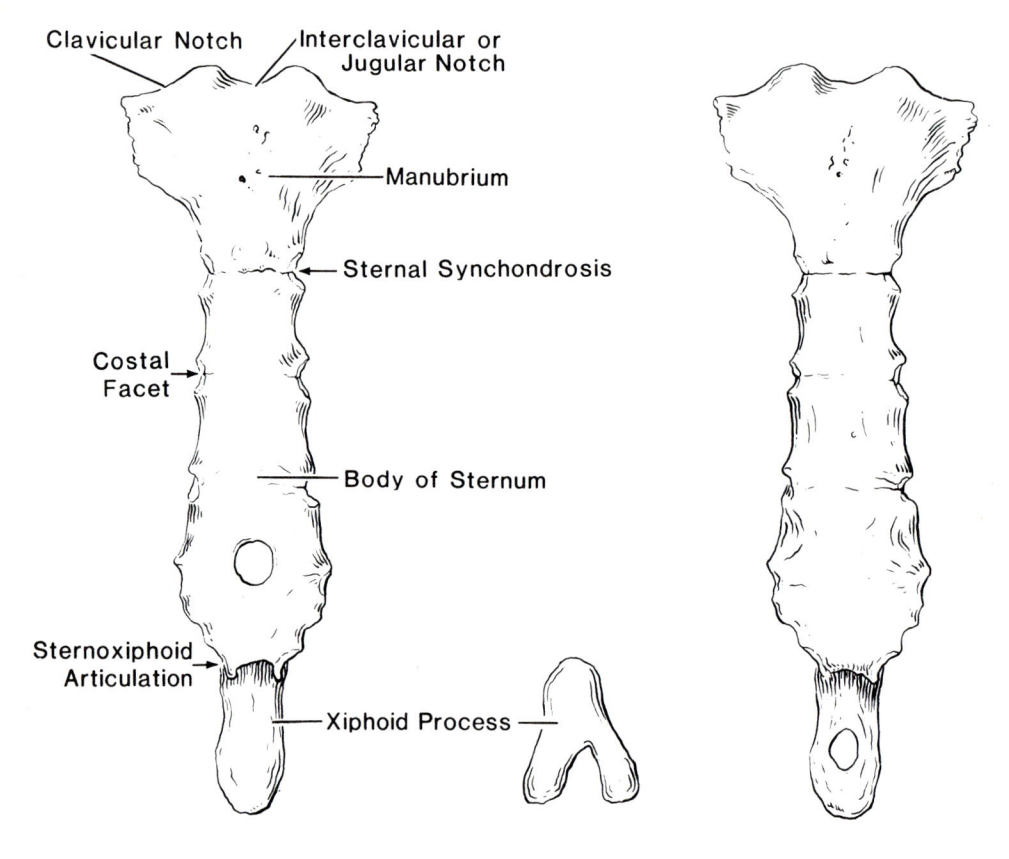

FIG. 46.
Variations in Sternum.
Suprasternal Bones.
Bifid Xiphoid Process.
Apertures in Body of Sternum and in Xiphoid Process.
Redrawn from various sources.

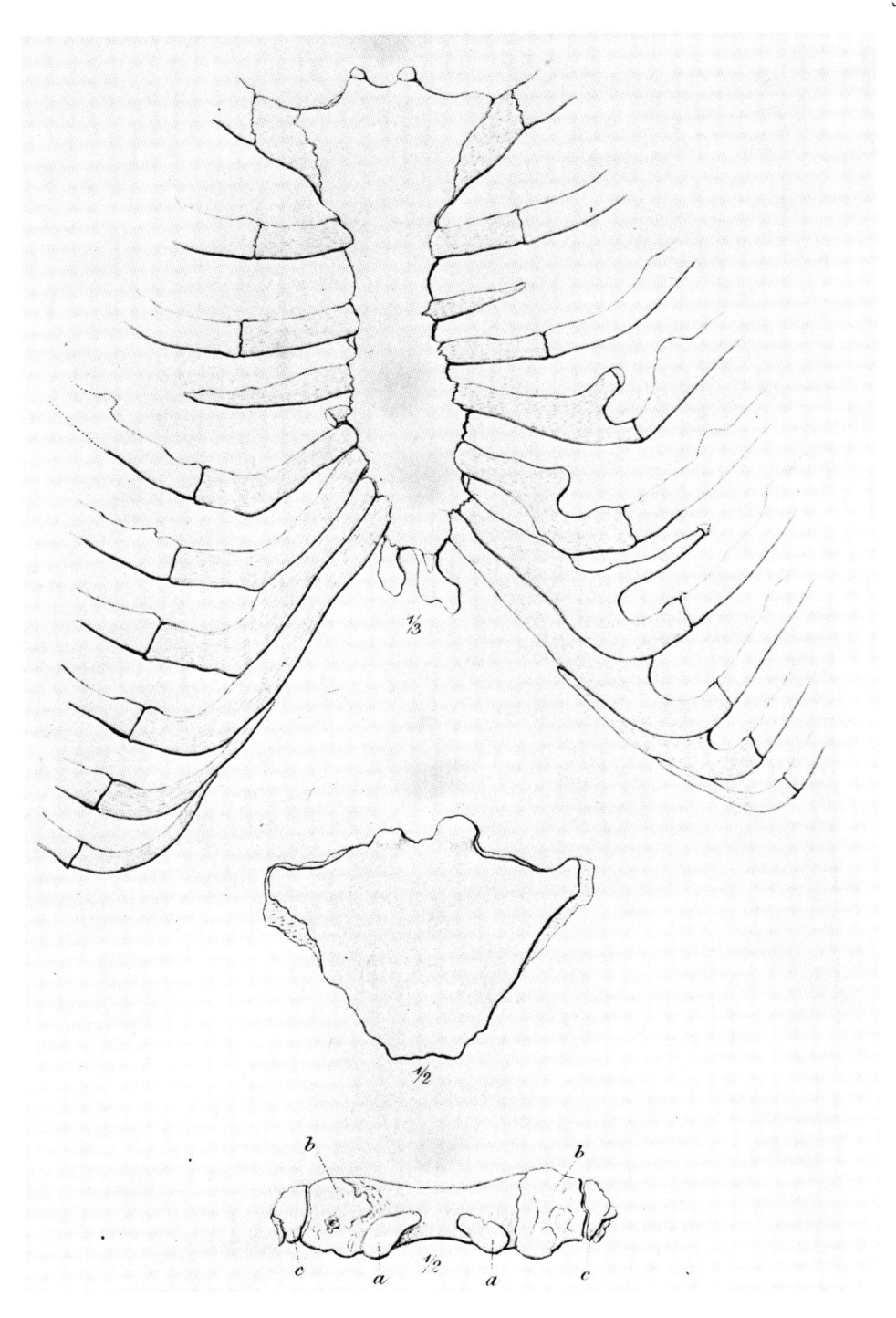

FIG. 47.
Suprasternal Bones.
Rib Anomalies.
Duplication of Xiphoid Process.

The suprasternal bones may articulate with the manubrium of the sternum or may become assimilated by the manubrium.

From Pansch, A. Über Anomalien am Thoraxskelete. *Arch. Anat. Physiol. Wissen Med.* 1875, pp. 552–564.

Human Anatomic Variation

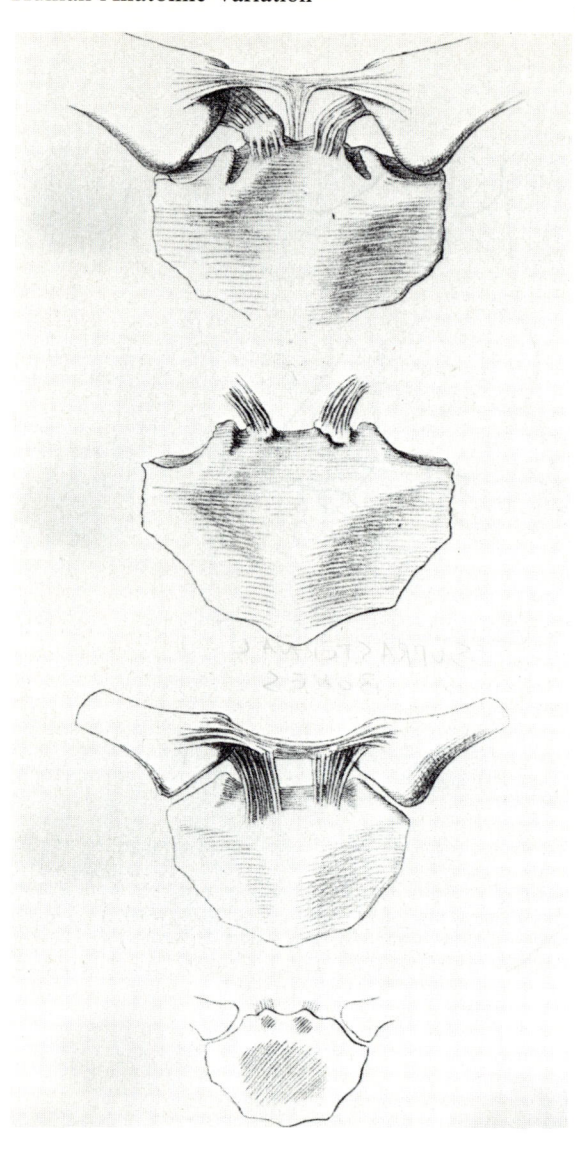

◀ FIG. 48.

Variations in Suprasternal Bones.

The drawings show variations in the suprasternal bones and in the ligaments associated with the bones or the region when the bones are absent.

From Carwardine, T. The suprasternal bones in man. *J. Anat. Physiol.* 27:232–234, 1893.

FIG. 49.

Fused First and Second Ribs.

C^1, First rib; C^2, second rib; l, ligament; c^1, c^2, cartilage; m, manubrium sterni.

The first and second ribs are fused at their distal end. The fused end of the first rib is joined to the manubrium through a ligament joined to cartilage. The extension of the fused end of the second rib is by cartilage.

From Aeby, C. Seltene Rippenanomalie des Menschen. *Arch. Anat. Physiol. Wissen Med.* 1868, pp. 68–71.

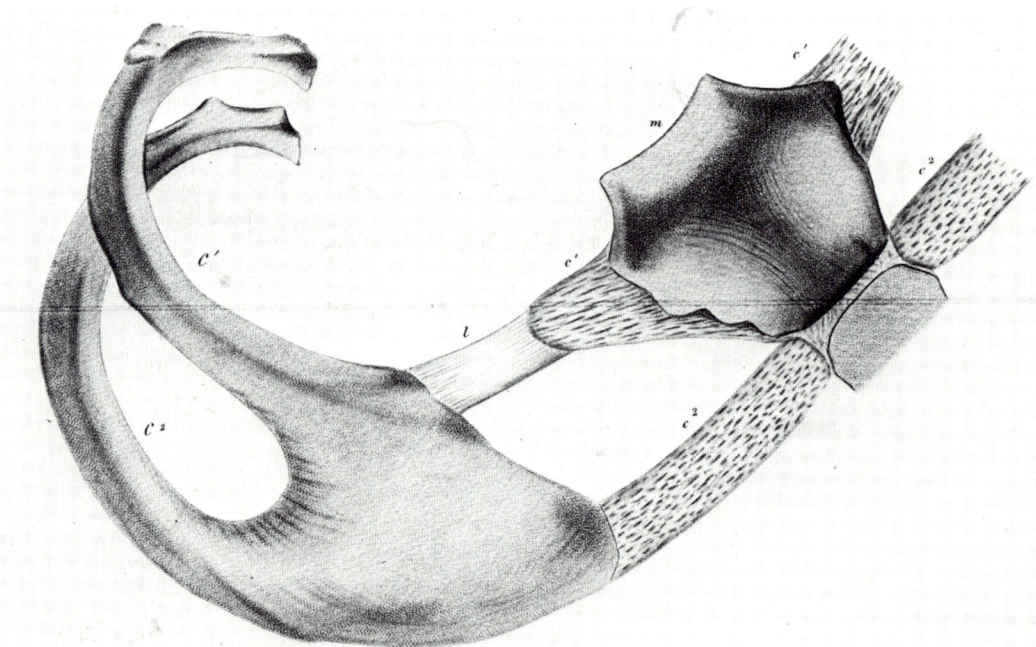

▼

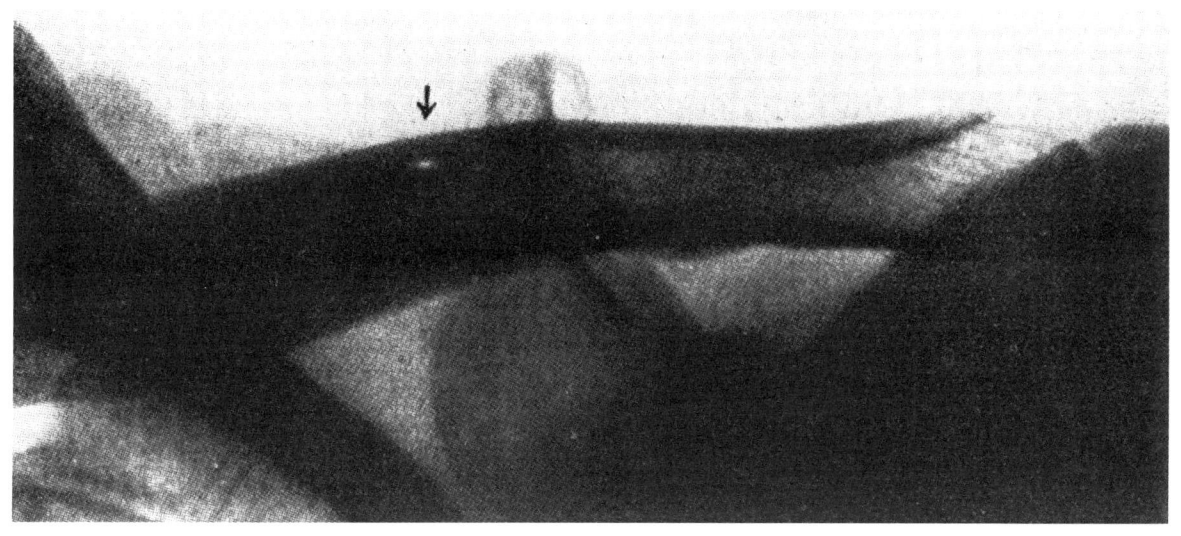

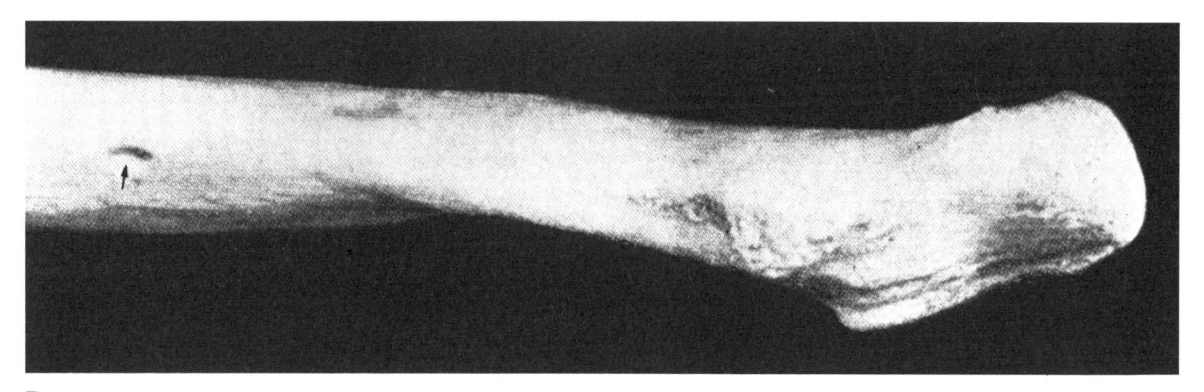

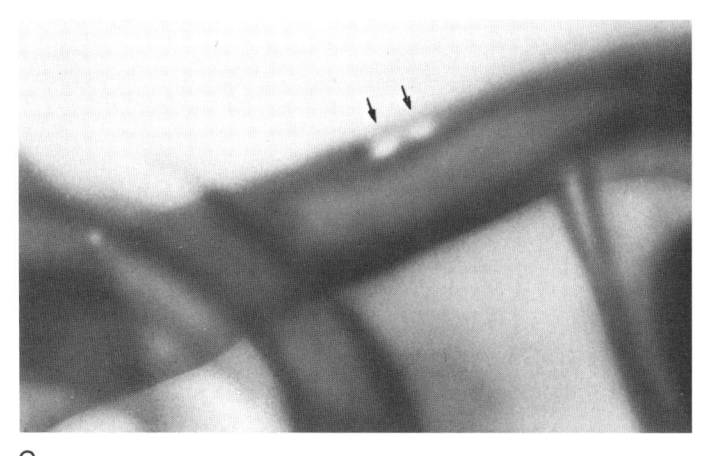

Skeleton

FIG. 50.
Clavicular Canals for Supraclavicular Nerves.

A, B: Canal for middle supraclavicular nerves (*single arrow*).
C: Double canal for middle supraclavicular nerves (*double arrows*).

From Skarby, H.G. Das Foramen nervi clavicularis im Röntgenbild. *Acta Radiol.* 17:397–402, 1936; and from Pahl, R. Doppelter Nervenkanal der Klavikula als diagnostische Fehlerquelle (Zugleich ein Beitrag über das Foramen nervi supraclavicularis.) *Fortschr. Röntgenstr.* 82:487–491, 1955.

Human Anatomic Variation

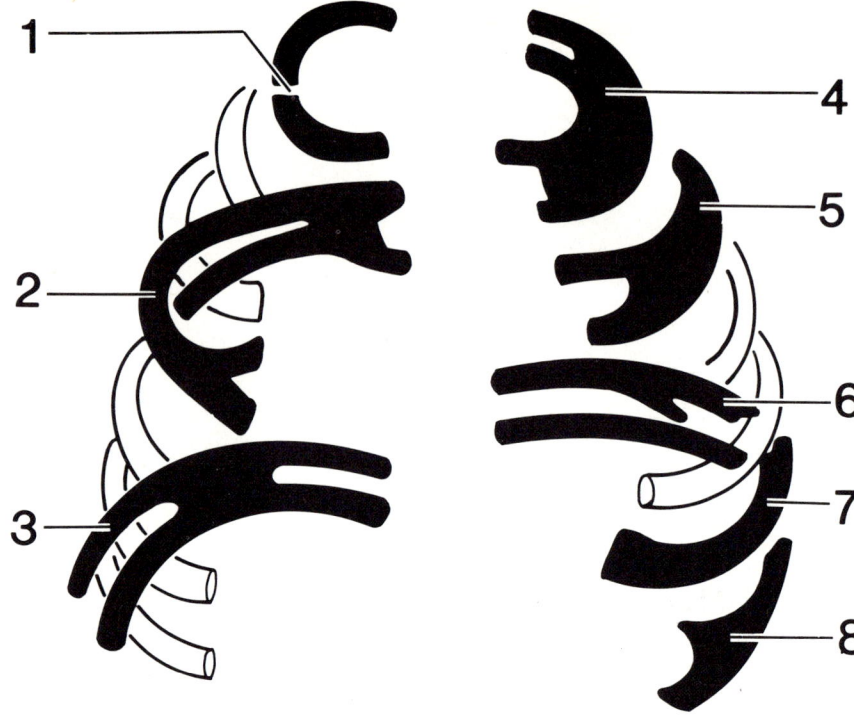

FIG. 51.
Radiographic Images of Some Rib Anomalies.
1, Discontinuity of the first rib; *2*, bridge formation posteriorly, forked rib anteriorly; *3*, costal bridge; *4*, bridge-shaped fusion; *5*, fusion dorsally; *6*, suggestion of costal bridging; *7*, bifurcation suggested; *9*, Luschka's bifurcated rib.

Redrawn from Zimmer, E.A. *A. Köhler and E.A. Zimmer Borderlands of the Normal and Early Pathologic in Skeletal Roentgenology,* 3rd American Ed., translated and edited by S.P. Wilk. Grune & Stratton, New York, 1968.

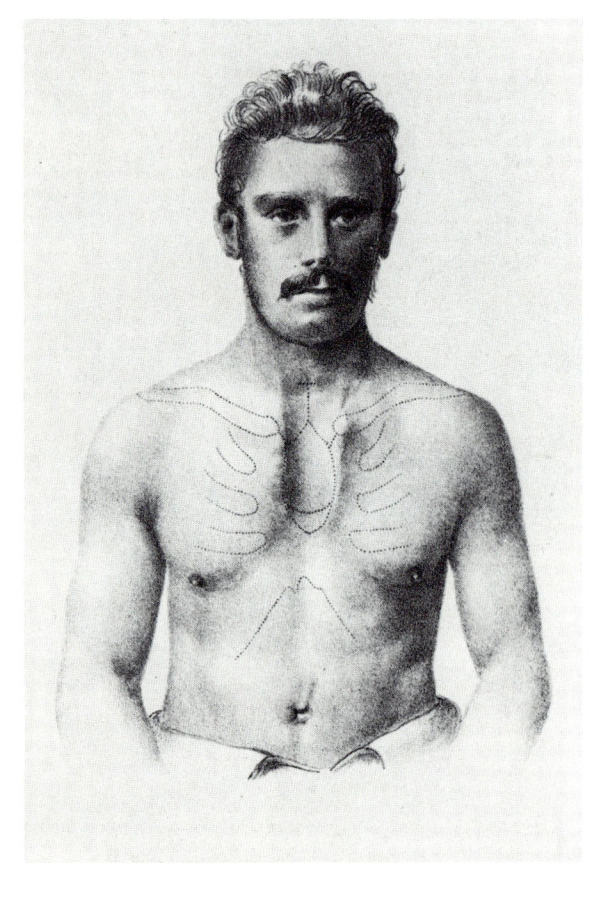

FIG. 52.
Fissura Sterna.
From Gibson, G.A. and H. Malet. Presternal fissure uncovering the base of the heart. *J. Anat. Physiol.* 14:1–9, 1879.

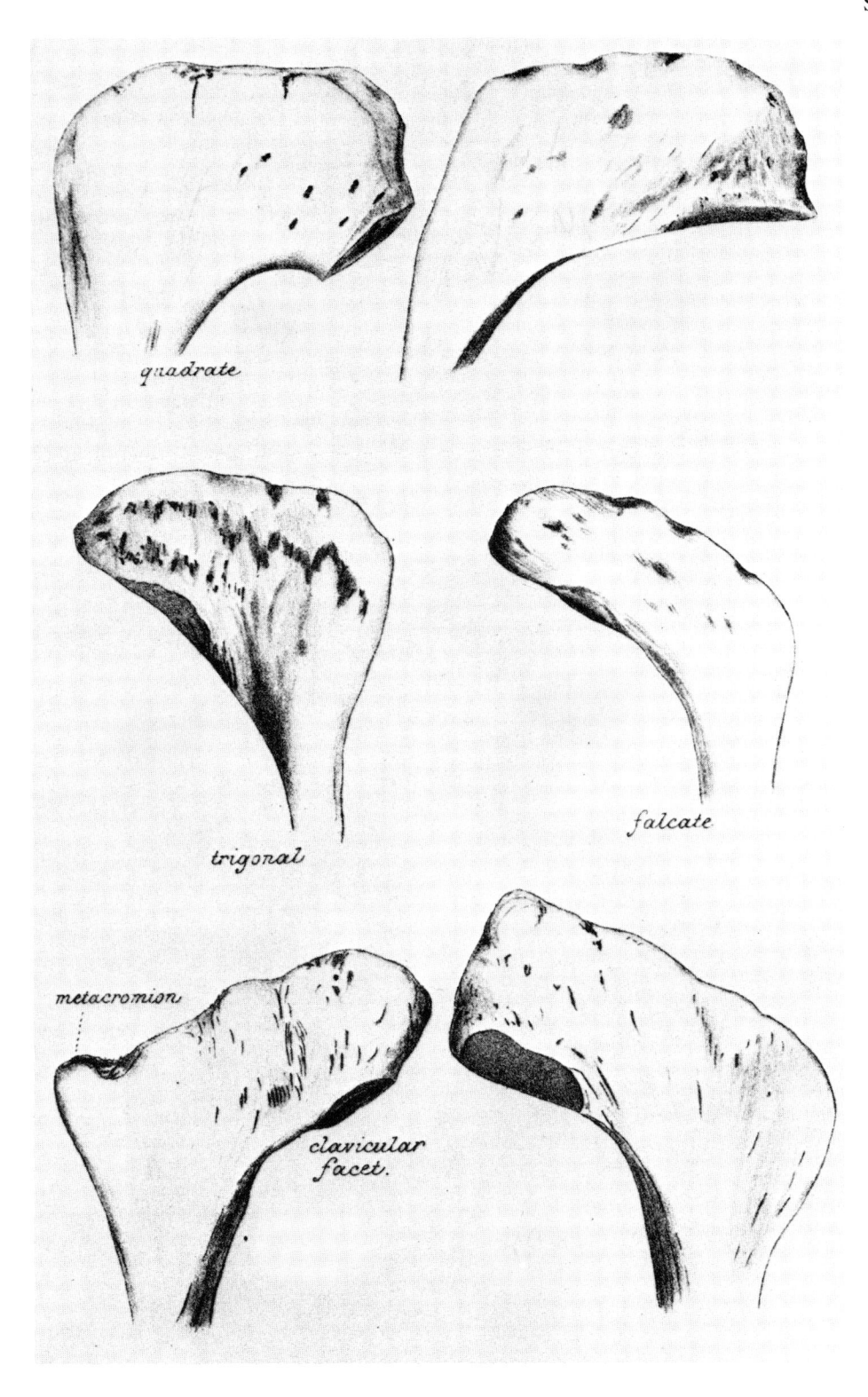

FIG. 53.
Various Types of Acromion Process.

From Macalister, A. Notes on acromion. *J. Anat. Physiol.* 27:245–251, 1893.

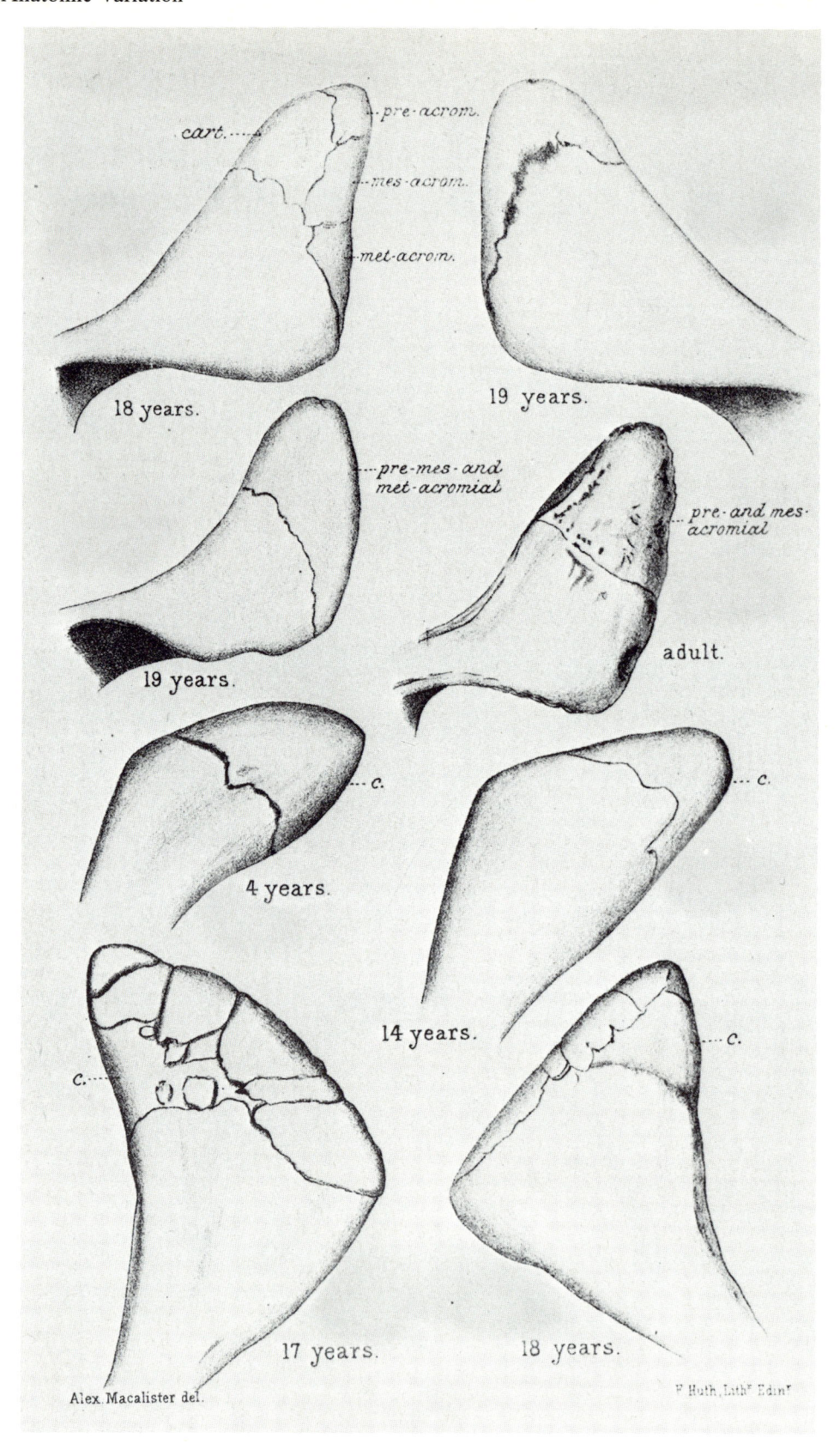

FIG. 54.
Ossification of Acromion Process.
c, Cartilaginous portion.

From Macalister, A. Notes on the acromion. *J. Anat. Physiol.* 27:245–251, 1893.

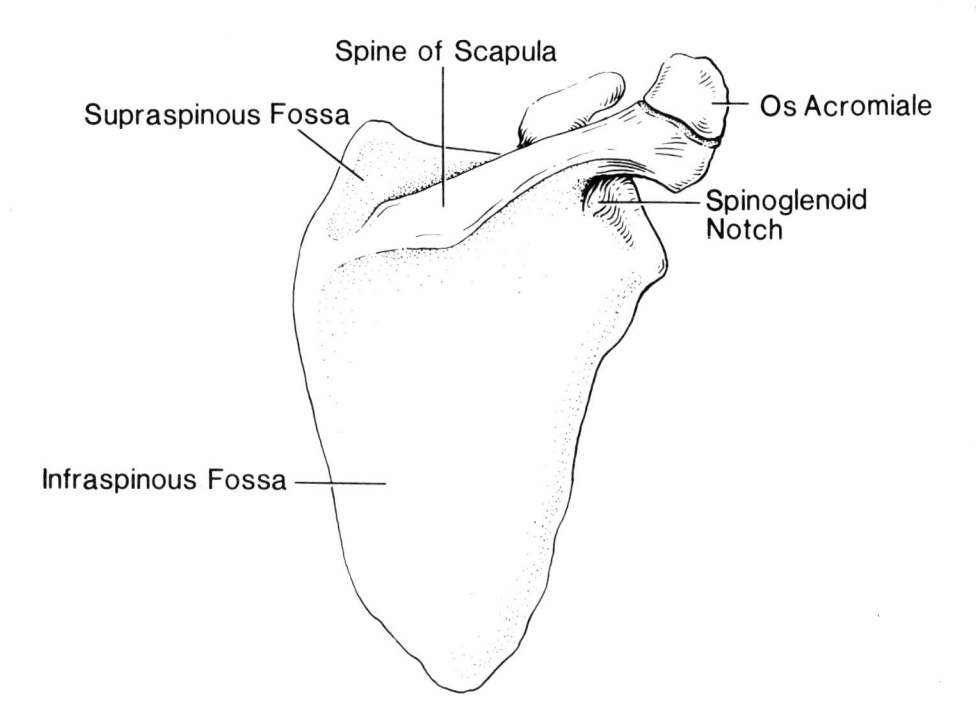

FIG. 55.
Os Acromiale.

Unfused acromial epiphysis is present in about 8% of specimens and it is found, more commonly, unilaterally. The drawing shows the dorsal aspect of the scapula.

Redrawn and modified from Zimmer, E.A. *A. Köhler and E.A. Zimmer Borderlands of the Normal and Early Pathologic in Skeletal Roentgenology*, 3rd American Ed., translated and edited by S.P. Wilk. Grune & Stratton, New York, 1968.

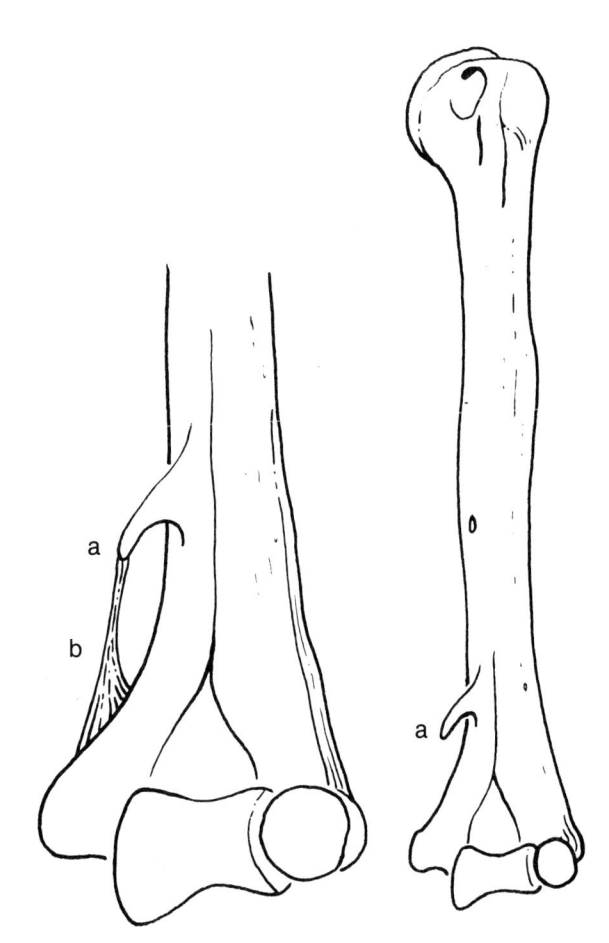

FIG. 56.
Supracondylar Process.

a,a, Supracondylar process; b, ligament completing the arch.

Authors' note: See Struthers, references 967 and 969; Calori, 144; Gantert and Alzheimer, 310; Ganz, 311; Gruber, 373; and Terry, 1004, 1005.

From Struthers, J. On some points in the abnormal anatomy of the arm. *Br. Foreign Medico-Chir. Rev.* 13:523–533, 1854.

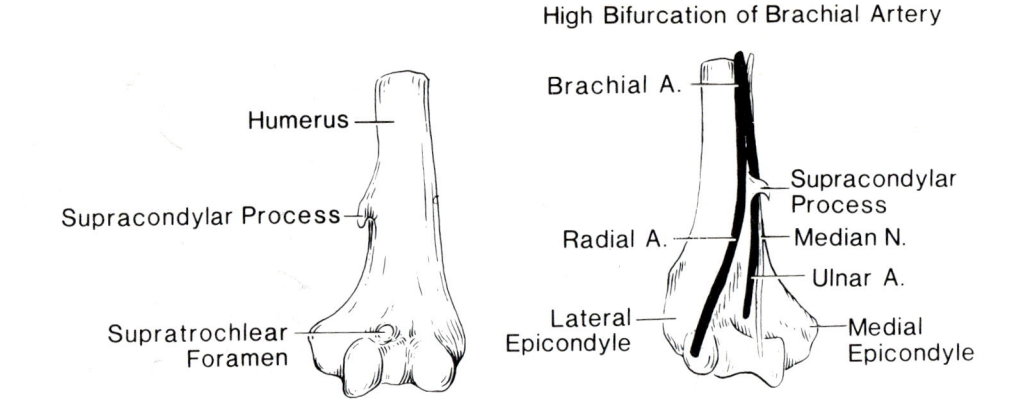

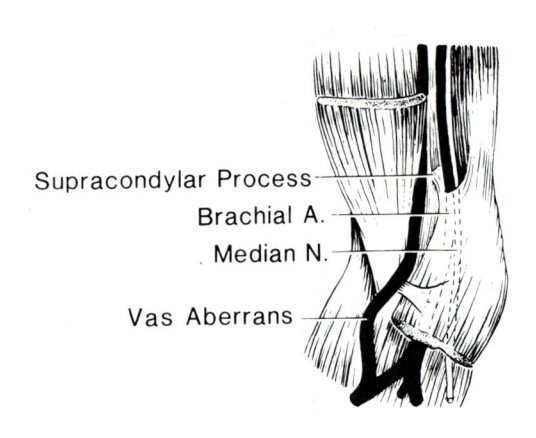

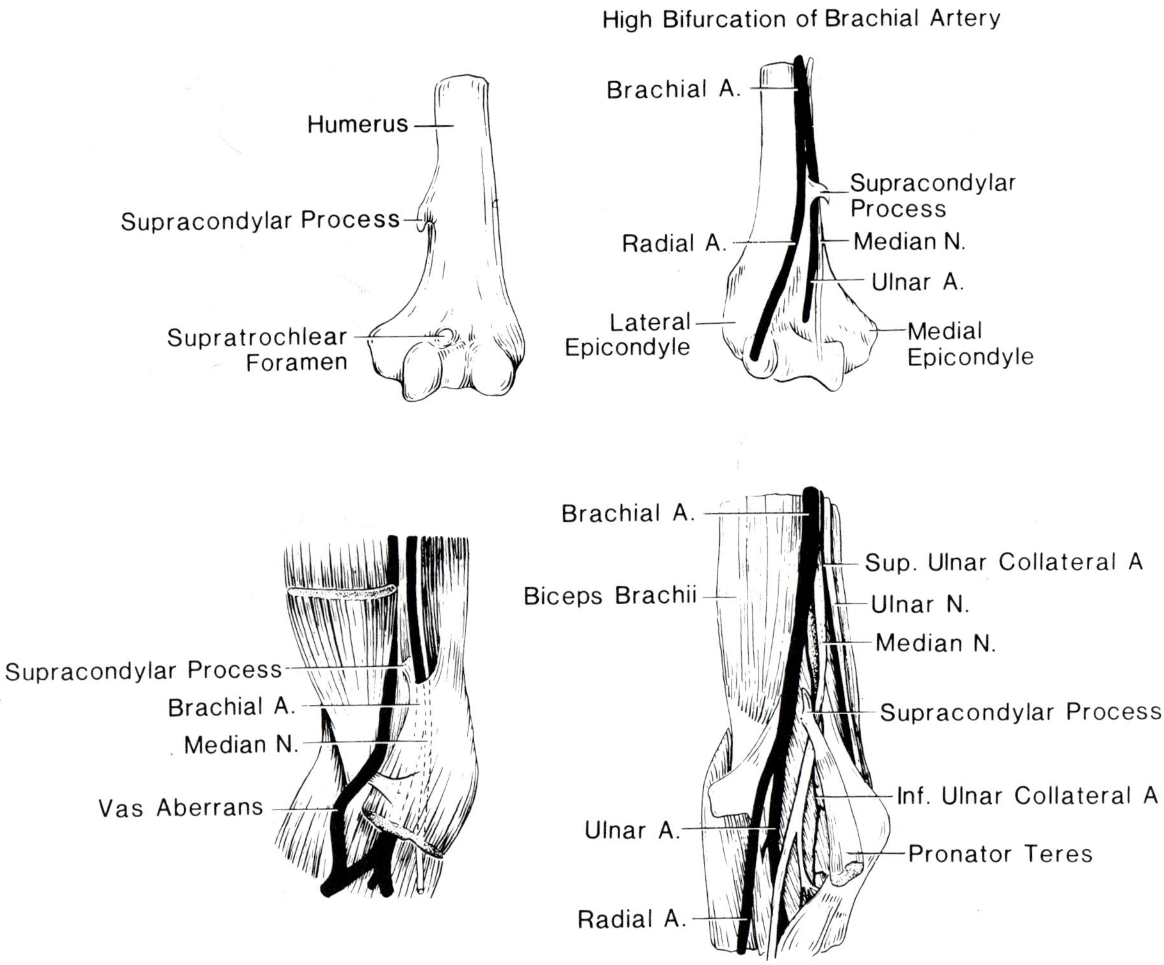

Human Anatomic Variation

FIG. 57.
Supracondylar Process.
Supratrochlear Foramen.

Present in 7 of 1000 subjects, a fibrous band joins the supracondylar process with the medial epicondyle forming a foramen through which the median nerve and brachial artery may pass.

A supratrochlear foramen when present may be closed in life by a membrane. It is more common on the left than the right and more common in females than males.

Authors' note: See also Struthers, references 967, 969; Calori, 144; Gantert and Alzheimer, 310; Ganz, 311; and Gruber, 373.

Redrawn from Terry, R.J. A study of the supracondyloid process in the living. *Am. J. Phys. Anthropol.* 4:129–139, 1921; Terry, R.J. New data on the incidence of the supracondyloid variation. *Am. J. Phys. Anthropol.* 9:265–270, 1926; and Trotter, M. Septal apertures in the humerus of American whites and Negroes. *Am. J. Phys. Anthropol.* 19:213, 1934.

Skeleton

FIG. 58.

Supracondylar Process.
(Supernumerary Belly of M. Biceps Brachii).
(Supernumerary Head of M. Pronator Teres).

Left arm. *1*, Supracondylar process; *2*, m. deltoideus; *3*, m. teres major; *4*, m. latissimus dorsi; *5*, m. biceps brachii: *5'*, long head; and *5"*, short head; *6*, m. triceps brachii; *7*, m. coracobrachialis; *8*, m. brachialis; *9*, prominence of lateral group of muscles; *10*, m. pronator teres; *a*, supernumerary belly of the short head of biceps brachii: α,β, lateral and medial insertion points; *b*, supernumerary portion of pronator teres; γ, small head of pronator teres arising from supracondylar process; *c*, brachial artery; *d*, median nerve; *e*, ulnar nerve; *, brachial fascial aperture of canal in medial bicipital sulcus and anterior medial cubital sulcus for brachial artery and median nerve.

From Gruber, W. Ein Nachtrag zur Kenntnis des Processus supracondyloideus (*internus*) humeri des Menschen. *Arch. Anat. Physiol. Wissen. Med.* 1865, pp. 367–375.

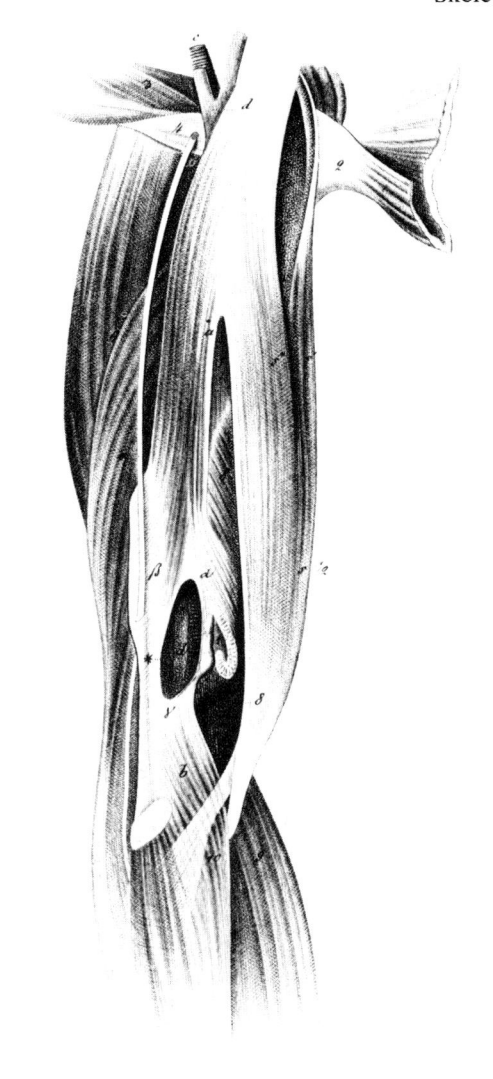

FIG. 59.

Sesamoid Bones of the Hand.

Sesamoid bones of the adult human hand. The sesamoid bones on the metacarpophalangeal joint are regularly present (100%). For those that are most commonly present (*circled black dots*) the percentages are given. More unusual locations, only occasionally found (*black dots*), are labeled *R*.

Authors' note: See also the study by R.F. Patterson, reference 772.

Redrawn from Degen, S.T. Über das Auftreten der Knochenkerne am Handskelett von der Geburt bis zur Reife. Mit einem Hinweis auf die Sesambeine der Hand. *Med. Klin.* 46:1330–1332, 1951.

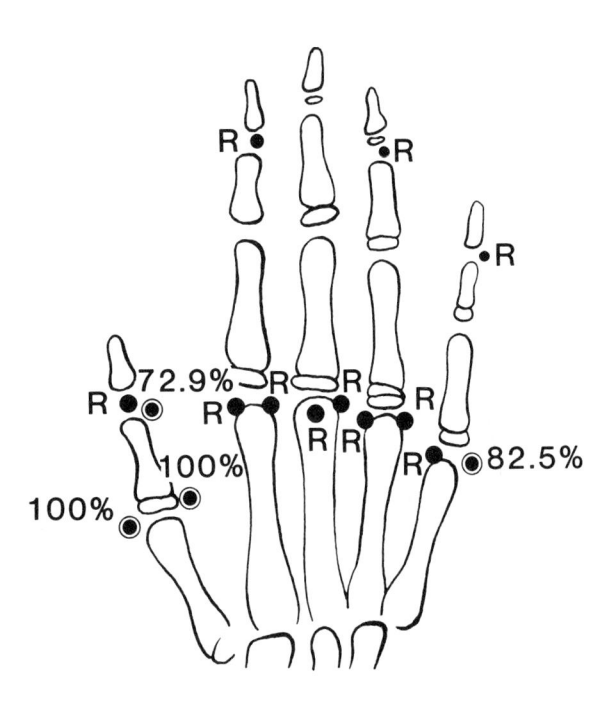

581

Human Anatomic Variation

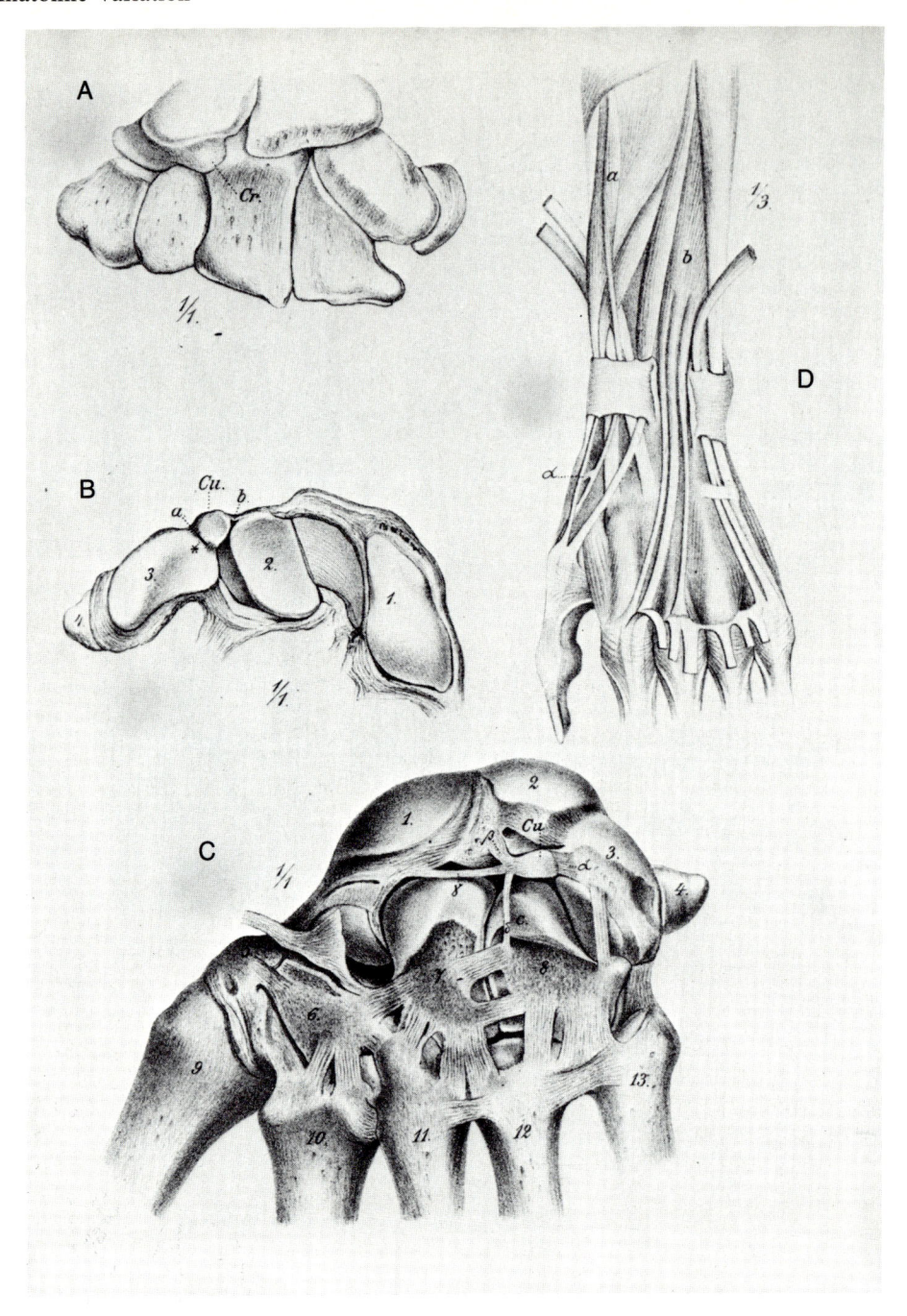

FIG. 60.
Os Centrale of Carpus.
(M. Radialis Digiti I. s. Pollicis.)

A: Os centrale, carpi radialis (Cr.).
B, C: 1, Scaphoid; 2, lunate; 3, triquetrum; 4, pisiforme; 5, trapezium; 6, trapezoid; 7, capitate; 8, hamate; 9–13, metacarpals, I through V; Cu, os centrale, carpi ulnari; a, capsule triquetrocentralis; b, articulation between lunate and centralis; α, dorsal ligament of triquetrocentralis articulation; β, dorsal lunatocentralis ligament; γ, scaphoidocentralis fibrous fascicle; δ, hamatocentralis fibrous fascicle.

D: a, M. radialis digiti primi s. l. s. pollicis s. extensor pollicis brevis accessorius; α, branch of its tendon of insertion to base of first metacarpal; b, three branches of mm. extensor digiti indicis and medii proprius.

From Gruber, W. Bericht über neue Funde des Os centrale carpi radiale (mihi) beim Menschen. Arch. Pathol. Anat. Physiol. Klin. Med. 98:396–402, 1884; Über das Os centrale carpi ulnare (mihi) bei dem Menschen und über das muthmaßliche homologe Carpalstück bei den Amphibien. Ibid. 98:402–408, 1884; and Neuer Musculus radialis digiti I. s. pollicis. Ibid. 98:422–424, 1884.

Skeleton

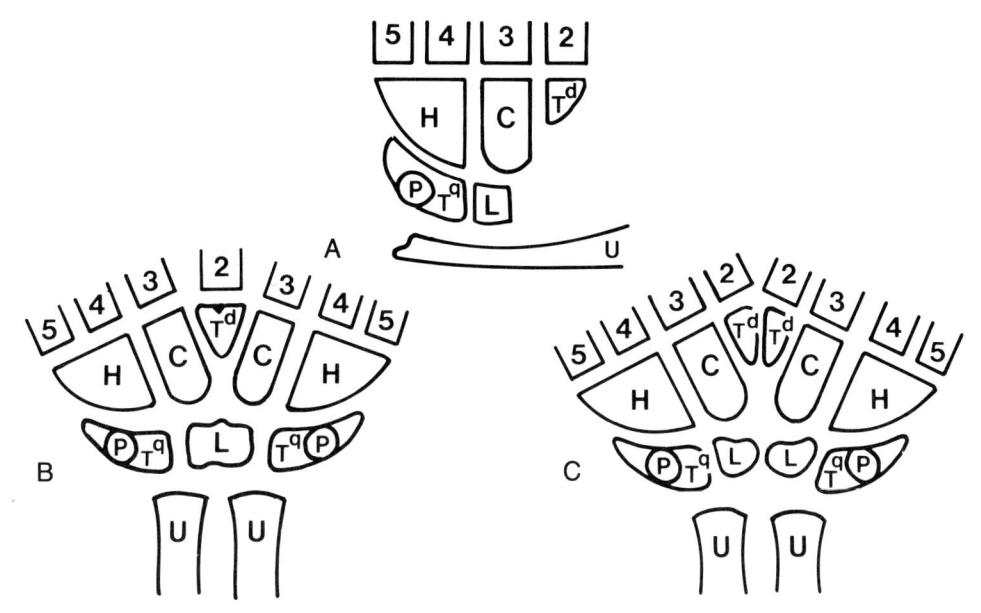

FIG. 61.
Extensive Carpal Anomalies.

A: Shows typical carpus in radial hemimelia (absence of the radial side of forearm and hand). The radius, scaphoid, trapezium, first metacarpal, and thumb are absent. The ulna (U), lunate (L), triquetrum (T^q), pisiforme (P), trapezoid (T^d), capitate (C), hamate (H), and second to fifth metacarpals and fingers are present. Note the radial deviation of the hand (manus vara).

B: Ulnar dimelia (duplication of ulnar side of limb), type I, with double radial hemimelia (neither ulna has a radius), and single index finger. Note that the same bones are absent as in the preceding case, but that the bones present in the preceding case are duplicated here and are in mirror-image formation.

C: Ulnar dimelia, type II, with double radial hemimelia, as above, and duplication of the index finger.

Redrawn from and used by permission of O'Rahilly, R. A survey of carpal and tarsal anomalies. *J. Bone Joint Surg. [Am.]* 35:626–642, 1953.

FIG. 62.
Carpal Accessoria.
Carpal Fusions.

A: Scheme of accessory carpal ossicles. Palmar aspect of the right hand; the more dorsally situated ossicles are shown in broken outline. From this diagram the names of the ossicles in a given region of the hand can be determined; the descriptions of these ossicles provided in O'Rahilly's text may then be used as a guide in the differential diagnosis of a given case.

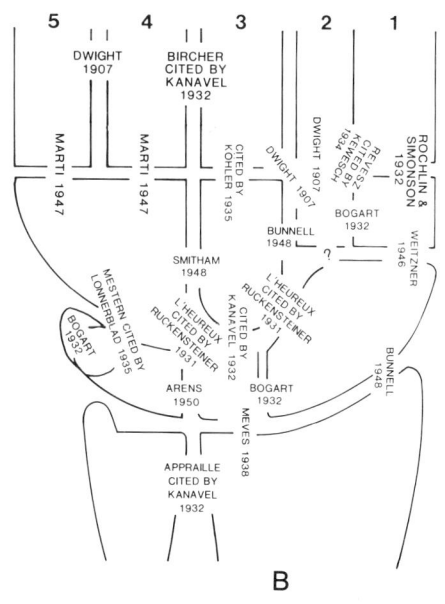

B: Scheme of right carpus, anterior aspect, showing some recent cases of carpal coherence. The name of the author reporting the case and the date of publication are placed across the fused bones in each case.

Redrawn from and used by permission of O'Rahilly, R. A survey of carpal and tarsal anomalies. *J. Bone Joint Surg. [Am.]* 35:626–642, 1953.

Human Anatomic Variation

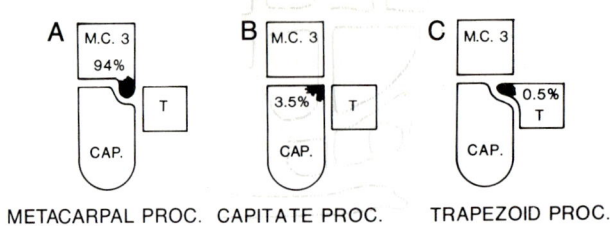

Metacarpostyloid Variations

A METACARPAL PROC. (94%)
B CAPITATE PROC. (3.5%)
C TRAPEZOID PROC. (0.5%)
D OS STYLOIDEUM (2%)
E "BILOCATION"

Calcaneonavicular Variations

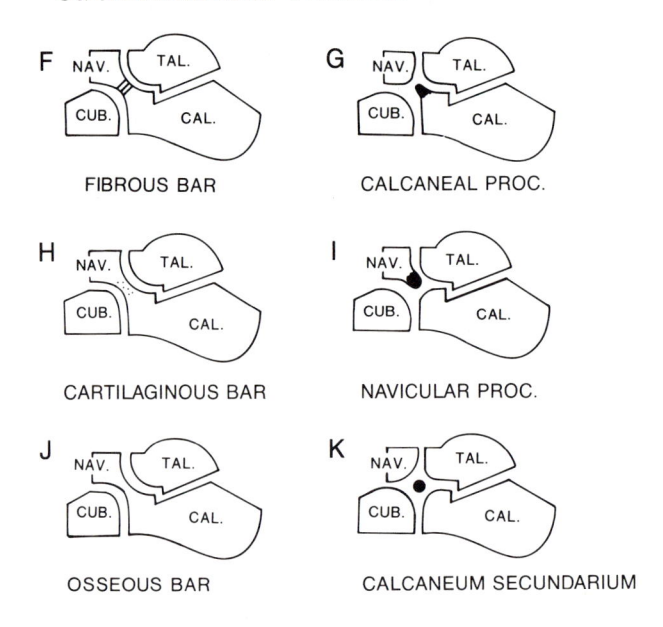

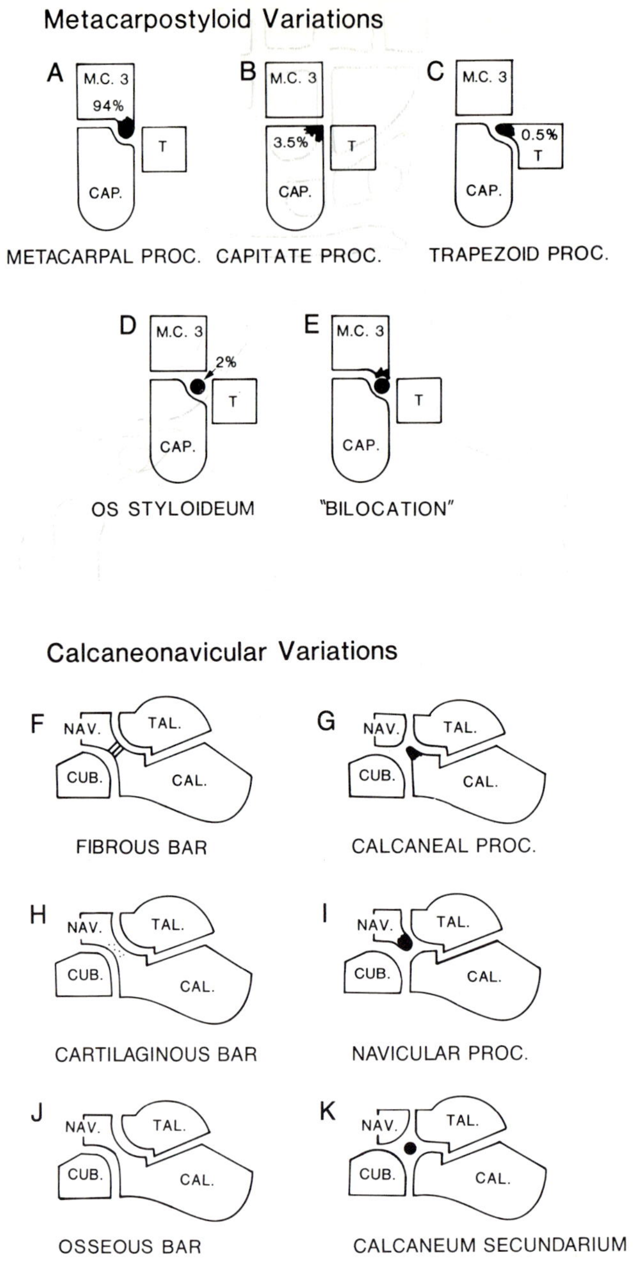

F FIBROUS BAR
G CALCANEAL PROC.
H CARTILAGINOUS BAR
I NAVICULAR PROC.
J OSSEOUS BAR
K CALCANEUM SECUNDARIUM

FIG. 63.
Metacarpostyloid Variations.
Calcaneonavicular Variations.

A–E: Variations of the processus styloideus of the third metacarpal bone. The third metacarpal (M.C.3) styloid process (black) is an integral part of the metacarpal in 94% of cases (A). It may be found, however, as (B) a portion of the capitate in 3.5%; as (C) a portion of the trapezoid (T) in 0.5%; or (D) as a free ossicle (os styloideum) in 2%. In some cases a separate ossicle may occur together with a distinct process (on the third metacarpal, for example, as in E) in the one hand.

A'–F': Variations at the site of the calcaneonavicular bar (modified from Seddon). On the left side the calcaneum and navicular are represented as united respectively by A', fibrous; B', cartilaginous; and C', bony tissue. On the right side are shown D', a prominent process on the calcaneum; E', a prominent process on the navicular; and F', a separate calcaneonavicular ossicle (calcaneum secundarium).

In a similar way a talocalcaneal bridge or a free ossicle (os sustentaculi) may occur between the talus and the sustentaculum tali of the calcaneum.

Redrawn from and used by permission of O'Rahilly, R. A survey of carpal and tarsal anomalies. J. Bone Joint Surg. [Am.] 35:626–642, 1953.

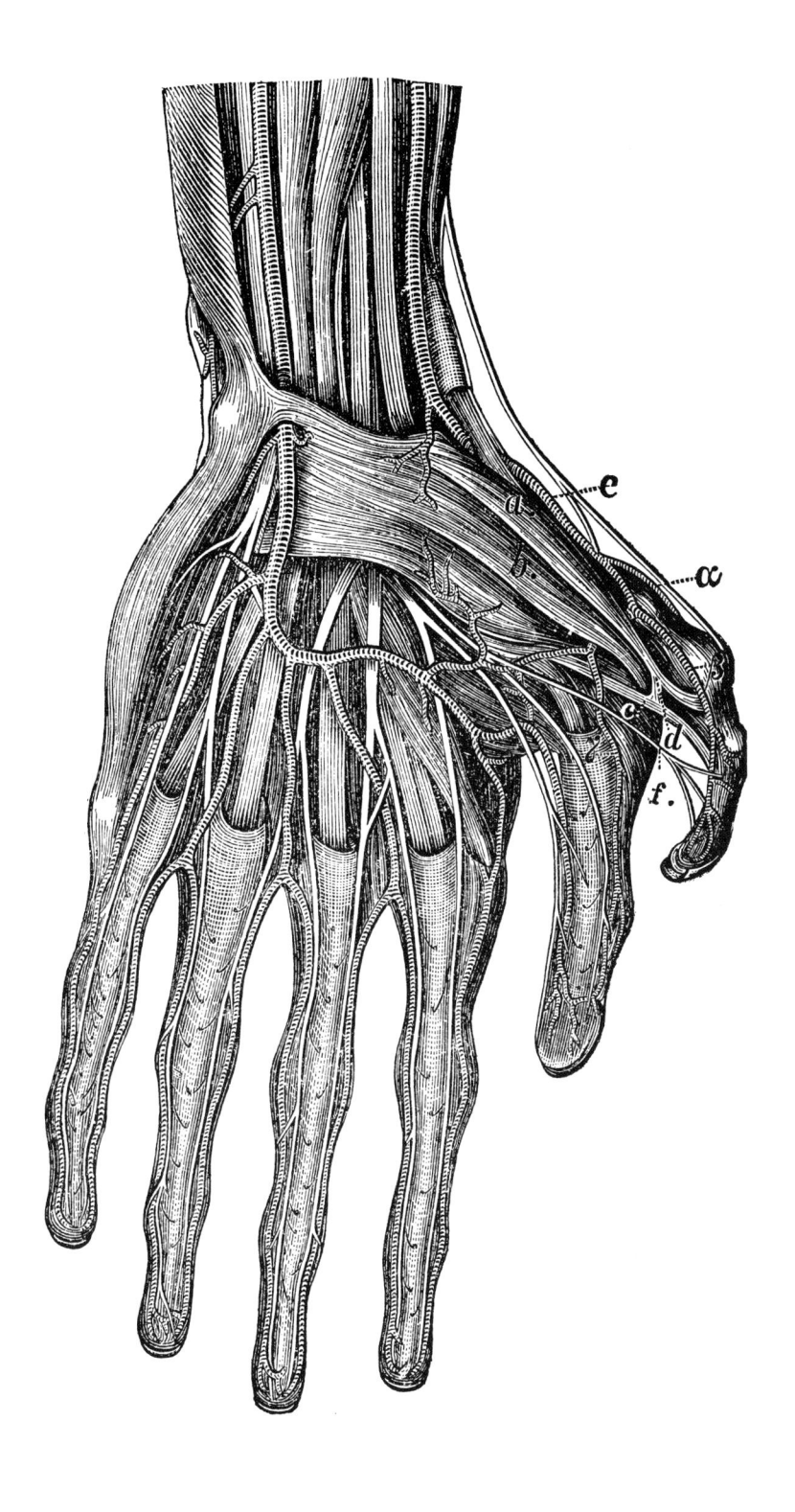

FIG. 64.
Supernumerary Thumb.
(Supernumerary M. Flexor Pollicis Brevis).
(Supernumerary Median Nerve Branch).

a, Supernumerary flexor pollicis brevis; b, flexor pollicis brevis; c, tendinous slip from flexor pollicis longus to supernumerary thumb; d, tendinous slip between supernumerary flexor pollicis brevis and flexor pollicis longus; e, supernumerary branch of radial artery: α, branch to dorsum of supernumerary thumb, β, branch to volar aspect of supernumerary thumb, and γ, normal volar branch; f, supernumerary branch from median nerve.

From Gruber, W. Zergliederung eines linken Armes mit Duplicität des Daumens. *Bull. Acad. Imp. Sci. St. Petersbourg* 17:24–31, 1872.

Human Anatomic Variation

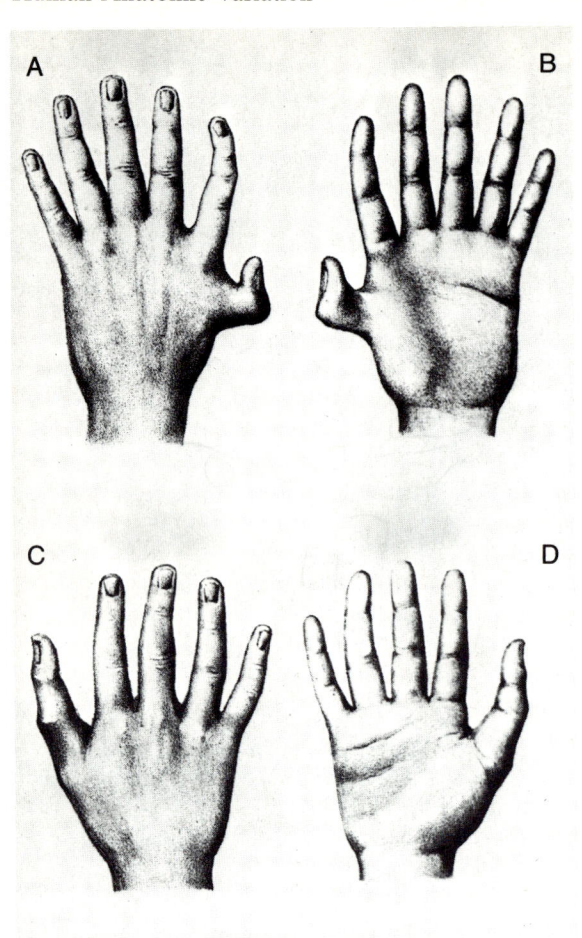

FIG. 65.
Three-Phalangeal Thumb.
Hexadactyly.

A: Hexadactyly, left hand, dorsal aspect.
B: Hexadactyly, left hand, palmar aspect.
C: Three-phalangeal thumb, right hand, dorsal aspect.
D: Three-phalangeal thumb, right hand, palmar aspect.

From Windle, B.C.A. Occurrence of an additional phalanx in the human pollex. *J. Anat. Physiol.* 26:100–116, 1892.

Septodactyly (manus) Octodactyly (pedis)

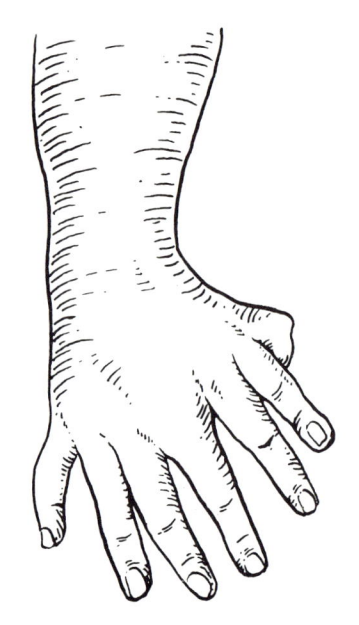

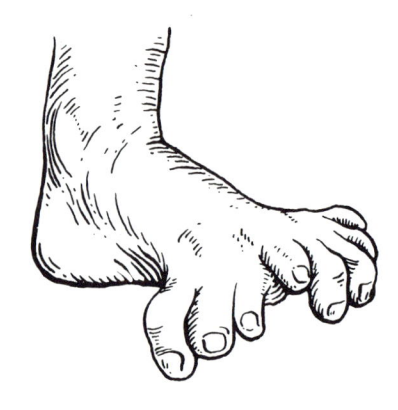

FIG. 66.
Septodactyly (Manus).
Octodactyly (Pedis).

Redrawn from Anatomie du vivant. *La Presse Medicale* No. 69. 30:1447, 1922.

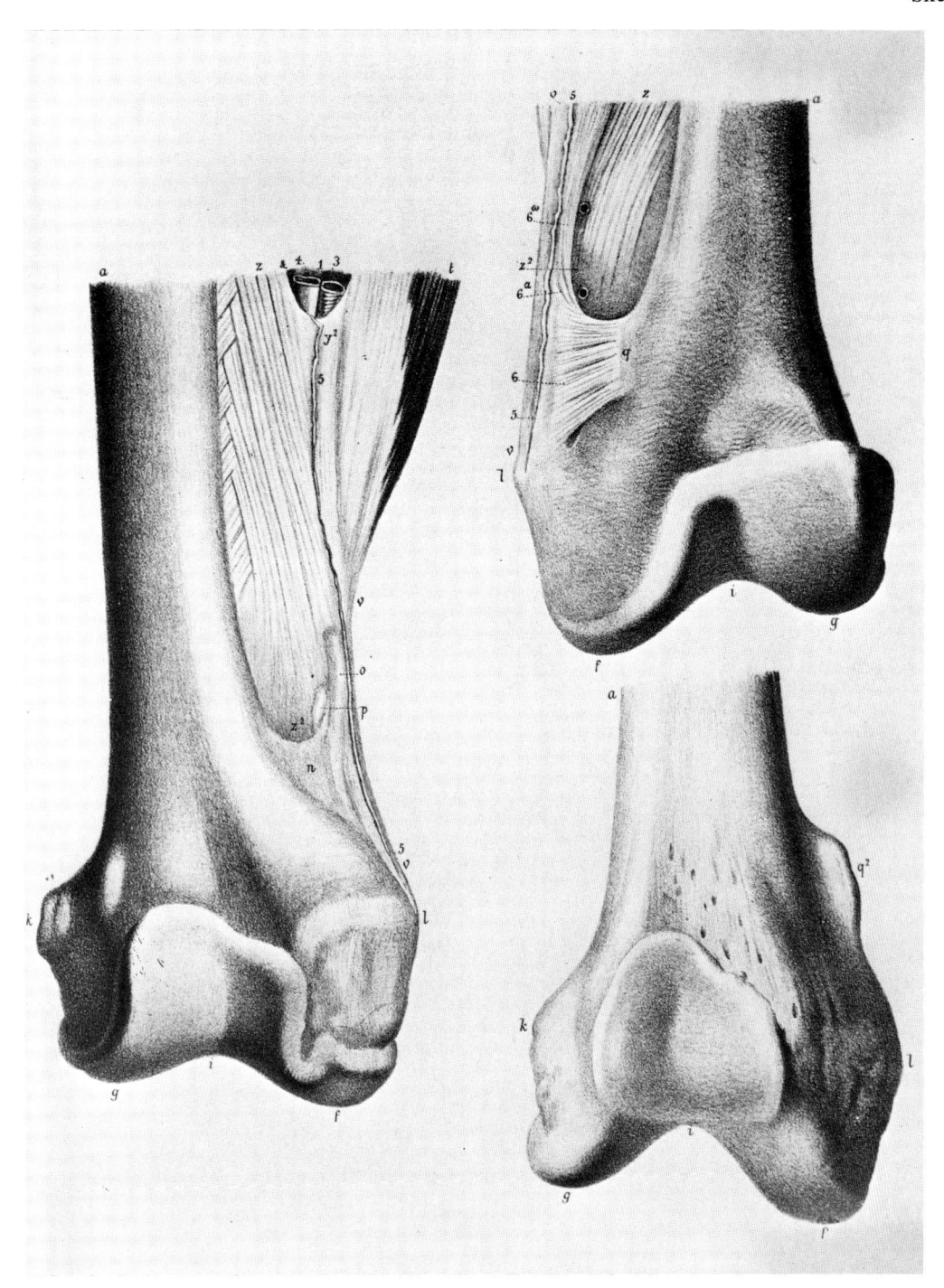

FIG. 67.
Medial Supracondylar Process of Femur.

a, Inferior portion of femur; f, medial condyle; g, lateral condyle; i, patellar facet; k, lateral condylar tubercle; l, medial condylar tubercle (adductor tubercle); n, medial supracondylar process, triangular portion; o, medial supracondylar process, long styloid portion; p, small styloid process; q, bony crest corresponding to the triangular portion; q^2, a large crest may be present; t, A portion of m. adductor magnus; v, insertion of m. adductor magnus; y^2, aponeurotic part of the tendon of insertion of m. adductor magnus; z,z^2, inferior portion of the medial intermuscular ligament adherent to the supracondyloid process; &,1, adductor canal ending at 2 through which courses the popliteal artery 3 and vein 4; 2, termination of the adductor canal; 3,4, popliteal artery and vein; 5, cut edge of the intermuscular ligament and a portion of the aponeurotic tendon of y^2; 6, transverse triangular ligament joining femur and styloid process; 6^a, 6^ω, superior end of the triangular ligament of 6 above continuing as the medial edge of the adductor canal.

From Calori, L. Intorno al processo sopracondiloideo interno del femore nei mammiferi e nell'uomo. *Mem. R. Accad. Sci. Istituto di Bologna S. 4* 4:585–592, 1882.

Human Anatomic Variation

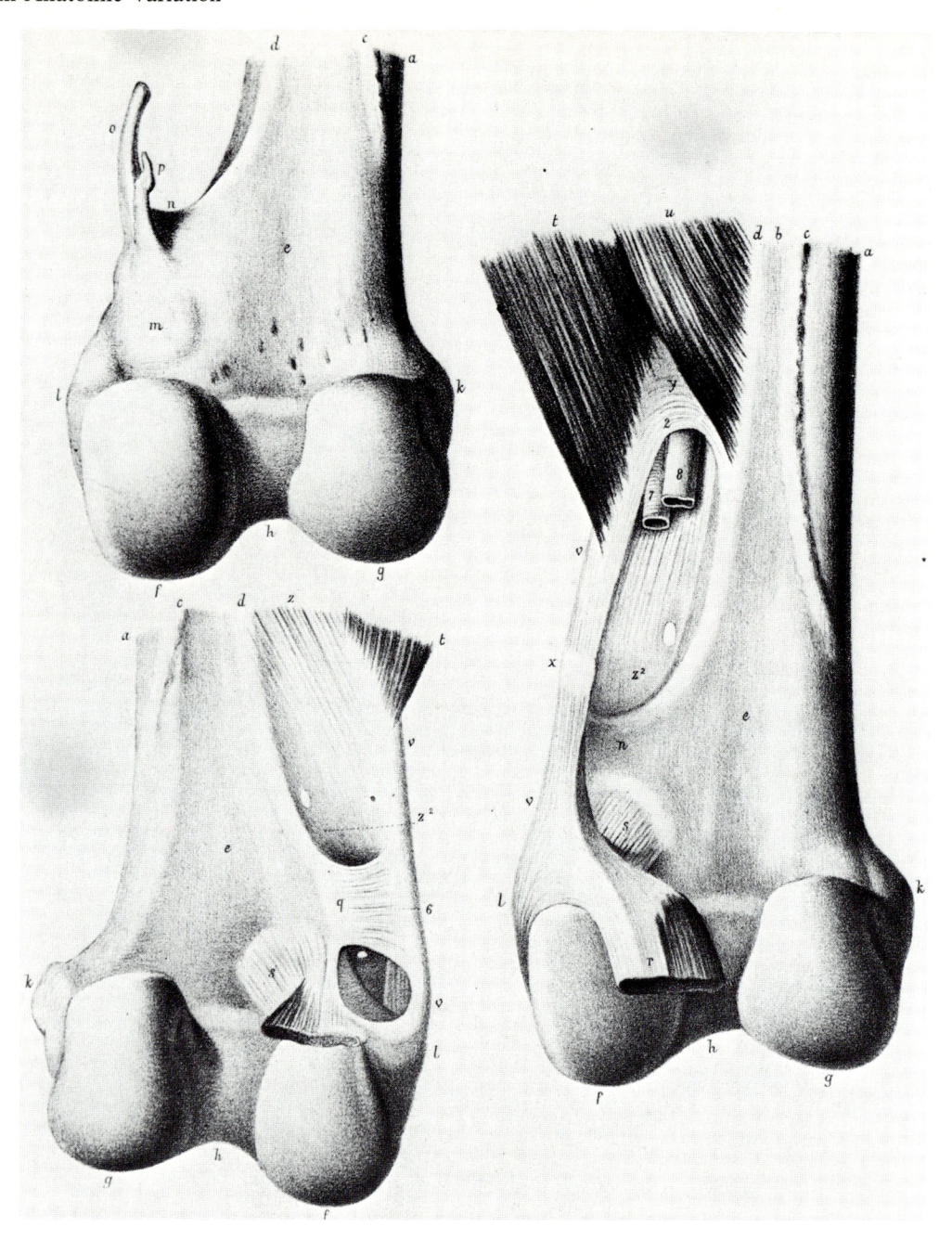

FIG. 68.
Medial Supracondylar Process of Femur.

a, Inferior portion of femur; *b*, linea aspera: *c*, lateral lip, and *d*, medial lip; *e*, popliteal floor; *f*, medial condyle; *g*, lateral condyle; *h*, intercondylar fissure or notch; *k*, lateral condylar tubercle; *l*, medial condylar tubercle; *m*, circular tubercle for attachment of medial head of gastrocnemius; *n,o*, medial supracondylar process: triangular element (*n*), and styloid process (*o*); *p*, small styloid process; *r*, tendon of insertion of medial head of gastrocnemius; *s*, cut edge of tendon of insertion of medial head of m. gastrocnemius; *t,u*, portion of adductor magnus; *v,v*, tendon of insertion of medial head of gastrocnemius; *x*, attachment of same tendon (*v,v*) to small styloid process; *y*, aponeurotic arch forming inferior extremity of adductor canal; z, z^2, inferior portion of the medial intermuscular ligament adherent to the supracondyloid process. 7, 8, popliteal artery and vein.

From Calori, L. Intorno al processo sopracondiloideo interno del femore nei mammiferi e nell'uomo. *Mem. R. Accad. Sci. Istituto di Bologna* S. 4 4:585–592, 1882.

FIG. 69.
Variations in Shapes of Patella Partita.
Radiographic images of patella partita. The last sketch is from a lateral film.
Redrawn from Zimmer, E.A. *A. Köhler and E.A. Zimmer Borderlands of the Normal and Early Pathologic in Skeletal Roentgenology,* 3rd American Ed., translated and edited by S.P. Wilk. Grune & Stratton, New York, 1968.

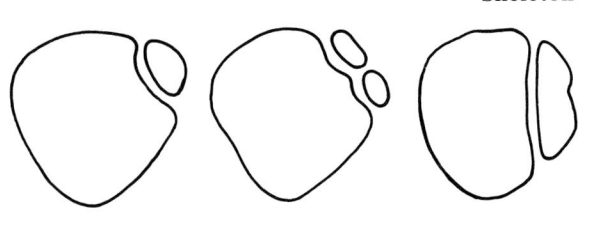

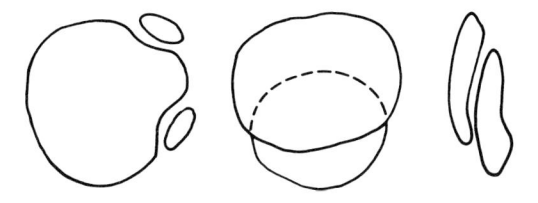

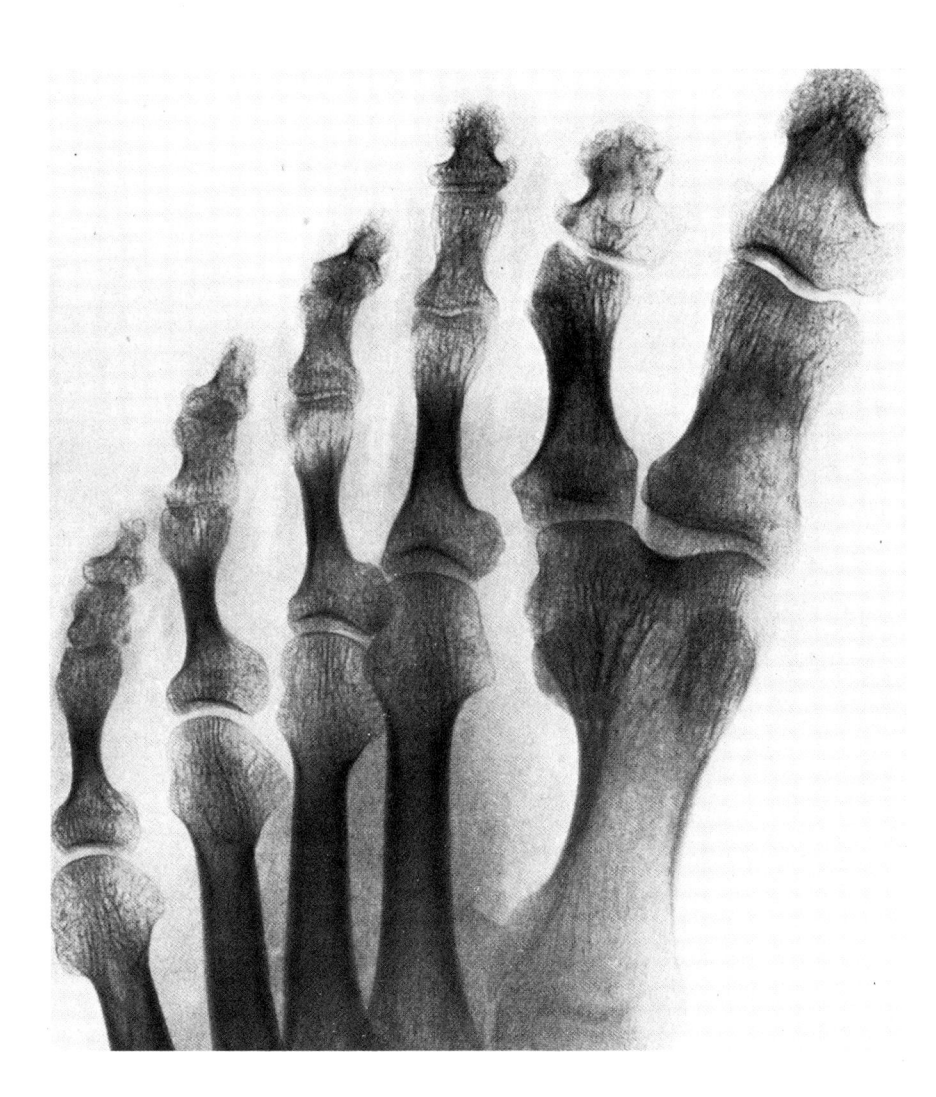

FIG. 70.
Duplication of the Great Toe.
From Zimmer, E.A. *A. Köhler and E.A. Zimmer Borderlands of the Normal and Early Pathologic in Skeletal Roentgenology,* 11th Ed., translated and edited by S.P. Wilk. Grune & Stratton, New York, 1968.

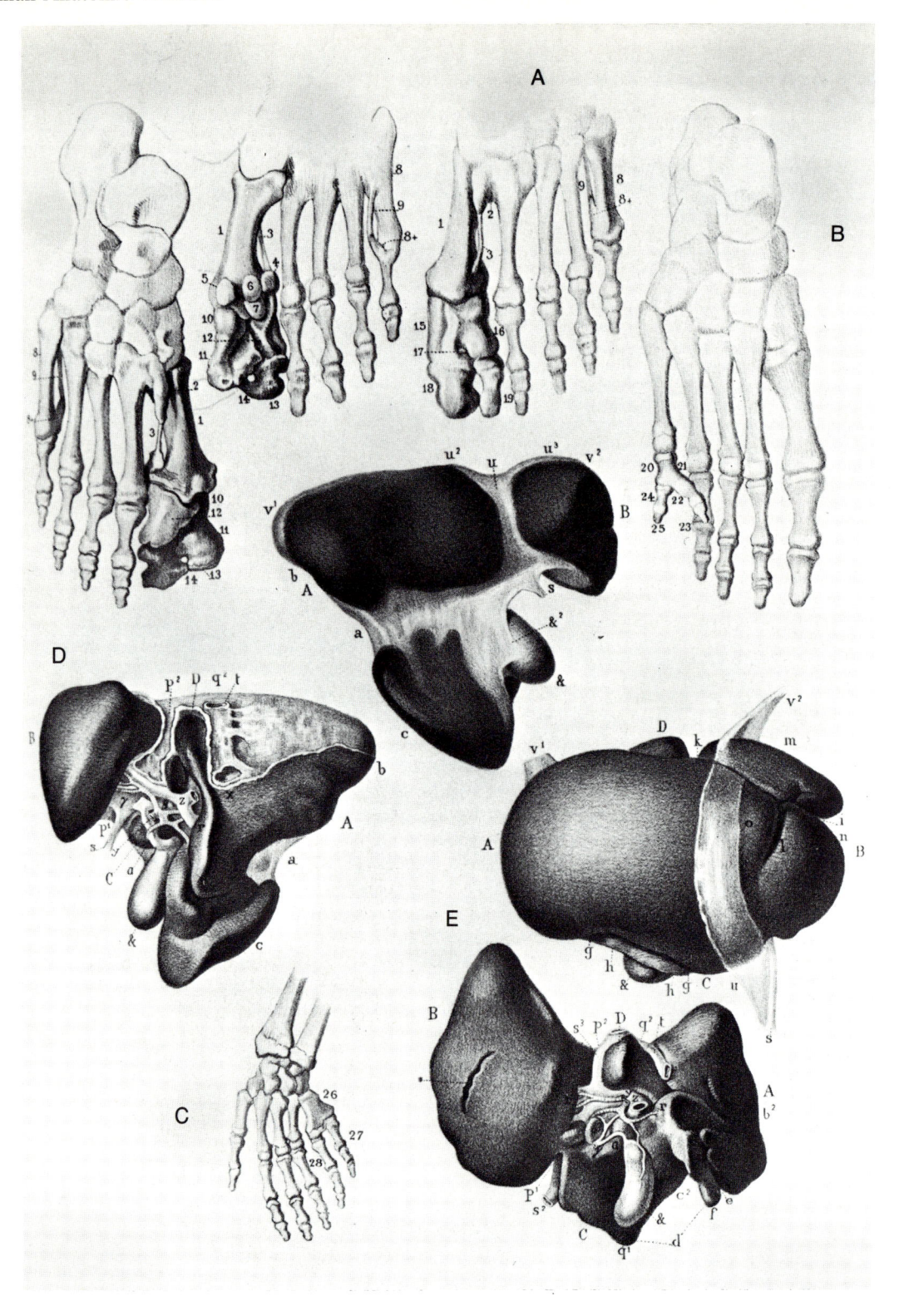

Human Anatomic Variation

FIG. 71.
Duplication of the Great Toe.
Duplication of the Little Toe.
Hexadactyly (Manus).
(Excessive Division of the Liver).

Anomalous digits of feet and hand in a ballerina.
A: Duplication of great toe.

B: Duplication of little toe.
C: Hexadactyly of the hand.
D, E: Excessive division of the liver.

From Calori, L. Sulla coesistenza di un'eccessiva divisione del fegato e di qualche dito soprannumerario nelle mani o nei piedi. *Mem. R. Accad. Sci. Istituto di Bologna* S. 4 2:335–343, 1880.

Skeleton

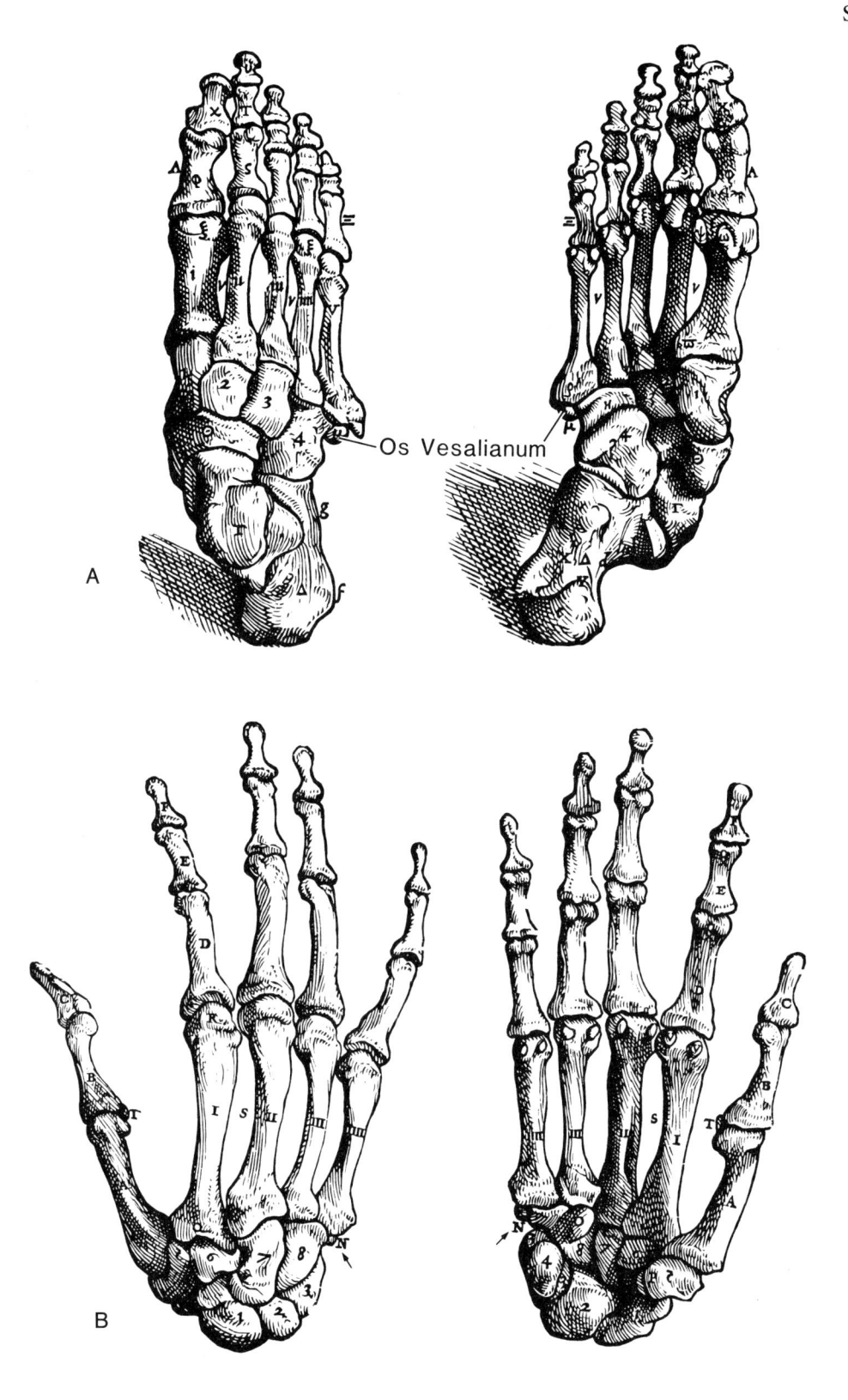

FIG. 72.
Os Vesalianum (Pedis).
Os Vesalianum (Manus).

A: Skeleton of foot showing os vesalianum.
B: Skeleton of hand showing os vesalianum, indicated by *arrows*.

Authors' note: The label "Os Vesalianum" and arrows have been added to the drawings for use in this Atlas.

From Vesalius, A. *Fabrica*, 1543. In Straus, W.L., Jr. and O. Temkin. Vesalius and the problem of variability. *Bull. Hist. Med.* 14:609–633, 1943.

Human Anatomic Variation

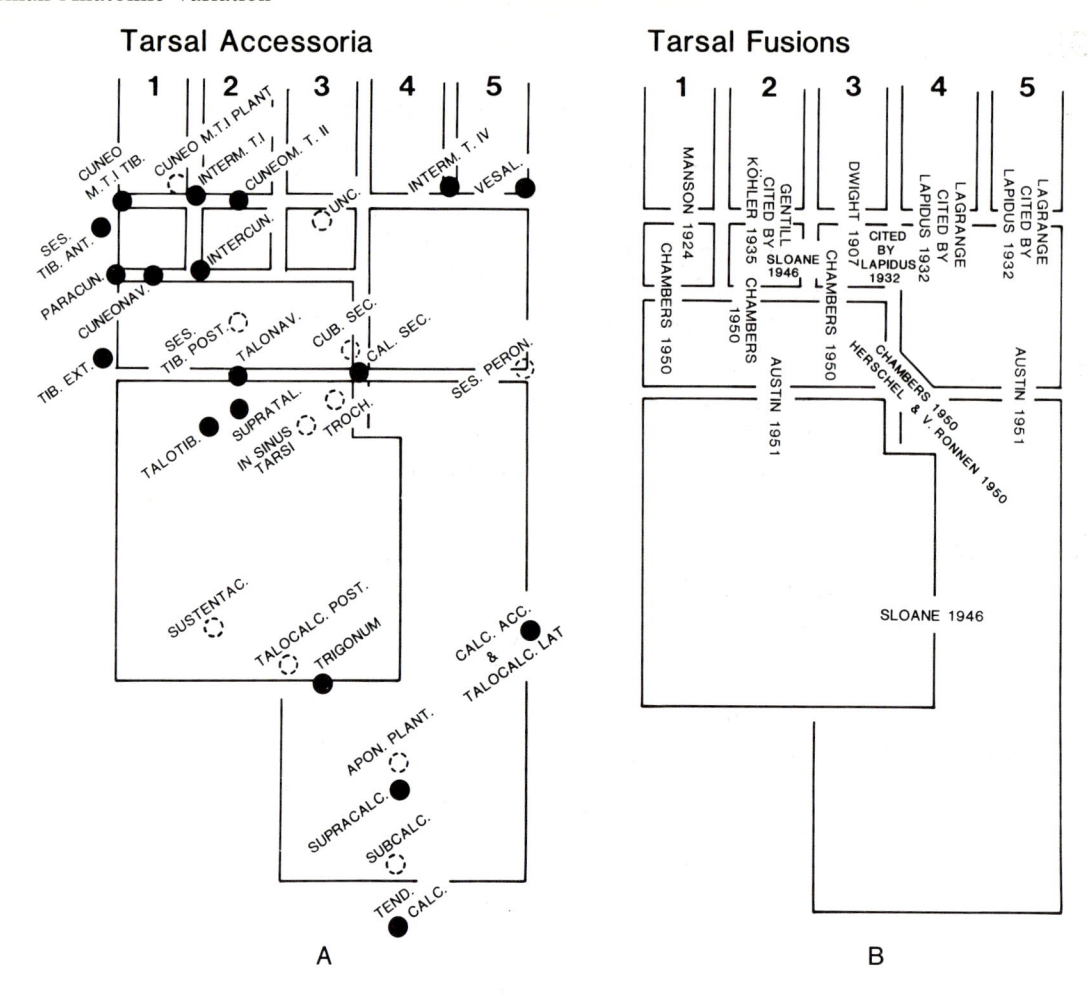

Tarsal Accessoria

Tarsal Fusions

FIG. 73.
Tarsal Accessoria.
Tarsal Fusions.

A: Scheme of accessory tarsal ossicles, dorsal aspect of the right foot; the more plantar-situated ossicles are shown in broken outline. From this diagram the names of the ossicles in a given region of the foot can be determined; the descriptions of these ossicles provided in O'Rahilly's text may be used as a guide in the differential diagnosis of a given case.

B: Scheme of the right tarsus, superior aspect, showing some recent cases of tarsal coherence. The name of the author and date of publication are placed across the fused bones in each case.

Redrawn from and used by permission of O'Rahilly, R. A survey of carpal and tarsal anomalies. J. Bone Joint Surg. [Am.] 35:626–642, 1953.

FIG. 74.
Supernumerary Tarsal Bones.

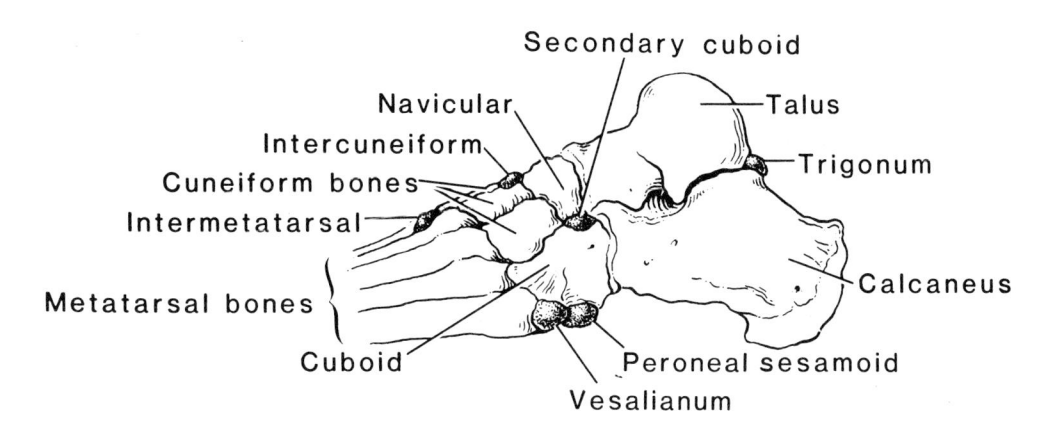

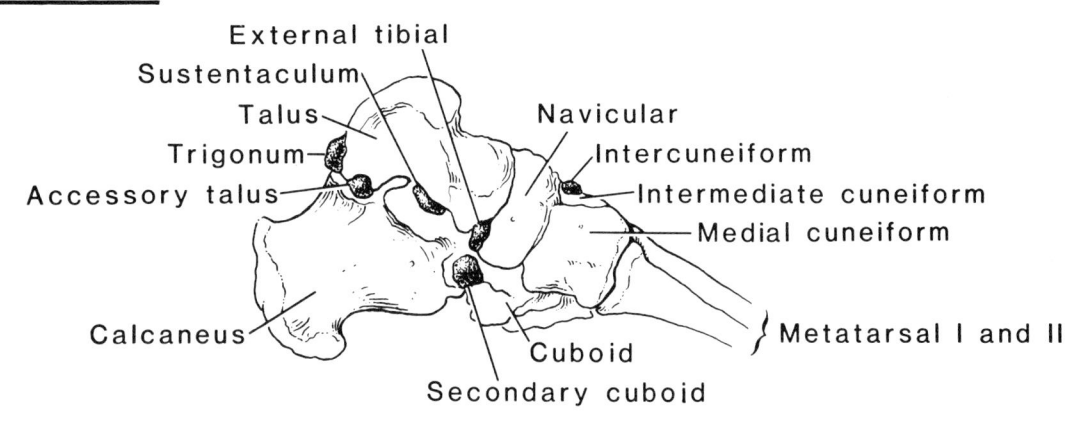

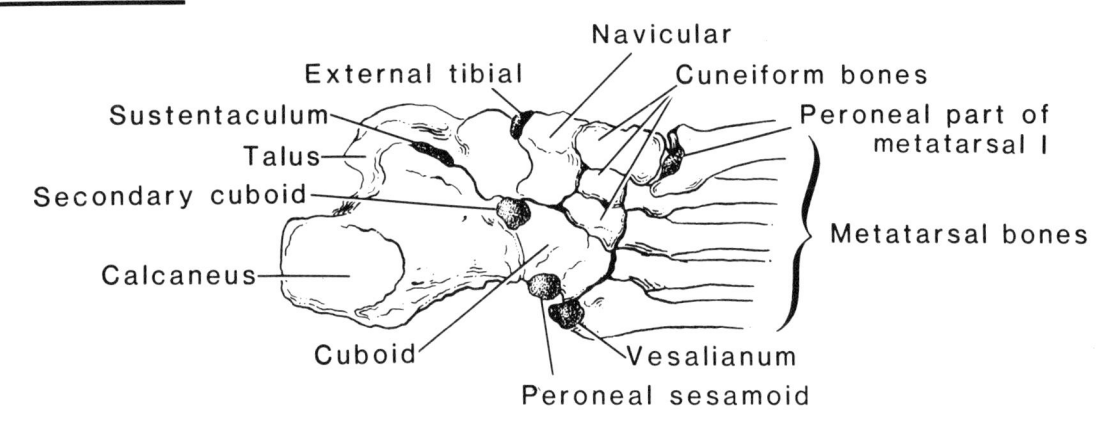

Redrawn from Pfitzner in *Morris' Human Anatomy*, 12th Ed., edited by B.J. Anson. The Blakiston Division, McGraw-Hill Book Co., New York, 1966.